Pathways to Astronomy

Fourth Edition

Stephen E. Schneider

Thomas T. Arny

LOOKING UP at the night sky is one pathway to astronomy. The beauty of the night sky, the glowing stars in their ancient constellations, invites us to wonder about our place in the universe. A small telescope shows even more remarkable sights, and further study reveals exotic and violent phenomena of surpassing beauty. The nine "**Looking Up**" figures on the following pages display a few of the amazing objects that fill the cosmos. Brief descriptions of each object list the Units where you can learn more about them.

LOOKING UP #1
Northern Circumpolar Constellations

For observers over most of the northern hemisphere, there are five constellations that are circumpolar, remaining visible all night long: Ursa Major (the Big Bear), Ursa Minor (the Little Bear), Cepheus (the King), Cassiopeia (the Queen), and Draco (the Dragon). The brightest stars in Ursa Major and Ursa Minor form two well-known asterisms: the Big and Little Dippers.

Delta Cephei
A pulsating variable star (Unit 64) at a distance of 980 ly.

Cassiopeia

Cepheus

~12 ly

M52
This is an open star cluster (Unit 69). Its distance is uncertain—perhaps 3000 to 5000 ly.

Draco

Polaris — The North Star
This star lies about 430 ly away, almost directly above the Earth's North Pole, making it an important aid for navigation (Unit 5).

Little Dipper

170,000 ly

M101
This spiral galaxy is ~27 million light years away from us (Unit 75).

Thuban
This was the north star when the pyramids were built in ancient Egypt (Unit 6).

Big Dipper

M81 and M82
Gravitational interactions between these two galaxies have triggered star formation (Unit 76).

Cassiopeia in 3-D

Earth — 55 ly, 230 ly, 550 ly, 100 ly, 410 ly

1 light year (ly) ≈ 10 trillion km ≈ 6 trillion miles

North

LOOKING UP #2
Ursa Major

Circling in the northern sky is the Big Dipper, part of the well-known constellation Ursa Major, the Big Bear. The Big Dipper is technically not a constellation, but just an asterism — a star grouping. It is easy to see in the early evening looking north from mid-March through mid-September. The Big Dipper can help you find the North Star, and with a telescope on a dark, clear night, you can find several other intriguing objects as shown below.

Over the course of a night, stars appear to rotate counterclockwise around the star Polaris, which remains nearly stationary because it lies almost directly above Earth's North Pole. Polaris is not especially bright, but you can easily find Polaris by extending a line from the two stars at the end of the bowl of the Big Dipper, the pointer stars, as shown by the dashed yellow line (Unit 13).

Pointer stars

Big Dipper

Location of the Hubble Deep Field (Unit 76)

Polaris

Little Dipper

~1.6 ly

M97 — The Owl
This planetary nebula (Unit 65) is ~2500 ly away.

Mizar and Alcor
If you look closely at it, you may notice that the middle star in the "handle" is actually two stars — Mizar and Alcor. Despite appearing close together in the sky, they are probably not in orbit around each other. However, with a small telescope, you can see that Mizar (the brighter of the star pair) has a faint companion star. This companion does in fact orbit Mizar. Moreover, each of Mizar's stars is itself a binary star, making Mizar a quadruple system (Unit 57).

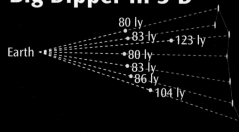

Big Dipper in 3-D

Earth — 80 ly, 83 ly, 123 ly, 80 ly, 83 ly, 86 ly, 104 ly

1 light year (ly) ≈ 10 trillion km ≈ 6 trillion miles

170,000 ly

M51
The Whirlpool Galaxy can be seen as a dim patch of light with a small telescope. M51 is about 37 million ly away from Earth (Unit 76).

LOOKING UP #3
M31 & Perseus

The galaxy M31 lies in the constellation Andromeda, near the constellations Perseus and Cassiopeia. It is about 2.5 million ly from us, the most distant object visible with the naked eye. Northern hemisphere viewers can see M31 in the evening sky from August through December.

Andromeda

M31 — Andromeda Galaxy (Unit 77)

~150,000 ly

~200 ly

The Double Cluster
If you scan with binoculars from M31 toward the space between Perseus and Cassiopeia, you will see the Double Cluster — two groups of massive, luminous but very distant stars. The Double Cluster is best seen with binoculars. The two clusters are about 7000 ly away and a few hundred light years apart (Unit 70).

Perseus

Algol
Algol, the "demon star," dims for about 10 hours every few days as its companion eclipses it (Unit 58).

California Nebula
An emission nebula (Unit 73) with a shape like the state of California.

Capella
The brightest star in the constellation Auriga, the Charioteer. A binary star (Unit 57).

Auriga

M45 — Pleiades

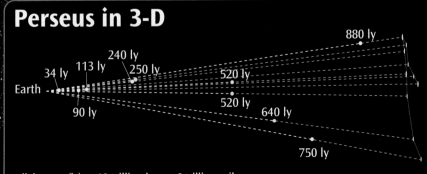

Perseus in 3-D

1 light year (ly) ≈ 10 trillion km ≈ 6 trillion miles

LOOKING UP #4
Summer Triangle

The Summer Triangle consists of the three bright stars Deneb, Vega, and Altair, the brightest stars in the constellations Cygnus, Lyra, and Aquila, respectively. They rise in the east shortly after sunset in late June and are visible throughout the northern summer and into late October (when they set in the west in the early evening). Vega looks the brightest to us, but Deneb produces the most light, only looking dimmer because it is so much farther from us.

Vega

Lyra

Epsilon Lyrae
A double, double star

Cygnus

M57 — Ring Nebula
This planetary nebula (Unit 65) is about 2300 ly distant. From its observed expansion rate it is estimated to be 7000 years old.

1 ly

Deneb
Deneb is a blue supergiant (Unit 67), one of the most luminous stars we can see. Deneb emits ~50,000 times more light than the Sun.

Altair

Albireo
Through a small telescope this star pair shows a strong color contrast between the orange red giant and blue main-sequence star (Unit 59). These stars may orbit each other every few hundred thousand years, but they are far enough apart that they may not be in orbit.

M27 — Dumbbell Nebula
Another planetary nebula (Unit 65), the Dumbbell is about 1200 ly distant and is about 2.5 ly in diameter.

~2.5 ly

The Summer Triangle in 3-D

Vega 25 ly
Earth
Altair 17 ly
Albireo 430 ly
Deneb 1400 ly

1 light year (ly) ≈ 10 trillion km ≈ 6 trillion miles

LOOKING UP #5
Taurus

Taurus, the Bull, is one of the constellations of the zodiac and one of the creatures hunted by Orion in mythology. Taurus is visible in the evening sky from November through March. The brightest star in Taurus is Aldebaran, the eye of the bull. The nebula and two star clusters highlighted below have been critical in the history of astronomy for understanding the distances and fates of stars.

M45 — Pleiades
This open star cluster (Unit 70) is easy to see with the naked eye and looks like a tiny dipper. It is about 400 ly from Earth.

~8 ly

Aldebaran
Aldebaran is a red giant star (Unit 63). It is about 67 ly away from Earth and has a diameter about 45 times larger than the Sun's. Although it appears to be part of the Hyades, it is less than half as distant.

~10 ly

M1 — Crab Nebula
The Crab Nebula is the remnant of a star that blew up in the year A.D. 1054 as a supernova. At its center is a pulsar (Unit 68). It is about 6500 ly away from us.

T Tauri
T Tauri is an erratically-varying pre-main-sequence star, prototype of a class of forming stars (Unit 61). It is about 600 ly distant.

Hyades
The "V" in Taurus is another nearby star cluster, measured to be 151 ly away by the *Hipparcos* satellite (Unit 54). It is easy to see its many stars with binoculars.

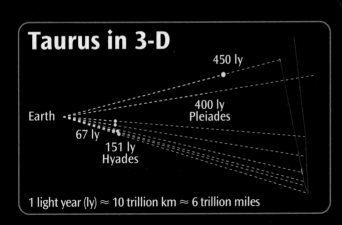

Taurus in 3-D
Earth
67 ly
151 ly Hyades
400 ly Pleiades
450 ly

1 light year (ly) ≈ 10 trillion km ≈ 6 trillion miles

LOOKING UP #6
Orion

Orion is easy to identify because of the three bright stars of his "belt." You can see Orion in the evening sky from November to April, and before dawn from August through September.

Betelgeuse
Betelgeuse is a red supergiant star (Unit 63) that has swelled to a size that is larger than the orbit of Mars. Its red color indicates that it is relatively cool for a star, about 3500 kelvin.

Horsehead Nebula
The horsehead shape is produced by dust in an interstellar cloud blocking background light (Unit 73).

M42 — Orion Nebula
The Orion Nebula is an active star-forming region rich with dust and gas (Units 61, 73).

Rigel
Rigel is a Blue Supergiant star (Unit 67). Its blue color indicates a surface temperature of about 10,000 kelvin.

Protoplanetary disk
This is the beginning of a star; our early Solar System may have looked like this (Unit 36).

Orion in 3-D

Earth — 250 ly, 640 ly, 690 ly, 740 ly, 650 ly, 860 ly, 1300 ly, 1340 ly

1 light year (ly) ≈ 10 trillion km ≈ 6 trillion miles

LOOKING UP #7
Sagittarius

Sagittarius marks the direction to the center of the Milky Way. It can be identified by its "teapot" shape, with the Milky Way seeming to rise like steam from the spout. From northern latitudes, the constellation is best seen July to September, when it is above the southern horizon in the evening. Many star-forming nebulae are visible in this region (Units 61, 73).

M16 — Eagle Nebula
This young star cluster and the hot gas around it lie about 7000 ly from Earth.

~70 ly

~1 ly

M17 — Swan Nebula

M8 — Lagoon Nebula

M20 — Trifid Nebula
The name Trifid was given because of the dark streaks that divide it into thirds. The distance of this nebula is uncertain, approximately 5000 ly away, making its size uncertain too.

~50 ly

M22
M22 is one of many globular clusters (Unit 70) concentrated toward the center of our Galaxy. Easy to see with binoculars, it is just barely visible to the naked eye. It is about 11,000 ly away from us.

~100 ly

The "teapot" of Sagittarius

Center of the Milky Way (Unit 74)

Sagittarius in 3-D

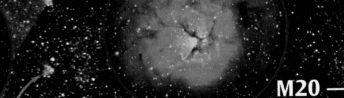

Earth — 78 ly, 97 ly, 230 ly, 240 ly, 122 ly, 350 ly, 88 ly, 143 ly

1 light year (ly) ≈ 10 trillion km ≈ 6 trillion miles

LOOKING UP #8
Centaurus and Crux, The Southern Cross

These constellations are best observed from the southern hemisphere. Northern hemisphere viewers can see Centaurus low in the southern sky during evenings in May – July, but the Southern Cross rises above the horizon only for viewers south of latitude ~25°N (Key West, South Texas, and Hawaii in the United States).

Proxima Centauri
This dim star is the nearest star to the Sun, 4.22 ly distant (Unit 54).

~50 ly

Alpha Centauri

Centaurus

Omega Centauri
This is the largest globular cluster (Unit 70) in the Milky Way, ~16,000 ly distant and containing millions of stars.

~200 ly

The Jewel Box
NGC 4755, an open star cluster (Unit 70) ~500 ly from us.

The Coal Sack
An interstellar dust cloud (Unit 73)

Crux
The Southern Cross

50,000 ly

Centaurus A
This active galaxy (Unit 78), about 11 million ly distant, is one of the brightest radio sources in the sky.

Eta Carinae
At over 100 times the mass of the Sun, this is one of the highest-mass stars known and doomed to die young (Unit 61). It is about 8000 ly distant.

Southern Cross in 3-D

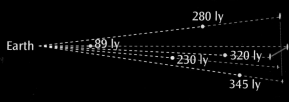

Earth ···· 89 ly ···· 280 ly
230 ly • 320 ly
345 ly

1 light year (ly) ≈ 10 trillion km ≈ 6 trillion miles

LOOKING UP #9
Southern Circumpolar Constellations

Most of the constellations in this part of the sky are dim, but observers in much of the southern hemisphere can see the Magellanic Clouds circling the south celestial pole throughout the night.

Crux
The Southern Cross

Musca

Hourglass Nebula
A planetary nebula (Unit 65) ~8000 ly distant

Apus

Octans
The constellation closest to the south celestial pole is named after a navigational instrument, the octant.

~0.5 ly

Chamaeleon

The South Celestial Pole
No bright stars lie near the south celestial pole (Unit 5), but the Southern Cross points toward it.

Thumbprint Nebula
A Bok globule (Unit 61) about 600 ly distant

Volans

Mensa

Hydrus

~1000 ly

Small Magellanic Cloud
A dwarf galaxy orbiting the Milky Way at a distance of ~200,000 ly (Unit 77).

Large Magellanic Cloud
A small galaxy orbiting the Milky Way at a distance of ~160,000 ly (Unit 77).

Tarantula Nebula
A star-formation region (Unit 60) in the Large Magellanic Cloud larger than any known in the Milky Way.

Pathways to Astronomy

Fourth Edition

Stephen E. Schneider
Professor of Astronomy
University of Massachusetts, Amherst

Thomas T. Arny
Professor of Astronomy, Emeritus
University of Massachusetts, Amherst

PATHWAYS TO ASTRONOMY, FOURTH EDITION

Published by McGraw-Hill Education, 2 Penn Plaza, New York, NY 10121. Copyright © 2015 by McGraw-Hill Education. All rights reserved. Printed in the United States of America. Previous editions © 2012, 2009, and 2007. No part of this publication may be reproduced or distributed in any form or by any means, or stored in a database or retrieval system, without the prior written consent of McGraw-Hill Education, including, but not limited to, in any network or other electronic storage or transmission, or broadcast for distance learning.

Some ancillaries, including electronic and print components, may not be available to customers outside the United States.

This book is printed on acid-free paper.

1 2 3 4 5 6 7 8 9 0 RMN/RMN 1 0 9 8 7 6 5 4

ISBN 978-1-259-25361-4
MHID 1-259-25361-9

All credits appearing on page or at the end of the book are considered to be an extension of the copyright page.

The Internet addresses listed in the text were accurate at the time of publication. The inclusion of a website does not indicate an endorsement by the authors or McGraw-Hill Education, and McGraw-Hill Education does not guarantee the accuracy of the information presented at these sites.

www.mhhe.com

To my father, who taught me the night sky when I was little.

—Steve

About the Authors

Stephen E. Schneider is a professor and Thomas T. Arny is an emeritus professor in the Astronomy Department at the University of Massachusetts at Amherst, which is part of the Five College Astronomy Department (comprising faculty from UMass and Amherst, Hampshire, Mt. Holyoke, and Smith Colleges). Both are recipients of their college's Outstanding Teacher Award, and they have collectively taught introductory astronomy for over 50 years to students with a wide variety of backgrounds.

Steve Schneider became interested in astronomy at the amateur level when he was a child. He studied astronomy as an undergraduate at Harvard and obtained his Ph.D. from Cornell. His dissertation work received the Trumpler Award of the Astronomical Society of the Pacific, and he was named a Presidential Young Investigator. In addition to teaching introductory astronomy, he works closely with science teachers, presenting workshops and special courses. He also loves to draw and paint.

Tom Arny received his undergraduate degree from Haverford College and his Ph.D. in astronomy from the University of Arizona. In addition to his interest in astronomy, he has a long-standing fascination with the natural world: weather (especially atmospheric optics such as rainbows), birds, wildflowers, and butterflies.

Brief Contents

Looking Up Illustrations i
Preface xxv

PART 1 THE COSMIC LANDSCAPE 1

Unit 1   Our Planetary Neighborhood 1
Unit 2   Beyond the Solar System 9
Unit 3   Astronomical Numbers 17
Unit 4   Scientific Foundations of Astronomy 25
Unit 5   The Night Sky 33
Unit 6   The Year 40
Unit 7   The Time of Day 51
Unit 8   Lunar Cycles 58
Unit 9   Calendars 69
Unit 10 Geometry of the Earth, Moon, and Sun 76
Unit 11 Planets: The Wandering Stars 84
Unit 12 The Beginnings of Modern Astronomy 93
Unit 13 Observing the Sky 100

PART 2 PROBING MATTER, LIGHT, AND THEIR INTERACTIONS 109

Unit 14 Astronomical Motion: Inertia, Mass, and Force 109
Unit 15 Force, Acceleration, and Interaction 114
Unit 16 The Universal Law of Gravity 120
Unit 17 Measuring a Body's Mass Using Orbital Motion 125
Unit 18 Orbital and Escape Velocities 129
Unit 19 Tides 135
Unit 20 Conservation Laws 141
Unit 21 The Dual Nature of Light and Matter 147
Unit 22 The Electromagnetic Spectrum 155
Unit 23 Thermal Radiation 162
Unit 24 Identifying Atoms by Their Spectra 168
Unit 25 The Doppler Shift 177
Unit 26 Special Relativity 181
Unit 27 General Relativity 189
Unit 28 Detecting Light—An Overview 196
Unit 29 Collecting Light 203
Unit 30 Focusing Light 210
Unit 31 Telescope Resolution 218
Unit 32 The Earth's Atmosphere and Space Observatories 224
Unit 33 Amateur Astronomy 232

PART 3 THE SOLAR SYSTEM 241

Unit 34 The Structure of the Solar System 241
Unit 35 The Origin of the Solar System 250
Unit 36 Other Planetary Systems 261
Unit 37 The Earth as a Terrestrial Planet 270
Unit 38 Earth's Atmosphere and Hydrosphere 281
Unit 39 Our Moon 292
Unit 40 Mercury 302
Unit 41 Venus 309
Unit 42 Mars 316
Unit 43 Asteroids 328
Unit 44 Comparative Planetology 336
Unit 45 Jupiter and Saturn 347
Unit 46 Uranus and Neptune 355
Unit 47 Satellite Systems and Rings 361
Unit 48 Ice Worlds, Pluto, and Beyond 371
Unit 49 Comets 382
Unit 50 Impacts on Earth 392

PART 4 STARS AND STELLAR EVOLUTION 402

Unit 51	The Sun, Our Star	402
Unit 52	The Sun's Source of Power	412
Unit 53	Solar Activity	420
Unit 54	Surveying the Stars	429
Unit 55	The Luminosities of Stars	440
Unit 56	The Temperatures and Compositions of Stars	446
Unit 57	The Masses of Orbiting Stars	454
Unit 58	The Sizes of Stars	459
Unit 59	The H-R Diagram	465
Unit 60	Overview of Stellar Evolution	473
Unit 61	Star Formation	481
Unit 62	Main-Sequence Stars	489
Unit 63	Giant Stars	496
Unit 64	Variable Stars	503
Unit 65	Mass Loss and Death of Low-Mass Stars	509
Unit 66	Exploding White Dwarfs	516
Unit 67	Old Age and Death of Massive Stars	522
Unit 68	Neutron Stars	532
Unit 69	Black Holes	539
Unit 70	Star Clusters	548

PART 5 GALAXIES AND THE UNIVERSE 557

Unit 71	Discovering the Milky Way	557
Unit 72	Stars of the Milky Way	564
Unit 73	Gas and Dust in the Milky Way	572
Unit 74	Mass and Motions in the Milky Way	580
Unit 75	A Universe of Galaxies	589
Unit 76	Types of Galaxies	598
Unit 77	Galaxy Clustering	609
Unit 78	Active Galactic Nuclei	618
Unit 79	Dark Matter	627
Unit 80	Cosmology	635
Unit 81	The Edges of the Universe	644
Unit 82	The Curvature and Expansion of the Universe	652
Unit 83	The Beginnings of the Universe	660
Unit 84	Dark Energy and the Fate of the Universe	671
Unit 85	Astrobiology	679
Unit 86	The Search for Life Elsewhere	688

Contents

Looking Up Illustrations   i
   #1 Northern Circumpolar Constellations   ii
   #2 Ursa Major   iii
   #3 M31 & Perseus   iv
   #4 Summer Triangle   v
   #5 Taurus   vi
   #6 Orion   vii
   #7 Sagittarius   viii
   #8 Centaurus and Crux, The Southern Cross   ix
   #9 Southern Circumpolar Constellations   x
Preface   xxv

PART 1   THE COSMIC LANDSCAPE   1

Unit 1 Our Planetary Neighborhood   1
- 1.1 The Earth   1
- 1.2 The Moon   2
- 1.3 The Planets   3
- 1.4 The Sun   4
- 1.5 The Solar System   5
- 1.6 The Astronomical Unit   6

Unit 2 Beyond the Solar System   9
- 2.1 Stellar Evolution   9
- 2.2 The Light-Year   10
- 2.3 The Milky Way Galaxy   11
- 2.4 Galaxy Clusters and Beyond   12
- 2.5 The Still-Unknown Universe   14

Unit 3 Astronomical Numbers   17
- 3.1 The Metric System   18
- 3.2 Scientific Notation   20
- 3.3 Special Units   21
- 3.4 Approximation   23

Unit 4 Scientific Foundations of Astronomy   25
- 4.1 The Scientific Method   25
- 4.2 The Nature of Matter   27
- 4.3 The Four Fundamental Forces   29
- 4.4 The Elementary Particles   30

Unit 5 The Night Sky   33
- 5.1 The Celestial Sphere   33
- 5.2 Constellations   34
- 5.3 Daily Motion   35
- 5.4 Latitude and Longitude   37
- 5.5 Celestial Coordinates   38

Unit 6 The Year   40
- 6.1 Annual Motion of the Sun   41
- 6.2 The Ecliptic and the Zodiac   42
- 6.3 The Seasons   43
- 6.4 The Ecliptic's Tilt   45
- 6.5 Solstices and Equinoxes   46
- 6.6 Precession   49

Unit 7 The Time of Day   51
- 7.1 The Day   51
- 7.2 Length of Daylight Hours   52
- 7.3 Time Zones   54
- 7.4 Daylight Saving Time   55
- 7.5 Leap Seconds   56

Unit 8 Lunar Cycles   58
- 8.1 Phases of the Moon   58
- 8.2 Eclipses   61
- 8.3 Eclipse Seasons   65
- 8.4 Moon Lore   66

Unit 9 Calendars   69
- 9.1 The Week   69
- 9.2 The Month   70
- 9.3 The Roman Calendar   71
- 9.4 The Leap Year   72
- 9.5 The Chronicling of Years   73

Unit 10 Geometry of the Earth, Moon, and Sun   76
- 10.1 The Shape of the Earth   76
- 10.2 Distance and Size of the Sun and Moon   77
- 10.3 The Size of the Earth   79
- 10.4 Measuring the Diameter of Astronomical Objects   80
- 10.5 The Moon Illusion   82

Unit 11	Planets: The Wandering Stars 84		**Unit 20**	Conservation Laws 141

Unit 11 — Planets: The Wandering Stars 84
- 11.1 Motions of the Planets 84
- 11.2 Early Ideas About Retrograde Motion 86
- 11.3 The Heliocentric Model 88
- 11.4 The Copernican Revolution 90

Unit 12 — The Beginnings of Modern Astronomy 93
- 12.1 Precision Astronomical Measurements 93
- 12.2 The Nature of Planetary Orbits 94
- 12.3 The First Telescopic Observations 97

Unit 13 — Observing the Sky 100
- 13.1 Learning the Constellations 100
- 13.2 Motions of the Stars 103
- 13.3 Motion of the Sun 104
- 13.4 Motions of the Moon and Planets 104
- 13.5 A Sundial: Orbital Effects on the Day 106

PART 2 PROBING MATTER, LIGHT, AND THEIR INTERACTIONS 109

Unit 14 — Astronomical Motion: Inertia, Mass, and Force 109
- 14.1 Inertia and Mass 109
- 14.2 The Law of Inertia 111
- 14.3 Forces and Weights 111
- 14.4 The Force in an Orbit 112

Unit 15 — Force, Acceleration, and Interaction 114
- 15.1 Acceleration 114
- 15.2 Newton's Second Law of Motion 116
- 15.3 Action and Reaction: Newton's Third Law of Motion 117

Unit 16 — The Universal Law of Gravity 120
- 16.1 Orbital Motion and Gravity 120
- 16.2 Newton's Universal Law of Gravity 121
- 16.3 Surface Gravity and Weight 122

Unit 17 — Measuring a Body's Mass Using Orbital Motion 125
- 17.1 Masses from Orbital Speeds 125
- 17.2 Kepler's Third Law Revisited 127

Unit 18 — Orbital and Escape Velocities 129
- 18.1 Circular Orbits 129
- 18.2 Escape Velocity 130
- 18.3 The Shapes of Orbits 132

Unit 19 — Tides 135
- 19.1 Cause of Tides 135
- 19.2 The Size of the Tidal Force 136
- 19.3 Solar Tides 138
- 19.4 Tidal Braking 139

Unit 20 — Conservation Laws 141
- 20.1 Conservation of Energy 141
- 20.2 Conservation of Mass (Almost) 144
- 20.3 Conservation of Angular Momentum 145

Unit 21 — The Dual Nature of Light and Matter 147
- 21.1 The Nature of Light 147
- 21.2 The Effect of Distance on Light 149
- 21.3 The Nature of Matter 150
- 21.4 The Interaction of Light and Matter 151

Unit 22 — The Electromagnetic Spectrum 155
- 22.1 Wavelengths and Frequencies 155
- 22.2 Energy Carried by Photons 157
- 22.3 White Light and the Color Spectrum 157
- 22.4 The Electromagnetic Spectrum 159

Unit 23 — Thermal Radiation 162
- 23.1 Blackbodies 162
- 23.2 Color, Luminosity, and Temperature 163
- 23.3 Measuring Temperature 164
- 23.4 Taking the Temperature of Astronomical Objects 165
- 23.5 The Stefan-Boltzmann Law 166

Unit 24 — Identifying Atoms by Their Spectra 168
- 24.1 The Spectrum of Hydrogen 168
- 24.2 Identifying Atoms by Their Light 170
- 24.3 Types of Spectra 173
- 24.4 Astronomical Spectra 174

Unit 25 — The Doppler Shift 177
- 25.1 Calculating the Doppler Shift 177
- 25.2 Astronomical Motions 179

Unit 26 — Special Relativity 181
- 26.1 Light from Moving Bodies 181
- 26.2 The Michelson-Morley Experiment 183
- 26.3 Einstein's Theory of Special Relativity 184
- 26.4 Special Relativity and Space Travel 186
- 26.5 The Twin Paradox 187

Unit 27 — General Relativity 189
- 27.1 The Principle of Equivalence 189
- 27.2 Gravity and the Curvature of Space 191
- 27.3 Gravitational Time Dilation 192
- 27.4 Gravitational Waves 194

Unit 28 — Detecting Light—An Overview 196
- 28.1 Technological Frontiers 196
- 28.2 Detecting Visible Light 197
- 28.3 Observing at Nonvisible Wavelengths 198
- 28.4 The Crab Nebula: A Case History 200

Unit 29	Collecting Light 203
29.1	Modern Observatories 203
29.2	Collecting Power 206
29.3	Filtering Light 207
29.4	Surface Brightness 208

Unit 30	Focusing Light 210
30.1	Refracting Telescopes 210
30.2	Reflecting Telescopes 212
30.3	Development of Larger Apertures 214
30.4	Color Dispersion 216

Unit 31	Telescope Resolution 218
31.1	Resolution and Diffraction 218
31.2	Calculating the Resolution of a Telescope 220
31.3	Interferometers 220

Unit 32	The Earth's Atmosphere and Space Observatories 224
32.1	Atmospheric Absorption 224
32.2	Atmospheric Scintillation 226
32.3	Atmospheric Refraction 228
32.4	Observatories in Space 229

Unit 33	Amateur Astronomy 232
33.1	The Human Eye 232
33.2	Your Eyes at Night 233
33.3	Choosing a Telescope 234
33.4	Completing the Telescope 235
33.5	Astronomical Photography 237

PART 3 THE SOLAR SYSTEM 241

Unit 34	The Structure of the Solar System 241
34.1	Components of the Solar System 241
34.2	Orbital Patterns in the Solar System 244
34.3	Compositions in the Solar System 246
34.4	Density and Composition 246

Unit 35	The Origin of the Solar System 250
35.1	The Age of the Solar System 250
35.2	Birth of the Solar System 252
35.3	From Dust Grains to Planetesimals 254
35.4	Formation of the Planets 256
35.5	Late-Stage Bombardment 258

Unit 36	Other Planetary Systems 261
36.1	Young Planetary Systems 261
36.2	Detecting Exoplanets by Doppler Shifts 262
36.3	Migrating Planets 265
36.4	New Exoplanet Detection Methods 266

Unit 37	The Earth as a Terrestrial Planet 270
37.1	Composition of the Earth 270
37.2	Heating of the Earth's Core 273
37.3	The Earth's Dynamic Surface 274
37.4	The Earth's Magnetic Field 278

Unit 38	Earth's Atmosphere and Hydrosphere 281
38.1	Structure of the Atmosphere 281
38.2	The Atmosphere, Light, and Global Warming 284
38.3	The History of the Atmosphere and Oceans 286
38.4	The Shaping Effects of Water 288
38.5	Air and Ocean Circulation: the Coriolis Effect 289

Unit 39	Our Moon 292
39.1	The Origin of the Moon 292
39.2	Surface Features 294
39.3	The Moon's Structure and History 297
39.4	The Absence of a Lunar Atmosphere 299
39.5	The Moon's Rotation and Orbit 300

Unit 40	Mercury 302
40.1	Mercury's Surface Features 302
40.2	Mercury's Interior 305
40.3	Mercury's Rotation 306
40.4	Mercury's Temperature and Atmosphere 307

Unit 41	Venus 309
41.1	The Venusian Atmosphere 309
41.2	The Surface and Interior of Venus 311
41.3	Rotation of Venus 314

Unit 42	Mars 316
42.1	The Major Features of Mars 316
42.2	A Blue Mars? 320
42.3	The Martian Atmosphere 324
42.4	The Martian Moons 326

Unit 43	Asteroids 328
43.1	The Discovery of Asteroids 329
43.2	Asteroid Sizes and Shapes 330
43.3	Asteroid Compositions and Origin 332
43.4	Asteroid Orbits 333

Unit 44	Comparative Planetology 336
44.1	The Role of Mass and Radius 337
44.2	The Role of Water and Biological Processes 339
44.3	The Role of Sunlight 341
44.4	The Outer Versus the Inner Solar System 342

Unit 45	Jupiter and Saturn 347
45.1	The Appearance of Jupiter and Saturn 348
45.2	The Giants' Interiors 349

45.3 Stormy Atmospheres 351
45.4 The Magnetic Fields 353

Unit 46 Uranus and Neptune 355
46.1 Discovery of Two New Planets 356
46.2 The Atmospheres of Uranus and Neptune 356
46.3 Oddly Tilted Axes 358

Unit 47 Satellite Systems and Rings 361
47.1 Satellite Systems 361
47.2 Satellite Properties 363
47.3 Ring Systems 366
47.4 Origin of Planetary Rings 368

Unit 48 Ice Worlds, Pluto, and Beyond 371
48.1 The Galilean Satellites 372
48.2 Saturn's Moon Titan 375
48.3 Neptune's Moon Triton 376
48.4 Pluto 377
48.5 The Trans-Neptunian Worlds 378

Unit 49 Comets 382
49.1 Comet Structure and Appearance 383
49.2 Composition of Comets 385
49.3 Origin of Comets 387
49.4 Meteor Showers 389

Unit 50 Impacts on Earth 392
50.1 Heating of Meteors 392
50.2 Meteorites 393
50.3 The Energy of Impacts 395
50.4 Impacts with the Earth 396
50.5 Mass Extinction Events 399

PART 4 STARS AND STELLAR EVOLUTION 402

Unit 51 The Sun, Our Star 402
51.1 The Surface of the Sun 402
51.2 Pressure Balance and the Sun's Interior 404
51.3 Energy Transport 406
51.4 The Solar Atmosphere 407
51.5 Solar Seismology 410

Unit 52 The Sun's Source of Power 412
52.1 The Mystery Behind Sunshine 412
52.2 The Conversion of Hydrogen into Helium 414
52.3 Solar Neutrinos 416
52.4 The Fusion Bottleneck 418

Unit 53 Solar Activity 420
53.1 Sunspots 420
53.2 Prominences and Flares 423
53.3 The Solar Cycle 425
53.4 The Solar Cycle and Terrestrial Climate 426

Unit 54 Surveying the Stars 429
54.1 Triangulation 429
54.2 Parallax 432
54.3 Calculating Parallaxes 434
54.4 Moving Stars 436
54.5 The Aberration of Starlight 437

Unit 55 The Luminosities of Stars 440
55.1 Luminosity 440
55.2 Measuring Luminosities Using the Inverse-Square Law 441
55.3 Distance by the Standard-Candles Method 442
55.4 The Magnitude System 443

Unit 56 The Temperatures and Compositions of Stars 446
56.1 Interactions of Photons and Matter in Stars 446
56.2 Stellar Surface Temperature 447
56.3 The Development of Spectral Classification 449
56.4 How Temperature Affects a Star's Spectrum 451
56.5 Spectral Classification Criteria 451

Unit 57 The Masses of Orbiting Stars 454
57.1 Types of Binary Stars 454
57.2 Measuring Stellar Masses with Binary Stars 455
57.3 The Center of Mass 456

Unit 58 The Sizes of Stars 459
58.1 The Angular Sizes of Stars 459
58.2 Using Eclipsing Binaries to Measure Stellar Diameters 461
58.3 Using the Stefan-Boltzmann Law to Calculate Stellar Radii 462

Unit 59 The H-R Diagram 465
59.1 Analyzing the H-R Diagram 466
59.2 The Mass–Luminosity Relation 469
59.3 Luminosity Classes 470

Unit 60 Overview of Stellar Evolution 473
60.1 Stellar Evolution: Models and Observations 473
60.2 The Evolution of a Star 474
60.3 The Stellar Evolution Cycle 477
60.4 Tracking Changes with the H-R Diagram 478

Unit 61 Star Formation 481
61.1 The Origin of Stars in Interstellar Clouds 481
61.2 Proto- and Pre-Main-Sequence Stars 482

61.3 Star Formation in the H-R Diagram 486	Unit 70 Star Clusters 548
61.4 Stellar Mass Limits 487	70.1 Types of Star Clusters 548
Unit 62 Main-Sequence Stars 489	70.2 Testing Stellar Evolution Theory 551
62.1 Mass and Core Temperature 489	70.3 The Initial Mass Function 553

- Unit 62 Main-Sequence Stars 489
  - 62.1 Mass and Core Temperature 489
  - 62.2 Structure of High-Mass and Low-Mass Stars 490
  - 62.3 Main-Sequence Lifetime of a Star 492
  - 62.4 Changes During the Main-Sequence Phase 494
- Unit 63 Giant Stars 496
  - 63.1 Restructuring Following the Main Sequence 496
  - 63.2 Helium Fusion 498
  - 63.3 Electron Degeneracy and the Helium Flash in Low-Mass Stars 499
  - 63.4 Helium Fusion in the H-R Diagram 500
- Unit 64 Variable Stars 503
  - 64.1 Classes of Variable Stars 503
  - 64.2 Yellow Giants and Pulsating Stars 505
  - 64.3 The Period–Luminosity Relation 507
- Unit 65 Mass Loss and Death of Low-Mass Stars 509
  - 65.1 The Fate of Stars Like the Sun 509
  - 65.2 Ejection of a Star's Outer Layers 511
  - 65.3 Planetary Nebulae 512
  - 65.4 White Dwarfs 514
- Unit 66 Exploding White Dwarfs 516
  - 66.1 Novae 516
  - 66.2 The Chandrasekhar Limit 518
  - 66.3 Supernovae of Type Ia 519
- Unit 67 Old Age and Death of Massive Stars 522
  - 67.1 The Fate of Massive Stars 522
  - 67.2 The Formation of Heavy Elements 524
  - 67.3 Core Collapse of Massive Stars 526
  - 67.4 Supernova Remnants 528
  - 67.5 Gamma-Ray Bursts and Hypernovae 529
- Unit 68 Neutron Stars 532
  - 68.1 Pulsars and the Discovery of Neutron Stars 532
  - 68.2 Emission From Neutron Stars 534
  - 68.3 High-Energy Pulsars 537
- Unit 69 Black Holes 539
  - 69.1 The Escape Velocity Limit 539
  - 69.2 Curving Space Out of Sight 541
  - 69.3 Observing Black Holes 544
  - 69.4 Hawking Radiation 545
  - 69.5 Small and Large Black Holes 546

PART V GALAXIES AND THE UNIVERSE 557

- Unit 71 Discovering the Milky Way 557
  - 71.1 The Shape of the Milky Way 558
  - 71.2 Star Counts and the Size of the Galaxy 559
  - 71.3 Globular Clusters and the Size of the Galaxy 560
  - 71.4 Galactic Structure and Contents 562
- Unit 72 Stars of the Milky Way 564
  - 72.1 Stellar Populations 564
  - 72.2 Formation of Our Galaxy 566
  - 72.3 Evolution Through Mergers 568
  - 72.4 The Future of the Milky Way 570
- Unit 73 Gas and Dust in the Milky Way 572
  - 73.1 The Interstellar Medium 572
  - 73.2 Interstellar Dust: Dimming and Reddening 574
  - 73.3 Radio Waves from Cold Interstellar Gas 576
  - 73.4 Heating and Cooling in the ISM 577
- Unit 74 Mass and Motions in the Milky Way 580
  - 74.1 The Mass of the Milky Way and the Number of its Stars 580
  - 74.2 The Galactic Center and Edge 583
  - 74.3 Density Waves and Spiral Arms 585
- Unit 75 A Universe of Galaxies 589
  - 75.1 Early Observations of Galaxies 589
  - 75.2 The Distances of Galaxies 592
  - 75.3 The Redshift and Hubble's Law 594
- Unit 76 Types of Galaxies 598
  - 76.1 Galaxy Classification 598
  - 76.2 Differences in Star and Gas Content 602
  - 76.3 The Evolution of Galaxies 603
  - 76.4 Galaxy Mergers and Changing Types 606
- Unit 77 Galaxy Clustering 609
  - 77.1 The Local Group 610
  - 77.2 Rich and Poor Galaxy Clusters 611
  - 77.3 Superclusters 613
  - 77.4 Large-Scale Structure 614
  - 77.5 Probing Intergalactic Space 616

Unit	Section	Title	Page
Unit 78		Active Galactic Nuclei	618
	78.1	Active Galaxies	618
	78.2	Quasars	620
	78.3	Cause of Activity in Galaxies	622
	78.4	Black Hole/Galaxy Interactions	624
Unit 79		Dark Matter	627
	79.1	Measuring the Mass of a Galaxy	627
	79.2	Dark Matter in Clusters of Galaxies	629
	79.3	Gravitational Lenses	630
	79.4	What Is Dark Matter?	632
Unit 80		Cosmology	635
	80.1	Evolving Concepts of the Universe	635
	80.2	The Recession of Galaxies	636
	80.3	The Meaning of Redshift	639
	80.4	The Age of the Universe	641
Unit 81		The Edges of the Universe	644
	81.1	Olbers' Paradox	644
	81.2	The Cosmic Microwave Background	647
	81.3	The Era of Galaxy Formation	649
Unit 82		The Curvature and Expansion of Universes	652
	82.1	Quantifying Curvature	652
	82.2	Curvature and Expansion	654
	82.3	The Density of the Universe	655
	82.4	A Cosmological Constant	657
Unit 83		The Beginnings of the Universe	660
	83.1	The Eras of the Universe	660
	83.2	The Origin of Helium	661
	83.3	Radiation, Matter, and Antimatter	664
	83.4	The Epoch of Inflation	665
	83.5	Cosmological Problems Solved by Inflation	668
Unit 84		Dark Energy and the Fate of the Universe	671
	84.1	The Type Ia Supernova Test	671
	84.2	An Accelerating Universe	672
	84.3	Dark Matter, Dark Energy, and the Cosmic Microwave Background	674
	84.4	A Runaway Universe?	675
	84.5	Other Universes?	676
Unit 85		Astrobiology	679
	85.1	The History of Life on Earth	679
	85.2	The Chemistry of Life	681
	85.3	The Origin of Life	683
	85.4	Life, Planets, and the Universe	686
Unit 86		The Search for Life Elsewhere	688
	86.1	The Search for Life on Mars	688
	86.2	Life on Other Planets?	690
	86.3	Are We Alone?	691
	86.4	SETI	694

Appendix

Scientific Notation   A-1

Solving Distance, Velocity, Time (d, V, t) Problems   A-1

Table 1	Physical and Astronomical Constants	A-2
Table 2	Metric Prefixes	A-2
Table 3	Conversion Between English and Metric Units	A-2
Table 4	Some Useful Formulas	A-2
Table 5	Physical Properties of the Planets	A-3
Table 6	Orbital Properties of the Planets	A-3
Table 7	Larger Satellites of the Planets	A-4
Table 8	Properties of Main-Sequence Stars	A-6
Table 9	The Brightest Stars	A-6
Table 10	The Nearest Stars	A-7
Table 11	Known and Suspected Members of the Local Group of Galaxies	A-8
Table 12	The Brightest Galaxies Beyond the Local Group	A-9
Table 13	The Messier Catalog	A-9
Table 14	A Cosmic Periodic Table of the Elements	A-12

Answers to Questions   AQ-1

Glossary   G-1

Credits   C-1

Index   I-1

Preface

APPROACH

There are many astronomy textbooks available today, but *Pathways to Astronomy* offers something different. . . .

Created by two veteran teachers of astronomy, both recipients of outstanding teaching awards, *Pathways* breaks down introductory astronomy into its component parts. The huge and fascinating field of astronomy is divided into 86 Units from which you can selectively choose topics according to your interests, while maintaining a natural flow of presentation.

One of the frustrations created by other current astronomy textbooks is that each chapter covers such a wide array of topics that it is difficult for students to absorb the large amount of material, and the texts are wed to such a specific order of presentation that it is difficult for the professor to link the chapter readings and review questions to his or her own particular approach to teaching the subject. Whether you are learning astronomy for the first time or teaching it for the tenth, *Pathways* offers greater flexibility for exploring astronomy in the way you want to.

The Unit structure allows the new learner and the veteran professor to relate the text more clearly to college lectures. Each Unit is small enough to be easily tackled on its own or read as an adjunct to the classroom lecture. For the faculty member who is designing a course to relate to current events in astronomy or a particular theme, the structure of *Pathways* makes it easier to assign reading and worked problems that are relevant to each topic. For the student of astronomy, *Pathways* makes it easier to digest each topic and to clearly relate each Unit to lecture material.

Each Unit of *Pathways to Astronomy* is like a mini-lecture on a single topic or closely related set of ideas. The same material covered in other introductory astronomy texts is included, but it is broken up into smaller, self-contained parts. This gives greater flexibility in selecting topics than is possible with the wide-ranging chapter in a traditional text that covers the same material as four or five *Pathways* Units.

Even though the Units are written to be as independent as possible, they still flow naturally from one to the next or even in alternative orders—different *Pathways*—through the book. Professors can select Units to fit their course needs and cover them in the order they prefer. They can choose individual Units that will be explored in lecture while assigning other Units for self-study. Or they can cover all the Units in full depth in a content-rich course. With the short length of Units, students can more easily digest the material covered in an individual Unit before moving on to the next Unit. And because the questions and problems are focused on a single topic, it is much easier to determine mastery of each topic.

The Unit format also provides an opportunity to take some extra steps beyond the ordinary text. The authors have included some material of special interest that introduces topics most introductory texts do not offer—for example, Units on calendar systems and special relativity. More advanced material within a particular Unit topic is also organized toward the end of the Unit so that the essentials are covered first—also providing flexibility for assigning readings.

Pathways to Astronomy makes it easy to tailor readings and exercises so they fit best within a course's structure. It also provides opportunities to travel down some fascinating paths to enhance a course or to provide additional reading for advanced students.

NEW TO THE FOURTH EDITION

New Book Elements In our efforts to help students understand and interpret this wide-ranging material, we have added several new elements to *Pathways*:

Learning Objectives: At the start of each Unit, a new list of learning objectives describes the most important skills and abilities that readers should strive for in studying that Unit. These identify specific actions (such as describing, explaining, comparing, and calculating) that demonstrate a good mastery of the material.

Mathematical Insight

New boxes in the margins explore the mathematics of the text more deeply

Mathematical Insights: These marginal notes provide mathematical details to clarify the discussion in the text or expand beyond it, and they provide worked examples of mathematical problems.

New map format: We have converted almost every global map of the Earth, Moon, and planets to a Mollweide equal-area projection. Data are frequently released in a wide variety of projections, which introduce different distortions. Reprocessing the data into a single format facilitates comparisons between planets, and the equal-area projection avoids distortions that exaggerate sizes in polar regions.

These marginal notes identify objects related to the text that can be seen in the Looking Up images at the front of the book—or with a small telescope.

Looking Up Icons: These marginal notes point out objects that can be seen in the Looking Up pages at the front of the book, helping to relate the textual descriptions to objects visible in the night sky.

End-of-Unit questions: We have revised and added new questions to each Unit.

Highlights of Changes In addition to our own monitoring of new and interesting results in the field, many readers and reviewers offered excellent suggestions for updates and improvements to *Pathways*. In all, more than 150 figures were added, updated, or replaced throughout the book to improve clarity and to include some of the best new images available. The Unit topics remain the same as the third edition, although a few Units were more extensively revised as noted below. The following is a partial list of changes, including the most interesting updates and additions:

Unit 4 Foundations of Astronomy: Revised text and new figures for particles and forces, with brief discussion of the Higgs particle.

Unit 6 The Year: Revised discussion of seasons, and added figures of the Karnak temple in Egypt and path of Sun as seen from different latitudes.

Unit 8 Lunar Cycles: Revised eclipse text and figures and new eclipse-path map.

Unit 10 Geometry of the Earth, Moon, and Sun: New figures illustrating Earth's curvature and ancient ideas of parallax.

Unit 19 Tides: Revised figures to clarify tidal interactions.

Unit 22 The Electromagnetic Spectrum: New figure with a multiwavelength comparison of M31.

Unit 23 Thermal Radiation: New figures with graphs illustrating the Wien and Stefan-Boltzmann Laws.

Unit 32 The Earth's Atmosphere and Space: New figures and images of atmospheric refraction.

Unit 33 Amateur Astronomy: Updated images and discussion of amateur astronomical photography.

Unit 34 The Structure of the Solar System: Revised discussion and table of chemical compositions in Solar System.

Unit 35 The Origin of the Solar System: Added graph of radioactive decay.

Unit 36 Other Planetary Systems: Extensively revised with new figures to illustrate exoplanet detection methods, planet migration, density determinations, and recent *Kepler* findings.

Unit 37 The Earth as a Terrestrial Planet: New figures and expanded discussion in section 3 on tectonics.

Unit 38 Earth's Atmosphere and Hydrosphere: Revised figures and text on Coriolis effect (section 5).

Unit 39 Our Moon: Reorganized section order, beginning now with the Moon's formation. Many new figures and images from *LRO* and *Lunar Prospector*.

Unit 41 Venus: Added recent findings from *Venus Express*, and reprocessed imagery from *Magellan*.

Unit 42 Mars: Unit extensively updated and reorganized, adding several new and reprocessed images, for clearer development. Early results and images from *Curiosity* included.

Unit 43 Asteroids: New comparison images including Vesta results from *Dawn*.

Unit 47 Satellite Systems and Rings: New *Cassini* images of Enceladus and rings.

Unit 49 Comets: Unit reorganized for clearer development with several revised diagrams and new images from *Stardust* and *STEREO*.

Unit 50 Impacts on Earth: Added Images from the Chelyabinsk meteor explosion.

Unit 51 The Sun, Our Star: New diagram of *Voyagers* nearing interstellar boundary.

Unit 54 Surveying the Stars: New image and discussion of finding Sun's distance from the transit of Venus.

Unit 59 The H-R Diagram: Some details removed from H-R diagrams to make them easier to understand.

Unit 68 Neutron Stars: Updated discussion of recent observations of nearby neutron stars and implications for their internal structure.

Unit 69 Black Holes: New section on Hawking radiation.

Unit 70 Star Clusters: New images of clusters showing them to the same scale.

Unit 74 Mass and Motions in the Milky Way: Added *Fermi* observations of gamma-ray bubbles and updated observations of orbits around central black hole.

Unit 79 Dark Matter: Added discussion of recent possible WIMP detections.

Unit 80 Cosmology: New image and discussion of tentative detections of galaxies at redshift greater than 10.

Unit 84 Dark Energy and the Fate of the Universe: New diagrams and updated cosmological parameters with results from *Planck*.

Unit 85 Astrobiology: Updated with recent microfossil findings.

Unit 86 The Search for Life Elsewhere: Added early *Curiosity* findings on Mars.

FEATURES

Looking Up Illustrations It can be challenging to link introductory astronomy to the sky around us. The nine "Looking Up" full-page art pieces at the front of the book provide another pathway to astronomy, connecting what we actually see when "looking up" at the night sky with the more theoretical side of astronomy. Each illustration displays a large-scale photograph of one or more constellations in the night sky. Each also contains close-up photographs and illustrations of some of the most interesting telescopic objects with cross-references to the text. Details are also given regarding the objects' distances from Earth, along with three-dimensional illustrations of some of the stars or other objects within the field of view. The Looking Up Illustrations begin on page ii, to make them easier to find.

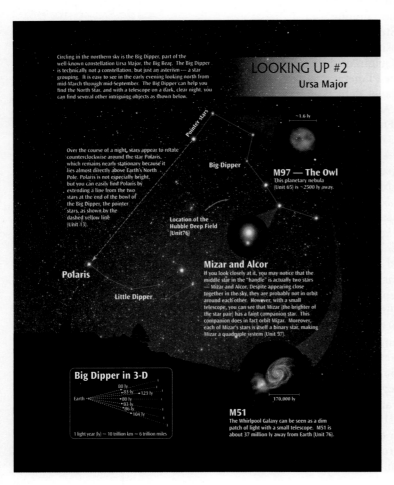

Star Chart A good star chart also helps link the study of astronomy to the night sky. *Pathways to Astronomy* offers a foldout star chart as well as seasonal star charts for Northern Hemisphere observers. These will help students to take that next step beyond the book—exploring the night sky.

Detailed Art *Pathways to Astronomy* has taken each illustration a step further than the norm. Many figures are annotated to describe the processes depicted within the illustration. Photos are often inserted next to the illustration for comparison so students can see the process in reality.

End-of-Unit Material The end-of-unit sections include hundreds of Key Points, Review Questions, Quantitative Problems, and Test Yourself questions to help students master the unit material. These elements allow students to apply what they have learned before moving on to another unit. The answers for all the Test Yourself questions are provided at the back of the text.

Clarification Point

Some widely held beliefs about astronomy are known to be incorrect.

Clarification Points Marginal notes call attention to common misunderstandings that many people new to astronomy may have. These are points of confusion that seem particularly difficult to overcome, so they deserve special attention.

Concept Questions Dozens of Concept Questions are scattered throughout the margins of the units. These questions are designed to invite students to think beyond the text, and to think about questions that have no easy answer. Many also make good discussion questions in class.

Mathematical Symbols To help students work with the math and avoid confusion, throughout the text we have chosen unique symbols for each physical quantity and used them consistently. Where tradition dictates that the same letter be used for different quantities (such as acceleration and semimajor axis), we have used distinctive fonts. A symbol glossary is included at the start of the glossary to help students keep track of the meaning of symbols in formulas.

Cosmic Periodic Table We have produced a periodic table geared toward astronomy. It is introduced in Unit 4 as part of the discussion of elements. This table is designed to convey information of interest to astronomers, such as relative abundances, condensation temperatures, mass excess (relative to iron-56), origins of elements, and radioactive lifetimes of unstable elements. The table has been revised with properties and names for some recently-measured elements.

Electronic Media Integration with the Text Icons for Interactives and Animations have been placed where additional understanding can be gained through an animation or interactive experience.

Interactives McGraw-Hill is proud to bring you an assortment of 23 outstanding Interactives like no other. Each Interactive is programmed in Flash for a stronger visual appeal. These Interactives offer a fresh, dynamic method to teach astronomy basics. Each Interactive allows users to manipulate parameters and gain a better understanding of topics such as blackbody radiation, the Bohr model, a "solar system builder," retrograde motion, cosmology, and the H-R diagram by watching the effects of these manipulations. Each Interactive includes an analysis tool (interactive model), a tutorial describing its function, content describing its principal themes, related exercises, and solutions to the exercises. Users can jump between these exercises and analysis tools with just a mouse click.

LEARNING RESOURCES

McGraw-Hill ConnectPlus®

Astronomy provides online presentation, assignment, and assessment solutions. It connects your students with the tools and resources they'll need to achieve success. With Connect Astronomy, you can deliver assignments, quizzes, and tests online. A robust set of questions and problems are presented and aligned with the textbook's learning goals. As an instructor, you can edit existing questions and author entirely new problems. Track individual student performance—by question, assignment, or in relation to the class overall—with detailed grade reports. Integrate grade reports easily with Learning Management Systems (LMS) such as WebCT and Blackboard—and much more. **ConnectPlus Astronomy** provides students with all the advantages of Connect Astronomy, plus 24/7 online access to an eBook. This media-rich version of the book is available through the McGraw-Hill Connect platform and allows seamless integration of text, media, and assessments. To learn more, visit **www.mcgrawhillconnect.com.**

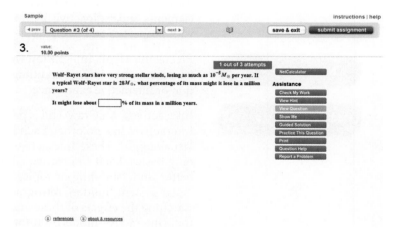

McGraw-Hill LearnSmart™

McGraw-Hill LearnSmart™ is available as a standalone product or as an integrated feature of McGraw-Hill Connect® Chemistry. It is an adaptive learning system designed to help students learn faster, study more efficiently, and retain more knowledge for greater success. LearnSmart assesses a student's knowledge

of course content through a series of adaptive questions. It pinpoints concepts the student does not understand and maps out a personalized study plan for success. This innovative study tool also has features that enable instructors to see exactly what students have accomplished and a built-in assessment tool for graded assignments. Visit the following site for a demonstration: **www.mhlearnsmart.com**

McGraw-Hill SmartBook™

Powered by the intelligent and adaptive LearnSmart engine, **SmartBook** is the first and only continuously adaptive reading experience available today. Distinguishing what students know from what they don't, and honing in on concepts they are most likely to forget, SmartBook personalizes content for each student. Reading is no longer a passive and linear experience but an engaging and dynamic one, where students are more likely to master and retain important concepts, coming to class better prepared.

SmartBook includes powerful reports that identify specific topics and learning objectives students need to study. These valuable reports also provide instructors insight into how students are progressing through textbook content and are useful for identifying class trends, focusing precious class time, providing personalized feedback to students, and tailoring assessment.

How does SmartBook work? Each SmartBook contains four components: Preview, Read, Practice, and Recharge. Starting with an initial preview of each chapter and key learning objectives, students read the material and are guided to topics over which they need the most practice based on their responses to a continuously adapting diagnostic. Read and practice continue until SmartBook directs students to recharge important material they are most likely to forget to ensure concept mastery and retention.

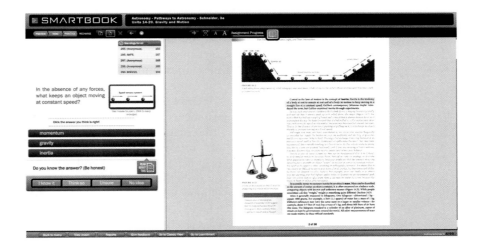

McGraw-Hill Create™

With **McGraw-Hill Create™,** you can easily rearrange chapters, combine material from other content sources, and quickly upload content you have written, for example, your course syllabus or teaching notes. Find the content you need in Create by searching through thousands of leading McGraw-Hill textbooks. Arrange your book to fit your teaching style. Create even allows you to personalize your book's appearance by selecting the cover and adding your name, school, and course information. Order a Create book and you'll receive a complimentary print review copy in 3–5 business days or a complimentary electronic review copy (eComp) via e-mail in minutes. Go to www.mcgrawhillcreate.com today and register to experience how McGraw-Hill Create empowers you to teach *your* students *your* way. **www.mcgrawhillcreate.com**

My Lectures—Tegrity®

McGraw-Hill Tegrity® records and distributes your class lecture with just a click of a button. Students can view anytime/anywhere via computer, iPod, or mobile device. It indexes as it records your PowerPoint® presentations and anything shown on your computer so students can use keywords to find exactly what they want to study. Tegrity is available as an integrated feature of McGraw-Hill Connect® Chemistry and as a standalone.

Classroom Performance System and Questions

The Classroom Performance System (CPS) brings interactivity into the classroom or lecture hall. CPS is a wireless response system that gives an instructor immediate feedback from every student in the class. Each CPS unit comes with up to 512 individual response pads and an appropriate number of corresponding receiver units. The wireless response pads are essentially remotes that are easy to use and engage students. The CPS system allows instructors to create their own questions or use the astronomy questions provided by McGraw-Hill.

For the Instructor

- **Test Bank** The electronic test bank offers a bank of questions that can be used for homework assignments or the preparation of exams. The test bank can be utilized to quickly create customized exams. It allows instructors to sort questions by format or level of difficulty; edit existing questions or add new ones; and scramble questions and answer keys for multiple versions of the same test.
- **McGraw-Hill Presentation Tools** Build instructional material wherever, whenever, and however you want! Accessed through *Pathways* Connect site, the Presentation Tools are an online digital library containing assets such as photos, artwork, animations, and other media types that can be used to create customized lectures, visually enhanced tests and quizzes, compelling course websites, or attractive printed support materials. Assets are copyrighted by McGraw-Hill Higher Education, but they can be used by instructors for classroom purposes. The visual resources in this collection include:
  - **Art** Full-color digital files of all illustrations in the book can be readily incorporated into lecture presentations, exams, or custom-made classroom materials. In addition, all files are preinserted into PowerPoint© slides for ease of lecture presentation.
  - **Photos** The photos collection contains digital files of photographs from the text, which can be reproduced for multiple classroom uses.
  - **Animations and Interactives** Numerous full-color animations and the astronomy interactives, illustrating important processes, are also provided. Harness the visual impact of concepts in motion by importing these files into classroom presentations or online course materials.

Also residing on your textbook's Connect website are:

- **PowerPoint Lecture Outlines** Ready-made presentations that combine art and lecture notes are provided for each unit of the text.
- **PowerPoint Slides** For instructors who prefer to create their lectures from scratch, all illustrations and photos are preinserted by chapter into blank PowerPoint slides.
- **Instructor's Manual** The Instructor's Manual is housed within the Connect site and can be accessed only by instructors. This manual includes hints for teaching with this text, additional thought and discussion questions, answers to end-of-unit material, and sample syllabi.

ACKNOWLEDGMENTS

Writing and revising a text such as *Pathways* is a collaboration with everyone who reads or uses it. We are deeply grateful to everyone who offered a suggestion, pointed out a mistake, or found a place where we might improve the content. Our sincere thanks to all the reviewers who have offered suggestions throughout the life of this book, especially those who offered their feedback in preparation of this fourth edition:

Lloyd Black *Rowan University*
Miles Blanton *Bowling Green State University*
Paul Butterworth *George Washington University*
Micol Christopher *Mt. San Antonio College*
James Cooney *University of Central Florida*
Robert Egler *North Carolina State University*
Paul Eskridge *Minnesota State University–Mankato*
Phillip Flower *Clemson University*
Steven Furlanetto *University of California–Los Angeles*
Janet Mclarty-Shroeder *Cerritos College*
Milan Mijic *California State University–Los Angeles*
Anatoly Miroshnichenko *University of North Carolina–Greensboro*
Jon Pedicino *College of the Redwoods*
Glenn Tiede *Bowling Green State University*
Jose Vazquez *New York University*

Special thanks to those who were instrumental in the preparation of LearnSmart and SmartBook for *Pathways to Astronomy* as well as to those who helped develop and enhance our online homework offerings in Connect.

Karen Castle, *Diablo Valley College*
Hugh H. Crowl *Bennington College*
Gregory Dolise, *Harrisburg Area Community College*
Beth Hufnagel *Anne Arundel Community College*
Patrick L. Koehn *Eastern Michigan University*
Susanna Lomant, *Georgia Perimeter College*
Thomas Pannuti, *Morehead State University*
Lara Arielle Phillips *University of Notre Dame*
Christopher Shope *Harrisburg Area Community College*
Michael D. Stage *Mount Holyoke College*
Christopher Taylor, *California State University, Sacramento*

Finally, the authors would like to thank the team at McGraw-Hill for all their assistance with updating *Pathways*, including Lora Neyens, April Southwood, Carrie Burger, Tara McDermott, Derek Elgin, Mary Reeg, and Pat Steele. Thank you to Louis Rubbo who revised and added new end-of-unit material. Thanks particularly to Beth Berry for a close reading of the revised manuscript and many corrections and suggestions for improvement.

Pathways to Astronomy

PART 1
UNIT 1

Our Planetary Neighborhood

1.1 The Earth
1.2 The Moon
1.3 The Planets
1.4 The Sun
1.5 The Solar System
1.6 The Astronomical Unit

Learning Objectives

Upon completing this Unit, you should be able to:
- List the kinds of objects that reside in our Solar System.
- Relate sizes of the Sun, planets, and other objects in the Solar System to the size of the Earth.
- Define the Astronomical Unit, and relate the distances between objects in the Solar System to their sizes.

Astronomy is the study of the universe: from the Earth itself to the most distant galaxy, from deciphering its nature at the beginning of time to predicting its eventual fate in the remote future. Within the vast space of the universe lie planets with dead volcanoes whose summits dwarf Mount Everest. There are stars a thousand times the size of the Sun—so large that in the Sun's place, they would swallow up the Earth. And there are galaxies—slowly whirling systems of billions of stars—so vast that the Earth is smaller by comparison to them than a single grain of sand is to the Earth itself. On this small planet within this immense cosmic landscape, life has evolved over billions of years, giving rise to creatures who seek to understand the nature of the universe.

Through astronomy we gain the ability to study places so remote that there is no possibility of our ever visiting them. We gain insights into alien environments unlike anything found on Earth. And our insights provide new perspectives on our home planet and about ourselves.

The vast variety of planets, their distances, and their sizes are almost unimaginable. We will explore the Earth and its neighboring planets in detail in Part III; but to gain some sense of scale, we begin in this Unit with a brief look at the Earth and the Earth's neighborhood—the Solar System. We proceed from the familiar to the less familiar, introducing some of the common terms astronomers use and examining how astronomers speak of the huge distances between the planets.

FIGURE 1.1
The planet Earth, our home, with blue oceans, white clouds, and multihued continents. This photo was taken by *Apollo 17* astronauts traveling to the Moon in 1972.

1.1 THE EARTH

We begin with the Earth, our home **planet** (Figure 1.1). This spinning ball of rock and metal, coated with a thin layer of gas and liquid, is huge by human standards, but it is one of the smaller bodies in the cosmic landscape. Nevertheless it is an appropriate place to start, because as the base from which we view the universe, it determines what we can see. We cannot travel to any but the nearest objects in our quest to understand the universe. Instead, we are like children who know their neighborhood well but for whom the larger world is still a mystery, known only from books and television.

But just as children use knowledge of their neighborhood to build their image of the world, so astronomers use their knowledge of Earth as a guide to distant worlds. The size of the Earth and features on it, for example, are useful reference points for appreciating the sizes of other objects. We will often refer to other planets in terms of their radii relative to the Earth's own radius of about 6371 kilometers

(3959 miles). Similarly, it is convenient to refer to other planets' masses in terms of the Earth's own mass, which is itself so large that it is difficult to imagine: 5,970,000,000,000,000,000,000,000 kilograms, or about 6 billion trillion tons.

Although few people realize it, we use the size of the Earth for defining the meter, the fundamental unit of the **metric system** (see Unit 3). The meter was originally defined to be one 10-millionth of the distance from the equator to the North Pole. The measurements made in the late 1700s were slightly off, but the Earth's circumference very nearly equals 40,000 kilometers (about 25,000 miles). When this system was introduced, it was a convenient way to measure the distances ships traveled on the Earth's oceans, and it offered a less arbitrary standard than the many other measurement systems used at the time.

The geological processes that occur on Earth provide another kind of measure—one for interpreting the processes that shape the other planets. When a volcano spews molten lava, it provides a hint that below the surface our planet is extremely hot. During the last century geologists discovered that this internal heat drives slow but powerful currents that shake our planet's crust, move continents, build mountains, and heave up volcanoes. We can carry over our understanding of such geological processes here on Earth to help us make hypotheses about the processes that create similar features on other planets. And when we discover different features on other worlds, we can use them to help us think about the Earth in new ways.

> The internal heat of planets comes from two main sources—heat left over from their formation and the decay of radioactive elements in their interior (Unit 35). Their ability to retain this heat depends mostly on their size.

1.2 THE MOON

Our nearest neighbor, the Moon, is a world profoundly different from the Earth. The Moon is our **satellite**, orbiting the Earth nearly 400,000 kilometers (about a quarter-million miles) away. A string stretched from the Earth to the Moon could wrap around the Earth almost 10 times. The Moon is much smaller, having a diameter only about one-quarter our planet's. The Moon also has symbolic significance for us—it marks the present limit of direct human exploration of space.

With the naked eye, the Moon appears to be a quiet, glowing orb (Figure 1.2A); but with a small telescope or binoculars, we can see that the Moon is somewhat like the Earth in having mountains and plains on its surface—yet utterly unlike the Earth as well (Figure 1.2B and C). Instead of white whirling clouds, green-covered hills, and blue oceans, we see an airless, pitted ball of rock. Instead of mountain ranges and volcanoes, the Moon's surface is peppered with circular craters blasted into the surface when bodies crashed into it. We see evidence of a history of steady pounding by objects ranging from the microscopic to the size of mountains, impacting at speeds more than 10 times faster than any rifle bullet. Some of the larger collisions carved out craters more than 100 kilometers (60 miles) in diameter, and innumerable smaller impacts pulverized surface rock to rubble and dust.

The Earth, so near to the Moon in space, must have suffered a similar pounding, yet looks utterly different. Why are these two worlds so different? Much of the explanation lies in their greatly different mass. The Moon's mass is only about 1/80th the Earth's, and its smaller bulk made the Moon less able to retain internal heat or an atmosphere after

FIGURE 1.2
(A) The Moon as we see it with unaided eyes. (B) The Moon through a small telescope. (C) Apollo 17 astronauts on the Moon's surface.

Concept Question 1

How do you imagine people's perceptions of the heavens might have changed after seeing the Moon through the first telescopes in the early 1600s?

its formation. With less internal heat, the Moon's interior is quiet. Heat-driven motion, so important in shaping and changing the Earth's surface over the eons, is nearly absent in the Moon. Without an atmosphere, the Moon is not protected from small impacting objects, which on Earth are vaporized through the heat of friction before reaching the ground. And without erosion caused by an atmosphere or modification of the surface caused by geological activity, the Moon's surface exhibits all of its old scars. Much of the record of Earth's past has been erased, but the Moon can help us reconstruct our planet's ancient history.

1.3 THE PLANETS

Beyond the Moon, orbiting the Sun as the Earth does, are seven other planets, sister bodies of Earth. To the unaided eye, the other planets are mere points of light whose positions shift slowly from night to night. But by observing them, first with Earth-based telescopes, then ultimately by remotely piloted spacecraft, we have learned that they are truly other worlds.

In order of increasing distance from the Sun, the eight planets are Mercury, Venus, Earth, Mars, Jupiter, Saturn, Uranus, and Neptune. These worlds have dramatically different appearances, as illustrated in Figure 1.3. The surfaces of the planets differ enormously from the landscape of Earth. Craters scar the airless surface of Mercury, while dense clouds of sulfuric acid droplets completely shroud Venus. Huge canyons, volcanoes, and deserts spread across the ruddy face of Mars.

The correct relative sizes of the planets are difficult to show in a single figure because the four inner planets are so much smaller than the four outer planets. All four outer planets have atmospheres so vast they could swallow the Earth whole. Jupiter has storm systems that are as big as the Earth. These giant planets all have ring systems, although Saturn's is the brightest by far. Uranus and Neptune are colored blue, not by water but by a deep layer of methane. Astronomers often indicate the sizes of the planets by comparing them to Earth's radius, which we abbreviate as $R_\oplus$ using the international symbol $\oplus$ for the Earth. The planets range in size from Jupiter, with a radius of 11 Earth radii (11 $R_\oplus$), down to Mercury, with a radius of slightly over one-third Earth's radius (0.38 $R_\oplus$).

Our two nearest neighboring planets, Venus and Mars, provide deeper insights into our own planet because they are so similar to, and yet so different from, the Earth. They are the closest to Earth in size, with radii of 0.95 and 0.53 $R_\oplus$, respectively. Their landscapes include features that resemble many seen on Earth—for example, they both have volcanoes and mountain ranges—so we conclude that geological processes like those that shaped the Earth must have occurred on both.

Despite their similarities in size and in distance from the Sun, Venus and Mars have atmospheres dramatically different from Earth's. On Venus we would be crushed and cooked by its intensely hot, dense atmosphere, whereas on Mars we would suffocate and freeze. And the other worlds astronomers have studied are even more alien. Every other planet and moon that we have found would also be hostile to human life. By studying the factors that make other planets so different from Earth, we may also begin to understand the potential consequences of our own impact on the Earth.

FIGURE 1.3

Images of the eight planets. The four inner planets are shown to the same scale in the upper panel, and the four much-larger outer planets are shown to the same scale in the lower panel, with the Earth for comparison.

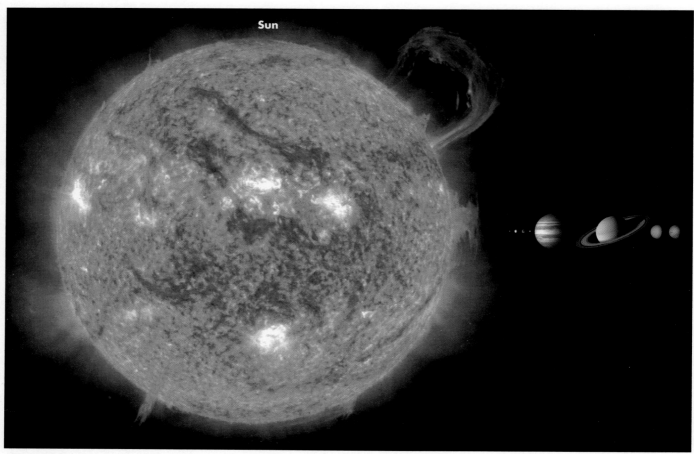

FIGURE 1.4
The Sun as viewed through a filter that allows its hot outer gases to be seen. The eight planets are shown to scale beside it for comparison. (The filter allows astronomers to see very hot helium gas.)

1.4 THE SUN

All of the planets of our Solar System are dwarfed by the Sun, whose immense gravity holds them in orbit. The Sun is a **star,** a huge ball of gas over 100 times the diameter of the Earth and over 300,000 times more massive. The difference in size between the Sun and the planets is illustrated in Figure 1.4.

The Sun, of course, differs from the planets in more than just size: It generates energy in its core by nuclear reactions that convert hydrogen into helium. The Sun is producing energy at a furious rate—more in just one second than all of the bombs and the energy ever generated on Earth during human history. The energy from the intensely hot core flows to the Sun's surface, which is more than a thousand times cooler than the core but is still hot enough to vaporize iron. From there the Sun's energy streams into space, illuminating and warming the planets' surfaces.

The energy the Sun can produce is enormous but nonetheless limited. The Sun's stream of energy has already lasted more than 4 billion years—enough time for life to form and evolve on Earth and for intelligent creatures who can marvel at these things to have come into being. However, much evidence indicates that the Sun will run out of fuel eventually, in about another 5 or 6 billion years. It will then fade away like a cooling ember. Thus, astronomy helps us to look deep into the past and far into the future as we consider phenomena of an enormous range in sizes and distances.

> **Concept Question 2**
>
> The Sun and all the planets are hotter in their interiors than on their surfaces. What different sources of energy can you think of that might produce their internal heat?

1.5 THE SOLAR SYSTEM

The Sun and the bodies orbiting it form the **Solar System,** bound together by the enormous gravity of the Sun. In addition to the eight planets, the Solar System is filled with a vast number of smaller bodies—satellites (moons) orbiting the planets, and asteroids and comets orbiting the Sun. However, the mass of all of the planets and other objects in the Solar System does not add up to even 1% of the Sun's mass.

The paths that the planets follow around the Sun (Figure 1.5) form a set of huge, approximately circular rings, one inside the other, with the Sun at their center. The orbits of the planets lie in nearly the same plane, so that the whole system is disk-shaped. The outermost planet—Neptune—lies about 4.5 billion kilometers (about 2.8 billion miles) from the Sun. The distances between planets in the outer Solar System are so much larger than in the inner Solar System that we have to plot these two regions separately to see them clearly.

Beyond Neptune's orbit are innumerable small objects orbiting in the "Kuiper belt," much as beyond Mars there are vast numbers of small bodies orbiting in the asteroid belt. The orbits of these objects are generally less circular and more tilted than the planets' orbits. This is especially true of comets (Unit 49) whose orbits carry them between the inner and outer parts of the Solar System.

Pluto is no longer classified as one of the major planets, but instead is classified as a "dwarf planet" or "plutoid" (Unit 48). Under this new definition, the asteroid Ceres (Unit 43) has also been reclassified as a dwarf planet. Eris, discovered in 2005, is the largest plutoid yet found, about 27% more massive than Pluto. The orbits of these three bodies are shown in Figure 1.5, and there are dozens more candidate dwarf planets orbiting in the Kuiper belt.

FIGURE 1.5
Sketch of the positions and orbits of the planets and dwarf planets in our Solar System on March 20, 2011. The orbit of Halley's comet along with another typical comet orbit are also shown. The approximate location of small bodies in the asteroid belt and Kuiper belt are indicated. To show the orbits to scale, the inner (left) and outer (right) Solar System are shown separately.

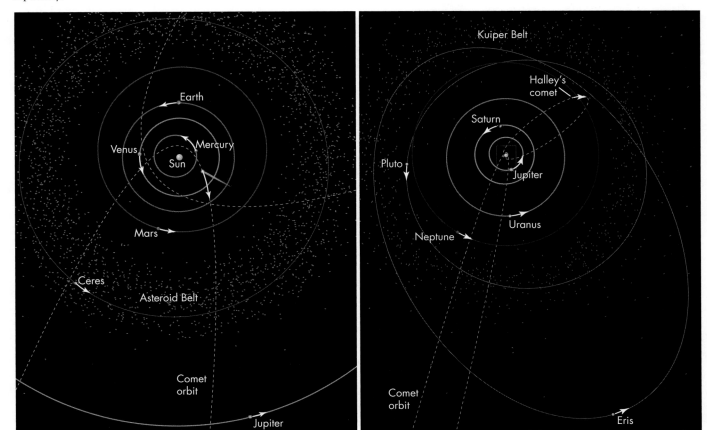

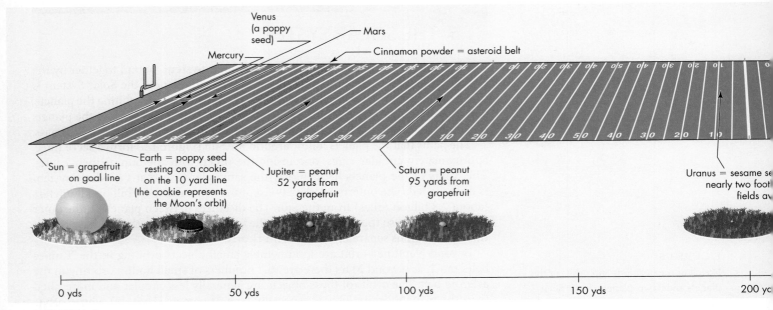

FIGURE 1.6
A scale model of the Solar System with the Sun the size of a grapefruit. At this scale all of the planets that are visible without a telescope fit within a single football field. Uranus and Neptune each require another whole field, while Pluto, Eris, and many other icy bodies are even farther away.

1.6 THE ASTRONOMICAL UNIT

The distances between objects in the Solar System are so vast that they are almost incomprehensible. One way to picture the distances involved is to construct a scale model of the Solar System. For example, if the Sun is represented by a grapefruit, Earth would be the size of a grain of sand or a poppy seed, orbiting about 9 meters (about 30 feet) away. This scale is illustrated in Figure 1.6, where the sizes and distances of the planets are shown on a string of U.S. football fields. In this scale model, the dwarf planets Pluto and Eris would be specks of dust orbiting more than 0.35 kilometer (0.22 mile) away.

Whenever possible, we use a **unit**—a quantity used for measurement—appropriate to the size of what we seek to measure. In earlier times people used units that were quite literally at hand, such as finger widths or the spread of a hand to measure a piece of cloth, and paces to measure the size of a field. In the same tradition, although on a much different scale, astronomers have adopted the Earth's distance from the Sun as a good unit for measuring the size of the Solar System. Astronomers call this distance the **astronomical unit,** abbreviated AU.

For several centuries astronomers did not know exactly how big an astronomical unit was. Even though they did not know how many kilometers were in an AU, they discovered ways to find out the distances to other planets in AU. For example, they could show that Mercury was 0.4 AU from the Sun, while Neptune is about 30 AU from the Sun.

Modern measurements show that the astronomical unit is about 150 million kilometers (about 93 million miles). Actually, as the Earth orbits the Sun, its distance varies from 147.1 to 152.1 million kilometers (91.4 to 94.5 million miles), just slightly different from a circle with a radius of 1 astronomical unit, as shown in figure 1.7. The modern definition of the astronomical unit is defined in terms of the average of these extremes, and has been very precisely measured as 149,597,870.7 kilometers.

The Solar System remains the limit to our exploration of the universe with spacecraft. Most of our probes have explored only the inner part of the Solar System, although we have sent a few to explore the outer planets. The *Voyager 1* and *2*

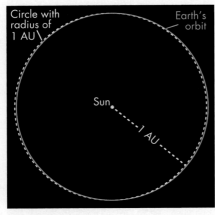

FIGURE 1.7
The Earth's orbit (shown in blue) is nearly a circle of radius 1 astronomical unit (dotted white circle), however the distance from the Sun is about 3% larger at its maximum as at its minimum.

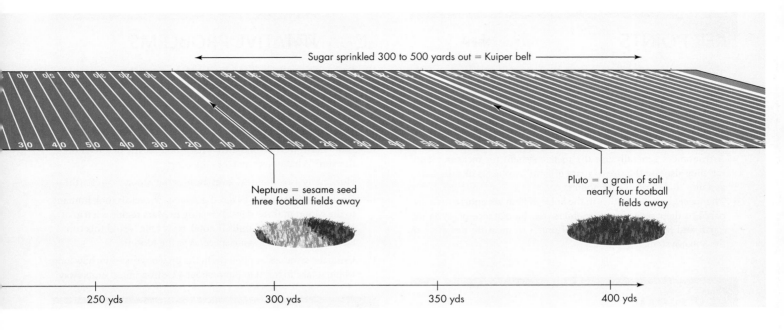

Concept Question 3

What do you suppose are the effects of the varying distance of the Earth from the Sun? How could we observe that?

spacecrafts launched in 1977 are our most distant explorers. In 1990 *Voyager 1*, having passed Pluto's orbit at more than 40 AU, looked back and made the image of the Solar System shown in Figure 1.8. From this distance the planets were just points of light and the Sun a bright star. The *Voyager 1* spacecraft is now more than three times as far from the Sun as Neptune, traveling about 3 AU farther away every year.

Even at their high speeds leaving the Solar System, the two *Voyager* spacecraft would take about 60,000 years to reach the nearest star. But exploration by spacecraft does not represent the limit of our vision. With telescopes and astrophysics, our view extends far beyond the Solar System to reveal that there are planets orbiting other stars, and there are billions of other stars with much to teach us about our own star.

FIGURE 1.8
(A) This view of the Solar System is based on a set of images made by the *Voyager 1* spacecraft when it was about 40 AU from the Sun and about 20 AU above Neptune's orbit. The images of the planets (mere dots because of their immense distance) and the Sun have been made a little bigger and brighter in this view for clarity. Mercury is lost in the Sun's glare; Mars happened to lie nearly in front of the Sun at the time the image was made, so it too is invisible. (B) A sketch of the orbits of the planets, showing approximately where each object was located when the image was made (February 1990). Notice the immense, empty spaces between the planets and the nearly concentric rings of the planets' orbits, which make a nearly flat system.

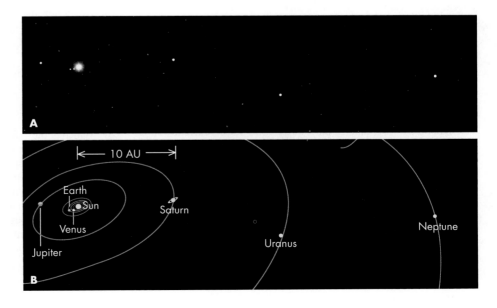

KEY POINTS

- The Earth is one planet in the Solar System, which contains the Sun, eight planets, and a huge number of smaller objects.
- The planets vary tremendously in size and appearance, some much larger than the Earth and some smaller.
- The Sun is the largest body by far in the Solar System, and the energy it radiates warms the surfaces of the planets.
- Astronomers generally use the metric system for measurements, but they also use the size or mass of familiar objects like the Earth as a unit to describe other bodies.
- The spaces between bodies in the Solar System are enormous compared to their sizes, so astronomers use the distance between the Earth and Sun (the "astronomical unit") to measure distances in the Solar System.

KEY TERMS

astronomical unit (AU), 6
metric system, 2
planet, 1
satellite, 2
Solar System, 5
star, 4
unit, 6

CONCEPT QUESTIONS

Concept Questions on the following topics are located in the margins. They invite thinking and discussion beyond the text.

1. The Moon and perception of the heavens. (p. 3)
2. Sources of internal heat. (p. 4)
3. Effect of distance from Sun on Earth. (p. 7)

REVIEW QUESTIONS

4. What are the eight planets in order of distance from the Sun?
5. What is a dwarf planet? Can you name two objects currently in this category?
6. What planets are most similar to Earth?
7. About how many times bigger in radius is the Sun than the Earth? How many times bigger in mass?
8. Besides the Sun and planets, what other kinds of objects are members of the Solar System?
9. What is an astronomical unit?

QUANTITATIVE PROBLEMS

10. If you use a volleyball as a model of the Earth, how big would 1 kilometer be on it? The volleyball has a circumference of 68 cm.
11. What would be the circumference and diameter (circumference = $\pi \times$ diameter) of a ball that would represent the Moon if the Earth were a volleyball? What kind of ball or object matches this size?
12. If the Earth were a volleyball, what would be the diameter of the Sun? What object matches this size?
13. How many astronomical units away is the Moon from Earth?
14. During the 1960s and 1970s, the *Apollo* spacecraft took humans to the Moon in three days. Traveling to Mars requires a trip of about 2 astronomical units in total. How long would this trip take, traveling at the same speed as to the Moon?
15. Using the same assumptions as in the previous question, how long would it take to travel to Pluto, about 40 astronomical units away?

TEST YOURSELF

16. Which of the following lists gives the sizes of the objects from smallest to largest?
    a. Moon, Earth, Pluto, Mars, Jupiter
    b. Pluto, Mars, Moon, Earth, Jupiter
    c. Jupiter, Pluto, Mars, Moon, Earth
    d. Moon, Mars, Jupiter, Earth, Pluto
    e. Pluto, Moon, Mars, Earth, Jupiter
17. Which of the following lists gives the distances of the objects from the Sun from smallest to largest?
    a. Ceres, Venus, Jupiter, Neptune, Eris
    b. Venus, Eris, Ceres, Neptune, Jupiter
    c. Eris, Ceres, Venus, Neptune, Jupiter
    d. Venus, Ceres, Jupiter, Neptune, Eris
    e. Ceres, Eris, Venus, Jupiter, Neptune
18. About how many times larger is the Earth's diameter compared to the Moon's?
    a. 2 times
    b. 4 times
    c. 10 times
    d. 25 times
    e. 100 times
19. About how many times larger is the Sun's diameter compared to Earth's?
    a. 2 times
    b. 4 times
    c. 10 times
    d. 25 times
    e. 100 times
20. Which of the following properties is primarily responsible for the difference in the surface appearances for the Moon and Earth?
    a. Over the history of the Solar System, the Moon has been struck more often than Earth.
    b. The Moon's mass is much smaller than Earth's.
    c. Earth has an abundant amount of liquid water while the Moon does not.
    d. Living organisms are found throughout the surface of Earth while none have been found on the Moon.

PART 1

UNIT 2

Beyond the Solar System

2.1 Stellar Evolution
2.2 The Light-Year
2.3 The Milky Way Galaxy
2.4 Galaxy Clusters and Beyond
2.5 The Still-Unknown Universe

Learning Objectives

Upon completing this Unit, you should be able to:
- List the range of masses of stars and describe the basic steps in their evolution.
- Describe the overall structure and content of the Milky Way.
- Explain how the light-year is calculated, and list its approximate size.
- Compare distances within and between galaxies, and describe larger structures.
- Describe the Big Bang and the "dark" components of the universe.

On a clear night you can see several thousand stars with your unaided eyes. If you are far from city lights, you may be able to see a pale band of light known as the Milky Way stretching across the sky. Astronomers use telescopes to probe much deeper. They have learned that the Sun is itself a star, just one among hundreds of billions of stars in the Milky Way. The Milky Way is in turn just one among hundreds of billions of galaxies in the known universe.

Some stars are similar to the Sun, and observations of them give us insights about the Sun and Solar System (Unit 1). Other stars are drastically different, and astronomers deduce that in some cases this is because the stars are in different stages of their lifetimes. We give a brief preview of a few aspects of stars in this Unit, returning for a deeper examination in Part IV.

Larger objects, such as galaxies, are held together by the force of gravity (Unit 4). Gravity reaches across the empty space between objects and can link them together into gravitationally bound systems. For example, gravity is what holds the planets in orbit around the Sun to make the Solar System. Gravity is also what pulls the hundreds of billions of stars within the Milky Way into orbit around the center of our Galaxy. To provide a context for our studies, we give an overview of galaxies and even larger gravitational systems in this Unit; we examine them in greater detail in Part V of the book.

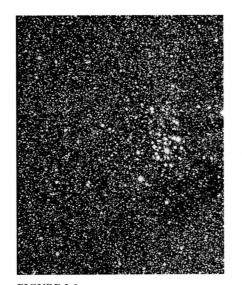

FIGURE 2.1
A deep image of a small portion of the Milky Way made with a telescope. The image reveals stars too faint to be seen individually with your eyes alone. Slight differences in the colors of stars are caused by the stars' temperatures.

2.1 STELLAR EVOLUTION

The night sky is filled with myriad stars, as seen in Figure 2.1. Some stars are much like the Sun, but others are thousands of times larger or smaller. The brightest stars produce over a billion times more light than the dimmest stars. Some stars are much hotter than the Sun and shine a dazzling blue-white, whereas others are cooler and glow a pale orange or red.

The Sun is a fairly typical star and its properties provide a convenient comparison for understanding other stars. Astronomers often describe stars' masses as ranging from 0.1 to 100 "solar masses," or 0.1 to 100 $M_\odot$ (the $\odot$ symbol is the international symbol for the Sun). The mass of a star appears to be the most critical factor in determining its difference from other stars. The stronger gravity of a more massive star drives up the rate of nuclear fusion dramatically, while smaller stars consume their fuel at a more leisurely pace.

In Part IV of this book we will see how astronomers have pieced together evidence that allows us to understand how stars are born, how they change as they age, and the dire fates they face when they run out of fuel. These changes, called **stellar evolution** by astronomers, are driven by the inexorable pull of gravity. Discovering the story of stars' lives has been one of the great triumphs of astrophysics and the scientific method during the last few centuries.

Deep images like that in Figure 2.2 reveal that stars intermingle with immense clouds of gas and dust. These are the sites of stellar birth. Within cold, dark clouds, gravity may draw the gas into dense clumps that collapse to form new stars, heating the gas around them until it glows. Stars eventually burn themselves out, but they do not disappear quietly. Stars like the Sun last billions of years before their final phases, when they tear themselves apart before fading away. The most massive stars die after just a few million years in titanic explosions, spraying matter outward to mix with the vast clouds of gas lying between the stars. This matter from exploded stars is ultimately recycled into new stars.

Making sense of the story of how stars are born and die has led to a greatly expanded vision of the kinds of objects that exist in the universe. We now know of a wide variety of objects that are beyond anything imagined a century ago. There are huge numbers of "failed" stars—objects that weren't quite massive enough to start the fusion of hydrogen to helium of the stars that shine in the night sky. And when stars die, they leave behind bizarre corpses—huge amounts of matter compacted into tiny objects such as white dwarfs, neutron stars, and black holes.

FIGURE 2.2
Hubble Space Telescope image of a cloud of gas and dust in the Milky Way. On the scale of this picture, the Solar System out to Neptune is about 1000 times smaller than the period ending this sentence.

The life stories of other stars tell us that when the Sun runs out of hydrogen fuel, it will undergo drastic changes as it restructures itself to use helium fuel instead. We can predict that the Sun will go through a phase in which it will expand and nearly swallow up the Earth. The enormous energy output during this phase will melt our planet entirely. The Sun will end this luminous phase by blowing away its own atmosphere and then slowly fading like a cooling ember after a campfire has burned out.

2.2 THE LIGHT-YEAR

While the astronomical unit works well for describing distances in our Solar System, distances between stars are so immense that the AU is an inappropriately small unit. The second nearest star (after the Sun!) is hundreds of thousands of AU away. A convenient way of describing such distances is the light-year.

Measuring a distance in terms of a unit of time may at first sound peculiar, but we do it all the time in everyday life. For example, we say that our town is a two-hour drive from the city, or our dorm is a five-minute walk from the library. In making such statements, we imply that we are traveling at a standard speed: freeway driving speed or a walking pace.

Astronomers have a superb speed standard: the speed of light in empty space. This is a constant of nature equal to 299,792,458 meters per second. We usually round this off to 300,000 kilometers per second (about 186,000 miles per second).

Mathematical Insight

To calculate the length of a light-year from light's speed in kilometers per second, you need to first find the number of seconds in a year. You can do this by multiplying the number of days in a year times the number of hours in a day times the number of minutes in an hour times the number of seconds in a minute.

Moving at this constant, universal speed, light in one year travels a distance defined to be 1 **light-year,** abbreviated as 1 ly. This works out to be about 10 trillion kilometers (6 trillion miles). As an example of the use of light-years, the star nearest our Sun is 4.2 light-years away. Although we achieve a major convenience in adopting such a huge distance for our scale unit, we should not lose sight of how truly immense such distances are. For example, if we were to count off the kilometers in a light-year, one every second, it would take us about 300,000 years!

We can use the light-year for setting the scale of the Milky Way Galaxy. In light-years, the visible disk is more than 100,000 light-years across, with the Sun orbiting roughly 25,000 light-years from the center. In the Sun's vicinity, stars are separated typically by a few light-years, but they are much more crowded toward the center and gradually thin out in the outer regions, so that there is no precisely defined outer edge. Throughout the disk of the Milky Way are scattered gas clouds with sizes of up to hundreds of light-years across. Some of these clouds contain more than a million times the Sun's mass, but they are so diffuse that their overall density is less than a billionth-trillionth of the density of the air we breathe.

2.3 THE MILKY WAY GALAXY

Concept Question 1

In what ways are the Solar System and Milky Way similar and different?

Our Sun and the stars we see at night are part of an immense system of stars called the Milky Way Galaxy. The **Milky Way Galaxy** is a collection of several hundred billion stars, along with billions of solar masses of gas and dust, gathered into a flattened disklike shape somewhat like the shape of the Solar System (Figure 2.3), but with a diameter roughly 100 million times larger. The Sun and other stars orbit around the center of the Milky Way, but our Galaxy is so vast that it takes the Sun several hundred million years to complete one trip around this immense disk. The center of the Milky Way looks bright because stars are more crowded together than they are out at the Sun's orbit. Nevertheless, the stars are still separated by huge distances and almost never collide.

In this huge swarm of stars and clouds, the Solar System is all but lost—like a single grain of sand on a vast beach—forcing us again to grapple with the problem of scale. Stars are almost unimaginably remote: even the nearest one to the Sun is about 40 trillion kilometers (25 trillion miles) away—about 10,000 times farther than Neptune. Such distances are so immense that analogy is often the only way to grasp them. For example, if we think of the Sun as a pinhead, the nearest star would be another pinhead about 60 kilometers (35 miles) away, and the space between them would be nearly empty. On this scale, the Sun is to the size of the Milky Way as a pinhead is to the size of the Sun itself.

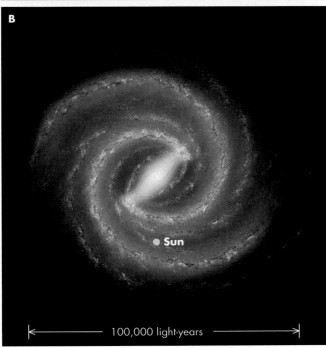

FIGURE 2.3

The Milky Way Galaxy. (A) The view from our position within the disk of the Galaxy made by plotting stars in the 2MASS star catalog. The brown-colored spots running horizontally across the image are caused by dust clouds that partially block the light from background stars. (B) The approximate structure of the Milky Way if it were seen from above, as mapped out by the Spitzer Space Telescope.

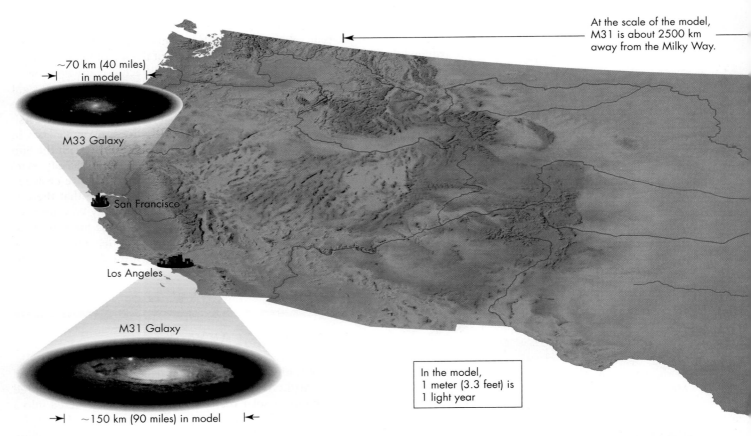

FIGURE 2.4
An approximate scale model of the Local Group in which 1 meter represents 1 light-year. At this scale the three largest galaxies in the Local Group are roughly as far from each other as Los Angeles, Chicago, and San Francisco, cities that, with their surrounding suburbs, are roughly the correct relative size of these galaxies. The entire Solar System is the size of a pinhead on this scale.

LOOKING UP

M31 is visible to the naked eye in the constellation Andromeda. See Looking Up #3 at the front of the book.

2.4 GALAXY CLUSTERS AND BEYOND

Having gained some sense of scale for the Milky Way, we now resume our exploration of the cosmic landscape, pushing out to the realm of other **galaxies** like the Milky Way, each an enormous island of stars, gas, and other matter. Here we find that just as stars are gravitationally bound into star clusters and galaxies, so galaxies are themselves bound into **galaxy groups** and **galaxy clusters**.

The Milky Way is part of a collection of galaxies called the **Local Group**. It is "local," of course, because it is the one we inhabit. The Local Group is a relatively small concentration of galaxies, containing only about four dozen galaxies, but it is still about 3 million light-years in diameter. The Milky Way is the second largest galaxy in the Local Group after the Andromeda Galaxy, also known as M31, the catalog number of a galaxy about 2.5 million light-years away from us.

To gain some perspective on the size of the Milky Way and Local Group, imagine that we built a model in which a light-year was scaled down to just 1 meter. At this scale Neptune's orbit would fit inside the head of a pin, and the Sun would be the size of a virus. The Milky Way Galaxy would be the size of a large city—about 100 kilometers (60 miles) across, while M31 would be another city about 2500 kilometers away (Figure 2.4). Given the enormous scale of this model, light crawls along about as fast as a plant might grow—just a meter per year; so that even at light speed, a trip from the Milky Way to M31 would take 2.5 million years!

The Milky Way is currently moving at about 90 kilometers per second (about 300,000 kilometers per hour, or 200,000 miles per hour) toward M31. Originally these two galaxies were moving away from each other, but the gravitational pull between them caused them to slow down, stop, and ultimately fall back toward

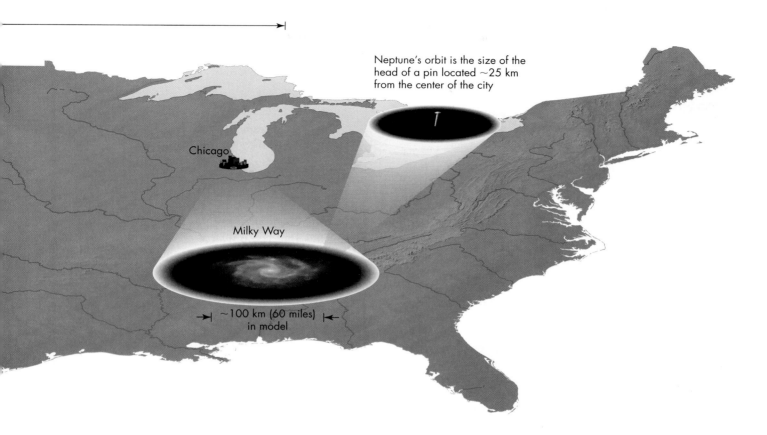

each other. In about 5 billion years the Milky Way and M31 will crash into each other, and the stars and gas in the Milky Way's disk will be flung into wild new orbits, probably sending our Solar System flying off into intergalactic space.

We have to adjust the scale of our thinking again to imagine even larger scales. The Local Group is one of dozens of galaxy groups surrounding a much larger assemblage of galaxies called the **Virgo Cluster,** like suburbs surrounding a major city. The Virgo Cluster is centered about 50 million light-years away and itself contains thousands of galaxies. The central region of the Virgo Cluster is shown in Figure 2.5.

The term *Virgo Supercluster* has sometimes been used to describe the enormous "metropolitan area" of the Virgo Cluster with its surrounding galaxy groups, but this collection of groups and clusters is just a modest example of an even larger scale of clustering. The term **supercluster** is now generally reserved for collections of many galaxy clusters (and their associated surrounding regions), gravitationally bound to one another in structures that span hundreds of millions of light-years.

The mind boggles at these enormous gravitational structures, and perhaps you have begun to wonder whether this hierarchy of structures extends ever upward. But structures of such vast size are about the largest objects we can see before we take the final jump in scale to the **universe** itself. Although for centuries our knowledge of the visible universe was confined by the limits of our telescopes and instruments, today we are reaching a fundamental limit: We can see only as far away as light has had time to travel in the age of the universe.

FIGURE 2.5
Photograph of the central region of the Virgo Cluster. The Milky Way and the entire Local Group have been slowed in their motion away from this galaxy cluster by its strong gravitational pull.

Concept Question 2

Suppose we extended the scale model in Figure 2.4. How far away would the Virgo Cluster be? How large would the observable universe (out to 13.8 billion light-years) be?

TABLE 2.1 Scales of the Universe

Object	Approximate Radius
Earth	6400 km (4000 miles) = $R_\oplus$
Moon's orbit	380,000 km ≈ 60 × $R_\oplus$
Sun	700,000 km = $R_\odot$ ≈ 109 × $R_\oplus$
Earth's orbit	150 million km = 1 AU ≈ 214 × $R_\odot$
Solar System to Neptune	4.5 billion km ≈ 30 AU
Nearest star	270,000 AU ≈ 4.2 light-years (ly)
Milky Way Galaxy	50,000 ly
Local Group	1.5 million ly
Local Supercluster	50 million ly
Visible Universe	13.8 billion ly

The best evidence today indicates that the universe is 13.8 billion years old, and therefore we cannot see any farther than light can have traveled in that time. The largest superclusters we see span nearly 1% of the visible universe, but as best we can determine the universe is relatively uniform over larger spans than this.

Within the 13.8 billion light-years we can see, there is no hint of an edge or change in the nature of the universe. This suggests that the universe must be far larger than what we can see. Current ideas of how the universe began predict that the universe is immensely larger, perhaps extending limitlessly. Regardless of our uncertainty about the known universe's overall size, we can observe that it has a well-ordered hierarchy of smaller structures. Small objects are clustered into larger systems, which are themselves clustered: satellites around planets, planets around stars, stars in galaxies, galaxies in clusters, clusters in superclusters, as shown in Table 2.1 and illustrated in Figure 2.6. This orderly structure originated as a result of the pull of gravity and the ways in which matter interacts on different scales.

2.5 THE STILL-UNKNOWN UNIVERSE

Over the last century, astronomers have uncovered evidence that the entire universe behaves in ways that were utterly unexpected a hundred years ago. Measurements show that galaxies are flying away from each other as a result of an all-encompassing explosion, called the **Big Bang.** Albert Einstein's general theory of relativity showed that this is the wrong way to interpret what we see, though—space itself is expanding, carrying the galaxies along with it. Einstein's theories (Units 26 and 27), now cornerstones of modern physics, have forced us to reevaluate our most basic notions of space and time, and we are discovering that there is much more to the universe than we can directly see through our telescopes.

Far more matter is found to fill the universe than what we can see in planets, stars, gas clouds, galaxies, and so on. Even after we correct for the amount of matter that is hard to detect, there is strong evidence that there is about 10 times more utterly invisible **dark matter** that is made of strange substances that *cannot* form stars or planets. Yet this matter appears to be the main component of every system the size of a galaxy or larger. The evidence that such matter exists comes from a variety of sources. The motion of stars in the outer parts of galaxies is so fast that they would fly off into intergalactic space unless much more mass is present than we can see. The same has been found true for galaxies in galaxy clusters. That is, the gravitational force that would be exerted by the observable stars and galaxies is far

Concept Question 3

The idea of dark matter may seem quite unusual, but can you think of any situations in which you have been able to find evidence that something was present without actually seeing it? Were you later proved correct?

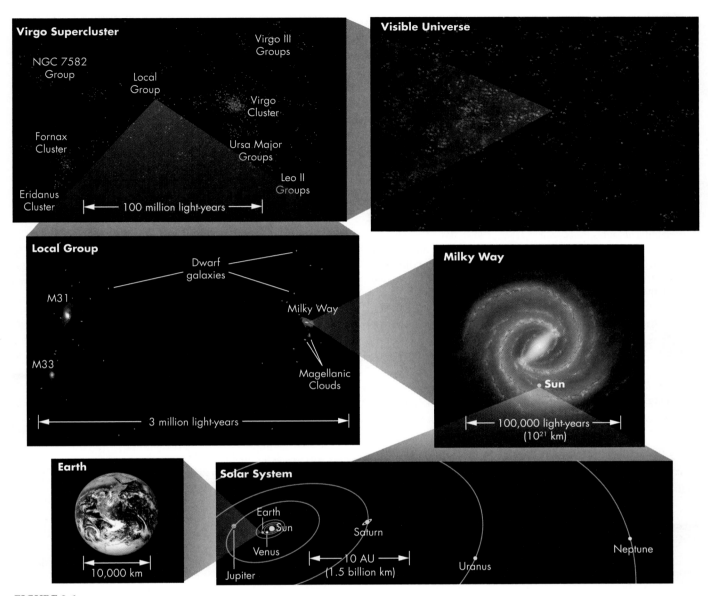

FIGURE 2.6
The Earth is but one of eight planets orbiting our star, the Sun. The Sun is but one of hundreds of billions of stars in our Galaxy, the Milky Way. The Milky Way is the second largest among dozens of galaxies in our "Local Group." The Local Group is one of the smaller "clusters" of galaxies that make up the "Virgo Supercluster." The universe is filled with millions of other superclusters stretching to the limits of our vision.

too small to hold the systems together. This is the subject of intense ongoing investigations, and it may be that particle physicists, exploring the submicroscopic world at scales far smaller than atoms, will uncover the nature of this strange kind of matter.

Over the last decade, astronomers have also accumulated evidence for an even stranger substance that has been dubbed **dark energy.** This is an energy that pervades every corner of space—even the emptiest vacuum. Einstein predicted the possibility of such an all-pervasive energy, but based on the evidence of his day, he abandoned the idea. New evidence suggests that not only does dark energy exist, it "outweighs" all of the visible and dark matter in the universe. If the latest measurements are correct, dark energy will drive the expansion of the universe faster and faster, perhaps in the extremely remote future causing space to expand so rapidly that it will tear apart the visible universe!

KEY POINTS

- The Sun is a fairly typical star, and we can learn about its past and future by studying other stars.
- Stars have a range of properties, some much brighter or dimmer than the Sun, some much more or less massive than the Sun.
- The Sun is one among the hundreds of billions of stars that make up the Milky Way Galaxy.
- Other galaxies are seen out to the remotest visible regions, often found in "clusters" or "superclusters."
- Astronomers often use the "light-year," the distance light travels in a year, to describe interstellar and intergalactic distances.
- The universe began with a "Big Bang" in which space began expanding at high speeds.
- Structures such as galaxies and clusters are held together by gravity, but there is evidence of much more matter in these structures than can be directly seen, which astronomers call "dark matter."

KEY TERMS

Big Bang, 14
dark energy, 15
dark matter, 14
galaxy, 12
galaxy cluster, 12
galaxy group, 12
light-year, 11

Local Group, 12
Milky Way Galaxy, 11
stellar evolution, 10
supercluster, 13
universe, 13
Virgo Cluster, 13

CONCEPT QUESTIONS

Concept Questions on the following topics are located in the margins. They invite thinking and discussion beyond the text.

1. Comparing the Solar System and Milky Way. (p. 11)
2. Making a scale model of the universe. (p. 14)
3. Detecting what you cannot see. (p. 14)

REVIEW QUESTIONS

4. What do astronomers mean by stellar "evolution"?
5. What is a galaxy?
6. Roughly how big across is the Milky Way Galaxy?
7. How is a light-year defined?
8. To what systems (in increasing order of size) does the Earth belong?
9. What do the terms *dark matter* and *dark energy* refer to?

QUANTITATIVE PROBLEMS

10. If the Milky Way were the size of a nickel (about 2 centimeters in diameter): (a) How big would the Local Group be? (b) How big would the Local Supercluster be? (c) How big would the visible universe be? (The data in Table 2.1 may help you here.)
11. If we detected radio signals of intelligent origin from the Andromeda Galaxy (M31) and immediately responded with a radio message of our own, how long would we have to wait for a reply?
12. How long does it take light to travel from the Sun to Earth?
13. If the Milky Way is moving away from the Virgo Cluster at 1000 kilometers per second, how long does it take for the distance between them to increase by 1 light-year?
14. The Milky Way is moving toward the larger galaxy M31 at about 100 kilometers per second. M31 is about 2.5 million light-years away. How long will it take before the Milky Way collides with M31 if it continues at this speed?
15. Some distant galaxies seen by the Hubble Space Telescope are 12 billion light-years away. Using the scale in Figure 2.4 (where the Milky Way is about the size of a large city), how far would that be in the model?

TEST YOURSELF

16. Which of the following is the best analogy? The Earth is to the Solar System as
    a. the Sun is to the Local Group.
    b. the Local Group is to the visible universe.
    c. the Sun is to the Milky Way.
    d. the Local Group is to the Milky Way.
    e. the Virgo Cluster is to the Local Group.
17. Which of the following astronomical systems is/are held together by gravity?
    a. The Sun   c. The Milky Way   e. All of them
    b. The Solar System   d. The Local Group
18. The light-year is a unit of
    a. time.   b. distance.   c. speed.   d. age.   e. All of these.
19. Rank the following systems from smallest to largest in size.
    a. Milky Way, Local Group, Virgo Cluster, Solar System
    b. Solar System, Local Group, Milky Way, Virgo Cluster
    c. Virgo Cluster, Local Group, Milky Way, Solar System
    d. Solar System, Milky Way, Local Group, Virgo Cluster
    e. Local Group, Solar System, Milky Way, Virgo Cluster
20. The Sun orbits the center of the Milky Way at a much higher speed than would be expected based on the mass of visible stars. What do astronomers think probably explains this discrepancy?
    a. Extrasolar planets
    b. A large number of undetected black holes
    c. Dark matter
    d. Dark energy
    e. An unobserved object that is gravitationally attracting the Sun

UNIT 3

Astronomical Numbers

3.1 The Metric System
3.2 Scientific Notation
3.3 Special Units
3.4 Approximation

Learning Objectives

Upon completing this Unit, you should be able to:
- Understand the meaning of exponential notation, and carry out calculations with exponential notation in multiplication and division.
- Identify the power of ten associated with commonly used metric prefixes.
- Describe the basic units in the MKS system and relate them to English system units.
- Convert a number into and out of scientific notation, and describe the purpose of this notation.
- Explain why astronomers use some non-metric units, and carry out a calculation to convert between different units.
- Define significant digits, and describe how results should be rounded off when carrying out calculations.

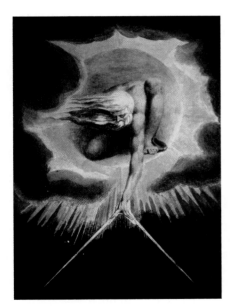

FIGURE 3.1
The "Ancient of Days" taking the measure of the universe. Etching by William Blake (ca. 1794).

Astronomy deals with a greater range of numbers than any other science. It is challenging to grasp, even figuratively, the vast range of the measurements needed to study the universe (Figure 3.1). To understand a planet that has a diameter of 13,000,000 meters, we also need to understand the interactions between atoms that are only 0.0000000001 meters in diameter. And as we discuss other aspects of the universe, we will be considering sizes, masses, brightness, energy, and times that span even greater ranges.

To deal with this vast array of numbers, astronomers use four different strategies: (1) the metric system, (2) scientific notation, (3) special units, and (4) approximation. We will look at each of these approaches in turn, but we can summarize these procedures briefly as follows:

The *metric system,* as opposed to the English system, allows easy conversions between larger and smaller **units** of measure, such as between meters and kilometers. To carry out many calculations, we must express measurements in fundamental units such as meters, seconds, and kilograms. Faced with numbers like 13,000,000 and 0.0000000001, astronomers use a way of abbreviating these numbers called *scientific notation.* This notation keeps track of the order of magnitude of a number separately from its precise value.

Metric units do not always prove convenient for interpreting the sizes of things, so sometimes we create special units with a clear physical meaning, such as the astronomical unit (Unit 1.6) or the light-year (Unit 2.2). Often the essential information we need about a number is just its approximate size, so astronomers round off values. This avoids focusing on a level of precision that is not important. We say, for example, that the astronomical unit, which stands for the Earth's distance from the Sun, is 150 million kilometers even though it is known to be 149,597,870 kilometers.

This mathematical background is important for understanding much of the material in astronomy, but it does not necessarily have to be mastered all at once. Return to this Unit whenever the numbers seem overwhelming.

FIGURE 3.2
The Mars *Climate Orbiter* crashed due to confusion about English and metric units.

It is often forgotten today that the English system also offers many intermediate size scales, but their relationships are complex. The inch is divided into 12 "lines," the foot into 12 inches, the yard into 3 feet, the "rod" into 5.5 yards, the "chain" into 4 rods, the "furlong" into 10 chains, the mile into 8 furlongs, and the "league" into 3 miles. The acre, still commonly used as a measure of area, is one furlong long by one chain wide. As if this is not enough to give one a headache, weights and volumes were divided in completely different ways!

Mathematical Insight

The trick to converting from one unit to another is to multiply by "1." One can be written in many ways. If a and b are equal, then $\frac{a}{b} = 1$. In the conversion of kilometers to meters, because 1 km = 10^3 m, we multiply by $\frac{10^3 \text{ m}}{1 \text{ km}}$. This lets us cancel out the kilometer units because they appear in both the numerator and the denominator.

3.1 THE METRIC SYSTEM

Before the metric system was introduced in 1791, widely varying systems of measurement were used in different countries, or even in different regions within a single country. For example, the Paris foot was 6.63% longer than the English foot, while the Spanish *vara* was 8.67% shorter than the English yard. Similarly, the French pound (*livre*) was 10.2% larger than the English pound, and many other weight measures were in common use even just within Europe. Confusion about measurements was widespread and required frequent conversion between different units.

That same confusion caused the loss of the Mars *Climate Orbiter* spacecraft in 1999. The company building navigation jets provided its thrust data in English units, but NASA mission controllers thought the units were metric. Firing the thrusters caused the craft to go off course and crash (Figure 3.2). Today the metric system has been widely adopted; the complex system of English units is still the primary choice only in the United States, Liberia, and Myanmar. In the end, the units used should not interfere with appreciating the important ideas of astronomy, but it is helpful to be familiar with the metric system for the calculations we will carry out.

The great advantage of the **metric system** over the English system is that different units are related by factors of 10, making it much simpler to convert between different units—1 kilometer is 1000 meters, whereas 1 mile is 5280 feet. We can usually convert between metric units by just "moving the decimal point" instead of multiplying or dividing by a conversion factor like 5280 feet per mile.

Each time we move the decimal point, we are performing the equivalent of multiplying or dividing by factors of 10. Mathematically, we can use **exponents** to indicate the *number* of times we multiply by 10. An exponent indicates how many times we multiply a quantity by itself. For example, there are 1000 meters in a kilometer. We can write 1000 as "10 to the third power" or "10 to the exponent 3" as follows:

$$1000 = 10 \times 10 \times 10 = 10^3.$$

When we convert meters to kilometers, we also move the decimal point over three places:

$$1 \text{ km} = 1. \text{ km} \times \frac{10^3 \text{ m}}{\text{km}} = 1000. \text{ m},$$

where we have used the abbreviations m for meters and km for kilometers, and we have explicitly included the decimal point for clarity.

When we multiply by 10^3, we move the decimal point to the right by 3 places. If we were to divide by 10^3, we would move the decimal point to the *left* by 3 places. This second case can also be written as multiplying by 10^{-3}. In other words "10 to the power negative 3" is the equivalent of one thousandth (1/1000). When we take 10 (or any number) to the "zeroth power" the result is always one: $10^0 = 1$.

Similarly, we can write 1 billion as

$$1{,}000{,}000{,}000. = 10 \times 10 \times 10 \times 10 \times 10 \times 10 \times 10 \times 10 \times 10 = 10^9,$$

or one millionth as

$$0.000001 = \frac{1}{1{,}000{,}000.} = \frac{1}{10 \times 10 \times 10 \times 10 \times 10 \times 10} = \frac{1}{10^6} = 10^{-6}.$$

In brief, rather than writing out all the zeros, we use the exponent to tell us the number of zeros.

In the metric system, **metric prefixes** identify various possible **powers of 10**. The prefix *kilo-*, for example, indicates 1000, while *milli-* indicates one thousandth, and *mega-* indicates 1 million. These prefixes can be added to any unit of measure to create a new unit that is closer to sizes we are interested in discussing: millimeter, kilogram, or megaton, for example.

TABLE 3.1 Metric Prefixes

Power of Ten	Exponential Notation	Metric Prefix	Abbreviation
septillion	10^{24}	yotta	Y
sextillion	10^{21}	zetta	Z
quintillion	10^{18}	exa	E
quadrillion	10^{15}	peta	P
trillion	10^{12}	tera	T
billion	$\mathbf{10^{9}}$	**giga**	**G**
million	$\mathbf{10^{6}}$	**mega**	**M**
thousand	$\mathbf{10^{3}}$	**kilo**	**k**
hundred	10^{2}	hecto	h
ten	10^{1}	deca	da
tenth	10^{-1}	deci	d
hundredth	$\mathbf{10^{-2}}$	**centi**	**c**
thousandth	$\mathbf{10^{-3}}$	**milli**	**m**
millionth	$\mathbf{10^{-6}}$	**micro**	**μ**
billionth	$\mathbf{10^{-9}}$	**nano**	**n**
trillionth	10^{-12}	pico	p
quadrillionth	10^{-15}	femto	f
quintillionth	10^{-18}	atto	a
sextillionth	10^{-21}	zepto	z
septillionth	10^{-24}	yocto	y

TABLE 3.2 MKS Units

Quantity	Metric Unit	MKS Equivalent	English Equivalent
Length	meter	m	3.28 feet
Mass	kilogram	kg	2.2 pounds (of mass)
Time	second	sec	(same)
Area	square meter	m^2	10.76 square feet
Volume	liter (L)	$10^{-3}\,m^3$	1.06 U.S. quarts
Speed	meters per second	m/sec	2.24 miles per hour
Acceleration	meters per square second	m/sec^2	3.28 feet per square second
Density	kilograms per liter	$10^3\,kg/m^3$	0.036 pounds per cubic inch
Force	newton (N)	$kg \cdot m/sec^2$	0.225 pounds (of force)
Energy	joule (J)	$kg \cdot m^2/sec^2$	0.000948 BTUs
Power	watt (W)	$kg \cdot m^2/sec^3$	0.00134 horsepower

Table 3.1 shows the standard metric prefixes along with their meanings in words and exponential notation. This is a complete listing of metric prefixes; in this book we use only the seven shown in boldface in the table. A nanometer, for example, is a unit we will use when discussing light waves. It is a billionth of a meter, or 10^{-9} m, and can be abbreviated nm. To convert 5 nanometers to meters, we would move the decimal point to the left by 9 places: 5. nm = 0.000000005 m. Thus, starting from a fundamental unit of measure like the meter, the metric prefixes give us a wide range of units that are appropriate in different contexts.

Any system of physical measurements requires three fundamental units—those describing length, mass, and time. In the metric system these are the meter (which is about 10% longer than the yard), the kilogram (which is about 2.2 pounds), and the second. This set of units defines the **MKS system,** which stands for meter, kilogram, and second. Units for measuring other quantities, such as force, energy, and power, can all be written in terms of these fundamental units. Some of the more common kinds of metric units we use in this book, and their nearest equivalent in the English system, are listed in Table 3.2.

One unit commonly used in this book, the liter, is a unit of volume, but is not quite as simple a combination of MKS units as the others. The liter is one thousandth of a cubic meter. The liter was the basis for the original definition of the kilogram—the mass of one liter of water—but because this was not very precise, it was redefined by a carefully preserved cylinder of platinum (Figure 3.3).

We will often be examining the **density** (mass per volume) of an object to try to understand its composition. For example, a planet with a density of 1 kilogram per liter has the density of water and is probably composed of ice, but one with a higher density must contain denser substances such as rock or iron. We will find places such as the centers of dying stars where the density is enormously larger, and regions of interstellar space where the density is a minuscule fraction of this.

The units in Table 3.2 can all employ metric prefixes, so the specifications for a car's power would be likely listed as, say, 200 kilowatts. To convert this to the English unit of horsepower, we could carry out the following calculation:

$$200\,kW = 200{,}000\,W \times 0.00134\,\text{horsepower}/W = 268\,\text{horsepower}.$$

Also be aware that in the English system, the term *pound* is used interchangeably for a mass (a measure of the amount of matter) as well as for the gravitational force with which the Earth pulls on that mass (see Unit 14.1 for details). As we will discover, the same mass weighs a different amount on different planets.

FIGURE 3.3
The prototype of the kilogram is a platinum-alloy cylinder stored under vacuum at the International Bureau of Weights and Measures in France.

Mathematical Insight

The conversion here is accomplished first by converting kilowatts to watts, multiplying by $1 = \frac{1000\,W}{1\,kW}$, and then converting units by once again multiplying by $1 = \frac{0.00134\,\text{horsepower}}{1\,W}$.

3.2 SCIENTIFIC NOTATION

When we make a calculation in scientific notation, we usually begin by converting all of the values into MKS units. For example, the mass of the Sun is 1,989,000,000,000,000,000,000,000,000,000 kilograms. There is no metric prefix close to the size of such an enormous number, so we use a more concise way to express such numbers called *scientific notation*.

Scientific notation combines the **powers-of-10** notation just described with the particular value of the number. We divide the number into a value between 1 and 10 and a power of 10 that when multiplied together yield the original number. Thus, we can write 600 (six hundred) as

$$600 = 6 \times 10 \times 10 = 6 \times 10^2.$$

We write 543,000 (five hundred forty-three thousand) as

$$543,000 = 5.43 \times 10 \times 10 \times 10 \times 10 \times 10 = 5.43 \times 10^5.$$

and 21 millionths becomes

$$\frac{21}{1,000,000} = 0.000021 = \frac{2.1}{100,000} = \frac{2.1}{10^5} = 2.1 \times 10^{-5}.$$

Once again the exponent indicates how we have moved the decimal point. The Sun's mass expressed this way is 1.989×10^{30} kg.

With scientific notation, multiplying and dividing very large numbers becomes easier. This is because when we multiply two powers of 10 we just add the exponents, whereas to divide we subtract them. For example,

$$10^2 \times 10^5 = (10 \times 10) \times (10 \times 10 \times 10 \times 10 \times 10) = 10^{2+5} = 10^7,$$

or as an example of division,

$$\frac{10^8}{10^5} = \frac{10 \times 10 \times 10 \times 10 \times 10 \times 10 \times 10 \times 10}{10 \times 10 \times 10 \times 10 \times 10} = 10 \times 10 \times 10 = 10^{8-5} = 10^3.$$

We can write this as a pair of general rules:

$$10^A \times 10^B = 10^{A+B} \quad \text{and} \quad 10^A/10^B = 10^{A-B}.$$

It is important to remember is that $10^A + 10^B$ does *not* equal 10^{A+B}. To add or subtract two numbers expressed in scientific notation, *first* convert both to the *same* power of 10, then add or subtract the values by which the power of 10 is multiplied.

As an illustration of the use of scientific notation, we can calculate the number of kilometers in a light-year. To do this, we multiply light's speed by the number of seconds in a year. A year is 365¼ days, each day having 24 hours, each hour 60 minutes, and each minute 60 seconds, so the total number of seconds in a year is

$$365.25 \text{ days} \times \frac{24 \text{ hours}}{1 \text{ day}} \times \frac{60 \text{ minutes}}{1 \text{ hour}} \times \frac{60 \text{ seconds}}{1 \text{ minute}}$$

$$= 31,557,600 \text{ seconds} \approx 3.156 \times 10^7 \text{ sec}.$$

The speed of light is 299,793 km/sec = 2.998×10^5 km/sec. Multiplying the speed by the time gives us the distance:

$$1 \text{ ly} = \text{speed of light} \times 1 \text{ year}$$
$$= 2.998 \times 10^5 \text{ km/sec} \times 3.156 \times 10^7 \text{ sec}$$
$$= 2.998 \times 3.156 \times 10^{5+7} \text{ km}$$
$$\approx 9.46 \times 10^{12} \text{ km},$$

or nearly 10 trillion kilometers.

Concept Question 1

How is the phrase "a six-figure salary" like scientific notation? Can you think of other ways we express powers of 10 in everyday usage?

Mathematical Insight

The conversion here is accomplished by multiplying by 1 written in three different ways. Make sure you can identify them.

3.3 SPECIAL UNITS

The enormous range of sizes, masses, and times studied in astronomy is illustrated in Figure 3.4, which provides several examples of objects on scales showing the power-of-ten value of meters, kilograms, and seconds. While we can always make measurements in terms of these MKS units, it is sometimes easier to grasp the meaning of a value using special units that are more familiar to us.

Few people would be able to understand if a man said he was 700 megaseconds old rather than saying 22 years. A "year" is a special unit that helps us interpret time spans more easily. There are many other places where astronomers have invented units to describe objects and phenomena in a way that is easier to understand. Usually these units make it easier to gain physical intuition, but they add to the number of units to learn, and they are not as easy to convert as metric units.

The **light-year,** the distance light travels in a year, is a good example. In principle we could use metric prefixes and a unit like "petameters" (Table 3.1), and write a light-year as 9.46 petameters or 9.46 Pm. However, the light-year has such a useful interpretation that we prefer to introduce it as another unit. Specifically, when we see an object 10 million light-years away, the light has taken 10 million years

Even the meter was originally a special unit based on the size of the Earth. The original plan was that the meter would be one 10-millionth of the distance from pole to equator. The original determination of the meter was slightly off, however, so that the equator-to-pole distance is today found to be 10,002 km, for a circumference of 40,008 km measured around the poles. More significantly, the Earth is not a perfect sphere, so the circumference measured at the equator bulges to 40,075 km.

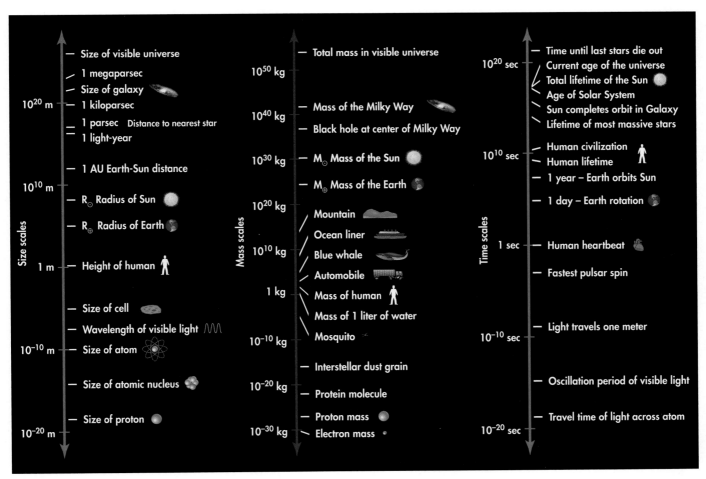

FIGURE 3.4
Scales showing relative sizes, masses, and times of special units and various objects and processes in the universe. Each is shown on a scale of powers of 10—ranging from the submicroscopic to the astronomical.

TABLE 3.3 Special Units

Quantity	Special Unit	Abbreviation	Metric Equivalent
Length	Earth's radius	$R_\oplus$	6.37×10^6 m
	Sun's radius	$R_\odot$	6.97×10^8 m
	Astronomical unit (Earth–Sun distance)	AU	1.50×10^{11} m
	Light-year (distance light travels in one year)	ly	9.46×10^{15} m
	Parsec (distance calculated by special geometric technique)	pc	3.09×10^{16} m
Mass	Earth's mass	$M_\oplus$	5.97×10^{24} kg
	Sun's mass	$M_\odot$	1.99×10^{30} kg
Time	Day (rotation period of Earth)	day	8.64×10^4 sec
	Year (orbital period of Earth)	yr	3.16×10^7 sec
	Sun's lifetime (estimated total time Sun will generate energy)	$t_\odot$	3.16×10^{17} sec
Speed	Speed of light (through empty space)	c	3.00×10^8 m/sec
Acceleration	Earth's surface gravity (the rate at which falling objects accelerate)	g	9.81 m/sec^2
Energy	Kiloton of TNT (energy released by a standard 1000 ton bomb)	kt	4.18×10^{12} joules
Power	Sun's luminosity (energy the Sun generates per second)	$L_\odot$	3.86×10^{26} watts

to reach us. That means that we are seeing the object *as it was 10 million years ago*. This becomes even more interesting as we look out billions of light-years, when the universe was just a fraction of its current age. We are literally able to look back toward the beginning of time, and the light-year unit helps us understand what, or rather "when," we are seeing.

A list of the special units we will use in this book is given in Table 3.3, and those describing sizes, masses, and times are shown in Figure 3.4. Many of these units are used to relate other objects to the more familiar Sun and Earth, such as the Sun's power output or the length of Earth's year. Some of the units were invented to make our calculations easier. For example, the astronomical unit simplifies the calculation of orbital parameters of objects in the Solar System (Unit 12) as well as of stars that orbit each other (Unit 17). The **parsec** was invented as a unit to simplify the formula for the primary method we have for finding stars' distances (Unit 54). Both the parsec and light-year are frequently used by astronomers when talking about stars and galaxies, so it is useful to be able to convert back and forth between them. A parsec is a little more than 3 times larger than a light year, so converting between them is similar to converting between meters and feet.

For many calculations these special units add an extra step to calculations, requiring us to convert the values to the MKS system. For example, suppose we wanted to calculate how fast the Earth moves as it orbits the Sun (Figure 3.5). The radius of its orbit is 1 astronomical unit, and it completes an orbit in 1 year. To carry out the calculation, first we need to know how far the the Earth travels in its orbit. Since the circumference of a circle is 2π times its radius and 1 AU is 1.50×10^{11} meters, we find that the Earth travels a distance

$$2 \times \pi \times 1 \text{ AU} = 6.283185 \times 1.50 \times 10^{11} \text{ m} = 9.42 \times 10^{11} \text{ m}$$

in each year, which is a time = 3.16×10^7 sec. The speed, V, of the Earth is the distance it travels divided by the time, which we calculate as follows:

$$V = \frac{\text{distance}}{\text{time}} = \frac{9.42 \times 10^{11} \text{ m}}{3.16 \times 10^7 \text{ sec}} = 2.98 \times 10^4 \frac{\text{m}}{\text{sec}}.$$

This value is nearly 30 kilometers per second, more than 100,000 kilometers (60,000 miles) per hour. This is about 100 times faster than a speeding bullet!

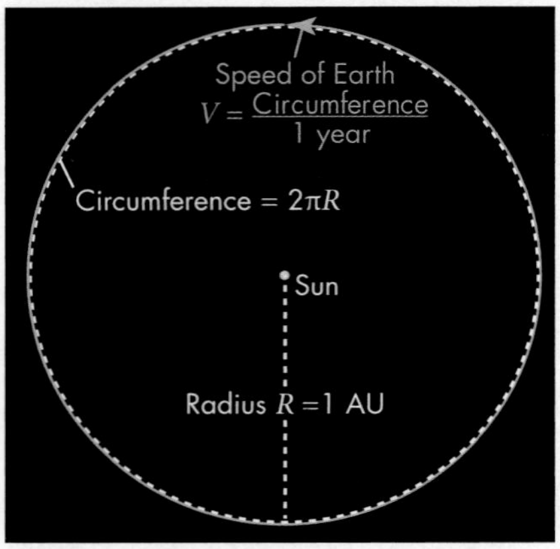

FIGURE 3.5
The speed of the Earth as it orbits the Sun is equal to the circumference of its orbit divided by the time it takes to complete the orbit (1 year).

3.4 APPROXIMATION

Astronomy is a science in which we often have to make uncertain estimates. Even though we can measure some numbers quite precisely, many others are subject to such uncertainty that there is no point in keeping track of more than the first couple of digits of a number.

For example, astronomers have determined that there are 3.261633 light-years in a parsec. However, we rarely know the distance of a star more precisely than within 10%. If a star's distance is listed as 10 parsecs, it is *not* correct to say the star is at a distance of 32.61633 light-years. The latter number gives the appearance of precision that does not exist. It would be more appropriate to convert 10 parsecs to 33 light-years or even 30 light-years—these better reflect the level of precision with which such distances are known.

Measurements are best reported showing only the number of digits that are accurately known. Values written as 374,000 or 0.00512 or 1.50×10^{11}, for example, are each reported to three **significant digits.** Having three significant digits means that the first two digits (37, 51, 15 in these three examples) are well determined, and the last one is fairly accurate, although it may have some uncertainty. Scientific notation is useful here for indicating when a trailing 0 is significant. This would be unclear if 1.50×10^{11} were written 150,000,000,000.

Any result from multiplying or dividing numbers in a formula can be only as accurate as the *least* accurate measurement used. For example, the nearest star, Proxima Centauri, is measured to be 1.3 parsecs away—to two digits of precision. We can multiply this by the number of light-years per parsec to get the distance in light-years:

$$1.3 \text{ pc} \times \frac{3.261633 \text{ ly}}{\text{pc}} = 4.2401229 \text{ ly} \approx 4.2 \text{ ly}.$$

We **round off** the result to 4.2 ly, with two digits to reflect the level of precision for this calculation.

Notice that because we are limited by the least precise number, there was really no need to use all of the digits 3.261633 in our calculation. If we rounded off our ly-to-pc conversion factor to 3.26, we would have gotten the same result:

$$1.3 \text{ pc} \times \frac{3.26 \text{ ly}}{\text{pc}} = 4.23 \text{ ly} \approx 4.2 \text{ ly}.$$

On the other hand, rounding off our conversion factor to just one digit, 3, we get:

$$1.3 \text{ pc} \times \frac{3 \text{ ly}}{\text{pc}} = 3.9 \text{ ly} \approx 4 \text{ ly}.$$

The result should be reported to only one digit because one of the numbers we used had only one significant digit. Finally, note that if we round the conversion factor to 3.3, the result is

$$1.3 \text{ pc} \times \frac{3.3 \text{ ly}}{\text{pc}} = 4.29 \text{ ly} \approx 4.3 \text{ ly},$$

which is rounded *up* because the digit (9) after the last significant digit, 2, is greater than or equal to 5. Unfortunately, when we carry out a calculation with several numbers that are approximate, we multiply the error, and so the result has a *greater* uncertainty than the least accurate of the numbers.

Good use of approximation is an art as well as a science. To be safe, when carrying out a calculation, use as many digits as are available for each value in the equation. However, with practice, you will find that you can round off some of the values to fewer digits with no significant effect on the accuracy of your result, making your calculations a little simpler.

Mathematical Insight

In the technical literature, astronomers often report numbers with a "plus or minus" range to indicate the uncertainty in a value, like 71 ± 6. Then if we multiplied the number by 2, we would find 142 ± 12, with both the value and the uncertainty doubling. This notation becomes cumbersome, especially when we start multiplying a set of numbers each of which has its own precision.

Concept Question 2

We encounter numbers frequently in everyday life, sometimes exact, sometimes indefinite. What's the most precise number you've encountered? What are some of the most imprecise numbers?

KEY POINTS

- Astronomers generally use the metric system.
- Metric prefixes provide a simple way of producing new units of measurement that are useful for different size objects.
- Scientific notation can be used for measurements of any size, and it offers certain conveniences for mathematical calculations.
- Astronomers use several nonmetric units, such as the light-year, that have useful physical meanings or facilitate comparisons.
- Some values are only approximately known in astronomy, and it is important to present values at a level of precision that reflects their accuracy.

KEY TERMS

density, 19
exponent, 18
light-year, 21
metric prefix, 18
metric system, 18
MKS system, 19
parsec, 22
powers of 10, 18
round off, 23
scientific notation, 20
significant digits, 23
unit, 17

CONCEPT QUESTIONS

Concept Questions on the following topics are located in the margins. They invite thinking and discussion beyond the text.

1. Exponentials in common parlance. (p. 20)
2. Precise and imprecise numbers. (p. 23)

REVIEW QUESTIONS

3. What are the advantages of the metric system?
4. What is meant by a positive exponent? a negative exponent? an exponent of zero?
5. How is a number expressed in scientific notation?
6. What nonmetric units do astronomers frequently use?
7. What are significant digits?
8. When calculations are carried out with several numbers of different precision, what is the precision of the result of the calculation?

QUANTITATIVE PROBLEMS

9. The radius of the Sun is 7×10^5 kilometers, and that of the Earth is about 6.4×10^3 kilometers. Show that the Sun's radius is approximately 100 times the Earth's radius.
10. Hypergiant stars can have radii as large as 1500 times that of the Sun. Estimate the radius of a hypergiant star in kilometers. How many AU is this equivalent to?
11. Using scientific notation, show that it takes sunlight about 8½ minutes to reach Earth from the Sun.
12. How many 1-kiloton bombs would need to be exploded to produce 3.86×10^{26} joules, the amount of energy emitted by the Sun in 1 second?
13. If I want to work out the distance to the Andromeda Galaxy, do I need to be concerned about whether the distance is from the center of the Milky Way or from the Sun's location? Why or why not?
14. Suppose two galaxies move away from each other at a speed of 6000 km sec and are 300 million (3×10^8) light-years apart. If their speed has remained constant, how long has it taken them to move that far apart? Express your answer in years.
15. The famous equation $E = mc^2$ tells us how much energy, E, is stored in a mass of m. Use this equation to calculate the energy stored in a 3.10 gram penny.
16. Using scientific notation, evaluate $(1.4 \times 10^9)^3 / (9.3 \times 10^8)^2$.
17. Napoleon Bonaparte is often said to have been short, but contemporaries described him as being of average or slightly above average height for his time. This mistake was apparently made because the height reported in Paris feet was misinterpreted. Bonaparte was reported to be about 5 feet 2 inches tall as reported by contemporaries and at his autopsy. If these were in Paris units, 6.63% longer than English inches and feet, how tall was he actually in (a) English units, and (b) metric units?

TEST YOURSELF

18. Which of the following is the correct method for calculating $10^5/10^3$?
    a. 10^{5+3}
    b. 10^{5-3}
    c. $10^{5\times3}$
    d. $10^{5/3}$
    e. $10^{(5+3)/2}$

19. Which of the following is *not* equivalent to 30 kilometers?
    a. 30,000,000 millimeters
    b. 3×10^6 cm
    c. 30,000 meters
    d. 3×10^3 meters
    e. 0.03 megameters

20. From the following, choose the best approximation for the sum of 3.14×10^{-1} and 6.86×10^4.
    a. 10.00×10^3
    b. 6.86×10^4
    c. 3.14×10^{-1}
    d. 1.00×10^3
    e. 3.72×10^3

PART 1
UNIT 4
Scientific Foundations of Astronomy

4.1 The Scientific Method
4.2 The Nature of Matter
4.3 The Four Fundamental Forces
4.4 The Elementary Particles

Learning Objectives

Upon completing this Unit, you should be able to:
- Describe the scientific method and the processes involved in it.
- Formulate a sample hypothesis, and explain the differences between a theory, a model, and an unscientific hypothesis.
- Describe the basic characteristics of atoms and the use of the periodic table.
- List the fundamental forces and the elementary particles in normal matter.
- Describe how particles interact depending on their mass and kinds of charge.

Our understanding of the universe has not come easily. Astronomy builds on many ideas from physics, chemistry, geology, mathematics, and biology, representing the intellectual work of thousands of men and women over thousands of years. Learning how to take experiments here on Earth to understand the far reaches of the universe is part of the science of astronomy.

It seems that whenever we try to understand *larger* objects, we need to know more about ever smaller elements of the *submicroscopic* world. To study planets we must understand atoms and molecules; to study stars we have to understand the nuclei of atoms; to study the universe we need to understand subatomic particles. And in studying the cosmos, we encounter extremes of temperature, size, gravity, and energy that no one could dream of testing in Earth-bound laboratories.

Remarkably, the discoveries we have made about the submicroscopic properties of matter in laboratories here on Earth have been found to apply to the rest of the universe in almost every instance. This is remarkable in itself, given that the Earth is such a tiny piece of the universe. Occasionally, in extrapolating our Earth-based findings to distant realms and extreme conditions, astronomers encounter unexpected differences from current theories. These differences often lead to new ideas for revising or extending our theories.

Scientific understanding about the nature of matter and the interactions that take place between particles of matter provides the foundation for all of astronomy. We examine these fundamental ideas in this Unit. First, though, we need to understand the methods that scientists use for testing ideas in new situations.

Concept Question 1

Make a hypothesis about the cause of some weather phenomenon you have observed. For example, what produces lightning or hail? What observations are you familiar with that support your hypothesis?

4.1 THE SCIENTIFIC METHOD

The **scientific method** is the procedure by which scientists construct their ideas about the universe and its contents, regardless of whether those ideas concern stars, planets, living things, or matter itself. In the scientific method, a scientist proposes an idea—a **hypothesis**—about some property of the universe and then tests the hypothesis by experiment. Ideally the experiment's results either support the hypothesis or refute it. The scientific method is sometimes portrayed as a

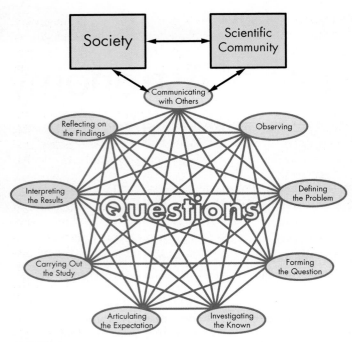

FIGURE 4.1
A diagram illustrating the scientific method as described by scientists. The figure is based on interviews with scientists carried out by Reiff, Harwood, and Phillipson of Indiana University. Far from being a regular step-by-step procedure, the scientific method was described by scientists as a set of different activities and processes, which might be visited and revisited in a wide variety of orders as different questions arose. This is illustrated in the figure by crisscrossing lines connecting nine processes identified from the interviews.

Concept Question 2

What kind of *unscientific* hypothesis (one that is impossible to disprove) can you think up about the cause of some weather phenomenon?

straightforward four- or five-step procedure, but in actuality it is far more complex, involving many social interactions, as depicted in Figure 4.1.

Astronomers face a special challenge in applying the scientific method because they cannot experiment with their subject matter directly. Astronomy is primarily an observational science, like geology or the study of human behavior; so the strength of astronomers' theories is tied to their ability to predict phenomena or circumstances not yet observed.

One of the essential requirements for a hypothesis to be regarded as scientific is that it must be clear how it can be disproved. In experimental sciences, this may be the prediction of the outcome of a laboratory experiment. For an observational science like astronomy, it might be a prediction that, for example, whenever we find a star with two particular properties, we will find that it also has a third property. To disprove the hypothesis, we need only demonstrate that the prediction is incorrect.

By requiring hypotheses to be disprovable, science not only provides a way of introducing new ideas; it also encourages scientists to test and retest older ideas from new perspectives. This openness to retesting should not be interpreted as vagueness or uncertainty—far from it. For an idea to stand up to this constant testing, it must be robust, providing us with much of the most rigorously tested information we possess.

Once a hypothesis has been thoroughly tested and verified, it may be termed a *theory* or *law*. When we use the word *theory* here, we do *not* mean that the idea is unproved or tentative, and the word *law* should not be interpreted as meaning the idea is beyond testing. Rather, theories and laws have achieved wide acceptance by successful testing over a long time. **Laws** are generally mathematical statements, whereas **theories** are generally expressed in words.

Sometimes complex ideas are described by "models." A **model** is a schematic representation of a real system that illustrates its most essential relationships and interactions. For example, the diagram in Figure 4.1 is a model of the scientific method. Usually astronomical models are expressed mathematically or geometrically, and often they oversimplify a complex set of relationships to try to identify the most important elements. For example, contrary models were developed to explain the motions of planets, one assuming that the Earth was the center of their motion, another that the Sun was. Both models can accurately predict the motions we see, but today we recognize the Sun-centered model as the more nearly correct of the two models. However, even it is not quite right, because in fact the bodies in the Solar System all interact with each other through the force of gravity in a more complex way than simply circling the Sun. And in some situations the Earth-centered model is more convenient for describing the motions we see—as when we speak of the "Sun setting" as opposed to saying that the Earth rotated so that the Sun became hidden below our local horizon.

When an idea has achieved the status of a theory or law, its ability to make accurate predictions does not end even when a new idea overtakes it. For example, in the late 1600s Isaac Newton proposed several mathematical relationships describing motion and gravity, known as Newton's laws. They are extremely successful mathematical models that are still used today despite the fact that they are now known to make small errors in the prediction of motions under extreme conditions beyond our common experience. This was shown in the early 1900s by Albert

Einstein. He proposed revisions to Newton's laws, known as special and general relativity, which explain how space and time are altered by motion and gravity. These are so well tested that they have risen to the level of well-accepted theories.

Successful application of the scientific method is no guarantee that its results will be believed. For example, even before 300 B.C.E., Greek scientists pursued several lines of inquiry demonstrating that the Earth is a sphere (Unit 10). Yet despite the evidence supporting this hypothesis, for thousands of years many people continued to believe the Earth to be flat. Today many people believe that it is possible to make a spaceship move faster than the speed of light, a belief that is depicted in many science fiction movies. However, Einstein's theory of special relativity has provided compelling evidence that this cannot happen.

We can be quite confident of the longest-standing theories because so many scientists have attempted to overturn them. Nevertheless, an exciting aspect of science is that it invites skepticism. At the forefront of the sciences, debate is very lively, with new hypotheses being proposed to explain new measurements, and challenges to accepted theories always encouraged. Individual scientists often dispute particular pieces of evidence. One astronomer may find the evidence supporting a hypothesis convincing, while another astronomer may think that the experiment was done incorrectly or the data were analyzed improperly. This is all part of the scientific process.

We need therefore to keep in mind that when we discuss ideas, they are never "proved," and sometimes they are not even widely accepted. This is especially true of ideas at the frontiers of our knowledge—for example, those dealing with the origin and structure of the universe or those dealing with black holes. However, the tentativeness of such ideas does not always stop astronomers from being positive about them, leading the Russian physicist Lev Landau to joke that astrophysicists are "often in error, but never in doubt." Therefore, keep in mind that some of the ideas we discuss in this book will be vastly improved on, or perhaps proved wrong, in the future. That is not a failing of science, however. It is its strength.

> Einstein's theories may someday also need revision; but it should be understood that, as with Newton's laws, their validity will not suddenly vanish. The many tests of the theories' predictions show that they work with enormous precision over a great range of circumstances.

4.2 THE NATURE OF MATTER

One of the most powerful ideas in astronomy is that we can apply to the universe at large what we learn about nature here on Earth. This is sometimes called the hypothesis of **universality.** Starting from this hypothesis, the step-by-step application of physical laws to the nature of matter drives us to many extraordinary conclusions. We will conclude that there are exotic things in the universe such as black holes, quasars, neutron stars, dark matter, and dark energy, as discussed in later Units. Yet the volume of space we have explored directly with space probes is just a thousandth-trillionth-trillionth-trillionth (10^{-39}) of the volume of the visible universe. How can we be sure that the properties of matter and physical laws do not change elsewhere in the universe?

Our studies of matter on Earth show that it is made up of tiny particles called **atoms,** and 92 kinds of these occur naturally. The simplest type, a hydrogen atom, is about one 10-billionth of a meter (10^{-10} m) in diameter, so that 10 million (10^7) hydrogen atoms could be put in a line across the diameter of the period at the end of this sentence. Despite this tiny size, atoms themselves have structure. Every atom has a small (about 1/10,000 the atom's diameter) central core called the **nucleus** that is surrounded by one or more even smaller lightweight particles called **electrons** (Figure 4.2). The nucleus is composed of two kinds of heavy particles called **protons** and **neutrons.**

Superficially, an atom resembles a miniature solar system, but different physical forces and behaviors are at play than the **gravitational force** between planets and

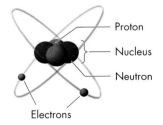

FIGURE 4.2
An atom can be modeled as a nucleus around which electrons orbit. The nucleus is itself composed of particles called protons and neutrons.

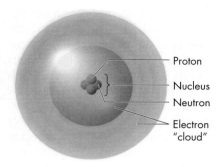

FIGURE 4.3
A more modern view of the structure of atoms attempts to illustrate the idea that the electrons do not travel in simple orbits, like planets, but instead in "orbital clouds."

The elements were created by several astronomical processes discussed later in the text. Some are radioactive and decay rapidly. Excluding these, only 79 stable elements occur naturally.

FIGURE 4.4
The periodic table of the elements displays properties of each type of atom. The full periodic table is shown on the inside back cover and displays astronomical properties of each kind of atom in addition to the basic chemical properties.

the Sun. At subatomic scales, particles do not behave in ways familiar to us—their motions are more wavelike (Unit 21). It is impossible to predict their precise position at any moment, so the location of the electron is better described in terms of an "orbital cloud" (Figure 4.3).

Experiments show that electrons repel each other, and protons repel each other, but electrons and protons *attract* each other. You can get some sense of the effects of electrical forces by rubbing a balloon on your hair in dry weather. The balloon collects electrons from your hair, while your hair is left with an excess of protons. The electric charge on the balloon attracts strands of your hair, while your hair stands on end because the strands of like-charged hair repel each other. This is the same force at play on an atomic level.

To model this idea, the proton is designated as having a "positive" **electric charge,** and an electron as having a "negative" electric charge. A neutron, as its name suggests, is neutral and has no charge. With this description, we can say that electric charges of the same sign repel each other (+ repels + ; − repels −) and opposites attract (+ attracts −). That attraction is what holds the electrons in their orbits around the nucleus of an atom.

The different kinds of atoms are displayed in the **periodic table** of the elements. The periodic table organizes the atoms in order of the number of protons, or **atomic number.** The table is organized so that atoms in the same column have similar chemical properties. The first part of the periodic table is shown in Figure 4.4, and the full table is shown inside the back cover. A wealth of additional information is shown for each atom, including the **atomic mass,** which is essentially the sum of the number of protons and neutrons in the atom's nucleus.

The light we see is also made up of elementary particles, called **photons,** which can be generated or affected by electrically charged particles such as electrons and protons. The characteristics of the light emitted by each kind of atom are distinct, and a careful analysis of photon energies allows us to determine the kind of matter that generated the light. Astronomers have used this fact to test the hypothesis of universality. In fact, by examining the characteristics of light emitted by very distant objects, we are also testing whether the universe was the same in the past when the light left those objects. We find that everywhere we look in the universe, matter has the same characteristics. The same kinds of atoms made of the same kinds of electrons, protons, and neutrons undergo the same kinds of chemical reactions as we observe here on Earth, just as they did "a long time ago in a galaxy far, far away."

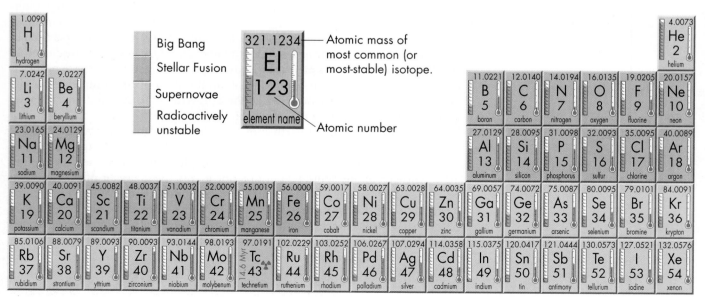

4.3 THE FOUR FUNDAMENTAL FORCES

When the model of the atomic nucleus was developed in the early 1900s, only electrical and gravitational forces were known. Actually, the electric force had been shown to be related to the magnetic force that makes a compass point north or holds magnets on the door of your refrigerator. Scientists discovered that moving electric charges generate fields of magnetic force, and moving magnets can generate electric forces. Scientists therefore refer to them jointly as the **electromagnetic force**. The electromagnetic force between two protons is 10^{36} times stronger than the gravitational force, so gravity plays very little role at subatomic scales.

There must be another force at work in an atomic nucleus, since gravity is too weak to overcome the electrical repulsion between the protons. This force, dubbed the **strong force**, binds protons and neutrons together. Experiments show that the strong force is about 100 times stronger than the electric repulsion between protons in a nucleus, but it acts only over extremely short distances.

Yet another force was needed to explain the **radioactive decay** of certain substances discovered in the late 1800s. In radioactive decay, an atomic nucleus splits into two smaller nuclei, with the release of radiation. The force that causes these changes is called the **weak force** because it is about a billion times weaker than the strong force between protons. The weak force can cause one kind of particle to turn into another through a process described further in the next section. The weak force is also a short-range force, with virtually no influence outside an atomic nucleus. Properties of all four forces are summarized in Table 4.1.

The strong and weak forces each have their own associated "charges." These are properties that respond to the particular force. The strong force has three kinds of charge that the strong force always drives toward balanced combinations, analogous to the way red, blue, and green light can be combined to make white light. This analogy has led the charge associated with the strong force to be called "color charge." Associated with the weak force is a "weak charge," which has a complex representation that depends on a pair of properties in different combinations.

If an object's charge is neutralized, then it will not respond to the force associated with that charge. Thus, the electromagnetic force keeps pulling electrons into an atom until their negative charges neutralize the positive charge in the nucleus. As a result, atoms are normally neutral and the net electromagnetic force of the Sun on our planet is negligible compared to the gravitational force. The same occurs with the strong force and the weak force. This is why gravity is the dominant force at astronomical size scales despite being so extremely weak compared to the other forces. Gravity remains because its "charge"—mass—does not come in opposites that can cancel each other's effects.

The strength of the weak force becomes about as strong as the electromagnetic force at very, very small distances—less than 10^{-18} (a millionth-trillionth) of a meter—only a thousandth of the size of a proton.

Concept Question 3

Unlike gravity or electromagnetism, which grow weaker at larger distances, the strong force is thought to grow stronger. Why then might it be that this force is virtually undetectable outside an atom's nucleus?

TABLE 4.1 Fundamental Forces

Force	Associated Property	Effect	Range	Carrier Particle	Relative Strength
Gravitational	Mass	All masses attract each other.	Infinite but weakens with distance	Graviton	10^{-36}
Electromagnetic	Electric charge	Opposites attract, likes repel.	Infinite but weakens with distance	Photon	1
Strong	Color charge	Three colors combine to make neutral combinations.	$\approx 10^{-15}$ meters (distance between protons in atomic nucleus)	Gluon	10^2
Weak	Weak charge	Massive particles decay to lower-mass particles.	$\approx 10^{-18}$ meters (1/1000 proton diameter)	W and Z	10^{-7}

4.4 THE ELEMENTARY PARTICLES

Early in the 1900s, it appeared that all matter could be explained in terms of just three subatomic particles: electrons, protons, and neutrons. However, that simple model began to unravel in the 1930s as more and more particles were found.

One of the first new kinds of particles found was the **neutrino**, given that name as a play on the word "new." The neutrino was needed to explain what happens when a neutron decays into a proton and electron (via the weak force). Physicists found that some energy seemed to disappear in the process, contrary to the law of energy conservation (Unit 20), and they hypothesized that a particle with no electric charge and extremely little mass had been emitted. These "ghostly" particles interact solely through the weak force, so weakly that they pass through the Earth like light through a window. Neutrinos were finally detected 20 years later, confirming the hypothesis and illustrating the power of the scientific method.

Over subsequent decades, physicists discovered hundreds more subatomic particles, with different masses, charges, and other properties. At first, it seemed that the subatomic world was inexplicably complex, but gradually patterns emerged that led to the development of a model based on just a few truly elementary particles. These particles, along with the four forces, are the basis of the **Standard Model** of particle physics. This model successfully explains how all of the subatomic particles form from various combinations of the elementary particles. Physicists continue to test and refine the Standard Model with particle accelerators, such as the Large Hadron Collider in Europe, the world's biggest accelerator.

Protons and neutrons are built of elementary particles called **quarks.** Two types of quarks, called the *up quark* (*u*) and *down quark* (*d*), combine in sets of three to make the proton (*uud*) and neutron (*ddu*). The strong force acting between quarks grows stronger with distance, making it impossible to remove a quark from nuclear particles to study in isolation. However, particle physicists can "see" the internal lumpy structure of protons or neutrons when they bombard them with high-energy electrons, and from this kind of experiment they can determine much about the characteristics of quarks.

Four elementary particles: up quarks, down quarks, electrons, and neutrinos make up nearly all of the matter in the known universe. There are two sets of more massive particles with properties that parallel those of these four particles, as illustrated in Figure 4.5. These second and third generations of matter are created at high temperatures and during high-energy collisions. However, the weak force causes them to decay rapidly into first generation particles. The other generations of quarks have been given fanciful names—*strange, charm, bottom,* and *top*—because they exhibit properties that we do not see in normal matter. The two other generations contain analogs of the electron called the *muon* and *tauon,* and these each have an associated neutrino, called the *muon neutrino* and the *tauon neutrino.*

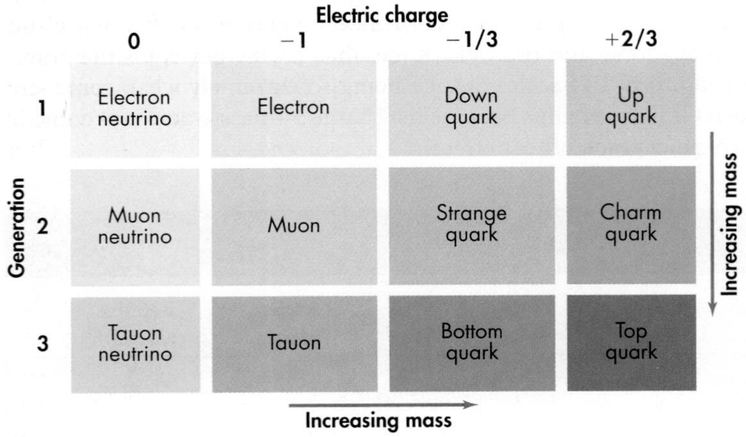

The four forces and the ways they interact with matter are illustrated in Figure 4.6. In the language of particle physics, each of the forces is "mediated," or carried, by a force particle. In Section 4.2 we discussed how photons (particles of light or electromagnetic radiation) interact with electric charges. Photons are the carriers of the electromagnetic force. A particle called a **gluon** (the name invented to suggest its strong gluelike properties) similarly carries the strong force, while particles called W

Each particle also has a corresponding "antiparticle" with opposite properties. When an antiparticle meets its partner, they annihilate each other, leaving nothing but energy. Antiparticles, and their role at the beginning of the universe, are discussed further in Unit 83.

Generation	Electric charge 0	−1	−1/3	+2/3
1	Electron neutrino	Electron	Down quark	Up quark
2	Muon neutrino	Muon	Strange quark	Charm quark
3	Tauon neutrino	Tauon	Bottom quark	Top quark

Increasing mass →

Increasing mass ↓

FIGURE 4.5
The elementary particles of the standard model. The general direction of increasing mass is shown, although the down quark appears to be more massive than the up quark. The particles are displayed here similar to the periodic table, with particles in each column having similar properties.

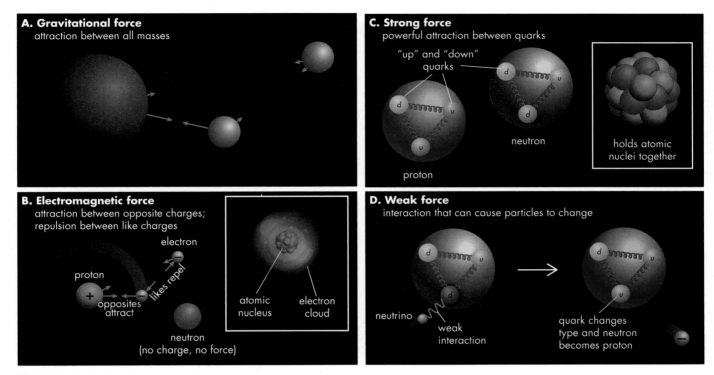

FIGURE 4.6
The four fundamental forces and their interactions with matter. (A) The force of gravity is present between all objects with mass. The force, represented by green arrows in the figure, is always attractive, but grows weaker with distance. (B) The electromagnetic force arises between particles with an electric charge. It causes electrons (negative charge) to be attracted to protons (positive charge) to form atoms. (C) Protons and neutrons are made of smaller particles called quarks, which are held together by the strong force. (D) The weak force causes some particles to change into others as they interact. The weak force causes radioactive decay and plays a critical role in energy formation in stars.

Physicists hypothesize that another particle—the graviton—carries the gravitational force, but this particle has not yet been detected.

Concept Question 4

What kinds of everyday objects can you think of that respond to one kind of force but not another?

and Z carry the weak force. The weak force has the unique ability to change one type of particle into another, as illustrated in Figure 4.6D. The carrier particles complete the Standard Model, except for one final particle, called the Higgs, which physicists think they detected at the Large Hadron Collider in 2012. The Higgs particle is generated by a field that is thought to fill space and interact with particles in a way that gives them their masses. The Higgs field is a little like a molasses that fills space and puts a greater drag on some kinds of particles, and that is why they appear heavier.

The science of astronomy uses what we have learned from particle physics and other sciences to interpret what we see. For example, the nuclear interactions inside stars generate neutrinos. In turn, astronomers studying neutrinos from the Sun found that the number of neutrinos detected did not match the number predicted. This forced physicists to revise their assumption that neutrinos had no mass (this is discussed further in Unit 52). In this way astronomy can sometimes drive forward new understandings in physics and other sciences.

Despite the successes of the Standard Model, physicists have not yet been able to develop a version that incorporates gravity. Recent advances in astronomy also suggest that there is a kind of matter not consistent with the Standard Model. Astronomers call this "dark matter" because it does not appear to interact with the electromagnetic force and therefore cannot emit light (Unit 79). Dark matter may interact through the weak force, and a number of hypotheses have proposed ways of extending the Standard Model to include it. Models of how the universe began expanding also suggest extensions to the Standard Model. Thus, as astronomers probe the largest scales in the universe, they are changing our understanding of the fundamental nature of the universe at the subatomic level.

KEY POINTS

- The scientific method is not a straightforward procedure, but a rich process where new ideas are developed and tested.
- For a hypothesis to be scientific it must be disprovable, therefore no scientific idea can ever be absolutely "proved."
- Hypotheses that stand up to intense testing may become known as laws or theories.
- The hypothesis of universality expresses the idea that matter and forces appear to be the same everywhere in the universe.
- Matter is made up of about 92 kinds of atoms, which are displayed in a periodic table that helps describe their properties.
- Atoms are in turn composed of electrons, protons, and neutrons, and protons and neutrons are themselves made of quarks.
- The elementary particles—quarks, electrons, and neutrinos—interact through just four forces: gravity, electromagnetism, and, at subatomic distances, the strong and weak forces.
- Neutrinos interact through the weak force, but not the strong or electromagnetic forces, so they are very difficult to detect.

KEY TERMS

atom, 27
atomic mass, 28
atomic number, 28
electric charge, 28
electromagnetic force, 29
electron, 27
gluon, 30
gravitational force, 27
hypothesis, 25
law, 26
model, 26
neutrino, 30
neutron, 27
nucleus, 27
periodic table, 28
photon, 28
proton, 27
quark, 30
radioactive decay, 29
scientific method, 25
Standard Model, 30
strong force, 29
theory, 26
universality, 27
weak force, 29

CONCEPT QUESTIONS

Concept Questions on the following topics are located in the margins. They invite thinking and discussion beyond the text.

1. Develop a hypothesis. (p. 25)
2. Develop an unscientific hypothesis. (p. 26)
3. The short range of the strong force. (p. 29)
4. Objects that do not respond to forces. (p. 31)

REVIEW QUESTIONS

5. What is meant by the scientific method?
6. Can a theory ever be proved to be true beyond any doubt?
7. What force(s) hold the Sun together?
8. What particles make up an atom? Which of these particles are used to determine its atomic number?
9. What force(s) hold an atom together?
10. What is a neutrino? What is a quark?

QUANTITATIVE PROBLEMS

11. If a proton were the size of a large apple (about 10 cm in diameter), how large would a hydrogen atom be?
12. Imagine that the Earth and the Moon were both positively charged by an amount proportional to their masses, so that their electrical repulsion canceled out their gravitational attraction. Using the relative strengths of gravity and the electromagnetic force, how much mass in the Earth would have to be positively charged?
13. Explain at a fundamental level what is happening when you take clothes out of a dryer and find them clinging to each other. What do you think is happening when you hear a crackling sound as you pull a sock and shirt apart?
14. It is estimated that a neutrino has to pass within a distance of about 10^{-24} meters of a quark for a high likelihood of interaction. If an atom has a radius of 10^{-10} meters and 12 nuclear particles in its nucleus, what is the probability that a neutrino passing through the atom will undergo an interaction? (Hint: Think of the atom as a target and the area around each quark as a bull's-eye.)

TEST YOURSELF

15. According to the Standard Model, which of the following is not an elementary particle?
    a. Muon   c. Charm quark   e. Neutrino
    b. Strange quark   d. Neutron
16. The best measure of an astronomical hypothesis's validity is
    a. its level of acceptance by scientists.
    b. the ease with which it can be understood by nonscientists.
    c. the sophistication of the mathematics it requires.
    d. its ability to make predictions about future observations.
    e. how well it describes what has been previously observed.
17. Rank the relative strength of the fundamental forces from weakest to strongest.
    a. Weak, Electromagnetic, Gravitational, Strong
    b. Weak, Electromagnetic, Strong, Gravitational
    c. Gravitational, Electromagnetic, Strong, Weak
    d. Gravitational, Weak, Electromagnetic, Strong
    e. Electromagnetic, Gravitational, Weak, Strong
18. Which of the following cannot be further subdivided?
    a. Hydrogen   c. Electron   e. Helium
    b. Proton   d. Iron
19. Which force has only one kind of "charge"?
    a. Gravity   c. Strong force   e. None of these
    b. Electromagnetism d. Weak force
20. What do physicists call the hypothesis that the laws of physics are the same everywhere?
    a. Periodic table   c. Universality   e. Nuclear theory
    b. Atomic theory   d. Standard Model

PART 1
UNIT 5

The Night Sky

5.1 The Celestial Sphere
5.2 Constellations
5.3 Daily Motion
5.4 Latitude and Longitude
5.5 Celestial Coordinates

Learning Objectives

Upon completing this Unit, you should be able to:
- Describe the celestial sphere and explain what makes it a useful model of the sky.
- List points of reference astronomers use in describing locations in the sky.
- Explain the daily motions of the sky and how they differ at different latitudes.
- Estimate the position of an object on the celestial sphere from its coordinates.

One of nature's spectacles is the night sky seen from a clear, dark location. Stars are scattered across the vault of the heavens in patterns that have guided people at night and inspired legends since time immemorial. From ancient records we know that the pattern of stars has changed little over the last several thousand years. Thus, the night sky affords us a direct link with our remote ancestors as they tried to understand the nature of the heavens. When you look up at the stars, you might be a shepherd in ancient Egypt, a hunter-gatherer on the African plains, a trader sailing along the coast of Persia, or even an airplane navigator in the early twentieth century.

Observations of astronomical phenomena are part of virtually every culture. Many astronomical phenomena—such as the rising and setting of stars, and the gradual seasonal changes in which stars can be seen each evening—are visible without any equipment. Sadly, many of the astronomical phenomena well known to ancient people are not familiar to us today because the bright lights of cities make it hard to see the sky and its rhythms. Therefore, if we are to appreciate the context of astronomical ideas, we need to first understand what our distant ancestors knew and what we ourselves can learn by watching the sky at night.

In this Unit we will examine the patterns we see in the night sky, then, in Unit 6, we will see how things change through the course of the year. Subsequent Units will examine the Sun, Moon, and planets in greater detail. Keep in mind that the stars we are describing are much like the Sun (Unit 2), but for the purpose of understanding the patterns and motions we see at night, we need not know what they really are.

5.1 THE CELESTIAL SPHERE

From a dark location on a clear night, thousands of stars are visible to the naked eye. The stars appear to be sprinkled on the inside of a huge dome that arches overhead (Figure 5.1). We call the line where the sky appears to meet the Earth the **horizon.**

Although the stars look as if they form a dome over us, the stars are actually at vastly different distances from us. For example, the nearest star is about 4 light-years away (Unit 2), while others that we can see are thousands of times more distant. And even the planets, the Sun, and the Moon, which is 100 million times closer than the nearest star, are still so distant that we are unable to get a direct sense of their true three-dimensional arrangement in space without precise measurements and careful deduction. Despite the great variety of distances, for the purposes of studying the patterns of the night sky with our naked eyes, we can treat all stars as if they are at the same distance from us (Figure 5.2A).

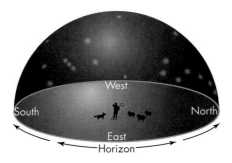

FIGURE 5.1
Our experience of the night sky. Stars seem to be placed in patterns upon a dome that stretches from horizon to horizon.

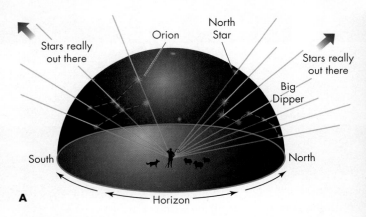
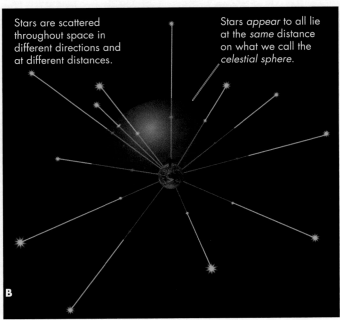

FIGURE 5.2
(A) Although the stars appear to lie on a hemisphere that meets the horizon, they are extremely far away. (B) Stars are scattered through space at very different distances in all directions, but for observations made from Earth we can model them as if they lie on the celestial sphere. Note: Sizes and distances are drastically exaggerated. If Earth was the size shown here, the nearest star would be 6000 miles away.

For clarity, in Figure 5.2 and many other figures throughout this book, the sizes of astronomical bodies are exaggerated compared to the distances between them.

The "dome" of stars that we see overhead represents only half of the sky. In fact, stars surround us in all directions—a starscape that we might imagine as a **celestial sphere,** with the Earth at its center, as depicted in Figure 5.2B. The celestial sphere is a model (Unit 4.1). It does not represent physical reality, but it serves as a useful way to visualize the arrangement and motions of celestial bodies.

At any moment as you look at the stars, another half of the celestial sphere lies below the horizon, hidden by the solid Earth beneath your feet. Only if we took a spaceship far enough away from the Earth would we see the whole panorama of stars surrounding us.

5.2 CONSTELLATIONS

The patterns formed by brighter stars sometimes suggest the shapes of animals, personages, or objects of cultural relevance. For example, the pattern of stars shown in Figure 5.3A looks a little like a lion, although other cultures saw a dragon or a sphinx. This **constellation**—a recognized grouping of stars—is known as Leo, a name that has been carried down through the centuries from the Latin word for "lion," and the connection with a lion dates back millennia earlier.

The International Astronomical Union now technically defines a constellation as an area of the sky containing one of these patterns of stars. The entire celestial sphere is divided into 88 constellations. Most of the larger constellations have ancient roots, but some were named more recently by cartographers making maps of the sky and filling in unnamed areas between the better-known constellations. Some groupings of stars have names that remain in common use even though they are not part of the 88 recognized constellations. These "unofficial" groupings of stars are called **asterisms.** For example, the pattern of stars that makes up the head of Leo is sometimes called the "Sickle." Similarly, the "Big Dipper" is an asterism within the constellation Ursa Major (Latin for "large bear").

Many constellations show no obvious resemblance to their namesakes. It takes some imagination to see a swan in the constellation Cygnus (Figure 5.3B), which is

Concept Question 1

If you lived on another planet in the Solar System, would you see the same constellations? What about from a planet orbiting another star?

FIGURE 5.3
The two constellations Leo (A) and Cygnus (B) with figures sketched in to help you visualize the animals they resemble.

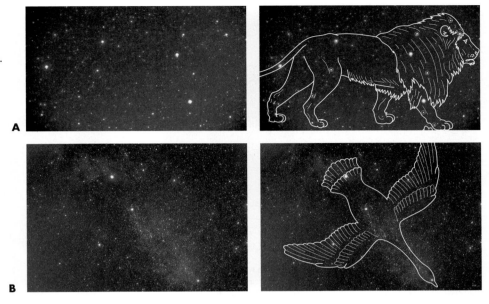

The constellation Orion is shown in Looking Up #6 at the front of the book.

one of the more recognizable constellations. In some cases, factors other than their shape may have played a role. The location of constellations like Ursa Major and Ursa Minor (large bear and small bear), always in the northern part of the sky, may have made early sky watchers think of real or legendary bears in northern lands. In some cases the names are connected with legends regarding other constellations, such as Canis Major and Canis Minor, the large and small dogs who were hunting companions of Orion—one of the more recognizable constellations. Some areas of the southern sky that are not visible from Europe were given names by early European navigators and explorers. This yielded names such as Telescopium, Microscopium, and Antlia, named respectively for the telescope, microscope, and the air pump!

Although most of the constellation names come to us from the classical world of ancient Greece and Rome, most star names come from Arabic words—Aldebaran, Betelgeuse, and Zubeneschamali, to name a few. Many of these star names describe the parts of a constellation; for example, the second brightest star in Leo is called Denebola, from Arabic words meaning "the lion's tail."

Keep in mind that stars in a constellation generally have no physical relation to one another. They simply happen to be in more or less the same direction in the sky. Also, all stars move through space; but as seen from Earth, they usually take tens of thousands of years to make any noticeable shift. Thus, we see today virtually the same pattern of stars that was seen by ancient peoples.

5.3 DAILY MOTION

If you watch stars near the western horizon, in as little as 10 minutes you will notice that they have moved closer to the horizon (Figure 5.4). Likewise, you can see stars rising above the eastern horizon. If you locate a constellation high in the sky and then come back to look at it after a few hours, you will find it has changed position dramatically. Everyone has noticed these phenomena for one star—the Sun—and the cause of stars' nightly motion is no different.

Our cycle of day and night and the motion of the Sun and other stars across the sky occur because, as the Earth spins, we face different parts of the celestial

FIGURE 5.4
Motion of stars near the western horizon.

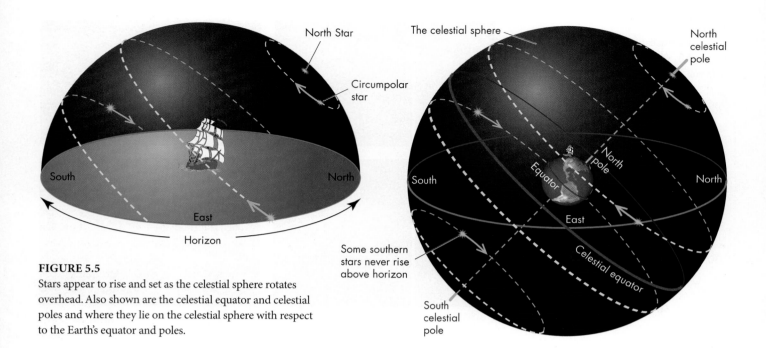

FIGURE 5.5
Stars appear to rise and set as the celestial sphere rotates overhead. Also shown are the celestial equator and celestial poles and where they lie on the celestial sphere with respect to the Earth's equator and poles.

Clarification Point

While it is true that the Sun, Moon, and planets shift relative to the stars on the celestial sphere, these motions are very slow. It is the Earth's rotation that causes these objects always to rise in the east and set in the west.

LOOKING UP

The region of the North Celestial Pole is shown in Looking Up #1 at the front of the book. Note that Polaris is *not* an especially bright star—a common erroneous belief. The region of the South Celestial Pole is shown in Looking Up #9.

Star rise and star set

sphere. If you watch the stars even more carefully, you will discover that not all stars rise and set, and near the northern and southern horizons some stars move nearly parallel to the horizon. How can we make sense of these motions?

From our perspective looking up at the sky, it seems as if the celestial sphere rotates around a stationary Earth (Figure 5.5) once each day. Ancient peoples had no compelling reasons to believe that the Earth spun, so they attributed the daily motion of the Sun and stars to the turning of the vast celestial sphere overhead. Today, of course, we know that it is not the celestial sphere that spins but the Earth; however, we still speak of the Sun and other stars "rising" and "setting."

As the celestial sphere turns overhead, two points on it do not move. These points are defined as the north and south **celestial poles** (Figure 5.5). The celestial poles lie exactly above the North and South Poles of the Earth, and just as our planet turns about a line running from its North Pole to its South Pole, so the celestial sphere seems to rotate around the celestial poles from our perspective on Earth. The star **Polaris** in the constellation Ursa Minor happens to lie close to the north celestial pole. This makes Polaris an important navigation aid, frequently called the "North Star," because it lies close to the direction of true north. There is no comparably bright star anywhere near the south celestial pole, so finding true south from the stars usually requires some triangulation between brighter constellations fairly far from the pole.

Another useful sky marker used by astronomers is the **celestial equator.** The celestial equator lies directly above the Earth's equator, just as the celestial poles lie above the Earth's poles, as Figure 5.5 shows. Stars on the celestial equator rise due east and set due west. Also, because the horizon cuts the celestial equator in half, stars there spend an equal amount of time above and below the horizon.

For people living in the Northern Hemisphere, stars north of the celestial equator spend a longer time above the horizon than below it, while the reverse is true for stars south of the celestial equator. Stars that are near enough the north celestial pole that they always remain above the horizon are said to be **circumpolar**. Likewise, there are some stars near the south celestial pole that always remain below the horizon (Figure 5.5). The reverse is true for observers south of the equator. The portion of the sky we can see depends on on our location on the Earth, as we discuss next.

5.4 LATITUDE AND LONGITUDE

To be more precise about what we can see in the sky, we need to be more exact about our location on the Earth. We can do this using **longitude,** defining our east–west position, and **latitude,** defining our north–south position. Each line of latitude is at a fixed distance from the equator, while lines of longitude run from pole to pole. The positions of these lines are measured not in a linear unit like kilometers, but instead by an angle in degrees (Figure 5.6). Ninety degrees (90°) makes up a right angle, and 360° makes up a full circle. Latitudes are measured north and south relative to the equator; for example, Boston, Massachusetts, and Rome, Italy, are both about 42°N, while Buenos Aires, Argentina, and Sydney, Australia, are both about 34°S.

The position east or west has no such natural reference line as the equator. By international agreement, the north–south line that runs through the Royal Observatory in Greenwich, England, is used to mark 0° longitude, and is known as the **prime meridian.** Using our sample cities above, Boston is 71°W, Rome is 12°E, Buenos Aires is 58°W, and Sydney is 151°E of this line.

When you move to different latitudes, different stars become visible, as illustrated in Figure 5.7. If you are standing at the North Pole, the star Polaris is almost straight overhead, at your **zenith,** while the south celestial pole is straight underneath your feet at your **nadir.** (Like many star names, both of these words have Arabic roots.) From the North Pole, the celestial equator is at your horizon. Therefore no stars south of the celestial equator are ever visible, and the stars that are visible all circle the sky parallel to the horizon. Standing on the equator, 90° from the North Pole, the celestial poles are on the horizon, and stars on the celestial equator pass straight overhead through your zenith as the Earth rotates (Figure 5.7).

Any two locations at the same latitude will see the same parts of the celestial sphere during a night, so Boston's sky is the same as Rome's. The difference in longitude means they do not see it at the same time. Because the Earth rotates by 15° each hour, and Boston is 83° west of Rome, the same star rises (or sets) in Boston about 5½ hours after it rises (or sets) in Rome.

The location of the celestial poles is particularly important for navigation by the stars because the celestial pole is at an angle above the horizon equal to your latitude. Observing from the Earth's North Pole, 90°N, the north celestial pole is 90° above the horizon. From the equator, the north celestial pole is just on the horizon—we could say 0° above it. Thus, if you can find the celestial pole, your latitude equals the angle at which it lies above the horizon.

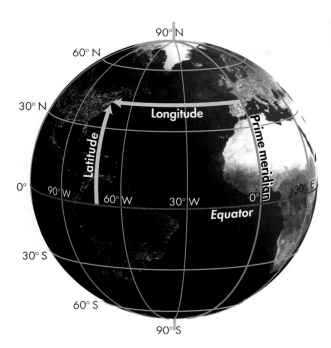

FIGURE 5.6
Latitudes and longitudes on the Earth. Latitudes are measured as the angle north or south of the equator. Longitudes are measured as the angle east or west of the prime meridian. Arrows indicate the latitude and longitude of Boston, Massachusetts.

Mathematical Insight

The Earth rotates 360° in 24 hours, so it rotates 360°/24 hr = 15°/hr. To rotate 83° takes 83° ÷ 15°/hr = 5.53 hr.

Concept Question 2

Can you think of a way to build a simple device to measure the angle of a star above the horizon?

FIGURE 5.7
The appearance of the sky from different latitudes. From the North Pole, stars appear to circle the sky at a fixed height above the horizon, and half the celestial sphere is never visible. At the equator, stars appear to rise and set perpendicular to the horizon, and all parts of the celestial sphere are visible at different times. At intermediate latitudes, the situation is between these extremes.

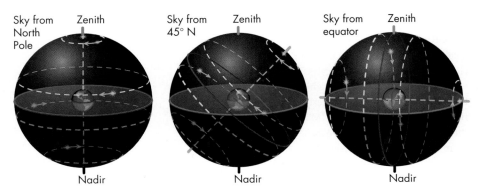

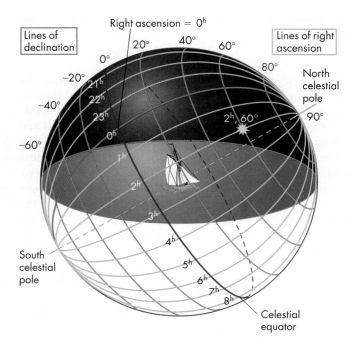

FIGURE 5.8
Locating a star according to right ascension and declination. Sky coordinates are similar to longitude and latitude, although the "celestial longitude" is divided into 24 hours of right ascension.

Concept Question 3

What is the approximate latitude of the boat in Figure 5.8? What different ways can you think of to estimate the latitude?

LOOKING UP

Compare this chart to the photograph in Looking Up #3 at the front of the book, and try to match up the brightest stars.

FIGURE 5.9
Small portion of a detailed star chart. Black circles are stars. Their size indicates their brightness—larger circles are brighter stars. Red ellipses are galaxies; the largest at the right is the Andromeda Galaxy (M31). The shaded blue area is the Milky Way.
From *Sky Atlas 2000.0*, 1st edition, by Wil Tirion. ©1981 Sky Publishing Corp. Reprinted with permission of the publisher.

5.5 CELESTIAL COORDINATES

The coordinate grid used by astronomers is similar to the longitude and latitude system used to describe positions on Earth. A star's location in the sky is described by a **right ascension,** defining its east–west position, and a **declination,** defining its north–south position. Right ascension (or **RA** for short) plays the same role as longitude, while declination (or **dec**) plays the same role as latitude.

The celestial sphere can thus be divided into a grid consisting of east–west lines parallel to the celestial equator, and north–south lines connecting one celestial pole to the other (Figure 5.8). Declination values run from $+90°$ to $-90°$ (from the north to the south celestial poles), with $0°$ at the celestial equator. Right ascension values can likewise be recorded in degrees, but they are commonly listed in "hours." Just as we can divide a circle into $360°$, we can divide it into 24 "hours." Each hour of RA equals $15°$; that is, $360° \div 24 = 15°$. The convenience of this system is that if a star at RA = 20^h is overhead now, a star at RA = 23^h will be overhead three hours from now. The right ascension of an object can be further refined to minutes (m) and seconds (s), just as we divide time intervals.

With a set of coordinate lines established, we can now locate astronomical objects in the sky the same way we can locate places on the Earth. Astronomers use star charts for this purpose, much as navigators use maps to find places on Earth. Part of a detailed star chart is shown in Figure 5.9. It shows the location of stars and other objects. It also gives some indication of the relative brightness of the stars by marking their positions with larger or smaller dots. Many charts also have information about the season and time of night at which the stars are visible. The foldout star chart in the back of this book labels the dates when different stars will be overhead at 8 P.M. in the evening—a common time to be outside viewing the stars.

In Figure 5.9 an oval-shaped object is located at right ascension 0 hours 43 minutes ($0^h\ 43^m$), declination $+41°$. This is a nearby galaxy called M31, located in the Local Group. Because its declination very nearly matches the latitude of Boston and Rome, M31 passes nearly through the zenith once each day as observed from both cities. From any location, if a star's declination matches your latitude, it will pass through your zenith.

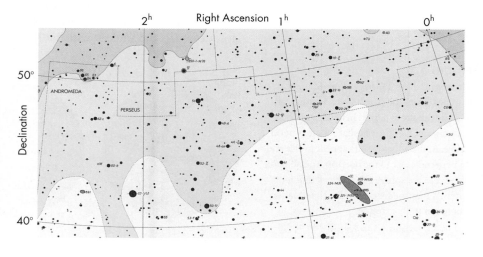

KEY POINTS

- Although stars are at different distances, they look as though they are located on a "celestial sphere" that surrounds the Earth.
- Names have been given to constellations of stars in the sky, usually based on objects or animals that their pattern suggests.
- From the surface of a spinning Earth, the celestial sphere appears to rotate around us, causing stars to "rise" and "set."
- The celestial sphere appears to rotate around celestial poles that lie straight above the Earth's poles; the celestial equator lies above Earth's equator.
- Positions on the Earth (and the celestial sphere) are defined by an east–west longitude (right ascension) coordinate and a north–south latitude (declination) coordinate.
- The apparent motion of the stars across the night sky depends on your latitude. A particular star will rise at different times when observed from different longitudes.

KEY TERMS

asterism, 34
celestial equator, 36
celestial pole, 36
celestial sphere, 34
circumpolar, 36
constellation, 34
declination (dec), 38
horizon, 33
latitude, 37
longitude, 37
nadir, 37
Polaris, 36
prime meridian, 37
right ascension (RA), 38
zenith, 37

CONCEPT QUESTIONS

Concept Questions on the following topics are located in the margins. They invite thinking and discussion beyond the text.

1. Constellations from other planets. (p. 34)
2. Measuring star angles. (p. 37)
3. Ways of estimating latitude. (p. 38)

REVIEW QUESTIONS

4. What is the celestial sphere?
5. What is a constellation? How does it differ from an asterism?
6. How are latitudes and longitudes defined on the Earth?
7. How are the celestial poles and celestial equator related to the Earth?
8. What is right ascension? declination?
9. Why is right ascension measured in hours, rather than degrees?

QUANTITATIVE PROBLEMS

10. What are your latitude and longitude? What are the latitude and longitude at the point on the Earth exactly opposite you (in the direction of your nadir)? What geographic features are located there?
11. If your observatory is located at a latitude of 45° N and a longitude of 90° W, what range of declinations are visible from your observatory? If you wanted to see the most stars, at what latitude should you build your observatory?
12. The constellations Ursa Major (the Great Bear), in the Northern Hemisphere, and Crux (the Southern Cross), in the Southern Hemisphere, can be used to locate the north and south celestial poles, respectively. Using the star charts at the end of the book can you show how this is done?
13. Buenos Aires and Sydney are both at latitude 34°S, but Buenos Aires is at longitude 58°W and Sydney at 151°E. Using the star chart at the back of the book, find a bright star that passes through the zenith of Buenos Aires and Sydney. How many hours after passing through the zenith in Buenos Aires will it pass through the zenith in Sydney?
14. How far is the celestial equator from your zenith if your latitude is 31° N? 15° S?
15. Using a protractor, draw a diagram of the Earth. Mark the poles, the equator, and your own latitude. With a ruler, find the line that is "tangent" to your location—that is, the line should touch the Earth's surface at your location while matching the surface at this point. This tangent line represents your horizon. Draw a line extending from pole to pole to indicate the Earth's axis, and extend both this line and your horizon line until they cross. Show that your latitude is the same as the angle of the celestial pole with respect to your horizon.
16. If you are observing the sky at a latitude of 39° N, what range of declinations do circumpolar stars fall within?

TEST YOURSELF

17. What makes the star Polaris special?
    a. It is the brightest star in the sky.
    b. The Earth's axis points nearly at it.
    c. It sits a few degrees above the northern horizon from any place on Earth.
    d. It is part of one of the most easily recognized constellations in the sky.
    e. All of the above.
18. Where must you be located if a star with a declination of +23° passes through your zenith?
    a. A longitude of 23° N
    b. A longitude of 67° N
    c. A longitude of 23° S
    d. A longitude of 67° S
    e. A star with a declination of +23° cannot pass through anyone's zenith.
19. If you are standing at the Earth's North Pole, which of the following will always be at the zenith?
    a. The celestial equator
    b. The Moon
    c. M31
    d. The north celestial pole
    e. The Sun
20. If you are standing at the Earth's equator, which of the following will always be at the zenith?
    a. The celestial equator
    b. The Moon
    c. M31
    d. The north celestial pole
    e. The Sun

UNIT 6

The Year

PART 1

6.1 Annual Motion of the Sun
6.2 The Ecliptic and the Zodiac
6.3 The Seasons
6.4 The Ecliptic's Tilt
6.5 Solstices and Equinoxes
6.6 Precession

Learning Objectives

Upon completing this Unit, you should be able to:
- Describe the annual motions of the Sun relative to the stars and your horizon.
- Recall the dates and names when the Sun reaches its extreme and midway points.
- Define the ecliptic and zodiac and relate them to the Earth's orbit and rotation axis.
- Explain how and why solar heating varies during the year depending on latitude.
- Describe the Earth's precession and its effects on our view of the Sun and sky.

For people long ago, observations of the heavens had more than just curiosity value. Because so many astronomical phenomena are cyclic—that is, they repeat at a regular interval—they can serve as timekeepers. The most basic of these cycles is the rhythm of day and night, as the celestial sphere appears to rotate about the Earth, as described in Unit 5. This cycle is not completely uniform—days and nights alternately lengthen and shorten over the year. This slow rhythm is tied to a gradual shift of the Sun's apparent position relative to the "fixed stars" on the celestial sphere. The shifting position of the Sun also leads to seasonal changes in the weather and temperature.

The motion of the Sun against the celestial sphere provides a means for tracking these changes predictably. We might imagine ancient peoples asking, When is it time to plant crops? Or move to the next location to ensure a ready supply of water? Or prepare for winter? Some societies built monumental structures to mark the changing position of the Sun. An example is the Mayan pyramid at Chichén Itzá. The pyramid was designed so that on the first day of spring or fall, shadow play creates the appearance of a serpent slithering down the staircase (Figure 6.1).

We will use much of the terminology for positions in the sky from Unit 5. Review that Unit if terms like *celestial sphere*, *constellation*, *zenith*, or *declination* are unfamiliar.

FIGURE 6.1
In late afternoon on the first day of spring and fall, the Sun casts a shadow that resembles a serpent slithering down the steps of the Mayan pyramid at Chichén Itzá. The head of the serpent is depicted in a sculpture at the base of the stairs.

6.1 ANNUAL MOTION OF THE SUN

As the Earth orbits the Sun, the stars that are visible each night change. The shift is so slow that it is difficult to appreciate from one night to the next, but in the span of a month the changes become obvious. Because these movements repeat after the Earth completes one orbit around the Sun—a **year**—they are called *annual motions*.

If you watch the sky each evening over several months, you will discover that new constellations appear in the eastern sky and old ones disappear from the western sky. For example, across most of North America, Europe, and Asia on an early July evening, the constellation Scorpius will be visible in the southern half of the sky. However, on December evenings the brilliant constellation Orion, the hunter, is visible instead.

The realization that different stars are visible at different times of the year was extremely important to early peoples because it provided a way to predict the changing of the seasons. A farmer might be tricked into planting too early by a short spell of warmer-than-normal weather in the late winter, but experience taught that each year the stars could reliably predict when spring was arriving. For example, if you live in the Northern Hemisphere and the early evening sky shows Leo in the south instead of Scorpius or Orion, it will soon be time to plant. Even in semitropical climates, where seasonal temperature differences are much smaller, planting with accuracy is also necessary to avoid crop damage due to annual flooding or dry periods.

The changing of constellations throughout the year is caused by the Earth's motion around the Sun. As the Earth **revolves** (orbits) around the Sun, the Sun's glare blocks our view of the part of the celestial sphere that lies in the direction of the Sun, making the stars that lie beyond the Sun invisible, as Figure 6.2 shows. For example, in early June, a line from the Earth to the Sun points toward the constellation Taurus, and its stars are completely lost in the Sun's glare. In the dusk after sunset, however, it is possible to see the neighboring constellation, Gemini, just above the western horizon. A month later, from the Earth's new position, the Sun lies in the direction of Gemini, causing this constellation to disappear in the Sun's glare. Looking to the west just after sunset, it is possible to see, just barely, the dim stars of the constellation Cancer above the horizon. A month after that, the Sun is in Cancer, and the constellation Leo is visible just above the horizon after sunset.

Month by month, the Sun hides one constellation after another. It is like sitting around a campfire and not being able to see the faces of the people on the far side. However, if we get up and walk around the fire, we can see faces that were previously hidden. Similarly, the Earth's motion allows us to see stars previously hidden in the Sun's glare.

Constellations by season (ANIMATION)

Ancient Egyptians watched for the first signs of the "dog star" Sirius rising just before dawn. They used this to predict when the Nile would flood, and it also marked the start of summer. This is the origin of the phrase "dog days" of summer.

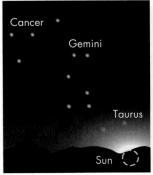  

FIGURE 6.2
The changing appearance of the evening sky over several months. In early June, the Sun appears to lie in the constellation Taurus and the constellation Gemini is visible in the west just after sunset. A month later the Sun is in Gemini, and a month after that it is in Cancer.

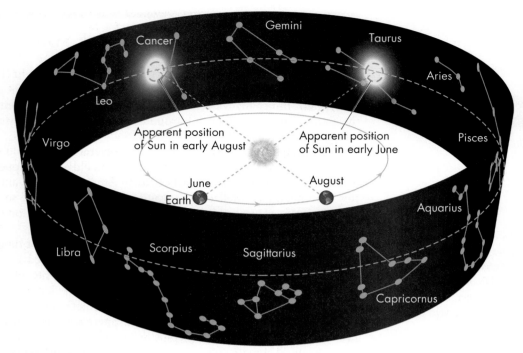

FIGURE 6.3
As the Earth orbits the Sun, the Sun appears to move around the celestial sphere through the background stars. The Sun's path is called the *ecliptic*. The Sun appears to lie in Taurus in June, in Cancer during August, in Virgo during October, and so forth. Note that the ecliptic is the extension of the Earth's orbital plane out to the celestial sphere. (Sizes and distances of objects are not to scale.)

6.2 THE ECLIPTIC AND THE ZODIAC

If we could mark on the celestial sphere the path traced by the Sun as it moves through the constellations, we would see that it moves around the celestial sphere, as illustrated in Figure 6.3. Astronomers call the path that the Sun traces across the celestial sphere the **ecliptic.** You can see in Figure 6.3 that the ecliptic is the extension of the Earth's orbit onto the celestial sphere, just as the celestial equator is the extension of the Earth's equator onto the celestial sphere.

The name *ecliptic* comes from the fact that only when the new or full moon crosses this line can an eclipse occur. (See Unit 8.)

The ecliptic passes through a dozen constellations, which are collectively called the **zodiac.** The word *zodiac* comes from Greek roots meaning "animals" (as in **zo**ology) and "circle" (as in **dia**meter). That is, *zodiac* refers to a circle of animals, which is what most of its constellations represent: Aries (ram), Taurus (bull), Gemini (twins), Cancer (crab), Leo (lion), Virgo (virgin), Libra (scale), Scorpius (scorpion), Sagittarius (archer), Capricornus (goat), Aquarius (water bearer), and Pisces (fish). Actually, the ecliptic passes through a thirteenth constellation, Ophiuchus (serpent holder), during the first half of December (between Scorpius and Sagittarius); but this constellation was not included in the zodiac, probably because of some uncertainty in ancient times about the precise path of the Sun and some vagueness about the boundaries of the constellations.

Concept Question 1

Can you think of an astronomical reason why the zodiac may have been divided into 12 signs rather than 13—or some other number entirely?

The names of some of the constellations of the zodiac may have originated in the seasons when the Sun passed through them. For example, rainy weather in much of Europe during winter was foretold by the Sun's appearance in the constellation Aquarius (the water bearer). Likewise, the harvest time was indicated by the Sun's appearance in Virgo (the virgin), a constellation often depicted as the goddess Proserpine, holding a sheaf of grain.

6.3 THE SEASONS

Many people mistakenly believe that we have seasons because the Earth's distance from the Sun changes. They suppose that summer occurs when we are closest to the Sun and winter when we are farthest away. It turns out, however, that the Earth is closest to the Sun in early January, when the Northern Hemisphere is coldest. Clearly then, seasons must have some other cause.

To see what does cause our seasons, we need to look at how our planet is oriented in space. As the Earth orbits the Sun, our planet also spins or **rotates.** That spin is around a line—the **rotation axis**—which we might imagine running through the Earth from its North Pole to its South Pole. The Earth's rotation axis is not perpendicular to its orbit around the Sun. Rather, it is tipped by 23.5° from the vertical, as shown in Figure 6.4A.

As our planet moves along its orbit, its rotation axis maintains nearly exactly the same tilt and direction, as illustrated in Figure 6.4B. That is, the Earth behaves much like a giant gyroscope or spinning top. In fact, every spinning object shows this tendency to maintain its orientation. This is why you need to put spin on a frisbee to keep it from flipping over when you throw it, and it is why a quarterback puts "spin" on a football. The constancy of our planet's tilt as we move around the Sun causes sunlight to fall more directly on the Northern Hemisphere for half of the year and more directly on the Southern Hemisphere for the other half of the year, as seen in Figure 6.4B. This in turn changes the amount of heat each hemisphere receives from the Sun.

A surface directly facing the Sun receives the most concentrated sunlight. If the surface receives the sunlight at an angle, the light is spread out over a larger area and therefore is less concentrated. You take advantage of this effect instinctively when you warm your hands at a fire by holding your palms flat toward the fire. You also may have experienced the high temperature of pavement or a beach around noon, when the Sun is shining most directly on it, whereas the same surface will be cooler in the late afternoon, even though it is not shaded, when sunlight strikes it more obliquely.

Clarification Point

Seasons are not caused by the Earth's distance from the Sun.

The constancy of Earth's tilt is a consequence of the *conservation of angular momentum*. (See Unit 20.)

Concept Question 2

When it is summer in the Northern Hemisphere, what is the season in the Southern Hemisphere? What does this demonstrate about possible causes of the seasons?

FIGURE 6.4
(A) The Earth's rotation axis is tilted 23.5° to its orbit around the Sun. (B) The Earth's rotation axis keeps nearly the same tilt and direction as it revolves (orbits) around the Sun. As a result, sunlight falls more directly on the Northern Hemisphere during half of the year and on the Southern Hemisphere during the other half of the year. (Sizes and distances are not to scale.)

Earth's rotation axis

Seasons

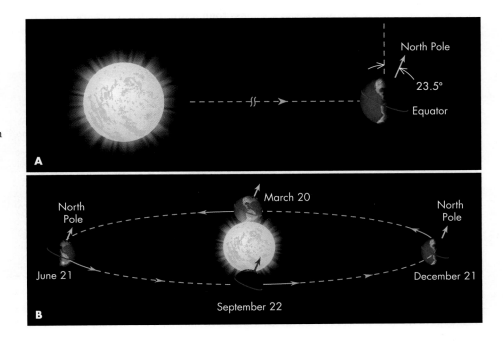

FIGURE 6.5

The portion of the Earth's surface directly facing the Sun receives more concentrated light (and thus more heat) than other parts of the Earth's surface. The same size "beam" of sunlight (carrying the same amount of energy) spreads out over a larger area where the surface is "tilted."

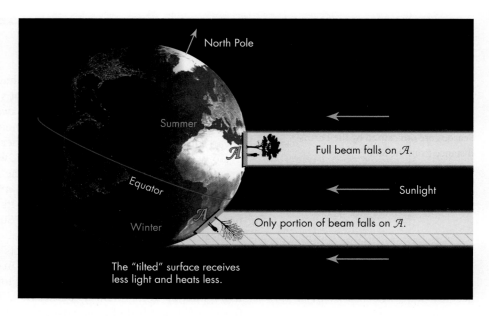

Concept Question 3

If the shape of the Earth's orbit were unaltered but its rotation axis were shifted so that it had no tilt with respect to the orbit, how would seasons be affected?

Clarification Point

Seasonal differences between the north and south are *not* caused by one hemisphere being closer to the Sun than the other hemisphere.

The effect of the Sun's angle relative to the ground is illustrated in Figure 6.5 for regions at similar latitudes in the Northern and Southern Hemispheres. An astronomer might compare these regions in terms of the energy received per square meter. The piece of land directly facing the Sun receives about 1300 watts of solar power over every square meter. Where the surface is tilted at an angle to the Sun's light, the same 1300 watts are spread out over a larger area on the ground, and each square meter of the Earth's surface receives only a fraction as much energy. The same change occurs in one spot over the course of a year because of the tilt of the Earth's axis, as shown in Figure 6.6, and this is the cause of the seasons.

An important point here is not to confuse the "directness" of the Sun's light with one hemisphere being closer to the Sun. It is true that the Northern Hemisphere of the Earth is a few thousand kilometers closer to the Sun than the Southern Hemisphere during the northern summer. However, the effect of this difference in distance is tiny. Compared to the millions of kilometers of distance to the Sun, this difference in distance between the two hemispheres changes the solar heating by less than a hundredth of one percent. By contrast, the differing angle at which the Sun shines on higher latitudes during the year changes the solar energy absorbed by the ground by a factor of two or more.

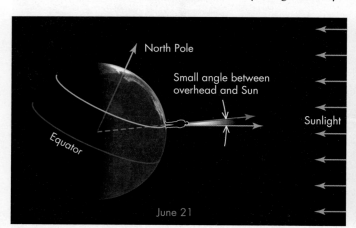
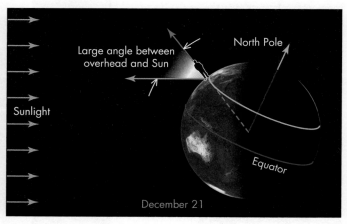

FIGURE 6.6

Why the Sun at noon is high in the sky in summer and low in the sky in winter. On June 21, from a latitude of 40°N, the noontime Sun is at an angle of just 16.5° (= 40° − 23.5°) from the zenith. On December 21, the Sun is at an angle of 63.5° (= 40° + 23.5°) from the zenith.

6.4 THE ECLIPTIC'S TILT

The tilt of the Earth's rotation axis not only causes heating differences, it makes the Sun appear to move north and south on the celestial sphere. Because of the 23.5° tilt of the Earth's axis, when the Earth is at the point of its orbit where the North Pole is most tipped toward the Sun, the Sun will pass straight overhead for someone at a latitude of 23.5°N. This means that the Sun lies 23.5° north of the celestial equator on about June 21. In other words its declination is +23.5° (see Unit 5.5). Likewise, on December 21 the Sun's declination is −23.5°. The Sun lies north of the celestial equator for half of the year and south of the celestial equator for the other half of the year.

On about March 20 and September 22, the Sun's path—the ecliptic—crosses the celestial equator. Just as the Earth's axis is tilted by 23.5° relative to its orbit, the ecliptic is tilted 23.5° relative to the celestial equator, as the sequence of sketches in Figure 6.7 shows. In terms of celestial coordinates (Unit 5.5), the Sun is at a declination of 0° on the March 20 and September 22, while it is at plus or minus 23.5° on June 21 and December 21, respectively. The Sun's position at the moment it crosses the celestial equator on the vernal (March) equinox is used to define 0 hours (or 0^h) for the right ascension system. On June 21 the Sun moves to a right ascension of about 6 hours, then 12 hours on September 22, then 18 hours on December 21, before returning to 0 hours of right ascension a year later (see Figure 6.7).

The Sun's position in the celestial sphere is shown in star charts by the curving line of the ecliptic—see the foldout star chart in the back of the book.

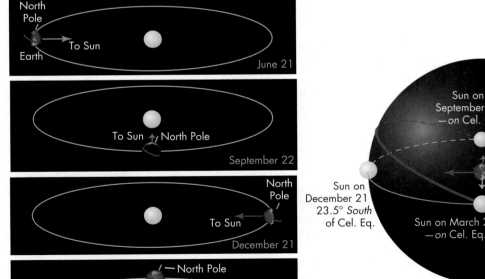

FIGURE 6.7
As the Earth orbits the Sun, the Sun's position with respect to the celestial equator changes. The Sun reaches 23.5° *north* of the celestial equator on June 21 but 23.5° *south* of the celestial equator on December 21. The Sun crosses the celestial equator on about March 20 and September 22 each year. The times when the Sun reaches its extremes are known as the solstices; the times when it crosses the celestial equator are the equinoxes. (The dates can sometimes vary because of the extra day inserted in leap years.)

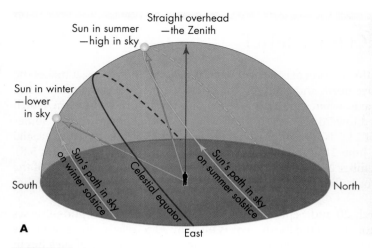
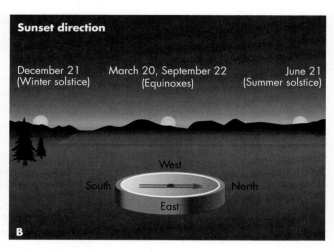

FIGURE 6.8
(A) The Sun's path on the sky changes during the year as it shifts north and south of the celestial equator. The paths here are illustrated for a latitude of 40°N. (B) The direction of the rising and setting Sun changes throughout the year. At the equinoxes, the rising and setting points are due east and due west. The sunset direction shifts slowly north from March until the summer solstice, after which it shifts back, reaching due east at the autumnal equinox. The sunset direction continues moving south until the winter solstice, then reverses direction again back to the north.

6.5 SOLSTICES AND EQUINOXES

The tilt of the ecliptic with respect to the celestial equator means that during the year, the Sun's path across the sky each day changes as illustrated in Figure 6.8A. This is most easily observed by watching where the Sun rises and sets. This is not due east or west, except when the Sun is crossing the celestial equator, on two days of the year that are called the **equinoxes.** The word comes from the Latin for "equal night," so named because the length of the night is approximately equal to the length of the day on those dates. The **vernal equinox** occurs near March 20, when the Sun is moving from the Southern Hemisphere of the celestial sphere into the Northern Hemisphere. Six months later the **autumnal equinox** occurs near September 22 as the Sun crosses the celestial equator on its way south. In the Northern Hemisphere, these dates mark the first day of spring and fall, respectively, but in the Southern Hemisphere this is reversed.

On every other day of the year the Sun rises either north or south of due east in a regular, predictable fashion as illustrated in Figure 6.8B. From the vernal equinox to the autumnal equinox (during the Northern Hemisphere spring and summer, and the Southern Hemisphere fall and winter), the Sun rises in the northeast and sets in the northwest. During the rest of the year, the Sun rises in the southeast and sets in the southwest.

The exact dates of the equinoxes and solstices vary slightly from year to year, mainly because of differences in the calendar due to leap years, but also because of slight variations in the Earth's orbit.

When the Sun reaches its farthest point north or south on the celestial sphere, 23.5° from the celestial equator, the Sun pauses in its north–south motion and changes direction. Accordingly, these times are called the **solstices,** meaning the Sun (*sol*) stops its northward or southward motion. The winter solstice is particularly celebrated by many northern cultures with holidays and festivals, often symbolizing rebirth as this date marks the beginning of longer days.

Ancient peoples all over the world used the northward and southward journeys of the Sun to track the seasons. They built a variety of structures to detect the limits of the Sun's motion. One of the best-known of these structures is Stonehenge, an ancient stone circle in England (Figure 6.9A). Although its exact use in ancient times is lost to us, it was laid out so that seasonal changes in the Sun's position

The Sun's seasonal motion

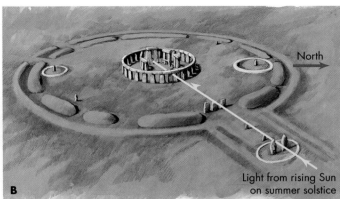

FIGURE 6.9
Stonehenge. (A) Massive stones were erected by ancient Britons more than 4000 years ago to mark the changing position of the Sun. (B) Diagram showing how an observer in the stone circle would see the rising Sun framed by a pair of standing stones on the summer solstice.

could be observed by noting through which stone arches the Sun rose or set. In particular, when the Sun reaches its farthest point north, an observer standing at the center of this circle of immense vertical stones would see the rising Sun framed by standing stones outside the main circle (Figure 6.9B).

Other cultures around the world built a variety of structures to track the Sun's motions. A Mayan pyramid that marks the equinoxes was noted in the chapter opening (Figure 6.1). Ancient Egyptians built their great Karnak temple complex at Luxor (ancient Thebes) along an axis aligned with the winter solstice sunrise (Figure 6.10A). An ancient observatory built about 2300 years ago in Peru was designed to track the Sun's progress throughout the year as the Sun shifted position. Sunrise positions could be measured against a series of towers constructed along a ridge to the east of the observatory (Fig. 6.10B).

The Sun's declination changes only gradually from one day to the next, so just like any other star, the Sun passes overhead at the latitude corresponding to its declination (see Unit 5.5). This means that for tropical regions, between latitudes 23.5°N and 23.5°S, the Sun passes straight overhead on two days during the year as the Sun moves south and then north again. The northern limit of tropical latitudes, 23.5°N, is called the **Tropic of Cancer** because the Sun reached its point

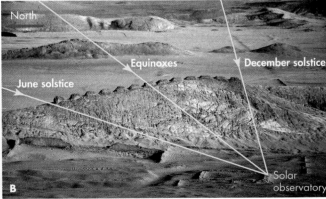

FIGURE 6.10
(A) The direction of the rising and setting Sun changes throughout the year. At the equinoxes, the rising and setting points are due east and due west. The sunrise direction shifts slowly north from March until the summer solstice, after which it shifts back, reaching due east at the autumnal equinox. The sunrise direction continues moving south until the winter solstice, then reverses direction again back to the north. (B) The oldest known astronomical observatory in the Americas is found in Chankillo, Peru. This ancient observatory marked the shifting position of sunrise with a series of 13 towers along a ridge built about 2300 years ago.

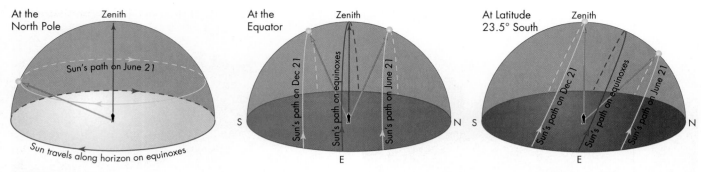

FIGURE 6.11

The path of the Sun in the sky differs depending on your latitude. These diagrams illustrate the Sun's path in the sky for three different latitudes: the North Pole (90° North); the equator (0°); and 23.5° South. At the North Pole, the Sun never sets for six months, but gradually spirals up from the horizon from the vernal equinox to the summer solstice, then spirals back down to the horizon at the autumnal equinox before it disappears for six months. At the equator, the Sun rises straight upward from the horizon, but reaches the zenith only on the equinoxes. At 23.5° South, the Sun reaches the zenith at noon only on December 21, the start of summer in the Southern Hemisphere.

Mathematical Insight

On June 21 at latitude 23.5°N, the Sun is straight overhead at noon—0° from the zenith. Because the relative angle of the Earth's surface depends on the difference of latitudes, at 40°N the Sun is 16.5° (= 40°N − 23.5°N) from the zenith. On December 21, when the Sun is overhead at 23.5°S, the difference in latitudes places the Sun 63.5° from the zenith.

Concept Question 4

During the course of the year, the sunset (and sunrise) position shifts. How should the amount of the shift depend on latitude?

farthest north when it was in the constellation Cancer at the time this term was defined. The southern limit is the called the **Tropic of Capricorn** after the constellation where the Sun was at its farthest south position. The Sun is no longer in these constellations when it is at its northern and southern extremes because of *precession* (Section 6.6).

Outside of the tropics, the Sun never passes straight overhead, but varies in its height above the horizon depending on latitude. This reaches an extreme in polar regions where the Sun does not rise during some part of the year, and does not set during another part of the year. At a latitude 23.5° from either pole—66.5°N, the **Arctic Circle,** and 66.5°S, the **Antarctic Circle**—the Sun stays up for 24 hours on the first day of summer, just touching the horizon at midnight, and stays below the horizon all day on the first day of winter.

The path of the Sun on the sky each day looks quite different in these different regions as illustrated in Figure 6.11. At the poles, the Sun circles the sky parallel to the horizon, while at the equator the Sun rises and sets perpendicular to the horizon. At intermediate latitudes, the amount of sunlight varies each day, depending how far you are from the equator and the distance of the Sun north or south of the celestial equator. (Differences in the length of the day are examined in more detail in Unit 7.2.) Thus, because of the Earth's tilted axis, in summer we receive more hours of sunlight in addition to having the Sun's light strike the surface more directly.

The seasons "officially" begin on the solstices and equinoxes, with northern spring running from the vernal equinox to the solstice in June. Even though the longest day is on the first day of summer, the hottest period of the year occurs roughly six weeks later, as shown for four cities in Table 6.1. The delay, known as the *lag of the seasons,* results from the oceans and land being slow to warm up in summer. Similarly, there is about a six-week lag after the shortest day of the year until the coldest period of the year.

TABLE 6.1 Monthly Average Temperatures in Four Cities (in Degrees Celsius)

City	Latitude	Jan.	Feb.	Mar.	April	May	June	July	Aug.	Sept.	Oct.	Nov.	Dec.
Buenos Aires	34°S	23.5	22.7	20.6	16.7	13.3	10.4	10.0	11.1	13.2	16.0	19.3	22.0
Boston	42°N	−2.2	−1.6	2.5	8.2	14.1	19.4	22.5	21.5	17.3	11.5	5.5	0.0
Rome	42°N	7.1	8.2	10.5	13.7	17.8	21.7	24.4	24.1	20.9	16.5	11.7	8.3
Sydney	34°S	22.1	22.1	21.0	18.4	15.3	12.9	12.0	13.2	15.3	17.7	19.5	21.2

6.6 PRECESSION

If you watch a spinning top, you will see that it "wobbles," often more extremely as it slows down. That it wobbles is another way of saying that its rotation axis slowly shifts direction (Figure 6.12). The spinning Earth wobbles too, in a motion called **precession.** Precession occurs very slowly for the Earth. A single "wobble" takes about 26,000 years, but it has both interesting and important consequences.

Currently the Earth's North Pole points very close to the star Polaris. But this is only temporary. When the Egyptian pyramids were built 4000 years ago, the "North Star" was Thuban (meaning "the star") in the constellation Draco (Figure 6.12). In the future the axis will continue shifting direction past Polaris, and it will not point close to any bright stars for thousands of years. In about 7000 years the south celestial pole will be very close to a star slightly brighter than Polaris in the constellation Vela. In that future time there will be a "South Star." In 12,000 years the rotation axis will have shifted so that the north celestial pole points fairly close to the bright star Vega. Then we will have a new, much brighter "North Star."

The changing direction of the Earth's pole does not alter the Earth's orbit, so the ecliptic and the constellations of the zodiac remain the same. However, it does change which constellation the Sun is in on the equinoxes and solstices. Several thousand years ago the Sun was in Cancer on the first day of summer, giving us the name "Tropic of Cancer" for the northernmost latitude where the Sun is ever directly overhead. Today it is in Gemini on the first day of summer. Because astronomers base the zero point of right ascension on the Sun's position at the vernal equinox (Section 6.5), celestial coordinates change a little bit every year.

Precession also slowly alters Earth's climate. At this time we are closest to the Sun during the northern winter. In about 13,000 years we will be farthest from the Sun during the northern winter. This will make seasons in the Northern Hemisphere more severe at that time. Precession is suspected to be one of the components that affect long-term changes in climate, which may have triggered past ice ages.

Precession is caused primarily by the Moon's gravitational pull trying to "straighten out" the direction of the Earth's spin.

LOOKING UP

Polaris and Thuban can be seen in Looking Up #1 at the front of the book.

You may be amused to learn that because of precession, horoscopes listed in newspapers are incorrect by about 1 month. If you thought your "sign" was Cancer, then you are probably "a Gemini"; if Gemini, then Taurus; etc.

FIGURE 6.12
Precession makes the Earth's rotation axis swing around, slowly tracing out a circle in the sky, somewhat like a spinning top.

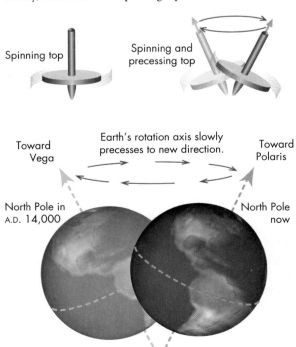

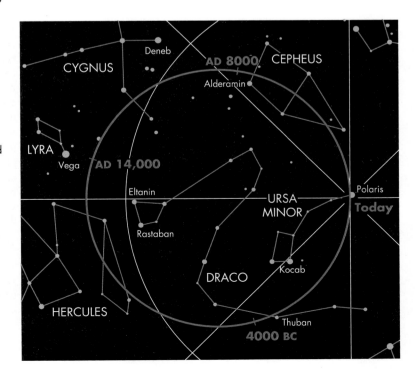

KEY POINTS

- The Sun appears to shift position among the stars during the course of the year as the Earth orbits it.
- The Sun's path on the celestial sphere is called the ecliptic, and the 12 constellations it moves through are known as the zodiac.
- The Earth's spin axis is tilted 23.5° relative to its orbit around the Sun, so the ecliptic is tilted relative to the celestial equator.
- Because of this tilt, the Northern Hemisphere receives sunlight at a more direct angle and for a longer period of time each day during half of the year. In the other six months the opposite is true.
- The changing amount of solar heating causes the seasons, and they are opposite in the Southern Hemisphere.
- The changing direction of sunrise and sunset throughout the year has been observed by peoples back to ancient times.
- The direction of the spin of the Earth's axis has been found to change very slowly over a 26,000-year period, an effect called precession.

KEY TERMS

Antarctic Circle, 48
Arctic Circle, 48
autumnal equinox, 46
ecliptic, 42
equinox, 46
precession, 49
revolve, 41
rotate, 43
rotation axis, 43
solstice, 46
Tropic of Cancer, 47
Tropic of Capricorn, 48
vernal equinox, 46
year, 41
zodiac, 42

CONCEPT QUESTIONS

Concept Questions on the following topics are located in the margins. They invite thinking and discussion beyond the text.

1. Why the zodiac is divided into 12 parts. (p. 42)
2. Differences in northern and southern seasons. (p. 43)
3. Seasons if the Earth's axis were not tilted. (p. 44)
4. The positions of sunset at different latitudes. (p. 48)

REVIEW QUESTIONS

5. What is the ecliptic? What is the zodiac?
6. What causes the seasons?
7. When it is winter in Australia, what season is it in the United States?
8. Where is the Sun located on the celestial sphere during the equinoxes and solstices?
9. Why is the summer solstice not the hottest day of the year?
10. How does the Sun's position on the horizon at sunset change through the course of the year?
11. What effect does precession of the Earth's rotation axis have on the Sun's location in the zodiac?

QUANTITATIVE PROBLEMS

12. Suppose the Earth's axis was not tilted, instead of its actual tilt of 23.5°. Where would the tropics and arctic regions be? How would seasons be different?
13. Suppose the Earth's axis were tilted by 50° instead of 23.5°. Where would the tropics and arctic regions be? How would seasons be different?
14. Suppose the Earth's axis were tilted by 90° instead of 23.5°. Where would the tropics and arctic regions be? How would the seasons be different?
15. Describe the motion you would see on the solstices and the equinoxes if you were observing the Sun from the Arctic Circle, at a latitude of 66.5°N.
16. If you wished to observe a star with a right ascension of 12^h, what would be the best time of year to observe it? What would be the best time to observe a star with a right ascension of 6^h? (Also see Unit 5.5.)

TEST YOURSELF

17. From what location on Earth will the Sun always rise due east and set due west?
    a. Everywhere
    b. Nowhere
    c. Either of the tropic lines
    d. Either of the poles
    e. The equator
18. On what day(s) of the year are nights longest at the equator?
    a. They are the same length throughout the year there.
    b. The solstices
    c. The equinoxes
    d. Around June 21
    e. Around December 21
19. During summer in either hemisphere the temperature is higher because the Sun
    a. is higher in the sky during the day.
    b. rises due east and sets due west.
    c. always passes through the zenith at noon.
    d. releases more energy.
    e. is closer to the Earth.
20. For someone in the Southern Hemisphere, which of the following is correct?
    a. The Sun rises in the west.
    b. The Sun rises in the southeast on December 21.
    c. Summer occurs when the Sun is rising lowest in the sky.
    d. The Sun is in the opposite sign of the zodiac than for an observer in the Northern Hemisphere.
    e. All of the above.
21. What is the slow shift of the position of the celestial poles?
    a. Solstice
    b. Ecliptic
    c. Precession
    d. Equinox
    e. Year

UNIT 7

The Time of Day

7.1 The Day
7.2 Length of Daylight Hours
7.3 Time Zones
7.4 Daylight Saving Time
7.5 Leap Seconds

Learning Objectives

Upon completing this Unit, you should be able to:
- Understand the complications in defining a day, and define a sidereal day.
- Explain the difference between apparent noon, and clock noon.
- Describe how the hours of daylight vary with latitude.
- Explain the function of time zones and the international dateline, and calculate time differences from a time zone map.
- Describe the purpose of daylight saving time and of the addition of leap seconds.

From before recorded history, people have used events in the heavens to mark the passage of time. The day was the time interval from sunrise to sunrise, and the time of day could be determined from how high the Sun was in the sky. As our ability to independently measure time has become more accurate, we have found that the apparent motions of the Sun across the sky are not as uniform as we once thought, and in this age of rapid travel and communications it no longer makes sense for each town to set its own time according to the Sun. With high-precision modern clocks we have even detected a gradual slowing of the Earth's spin!

Each of these adjustments to our understanding of how to keep time provides an insight into the workings of astronomy. In this Unit we explore the motions of the Sun in detail. We use some basic ideas of day and night and celestial coordinates presented in Unit 5 and of the apparent motion of the Sun against the stars from Unit 6. If these ideas are unfamiliar, you may want to review those Units first.

7.1 THE DAY

The length of the day is set by the Earth's rotation speed on its axis. One day corresponds to one rotation. However, we must be careful how we measure our planet's rotation. For example, we might use the time from one sunrise to the next to define a day. That was a traditional definition that, after all, is what establishes the day–night cycle around which we structure our activities. However, the time from sunrise to sunrise changes steadily throughout the year as a result of the seasonal change in the number of daylight hours. A better time marker is the time it takes the Sun to move from its highest point in the sky on one day, what we technically call **apparent noon,** to its highest point in the sky on the next day—a time interval that we call the **solar day.**

Traditionally, we divide a day into "**A.M.**" and "**P.M.**," which stand for **ante meridian** and **post meridian,** respectively. The **meridian** is a line that divides the eastern and western halves of the sky. The meridian extends from the point on the horizon due north to the point due south and passes directly through the zenith, the point exactly overhead. As the Sun moves across the sky (Figure 7.1), it crosses the meridian at apparent noon. Before (*ante*) noon is thus A.M., while after (*post*) noon is P.M. Apparent solar time is what a sundial measures, and during the year this time may

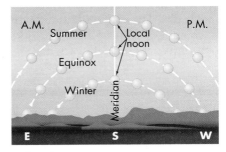

FIGURE 7.1
The Sun rises in the east, crosses the meridian at local noon, then sets in the west. This figure depicts the path of the Sun seen from the Northern Hemisphere on the equinoxes and the solstices.

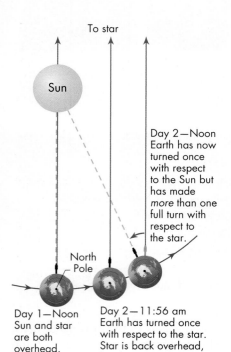

FIGURE 7.2
The length of the day measured with respect to the stars is not the same as the length measured with respect to the Sun. The Earth's orbital motion around the Sun makes it necessary for the Earth to rotate slightly more before the Sun will be back overhead. (Motion is exaggerated for clarity.)

> **Mathematical Insight**
>
> In 24 hours the Earth rotates 360°. 24 hours contain 1440 (= 24 × 60) minutes. For the Earth to rotate about 1° extra so that the same side is facing the Sun again, takes about 1440/360 = 4 minutes.

Hours of daylight

be ahead of or behind clock time by as much as a quarter of an hour. This variation arises from the Earth's non-circular orbit and the tilt of its axis (see Unit 13). Therefore, even though the Sun's position determines the day–night cycle, it is not a stable reference for measuring Earth's spin.

We can avoid most of the variation in the day's length if, instead of using the Sun, we use a star as our reference. For example, if we pick a star that crosses our meridian at a given moment and measure the time it takes for that same star to return to the meridian again, we will find that this time interval repeats quite precisely. However, this interval is not 24 hours, but about 23 hours, 56 minutes, and 4.0905 seconds. This day length, measured with respect to the stars, is called a **sidereal day,** which is divided into correspondingly shorter hours, minutes, and seconds of **sidereal time.**

Astronomers find that a clock set to run at this speed is very useful. It is not just that the sidereal day is much more stable. Another reason is that, at a given location, any particular star will always rise at the same sidereal time. To avoid the nuisance of A.M. and P.M., sidereal time is measured on a 24-hour basis. For example, the bright star Procyon in the constellation Canis Minor (small dog) rises at about 10 P.M. in November but at about 8 P.M. in December and 6 P.M. in January by solar time. However, on a clock keeping sidereal time, it always rises at the same time at a given location: about 01:30 by the sidereal clock.

Why is the sidereal day shorter than the solar day? We can see the reason by looking at Figure 7.2, where we measure the interval between successive apparent noons—a solar day. Let us imagine that at the same time as we are watching the Sun, we can also watch a star, and that we measure the time interval between its passages across the meridian—a sidereal day.

As we wait for the Sun and star to move back across the meridian, the Earth moves along its orbit. The distance the Earth moves in one day is so small compared with the star's distance that we see the star in essentially the same direction as on the previous day. However, we see the Sun in a measurably different direction, as Figure 7.2 illustrates. The Earth must rotate a bit more before the Sun is again on the meridian. That extra rotation, needed to compensate for the Earth's orbital motion, makes the solar day slightly longer than the sidereal day.

It is easy to figure out how much longer, on the average, the solar day must be. Because it takes us 365¼ days to orbit the Sun and because there are 360° in a circle, the Earth moves approximately 1° per day in its orbit around the Sun. That means that for the Sun to reach its noon position, the Earth must rotate approximately 1° past its position on the previous day. Another way of thinking about this is that the Sun is slowly moving eastward across the sky through the stars at the same time as the Earth is rotating. Therefore, in a given "day," the Earth must rotate a bit more to keep pace with the Sun than it would to keep pace with the stars.

The extra 4 minutes the Earth must rotate to face the Sun is relative to an average day length over the year called the **mean solar day.** The amount of time from one apparent noon to the next differs by up to about 30 seconds. Since our clocks run at a constant rate, the time of apparent noon shifts on our clocks each day.

7.2 LENGTH OF DAYLIGHT HOURS

Although each day lasts 24 hours, the number of hours of daylight, or the amount of time the Sun is above the horizon, changes dramatically throughout the year unless you are close to the equator. For example, in northern middle latitudes, including most of the United States, southern Canada, and Europe, summer has about 15 hours of daylight and only 9 hours of night. In the winter, the reverse is true. Anywhere north of the **Arctic Circle** at a latitude of 66.5° (90° − 23.5°), the Sun remains above the horizon for 24 hours on the summer solstice as shown in

FIGURE 7.3
Sequence of 24 pictures of the Sun taken from a spot close to the Arctic Circle near the summer solstice. The pictures were taken 1 hour apart and show the Sun circling the sky but never setting.

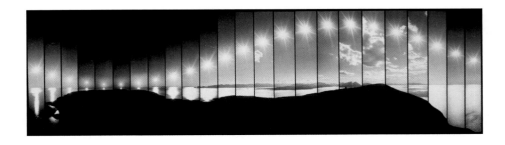

Clarification Point

Many people believe that the Sun is straight overhead every day at noon. Outside of the tropics (latitudes within 23.5° of the equator) the Sun is *never* straight overhead, and even within the tropics the Sun passes overhead around noon on only one or two days each year.

Figure 7.3, and it remains below the horizon the entire day on the winter solstice. On the equator, day and night are each 12 hours every day.

This variation in the number of daylight hours is caused by the Earth's tilted rotation axis. Remember that as the Earth moves around the Sun, its rotation axis points in very nearly a fixed direction in space. As a result, the Sun shines more directly on the Northern Hemisphere during its summer and at a more oblique angle during its winter. The result (as you can see in Figure 7.4) is that a large fraction of the Northern Hemisphere is illuminated by sunlight at any time in the summer, but a small fraction is illuminated in the winter. So as rotation carries us around the Earth's axis, only a relatively few hours of a summer day are unlit, but a relatively large number of winter hours are dark. On the first days of spring and autumn (the equinoxes), the hemispheres are equally lit, so that day and night are of equal length everywhere on Earth.

If we change our perspective and look out from the Earth, we see that during the summer the Sun's path is high in the sky, so that the Sun spends a larger portion of the day above the horizon (Figure 7.1). This gives us not only more heat but also more hours of daylight. On the other hand, in winter the Sun's path across the sky is much shorter, giving us less heat and fewer hours of light (also see Unit 6).

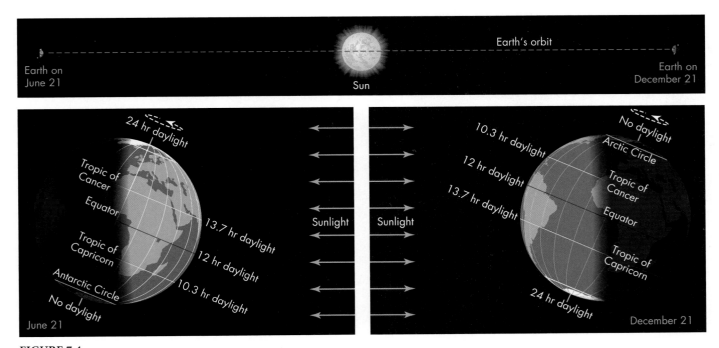

FIGURE 7.4
The tilt of the Earth affects the number of daylight hours. Locations near the equator always receive about 12 hours of daylight, but locations toward the poles have more hours of darkness in winter than in summer. In fact, above latitudes 66.5° the Sun never sets for part of the year (the *midnight sun* phenomenon) and never rises for another part of the year. At the equinoxes, all parts of the Earth receive the same number of hours of light and dark. (Sizes and separation of the Earth and Sun are not to scale.)

7.3 TIME ZONES

Because the Sun is our basic timekeeping reference, most people like to measure time so that the Sun is highest in the sky at about noon. As a result, clocks in different parts of the world are set to different times so that the local clock time approximately reflects the position of the Sun in the sky. Because the Earth is round, the Sun can't be "highest" everywhere at the same time, so it can't be noon everywhere at the same time.

By the late 1800s, with the increasing speed of travel and communications, it became confusing for each city to maintain its own time according to the position of the Sun in the sky. By international agreement, the Earth was therefore divided into 24 major **time zones,** centered every 15° of longitude, in which the time differs by one hour from one zone to the next. With this system, clocks in a time zone all read the same, and they are at most a half hour ahead of or behind what they would be if the time were measured locally. Many regions use local geographic features or political borders to define the boundaries between time zones rather than strictly following the longitude limits (see Figure 7.5). Authorities in a few countries and regions did not adopt this agreement, choosing instead to maintain a time standard that was closer to local time. For example, Newfoundland, India, Nepal, and portions of Australia are offset by 30- or 15-minute differences from the international standard.

Across the lower 48 United States, the time zones are, from east to west, Eastern, Central, Mountain, and Pacific. Within each zone the time is the same everywhere and is called **standard time.** In the eastern zone the time is denoted Eastern Standard Time (EST), in the central zone it is denoted Central Standard Time (CST), and so on. If you travel across the United States, you reset your watch as you cross from one time zone to another, adding one hour for each time zone as you go east, and subtracting one hour for each time zone as you go west.

If you travel through many time zones, you may need to make such a large time correction that you shift your watch past midnight. For example, if you could travel westward quickly enough that little time elapsed, setting your watch back each time

> **Concept Question 1**
>
> In Figure 7.5 the international date line has a very complex shape. If you were sailing north through the -11^h time zone, how would your time and date change?

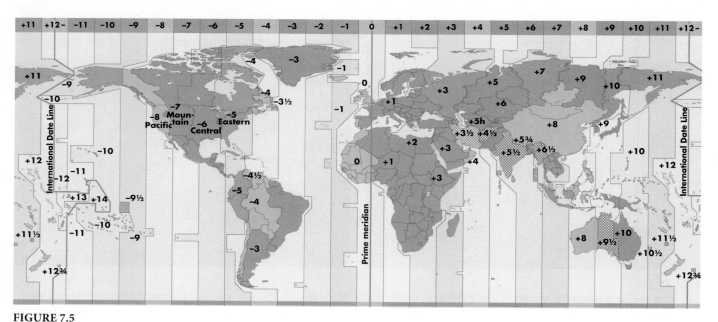

FIGURE 7.5

Time zones of the world and the international date line. Local time = universal time + numbers on top of the chart.

Concept Question 2

In Jules Verne's story *Around the World in 80 Days*, the travelers discover that although they have experienced 80 days and nights, they still have one more day before 80 days will have passed in London. How can this be?

you crossed a time zone, you could end up at your starting point with your watch turned backward by 24 hours. But you would not have traveled back in time!

When you cross longitude 180° (roughly down the middle of the Pacific Ocean), you add a day to the calendar if you are traveling west and subtract a day if you are traveling east. For example, you could celebrate the New Year in Japan and take a flight after midnight to Hawaii, where it would still be the day before, so you could celebrate the New Year that night too! The precise location where the day shifts is called the **international date line** (Figure 7.5). It generally follows 180° longitude but bends around extreme eastern Siberia and some island groups to ensure that they keep the same calendar time as their neighbors.

The nuisance of having different times at different locations can be avoided by using **universal time,** abbreviated as **UT.** Universal time is the time kept in the time zone containing the longitude zero, which passes through Greenwich, England. By using UT, which is based on a 24-hour system to avoid confusion between A.M. and P.M., two astronomers at remote locations can make a measurement at the same time without worrying about what time zones they are in.

7.4 DAYLIGHT SAVING TIME

In many parts of the world, people set clocks ahead of standard time during the summer months and then back again to standard time during the winter months (Figure 7.6). This has the effect of shifting sunrise and sunset to later hours during the day, thereby creating more hours of daylight during the time when most people are awake. Time kept in this fashion is called **daylight saving time** in the United States. In some other parts of the world it is called "Summer Time."

Daylight saving time was originally established during World War I as a way to save energy. By setting clocks ahead, less artificial light was needed in the evening hours. In effect, it is a method to get everyone to wake up and go to work an hour earlier than they would normally and take advantage of the earlier rising of the

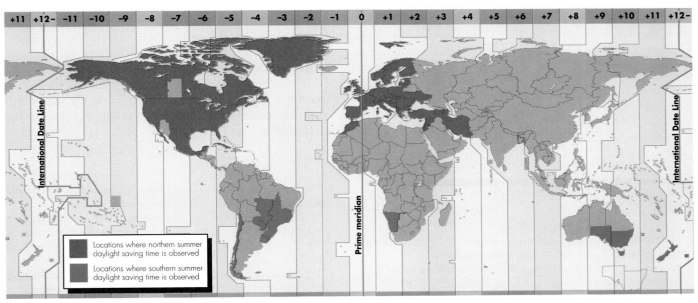

FIGURE 7.6
Regions where daylight saving time is observed for some portion of the year.

Concept Question 3

If daylight saving time saves energy, why not just have people come to work earlier when the Sun rises earlier and later when it rises later?

Sun. Clocks are not set ahead permanently because in some parts of the country the winter sunrise is so late that daylight saving time would require people to wake and go to work in the dark. In the United States daylight saving time runs from the second Sunday in March to the first Sunday in November. In Europe it runs from the last Sunday in March to the last Sunday in October, whereas in Australia it runs from October to April. Many other countries, and even some states within the United States, follow different rules, while most tropical countries keep their clocks fixed on standard time.

In recent years, a number of countries have stopped the practice of switching to daylight saving time in the summer. Almost all of Asia, Australia, and South America used to switch time during the summer, but now the practice is confined mostly to North America and Europe.

7.5 LEAP SECONDS

Highly accurate time measurements have allowed us to detect a gradual slowing of the Earth's spin in recent decades. The second was defined as 1 part in 86,400 ($24 \times 60 \times 60$) of a day, based on astronomical measurements made more than a century ago. With modern atomic clocks it has been determined that the length of the mean solar day is now about 86,400.002 seconds, and it is increasing by about 0.0014 second each century. Over a year the 0.002 second add up to most of a second: $365 \times 0.002 \approx 0.7$ second each year. This means that an accurate clock set at midnight on New Year's Eve would signal the beginning of the next new year almost 1 second too early. By the year 2100, an accurate clock will be off by 1.2 seconds after a year.

A strong earthquake in Chile in March 2010 sped up the Earth's rotation slightly, shortening the day by about 1 *millionth* of a second.

This time change might sound insignificant, but over long periods its effects accumulate. Fossil records from corals that put down visible layers each day indicate that 400 million years ago the Earth's year had about 400 days. The Earth took the same amount of time to orbit the Sun, but the Earth took only about 22 hours to spin on its axis.

Note that atomic clocks are *not* radioactive. They use an internal vibration frequency of cesium atoms to determine precise time intervals.

With the precise timing needed today for technological uses such as the Global Positioning System (GPS), even millisecond errors are critical. To keep our clocks in agreement with the Sun and stars, we now adjust atomic clocks with a **leap second** every year or two. This is coordinated worldwide so that clocks everywhere remain in agreement.

The slowing of the Earth is a consequence of the interaction of the spinning Earth with ocean tides, which are held relatively stationary by the Moon's gravity. In effect, the eastern coasts of the continents "run into" high tide as the solid Earth spins beneath the tide, and this creates a **tidal braking** force that is slowing the Earth. Tides are discussed further in Unit 19.

KEY POINTS

- The Earth's spin is very regular with respect to the stars, taking about 23 hours 56 minutes to complete one sidereal day.
- Because of Earth's orbital motion, the rotation relative to the Sun takes about 4 minutes longer and is less regular.
- The Sun crosses your meridian at local noon, but at clock times that may vary by over a quarter hour at different times of year.
- The time from sunrise to sunset varies depending on latitude and season, with more extremes farther from the equator.
- In the tropics, the Sun can pass overhead one or two days each year; in polar regions, the Sun is sometimes above the horizon for 24 hours, and below the horizon for 24 hours six months later.
- Time zones help standardize times to the closest hour, although these are often modified for local reasons or to "save daylight."
- The Earth's spin is gradually slowing, causing days to grow longer by about 1 thousandth of a second in the last century.

KEY TERMS

A.M. (ante meridian), 53
apparent noon, 53
Arctic Circle, 54
daylight saving time, 57
international date line, 57
leap second, 58
mean solar day,
meridian, 53

P.M. (post meridian), 53
sidereal day, 54
sidereal time, 54
solar day, 53
standard time, 56
tidal braking, 58
time zone, 56
universal time (UT), 57

CONCEPT QUESTIONS

Concept Questions on the following topics are located in the margins. They invite thinking and discussion beyond the text.

1. The international date line. (p. 54)
2. Days passed when circling the globe. (p. 55)
3. Why use daylight saving time. (p. 56)

REVIEW QUESTIONS

4. Where is the meridian located?
5. How is the sidereal day defined? Why do the sidereal and solar days differ in length?
6. What is universal time?
7. What are the advantages and disadvantages of time zones?
8. How does time shift across the international date line?
9. Why are "leap seconds" added to our clocks every few years?

QUANTITATIVE PROBLEMS

10. Philadelphia is 5° east of Pittsburgh. If both cities synchronize their clocks with apparent noon—that is, all the local clocks read noon when the Sun passes through it highest point in the sky—by how many minutes do the clocks in the two cities differ from one other?
11. If Russia had all clocks set to Moscow time (time zone +3 in Figure 7.4) instead of using many time zones, what time would the Sun rise at the easternmost tip of Siberia on the equinoxes? Examining Figure 7.4, what places in the world have the Sun crossing the meridian earliest according to local time? What places have it latest?
12. One city is at a longitude of 21° E. A second city is at 104° W. How many hours apart will a star cross the meridian in the first city and then in the second? If the time zone of each city is based on the closest longitude that is a multiple of 15°, and the star passes overhead in the first city exactly at midnight, at what time (by the clock) will it pass over the second city?
13. Suppose the Earth's spin slowed down until there were just 180 days in a year. Compare the length of a sidereal and a solar day in this new situation. (Do not redefine units of time—just express them in terms of our current hours, minutes, and seconds.)
14. It is thought that the angle of the Earth's axis varies by a few degrees over tens of thousands of years. If the angle were 20° instead of 23.5°, what would be the angle of the noontime Sun from the horizon at your own latitude on the solstices?
15. What is your longitude, and what is the longitude of the center of your time zone? (This should be a multiple of 15°.) Calculate the time at which the Sun should, on average, cross your own meridian, assuming that the Sun crosses the meridian at exactly 12:00 at the center of your time zone.

TEST YOURSELF

16. Earth's orbit around the Sun has what effect on the rising time of non-circumpolar stars?
    a. Nothing—stars rise the same time every night.
    b. They rise several minutes earlier each night.
    c. They rise several minutes later each night.
    d. They rise a little farther south or north each night, depending on the season.
    e. Nothing—only the Sun rises, not stars.
17. In which of the following locations can the length of daylight range from zero to 24 hours?
    a. Only on the equator
    b. At latitudes closer than 23.5° to the equator
    c. At latitudes between 23.5° and 66.5° north or south
    d. At latitudes greater than 66.5° north or south
    e. Nowhere on Earth
18. In which of the following locations is the length of daylight 12 hours throughout the year?
    a. Only on the equator
    b. At latitudes closer than 23.5° to the equator
    c. At latitudes between 23.5° and 66.5° north or south
    d. At latitudes greater than 66.5° north or south
    e. Nowhere on Earth
19. Daylight saving time
    a. shifts daylight from summer days into the winter.
    b. corrects clocks for errors caused by the Earth's tilted axis.
    c. results in the Sun crossing the meridian around 1 P.M.
    d. puts clocks in agreement with the Sun's position in the sky.
    e. All of the above.
20. If the Earth began orbiting the Sun twice as fast as it does now, the length of a solar day would
    a. stay the same.
    b. become twice as long.
    c. become slightly longer.
    d. become half as long.
    e. become slightly shorter.

PART 1 UNIT 8

Lunar Cycles

8.1 Phases of the Moon
8.2 Eclipses
8.3 Eclipse Seasons
8.4 Moon Lore

Learning Objectives

Upon completing this Unit, you should be able to:
- Identify the phases of the Moon, and explain their geometry.
- Predict when and where the Moon will be in the sky depending on its phase.
- Describe when and where solar and lunar eclipses are visible, the differences between partial and total eclipses, and the visual phenomena associated with each.
- Explain why eclipses do not occur every month and patterns in their occurrence.

The Moon is one of the loveliest of astronomical sights; its beauty is extolled in literature, poetry, song, and art. The Moon is forever changing shape, brightness, position, and the times when it is visible. Sometimes it is so bright that it can illuminate the night well enough to hike or even read by its light, whereas at other times it provides little light or is altogether absent. The pattern of these changes may not be obvious unless you spend many successive nights studying the Moon. The Moon's changes follow a regular, clocklike 29½-day cycle. This is the origin of our time period of a **month**—the term coming from the word *Moon*.

The Moon rises in the east and sets in the west because of the Earth's rotation (Unit 5). Also, like the Sun, the Moon shifts its position across the background stars from west to east (Unit 6), but about 12 times more rapidly. You can see this motion in as little as 10 minutes when the Moon happens to lie close to a bright star. The Moon is the most quickly shifting of the astronomical bodies that we can regularly see with the naked eye.

8.1 PHASES OF THE MOON

If you spot the Moon in the west shortly after sunset one evening, on subsequent nights it will have shifted progressively farther east, and in about two weeks it will be in the eastern sky as the Sun sets. These changes occur as the Moon orbits the Earth, causing the Moon to rise and set about 50 minutes later each day.

One of the most striking features of the Moon is that its shape seems to change from night to night in what is called the cycle of lunar **phases.** This sequence is shown in Figure 8.1 by a series of photographs of the Moon taken over the course of a month. Starting from when the Moon appears as a thin **crescent** just after sunset (upper left in Figure 8.1), more of it becomes illuminated each night—it

FIGURE 8.1
A series of images of the Moon taken almost every night for one month. The Moon begins the month as a thin crescent (upper left), grows to a full moon (start of 2nd row), then shrinks back to a crescent.

FIGURE 8.2

The cause of the Moon's phases. As the Moon orbits the Earth, we see different portions of its illuminated half, which faces the Sun. This causes the Moon to cycle from new to full and back again. (Sizes and distances of objects are not to scale.)

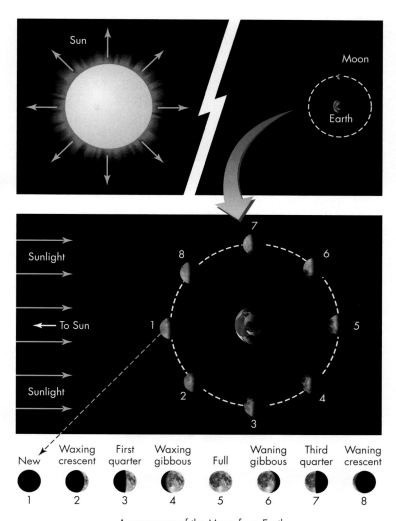

TABLE 8.1	The Lunar Cycle		
Position in Orbit (Fig. 8.2)	Day	Phase	Location and Time for Viewing
1	0.0	New	In same direction as the Sun
2	0.0–7.4	Waxing crescent	Near the Sun in early evening sky
3	7.4	First quarter	90° from the Sun in evening sky
4	7.4–14.7	Waxing gibbous	Far from the Sun in evening sky
5	14.7	Full	Opposite the Sun, rising at sunset
6	14.7–22.1	Waning gibbous	Far from the Sun in morning sky
7	22.1	Third quarter	90° from the Sun in morning sky
8	22.1–29.5	Waning crescent	Near the Sun in predawn sky
1	29.5	New	In same direction as the Sun

Lunar phases

Clarification Point

Phase depends on the place the Moon is in its orbit, which changes relatively little in a day. Therefore the phase you see tonight is almost exactly the same for everyone else on Earth on the same date.

waxes—until after about two weeks it appears as a fully illuminated disk. Then it steadily decreases—it **wanes**—back to a thin crescent again, and finally disappears for a day or two before beginning the cycle again. In addition to *crescent,* several descriptive names are used for the phases seen during the period of approximately 29½ days of the **lunar month.** When we see more than half of the Moon lit, it is **gibbous;** when completely illuminated, it is **full;** and when completely dark, it is **new.**

The half-lit Moon is also given a special name. Because this phase occurs one-quarter or three-quarters of the way through the lunar cycle, it is called a **first-quarter moon** or **third-quarter moon.** The **quarter,** new, and full phases refer to particular moments in the orbit of the Moon around the Earth—although we often refer to the Moon being full, for example, if it is within about a day of that phase. A new moon occurs when the Moon lies as near as possible to the line between us and the Sun. A full moon occurs when the Moon is on the other side of the Earth from the Sun, opposite it in the sky. The quarter moon occurs when the Moon is 90° away from the Sun in the sky.

All of these terms can become confusing, so they are summarized in Table 8.1 and illustrated in Figure 8.2. The times of day and night when you can see the Moon also change throughout the lunar cycle. This is described in the last column of Table 8.1 and illustrated graphically in Figure 8.3.

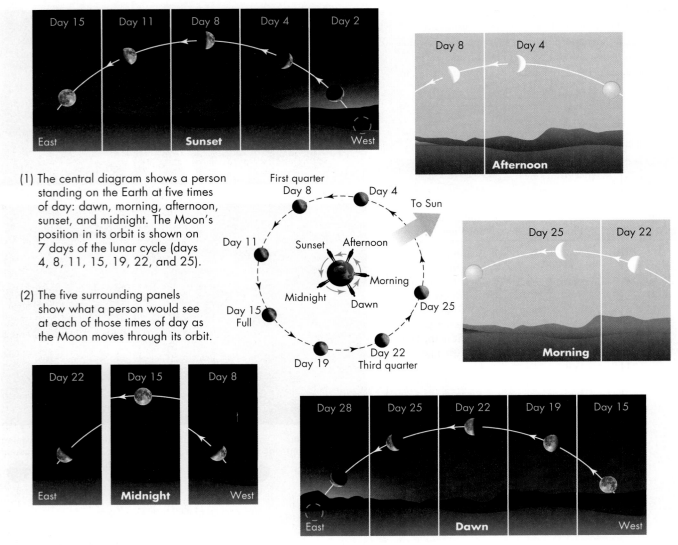

FIGURE 8.3
When the Moon is visible. This figure shows where to look for the Moon and how it appears at different times of day as it goes through its monthly cycle of phases. The central sketch shows the Earth and Moon and where the Moon is in its orbit at different phases, indicated here by the number of days since the new moon. The yellow arrow points toward the Sun.

Clarification Point

Moon phases are *not* caused by the Earth's shadow.

Many people mistakenly assume that the changes in the Moon's shape are caused by the Earth's shadow falling on the Moon. However, you can deduce that this cannot be the explanation because crescent phases occur when the Moon and Sun lie approximately in the same direction in the sky, and the Earth's shadow must therefore point away from the Moon. In fact, we see the Moon's shape change because as the Moon moves around us, we see different amounts of its illuminated half. For example, when the Moon lies approximately opposite the Sun in the sky, the side of the Moon toward the Earth is fully lit. On the other hand, when the Moon lies approximately between us and the Sun, its fully lit side is turned nearly completely away from us, and therefore we glimpse at most a sliver of its illuminated side, as illustrated in Figure 8.3.

Concept Question 1

If the Earth did not spin, how often would the Moon rise? In what directions would it rise and set?

Another confusing point is that the Moon's orbit around the Earth is actually about 27.3 days, not the 29.5 days of the lunar cycle. This is also called the **sidereal month** (from the Latin *siderea* for "starry") because it is the time the Moon takes to return to the same position relative to the stars. The difference results from the Earth orbiting the Sun as the Moon orbits the Earth. As illustrated in Figure 8.4, after 27.3 days the

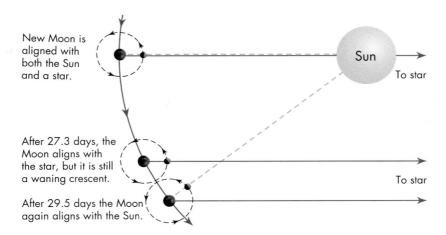

FIGURE 8.4
The *sidereal* month is the time the Moon takes to complete an orbit relative to the distant stars. This is about 27.3 days, less than the lunar month because as the Moon is orbiting the Earth, the Earth is orbiting the Sun. It takes about two additional days for the Moon to come back in alignment with the Sun.

Moon is aligned with the same distant stars again. However, the Earth has reached a new position in its orbit. From our perspective, the Sun has shifted into the next constellation of the zodiac, so the Moon must orbit about two more days to catch up to the Sun and reach the new phase again. This is similar to the reason why the *sidereal* day is shorter than the solar day (Unit 7).

8.2 ECLIPSES

An **eclipse** occurs when the Moon lies exactly between the Earth and the Sun, or when the Earth lies exactly between the Sun and the Moon, so that all three bodies lie on a straight line. When this happens the Moon will cast a shadow on the Earth or vice versa. A **solar eclipse** occurs whenever the Moon passes directly between the Sun and the Earth and blocks our view of the Sun, as depicted in Figure 8.5. A **lunar eclipse** occurs when the Moon passes through the Earth's shadow as it moves through the portion of its orbit opposite the Sun, as shown in Figure 8.6. Thus, solar eclipses can occur *only* when the Moon is new, and lunar eclipses *only* when the Moon is full.

FIGURE 8.5
A solar eclipse occurs when the Moon passes between the Sun and the Earth so that the Moon's shadow touches the Earth. The photo inset shows what a solar eclipse looks like from Earth, from the center of the Moon's shadow, where the Sun is completely covered.

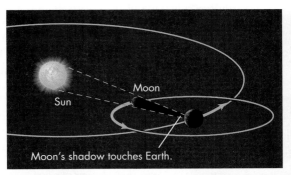

FIGURE 8.6
A lunar eclipse occurs when the Earth passes between the Sun and the Moon, causing the Earth's shadow to fall on the Moon. Some sunlight leaks through the Earth's atmosphere, casting a deep reddish light on the Moon. The photo inset shows what the eclipse looks like from Earth.

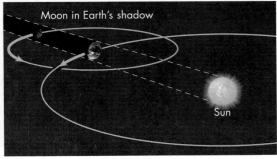

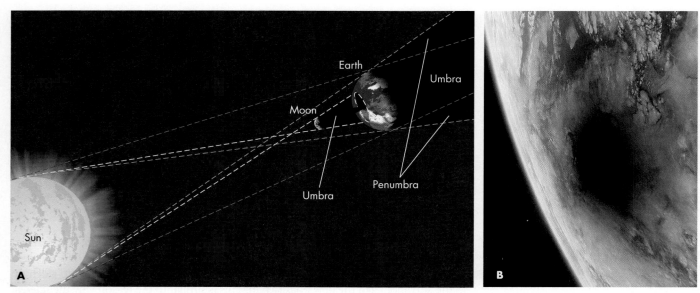

FIGURE 8.7
(A) The shadow of the Earth contains regions where the Sun is completely blocked (the umbra) and regions where sunlight is only partially blocked (the penumbra). (B) A photograph of the Earth from the MIR space station made during a solar eclipse. The Moon's umbra is almost black, while the surrounding region of the penumbra goes from black to fully lit, depending on how much of the Sun is covered by the Moon.

Eclipses generally take a few hours from start to finish, but most go almost unnoticed because the Sun's light is only partially blocked. The shadow cast by the Moon or by the Earth contains a central region where the Sun's light is blocked completely, the **umbra,** surrounded by a region where the light is blocked only partially, the **penumbra** (Figure 8.7). When we are in the Moon's penumbra or when only part of the Moon is in Earth's umbra, we call it a **partial eclipse.** These occur about twice as often as **total eclipses**, where the Moon completely covers the Sun, or the Moon is completely within the Earth's umbra.

Total eclipses are beautiful and marvelous events. During a total lunar eclipse, the Earth's shadow gradually spreads across the full Moon's face, cutting an ever deeper dark semicircle out of it. The shadow takes about an hour to completely cover the Moon and produce totality. At totality, the Moon generally appears a deep ruddy color, almost as if dipped in blood. Sometimes it even disappears. After totality, the Moon again becomes lit, bit by bit, reverting to its unsullied, silvery light.

A little light falls on the Moon even at totality because the Earth's atmosphere bends some sunlight into the shadow, as shown in Figure 8.8. The light reaching the Moon is red because interactions with air molecules remove the blue light as it passes through our atmosphere and is bent, exactly as happens when we see the setting Sun. During a lunar eclipse, if you were looking at the Earth from the Moon, the Earth would appear as a fiery ring of red—in effect you would be seeing sunsets from all over the Earth simultaneously!

Some portion of the Moon passes through the penumbra of the Earth's shadow about twice as often as it completely enters the umbra, producing a total lunar eclipse. In partial lunar eclipses, only part of the Moon enters the umbra. The weakest types of partial eclipses are called **penumbral eclipses** because no part of the Moon enters the umbra. They are often so minor that they are not marked on astronomical calendars, although someone on the Moon might see the Earth partially eclipsing the Sun.

Lunar eclipses are visible if you are anywhere on the night side of the Earth when the eclipse is occurring. It is far rarer to see a total solar eclipse because the Moon's shadow on the Earth is quite small. In fact, you are unlikely to ever see a

Concept Question 2

If you were standing on the Moon during a total lunar eclipse, what sequence of events would you observe?

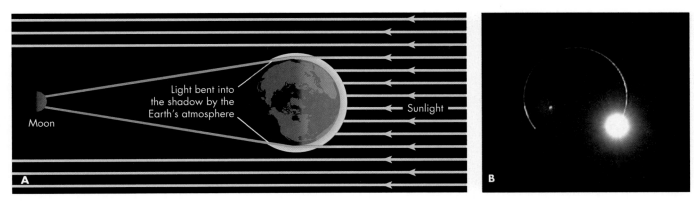

FIGURE 8.8

(A) As sunlight falls on the Earth, some passes through the Earth's atmosphere and is slightly bent so that it ends up in the Earth's shadow. In its passage through our atmosphere, most of the blue light is removed, leaving only the red. That red light then falls on the Moon, giving it its ruddy color at totality. (B) The Japanese satellite *Kaguya* captured this picture of the Earth from the Moon in 2009 when the Earth partially blocked the Sun's light. A small portion of the Sun is visible as the bright spot in the image while sunlight bent by the Earth's atmosphere outlines the Earth. (The bottom portion of the Earth is cut off in this image by the Moon's horizon.)

total solar eclipse in your lifetime unless you travel to see it, because on average they occur in any location only once every several centuries.

A total solar eclipse begins with a small black "bite" taken out of the Sun's edge as the Moon cuts across its disk (Figure 8.9A). Over the next hour or so, the Moon gradually covers more and more of the Sun. While the Sun is only partially covered, you must be careful when viewing it, so you don't injure your eyes.

If you are fortunate enough to be at a location where the eclipse is total, you will see one of the most amazing sights in nature. As the time when the Moon's disk completely covers the Sun (totality) approaches, the landscape takes on an eerie light. Shadows become incredibly sharp and black: even individual hairs on your head cast crisp shadows. Sunlight filtering through leaves creates tiny bright crescents on the ground. Seconds before totality, pale ripples of light sweep across the ground, and to the west the deep purple shadow of the Moon hurtles toward you at more than 1000 miles an hour. In one heartbeat you are plunged into darkness. Overhead, stars become visible. Perhaps a solar prominence—a tiny, glowing, red flamelike cloud in the Sun's atmosphere—may protrude beyond the Moon's black disk (Figure 8.9C). The corona of the Sun—its outer atmosphere—gleams with a steely light around the Moon's black disk (Figure 8.9D). Birds call as if it were evening. A deep chill descends, because for a few minutes the Sun's warmth is blocked by the Moon. The horizon takes on sunset colors: the deep blue of twilight with

FIGURE 8.9

Pictures of a total solar eclipse in 2010. (A) An hour before totality, the Moon only partially eclipses the Sun. (B) About 5 minutes before totality. (C) With the bright part of the Sun covered, the Sun's glowing pink atmosphere becomes visible. (D) Faint hot gases form a corona around the Sun. (E) As the Moon slides off the Sun, the first glimpse of the bright portion of the Sun makes a "diamond ring," while thin clouds in Earth's atmosphere diffract the light into a colored glow around the Sun.

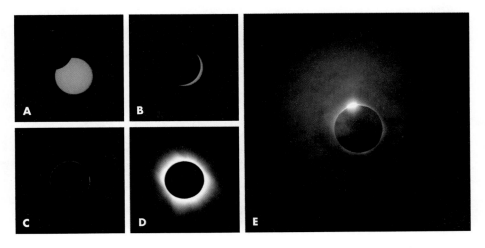

Concept Question 3

The Moon's distance from the Earth is measured to be gradually increasing. When the Moon is much farther away than it is now, what will happen to the frequency of the various kinds of eclipses?

perhaps a distant cloud in our atmosphere glowing orange. As the Moon continues in its orbit, it begins to uncover the Sun, and in the first moments after totality, the partially eclipsed Sun looks a little like a diamond ring (Figure 8.9E). Now the cycle continues in reverse. The sky rapidly brightens, and the shadow of the Moon may be glimpsed on distant clouds or mountains racing away to the east. If you ever have the chance to see a total eclipse, do it!

Given the remarkable nature of these events, it is not surprising that early people recorded them and sought (successfully) to predict them. Ancient Babylonian and Chinese astronomers kept records stretching back centuries and, based on the patterns they found, were able to predict future eclipses. In more recent times, astronomers have used total solar eclipses to study the outer layers of the Sun, which are normally hidden from sight by the enormously brighter central region of the Sun.

Because the Moon is so small compared with the Earth, its shadow is small, so you can see a total solar eclipse only from within a narrow band on the Earth, as illustrated in Figure 8.10. Thus, you are much more likely to see a total lunar eclipse, which is visible anywhere that the Moon is above the horizon at the time of the eclipse.

FIGURE 8.10

(A) Sketch of how the Moon's shadow travels across the Earth. (B) Location of recent and upcoming total solar eclipses through 2035.

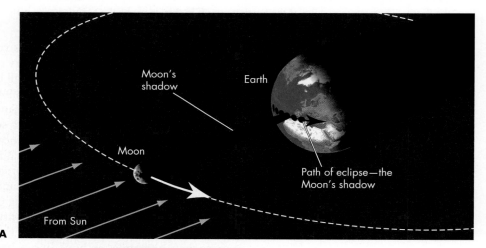

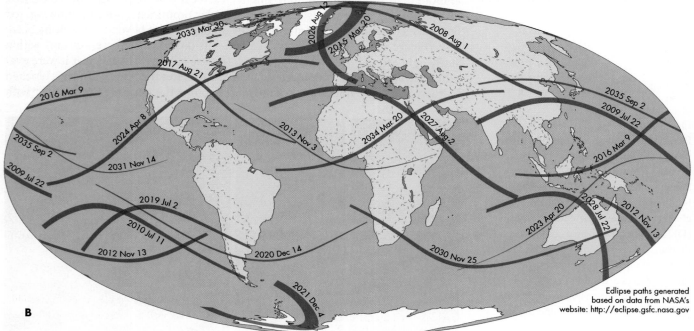

The Moon's orbit around the Earth is not perfectly circular, so sometimes the Moon is far enough away that the umbra does not quite reach the Earth. When this happens, the Moon will appear smaller than the Sun (Figure 8.11), so even if we are in a position where the Sun and Moon are exactly aligned, we can still see the Sun's surface in a ring, or *annulus,* around the Moon. This type of partial eclipse is called an **annular eclipse.**

8.3 ECLIPSE SEASONS

You may wonder why we do not have eclipses every month. The answer is that the Moon's orbit is tipped by about 5° with respect to the Earth's orbit around the Sun (Figure 8.12). Because of this tilt, even if the Moon is new its shadow may pass above or below Earth. As a result, no eclipse occurs. Similarly, when the Moon is full, it may pass above or below the Earth's shadow so that again no eclipse occurs. Eclipses can occur only when the Moon is within about 1° of crossing the ecliptic—the plane of the Earth's orbit around the Sun. This usually occurs only twice during the year (Figure 8.12A).

For example, if a solar eclipse occurs on a given day in May, a lunar eclipse is likely to occur roughly two weeks later or earlier (the interval between full

FIGURE 8.11
An annular eclipse of the Sun in 1992 occurring near sunset. The Moon is at a distant point in its orbit, so it appears smaller and cannot block the Sun entirely.

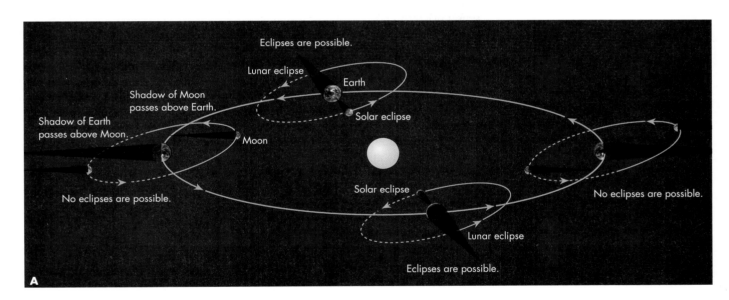

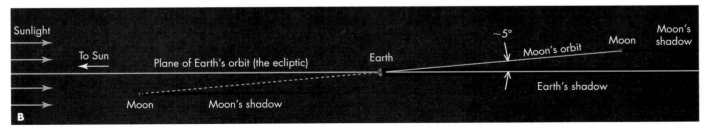

FIGURE 8.12
(A) The Moon's orbit keeps approximately the same orientation as the Earth orbits the Sun. Because of its orbital tilt, the Moon generally is either above or below the Earth's orbit. Thus, the Moon's shadow rarely hits the Earth, and the Earth's shadow rarely hits the Moon. Eclipse seasons occur when the Earth is in either of two places in its orbit about six months apart. These occur when the Moon's orbital plane, if extended, intersects the Sun. (B) The Earth and Moon are drawn to correct relative size and separation with their orbits seen here edge-on. Note how thin their shadows are.

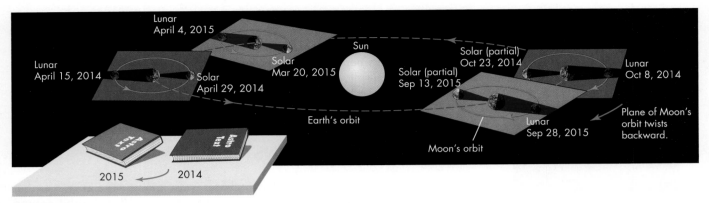

FIGURE 8.13
Precession of the Moon's orbit causes eclipses to come at different dates in successive years. Notice its similarity to twisting a tilted book that has one edge resting on a table. Sizes of objects and their separations are not to scale.

The 18-year, 11-day, 8-hour period, called the *saros*, between similar eclipses was discovered by ancient astronomers. For a solar eclipse, the eight additional hours shift the location where the eclipse can be seen about one-third of the way around the globe. However, after three of these periods (about 54 years, 31 days), a similar eclipse occurs near the original location.

and new moons), and it is even possible to have two partial solar eclipses on two new moons in a row. Half a year later, in November, the Moon will again be close to the ecliptic when it is new or full, and another set of eclipses may occur.

The times of year at which eclipses can occur gradually shift from year to year because the plane of the Moon's orbit "wobbles." This wobble, called *precession* (see also Unit 6.6), slowly changes the plane of the Moon's orbit. This orbital precession makes the dates of the eclipse seasons shift by about 20 days each year. Thus, in 2015, eclipses occur about three weeks earlier, on average, than in 2014, as illustrated in Figure 8.13.

The Moon's orbital plane swings back to the same orientation after 18.6 years. A full (or new) moon occurs during the eclipse season with almost exactly the same configuration of Moon, Earth, and Sun after 18 years, 11 days, and 8 hours. This can be seen in Figure 8.9, where similarly shaped shadow paths occur about 18 years apart.

During each eclipse season there is always at least one solar and one lunar eclipse, and sometimes two of either. As a result there are always at least two solar and two lunar eclipses each year. Because eclipse seasons occur a little less than six months apart, there can be as many as five solar or lunar eclipses in a year. Most of these eclipses are partial, however, and they may not be visible to an observer at a given location, since the eclipse may be visible only from another part of the Earth.

8.4 MOON LORE

The Moon figures prominently in folklore around the world. Anyone who spends time far from the glow of city lights will have noted how different a moonlit night is from the pitch dark when the Moon is below the horizon. This clearly affects animal behavior—predators and prey using the extra light or the cover of darkness to their own advantage. However, most stories concerning the Moon's powers are false.

For example, people often claim that the full moon triggers antisocial behavior—hence the term *lunacy*. Such supposed correlations with lunar phase do not stand up under closer scrutiny. Careful examination of the data shows that automobile accidents, murders, admissions to clinics, and so forth do not correlate with lunar phase. It is true that the extra light at night around the time of the full moon makes it easier to see and do things outside, so we may simply be more aware of what happens then.

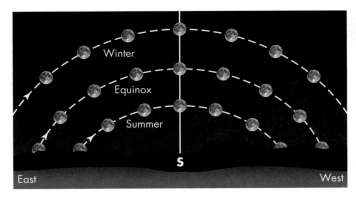

FIGURE 8.14
Path of the full moon through the sky during different seasons. It reaches its highest point in the sky at midnight, but it is lower in the sky in summer than in winter.

TABLE 8.2 Names Used for Full Moons	
Month	Name of Full Moon
January	Old moon
February	Hunger moon
March	Sap or crow moon
April	Egg or grass moon
May	Planting moon
June	Rose or flower moon
July	Thunder or hay moon
August	Grain or green corn moon
September	Harvest moon
October	Hunter's moon
November	Frost or beaver moon
December	Long night moon

Concept Question 4

What natural phenomena that occur on Earth do you think might be linked to the lunar cycle? How could you test your idea?

The position of the full moon in the sky changes during the year. The full moon is almost opposite the Sun, so in winter the full moon is located nearly where the Sun is on the celestial sphere in summer, and vice versa. Thus, the full moon traces a path through the sky similar to that of the Sun six months earlier or later, so the full moon reaches a much higher point in the sky in the winter than in the summer (compare Figure 8.14 to Figure 7.1).

Special names are sometimes given to the full moon at different times of the year. Some examples from North American folklore are listed in Table 8.2. The full moon nearest the time of the autumnal equinox is often called the harvest moon because, as it rises in the east at sunset, the light from the harvest moon helps farmers see to get the crops in. Full moons in other months also have special names, but only the names "harvest moon" and "hunter's moon" (for a full moon in the October hunting season) are widely used.

It is unclear what the origin is of the phrase *once in a blue moon,* indicating a rare event. One suggestion is that this phrase refers to months with two full moons. It is unusual to have two full moons in one month because the cycle of phases is 29.5 days; therefore, unless the Moon is full on the first day of the month, the next full moon will fall in the following month. The odds of the Moon being full on the first of the month are about 1 in 30, so two full moons in a given month happen about every 2½ years. On some calendars the symbol for a second full moon in a month is printed in blue ink. This printing practice appears to be more recent than the earliest uses of the phrase, however, so a "blue moon" must have meant something else originally.

It is possible, in fact, that on rare occasions the Moon may look blue. This odd coloration can be caused by particles in the Earth's atmosphere. Normally our atmosphere filters out blue colors and allows red to pass through, which is why the Sun looks red when it is low in the sky. However, if the atmosphere contains particles whose size falls within a narrow range, the reverse may occur. Airborne particles from volcanic eruptions or in the smoke from forest fires occasionally are just the right size to filter out red light, allowing mainly the blue colors of the spectrum to pass through. Under these unusual circumstances, we may therefore see a "blue moon."

KEY POINTS

- As the Moon orbits the Earth, we see different portions of its illuminated half creating the lunar cycle of phases.
- The phases run from "new," when the Moon is in the same direction as the Sun, through waxing crescent, first quarter, waxing gibbous, full, waning gibbous, third quarter, and waning crescent.
- The angle between the Moon and Sun determines the phase, so the time of rising and setting can be predicted from the phase.
- A solar eclipse occurs when the Moon's shadow passes over the Earth; a lunar eclipse when the Moon passes through the Earth's shadow.
- Total eclipses occur when sunlight is completely blocked.
- During a total lunar eclipse, the Moon appears red because of light bent into the shadow by the Earth's atmosphere.
- The Moon's orbit is tilted relative to the ecliptic, so in most months the shadows pass "above" or "below" the Moon or Earth.
- The dates when eclipses can occur gradually shift because the tilt of the Moon's orbit gradually precesses over an 18-year period.
- When the Moon is full, it is opposite the Sun, so it is highest in the sky at the winter solstice and lowest at the summer solstice.

KEY TERMS

annular eclipse, 65
crescent moon, 58
eclipse, 61
first-quarter moon, 59
full moon, 59
gibbous, 59
lunar eclipse, 61
lunar month, 59
month, 58
new moon, 59
partial eclipse, 62
penumbra, 62
penumbral eclipse, 62
phase, 58
quarter moon, 59
sidereal month, 60
solar eclipse, 61
third-quarter moon, 59
total eclipse, 62
umbra, 62
wane, 58
wax, 59

CONCEPT QUESTIONS

Concept Questions on the following topics are located in the margins. They invite thinking and discussion beyond the text.

1. Moonrise if the Earth did not spin. (p. 60)
2. A total lunar eclipse observed from the Moon. (p. 62)
3. Effect of Moon's distance on types of eclipses. (p. 64)
4. Testing effects of lunar cycle. (p. 67)

REVIEW QUESTIONS

5. How long does it take the Moon to go through a cycle of phases?
6. What are the names of the phases, and when do they occur?
7. At what time does a first-quarter moon rise? At what time does it set?
8. How is phase related to the angle between the Moon and Sun?
9. What is the phase of the Moon during a total solar eclipse? during a total lunar eclipse?
10. Why don't we have a lunar and a solar eclipse every month?
11. Why does the Moon look red during a total lunar eclipse?
12. What parts of the Sun are visible during a total solar eclipse?

QUANTITATIVE PROBLEMS

13. How long does someone have to wait from the time of a third-quarter moon until the next full moon?
14. Calculate how much later the Moon rises each day.
15. Show that the ratio of the Moon's diameter to the Sun's diameter is very similar to the ratio of the Moon's distance from the Earth to the Sun's distance from the Earth. (Sizes and distances needed to solve this are in the tables in the appendix.)
16. A total solar eclipse occurred over Europe on August 11, 1999. Based on the 18 year, 11 day, 8 hour period (6585.2313 days) between repetitions of similar eclipses, when and where will the next similar total eclipse occur?
17. If Figure 8.12B were expanded to include the Sun at the same scale, how big would the Sun be? How far away would it be? (Sizes and distances needed to solve this are in Appendix Tables 1 and 7.)

TEST YOURSELF

18. If you see the Moon rising at noon, its phase must be about
    a. full.
    b. new.
    c. first-quarter.
    d. third-quarter.
19. Suppose you observe a solar eclipse shortly before sunset. The phase of the Moon must be
    a. full.
    b. new.
    c. first-quarter.
    d. third-quarter.
20. During a lunar eclipse,
    a. the Earth's shadow falls on the Sun.
    b. the Moon's shadow falls on the Earth.
    c. the Sun's shadow falls on the Moon.
    d. the Earth's shadow falls on the Moon.
    e. the Earth stops turning.
21. Each day, the Moon rises about
    a. the same time.
    b. an hour earlier.
    c. an hour later.
    d. earlier or later depending on the season.
    e. None of the above because the Moon is always above the horizon.

PART 1 UNIT 9

Calendars

9.1 The Week
9.2 The Month
9.3 The Roman Calendar
9.4 The Leap Year
9.5 The Chronicling of Years

Learning Objectives

Upon completing this Unit, you should be able to:
- Explain the historical and astronomical origins of the week and month.
- Compare how calendar systems used in different cultures and religions differ from the Gregorian calendar that is in common use in the world today.
- Explain the purpose of the leap year and the revisions in the Gregorian calendar.
- Compare how different cultures have counted the passage of years.

Different cultures record the passage of time in many ways. At first it may seem bewildering that there are so many approaches, but they are based on just a few astronomical cycles.

We group days into weeks, months, and years to track time periods, which reflect natural cycles that shape our lives. A problem arises, though. These various time periods are not simple multiples of one another. The lunar cycle is about 29.53 days, and the year (from vernal equinox to vernal equinox) is about 365.24 days; so the number of days in the month or year is not a whole number, and the number of weeks in a month, or months in a year, is not a whole number. As a result, there is no obvious or best way to record the passage of time.

For example, is it more important that spring arrives on the same date, that the Moon is new on the first day of the month, or that the method of counting days is simple? This question has been answered in different ways by different cultures. Some developed complex systems for keeping their calendars connected with the astronomical cycles, while others ignored the incompatibilities and let their calendars lose that connection. We concentrate in this Unit on the schemes that different cultures developed for keeping track of the Sun's and Moon's motions. These motions themselves are discussed in detail in Unit 6 (The Year) and Unit 8 (Lunar Cycles).

9.1 THE WEEK

The origin of the 7-day week is uncertain. Seven days is approximately a quarter of the lunar cycle, but there is no simple interval of days that will keep alignment with the 29.53-day lunar month. The 7-day week was used by the Babylonians and Sumerians, although other cultures used anywhere from 6 to 10 days. In ancient Rome an 8-day week (marking the interval between market days) was used (Figure 9.1); in more recent history, after the French Revolution, a 10-day week was briefly adopted in France.

It is possible that there are seven days in the week to recognize the seven visible objects that move across the sky with respect to the stars: the Sun, the Moon, and the planets Mercury, Venus, Mars, Jupiter, and Saturn. We can see the names of some of these bodies in our English day names (*Sunday*, *Monday*, and *Saturday*). The influence is even clearer in the romance languages, which have their roots in

FIGURE 9.1
A portion of an ancient Roman calendar. Note the Roman numerals indicating the number of days in the months at the bottom of each column. The letters A through H indicate the 8 days of the week.

> **Concept Question 1**
> Where else do the names of ancient gods show up in everyday life? What qualities do the names convey?

Latin. An example is Spanish, with the day names *lunes, martes, miércoles, jueves,* and *viernes* relating to the Moon, Mars, Mercury, Jupiter, and Venus.

Some English day names come to us through the names of Germanic gods, many of whom have a direct parallel with the Greco-Roman gods after whom the planets are named. For example, Tuesday is named for Tiw, a god of war like Mars (giving Spanish *martes*). Wednesday is named for Woden, the chief god of Germanic peoples, identified with Mercury (*miércoles*). Thursday is named for Thor, the thunder god. He had powers like those of Jupiter, who was also called Jove (*jueves*). Friday is named for Freya, a love goddess like Venus (*viernes*).

Despite the astronomical connections to days' names, the week may simply reflect a human cycle: reasonable periods for work and rest, for purchasing goods and social gatherings. The seven-day week became dominant worldwide with the spread of Jewish, Christian, and Islamic cultures, all of which base their week on the biblical account of the creation of the world in seven days, as well as several other cultures that used seven days but without a biblical origin. Longer time periods such as the month and year have a clearer connection to astronomical cycles.

9.2 THE MONTH

The month is the next largest unit that is used by nearly all cultures. This time interval, and its name, derives from the Moon's cycle of phases. The interval between new moons averages 29.53059 … days, and because the year has about 365¼ days in it, there are about 12 lunar cycles per year.

However, 12 lunar cycles end up 11 days short of a full year, so there is no way to build a simple calendar that reflects both cycles. This has led to several different calendar systems used by different cultures and religions.

The commonly used calendar system maintains the "month," but it has no relationship to the phases of the Moon. It is an interval of 28 to 31 days, 12 of which add up to a year. This system is really just a solar calendar, maintaining the shorter period we call a month only as a timekeeping convenience. The history of this system is quite complicated, and we will return to it in Section 9.3.

Other calendar systems have retained a much closer connection to the Moon. For example, the Islamic calendar system is purely lunar, defining a year as 12 lunar cycles (Figure 9.2). However, this means dates shift relative to the seasonal position of the Sun by almost 11 days each year. For example, the holy month of Ramadan began on September 24 in 2006 but on September 13 in 2007 and September 2 in 2008. This system is a logical development for a culture that arose in the Middle East, where seasonal differences are not especially pronounced and where night travel by moonlight through the desert was commonplace.

The Chinese and Jewish calendars are called **lunisolar calendars** because they maintain a connection to the cycles of both the Sun and the Moon by adding a 13th month about every three years. The extra month is added whenever the calendar gets too far out of alignment with the seasonal year, so some years have about 354 ($\approx 12 \times 29.53$) days while others have about 384 ($\approx 13 \times 29.53$) days. For example, the Jewish New Year, Rosh Hashanah, occurred on September 30,

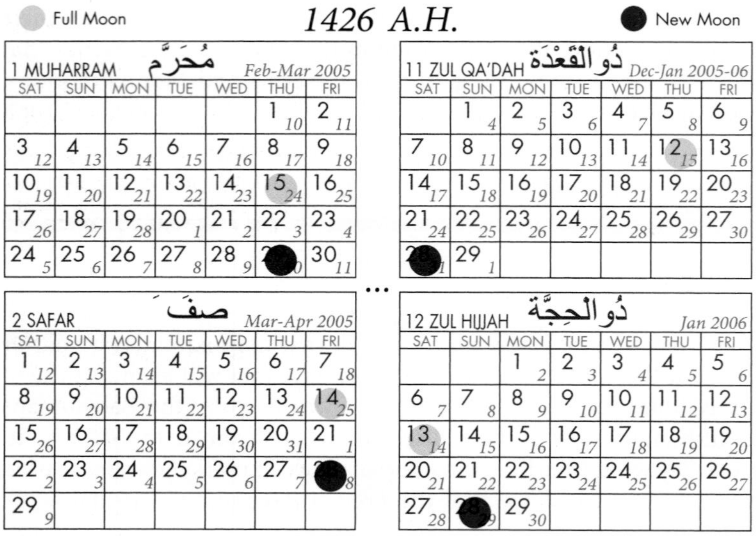

FIGURE 9.2

An Islamic calendar, showing the first two and last two months of the year 1426. Corresponding dates in the common calendar are shown in green (in 2005–2006 C.E.). Each month begins when a crescent moon is first visible, so the new moon comes about 2 days before the end of the month.

2008, then shifted to September 19 and September 9 in the subsequent two years. It shifted back to September 29 in 2011 after a year in which a 13th month was added in March of the common calendar.

The Chinese calendar is similar, but the month names are based on the segment of the ecliptic where the Sun is located when the Moon is new (Figure 9.3). This is somewhat similar to naming the months according to the constellation of the zodiac that the Sun is in at each new moon. When two new moons occur while the Sun is in the same segment of the ecliptic, there can be two of the same named month in a row. This can happen, for example, when a new moon occurs just as the Sun enters the segment of the ecliptic designated for "August" and then another new moon occurs just before the Sun leaves that segment. In that year there will be a second August. Using this procedure adds an extra month once every three or so years at varying times during the year. This shifts the date of the Chinese New Year in a pattern similar to that for Rosh Hashanah.

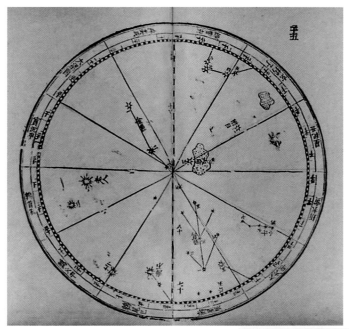

FIGURE 9.3
A Chinese calendar dividing up the year by days and position of the Sun along its path through the stars.

9.3 THE ROMAN CALENDAR

Our commonly used calendar is based on one developed by the Romans about 200 B.C.E. (before the Common Era—see Section 9.5). In fact, the word *calendar* is itself of Roman origin. There is some controversy about how the original Roman calendar was organized. It appears to have begun with only 10 named months and a gap during the winter, and it probably began on the first day of spring (the vernal equinox) rather than in January.

Many of the names of our months reflect that early calendar system. For example, if the year began in March, then September, October, November, and December would be the 7th (Sept.), 8th (Oct.), 9th (Nov.), and 10th (Dec.) months, each word's root reflecting the Latin name for the number of the month. Origins of the names of the months, some of which are less certain, are listed in Table 9.1.

Because the Roman calendar was not required to conform to the cycles of astronomical phenomena, it became a form of political patronage. The priests who regulated the calendar would add days and even months to please one group, and take days off to punish another. Think about the possibilities when rent and bills

TABLE 9.1 Origins of the Names of the Months

Month	Origin of the Name
January	Janus (gate), the two-faced god looking to the past and future; hence, beginnings
February	*Februa* ("expiatory offerings")
March	The god Mars
April	Etruscan *apru* (April), probably shortened from the Greek Aphrodite, goddess of love and earlier of the underworld
May	Maia's month; Maia ("she who is great"), the eldest of the Pleiades and the mother of Hermes by Zeus
June	Junius, an old Roman noble family (from Juno, wife and sister of Jupiter, equal to Greek Hera)
July	Julius Caesar (*Julius* means "descended from Jupiter"; the *Ju-* of June and July are the same: Jupiter, "Sky-father")
August	Augustus Caesar (*augustus* means "sacred" or "grand")
September–December	"Seventh month" to "tenth month" (the *-ember* may come from the same root as month)

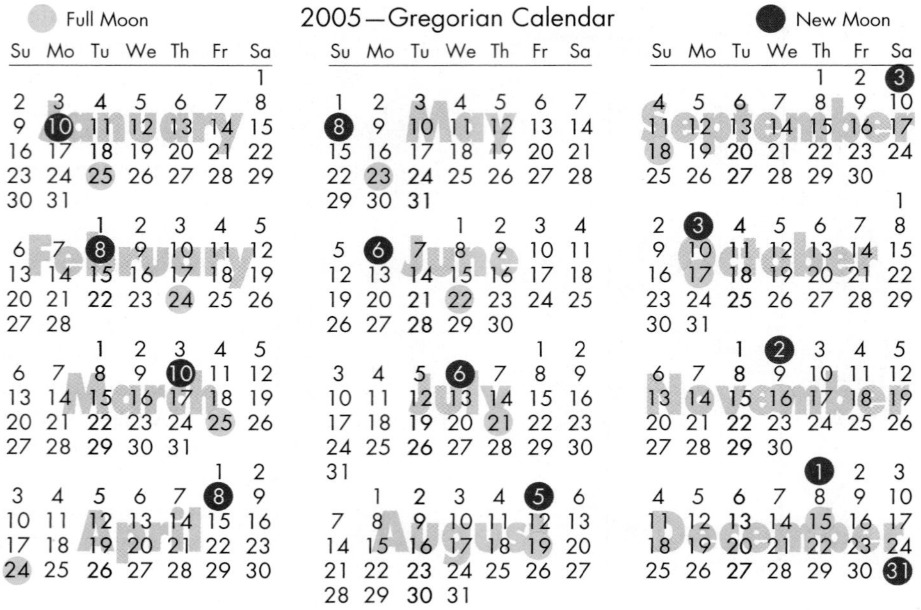

FIGURE 9.4

Sample modern calendar for the year 2005. Note that the dates of the full and new moon do not stay in the same parts of the month during the year.

are due! Such confusion resulted from these abuses that in 46 B.C.E. Julius Caesar asked the astronomer Sosigenes to design a calendar that would fit the astronomical events better and give less room for the priests and politicians to manipulate it. The resulting calendar, known as the **Julian calendar,** consisted of 12 months, alternating in order between 31 or 30 days, except February, which is discussed further in Section 9.4.

The Julian calendar barely survived Caesar before the politicians were at it again. First the name of the month *Quintilis* (originally the fifth month of the Roman year) was changed to *Julio* to honor Julius Caesar—hence our *July*. Next, on the death of Julius Caesar's successor, Augustus Caesar, an able and highly respected leader, the following month, *Sextilis*, was renamed in his honor—hence the name *August*. However, the order of months originally made August 30 days long, and it would have been impolitic for Augustus's month to be a day shorter than Julius's; so August was changed to have 31 days, and the numbers of days in all the following months were changed to reestablish the 30/31 alternation. Unfortunately, this used up one more day than the year allowed. Poor February, already one day short, was trimmed by a second day, leaving it with only 28 days. With only minor modifications, this is the calendar we use today (Figure 9.4).

9.4 THE LEAP YEAR

The ancient Egyptians knew that the year is not exactly 365 days long. It takes about 365¼ days for the Earth to complete an orbit around the Sun. Because we can't have fractions of a day in the calendar, a calendar based on a year of 365 days will come up 1 day short every 4 years.

Does a quarter of a day really matter? Consider that the seasons are set by the orientation of the Earth's rotation axis with respect to the Sun, not by how many days have elapsed. We therefore want to make sure that we start each year when the Earth has the same orientation. Otherwise the seasons get out of step with the calendar. For example, because in 4 years you will lose 1 day, in 120 years you will lose a month, and in 360 years you will lose an entire season. With a 365-day year, in a little over three centuries April would begin in what is now January.

The vernal equinox and some other astronomical events occur at one-year intervals, but the insertion of a leap day causes their calendar date to shift from year to year.

This problem is corrected by the **leap year,** a device introduced with the Julian calendar to keep the calendar in step with the seasons. The leap year corrects for the quarter day by adding a day to the calendar every fourth year. The extra day is added to February, which alternates between 28 and 29 days. The tradition of making adjustments to February may date back to when February was the last month, and year-end adjustments were made to keep the calendar in agreement with the vernal equinox. The civil calendar of India today has a similar system of leap years. It has a set of 12 months of 30 or 31 days that begins at the vernal equinox, and the first month of the year increases from 30 to 31 days during a leap year.

Unfortunately, the year is actually a little shorter than 365¼ days, so the leap year corrects by a tiny bit too much. To address this problem, the Julian calendar was modified in 1582 at the direction of Pope Gregory XIII. The common calendar we use today is therefore known as the **Gregorian calendar.** The change this calendar system introduced was to eliminate leap years in centuries that are not divisible by 400. Thus, 1700, 1800, and 1900 were not leap years, but 1600 and 2000 were.

The inauguration of the Gregorian calendar in 1582 was not a smooth transition. Because this calendar was adopted roughly 1600 years after the introduction of the Julian calendar, the accumulated error in the relationship between the calendar and the seasons came to about 10 days. To bring the calendar back into synchrony with the seasons, Pope Gregory XIII simply eliminated 10 days from the year 1582 so that the day after October 4 became October 15.

Although the changeover was accepted in much of Europe, non-Catholic countries such as Protestant England refused to abide by the Pope's edict. The change was not made in England until 1752. In Russia the change was not made until the revolution in the early 1900s. Other countries (Greece and Turkey, for example) changed in the 1920s.

Concept Question 2

If you were to create a calendar completely on your own, what would it look like?

9.5 THE CHRONICLING OF YEARS

The common scheme for numbering years began over a millennium ago in the year 532. But why "532"? The numbering of years by different cultures often relates to important religious dates, or the estimated starting date of a nation, or the years since an important ruler came to power. Sometimes these numbering systems were developed decades or even centuries after the event whose anniversary they are intended to mark, so the numbering is somewhat uncertain.

For example, until 532, throughout Europe the year was still commonly counted from the supposed founding of Rome (in 753 B.C.E., according to the common calendar), which made 532 the year 1285. Others, though, marked 532 as the year 248, reckoning from the year that Diocletian became emperor of Rome. Dionysius Exiguus, a monk of this era, carried out calculations for determining the date of the Christian holiday of Easter, which is related to the date of the full moon after the vernal equinox. The monk published a table of Easter dates beginning in the year 532 "*anno domini*" (or **A.D.,** meaning "in the year of the Lord," *not* "after death") to honor Jesus's life instead of Rome or its emperors. However, the monk was a better astronomer than historian. Christian scholars even at the time pointed out that this could not mark the years since the birth of Jesus because of obvious inconsistencies with other historical dates. But the table enjoyed wide circulation, and the idea of marking years this way became popular in Christian countries. So despite its historical inaccuracy, it became widely used and has become the common numbering system used today.

Based on various biblical references to Roman events and rulers, the date of Jesus's birth would have to be between 4 and 10 B.C.E.

For the years that preceded A.D. 1, the convention became to write them as 1, 2, 3 ... **B.C.,** standing for "before Christ." There is no year 0, which complicates mathematical treatment of years and introduces confusion about when centuries and millennia begin. In fact, the twenty-first century and the third millennium began on January 1, *2001*, because a decade, century, or millennium is complete only after its 10th, 100th, or 1000th year is completed. December 31, 2000, marked such a completion. Hence, starting from A.D. 1, the transition from A.D. 1999 to A.D. 2000 marked the completion of only 1999 years, not two full millennia.

Recently two different abbreviations have begun to replace A.D. and B.C. They are **C.E.** and **B.C.E.,** which stand, respectively, for "Common Era" and "before the Common Era." "Common Era" refers to our present calendar, which is used nearly worldwide for most business purposes, and thereby avoids reference to a particular religion. This also avoids the problem that the numbering was incorrect in the first place. Yet another abbreviation—**B.P.,** which stands for "before the present (era)"—is used, especially in archaeological, paleontological, and geological works. B.P., used for dates determined by analyzing radioactive carbon, takes 1950 C.E. as its base year.

The Jewish calendar is dated from the biblical creation of the world, so Rosh Hashanah in 2000 began the year 5761. The Islamic calendar is dated from Mohammed's emigration to Medina in 622 C.E. Because the Islamic year is shorter than the solar year, its numbering increases by one year relative to the common system every 33 or 34 years. The Islamic year 1421 began April 6, 2000.

The Chinese year-numbering system is based on cycles of 60 years, broken into five sequences of 12 years, each year of which is associated with a particular animal—Rat, Ox, Tiger, Rabbit, Dragon, Snake, Horse, Sheep, Monkey, Chicken, Dog, Pig. Every two years are associated with a particular element—Wood, Fire, Earth, Metal, Water—and the combination of animal and element characteristics repeats after 60 years. We are currently in the 79th of these 60-year cycles; year 1 of this cycle was in 1984, associated with the Rat and Wood. According to tradition this system originated in 2697 B.C.E. Continuous chronologies have been maintained since about the ninth century B.C.E.

Only a few calendar systems have survived in relatively common use. Other cultures had systems that used additional astronomical cycles. For example, the Maya included the motions of Venus in their time reckoning, making their calendar quite complex (Figure 9.5). The Mayan calendar involved 20-day "weeks" and a complex series of multiples of these "weeks" making up time periods of approximately 1 year, 20 years, 400 years, and 5000 years. One of these "long-count" periods of about 5000 years ended in 2012. Although Mayan scholars do not think this had a significant meaning in Mayan religion, some sensationalist claims were made that various cataclysms would occur in 2012, such as a collision with a "rogue planet," an outburst from the Sun, or a peculiar alignment with the Milky Way. These had no astronomical foundation.

Concept Question 3

In recent decades, doomsday claims have been made related to a variety of astronomical and calendrical events, such as Comet Kohoutek in 1974, Comet Hale-Bopp in 1996, and the "Y2K millennium" bug. Why do you suppose that claims of the end of the world are so popular?

FIGURE 9.5
A portion of a Mayan calendar, which is broken up into 20-day "weeks."

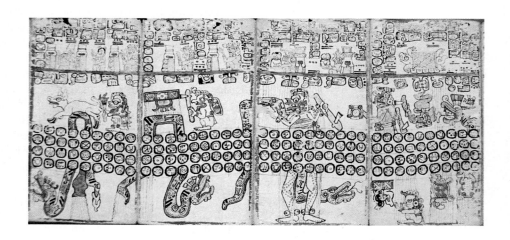

KEY POINTS

- A week of seven days may reflect quarters of the lunar cycle.
- Weekday names are related to the Sun, Moon, and planets.
- The Islamic calendar is strictly lunar with 12 months each year, making it only 354 days long.
- Chinese and Jewish calendars add a 13th lunar month every few years to maintain closer correspondence with the solar year.
- The common calendar in widespread use today dates back to ancient Rome and maintains no connection with Moon phases.
- The common calendar is known as the Gregorian calendar; it adds a leap year every 4 years (except in some centuries) to keep within about half a day of the solar year.
- Numbering of years has been done in many different ways by different cultures.
- The most commonly used numbering scheme was intended to correspond to the Christian religion, but was miscalculated.
- Claims of a Mayan calendar cataclysm have no astronomical basis.

KEY TERMS

A.D. (*anno domini*), 73
B.C. (before Christ), 73
B.C.E. (before the Common Era), 74
B.P. (before the present), 74
C.E. (Common Era), 74
Gregorian calendar, 73
Julian calendar, 72
leap year, 73
lunisolar calendar, 70

CONCEPT QUESTIONS

Concept Questions on the following topics are located in the margins. They invite thinking and discussion beyond the text.

1. Names of gods in everyday usage. (p. 70)
2. Constructing a better calendar. (p. 73)
3. Popularity of doomsday claims. (p. 74)

REVIEW QUESTIONS

4. What connection does the seven-day week have to astronomical phenomena?
5. What is the difference between lunar, lunisolar, and solar calendars?
6. What is the purpose of having leap years?
7. What was the purpose of the change to the Gregorian calendar system in recent centuries?
8. What establishes the initial year in the various calendar systems?
9. When and why did the year-numbering of the common calendar system begin to be used?

QUANTITATIVE PROBLEMS

10. If the same year-numbering systems were used today as were used in Europe before 532 C.E., what year would we currently be in?
11. What year will 2020 C.E. be in the Chinese, Islamic, and Jewish calendar systems?
12. Suppose a 12-month lunar calendar year (such as in the Islamic calendar) began on January 1. How many solar years would pass before the lunar calendar year would again begin within one week of January 1? How many lunar calendar years would have elapsed?
13. The first day of the month in the Islamic calendar begins when the crescent moon is first spotted. Usually the earliest this is possible is about 16 hours after the new moon. How many degrees has the Moon moved since it was new at this time? Approximately how long after the Sun sets will the Moon set if it is just 16 hours past new?
14. Under the Julian calendar system, the year averages to 365.25 days after 4 years. In the Gregorian system, determine the average length of the year after 400 years. If a country needed to convert from the Julian to the Gregorian calendar now, how many days would it need to remove from the calendar?
15. Suppose the year were 365.170 days long. How would you design a pattern of leap years to make the average year length exactly match this value within a span of 1000 years?

TEST YOURSELF

16. According to the Gregorian calendar, which of the following years is *not* a leap year?
    a. 1976
    b. 1900
    c. 2000
    d. 2016
    e. 2400
17. Suppose that the length of the year were 365.2 days instead of 365.25 days. How often would we have leap year? Every
    a. 2 years.
    b. 5 years.
    c. 10 years.
    d. 20 years.
    e. 50 years.
18. Why does the Chinese New Year occur on different dates in the Western calendar each year?
    a. The Chinese year always starts when the Moon is new.
    b. The precession of the Earth's axis shifts it a little each year.
    c. The date is based on when Mars undergoes retrograde motion.
    d. The date is chosen each year for when a positive horoscope is cast.
    e. The Chinese calendar slips by a quarter day each year because it has no leap days.
19. Why was the Roman calendar system of a leap year every 4 years changed in the 1500s?
    a. The year is not exactly 365.25 days long.
    b. The Earth had slowed down its spin since ancient Roman times.
    c. Precession of the Earth had changed the length of the year.
    d. The lunar phases had gone out of phase with our months.
20. If we were to stop the practice of including leap years, then the date of the vernal equinox would
    a. slowly shift toward earlier dates in the calendar year.
    b. stay the same.
    c. jump forward by 10 days each year.
    d. slowly shift toward later dates in the calendar year.

PART 1 UNIT 10

Geometry of the Earth, Moon, and Sun

10.1 The Shape of the Earth
10.2 Distance and Size of the Sun and Moon
10.3 The Size of the Earth
10.4 Measuring the Diameter of Astronomical Objects
10.5 The Moon Illusion

Learning Objectives

Upon completing this Unit, you should be able to:
- Show that the Earth is round and how its size can be found from basic observations.
- Explain how ancients estimated the sizes and distances of the Moon and Sun.
- Discuss why ancients rejected the idea that the Earth orbited the Sun.
- Define angular measurements and calculate angular and linear sizes using distances.
- Describe the ways that angular sizes are often perceived incorrectly.

It is possible to draw several important conclusions about the sizes of and distances between the Earth, Moon, and Sun based on careful but simple measurements. As far as we know, the Greek astronomers of classical times were the first to do this. The values they determined were not always precise, but they were surprisingly good. They demonstrated that you do not need telescopes, high-speed communications, computers, or satellites to determine, for example, that the Earth is a sphere with a circumference of about 40,000 kilometers.

We study ancient ideas of the heavens not so much for what they tell us about the cosmos but to learn how observation, geometry, and careful reasoning can lead us to a deeper understanding of the universe. Remarkably, we can use some of the basic ideas about the orbit, phases, and eclipses of the Moon from Unit 8 to derive distances and sizes for both the Moon and the Sun. Similar kinds of geometric ideas are still widely used today to measure the sizes of astronomical objects.

10.1 THE SHAPE OF THE EARTH

Clarification Point

It is a modern myth that medieval scholars believed the Earth was flat and that Christopher Columbus's crew feared "sailing off the edge of the Earth."

Concept Question 1

A few people today still believe the Earth is flat. What "proof" would you offer them that it is round? What argument do you think would be most persuasive?

The ancient Greeks knew that the Earth is round. As long ago as about 500 B.C.E., the mathematician **Pythagoras** (about 560–480 B.C.E.) was teaching that the Earth is spherical, but the reason for his support of this idea was as much mystical as rational. Like many of the ancient philosophers, he believed that the sphere was the perfect shape and that the gods would therefore have utilized that perfect form in creating the Earth.

By 300 B.C.E., however, **Aristotle** (384–322 B.C.E.) was presenting arguments for the Earth's spherical shape that were based on simple naked-eye observations that anyone could make. Such reliance on direct observation was the first step toward generating scientifically testable hypotheses about the contents and workings of the universe. For instance, Aristotle noted that if you look at a lunar eclipse when the Earth's shadow falls on the Moon, the shadow can be clearly seen as curved, as Figure 10.1A shows. As he wrote in his treatise "On the Heavens,"

The shapes that the Moon itself each month shows are of every kind—straight, gibbous, and concave—but in eclipses the outline is always curved;

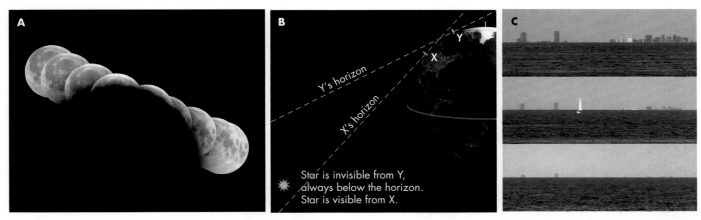

FIGURE 10.1

Demonstrations that the Earth is spherical. (A) A sequence of photographs during a partial lunar eclipse shows that the edge of the Earth's shadow on the Moon is always a portion of a circle, showing that the Earth must be round. (B) As a traveler moves from north to south on the Earth, different stars become visible. Some disappear below the northern horizon, whereas others, previously hidden, become visible above the southern horizon. This variation would not occur on a flat Earth. (C) A sequence of photos taken from a boat traveling away from Boston. Note that the tops of the buildings remain visible as the bottom parts and shorter buildings disappear over the horizon.

and, since it is the interposition of the Earth that makes the eclipse, the form of this line will be caused by the form of the Earth's surface, which is therefore spherical.

Another of Aristotle's arguments that the Earth is spherical was based on the observation that a traveler who moves south will see stars that were previously hidden below the southern horizon, as illustrated in Figure 10.1B. For example, the bright southern star Canopus is easily seen in Miami but is always below the horizon in Boston. The Earth's curvature can also be seen by looking across the ocean at a receding ship or from the ship toward the land. The top of these objects remain visible at large distances while the base becomes hidden below the horizon (Figure 10.1C). This could not happen on a flat Earth.

10.2 DISTANCE AND SIZE OF THE SUN AND MOON

One of the most remarkable ancient Greek astronomers was **Aristarchus** of Samos (about 310–230 B.C.E). He was able to estimate the size and distance of the Moon and Sun. Even though his values were not highly accurate, they gave at least the correct sense of which was larger and which was smaller than the Earth, as well as approximate indications of their distances from Earth. Few of his writings survive intact, but references to his works by later astronomers allow us to reconstruct many of his findings. For example, by comparing the size of the Earth's shadow on the Moon during a lunar eclipse to the size of the Moon's disk, as can be estimated in Figure 10.1A, Aristarchus calculated that the Moon's diameter was about one-third of the Earth's. His estimate that the Moon's diameter is about 0.33 that of the Earth's is not far from the correct value of about 0.27.

Aristarchus then estimated the distance to the Moon by noting that when the Moon passes through the middle of the Earth's shadow during a lunar eclipse, it takes about three hours to complete its passage through the shadow. On the other hand, the Moon takes about 660 hours to complete its passage around the zodiac.

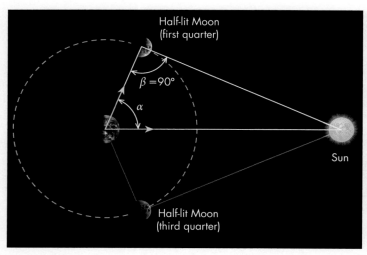
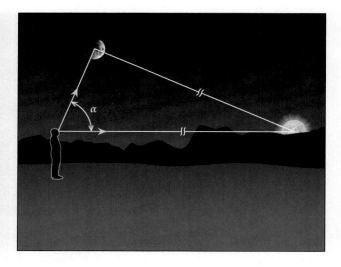

FIGURE 10.2

Aristarchus estimated the relative distance of the Sun and Moon by observing the angle α between the Sun and the Moon when the Moon is exactly half lit. Angle β must be 90° for the Moon to be half lit. Knowing the angle α, he could then set the scale of the triangle and thus the relative lengths of the sides. (Sizes and distances are not to scale.)

Mathematical Insight

If the circumference of the Moon's orbit is 220 times Earth's diameter, it is $440 \times R_\oplus$. The circumference of a circle is 2π times its radius, so the Moon's orbital radius is $440 \times R_\oplus/2\pi \approx 70 \times R_\oplus$.

Even with modern instruments, it is impossible to measure the angle between the quarter Moon and the Sun well enough to obtain the Sun's distance accurately. If the Sun were as close as about 10 times the Moon's distance, though, there would be about a day less between new Moon and first quarter than between first-quarter and full moon, which would be fairly easy to detect.

Concept Question 2

Can you think of other instances where a hypothesis was correct but the evidence initially appeared to contradict it?

If the Earth's shadow is about the same size as the Earth itself, then the entire circumference of the Moon's orbit is about 220 (= 660/3) times bigger than the Earth's diameter, and its distance is about 70 times the Earth's radius. This is not far from the actual distance of the Moon, which is about 60 Earth radii away from our planet.

Aristarchus also calculated the Sun to be about 20 times farther away from the Earth than the Moon is. He carried out this calculation by measuring the angle between the Sun and the Moon when the Moon was exactly half lit at the first- and third-quarter phases (Figure 10.2). From the apparent size in the sky of the Sun and the Moon and their relative distances, and his calculation of the Moon's diameter relative to the Earth's, Aristarchus deduced that the Sun's diameter was about seven times that of the Earth (see Section 10.4). This is far smaller than we know it is today—the Sun is more than 100 times larger than the Earth. Nevertheless, it was the first demonstration that the Sun is much bigger than the Earth and much more distant than the Moon.

It was perhaps his recognition of the large size of the Sun that led Aristarchus to propose the revolutionary idea that the Earth revolves around the Sun rather than the reverse. Aristarchus was, of course, correct; but his idea was too radical, and another 2000 years passed before scientists became convinced of its correctness. However, this was not mere stubbornness. There was a good reason not to conclude that the Earth moves around the Sun. If it does, the positions of stars should change during the year as the Earth moves from one side to the other within the celestial sphere. Looking at Figure 10.3A, you can see that the nearby star should appear to lie at a different angle from, and in a different position with respect to, each of the more distant stars in January and in July.

This shift in a star's apparent position resulting from the Earth's motion around the Sun is called **parallax**, and Aristarchus's critics were right in supposing that it should occur. Actually, many astronomers of the time assumed the stars were all at the same distance on the celestial sphere, but even so there would be a parallax effect. If the Earth moved around inside this hypothesized celestial sphere as in Figure 10.3B, we would see changes in the stars' apparent separation—just as we can tell how we are moving in the dark by looking at nearby lights. Because no parallax was seen, astronomers rejected Aristarchus's idea. What they failed to realize was that because of the immense distance to the stars, the parallax is so tiny that

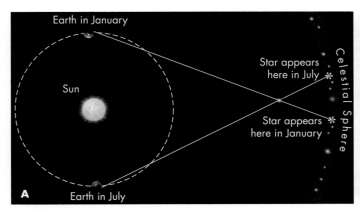
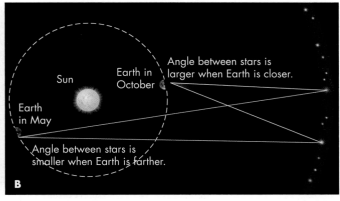

FIGURE 10.3
Most ancient astronomers argued against the idea of the Earth revolving around the Sun because: (A) if some stars are nearer than other stars, we would see their positions appearing to shift relative to their neighbors (stellar parallax) as the Earth moved around the Sun; and (B) even if all the stars lay at the same distance (on the celestial sphere), as Earth orbited we would sometimes be closer to the stars and sometimes farther, so the angular size of constellations would change. Neither effect was seen because stars are so tremendously distant.

we need a powerful telescope and very precise equipment to measure it. In fact, parallax of stars was not successfully measured until 1838 (see Unit 54). Thus, an idea may be rejected for reasons that are logically correct but are flawed because of limited data accuracy.

10.3 THE SIZE OF THE EARTH

Even though few astronomers accepted Aristarchus's idea that the Earth moves around the Sun, they did accept his estimates of the sizes and distances of the Sun and Moon. However, these measurements were all scaled to the size of the Earth, whose size was not known. The Moon might have one-third the diameter of the Earth, but was that 1000 kilometers or 100,000 kilometers?

Eratosthenes (276–195 B.C.E), head of the famous library at Alexandria in Egypt, made the first measurement of the Earth's size. He obtained a value for its circumference that very nearly matches the presently measured value. Eratosthenes's demonstration is one of the most elegant ever performed because it so superbly illustrates how scientific inference links observation and logic.

Eratosthenes, a geographer as well as an astronomer, heard that to the south, in the Egyptian town of Syene (the present city of Aswân), the Sun would be directly overhead at noon on the summer solstice and thus it cast no shadow. Proof of this was that on the solstice the Sun lit the bottom of a deep well there. However, on that same day, the Sun did cast a shadow in Alexandria, where he lived. Knowing the distance between Alexandria and Syene and appreciating the power of geometry, Eratosthenes realized he could deduce the size (circumference) of the Earth. He analyzed the problem as follows: Because the Sun is far away from the Earth, its light travels in parallel rays toward the Earth. Thus, two rays of sunlight, one hitting Alexandria and the other shining down the well, are parallel lines, as depicted in Figure 10.4.

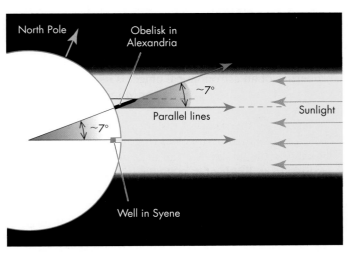

FIGURE 10.4
Eratosthenes's calculation of the circumference of the Earth. The Sun is directly overhead on the summer solstice at Syene in southern Egypt. On that same day, Eratosthenes found the Sun to be 1/50th of a circle (about 7°) from the vertical in Alexandria, in northern Egypt. Eratosthenes deduced that the angle between two verticals placed in northern and southern Egypt must be 1/50th of the circumference of the Earth.

With our modern perspective, we would say that the Sun was straight overhead in Syene because its latitude is close to 23.5°N, whereas Alexandria lies at about 31°N, where the Sun never passes through the zenith.

Mathematical Insight

The "stadium" is an ancient unit of measure used by many cultures, and with different lengths, so the exact value used by Eratosthenes is uncertain and may have ranged from about 0.15 to about 0.21 km. The modern use of the word *stadium* comes from the use of the term to refer to a racetrack one stadium long.

Since the Sun is straight above Syene at noon on the summer solstice, if the Sun's rays continued downward through Syene, they would go straight to the center of the Earth. Because of the curved surface of the Earth, a vertical line down to the center of the Earth from Alexandria is at an angle to the rays of sunlight. From geometry we know that the corresponding angles formed where a single line crosses two parallel lines are equal, so the line passing vertically through the Earth's surface in Alexandria makes the same angle with all of the Sun's parallel rays. As a result of this, the angle between vertical and the Sun's rays is the same as the angle of latitude between the two cities as shown in Figure 10.4.

This angle can be measured with sticks and a protractor (or their ancient equivalent) and is the angle between the direction to the Sun and a straight stick pointing vertically upward (see Figure 10.4). Eratosthenes found this angle to be about 1/50th of a circle, or roughly 7°. Therefore, the angle formed by a line from Alexandria to the Earth's center and a line from the well to the Earth's center must also be 1/50th of a circle.

Eratosthenes reasoned that if the angle between Alexandria and Syene is 1/50th of a circle, the distance between these cities must be 1/50th the circumference of the Earth. Because he knew the distance between Alexandria and the well to be 5000 stadia (where a stadium is about 0.16 kilometer), the distance around the entire Earth must be 5000 stadia × 50, or 250,000, stadia. When expressed in modern units, this is roughly 40,000 kilometers, or a diameter of about 13,000 kilometers, which is approximately the size of the Earth as we know it today.

By modern standards, there is absolutely nothing wrong with this technique. You can use it yourself to measure the size of the Earth. Eratosthenes's success was a triumph of logic and the scientific technique. The method required that he assume the Sun was so far away that its light reached Earth along parallel lines. That assumption, however, was supported by the earlier set of measurements made by Aristarchus, showing that the Sun is much more distant than the Moon (Section 10.2).

10.4 MEASURING THE DIAMETER OF ASTRONOMICAL OBJECTS

One of the most basic properties of an astronomical object is its size. Because we cannot stretch a tape measure across the disk of the Sun or Moon, how do we know the size of such heavenly bodies?

The basic method for measuring the size of a distant object was worked out long ago and is still used today. This method involves first measuring how big an object *looks*, a quantity called its **angular size**. We can measure an object's angular size by drawing imaginary lines to each side of it, as shown in Figure 10.5, and then measuring the angle between the lines using a protractor.

As an astronomical example, we can measure the angular size of the Moon, or equivalently, because it is a round object, its *angular diameter*. The Moon's angular

FIGURE 10.5

The angle that an object covers from an observer's point of view is called its angular size. Note that a larger object at a larger distance may have the same angular size as a nearer, smaller object.

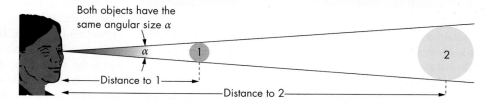

A

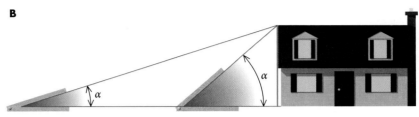

B

FIGURE 10.6
(A) A scale drawing of an observation of the Moon shows that it has the angular size of a 1.5-mm object held 172 mm from your eye. This is equivalent to about the head of a pin held about 7 inches from your eye. (B) Angular size grows smaller in inverse proportion as the distance becomes larger.

size is small, and the Moon moves on the sky, so it is not practical to aim lines on each side. Instead, we can determine its angular size by the following procedure: Hold up a ruler at arm's length in front of the Moon. If you do this, you might find that the Moon is about 6 mm across on the ruler. Next measure the distance from your eye to the ruler; about 688 mm in this example. Drawing this to scale, dividing the sizes by 4 to fit across a page, yields Figure 10.6A. Measuring the angle between the lines with a protractor, we find that the Moon's angular size is about a half degree (0.5°).

Notice that an object's angular size depends on its distance (Figure 10.6B). For example, if you move closer to an object, its angular size increases. Therefore,

The angular size of an object depends inversely on its distance.

From its angular size and distance, we can find an object's *linear size*—that is, its size in units of length such as meters. To work out a mathematical formula relating angular size and linear size, imagine that you are at the center of a circle whose radius is d, the distance to the body. The circle passes through the object, as illustrated in Figure 10.7. Let ℓ be the linear size of the body. Draw lines from the center to each end of ℓ, letting the angle between the lines be α (the Greek letter alpha), the object's angular size.

We now determine the object's linear size, ℓ, by forming the following proportion: ℓ is to the circumference of the circle as α is to the total number of degrees around the circle, which we know is 360. Thus,

$$\frac{\ell}{\text{circumference}} = \frac{\alpha}{360°}.$$

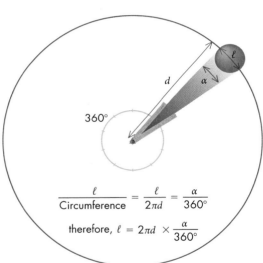

FIGURE 10.7
How to determine linear size from angular size.

ℓ = linear size
α = angular size
d = distance

However, we know from geometry that the circle's circumference is $2\pi \times d$. Thus, we can solve for ℓ by multiplying both sides by the circumference:

$$\ell = \text{circumference} \times \frac{\alpha}{360°} = 2\pi \times d \times \frac{\alpha}{360°}.$$

Dividing 360° by 2π gives an important formula relating angular size, linear size, and distance:

$$\ell = d \times \frac{\alpha}{57.3°}.$$

This is known as the *angular size formula*, or sometimes the *small-angle formula*, because it is most accurate for small angles. As an example of how we might apply this formula, we use it to measure the Moon's diameter. We stated previously that the Moon's angular size is about 0.5°. Its distance is about 384,000 kilometers. Thus, its diameter is

$$\ell = 384{,}000 \text{ km} \times \frac{0.5°}{57.3°} = 3400 \text{ km}.$$

So the Moon's diameter is a little more than one-quarter of the Earth's.

Mathematical Insight

The angular size formula has an error of only about 1% for objects 28° across, and less than 0.1% at angles smaller than about 9°.

10.5 THE MOON ILLUSION

Your perception of angular size can be tricked by the context of your observations. One of the best known examples of this is the Moon. To most people the Moon appears to be much larger when they see it rising or setting than when it is high in the sky. But if you measure the Moon's angular diameter carefully, you will find it to be slightly *smaller* when it is near the horizon than when it is nearly overhead. In fact, when the Moon is nearest its zenith, we are slightly closer to it. The difference in distance occurs because, as the Earth spins, we are carried closer to and farther from the Moon by a distance about equal to the Earth's radius, and when we see the Moon overhead, we are on the part of the Earth closest to the Moon.

The mental causes of this misperception of sizes, known as the **Moon illusion**, are still debated by psychologists. It appears to be an optical illusion caused, at least in part, by the observer's comparing the Moon with objects seen near it on the horizon, such as distant hills and buildings. You know those objects are big even though their distance makes them appear small. Therefore, you unconsciously magnify both them and the Moon, making the Moon seem larger. You can verify this sense of illusory magnification by looking at the Moon through a narrow tube that blocks out objects near it on the skyline. You can also hold up a small coin at arm's length and compare its size to the Moon's. Observed these ways, you will find that the Moon appears roughly the same size regardless of where it is on the sky.

Figure 10.8 shows a similar effect. Because you know that the rails are really parallel, your brain ignores the apparent convergence of the railroad tracks and mentally spreads the rails apart. That is, your brain provides the same kind of enlargement to the circle near the rails' convergence point as it does to the rails, causing you to perceive the middle circle as larger than the one near the bottom of the image, even though they are the same size.

Clarification Point

Aristotle suggested a now popular belief that the atmosphere magnifies the Moon when it is near the horizon. However, the Moon's vertical extent is actually shrunk by bending of light near the horizon, while its width is unaffected.

FIGURE 10.8
Circles along converging rails illustrate how your perception may be fooled. The bottom circle looks smaller than the circle on the horizon but is in fact the same size. Similarly, the identical circle high in the sky looks smaller than the circle on the horizon.

KEY POINTS

- The Earth's shape has been known to be spherical for thousands of years from observations of eclipses and the disappearance of southern stars below the horizon as one travels northward.
- The ancient Greek astronomer Aristarchus used lunar eclipses to show that the Moon is about a third the size of the Earth and that it orbits about 70 Earth radii away from us.
- He also showed that the Sun is much farther from us than the Moon, and that the Sun is many times larger than the Earth.
- Aristarchus also proposed that the Earth orbits the Sun, but failure to detect parallax led others to doubt this hypothesis.
- The ancient Greek astronomer Eratosthenes was able to find the size of the Earth by comparing the angle of sunlight shining on separate parts of the Earth's surface.

- The angular size of an object depends on its linear size divided by its distance; knowing two of these quantities lets us find the third.
- Mental perception of angular size can be tricked by optical illusions, such as the Moon looking larger near the horizon.

KEY TERMS

angular size, 80
Aristarchus, 77
Aristotle, 76
Eratosthenes, 79
Moon illusion, 82
parallax, 78
Pythagoras, 76

CONCEPT QUESTIONS

Concept Questions on the following topics are located in the margins. They invite thinking and discussion beyond the text.

1. Best evidence that the Earth is not flat. (p. 76)
2. Correct hypotheses that initially appear false. (p. 78)

REVIEW QUESTIONS

3. What evidence did ancient astronomers have that the Earth is round?
4. How was the circumference of the Earth determined by Eratosthenes?
5. How was the size of the Moon estimated by ancient observations?
6. Why was Aristarchus's estimate of the distance to the Sun uncertain?
7. What is the difference between *angular size* and *linear size*?
8. If you triple your distance from an object, what happens to its angular size?
9. Why does the Moon look larger when it is near the horizon?

QUANTITATIVE PROBLEMS

10. The smallest separation between Earth and Jupiter is 588 million km while the largest separation is 968 million km. Given that Jupiter has a physical diameter of 140,000 km, what is the change in the angular size of Jupiter throughout the year?
11. Suppose you are an alien living on the fictitious warlike planet Myrmidon and you want to measure its size. The Sun is shining directly down a missile silo 1000 miles to your south, while at your location the Sun is 36° from straight overhead. What is the circumference of Myrmidon? What is its radius?
12. A cloud directly above you is about 10° across. From the weather report you know that the cloud is at an altitude of 2200 m. How wide is the cloud?
13. The Andromeda Galaxy, M31, is in many ways similar to our own galaxy but only slightly larger. The linear diameter of the Andromeda Galaxy along its longest axis is 140,000 light-years, but from our perspective, the Andromeda Galaxy has a maximum angular diameter of 3.18°. How far away is the Andromeda Galaxy?
14. Suppose you see a person whom you know is 1.8 m (6.0 ft) tall, standing at a distance where he appears to have an angular height of 2°. How far away is he?
15. Suppose the Moon is exactly 0.500° wide when it is near the horizon. If later in the night you saw it straight overhead, what would its angular diameter be? (Assume the Earth's radius is 6378 km and the Moon's distance at the first observation is 398,000 km.)
16. Aristarchus's critics argued that parallax was not observed and therefore the Earth could not be in motion. In fact we know now that the stars are very far away and parallax motion is very small. Let's examine this by estimating the change in angular size of the angular separation between two stars 10 light-years away from the Sun. The two stars are 3 light-years apart. Calculate the angular separation of the two stars when the Earth is closest to them in its orbit and when it is farthest away, on the opposite point on its orbit. If human vision (without the aid of a telescope) can at best distinguish a difference of about 0.02°, would the change in angular separation be detectable to the naked eye?

TEST YOURSELF

17. The circular shape of the Earth's shadow on the Moon led early astronomers to conclude that
    a. the Earth is a sphere.
    b. the Earth is at the center of the Solar System.
    c. the Earth must be at rest.
    d. the Moon must orbit the Sun.
    e. the Moon is a sphere.
18. The ancient Greeks are *not* credited with
    a. measuring the size of the Earth.
    b. finding the relative sizes of the Earth and Moon.
    c. determining that the Earth is round.
    d. detecting the parallax of stars.
    e. suggesting that the Sun is at the center of the Solar System.
19. If the Moon were moved to a distance that is one-fourth of its current value, its angular size would be
    a. twice as big.
    b. half as big.
    c. four times as big.
    d. one-quarter as big.
    e. the same as it is now.
20. Earth's distance from the Sun varies as it orbits the Sun. Consequently,
    a. the Sun has a larger angular size when Earth is closer to the Sun.
    b. the Sun's angular size does not change throughout the year.
    c. the Sun has a smaller angular size when Earth is closer to the Sun.
    d. the Sun has an oval shape as viewed from Earth.

PART 1

UNIT 11

Planets: The Wandering Stars

11.1 Motions of the Planets
11.2 Early Ideas About Retrograde Motion
11.3 The Heliocentric Model
11.4 The Copernican Revolution

Learning Objectives

Upon completing this Unit, you should be able to:
- Describe the patterns of motion in the sky that the planets exhibit.
- Contrast the geocentric and heliocentric models' explanation of retrograde motion.
- Define terms associated with planetary positions, such as opposition, conjunction, and greatest elongation.
- Explain how Copernicus estimated the distances of the planets from the Sun.

Five of the brightest "stars" in the sky do not remain in the same position on the celestial sphere, but instead gradually shift from night to night. The motions of these objects are quite complex and difficult to predict. The Greeks gave them the name *planetai,* meaning "wanderers," from which our word planet comes.

Until the invention of telescopes, **planets** were thought of as special stars, free to roam about the sky, unlike the thousands of other stars visible on the celestial sphere. This freedom of movement suggested they had special powers, and perhaps this inspired their association with the gods Mercury, Venus, Mars, Jupiter, and Saturn. Careful recording of the positions of the planets revealed that they follow regular, if complex, patterns. As astronomers sought to understand these motions, they eventually realized that the motions made more sense if the Earth itself was a planet too. In the present day when we accept as common knowledge that the Earth is a planet, it is hard to imagine how bizarre it must have sounded when it was first proposed that Earth was a "wandering star."

This Unit focuses on the interpretation of the motion of the planets and the discovery that the planets orbit the Sun, spanning thousands of years of astronomical history. If you are unfamiliar with the basic structure of the Solar System, it may help to review Unit 1. The motion of the Earth around the Sun (and the apparent motion of the Sun among the stars) is described in Unit 6.

11.1 MOTIONS OF THE PLANETS

Planets move against the background of stars because of a combination of the Earth's and their own orbital motion around the Sun. One of the more important features of this motion is that the planets always remain within a narrow band on either side of the ecliptic, within the constellations of the zodiac, as the Sun does (Unit 6). The planets remain close to the ecliptic because their orbits, including that of the Earth, all lie in nearly the same plane (Figure 11.1A). Like the Sun and the Moon, the planets shift north and south of the celestial equator because the ecliptic is tilted by 23.5° (Figure 11.1B).

Normally the planets are seen to move from west to east past the stars. This does not mean that the planets rise in the west and set in the east. As seen from the

Concept Question 1

If you see a bright "star" in the sky, how can you tell whether it is a star or a planet (such as Venus) without using a telescope?

Unit 11 Planets: The Wandering Stars

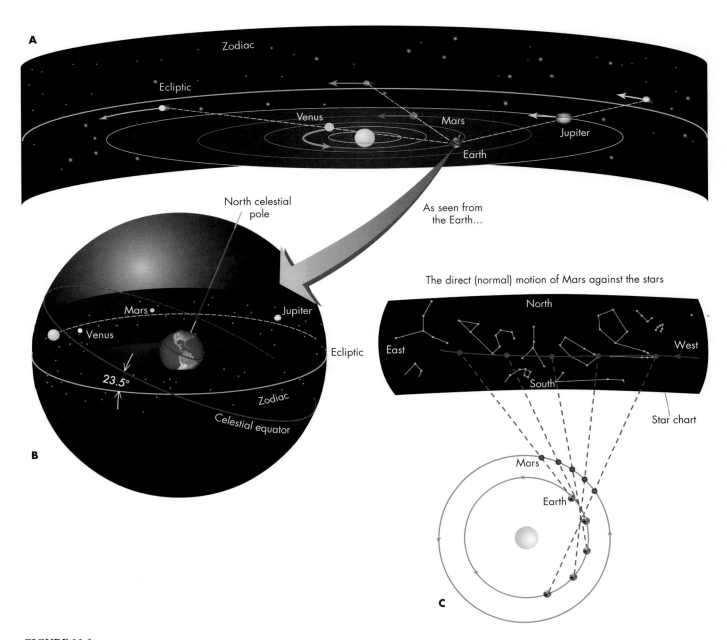

FIGURE 11.1
(A) The planets move through the zodiac because they all orbit the Sun in approximately the same plane as the Earth. (B) As viewed from the Earth, the ecliptic is tilted 23.5° with respect to the celestial equator, so the planets move north and south on the celestial sphere as they orbit the Sun. (C) Tracking the motion of a planet over several months, it generally moves eastward against the background stars. Note: Star maps usually have east on the left and west on the right, so that they depict the sky when looking upward, whereas maps of the Earth face downward so east and west are reversed.

Pluto, Eris, and a number of other small Solar System bodies sometimes move outside the zodiac.

surface of Earth, planets always rise in the east and set in the west because their movements across the celestial sphere are far slower than the effect of the Earth's rotation. The motion of a planet relative to constellations of the zodiac can be seen by marking off the planet's position on a star chart over a period of several months. Figure 11.1C illustrates such a plot and shows that planets normally move eastward through the stars as a result of their orbital motion around the Sun.

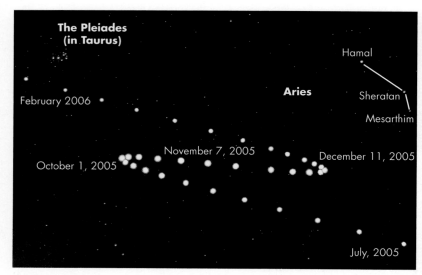

FIGURE 11.2
Images showing the retrograde motion of Mars in late 2005. Pictures were taken over an eight-month period, approximately one week apart, as Mars moved through the constellations Aries and Taurus.

Sometimes, however, this movement is interrupted. Periodically each planet moves west with respect to the stars, a condition known as **retrograde motion.** The word *retrograde* means "backward moving," and when a planet is in retrograde motion, its path through the stars zigzags or loops backward for a month or more.

Figure 11.2 illustrates this peculiar motion in a time-lapse sequence of photographs of Mars taken over about eight months. Mars begins its journey through this picture in July 2005 on the lower right side of the image, then while it steadily brightens, it changes direction in October 2005. After about two months traveling backward among the stars, it reverts to its original direction, meanwhile fading in brightness. All planets undergo retrograde motion when they lie in the same direction as the Earth from the Sun. This strange behavior became one of the major puzzles in astronomy for almost 2000 years.

11.2 EARLY IDEAS ABOUT RETROGRADE MOTION

One of the most basic of observations of the sky is that everything seems to spin around the Earth from east to west. From earliest times, this led to the model that the Earth was at the center of the universe and the planets and stars moved around it. An Earth-centered explanation of this type is called a **geocentric model.**

Figure 11.3 shows a typical geocentric model based on the work of the Greek astronomer Eudoxus, who lived about 400–347 B.C.E. The Sun, Moon, and planets all move around the Earth at different speeds than the stars and are pictured as being located on different transparent (crystalline) spheres that turned at different speeds. The bodies that move fastest relative to the stars are those that are nearest to

> **Clarification Point**
>
> The normal and retrograde motions of planets are very slow compared to the Earth's spin, so they always still rise in the east and set in the west.

FIGURE 11.3
A cutaway view of the geocentric model of the Solar System according to Eudoxus. The celestial bodies were pictured as being carried around the Earth on transparent spheres. The spheres were imagined to spin around Earth once each day, but at slightly different rates so that the Sun, Moon, and planets moved at slightly different rates relative to the stars.

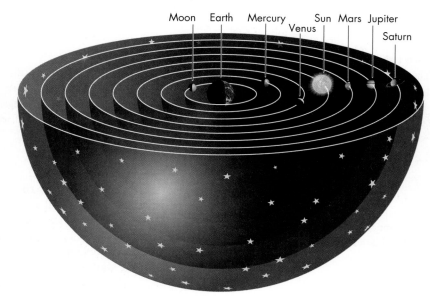

the Earth. Thus, the Moon completes its orbit most rapidly and is nearest, whereas Saturn, whose path through the stars takes roughly 29 years to complete, is located the farthest out. By tipping the rotation axis of each sphere slightly, Eudoxus was able to give a satisfactory explanation of the normal motions of the heavenly bodies. However, this model does not explain retrograde motion, unless one believes that the crystalline spheres sometimes stop turning, reverse direction, stop again, and then resume their original motion. This idea is clumsy and unappealing.

Eudoxus was able to develop a more complex model to explain retrograde motion by requiring that each planet's crystalline sphere should itself be attached to another crystalline sphere that rotated at a different angle. By combining steady rotations of both spheres, he was able to make zigzag paths that roughly resembled retrograde motion. Over the next five centuries, astronomers proposed variants of this complex model, leading to a very successful model that could predict the planets' positions with good accuracy, developed by Claudius **Ptolemy,** the great astronomer of Roman times, who lived about 100 to 170 C.E.

Ptolemy lived in Alexandria, Egypt, which at that time was one of the intellectual centers of the world, in part because of its magnificent library. Ptolemy fashioned a model of planetary motions in which each planet moved on one small circle, which in turn moved on a larger one (Figure 11.4). The small circle, called an **epicycle,** was supposed to be carried along on the large circle like a Frisbee spinning on the rim of a bicycle wheel or the children's toy called a "Spirograph."

According to Ptolemy's model, the general eastward motion of the planet relative to the stars is caused by a slow rotation of the large circle (the bicycle wheel, in our analogy) relative to the celestial sphere. Retrograde motion occurs when the epicycle carries the planet in a reverse direction (caused by the rotation of the Frisbee, in our analogy) relative to the celestial sphere. Thus, epicycles can account for retrograde motion, and they even suggested that planets are nearer the Earth at this time (see Figure 11.4), which could explain why they are brighter.

In order for this model to make accurate predictions, Ptolemy developed additional "devices" to describe each planet's motion. The large circle that each epicycle moved along was set off center from the Earth, and it turned not at a constant speed around the Earth, but around another point in space. This made the model more accurate, but at the cost of increased complexity. Each planet required many parameters: the sizes of the circle and epicycle, the offset from the Earth, and rates of turning of the circle and epicycle. Ptolemy's model survived until the 1500s, thanks to its success at predicting the positions of the planets. However, there is something unsatisfying about such a complex model—the motion of each planet requires many different parameters that seemed arbitrary.

Simplicity is an important element of scientific theory. As the medieval British philosopher William of Occam wrote in the 1300s, "Entities must not be unnecessarily multiplied." This expresses the idea that a scientific model that requires many parts or parameters to explain a phenomenon is less desirable than a simpler model that achieves the same result, a principle known as **Occam's razor.** It expresses a situation often encountered in science: When we begin with incorrect assumptions, as new evidence accumulates, we will often have to add more and more complexity to our models to explain discrepancies with the evidence.

Ptolemy's model (and much of what we know about Greek and Roman civilization) was preserved through medieval times from about 700 to 1200 C.E. by Islamic scholars around the southern edge of the Mediterranean. They expanded on Ptolemy's work and developed more advanced mathematical techniques using algebra (*algebra* is an Arabic word) and Arabic numerals, as well as making many detailed measurements of the motions of the Sun, Moon, and planets. With improved measurements, discrepancies became apparent. Astronomers found that they could add additional small epicycles to each planet's motion to correct for errors in Ptolemy's model, but to some it seemed there must be a better model.

Ptolemy's model

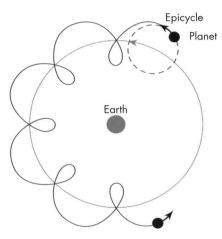

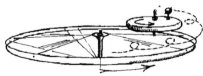

FIGURE 11.4
To explain the retrograde motion of the planets in the geocentric model, Ptolemy employed epicycles. The planets were pictured as being attached to their transparent sphere by a smaller sphere that rotated at a different rate, producing a looping motion. Epicycles are a bit like a bicycle wheel with a spinning Frisbee attached to its rim.

Although we call our numerals "Arabic," they were borrowed by Arab traders from the civilizations farther east in what is now India.

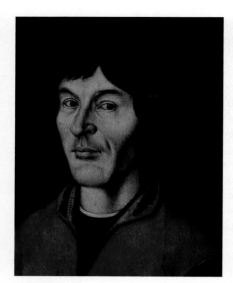

FIGURE 11.5
Nicolaus Copernicus (1473–1543).

11.3 THE HELIOCENTRIC MODEL

The man who began the demolition of the geocentric model and started a revolution in astronomical thinking was a Polish physician and lawyer by the name of Nicolaus **Copernicus** (1473–1543). Copernicus (Figure 11.5) wrote in about 1510 that he was dissatisfied with one of the many devices that had been added to make the Ptolemaic model work. The mathematical device (known as an equant) made the planets speed up and slow down as they circled the Earth, and to Copernicus this violated the original simplicity of the epicyclic model. He decided to reconsider the possibility that the Earth moves around the Sun, as proposed by Aristarchus almost 2000 years earlier (Unit 10.2), to see if that could simplify the model.

A **heliocentric model** places the Sun (*helios* in Greek) at the center of a system of planets, and the Earth is just one of the planets. The idea that the Earth moves seemed to go against common sense, but Copernicus showed that such a model offered a far simpler explanation of retrograde motion. In fact, if the planets orbit the Sun, retrograde motion becomes a simple consequence of one planet on a smaller, faster orbit overtaking and passing another on a larger, slower orbit.

To understand why retrograde motion occurs in a heliocentric system, examine Figure 11.6A. Here we see the Earth and Mars moving around the Sun. The Earth completes its orbit, circling the Sun, in 1 year, whereas Mars takes 1.88 years to complete an orbit, with the Earth overtaking and passing Mars every 780 days. Lines drawn from the Earth to Mars show that Mars appears to change its direction of motion against the background stars each time the Earth overtakes it. A similar phenomenon occurs when you drive on a highway and pass a slower car. Both cars are, of course, moving in the same direction. However, as you pass the slower car, it looks as if it is moving backward against the stationary objects behind it.

Retrograde motion occurs around the time when Mars is directly opposite the Sun in the sky, which we call **opposition** (Figure 11.6B). When Mars is at opposition, it is also nearest Earth, as well as being brightest and easiest to see.

FIGURE 11.6
(A) We see retrograde motion because as Earth approaches and passes Mars, Mars appears to move backward against the background stars. (B) Retrograde motion occurs when a planet is in opposition or inferior conjunction, so that the Earth is either passing the planet or the planet is passing Earth. The diagram shows several other planetary configurations as well.

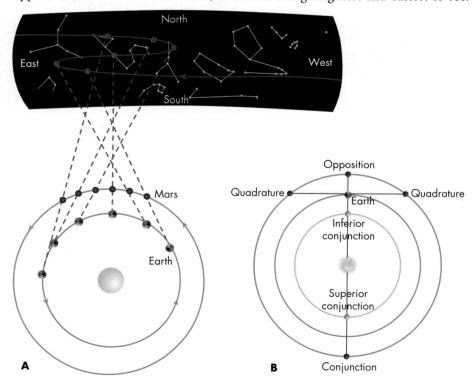

Retrograde motion

Retrograde motion of Mars

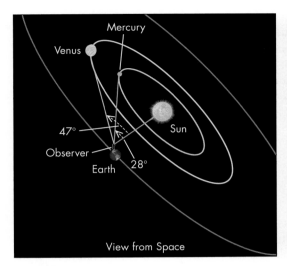
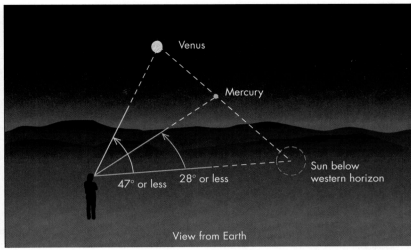

FIGURE 11.7
The greatest elongations of Mercury and Venus, and the Evening Star phenomenon. The diagram on the left shows that Mercury and Venus can *never* appear more than 28° and 47°, respectively, from the Sun. Because their (and the Earth's) orbits are not perfectly circular, the greatest elongation in each orbit varies, averaging 22.5° for Mercury and 46° for Venus.

Jupiter and Saturn (as well as Uranus, Neptune, and other bodies in the outer Solar System) exhibit retrograde motion around the time of opposition, but Mercury and Venus are different. Because they are nearer the Sun than the Earth, they can never be in opposition. Instead they appear to reverse direction against the stars as they pass between the Earth and the Sun. Astronomers use the term **conjunction** to refer to when a planet lies in the same direction as the Sun. All of the planets come into conjunction when they are on the far side of the Sun (often called **superior conjunction**) as illustrated in Figure 11.6B. Venus and Mercury undergo retrograde motion around the time when they are passing the Earth in **inferior conjunction** between the Earth and the Sun.

Copernicus found that he could determine the relative sizes of the planets' orbits from information about the planets' configurations. For example, Venus and Mercury never get very far from the Sun, as shown in Figure 11.7. Mercury averages about 22.5° from the Sun when it is at greatest elongation, and Venus averages about 46° from the Sun, as seen from Earth. The size of this largest angular separation from the Sun, or **greatest elongation,** is determined by the sizes of these planets' orbits relative to the Earth's orbit. By assuming that Venus, Mercury, and the Earth orbit the Sun in concentric circles, we can make a geometric construction (or use trigonometry) to find the sizes of the orbits, as shown in Figure 11.8.

Mathematical Insight

To apply trigonometry to solve this problem, we note that the *sine* of α is equal to the distance between the Sun and the planet divided by the distance between the Sun and Earth (1 AU). Therefore, Venus's orbit has a radius of $1 \text{ AU} \times \sin(46°) = 0.72 \text{ AU}$

FIGURE 11.8
How Copernicus calculated the distance to Mercury and Venus. When a planet interior to the Earth appears farthest from the Sun, the planet's angle on the sky away from the Sun, α, can be measured. You can see from the figure that at this time the direction from the planet to the Sun makes an angle of 90° to our line of sight. The planet's distance from the Sun can then be calculated with geometry, if one knows the measured value of angle α and the fact that the Earth–Sun distance is 1 AU.

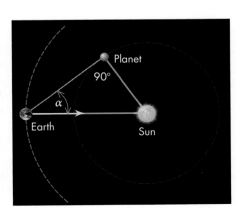

FIGURE 11.9

The distance to an outer planet can be found by determining the interval of time from when the planet is opposite the Sun in the sky (the planet rises at sunset) to when the planet is 90° away from the Sun in the sky. From that time interval we can determine what fraction of each of their orbits the Earth and the planet moved through in that time. Multiplying those fractions by 360° gives the angles the planet and the Earth moved. The difference between those angles gives angle β. Finally, using geometry and the value of angle β, the planet's distance from the Sun can be calculated.

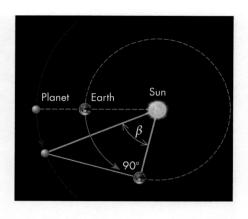

TABLE 11.1 Copernicus's Sizes of Planetary Orbits

Planet	Radius of Orbit in AU	
	Copernicus	Modern
Mercury	0.38	0.39
Venus	0.72	0.72
Earth	1.00	1.00
Mars	1.52	1.52
Jupiter	5.22	5.20
Saturn	9.17	9.54

Determining the sizes of the outer planets' orbits is more complicated. For this, Copernicus noted how long the planets took to move from opposition (180° from the Sun) to **quadrature** (90° from the Sun), as shown in Figure 11.9. For a distant planet like Saturn, the time between opposition and quadrature is nearly a quarter of the orbit. However, for a planet like Mars, whose orbit is not much bigger than the Earth's, the portion of the orbit between opposition and quadrature is a fairly small fraction of the total orbit. Again, a geometric construction can be used to establish the relative size of the planets' orbits.

Remarkably, by using the heliocentric model, Copernicus not only could provide a simpler explanation of retrograde motion, he could even calculate the relative sizes of all of the planets' orbits (Table 11.1). He could not determine how far away the Earth was from the Sun, but he could calculate that Jupiter was about five times farther from the Sun than the Earth and that Mercury was about one-third as far. The distances found in this manner were expressed in terms of the Earth's distance from the Sun—the astronomical unit—whose value was not accurately known until several hundred years later. Table 11.1 illustrates that Copernicus's calculations agree remarkably well with modern values. These distances became an important starting place for astronomers who came later.

11.4 THE COPERNICAN REVOLUTION

Copernicus described his model of a Sun-centered universe in one of the most influential scientific books of all time, *De revolutionibus orbium coelestium* (On the Revolutions of the Celestial Orbs). Because his ideas ran counter to beliefs of the day, they were met with hostility. Copernicus quietly circulated manuscripts of his ideas for about 30 years before allowing the book to be published in 1543 (Fig. 11.10). Reportedly he saw the first copy while on his deathbed.

Some criticism of Copernicus's work was justified because his model could not predict the positions of the planets any better than Ptolemy's. This lack of success arose largely because Copernicus assumed that the planets orbited in circles at a constant speed, an idea that was disproved only in the next century. To make his model more accurate, Copernicus even resorted to the use of small epicycles to adjust his predictions of the planets' positions. Furthermore, his model again raised the question of why no stellar parallax could be seen, which was an objection raised to the heliocentric model when Aristarchus first proposed it (Unit 10).

There were counterarguments to the objections, however. The lack of parallax could be explained if the stars were very distant. On a technical level, Copernicus's model did not require large epicycles to correct his positions because the circular orbits were not bad approximations of the planets' orbits. In addition, his model

FIGURE 11.10
The title page and a diagram showing the heliocentric system of the planets from the first edition of *De revolutionibus orbium coelestium*. The text on the title page advertises that the reader will find admirable new hypotheses about the planets, and warns at the bottom that the reader needs training in geometry.

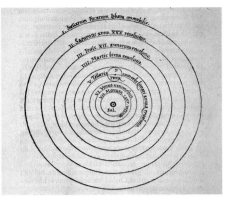

Concept Question 2

Imagine a planet far beyond Saturn. What would its retrograde motion look like? How does the idea of parallax compare to the idea of retrograde motion?

did not need to invoke what seemed to be chance "coincidences" in the epicyclic model, such as the "crystalline spheres" of Mercury and Venus just happening to rotate at the same speed as the Sun's "sphere" to keep these planets near the Sun. So even though Copernicus's model predicted the positions of planets no better than Ptolemy's model, it offered a simpler explanation of their motions.

The main objection to Copernicus's views of planetary motion was that they ran counter to the widely accepted teachings of the ancient Greek philosopher Aristotle, whose views were supported by religious doctrine. After all, if the Earth is just one of the planets, it holds no special place in the universe. Also, at that time it was not understood how the Earth could be spinning and moving at tremendous speeds without our knowing it. When we observe the sky, it looks as if it moves around us, and we do not detect any sensations caused by the Earth's motion. This mixture of rational and irrational objections made even many scientists slow to accept the Copernican view.

However, conditions were favorable for such new ideas. The cultural Renaissance in Europe was at its height; the Protestant Reformation had just begun; Europeans were sailing to the new world; and there was a growing acceptance of the immensity of the universe. In such an intellectually stimulating environment, new ideas flourished and found a more receptive climate than in earlier times. In the late 1500s the English astronomer Thomas Digges argued in favor of the Copernican model, proposing that the stars were other suns at enormous and various distances. The Italian philosopher and monk Giordano Bruno went so far as to claim that the planets around these other stars were inhabited.

The aesthetic appeal of the heliocentric system led to a growing acceptance of the model despite church efforts to suppress it. Bruno was burned at the stake in 1600 for his heretical ideas, but others continued their quest for a deeper understanding of the workings of the cosmos. Are the stars other suns? How can we reconcile motion of the Earth with our everyday experience that it seems stationary? What keeps the planets in their orbits? The steady work of science was on the verge of answering these and many other challenging questions.

Giordano Bruno was accused of a number of heresies, and some argue that his advocacy of the Copernican system was not central to his conviction. However, it is worth noting that one of the inquisitors who sentenced him to death also ordered Galileo Galilei not to defend the Copernican model. The head inquisitor, Cardinal Ballarmine, was declared a saint in 1930.

KEY POINTS

- The planets look like bright "wandering stars," moving among the constellations of the zodiac.
- The planets normally move from west to east among the stars, like the Sun and the Moon, but sometimes they change direction for a period of months.
- Early astronomers believed the planets orbited the Earth.
- In the 2nd century C.E., Ptolemy made a model of retrograde motion proposing that planets move along epicycles (loops) as they orbited the Earth.
- In the 16th century C.E., Copernicus showed that if the planets and Earth orbit the Sun, retrograde motion appears to occur when planets pass each other in their orbits.
- Copernicus was able to derive the distances of the planets by using geometry.

KEY TERMS

conjunction, 89
Copernicus, 88
epicycle, 87
geocentric model, 86
greatest elongation, 89
heliocentric model, 88
inferior conjunction, 89
Occam's razor, 87
opposition, 88
planet, 84
Ptolemy, 87
quadrature, 90
retrograde motion, 86
superior conjunction, 89

CONCEPT QUESTIONS

Concept Questions on the following topics are located in the margins. They invite thinking and discussion beyond the text.

1. Distinguishing a bright star from a planet. (p. 84)
2. Retrograde motion versus parallax. (p. 91)

REVIEW QUESTIONS

3. What is retrograde motion?
4. List the arguments in favor of and against Copernicus's model of a Sun-centered universe.
5. What is the best location for Mercury to appear in the sky for us to observe it?
6. What is the best location for Mars to appear in the sky for us to observe it?
7. Why is the time it takes Earth to overtake and pass Mars different from Mars's orbital period?
8. How did Copernicus estimate the planets' distances from the Sun?
9. When a planet is executing retrograde motion it appears brighter than usual from Earth. Explain why this observation does not allow us to eliminate Ptolemy's geocentric model in favor of the heliocentric Copernican model.

QUANTITATIVE PROBLEMS

10. If we discover a new planet in the Solar System and observe that it takes about 366 days between oppositions, what could we say about it?
11. The smallest separation between Earth and Jupiter is 588 million km while the largest separation is 968 million km. Given that Jupiter has a physical diameter of 140,000 km, what is the change in the angular size of Jupiter throughout the year?
12. When Mars is in conjunction, within an angle of about 5° of the Sun, we lose radio contact with our spacecraft there. Draw a scale diagram of the orbits of Earth and Mars and, given that the Earth moves about 1° per day and Mars about ½° per day, estimate approximately how long this period of communication problems lasts.
13. Make a geometric construction (or use trigonometry) to show that Venus's orbit is about 0.72 times the size of Earth's, based on the fact that Venus's greatest elongation is 47°.
14. Redraw Figure 11.6 to show that retrograde motion can also be seen for a planet closer to the Sun than the Earth. Near what position would we expect to see retrograde motion for such a planet? What would an observer on this planet see of the Earth's motion?
15. Mars spends, on average, about 79 days (11.5% of its orbit) moving from opposition to quadrature. Make a geometric construction like Figure 11.9 (or use trigonometry) to show that Mars's orbit is about 1.52 times the size of Earth's.

TEST YOURSELF

16. Which of the following is a place where you would never expect to see one of the five bright planets in the sky?
    a. In the western sky after sunset
    b. In the zodiac constellation Scorpius
    c. Close to the North Star
    d. High in the sky at midnight
    e. Close to the Moon
17. Over the course of many nights, a planet in retrograde motion will
    a. shift west to east on the celestial sphere.
    b. shift east to west on the celestial sphere.
    c. remain fixed on the celestial sphere.
    d. move randomly about the celestial sphere.
18. Which of the following planets cannot be seen at midnight?
    a. Mercury
    b. Mars
    c. Jupiter
    d. Saturn
19. Occam's razor is used to
    a. discriminate between models based on their simplicity.
    b. measure the angle between the Sun and planets.
    c. explain why planets change the direction of their orbits.
    d. describe the path that the planets follow across the ecliptic.
    e. execute heretics.
20. At the time it was proposed, what was most convincing about Copernicus's model of the Solar System versus earlier models?
    a. It was just obviously true once it had been stated.
    b. It was much more accurate at predicting planets' positions.
    c. It explained retrograde motion more simply and naturally.
    d. It had the strong backing of religious leaders.
21. Retrograde motion is discernible by watching a planet over the course of
    a. a few minutes.
    b. a few hours.
    c. many nights.
    d. many years.

PART 1
UNIT 12
The Beginnings of Modern Astronomy

12.1 Precision Astronomical Measurements
12.2 The Nature of Planetary Orbits
12.3 The First Telescopic Observations

Learning Objectives

Upon completing this Unit, you should be able to:
- Identify the contributions of Brahe, Kepler, and Galileo to astronomy.
- Describe the properties of ellipses and how they relate to planetary orbits.
- List Kepler's three laws and carry out calculations using each of them.
- Discuss why Galileo's first telescopic observations were so revolutionary.

A story is told that in medieval Europe, a gathering of scholars debated how many teeth a horse has—according to Aristotle versus the Bible. The debate went on for hours with no apparent resolution. A young novitiate proposed going out to the stable to look in a horse's mouth and count. After being castigated for his unscholarly proposal, he was ejected from the assembly!

This story may be no more than legend, but it illustrates a difference in scientific approach before about 1600. Scholars used to believe that one should look for answers in the writings of great authorities of the past. Scientists today take it for granted that if they have questions or disagreements, they should conduct experiments and make new observations to settle the issue. In the century following Copernicus's proposal that the Earth orbited the Sun (Unit 11), new observations finally settled the issue, as we examine in this Unit.

12.1 PRECISION ASTRONOMICAL MEASUREMENTS

Early first steps in solving the problem of astronomical motion were made by the sixteenth-century astronomer Tycho **Brahe** (1546–1601), pronounced *TEE-koh BRAH-hee*. Born into the Danish nobility, Brahe used his position and wealth to indulge his passion for study of the heavens, a passion based in part on his professed belief that God placed the planets in the heavens to be used as signs to mankind of events on Earth. Driven by this interest in the skies, he designed, had built, and used instruments that permitted the most accurate pretelescopic measurements ever made. The instruments he used were similar to giant protractors that he could turn to measure positions in the sky with high precision (Figure 12.1). His instruments allowed him to measure the positions of planets to about 1 **arc minute,** or 1/60 of a degree. This is 30 times smaller than the angular size of the Moon, and about the limit of human vision.

Brahe did more than just record planetary positions. In 1572, when he saw an exploding star (what we now call a supernova—Unit 66), he showed that this bright new object maintained a fixed position in the celestial sphere and therefore had to be far beyond the supposed spheres that move the planets. When a bright comet appeared in 1577, he showed that it lay far beyond the Moon, not within the Earth's atmosphere as taught by the ancients. These observations suggested that the heavens were more changeable and more complex than previously believed.

FIGURE 12.1
Painting of Tycho Brahe (1546–1601) and an assistant making measurements from his observatory in Uraniborg.

Concept Question 1

Tycho Brahe's time measurements would have had to be precise to within a few seconds to achieve an accuracy of 1 arc minute for positions relative to the horizon. Can you think of ways to measure positions on the celestial sphere that might not require such accurate timing?

Although Brahe could see the virtues of the simplicity offered by the Copernican model, he was unconvinced of its validity because even his instruments were unable to detect stellar parallax—the apparent shift in star positions expected if the Earth moves around the Sun (Unit 10). Therefore, he offered a compromise model in which all of the planets except the Earth went around the Sun, while the Sun, as in earlier models, circled the Earth. Brahe was the last of the great astronomers to hold that the Earth was at the center of the universe. Brahe's model won favor with those who sought to maintain the idea of a geocentric universe. It was difficult to disprove because it was essentially the same as the Copernican model, except for the change of perspective about whether the Earth moved around the Sun or vice versa. Others, though, began to find additional reasons for doubting that the Earth was so central to the rest of the universe.

12.2 THE NATURE OF PLANETARY ORBITS

A few years before he died, Tycho Brahe hired a younger assistant, Johannes **Kepler** (1571–1630). They did not get along well, and Kepler (Figure 12.2) was fired and rehired during their brief association. Kepler had a strong grasp of geometry and unusual ideas. For example, he searched for a fundamental relationship between the spacing between the planets and how tightly various geometrical figures could nest inside each other—such as a sphere inside a cube. Kepler did not think highly of Brahe's planetary model. He described Brahe's planets orbiting the Sun which orbits the Earth as a "pretzel."

When Brahe died, his observational data passed into Kepler's hands. Brahe's family accused Kepler of stealing them, but their real value was apparent only after years of Kepler's painstaking work. He was able to use the data from multiple orbits of Mars to plot out the path Mars took through space. Whereas all previous investigators had assumed planets moved on circles, Kepler showed—with Brahe's superb data—that Mars moved not along a circular path but rather along an **ellipse**.

An ellipse can be drawn with a pencil inserted in a loop of string that is hooked around two thumbtacks. If you move the pencil while keeping it tight against the string, as shown in Figure 12.3A, you will draw an ellipse. Each point marked by a tack is called a **focus** of the ellipse. The shape of an ellipse is described by its long and short dimensions, called its *major axis* and *minor axis* (Figure 12.3B) respectively. For most calculations the most important value is half the major axis length, or the **semimajor axis**.

Kepler's careful reconstruction of Mars's orbit showed that the Sun is located *not* at the center of the ellipse but off-center, at one focus of the ellipse (Figure 12.3C).

FIGURE 12.2
Johannes Kepler (1571–1630).

Kepler's laws

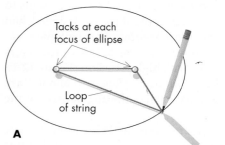
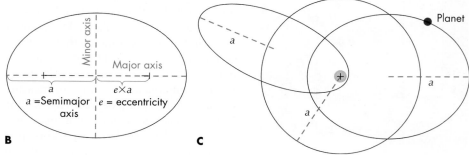

FIGURE 12.3
(A) Drawing an ellipse. (B) The major and minor axes. The semimajor axis, *a*, is half of the major axis. The amount by which the Sun is off-center in the ellipse determines the eccentricity, *e*, of the ellipse. (C) The Sun lies at one focus of the ellipse. Three different orbits are shown that have the same size semimajor axis but differing eccentricities.

It may help you remember the meaning of the values of eccentricity to note that for eccentricity zero ($e = 0$) the numeral looks like a circle, while for eccentricity one ($e = 1$) the numeral looks like an ellipse stretched into a line.

Kepler's second law

TABLE 12.1 Kepler's Third Law

Planet	a (AUs)	P (years)	a^3	P^2
Mercury	0.387	0.241	0.058	0.058
Venus	0.723	0.615	0.378	0.378
Earth	1.0	1.0	1.0	1.0
Mars	1.524	1.881	3.54	3.54
Jupiter	5.20	11.86	141	141
Saturn	9.54	29.46	868	868

P = orbital period, measured in years
a = semimajor axis, measured in AU

Astronomers use the term **eccentricity** to describe how round or "stretched out" an ellipse is. Mathematically, the eccentricity e indicates how far the Sun is from the center of the ellipse as a fraction of the semimajor axis (Figure 12.3C). If the Sun is exactly at the center, the two foci are on top of each other, so the eccentricity is zero. In this case, the orbit is a circle and the semimajor axis is the same as the radius of the circle. As the eccentricity of the orbit increases toward 1, the position of the Sun becomes increasingly offset from the center. Figure 12.3C illustrates ellipses with several different eccentricities but the same value for the semimajor axis.

Kepler found further that Mars moves faster in its orbit when it is closer to the Sun. He found a way to describe this geometrically by imagining the motion of a line joining the planet to the Sun: The planet's speed changes so that the line sweeps out equal areas in equal time intervals, as illustrated by the shaded areas in Figure 12.4. For the areas to be equal, the distance traveled along the orbit in a given time must grow larger the closer the planet is to the Sun. With the elliptical shape and speed variations of the orbit now established, Kepler was able to obtain excellent agreement between the calculated and the observed positions not only of Mars but also of the other planets.

Kepler also discovered a relationship between the size of a planet's orbit and how long it takes to orbit the Sun—its orbital **period.** Kepler determined that the period (P) depends on the semimajor axis of the orbit (a) according to a mathematical formula. The period increases as the semimajor axis increases, but not as a simple proportionality. He found that the square of the orbital period measured in years equals the cube of the semimajor axis measured in astronomical units (Unit 1.6). This is demonstrated in Table 12.1.

Kepler's discoveries of the nature of planetary motions are expressed in what are known today as **Kepler's three laws:**

I. **Planets move in elliptical orbits with the Sun at one focus of the ellipse.**

II. **The orbital speed of a planet varies so that a line joining the Sun and the planet will sweep over equal areas in equal time intervals.**

III. **The amount of time a planet takes to orbit the Sun, P, is related to its orbit's size, a: Mathematically,**

$$P^2 = a^3$$

where P is measured in years, and a is measured in astronomical units.

Kepler's three laws describe the essential features of planetary motion around the Sun. They describe not only the shape of a planet's path but also its speed and distance from the Sun. The three laws are illustrated in Figure 12.4.

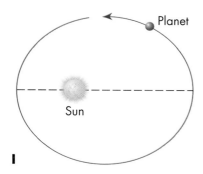
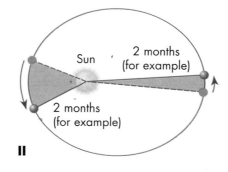
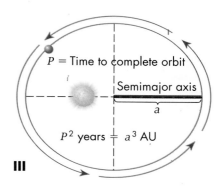

FIGURE 12.4
Kepler's three laws. (I) A planet moves in an elliptical orbit with the Sun at one focus. (II) A planet moves so that a line from it to the Sun sweeps out equal areas in equal times. Thus, the planet moves fastest when nearest the Sun. For the drawing a two-month interval is chosen. (III) The square of a planet's orbital period (in years) equals the cube of the semimajor axis of its orbit (in AU).

Kepler's second law is a consequence of the conservation of angular momentum, which we will discuss further in Unit 20.

Concept Question 2

Kepler spent much of his life searching for connections between pure geometry and nature. Where have you encountered geometric patterns in nature?

Kepler's third law

Mathematical Insights

Taking the square or cube root of a number can be done with a scientific calculator. If there is not a special key for taking roots, you can usually use an x^y key to raise the number to the power 0.5 for a square root or to the power 0.3333... for a cube root. More generally, the reverse of raising a number to a power is to raise it to the inverse of that power.

Kepler's laws are *empirical* findings—that is, they are determined directly from measurements rather than from a theoretical idea about what causes the planets to move. They remain valid today, even though we no longer agree with many of Kepler's own interpretations of their causes. Later scientists discovered that these laws can all be derived from more fundamental laws about the nature of motion and gravity. Describing the heavens with mathematical laws revolutionized our way of thinking about the universe. Without such mathematical formulations of physical laws, much of our technological society would be impossible. Kepler's laws were a major breakthrough in our quest to understand the universe.

It is perhaps ironic that such mathematical laws should come from Kepler, because so much of his work is tinged with mysticism. For example, Kepler's third law evolved from his attempts to link planetary motion to music, using the mathematical relations known to exist between different notes of the musical scale. Kepler even attempted to compose "music of the spheres" based on such a supposed link. Nevertheless, despite such excursions into these nonastronomical matters, Kepler's discoveries remain the foundation for our understanding of how planets move.

The third law has implications for the relative speeds of planets whose orbits are at different distances from the Sun. Because the third law states that $P^2 = a^3$, a planet far from the Sun (larger a) has a longer orbital period (P) than one near the Sun. If the third law had stated that $P = a$, then the speeds of the planets would all have been the same, because the distance traveled in the orbit is proportional to the semimajor axis. However, the third law implies that planets nearer the Sun move faster than outer planets. For example, the Earth takes 1 year to complete its orbit; but Jupiter, whose orbit is about 5 times larger than the Earth's, takes about 12 years. Thus, Jupiter moves about 0.42 (= 5/12) times as fast as Earth. This difference in speed is the source of retrograde motion—as Earth overtakes and passes Jupiter, we perceive Jupiter as moving backward, as discussed in Unit 11.

Kepler's third law allows us to predict the period of any body orbiting the Sun if we know its semimajor axis. For example, Pluto was discovered in 1930 at a distance of about 40 AU from the Sun. Assuming that this is approximately its semimajor axis, a, we can determine the period of its orbit from Kepler's third law as follows. The cube of the semimajor axis (40^3) must equal the square of the period P in years, so

$$a^3 = 40^3 = 64{,}000 = P^2.$$

Taking the square root of 64,000 gives us $P \approx 250$ years. That is, we can predict that Pluto will take about 250 years to complete an orbit, which is longer than we have known of its existence. Modern detailed measurements of its orbit show that this initial estimate was quite good: Pluto's orbital period is 248.1 years, and its semimajor axis is 39.48 AU.

It is also possible to find the semimajor axis of a body orbiting the Sun from its orbital period. For example, in 1705 the English astronomer Sir Edmond Halley realized that a bright comet had been reported every 76 years in historical records. Known today as Comet Halley, we now know that this object becomes visible only in the part of its orbit when it is close to the Sun. Its semimajor axis can be calculated by taking the square of its period (76^2) and equating that to its semimajor axis cubed:

$$P^2 = 76^2 = 5800 = a^3.$$

Taking the cube root of 5800, we find that $a \approx 18$ AU. In this case, the comet must have a highly elliptical orbit, because it is observed to come within 0.6 AU of the Sun. Based on Kepler's first law, the Sun must be at one focus of an ellipse with a major axis of 36 AU (twice a). Therefore, Comet Halley must travel out to beyond 35 AU—nearly as far as Pluto's orbit—where it is so dimly lit that it cannot be seen from Earth.

By long-standing tradition, Galilei is usually referred to by his first name, Galileo. The same used to be true of Tycho Brahe, but the more standard reference by last name is growing more common.

FIGURE 12.5
Galileo Galilei (1564–1642).

Kepler introduced the word *satellite*. When he saw the moons of Jupiter with a small telescope, their motion around the planet made him think of attendants or bodyguards—*satelles* in Latin.

12.3 THE FIRST TELESCOPIC OBSERVATIONS

At about the same time that Tycho Brahe and Johannes Kepler were striving to understand the motion of heavenly bodies, the Italian scientist **Galileo** Galilei (1564–1642) was likewise trying to understand the heavens. However, his approach was dramatically different.

Galileo (Figure 12.5) was interested not just in celestial motion but in all aspects of motion. He studied falling bodies, swinging weights hung on strings, and so on. In addition, he used the newly invented telescope to study the heavens and interpret his findings, and what he found was astonishing.

In looking at the Moon, Galileo saw that its surface had mountains and was in that sense similar to the surface of the Earth. Therefore, he concluded that the Moon was not some mysterious ethereal body but a ball of rock. He discovered that the Sun had dark spots (now known as **sunspots**) on its surface. He noticed that the position of the spots changed from day to day. This showed that the Sun not only had blemishes and was not a perfect celestial orb, but also that it changed. These observations conflicted with earlier beliefs that the heavens were perfect and unchanging. In fact, by observing the changing position of the spots from day to day, Galileo deduced that the Sun rotated.

These observations gave evidence that the Sun and Moon are physical bodies. This was a problem for geocentric theories. To explain the rapid motion of all the objects in the sky around the Earth (once every 24 hours or so), early astronomers supposed the heavenly bodies were made of some light substance that could move at extremely high speeds—perhaps like the beam of a searchlight. It was harder to conceive of mountains of matter swinging around the Earth at such speeds.

Galileo looked at Jupiter and saw four smaller objects orbiting it, which he concluded were moons of the planet. These bodies, known today in his honor as the **Galilean satellites** (Unit 48), proved that at least some bodies in the heavens do not orbit the Earth. Seeing these satellites remain in orbit around Jupiter also refuted the notion that if the Earth moved, the Moon would fall behind it. But perhaps even more important, the motion of these bodies raised the crucial problem of what held them in orbit.

Galileo also looked at Saturn and discovered that it did not appear as a perfectly round disk but that it appeared to have "satellites" that remained stationary on either side of the planet—these were actually Saturn's rings, but his telescope was too small and too crudely made (inferior to today's inexpensive binoculars) to show them clearly. When Galileo examined the faint band of light on the celestial

FIGURE 12.6
Drawings from Galileo's 1610 book *Sidereus Nuncius* (*Starry Messenger*). (A) A sketch of the Moon seen through his telescope, showing mountainous features. (B) A series of diagrams of Jupiter and its moons, seen shifting from night to night as they orbited Jupiter.

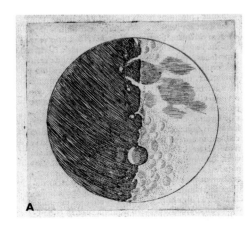
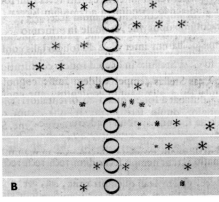

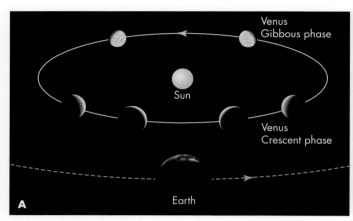
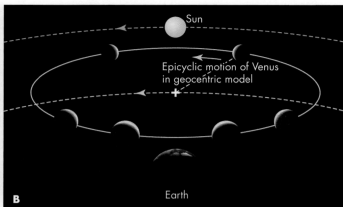

FIGURE 12.7
As Venus orbits the Sun, it goes through a cycle of phases (A). The phase and its position with respect to the Sun show conclusively that Venus cannot be orbiting the Earth. The gibbous phases Galileo observed occur for the heliocentric model but cannot happen in the Earth-centered Ptolemaic model, as illustrated in (B).

FIGURE 12.8
Images of Venus made with a small telescope in 2004 show it changing from a gibbous phase on the far side of the Sun to a crescent phase as it passes between the Earth and Sun.

Phases of Venus

Concept Question 3

Brahe argued that the Sun orbited the Earth but that the other planets orbited the Sun. Could Brahe's model explain the phases of Venus as observed by Galileo? Why or why not?

sphere that we call the Milky Way, he saw that it was populated with an uncountable number of stars. This observation, demonstrating that there were far more stars than anyone had imagined, shook the complacency of those who believed in a simple Earth-centered universe.

Galileo's observations of Venus were perhaps the most decisive in ruling out Ptolemy's geocentric model (Unit 11). Galileo observed that Venus went through a cycle of phases like the Moon, as illustrated in Figure 12.7. Consider what the geocentric model predicts: Venus should be on a crystalline sphere inside the Sun's own crystalline sphere. Because it never reaches an angle larger than 47° from the Sun, sunlight shining on Venus would always illuminate it from behind, like a crescent moon.

By contrast, the heliocentric model predicted that Venus would cross both in front of and behind the Sun from our perspective. When Venus is on the far side of the Sun, its angular size is small (because of the large distance), but we see the side of the planet illuminated by the Sun. Therefore, its phase is nearly full. On the other hand, when it is passing between the Earth and the Sun, it will be angularly large, but we are facing mostly the dark side of the planet, with only a thin crescent illuminated (Figure 12.8). This behavior was exactly what Galileo found, leaving no doubt that Venus orbited the Sun.

Galileo's probings into the nature of the universe led him into trouble with religious "law." He was a vocal supporter of the Copernican view of a Sun-centered universe and wrote and circulated his views widely and somewhat tactlessly. He presented his arguments as a dialogue between a wise teacher and a nonbeliever in the Copernican system whom he called *Simplicio* and who, according to his detractors, was patterned after the pope. Although the pope was actually a friend of Galileo, conservative churchmen urged that Galileo should be brought before the Inquisition because his views that the Earth moved were counter to the teachings of the Catholic Church. Considering that his trial took place when the papacy was attempting to stamp out heresy, Galileo escaped relatively lightly. He was forced to recant his "heresy" and was put under house arrest for the remainder of his life. Only in 1992 did the Catholic Church admit it had erred in condemning Galileo for his ideas.

Galileo's contributions to science would be honored even had he not made so many important observational discoveries. Beyond his revolutionary work in astronomy, he carried out fundamental experiments in physics that provided the beginnings of our understanding of the nature of motion and gravity. He was one of the principal founders of the experimental method for studying scientific problems, and his methods provided the direction for a new approach toward studying the cosmos, as we will begin exploring in Unit 14.

KEY POINTS

- Accurate measurements of the positions of planets made in the late 1500s by Tycho Brahe allowed Johannes Kepler to examine the orbits of the planets in greater detail than ever before.
- Kepler's first law: planets orbit the Sun along ellipses with the Sun at one focus.
- Kepler's second law: planets speed up as they get closer to the Sun along their orbit, "sweeping out equal areas in equal times."
- Kepler's third law: the cube of the size of an orbit is proportional to the square of the time taken to complete the orbit.
- Other bodies orbiting the Sun follow the same laws.
- Galileo Galilei was the first to employ a telescope to study celestial objects, discovering many things contradictory to earlier beliefs.
- The Moon and Sun have irregular surfaces, more like Earthly bodies than the expected heavenly perfection.
- Jupiter has satellites orbiting it, much as Earth has its own Moon.
- Venus passes through a full set of phases, including crescent and gibbous, indicating that it must orbit the Sun.

KEY TERMS

arc minute, 93
Brahe, 93
eccentricity, 95
ellipse, 94
focus, 94
Galilean satellite, 97
Galileo, 97
Kepler, 94
Kepler's three laws, 95
period, 95
semimajor axis, 94
sunspot, 97

CONCEPT QUESTIONS

Concept Questions on the following topics are located in the margins. They invite thinking and discussion beyond the text.

1. Making accurate measurements on the celestial sphere. (p. 94)
2. Geometric patterns in nature. (p. 96)
3. Brahe's model of the Solar System. (p. 98)

REVIEW QUESTIONS

4. How were Tycho Brahe's observations critical to the development of astronomy?
5. What is the shape of a planet's orbit? Where is the Sun located relative to this shape?
6. How does a planet's speed change as it orbits the Sun?
7. What is the relationship between the size of a planet's orbit and the length of time it takes to complete an orbit?
8. How did Galileo's observations help to rule out a geocentric model of the Solar System?
9. Which planets can we see through a telescope as a crescent? When does this happen?

QUANTITATIVE PROBLEMS

10. Ceres has an orbital period of 4.60 years. What is Ceres' semimajor axis? Where in the Solar System does this place Ceres?
11. Sedna, an object in the outer Solar System, has a semimajor axis of 526 AU. What is its orbital period? How does this compare to Pluto's orbital period of 248.1 years?
12. Suppose an asteroid is discovered with an orbit that brings it as close as 0.5 AU from the Sun and as far away as 5.5 AU. What is the semimajor axis of its orbit? What is its orbital period?
13. If a comet orbits the Sun and reaches 1 AU at its closest approach to the Sun, and its orbital period is 27 years, what is its maximum distance from the Sun?
14. For an asteroid in a circular orbit at 4 AU, calculate its orbital speed. Compare this Earth's orbital speed of 29.8 km/sec.
15. A comet is discovered with an orbit that extends from Venus's orbit to Neptune's orbit. Use Kepler's second law to estimate how many times faster the comet is moving when it is closest to the Sun compared to when it is farthest from the Sun. (Hint: The area swept out over a short period of time when the planet is at these two extremes can be approximated by a triangle, and the area of a triangle equals one-half of its base times its height.)

TEST YOURSELF

16. Galileo's observations of the changing phases of Venus supported the view that
    a. the Sun orbits the Earth.
    b. the universe is infinite.
    c. the Earth is a sphere.
    d. the Moon orbits the Sun.
    e. Venus orbits the Sun.
17. Kepler's third law
    a. relates the duration of a planet's orbit to the size of its orbit.
    b. relates a body's mass to its gravitational attraction.
    c. allowed him to predict when eclipses occur.
    d. allowed him to measure the distance to nearby stars.
    e. showed that the Sun is much farther away than the Moon.
18. Which of the following did the models of Copernicus and Kepler have in common?
    a. Planets move in elliptical orbits.
    b. The inner planets move faster than the outer planets.
    c. The motions of the planets are uniform.
    d. The planets move on epicycles.
    e. The Sun is located at the exact center of the Earth's orbit.
19. Which of Kepler's laws explains why the Sun has a slightly larger apparent (angular) size in January than in July?
    a. First law   b. Second law   c. Third law
    d. None of them; this is unrelated to Kepler's laws.
20. Which of Kepler's laws explains why comets on high eccentricity orbits spend most of their orbital period far from the Sun?
    a. First law   b. Second law   c. Third law
    d. None of them; this is unrelated to Kepler's laws.
21. Galileo's discovery of Jupiter's moons was surprising because
    a. there were 4 moons, not 1.   c. they did not orbit Earth.
    b. some outweigh our Moon.   d. they had retrograde orbits.

UNIT 13

Observing the Sky

13.1 Learning the Constellations
13.2 Motions of the Stars
13.3 Motion of the Sun
13.4 Motions of the Moon and Planets
13.5 A Sundial: Orbital Effects on the Day

Learning Objectives

Upon completing this Unit, you should be able to:
- Estimate angular sizes and directions on the sky and relative to the horizon.
- Understand how to find constellations and planets using star charts, and demonstrate how stars change position over the course of a night.
- Demonstrate how the Sun's position shifts during the year.
- Explain the differences between sundial time and clock time.

Experiencing the astronomical changes that occur each night, or over several days or months, will give you a much deeper appreciation of the cosmos. This Unit offers a number of projects for exploring the sky and connecting your observations to the models and theories discussed in earlier Units. These are noted in connection with each topic.

There are many important aspects of the universe you can learn by simply watching the sky. You do not need fancy equipment. Just a star atlas or a computer with planetarium software can help you get oriented as you look at the sky. Best of all, stargazing is fun and connects you to cultural traditions reaching into prehistory as well as current scientific research. If you find astronomical observation rewarding, there are opportunities to engage in more serious pursuits, such as searching for new comets or monitoring variable stars. These possibilities are discussed further in Unit 33.

13.1 LEARNING THE CONSTELLATIONS

One of the best ways to get started in your studies of the sky is to learn the constellations (Unit 5). All you need is a star chart, a dim flashlight, and a place that is dark and has an unobstructed view of the night sky. Star charts are located in the back of this book. The charts give instructions on how to hold them so that for the date and time at which you are observing. A planetarium software program can generate charts for a specific time and location.

Start by determining which way is north, using a compass if necessary. Then try to locate a few of the brighter stars, matching them up with your chart. Next, try to identify a few star patterns. Focus at first on just some brighter ones. For example, if you live at midlatitude in the Northern Hemisphere (as in most of the United States, Canada, Europe, and Asia), the Big Dipper—the asterism that is part of the constellation Ursa Major—is a good group to start with because it is circumpolar for anyone living at a latitude north of about 35°N.

As you attempt to find and identify stars, your spread hand held at arm's length makes a useful scale. For most people, a fully spread hand at arm's length covers an angular size of about 20° of sky, or about the length of the Big Dipper from tip of handle to bowl, as shown in Figure 13.1A. For smaller distances, you can use finger widths: held at arm's length, your thumbnail is about 2° wide, and the tip of your little finger is about 1° wide (Figure 13.1B).

LOOKING UP

The Big Dipper can be seen in Looking Up #1 and #2 at the front of the book.

Mathematical Insights

The angular size of an outstretched hand is usually similar because (by the angular size formula in Unit 10.4) the angle (α) depends on the ratio of the size of the hand (ℓ) to the length of the arm (d), and ℓ and d are proportional in most people.

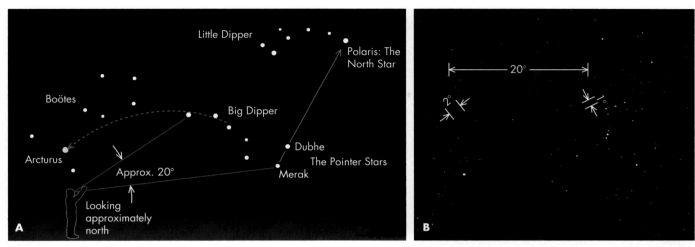

FIGURE 13.1
(A) The Big Dipper, part of the constellation Ursa Major, the Great Bear. A line through the two pointer stars points toward Polaris. The Big Dipper spans about 20° of the sky. The sky is shown approximately as it looks in mid-September at about 8 P.M. from mid-northern latitudes. (B) You can estimate angular separations on the sky using your hand stretched out at arm's length in front of you. Your handspread is about 20°, your thumb is about 2° wide, and the tip of your little finger is about 1° wide.

The Big Dipper not only is easy to spot, but is also an excellent signpost to other asterisms and stars. For example, the two stars at the end of its "bowl" away from the "handle" (Figure 13.1A) are called the "pointers" because they point, roughly, to the North Star, Polaris, about 30° or 1½ handspreads away. Because Polaris lies nearly above the Earth's North Pole, it is useful in orienting yourself to compass directions. Polaris marks the end of the handle of the Little Dipper, an asterism that is part of the constellation Ursa Minor, the Little Bear. If you extend the arc formed by the stars in the handle of the Big Dipper, you will find a path that curves to the bright star Arcturus ("follow the arc to Arcturus"). Arcturus is also about 1½ handspreads away from the Big Dipper in the constellation Boötes (*boh-OH-teez*).

Estimating angles with your hand makes it easy to point out stars to other people. For example, you can say that a star is half a handspread away from the Moon and at the 4 o'clock position, as illustrated in Figure 13.2A. A more general method for locating a star is to measure its altitude and azimuth, as shown in Figure 13.2B. The star's **altitude** is its angle above the horizon, while its **azimuth** is defined as the angle measured eastward from true north to the point on the horizon below the star.

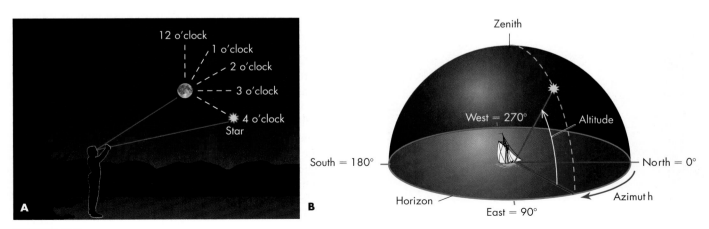

FIGURE 13.2
(A) Describing the location of stars by clock position. The star is half a handspread from the Moon and at the 4 o'clock position. (B) A star's position can be indicated by its altitude above the horizon and its azimuth measured eastward around the horizon from true north.

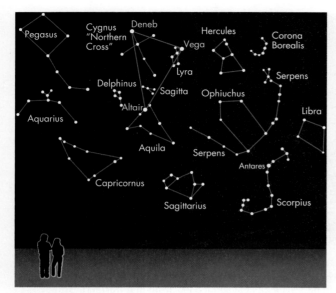

FIGURE 13.3
Dominating the night sky in July, August, and September are the three bright stars Vega, Altair, and Deneb, which form the Summer Triangle. This sketch shows almost half of the whole sky looking south (from mid-northern latitudes) at about 8 P.M. in early September.

A photograph of the summer triangle is shown in Looking Up #4 at the front of the book.

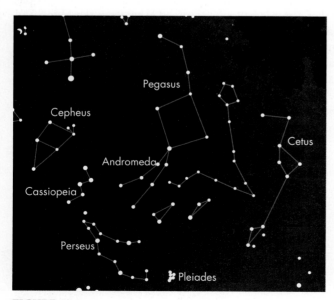

FIGURE 13.4
Perseus, Andromeda, Cassiopeia, Cepheus, Cetus, and Pegasus. The sky is drawn as it looks in mid-November at about 8 P.M., looking upward from mid-northern latitudes.

Learning to recognize some of the brightest stars is another way to locate constellations. A good example is the asterism known as the Summer Triangle, which spans three constellations. It consists of three bright stars conspicuous in evenings most places from July to November: Deneb (in Cygnus, the Swan), Altair (in Aquila, the Eagle), and Vega (in Lyra, the Harp), shown in Figure 13.3. The three stars almost form an isosceles triangle, with Deneb and Altair 38° apart, while Deneb and Vega are 23° apart. To Polynesians, these were known as the Navigator's Triangle because of their importance for traveling between Pacific islands. A modern invention is that they mark out a "V" for summer Vacation.

Once you recognize a few constellations, you may find that learning the stories behind them will help you remember their shapes and locations. It has been suggested that many such stories were created as aids to memory, especially important when familiarity with the stars could be literally a matter of life or death to a farmer or a navigator. Scientists have even shown that baby birds learn to recognize star patterns and movements and use them to navigate safely—unguided by their parents—across thousands of miles of ocean to their winter homes.

The native inhabitants of North America had a story about the Big Dipper. Its bowl represented a huge bear, and the handle represented three warriors in pursuit of the bear. They had wounded it, and it was bleeding. The red color of autumn leaves was said to be caused by the bear's blood dripping on them when the constellation lies low in the sky during the evening hours of the autumn months.

Stories are also told that connect multiple constellations. For example, if you follow the pointer stars in the Big Dipper past the Little Dipper and Polaris, you will come to a set of constellations tied together by an ancient Greek myth. The constellations are shown in Figure 13.4, as they might be seen in a northern autumn sky. Their story goes as follows:

In ancient days there lived a queen of Ethiopia, Cassiopeia, who was very beautiful but also very vain. She and king Cepheus, her husband, and their daughter, Andromeda, lived happily until one day the queen boasted that she was more beautiful than the daughters of Nereus, a sea god. In punishment for such pride, the sea god Neptune sent a sea monster, Cetus, to ravage the kingdom. To save his people and appease the gods, Cepheus was instructed to tie his daughter, Andromeda, to a rock for the monster to devour. Meanwhile, Perseus was returning home from a quest in which he slew the snake-haired Gorgon, Medusa. Upon Medusa's death, her blood dripped into the sea and turned into the flying winged horse, Pegasus. Perseus saw the maiden's peril and flew to her rescue, slaying the monster. They all lived as happily ever after as most mythological families.

There are many other stories about constellations, but the one just described may give you some sense of those that have been handed down over thousands of years of written and oral history. Explore these stories as you learn the constellations: They will help you remember the relative locations in the sky of the various constellations.

13.2 MOTIONS OF THE STARS

Many people are surprised when they are told that the stars rise and set and move across the sky in much the same way that the Sun does (Unit 5). However, it is easy to show that stars move.

Use a tripod or other object on which you can affix a ruler in some manner. A ruler taped to a camera tripod would be ideal. Find a bright star and sight along the ruler toward a star near the horizon, as sketched in Figure 13.5. If you now wait about 15 minutes and again sight along the stick, you will see that the ruler no longer points to the star. You can do this experiment indoors if you can see a star through a window. Set up a tripod or other object close to the window to help you establish a sightline toward the star. Sighting over the top of your tripod, use a grease pencil or a piece of tape to mark the point on the glass where the star appears to be. Again, after about 15 minutes, the star's motion will be clearly visible.

The same motion can be demonstrated in just a few moments with any small telescope mounted on a tripod. If the telescope is set to point at a star, the star will steadily drift across the field of view. The image seen through a telescope may be reversed, but you can determine what direction the star is moving by seeing which way the telescope has to be adjusted to bring the star back to the center of the field of view.

This experiment will not work if you look for motion of the North Star! The North Star, Polaris, lies less than 1° from the north celestial pole. Because of its position there, it is the only moderately bright star in the northern sky that shows no obvious motion during the night. Its relatively fixed position is illustrated by the time exposure in Figure 13.6, showing the stars' apparent rotation around the north celestial pole. The counterclockwise motion of the stars around the celestial pole can be observed by noting the "o'clock" position of one of the bright stars around Polaris (Figure 13.2A). If you come back in about 2 hours, the position will be about an hour "earlier" on the imaginary clock.

In all of these experiments the stars are moving to the west of their original position. What is meant by "west" on the sky often appears to be quite different from what we might think of as west on the ground. A star's westward motion may have it moving up, down, or at various angles to the horizon. This is particularly true in Figure 13.6, where westward motion is to the left for stars above Polaris and to the right for stars below it.

> **Clarification Point**
>
> Remember that most objects' motions we see in the sky during a night are actually due to the Earth spinning.

> **Concept Question 1**
>
> If you observe the motion of a star rising in the east, how does the angle between its path and the horizon depend on your latitude?

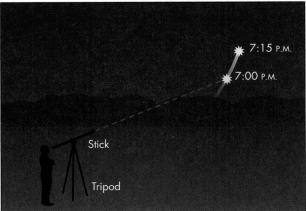

FIGURE 13.5
A sketch illustrating how to observe the motion of the stars across the sky by sighting along a stick.

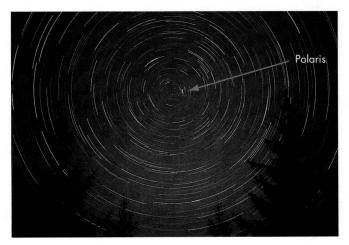

FIGURE 13.6
A time exposure showing how Polaris remains essentially fixed while the sky appears to pivot around it.

FIGURE 13.7
A pair of photographs taken eight days apart from the same location. The pictures were taken just before sunset, close to the date of the autumnal equinox. The sunset position changed by more than 4° during this time—just about the size of the Sun's apparent diameter each day. The width of the outstretched thumb in the bottom picture helps indicate a scale of about 2°.

13.3 MOTION OF THE SUN

Just as for other stars, the rising and setting of the Sun teaches us about the Earth's daily rotation. However, by observing the Sun for about a week we can also detect the effects of the Earth's tilted axis of rotation. You can observe the Sun's shift north and south on the celestial sphere based on its location at sunset. If you observe the Sun setting from the same location for even a few nights, you can begin to trace the patterns that led ancient peoples to build remarkable structures like Stonehenge.

Find a spot, perhaps on a hill or out a window facing west, where you can see the western horizon in the evening. Take a photograph or make a sketch of the western horizon, noting hills, buildings, or trees that might serve as reference marks. Use your hand and fingers to estimate the angular size of features on the horizon, as discussed in Section 13.1. From your viewing spot, watch the sunset and mark on your sketch where the Sun goes down, noting the date and time.

Make observations for as many consecutive days as you can. You will discover that the Sun's position changes by an obvious amount in a single day near the equinoxes, but much more slowly near the solstices (Unit 6). Observe how the times of sunset change as the Sun's position changes. You might try taking photographs of the sunset like those shown in Figure 13.7. It is extremely important, however, never to look directly at the Sun, especially through any kind of magnifying lenses, because doing so can damage your eyes.

Although the Sun is often dimmed enough at sunset for safe viewing, it is not always so. The only way to be certain it is safe to photograph the Sun at sunset is to use a camera that shows you the image of the Sun indirectly on a display screen, such as most digital cameras.

13.4 MOTIONS OF THE MOON AND PLANETS

Determine from Table 13.1 when the Moon will be a few days past new so it will be visible in the early evening. Go outside shortly after sunset and look for the Moon in the west near where the Sun went down. If you have a clear, dark sky, you may even be able to see the dark side of the Moon, lit up by reflected light from the Earth as shown in Figure 13.8.

On the foldout star chart in the back of the book, you will find a list of dates along the top. At 8 P.M. the stars below this date lie along the meridian—the north–south line running overhead. The stars below the next date to the left are overhead an hour later, and to the right an hour earlier. Once the sky is dark, you will be able to see most of the stars six hours to the right and six hours to the left of the ones that are overhead. As the sky darkens, locate the brighter stars near the Moon, and then mark on the chart where the Moon is with respect to those stars. Finally, sketch the Moon's shape.

Concept Question 2

When the Moon is in its crescent phase, what phase would the Earth appear to be in for someone on the Moon? How bright would "earthlight" be on the Moon, compared to moonlight on Earth?

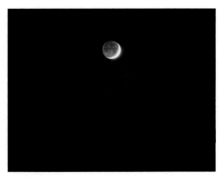

FIGURE 13.8
This photograph shows the waxing crescent Moon a few days after the New Moon. The bright crescent portion is lit by the Sun, but the dark side is also visible, illuminated by light reflected off the Earth.

Morning and Evening Stars

Mathematical Insight

Mathematically it can be shown that the synodic period of a planet (S) is related to its orbital period (P) and the Earth's orbital period (1 yr) by the formula $1/S = |1 - 1/P|$, where P and S are both measured in years.

The dates of the New Moon and the locations of the five bright planets are given each month. Dates of total solar eclipses are noted with a black circle; total lunar eclipses with a red circle. The table is continued on the foldout star chart in the back of the book. The abbreviation for the constellation where the planet can be found is listed in green when the planet is primarily in the morning sky, in blue for the evening sky, and in black when opposite the Sun. "Sun" is listed when the planet is nearly in line with the Sun and difficult to see. For Venus, the month when it is at its greatest elongation from the Sun is marked by an asterisk. Mercury is always fairly close to the Sun and can be seen only shortly after sunset or shortly before sunrise. The date of its greatest elongation is given, and the best opportunity for seeing it is generally within about one week of this date.

Repeat this process for the next four or five nights. The Moon will set about 50 minutes later each evening; note these times and adjust your observing time accordingly. After watching for a few nights, mark out the Moon's path on the star chart. Ideally, you might want to follow the Moon's track for about two weeks, although as the Moon reaches its third quarter (three weeks after new moon), you will have to stay up late because the third-quarter moon does not rise until about midnight. Early risers who get up before dawn can watch the Moon become a waning crescent as it approaches the new phase.

You can also use a star chart to study the motion of the planets. If it is visible, Venus is a good choice because it is bright and moves rapidly across the sky. It is often called the **Evening Star** because it stands out as the brightest "star" in the evening sky. However, Venus spends half its time as the **Morning Star** and is sometimes too close to the Sun to be seen easily, so it is not always a convenient target for observation. The locations where Venus and other planets are visible each month are given in Table 13.1. The table indicates whether the planets are more easily observed in the evening sky (blue) or morning sky (green), and when they are close to the Sun. Mercury is always quite challenging to see, requiring a clear view near to where the Sun has just set or is about to rise. The dates when it reaches its greatest angle from the Sun are indicated in the table, and you will have the best chance of finding it within a week of those dates.

If you watch the planets long enough, you can see how closely they follow the ecliptic—shown in the foldout star chart by a curving line that crosses both sides of the equator. Because the outer planets move relatively slowly across the sky, you should space out your observations, perhaps marking positions once a week rather than every night. You will also be able to see whether they are making direct or retrograde motion against the background stars. Retrograde motion occurs for a month or more around the time the planets are closest to Earth in their orbits. This occurs for the outer planets for the months where the constellation is marked in black in Table 13.1.

The interval between these periods of retrograde motion (or any other successive planetary configurations such as opposition or conjunction) is called the **synodic period**. The synodic period differs from the planet's orbital period because both the Earth and the other planets move around the Sun. Therefore, the interval between oppositions is neither an Earth year nor the other planet's orbital period. For example, the Earth takes over two years (780 days) to catch up to and overtake Mars after an opposition. The Earth overtakes the slower-moving, more distant planets more quickly, so Saturn's synodic period, for example, is just 378 days.

TABLE 13.1 Moon and Planet Finder

	Month	New Moon & Eclipses	Mercury	Venus	Mars	Jupiter	Saturn		Month	New Moon & Eclipses	Mercury	Venus	Mars	Jupiter	Saturn
2014	Jan	1, 30	31	Sun	Vir	Gem	Lib	**2015**	Jan	20	14	Cap	Aqr	Leo	Sco
	Feb		Sun	Sgr	Vir	Gem	Lib		Feb	18	24	Aqr	Psc	Cnc	Sco
	Mar	1, 30	14	Cap*	Vir	Gem	Lib		Mar	●20	Sun	Psc	Psc	Cnc	Sco
	Apr	●15, 29	Sun	Aqr	Vir	Gem	Lib		Apr	●4, 18	Sun	Tau	Ari	Cnc	Sco
	May	27	25	Psc	Vir	Gem	Lib		May	18	7	Gem	Tau	Cnc	Sco
	Jun	26	Sun	Ari	Vir	Gem	Lib		Jun	16	24	Cnc*	Sun	Leo	Lib
	Jul	26	12	Tau	Vir	Sun	Lib		Jul	16	Sun	Leo	Gem	Leo	Lib
	Aug	25	Sun	Cnc	Lib	Cnc	Lib		Aug	14	Sun	Sun	Cnc	Sun	Lib
	Sep	24	21	Leo	Sco	Cnc	Lib		Sep	13, ●28	4	Cnc	Leo	Leo	Lib
	Oct	●8, 23	Sun	Sun	Oph	Leo	Lib		Oct	13	16	Leo*	Leo	Leo	Sco
	Nov	22	1	Lib	Sgr	Leo	Sun		Nov	11	Sun	Vir	Vir	Leo	Sun
	Dec	22	Sun	Sgr	Cap	Leo	Lib		Dec	11	29	Lib	Vir	Leo	Oph

13.5 A SUNDIAL: ORBITAL EFFECTS ON THE DAY

The observations described in the preceding sections reveal many basic features of the sky and planetary motion. Even more sophisticated information about the nature of Earth's orbit can be learned from careful observations of a sundial over the course of a year.

A sundial can be as simple as a flagpole or any other fixed tall pole where you can mark the shadow cast by the top of the pole. If we were to measure the length of the solar day from noon to noon with a stopwatch, we would discover that it is in general *not* exactly 24 hours. We can do this with a flagpole by marking when the shadow lies exactly along a north–south line. The time between successive noons varies by as much as half a minute at different times during the year.

Our clocks do not change speed during the year, of course, but instead use the average day length during a year. That average day length is called the **mean solar day,** which has, by definition, 24 hours of clock time. Therefore, even if you live on the central longitude of your time zone (Unit 7), the Sun will not normally lie along the meridian when your clock indicates noon. This is illustrated in Figure 13.9A, which shows photographs that were taken at the same time on many days over the course of a year. The Sun moves north and south because of the Earth's tilted axis, but it also makes a figure-eight pattern, called an **analemma.**

The difference in length between the mean solar day and the true solar day accumulates to a difference of over 16 minutes between clock time and time based on the position of the Sun at different times of year. This difference is described by the **equation of time,** which is shown graphically in Figure 13.9B. The equation of time gives the correction needed on a sundial if it is to give the same time as your watch.

The variation in the solar day arises because of two effects: The Earth's orbit is not circular, and the Earth's axis is tilted with respect to the orbit. Both of these have similarly sized effects on the length of the day; but they follow different patterns that are offset in time, which makes the equation of time quite complicated.

The effect of the Earth's elliptical orbit on the day is illustrated in Figure 13.10. In January, when the Earth is closest to the Sun, the Earth sweeps through a larger

Concept Question 3

What differences would you see between a sundial in the Northern Hemisphere and one in the Southern Hemisphere?

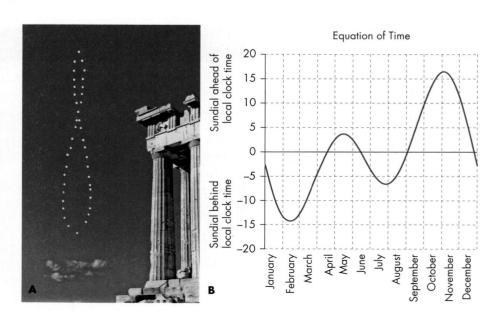

FIGURE 13.9
(A) A series of photographs of the Sun at noon (by the clock) throughout a year shows that the Sun is sometimes a little east and sometimes a little west of the meridian. (B) A graph showing the equation of time, which is the correction that must be applied to sundial time to determine mean solar time.

FIGURE 13.10

As the Earth moves around the Sun, its orbital speed changes as a result of Kepler's second law of motion. For example, the Earth moves faster in January when it is near the Sun than in July when it is far from the Sun. Therefore, in 24 hours the Earth moves farther along its orbit in January than in July. As a result, the Earth must turn slightly more in January to bring the Sun back to overhead. This makes the interval between successive noons longer in January than in July and means they are not exactly 24 hours. For that reason, time is kept using a "mean Sun" that moves across the sky at the real Sun's average rate. (The extremes of the Earth's distance from the Sun have been greatly exaggerated in this figure.)

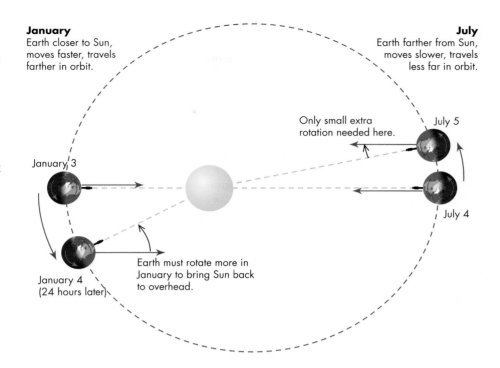

Mathematical Insights

The difference between sundial and clock time may seem just a curiosity today; but for a navigator using the Sun to determine a ship's longitude, a 15 minute offset could be disastrous. 15 minutes is 1/4 of an hour, or 1/96 of 24 hours. At the equator, this means you might be 1/96 of the Earth's circumference east or west of your intended position. This is a potential error of $\sim 40{,}000$ km$/96 \approx 420$ km.

angle in its orbit (Kepler's second law; Unit 12.2), and the Sun appears to shift farther eastward on the celestial sphere than it does on average. This in turn means the Earth has to rotate a bit farther to face the Sun again, with the interval between successive noons becoming about 10 seconds longer than average. This effect adds in the same direction for about half the year, and consequently solar time falls farther and farther behind clock time. The reverse applies in the months of April through September, when the Earth is farther from the Sun. During this half of the year, solar time gets ahead of clock time.

The effect of the Earth's tilted axis on the day is easiest to understand if we think of how the Sun shifts position on the celestial sphere. Suppose the Earth's orbit were perfectly circular and the Sun shifted by the same distance along the ecliptic each day. In this case, the Sun would make its most rapid progress in an eastward direction among the stars when it was moving due east rather than when part of its motion was northward or southward. Because the ecliptic is tilted, the Sun moves due east among the stars, parallel to the celestial equator, only during the two solstices. It makes the least eastward progress when its motion north or south is largest at the equinoxes. Thus, the Earth has to turn farther when the Sun's apparent position has shifted farther eastward, and this effect can lengthen the day by up to about 20 seconds at the solstices, shortening it by the same amount near the equinoxes.

Because both effects reach maxima near the end of the year (winter solstice and closest approach to the Sun), the longest day of the year is actually a couple of days after the December solstice—about 30 seconds longer than average. This is not the same as the length of daylight hours, which are shortest (in the Northern Hemisphere) at this time of year, as discussed in Unit 7.

The elliptical orbit of the Earth and its changing speed also explains why the times between equinoxes and solstices are not the same throughout the year. The average dates for the start of each season are March 20, June 21, September 22, and December 21. This gives 93 days each for northern spring and summer, but just 90 and 89 days for autumn and winter. This difference was a puzzle to ancient astronomers. It was finally understood when Kepler (Unit 12) discovered that planets travel on elliptical orbits with a varying speed.

KEY POINTS

- Stories about the constellations may have arisen to help people memorize them.
- You can use your hand to measure angular sizes and separations.
- Star positions relative to the local horizon are measured by angles of azimuth and altitude.
- The shift of the sunset position can be seen in as little as a few days around the equinoxes.
- Mercury is visible only for a week or two close to greatest elongations; the other planets move more slowly through the zodiac.
- Time measured on a sundial can vary by more than a quarter hour, depending on the time of year, as specified by the "equation of time."
- The tilt of the Earth's axis and the ellipticity of its orbit make the time from noon to noon vary throughout the year.

KEY TERMS

altitude, 101
analemma, 106
azimuth, 101
equation of time, 106
Evening Star, 105
mean solar day, 106
Morning Star, 105
synodic period, 105

CONCEPT QUESTIONS

Concept Questions on the following topics are located in the margins. They invite thinking and discussion beyond the text.

1. Latitude and angle of star rise. (p. 103)
2. Phase of Earth from crescent Moon. (p. 104)
3. Sundial in Southern versus Northern Hemisphere. (p. 106)

REVIEW QUESTIONS

4. How can you use your hand to measure angles between objects in the sky?
5. What methods can you use for describing positions in the sky?
6. How can you demonstrate that stars change their position throughout the night?
7. Of the Sun, Moon, and planets, which move most quickly, and which move most slowly, relative to the stars?
8. Which planets appear as "Morning" or "Evening" stars?
9. Why does sundial time differ from clock time?

QUANTITATIVE PROBLEMS

10. Use your hand and fingers to estimate the angular size of at least three constellations. Sketch the constellations to scale with your measurements.

11. Sailors have "handy" rules for estimating the time until sunset. Approximately how many minutes before sunset is the Sun "one finger" above the horizon at the equator? At 45° N or S latitude?
12. Demonstrate through the use of a drawing, that the altitude of the North Star as seen by anyone in the northern hemisphere, is equal to the observer's latitude.
13. Calculate a precise value for how many degrees the Moon moves each day, keeping in mind that to complete one lunar month of 29.53 days, the Moon must once again align with the Sun, which also has shifted along the ecliptic during that time.
14. Even at greatest elongation, Mercury will not always be visible high above the horizon after sunset or before sunrise. This is primarily because the angle between the ecliptic and the horizon is shallow sometimes during the year. In that case, Mercury is far from the Sun but still close to the horizon. When the ecliptic is more nearly perpendicular to the horizon in your viewing location, Mercury will have a higher altitude. By examining the foldout star chart, estimate at what times of year the ecliptic will be at the best angle for you to observe Mercury in the evening sky. Is there an upcoming greatest elongation that is favorable for you?
15. If you mark the position of the shadow cast by the top of a flagpole at the same clock time every day throughout the year, you will find that the marks trace out a figure-eight pattern. Explain why this happens by referring to the equation of time.

TEST YOURSELF

16. If a Northern-Hemisphere observer sees a star just above the north point on the horizon, a few minutes later the same star
    a. will have moved to the east.
    b. will have moved to the west.
    c. will not have moved.
    d. will have moved straight up from the horizon.
    e. will have set.
17. Suppose you see Venus at greatest elongation (farthest from the Sun) in the evening tonight. About how long will you have to wait to see Venus at greatest elongation in the evening again?
    a. Venus's orbital period (225 days)
    b. Venus's rotation period (243 days)
    c. Earth's orbital period (365 days)
    d. Venus's synodic period (584 days)
18. The equation of time describes differences between clock time and sundial time caused by
    a. wobbling of the Earth's axis.
    b. the tilt of the Earth's axis relative to its orbit.
    c. tugs on Earth's orbit by other planets.
    d. changes in Earth's speed as it orbits the Sun.
    e. Both (b) and (d) are true.
19. If someone in the northern hemisphere notices a star rising due east, when that star reaches its highest point in the sky it will be
    a. high in the northern sky.
    b. high in the southern sky.
    c. directly overhead.
    d. it depends on the time of year the star is observed.

Astronomical Motion: Inertia, Mass, and Force

14.1 Inertia and Mass
14.2 The Law of Inertia
14.3 Forces and Weights
14.4 The Force in an Orbit

Learning Objectives

Upon completing this Unit, you should be able to:
- Define and provide examples of inertia, mass, force, speed, and velocity.
- State the law of inertia, and give examples of its application in terms of both speed and direction.
- Explain the difference between mass and weight, providing examples.
- Describe the nature of the force required for circular motion.

Astronomers of antiquity did not make the connection between gravity and astronomical motion that we recognize today. Since Aristotle, it was believed that without some continuing application of power, all objects would come to rest. Scientists puzzled over what kept the planets moving in their orbits, and they wondered why, if the Earth moved, people did not simply fly off into space.

The solutions to these mysteries began with a series of careful experiments conducted in the early 1600s by Galileo Galilei (Unit 12). Apart from his famous—and perhaps fictitious—demonstration of weights dropped from the Leaning Tower of Pisa, Galileo experimented with projectiles and with balls rolling down planks. The behavior of balls rolling down planks sounds far removed from the behavior of planets, but these basic experiments led him to recognize several properties of motion. A new understanding of forces and motion was essential to make Copernicus's heliocentric model of the Solar System plausible.

In this Unit we begin examining the framework of ideas that allowed us to understand how Earth can be in orbit, as is implied by the heliocentric model developed in Unit 12. Further physical concepts are explored in Unit 15 leading to the synthesis of ideas in the law of gravity in Unit 16.

14.1 INERTIA AND MASS

Much of our modern understanding of physics was developed by a scientist who was born the year Galileo died, **Isaac Newton** (1642–1727). Newton (Figure 14.1) is arguably the greatest scientist of all time. He made astounding contributions to mathematics, physics, and astronomy. Moreover, he pioneered the modern studies of motion, optics, and gravity. Many of these ideas were conceived when he was 23, forced to stay home from college because the plague was ravaging England. Newton was a fascinating individual. He was deeply religious and wrote prolifically on theological matters as well as science.

In his attempts to understand the motion of the Moon, Newton did much more than deduce the law of gravity. He realized that the mathematical methods he needed did not exist, so he invented calculus! What is especially remarkable about

FIGURE 14.1
Isaac Newton (1642–1727).

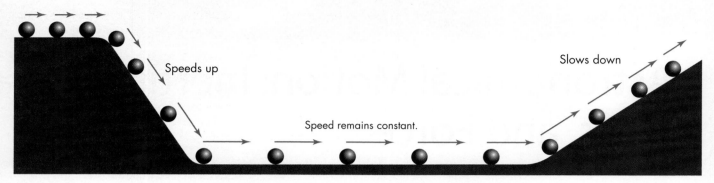

FIGURE 14.2
A ball rolling down a slope speeds up. A ball rolling up a slope slows down. A ball rolling on a flat surface rolls at a constant speed if no forces (such as friction) act on it.

Newton's work is that the discoveries he made in the seventeenth century form the basis for calculating the trajectories of spacecraft today. The physical laws Newton deduced took form as he carefully examined basic concepts like mass and force and developed precise definitions that could be expressed mathematically.

Central to the laws of motion is the concept of **inertia**. Inertia is the tendency of a body at rest to remain at rest and of a body in motion to keep moving in a straight line at a constant speed. Galileo's contemporary Johannes Kepler introduced the term, but Galileo examined inertia through experiments.

In one such experiment, Galileo rolled a ball down a sloping board repeatedly and noticed that it always sped up as it rolled down the slope (Figure 14.2). He next rolled the ball up a sloping board and noticed that it always slowed down as it approached the top. He hypothesized that if a ball rolled on a flat surface and there was no friction, its speed would neither increase nor decrease but remain constant. That is, in the absence of any force pushing or pulling on it, inertia keeps an object already in motion moving at a fixed speed.

Although you may not have considered it, we encounter inertia frequently in everyday life. Apply the brakes of your car suddenly, and the bag of groceries beside you tips over. Why is that? The bag's inertia keeps it moving forward at its previous speed until it hits the dashboard or spills onto the floor. You may have experienced this yourself standing in a bus or train. As the vehicle starts to move, you feel as if you are pushed backward, and if you are carrying a heavy suitcase, you may discover that its extra inertia causes you to lose your balance.

Inertia is one of those properties that are so fundamental that it is difficult to find simpler words to describe them. Instead we rely on analogy or describe what happens in various situations. Someone might say that the amount of inertia describes how strongly an object "wants" to remain at rest or in constant motion. We need to be cautious when speaking so colloquially, however. An object that is very heavy or difficult to move may have a lot of inertia, but heaviness and ability to move can depend on other factors. For example, your own body or an object you are carrying may feel lighter under water or heavier on an amusement park ride. Or a boat that weighs more than a car may be easier to move because the water it floats in offers little resistance.

In scientific terms we measure inertia by an object's **mass**. Mass can be described as the amount of matter an object contains. It is often measured on a balance scale, comparing objects with known and unknown masses (Figure 14.3). While people sometimes call this "weight," weight is something quite different (Section 14.3).

Mass is generally measured in kilograms. One kilogram—abbreviated 1 kg—equals 1000 grams. For example, a liter (1.1 quarts) of water has a mass of 1 kg. Different substances may have the same mass in a larger or smaller volume—for example, about 1/3 liter of rock has a mass of 1 kg, and about 800 liters of air have this mass. The kilogram standard is a cylinder of an alloy of platinum, copies of which are kept by governments around the world. All other measurements of mass are made relative to these official standards.

FIGURE 14.3
A balance determines an object's mass by comparing it to objects of known mass.

Concept Question 1

Comparisons of kilogram standards around the world suggest that the original has lost about 50 micrograms after a century. What might have caused such a change?

14.2 THE LAW OF INERTIA

Newton recognized the special importance of inertia and helped clarify various aspects of it. He described inertia as "a power of resisting, by which every body . . . [endeavors] to persevere in its present state, whether it be of rest, or of moving uniformly forward in a right line." This is encapsulated in what is now called **Newton's first law of motion** (sometimes referred to simply as the *law of inertia*). The law can be stated as follows:

> **I. A body continues in a state of rest, or in uniform motion in a straight line at a constant speed, unless made to change that state by forces acting on it.**

An essential point here is that inertia causes an object to resist changes in either speed or direction. This is again exemplified by groceries on a car seat. If the car turns a corner at a constant speed, the grocery bag will slide sideways on the seat and perhaps tip over. Its inertia keeps it going in the same direction as before unless you apply a force to it. To prevent the bag from continuing with its former speed and direction, you need to apply a force on it. Whether you are stopping or turning, you must reach over and hold the bag to keep it from falling over.

Because the speed and direction of motion are both important, scientists use a quantity that incorporates both: **velocity**. A velocity might be written as 100 kilometers per hour (kph), or 60 mph, to the northeast. A body's velocity changes if either its speed or its direction changes. This lets us simplify Newton's first law:

> **I. A body maintains a constant velocity unless forces act on it.**

In space we have to define a velocity in three dimensions, but the same rule applies. A spacecraft, a planet, or an entire galaxy will continue to move through space along a straight line at a constant speed unless some force acts upon it.

Actually, Newton was preceded in stating the law of inertia by the seventeenth-century Dutch scientist Christian Huygens. However, Newton went on to develop additional physical laws and—more important for astronomy—showed how to apply them to the universe.

Concept Question 2

If you watch passengers on a train that is going around a curve to the right, the passengers all lean to the right also. What is going on here, and how does it relate to inertia?

14.3 FORCES AND WEIGHTS

In effect, Newton's first law defines what a **force** is—anything that can cause a body to change velocity. A force is often a push or a pull, or it can be something that resists motion, such as friction. Like Earth's gravity as it pulls a skydiver downward, forces do not necessarily require direct physical contact.

In some situations forces may be applied to an object, yet there is no **net force**—that is, the forces acting on the object cancel each other out. For example, if a box is at rest on a table and pushed equally by two opposing forces, the forces are balanced. Therefore the box in Figure 14.4 experiences no net force when people push it equally from opposite sides, and it does not move. That same box is also being pulled downward by gravity, but the table top is pushing upward with an equal force, so there is no net force up or down either.

The force pulling the box down is also known as its weight, which is *not* the same as its mass. Because an object's mass is the amount of matter in the object, its mass is a fixed quantity. An object's **weight**, however, is a measure of the external forces acting on it and therefore depends on the object's environment.

Your weight on the Moon will be different from your weight on Earth because the strength of the Moon's gravitational pull on you is different from the Earth's. Your

FIGURE 14.4
Forces in opposite directions can cancel each other, resulting in no net force.

Clarification Point

Mass is a property of matter. Weight is a *force* on an object.

> **Concept Question 3**
>
> Would a balance scale give a correct reading for a mass if it were used under water?

> Some scientists prefer to say that weight refers only to the gravitational pull on you, and that other effects (like buoyancy or orbiting in a spacecraft) change only your *apparent* weight.

weight can also vary on Earth because of other forces acting on you. For example, the buoyancy of water may make you feel lighter or even weightless. Weight can be measured by a bathroom "spring" scale—you compress the spring more or less depending on your weight. By contrast, mass is usually measured with a balance that is designed so both the unknown and the reference masses experience the same forces. That way, the same masses that balance on Earth will balance on the Moon.

In an elevator, as it first starts rising, you may feel heavier—and in fact if you took a bathroom scale into the elevator, you would discover that your weight does increase momentarily. On the other hand, when an elevator starts downward, you will feel momentarily lighter, and in an orbiting space capsule, astronauts feel weightless. We experience changes in weight as a result of the gravitational force on us, other forces acting on us, or the way our surroundings move; but no matter where we are, we have the same mass.

14.4 THE FORCE IN AN ORBIT

Newton's first law may not sound impressive at first, but it carries an idea that is crucial in astronomy: If a body is changing speed or moving along a curved path, some net force must be acting on it. The fact that the Earth and other planets follow curved paths through space is a clue that they are being acted on by a force. But what force is it and where is it coming from?

A simple experiment helps to make clearer what forces must be present in orbital motion. If we tie a mass to a string and swing it in a circle, Newton's first law tells us that the mass's inertia will carry it in a straight line if no forces are acting on it. What force, then, is acting on the circling mass? The force is the one exerted by the string, preventing the mass from moving in a straight line and keeping it turning in a circle. We can feel that force as a tug on our hand from the string, and we can see its importance if the string breaks or we suddenly let go of it. With the force no longer acting on it, the mass flies off in a straight line, demonstrating the first law, as illustrated in Figure 14.5.

> **Concept Question 4**
>
> How do you think you could measure the mass of an object if you were in a weightless environment, like the International Space Station?

If we watch the mass in the last example after the string breaks, we will notice that its path is not completely straight. If we release it parallel to the ground, we will notice that the trajectory of the mass begins curving down toward the ground. The Earth must therefore be exerting a force on the object—the force of gravity.

Newton's great insight was to translate this example to an astronomical setting. The Moon's curved path around the Earth implies that Earth must be pulling on the Moon. Likewise, the planets' orbits imply that they feel a pull from the Sun. Gravity, he realized, is not just some phenomenon on Earth's surface—it must be a force that tugs on distant objects through space like an invisible string.

FIGURE 14.5
For a mass on a string to travel in a circle, a force (green arrow) must act along the string to overcome inertia. Without that force, inertia makes the mass move in a straight line.

Unit 14 Astronomical Motion: Inertia, Mass, and Force

KEY POINTS

- Bodies resist changes to their state of motion because of their inertia.
- The amount of inertia of an object is given by its mass, which is measured in kilograms.
- The term *velocity* is used to specify speed in a particular direction.
- Newton's first law: A body maintains a constant velocity unless forces act on it (law of inertia).
- Anything that causes an object to change speed or direction is a force; a planet on a curved orbit must be experiencing a force.
- Mass is a property indicating the amount of matter in an object.
- Weight depends on both the mass of an object and the forces that are acting upon it.

KEY TERMS

force, 111
inertia, 110
mass, 110
net force, 111
Newton, Isaac, 109
Newton's first law of motion, 111
velocity, 111
weight, 111

CONCEPT QUESTIONS

Concept Questions on the following topics are located in the margins. They invite thinking and discussion beyond the text.

1. Variations in the kilogram standard. (p. 110)
2. Why passengers lean when a train rounds a corner. (p. 111)
3. Using a balance under water. (p. 112)
4. Measuring mass in a weightless environment. (p. 112)

REVIEW QUESTIONS

5. What is meant by *inertia*?
6. What is the difference between mass and weight?
7. Why does an object sliding on the floor come to a stop? Does this violate the law of inertia?
8. What does Newton's first law of motion tell you about the difference between motion in a straight line and motion along a curve?
9. When a ball slows down while it is rolling up a slope, what force is changing the ball's velocity?
10. Why is it necessary to specify direction when speaking of velocity or force, but not when considering other quantities such as mass?
11. Is an object accelerating if it is moving at a constant speed in a circular orbit?

QUANTITATIVE PROBLEMS

12. Picture a soccer ball rolling across the ground at 1 meter per second (1 m/sec). One player deflects the ball so that it turns by a small angle but still moves at the same speed, then a second player brings the ball to a stop. Draw a diagram of this with arrows indicating the relative direction and magnitudes of the velocities of the ball and of the forces that the two players apply to the ball.
13. Two books, one a thin ¼-kg paperback and the other a thick 1-kg dictionary, are placed side-by-side on the same table. Is the force by the table on each book the same (because it is the same surface) or different (for some other reason)? Explain your answer.
14. In an amusement park ride called The Rotor, you are spun in a cylinder and then held against the wall as a result of the spin when the floor drops away. People sometimes describe that effect as being due to "centrifugal force." What is really holding you against the wall of the spinning cylinder? Draw a diagram showing the direction of all forces being applied to you. What force is keeping you from sliding down the wall, countering the force of gravity? How big is that force?
15. Consider the situation shown in Figure 14.5. If the mass of the ball were twice as large, do you think this would require a greater or lesser force applied by the string? What if the same mass were swung around twice as fast? (A qualitative result is sufficient for this problem, but try to guess how much bigger by performing the experiment if possible.)

TEST YOURSELF

16. Which of the following demonstrate(s) the property of inertia?
    a. A hockey puck sliding across an ice rink
    b. A skateboarder coasting along a level sidewalk
    c. A coffee mug sitting on a table
    d. Whipping a tablecloth out from under the dishes on a table
    e. All of the above.
17. If an object moves along a curved path at a constant speed, you can infer that
    a. a force is acting on it.   d. Both (a) and (b) are true.
    b. its velocity is not changing. e. None of these is true.
    c. no forces are acting on it.
18. The mass of a 5-kg bowling ball would be _____ if it were located in deep space, far from any star or planet.
    a. zero   c. slightly smaller
    b. much smaller   d. the same
19. What physical quantity is used to determine the amount of inertia an object has?
    a. Force   b. Mass   c. Weight   d. Velocity
20. You observe a spacecraft moving with a constant velocity. Which of the following must be true about forces acting on the spacecraft?
    a. No forces can be acting on it.
    b. A steady force must be acting in its direction of motion.
    c. Forces may be acting on it, but the net force must be zero.
    d. Gravity must be forcing the spacecraft into a circular orbit.

Force, Acceleration, and Interaction

15.1 Acceleration
15.2 Newton's Second Law of Motion
15.3 Action and Reaction: Newton's Third Law of Motion

Learning Objectives

Upon completing this Unit, you should be able to:
- Define acceleration, and calculate it from a changing velocity.
- Carry out a vector addition graphically.
- State Newton's second and third laws of motion, giving practical examples of each.
- Calculate an acceleration based on the force applied to a mass.
- Describe and explain the motions of objects of unequal mass that orbit each other.

In Unit 14 we looked at the idea of inertia and how a force acting on a mass causes its speed and/or direction (its velocity) to change. But exactly how much change in motion will a particular force produce? To answer that question, we need to define carefully what we mean by changes in motion, or *accelerations*.

Isaac Newton again clarified these ideas so that they could be treated mathematically. The ideas encapsulated in two further laws are simple to write down, but their implications are enormous. With this carefully constructed understanding of ideas such as mass, force, velocity, and acceleration, it becomes possible to track and predict the motions of all kinds of objects.

15.1 ACCELERATION

Mathematical Insight

To convert between km/hr and m/sec, remember that there are 1000 meters in a kilometer and 3600 seconds in an hour. Therefore:

$$10 \, \frac{m}{sec} \times \frac{3600}{3600} = \frac{36{,}000 \, m}{3600 \, sec} = 36 \, \frac{km}{hr}$$

Motion of an object is a change in its position, which we characterize both by the direction of the object and by its speed. For example, a car is moving east at 10 meters per second (this is the same as 36 km per hour, or about 22 mph). A quantity that has both a size and a direction is known as a **vector**, which is generally depicted by an arrow to indicate direction, the length of which indicates the size.

If the car's speed and direction remain constant, we say it has a constant velocity. If the car changes either its speed or direction, it is no longer moving uniformly, as depicted in Figure 15.1. A change in velocity occurring during a period of time is defined as an **acceleration**. A common experience of acceleration is when we step

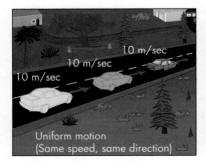
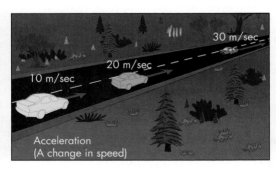
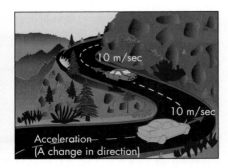

FIGURE 15.1
Views of a car in uniform motion and accelerating. (A) Uniform motion implies no change in speed or direction. The car moves in a straight line at a constant speed. If an object's speed (B) or direction (C) changes, the object undergoes an acceleration.

> **Clarification Point**
>
> An acceleration is any change of speed or direction, not just an increase in speed.

on the accelerator in a car and it speeds up. Although in everyday usage *acceleration* implies an increase in speed, scientifically *any* change in speed is an acceleration, so a car also "accelerates" when we apply the brakes and it slows down.

The acceleration of an object is defined as its change in velocity divided by the time taken to change it. This can be written mathematically as

$$\text{acceleration} = \frac{\text{change in velocity}}{\text{change in time}}$$

or, using symbols,

$$a = \frac{\Delta V}{\Delta t}.$$

Suppose a car is traveling at 10 meters per second eastward along a straight road. If the car increases its speed to 30 m/sec over 5 seconds (Figure 15.1B), we would say that its change of velocity is $\Delta V = 30$ m/sec $- 10$ m/sec $= 20$ m/sec, while the time change is $\Delta t = 5$ sec. The acceleration is therefore

$$a = \frac{20 \text{ m/sec}}{5 \text{ sec}} = 4 \frac{\text{m}}{\text{sec}^2}.$$

Acceleration is usually written in units of "m/sec^2," and we might say that the car's acceleration during this 5-second interval is "four meters per second squared" or "four meters per second per second." What this means is that the speed changes, on average, by 4 m/sec during each second. After 1 second of this acceleration, the car sped up by 4 m/sec and was traveling at 14 m/sec. After 2 seconds, it was traveling at 18 m/sec. After 3 seconds, it was traveling at 22 m/sec, and so forth.

Accelerations have direction and size, and so do vectors. The direction of an acceleration is the same direction as the velocity difference. To find this direction, we need to do vector math. We illustrate this in Figure 15.2A for the previous example of an accelerating car. The orange arrow represents 10 m/sec to the east and the red arrow is three times longer, representing 30 m/sec to the east. The velocity difference vector is the vector you would have to add to the initial velocity to give you the final velocity. The result (black arrow) is a velocity change of 20 m/sec to the east.

If you braked from 30 m/sec down to 20 m/sec (Figure 15.2B), the velocity difference is $\Delta V = 10$ m/sec in the westward direction. Depending on how quickly the car slows, the acceleration we would feel inside the car might be large or small. If the braking is done gradually over 5 seconds, the acceleration is $a = (10 \text{ m/sec})/5 \text{ sec} = 2$ m/sec^2—a relatively gentle acceleration. But if the braking takes just 1 second, the acceleration would be $a = (10 \text{ m/sec})/1 \text{ sec} = 10$ m/sec^2—which is quite strong.

Suppose a driver going 50 m/sec eastward swerves to the right to avoid an obstacle. The driver has made a change in velocity of 5 m/sec southward as illustrated by the black ΔV vector in Figure 15.2C. If the car swerved in the space of 0.5 second, the acceleration would be 10 m/sec^2 southward. Although this is as much acceleration as the example above, the speed barely changes, as can be seen by the length of the vectors. The motion to the right adds a component of velocity perpendicular to the original motion, but it does not change the overall speed very much.

> **Mathematical Insights**
>
> A change in velocity of 5 m/sec perpendicular to an initial velocity of 50 m/sec forms a right triangle (Figure 15.2C). The final speed can be found from the Pythagorean formula: $\sqrt{50^2 + 5^2} = \sqrt{2525} = 50.25$.

FIGURE 15.2
Velocity vector math. The diagrams show the change-in-velocity vector ΔV (in black) when the initial velocity V_1 (A) increases and remains in the same direction or (B) decreases but remains in the same direction, (C) experiences a small change perpendicular to the initial velocity, and (D, E) undergoes large changes in direction and speed.

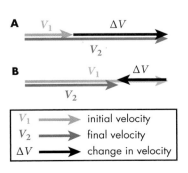
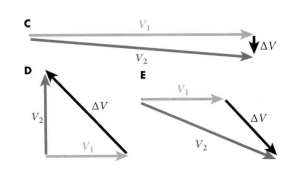

> **Clarification Point**
>
> Circular movement can only occur with a continuous acceleration.

If a car drives around a circular track at a fixed speed, its direction of travel is changing steadily, so it is undergoing constant acceleration. Sitting inside the car, you would experience this as a constant push toward the outside of the curve as your inertia (Unit 14) attempts to keep you moving in a straight line. The same is true for a circular orbit of a planet.

15.2 NEWTON'S SECOND LAW OF MOTION

How do we produce acceleration? Newton realized that for a body to accelerate, a force must act on it. When you press the accelerator pedal in a car, you make the engine run faster and transmit a force to the car by turning the wheels faster. When you steer the car to the right, the tires transmit a sideways force to the car that changes its direction of motion. For example, to accelerate—change the direction of—a mass whirling on a string, we must constantly exert a pull on the string (Unit 14.4). Similarly, to accelerate a shopping cart, we must exert a force on it.

Simple experiments can show that the acceleration we get is proportional to the force we apply, and in the same direction as the force. Forces are expressed in the MKS unit **newtons,** so applying more newtons of force produces a larger acceleration. For example, if we push a shopping cart gently, its acceleration is slight. If we push harder, its acceleration is greater. But experience shows us that more than just force is at work here. For a given push, the amount of acceleration also depends on how full the cart is. A lightly loaded cart may scoot away under a slight push, but a heavily loaded cart hardly budges given the same push, as illustrated in Figure 15.3. Thus, the acceleration produced by a given force also depends on the amount of matter being accelerated.

With an understanding of how to measure the acceleration of an object, we are now prepared to write an equation that can describe how forces affect the motions of any object in the universe. This is **Newton's second law of motion,** and it is surprisingly simple. Mathematically, in its most familiar form, the law states this:

$$F = m \times a.$$

In words,

> **II. The force (F) acting on an object equals the product of its acceleration (a) and its mass (m).**

As it is written here, the formula would be most useful if we wanted to determine what force must be acting on an object of known mass that is undergoing a

> **Mathematical Insights**
>
> In the MKS system (Unit 3), the unit of force is appropriately enough called the newton, which is the force needed to accelerate a 1-kg mass to a speed of 1 m/sec in 1 second (in other words, 1 newton = 1 kg·m/sec^2). This is about one-tenth the downward force of a 1-kg mass in Earth's gravitational field, and about a quarter pound of force in English units.

a = acceleration (in m/sec^2)
F = force (in newtons or kg·m/sec^2)
m = mass (in kg)

FIGURE 15.3
The same force applied to a loaded shopping cart will not make it accelerate as much as an empty cart.

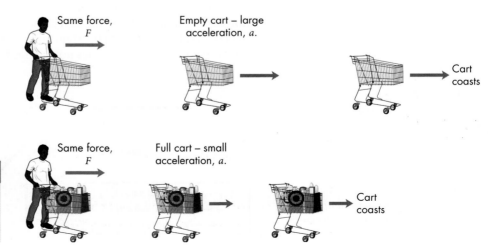

> **Concept Question 1**
>
> In what situations in everyday life have you experienced Newton's second law?

measured acceleration. Much more often we are interested in predicting how we will change the velocity of an object (accelerate it) when applying a known force. Thus, a useful way of thinking about Newton's formula is this:

$$a = F/m.$$

In words,

> II. **The amount of acceleration (a) that a body will experience is equal to the force (F) applied divided by the body's mass (m).**

Another way of saying this is that the acceleration grows in proportion to the force applied, but is inversely proportional to the mass of the object. If the force is doubled, the acceleration doubles; if the mass doubles, the acceleration is half.

Incidentally, we can also write Newton's second law as $m = F/a$. This form is useful for measuring the mass of an object even if it is floating, weightless, in space. This is the way astronauts measure their mass while in orbit. A known force is applied, and the resulting acceleration is measured; these numbers are "plugged into" the equation, and the mass is calculated.

> **Mathematical Insight**
>
> Mathematically, one quantity is proportional to another if when one changes by a certain factor, the other changes by the same factor (when all else is kept the same). Proportionalities are a good way of examining how an equation works in different situations.

15.3 ACTION AND REACTION: NEWTON'S THIRD LAW OF MOTION

Newton's second law indicates the response of an object to a force, but what happens to the object applying the force? This is explained by a final law of motion, **Newton's third law of motion**, sometimes called the *law of action–reaction:*

> III. **When two bodies interact, they create equal and opposite forces on each other.**

This law is somewhat counterintuitive because we usually think of one object supplying a force and another receiving it; but the force is felt by both with equal intensity. Two skateboarders side by side serve as an example of the third law. In Figure 15.4A, if X pushes on Y, both move. According to Newton's law, when X exerts a force on Y, Y exerts a force on X, so that both accelerate.

Suppose, though, that one of the skateboarders has a much larger mass than the other (Figure 15.4B). No matter which of the skateboarders pushes on the other, the one with the smaller mass will move away faster. The force F acting on each is the same, but because of Newton's second law the resulting accelerations differ because $a = F/m$. The skateboarder with the smaller mass m will have the larger acceleration because m appears in the denominator. A larger force would have to be applied to the heavier skateboarder to make him or her accelerate as much as the lighter skateboarder.

The gravitational force between the Earth and the Moon affords an astronomical example of Newton's second and third laws and at the same time leads us a step closer to

> **Concept Question 2**
>
> Suppose a skateboarder pushed off a wall instead of off another skateboarder. How does Newton's third law apply? What accelerates in reaction to the push?

FIGURE 15.4
Skateboarders illustrate Newton's third law of motion. (A) When X pushes on Y, an equal push is given to X by Y. (B) When X pushes on a much heavier person Z, an equal and opposite push is given to each, but Z moves off at a slower speed.

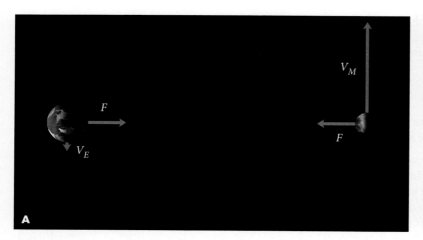

FIGURE 15.5
(A) The Moon exerts a force on the Earth equal and opposite to the force that the Earth exerts on the Moon. (Sizes and separations are not shown to scale.) (B) A hammer thrower feels an equal and opposite force matching the force he exerts on the swinging hammer. In both instances the more massive object (the Earth and the man) accelerates less according to Newton's second law of motion.

understanding gravity. According to Newton's third law, the gravitational force of the Moon on the Earth must be exactly the same as the gravitational force of the Earth on the Moon. Why, then, does the Moon orbit the Earth and not the other way around? The answer is the same as for the skateboarders: Because of Newton's second law, $a = F/m$. Thus, even though the Earth and Moon exert precisely equal forces upon each other, the Earth accelerates about 81 times less because it is 81 times more massive than the Moon. Because the Moon's acceleration is so much larger than the Earth's, the Moon does most of the moving (Figure 15.5A). In fact, however, the Earth does move a little bit as the Moon orbits it, much as you must lean back and move in a circle yourself if you swing a heavy weight around you (Figure 15.5B).

The same is true of the planets orbiting the Sun. Because the Sun is so much more massive than all of the planets, it wobbles by only a fraction of its own radius as the planets orbit it. Astronomers have been able to detect hundreds of other stars wobbling in this way, giving us evidence that planets orbit other stars (Unit 36).

The reaction force is what allows a spacecraft to accelerate. A rocket engine works by accelerating its propellant to a high speed and firing it out the back of the ship. The ship feels an equal force in the opposite direction. As illustrated in Figure 15.6, the ignited rocket fuel exerts as big a force on itself as on the rocket ship.

Astonishingly, Newton's three laws of motion allow scientists to plan and predict virtually all features of a body's motion. With $a = F/m$ and with knowledge of the masses and the forces in action, engineers and scientists can, for example, target a spacecraft safely between Saturn and its rings by setting off its thrusters to produce forces on the spacecraft to change its velocity (speed and direction) by a predictable amount even though it is millions of miles away from its target and hundreds of millions of miles away from the Earth.

> **Concept Question 3**
>
> If you were an astronaut who accidentally drifted away from your spaceship along with your toolbox, how could you use the tools to get back to the ship?

> **Clarification Point**
>
> A rocket accelerates forward as a reaction to pushing rocket fuel out the back, accelerated to high speed. It is *not* caused by the rocket flame "pushing against" the ground or air.

FIGURE 15.6
A rocket accelerates according to Newton's third law. Ignited propellants are pushed out the back of the rocket at high speed. They apply an equal and opposite force on the rocket, accelerating the ship forward.

Unit 15 Force, Acceleration, and Interaction

KEY POINTS

- An acceleration is a change in velocity occurring during a period of time.
- Accelerations, velocities, and forces are vector quantities because they have both size and direction.
- The direction of an object's acceleration can be found from vector math comparing its initial and final velocities.
- Newton's second law: The amount of acceleration that a body will experience is equal to the force applied divided by the body's mass.
- Newton's third law: When two bodies interact, they create equal and opposite forces on each other.
- Although forces in an interaction are equal, the more massive object will experience less acceleration due to Newton's second law.

KEY TERMS

acceleration, 114
newton, 116
Newton's second law of motion, 116
Newton's third law of motion, 117
vector, 114

CONCEPT QUESTIONS

Concept Questions on the following topics are located in the margins. They invite thinking and discussion beyond the text.

1. Newton's second law in everyday life. (p. 116)
2. Reaction force when pushing against a wall. (p. 117)
3. Using tools to get back to your spaceship. (p. 118)

REVIEW QUESTIONS

4. How does acceleration differ from velocity? from force?
5. What is happening when an object has a velocity in the positive X direction and an acceleration in the negative X direction at the same time?
6. What is Newton's second law?
7. Under what circumstances can an object experience a force yet maintain an unchanging speed?
8. How can objects move if every force is associated with an equal and opposite force?
9. If you exert a gravitational force on the Earth equal to the force Earth applies on you, why do *you* fall instead of the Earth?

QUANTITATIVE PROBLEMS

10. After driving behind a car that is traveling at 90 km/hr, you decide to pass the car by changing your speed to 110 km/hr. What was your average acceleration if it takes 3 seconds to change your speed?

11. In metric units, forces are measured in "newtons," which can also be written as kilograms × meters/second^2. What is the acceleration experienced by a 100-kg mass if a 100-newton force is applied to it? What is the acceleration experienced by a 200-kg mass if the same force is applied to it?
12. A force of 1 newton acting for 1 second on a 1-kilogram-mass object will change the object's velocity by 1 m/sec. For example, suppose an object with a mass of 20 kg is moving at a rate of 6 m/sec in a straight line. A force of 30 newtons acting opposite the direction of motion of the object for 4 sec will bring the object to a halt.
    a. What would happen if the force continued after 4 sec?
    b. What would be the resulting motion of the object if the force acts in the opposite direction?
    c. What if the same force acted perpendicular to the motion?
13. How long should a 4-newton force be applied to a 6-kg bowling ball to get the ball rolling at 5 m/sec? If a greater force is applied, does it take less or more time to reach the same velocity? Explain.
14. If the bowling ball from the previous problem is now rolling uniformly at 5 m/sec, describe the direction a force must be applied to get the ball to stop moving. If you want the ball to be at rest in 1 s, what magnitude of a net force must be applied?
15. Suppose that a 1000-kg rocket's thrusters give it an acceleration of 20 m/sec^2 when it is launched from Earth's surface. If Earth's gravity applies a downward force of approximately 10,000 newtons on the rocket, what is the force of the thrusters on the rocket? If the same rocket were launched from the surface of another planet, would the acceleration of the rocket be greater or less than the acceleration experienced by the rocket taking off from the Earth? Explain your answer.

TEST YOURSELF

16. A rocket blasts propellant out of its thrusters and "lifts off," heading into space. What provided the force to lift the rocket?
    a. The propellant pushing against air molecules in the atmosphere
    b. The propellant heating and expanding the air beneath the rocket, and so pushing the rocket up
    c. The action of the propellant accelerating down, giving a reaction force to the rocket
    d. The propellant reversing direction as it strikes the ground below the rocket, then bouncing back and pushing the rocket up
17. Which of the following cases does *not* describe an acceleration? The space shuttle
    a. lifting off from Earth   c. coming to a stop upon landing
    b. orbiting Earth   d. being towed at a steady pace
18. If you apply the same force to two carts, the first with a mass of 100 kg, the second with a mass of 10 kg, the acceleration of the 100-kg cart will be _____ the acceleration of the 10-kg cart.
    a. 10 times larger than   c. the same as
    b. 10 times smaller than
19. You give a shopping cart a shove to the east. It rolls for a while, but eventually it slows down and comes to a stop. What was the direction of the net force acting on it after you shoved it?
    a. To the east   b. To the west   c. Downward

The Universal Law of Gravity

16.1 Orbital Motion and Gravity
16.2 Newton's Universal Law of Gravity
16.3 Surface Gravity and Weight

Learning Objectives

Upon completing this Unit, you should be able to:
- Explain how orbital motion and falling can be caused by the same force.
- Write out the law of gravity and explain the meaning of each term.
- Give examples of how the force of gravity varies with distance.
- Carry out a calculation with the law of gravity to find an acceleration.
- Explain why different masses all fall at the same (accelerating) rate.

Isaac Newton made sense of motion with his three laws, as we explored in Units 14 and 15, but his larger goal was to understand the mysterious force of gravity. Newton was not the first person to attempt to discover and define the force that holds planets in orbit around the Sun. Nearly 100 years earlier, Kepler recognized that some force must hold the planets in their orbits and proposed that something similar to magnetism might be responsible. Newton was not even the first person to suggest that gravity is responsible. Other members of the Royal Society in England speculated about gravity's role, but none was able to present a convincing case.

According to one story, Newton had a flash of inspiration when he saw an apple falling from a tree. The apple falling down to the Earth's surface made him speculate whether Earth's gravity might extend to the Moon. Newton realized that if the Earth's gravitational pull reached all the way to the Moon, it could provide the force that keeps the Moon circling the Earth—like a string pulling on a twirling mass.

16.1 ORBITAL MOTION AND GRAVITY

Much of Newton's work is very mathematical, but as part of his discussion of orbital motion he described a thought experiment to demonstrate how a body can move in orbit. Thought experiments are not actually performed; rather, they serve as a way to think about problems. In Newton's thought experiment, we imagine a cannon on a mountain peak firing a projectile (Figure 16.1A). From our everyday experience, we know that whenever a body is thrown horizontally, gravity pulls it downward so that its path is an arc. Moreover, the faster we throw the body, the farther it travels before striking the ground.

Now imagine increasing the projectile's speed more and more, allowing it to travel ever farther. As the distance traveled by the ball becomes very large, we see that the Earth's surface curves

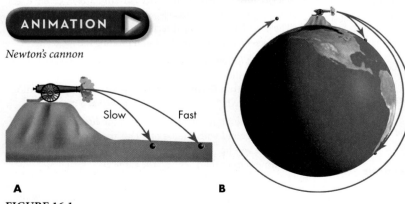

FIGURE 16.1
(A) A cannon on a mountain peak fires a projectile. The faster the projectile is fired, the farther it travels before hitting the ground. (B) At a sufficiently high speed, the projectile travels so far that the Earth's surface curves out from under it as fast as it falls, and the projectile is in orbit.

Mathematical Insights

If we increase the speed of the cannonball beyond a circular orbit, it will begin to swing farther and farther out from the Earth in an elliptical path. If its speed is great enough, it will escape from the Earth and never return (Unit 18).

Gravity variations

Concept Question 1

Why do you suppose most spacecraft are launched to the east? Why are they generally launched from near the equator?

Force of gravity

Concept Question 2

Does the saying "The bigger they are, the harder they fall" make sense given Newton's laws and the law of gravity?

away below the projectile (Figure 16.1B). Therefore, if the projectile moves at the right speed, its curvature downward will match the curvature of the Earth's surface, and the projectile will never hit the ground. Such is the nature of orbital motion and how the Moon orbits the Earth. The Moon is "falling," but because of its sideways motion it always misses the Earth. This does not answer the question of *how* the Moon got its sideways motion (presumably it was not fired out of a cannon!); but once set up with the right velocity, it can continue "falling" forever.

We can phrase this thought experiment more specifically using Newton's first law of motion. According to that law, in the absence of forces the projectile would travel in a straight line at constant speed. But because a force—gravity—is acting on the projectile, its path is not straight but curved. Moreover, the law helps us understand that the projectile keeps moving because it has inertia.

Notice that in this discussion we used no formulas. All we needed was Newton's first law and the idea that gravity supplies the force. To make further progress—for example, to determine how rapidly the projectile must move to be in orbit—we need a mathematical formulation.

16.2 NEWTON'S UNIVERSAL LAW OF GRAVITY

Years after developing his initial ideas, Newton published his law of gravity in 1687 in his *Philosophiae Naturalis Principia Mathematica* (Mathematical Principles of Natural Philosophy), one of the milestones in the history of science. He demonstrated the properties that gravity must have if it is to control planetary motion. Moreover, Newton went on to derive the **law of gravity** in mathematical form, allowing astronomers to predict the positions and motions of the planets and other astronomical bodies.

The law of gravity must describe the force that acts in many circumstances—a falling apple, an orbiting satellite, or planets circling the Sun—with a single equation. We can begin to see what form this equation must have if we consider all of these different situations. First, it seems clear that gravity must depend on mass, because larger bodies, like the Sun, produce a stronger force than smaller bodies, like the Earth or an apple. And the pull an object exerts must also depend on the mass of the object being pulled—for example, the Earth must pull on the Moon with a far larger force than it exerts on the apple, in order to overcome the Moon's larger inertia. Newton determined that the only way to explain these diverse situations and make the law of gravity consistent with his laws of motion was that the gravitational force between two bodies must depend on the product of their masses. Finally, Newton determined that the force must grow weaker with distance to explain the slower speeds of the outer planets in their orbits about the Sun (Kepler's third law, Unit 12.2).

Newton analyzed these issues and concluded the following:

> **Every mass exerts a force of attraction on every other mass. The strength of the force is directly proportional to the product of the masses divided by the square of the distance between them.**

An important note about the distance used in this calculation: it is the distance between the centers, or technically the **centers of mass,** of the two objects. For a spherical object the center of mass is at the center, but for more complicated shapes it is the balance point for the mass distribution—for your body, this is roughly at the center of your pelvis. If you are standing on the Earth, the distance used to calculate the gravitational force exerted on you is approximately the radius of the Earth, or about 6380 km. Your body is too small to add significantly to this.

We can write this extremely important result in a shorthand mathematical manner by defining several algebraic variables. Let M and m be the masses of the

FIGURE 16.2
Gravity produces a force of attraction (green arrows) between bodies. The strength of the force depends on the product of their masses, M and m, divided by the square of their separation, d^2. G is the universal gravitational constant.

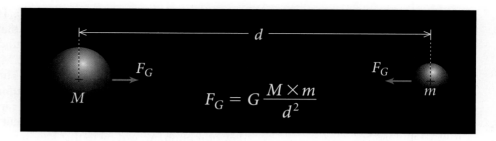

two bodies (Figure 16.2), and let the separation between their centers be d. Then the strength of the gravitational force between them, F_G, is

$$F_G = G \frac{M \times m}{d^2}.$$

The factor **G** is a constant, a conversion factor, that lets us translate from the units of mass and distance on the right side of the equation to units of force for F_G. The value of G is found by measuring the force between two bodies of known mass and separation—for example, two large lead masses in a laboratory—which was first done by Henry Cavendish in 1797. He placed two masses (red spheres in Figure 16.3) at either end of a rod suspended from a cable, then moved two large masses (purple spheres) close to them, causing the cable to twist, requiring a known force.

F_G = Force of gravity
G = Newton's gravitational constant
M, m = Masses of objects
d = Distance between their centers

The numerical value for G depends on the units chosen to measure M, m, d, and F_G. For example, if M and m are measured in kilograms, d in meters, and F_G in newtons, the MKS unit of force (Unit 15), then

$$G = 6.67 \times 10^{-11} \text{ newtons} \cdot \text{meters}^2/\text{kilogram}^2.$$

As long as we make measurements using MKS units, G has this same value whether we are dealing with stars, planets, or apples.

Writing the law of gravity as an equation helps us see several important points. If either M or m increases, and the other factors remain the same, the force increases by the same amount. We call this a *direct proportionality*. On the other hand, if d (the distance between the objects) increases, the force gets weaker. In fact, it weakens as the square of the distance. That is, if the distance between two masses is doubled, the gravitational force between them decreases by a factor of four, not two. We call this an *inverse-square proportionality*.

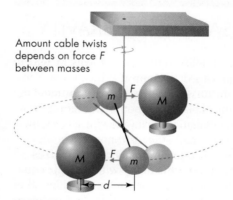

FIGURE 16.3
Cavendish's experiment that measured the gravitational constant G, by measuring the twist caused by a pair of large masses.

16.3 SURFACE GRAVITY AND WEIGHT

Several scientists during the 1500s, most famously Galileo, observed that balls with different weights dropped together from the same height strike the ground simultaneously. This seems counterintuitive because we feel a greater force downward when we heft a large mass than a light one, but gravity accelerates both in precisely the same way. A feather falls more slowly than a hammer on Earth because air resistance reduces the net force on the feather, but in a vacuum they fall at the same rate, as demonstrated by astronauts on the Moon (Figure 16.4).

The Earth *does* pull with a larger force the more massive the object, but this is counterbalanced by the object's greater inertia. Gravity provides just the right force to accelerate all objects at the same rate on our planet's surface. We can show this mathematically using Newton's laws.

Consider the gravitational acceleration on an object with mass m dropped near the surface of Earth. The Earth has a mass $M_\oplus$ and a radius $R_\oplus$. The distance between the center of the object and the center of the Earth is approximately $R_\oplus$, so we can use $R_\oplus$ as a close approximation to the actual distance. We can use Newton's

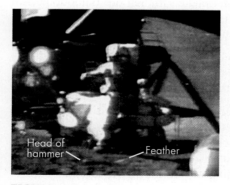

FIGURE 16.4
Frame from video showing a hammer and feather striking the ground at the same time after being dropped on the airless Moon.

second law, $a = F/m$, to calculate the acceleration due to the Earth's gravity. This acceleration is usually written as **g**. Using Newton's universal law of gravity, we find

$$g = \frac{F_G}{m} = G\frac{m \times M_\oplus}{m \times R_\oplus^2} = G\frac{M_\oplus}{R_\oplus^2}.$$

This is often simply called the Earth's **surface gravity.**

Note that the mass of the falling object does not matter. The Earth's surface gravity, g, depends only on the mass and radius of the Earth and the gravitational constant G. If we "plug in" the values for these quantities, we can find the acceleration that all objects experience at the Earth's surface:

$$g = 6.67 \times 10^{-11} \frac{\text{newton} \cdot \text{m}^2}{\text{kg}^2} \times \frac{5.97 \times 10^{24}\,\text{kg}}{(6.37 \times 10^6\,\text{m})^2}$$
$$= 9.81 \frac{\text{newton}}{\text{kg}} = 9.81 \frac{\text{m}}{\text{sec}^2}.$$

In the last step, we have used the definition of a newton: 1 newton = $1\,\text{kg}\cdot\text{m}/\text{sec}^2$.

Therefore, all objects at the Earth's surface accelerate at this same rate. After falling for 1 second, they have a downward velocity of 9.81 m/sec (about 35 kph, or 22 mph); after 2 seconds, 19.62 m/sec (70 kph, 44 mph); after 3 seconds, 29.43 m/sec (105 kph, 66 mph)—increasing by 9.81 m/sec each second. This acceleration is illustrated in Figure 16.5 for a ball dropped from the Leaning Tower of Pisa. The ball would take over 3 seconds to reach the ground 56 m below, by which time it would be falling at about 33 m/sec (120 kph, or 75 mph).

The value of "g" on Earth's surface is often used in describing other rates of acceleration. For example, you may experience up to about 5 g's on an amusement park ride or race car, and automobile airbags are triggered by a negative acceleration (*deceleration*) of about 10 g's.

We can also compare the rates of acceleration experienced at the surface of other astronomical bodies. The Moon's surface gravity is about 0.17 g—that is, on the Moon you would weigh about 17% of what you weigh on the Earth. Figure 16.6 shows an astronaut jumping up about a meter from a stand while wearing a spacesuit and life support system with a combined mass of about 90 kg (about 200 pounds). On the Moon your mass remains the same, but your weight is only 17% of that on the Earth.

The force of gravity you feel is strongest when you stand on the surface of a body because your distance from its center is as small as possible. The force weakens with increasing distance, but it never disappears. The gravitational attraction of a body reaches across the entire universe. The Earth's gravity not only holds you onto its surface but also exerts the force that holds the Moon in orbit around the Earth. Earth's pull extends even to distant stars and galaxies, although its pull at such distances is minuscule and just one among the forces from countless other objects.

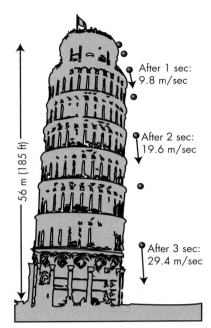

FIGURE 16.5
A ball dropped from the Leaning Tower of Pisa will take more than 3 seconds to reach the ground. At 1 g acceleration, after each second it will be falling 9.8 m/sec faster, striking the ground at a speed of about 33 m/sec.

Clarification Point

Astronauts orbiting the Earth appear weightless because they and their ship are both "falling" at the same rate, so the floor of the ship drops away just as fast as the astronaut falls. In fact, the Earth's gravitational pull is only slightly less than on the Earth's surface.

FIGURE 16.6
Apollo 16 astronaut John Young making a jumping salute. Despite a total mass of over 170 kg in his space suit, he easily jumps up in the Moon's weak gravity.

Concept Question 3

In Figure 16.6, can you locate the astronaut's shadow? How might you use the shadow to determine how high he is jumping?

KEY POINTS

- An object in orbit is falling, but it has a large enough horizontal velocity to never hit the surface of the object it is orbiting.
- Gravity is a pull felt between all objects with mass.
- The force of gravity between two bodies is proportional to the product of their masses and inversely proportional to the square of the distance between them.
- Objects near the surface of the Earth all accelerate downward at the same rate, one "g"—about 9.8 m/sec^2.
- The surface gravity felt on other bodies (such as the Moon) differs from that on the Earth.
- The force of gravity grows weaker with distance, but it never drops to zero.

KEY TERMS

center of mass, 121
g (acceleration due to gravity at Earth's surface), 123
G (gravitational constant), 122
law of gravity, 121
surface gravity, 123

CONCEPT QUESTIONS

Concept Questions on the following topics are located in the margins. They invite thinking and discussion beyond the text.

1. Best direction for launching a rocket into orbit. (p. 121)
2. The truth of "the harder they fall." (p. 121)
3. Estimating height from a shadow. (p. 123)

REVIEW QUESTIONS

4. What are the quantities that determine the force of gravity between two objects?
5. In what sense is gravity an "inverse-square law"?
6. What is the function of the gravitational constant G?
7. The force of gravity holds the Moon in orbit, while the same force causes an apple released from a tree to fall to the ground. Why does the same force have different effects on these two objects?
8. What is surface gravity?
9. How could it be possible that an object on a planet more massive than the Earth could have the same weight on that planet's surface as on the Earth?

QUANTITATIVE PROBLEMS

10. Given that the mass of Mars is 6.4×10^{23} kg, how influential (gravitationally speaking) is Mars on you when the red planet is at its smallest separation from Earth, which is 56 million km? How does this value compare to the gravitational force between you and a person standing 1 m away?
11. What would be the weight of an object that weighs 180 newtons on the Earth if it were moved to a planet with twice the mass of the Earth and a radius twice the radius of the Earth?
12. The Moon's radius is 1.74×10^6 m; its mass is 7.35×10^{22} kg. Find the surface gravity on the Moon from these values.
13. The Sun's radius is 6.97×10^8 m, while its mass is 1.99×10^{30} kg. Using these values find the surface gravity on the Sun. Without changing its mass, what would the Sun's radius have to change to in order to have the same surface gravity as Earth?
14. The mass of the planet Jupiter is 1.90×10^{27} kg, and the minimum distance from us (on Earth) to Jupiter is 6.30×10^{11} m. The mass of a loaded 18-wheeler is 35,000 kg.
    a. What is the gravitational force of Jupiter on you?
    b. What is the force of the 18-wheeler on you if you are standing 2 meters away from its center of mass?
15. Using the data in Appendix Table 5, find the surface gravity at the cloud-tops of Uranus.

TEST YOURSELF

16. The Earth's mass is about 80 times larger than the Moon's. What is the ratio of the gravitational force the Earth exerts on the Moon to the gravitational force the Moon exerts on the Earth?
    a. 80 to 1   c. 1 to 1   e. 1 to 80^2
    b. 1 to 80   d. 80^2 to 1
17. Astronauts inside the International Space Station appear weightless because
    a. there is no gravity in space.
    b. the space station is far from Earth, therefore the gravitational force is so small it barely affects the astronauts.
    c. the gravitational force from the Sun counter-balances the gravitational force by Earth.
    d. both the space station and the astronauts are orbiting Earth, therefore, the astronauts appear weightless because both objects are falling together toward Earth.
18. Gravity
    a. is the result of the pressure of the atmosphere on us.
    b. occurs between objects that are touching each other (or that are both touching the atmosphere).
    c. is the force larger objects exert on smaller ones.
    d. is the attraction between all objects that have mass.
    e. is caused only by planets and the Sun.
19. Why is it that on Earth a feather will fall at a slower rate than a rock?
    a. Earthen materials are more strongly attracted to Earth.
    b. The gravitational force on the rock is larger because it is more massive.
    c. Air resistance reduces the net downward force on the feather.
    d. Gravity exerts a stronger force on denser materials.
    e. They actually do fall at the same rate even near the surface of Earth.
20. At the distance of Jupiter, about 5 AU from the Sun, how strong is the Sun's gravitational force on a spaceship compared to when the spaceship was at Earth's distance from the Sun?
    a. about 25 times weaker   d. about 5 times stronger
    b. about 25 times stronger   e. the same
    c. about 5 times weaker

Measuring a Body's Mass Using Orbital Motion

17.1 Masses from Orbital Speeds
17.2 Kepler's Third Law Revisited

Learning Objectives

Upon completing this Unit, you should be able to:
- Describe a centripetal force and define it mathematically.
- Identify situations where separations and velocities can be used to determine a mass, and apply the gravitational formula to find the mass.
- Explain how Newton's law of gravitation relates to Kepler's third law.
- Apply Newton's revised version of Kepler's third law to find the gravitational mass.

Newton's law of gravity (Unit 16) not only describes the motions of the Moon and planets and explains Kepler's laws (Unit 12), it applies to situations Newton never anticipated. This is a mark of the best theories, that they reach beyond what was previously known to whole new situations.

Because the force of gravity depends on the masses of the bodies, if we can independently measure the force, we can determine what masses must be present. Determining what forces are present might at first seem an even more difficult task than estimating the mass of an object from its size and composition. After all, you cannot connect a spring to the Earth to see how much the Sun is pulling on it. However, Newton realized that his laws of motion (Units 14 and 15) could reveal what forces are present from the size and period of an orbit.

Newton had to develop new mathematical tools to solve problems like this, and this was one of his motivations for developing calculus. The resulting relationships allow astronomers to work backward from the properties of an orbit to find the masses of the orbiting objects. We can find the mass of Jupiter from the orbits of its satellites, or the mass of an unseen planet or black hole in orbit around a distant star, and even determine that mysterious "dark matter" must be present in galaxies.

17.1 MASSES FROM ORBITAL SPEEDS

Consider a small body in a circular orbit around a large body. To simplify our calculation we will suppose that the small body has such a small mass that it only negligibly shifts the position of the more massive body. We can therefore treat the more massive object as essentially stationary. These restrictions are met to high precision in many astronomical systems, such as the Earth's motion around the Sun and the Sun's motion around the center of the Milky Way. These assumptions simplify the mathematics, but the results turn out to correct for the more complex case when the masses of the orbiting bodies are not so dissimilar.

From Newton's first law, we know that there must be a force acting on a body that moves along a circular path. This **centripetal force** must be applied to any body moving in a circle, whether it is a car rounding a curve, a mass swung on a string, or the Earth orbiting the Sun.

Concept Question 1

If you sit on a spinning amusement park ride, you will feel a "centrifugal force" throwing you outward. Why do you suppose that this is usually called a fictitious force?

FIGURE 17.1
(A) The centripetal force, F_C, depends on the mass and speed at which an object swings in a circle as well as the object's distance from the center. (B) The gravitational force between the Sun and the Earth holds the Earth in its nearly circular orbit.

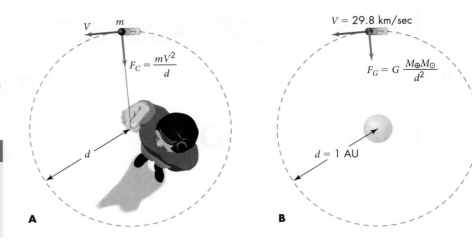

Concept Question 2

If you are rounding a corner in a car, what does the centripetal force equation tell you about what conditions might make the car skid?

F_C = Centripetal force
m = Mass moving on circular path
V = Velocity of circular motion
d = Distance from center of circular motion

Step-by-step derivation:
$$F_G = F_C$$
$$G\frac{m \times M}{d^2} = \frac{m \times V^2}{d}$$
$$G\frac{M}{d} = V^2$$
$$\frac{d}{G} \times G\frac{M}{d} = \frac{d}{G} \times V^2$$
$$M = \frac{d \times V^2}{G}$$

where M is the mass of the object being orbited, d is the distance to and V is the speed of the orbiting body.

Mathematical Insight

Step-by-step derivation of the Earth's orbital speed from its orbital distance d (1 AU) and period P (1 yr):
$V = 2\pi \times d/P = 6.28 \times 1.50 \times 10^{11}$ m $/(3.16 \times 10^7$ sec$) = 2.98 \times 10^4$ m/sec $= 29.8$ km/sec.

Newton used calculus (which he invented to help solve this type of problem) to determine the acceleration a body undergoes when it travels in a circle, as illustrated in Figure 17.1A. He derived the following equation for the centripetal force F_C on a mass m moving with a velocity V at a distance d from the center of the circle:

$$F_C = \frac{m \times V^2}{d}.$$

Without going into the details of the derivation of this formula, we can still understand why it has the dependencies it does. The *size* of the change in velocity at any moment and the *rate* at which it changes are both proportional to the velocity. Hence, the result depends on V^2. The turns become tighter and the acceleration is greater if the circle is smaller, which gives an inverse dependence on the distance from the center. Finally, the m in the equation comes from Newton's second law, reflecting the fact that a larger mass takes a greater force to turn.

Using this equation, we can find the mass of a star if we know the speed and radius of a planet's orbit around it. Let the star's mass be M and the planet's mass be m, with the planet's mass assumed to be much smaller than M. Assume that the planet moves in a circular orbit at a distance d from the Sun with a velocity V. The gravitational attraction between the star and the planet provides the force that deflects the planet from its tendency to move in a straight line, creating the force needed to produce the observed centripetal acceleration.

For an object in a circular orbit, the gravitational force F_G must equal the centripetal force F_C. If we set $F_G = F_C$ and carry out the algebra (see box to left), we find that the mass is related to the size and speed of the orbit as follows:

$$M = \frac{d \times V^2}{G}.$$

Therefore, we can determine the mass of an object if we know the speed and distance of another body orbiting it. And we do not even need to know the mass of the orbiting body!

For example, the orbit of the Earth around the Sun (Figure 17.1B) allows us to determine the Sun's mass. The Earth's orbital velocity V is 29.8 km/sec, or 2.98×10^4 m/sec. Using this value together with the Earth–Sun distance (1 AU = 1.50×10^{11} m), we find that the Sun's mass must be:

$$M_\odot = \frac{1.50 \times 10^{11} \text{ m} \times (2.98 \times 10^4 \text{ m/sec})^2}{6.67 \times 10^{-11} \text{ m}^3/\text{sec}^2 \cdot \text{kg}} = 2.0 \times 10^{30} \text{ kg}.$$

This same method can be used to calculate, for example, the mass of the Earth from an orbiting satellite, or the mass of a galaxy from an orbiting star. This is an especially convenient way of finding masses, because astronomers have methods

17.2 KEPLER'S THIRD LAW REVISITED

In Section 17.1 we could have written the expression for M in a slightly different way: by expressing the velocity, V, as the ratio of the orbital circumference ($2\pi d$) and the period, P. If we were to do that, we would end up with:

$$M = \frac{4\pi^2}{G} \times \frac{d^3}{P^2}.$$

This expression bears a certain resemblance to Kepler's third law, $P^2 = a^3$, where P is measured in years and a in astronomical units. As a reminder that these values need to be measured in these units, we sometimes write P_{yr} and a_{AU}.

The resemblance to Kepler's third law is a little clearer if we rewrite it as:

$$1 = \frac{a_{AU}^3}{P_{yr}^2}.$$

One difference from Newton's formula is that instead of measuring distances in meters, periods in seconds, and masses in kilograms, Kepler used the Earth's orbital radius and period, essentially folding the Sun's mass and the constants (4, π, and G) into one value.

We know that Kepler's third law also applies to elliptical orbits, not just circular orbits, and a more detailed analysis using Newton's laws of motion shows this is true in general. **Newton's version of Kepler's third law** applies not just to objects orbiting the Sun but to any two objects orbiting each other (Figure 17.2). It applies even if the two bodies have similar masses, and even if the orbit is elliptical. Newton showed that the sum of the two bodies' masses obeys the following law:

$$M_A + M_B = \frac{a_{AU}^3}{P_{yr}^2}$$

where the masses of the two orbiting bodies, M_A and M_B, are in units of the Sun's mass $M_\odot$.

Newton's version of the law matches Kepler's version within the Solar System because the Sun's mass is so much larger than anything else in the Solar System. As a result, the sum of the Sun's mass and any other object's mass is not much different from the Sun's mass alone. For example, the Earth's mass is just 0.000003 $M_\odot$, so when it is added to the Sun's mass, the left side of the equation becomes 1.000003. It requires very precise measurements to detect the small difference of this value from 1. The only planet for which it is not too difficult to detect the difference is Jupiter, which has a mass of about 0.001 $M_\odot$.

The modified form of Kepler's law is especially important because we can use it to measure the masses of stars that are orbiting each other (Unit 57). Whenever we can measure the orbital period, P, of two stars in years, along with the semimajor axis of their orbit, a, in AU, we can determine the sum of the two stars' masses.

With these equations, gravity becomes a powerful tool for determining the masses of astronomical bodies. We shall use this method many times throughout our study of the universe.

Concept Question 3

Might alien astronomers living in a different system of planets orbiting a star have derived Kepler's third law? How would it differ from our version?

Mathematical Insight

Astronomers and other scientists often use subscripts as reminders about the units to use or the meaning of symbols in an equation. The subscripts do not otherwise affect how we carry out the arithmetic.

$M_A + M_B$ = Total mass of orbiting bodies in solar masses

a_{AU} = Semimajor axis of orbit in AU

P_{yr} = Period of orbit in years

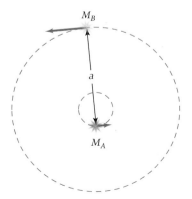

FIGURE 17.2
A pair of stars in orbit around each other. The sum of their masses can be determined by the modified form of Kepler's third law.

KEY POINTS

- Because gravity depends on an object's mass, orbital motions can be used to determine the mass of an object.
- The centripetal force is the force needed to make an object curve along a path with a given radius at a given speed.
- By equating the gravitational force to the centripetal force needed to make a body follow the curved path of its orbit, we can relate the mass of an object to the distance and speed of orbiting bodies.
- Newton derived a modified version of Kepler's third law that is applicable to orbits around bodies other than the Sun; he showed that the relationship depends on the sum of the bodies' masses.

KEY TERMS

centripetal force, 125

Newton's version of Kepler's third law, 127

CONCEPT QUESTIONS

Concept Questions on the following topics are located in the margins. They invite thinking and discussion beyond the text.

1. Fictitious centrifugal force. (p. 125)
2. Centripetal forces and skidding in a car. (p. 126)
3. Kepler's third law in a different solar system. (p. 127)

REVIEW QUESTIONS

4. What is a centripetal force?
5. In what situations have you experienced a centripetal force?
6. How does a centripetal force depend on the speed and radius of the circular motion?
7. Why is it generally unnecessary to know the mass of a planet if we want to measure the mass of the star it orbits?
8. What are the differences between Kepler's third law and Newton's version of this law?

QUANTITATIVE PROBLEMS

9. Given that Neptune is about 30 AU from the Sun, how many years does it take to complete one orbit around the Sun? Approximately how many orbits has Neptune completed since its discovery in 1846?
10. Derive the mass of the Earth from the Moon's orbital period of 27.3 days and orbital distance of 3.84×10^8 m.
11. Communication satellites are typically placed in what are called geostationary orbits. These are orbits where the orbital period of the satellite matches Earth's rotational period, allowing the satellite to stay over a fixed point on Earth. Calculate the altitude (distance above Earth's surface) of a geostationary orbit and compare this value to the altitude of the International Space Station, which is 360 km.
12. A pair of identical stars is separated by 25.0 AU, and the stars orbit each other with a period of 62.0 years. What are the masses of the stars?
13. The Sun orbits the center of the Milky Way Galaxy at about 220 km/sec at a distance from the center of about 2.6×10^{20} meters (about 28,000 light-years). What is the mass of the Milky Way inside the Sun's orbit? (Note: In this problem, we assume that the Galaxy can be treated as a sphere of matter. Strictly speaking, this is not precisely correct, but the more elaborate math needed to calculate the problem properly ends up giving almost the same answer.)
14. The orbital speed of stars orbiting 14,000, 56,000, and 84,000 light-years from the center of the Milky Way are also found to be 220 km/sec, as in the previous problem. What is the mass out to each of these radii? If all the mass had been concentrated in the very center of the Milky Way, what orbital speeds would you have expected at these distances from the center (assuming the speed was 220 km/sec at the Sun's distance)?
15. Jupiter's orbital period is 11.8622 years and its semimajor axis is 5.203 AU. Show that this implies a mass for Jupiter of approximately 1/1000 the Sun's mass. If Jupiter had a completely negligible mass, but orbited at the same distance from the Sun, what would its orbital period be?
16. In searching for planets orbiting other stars, you notice a star that looks just like the Sun (and presumably has the same mass as the Sun) whose position is slightly shifting back and forth due to an unseen orbiting planet. It is determined that the star is moving in a circular orbit with a period of 14 days. What is the semimajor axis for the planet? Compare this value to Mercury's semimajor axis of 0.387 AU.

TEST YOURSELF

17. If you are riding a merry-go-round and experience a centripetal acceleration of 0.1g, and the merry-go-round starts spinning twice as fast, how big will your acceleration be?
    a. 0.05 g
    b. 0.1 g
    c. 0.2 g
    d. 0.3 g
    e. 0.4 g
18. Suppose astronomers discover a dim star orbiting a distant bright star. What properties do they need to measure in order to find the mass of the bright star?
    a. The mass of the dim star
    b. How far the dim star orbits from the bright star
    c. The orbital period of the dim star
    d. All of the above.
19. If you see two bodies in orbit around one another, separated by 1 AU and taking 1 yr to complete an orbit, which of the following *cannot* be true?
    a. One of the bodies has a mass of about 1 $M_\odot$.
    b. The two bodies have equal masses.
    c. One of the bodies is a planet like the Earth.
    d. One of the bodies has a mass of about 2 $M_\odot$.
20. In the distant past, it was thought that the Moon orbited four times closer to the Earth. What would its speed have been in its orbit at this distance, compared to its currently measured speed?
    a. 4 times slower
    b. 4 times faster
    c. 2 times slower
    d. 2 times faster
    e. The same

Orbital and Escape Velocities

18.1 Circular Orbits
18.2 Escape Velocity
18.3 The Shapes of Orbits

Learning Objectives

Upon completing this Unit, you should be able to:
- Describe how orbital and escape velocities depend on distance and mass.
- Explain how an object can escape a planet if gravity extends to infinite distance.
- Calculate the orbital and escape velocities at a given distance from a massive body.
- Describe the possible shapes of an orbit and how they depend on speed.

What goes up does not always come down. Without friction to slow it, a satellite around a planet, or a planet around the Sun, can remain in orbit essentially forever. Newton's idea of a cannonball circling the Earth due to gravity's pull (Unit 16) is not far removed from the satellites of today. Cannonballs and satellites can travel along orbits with many shapes other than a circle. As Kepler's first law (Unit 12) indicates, orbits can also have elliptical shapes. In fact, if a cannonball or satellite is given a high enough speed, it will continue to travel outward forever.

We explore the mathematics of orbits in greater detail in this Unit. Basics of the law of gravity are examined in Unit 16, and of circular motion in Unit 17.

18.1 CIRCULAR ORBITS

Orbital velocity

V_{circ} = Velocity of a circular orbit
G = Newton's gravitational constant
M = Mass of body being orbited
R = Radius of orbit (from center of body being orbited)

Most spacecraft orbit the Earth only as far up as necessary to avoid friction from the Earth's atmosphere. The international space station, for example, orbits at about 360 km (220 miles) above the Earth's surface. Even at that height, drag from the very tenuous atmosphere there causes it to drop about 50 m closer to Earth each day. Periodic visits by the space shuttle bring not only food and supplies, but an opportunity to boost the station back to higher altitudes.

Compared to the Earth's radius of about 6400 km, the space station's, and most other satellites', orbits are less than 10% farther from the Earth's center than the Earth's own surface. If we calculate the balance between gravity and centripetal force (Unit 17), we find that an object in a circular orbit must have an **orbital velocity** of

$$V_{circ} = \sqrt{\frac{GM}{R}}$$

where M is the mass of the planet (or other object) being orbited and R is the orbital distance from the *center* of the planet. Without deriving this equation, we can see that it makes sense because the orbital speed is larger for a more massive planet, and it grows slower at larger radii, where gravity is weaker.

For an orbit just above the Earth's surface (Figure 18.1) the circular orbital speed is

$$V_{circ} = \sqrt{\frac{GM_\oplus}{R_\oplus}} = \sqrt{\frac{6.67 \times 10^{-11} \, \text{m}^3/\text{sec}^2 \cdot \text{kg} \times 5.97 \times 10^{24} \, \text{kg}}{6.37 \times 10^6 \, \text{m}}}$$
$$= \sqrt{6.25 \times 10^7 \, \text{m}^2/\text{sec}^2} = 7900 \, \text{m/sec}$$

or a little under 8 km/sec. Notice that for a satellite orbiting at larger radii, the circular velocity becomes slower.

129

FIGURE 18.1

A circular orbit is slower at larger distances from the Earth. A rocket has to fire its thrusters to speed up to travel out to the slower orbit. At twice the Earth's radius, the circular velocity is only 5.6 km/sec, but a rocket on an elliptical orbit traveling out to this radius is going even slower. Unless it fires its thrusters again, it will complete its elliptical orbit, "falling" back close to the Earth again. If it can match the circular orbit velocity at the new radius, it can remain in orbit there.

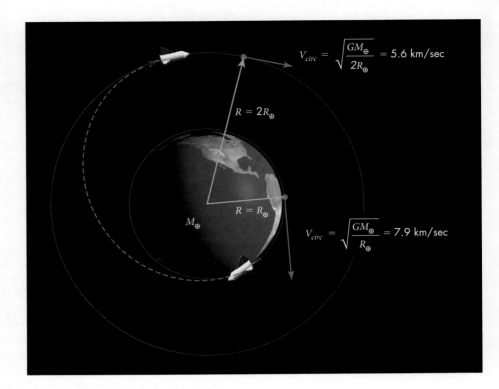

Concept Question 1

If you were commanding an orbiting space shuttle and wanted to reach a space station several thousand kilometers ahead of you, how would you maneuver to reach it?

The slower speed in higher orbits leads to an oddity—to "slow down" its orbit, a rocket must "speed up." For example, to travel out to the larger, slower orbit in Figure 18.1, the rocket must fire its thrusters to speed itself up. That puts it into an elliptical orbit that carries it out to the larger radius. When it reaches the larger orbital radius, it is traveling more slowly than the outer circular orbit speed. It will fall back along its elliptical orbit unless it again fires its rockets to speed up a second time and match the circular orbit speed at this radius.

18.2 ESCAPE VELOCITY

To overcome a planet's gravitational force and escape into space, a rocket must achieve a critical speed known as the **escape velocity.** Escape velocity is the speed at which an object needs to move away from a body so as not to be drawn back by its gravitational attraction. We can understand how such a speed might exist if we think about throwing an object into the air. The faster the object is tossed upward, the higher it goes and the longer it takes to fall back. Escape velocity is the speed an object needs so that it will never fall back, as depicted in Figure 18.2. Thus, escape velocity is of great importance in space travel if craft are to move away from one body and not be drawn back to it. However, escape velocity is also important for understanding many astronomical phenomena, such as whether a planet has an atmosphere and the nature of black holes.

The escape velocity can be phrased in terms of the energy needed to escape, as discussed further in Unit 20.

The escape velocity, V_{esc}, for a spherical body such as a planet can be found from the law of gravity and Newton's laws of motion. It is given by the following formula:

$$V_{esc} = \sqrt{\frac{2GM}{R}}.$$

V_{esc} = Velocity needed to escape from gravitational pull of a body
G = Newton's gravitational constant
M = Mass of body
R = Starting distance from center of body

Note how similar this formula is to the one for circular orbit speed. The escape velocity is only $\sqrt{2}$ (≈ 1.414) times larger than the circular orbit speed at any radius. Thus, if we know one, we can multiply or divide by $\sqrt{2}$ to find the other.

FIGURE 18.2
Escape velocity is the speed an object must have to overcome the gravitational force of a body such as a planet or star and not fall back. In general, the larger the mass of the planet or star and the smaller the distance from its center, the greater the escape velocity will be.

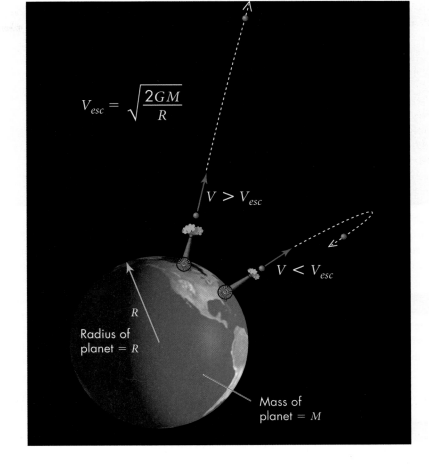

Escape velocity

Escape velocity

Mathematical Insights

Multiplying $\sqrt{2}$ times the speed of 7.9 km/sec calculated in Section 18.1 gives the escape velocity from Earth's surface: 11.2 km/sec.

Clarification Point

The escape velocity from a planet does *not* depend on the mass of the escaping object.

Notice in the equation for V_{esc} that if two planets of the same radius are compared, the one with the larger mass will have the larger escape velocity. On the other hand, if two planets of the same mass are compared, the one with the smaller radius will have the greater escape velocity.

Notice also that the escape velocity does *not* depend on the mass of the spacecraft, and neither does the orbital velocity. This is a common point of confusion. These speeds are independent of the mass of the spacecraft for essentially the same reason bodies of different masses fall at the same rate, as Galileo demonstrated. It does, however, require a greater force to accelerate a larger mass to the same speed, so more power is needed to launch a heavier spacecraft, as common sense would suggest.

To illustrate the use of the formula, we calculate the escape velocity from the Moon. The Moon's mass is 7.35×10^{22} kilograms. Its radius is 1.74×10^6 meters. We insert these values in the formula for escape velocity and find:

$$V_{esc}(\text{Moon}) = \sqrt{\frac{2 \times 6.67 \times 10^{-11} \text{ m}^3/\text{sec}^2 \cdot \text{kg} \times 7.35 \times 10^{22} \text{ kg}}{1.74 \times 10^6 \text{ m}}}$$
$$= 2.4 \times 10^3 \text{ m/sec}$$
$$= 2.4 \text{ km/sec}.$$

This low escape velocity compared to the Earth's 11.2 km/sec means it is much easier to blast a rocket off the Moon than off the Earth.

The low escape velocity from the Moon is also one of the primary reasons the Moon lacks an atmosphere (Figure 18.3). The average speed of molecules in a gas

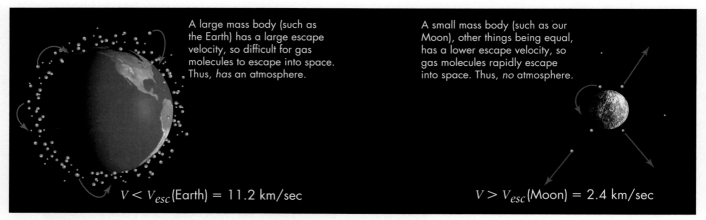

FIGURE 18.3
Atmospheres and escape velocities. Molecules in the Earth's atmosphere are represented by small blue dots (not to scale!). At Earth's temperature, almost no molecules have a speed great enough to escape Earth's gravity; however at the same speed, they will escape from the Moon. This is one of the factors that leads to the absence of an atmosphere on small planets and satellites.

> **Mathematical Insights**
>
> The average speed of molecules in a gas is proportional to the square root of the temperature.

> **Concept Question 2**
>
> Can a body have a low density but a high escape velocity?

at the Earth's and Moon's temperatures is typically about 0.5 km/sec. However, individual molecules may randomly travel faster, and the higher the temperature of a gas, the more molecules will be traveling above a given speed. It has been found that if the escape velocity is not at least 10 times larger than the average molecular speed of a gas, a planet does not generally retain the gas. By contrast, the Sun's escape velocity is 617 km/sec; thus it is able to hang on to the gas in its atmosphere even though it is heated to a high temperature.

The escape velocity formula applies not only to planets, but to any massive body. For example, a black hole (Unit 69) is a body, such as a collapsed star, whose radius is so small that the escape velocity exceeds the speed of light. We can even consider whether the entire universe is expanding at a speed greater than its "escape velocity" and will expand forever or collapse back on itself (Unit 82).

18.3 THE SHAPES OF ORBITS

Circular and escape velocities are just two possibilities out of an array of possible motions that bodies follow in a gravitational field. At speeds slower than the escape velocity, objects follow elliptical orbits, but at higher velocities the orbits are described by other geometric shapes.

The many possible trajectories of a spacecraft are illustrated in Figure 18.4. This figure illustrates what might happen to a spacecraft in circular orbit (white circle) around the Earth if it fired its engines to make its speed slower or faster. If the spacecraft fires thrusters to slow itself down, it will now follow an elliptical orbit with the center of the Earth at the farther focus of the ellipse (Unit 12.2). If it slows down too much, its elliptical orbit may cross the surface of the Earth, and the spacecraft will crash unless further maneuvers are made.

If the spacecraft fires its engines to speed up, it will also enter an elliptical orbit—up to a limit. This time the orbit follows an ellipse with the Earth at the nearer focus of the ellipse. The more the speed is increased, the greater the eccentricity of the ellipse (Unit 12.2). As the speed gets close to the escape velocity, the elliptical orbits will become more elongated, carrying the spacecraft far from the Earth before it "falls back" along its elliptical trajectory.

FIGURE 18.4
From a circular orbit around the Earth (shown in white), if a spacecraft fires its engines, it can end up in a variety of different trajectories. If it slows down, it will end up in an elliptical orbit (red dashed lines), possibly even colliding with Earth. The more it speeds up, the more stretched out its elliptical orbit will become (orange). If it speeds up to the escape velocity, it will end up on a "parabolic" trajectory (yellow) that never returns to Earth. Faster still and it follows a "hyperbolic" path (blue).

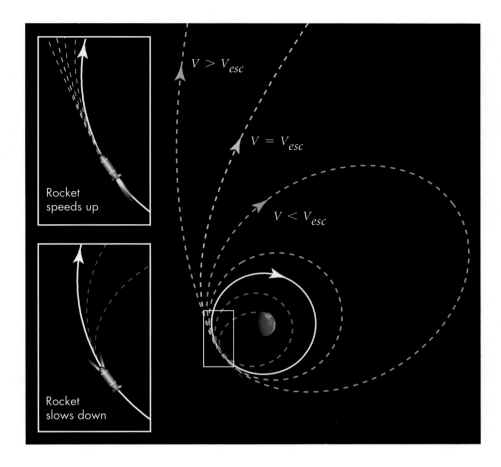

Concept Question 3

What would be the best way to orient the orbit of a launching platform around Earth for travel to other parts of the Solar System? Where would you want to be in the orbit to get the maximum boost from the Earth's own orbit?

At the escape velocity, the trajectory follows a mathematical curve called a **parabola.** At even higher speeds the trajectory will follow a flatter curve called a **hyperbola.** These bowl-shaped curves are illustrated in Figure 18.4 in yellow and blue. The spacecraft would never return once it began to travel on this trajectory unless it slowed itself down again relative to the Earth. Parabolic and hyperbolic trajectories are followed by some other objects in nature, such as some comets that enter the Solar System at such high speeds that they pass through and then travel outward never to be seen again.

Aerospace engineers have considered building a launching platform in high Earth orbit, similar to the idea illustrated in Figure 18.4. If spacecraft were to be launched from such a platform, the escape velocity would be smaller because the distance from the center of the Earth, R, would be greater. Moreover, because they would already be traveling at orbital velocity, they would need to be boosted only by the difference between escape and circular velocities at that distance.

Even though launching from an orbiting space platform would make the escape from Earth's pull easier, that is only half the challenge. As a spacecraft moves beyond Earth's immediate vicinity, the effects of the Sun's gravity become more apparent. Even at Earth's orbital distance, the Sun still exerts a substantial force. At 1 AU, the escape velocity from the Sun is about 42 km/sec. For probes to reach the outer parts of the Solar System therefore requires accelerating spacecraft to about 12 km/sec faster than Earth's orbital speed of 30 km/sec.

It is almost as difficult to send a space probe to Mercury, even though it is closer to the Sun. The spacecraft must slow down by about 8 km/sec relative to Earth's orbit in order to make its way so close to the Sun.

KEY POINTS

- A circular orbit around a planet (or other body) has a speed that depends on the square root of the ratio of the mass of the planet and the radius of the orbit.
- Objects can travel away from a planet forever if their speed exceeds the planet's escape velocity.
- The escape velocity is about 41% larger than the circular orbital velocity.
- A planet cannot hold on to an atmosphere unless the escape velocity is many times larger than the average speed of atmospheric molecules.
- The orbital (or escape) velocity depends on the mass of the body being orbited (or escaped from), not the mass of the orbiting object.
- If an object reaches or exceeds the escape velocity, it travels along a parabolic or hyperbolic path; if slower, a closed elliptical path.

KEY TERMS

escape velocity, 130   orbital velocity, 129
hyperbola, 133   parabola, 133

CONCEPT QUESTIONS

Concept Questions on the following topics are located in the margins. They invite thinking and discussion beyond the text.

1. Maneuvering a shuttle to the space station. (p. 130)
2. Density's relationship to escape velocity. (p. 132)
3. Best orbit for a space launch platform. (p. 133)

REVIEW QUESTIONS

4. How does the speed for a circular orbit depend on the distance from a planet? How does it depend on the mass of the planet?
5. What must be done to switch from a smaller circular orbit to a larger one? from a larger to a smaller one?
6. What is meant by *escape velocity*?
7. What is the relationship between escape velocity and circular velocity?
8. If you are orbiting at a fixed distance from an object, does the orbital or escape velocity depend on the radius of the object?
9. What shape trajectories do objects follow if they are traveling at the escape velocity or higher?

QUANTITATIVE PROBLEMS

10. Calculate the orbital speed for a satellite 1000 km above the Earth's surface, using the fact that $M_\oplus = 5.97 \times 10^{24}$ kg and $R_\oplus = 6.37 \times 10^6$ m.

11. How small would the Sun ($M_\odot = 1.99 \times 10^{30}$ kg) have to shrink to make its escape velocity equal to the speed of light ($c = 3.0 \times 10^8$ m/sec)?

12. On average the International Space Station is 360 km above the surface of Earth. By what percent is the escape velocity from Earth reduced by launching from the ISS rather than the surface of Earth?

13. The escape velocity of the Sun can be considered to be the escape velocity of the Solar System.
    a. Calculate the escape velocity of the Sun at its surface.
    b. Calculate the escape velocity of the Sun at the orbit of the Earth.

14. What is the escape velocity from the surface of an asteroid with a radius of 50 km and a mass of 1.0×10^{18} kg? (These are the approximate values for the asteroid Hekate.) If very good pitchers can throw a fastball with a speed of 162 kph (or 101 mph), could they throw the ball off the asteroid?

15. What is the escape velocity from a galaxy at a radius of 50,000 ly if the mass of the galaxy is 10^{12} times the Sun's mass?

16. Which body has a larger escape velocity, Mars or Saturn?

$M_{Mars} = 0.1\, M_\oplus$ $R_{Mars} = 0.5\, R_\oplus$
$M_{Saturn} = 95\, M_\oplus$ $R_{Saturn} = 9.4\, R_\oplus$

TEST YOURSELF

17. The Apollo missions required powerful rockets to liftoff from Earth but less powerful rockets to liftoff from the Moon because
    a. the Moon's radius is smaller.
    b. the Moon's mass is smaller.
    c. the Moon is far from Earth.
    d. the Moon has no atmosphere.
    e. All of the choices are correct.

18. Suppose the Sun suddenly collapsed in on itself, dropping to half its current radius. How big would the escape velocity from the surface be, compared to its current value?
    a. ½ as big
    b. 2 times bigger
    c. $\sqrt{2}$ times bigger
    d. 4 times bigger
    e. The same

19. The escape velocity from Earth for the very massive Saturn V rockets was _____ the escape velocity for the much less massive Atlas rockets.
    a. much greater than
    b. slightly greater than
    c. the same as
    d. slightly less than
    e. much less than

20. What do astronomers mean when they say a rocket has reached the escape velocity of Earth?
    a. The rocket has a high enough speed to travel away from the Earth forever without ever firing its rocket engine again.
    b. The rocket is going fast enough to reach another planet, but it would fall back to Earth otherwise.
    c. The rocket is moving faster than the Earth's orbital velocity around the Sun.
    d. The rocket has passed outside of the Earth's atmosphere, so it no longer feels the pull of Earth's gravity.

Tides

19.1 Cause of Tides
19.2 The Size of the Tidal Force
19.3 Solar Tides
19.4 Tidal Braking

Learning Objectives

Upon completing this Unit, you should be able to:
- Explain why gravitational forces raise tides on near and far sides of a body.
- Describe ocean tides on Earth, their cause, and timing with respect to the Moon.
- Explain how and why the strength of ocean tides changes with the Moon's phase.
- Explain how tides are causing the Earth's spin to slow and months to lengthen.

Anyone who has spent even a few hours by the ocean knows that the water's level rises and falls during the day. A blanket set on the sand 10 feet from the water's edge may be under water an hour later, or an anchored boat may be left high and dry. These regular changes in the height of the ocean are called the **tides**, which are caused primarily by the Moon.

In fact, tides occur everywhere bodies interact gravitationally: between planets and their satellites; between stars that orbit each other; and between neighboring galaxies. Tidal interactions are a consequence of the way gravitational forces vary with distance (Unit 16). In this Unit we focus on Earth–Moon tidal interaction and relate tides to the phase of the Moon (Unit 8), but a similar analysis might be applied to a pair of stars or a pair of galaxies.

19.1 CAUSE OF TIDES

Up till now we have treated the gravitational pull between interacting bodies as if the force was everywhere the same on each body, but in fact gravity pulls with different strengths and directions on every atom within each body. The force of gravity weakens with distance because of the $1/d^2$ dependence of gravity. Hence, the Moon's attraction is stronger than average on the side of the Earth nearer the Moon and weaker on the far side (Figure 19.1A). Having forces of different strength pulling on opposite sides of a body produces a stretching effect. Astronomers call this stretching effect a "differential gravitational force" or simply a **tidal force.**

The tidal force draws water in the oceans into **tidal bulges** on the side of the Earth facing the Moon as well as on the Earth's far side, where the solid Earth is, in effect, pulled away from the oceans as illustrated in Figure 19.1A. The same occurs when two galaxies interact gravitationally, as shown in Figure 19.1B. The galaxy on the right in this image pulls out long "tidal tails" from the galaxy on the left, both toward and away from it. For the Earth, the material being pulled into tides is water; for the galaxy, it is stars orbiting the galaxy. While the physical process is similar, in one case the tides occur in hours over thousands of kilometers, while in the other it is hundreds of millions of years and thousands of light-years.

The tidal bulge on the far side of the Earth (or a galaxy) may seem counterintuitive. Consider the following analogy: Imagine that you are being lifted up by hundreds of strings tied to all parts of you and your clothing. Suppose the strings tied to the hair on your head are being pulled a bit harder than the rest—your hair will end up being pulled upward until it stands on end. And suppose that the strings tied to

Origin of tides

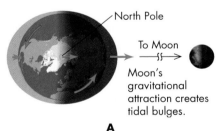

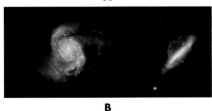

FIGURE 19.1

Tidal deformations. (A) The difference in the Moon's gravitational pull between the near and far sides of the Earth creates tidal bulges on the Earth. (B) Image from the Hubble Space Telescope of two galaxies. Tidal forces on the stars orbiting the galaxy on the left cause their orbits to be stretched toward and away from the companion galaxy on the right.

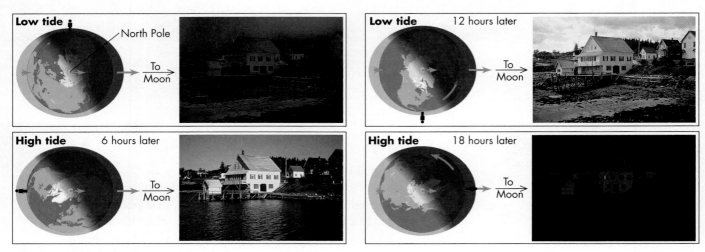

FIGURE 19.2
As the Earth rotates, it carries us (small black figure) through the tidal bulges. Because there are two bulges where the water is high and two regions where the water is low, we get two high tides and two low tides each day at most coastal locations. The images on the right side of each panel illustrate high and low tides on the coast of Maine. (Note that the size of the tidal bulge in the diagrams is greatly exaggerated for illustrative purposes.)

your feet and shoelaces are pulled a bit less strongly—your feet and shoelaces will dangle downward, away from the direction in which you are being pulled. This stretching effect is produced by the difference of forces across your body.

The oceans' tidal bulges are drawn toward the points facing and opposite the Moon, but the Earth spins. The Earth's rotation therefore carries us first into one bulge and then the next. As we enter the bulge, if we are on the ocean, the water level rises, and as we leave the bulge, the level falls. Because there are two bulges, we are carried into high water twice a day; these are the twice-daily high tides. Between the times of high water, as we move out of the bulge, the water level drops, making two low tides each day (Figure 19.2).

This simple picture becomes more complicated when the tidal bulge reaches shore. In most locations the tidal bulge has a depth of about 2 meters (6 feet), but it may reach 10 meters (30 feet) or more in some long narrow bays (as you can see in the photographs of high and low tides along the coast of Maine—Figure 19.2) and may even rush upriver as a *tidal bore*—a cresting wave that flows upstream. On some rivers, surfers ride the bore upstream on the rising tide. The rising tidewater in these regions is funneled into a narrower region, "piling" up the water to greater depth.

The motion of the Moon in its orbit makes the tidal bulge shift slightly from day to day. Thus, low tides come about 50 minutes later each day, the same delay as for moonrise (see Unit 7).

Clarification Point

Many people mistakenly think that the Moon's gravity makes a tidal bulge only on the side of Earth facing the Moon or that the second high tide is caused by the Sun. One way to understand that the Moon causes the far bulge is that the Moon's gravity also pulls the Earth away from water on the side opposite the Moon.

Concept Question 1

One kind of alternative energy source is tidal power. How could you extract energy out of the tides?

19.2 THE SIZE OF THE TIDAL FORCE

How big is the tidal force? It turns out to be quite small. Consider 1 kg of water at the average distance between the Earth's and Moon's centers: 3.84×10^8 m. Because the Moon's mass is 7.35×10^{22} kg, the gravitational force of the Moon on this kilogram of water will be:

$$F_M = G\frac{M \times m}{d^2} = 6.67 \times 10^{-11} \text{ newton} \cdot \text{m}^2/\text{kg}^2 \frac{7.35 \times 10^{22} \text{ kg} \times 1 \text{ kg}}{(3.84 \times 10^8 \text{ m})^2}$$

$$= 3.33 \times 10^{-5} \text{ newton}.$$

This is the average force felt by each kilogram of water on Earth, but the size (and direction) of the force varies, as illustrated in Figure 19.3.

A kilogram of water on the side of the Earth nearest the Moon is 6370 km (the Earth's radius, $R_\oplus$) closer to the Moon, so it feels a slightly larger force:

$$F_{M,\text{near}} = G \frac{M \times m}{(d - R_\oplus)^2}$$

$$= 6.67 \times 10^{-11} \text{ newton} \cdot \text{m}^2/\text{kg}^2 \frac{7.35 \times 10^{22} \text{ kg} \times 1 \text{ kg}}{(3.84 \times 10^8 \text{ m} - 6.37 \times 10^6 \text{ m})^2}$$

$$= 3.44 \times 10^{-5} \text{ newton}.$$

The *difference* of these two forces, 1.1×10^{-6} newton, is the tidal force. If you carry out the same calculation for water on the side of the Earth farthest from the Moon, you will get an almost identical result, except in the opposite direction. This is the force that the kilogram of water feels relative to the solid Earth, pulling it in a direction away from the surface. The same kilogram of water feels a force of 9.8 newtons "downward" due to the Earth's gravity. This is almost 10 million times stronger, so when the Moon is overhead (or on the far side of the Earth below your feet), you weigh about one ten-millionth less than when the Moon is near the horizon.

Such a tiny force is barely measurable, so why do the oceans change in height so dramatically? The answer is that water flows, and the tidal force can create forces that pull the water along the surface. For example, in Figure 19.3, water at point A feels a net force that causes it to flow toward the point directly beneath the Moon. On the far side of the Earth from the Moon, there are similar net tidal forces along the surface causing water to flow toward the point farthest from the Moon. Here again, the flow is not a strong current of water traveling thousands of kilometers, but mostly small shifts in the position of water in oceans all over the world creating the excess we see in the tidal bulges.

> **Mathematical Insights**
>
> To subtract the average force vector, we add a vector of the same size but the opposite direction. This vector's tail is placed at the head of the local force vector. The final vector is from the tail of the local vector to the head of the negative vector. See Unit 15.1 for more about vector math.

FIGURE 19.3
In the upper panel, arrows schematically show the Moon's gravitational force acting on a kilogram of water at several points on the Earth. The tidal force experienced on the Earth is the difference between the local force and the average force as measured at the center of the Earth at point C. To find the magnitude and direction of the tidal force, the force vector at point C must be subtracted from the individual force vectors as illustrated in the lower panel. (The Moon is shown much closer to Earth than it really is to make the angles clearer.)

> **Concept Question 2**
>
> The Moon does not generally lie directly above the Earth's equator, but may be as much as about 30° north or south of the equator. How can this cause the two high tides in a day to have different strengths?

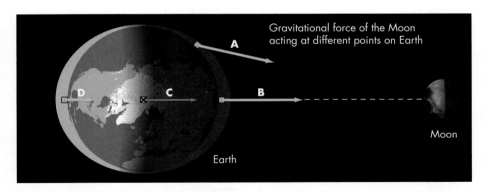

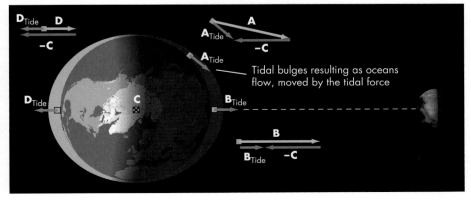

19.3 SOLAR TIDES

The Sun also creates tides on the Earth. Yet, despite the fact that the Sun is much more massive than the Moon and exerts a larger gravitational force on the Earth, it is also much farther away; so the *differential* force from one side of the Earth to the other is smaller. If you carry out a calculation like that in Section 19.2, you will find that the difference of the force on a kilogram of water on the side facing the Sun is about half what we found for the Moon.

The result is that the Sun's tidal effect on the Earth is only about one-half the Moon's, so the Moon's position always determine when tides will be high or low. Nevertheless, it is easy to see the Sun's effect when the two tides cooperate in **spring tides,** which are unusually large tides that occur at new and full moons. At these times the lunar and solar tidal forces work together, adding their tidal bulges, as illustrated in Figure 19.4A. Notice that spring tides have nothing to do with the seasons; rather they refer to the "springing up" of the water at new and full moons.

It may seem odd that spring tides occur at both new and full moons because the Moon and Sun pull together when the Moon is new but in opposite directions when it is full. However, the Sun and the Moon each create two tidal bulges, and the bulges combine regardless of whether the Sun and Moon are on the same or opposite sides of the Earth. On the other hand, at first and third quarters, the Sun's and the Moon's tidal forces work at cross-purposes, creating tidal bulges at right angles to one another, as shown in Figure 19.4B. The **neap tides** that result show the smallest change between high and low tides.

The strong dependence of the tidal force on distance means that when bodies are separated by very small distances, tides can become so large that they can literally pull a body apart. This may be how some planets ended up with rings, as we will consider in Unit 47.

> **Clarification Point**
>
> Spring tides occur at new moon and full moon and can occur in any season, not just the spring.

FIGURE 19.4
The Sun's gravity creates tides, too, though its effect is only about half that of the Moon. (A) The Sun and Moon each create tidal bulges on the Earth. When the Sun and Moon are in line (when the Moon is new or full), their tidal forces add together to make larger-than-average tides. (B) When the Sun and Moon are at 90° as seen from Earth (when the Moon is at first or third quarter), their tidal bulges are at right angles and partially nullify each other, creating smaller-than-average tidal changes. (Sizes and separations are not to scale.)

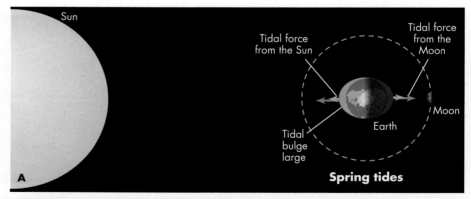

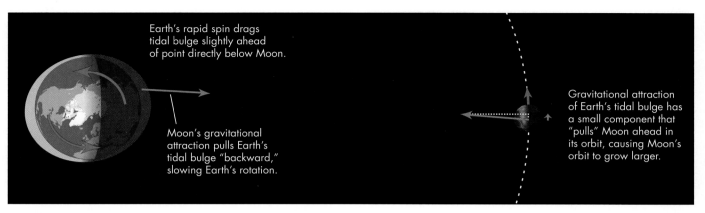

FIGURE 19.5
Tidal braking slows the Earth's rotation and speeds up the Moon's motion in its orbit. Friction between the oceans and Earth's solid crust gradually slows the Earth's rotation. At the same time, the friction "drags" the bulges of water "ahead" of the Earth–Moon line, producing a reaction force on the Moon that causes it to orbit farther from the Earth. (The actual separation of the Earth and Moon is much greater than depicted in this figure, and the shift of the tidal bulge is smaller.)

19.4 TIDAL BRAKING

Over a long period of time the effects of tides tend to slow down an object's rotation, a phenomenon known as **tidal braking.** Figure 19.5 shows how the Moon is currently tidally braking the Earth. As the Earth spins, friction between the ocean and the solid Earth drags the tidal bulge ahead of the imaginary line joining the Earth and Moon. The Moon's gravity pulls on the bulge, as shown by the arrow in the figure, and holds it back. The resulting drag is transmitted through the ocean to the Earth, slowing its rotation the way your hand placed on a spinning bicycle wheel slows the wheel. The result of this slowing of our planet's spin is that the day lengthens by about 0.002 seconds each century. At this rate, after about 100 billion years, the Earth's spin would slow to the point where it was locked into synchrony with the Moon's orbital period, although the Sun will not live long enough for this to happen.

Similar tidal effects have slowed the rotation of most of the moons in the Solar System to the point where one side of the moon always faces the planet. The Sun has probably caused tidal braking of Mercury and Venus, which experience stronger solar tides than the Earth because they are closer to the Sun. These effects are explored further in Part Three of this book. The effects of tidal braking may have far-reaching consequences. For example, some scientists speculate that life may be unlikely to form on a planet that always has the same face to its star.

The tidal braking that the Moon applies to the Earth leads to a reaction force on the Moon itself. Because the Earth's spin pulls the tidal bulge a little bit ahead of the Moon, the bulge exerts a pull that is a little off-center from the Moon's orbit. This produces a small "torque" on the Moon, as illustrated in Figure 19.5. Note that the tidal bulge on the far side of the Earth exerts a torque in the opposite direction, but it is smaller because of gravity's weakening strength at larger distances. This is not just a hypothetical result. Apollo astronauts placed special reflectors on the Moon, and by bouncing laser light from them and recording the round-trip travel time, we have determined that the Moon is moving away from the Earth at 4 cm (about 1.5 inches) per year.

Concept Question 3

Because of continental drift, all the continents were grouped into one large landmass several hundred million years ago. Would you expect more or less tidal braking back then?

KEY POINTS

- Because the gravitational force declines with distance, there are differences in the strength of the force on different parts of an astronomical body.
- After subtracting the average force on an astronomical body, the remaining component of force is known as a tidal force.
- The direction and size of tidal forces can be found using vector math, subtracting the average force from the forces at each point.
- The tidal force of the Moon on the Earth gives rise to tidal bulges of ocean water on the sides of the Earth facing and away from the Moon.
- As Earth rotates through the tidal bulges we experience two high tides and two low tides each day.
- The Sun's tidal force on the Earth is smaller than the Moon's because the tidal force depends very strongly on the distance.
- The tides raised by the Moon on the Earth are gradually slowing Earth's spin and making the Moon's orbit slowly grow larger.

KEY TERMS

neap tide, 138
spring tide, 138
tidal braking, 139
tidal bulge, 135
tidal force, 135
tides, 135

CONCEPT QUESTIONS

Concept Questions on the following topics are located in the margins. They invite thinking and discussion beyond the text.

1. Extracting energy from tides. (p. 136)
2. The Moon's "latitude" and the strength of tides. (p. 137)
3. Continental drift and tidal braking. (p. 139)

REVIEW QUESTIONS

4. What is a tidal force?
5. Why is there a tidal bulge on the side of the Earth opposite the Moon?
6. If the same side of the Earth faced the Moon at all times, how would this affect the tides?
7. If there were no coastal effects, would you expect the tides to be larger near the poles or near the equator?
8. Why are the tides caused by the Moon larger in effect than the tides caused by the Sun?
9. What are spring and neap tides?

QUANTITATIVE PROBLEMS

10. At what times would you expect high and low tides when the Moon is full? Draw a diagram to support your result.

11. Estimate the tidal force between your feet and your head when you are standing on Earth. Assume that both your feet and head have a mass of approximately 5 kg and that you are 2 m tall.
12. Carry out the same calculation as in Section 19.2 for the tidal force on the far side of the Earth. Show that this tidal force has approximately the same strength as the tidal force on the near side but points in the opposite direction.
13. Copy Figure 19.3 and add a point E on the side of Earth opposite point A. Draw a vector of appropriate length and direction to represent the force at this point, and show that after subtracting off the average force, the tidal force is in the direction toward the bulge on the side of Earth opposite the Moon.
14. If the Moon were massive enough, it could pull a kilogram of water off of the Earth's surface with its tidal forces. How massive would it have to be for this to happen?
15. Demonstrate that the Sun's tidal effects are approximately half that of the Moon's by finding the tidal force on 1 kg of water located on Earth. The Sun's mass is 1.99×10^{30} kg, and the distance between the Earth and Sun is 1.50×10^{11} m.
16. Calculate the tidal force of Mars on 1 kg of water on the Earth. The mass of Mars is 6.42×10^{23} kg, and the minimum distance between the Earth and Mars is 7.78×10^{10} m.

TEST YOURSELF

17. As a result of the Moon's gravitational pull, when would you weigh less?
    a. Whenever it is high tide locally
    b. Whenever it is low tide locally
    c. Only when the Moon is overhead
    d. Only if you were near one of the Earth's poles
    e. Weight never changes as a result of the Moon's gravity.
18. Low tide during the new moon occurs at about
    a. midnight.
    b. 6 P.M.
    c. midnight and noon.
    d. noon.
    e. 6 A.M. and 6 P.M.
19. What phase(s) is the Moon in when there are high tides at 6:00 P.M.?
    a. First quarter
    b. First or third quarter
    c. New moon
    d. New or full moon
    e. There is no relationship between the lunar phase and ocean tides.
20. The length of a day is increasing because
    a. Earth's tidal bulge is not aligned with the Moon.
    b. Earth is slowly moving away from the Sun.
    c. the Moon's gravitational pull is weakening.
    d. the Moon's tidal effects are greater than the Sun's.
21. Which of the following is caused by the Moon's tidal force?
    a. The rising and sinking of the ocean
    b. The slowing of the Earth's spin
    c. The gradual increase in the size of the Moon's orbit
    d. All of the above.

Conservation Laws

20.1 Conservation of Energy

20.2 Conservation of Mass (Almost)

20.3 Conservation of Angular Momentum

Learning Objectives

Upon completing this Unit, you should be able to:
- State the law of conservation of energy and explain how it is used.
- Give the mathematical formulas for kinetic and potential energies, and carry out a calculation giving an example where one is converted to the other.
- Define angular momentum, and illustrate how its conservation can be used to explain changes in rotation speed.

Newton's laws of motion and gravity (Units 14–16) are but one set of laws important for understanding the universe. Another set of laws, the conservation laws, are also very powerful. Some physical properties of matter or a system remain the same—are conserved—in almost all circumstances. For example, in cooking (and other chemical processes) the numbers of each type of atom are conserved; the atoms may combine with each other in different kinds of molecules, but no atoms are created or destroyed. Knowing this allows us to make a number of predictions about what is possible when we cook a certain combination of ingredients. It may not teach us how to bake a cake, but it can help us predict the consequences of leaving out certain ingredients!

Conservation laws identify fundamental properties that do not change under almost any condition. We will discuss three such laws here: conservation of energy, conservation of mass, and conservation of angular momentum. The power of these conservation laws is that, if we can measure the energy, the mass, or the angular momentum of a system at one time, we know that all subsequent changes in the system can occur only in ways that conserve all three quantities. With these laws we can predict how fast a collapsing star will spin or the explosive energy an asteroid will have if it strikes the Earth.

20.1 CONSERVATION OF ENERGY

Energy is a familiar term from everyday conversation, but that usage is not always the same as is meant in the sciences. **Energy** can come in many forms, but in general it can be described as the ability to generate motion. Thus, a moving ball may be able to hit a stationary ball and set it in motion. The moving ball therefore has energy. Energy can also be present where there is no motion. When you pick up a ball off the ground and hold it at rest in your hand, the ball is stationary, but if you let go of the ball, it will be pulled downward by gravity and set into motion. Thus, picking it up gave it a different form of energy that would allow it to generate motion.

Energy is neither created nor destroyed—it just changes forms. This can be described by the law of conservation of energy, which states the following:

The energy in a closed system may change form, but the total amount of energy does not change as a result of any process.

The idea of a *closed system* is important: It means that energy is not exchanged with anything outside the system. For example, the Earth is not a closed system, because it

Although energy is conserved, not all forms of energy are useful to us. For example, a machine may produce heat energy that is waste for the intended purpose.

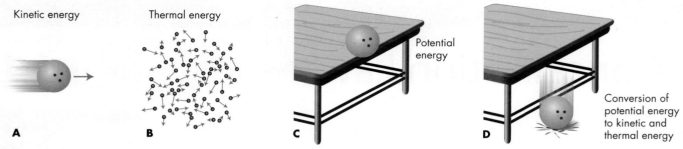

FIGURE 20.1
Energy can be found in many forms. (A) A moving body has kinetic energy. (B) The energy we associate with heat is produced by the motions of particles within the substance. (C) Energy can also be found in a "potential" form, such as a bowling ball at the edge of a table. (D) If the bowling ball falls off the edge of the table, it produces kinetic energy as it falls to the floor.

E_K = Kinetic energy
m = Mass of moving object
V = Speed of moving object

E_G = Gravitational potential energy
m, M = Masses of objects
d = Distances between objects' centers
G = Newton's gravitational constant

receives energy from the Sun and radiates heat energy into space. Very few systems remain isolated at all times, but often we can examine an individual process—energy conversion during a roller coaster ride, for example—as if it occurred in isolation.

To appreciate the power of this law, we need to look more closely at how we measure energy as well as some of the forms it may take. Energy is interesting in part because it can have so many forms. The most obvious form is the energy of an object that is already in motion, or **kinetic energy.** A simple example of kinetic energy is the energy of a moving car or a thrown ball (Figure 20.1A). Physicists measure energy in **joules,** the unit of energy in the MKS system (see Unit 3), named after the nineteenth-century British physicist James Joule, who helped verify the law of conservation of energy. The kinetic energy of a body of mass m moving with a speed V can be expressed mathematically as

$$E_K = \tfrac{1}{2} m \times V^2.$$

If m is measured in kilograms and V in meters per second, the calculation will give the energy in joules.

Another form of energy is heat, or **thermal energy.** Heat is also energy of motion, but at the atomic or molecular level. That is, the hotter something is, the more rapidly its atoms and molecules move within it. Sometimes that motion is a vibration. At other times it is a random careening of particles (Figure 20.1B).

Yet another common form of energy is **potential energy.** Potential energy is energy that has not been liberated but is available. For example, water stored behind a dam, and a bowling ball poised on the edge of a table, both have potential energy (Figure 20.1C). If the dam bursts or the ball rolls off the table, the potential energy can be released and converted into motion. These examples of potential energy result from the gravitational attraction between the Earth and the water or the bowling ball. To lift an object up against Earth's gravity takes energy, and that energy is stored until the object falls back to the Earth. The **gravitational potential energy** can be written mathematically as

$$E_G = -G \frac{m \times M}{d}$$

where G, m, M, and d are all defined as in the law of gravity (Unit 16). Again, if all the quantities are measured in MKS units, the resulting energy will be in joules. The form of this equation is what you might expect—potential energy has a bigger magnitude for larger masses, and it grows weaker with distance as gravity grows weaker. A curious aspect of this equation, though, is the minus sign. We discuss this further a little later.

The usefulness of the law of conservation of energy is that when you add together all the forms of energy at the beginning of a process, their sum must equal the total amount of energy in all its forms at the end. As an example, let's consider the bowling ball on the table edge. If it is just sitting there, it initially has only potential energy. If it falls from the table, that potential energy is converted to energy of motion as the ball falls. Upon hitting the ground it may look as if the energy has disappeared: The ball is at rest again so that it no longer has energy of motion. Likewise, it has given up its potential energy. However, its energy has

Concept Question 1

How will the temperature of a gas cloud change as gravity causes it to collapse?

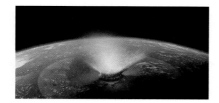

FIGURE 20.2
An asteroid striking the Earth converts the kinetic energy of its orbital motion around the Sun into many different forms, such as heat, light, waves, motion of the ground, and breaking of molecular bonds.

Concept Question 2

Considering your own body and all of the things you have used, what forms of energy have you used today?

FIGURE 20.3
The cars of a roller coaster exchange kinetic and potential energy as they coast up and down the track. Potential energy is given to the cars at the start of the ride by pulling them up to the highest point on the track.

not disappeared. On impact, the ball shakes the ground and gives its energy of motion to the atoms in the ground. Their motion produces heat and sound (vibrations of atoms in the air), so that the ball's energy has changed form yet again (Figure 20.1D). If you pick up the bowling ball and set it back on the table, in lifting the ball you will expend the same amount of **chemical energy**—potential energy in the electrical bonds of molecules such as fat and sugar stored in your body—as the potential energy that the bowling ball gains.

An astronomical example that illustrates the conservation of energy and its conversion to different forms is an asteroid impacting the Earth (Figure 20.2). As it approaches the Earth, the asteroid has energy of motion. During its passage through the atmosphere, a small amount of the energy of motion is converted to heat and light as it tears through the atmosphere. The asteroid's kinetic energy drives the asteroid deep into the solid crust of the Earth; as it slows, it transfers its kinetic energy by compressing and heating the rock at the point of impact. This violent compression is transferred to surrounding rock, expanding outward in all directions. In a fraction of a second, so much heat is generated that the asteroid and surrounding rock melt or vaporize. Some rock is blasted up and out, giving it kinetic energy, and seismic waves (Unit 37) travel through the Earth, shaking the ground thousands of kilometers away. Most of the asteroid's kinetic energy ends up as thermal energy, which is eventually radiated away into space (Unit 23). Some of the energy remains as potential energy—stored in chemical bonds created by reactions in the heat of the blast or stored as gravitational potential energy in rock pushed upward in the region surrounding the blast crater. Thus, the energy changes form, but the total energy at the beginning equals the total energy at the end.

Galileo's experiments showing that a ball slows down as it goes up a ramp or speeds up as it goes down (Unit 14) can also be understood in terms of conservation of energy. As the ball rolls up a ramp, its kinetic energy decreases as its gravitational potential energy increases. This is much like a roller coaster (Figure 20.3). Energy is expended pulling the cars to the top of the track, then they coast, trading potential and kinetic energy back and forth as they move up and down.

Extending this idea, we can phrase the question of how fast a rocket must travel to leave the Earth (its *escape velocity*, Unit 18) in terms of the kinetic energy it must have to completely overcome the Earth's gravitational potential energy. The rocket slows, and as it reaches an infinite distance where the gravitational pull approaches zero, its kinetic energy also drops to zero. If a rocket has enough (positive) kinetic energy to overcome the (negative) gravitational potential energy, it will escape from the Earth's surface. In fact, we can recover the formula for escape velocity by setting $E_K + E_G = 0$.

Mathematical Insights

Assigning a negative value to gravitational potential energy may seem strange, but the alternatives are perhaps stranger. If we assigned "zero energy" to when two objects were put close together, we would have to say that the potential energy was very large when the objects were barely feeling each other's gravity—and the gravitational potential energy would still become negative if the objects were put even closer together.

This helps explain why gravitational potential energy (or potential energy based on any attractive force) has a negative value. When two objects are very far apart, the force of gravity becomes negligible, so it is natural to describe the gravitational potential energy as near zero. If the two objects fall together, their kinetic energy increases—subtracting energy from the gravitational potential energy. If we start with nearly zero joules of gravitational potential energy and subtract thousands of joules, E_G must become more negative as gravity pulls objects closer together.

Curiously, on astronomical scales many objects in the universe, including perhaps the entire universe itself, exhibit a near balance between the gravitational potential and kinetic energies. Both kinds of energy may be individually large, but their sum is usually close to zero.

20.2 CONSERVATION OF MASS (ALMOST)

Even before the discovery of the conservation of energy, scientists had proposed another fundamental conservation law: the conservation of mass. This was thought to be precisely true, until a very interesting exception was discovered by Albert Einstein in the 1900s. In fact, it is so close to being true in everyday circumstances that for almost all practical purposes it can be treated as a conservation law.

In the 1700s, chemical conversions and interactions that occurred between substances were not fully understood. However, a number of scientists had noted that the total mass of the material at the end of any chemical experiment was the same as at the beginning. The more carefully all of the products of the experiment were collected and measured, the more precisely this rule was found to be obeyed.

By the early 1800s the English chemist and physicist James Dalton realized that this behavior of matter could be explained if matter was made up of atoms—submicroscopic particles that had specific unchanging properties and masses (Unit 4). The different kinds of atoms are listed in the Periodic Table of the Elements, shown in the back inside cover of this book. These atoms can be arranged in different combinations or molecules, but they cannot be destroyed or altered. Even though the matter might appear to change form drastically, in fact the same number of each type of atom is present at the end of an experiment as at the beginning. As a result, the mass is also conserved.

The notion of the conservation of matter began to be altered in the 1900s. It was discovered that atoms themselves are made of smaller subatomic particles, such as electrons, protons, and neutrons (Unit 4). In some circumstances those particles can be pulled apart—allowing one type of atom to be converted into another. The interactions that cause these kinds of conversions generally require extremely high energies and are quite rare at Earthly temperatures. However, they are not so rare at the temperatures found inside stars.

At about the same time Albert Einstein showed that mass and energy can be converted into one another. For example, the subatomic particles that make up atoms of helium have a lower mass when combined to form atoms of carbon. The loss in mass is balanced by a release of energy. The conversion of atoms from one type to another proves to be the source of energy of the Sun and other stars (Unit 52).

One way of understanding this relationship between energy and mass is to think of mass as a kind of extremely compact form of potential energy. In fact, Einstein showed that mass and energy need to be treated similarly. He showed, for example, that gravity depends not just on the mass of particles but on their energy as well. We tend to think of mass and energy as being very different quantities, partly because the energies we deal with in everyday life are minuscule compared to the potential energy bundled up in matter. To be exact, physicists today refer to conservation of "mass-energy." In most circumstances, however, mass and energy are separately conserved to a very high degree of precision.

Concept Question 3

What kind of energy is nuclear energy? What is its source?

20.3 CONSERVATION OF ANGULAR MOMENTUM

You have perhaps seen water going down a drain and noticed that as the water moves inward toward the drain hole, it spirals faster and faster. Likewise, ice skaters who go into a spin with outstretched arms and then fold their arms inward spin even faster. These are both examples of the conservation of angular momentum.

Angular momentum is a measure of the tendency of a spinning object to keep spinning with its rotation axis pointing in the same direction unless acted on by an external "rotational force," called a **torque,** that works to alter the rotation. The distinction between a force and a torque is that some forces act in directions that do not affect rotation, while the strength of a torque is measured by its ability to increase or decrease rotation. If a mass m is moving around a rotation axis at a distance R with a velocity V, then mathematically its angular momentum is $m \times V \times R$. The law of conservation of angular momentum states this:

If no external force acts to change the spin rate, then $m \times V \times R$ remains constant.

How does this explain the faster spin of the ice skater whose arms are pulled in? If $m \times V \times R$ is constant, and if the skater makes R (the length of the extended arms) shorter, then V must get larger (Figure 20.4). Notice that when skaters are pulling their own arms in, no *external* forces are involved. Likewise, as the water flowing toward a drain circles nearer the drain hole, its R decreases; therefore its V of rotation must increase.

We have already encountered an astronomical example of this conservation law in Kepler's second law of planetary motion. As a planet moves closer to the Sun, its orbital velocity must increase to keep $m \times V \times R$ constant. Because nearly all objects have some intrinsic amount of rotation, if any process causes them to change size, they will change their spin rate in response.

Another example of the conservation of angular momentum is tidal braking (Unit 19.4). As the Earth's spin slows due to its interaction with ocean tides, its angular momentum is decreasing. However, the lost angular momentum is transferred to the Moon, whose angular momentum increases as its orbital radius grows larger. Using the law of conservation of angular momentum, we can deduce the rate at which the Moon must be moving away based on the rate of slowing of the Earth's spin, and this is confirmed by experiment.

---

m = Mass of rotating object
V = Speed of rotation around axis
R = Radius from axis of rotation

Concept Question 4

Suppose you were in a spinning spacecraft. How could you use a jet thruster to stop your spin? If you can slow your spin, in what sense is angular momentum conserved?

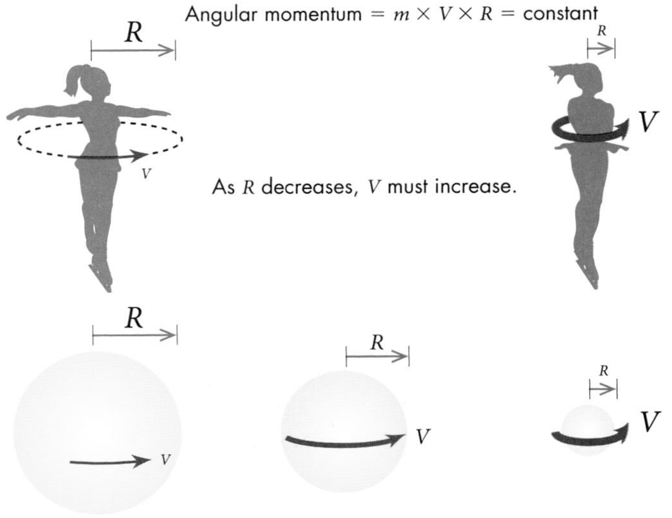

FIGURE 20.4
Conservation of angular momentum spins up a skater and a spinning star. Angular momentum is proportional to a body's mass, m, times its rotation speed, V, times its radius, R. Conservation of angular momentum requires that mathematically the product $m \times V \times R$ remains constant. Thus, if a given mass changes its R, its V must also change. In particular, if R decreases, V must increase, as the ice skater demonstrates by spinning faster as she draws her arms in. Similarly, a rotating star spins faster if its radius shrinks.

KEY POINTS

- Conservation laws describe physical properties that remain constant no matter what happens.
- Energy is conserved. It comes in many forms, but may be described as a property that quantifies the ability to create motion.
- Energy comes generally in two forms: kinetic (the energy of motion) and potential (a stored form of energy).
- Mass is very nearly conserved, but in subatomic interactions there can be important exceptions to this.
- Mass is essentially a kind of energy, and each can be converted into the other, so the combined mass and energy is conserved.
- Angular momentum is another conserved quantity, describing the amount of spin of an object or system, and depends on the product of mass, size, and rotation speed.
- When stars or other astronomical bodies contract or expand, their speed of rotation must change to conserve angular momentum.

KEY TERMS

angular momentum, 145
chemical energy, 143
conservation law, 141
energy, 141
gravitational potential energy, 142
joule, 142
kinetic energy, 142
potential energy, 142
thermal energy, 142
torque, 145

CONCEPT QUESTIONS

Concept Questions on the following topics are located in the margins. They invite thinking and discussion beyond the text.

1. Temperature of a collapsing cloud. (p. 143)
2. Kinds of energy you use. (p. 143)
3. Nuclear energy. (p. 144)
4. Changing angular momentum of a spacecraft. (p. 145)

REVIEW QUESTIONS

5. What is a conservation law?
6. An object slows to a halt as it rolls uphill. What happened to the kinetic energy?
7. What are the different forms of energy?
8. In what situations is mass conserved, and what happens to it when it is not conserved?
9. How does the conservation of angular momentum explain why a planet moves faster when it is closer to the Sun in its orbit?
10. What is a torque?

QUANTITATIVE PROBLEMS

11. Some professional baseball pitchers can throw a 0.145-kg baseball at 160 km/hr (about 100 mph). What is the kinetic energy of the ball? How much larger is the gravitational potential energy compared to kinetic energy?
12. What kinetic energy must a 10,000-kg rocket have to escape from the surface of the Earth?
13. Show that if the gravitational potential energy and kinetic energy add up to zero for an object on the Earth's surface, you can derive the escape velocity formula given in Unit 18.
14. Suppose that a comet with a mass of 10^9 kg strikes the Earth at a velocity of 40 km/sec. How much energy would be released by the impact? Convert this energy into the equivalent of "kilotons of TNT" (kt), where 1 kt $\approx 4.2 \times 10^{12}$ joules.
15. Astronomers estimate that about 100 metric tons (1×10^5 kg) of meteorites strike the Earth every day.
    a. If the meteorites increased the mass of the Earth by 1% without adding any angular momentum, how long would the day become?
    b. The Earth's mass is 5.97×10^{24} kg. How many years would it take to increase the Earth's mass by 1% at this rate?
16. Suppose that a star collapses inward until it is reduced to one-tenth of its original size while losing no mass. (Note: We are treating the star as if it rotates like a solid, uniform ball, for the purposes of this problem.)
    a. How many times faster will its surface be moving?
    b. If initially it was spinning around once every 100 hours, how long will it take to spin around after it collapses?

TEST YOURSELF

17. A satellite orbiting a planet in a circular path at a constant speed has what kind of energy?
    a. Just gravitational potential energy
    b. Just kinetic energy
    c. Both gravitational potential and kinetic energy
    d. Neither gravitational potential nor kinetic energy
18. Suppose a comet has an elliptical orbit that brings it twice as close to the Sun at its closest approach as when it is at its largest separation. If its speed is 10 km/sec at its largest separation, what will its speed be when it is closest?
    a. 5 km/sec   c. 20 km/sec
    b. 10 km/sec   d. 40 km/sec
19. How many times more chemical energy must a bicyclist expend to reach a speed of 40 kph compared to another bicyclist of the same mass to reach a speed of 20 kph?
    a. 40   c. 10   e. 2
    b. 20   d. 4
20. Several asteroids are heading toward a collision with Earth. Based on its mass and expected collision speed, which of the following is the most dangerous?
    a. Mass of 1 ton; speed 1000 km/sec
    b. Mass of 10 tons; speed of 100 km/sec
    c. Mass of 100 tons; speed of 10 km/sec
    d. Mass of 1000 tons; speed of 1 km/sec

The Dual Nature of Light and Matter

21.1 The Nature of Light
21.2 The Effect of Distance on Light
21.3 The Nature of Matter
21.4 The Interaction of Light and Matter

Learning Objectives

Upon completing this Unit, you should be able to:
- Describe how light sometimes behaves like waves and sometimes like particles.
- Recall the speed of light and the meaning of light being an electromagnetic wave.
- Define and explain the inverse square law for light.
- Describe how light interacts with matter and how light is used to identify elements.

Our home planet is so far from other astronomical bodies that, with few exceptions, we cannot bring samples from them to study in our laboratories. And even for objects within the range of our spacecraft, physical conditions may make it impossible to send probes to make direct measurements. For example, if we want to know how hot the Sun is, we cannot stick a thermometer into it. Nor can we directly sample the atmosphere of a distant star. However, we can sample such remote bodies indirectly by analyzing their light.

At first this might appear to be an enormous limitation—allowing us to say something about only the direction to an object and the quantity of light it produces. However, if we can understand the energetics of light and the way it interacts with matter, we can also learn what the body is made of and its temperature. Light, therefore, is our key to studying the universe; but to use this key, we need to understand light's unusual nature. We introduced some basic ideas about light and matter in Unit 4, but in this Unit we will explore how they both behave in some ways like waves and in some ways like particles. In the following Units, we will delve into aspects of light that allow astronomers to explore various aspects of remote objects.

21.1 THE NATURE OF LIGHT

Light is radiant energy; that is, it is energy that can travel through space from one point to another without the need for a direct physical link. Light is, therefore, very different from, for example, sound. Sound can reach us only if it is carried by a medium such as air or water. Light can reach us even across empty space. In empty space we can see the burst of light of an explosion, but we will hear no sound from it at all.

Light's capacity to travel through the vacuum of space is paralleled by another special property: its high speed. In fact, the speed of light is an upper limit to all motion (Unit 26). In empty space, light travels at the incredible speed of 299,792,458 meters per second—a speed that would take an object seven times around the Earth in 1 second! The speed of light in empty space is a universal constant and is denoted by c, and although it is very precisely measured, for most purposes it can be rounded off to 300,000 km/sec.

Observation and experimentation on light throughout the last few centuries produced two different models of what light is and how it works. According to one

---

Light moves slower when it travels through matter. The speed at which light travels through glass is about 60% of its speed in a vacuum; through water, about 75%; through air, about 99.97%.

Clarification Point

Water waves and sound waves require a medium, water or air, to propagate. Light waves can travel through a total vacuum.

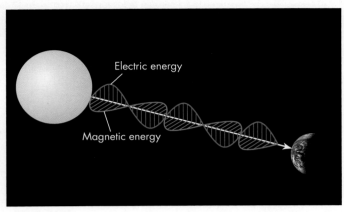

FIGURE 21.1
A wave of electromagnetic energy moves through empty space at the speed of light, 299,792.5 kilometers per second. The wave carries itself along by continually changing its electric energy into magnetic energy, and vice versa.

The joule (abbreviation J) is the MKS unit of energy (Unit 3).

Photons from a light source

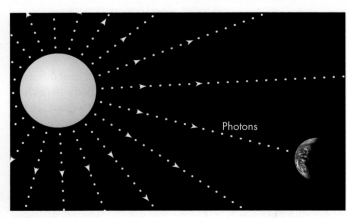

FIGURE 21.2
Photons—particles of electromagnetic energy—stream away from a light source at the speed of light.

model, light is a wave, somewhat like sound or water waves, but made up of a mix of electric and magnetic energy. The ability of such a wave to travel through empty space comes from the interrelatedness of electricity and magnetism.

You can see this relationship between electricity and magnetism in everyday life. For example, when you start your car, turning the ignition key sends an electric current from the battery to the starter. There the current generates a magnetic force that rotates the engine. Similarly, when you pull the cord on a lawn mower, you spin a magnet, generating an electric current that creates the spark to ignite the gas in the engine.

Light can travel through empty space because a small disturbance of an electric field creates a magnetic disturbance in the adjacent space, which in turn creates a new electric disturbance in the space adjacent to it, and so on, as illustrated in Figure 21.1. In this fashion, light can move wavelike through empty space, "carrying itself by its own bootstraps." Thus, we say that light is an **electromagnetic wave.** When such a wave encounters the electrons in matter, it may interact with them, a little bit like the way a water wave makes a boat rock. It is from such tiny disturbances in our eyes or in an antenna that we can detect electromagnetic waves.

Many of the early experiments on light demonstrated this distinctly wavelike behavior. For example, sources of light can **interfere** with each other like the interacting waves in an ocean, reinforcing and canceling each other's effects in complex patterns. And like ocean waves passing through a narrow inlet into a bay, light passing through an opening does not travel in a straight line, but spreads out. This is an important consideration in the design of telescopes, because it limits the sharpness of images a telescope can produce (Unit 31).

The wave model of light works well to explain many phenomena—for example, the length between waves determines the color of light (Unit 22). The wave model fails, though, to explain other properties of light. If light were like a water wave, for example, we could deliver any amount of electromagnetic energy by varying the intensity of the electromagnetic wave. However, we find something quite different for light. For example, the red light we see coming from a stoplight is made up of little parcels containing 3×10^{-19} joules of energy. We cannot get, for instance, half this much energy or 1.7 times this much energy from red light, just whole multiples of this much energy. We can vary the energy in a water wave by making the waves taller or shorter, but light behaves as if, in the analogy to water waves, we can make waves in a bathtub that are 3, 6, or 9 centimeters tall, but no size in between. To describe this property, we say that the energy of light is **quantized;** that is, it comes in discrete packets.

Based on this property, light is better described as being made of particles. These particles are called **photons** (Figure 21.2). You can have 1 or 2 or 3 photons, but not a half a photon or 1.7 photons. In this model, the photons are packets of energy. When they enter your eye, they produce the sensation of light by striking your retina like microscopic bullets, releasing their energy and causing a chemical change in your photoreceptor cells (Unit 33). In empty space, photons move in a straight line at the speed of light.

For centuries, scientists argued over whether light consisted of waves or particles, pointing out experiments in which the opposing model could not give an adequate explanation. By the early 20th century, however, it became clear that neither

Concept Question 1

If you start a water wave from one side of a bathtub (or a sink or pool), do any water molecules travel from one side of the tub to the other? How can you test your idea?

model completely describes light. At the submicroscopic scales in which it operates, light has a more complex dual nature that we can describe only partially with our wave and particle models. So today scientists describe light as having a **wave–particle duality.** What light and other particles do at submicroscopic levels is both wavelike and particle-like and neither. It can be described very precisely using mathematics, but it is difficult to describe based on what is familiar to us. We should not think of this wave–particle duality as peculiar, however, because it is the reality that underlies everything we experience.

In practice, scientists use whichever model—wave or particle—best describes or explains a particular phenomenon. For example, the way light reflects off a mirror is easily understood if you imagine photons striking the mirror and bouncing back just the way a ball rebounds when thrown at a wall. On the other hand, the focusing of light by a lens is better explained by the wave model.

21.2 THE EFFECT OF DISTANCE ON LIGHT

An individual photon traveling through space maintains its speed and energy; however, we also know that objects appear dimmer when they are farther away. This dimming is the result of the light's spreading out over a larger area as the photons move away from the source of light—much as the height of a wave produced by a rock tossed into a pond grows shorter as the wave ring expands.

This spreading out of photons explains why the planets in the inner Solar System are warmer than the outer planets (Unit 34), even though they receive photons that have the same energy. For example, photons traveling away from the Sun become spread farther apart, as illustrated in Figure 21.3. Suppose we traced the path of 36 photons that were spread out over a spherical surface area of 1 AU^2 at Earth's distance. At 2 AU they would be spread out over a surface area of 4 AU^2, at 3 AU they would be spread out over a surface area of 9 AU^2, and so on. The number of photons reaching each square AU of surface area is decreasing as a function of the distance *squared*.

We can make this more mathematically precise by noting that at the distance of the Earth from the Sun, the photons are spread over the area of a sphere that has a radius of 1 astronomical unit (the Earth–Sun distance). The surface area of this sphere, given by the geometric formula $4\pi \times (1 \text{ AU})^2$, is about 12 square AU. At Jupiter's distance from the Sun—about 5 AU—the light is spread out over a sphere with a surface area of $4\pi \times (5 \text{ AU})^2$, which is an area 25 times larger than at the Earth's distance. Thus, each square meter of the Earth's surface receives 25 times more light than each square meter of Jupiter's surface. As a result, the Sun looks 25 times dimmer at Jupiter's distance than at the Earth's distance.

Astronomers describe this change in the amount of light as a change in its **brightness.** At a distance d from the source, the light is spread out over an area of $4\pi d^2$, so we can write this relationship in mathematical form as

$$\text{brightness} = \frac{\text{total light output}}{4\pi d^2}.$$

This relationship is known as the **inverse-square law** because the light grows dimmer in proportion to the square of the distance. The inverse-square law is important for determining how much energy is received at different distances from a source of light. The relationship also allows us to calculate the distances to many stars (Unit 55).

Inverse-square law

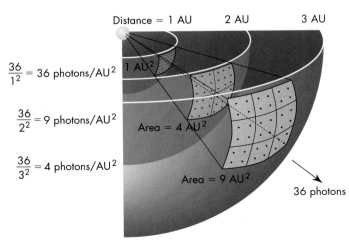

FIGURE 21.3
Photons become spread over successively larger areas as the distance from the source of light increases. The number of photons in the same area decreases as a function of the inverse square of the distance.

21.3 THE NATURE OF MATTER

Just as scientists have puzzled over the nature of light, they have also puzzled over the nature of matter. Like so many of our ideas about the nature of the universe, our ideas about matter date back to the ancient Greeks. For example, Leucippus and his student Democritus, who lived around the fifth century B.C.E. in Greece, taught that matter is composed of tiny indivisible particles. They called these particles *atoms,* which means "uncuttable" in Greek. They argued, for example, that the water in a tub could be subdivided into smaller and smaller pieces—droplets—only down to some finite size for the smallest particle of water.

Our current model for the nature of atoms dates back to the early 1900s and the work of the New Zealand physicist Ernest Rutherford. He showed that an atom can be thought of as a dense core called a *nucleus* around which smaller particles called *electrons* orbit (Figure 21.4). The force that holds the electrons in orbit is the electrical attraction between the positively charged protons in the nucleus of the atom and the negative electric charge of the electrons. The orbits of electrons are extremely small. For example, the diameter of the smallest electron orbit in a hydrogen atom is only about 10^{-10} (1 ten-billionth) meter. The nucleus of the atom, where the protons and neutrons are packed together, is less than one ten-thousandth as big as the electron's orbit.

Atoms come in about a hundred different types of what are called chemical **elements.** The main characteristics of a chemical element are determined by the number of protons in its nucleus. For example, hydrogen atoms contain 1 proton; helium atoms contain 2 protons; carbon, 6; oxygen, 8; and iron, 26. A complete list of the elements is given in the Periodic Table of the Elements shown on the inside back cover of the book (see also Unit 4). The **atomic number** of each element indicates the number of protons in its nucleus.

Concept Question 2

Based on the forces present in an atomic nucleus, why do you suppose there are no stable elements with more than about 92 protons in the nucleus?

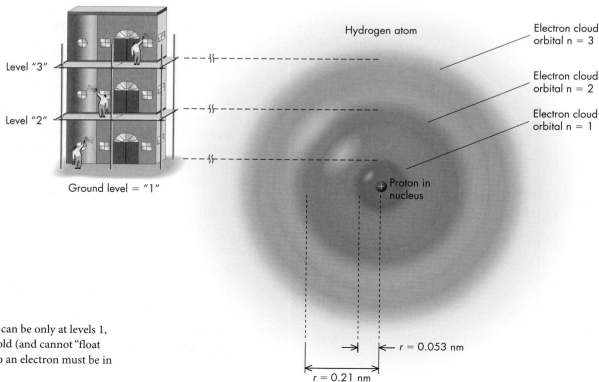

FIGURE 21.4
Just as the painters can be only at levels 1, 2, 3, ... of the scaffold (and cannot "float in between"), so too an electron must be in orbital 1, 2, 3, ...

Although the identity of an element is determined by the number of protons in its nucleus, the chemical properties of each element are determined by the electrons orbiting the nucleus. The number of electrons normally equals the number of protons, so that each atom normally has an equal number of positive and negative electrical charges and is therefore electrically neutral.

The structure of an atom with electrons orbiting an atomic nucleus superficially resembles planets orbiting the Sun. The nucleus of the atom is similar to the Sun in that it contains almost all of the mass of the system and provides the primary source of attraction for the orbiting bodies. However, the resemblance ends there.

Unusual effects operate at an atomic level that have no counterpart in larger systems. One of the first clues of this was a puzzle facing Rutherford and other physicists studying atoms in the early 1900s. Experiments had shown that electrical charges circling about produce electromagnetic radiation. Yet electrons in atoms can remain "in orbit" for indefinitely long times without emitting any radiation. This is a good thing! If electrons constantly emitted energy, they would spiral into the nucleus, and atoms would collapse in on themselves, so matter as we know it would not exist.

Electrons in an atom do not radiate because they are not simply particles. Like light, they too have a wave–particle duality. Their wave nature is not another kind of electromagnetic wave, and it is not related to the fact that an electron has an electric charge; rather, it concerns fluctuations of "electron-ness." In the early 20th century it was discovered that all matter behaves like a wave in certain respects, although usually this is apparent only for the individual subatomic particles. The existence and position of such particles can fluctuate according to rules that are described by the area of physics called *quantum mechanics*.

One of the most important consequences of electrons' duality is that their orbits may have only certain quantized sizes. Although a planet may orbit a star at any distance, an electron may orbit an atomic nucleus only at certain distances—much as when you climb stairs, you can gain only certain discrete heights. The reason for this peculiar steplike character to electrons' orbits was worked out by the French physicist Louis de Broglie (pronounced *de-BROY*) in the early 1900s. De Broglie showed that the electron's wave nature forces the electron to move only in orbits whose circumference is a whole number of waves. If it moved in an orbit whose circumference was ½ or 1½ times the length of a wave, the crest of a subsequent wave would land where the trough of the wave had been previously as it circled the nucleus, and the electron wave would "cancel" itself out. This property prevents the electron from being pulled into the nucleus because the smallest orbit, or **ground state,** is one electron wave long.

> In quantum mechanics the location of a particle is described in terms of a "wave function" that describes the probability of finding it at any position.

The wave nature of the electron "smears out" the position of the electrons. Just as a water wave does not exist at a single point, the electron wave cannot be defined as a single point. As a result, although we have described the electrons as orbiting like tiny particles around the nucleus, a better description is that they exist as a three-dimensional electron wave or **orbital.** These discrete, yet smeared-out, orbitals are illustrated as green "clouds" in Figure 21.4.

21.4 THE INTERACTION OF LIGHT AND MATTER

The structure of atoms determines not only their chemical properties but also the way they interact with light. The difference in these interactions for different elements is what allows astronomers to deduce whether an astronomical body—a star or a planet—contains iron, hydrogen, or whatever chemicals happen to be present. Therefore, understanding the structure of atoms ultimately helps us understand the nature of stars.

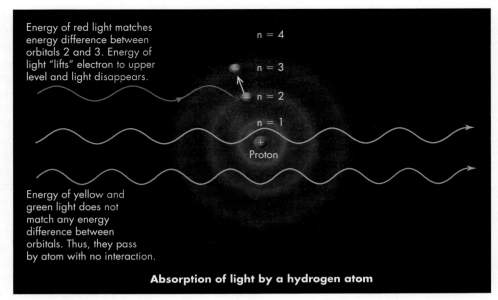

FIGURE 21.5
An atom can absorb a photon if the photon's energy matches the energy required for one of the atom's electrons to jump to a higher orbital.

Absorption of electromagnetic waves

Concept Question 3

Our eyes are only sensitive to a certain range of energies. From an evolutionary standpoint, what factors might have helped determine the energies of photons we sense?

Electrons can shift from one orbital to another as they interact with light. This shifting changes their energy, as can be understood by a simple analogy. The electrical attraction between the nucleus and the electron creates a force between them like a spring. If the electron increases its distance from the nucleus, the spring must stretch. The atom requires energy to accomplish this "stretch." Similarly, if the electron moves closer to the nucleus, the spring contracts and the atom must give up, or emit, energy. We perceive that emitted energy as electromagnetic waves.

The process in which light's energy is stored in an atom as energy is called **absorption** (Figure 21.5). Absorption lifts an electron from a lower to a higher orbital and excites the atom by increasing the electron's energy. Absorption may even **ionize** the atom—this occurs if the electron gains so much energy that it leaves orbit completely. Absorption is important in understanding such diverse phenomena as the temperature of a planet and the identification of star types, as we will discover in later Units.

We can describe this interaction in two ways corresponding to the wave–particle duality of nature at submicroscopic scales:

Particles: An electron is orbiting an atomic nucleus at a particular distance. It is struck by a photon that contains the exact amount of energy needed to knock the electron into the next larger possible orbit. The electron absorbs the photon and jumps into the new, larger orbit.

Waves: An atom has an electron orbital of a particular wave configuration. An electromagnetic wave comes by with electric and magnetic fields oscillating at just the exact rate so that the orbital interacts with the wave, absorbing energy from it, and begins to vibrate in a new, more energetic mode in a larger orbital.

Neither description is exactly right because matter at these scales is both a particle *and* a wave—a duality that is foreign to our everyday experience.

In some ways, the particle description is clearer. Discrete particles interact in a clear and sequential way. But it may lead one to wonder how the electron "knows"

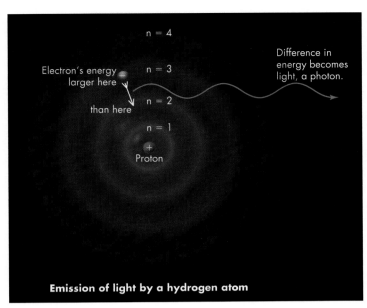

FIGURE 21.6
Energy is released when an electron drops from a higher to a lower orbital, causing the atom to emit a photon.

Emission of electromagnetic waves

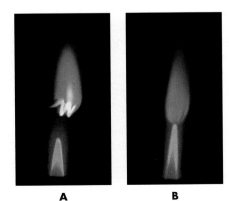

FIGURE 21.7
When different elements are heated in a flame, they emit different colors of light. For example, (A) copper appears green, while (B) strontium appears red.

that it should not react when a photon with slightly less or slightly more energy hits it. The wave picture is more challenging in other ways, but the interaction between a photon and an electron may be clearer. The electron orbital responds to the electromagnetic oscillations, much as the string of a musical instrument may begin vibrating in response to a sound wave of the right pitch, but not to all pitches.

We have seen that when an electron moves from one orbital to another, the energy of the atom changes. If the electron is in its innermost orbital, closest to the nucleus, the atom's energy is at its lowest possible energy, or ground state. If one or more electrons is shifted outward into a larger orbital, the atom's energy is increased. Such an atom is said to be in an **excited state.**

The law of energy conservation requires that energy does not appear or disappear; it just changes form (Unit 20). According to this principle, when an atom loses energy, exactly the same amount of energy must reappear in some other form, such as an electromagnetic wave.

How is electromagnetic radiation created? When an electron drops from one orbital to another, it alters the electric energy of the atom. As we described earlier, such an electrical disturbance generates a magnetic disturbance, which in turn generates a new electrical disturbance. Thus, the energy released when an electron drops from a higher to a lower orbital becomes an electromagnetic wave, a process called **emission** (Figure 21.6).

For our purposes, we can simplify the discussion of electron behavior by focusing on just the energies that are involved. When an electron moves between orbitals—an **electronic transition**—the energy change depends on the particular orbital and on the kind of atom. The quantum levels also mean that a photon with an energy that does not match the energy difference to a higher level *cannot* be absorbed, as illustrated by the yellow and green photons in Figure 21.5.

The possible energy levels in an atom are quite complicated to calculate. An atom with six protons does not simply have electron orbitals that interact with photons six times more energetic than those that interact with hydrogen, or any other such simple relationship. The electron orbital clouds interact with each other in complex ways, giving each element a rich variety of unique atomic properties, which we examine further in Unit 24.

Emission plays an important role in many astronomical phenomena. The aurora (northern lights) is an example of emission by atoms in the Earth's upper atmosphere. Gas in the Sun's outer atmosphere and clouds of gas in deep space also emit light as electrons drop to lower energy orbitals. The light from the flame of a candle is another example of emission, and when different elements are burned, they produce different colors because of their different electron energy levels (Figure 21.7).

Whether you picture emission and absorption as photons and electrons knocking about or as waves exciting vibrations, the important thing to keep track of is how much energy has been transferred. You may find it helpful in understanding emission and absorption if you think of an analogy. Absorption is a bit like drawing an arrow back, preparing to shoot it from a bow. Emission is like the arrow being shot. In one case, energy of your muscles is transferred to and stored in the flexed bow. In the other, the stored energy is released as the arrow takes flight.

KEY POINTS

- Light travels through space at the highest possible speed in nature, about 300,000 km/sec.
- Light is an electromagnetic wave, but it is also composed of quantum particles of energy called photons.
- The behavior at subatomic scales, with simultaneous wavelike and particle-like aspects, is called wave–particle duality.
- As light travels away from its source, it becomes spread out, dropping in intensity as the inverse square of the distance.
- Matter particles also behave like waves, and this is most clearly seen in electrons, which occupy quantized orbitals in atoms.
- Electrons shift between two orbitals if they absorb or emit a photon with the energy equal to the difference in energy between the orbitals.

KEY TERMS

absorption, 152
atomic number, 150
brightness, 149
electromagnetic wave, 148
electronic transition, 153
element, 150
emission, 153
excited state, 153
ground state, 151
interfere, 148
inverse-square law, 149
ionize, 152
light, 147
orbital, 151
photon, 148
quantized, 148
wave–particle duality, 149

CONCEPT QUESTIONS

Concept Questions on the following topics are located in the margins. They invite thinking and discussion beyond the text.

1. Motion of matter in a wave. (p. 149)
2. Maximum number of protons in nucleus. (p. 150)
3. Evolutionary factors for sensing photons. (p. 152)

REVIEW QUESTIONS

4. Why is light called an *electromagnetic wave*?
5. What is a photon? What is meant by duality?
6. How is the brightness of an object related to the total light output of the object and its distance?
7. What is the inverse-square law?
8. What determines whether a particular photon can interact with an atom?
9. What is the difference between emission and absorption in terms of what happens to an electron in an atom?

QUANTITATIVE PROBLEMS

10. Calculate how long it takes light to travel the 150 million kilometers from the Earth to the Sun.
11. Suppose you are operating a remote-control spacecraft that is attempting to fly through the rings of Saturn, which are composed of icy chunks of rock. Calculate the round-trip communication time, which includes a downlink of an image and uplink steering command to the spacecraft. Is this a reasonable approach to navigating through Saturn's rings?
12. How much brighter does the Sun appear from Earth when their separation is the smallest (0.983 AU) compared to when they are at their largest separation (1.017 AU)?
13. The Sun is about 10 billion times brighter than the next brightest star, Sirius (as seen from Earth). How far would we have to be from the Sun for the Sun to be as bright as Sirius? (The distance from the Earth to the Sun is 1.00 AU.)
14. Suppose there were an atom with three excited states with energies 1, 3, and 7×10^{-19} joules above the ground state. If an electron began in the highest energy level, determine all the different ways the electron could descend to the ground state and the energies of the photons it would emit in each case.
15. For the atom in the preceding problem, if the electron were in the ground state, what energy photons could it absorb? What if it were in the first excited state?

TEST YOURSELF

16. Light is an electromagnetic wave that
    a. is made of magnetized electrons moving back and forth.
    b. can be generated by the motions of electrical charges.
    c. travels through water in microscopic oscillations.
    d. carries no energy.
17. When a photon interacts with an atom, what changes occur in the atom?
    a. The atomic number increases.
    b. The nucleus begins to glow.
    c. An electron changes its orbital.
    d. The photon becomes trapped in orbit.
18. Describing an atom's orbitals as being *quantized* means that the
    a. atom is made of individual electrons, protons, and neutrons.
    b. electron can oscillate only with certain discrete patterns.
    c. number of electrons equals the number of protons.
    d. number of electrons in an atom is a whole number.
    e. electron's electric charge has a single fixed value.
19. Suppose we receive 1000 photons each second from a star. If the same star was 10 times closer, how many photons would we receive each second?
    a. 10   b. 100   c. 1000   d. 10,000   e. 100,000
20. When an atom becomes ionized it
    a. has more protons than electrons.
    b. has an electron in an excited state.
    c. is emitting electromagnetic radiation.
    d. has an extra neutron in its nucleus.

The Electromagnetic Spectrum

22.1 Wavelengths and Frequencies
22.2 Energy Carried by Photons
22.3 White Light and the Color Spectrum
22.4 The Electromagnetic Spectrum

Learning Objectives

Upon completing this Unit, you should be able to:
- Explain how wavelength and frequency are related and calculate one from the other.
- Describe and calculate the energy carried by a photon for different wavelengths.
- Explain what color and white light correspond to in terms of wavelengths of light.
- Name the different bands of light and order them by wavelength.

There is more to light than what meets the eye. Our eyes can detect some electromagnetic waves, but there are others we cannot see. Modern technologies make us familiar with many of these other kinds of electromagnetic waves, such as radio waves, X-rays, and ultraviolet light. They differ from visible light only in the rate at which the electromagnetic waves oscillate.

We examined in Unit 21 how light can alternatively be described as waves or photons. There is a direct relationship between the energy of these photons and the rate at which the wave oscillates. X-rays are rapidly oscillating waves; equivalently, we might call them highly energetic photons. Radio waves are slowly oscillating waves—or low-energy photons. This entire range of waves is called the **electromagnetic spectrum.**

22.1 WAVELENGTHS AND FREQUENCIES

Regardless of whether we consider light to be a wave or a stream of photon particles, our eyes perceive one of its most fundamental properties as color. Human beings can see colors ranging from deep red through orange and yellow into green, blue, and violet. The colors to which the human eye is sensitive define what is called the **visible spectrum.** But what property of photons or electromagnetic waves corresponds to light's different colors?

According to the wave interpretation of light, light's color is determined by the light's **wavelength,** which is the spacing between wave crests (Figure 22.1). That is, instead of describing a quality, light's *color,* we can specify a quantity, its *wavelength,*

FIGURE 22.1
The distance between crests defines the wavelength, λ, for any kind of wave, be it water or electromagnetic.

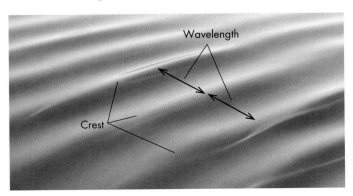
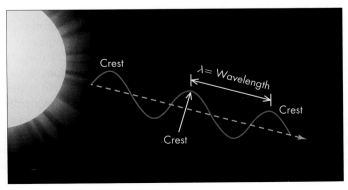

FIGURE 22.2
The frequency at which you bob up and down while floating in the water depends on the wavelength of the wave. Short-wavelength waves move past you quickly, so you bob up and down at a high frequency (A–B). A long-wavelength wave (C) takes longer to pass by, even though the speed of the wave crest is the same, so the frequency is slower.

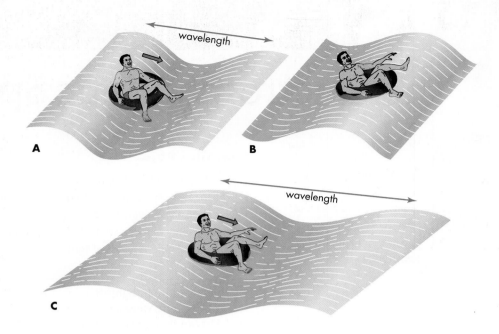

denoted by the Greek letter lambda λ. The wavelengths of visible light are very small—roughly the size of a bacterium. For example, the wavelength of deep-red light is about 7×10^{-7} meter, or 700 nanometers (nm). The wavelength of violet light is about 4×10^{-7} meter, or 400 nm. Intermediate colors have intermediate wavelengths.

Sometimes it is useful to describe electromagnetic waves by their frequency rather than their wavelength. **Frequency** is the number of wave crests that pass a given point in 1 second. The number of waves or "cycles" per second is indicated with the MKS unit **hertz** (abbreviated as Hz), named after Heinrich Hertz, who pioneered the broadcast and reception of radio waves in the late 1800s. In equations the frequency is usually denoted by the Greek letter nu ν.

The frequency and wavelength of a wave have a simple relationship to each other. When the wavelength is longer, the frequency is lower (slower); when the wavelength is shorter, the frequency is higher (faster). If you float in ocean waves, as illustrated in Figure 22.2, you can experience this in terms of the rate at which you bob up and down (frequency) as waves pass by.

The relationship between the frequency and wavelength of a wave is determined by the wave speed, because in one vibration a wave must travel a distance equal to one wavelength. Thus, if we multiply the wavelength λ by the number of waves per second, ν, the product is the speed of light:

$$\lambda \times \nu = c.$$

Because all light waves travel at the same speed, c, through empty space, the wavelength determines the frequency, and vice versa. We will generally use λ to characterize electromagnetic waves, but ν is just as good. For example, yellow light vibrates about 500 trillion times a second (5×10^{14} Hz or 500 THz); red light vibrates a little slower; blue light a little faster.

Be sure to remember that *longer wavelengths of visible light correspond to redder colors*. Equivalently, *lower frequencies correspond to redder colors*. In the opposite case, astronomers commonly say that *shorter wavelengths and higher frequencies are bluer* (although it would be more accurate to say that they are more violet). We will use this information many times as we interpret the significance of different colors of light coming from astronomical objects. And as we shall see next, these color differences have a deeper significance in terms of energies.

Clarification Point

A common mistake is to think that a rapid frequency corresponds to a higher speed. In fact, the waves travel at the same speed, but more crests go by if the frequency is higher.

λ = wavelength of wave
ν = frequency of wave
c = speed of light

Light travels slower through transparent materials than it does through empty space, and different wavelengths of light are slowed differently. In materials like glass, water, or air, blue light travels slightly slower than red light.

22.2 ENERGY CARRIED BY PHOTONS

The warmth we feel on our face from a beam of sunlight demonstrates that light carries energy, but not all wavelengths carry the same amount of energy. It turns out that the amount of energy E carried by a photon depends on its wavelength λ. Each photon of wavelength λ carries an energy E given by

$$E = \frac{h \times c}{\lambda}.$$

E = energy of photon
h = Planck's constant
c = speed of light
λ = wavelength

The quantity h is called *Planck's constant*, a constant of nature that describes the fundamental relationship between energies and waves in the submicroscopic realm of quantum mechanics. This equation can also be written in terms of the frequency of a photon. From the equation relating wavelength and frequency to the speed of light, we find that $\nu = c/\lambda$, so the energy of a photon is equivalently $E = h \times \nu$.

Thus, the wavelength of a photon determines how much energy it carries. Long-wavelength, low-frequency photons contain less energy. In other words,

A photon carries an amount of energy proportional to its frequency and inversely proportional to its wavelength.

Mathematical Insights

In MKS units, Planck's constant is:
$h = 6.63 \times 10^{-34}$ joule·second, and
$h \times c = 1.99 \times 10^{-25}$ joule·meter.
So for a blue-green photon with a wavelength of 500 nm,

$$E = \frac{h \times c}{\lambda} = \frac{1.99 \times 10^{-25} \text{ joule} \cdot \text{m}}{500 \text{ nm}}$$

$$= \frac{1.99 \times 10^{-25} \text{ joule} \cdot \text{m}}{5.0 \times 10^{-7} \text{ m}}$$

$$= 4.0 \times 10^{-19} \text{ joule}.$$

For example, a blue-green photon with a wavelength of 500 nm has an energy of about 4×10^{-19} joule (see calculation at left). A deep-red photon with a wavelength of 700 nm will have an energy 1.4 ($= 700/500$) times lower, or 2.8×10^{-19} joule; a violet photon at 400 nm will have an energy 1.25 ($= 500/400$) times higher, or 5×10^{-19} joule.

Thousands of photons must enter our eyes each second for us to be able to see the dimmest stars visible to us. This adds up to about 10^{-15} joule per second, or 10^{-15} watt (1 joule/sec is defined as 1 watt). Note that any number of red photons will always look red, even if their total energy exceeds the energy from a smaller number of violet photons. The receptors in the eye react according to the energy of each photon. The same kind of "red" receptors will react more strongly as the number of red photons climbs, but a different kind of receptor is sensitive to violet photons. The way the eye works is examined further in Unit 33.

Concept Question 1

It might help you remember the connection between energy, wavelength, and frequency to make waves on a rope. Tie one end to a doorknob and shake the other end. To make short-wavelength waves along the rope, how rapidly (frequently) do you shake the rope? Does it take more of your energy to make short waves or long waves?

22.3 WHITE LIGHT AND THE COLOR SPECTRUM

We perceive many colors of light, which we can describe by the wavelength of the electromagnetic waves or the energy of its photons. However, some light seems to have no color, and not all photons have wavelengths within the range we have specified. What causes us to perceive a lack of color? How do we sense wavelengths outside the range of our eyes?

The Sun high in the sky and an ordinary lightbulb appear to have no dominant color. Light from such sources is called **white light**. White light is not a special color of light; rather it is a mixture of all colors. That is, the sunlight we see is made up of all the wavelengths of visible light—a blend of red, yellow, green, blue, and so on—and our eyes perceive the combination of all these as white. Newton demonstrated this property of sunlight by a simple but elegant experiment. He passed sunlight through a prism (Figure 22.3A) so that the light was spread out into the visible spectrum (which we see as a rainbow of colors). He then recombined the separated colors with a lens and reformed the beam of white light.

You can see how colors of light mix if you look at a color television screen with a magnifying glass. You will notice that the screen is covered with tiny red, green,

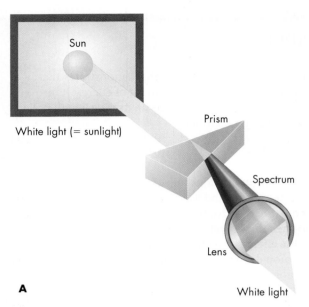

FIGURE 22.3
(A) A prism spreads "white" light into its component colors (a spectrum). Combining the colors again with a lens makes the light "white" again.
(B) Spectra in everyday life.

Concept Question 2

Why do you suppose that when you mix together red, green, and blue paint, you get a dark gray or brown color instead of white?

and blue dots. In a red object, only the red spots are lit. In a blue object, only the blue spots are illuminated. In a white object, all three color spots are lit, and your brain mixes these three colors to form white. Other colors are made by appropriate blending of red, green, and blue.

You may have noticed that photos taken in artificial light look much more yellow than the scene you remember, or outdoor photographs may look bluer than you remember. That is because most lightbulbs are much more yellow than sunlight, but we adjust to the difference. We generally see a white piece of paper as white even under different-color lighting conditions. Presumably because our senses have evolved to make us aware of differences in our surroundings, we ignore the ambient "color" of sunlight or a lightbulb, just as in time we learn to ignore a steady background sound or smell.

Our eyes respond to photons with energies ranging from about 2.8 to 5×10^{-19} joule, but are there photons with higher or lower energies? Studies of spectra reveal that light extends beyond what is visible. This was first seen in 1800 when Sir William Herschel (discoverer of the planet Uranus) showed that heat radiation, such as you feel from the Sun or from a warm radiator, though invisible, is related to visible light.

Herschel was trying to measure heat radiated by astronomical sources. He projected the Sun's spectrum onto a tabletop and placed a thermometer in each color to measure its energy. He was surprised that when he put a thermometer just off the red end of the visible spectrum, the thermometer registered an elevated temperature there, just as it did in the red part of the spectrum. He concluded that some form of invisible energy detectable as heat existed beyond the red end of the spectrum, and he therefore called it **infrared.** Even though your eyes cannot see infrared light, when it strikes your skin it deposits energy that warms the skin, and the nerves in your skin can feel the heat.

Just a year later, in 1801, the German scientist Johann Ritter was studying the light sensitivity of chemicals that were later to become the basis for photography. When he shone light through a prism onto one of these chemicals, he discovered

Although humans cannot see infrared radiation, several kinds of snakes, including the rattlesnake, have special infrared sensors located just below their eyes. These allow the snake to "see" in what for us would be total darkness, helping it to find warm-blooded prey such as rats.

Concept Question 3

Our eyes are insensitive to infrared light, but there are "infrared night-vision goggles." How do you suppose we can see infrared light with these?

that there was an intense photochemical reaction beyond the last visible portion of the spectrum at its violet end—**ultraviolet** radiation.

Infrared and ultraviolet radiation differ in no physical way from visible light except in their wavelength. Infrared has longer wavelengths and ultraviolet shorter wavelengths than visible light. Both can be described as waves or photons of electromagnetic energy, like visible light. They just cannot be seen by the human eye.

22.4 THE ELECTROMAGNETIC SPECTRUM

White light and the enormous variety of colors we can see represent just some of the richness possible for electromagnetic waves. Even infrared and ultraviolet light merely hint at a much broader spectrum. Today there are instruments capable of detecting electromagnetic waves with wavelengths anywhere from thousands of kilometers long down to 10^{-18} meter or less. Ordinary visible light falls in a narrow section in the middle of a broad spectral range (see Figure 22.4). Objects may emit and absorb light in every part of the electromagnetic spectrum, creating "colors" completely outside the range of our eyes. Viewed with our eyes alone, an object may appear to be dark; but it may be the source of an assortment of different-wavelength photons when viewed with a camera that is sensitive to a different part of the electromagnetic spectrum.

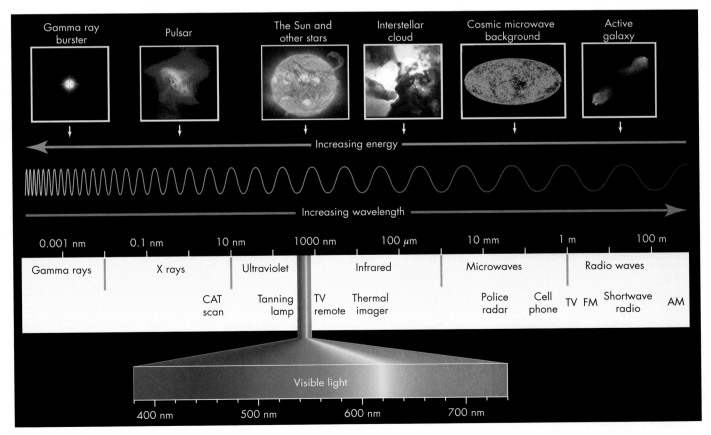

FIGURE 22.4
The electromagnetic spectrum. Examples are given of astronomical objects that are studied at the various wavelengths, as well as technologies operating at these wavelengths. Note that most astronomical sources will emit some electromagnetic radiation at other wavelengths as well.

> **Concept Question 4**
>
> Why do you suppose that the centers of some filled pastries grow hot in a microwave oven while their outer portions do not grow as hot?

> **Clarification Point**
>
> Radio waves are frequently confused with sound waves, perhaps because these light waves are often used to transmit signals that contain sound. Radio waves are another wavelength of light and travel at the speed of light.

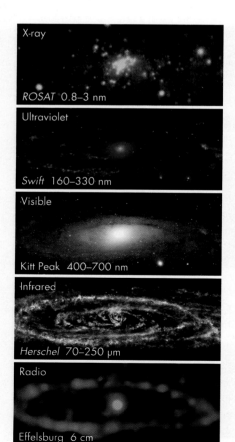

FIGURE 22.5
A series of images of the Andromeda galaxy (M31) at wavelengths from X rays to radio waves.

James Clerk Maxwell, a Scottish physicist, predicted the existence of **radio waves** in the mid-1800s. It was some 20 years later, however, before Heinrich Hertz produced them experimentally in 1888, and another 50 years had to pass before Karl Jansky discovered naturally occurring radio waves coming from cosmic sources.

Radio waves range in length from millimeters to hundreds of meters and longer, making them much longer than visible and infrared waves. An AM "radio" is a receiver for electromagnetic waves that are hundreds of meters long; FM radio corresponds to wavelengths of a few meters. Actually radio wavelengths are more commonly listed by their frequency. You can see this on a radio dial, where you tune in a station by its frequency in kilohertz or megahertz (thousands or millions of waves per second) rather than its wavelength.

Today we can generate radio waves and use them in many ways, ranging from communication to radar to cooking. Given the enormous span of wavelengths and uses, the wide range of radio waves is often subdivided, with the shorter wavelengths being called **microwaves.** These are waves of about 1 meter or less, although there is no definite standard on what the boundaries are between infrared, microwave, and radio. Most microwave ovens, for example, generate electromagnetic waves with wavelengths of about 12 cm.

The strong connection between radio waves and their use for communication systems confuses many people into thinking that radio waves are a kind of sound wave. Not so! Radio waves are a kind of light, traveling at the same enormous speed as all forms of electromagnetic radiation. If radio waves traveled anywhere near as slowly as sound, you would have to wait tens of seconds for every response when calling on a cell phone to a friend just a few kilometers away, and hours when calling across the country!

Many kinds of astronomical objects such as forming stars, exploding stars, active galaxies, and interstellar gas clouds generate these low-energy radio waves by natural processes. Understanding this radiation reveals information about the temperature of the gas, the molecules present, magnetic fields in the region, and a wide variety of other factors. Astronomers detect this radiation with radio telescopes to study the kinds of astrophysical processes and conditions that are different from those that generate visible light.

Over the last several decades, astronomers have also probed very short-wavelength regions called the **X-ray** and **gamma-ray** parts of the spectrum. X-ray wavelengths are even shorter than those of visible and ultraviolet light, typically between 0.01 and 10 nanometers, and gamma-ray wavelengths are shorter yet (see Figure 22.4). The fact that X-rays and gamma rays have such short wavelengths implies that they carry a great deal of energy and are generated only in regions with extremely high temperatures or high-energy reactions taking place. These wavelengths therefore reveal to astronomers some of the universe's most exotic objects and processes, such as hot gas falling into black holes.

The high energy of these photons allows them to penetrate many objects that are opaque to visible light. One familiar example is the ability of X-rays to penetrate soft tissue, which allows us to make diagnostic medical X-ray images. The X-ray photons that are not blocked by bone or other dense material travel through and expose a photographic negative—much like Ritter's 1801 experiment with ultraviolet light. Note that because X-rays are outside the range that our eyes can see, there is no such thing as a pair of "X-ray glasses" that lets you see through objects. If you looked through a pair of glasses that transmitted only X-rays, you would see . . . *nothing!*

Despite the enormous variety of electromagnetic waves, they are all the same physical phenomenon: the vibration of electric and magnetic energy traveling at the speed of light. The essential difference between kinds of electromagnetic radiation is merely their wavelength—or frequency—or energy. However, the different appearance of an astronomical object at each wavelength (Figure 22.5), reveals important clues about the physical processes taking place.

KEY POINTS

- Light waves have a wavelength that corresponds to the distance between wave crests, and a frequency that indicates the rate at which wave crests pass a point.
- The product of the wavelength and frequency equals the speed of the wave.
- The energy of photons in a light wave is directly proportional to the frequency and inversely proportional to the wavelength.
- Planck's constant gives the conversion from frequency to energy for a photon.
- The colors we perceive are determined by the light's wavelength; red is the longest wavelength we can see and blue is the shortest.
- We perceive the combination of all colors as white light.
- Visible light is just a small portion of the electromagnetic spectrum, which includes wavelengths millions of times longer and millions of times shorter.
- Gamma rays, X-rays, ultraviolet, infrared, microwave, and radio waves are all different wavelengths of light; all are also photons and travel at the same speed as visible light.

KEY TERMS

electromagnetic spectrum, 155
frequency, 156
gamma ray, 160
hertz, 156
infrared, 158
microwaves, 160
radio waves, 160
ultraviolet, 159
visible spectrum, 155
wavelength, 155
white light, 157
X-ray, 160

CONCEPT QUESTIONS

Concept Questions on the following topics are located in the margins. They invite thinking and discussion beyond the text.

1. Modeling the relationship of frequency to energy. (p. 157)
2. Mixing paint colors instead of light. (p. 158)
3. How infrared night-vision goggles work. (p. 159)
4. Why microwaves can heat the interior of an object. (p. 160)

REVIEW QUESTIONS

5. How are color and wavelength related?
6. What is meant by the *electromagnetic spectrum*?
7. Why is it more dangerous to be exposed to a weak X-ray source than to a strong infrared source?
8. Does all electromagnetic radiation travel at the same speed?
9. What are the bands of the electromagnetic spectrum from short to long wavelengths?
10. In what ways is it possible and impossible for people to personally detect energy outside the visible spectrum?

QUANTITATIVE PROBLEMS

11. A typical frequency used for cell phone communication is 800 megahertz. What is its wavelength? How does the wavelength compare to the size of a standard cell phone?
12. Microwave ovens operate at a 12-cm wavelength by making water molecules oscillate at the frequency of the electromagnetic radiation. What frequency is this? Explain at the atomic level why some substances heat up and others do not. How does this differ from what happens in conventional ovens at the atomic level?
13. A hydrogen atom has a diameter of about 0.1 nanometers. What is the frequency and energy of a photon that has a wavelength equal to the diameter of the hydrogen atom?
14. What is the wavelength of a photon that has 1000 times as much energy as a photon with a wavelength of 500 nm? Would it be more dangerous to be exposed to 1 million photons with a wavelength of 500 nm, or 1000 photons with 1000 times the energy of a 500-nm photon? Why?
15. Suppose the energy of an electron in its atomic orbital is 10^{-18} joule above the ground state. What wavelength of light would have photons matching this energy difference? Which band of electromagnetic radiation is this in?
16. a. What is the energy (in joules) of an extremely powerful gamma ray with a wavelength of 10^{-7} nanometer?
    b. How many of these gamma-ray photons would add up to the same amount of energy as a housefly in flight, which has a kinetic energy of about one millionth of a joule as it flies along?

TEST YOURSELF

17. Which type of electromagnetic radiation has the longest wavelength?
    a. Ultraviolet   c. X-ray   e. Radio
    b. Visible   d. Infrared
18. Which kind of photon has the lowest energy?
    a. Ultraviolet   c. X-ray   e. Radio
    b. Visible   d. Infrared
19. When a source of light is moving away from us at high speed, its light may be Doppler-shifted (Unit 25) to longer wavelengths. If the frequency drops to half its former value, the energy of the photons will
    a. drop by half too.   b. stay the same.   c. double.
20. Radio waves travel _____ visible light.
    a. faster than
    b. slower than
    c. at the same speed as
21. When does light look white?
    a. When the photons are so energetic that they excite all of the sensors in our eyes
    b. When all of the colors are present in nearly equal strength
    c. When the light has a wavelength of 0 (zero), so its color cannot be determined
    d. When the photons are too small to be seen individually

Thermal Radiation

23.1 Blackbodies
23.2 Color, Luminosity, and Temperature
23.3 Measuring Temperature
23.4 Taking the Temperature of Astronomical Objects
23.5 The Stefan-Boltzmann Law

Learning Objectives

Upon completing this Unit, you should be able to:
- Define a blackbody and explain how the color and brightness of thermal emission from an object depends on its temperature.
- Explain the value of the Kelvin temperature scale, and how other scales relate to it.
- Carry out calculations using Wien's Law and the Stefan-Boltzmann Law.

When a body becomes hotter, the atoms and molecules inside it vibrate more rapidly, undergo more collisions, and generally interact more energetically. We saw in Unit 22 that the wavelength of light is inversely related to the energy of its photons. It is not surprising, then, that higher temperatures are associated with radiation at shorter wavelengths. When a chunk of charcoal becomes hot, it may glow visibly. This kind of electromagnetic emission is called **thermal radiation.**

Thermal radiation is produced by many objects that you might not think of as "glowing." Your body or an animal's body produces infrared thermal radiation, as shown in Figure 23.1. On the other hand, some objects, like the collapsed remnants of stars, are so hot that their thermal radiation is mostly at X-ray wavelengths. Not all kinds of materials produce thermal radiation—in particular, low-density gas produces emission at the individual wavelengths associated with electrons changing orbitals, as we discussed in Unit 21. Most denser materials produce thermal emission that looks similar no matter their chemical composition.

Wherever thermal radiation occurs, it follows two simple rules: (1) The photons have higher energy when the material is hotter, and (2) the total amount of radiation produced increases rapidly as the temperature climbs. These two rules have mathematical formulations that allow us to learn about the physical conditions in material from the thermal radiation it emits. These provide important tools for studying objects throughout the universe.

FIGURE 23.1
A false-color image of the infrared thermal radiation from a dog. The dog's eyes are warmest and produce the most light, while its nose is cold and emits little light.

Blackbody radiation

23.1 BLACKBODIES

A **blackbody** is a theoretical object that absorbs all the radiation falling on it. Because such a body reflects no light, it looks black to us when it is cold, hence its name. For an object to appear black, it must be capable of absorbing all wavelengths of electromagnetic radiation. This implies that the substance is capable of undergoing energy transitions corresponding to all wavelengths. This is *not* true of a low-density gas, as we will examine in Unit 24, but it is true or nearly true for a wide range of denser materials.

When blackbodies are heated, they can also *radiate* at all wavelengths. This is likewise because transitions of all wavelengths are possible. Thus, blackbodies are both excellent absorbers and excellent emitters. Moreover, the intensity of their radiation changes smoothly from one wavelength to the next with no gaps or narrow peaks of brightness. Many objects—a piece of charcoal, an electric stove burner, the Sun, the Earth—are near enough to being blackbodies that their radiation looks very similar to that of the idealized blackbody.

Concept Question 1

Suppose you want to design a paint that will keep a building cool in bright sunlight. At what wavelengths would you want it to absorb light well, and at what wavelengths would you want it to reflect light well?

By contrast, gases (unless compressed to a high density) are generally not blackbodies and do not produce thermal emission. The wavelengths emitted by a gas are determined by its composition and the electron energy-level transitions that are possible for the kinds of atom and molecules present in the gas.

What makes a dense material different? Within it, the orbitals of each atom's electrons are disturbed by neighboring atoms. The possible energies of the electrons are no longer precisely defined as they are for isolated atoms, so an incoming photon is likely to encounter an atom that can absorb its energy. Solid materials often retain some characteristics of their component atoms and molecules, which is why the things we see are not all black. For example, a leaf appears green because it absorbs most wavelengths *except* those with green colors.

As a dense gas or solid becomes hotter, its similarity to a blackbody increases. This is because the vibrations and energetic collisions associated with higher temperatures cause the electrons' orbitals to be disturbed even more, further "smearing out" the range of energies—and wavelengths—they can absorb or emit.

23.2 COLOR, LUMINOSITY, AND TEMPERATURE

One important property we can infer about an object from the light it emits is its temperature. Temperature gives rise to subtle differences in the colors of stars at night. And by observing beyond the wavelengths of visible radiation, we can detect objects both much cooler and much hotter than stars.

A hot object emits light, as you can easily see if you turn on the burner of an electric stove. And the color of the light is related to the temperature of the burner. As the burner grows hotter, at first it emits infrared light that you might sense as heat; then it shifts to red, then orange (Figure 23.2). This relation between color and temperature allows astronomers to measure the temperature of stars and other astronomical objects from the wavelengths of light they emit. They can make this measurement using a relation called **Wien's law**, named for the German physicist Wilhelm Wien (pronounced *Veen*), who discovered it in the 1890s.

Wien's law states that the wavelength at which an object radiates most strongly is inversely proportional to the object's temperature:

The hotter a body is, the more strongly it will radiate at short wavelengths.

We can see this effect illustrated in Figure 23.3, where we plot the amount of energy radiated at each wavelength (color) for an object at three different temperatures. Notice that the hotter body has its most intense emission (highest point) at a shorter wavelength than the cooler body. This is what gives it a different color.

FIGURE 23.2
An electric stove burner illustrates the relationship between color and temperature.

FIGURE 23.3
As an object is heated, the wavelength at which it radiates most strongly, λ_{max}, shifts to shorter wavelengths, a relation known as *Wien's law*. Note also that as the object's temperature rises, the amount of energy radiated increases at *all* wavelengths.

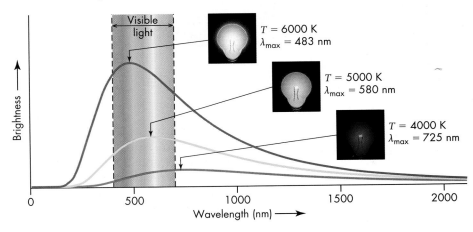

> **Clarification Point**
>
> The amount of light an object emits can be described in different ways. Luminosity is the total light output. Brightness is the light observed, which depends on the distance (Unit 21.2). The amount of light emitted per square meter is sometimes called a "surface brightness."

Note also that for objects of the same size, the hotter object emits more energy at every wavelength than the cooler one does. This is obvious looking at a stove burner. As it grows hotter, going from dull red to orange, it also grows much brighter. This is also the basis for the dimmer switches on lights: Changing the amount of electricity going to the lamp changes the temperature of the filament in the lightbulb and thereby its brightness. The light output of an object is known as its **luminosity**. What is observed is that even if the filament rises only a little bit in temperature, its luminosity increases substantially. Thus:

The luminosity of a hot body rises rapidly with temperature.

A mathematical form of this principle is formulated as the *Stefan-Boltzmann law*, which we examine in more detail in Section 23.5.

Before we can write out these mathematical laws in a form that allows us to carry out specific calculations, we must decide what temperature scale we wish to use. To make this choice, it will be helpful if we briefly discuss what temperature measures.

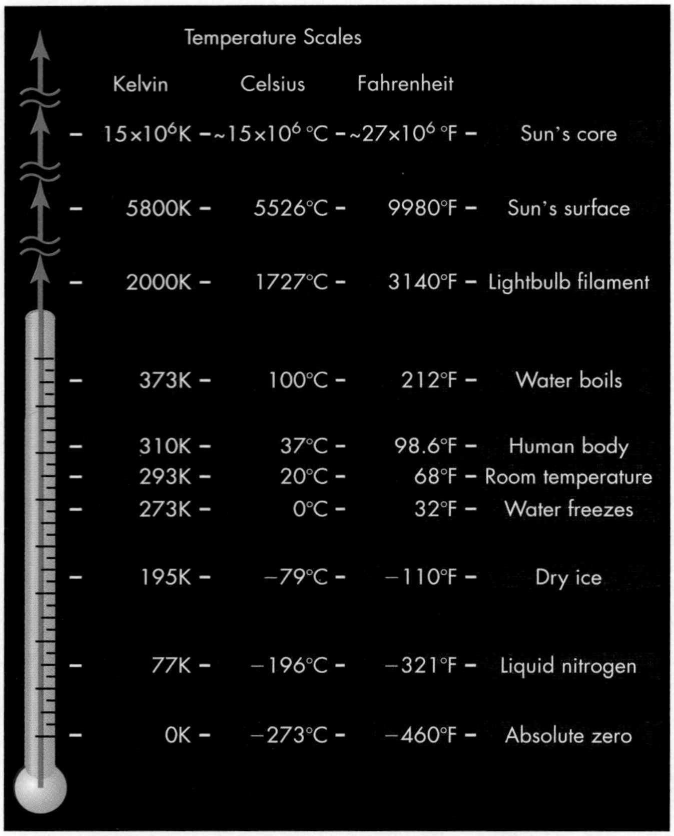

23.3 MEASURING TEMPERATURE

An object's temperature is directly related to its energy content and to the speed of its molecular motions. That is, the hotter the object, the faster its atoms and molecules move and the more energy it possesses. Similarly, if the body is cooled sufficiently, molecular motion within it slows to a virtual halt and its energy approaches zero.

In the late 1800s the British scientist Lord Kelvin devised a temperature scale based on these properties (Figure 23.4). At zero on the **Kelvin** scale, molecular motion ceases and objects have no heat energy, which is known as **absolute zero**. Notice that negative temperatures therefore have no meaning on the Kelvin scale because there cannot be less motion than none.

It is simple to convert from the Kelvin to the Celsius scale because Kelvin made the steps (kelvins) on his scale the same size. The temperature in kelvins equals 273.15 plus the Celsius temperature. For example, the freezing and boiling points of water, 0°C and 100°C, are 273.15 K and 373.15 K, respectively. Room temperature is about 300 K. Converting to the Fahrenheit scale is a little more complicated because each degree is 5/9 the size of a Celsius degree. The formula for the temperature in Fahrenheit degrees is $T_{°F} = 9/5 \, T_K - 459.4$, where T_K is the temperature in kelvins.

The Kelvin scale is useful in astronomy and other sciences because the temperature in kelvins is directly proportional to an object's **thermal energy** (the energy due to the internal vibrations and motions of the atoms within the object). Thus, when the thermal energy doubles, the value of the temperature in kelvins also doubles.

The Celsius and Fahrenheit scales chose values for "zero" that are above absolute zero, so at lower temperatures they go negative. When the temperature of an object doubles in the Celsius and Fahrenheit scales, it is *not* true that its thermal energy doubles. Because of its direct relation to so many physical processes, we will use the Kelvin scale in most of the remainder of this book. With a temperature scale chosen, we can now learn how we might measure a star's temperature.

Temperature Scales

Kelvin	Celsius	Fahrenheit	
15×10^6 K	~15×10^6 °C	~27×10^6 °F	Sun's core
5800 K	5526 °C	9980 °F	Sun's surface
2000 K	1727 °C	3140 °F	Lightbulb filament
373 K	100 °C	212 °F	Water boils
310 K	37 °C	98.6 °F	Human body
293 K	20 °C	68 °F	Room temperature
273 K	0 °C	32 °F	Water freezes
195 K	−79 °C	−110 °F	Dry ice
77 K	−196 °C	−321 °F	Liquid nitrogen
0 K	−273 °C	−460 °F	Absolute zero

FIGURE 23.4
Temperatures in Kelvin (K) and on the Celsius and Fahrenheit scales.

> Note that temperatures on the Kelvin scale are not given in "degrees" but are simply called "kelvins" or K.

23.4 TAKING THE TEMPERATURE OF ASTRONOMICAL OBJECTS

Wien's law allows us to measure how hot something is from the color of light it radiates most strongly. To measure a distant body's temperature using Wien's law, we proceed as follows. First we measure the body's brightness at many different wavelengths to find the particular wavelength at which it is brightest (that is, its wavelength of maximum emission). Then we use the law to calculate the body's temperature. To see how this is done, however, we need a mathematical expression for the law.

If we let T be the body's temperature measured in kelvins, and we let λ_{max} be the wavelength in nanometers at which the body radiates most strongly (Figures 23.3 and 23.5), Wien's law can be written in this form:

$$T = \frac{2.9 \times 10^6 \text{ K} \cdot \text{nm}}{\lambda_{max}}.$$

The subscript *max* on λ indicates that it is the wavelength of maximum emission.

As an example, we can measure the Sun's temperature. The Sun turns out to radiate most strongly at a wavelength of about 500 nanometers. Substituting $\lambda_{max} = 500$ nanometers, we find

$$T_\odot = \frac{2.9 \times 10^6 \text{ K} \cdot \text{nm}}{500 \text{ nm}} = 5800 \text{ K}.$$

The Sun is not a perfect blackbody; nevertheless, this result is within about 20 K of the value found by much more sophisticated analyses.

You might note that the wavelength at which the Sun radiates most strongly corresponds to a blue-green color, yet the Sun looks yellow-white to us. The reason we see it as whitish is related to how our eyes perceive color. Physiologists have found that the human eye interprets sunlight (and light from all extremely hot bodies) as whitish, with only tints of color. Such hot bodies emit a significant amount of light at all visible wavelengths, not just at the color corresponding to the wavelength of maximum emission. Thus, cool stars look white tinged with red, whereas very hot stars look white tinged with blue.

Wien's law works quite accurately for most stars and planets or almost any material that has atoms packed tightly together. It is important to remember, though, that the law tells us only about thermal radiation *emitted* by an object. An object may simultaneously be emitting thermal radiation and *reflecting* light from another source. For example, the red color of an apple and the green color of a lime have nothing to do with their temperatures. They both do emit thermal radiation, but if they are at normal room temperature, the radiation they emit will be primarily in the infrared range (Figure 23.6).

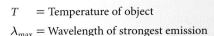

T = Temperature of object
λ_{max} = Wavelength of strongest emission

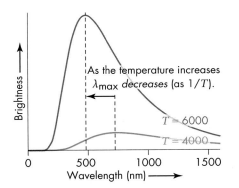

FIGURE 23.5
Wien's law. Hotter objects emit radiation that peaks at a shorter wavelength in inverse proportion to the temperature.

FIGURE 23.6
Images of two cups containing hot and cold water, at visible (left) and infrared (right) wavelengths. At visible wavelengths we see light reflected from the surface, the colors resulting from the wavelengths *not* absorbed by molecules in the plastic. The infrared picture is shown in false colors according to the inset scale. Hot liquid in the cup on the left makes it glow at shorter infrared wavelengths than the surrounding area or the cold liquid in the other cup.

23.5 THE STEFAN-BOLTZMANN LAW

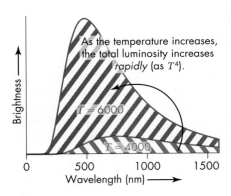

FIGURE 23.7
The Stefan-Boltzmann law. Hotter objects emit a larger total amount of radiation that grows in proportion to the temperature to the fourth power.

In the late 1800s two Austrian scientists, Josef Stefan and Ludwig Boltzmann, showed that the total amount of light emitted by an object increases as its temperature rises. The total light output can be thought of as adding up all of the light emitted at all wavelengths, as represented by the curves in Figure 23.7.

The **Stefan-Boltzmann law,** as their discovery is now called, states that a body of temperature T (measured in kelvins) radiates an amount of energy each second equal to σT^4 per square meter. The quantity σ is called the **Stefan-Boltzmann constant,** and its value is 5.67×10^{-8} watt per meter^2 per kelvin^4. Like Wien's law, this law is precisely true for blackbodies, but it is also quite accurate for many different kinds of materials, both solids and gases, as long as the material is relatively dense. It works well for embers of coal, a lightbulb filament, or the surface of a star.

The Stefan-Boltzmann law shows mathematically what we observe as rapid brightening as an object gets hotter. If the temperature doubles, for example, the light output increases by a factor of 16. We can show this mathematically for material heated to one temperature that's twice another. For example, suppose a lightbulb filament is heated to 1000 K. The amount of light it generates is

$$\sigma \times (1000\ \text{K})^4 = (5.67 \times 10^{-8}\ \text{watt/m}^2/\text{K}^4) \times (10^3)^4 \times \text{K}^4$$
$$= 5.67 \times 10^{-8} \times 10^{12}\ \text{watt/m}^2$$
$$= 5.67 \times 10^4\ \text{watt/m}^2$$

On the other hand, if its temperature is raised to 2000 K, the luminosity is

$$\sigma \times (2000\ \text{K})^4 = (5.67 \times 10^{-8}\ \text{watt/m}^2/\text{K}^4) \times (2 \times 10^3)^4 \times \text{K}^4$$
$$= 5.67 \times 10^{-8} \times 16 \times 10^{12}\ \text{watt/m}^2$$
$$= 90.72 \times 10^4\ \text{watt/m}^2$$

Comparing the second line in each of these calculations, you can see that the only difference is the factor of 16, which comes from taking 2 to the fourth power: $2^4 = 2 \times 2 \times 2 \times 2 = 16$. It would not matter what pair of temperatures we put into the equation: As long as one is *two times hotter* than the other, the amount of light generated by each square meter of the hotter object will always be *16 times more luminous* than the cooler one.

The other thing to notice from these calculations is the enormous wattages generated by material at these high temperatures. A square meter (about 10 square feet) of material heated to 2000 K generates over 900,000 watts of luminous power! A lightbulb filament is typically about 2000 to 2500 K. If the total surface area of the filament is one square centimeter, that's equivalent to 10^{-4} m^2. Such a filament would produce about 91×10^4 watt/m$^2 \times 10^{-4}$ m$^2 = 91$ watts of thermal energy.

For a 2000-K filament, Wien's law tells us that most of the emission of lightbulbs is at infrared wavelengths we cannot see. Typically, less than 10% of the power is produced as visible light. By raising the square-centimeter filament's temperature to 2500 K, we would now produce 221 watts. In addition, Wien's law tells us that if we raised the temperature, we would shift the peak of the emission closer to optical wavelengths, so the amount of visible light would increase even more.

The Sun's surface is about 5800 K, so compared to the 1000-K thermal emission we just calculated, the Sun's surface must generate $5.8^4 = 5.8 \times 5.8 \times 5.8 \times 5.8 = 1130$ times more power. This is over 64 million watts per square meter emitted by every square meter over the entire surface of this enormous body! When you look at photographs of the Sun's surface (Figure 23.8), you will see some regions that are darker and some lighter. These are regions with slightly different temperatures. The strong dependence of luminosity on temperature means that a region that is at 5000 K will look dimmer than the surrounding regions.

Concept Question 2

Why do you suppose a lightbulb filament is usually a thin winding wire instead of simply being thicker and straight? How might this thinking apply to the thermal radiation from a rocky asteroid versus the same amount of rock broken up into small pieces?

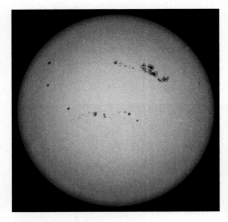

FIGURE 23.8
A portion of the Sun's surface. The darker spots in the picture are cooler than surrounding regions, so they look dark by contrast. However, they are generating light.

KEY POINTS

- A material that can absorb or emit perfectly at all wavelengths is known as a blackbody; such materials produce thermal emission with well-defined properties.
- Many materials, particularly solids and dense gases, produce thermal emission similar to blackbodies.
- Wien's law: the wavelength of strongest emission for a blackbody depends on the inverse of the temperature.
- Stefan-Boltzmann law: the total light output, or luminosity, of a blackbody rises as the temperature to the fourth power.
- The Kelvin temperature scale is the only scale that sets zero temperature to when there is no thermal energy.

KEY TERMS

absolute zero, 164
blackbody, 162
Kelvin, 164
luminosity, 164
Stefan-Boltzmann constant, 166
Stefan-Boltzmann law, 166
thermal energy, 164
thermal radiation, 162
Wien's law, 163

CONCEPT QUESTIONS

Concept Questions on the following topics are located in the margins. They invite thinking and discussion beyond the text.

1. What characteristics of paint keep a building cooler. (p. 163)
2. Emission from many small objects versus single large one. (p. 166)

REVIEW QUESTIONS

3. How are color and temperature related?
4. Under what circumstances can an object that looks yellow be hotter than one that looks red?
5. What are the advantages of the Kelvin temperature scale?
6. What is Wien's law?
7. What is the Stefan-Boltzmann law?
8. What could explain how you might have two lightbulbs of equal brightness, both producing thermal radiation, but one bluer than the other?

QUANTITATIVE PROBLEMS

9. What is the temperature range for objects whose wavelength at maximum falls within the visible spectrum, 400 to 700 nm?
10. A lightbulb radiates most strongly at a wavelength of about 3000 nanometers. How hot is its filament?
11. How hot would a source need to be to emit its maximum radiation in the FM range of the radio dial (100 MHz)? Why don't common objects interfere with our radio reception?
12. By measuring relic radiation left over from the Big Bang, astronomers have measured the average temperature of the Universe to be 2.73 K. At what wavelength does this background radiation emit most strongly? What part of the electromagnetic spectrum does this wavelength lie in?
13. In order to forge copper it requires heating the metal to a temperature of 1200 K. How much more thermal radiation does a piece of copper produce when heated to its forging temperature as compared to room temperature?
14. The total light an object emits depends on its temperature and its surface area. How large would a star with a surface temperature of 3000 K need to be to have the same luminosity as a star with a surface temperature of 6000 K? (Phrase your answer in terms of the size of the 6000-K star.)
15. What is the ratio of the luminosity per square meter coming from a part of the Sun that is at 5800 K versus a sunspot where the temperature is 5000 K?
16. Find the temperature in degrees Celsius and Fahrenheit for (a) a star whose surface is at 10,000 kelvins, and (b) the 3-kelvin temperature of deep space far from any stars.

TEST YOURSELF

17. Suppose we doubled the thermal energy of a rock that had a temperature of 7°C = 45°F = 280 K. What would its new temperature be?
    a. 14°C
    b. 90°F
    c. 560 K
    d. All of the above.
18. If the temperature of an object were halved, the wavelength where it emits the most amount of light will be
    a. 4 times longer.
    b. 2 times longer.
    c. the same.
    d. 2 times shorter.
    e. 4 times shorter.
19. If the temperature of an object doubles, the total amount of its thermal radiation will be
    a. the same.
    b. 2 times larger.
    c. 4 times larger.
    d. 8 times larger.
    e. 16 times larger.
20. Star A has a wavelength of maximum emission at 500 nm, while star B has a wavelength of maximum emission at 600 nm. Which of the following can we conclude?
    a. Star A is more luminous than star B.
    b. Star B is more luminous than star A.
    c. Star A is hotter than star B.
    d. Star B is hotter than star A.
21. The color of a star indicates its
    a. size.
    b. luminosity.
    c. temperature.
    d. composition.

PART 2

UNIT 24

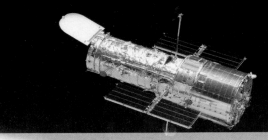

Identifying Atoms by Their Spectra

24.1 The Spectrum of Hydrogen
24.2 Identifying Atoms by Their Light
24.3 Types of Spectra
24.4 Astronomical Spectra

Learning Objectives

Upon completing this Unit, you should be able to:
- Describe the spectrum of hydrogen and explain how the visible lines arise.
- Discuss how spectra differ according to composition, temperature, and density.
- List Kirchoff's laws and describe the conditions that produce each type of spectrum.
- Describe the different properties that may alter the strength of a spectral line.

The keys to determining the composition and conditions of an astronomical body are the wavelengths of light it absorbs and emits. The technique used to capture and analyze the light from an astronomical body is called **spectroscopy.** In spectroscopy, electromagnetic radiation that is emitted or reflected by the object being studied is collected and spread into its component colors to form a spectrum (Unit 22). For visible wavelengths this is done by passing the light through a prism or through a grating consisting of numerous tiny, parallel lines (Figure 24.1).

We examined in Unit 21 how light and matter interact. Because photons of particular wavelengths are emitted and absorbed by atoms as electrons shift between orbitals, the spectrum of light will bear an imprint of the kinds of atoms the light interacted with. This allows astronomers to search for the "signatures" of various atoms by measuring how much light is present at each wavelength.

Spectroscopy is such an important tool for astronomers that we will examine it at a fairly high level of detail. Specifically, why does an atom produce a unique spectral signature? To understand that, we need to recall how light is produced.

24.1 THE SPECTRUM OF HYDROGEN

When an electron moves from one orbital to another, the atom's energy changes by an amount equal to the difference in energy between the two orbitals. Because the

FIGURE 24.1
Sketch of a spectroscope and how it forms a spectrum. Either a prism or a grating may be used to spread the light into its component colors.

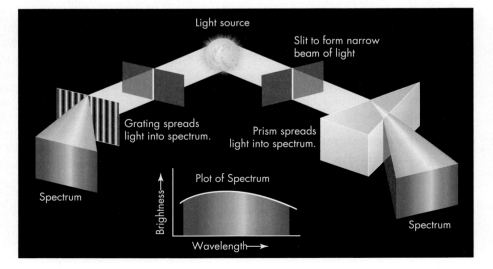

The Bohr atom

E_n = energy of atom in state n
n = level starting at ground state
$n = 1$

Scientists often use a nonstandard unit called an electron-volt (1 eV = 1.602 × 10^{-19} joules) for atomic energies. It is the amount of energy an electron gains when accelerated by an electric potential of 1 volt. In these units,
E_n(hydrogen) = -13.6 eV/n^2.

Concept Question 1

In what ways is the hydrogen energy level formula similar to the formula describing gravitational potential energy (Unit 20)?

FIGURE 24.2
Emission of light from a hydrogen atom. The energy of an electron dropping from an upper to a lower orbital is converted to light. The light's color depends on the orbitals involved.

atom's energy is determined by what orbitals its electrons move in, orbitals are often referred to as **energy levels**. As we saw in Unit 21, the orbitals of electrons cannot have just any energy: the wavelike nature of matter allows them to have only discrete energy levels. The allowed energy levels are very difficult to calculate for all but the simplest atom, hydrogen, which consists of a single electron orbiting a proton.

In the early 1900s, Danish physicist Niels Bohr developed a model of the hydrogen atom that gives its possible energy levels. The **Bohr model** of hydrogen describes the electron as orbiting at a set of discrete radii. If we number the energy levels starting from the *ground state* as $n = 1, 2, 3$, and so on, then the electron's orbital radius is $0.053 \times n^2$ nanometers. (A nanometer is 10^{-9} meter; see Unit 3.) The energy of level n can be described by the formula:

$$E_n(\text{hydrogen}) = -2.180 \times 10^{-18} \text{ joule}/n^2.$$

As n gets large, the electron is far from the nucleus, where the electrical attraction is weak, and it is natural to describe the energy of the orbital as being close to zero. As the electron is drawn in closer to the nucleus, it drops to a lower energy and emits a photon. The energy levels must therefore be less than zero, which is why there is a minus sign in the equation.

Suppose we look at an electron shifting from orbital 3 to orbital 2, as shown in Figure 24.2A. The wavelength of the emitted light can be calculated from the energy difference ΔE of the levels and the relation between energy and wavelength: $\Delta E = hc/\lambda$ (Unit 22). Using the energy-level formula we find:

$$\lambda = \frac{h \times c}{\Delta E} = \frac{1.987 \times 10^{-25} \text{ joule} \cdot \text{m}}{2.180 \times 10^{-18} \text{ joule} \left(\frac{1}{2^2} - \frac{1}{3^2}\right)}$$

$$= \frac{9.115 \times 10^{-8} \text{ m}}{(1/4 - 1/9)} = 6.563 \times 10^{-7} \text{m}.$$

This wavelength of about 656 nanometers is in the visible part of the spectrum, a red color. An electron dropping from orbital 3 to orbital 2 in a hydrogen atom always produces light of this wavelength. In fact, in astronomical images where there is glowing red-colored gas, it is almost certainly produced by hydrogen atoms undergoing this transition.

If, instead, the electron moves between orbital 4 and orbital 2 (Figure 24.2B), there will be a different change in energy because orbital 4 has higher energy than orbital 3. The larger energy change corresponds to a shorter-wavelength photon. A calculation of its energy change leads in this case to a wavelength of 486 nanometers, a turquoise blue color.

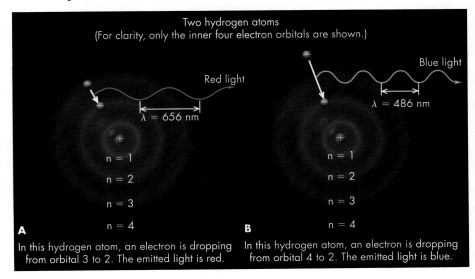

A. In this hydrogen atom, an electron is dropping from orbital 3 to 2. The emitted light is red.

B. In this hydrogen atom, an electron is dropping from orbital 4 to 2. The emitted light is blue.

24.2 IDENTIFYING ATOMS BY THEIR LIGHT

Suppose we heat up a gas. Heating speeds up the atoms in the gas, causing more forceful and frequent collisions that can knock an atom's electrons into more energetic orbitals. At the same time, the electrical attraction between the nucleus and the electron draws the electron back down to lower energy levels. As the electron shifts downward, the atom's energy decreases, and the energy that is lost appears as a photon.

If we spread the light from a hot gas into a spectrum, we will see that in general the spectrum contains light at only certain wavelengths. In a gas made up of hydrogen atoms, we will see light from atoms undergoing changes between many different energy levels simultaneously, as depicted in Figure 24.3. Thus, hydrogen gas produces a spectrum containing the red 656-nanometer and turquoise 486-nanometer lines, as well as violet light corresponding to electrons dropping from orbitals 5 to 2, and from higher energy levels down to the $n = 2$ energy level. This is known as an **emission-line spectrum.** This set of lines is the "signature" of hydrogen, unique to this element, and it allows astronomers to determine whether an astronomical object contains hydrogen.

The transitions between the $n = 2$ energy level and higher levels are known as **Balmer lines.** They are named for the Swiss mathematician who discovered the numerical relationship between their wavelengths in the late 1800s even before the quantum nature of matter was understood. These lines are particularly important in astronomy because hydrogen is such a common element and the wavelengths

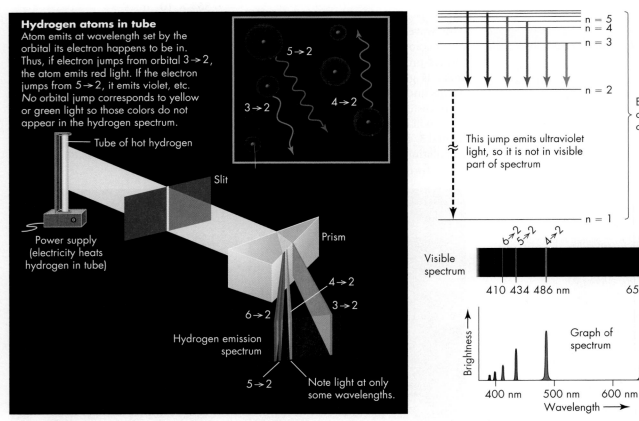

FIGURE 24.3
The spectrum of hydrogen in the visible wavelength range.

> **Concept Question 2**
>
> Can an atom in a highly excited state generate a low-energy photon? Can an atom in a low energy level produce a photon with high energy?

are in the visible part of the spectrum. Spectral lines of hydrogen are produced by transitions between other pairs of energy levels, but these lie outside the wavelength range that our eyes can detect. Transitions to the ground state are in the ultraviolet part of the spectrum, while transitions between higher levels are in the infrared or even longer wavelengths.

If instead of hydrogen we look at heated helium, we see a very different spectrum. The reason is simple. Helium has two electrons instead of one, and they interact with each other. This leads to a more complex set of electron energy levels and thus a different set of spectral lines, as you can see in Figure 24.4. The spectrum thus becomes a means of identifying what atoms are present in a gas.

In fact, helium was first discovered in 1868 by spectroscopic observations of the Sun during an eclipse. Astronomers recognized that the wavelength of the strong yellow line in its spectrum had no known counterpart, and they gave the name *helium* to the unknown element after the Greek word for the Sun, *helios*. Helium was not discovered by chemists on Earth until the 1890s.

For each element present in a gas we will see specific sets of spectral lines. We will see no light at most other colors because there are no electron orbital transitions corresponding to those energies. Therefore, the spectrum shows a set of colored lines separated by wide, dark gaps. For elements other than hydrogen, there is no simple mathematical formula we can write down to describe the energy levels, but the patterns of spectral lines have been determined from laboratory studies.

It is also possible to identify atoms in a gas from the wavelengths at which they *absorb* light. Light can be absorbed by an atom only if the energy of its wavelength corresponds to the energy difference between two energy levels in the atom. If the wavelength does not match, or the atom is not in the right starting state, the

> Transitions between energy levels must obey certain rules, for example angular momentum must be conserved (Unit 20.3). As a result, transitions are not possible between all pairs of energy levels.

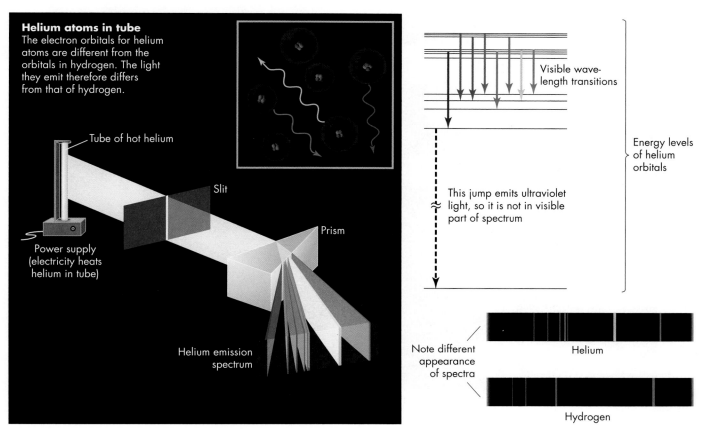

FIGURE 24.4
The spectrum of helium in the visible wavelength range.

photon cannot be absorbed. The photon will simply move past the atom, leaving itself and the atom unaffected.

For example, suppose we shine a beam of light that initially contains all the colors of the visible spectrum through a cloud full of hydrogen atoms. If we examine the spectrum of the light after it has passed through the cloud, we will find that certain wavelengths of the light are missing from the spectrum (Figure 24.5). In particular, the spectrum will contain gaps that appear as dark lines at 656 nanometers and 486 nanometers, precisely the wavelengths at which the hydrogen atoms emit. The **absorption-line spectrum** is, in effect, the opposite of the emission-line spectrum, because an atom's possible absorption lines have exactly the same wavelengths as its possible emission lines.

The gaps in the spectrum are created when photons at 656 nanometers and 486 nanometers interact with hydrogen atoms' electrons, lifting them from orbital 2 to 3 or orbital 2 to 4, respectively. Light at other wavelengths in this range has no effect on the atoms. Thus, we can tell that hydrogen is present from either its emission or its absorption spectral lines.

In our discussion we have considered light emitted and absorbed by individual atoms in a gas. If the atoms are linked to one another to form molecules, such as water or carbon dioxide, the new configuration of the electron orbitals results in an entirely different set of emission and absorption lines, so the molecules can also be identified from spectroscopy. In fact, even solid objects may imprint spectral lines on light that interacts with them. For example, when light from the Sun

Concept Question 3

If hydrogen atoms must be in the second energy level to absorb Balmer-line photons, how might hydrogen's ability to absorb these photons be affected by temperature?

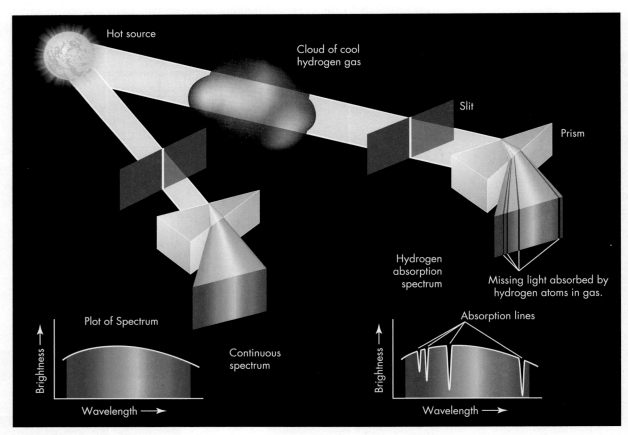

FIGURE 24.5

A hot, dense substance produces a continuous spectrum. Atoms in an intervening gas cloud absorb only those wavelengths whose energy equals the energy difference between their electron orbitals. The absorbed energy lifts the electrons to upper orbitals. The lost light makes the spectrum darker at the wavelengths where it is absorbed.

> **Concept Question 4**
>
> How could you use a spectrum produced from sunlight reflected off the atmosphere of Venus to determine something about Venus's atmospheric composition?

reflects from an asteroid, spectral features appear that were not present in the original sunlight. This gives astronomers information about the surface composition of bodies too cool to emit significant light of their own.

We conclude that in general we can identify the kinds of atoms or molecules that are present by examining either the bright or the dark spectral lines. Gaps in the spectrum at 656 nanometers and 486 nanometers imply that hydrogen is present. Similar gaps at other wavelengths would show that other elements are present. By matching the observed gaps to a directory of absorption lines, we can identify the atoms and molecules that are present. This is the fundamental way astronomers determine the chemical composition of astronomical bodies.

24.3 TYPES OF SPECTRA

The basic forms of spectra we have discussed can be categorized into three classes, and each implies a set of physical conditions. This classification system was first formulated by the German physicist Gustav Kirchhoff in the mid-1800s, so it is usually known as **Kirchhoff's laws**:

1. Some sources emit light in such a way that the intensity changes smoothly with wavelength and all colors are present, as we saw for thermal emission in Unit 23. We say such a light has a **continuous spectrum** (Figure 24.6). For a source to emit a continuous spectrum, its atoms must in general be packed so closely that the electron orbitals of each atom are influenced by the presence of neighboring atoms. When an atom is interacting with other atoms, the electron energy levels become altered by various amounts so that each atom can interact with photons of different energies, allowing photons of all wavelengths to be emitted or absorbed. Such conditions are typical of solid or dense materials such as the heated filament of an incandescent lightbulb, a glowing piece of charcoal, or the denser gas in the atmosphere of a star.

2. Conversely, an emission-line spectrum (Figure 24.7) implies that the atoms or molecules are well separated and the atoms are hot enough to have excited the electrons into higher energy levels. Isolated atoms or molecules all behave essentially identically, emitting light only at the specific wavelengths associated with their particular set of energy levels. Emission-line spectra are usually produced by hot, low-density gas, such as that in a fluorescent tube, the aurora, or interstellar gas clouds.

3. An absorption-line spectrum is produced when light from a hot, dense body (as in law 1 above) passes through cooler gas between it and the observer. In this case, nearly all the colors are present, but light is either missing or much dimmer at the specific wavelengths corresponding to energy transitions in the gas atoms (Figure 24.8). Absorption-line spectra are seen, for example, in most stars' spectra, produced by cooler, low-density gas lying above hotter layers (which produce a continuous spectrum). Cool interstellar gas clouds between a star and us can also produce absorption lines in a star's spectrum.

Kirchhoff's laws give us a starting point for interpreting the physical conditions that need to be present to produce each kind of spectrum we have discussed. If we see absorption lines, there must be a hotter, denser material shining through a cooler, low-density gas. If we see emission lines, there must be a hot gas. Of course, *hot* and *cool* are relative terms. If an atom or molecule has an excited state that requires very little energy to reach, we might see emission lines at what would be considered a low temperature on Earth. And when we see absorption lines in the atmosphere of a star, the gas doing the absorbing may be hot by our standards; it just needs to be cooler than the surface of the star shining through it.

FIGURE 24.6
Continuous spectrum. Normally produced by a hot solid or hot dense gas.

FIGURE 24.7
Emission-line spectrum. Generally produced by a hot, low-density gas.

FIGURE 24.8
Absorption-line spectrum. This spectrum results when light with a continuous spectrum passes through a cool gas.

24.4 ASTRONOMICAL SPECTRA

The spectrum of the Sun is shown in Figure 24.9. The spectrum is shown in two different styles. Figure 24.9A is an image of the light after it has been spread out by a prism or grating. Astronomers often find it more useful to plot a line graph of the brightness of the light at each wavelength as in Figure 24.9B.

The Sun's spectrum is typical of most stars. It shows a continuous spectrum interrupted by narrow absorption lines. In the graph (Figure 24.9B), the absorption lines are seen as sharp dips in the brightness. Astronomers can often identify which elements produce these absorption lines from large directories of spectral lines. By matching the wavelength of the line of interest to a line in the directory, astronomers can determine what kind of atom or molecule created the line.

Some of the lines are faint and hard to see, whereas other lines are obvious and strong. The strength or weakness of a given line depends on the number of atoms or molecules absorbing (or emitting, if we are looking at an emission line) at that wavelength. Complicating the interpretation, the number of atoms or molecules that can absorb or emit depends not just on how many of them are present but also on their temperature and "transition probabilities." Some transitions are more easily excited than others because of internal properties of an atomic element, and these can often be determined by laboratory experiments. Knowing these probabilities and the temperature, however, astronomers can deduce from the strength of emission and absorption lines the relative quantity of each atom.

We can see from the spectral lines in Figure 24.9 that the Sun contains hydrogen from the features at 486 nm and 656 nm discussed in Section 24.1. In fact, when a detailed calculation is made of the strength of the lines, it turns out that roughly 90% of the atoms in the Sun are hydrogen. When the Sun's spectrum is examined in detail, thousands of absorption lines from nearly every element are found. Table 24.1 shows a partial list of the Sun's composition based on an analysis of the absorption lines. Analyses of other stars and other astronomical objects show that the composition of the Sun is quite typical. The first eight elements listed in Table 24.1 are the eight most abundant elements throughout the universe, accounting for over 99.8% of all atomic matter.

Table 24.1 also shows the energy needed to ionize each element. The energy listed is the amount needed to remove an electron when the atom is in its ground state. For a star such as the Sun, many of the atoms are already in an excited state,

> The relative number of atoms of each type is also indicated in the periodic table inside the back cover of this book.

Mathematical Insight

Because the 10% of atoms in the Sun besides hydrogen are more massive than hydrogen, they contribute about 29% of the overall mass, and hydrogen contributes "only" about 71% of the Sun's mass.

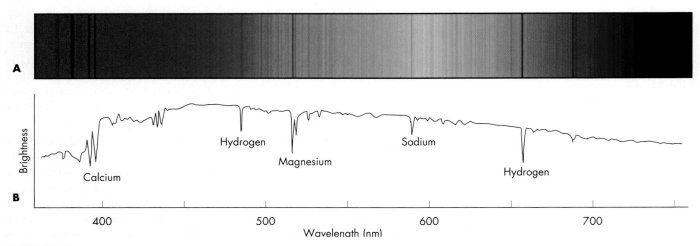

FIGURE 24.9

(A) The spectrum of the Sun. Note the narrow dark absorption lines. (B) A graphical representation of the Sun's spectrum. Several of the absorption lines are identified according to the element in the Sun's atmosphere that is causing the absorption.

Mathematical Insight

The ionization energy for hydrogen in Table 24.1 can be converted to a wavelength using the formula in Unit 22.2:

$$\lambda = \frac{h \times c}{E}$$
$$= \frac{6.626 \times 10^{-34} \text{J·sec} \times 2.998 \times 10^8 \text{m/sec}}{2.179 \times 10^{-18} \text{J}}$$
$$= 9.116 \times 10^{-8} \text{ m} = 91.2 \text{ nm}.$$

Wavelengths of light shorter than this will ionize hydrogen from its ground state.

TABLE 24.1 Composition of a Typical Star, Our Sun

Element	Number of Protons (Atomic Number)	Relative Number of Atoms	Percentage by Mass	Energy to Remove Outermost Electron
Hydrogen	1	10^{12}	71.1%	2.179×10^{-18} J
Helium	2	9.64×10^{10}	27.4%	3.940×10^{-18} J
Carbon	6	2.88×10^8	0.25%	1.804×10^{-18} J
Nitrogen	7	7.94×10^7	0.08%	2.328×10^{-18} J
Oxygen	8	5.75×10^8	0.65%	2.182×10^{-18} J
Neon	10	8.91×10^7	0.13%	3.454×10^{-18} J
Silicon	14	4.07×10^7	0.06%	1.306×10^{-18} J
Iron	26	3.47×10^7	0.14%	1.266×10^{-18} J
Gold	79	8	0.00000011%	1.478×10^{-18} J
Uranium	92	0.4	<0.000000007%	0.992×10^{-18} J

Note: This table lists eight of the most common elements along with gold and uranium to illustrate how extremely rare they are. Data on relative number of atoms drawn from Lodders (2003) *The Astrophysical Journal*, vol. 591, pp. 1220–1247.

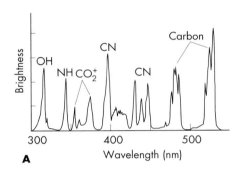
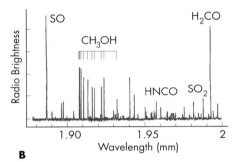
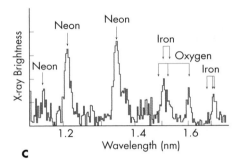

FIGURE 24.10
Emission-line spectra at a variety of different wavelengths. (A) A spectrum of a comet at visible and ultraviolet wavelengths. (B) A microwave spectrum of a cold interstellar cloud. (C) An X-ray spectrum of hot gas from an exploding star. Many of the atoms and molecules responsible for the emission lines are identified.

so less energy is required to ionize the atom. For example, hydrogen atoms in the $n = 2$ level require only one-fourth as much energy to be ionized as is listed in the table. This is the energy of a 365-nm photon. Ultraviolet photons can ionize many kinds of atoms in a star's atmosphere.

When an atom can be ionized by photons, the effect on a spectrum is different from the narrow absorption lines we have seen so far. *Any* photon that exceeds the **ionization energy** threshold can be absorbed by the atom. As a result, we do not see a narrow absorption line; instead, all photons at wavelengths shorter than the threshold can be absorbed. Stars' spectra generally show a sharp drop in light output at ultraviolet wavelengths because of the photons' ability to ionize atoms.

Spectroscopy is not limited to stars or to visible wavelengths of light. Figure 24.10 shows spectra of a comet, an interstellar gas cloud, and an exploding star. All three spectra show emission lines, but in very different parts of the electromagnetic spectrum. The best region to examine for spectral lines often depends on the temperature of the gas. A radio spectrum is well suited to a cold interstellar cloud, while X-ray observations are appropriate for a hot exploding star.

In both the comet and the interstellar cloud, the temperatures are low enough for molecules—some quite complex—to have formed. By contrast, the wavelengths of the emission lines seen in the X-ray spectrum tell us that these atoms are highly ionized. In fact, the gas is so hot that 7 electrons have been stripped from the neon atoms, 9 from the oxygen, and 16 from the iron atoms!

KEY POINTS

- The identity of atoms and molecules can be determined from the wavelengths of their spectral lines.
- Hydrogen has a set of energy levels that can be calculated fairly simply, but other atoms and molecules have complex spectra often determined through laboratory measurements.
- Hydrogen's Balmer lines, which are seen at visible wavelengths, occur between its second energy level and higher levels.
- If a continuous spectrum of light shines through a cool cloud of gas, dark absorption lines occur at the same wavelengths where emission lines might be seen from a hot gas.
- Kirchhoff identified the physical conditions giving rise to three kinds of spectra: continuous, emission-line, and absorption-line.
- From detailed analyses of the spectra of astronomical objects, we can determine their composition.
- The strength of spectral lines depends on the abundance of the particular atom or molecule and the temperature.

KEY TERMS

absorption-line spectrum, 172
Balmer lines, 170
Bohr model, 169
continuous spectrum, 173
emission-line spectrum, 170
energy level, 169
ionization energy, 175
Kirchhoff's laws, 173
spectroscopy, 168

CONCEPT QUESTIONS

Concept Questions on the following topics are located in the margins. They invite thinking and discussion beyond the text.

1. Comparing atomic energy to gravitational energy. (p. 169)
2. Photon energies versus excitation state. (p. 171)
3. Temperature and the absorption of Balmer lines. (p. 172)
4. Determining composition from reflected sunlight. (p. 173)

REVIEW QUESTIONS

5. What can astronomers learn about the chemical and physical properties of astronomical objects from their spectra?
6. Why don't atoms emit a continuous spectrum?
7. How do astronomers estimate the relative abundance of different elements from a spectrum?
8. How do the wavelengths of an element's emission lines relate to the wavelengths of its absorption lines?
9. What elements does spectroscopy indicate are the most common in the Sun's atmosphere?
10. How is it possible for the same cloud of gas to provide an emission spectrum to one observer and an absorption spectrum to a different observer?
11. If you were to look at the spectrum of a candle flame, what sort of spectrum would you expect to see: absorption, emission, or continuous? Why?

QUANTITATIVE PROBLEMS

12. In the formula for hydrogen's energy levels, find the energy difference between levels $n = 1$ and $n = 2$, and determine the wavelength of a photon with this energy. What part of the electromagnetic spectrum is this?
13. Using Wien's law (Unit 23.4), find the temperature of a star that has its wavelength of maximum emission at the wavelength needed to raise an electron from the ground state to the $n = 2$ state.
14. Using the formula for hydrogen's energy levels, calculate the energy of a photon emitted when hydrogen goes from level 143 to level 142 (a transition that has been observed by astronomers). What is the wavelength of this spectral line? What part of the electromagnetic spectrum is this?
15. Explain the physical meaning of a transition from $n = 1$ to $n =$ infinity. For hydrogen, calculate the energy associated with this transition (recall that anything divided by an infinite number is basically 0).
16. From the data in Table 24.1, what wavelength photon can ionize helium? What wavelength can ionize uranium?

TEST YOURSELF

17. An astronomer finds that the visible spectrum of a mysterious object shows bright emission lines. What can she conclude about the source?
    a. It contains cold gas.
    b. It is a hot, solid body.
    c. It has many different light sources, one for each emission line.
    d. It contains hot, relatively tenuous gas.
    e. It contains hot, tightly compacted gas.
18. When sunlight is passed through a spectroscope an absorption spectrum is observed because the Sun
    a. is made almost entirely of hot, low-density gas.
    b. is made almost entirely of cool, low-density gas.
    c. has a warm interior that shines through hotter, high density gas.
    d. has a hot interior that shines through cooler, low-density gas.
19. Suppose we detect the 656-nm wavelength of the $n = 3$ to $n = 2$ transition of hydrogen from a 5000-K gas cloud. If the cloud was heated to 10,000 K, what would the wavelength of the $n = 3$ to $n = 2$ transition now be?
    a. 278 nm
    b. 1312 nm
    c. 5656 nm
    d. 658 nm
    e. 656 nm
20. What do astronomers need to know before they can estimate the relative amount of different elements in a star's atmosphere?
    a. The number of absorption lines from each element
    b. The strength of absorption from each element's lines
    c. The temperature of the star's atmosphere
    d. Both a and b
    e. Both b and c

The Doppler Shift

25.1 Calculating the Doppler Shift
25.2 Astronomical Motions

Learning Objectives

Upon completing this Unit, you should be able to:
- Describe the Doppler effect and how it depends on the direction of motion.
- Calculate the Doppler shift from a velocity and vice versa.
- Describe Doppler effects that arise from rotation and internal motions of an object.

When we observe light from an object that is moving toward or away from us, we will find that the wavelengths we receive from it are altered by its motion. If it moves toward us, its wavelengths will be shorter. If it moves away from us, its wavelengths will be longer. Furthermore, the faster the source moves, the greater those changes in wavelength will be. This change in wavelength caused by motion is called the **Doppler shift,** and it is a powerful tool for measuring the speed of approach or recession of astronomical objects. Not all motion produces a Doppler shift. Motion perpendicular to the line of sight creates no Doppler shift because the source is neither approaching nor moving away from us. Astronomers use this shift to measure the motions and rotation of a wide variety of astronomical objects. The whole universe is in motion, and the motions reveal the presence of unseen planets, massive black holes, and dark matter.

The Doppler effect

Doppler shift

25.1 CALCULATING THE DOPPLER SHIFT

The Doppler shift occurs for all kinds of waves. You may have heard the Doppler shift of sound waves from a car as it passes you on a highway or a racetrack: The sound of the engine drops from a high-pitched whine to a low-pitched hum (corresponding to a shift from short to longer wavelengths of the sound) as the approaching vehicle passes you and then moves away (Figure 25.1A). The same thing happens with electromagnetic waves, although the change in wavelength at highway speeds is too subtle for our eyes to detect. However, electronic devices can detect the shift as when, for example, the Doppler shift of a radar beam that bounces off your car reveals to a law enforcement officer how fast your car is moving (Figure 25.1B).

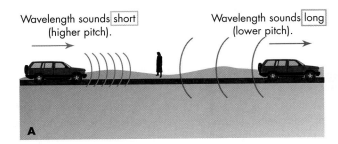
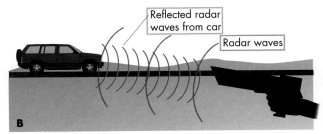

FIGURE 25.1
The Doppler shift: Waves shorten as a source approaches and lengthen as it recedes. (A) The Doppler shift of sound waves from a passing car. (B) The Doppler shift of radar waves in a speed trap.

FIGURE 25.2
(A) A stationary source of light sends out waves whose crests are equally spaced in all directions. (B) A source of light moving away from you will increase the distance the next wave must travel to reach you by $\Delta\lambda$, so the distance between wave crests grows longer by that amount. The position of the light source when it emitted the first wave crest is marked 1; when it emitted the next wave crest is marked 2. (C) Waves travel out spherically. Wave crest 1 was emitted when the light source was at position 1; wave crest 2 from position 2; and so on. The observed wavelength depends on the component of a source's motion that is along the observer's line of sight.

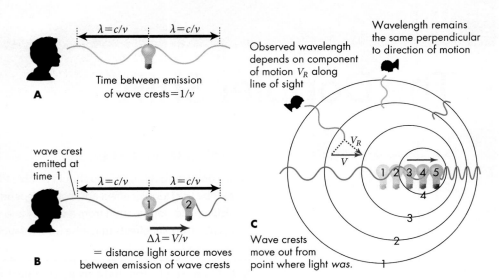

Mathematical Insights

$\Delta\lambda = V_R \times \Delta t = V_R \times \frac{1}{\nu} = V_R \times \frac{\lambda}{c}$

Multiply both sides by c/λ and rearrange to get the Doppler formula.

V_R = Radial velocity (toward or away from observer)
$\Delta\lambda$ = Change from rest wavelength
λ = Rest or normal wavelength
c = Speed of light

Concept Question 1

Wien's law indicates that the wavelength of light is longer from a cooler object. How could we tell the difference between this and an object that is moving away from us?

We can understand the Doppler shift by thinking about the spacing between wave crests from a light source. Suppose the source emits light at a particular frequency, like a spectral line from a hot gas (Unit 24). For a frequency ν, the time interval between wave crests departing the light source is, by definition, the inverse of the frequency, $1/\nu$. If the light source is not moving, the crests reach us with the same frequency, and the wave crests are separated by the normal or "rest" wavelength $\lambda = c/\nu$ (Unit 22), as illustrated in Figure 25.2A.

Now consider if the object is moving directly (or "radially") away from us at a speed V_R, known as the **radial velocity.** The light source still generates wave crests at the same frequency as when it was stationary, but in the time between the generation of one wave crest and the next, the source has moved farther away, as illustrated in Figure 25.2B. The resulting wavelength we observe, λ_{obs}, is therefore longer than the rest wavelength λ.

The difference between λ_{obs} and λ is written $\Delta\lambda$ (called "delta lambda" or "the change in λ"). For example, suppose a spectral line is generated at 656 nm (λ), but the line instead appears to be at 658 nm (λ_{obs}). The change in wavelength is

$$\Delta\lambda = \lambda_{obs} - \lambda = 658 \text{ nm} - 656 \text{ nm} = 2 \text{ nm}.$$

The change in wavelength is equal to the distance the light source travels in the time between emitting wave crests, $1/\nu$. This distance is the product of the speed and time, so $\Delta\lambda = V_R \times 1/\nu = V_R \times \lambda/c$. Multiplying both sides of this equation by c/λ gives us the Doppler formula for finding the light source's radial velocity:

$$V_R = \frac{\Delta\lambda}{\lambda} \times c.$$

Note how an observer to the right of the light source in Figure 25.2B would see the wavelength shortened by $\Delta\lambda$. By convention, $\Delta\lambda$ and V_R are negative if the object is approaching us (so that the observed wavelength, λ_{obs}, is shorter than the normal rest wavelength) and positive if it is moving away from us. In the previous example, the speed is

$$V_R = \frac{2 \text{ nm}}{656 \text{ nm}} \times c = 0.003 \times 300{,}000 \text{ km/sec} = 900 \text{ km/sec}$$

and the object must be moving away because the wavelength has grown longer.

Doppler shift measurements can be made at any wavelength of the electromagnetic spectrum. But regardless of the wavelength region observed, astronomers refer to shifts that *increase* the measured wavelength of the radiation as **redshifts** and those that *reduce* the wavelength as **blueshifts.** This harkens back to the early days of spectroscopy when only the visible portion of the electromagnetic spectrum was observable.

> **Clarification Point**
>
> A photon's speed cannot be boosted or reduced by motion of the source; only its wavelength (and therefore its energy) is altered.

Light waves do not just move in two directions, of course. A wave crest expands out spherically around the position from which it was emitted. When the light source is moving, the wave crests make a series of shells that are packed tightest in the direction in which the source is moving. This is illustrated in Figure 25.2C. Note how an observer who sees light coming from a direction perpendicular to the direction of motion sees no change in wavelength. Other observers see the wavelength lengthened or shortened according to the component of the light source's velocity that is along their line of sight.

25.2 ASTRONOMICAL MOTIONS

Observations of all sorts of astronomical sources reveal that they have redshifts and blueshifts. Based on the motions of thousands of nearby stars, the Solar System is moving at about 19 kilometers per second toward the constellation Hercules. We find that the stars in our Galaxy are orbiting at over 200 kilometers per second, and our Galaxy is moving at even larger speeds relative to other nearby galaxies.

We can discover motions within an object through detailed examination of its spectral lines. If the atoms within a gas are moving because of thermal motions or turbulence, the spectral lines will be spread over a wider range of wavelengths because the gas atoms will have different radial velocities. This is known as **Doppler broadening**. If a star is rotating, one side of the star's surface moves toward us, while the other moves away. This causes broadening of spectral lines with a particular shape, as illustrated in Figure 25.3. Similarly, we can measure the rotation speeds of an entire galaxy by measuring the broadening of spectral lines.

Within the Solar System, the rotation of objects allows astronomers to carry out radar measurements that can give detailed "pictures" of the surface of various planets, moons, and asteroids. Transmitters can send a powerful pulse of radio waves toward an asteroid that appears as just a point of light in the largest telescopes. The waves bounce off the asteroid and return to Earth with different time delays depending on how far each point is from the Earth, and they have different Doppler shifts across the asteroid because of its rotation. This gives us a two-dimensional image that is not exactly like a photographic view, but reveals the structure and surface features of the object. A number of asteroids (Unit 43) that have passed near to the Earth have been imaged in this way (Figure 25.4).

> **Concept Question 2**
>
> Some spacecrafts map the surface of planets using radar from orbit. How will the Doppler shift of the return signal depend on the direction from the spacecraft?

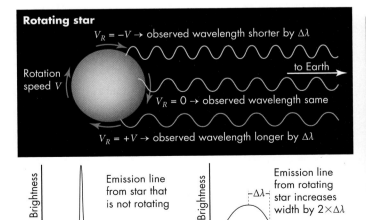

FIGURE 25.3
A rotating object has different Doppler shifts over its surface, so the spectral line (illustrated at bottom) becomes broader.

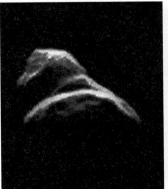 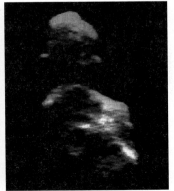

FIGURE 25.4
Two Doppler images of the asteroid Toutatis. The time delay and wavelength of radar beams bounced off a rotating object's surface provide a picture of its structure. These two views show that the asteroid is shaped something like a bowling pin.

KEY POINTS

- The wavelength of light we observe from an object shows a Doppler shift if it moves toward or away from us.
- The wavelength is lengthened (redshifted) if the source is moving away from us, shortened (blueshifted) if it is moving toward us.
- The size of the wavelength shift is proportional to the radial velocity of the object, and is given a negative value if approaching.
- The radial velocity is the component of motion along the observer's line of sight, so any portion of an object's motion that is perpendicular to the line of sight will produce no Doppler shift.
- The entire Solar System is moving through space relative to nearby stars, and our Galaxy is moving relative to other galaxies.
- If a gas has internal motions, the spectral lines will be Doppler broadened.
- Rotating objects also have broadened spectral lines, from which the rotation speed can be determined.
- Using radar, it is possible to map the surface of rotating objects by measuring the Doppler shift and time delay of the return signal.

KEY TERMS

blueshift, 178
Doppler broadening, 179
Doppler shift, 177
radial velocity, 178
redshift, 178

CONCEPT QUESTIONS

Concept Questions on the following topics are located in the margins. They invite thinking and discussion beyond the text.

1. Temperature difference versus redshift. (p. 178)
2. Radar imaging from orbit. (p. 179)

REVIEW QUESTIONS

3. What is the Doppler shift?
4. What is the difference between an object's velocity and its radial velocity?
5. Do we always see a Doppler shift when an object is moving?
6. How could two objects be moving at different velocities but show the same redshift?
7. If you made a spectrum of a star and it was redshifted by 1 nm at the red end of the spectrum, how much would it be redshifted at the violet end?
8. What is the difference between a redshift and a blueshift?
9. What motions change the shape of an object's spectral lines?

QUANTITATIVE PROBLEMS

10. About how fast would you have to be driving toward a red stoplight for it to appear green?
11. The $n = 3$ to $n = 2$ transition of hydrogen is normally at 656.3 nm. If it is detected from a star at 655.5 nm, first argue if the star is approaching or receding, then calculate the radial velocity of the star.
12. Suppose that a star is rotating at a speed of 50 km/sec at its equator. Its spectral lines will be spread out because one side of the star will be redshifted and the other side will be blueshifted. For the 656.3-nm line of hydrogen, over how wide a range in wavelengths will the spectral line be spread?
13. Hydrogen produces a spectral line at 656.3 nm. Oxygen produces a spectral line at 645.5 nm. If you detected a galaxy with a spectral line at 655.0 nm, what Doppler shift would you derive if you assumed it was produced by hydrogen? by oxygen? How could you determine which was correct?
14. Astronomers study planets in the Solar System with radar. Suppose you were observing Venus, which orbits the Sun at 35 km/sec, from the Earth, which orbits at 30 km/sec. Draw a diagram of the two planets' orbits (Venus orbits at about 0.7 AU) and use your diagram and vectors representing the velocities to estimate the radial velocity we would see when Venus is at its greatest elongations and when it is in conjunction with the Sun. What would be the change in wavelength for a 10-cm radar signal bounced off Venus at these different locations?
15. At its equator Mercury rotates at 3.0 m/sec. If an astronomer sent a radar signal at precisely 10 cm to bounce off of Mercury, what would be the breadth of the return signal?
16. Hydrogen produces spectral lines at radio wavelengths, notably at 21.1 cm. If a galaxy is moving away from us at 10% of the speed of light, at what wavelength will we detect this line? Convert this into a frequency.

TEST YOURSELF

17. If an object's spectral lines are shifted to shorter wavelengths, the object is
    a. moving away from us.
    b. moving toward us.
    c. very hot.
    d. very cold.
    e. emitting X-rays.
18. What properties of stars can be determined from their spectra?
    a. Chemical composition
    b. Surface temperature
    c. Radial velocity
    d. Rotation speed
    e. All of the above.
19. The wavelength of a radio emission line from a galaxy normally seen at 100 cm is detected at 105 cm. This implies that the galaxy is moving
    a. toward us at 5% of the speed of light.
    b. away from us at 5% of the speed of light.
    c. toward us at 5 centimeter per second.
    d. away from us at 5 centimeter per second.
20. Suppose that observations of a star revealed that the wavelengths of its spectral lines decreased for several weeks, then increased for several weeks, then repeated this pattern. Which of the following could best explain this behavior?
    a. The star is spinning.
    b. The star is in orbit with another object.
    c. The star is passing the Sun.
    d. This is an effect of Earth's own orbital motion.

Special Relativity

26.1 Light from Moving Bodies
26.2 The Michelson-Morley Experiment
26.3 Einstein's Theory of Special Relativity
26.4 Special Relativity and Space Travel
26.5 The Twin Paradox

Learning Objectives

Upon completing this Unit, you should be able to:
- Show how we "normally" expect velocities to add, and how light disobeys this.
- Describe how light's behavior raised questions about the nature of space and time.
- Calculate the Lorentz factor for a particular speed, and explain what it measures.
- Discuss the twin paradox, and explain it using special relativity.

The history of astronomical study of the Solar System progressed from remote observation to direct exploration of individual bodies with space probes. The same will never be true of our study of the stars. Today's fastest space probes would take hundreds of generations to reach other stars. Even supposing a space probe were already orbiting the nearest star beyond the Solar System, Proxima Centauri, the distance is so vast that if we sent out a message saying "Are you there?" today, we would have to wait nine years to hear back "Yes." The Solar System is isolated not just by distance, but by the time it takes signals to travel over these immense distances.

In attempting to understand how light moves, **Albert Einstein** discovered the surprising result that *nothing* can travel through space faster than 299,792 kilometers per second—the speed of light. His discovery, called **special relativity**, has changed some of our most basic ideas about the nature of space and time. One of the central discoveries of special relativity is that motion fundamentally alters the way space and time behave. It also has implications for the possibility of ever traveling to distant stars.

The universe is very different from the popular vision of science fiction movies. There can be no dialogues at interstellar distances, and high-speed travelers will find themselves traveling rapidly into their home planet's future. Yet, as strange as special relativity seems, it is now as solidly confirmed as that the Earth is not flat.

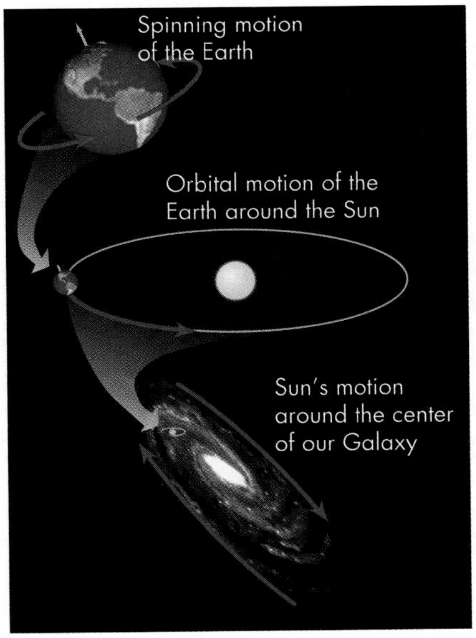

26.1 LIGHT FROM MOVING BODIES

The universe is full of motion at every scale (Figure 26.1). Once the age-old notion of the Earth being fixed at the center of things crumbled, it soon became clear that the Sun cannot be assumed to be stationary either. We now know that all stars move relative to one another as they orbit within our Galaxy, and that galaxies move at even greater speeds relative to one another.

Astronomers began detecting the motions of stars in the 1800s both by their slowly shifting positions in the sky and their Doppler shifts (Unit 25). These motions were measured relative to our local **rest frame**—a coordinate system tied to the Solar System's position and motion, in which it appears to us that we are at rest. In our rest frame we observe, for example, that nearly all stars are moving at tens or hundreds of kilometers per second in a wide variety of directions. However, it is important to recognize that an alien living in one of these other star systems could define its own rest frame. To this alien, it would appear that the Sun is moving through space. We might ask: Is there some way to decide which is stationary and which is not?

FIGURE 26.1
Motions are relative. We may think an object on the surface of Earth is stationary, but the surface is spinning, the Earth is orbiting the Sun, and the Sun is orbiting the center of our Galaxy.

Space itself might have its own rest frame. For example, light waves traveling through space might be similar to sound waves traveling through the air. In air, sound waves are carried along faster in the direction that the wind is blowing. By analogy, we might hypothesize that we should measure light traveling at different speeds in different directions depending on the Solar System's motion relative to the rest frame of space.

If there is no fundamental rest frame, we might predict light should behave more like a thrown ball or javelin, with photons receiving a boost or reduction in their speed depending on the motion of the emitter. For example, you can throw a baseball to first base faster if you are running toward the base rather than away from it. Similarly, you will find that you can throw a javelin faster and therefore farther if you are running forward when you release it (Figure 26.2). If you were to run *backward*, your throw would be weak and short even if you swing your arm just as hard. This is because the speed of the javelin (or baseball) relative to the ground is the vector sum of the speed you run and the speed you throw. Running backward therefore subtracts from the overall speed

Using the term we introduced at the beginning of this section, the javelin is traveling faster or slower in the target's rest frame depending on whether you move toward the target or away from it. In your own rest frame, the javelin moves just as fast (you throw it just as hard), but the difference in your rest frames combines with the speed of the javelin to produce the overall speed.

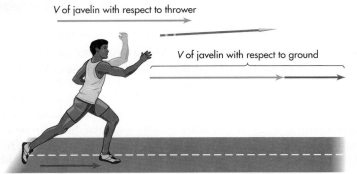

FIGURE 26.2
The speed of a javelin throw with respect to the ground is found by adding the velocity of the thrower across the ground to the speed with which the javelin is thrown.

The addition of relative speeds (you plus javelin) in this fashion is an example of what is sometimes called **Galilean relativity**. Galilean relativity gets its name because Galileo was one of the first scientists to recognize how motions add to and subtract from one another, and how the measurements of these motions depend on the rest frame of the observer.

Concept Question 1

What are some other examples of situations where different speeds add and subtract from each other?

If Galilean relativity applied to photons from a star moving toward us, we would expect the photons to have an increased speed toward us and to reach us in a shorter time. Likewise, if the star were moving away from us, we would expect the photons to have a decreased speed and take longer to reach us.

A difference in light speed would be obvious for one star in orbit around another star. Astronomers first detected stars that orbit each other in the early 1800s. An orbiting star will move toward us during one part of its orbit, and away from us half an orbit later (Figure 26.3). The speed of the motion is readily measured from the star's changing Doppler shift. If the photons traveled faster when the star was moving toward us, they would overtake photons that left the star earlier when it was moving away from us. The result would be that we would receive photons from the star when it was in different parts of its orbit in a scrambled order. Depending on the time delays, we might even see the star in several different parts of its orbit simultaneously!

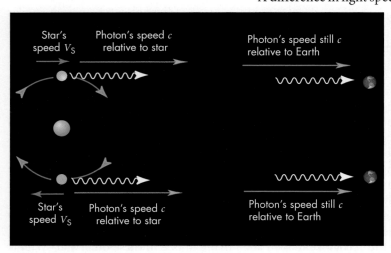

FIGURE 26.3
Light from a star in orbit reaches Earth at the same speed whether the star is moving toward or away from us.

The light we see from binary stars does nothing of the sort; we see no oddities in the stars' positions that would be expected if light's speed could change. Light therefore must move through space at the same speed no matter what the speed of the star. This resembles what we described for sound waves, or as another example, the waves a boat makes on a lake. The boat may move toward or away from us, but the waves travel through the water at the same speed.

In the 1800s scientists concluded that space must have its own rest frame something like the air or water in these examples. They believed space must contain some substance through which light waves always traveled at the same speed. They called this hypothesized substance the **aether**. However, as we will see in the next section, the behavior of light in other situations cannot be explained by an aether. Light's behavior requires instead a completely new understanding of space and time.

26.2 THE MICHELSON-MORLEY EXPERIMENT

By the late 1800s physicists were able to measure the speed of light with such high precision that it began to look as if it might be possible to detect the absolute motion of the Earth as it traveled through the hypothesized aether. As the Earth moved through the aether, it would appear to us that the aether was "flowing" past us. This flowing aether should affect the motions of photons moving through it, much as flowing water affects the motion of boats traveling through it. In other words, the photons should travel at different speeds, depending on the direction of their motion relative to the flow.

FIGURE 26.4
Photograph of the experimental apparatus devised by Albert Michelson and Edward Morley. An optical assembly on a granite slab floated on a pool of mercury, allowing extremely sensitive measurements of the speed of light in perpendicular directions illustrated by the red lines. No speed differences were found, no matter what direction the light traveled.

American scientists Albert Michelson and Edward Morley designed an apparatus in 1887 that could compare the speed of light moving in two perpendicular directions simultaneously (Figure 26.4). This was an extremely sensitive apparatus built on a massive granite slab that floated in a pool of mercury, which allowed them to rotate the slab smoothly while isolating the apparatus from any vibrations. They reasoned that if the slab was oriented in the right direction, then along one path through the apparatus the light would travel parallel to the flow of the aether. Along the other path the light would travel perpendicular to the aether. Therefore, the speed of the light would be different along the different paths.

However, when Michelson and Morley conducted measurements using their apparatus, they detected no difference in the speed of light along the perpendicular paths, no matter how they oriented the apparatus. They realized that this could just be bad luck during the first experiment—the combination of the Earth's motion around the Sun and the Sun's motion through space might just happen to make the Earth stationary relative to the aether at that point in its orbit. So they repeated their experiment many times over a year. At some point they should have easily detected the motion of Earth relative to the aether, but they found no sign of any relative motion at all.

The Michelson-Morley experiment has been called the most famous "failed" experiment in history, because it led to a revolution in physics. The results implied that there was no aether regulating the speed of light. However, if light was not moving relative to the aether, then physicists could not explain the constancy of the speed of light reaching us from sources moving at different speeds toward or away from us as we saw with the binary stars.

All experiments made then and now show that no matter how fast the source generating the light is moving, and no matter how fast the observer measuring the light is moving, the speed of light is always measured to be $c = 299{,}792{,}458$ meters per second. How can this be?

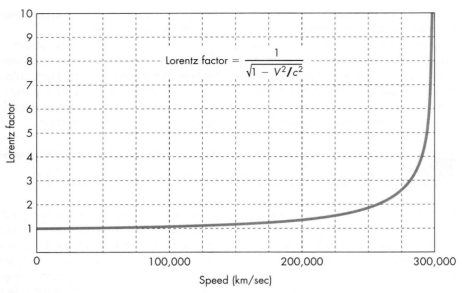

FIGURE 26.5
The Lorentz factor expresses the amount by which objects appear to compress in the direction of their motion. It is also the factor by which a moving clock appears to slow down, and the factor by which a moving object's mass appears to increase.

γ = Lorentz factor
V = speed relative to observer
c = speed of light (3.00×10^8 m/sec)

The idea of a contraction caused by motion was first proposed by Irish physicist George Fitzgerald. The Dutch physicist Hendrik Lorentz developed a model for explaining how the contraction might arise.

FIGURE 26.6
Albert Einstein (1879–1955).

One explanation offered in the late 1800s was that motion through space somehow causes matter to contract in the direction of motion through the aether. If matter contracts when it is moving, this would change our perception of length, so that we might be tricked into thinking the speed of light had not changed. For example, when Michelson and Morley were searching for a difference in the speed of light in two perpendicular directions, if the apparatus were compressed in the direction in which it was moving through space but not in the perpendicular direction, the path the light traveled would be different in the two directions. Such a contraction could cancel out the effect that Michelson and Morley were searching for. The factor by which the apparatus would need to contract is known today as the **Lorentz factor**. The Lorentz factor, usually denoted by the Greek letter gamma (γ), hypothesizes that an object's length shortens in the direction in which it is moving by a factor equal to

$$\gamma = \frac{1}{\sqrt{1 - V^2/c^2}}$$

where V is the speed of the object and c is the speed of light. The value of γ for different speeds V is plotted in Figure 26.5.

The Lorentz factor is close to 1 at small speeds. At 30 km/sec, the speed at which the Earth orbits the Sun, the Earth would contract by only a few centimeters in its direction of motion. However, at high speeds the contraction factor becomes very large, growing to infinity if the speed V were to reach the speed of light. The Lorentz contraction factor explained the Michelson-Morley results, but it could not explain the results of a variety of other experiments. It contains, however, an important idea that grew into a whole new concept about the nature of motion.

26.3 EINSTEIN'S THEORY OF SPECIAL RELATIVITY

In 1905 a 26-year-old graduate student named Albert Einstein (Figure 26.6) took on the problem of the seemingly inexplicable measurements of the speed of light. He was completing his physics degree while working in the Swiss patent office and supporting a new family. Yet in that one year alone, he completed his doctorate degree and wrote four papers in several areas of physics. Physicists widely agree that three of these papers were each worthy of a Nobel Prize! Einstein was little known at the time and had few colleagues with whom to discuss his ideas; but nonetheless in one of these papers he came up with a brilliant new approach to the question of the motion of light.

Einstein began by concentrating on the findings that light travels at the same speed no matter what the speed of its source or of the observer measuring the light. Even though many experiments had come to this conclusion, most physicists

TABLE 26.1 The Lorentz Factor at High Speeds

Speed	Lorentz Factor
0.87 c	2.0
0.97 c	4.1
0.99 c	7.1
0.999 c	22.4
0.9999 c	70.7
0.99999 c	223.6

had assumed it was an impossibility and so were seeking other explanations—such as errors in the experiments or the Lorentz contraction. Einstein, instead of thinking that a constant speed of light for all observers was an impossibility, accepted this finding as correct and proceeded to work out its consequences. He found that this led inevitably not only to a Lorentz-like contraction of space, *but also to a stretching of time by the same factor*. Even more important, he found that this contraction was not relative to some imagined aether filling space, but that these effects depended on just the relative motions of any two objects.

These alterations of space and time affect everything we see that moves relative to us. If a rocket moves by us at high speed, we will see it squashed in its direction of motion by a Lorentz factor that may be very large, as shown in Table 26.1. If we could watch a clock tick or measure the rate of an astronaut's heartbeat in the spaceship, we would discover that all these processes occur more slowly by the Lorentz factor. This is illustrated in Figure 26.7 for a spaceship traveling by the Earth at 87% of the speed of light, so its Lorentz factor is 2. When we observe the spaceship, it appears half as long, and each tick of its clock takes 2 seconds.

What is even more remarkable is that the mathematics Einstein worked out showed that the situation is exactly symmetrical. The astronaut moving by us at high speed will sense herself being the one who is stationary and will see us on the Earth as moving by her at high speed. She will see us and the Earth contracted in the direction of "our" motion, and she will see our clocks and our hearts running slowly (second panel of Figure 26.7). So what two observers in relative motion see is parallel—each would conclude that the other was the one undergoing the distortions of space and time.

An especially important feature of Einstein's work is the behavior of light. Suppose we or the passing astronaut shine a light beam at each other. We will each measure that the light is moving past us at the same speed, c. However, as we watch each other making these measurements, we will each think that the other is measuring a shorter distance and using a clock that runs too slow.

Special relativity is far-reaching in its implications. The theory is the basis for Einstein's discovery of the relationship between energy and mass: $E = m \times c^2$, which is explored further in Unit 52. Other fundamental quantities also change for moving objects. For instance, a moving object appears to grow more massive by the same Lorentz factor that describes how lengths grow shorter and time slows. This apparent higher mass is known as the **relativistic mass**.

These effects mean that nothing can reach the speed of light in our rest frame, let alone exceed it. The rocket ship traveling by us can fire its rockets and accelerate forever—but every time it goes a little faster, its time slows down more, and the rocket exhaust comes out more slowly and does not travel as far because of the contraction in length of the speeding ship. From our perspective the ship's mass appears to grow heavier and heavier, so it picks up less and less speed. The ship may approach the speed of light, but because the Lorentz factor goes to infinity, it would require infinite energy to reach the speed of light.

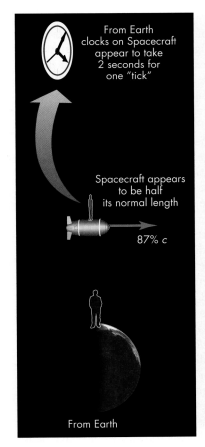

FIGURE 26.7
When a spacecraft travels by us, its length is contracted by the Lorentz factor and clocks on board run more slowly by the same factor. If the spacecraft is moving at 87% of the speed of light relative to Earth, it has a Lorentz factor of $\gamma = 2.0$. As a result, the spacecraft looks 2.0 times shorter, and each tick of its clock appears to take 2.0 seconds. From on board the spacecraft it appears that the Earth is compressed and time on Earth is running slowly.

FIGURE 26.8
From the rest frame of Earth and a star 100 light-years away (top), it appears that a spaceship traveling at 99.5% of c is shortened to about one-tenth of its original length. From the spaceship (bottom), it appears that the Earth and star are both moving at 99.5% of c. They and the distance between them are shortened to about one-tenth of their original length.

Mathematical Insights

You can find the speed V that gives a Lorentz factor $\gamma = 10$ as follows:

$$\gamma = 10 = \frac{1}{\sqrt{1-V^2/c^2}}$$

Square and invert both sides:
$1 - V^2/c^2 = 1/100$
$V^2 = (1 - 1/100)\, c^2 = 0.99\, c^2$
$V = \sqrt{0.99\, c^2} = 0.995\, c$

Mathematical Insights

At a speed of $V = 0.995\,c$, in 100 years a spaceship travels $0.995 \times 100 = 99.5$ light-years—nearly 100 light-years.

26.4 SPECIAL RELATIVITY AND SPACE TRAVEL

The theory of special relativity may seem strange, but it has been tested for a century by high-precision experiments. There is not a single verified contradiction to it, and its predictions about the slowing of time have been verified directly.

Special relativity is also about more than just perceived differences in space, time, and mass. For example, when atomic clocks (the most accurate clocks, used for establishing time worldwide) are flown on airplanes, it is found that their travel leaves them a little bit slow relative to the network of fixed clocks maintained around the world. The moving clocks must be readjusted after each trip. The Lorentz factor for traveling at airplane speeds of about 1000 kilometers per hour (600 mph) is just $\gamma = 1.0000000000004$, so every tick of the clock on an airplane takes about 4 ten-trillionths of a second longer than a tick of the clock in the ground-based network; but after several hours of flying the effect is measurable.

More intriguing is what happens at such high speed that the Lorentz factor is large. Experiments have demonstrated, for example, that a subatomic particle called a *muon* (Unit 4) normally has a lifetime of only about 2 millionths of a second before it decays. However, when muons are traveling at 99.5% of the speed of light (and therefore have a Lorentz factor of about 10), they live about 20 millionths of a second. This much longer lifetime allows them to travel distances that would be impossible for them within their normal lifetimes.

When speaking of microseconds, this change seems minor; but the same factor would apply for human space travel at 99.5% of the speed of light. If a spaceship could be built that traveled that fast, time would effectively run ten times more slowly on board compared with time here on Earth. If astronauts had food and air supplies for a 10-year trip, the Lorentz factor of $\gamma \approx 10$ would mean they could travel for $\gamma \times 10$ years ≈ 100 years in Earth time. At their speed relative to Earth of $0.995\,c$ they would be able to reach a distance of almost 100 light-years (Figure 26.8 top) before they ran out of supplies. At speeds even closer to c the Lorentz factor becomes even larger and the potential distances greater.

From the perspective of astronauts on a craft traveling at $0.995\,c$, it would not seem that time was passing any more slowly than normal. Nor would they feel that they or their ship was foreshortened or more massive. From their perspective the Earth and the star they are visiting *and the distance between the two* are contracted by the Lorentz factor (Figure 26.8 bottom). The distance would look like 100 light-years when they were in the rest frame of Earth and the distant star; but at their high speed, the distance would look only one-tenth as large!

Theoretically, at a high enough speed, time slows down so much that you could travel a million light-years away within your lifetime. This is far beyond current technologies, but it is also important to realize that such travel would have major challenges beyond simply reaching such high speeds. From the perspective of your spaceship traveling among the stars at near the speed of light, every atom and every dust particle in space along the ship's path has a relativistic mass increased by the Lorentz factor, and it is heading toward the ship at near the speed of light! This can impart to a pebble the impacting force of a ship-destroying asteroid.

Also, as intriguing as these possibilities are, they offer no time savings for the rest of us back on Earth. Consider again the astronauts traveling at 99.5% of the speed of light for 10 years to visit a distant star. If they then turned around and came home in another 10 years (by their reckoning), they would find that 200 years had passed on Earth. Everyone they knew would have grown old and died. They would likely be younger than their great-great-great-great-grandchildren!

26.5 THE TWIN PARADOX

Something may seem wrong about the description of space travel in the previous section, because when we first introduced relativity we noted that the time stretching appears symmetrical. While from the rest frame of the Earth it appears as though the astronauts' time is running slowly, from the rest frame of the astronauts it appears that time on Earth is running slowly. This is sometimes presented as the **twin paradox:** If one of the astronauts left a twin back on Earth, how can we say that one ages more than the other?

The explanation of this seeming paradox is that the situations are not actually symmetrical. Astronauts aboard a spacecraft experience accelerations as they first speed up their ship and then slow down at the star they visit. They turn their ship around and experience more accelerations as they again speed up and later slow down when they reach Earth again. The people remaining back on Earth experience none of these accelerations because they always remain in the same rest frame, and so the progression of time remains constant on Earth. By contrast, the astronauts' rest frame keeps changing, and in the end they return to the rest frame of the Earth.

Imagine, for example, that one of the astronauts sends messages once each day to her twin back on Earth, and meanwhile the twin on Earth sends messages once each day to his astronaut sister on the ship. As they part, they each receive the other's messages much more slowly—both because of the Lorentz factor *and* because the separation between the ship and Earth is growing larger and the messages take longer to reach their recipients. When the astronaut twin reaches the distant star, she sends her 10th-year message. Until this time, the situations are symmetrical. Both have received only a small fraction of each other's messages because of the Lorentz factor and growing separation.

The astronaut twin turns around and starts back, receiving the messages from her brother that were sent years earlier and have been on their way to her across the 100-light-year gap. They come more quickly now because she is approaching Earth, cutting the distance each message has to travel. She reads about her brother getting older and older, now seemingly very rapidly. In the meantime, her brother back on Earth is still reading her messages from the outgoing trip. He will die before the arrival of the 10th-year message announcing that the ship had reached the other star. At the speed of light in Earth's rest frame, that 10th-year message would take 100 years to reach Earth. In fact, the spaceship, traveling at 99.5% of the speed of its messages beamed back toward Earth, arrives just shortly after the message announcing they had reached the star.

In the movies, space travel is fast and everyone ages at the same rate. In reality, traveling at high speeds means that the travelers must leave behind not only their homes but their own times and the people in them.

Concept Question 2

How do science fiction stories you have seen or read treat motion at speeds approaching *c*? Do any of these stories correctly depict the effects of special relativity?

Concept Question 3

With everyone's time running at different speeds, can you imagine a story line that would make a good movie if it portrayed space travel accurately?

KEY POINTS

- At speeds much less than the speed of light, velocities appear to add by simple vector addition, as found by Galileo.
- Light's speed has been measured to be the same for all observers, no matter what their speed relative to the light source.
- Einstein showed that the constancy of the speed of light can occur only if lengths of moving objects contract by the Lorentz factor and their times dilate by the Lorentz factor.
- Observers you see as moving will see you as moving and your space as contracted and time dilated while their own is normal.
- The Lorentz factor rises to infinity at the speed of light, so it would require infinite energy to accelerate any object up to c.
- From a spaceship traveling close to c, "stationary" objects appear flattened and the distances between them are small.
- If two observers with relative motion come back together to compare their clocks, the one who undergoes accelerations will find that their clock ran slower.

KEY TERMS

aether, 183
Einstein, Albert, 181
Galilean relativity, 182
Lorentz factor, 184
relativistic mass, 185
rest frame, 181
special relativity, 181
twin paradox, 187

CONCEPT QUESTIONS

Concept Questions on the following topics are located in the margins. They invite thinking and discussion beyond the text.

1. Situations with relative velocities. (p. 182)
2. Treatment of high-speed motion in stories. (p. 187)
3. Devising a movie plot that is relativistically correct. (p. 187)

REVIEW QUESTIONS

4. What is Galilean relativity? When is it reasonable to use this concept?
5. What did the Michelson-Morley experiment fail to detect?
6. What quantity is always constant in special relativity?
7. What is the Lorentz factor?
8. How are length, time, and mass altered according to special relativity?
9. Why does special relativity rule out the possibility that a spacecraft could travel at the speed of light?
10. If it is impossible to travel faster than light, why is it possible (in principle) for you to travel millions of light-years away from Earth within your own lifetime?

QUANTITATIVE PROBLEMS

11. For the javelin thrown in Figure 26.2, would the effect of special relativity result in a speed for the javelin slower than Galilean relativity would predict, or faster? Explain.
12. *Voyager 1*, the farthest and one of the fastest artificial satellites, is now traveling at a speed of 17.2 km/sec relative to the Sun. What fraction of the speed of light is *Voyager 1* traveling? Approximately what is the Lorentz factor associated with this speed?
13. Assuming that the Lorentz factor becomes significant if it affects measured lengths and times at the 10% level (a Lorentz factor value of 1.1), how fast must the observed speed of an object be to reach this threshold?
14. What speed would give a Lorentz factor of about 1 million (10^6)?
15. The Earth is traveling at about 30 km/sec as it orbits the Sun.
    a. How much more slowly does time run on the Earth than in the Sun's rest frame?
    b. Why will the time be measured as slower on the Earth than in the Sun's rest frame? (Why is it *not* "just relative"?)
16. Suppose a 1-gram paperclip struck a spaceship at 99% of the speed of light. Special relativity indicates that the paperclip's energy goes from $\gamma \times mc^2$ and drops down to its "rest mass energy" of mc^2 once it stops moving.
    a. How much energy is released in the collision?
    b. How much energy is this in kilotons of TNT?
17. How fast must a spacecraft travel to reach the nearest star system (4.3 ly away) in 1 year "ship time"?

TEST YOURSELF

18. When a spaceship is traveling at 99% of the speed of light (Lorentz factor 7), an astronaut on board the ship will find that
    a. everything in the ship weighs seven times more.
    b. the ship has shrunk to one-seventh its original length.
    c. everyone on the ship talks seven times slower than normal.
    d. All of the above.
    e. None of the above. Everything seems normal to the astronaut on board.
19. Suppose an alien spacecraft approaches the Earth at half the speed of light (0.5 c). When the aliens turn on their searchlights, what speed will we measure the light coming off their spacecraft?
    a. 0.25 c b. 0.5 c c. 1.0 c d. 1.5 c e. 2.0 c
20. If Bob travels at close to the speed of light to another star and then returns, he will find that his twin sister Alice, who remained on Earth, is
    a. younger than him.
    b. older than him.
    c. the same age as him.
    d. He cannot return to Earth because it would violate relativity.
21. The Michelson-Morley experiment showed that the speed of light
    a. is slightly smaller in the direction of Earth's motion.
    b. is the same no matter what motion the observer has.
    c. is slightly larger in the direction of Earth's motion.
    d. depends on the wavelength of the light.

General Relativity

- **27.1** The Principle of Equivalence
- **27.2** Gravity and the Curvature of Space
- **27.3** Gravitational Time Dilation
- **27.4** Gravitational Waves

Learning Objectives

Upon completing this Unit, you should be able to:
- Define and provide examples of the principle of equivalence.
- Explain how gravity affects light and time according to general relativity.
- Calculate the time dilation in a region from the escape velocity.
- Give examples of general relativistic effects that have been detected astronomically.

Despite the success of the theory of special relativity (Unit 26), Einstein saw that his theory was not yet complete. Special relativity shows how observers in different rest frames measure space and time in different ways. There is a problem, however, when we try to think about special relativity in the presence of gravity.

Einstein (and Newton centuries before him) had pictured a rest frame as being like a spaceship that moves at a constant speed in a straight line. Its uniform motion shows that no forces act on it (Unit 14). If we moved in a perfectly straight line at a steady speed through the Solar System, however, the gravitational pull of different objects would tug us first in one direction then another inside our ship. Is constant-velocity motion truly in a rest frame, Einstein wondered, or is the more natural motion that of an unpowered ship, curving freely between the planets?

These ideas led Einstein to a new understanding of gravity, which he published in 1915. The mathematics of this approach are far more complex than for special relativity, but the underlying ideas we will discuss in this Unit give us a new way of understanding gravity. Einstein's theory of gravity, known as **general relativity,** is now extremely well supported by astronomical observations and laboratory experiments. In fact, GPS (Global Positioning System) units would not work if they did not make corrections for both special and general relativistic effects between orbiting satellites and the user on the ground. General relativity has supplanted Newton's law of gravitation (Unit 16), leading many to think that Albert Einstein is the greatest scientist of all time.

27.1 THE PRINCIPLE OF EQUIVALENCE

To understand the nature of the conflict between special relativity and gravity, we might ask: How do astronauts determine that they are remaining in a particular rest frame? Besides looking out the window and comparing their motion to neighboring objects, they can tell that they are not changing their rest frame—that is, not changing velocity—because they experience no acceleration. We saw in Unit 26.5 how in the "twin paradox," the traveler who experienced accelerations and returned home was the one who was younger than her twin. This is because effects such as time dilation are completely parallel for observers in different rest frames until one changes frames to join the other observer in his or her rest frame.

Gravity also changes the nature of motion and an observer's sensation of acceleration. For example, people standing on the surface of the Earth feel a constant acceleration downward even though they are not moving. On the International Space Station, on the other hand, astronauts inside the station feel weightless, as

FIGURE 27.1
The principle of equivalence. (A) Inside a free-falling elevator, you feel weightless, just as you would feel (B) in a spaceship that is drifting far from any sources of gravity. (C) If you are stationary in the gravitational field of the Earth, objects fall in a way that is equivalent to (D) being aboard a spaceship accelerating at 1 g.

Concept Question 1

When parachutists jump from an airplane, are they in free fall according to physicists' definition?

though no forces are acting on them, even though their orbit causes them to constantly change direction. If there were no windows, the astronauts could easily believe that they were traveling in a straight line, far from any massive objects. Thus, gravity creates accelerations without motion and changes in velocity where we feel no acceleration. Gravity has the unique ability to change the direction or speed of observers without them feeling a force being applied.

Imagine that you had the bad fortune to be inside an elevator whose cable broke (Figure 27.1A). As it was falling, you would find yourself floating inside the elevator, like an astronaut in a space capsule far from any massive objects (Figure 27.1B). You float in the elevator because the floor is accelerating downward at the same rate as you do. If you let go of an apple, it will float beside you. Physicists call this situation **free fall**, when no forces resist motion in a gravitational field. Similarly, the Moon or a satellite in orbit around Earth are in free fall, but their sideways motion keeps them from ever striking the Earth's surface (Unit 16). Astronauts train for weightlessness aboard airplanes in a similar way as the falling elevator. The airplane travels along a free-fall trajectory for as long as a minute, during which time the astronauts feel no force from their surroundings and appear to float.

Now suppose you are on a stationary elevator in Earth's gravitational field (Figure 27.1C). You feel held down to the floor, and if you drop an apple, it accelerates down toward the floor at the acceleration of gravity $g = 9.8$ meters/second per second. This is the same thing you would experience in a spaceship that was accelerating at 9.8 m/sec^2 (Figure 27.1D). For the spaceship we might say that the apple remains stationary as the floor of the ship accelerates upward to meet it, but for passengers the experience would be the same as if they were standing on Earth's surface.

Einstein had the remarkable insight that gravity actually alters the nature of space. When you are near a massive object, it causes space to, in effect, flow past you. Thus, when you let go of an apple, it moves downward as if you were moving upward through space, so what you observe is the same as letting go of an apple in an accelerating spaceship. The idea that gravity is the same as being in an accelerating frame of reference is known as the **principle of equivalence**.

Einstein spent several years searching for a way to describe how gravity alters space and time, finally developing general relativity in 1915. Einstein constructed this theory with the idea that the laws of special relativity must remain correct for any rest frame in free fall. Einstein reasoned that an observer in free fall feels no forces, so this is the most natural frame. Standing on the surface of the Earth, we always feel a force pulling on us. We may think that we and our surroundings are stationary, but space is flowing by us like a steady wind dragging us downward.

27.2 GRAVITY AND THE CURVATURE OF SPACE

Newton described gravity as a force that masses exert on each other at a distance, causing the masses to accelerate toward each other. By contrast, Einstein described space itself as being altered by masses so that an object follows a curved path in much the way a golf ball follows a curved path or accelerates as it rolls along the hills and slopes of a putting green.

One prediction of the principle of equivalence is that the path of light will curve due to gravity. We can illustrate this by a thought experiment: shining a laser at a target across a laboratory (Figure 27.2). By the principle of equivalence, the laboratory can be treated as if it is a spaceship accelerating at 9.8 m/sec^2. At the moment the light leaves the laser, it is aimed at the target, but the acceleration of the lab steadily changes the lab's rest frame, so if the path of the light is straight in the original frame, it must curve in the accelerating frame.

The flow of space toward massive objects can be pictured by an analogy. Imagine a large artificial lake in which there is a drain in the middle that draws out water at a steady rate. If you sit in a boat on the lake, the outflow of water draws you toward the drain, with a stronger pull the closer you get to the drain. This is analogous to how gravity exerts a stronger pull the closer you get to a massive object. If you were to travel across the lake in a motorboat, holding your steering wheel in a fixed direction, you would discover that your path was curved by the flow of water toward the drain. A photon curving as it passes a massive object is like the motorboat traveling by the drain. The water in the analogy represents space, and it illustrates an important idea of general relativity: space itself can have motion. While the notion of "nothing" having motion seems strange, it is one of the crucial ideas for understanding such things as black holes and the expansion of the universe.

Starlight passing close to the Sun, where the gravity is strongest, is deflected by the greatest amount (Figure 27.3). A total solar eclipse in 1919 gave astronomers their first opportunity to measure this deflection. Astronomers measured positions of stars near the Sun during the eclipse and discovered that the stars were deflected by the amount predicted by Einstein's theory, as illustrated in the inset to Figure 27.3. The measurements were somewhat uncertain, but they were convincing enough to make Einstein a celebrity. He had overturned Newton's law.

Astronomers have subsequently made measurements at radio wavelengths that do not require an eclipse and confirm Einstein's prediction with high precision. They have also detected the deflection of light around many other bodies. The

> Not only does space "flow" toward massive objects, moving masses pull nearby space along with them in a phenomenon known as "frame dragging." This effect has recently been measured with satellites orbiting Earth, but the effect is most dramatic around black holes, as discussed in Unit 69.

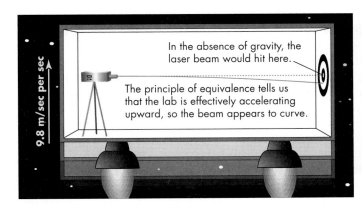

FIGURE 27.2
Gravitational curvature of the path of light. By the principle of equivalence, light must bend in the accelerating frame caused by gravity.

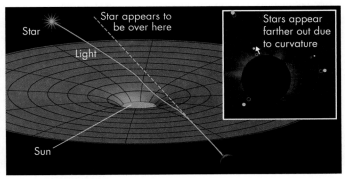

FIGURE 27.3
The "flow" of space toward a massive object like the Sun deflects the path of starlight passing near it. The prediction of shifted positions was confirmed during a solar eclipse in 1919.

deflections can sometimes even focus and brighten the light from a background object, which astronomers refer to as **gravitational lensing**. This allows astronomers to detect objects that might not otherwise be visible, such as planets around other stars (Unit 36) or dim stars in the outer part of our own Galaxy (Unit 79).

The curving path of light forces us to rethink what we mean by straight lines. If we were to sight along a laser beam between two points, it would look straight to us despite gravitational forces because all light would be curved the same way. A straightedge held up next to the path would seem to confirm that the path was straight, because the straightedge would also be affected by gravity in the same way. The curving path of light is not apparent to the eye because we and the light that enters our eyes all experience the effects of gravity. General relativity provides a way of calculating the **curvature of space** produced by masses, and Einstein showed further it is not just mass, but energy as well, that produces this curvature.

27.3 GRAVITATIONAL TIME DILATION

One of the more remarkable aspects of general relativity is that it predicts that time runs slower where gravity is stronger. This has been confirmed directly by experiments comparing clocks on the surface of the Earth with those on spacecraft far above the Earth, where the gravitational pull is weaker.

We can use the principle of equivalence to understand the difference in the rates of clocks due to gravity. Imagine an elevator in free fall, falling toward the surface of Earth, as pictured in Figure 27.4. Suppose the elevator is falling past a tower on which clocks are mounted. When the elevator is high up, its velocity downward is small, so according to special relativity there is a small Lorentz factor between the falling elevator and the clock next to it on the tower. As the elevator continues to fall, the relative velocity grows larger, so the Lorentz factor grows larger too. This means the clocks lower on the tower run at slower and slower rates relative to a clock in the free-falling elevator.

The principle of equivalence tells us that an observer in free fall measures time normally, so it is the "stationary" clocks on the tower that must be running slower the deeper they are in the Earth's gravitational field. Imagine now an almost infinitely tall tower with the free-falling elevator dropped from the top of the tower. The gravity is extremely weak at the top of the tower so far away from Earth, and its pull at first barely causes the elevator to accelerate, so the clocks at this great distance run at the same rate. By the time the elevator falls to the surface of Earth it speeds up to the Earth's escape velocity (Unit 18). This can be understood because the elevator would have to be launched at the escape velocity to travel out infinitely far from the Earth, so conservation of energy (Unit 20) tells us it must be the same velocity the elevator would have if it fell back down from infinite distance.

We conclude that the free-falling observer will see clocks slowed by the Lorentz factor corresponding to the local escape velocity. We can calculate a **gravitational time dilation** γ_G that tells us how much slower clocks run in a gravitational field based on the escape velocity:

$$\gamma_G = \frac{1}{\sqrt{1 - V_{esc}^2/c^2}}$$

where V_{esc} is the velocity needed to escape gravity's pull and c is the speed of light.

The distortion of time by gravity is also visible in the way gravity affects the wavelength of light emitted in a region where gravity is strong. Physical processes, including the frequency of vibrations in atoms, all occur more slowly where gravity is strong. When light is emitted from the surface of a dense type of star known as a white dwarf (Unit 65.4), for example, the spectral lines are all found to be

FIGURE 27.4
A free-falling observer will see clocks running slower the farther down they fall, because the relative speed grows larger.

γ_G = gravitational time dilation
V_{esc} = velocity to escape gravity's pull
c = speed of light

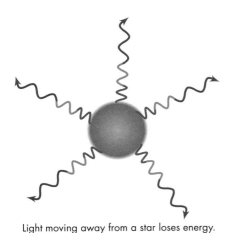
Light moving away from a star loses energy.

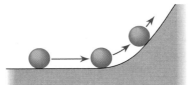
Ball rolling up a hill loses energy and slows down.

FIGURE 27.5
Time runs slower where the gravitational field is stronger, so electromagnetic waves emitted at a particular frequency from the surface of a star will have a lower frequency (longer wavelength) at larger distances. The waves lose energy as they travel away from the star, similar to how a ball loses energy rolling up a hill.

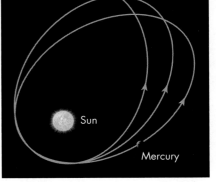

FIGURE 27.6
The orientation of Mercury's orbit shifts slightly each orbit by an amount predicted precisely by general relativity.

at lower frequencies (longer wavelengths) than we would normally expect. This effect is called a **gravitational redshift,** because it shifts visible light toward the red end of the spectrum. For example, hydrogen on the surface of a white dwarf may emit a spectral line of a known frequency; but when that light reaches a distant observer whose clock is running faster, the waves have a lower frequency, as illustrated in Figure 27.5. Physics experiments have directly measured the gravitational redshift even between the top and bottom of a building.

This stretching of time means that if one astronaut stayed in a spaceship and watched her twin brother travel down to the surface of a white dwarf, the twin on the surface of the white dwarf would appear to be moving, speaking, and breathing more slowly than normal because of the strong gravitational field. On the other hand, to the twin on the surface of the white dwarf, it would appear that his sister on the spaceship was moving, speaking, and breathing faster than normal. When he returned to the spacecraft, he would discover that less time had passed according to his clocks than had passed according to the clocks on the spaceship. He would be slightly younger than his sister.

When Einstein introduced general relativity, he showed that the stretching of space and time that it predicted solved a long-standing mystery about a small deviation in the orbit of Mercury. Mercury's orbit does not repeat itself precisely as predicted by Kepler's laws (Unit 12), but instead shifts direction slightly with each orbit, as illustrated in Figure 27.6. For decades astronomers had tried unsuccessfully to explain this deviation. General relativity predicts that the space near the Sun—deeper in the Sun's gravitational field—is stretched more than the space far away. In addition, general relativity predicts that time runs slower where gravity is stronger. Because Mercury's orbit is elliptical, when it is closest to the Sun, it is traveling through stretched-out space at a rate that is slightly too slow. As a result it does not travel as far as expected, making the orientation of the orbital ellipse to shift slightly each orbit.

The degree by which time slows depends on the strength of the gravitational field and the escape velocity. For objects known as black holes, the escape velocity reaches the speed of light and the gravitational time dilation becomes infinitely large. To an outside observer, time appears to stop at the edge of a black hole; while far from any masses, time runs at its normal rate. Black holes have many other unusual properties, and we will return to examine them in Unit 69.

Concept Question 2

What differences could you look for to distinguish between a gravitational redshift from the surface of a star and a Doppler redshift because the star is moving away from us?

Concept Question 3

What are the similarities and differences between the twin paradox of special relativity and the situation in which one twin travels deep into a strong gravitational field?

Clarification Point

In both special and general relativity, observers who experience the largest accelerations discover that their clocks have run slowest when they get together with other observers.

27.4 GRAVITATIONAL WAVES

Another prediction of Einstein's theory of general relativity is that when one object orbits another, the motion of the two objects generates **gravitational waves.** Just as ripples spread away from a stone tossed into a pond, so too gravitational waves spread across space, stretching and distorting the space and time through which the waves move (Figure 27.7). These waves have not yet been directly detected, but they could potentially provide a way of probing events such as the formation of a black hole, or collisions between extremely dense stars called neutron stars, which are the collapsed remnants of massive stars (Unit 68).

U.S. scientists have recently built highly sensitive detectors, called the Laser Interferometer Gravitational-wave Observatory. LIGO consists of vacuum tunnels, 4 kilometers long, in which light is reflected back and forth between mirrors. Separate stations are located in the states of Louisiana and Washington so that a passing gravitational wave can be discerned from local sources of vibration, such as seismic noise. The U.S. observatories have teamed up with other gravitational-wave observatories in Europe to further improve their ability to detect these waves. If a gravitational wave of sufficient strength passes through the detectors, it will shift the distance between the mirrors by a fraction of the diameter of an atomic nucleus in a predictable way, slightly altering the light patterns between the mirrors. The detectors have begun operating, and their sensitivity is being steadily improved. They are expected to eventually have the sensitivity to measure the passing distortions of space and time caused by distant, but violent astronomical events.

Although gravitational waves have not yet been directly detected, indirect evidence of such waves has been observed from rapidly orbiting neutron stars. Such a detection is possible because as the waves carry gravitational energy away from the orbiting stars, the stars spiral inward toward each other. Exactly such behavior was detected in 1974 by the U.S. astronomers Joseph Taylor and Russell Hulse, who discovered two neutron stars in a tight orbit about each other. Their observations showed that as the neutron stars orbit each other, they are gradually losing energy and the two stars are spiraling toward each other. Taylor and Hulse were able to measure this inward motion with high precision and showed that the rate of orbital decay exactly matches the rate of predicted energy loss in the form of gravitational waves. This has become one of the strong tests supporting the theory of general relativity. For this discovery, Taylor and Hulse were awarded the 1993 Nobel Prize in physics.

Clarification Point

Einstein's theory of general relativity shows that Newton's law of gravity is incorrect, but the differences become large only for extremely strong gravitational fields. In most situations Newton's law remains sufficient.

FIGURE 27.7
Gravitational waves generated by a rapidly orbiting pair of neutron stars. Such waves are "ripples" in space—and like ripples caused by a disturbance in water, they can make distant objects move slightly as the waves pass by.

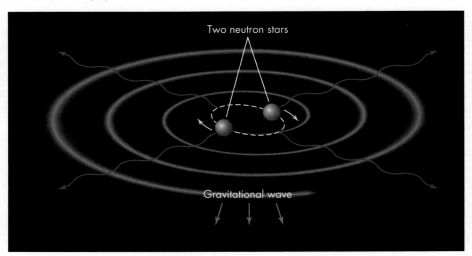

KEY POINTS

- General relativity is a theory of gravity, describing how masses "curve" space or, equivalently, cause space to "flow" toward them.
- Remaining stationary in a gravitational field is equivalent to accelerating through empty space (principle of equivalence).
- The path of light through space is curved by gravity, which was observed during a solar eclipse, confirming Einstein's theory.
- Time slows down in a gravitational field by an amount equivalent to special relativity's Lorentz factor for the local escape velocity.
- General relativity explains why Mercury's orbit gradually shifts its orientation.
- Orbiting masses produce gravitational waves that cause the masses to gradually lose energy and spiral together.
- Energy loss from gravitational waves has been detected, and new observatories are being designed to detect the waves directly.

KEY TERMS

curvature of space, 192
free fall, 190
general relativity, 189
gravitational lensing, 192
gravitational redshift, 193
gravitational time dilation, 192
gravitational wave, 194
principle of equivalence, 190

CONCEPT QUESTIONS

Concept Questions on the following topics are located in the margins. They invite thinking and discussion beyond the text.

1. Free fall by parachutists. (p. 190)
2. Gravitational versus Doppler redshifts. (p. 193)
3. Twin paradoxes in special and general relativity. (p. 193)

REVIEW QUESTIONS

4. What is the principle of equivalence?
5. How do physicists define free fall?
6. How does general relativity describe gravity differently from Newton's law of gravity?
7. What effect does the Sun's gravity have on starlight? What have astronomers observed to confirm this?
8. In what way is time affected by gravity?
9. What are gravitational waves? Have they been detected?

QUANTITATIVE PROBLEMS

10. The discrepancy in Mercury's orbit explained by general relativity is a shift in the orientation of the long axis of the orbital ellipse of 43 arc seconds per century. (An arc second is 1/3600 of a degree.) Mercury's orbital period is 88 days, so how much does the orientation change after a single orbit?
11. The escape velocity from Jupiter is about 60 km/sec while for Mercury it is 4 km/sec. Find the gravitational time dilation on the surface of these planets.
12. The neutron star system discovered by Taylor and Hulse takes about 8 hours to complete an orbit, but the period is shortening by about 80 microseconds per year. If it continues shortening at this rate, how long is it until the neutron stars collide?
13. Earth's escape velocity is 11 km/sec. How much slower do clocks run on Earth's surface than in deep space, far away from any massive object?
14. What is the gravitational time dilation for the Sun's escape velocity of 618 km/sec? If the frequency of light corresponding to a 500-nm photon is slowed by this amount, what wavelength will this photon have far from the Sun?
15. Find the escape velocity from a neutron star with the mass of the Sun and a radius of 10 km. What is the gravitational time dilation at the surface of the neutron star?
16. Suppose that the gravitational time dilation on the surface of a star is 1.1. What would be the escape velocity from the star?
17. What is the maximum delay between detections of a gravitational wave that passes through each of the two LIGO observatories? The observatories are separated by 3,000 km and gravitational waves travel at the speed of light.
18. For a satellite in a low circular orbit around the Earth, calculate and compare the Lorentz factor associated with its orbital velocity (Unit 18) with the gravitational time dilation.

TEST YOURSELF

19. What is "equivalent" in the principle of equivalence?
    a. Space and time
    b. Matter and energy
    c. Gravity and acceleration
    d. Forward and reverse directions of time
    e. All frames of reference
20. Among the following locations, where do clocks run slowest?
    a. On the surface of Earth
    b. On the International Space Station
    c. On the surface of the Sun
    d. In deep space far from any massive objects
    e. On the surface of a massive neutron star
21. Observations of the 1919 total solar eclipse were used to confirm what general relativistic prediction?
    a. Time dilation near a massive object
    b. Mercury's deviations from a Keplerian orbit
    c. The Sun emits gravitational waves
    d. Deflection of starlight as the light passes a massive object
22. The television signals that escape from Earth's surface and travel outward through space are at
    a. slightly lower frequencies than what they were emitted at.
    b. much lower frequencies than what they are emitted at.
    c. the same frequencies they are emitted at.
    d. much higher frequencies than what they are emitted at.
    e. slightly higher frequencies than what they are emitted at.

Detecting Light—An Overview

28.1 Technological Frontiers
28.2 Detecting Visible Light
28.3 Observing at Nonvisible Wavelengths
28.4 The Crab Nebula: A Case History

Learning Objectives

Upon completing this Unit, you should be able to:
- Describe the basic functions and limitations of telescopes.
- Explain how a CCD works.
- Interpret a false-color image and describe the different kinds astronomers use.
- Discuss what new technologies allow us to learn about astronomical objects.

Like all scientists, astronomers rely heavily on observations to guide them in proposing hypotheses and in testing theories already developed. Unlike most scientists, however, astronomers usually cannot actively probe the objects they study. Instead, they must perform their observations passively, collecting whatever light and other forms of radiation have been emitted by the bodies they seek to study.

The quest to understand the cosmos ultimately depends on the quality and detail of the data we can collect. Sometimes that quality and detail are limited by the available technology; sometimes they are constrained by the physical properties of the electromagnetic radiation that we are attempting to detect. This Unit provides an overview of the instruments and techniques used for astronomical observation. For a more in-depth examination, Units 29–32 explore the physical principles of telescopes, while Unit 33 explores some practical aspects of making your own observations of the night sky.

28.1 TECHNOLOGICAL FRONTIERS

Astronomy is the most ancient of the sciences, but many of its most exciting discoveries are very recent. For example, in the long history of astronomy, only in the last 200 years have astronomers known that "light"—electromagnetic radiation—extends beyond the visible spectrum; and only in the last half century have instruments been developed to study these other bands of electromagnetic radiation. And at visual wavelengths, telescopes have progressed to achieve remarkable precision and size (Figure 28.1).

The design and construction of instruments to collect and detect electromagnetic radiation of all wavelengths is a branch of astronomy called **instrumentation.** The amazing images and data that have changed many of our ideas about the nature of the cosmos in recent decades are due to the remarkable strides made in instrumentation.

The most recognizable tool of astronomical technology is the **telescope.** Telescopes allow astronomers to observe things that are not visible to the naked eye either because they are too dim or because they emit radiation outside the visible range of the electromagnetic spectrum. Telescopes

FIGURE 28.1
The twin Keck telescopes on the summit of Mauna Kea, Hawaii, during a total eclipse of the Sun. These are just slightly smaller than the largest individual optical telescope in the world, which was built in the Canary Islands in 2008 using a similar design.

are basically devices for collecting a large amount of electromagnetic radiation, focusing this light onto detectors, and magnifying the images to resolve fine detail. Astronomers strive to build bigger telescopes because the amount of light that can be collected depends on the size of the telescope's **aperture,** or collecting area (Unit 29). Collecting more light allows us to see dimmer and more distant objects. If we collect enough light, it is possible to examine the spectrum of radiation from a source, which reveals details of its physical and chemical nature (Unit 24).

A variety of telescope designs use different techniques to take the light that enters the telescope aperture and concentrate or **focus** it (Unit 30). Many of the principles involved in focusing light are similar whether we are dealing with radio waves or X-rays; but the construction details differ enormously because the precision of the construction is determined by the wavelength of the light being observed. A telescope's **resolution**—its ability to detect fine detail—is a function of the size of the telescope relative to the wavelength of the radiation it detects (Unit 31). This allows radio telescopes to be built with rougher surfaces; however, they have to be built much larger to achieve a comparable resolution.

The Earth's own atmosphere presents challenges for observing electromagnetic radiation. The atmosphere absorbs some wavelengths and alters the path of light in ways that badly degrade the quality of images. For these reasons, a number of space telescopes have been developed in recent years and placed in orbit above the atmosphere (Unit 32). However, putting telescopes in space presents significant challenges—and expense—and one of the goals of modern instrumentation is to find ways to compensate for the atmosphere for ground-based telescopes and to improve the resolution of telescopes with new designs.

28.2 DETECTING VISIBLE LIGHT

Modern telescopes bear little resemblance to the long tubes depicted in cartoons (Figure 28.2). Moreover, professional astronomers almost never sit at the eyepiece of a telescope. And few astronomers ever wear lab coats! Not only are astronomers not needed close to a telescope, but the heat from their bodies can ruin the image quality. With the major research telescopes of today, an astronomer is more likely to be dressed in jeans and a T-shirt, operating the telescope from a computer terminal that may be thousands of miles away. She or he will examine the data being collected, along with a wide array of data about the telescope's performance and weather conditions, and attempt to gauge whether enough data of sufficient quality have been collected to move on to the next set of observations. To make the most efficient use of the instruments, it is becoming more common for telescopes to be operated, and the data checked, by computer programs. A team of astronomers develops computer programs to observe the targeted sources efficiently, then waits for the requested data but has no immediate control over the telescope at all.

Before astronomical photography became widespread in the late 1800s, astronomers did stare through telescope eyepieces, and they wrote down data or made sketches of the objects being observed. However, despite its marvelous abilities, the human eye has difficulty detecting very faint light (Unit 33). Many astronomical objects are too dim for their few photons to create a sensible effect on the eye. For example, even if you were to look at a galaxy through the largest telescope available today, the light is so spread out that it would appear merely as dim smudges to your eye. Only by storing galaxies' light, sometimes for many hours, can we discern the detailed structure of the galaxies. Thus, to see very faint objects, astronomers must use detectors that can store light in some manner.

Until the 1980s, astronomers generally used photographic film to record the light from the objects they were studying. Astronomers today, however, almost

"It's somewhere between a nova and a supernova -- probably a pretty good nova."

FIGURE 28.2
Astronomers discussing the nature of their discovery? ScienceCartoonsPlus.com

FIGURE 28.3

Simplified diagram of a CCD. Photons striking the photoelectric layer free an electron (e). A positive voltage applied to one set of electrodes attracts the electrons and holds them in place under each pixel. During the readout phase, the voltages are changed to "push" the collected electrons along the readout electrodes.

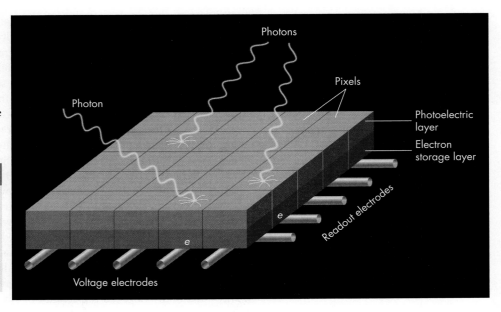

Concept Question 1

Many CCDs are refrigerated to very low temperatures. How do you suppose keeping the temperature low might make the image quality better? (Consider thermal effects on atoms within the CCD.)

Albert Einstein explained the photoelectric effect in 1905, providing one of the foundations for understanding the particle-like nature of light. His work showed that light is composed of discrete photons with energies that depend on their wavelength. His only Nobel Prize was awarded for this work, though he is even better known for his development of relativity theory, explaining the nature of space, time, and gravity (Units 26 and 27).

Long-wavelength radio waves cause electrons in an antenna to oscillate in response, and the small electric currents of these moving electrons are detected and amplified.

always use an electronic detector similar to those in digital cameras, called a **CCD,** which stands for "charge-coupled device." These devices use the **photoelectric effect,** in which a photon that has sufficient energy causes an electron to become unbound from the atom it is orbiting. Thus, when the incoming light strikes the "semiconductor" surface of the CCD, it frees electrons to move within the material. The surface is divided into thousands of little squares, or **pixels** (Figure 28.3). Electrical voltages on the surface pull the freed electrons into the nearest pixel. Here they are temporarily stored before being "read out" by the electronics. Then the whole chip is cleared and returned to its original state. The number of freed electrons in each pixel is proportional to the number of photons hitting it (that is, proportional to the intensity of the light). An electronic device coupled to a computer then scans the CCD, counting the number of electrons in each pixel and generating a picture.

Such cameras are extremely efficient, generally recording more than 75% of the photons striking them. This allows astronomers to record images much faster than with film, which might detect only 10% of the photons. Furthermore, unlike film, CCDs are reusable, and the digital images can be processed rapidly by a computer. Astronomers work with the digital images to correct them for instrumental effects, remove extraneous light, and determine the exact amount of light striking different parts of the image.

28.3 OBSERVING AT NONVISIBLE WAVELENGTHS

Many astronomical objects radiate electromagnetic energy at wavelengths outside the range of visible light, and so astronomers have devised new ways to observe such objects. For example, cold clouds of gas in interstellar space emit little visible light but large amounts of radio energy. To observe them, astronomers use radio telescopes, which employ technologies like those used for detecting television or radio station transmissions, but which are far more sensitive.

At infrared and shorter wavelengths astronomers generally use CCDs or similar technologies to detect photons. For example, dust clouds in space are too cold to emit detectable visible light, but they do radiate infrared photons. Normal CCDs are much more sensitive to infrared photons than photographic film (with wavelengths up to about twice the wavelength of optical photons). But longer-wavelength photons do not have enough energy to free electrons in the photoelectric material.

New photoelectric materials have been developed for this purpose. These detectors must be kept very cold, however, because heat from the instrument itself or any surrounding equipment will generate photons of these energies, which will overwhelm the photons from the astronomical source of interest.

Astronomers use X-ray and gamma-ray telescopes to observe, for example, the hot gas that accumulates around black holes. A single X-ray photon has sufficient energy to free many electrons within the CCD material used, and thus the number of electrons freed can be used as a measure of the energy of the incoming X-ray photon.

These technologies improved dramatically in the last few decades. An astronomer spending several years using cutting-edge infrared detectors around 1980 to collect data for a Ph.D. dissertation could with today's instruments collect those same data in five minutes! Astronomy at other wavelengths is likewise advancing rapidly, and astronomers are learning to interpret an ever-growing array of data as these other areas begin to "catch up" with the much better established field of visible-wavelength astronomy.

One of the challenges in interpreting data collected at nonvisible wavelengths is how to display it. Because our eyes cannot see these other wavelengths, astronomers must devise ways to depict what such instruments record. One way is with **false-color images,** as shown in Figure 28.4. In a false-color image, colors are used to represent information from other wavelengths in a way that takes advantage of our eyes' ability to discern fine differences in color. In a common type of false-color image, colors are used to represent the intensity of radiation that we cannot see—often using colors as they are shown in a geographic map, with reds and oranges representing the high levels (like mountains) and blues and purples representing low areas (like oceans). For example, in Figure 28.4A (a radio image of a galaxy and the jet of hot gas spurting from its core), astronomers color the regions emitting the most intense radio energy red; they color areas emitting somewhat less energy yellow and the faintest areas blue.

Figure 28.4B shows a different kind of false-color image (of the gas shell ejected by an exploding star). It is an X-ray image where different energy X-ray photons have been assigned colors analogously to how we see colors in the visible spectrum—blue for high energies, green for medium, and red for lower energies.

A false-color image may be coded to depict data in many different ways. Colors may be used to represent different wavelength ranges—for example, red representing radio, yellow representing infrared, and blue representing X-ray emission. Or the colors may be coded to indicate derived quantities like the temperature of gas, the strength of magnetic fields, or Doppler shift values. These techniques of visualizing data help astronomers to "see" connections between processes going on within an astronomical object.

FIGURE 28.4

(A) False-color picture of a radio galaxy. We can't see radio waves, so colors are used to represent their brightness—red brightest, blue dimmest. (B) False-color X-ray picture of Cas A, an exploding star.

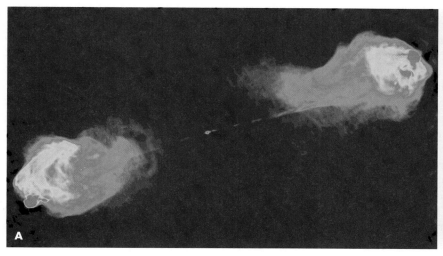
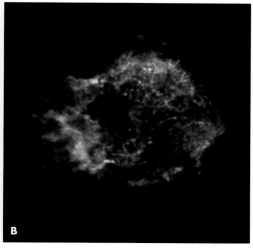

28.4 THE CRAB NEBULA: A CASE HISTORY

The history of observational astronomy is illustrated by the story of the discovery and gradual piecing together of the astronomical mystery of an exploding star, first seen nearly 1000 years ago. Although the story began with naked-eye observations, it continues with observations made with telescopes on the ground and in space. Moreover, the story illustrates how astronomers have come to rely on observing radiation at many wavelengths, not just visible light.

On July 4, 1054 C.E., just after sunset, astronomers in ancient China and other east Asian countries noticed a brilliant star near the crescent moon in a part of the sky where no bright star had previously been seen. They wrote of this event, "In the last year of the period *Chih-ho,* a guest star appeared. After more than a year it became invisible." The star was so bright that for several weeks it was visible during the day; but it gradually faded away, and after more than a year it grew too faint to see even at night and was forgotten.

Curiously, no records of this spectacular event have been found in European archives. It is difficult to imagine that such a spectacular event, visible for over a year from most of the Earth, went unnoticed by anyone in Europe. Some scholars speculate that written records of this event might have been suppressed because they ran counter to religious beliefs of the medieval church.

Nearly 700 years later, in 1731, John Bevis, a British physician and amateur astronomer, noticed with his telescope a faint patch of light in the constellation Taurus. In the late 1700s the French astronomer Charles Messier, who was hunting for comets, developed a catalog of similar dim patches of light that might be confused with comets. He listed the object in Taurus first, and so it became known as "Messier 1" or M1.

In 1844 William Parsons (also known as Lord Rosse), an Irish astronomer and telescope builder, noticed that M1 contained filaments that to his eye resembled a crab (Figure 28.5A). He therefore named it the "Crab Nebula." Modern photographs (Figure 28.5B) look quite different from Rosse's sketch, but perhaps still resemble a crab—for someone with a generous imagination!

In 1921 John Duncan, an American astronomer, compared two photographs of the nebula taken 12 years apart and noticed that it had increased slightly in diameter. He therefore deduced that the nebula was expanding. At the same time, several other astronomers came across the ancient Chinese records and noticed the coincidence in position of the nebula with the report of the exploding star. Then, seven years later, Edwin Hubble, at Mount Wilson Observatory in California, measured the increase of size more accurately and calculated from the rate of expansion that the nebula was about 900 years old—roughly matching the date recorded by the Chinese astronomers. The speed of the nebula's expansion calculated from these observations was astonishingly large—more than 1000 kilometers per second. Astronomers realized that the Chinese astronomers had witnessed an explosion that had sent material flying violently outward at almost unimaginable speeds.

Since the second half of the 20th century, astronomers have examined the Crab Nebula with telescopes using virtually all wavelength bands and, in doing so, have added yet more to their understanding of a star's demise. For example, the nebula is a powerful source of both radio waves (Figure 28.5C) and X-ray radiation (Figure 28.5D). These observations have led astronomers to the conclusion that this type of *supernova* explosion occurs after massive stars use up their fuel—teaching us that stars may die violent deaths (Units 66, 67).

The assortment of modern observations at many wavelengths has given astronomers one of their most detailed understandings of the death throes of a

---

Concept Question 2

How might a bright new star run counter to religious beliefs in the middle ages? What other kinds of astronomical events might be upsetting to religious beliefs?

LOOKING UP

M1 can be seen with a small telescope in the constellation Taurus. See Looking Up #5 at the front of the book.

Messier's catalog contains more than 100 of the brightest "nebulae" (Latin for "clouds") in the sky. The complete catalog, including coordinates, is given in Appendix Table 12. A number of the nebulae are also identified on the Looking Up illustrations at the front of the book.

FIGURE 28.5
(A) Lord Rosse's 1844 drawing of the "Crab Nebula." (B) Visible-light photograph of the Crab Nebula. (C) Radio image of the Crab Nebula. (D) X-ray image in false color of the Crab Nebula. X-rays are shown as blue/white while optical light is shown as red. The flattened round shape of the glowing gas suggests a spinning disk.

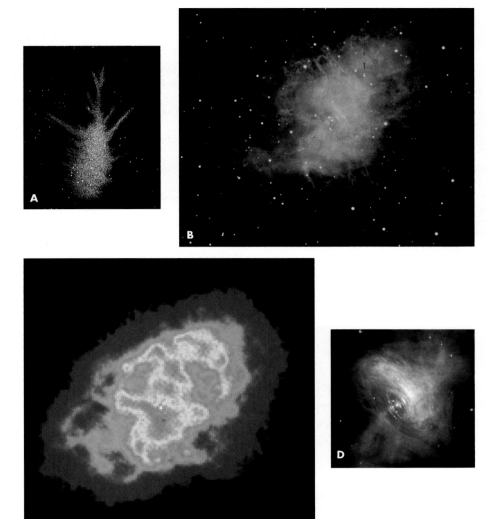

Concept Question 3

Based on what you know about the interactions of light and matter, what kinds of information might you hope to learn by studying radiation in the different electromagnetic bands?

star. High-speed radio-wavelength observations in 1968 exhibited rapid pulsations, from which astronomers deduced that the "remains" of the original star are spinning incredibly fast—about 30 revolutions per second. This in turn provides an essential clue that the star has collapsed to such a small size that it is as dense as the nucleus of an atom—an object called a *neutron star* (Unit 68). The X-ray observations reveal details of high-energy particles being ejected at near the speed of light from the dead star and trapped in rings around the star by its strong magnetic fields.

By observing at a variety of wavelengths, astronomers deduced far more than they could have from observations at visible wavelengths alone. And understanding the violent nature of the collapse of the central star suggested further types of observations in future supernova explosions. Thus, when a relatively nearby supernova explosion was detected in 1984, astronomers were able to recover information from physics experiments on neutrinos, revealing that the collapsing star generated an enormous output of these "ghostly" subatomic particles. Neutrino astronomy is still in its infancy, but it is becoming a standard technique for studying astronomical objects. And looking even further into the future, we may someday have gravitational wave observatories that could detect the ripples in space and time that general relativity (Unit 27) predicts these vast explosions could generate.

KEY POINTS

- Modern observations are more sensitive because today's larger telescopes collect more light, and new technologies are more sensitive to electromagnetic radiation.
- The ability of a telescope to resolve fine detail is a function of its size relative to the wavelength of the light it detects.
- The Earth's atmosphere can blur light passing through it, and not all wavelengths of radiation can penetrate the atmosphere.
- Photography has been replaced almost entirely with detectors based on the photoelectric effect.
- Data at non-visible wavelengths are generally presented using "false colors" to depict various physical properties.
- The explosion leading to the Crab Nebula was detected in 1054 C.E. in China, but despite being visible in daylight was not reported in Europe.
- Modern photographic, then multiwavelength, observations have revealed that the Crab Nebula is the remnant of a stellar explosion.

KEY TERMS

aperture, 197
CCD, 198
false-color image, 199
focus, 197
instrumentation, 196
photoelectric effect, 198
pixel, 198
resolution, 197
telescope, 196

CONCEPT QUESTIONS

Concept Questions on the following topics are located in the margins. They invite thinking and discussion beyond the text.

1. Thermal effects in CCDs. (p. 198)
2. The missing reports of the Crab Nebula explosion. (p. 200)
3. Information from different wavelengths. (p. 201)

REVIEW QUESTIONS

4. What advantages does a large telescope have over a small one?
5. What are the advantages of having a telescope in space? Is this true at all wavelengths?
6. How does a CCD work?
7. What is a false-color image?
8. How have astronomical techniques changed over the last century?
9. How did astronomers determine that the "Crab Nebula" was produced by an explosion almost a thousand years ago?

QUANTITATIVE PROBLEMS

10. Suppose that you are using an older CCD camera that collects only 40% of the light that would otherwise be gathered. How would this affect the magnification of the image you are taking? How would this affect the brightness of the image?
11. The Crab Nebula is about 6,500 ly from Earth. In what year did the supernova event that formed the Crab Nebula occur?
12. One of the 10-m diameter Keck telescopes can observe 500-nm visible-wavelength photons. If a radio telescope observing 10-cm photons was built to be proportionally as large relative to the photons it was observing, what would its diameter be? What about a gamma-ray telescope observing 0.01-nm photons?
13. The diameter of your eye's pupil is about 8 mm, while the Arecibo radio telescope has a diameter of about 300 m. For what wavelengths of light would the Arecibo radio telescope have the same resolution as your eye does to 500-nm visible light?
14. To provide clear images, the surface of a telescope mirror should be smooth to within 1/10 of a wavelength of the light being collected. What size is this for a visible-wavelength telescope? for a radio telescope? for a gamma-ray telescope? What physical objects match these sizes?
15. Small knots of gas in the Crab Nebula are observed to be moving away from the position of the dead star at its center. Photographs taken in 1940 and 1990 show that knots originally located 150 arc seconds from the center have moved 8 to 9 arc seconds farther out. From this information, estimate the range of dates when the explosion might have occurred.

TEST YOURSELF

16. Which of the following is *not* a reason astronomers seek to build larger telescopes? Larger telescopes
    a. can resolve greater detail.
    b. collect more photons.
    c. can detect fainter objects.
    d. are less affected by the Earth's atmosphere.
17. Which of the following is an *advantage* of a CCD over photographic film? CCDs
    a. can collect light for a long time.
    b. record a greater percentage of the photons striking them.
    c. have better resolution than photographic film.
    d. sometimes require cooling to very low temperatures.
    e. All of the above.
18. A CCD is
    a. an instrument for detecting radio waves.
    b. a device for making images of electromagnetic radiation.
    c. a linked set of telescopes that together produce one image.
    d. a kind of visible-light telescope with corrective optics.
19. False-color images are often used because
    a. human eyes only detect visible light.
    b. not enough light is collected to see the true image.
    c. all astronomical objects are nearly white in color.
    d. very few astronomical objects radiate in visible light.

Collecting Light

29.1 Modern Observatories
29.2 Collecting Power
29.3 Filtering Light
29.4 Surface Brightness

Learning Objectives

Upon completing this Unit, you should be able to:
- Explain the purpose of building large telescopes and the importance of coherence.
- Calculate and compare the collecting power of telescopes.
- Explain how and why astronomical filters are used.
- Define surface brightness and describe how telescopes affect it

Because most astronomical objects are so remote, their radiation is extremely faint by the time it reaches us. For example, to collect enough light to study the most remote galaxies, astronomers use telescopes with mirrors the size of swimming pools, and it may be necessary to point a telescope for hours at a spot so small that it would require thousands of years to look at every object in the sky this way. Therefore astronomers have to be selective in their approach—choosing objects to study that may help to confirm or reject a detailed hypothesis or to refine a theory.

It is usually not enough to simply determine that an object is present at a particular position in the sky. Special instruments are often used to extract the information desired from the radiation—instruments that can measure the brightness at different wavelengths (Unit 24). These instruments require much more light than is needed simply to detect an object, and this in turn requires large telescopes to keep the required observing time within practical limits. Astronomers build telescopes to examine all wavelength ranges (Unit 22), but some wavelength ranges are blocked by the Earth's atmosphere, so space-based telescopes are necessary (Unit 32). The largest telescopes are all built on the Earth's surface because it is extremely expensive to launch telescopes into space. In this Unit we will examine the design of some of the largest modern telescopes and the advantages of size.

29.1 MODERN OBSERVATORIES

Telescopes are essentially large "photon buckets" used to collect light over a large area and then concentrate the light for use by a high-sensitivity detector. The challenge is to build a telescope as large as possible but with every surface and separation built to a very exacting standard—to within about 1/10 of a wavelength of the electromagnetic radiation being observed. This high precision is required to keep the light waves **coherent**: The electromagnetic waves passing through different parts of lenses or bouncing off different parts of mirrors must all arrive at the detector with their wave crests in unison to achieve the maximum strength, as illustrated in Figure 29.1. If the waves arrive out of unison, they will partially cancel each other—somewhat like how the rowers in a large racing shell must row in unison for maximum effect.

The telescopes and the associated equipment that astronomers use are expensive. Therefore, the largest telescopes are often national or international facilities. For example, the United States operates the National Optical Astronomical Observatory and the National Radio Astronomy Observatory, Europe operates the European Southern Observatory, and dozens of other countries each operate many different telescopes. They are open to use by astronomers from around the world

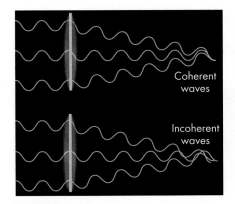

FIGURE 29.1
Optical elements such as lenses must bring light waves together coherently, as illustrated at top. Otherwise the waves may partially cancel each other out.

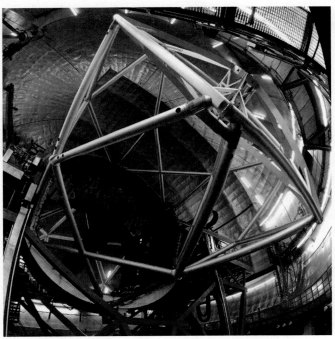

FIGURE 29.2
"Fisheye" view of the world's largest segmented-mirror telescope, in the Canary Islands. The mirror is 10.4 meters (34 feet) across.

FIGURE 29.3
The largest telescope mirrors yet built are 8.4 meters (about 27 feet) in diameter. Two of these huge mirrors are used together on the Large Binocular Telescope on Mount Graham, Arizona.

through a competitive review of observing proposals. Many colleges and universities have their own research telescopes (in addition to smaller ones near campus for instructional purposes). Altogether, several thousand observatories exist around the world, and observatories can be found on every continent. There are even telescopes at the South Pole in Antarctica. These take advantage of the long night and extreme dryness of the bitterly cold antarctic air to view wavelengths that are absorbed by water vapor, such as infrared light.

The optical telescope with the largest "collecting area" is currently the 10.4-meter diameter Gran Telescopio Canarias in the Canary Islands, Spain (Figure 29.2), which is made up of 36 mirror segments that are positioned by lasers to keep the light waves coherent. The largest single-piece mirrors are 8.4 meters in diameter. A pair of these are in the Large Binocular Telescope in Arizona (Figure 29.3), doubling the collecting area. The VLT (Very Large Telescope) takes this a step further, grouping together four 8.2-meter telescopes (Figure 29.4). This increases the collecting area and improves the resolution (Unit 31). The VLT is operated by a consortium of European countries and Chile and is located in the extremely dry, high-altitude mountains of northern Chile.

FIGURE 29.4
The four 8.2-meter diameter telescopes of the VLT in Cerro Paranal, Chile, can work in unison or independently.

FIGURE 29.5
Designs for a 39-meter telescope planned for completion in 2018 by the European Southern Observatory.

To observe the tiny trickle of light from the faintest objects, several 25- to 40-meter diameter telescopes are currently being designed. Three targeted for completion by the early 2020s are the 39-meter-diameter European Extremely Large Telescope (Figure 29.5) and the 25-meter Giant Magellan Telescope, both in Chile, and the 30-Meter Telescope on Mauna Kea, Hawaii. Such telescopes may allow us to detect reflected light from an Earth-size planet around another star, or to study the very first stages of star and galaxy formation shortly after the universe began. Building such a large structure precise to about 50 nanometers is extremely challenging, but it appears feasible using a segmented mirror.

The largest telescopes of all are generally radio telescopes. Radio astronomers must collect very large numbers of these lowest-energy photons to produce a signal powerful enough to be detected. Fortunately, because the wavelengths are much longer than visible light, the optical elements do not require such precise positioning. To observe centimeter wavelength, a precision of millimeters is sufficient.

One such huge radio telescope is located in Arecibo, Puerto Rico (Figure 29.6). The Arecibo radio telescope is more than 300 meters (1000 feet) in diameter. The huge "dish" is fixed in place in a natural basin and consequently has limited ability to look in different directions. It depends in large part on the rotation of the Earth to observe different parts of the sky. Fully steerable radio telescopes of 100 meters in diameter are located in West Virginia and in Bonn, Germany. At microwave wavelengths a 50-meter-diameter Large Millimeter Telescope is currently being completed in Mexico (Figure 29.7) using the segmented-mirror design of large optical telescopes. Radio astronomers are also thinking about ways to collect signals from very faint sources, and plans for a "Square Kilometer Array," which would have a collecting area of 1 kilometer square, are currently being tested. In the more distant future, astronomers would like to build a radio telescope on the far side of the Moon, where it would be shielded from most human-made radio-wavelength transmissions.

Concept Question 1

The "dish" of the Arecibo telescope is fastened to the ground. How do you suppose it is possible to look in any direction other than straight overhead with this telescope?

FIGURE 29.6
The 300-meter diameter Arecibo radio telescope in Puerto Rico.

FIGURE 29.7
The 50-meter Large Millimeter Telescope in Mexico.

FIGURE 29.8
A large aperture collects more light (photons) than a small one, leading to a brighter image.

Clarification Point
A larger aperture does not necessarily indicate that the magnification is higher.

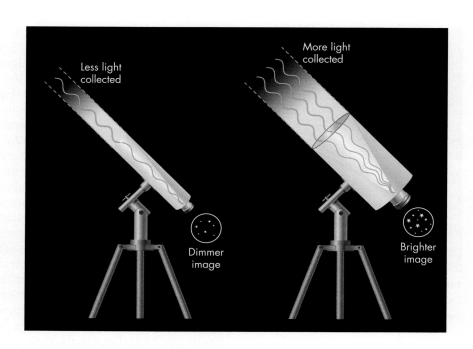

29.2 COLLECTING POWER

For us to see an object, light or photons from it must enter our eyes. How bright the object appears to us depends on the number of its photons that enter our eye per second. Astronomers can increase this amount by collecting photons with a telescope, which then funnels the photons to our eye. The bigger the telescope's **collecting area,** the more photons it gathers, as shown in Figure 29.8. The result is a brighter image, which allows us to see dim stars that are invisible in telescopes with smaller collecting areas. The area of a circular collector of diameter D is

$$\text{collecting area} = \frac{\pi}{4} \times D^2$$

Concept Question 2

Based on the area formula, if a 10-inch pizza is enough for one person, how big a pizza would you need for two? for four?

(or equivalently, $\pi \times \text{radius}^2$). As a result, a relatively small increase in the radius of the collecting area gives a larger increase in the number of photons caught. For example, doubling the radius of a lens or mirror increases its collecting area by a factor of 4. Because the collecting area is so important to a telescope's performance, astronomers usually describe a telescope by the diameter of its **aperture**—the area over which it collects photons. Thus, the "10-meter" Keck Telescopes in Hawaii have mirrors spanning 10 meters (roughly 30 feet) in diameter.

When your eye's pupil is completely dilated, it has an aperture with a diameter of about 8 mm. The Hubble Space Telescope has an aperture with a diameter of 2.4 m. Comparing the size of the collection area for the HST with that of your eye's pupil, we get:

$$\frac{\frac{\pi}{4} \times (2.4\,\text{m})^2}{\frac{\pi}{4} \times (8\,\text{mm})^2} = \left(\frac{2400\,\text{mm}}{8\,\text{mm}}\right)^2 = 300^2 = 90{,}000.$$

Clarification Point

Many people think the Hubble Space Telescope is the largest telescope ever built, probably because of its prominence in reporting new discoveries. However, there are dozens of larger ground-based telescopes.

Thus, the Hubble Space Telescope collects 90,000 times more light per second than your eye. Carrying out the same calculation for the Keck telescope shows that it collects more than 1.5 million times more light than your eye, and 17 times more light than the Hubble Space Telescope. The advantage of the Hubble, of course, is that it is free of the atmospheric distortions that affect light reaching ground-based telescopes (Unit 32).

FIGURE 29.9

Graphs of the transmission for several of the more common filters. Astronomers use color filters to allow a narrow range of wavelengths to reach detectors. At right, a V filter is illustrated. Light with a wavelength of about 530 nm passes through the filter at almost full strength, while light at nearby wavelengths is partially diminished. The filter blocks other wavelengths.

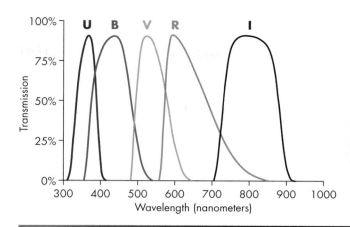

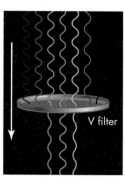

Concept Question 3

U and I filters look almost black, even if they transmit nearly perfectly. Why is this?

Different kinds of film, CCDs, and lenses may behave differently, depending on the wavelength of the light. By narrowing the range of wavelengths observed, filters make it easier to compare and combine data gathered with different telescopes and detectors and under different atmospheric conditions.

29.3 FILTERING LIGHT

Telescopes generally collect a wide range of wavelengths. For example, most visible-wavelength telescopes also work into the infrared and ultraviolet. However, astronomers rarely collect all of the incoming light, but instead **filter** out all but a narrow range of wavelengths, selected to learn about the colors of the particular objects they are studying.

Filters are often thin pieces of colored glass. The filter glass is designed to reject all wavelengths but those in the range of interest and to be highly transparent within that range. At visual and neighboring wavelengths astronomers commonly use the following set of filters:

  U, or ultraviolet, transmission strongest at 370 nm
  B, or blue, transmission strongest at 420 nm
  V, or "visual," transmission strongest at 530 nm
  R, or red, transmission strongest at 600 nm
  I, or infrared, transmission strongest at 800 nm

The transmission curves for filters of these types are plotted in Figure 29.9. Astronomers also use specialized filters to select narrow ranges of wavelength to examine the light from a single spectral line.

By observing stars and other objects through filters (Figure 29.10), astronomers can measure how much red light or blue light an object emits, which can be used to determine its temperature (Unit 23). In addition, filters help to reject unwanted light—when we observe an object known to emit most strongly at one wavelength, a filter can be chosen to admit light at that wavelength and no others.

Note that a filter does not "color" the light passing through it. When you look through a blue filter, you are seeing just the blue photons from each object, not changing the wavelength of other color photons. Most objects produce or reflect some photons of every wavelength, so a yellow object will not necessarily appear black through a blue filter.

FIGURE 29.10

A star field through B, V, and R filters, and combined to make a color image. Note the star at the bottom that is brighter in the R filter than in V or B, and the emission from gas at the top only seen in the R image (probably from a spectral line of hydrogen).

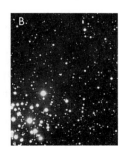   

Always look well away from the Sun. It can damage your eyes permanently. It should never be looked at directly except with specialized equipment designed specifically for the purpose.

29.4 SURFACE BRIGHTNESS

Look at a star with your naked eyes. Next, look at the same star through binoculars or a small telescope; you will notice it appears brighter. If, however, during the day you look at birds or trees through the same telescope, they will appear larger, but not brighter. Now, with one eye, try looking at the blue sky through a telescope, and look at the sky with the other eye unaided. The sky will appear the same or perhaps even brighter with the unaided eye. What these exercises demonstrate is that a telescope does not increase the "surface brightness" of an object.

We can express the **surface brightness** of the sky, a planet, or a distant galaxy in terms of the number of photons or total watts received per square degree over the surface of the object. With a telescope you will collect more photons from each square degree than with your naked-eye observation; but the telescope also magnifies the image, spreading its photons out by a compensating amount. Hence, if we use a telescope with an aperture big enough to collect 100 times more photons than your naked eye can collect, the magnification of the image will spread the image's light over an area at least 100 times larger. As a result, the amount of light reaching one square millimeter at the back of your eye, or a CCD detector at the back of a telescope, will be no greater however big a telescope is used. This is why the sky does not appear any brighter through a telescope.

But why do stars appear brighter when magnified? Stars are so far away that they look like points of light to your eye. Even when they are magnified by a telescope to cover an area 100,000 times larger, and we are collecting 100,000 times more photons from the star, the magnified image of the star still remains imperceptibly small. The image size is smaller than the pixel size of a CCD or of a cell in your eye, either of which is then further limited by telescope diffraction (Unit 31) and by the blurring of the Earth's atmosphere (Unit 32). As a result, the increased numbers of photons appear still to be coming from one point, making the star look brighter, but not discernibly bigger.

When you look through a telescope at a dim astronomical object, such as a nebula or galaxy, it almost never matches its appearance in photographs (Figure 29.11). Photographs exhibit low surface brightness details you could never see with your eyes. If magnification cannot make an object appear brighter, how can photographs look so bright? The answer is that these photographs collect more photons in each square degree by making long exposures. Unlike our eyes, a CCD or photographic film can accumulate the photons over a long period of time. Even if very few photons arrive each second from a low surface brightness region, by training our CCD or open camera shutter on the region for hours, we will gather many photons. At the end of our collection period, their cumulative numbers will cause the area to appear brighter.

Surface brightness presents some practical considerations for astronomical observations. To get a detailed image of the outer regions of a galaxy, which have a low surface brightness, it requires long integration times, no matter what the size of the telescope. Alternatively, if you are trying to photograph the sunlit Moon, exposures must be as short as a daylight snapshot on the Earth. Automatic exposures on cameras often are confused by the large amount of black in a night-sky image, and overexpose the Moon.

FIGURE 29.11
The galaxy M31 (see Unit 75) looks like a faint fuzzy spot of light with the naked eye or through a telescope, although much magnified through the telescope (top). A long photographic exposure (bottom) brings out features and colors that cannot be seen directly through a telescope.

KEY POINTS

- Telescopes are designed to capture as much light as possible in order to study the faintest and most distant objects.
- Optics and spacings in a telescope need to be accurate to a fraction of a wavelength for light to remain coherent.
- Single mirrors more than 8 meters in diameter have been constructed, but the largest telescopes use segmented mirrors.
- Wavelengths other than visible and radio are blocked by the atmosphere to various degrees, requiring space telescopes.
- The collecting area of a telescope is proportional to the area of its aperture, which depends on the square of the diameter.
- Astronomers use color filters to collect light over a narrow range of wavelengths to study some physical properties of an object.
- An object's "surface brightness" does not increase through larger telescopes, so long-duration observations are needed to see low-surface-brightness features.

KEY TERMS

aperture, 206
coherent, 203
collecting area, 206
filter, 207
surface brightness, 208

CONCEPT QUESTIONS

Concept Questions on the following topics are located in the margins. They invite thinking and discussion beyond the text.

1. Aiming the Arecibo telescope. (p. 205)
2. Ordering the right size pizza. (p. 206)
3. Why ultraviolet and infrared filters look dark. (p. 207)

REVIEW QUESTIONS

4. What kinds of telescopes are ground-based? Why not all types?
5. How do modern telescopes differ from earlier telescopes?
6. What is meant by the term *collecting area*?
7. When the diameter of one telescope is twice that of another, how does its collecting area compare?
8. How does a larger telescope make it easier to see faint objects?
9. Why do astronomers use filters instead of collecting all wavelengths of light?
10. Why would you be unable to detect spectral lines of hydrogen if you observed them through a V filter?

QUANTITATIVE PROBLEMS

11. How thick are the pages of this book in nanometers? Find this by measuring the thickness of many sheets and dividing. How much bigger is this in comparison with the precision that an optical telescope operating at 500 nm must maintain?

12. The smallest binoculars have an aperture diameter of approximately 1 inch, while the largest commercially available telescopes have apertures of about 50 inches. How many times fainter a star should you be able to see using the larger telescope in comparison to the binoculars?
13. Compare the collecting area of one of the 36 segments of the 10.4-m Gran Telescopio Canarias to the collecting area of the 2.4-m Hubble Space Telescope. Which is larger? By what factor?
14. A group of stars is observed through several different filters, and some stars are seen to be brightest when observed through either the V (530 nm) or R (600 nm) filter. Estimate the range of temperatures for these stars as follows:
    a. Assuming a star is brightest in the middle of the R filter, use Wien's law (Unit 23.4) to estimate its temperature.
    b. If a star is brighter through the V filter, use Wien's law to estimate the range of temperatures it might have.
15. Suppose you used a telescope to detect light from stars that is spread fairly evenly across 300- to 900-nm wavelengths. If you put a V filter in front of your detector, about how many times longer will you have to observe to collect as many photons? (Estimate this approximately, based on Figure 29.9.)
16. The power of the light reaching Earth from a "first magnitude" star is approximately 10^{-6} watts (joules per second) over each square meter of a telescope's collecting area. Assuming that this light is composed of photons of wavelength 500 nm, about how many photons are arriving in a square meter each second? How many photons enter your eye each second if your pupil is 8 mm in diameter?
17. At a radio wavelength around 30 cm, the Arecibo radio telescope can collect about 3×10^{-14} watts (joules per second) of radiation from the Crab Nebula. About how many photons is it collecting each second? At visible wavelengths, about 10^{-9} watts per square meter reach us from the Crab Nebula.

TEST YOURSELF

18. A person's pupil can range in size from about 3 mm to 9 mm in diameter, so the collecting area of a fully dilated pupil is ___ times greater than a non-dilated one.
    a. 3   b. 6   c. 9   d. 12   e. 18
19. What is a good reason to build larger telescopes?
    a. They permit us to detect dimmer stars.
    b. Larger telescopes are more durable.
    c. The greater the size, the more wavelengths they can let in.
    d. They make the light more coherent.
    e. National pride
20. The purpose of a blue filter is to
    a. shift the wavelength of the light to make it easier to detect.
    b. eliminate the color of the sky from photographs.
    c. let through only the blue wavelengths of light.
    d. allow a telescope to collect more light.
21. Suppose just as many photons were detected from an object when observed with a B filter as with no filter. Which of the following might explain this? The object is
    a. so small that its light can go through the filter.
    b. very hot so that it emits blue colored light.
    c. emitting only an emission line at a blue wavelength.
    d. producing uniform radiation at all wavelengths.

Focusing Light

30.1 Refracting Telescopes
30.2 Reflecting Telescopes
30.3 The Development of Larger Apertures
30.4 Color Dispersion

Learning Objectives

Upon completing this Unit, you should be able to:
- Define refraction and color dispersion and explain what causes them.
- Describe the focusing mechanisms for refracting and reflecting telescopes, and what differences there are in telescopes at different wavelengths.
- Discuss why reflectors have become the primary telescope type used for research.

In addition to a telescope's role as a large "photon bucket" (Unit 29) telescopes are designed to distinguish the direction from which light arrives and to **focus** the light. Focusing light is the process of bringing the electromagnetic waves together to form the strongest and most detailed signal possible.

Galileo Galilei is sometimes credited with the invention of the telescope, but the Dutch spectacle maker Hans Lippershey built one before him in 1608. News of the invention spread rapidly, and by the summer of 1609, Thomas Harriott, an English mathematician and scientist, had used a telescope to study the Moon. Just a few months later, Galileo, quickly grasping the idea behind the design of the telescope, built a telescope of better quality than any previous, capable of magnifying images by about a factor of 20. With this telescope he not only observed the Moon but discovered Jupiter's moons, Venus's phases, and the vast numbers of stars in the Milky Way. Most important, he published and interpreted his findings (Unit 12.3).

The competition to build better telescopes has continued nonstop to this day, and now includes specialized telescopes for every wavelength range. Astronomers have not only sought to build bigger telescopes that can detect dimmer objects (Unit 29), they have developed improved designs for focusing the light to provide more detailed images, as well as working at wavelengths across the electromagnetic spectrum (Unit 22). Combined with new developments in instrumentation (Unit 28), telescopes today are orders of magnitude more sensitive than telescopes just a few decades ago. The methods used for focusing light fall into two broad categories, employing either a lens or a mirror to concentrate the light that enters the main aperture of the telescope. Each of these methods has advantages and disadvantages, as we discuss next.

30.1 REFRACTING TELESCOPES

The first telescopes used **lenses** to focus the light. Transparent substances like the glass of a lens can alter the path of electromagnetic waves. This effect is called **refraction**. Eyeglasses and the lenses inside our eyes use refraction to focus light.

Telescopes are called **refractors** if they use a lens as the primary means to focus the light that enters the aperture. The lens of a refractor focuses the light by bending the rays, as shown in Figure 30.1. A lens that is just the right shape can collect light over the entire aperture and bend it so that it arrives at a focus behind the lens, creating an image. The figure depicts the light from two stars, a yellow star straight in front of the telescope and a red star above it. The light from each star is focused to a point, making an image of the sky over a region called the **focal plane**. This is where photographic film or a CCD might be placed, although often there

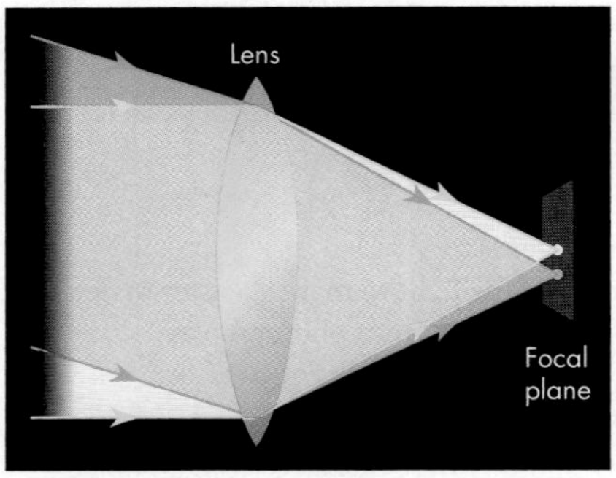

FIGURE 30.1
Light is shown arriving from a yellow star straight in front of the lens and from a red star above it. Each star's light is brought to a focus at the focal plane.

FIGURE 30.2

Refraction of light in water. Notice how the ruler appears to be bent.

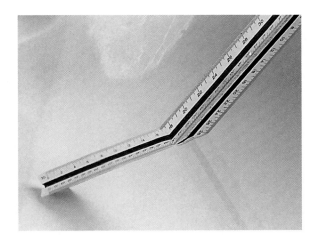

Concept Question 1

If you are scuba diving and you look up at the surface of the water overhead, you will see light from the sky in a region only out to about 45° away from the zenith. How might refraction explain this?

Ocean waves travel slower where the water grows shallower, so they refract like light waves. Shallow water can often be deduced where waves change direction.

FIGURE 30.3

Cause of refraction. (A) Light entering the dense medium is slowed, while the portion still in the less dense medium proceeds at its original speed. (B) A similar effect occurs when you walk hand-in-hand with someone who walks more slowly than you do.

are additional lenses in front of the detector to provide extra magnification or to correct for distortions produced by a single lens.

Lenses can bend light because when light moves from one transparent material into another at an angle, the direction in which the light travels is bent. Bending occurs when light from a distant star enters the Earth's atmosphere, or when that light travels from air into water or glass. The bending is generally stronger with a greater difference in density of the materials. For example, glass is slightly more dense than water, and light entering glass from air is bent slightly more than when it enters water. However, a clear piece of glass under water may be hard to see because of the small difference in refracting properties.

You can easily see the effects of refraction by placing a ruler in a glass of water and noticing that the ruler appears bent, as shown in Figure 30.2. The ruler in water also illustrates an important property of refraction. If you look along a ruler placed partly in the water and change its tilt, you will see that the amount of bending (refraction) changes. Exactly vertical rays are not bent at all, nearly vertical rays are bent only a little, and rays entering at a grazing angle are bent most.

Light is bent because its speed changes as it enters matter, generally slowing in denser material. This decrease in the speed of light arises from its interaction with the atoms through which the electromagnetic waves move. To understand how this reduction in light's speed makes it bend, imagine a light wave approaching a slab of material. The part of the wave that enters the material first is slowed while the part remaining outside is unaffected, as depicted in Figure 30.3A. Imagine what would happen if the right side of your car went off the road into mud. As that side of the car slowed, the car would swerve to the right. Or if two people are walking hand-in-hand (as sketched in Figure 30.3B), if the one on the right slows

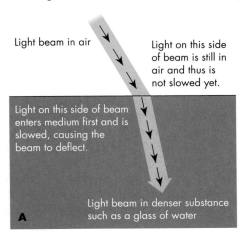
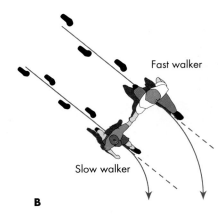

FIGURE 30.4
The largest refracting telescope ever built. Completed in 1897 for the University of Chicago's Yerkes Observatory in Williams Bay, Wisconsin, this refractor has a lens approximately 1 meter (40 inches) in diameter. The photograph shows Albert Einstein visiting the telescope in 1921.

Large lenses are used for some microwave applications. To our eyes these may look like large pieces of opaque plastic, but they are transparent to microwaves and refract them.

down, the direction of their motion turns to the right. So, too, if one portion of a light wave moves more slowly than another, the light's path will bend.

Figure 30.4 shows a photograph of the world's largest refractor, the 1-meter (40-inch) diameter Yerkes telescope of the University of Chicago. This telescope, completed in 1897, was one of the last large-lens telescopes ever built. Building a telescope with such a large lens presents serious structural problems. The Yerkes lens has a mass of more than 200 kilograms (450 pounds). This massive piece of glass has to be supported by its edges, where the glass is thinnest, so the lens flexes slightly, causing image distortions. In addition, the large mass of glass is located at the end of a long telescope tube, which must be even more massive—about 20 tons—to keep it from flexing. Building larger refractors would require even more massive telescopes with even greater problems of structural flexing. To build larger-aperture telescopes, an alternative approach was needed.

30.2 REFLECTING TELESCOPES

Not long after the invention of telescopes, people realized that mirrors could be used to bring light to a focus; but a practical design to do this was not worked out until 1670 by Isaac Newton. His first telescope used mirrors made of metal and did not provide images as sharp as refractors, but technological improvements over the subsequent centuries allowed mirrors to surpass lenses. Today almost all research telescopes, whether used for visible, radio, or X-ray wavelengths, are *reflecting telescopes*—**reflectors** for short—employing mirrors as the primary element to collect and focus the light.

As Figure 30.5 illustrates, a curved mirror can focus light rays reflected from it. The figure depicts the light from two stars, a yellow star straight in front of the telescope and a red star above the yellow star. Each star's light is focused at a point on the focal plane, but this is now in front of the mirror, so the detector blocks some of the light from reaching the **primary mirror.** This is usually only a small fraction of the total aperture. If the region blocked has one-third the diameter of

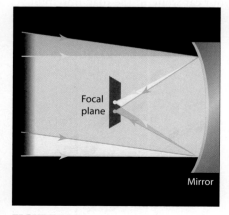

FIGURE 30.5
A curved mirror focuses light from a yellow star straight in front of it and a red star above that.

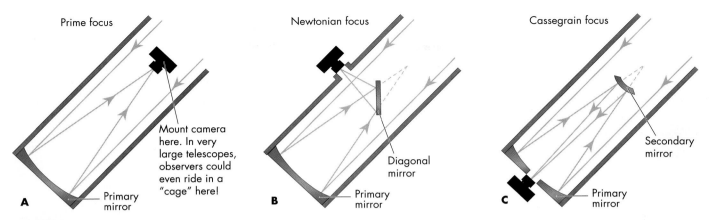

FIGURE 30.6
Sketches of different focus arrangements for reflectors.

the primary (Figure 30.6A), it will block only one-ninth of the light, because the collecting area (and blockage) is proportional to the diameter squared (Unit 29). Some large telescopes were big enough that they could even be built with a "cage" in which an astronomer rode inside the telescope to change photographic plates at the focal plane or "prime focus." Today, however, it is recognized that the heat from a human body can cause the air to refract light and affect images.

A variety of designs have been developed to move the focus outside the path of the incoming light. Newton's solution was to use a **secondary mirror** to deflect the light out to the side (Figure 30.6B). This is today known as a **Newtonian telescope,** and it remains a popular design for smaller reflectors. Most modern research telescopes, whether optical or radio, use a secondary mirror that reflects the light back through a hole cut in the middle of the main mirror. This is called a **Cassegrain telescope** (Figure 30.6C), named for the French sculptor Guillaume Cassegrain, who designed it in 1672.

It is important that the electromagnetic waves are coherent (Unit 29), that is, they must arrive at the detector in unison, so the shape of the reflector is critical. The mirror needs to be designed so that after electromagnetic waves bounce off the mirror, the crests of the waves come together at the focus all at the same time. By achieving this simultaneous arrival, the strength of the electromagnetic vibration at the focus will be the concentrated strength of the entire wave collected over the whole aperture. If the mirror shape is not correct, not all portions of the crest of the wave will arrive at the same time, and they will not combine to the maximum strength. As a result, the signal will be weaker at the detector, and the images will be blurry.

To keep the waves coherent, the smoothness of the surface is also essential. The first metal reflectors were difficult to polish. It was not until the late 1800s, when the technology of placing a thin layer of metal on glass was developed, that reflectors began to replace refractors. Glass can be shaped and polished smooth to better than one-tenth of the wavelength of the visible light, which is essential; otherwise the light will scatter in many directions.

A radio telescope designed to study 10-centimeter-wavelength light, by contrast, needs to be smooth only at the level of 1 centimeter, so it can be built of metal plates bolted together. At the other extreme, X-ray and gamma-ray photons have wavelengths that may be smaller than the size of an atom. X-ray photons encountering an optical telescope mirror would be scattered in many directions or absorbed by the mirror. However, they will reflect off a surface if they encounter it at a very shallow angle. (You can demonstrate this principle to yourself with visible light by tilting a rough surface so light hits it at a glancing angle.) The mirror designs for X-ray telescopes therefore look completely different from optical

> When the Hubble Space Telescope was first put into service, it was discovered that the surface had been polished to the wrong shape in a manufacturing error. Fortunately, it was possible to insert instruments with lenses to correct for the shape error, much as you can wear glasses to correct for focusing problems caused by the shape of your eye.

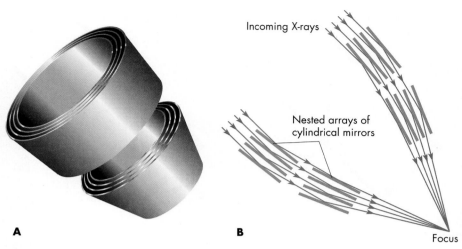

FIGURE 30.7
Grazing incidence optics in an X-ray telescope. (A) The mirrors are a series of nested cylinders. (B) This cross-sectional view shows how the incoming photons are reflected through shallow angles by each mirror to bring them to a focus.

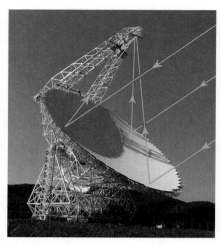

FIGURE 30.8
The 100-meter diameter radio telescope in Green Bank, West Virginia. The main dish reflects light off-axis so the secondary mirror does not block the telescope's aperture.

designs, but they achieve the same result of funneling the photons toward the detector, as illustrated in Figure 30.7.

Refracting telescopes have the advantage that the focus is behind the lens. This simplifies the design, and there is no need for a secondary mirror or other equipment that partially blocks the telescope aperture. Image quality is degraded by anything that blocks part of the aperture. For example, the "spikes" seen on stars in most astronomical images are caused by the support structures that hold the secondary mirror in place in a reflector. (This is examined further in Unit 31.)

One way to eliminate the problem of having a secondary mirror blocking the primary is to construct a reflector with an *off-axis* focus. The primary mirror reflects the light at an angle, bringing it to focus at a point outside the field of view of the aperture. The German-English astronomer Sir William Herschel built a telescope of this design in the 1780s. However, because the primary mirror needs to be asymmetrical, such telescopes are substantially more difficult to build and are relatively uncommon. The 100-meter (330-foot) diameter Green Bank Telescope (GBT), built using this idea, is shown in Figure 30.8.

30.3 DEVELOPMENT OF LARGER APERTURES

After 1900 almost all telescopes built for research were reflectors. This was primarily because of the expense of building larger lenses and the technological problems of supporting such massive pieces of glass (Section 30.1). Lenses are still frequently used as part of the optics of many research telescopes, but a telescope is still called a *reflector* if a mirror does the primary work of focusing the light that enters the aperture.

As a telescope moves, its mirrors and lenses must keep their same precise shapes and relative positions if the images are to be sharp. This is one of the most technically demanding parts of building a large telescope, because large pieces of glass or metal bend slightly when their positions shift. Unlike lenses, however, a mirror can be supported from behind, thereby helping to hold the glass in shape.

Development of large reflectors proceeded rapidly during the early 1900s, greatly surpassing the apertures of the largest refractors. A 5-meter (17-foot) diameter

FIGURE 30.9
(A) One of the twin Keck telescopes. The 36 mirrors cover an area 10 meters (about 33 feet) in diameter. (B) One of the two Gemini telescopes, located on Mauna Kea, Hawaii; the other one is in Chile. The mirror is 8.1 meters (26 feet) in diameter. The yellow lines show the light path through the Cassegrain design of the telescope.

Computer controls have also helped to reduce the weight of telescopes. Rather than building a telescope with an angled axis to mechanically compensate for the Earth's rotation, telescopes are now built with the weight more directly supported, and a computer steers the telescope.

mirror was completed in 1948 at the Palomar observatory in California. However, the mirror alone weighed nearly 15 tons, the whole telescope nearly 500 tons. Building larger telescopes along similar lines was found to be impractical because the weight of glass and supporting structures grew too rapidly with size, so the Palomar telescope remained the largest working telescope for more than 40 years.

Innovations over the last two decades have provided new ways for building telescopes with larger collecting areas. One way is to collect and focus the light with several smaller mirrors and then align each one individually instead of using a single large mirror. Mirrors designed this way are called *segmented mirrors*. Currently two of the largest such segmented mirrors are in the twin 10-meter (33-foot) Keck telescopes (Figure 30.9A), operated by the California Institute of Technology and the University of California and located on the 4200-meter (14,000-foot) volcanic peak Mauna Kea in Hawaii. The total weight of the glass in this design is about the same as in the Palomar 5-meter telescope, but the total telescope weight is only about half as much. Each Keck telescope consists of 36 separate mirrors (visible in Figure 30.9A) that are kept aligned by lasers that measure precisely the tilt and position of each mirror. If any misalignment is detected, tiny motors shift the offending mirror segment to keep the image sharply in focus. The same principle is being used to build large new radio telescopes.

Another approach for building large optical telescopes is to build mirrors much thinner and lighter than those in earlier, smaller telescopes. For many years telescope mirrors were made thicker as they were made larger, to keep them stiff. By making a back structure that uses sensors and tiny motors to compensate for bending as the mirror is tilted, much as described for the Keck telescopes, a thin mirror can be kept precisely shaped. New kinds of glass have also been developed that are very stable against changes in shape as the temperature changes during the night. A number of thin 8-meter (26-foot) glass mirrors have been built in recent years such as the Gemini 8.1-meter telescope shown in Figure 30.9B. These achieve overall weights of glass and telescope that are similar to those of the Palomar 5-meter telescope. Four 8.2-meter telescopes are used in unison at the VLT (Very Large Telescope) of the European Southern Observatory in Chile, giving the combination of the four the largest total collecting area of any current visible-wavelength observatory.

FIGURE 30.10
(A) Glass refracts blue light more than red light, producing a spectrum. (B) The dispersion of light caused by glass can result in different wavelengths focusing differently. The result can appear in images as fringes of color.

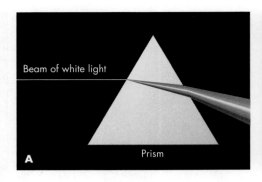
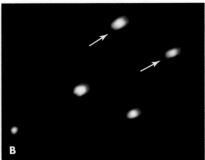

30.4 COLOR DISPERSION

Even though refractors do not have any blockage in their aperture, they have some significant disadvantages because of the properties of glass. The lenses may be as opaque as a chunk of concrete to shorter-wavelength ultraviolet photons. Mirrors, by contrast, are generally much less selective in the wavelengths of electromagnetic radiation they reflect, allowing much broader wavelength coverage with a single telescope. Mirrors also reflect light at the same angle, independent of wavelength, but the same is not true of lenses. The angle by which glass bends light depends on the light's wavelength.

Thus, as we saw in Unit 24, when white light—which is a mix of all colors—passes through a refracting material at an angle, it is spread into a spectrum, or rainbow. This happens to light passing through the glass of a prism (Figure 30.10A) or through drops of water in a rainstorm. The ability of glass to spread light into its component colors is enormously useful for carrying out spectroscopy on the light from astronomical sources. Astronomers can use a large prism to split the light of all of the astronomical sources within the telescope's field of view into spectra all at once. Stars and other objects may display slight or large differences in the characteristics of their spectra even though their color and appearance otherwise look the same. This gives astronomers an important tool for classifying objects and discovering new kinds of sources, but this same property of glass can cause problems for refractors.

Glass causes light to split into different colors through a process called **dispersion.** Different wavelengths of light travel at slightly different speeds in most materials and are therefore refracted by different amounts. Shorter wavelengths of light are generally refracted more strongly. The dispersion caused by glass is one reason astronomical mirrors have reflective aluminum coating on top of the glass instead of underneath it (as in a conventional mirror). However, this has the disadvantage that telescope mirrors can be easily marred and must be recoated over time as the metal becomes tarnished by exposure to air.

The dispersion of refracting materials poses a problem for making a sharp image. Light passing through the edge of a lens becomes dispersed, causing the shorter wavelengths of light from a star to focus at a different position from the longer wavelength. And even if the different colors are aligned, the red wavelengths of light may be out of focus when the blue wavelengths are sharp, creating images fringed with color. This error between different colors (*chroma* in Greek) is called **chromatic aberration,** and an example is shown in Figure 30.10B. Because of this problem, research instruments avoid lenses if mirrors can achieve the same result, although mirrors are not always practical.

Concept Question 2

Suppose you were examining a pulse of radio signals from a distant civilization far away across interstellar space. Knowing that the ionized gas in interstellar space causes dispersion of radio waves, what effect would you expect this to have on the signal?

Concept Question 3

Most refractors have lenses that bring light to focus over a large distance, making the telescopes very long. Why would a lens that focuses light within a short distance be likely to have more problems with chromatic aberration?

KEY POINTS

- The optics of a telescope are designed to redirect light to a focus.
- Substances that are transparent to electromagnetic radiation may nevertheless change its direction or refract it.
- Lenses use refraction to bring light to a focus, and a telescope using a lens as the primary means of focusing is called a refractor.
- Curved mirrors can also focus light, and a telescope using a mirror as the primary means of focusing is called a reflector.
- Reflectors focus light in front of the mirror, so there are many designs to get the light to a more practical position for observation.
- Problems with the weight of optics set limits on the size of telescope apertures of both reflectors and refractors.
- New technologies have allowed new large reflectors with lighter mirrors whose shape is maintained by active controls.
- Lenses usually bend different-wavelength light by different amounts, which can cause chromatic aberration.

KEY TERMS

Cassegrain telescope, 213
chromatic aberration, 216
dispersion, 216
focal plane, 210
focus, 210
lens, 210
Newtonian telescope, 213
primary mirror, 212
reflector, 212
refraction, 210
refractor, 210
secondary mirror, 213

CONCEPT QUESTIONS

Concept Questions on the following topics are located in the margins. They invite thinking and discussion beyond the text.

1. The sky as seen from under water. (p. 211)
2. Dispersion of extraterrestrial radio signals. (p. 216)
3. Chromatic aberration for short focal length lenses. (p. 216)

REVIEW QUESTIONS

4. What do we mean by the phrase *to focus light*?
5. What is the difference between reflecting and refracting telescopes?
6. Why do reflecting telescopes usually have a secondary mirror?
7. Why have the large telescopes built in the last century all been reflectors, instead of refractors?
8. What technologies are needed to allow the giant reflector mirrors of today to maintain a sharp focus as they are aimed in different directions?
9. How is dispersion of light useful in astronomy?

QUANTITATIVE PROBLEMS

10. An image will appear upside down when we look through a simple refracting telescope, as can be seen from Figure 30.1. Trace the light through a similar diagram to show what orientation objects would have through a Cassegrain telescope. Assume the Cassegrain's secondary mirror is flat.
11. A particular amateur astronomy reflector has a primary mirror with a diameter of 28 cm, and secondary mirror with a diameter of 9.5 cm. What fraction of the collecting area does the secondary mirror block? What diameter telescope has the same collecting area if no part of its aperture is blocked?
12. The front surface of your eye (the cornea) does most of the focusing of light entering your eye; the back wall of the eye is like a focal plane. Draw a diagram similar to Figure 30.1 to show how your eye focuses the light rays from a distant star straight in front of you. Now draw light rays for a nearer point of light (a few times farther away from your cornea than the cornea is from the back of the eye). The light rays will be bent by the cornea by about the same angle as the distant starlight, so do they now focus closer to or farther from the cornea?
13. A nearsighted person is able to focus on nearby objects, but not distant ones. Based on the previous problem, what does this tell you about the shape of the cornea for a nearsighted person? Will a nearsighted person focus red or blue colors better?
14. After the Palomar 5-m telescope was completed, astronomers estimated that the mass of larger telescopes would increase as the (mirror diameter)^3 or (mirror diameter)^4 because of the greater mass of glass and support needed to keep the entire telescope stiff. If the 10-m Keck telescopes followed this rule, how many times more massive would they have been than the Palomar telescope?
15. The *index of refraction* measures the factor by which light slows down in a substance. The index of refraction of water is 1.33. If you sent a pulse of light through water and a pulse of light through air (index of refraction of 1.00), how much longer would it take the light to travel 1 km through the water? (The speed of light in air is 299,800 km/sec.)

TEST YOURSELF

16. Which of the following is not a disadvantage of a reflecting telescope?
    a. The secondary mirror blocks some of the collecting area of the primary mirror.
    b. The support structure for the secondary introduces "spikes" on some images.
    c. The primary mirror is supported from behind the mirror.
    d. The reflective metal is exposed to the air allowing it to be marred.
17. The 5-meter Palomar reflector remained the largest optical telescope for more than 40 years. What did astronomers discover that led to larger telescopes?
    a. How to make lightweight refractors
    b. Active control systems to maintain mirror shape
    c. The first distant quasars that required a larger telescope
    d. A process for forming mirrors out of metal
18. How does the speed of light in glass compare to the speed of light in empty space? It is
    a. slower in glass.   b. slower in space.   c. the same.
19. What process causes light to be spread into a spectrum?
    a. Reflection   b. Refraction
    c. Dispersion   d. Focusing

Telescope Resolution

31.1 Resolution and Diffraction
31.2 Calculating the Resolution of a Telescope
31.3 Interferometers

Learning Objectives

Upon completing this chapter, you should be able to:
- Explain the effects of diffraction on the resolution of a telescope.
- Calculate the resolution of a telescope from its diameter and wavelength observed.
- Describe how an interferometer works and what determines its resolution.

People often ask how high a magnification a telescope can produce. In reality, any telescope can deliver as high a magnification as we want with the right eyepiece. What is more important to understand is the **resolution** of a telescope—the level of detail it is capable of revealing. A highly magnified blurry image is of little use.

Resolution is determined in part by the quality of the optics; but even with perfectly made lenses and mirrors, there is a fundamental limit on resolution imposed by the aperture size of a telescope (Unit 29). The larger the aperture, the finer the detail it is capable of detecting. This is the other reason astronomers strive to build larger telescopes—so they can resolve increasingly finer structural details of the objects they observe.

In this Unit we will examine the limitations imposed on the clarity of telescopic images by the interaction of light with a telescope's aperture, along with some "tricks" astronomers use to achieve the sharpest possible images. Improved resolution has led to many major astronomical discoveries, because until we can see an object clearly, it is difficult to interpret what it is.

31.1 RESOLUTION AND DIFFRACTION

If you mark two black dots close together on a piece of paper and look at them from the other side of the room, at a great enough distance your eyes will no longer be able to see them as separate spots. Similarly, stars that lie very close together, or markings on a planet, may not be clearly distinguishable. The smallest separation between features that a telescope is able to distinguish is a measure of the telescope's resolution.

Resolution is fundamentally limited by the wave nature of light. Whenever waves pass through an opening, they are bent at the edges of the opening by a phenomenon called **diffraction.** Figure 31.1 shows how water waves, all initially parallel to each other, are diffracted after they pass through a narrow opening. The central portion of the set of waves travels straight through the opening, but weaker waves radiate away from the edges of the opening.

Light waves are similarly diffracted at the edge of a telescope's aperture. The result is that the light waves from a star are shifted into slightly different directions, smearing out the light. Figure 31.2A shows an image of

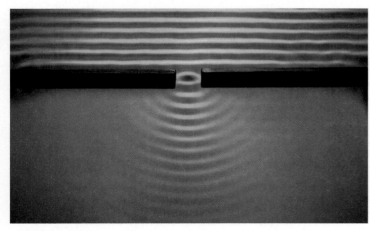

FIGURE 31.1
Water ripples originating from the top in this image pass through a narrow opening in a wall. The waves are diffracted at the edges of the opening and spread out in different directions than they were originally moving.

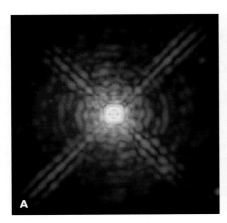 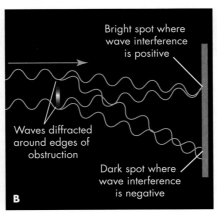 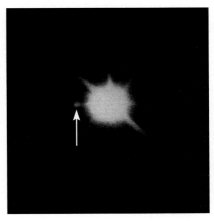

FIGURE 31.2
Diffraction and interference. (A) Image of a star made with the Hubble Space Telescope. Light coming from a single point is spread out into a very faint, complex diffraction pattern. (B) Diffraction at the edge of the telescope aperture or any internal structures bends waves into many directions, and these waves interfere with each other. Two pairs of waves are shown, one where the wave crests align and the waves add in intensity, and another where the waves oscillate in opposite directions, canceling each other out.

FIGURE 31.3
The bright star Sirius and its faint companion star (arrow). The companion orbits the brighter star with a period of about 50 years. The bright star appears bigger, but both stars are points of light.

a star from the Hubble Space Telescope. The image has been enhanced to bring out the very faint **diffraction pattern** produced when a star is imaged. Diffraction spreads the star's image into a blur, even though the light is coming from a single point within this image. A small percentage of the light is diffracted at larger angles and produces the complex outer parts of the diffraction pattern.

The complex set of rings and spikes seen in the diffraction pattern is caused by interference between waves coming through different parts of the aperture or diffracting from the edges of any other structures within the telescope's aperture. **Interference** occurs when light waves add to or subtract from each other. The diffracted light travels through the telescope along slightly different paths, offset from its original direction. Light diffracted by different edges of the aperture and following slightly longer or shorter paths may arrive at the detector with the wave crests adding together or canceling each other out (Figure 31.2B). Where they add together, we see a brighter spot in the diffraction pattern. Where the interference causes the waves to arrive with the crests of one wave adding to the troughs of another wave, they cancel one another, and we see a darker spot in the diffraction pattern. The diffraction rings are produced by interactions between light waves diffracted from different parts of the edges of the aperture. The X-shaped pattern surrounding the star is caused by light diffracted by the support structure holding a secondary mirror in the telescope.

The diffraction pattern can make brighter stars appear larger in an astronomical image even though the light from all stars is essentially coming from a single point. This is especially true when images are optimized to examine fainter objects. When the image is brightened, the inner portion of the diffraction pattern around a bright star may become completely "saturated" (completely white or bright). For the brightest stars, the diffraction pattern will be saturated even farther from the star. Detecting a faint star near a bright star can be difficult because a dim object may become lost within the bright star's diffraction pattern even if the separation is relatively large. For example, the image in Figure 31.3 has just enough resolution to see a faint companion of the bright star Sirius. With just slightly poorer resolution, the companion star could not be seen. This makes it extremely difficult to image faint objects such as planets around other stars.

> **Clarification Point**
>
> The apparent sizes of stars and spikes extending from them in telescopic images are usually caused by properties of the telescope, *not* by the stars.

One way to describe the resolution of a telescope is by the closest pair of stars that can still be discerned from each other. If the stars are separated by too small an angle, diffraction by the telescope aperture (or poor optics) will overlap the light from the stars to such an extent that they are indistinct from each other.

31.2 CALCULATING THE RESOLUTION OF A TELESCOPE

The best possible resolution of a telescope depends on both the wavelength of the radiation being observed and the diameter of the lens or mirror that collects the light. Resolution can be improved by using a larger mirror or lens, and it can also be improved by observing at shorter wavelengths.

Detailed calculations show that if two points of light are separated by an angle α (measured in arc seconds) and are observed at a wavelength λ (measured in nanometers), they cannot be seen as separate sources unless they are observed through a telescope whose diameter D (in centimeters) is larger than

$$D_{cm} > 0.02\, \lambda_{nm}/\alpha_{arcsec},$$

where the subscripts cm, nm, and arcsec have been added as reminders of the units that must be used for measuring each term. Notice that the diameter needed to resolve two sources increases as the sources get closer together.

Alternatively, we can express the smallest separation angle a telescope is capable of resolving by rearranging the equation terms:

$$\alpha_{arcsec} > 0.02\, \lambda_{nm}/D_{cm}.$$

For example, a telescope of diameter 100 centimeters (about 40 inches) observing visible light ($\lambda \approx 500$ nanometers) will be able to resolve two stars as little as 0.1 arc seconds apart.

Radio telescopes, despite their enormous size, observe very long wavelengths of light. In consequence, they usually have much poorer resolution than optical telescopes. The Arecibo radio telescope has a diameter of 300 m ($D = 30{,}000$ cm). Observing a radio wavelength of 10 cm ($\lambda = 10^8$ nm), the smallest resolvable angle at Arecibo is

$$0.02 \times 10^8/30{,}000 \approx 70 \text{ arcsec}.$$

The unaided human eye (pupil diameter 0.8 cm) observing visible wavelengths (500 nm) can resolve

$$0.02 \times 500/0.8 \approx 13 \text{ arcsec}.$$

Thus, the human eye can resolve finer detail than the largest single radio telescope in the world! Of course, if our eyes were sensitive to radio wavelengths, they would have extremely poor resolution.

31.3 INTERFEROMETERS

The limitation on resolution caused by diffraction is a major impediment to studying distant objects, where the angular size becomes very small. There are structural and financial limits to building larger and larger telescopes, so how can we achieve better resolution?

An answer to this problem begins to become apparent when we realize that a telescope aperture does not need to be continuous. If you take binoculars or a small telescope and put narrow strips of black tape in an X across the front of them—leaving gaps where light can pass through—the image you see through the binoculars does not have an X through it. The separate parts of the lens where light can still pass will focus the light perfectly well. Similarly, if you blacked out portions of the mirror on a reflecting telescope, the telescope would still function. The interesting

---

An arc second is a unit of angle and is equal to 1/3600 of a degree.

α_{arcsec} = smallest resolvable angle (in arc seconds)

λ_{nm} = wavelength of observation (in nanometers)

D_{cm} = diameter of telescope aperture (in centimeters)

Concept Question 1

Most people would say that they can resolve greater detail in bright light, when their pupils are very small, than in dim light. What other factors might account for this impression?

If you try putting black tape over a lens, do not stick the tape to the glass of the lens! Some lenses have special coatings or might be difficult to clean.

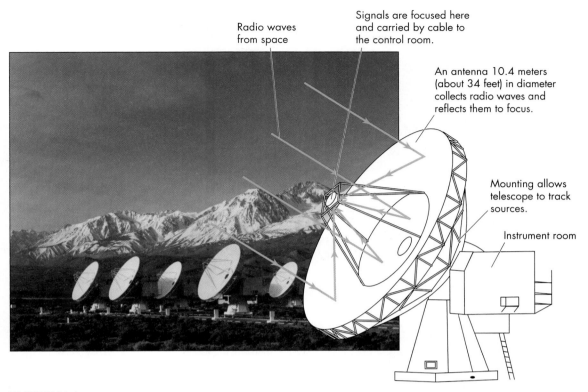

FIGURE 31.4
Photograph of the Owens Valley Millimeter Wavelength radio telescope, operated by the California Institute of Technology. In the background you can see the Sierra Nevada mountain range. The "telescope" is an array of six separate dishes that collect the radio waves. The captured radiation is then combined by computer to increase the resolution of the instrument.

thing is that as long as portions near the edges of the mirror remain uncovered, the resolution remains about the same. Less light may be collected, and the diffraction pattern would grow more complex, but the resolution would be little affected.

The pieces of the aperture actually do not need to be connected at all. As long as we can determine the relative positions of the waves precisely enough, we can add the light waves together so that they are in sync. More remarkable still, it is possible to record the electromagnetic wave from the separate "pieces" and later join together the recorded signals to form the equivalent of a single extended telescope aperture. This process is now regularly carried out at radio wavelengths (Figure 31.4) to join the signals from many separate telescopes. Major radio telescope arrays are located in many countries, the largest in steady use being the VLBA (Very Long Baseline Array) that spans the United States from the Caribbean to Hawaii. An international consortium is currently building a millimeter-wavelength array, in the high arid Atacama desert of Chile, known as ALMA (Atacama Large Millimeter Array). Radio astronomers sometimes arrange to make observations with even larger arrays by linking observations from radio telescopes around the world, and even space-borne radio telescopes, to achieve a resolution the equivalent of a single telescope that has an aperture more than 10,000 kilometers across!

Astronomers call a combined set of telescopes like this an **interferometer**, named after the principle of interference discussed earlier. To achieve the equivalent of a single large aperture, the same wave crest must be combined from the individual telescopes so that the interference between waves does not cause them to subtract from each other. Because the telescopes may be widely separated,

> A new design called a "nulling interferometer" attempts to cancel the signal in one direction to eliminate the light from a bright star, for example. Light from dimmer neighboring objects (planets, for example) enters the interferometer at slightly different angles, so the waves do not completely cancel each other.

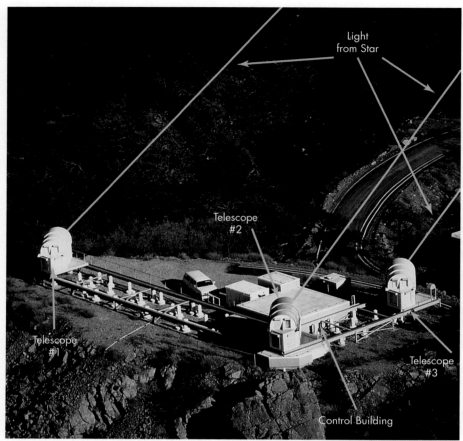

FIGURE 31.5

Photograph of an infrared and optical wavelength interferometer (IOTA) on Mt. Hopkins, Arizona. Light from the object of interest is collected by the three telescopes and sent to a control room. Computers there combine the light and reconstruct an image of the object.

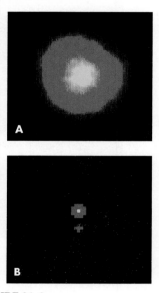

FIGURE 31.6

(A) A star observed with an ordinary telescope. (B) The same star observed with an interferometer. The higher resolution of the interferometer reveals that the "star" is actually two stars in orbit around one another.

the same wave crest will arrive at the different telescopes at different times. The electromagnetic waves from each telescope must be delayed by corresponding amounts of time so that the combined wave generates the strongest possible coherent signal. This indicates that the waves are all being brought together in unison—similar to the function of properly shaped lenses and mirrors in an individual telescope.

The result of this process is the ability to produce images in which the resolution is set not by the size of the individual telescopes but rather by their separation. The twin Keck telescopes and the four telescopes of the VLT are designed to be used in this way. A photograph of an interferometer using three telescopes is shown in Figure 31.5. For example, if the mirrors are 100 meters apart, the interferometer has the same resolution as a telescope 100 meters in diameter.

The high resolution of interferometers is far beyond what can be obtained in other ways. Figure 31.6A shows a view of two closely spaced stars as observed with a single telescope. Their images are blended as a result of diffraction. Figure 31.6B shows the same stars observed with an interferometer after the image has been processed by a computer. The two stars can now be easily distinguished: The separate mirrors produce the resolution of a single mirror with a diameter equal to the spacing between the individual telescopes.

Concept Question 2

Interferometers can provide superb resolution, but most optical observations are still made with individual telescopes. What would make optical interferometry more challenging than radio interferometry?

KEY POINTS

- A telescope's resolution is a measure of its ability to discriminate between objects and features that are close together.
- Resolution is limited by diffraction effects from the edge of the aperture and any obstacles along the optical path.
- Diffraction can produce complex patterns of interference that make it difficult to detect faint sources near bright ones.
- A telescope with no obstructions in its aperture still has a fundamental limit to its resolution, due to its aperture size.
- The smallest resolvable angle is proportional to the wavelength of light divided by the size of the aperture.
- Telescopes can work together in interferometers to achieve a resolution comparable to that of a telescope with a diameter equal to the *separation* between the telescopes.

KEY TERMS

diffraction, 218
diffraction pattern, 219
interference, 219
interferometer, 221
resolution, 218

CONCEPT QUESTIONS

Concept Questions on the following topics are located in the margins. They invite thinking and discussion beyond the text.

1. Bright light, pupil size, and resolution. (p. 220)
2. Difficulty of optical interferometry. (p. 222)

REVIEW QUESTIONS

3. What happens to a wave after it is diffracted?
4. What is meant by the *resolution* of a telescope?
5. If the optics could be made perfectly, why would there still be a limit on a telescope's resolution?
6. For telescopes operating at different wavelengths, how big must their apertures be to achieve the same resolution?
7. What effects do blockages within a telescope's aperture have on the quality of the image it produces?
8. What is an interferometer?
9. What advantages do we gain using an interferometer?

QUANTITATIVE PROBLEMS

10. If you are trying to resolve a double star with a 10-cm-diameter telescope, should a blue filter or a red filter make it easier to tell the stars apart? What resolution do you expect with each?
11. When the 2.4-m diameter Hubble Space Telescope (HST) is observing 500 nm visible light:
    a. What is the smallest resolvable angle it can resolve?
    b. Considering that the Moon is 384,400 km away, what is the smallest object HST could resolve on the Moon (see Unit 10.4)? Could the HST resolve the Apollo landing module?
12. What is the theoretical resolution of your eye to 500-nm light in bright light when the pupil is only 3 mm across? How many times brighter would the light have to be to get the same number of photons entering the eye as when the pupil is dilated to 8 mm?
13. Find the smallest resolvable angle of the 100-m diameter Green Bank Telescope when it is observing 20-cm wavelength radio waves. What is its resolution when observing at 3-mm wavelength?
14. Suppose that for an interferometer to work properly, the electromagnetic waves must arrive "in sync" to 1% of the period of the wave. What time difference is this for a 20-cm wavelength radio wave? What time difference is this for a 500-nm visible-wavelength wave? (Recall that the period of the wave is the inverse of its frequency ν.)
15. The Very Large Array consists of 27 radio dishes, each 25 m in diameter, connected together as an interferometer. Compare the resolution of a single VLA dish to the resolution of the VLA working as an interferometer assuming its maximum baseline of 36 km and observing 7-mm microwave signals.
16. Suppose you wanted to build a telescope capable of resolving a planet the size of Earth that is 10 light-years distant.
    a. Calculate the angular diameter of Earth if it were that far away (see Unit 10.4). To see any detail on the surface, we would want to see an angular size about 10 times smaller than this.
    b. Calculate the diameter of a telescope that would be needed to resolve this angular size if observing at 500 nm. Is this feasible?

TEST YOURSELF

17. A telescope's resolution quantifies its ability to
    a. measure the temperature of a source.
    b. detect dim sources.
    c. see finer details in sources.
    d. detect distant objects.
    e. increase the apparent size of the object.
18. Astronomers use interferometers to
    a. observe extremely dim sources.
    b. measure the speed of remote objects.
    c. detect radiation that otherwise cannot pass through our atmosphere.
    d. improve the ability to see fine details in sources.
    e. measure accurately the composition of sources.
19. If you are observing with a reflecting telescope, which of the following would increase the resolution?
    a. Switch to a telescope that has a bigger mirror.
    b. Observe at longer wavelengths.
    c. Use a refracting telescope with a lens the same diameter as the mirror in your reflector.
    d. Switch to a telescope that uses a more-reflective material.
    e. Any of the above would increase the resolution.
20. Which of the following wavelength and telescope diameter combinations should give the best resolution?
    a. 500 nm with a 5-m-diameter telescope
    b. 5 cm with a 5-km-diameter interferometer array
    c. 5 mm with a 50-m-diameter radio dish

The Earth's Atmosphere and Space Observatories

- **32.1** Atmospheric Absorption
- **32.2** Atmospheric Scintillation
- **32.3** Atmospheric Refraction
- **32.4** Observatories in Space

Learning Objectives

Upon completing this Unit, you should be able to:
- Identify the wavelength bands where the atmosphere is opaque and its causes.
- Explain what causes stars to "twinkle" and how astronomers attempt to correct this.
- Describe atmospheric refraction and its effect on the positions of stars.
- Discuss the advantages and disadvantages of placing telescopes in space.

The Earth's atmosphere presents a variety of challenges for carrying out astronomical observations across the electromagnetic spectrum (Unit 22). Although our atmosphere is transparent at visible wavelengths, it is not transparent at all wavelengths. Moreover, the atmosphere distorts the radiation that does pass through it. For example, sometimes it acts like an imperfect lens, distorting light as it moves from space into progressively denser layers of air. To complicate matters further, these effects vary with wavelength and weather conditions.

The atmosphere makes it challenging to build telescopes that work well. Some wavelengths cannot penetrate the Earth's atmosphere, so they cannot be observed except from space. For other wavelengths, conditions may be better in space, but observations from the ground are possible too. In these cases, there are trade-offs to consider. Space telescopes are generally much more expensive to build, launch, and maintain than ground-based telescopes. Also, limitations on the weight that can be launched into space set limits on the size of a space telescope. For such reasons, astronomers have explored many technological innovations, which by compensating for the distorting effects of the Earth's atmosphere will improve observations made from the ground.

32.1 ATMOSPHERIC ABSORPTION

Atmospheric absorption

Telescopes operating at many wavelengths face a major obstacle: Most of the radiation they seek to measure cannot penetrate the Earth's atmosphere. Gases in the Earth's atmosphere absorb electromagnetic radiation, and the amount of this absorption varies greatly with wavelength. For example, atmospheric gases affect visible light relatively little, so our atmosphere is nearly completely transparent to the wavelengths we see with our eyes. On the other hand, some of the gases strongly absorb infrared radiation while others absorb ultraviolet radiation.

The transparency of the atmosphere to visible light compared with its nontransparency to infrared and ultraviolet radiation creates what is called an **atmospheric window**. An atmospheric window is a wavelength region in which electromagnetic energy comes through our atmosphere easily compared with other nearby wavelengths (Figure 32.1). For example, there is a large window at radio wavelengths that makes most ground-based radio observations practical.

Unit 32 The Earth's Atmosphere and Space Observatories

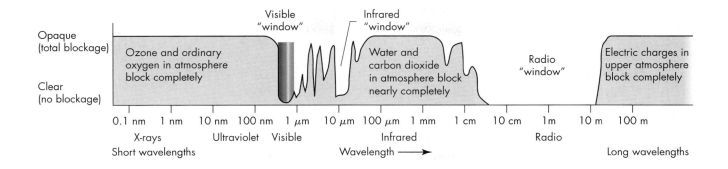

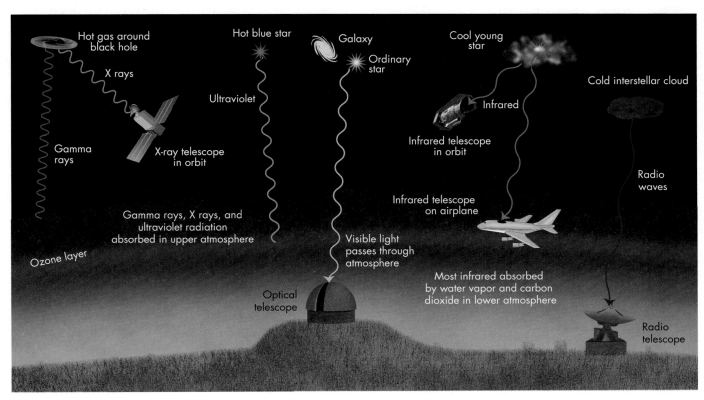

FIGURE 32.1

Atmospheric absorption. Wavelength regions where the atmosphere is essentially transparent, such as the visible spectrum, are called *atmospheric windows*. (Wavelengths and atmosphere are not drawn to scale.)

Concept Question 1

Some telescopes operate from high-flying airplanes. What wavelength ranges would this be most useful for accessing? What potential challenges would arise when observing from an airplane?

Other wavelength ranges present greater difficulties. Water (H_2O) and carbon dioxide (CO_2) absorb most infrared wavelengths, although there are some narrow windows within the range. This blockage by molecules in the Earth's atmosphere makes it difficult to detect emission from water or carbon dioxide located elsewhere in the cosmos. At shorter wavelengths than visible light, ozone (O_3) and ordinary oxygen (O_2) strongly absorb ultraviolet radiation, while oxygen and nitrogen absorb X-rays and gamma radiation. As a result, only from space can we observe the high-energy photons from some of the most violent processes occurring in the cosmos.

Because the atmospheric window at visible wavelengths permits us to observe the Sun and other stars, we can do much astronomy from the ground. But the blockage at other wavelengths is beneficial, too. For example, because our atmosphere blocks wavelengths shorter than visible light, it protects us from high-energy photons that are dangerous to living organisms. Similarly, because our

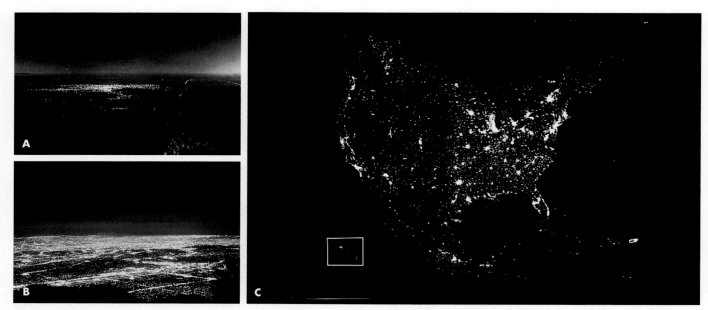

FIGURE 32.2
Photographs illustrating light pollution. (A) Los Angeles basin viewed from Mount Wilson Observatory in 1908. (B) Los Angeles at night in 1988. (C) Notice the pattern of the interstate highway system visible in the satellite picture of North America at night.

atmosphere is not transparent to infrared wavelengths, it traps infrared radiation emitted by the warmed ground at night, retaining much of its heat and preventing the oceans and us from freezing.

In recent decades astronomers have had to contend with another factor that limits their observations from the ground: **light pollution.** Most inhabited areas are peppered with nighttime lighting such as streetlights and outdoor advertising displays (Figure 32.2A and B). Although some such lighting may increase safety, much of it is wasted energy, illuminating unessential areas and spilling light upward into the sky, where it serves no purpose. Figure 32.2C shows a satellite view of North America at night and illustrates the wasted energy created by light pollution. Such stray light can interfere with astronomical observations. Many early observatories were built in cities to make them more accessible, but they have become essentially useless for research because of light pollution. In some places astronomers have persuaded regional planning bodies to develop lighting codes to minimize light pollution. Light pollution not only wastes energy and interferes with astronomy; it also destroys a part of our heritage—the ability to see stars at night. The night sky is a beautiful sight, and it is shameful to deprive people of it.

Get involved in your community by letting people know that they can save money by using light fixtures that have reflectors to keep the light shining downward. They are paying for the electricity that a poorly designed fixture uses to light up the night sky! See the International Dark-Sky Association website for more details.

32.2 ATMOSPHERIC SCINTILLATION

Even from a location where there is no light pollution and the air is clear, the atmosphere still causes significant problems for observations at visible wavelengths. Light's path bends as it travels through materials of different density. This is the same principle of refraction that allows lenses to work (Unit 30), but even pockets of air of different density can alter the path of light.

If you have ever noticed a star twinkling in the night sky, then you have observed atmospheric **scintillation.** This is caused by various pockets of warmer and cooler air that have slightly different densities. Each acts as a lens on the light

Why stars twinkle

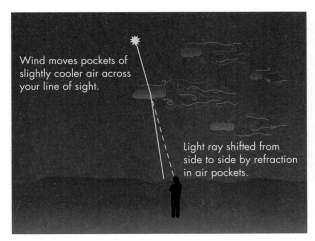

FIGURE 32.3
Twinkling of stars (seeing) is caused by moving atmospheric irregularities that refract light in random directions.

Concept Question 2

One way you can often tell a planet from a bright star without using a telescope is by how much they twinkle. Planets tend to shine more steadily than a star in the same position. Why do you suppose there is a difference?

traveling through it. As hot air rises or as winds and turbulence stir the air, these pockets will cross in front of a star, magnifying and demagnifying, and bending the path of the light first one way and then another. The color dispersion of these "lenses" will even make the star's color appear to vary. These effects are particularly apparent when we view a star near the horizon, where the light travels a longer path through the atmosphere. Depending on atmospheric conditions, a star may flicker and shift more or less, conditions astronomers call, respectively, bad or good **seeing.**

With bad seeing, the starlight you see at any instant is a blend of light from many slightly different directions, which smears the star's image and makes it dance (Figure 32.3). You can see a similar effect if you look down through the water at something on the bottom of a swimming pool. If the surface of the water has even slight disturbances, a pebble or coin on the bottom seems to dance around. Scintillation in our atmosphere limits the ability of Earth-bound observers to see fine details in astronomical objects. The dancing image of a star or planet distorts its picture when recorded by a camera or other device.

Until recently, ground-based astronomers had to submit to the distortions of seeing, but now they can partially compensate for such distortions in several ways. One technique involves observing a bright star simultaneously with the object of interest. Extremely rapid measurements are made of the bright star's shifting position, then tiny motors make compensating adjustments to the tilt of a secondary mirror to keep the bright star's position the same. Because the pockets of air in the atmosphere are fairly large, faint objects in the vicinity of the bright star are also kept stationary, and the image remains sharp. Astronomers are learning to make even more rapid adjustments for atmospheric effects, eliminating many of the distortions of atmospheric seeing. This technique, called **adaptive optics,** has already given astronomers dramatically improved views through the turbulence of our atmosphere. Because bright stars are not present next to all of the objects we might want to observe, astronomers have developed a technique using a powerful laser beam to create an artificial star high in the Earth's atmosphere, as shown in Figure 32.4A. The difference in the sharpness of an image of a distant galaxy is shown in Figure 32.4B with the adaptive optics system turned on and with it off.

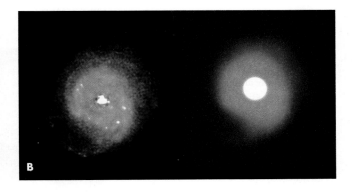

FIGURE 32.4
(A) A laser beam creates an artificial star whose image serves as a reference to eliminate the atmosphere's distortion of real stars (Starfire Optical Range of the Phillips Laboratory at Kirtland Air Force Base in New Mexico). (B) This pair of images illustrates the difference in image quality for a galaxy with adaptive optics turned on and off (images made at the Canada-France-Hawaii Telescope on Mauna Kea).

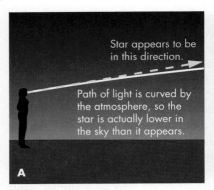
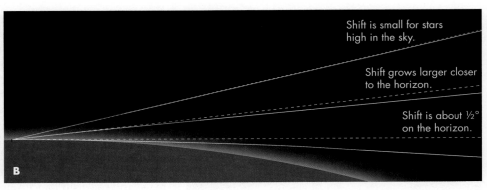

FIGURE 32.5
(A) Atmospheric refraction makes the Sun and stars look slightly higher in the sky than they really are. (B) Refraction is stronger for objects nearer the horizon, and objects seen at the horizon would actually be about half a degree below the horizon but for the atmosphere.

32.3 ATMOSPHERIC REFRACTION

Refraction by the atmosphere causes a larger, yet less obvious, effect. As light enters the atmosphere from space, its overall path is bent (Figure 32.5A). The bending is so large near the horizon that we can see stars that are actually below the horizon! Just as for the refraction of light by a lens (Unit 30), light coming nearly straight down into the atmosphere is little affected. Light entering the atmosphere at a steeper angle is bent by a larger amount (Figure 32.5B).

At the horizon, the effect of the bending is about half a degree. As a result, when the Sun or Moon appears to be just touching the horizon, it is in fact below the horizon—if the Earth had no atmosphere, it would already be out of view. Thus, atmospheric refraction alters the time at which the Sun seems to rise or set. It also distorts the shape of the Sun because of differences in the amount of bending from the bottom to the top of the Sun's disk (Figure 32.6).

By "lifting" the Sun's image above the horizon, even though the Sun has set, refraction slightly extends the length of daylight hours. As a result, the day of the year when the Sun appears to be above the horizon for exactly 12 hours is not the equinox, but rather a few days before the first day of spring and a few days after the first day of autumn. This depends on latitude and atmospheric conditions, but it turns out that near latitude 40°N, on St. Patrick's Day (March 17) the day usually has 12 hours between sunrise and sunset, whereas the actual equinox occurs on about March 21.

Concept Question 3

How do the refraction and absorption of light by the atmosphere relate to the color of the Moon during a total lunar eclipse?

Concept Question 4

Why do stars twinkle more when they are near the horizon than when they are near the zenith?

FIGURE 32.6
The Sun grows increasingly more flattened as it sets because atmospheric refraction "lifts" its lower edge more than its upper edge. The red line below the Sun is its reflection off water near the horizon.

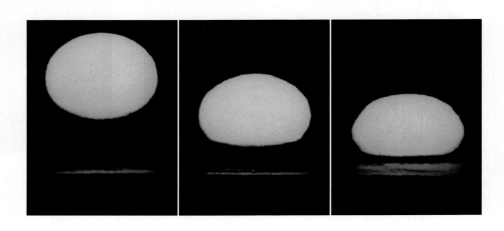

32.4 OBSERVATORIES IN SPACE

Figure 32.7 shows several of the major telescopes that have been launched in recent decades. Some, like the Hubble Space Telescope, which was launched in 1990, have operated for many years thanks to their low orbit around Earth that allows servicing missions with the space shuttle to repair problems and upgrade equipment. Most other space telescopes, like the Spitzer Infrared Space Telescope, have a shorter lifetime because the detectors require coolants such as liquid helium to keep them close to a temperature of absolute zero. When the supply of liquid helium was used up after almost 6 years, it could not be resupplied because the telescope was placed in deep space millions of kilometers from Earth. It was positioned far from Earth so that the Earth's own infrared radiation would not interfere with observations. By contrast, the Hubble Space Telescope orbits about 500 km (300 miles) above the Earth's surface.

More than 50 space telescopes have been launched by countries around the world to examine all different parts of the electromagnetic spectrum. Many of these missions were brief and were designed to give us our first clues about cosmic radiation outside of the Earth's atmospheric windows. With each mission we have developed a clearer idea of the characteristics of electromagnetic emission, leading to improved designs. Telescopes like the Chandra X-ray telescope, launched in

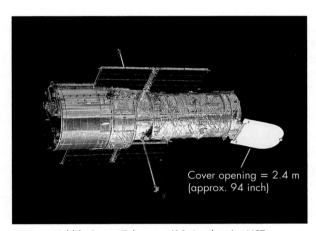

Hubble Space Telescope (13.6 m long) – HST

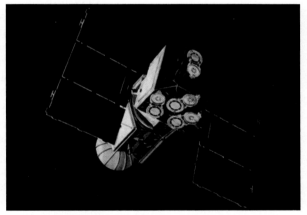

Extreme Ultraviolet Explorer – EUVE

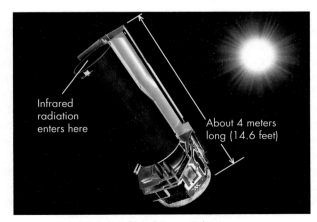

Spitzer Infrared Space Telescope

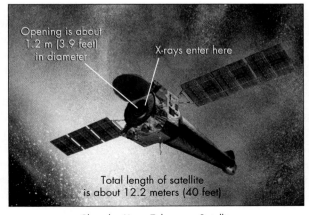

Chandra X-ray Telescope Satellite

FIGURE 32.7
Photograph of the Hubble Space Telescope and drawings of some other space observatories: Spitzer, EUVE, and Chandra.

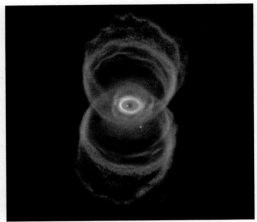

Sombrero Galaxy
A system of billions of stars and dark dusty interstellar matter

Hourglass Nebula
A dying star

FIGURE 32.8
The Hubble Space Telescope has provided some of the most detailed images of thousands of astronomical objects. Two examples are shown here, and dozens of others are located throughout this book.

1999, provided some of the first images of X-ray emissions, with resolutions comparable to visible-wavelength images, while other X-ray telescopes have focused on spectroscopy or high-speed timing for transient events.

Of the many orbiting telescopes used by astronomers, the Hubble Space Telescope (HST) is the most ambitious to date. The HST is designed to observe at visible, infrared, and ultraviolet wavelengths and has a mirror 2.4 meters (about 8 feet) in diameter. Its instruments allow it to take both pictures and spectra of astronomical objects. Although the HST initially had a number of problems, astronauts have repaired the major defects; and astronomers are now delighted with the clarity of its images (Figure 32.8). These images reveal details difficult to see with telescopes on the ground because such telescopes must peer through the blurring effects of our atmosphere.

Despite the freedom from atmospheric blurring and absorption that space observatories enjoy, much astronomical work will be done from the ground for the foreseeable future. Ground-based telescopes can be built much larger than orbiting telescopes. Moreover, equipment problems on the ground can be corrected without the expense, delay, danger, and complexity of a space mission.

Because huge telescopes in space or even on the Moon will remain only dreams for years to come, astronomers choose with care the location of ground-based observatories. Sites are picked to minimize clouds and the inevitable distortions and absorption of even clear air. Thus, nearly all observatories are in dry, relatively cloud-free regions of the world, such as the American Southwest, the Chilean desert, Australia, and a few islands, such as Hawaii and the Canary Islands off the coast of Africa. Moreover, astronomers try to locate observatories on mountain peaks to get them above the haze that often develops close to the ground in such dry locales and to improve the seeing.

Steady improvements in adaptive optics have made the sharpness of images from ground-based telescopes competitive with the HST. New designs promise to do even better, so plans for new space telescopes are mostly being targeted for wavelengths inaccessible from the ground because the electromagnetic waves cannot penetrate the Earth's atmosphere.

> Probably the most ambitious new space telescope currently planned is the James Webb Space Telescope, an infrared space telescope with a 6.5-meter diameter, targeted for launch in 2018.

KEY POINTS

- Earth's atmosphere blocks radiation at most wavelengths except for "windows" at visible and radio wavelengths.
- Light pollution around populated areas makes the night sky so bright that astronomical research is possible only in remote spots.
- Moving pockets of warmer or cooler air in the atmosphere bend light passing through it, making stars twinkle or scintillate.
- Astronomers can compensate for some atmospheric scintillation by using adaptive optics.
- Refraction of light by the Earth's atmosphere shifts the position of stars by a larger amount the nearer you look to the horizon.
- Dozens of space telescopes have been put in service, most to observe wavelengths blocked by the Earth's atmosphere.

KEY TERMS

adaptive optics, 227
atmospheric window, 224
light pollution, 226
scintillation, 226
seeing, 227

CONCEPT QUESTIONS

Concept Questions on the following topics are located in the margins. They invite thinking and discussion beyond the text.

1. Advantages and challenges of airborne telescopes. (p. 225)
2. Why planets do not twinkle. (p. 227)
3. Causes of Moon's color during a total lunar eclipse. (p. 228)
4. Twinkling stars at horizon and zenith. (p. 228)

REVIEW QUESTIONS

5. What is an atmospheric window?
6. What do astronomers mean by *seeing*?
7. How can light pollution be reduced?
8. Why does the Sun look flattened when it is close to the horizon?
9. What are the advantages and disadvantages of putting telescopes in space? on a mountaintop?
10. How are ground-based images made sharper by adaptive optics?

QUANTITATIVE PROBLEMS

11. Suppose that in the atmosphere there are pockets of air of different density that are 10 meters across. If winds move these pockets of air at a speed of 30 kph, for how long is an image through a telescope likely to remain stable? How would having many of these pockets along your line of sight change your time estimate?

12. A telescope in low orbit around the Earth takes about 90 minutes to circle the Earth. If the telescope remained oriented in the same way relative to the Earth, by how many arc seconds would its direction change relative to the stars during a 1-second exposure? (Actually, space telescopes are stabilized by gyroscopes that maintain their orientation relative to the stars.)

13. If the refraction of the atmosphere causes the Sun's position to shift up by 0.5° at the horizon, calculate how long the day would be on the equinox for someone living at the equator. Would that same day be longer or shorter for someone living at latitude 45°? Explain using a diagram.

14. Based on the graph in Figure 32.1, estimate the maximum energy and frequency for a photon that can be detected by a ground-based telescope?

15. The orbiting Chandra X-ray telescope, observing at about 0.1 nm, can collect about 2×10^{-13} watts (joules per second) of radiation from the Crab Nebula. About how many photons is it collecting each second?

16. If the cost of electricity is $0.10 per hour to run a 1000-watt lightbulb, estimate the total annual cost for a town running 200 streetlights, each consuming 500 watts. If a reflector is placed over each lamp so that twice as much light shone downward, how much money can the town save each year by switching to lower-wattage bulbs?

TEST YOURSELF

17. Which type of electromagnetic radiation must be observed using a space-based telescope?
    a. Radio
    b. Infrared
    c. Visible
    d. X-rays

18. Which of the following explains the advantage of the Hubble Space Telescope compared with ground-based telescopes?
    a. It is the largest reflector ever made.
    b. It can detect X-rays from space.
    c. It is not affected by atmospheric scintillation.
    d. It is closer to the objects it is imaging.
    e. All of the above

19. The purpose of adaptive optics is to make telescopes
    a. more flexible so they can fit in smaller buildings.
    b. look in several directions without having to move the primary mirror.
    c. capture more photons within their aperture.
    d. adjust for the distortions caused by the Earth's atmosphere.
    e. All of the above

20. Atmospheric refraction makes the Sun look
    a. larger in diameter.
    b. brighter.
    c. bluer in color.
    d. higher in the sky.

21. Astronomers have discussed building a radio telescope on the far side of the Moon. What would be the main advantage of a radio telescope there compared to one on Earth?
    a. To avoid the blockage of radio waves by Earth's atmosphere.
    b. Because the lower gravity would make it easier to build.
    c. To avoid the "light pollution" from our radio emissions.
    d. Because you can see much more of the sky from the Moon.
    e. All of the above are of about equal advantage.

Amateur Astronomy

33.1 The Human Eye
33.2 Your Eyes at Night
33.3 Choosing a Telescope
33.4 Completing the Telescope
33.5 Astronomical Photography

Learning Objectives

Upon completing this Unit, you should be able to:
- Explain how the eye works, relating its various parts to the equivalent parts of a telescope and detector.
- Discuss the features and trade-offs of various amateur telescope designs.
- Understand how eyepiece focal length determines the magnification of a telescope.
- Take some basic astronomical photographs.

Anyone with access to a modern small telescope has better equipment than Galileo ever had. With such equipment and a dark sky, you can become an amateur astronomer. The pleasures of the hobby can range from the aesthetic satisfaction of taking a lovely photograph to the thrill of discovering a new comet or an exploding star. Much of the enjoyment of the night sky comes from learning the constellations (Unit 5) and studying the changing positions of the Sun, Moon, and planets (Unit 13). Binoculars will let you begin to appreciate many of the more remote objects of the night sky, and a small telescope will let you see even more. Amateur astronomy also provides an excellent opportunity to put much of the knowledge about telescopes in Units 28–31 into practice. Here we offer a few suggestions for taking your interest in astronomy a step further.

33.1 THE HUMAN EYE

The human eye is a remarkable device. It not only detects visible-wavelength photons, but can discriminate between different wavelengths. It has a self-adjusting aperture and can adjust the sensitivity of its receptor cells by nearly a factor of a million. Some parts of the eye allow for high resolution and color discrimination, while other portions permit us to detect extremely low light levels. Understanding how the eye works will help you get the most from your observations of the night sky.

Your eye (Figure 33.1) contains many of the same elements as a telescope. The eye's **iris** can expand or contract to adjust the aperture, or **pupil**, to allow in more or less light. Light passing through the pupil is focused on the back surface called the **retina**. The shape of a transparent layer of your eye called the **cornea** does most of the focusing of light entering your eye, but an internal lens can be adjusted to allow the eye to focus at different distances.

The light focused on the retina can trigger a response in light-sensitive nerve cells called **rods** and **cones**, so named because of their shapes. There are more than 100 million of these "detectors"—akin to the pixels of a CCD—that send signals to your brain, providing your sense of vision. The rods and cones all use **photopigments**—proteins that change form when they absorb light, causing a sequence of chemical reactions. Your eye uses four kinds of photopigments that respond to different wavelengths of light (Figure 33.2). The rods use a photopigment that has a peak sensitivity at a blue-green wavelength. There are three kinds of cones, each using one kind of photopigment that is most sensitive to blue, green, or red light.

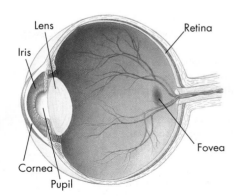

FIGURE 33.1
Structure of the human eye. The transparent cornea and lens focus light onto the retina at the back of the eye, which contains the rods and cones. The center of each eye's visual field is focused on the fovea. Focusing is accomplished by contraction and relaxation of the muscles that adjust the curvature of the lens.

FIGURE 33.2
The spectral sensitivities of the photopigments in rods and three types of cones.

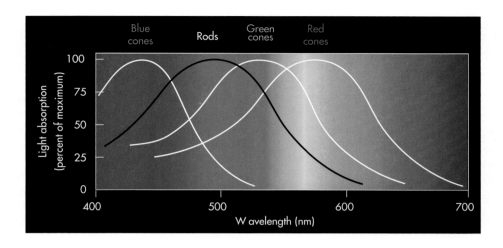

Concept Question 1

To "see" your blind spot, put a large black dot near one end of a piece of paper, then from the dot draw a straight line that is about 20 cm (8 inches) long. Close one eye and hold the paper in front of you with the line horizontal and the dot at the right side for your right eye (or vice versa). Look first at the dot and then slowly "slide" your center of vision along the line. You will find that at a particular angle the dot seems to disappear. Why do you suppose we are unaware of our own blind spots?

The cones and rods are not distributed uniformly in the eye. Instead, the cones are concentrated in a region called the **fovea** (Figure 33.1) in the center of your field of vision. By comparing the response of the three kinds of cones, your brain assigns a color. Note that a 500-nm photon can excite any of the photopigments, so your brain assigns a perceived color based on the relative strength of the reaction by different cones. Digital cameras employ a similar strategy to produce color images. Alternate pixels are behind one of three different color *filters* (Unit 29.3), mimicking the way the eye works.

Your color perception is strongly concentrated in the middle of your field of vision, so your color discrimination is best when you look straight at an object, and it drops sharply away from there. There are few cones beyond 10° from the center of your vision, and almost none beyond 40°. You may think you see colors in your peripheral vision, but you get this impression only because your eye constantly glances in different directions to fill in detail, and your brain remembers the colors from when you saw an object earlier. Your brain also compares the images from both eyes to estimate the distances of objects. In addition, it tries to fill in missing information from parts of the eye where there are no cones and rods—such as the blind spot about 15° to the outside of the center of your field of vision in each eye, where the nerves come together to carry the signals to your brain.

33.2 YOUR EYES AT NIGHT

You will discover that the longer you stay outside in dim light, the more sensitive your eyes will become and the fainter the stars you will be able to see. This is the result of physiological changes in your eye referred to as **dark adaption.**

The simplest change in your eye occurring in dim light is that the pupil opens wider. This is easy to verify by looking at yourself in a mirror in a dimly lit room. In bright sunlight, your pupil can shrink to a diameter of as little as 2 millimeters, but in total darkness, its diameter may expand to about 8 millimeters, which allows more light to enter your eye. A change from 2 to 8 millimeters is a factor of 4 in diameter and therefore a factor of 16 in the amount of light admitted to the eye.

The maximum size of the pupil tends to decrease steadily with age to about 5 or 6 mm.

Your eyes undergo another change in the dark. The rods can build up the level of photopigments within them until they are about 1 million times more sensitive to light than they are in full daylight. The process takes about 20 minutes to get fully established but is undone by even a few seconds' exposure to bright light, so once you are dark-adapted, you should stay away from bright lights. Smoking, alcohol, and other drugs also reduce your ability to adapt to the dark by reducing the iris's ability to expand and by interfering with the chemistry of the rods, cones, and other nerve cells. Getting enough vitamin A is important too, because this is an essential chemical for the photopigments in your eye.

The cones have much more limited ability to increase their sensitivity than the rods; so as light levels decrease, your vision relies more on the rods. This has several effects: your ability to discriminate between colors declines or disappears completely; the color of light you are most sensitive to shifts to blue wavelengths, a phenomenon known as the *Purkinje effect;* and the position of your best sensitivity to light shifts outside of the perceived center of your field of vision in the fovea.

The Purkinje effect can change your perception of the relative brightness of different-colored objects. In full daylight, the eye's overall response is strongest to yellow-green colors. At low light levels, you become most sensitive to the blue-green colors that the rods' photopigment responds to. This might be the result of natural selection: because the average color of starlight is bluer than sunlight, eyes responsive to blue at night should aid survival. As a result of this effect, your perception of the relative brightness of two differently colored stars will change as your eyes adapt to the dark—an important consideration if you are trying to study the changes in brightness of **variable stars** (Unit 64) by comparing them to neighboring stars. It is certainly the case that night-flying insects see blue light better than yellow. That is why bug zappers use a blue light to attract insects, and also why a yellow lightbulb is often used for outdoor night lighting to be less attractive to insects.

Within a circle about 1° across in the center of your field of vision, the fovea contains no rods at all. The number of rods relative to cones climbs steadily in the surrounding area of the retina. You can apply this information practically with the following observational technique for viewing dim objects. Look at a point about 5° away from the dim object you are trying to see (about the length of your outstretched thumb). By doing so you gain sensitivity to faint objects. This technique, called **averted vision,** places the focused image of the object you are trying to see on a part of the iris with a high density of rods. This is not easy to do at first, but with practice you will be able to see some extremely faint objects that are "invisible" when you look straight at them.

> To help keep their eyes adapted to the dark, astronomers use flashlights covered with a red filter to look at star charts, change eyepieces, and engage in other activities while observing the night sky.

33.3 CHOOSING A TELESCOPE

Although your eyes can see remarkably faint stars and even a few galaxies, a small telescope will greatly increase the number of objects you can observe and will offer you far better and more interesting views of the Moon, planets, and even more remote objects. The basics discussed in previous Units about the different kinds of telescopes and how they work (Unit 30) apply to small telescopes as well as research telescopes. However, there is an even wider range of telescope designs available for amateur astronomy, so selecting the best one for your needs can be confusing.

A good starting size is a 100- to 150-mm (4- to 6-inch) aperture telescope. When a telescope is referred to by size, as in "a 100-mm telescope," the size refers to the diameter of the primary mirror or lens (Unit 29). With a 100-mm telescope, you can easily study our Moon, the moons of Jupiter, the rings of Saturn, and most of the star clusters, nebulae, and galaxies in the Messier Catalog (Appendix Table 13). A relatively inexpensive design is a Newtonian reflector (Unit 30.2), although a wide variety of other designs are available with various advantages and disadvantages.

A well-manufactured 100-mm telescope can provide just about as sharp a view as you can get. The resolution of a telescope is limited by diffraction, and for a 100-mm aperture this limit is about 1 arc second (Unit 31.2). This is about the same as the smearing typically caused by atmospheric turbulence or "seeing" (Unit 32.2) on a good night from most locations. A larger telescope will offer relatively little

> **Concept Question 2**
>
> Suppose a star and a galaxy are both listed as being of equal total brightness. Which would be easier to observe through a telescope?

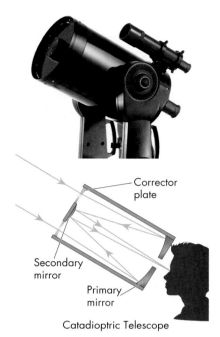

FIGURE 33.3
Photograph and diagram of a catadioptric telescope. This type of telescope has a corrector plate—a large lens over the main aperture—that works in combination with the primary mirror to focus the light. A Schmidt design with a Cassegrain focus is illustrated. A Maksutov design, another popular type, uses a different shape for the corrector plate.

Clarification Point

Many people believe that a telescope's magnification is its most important feature, but sharp images and good contrast are more critical for most observing experiences.

improvement in the detail it is possible to see, unless you are observing under exceptional atmospheric conditions.

Refractors are also an excellent choice, but they tend to be more expensive for the same size aperture. Because glass causes color dispersion (Unit 30.4), you should avoid inexpensive refractors that have a single-element primary lens. All better-quality refractors use two or three lenses designed to correct each other's color dispersion. Good-quality refractors have what are called "achromatic" lenses and the best have "apochromatic" lenses—indicating that the color problems of the lens have been corrected to different degrees. Refractors are particularly favored for bright objects such as the Moon and planets, because they can give the crispest and highest-contrast images. Their high contrast is a result of having a clear aperture without secondary mirrors or internal support structures that would scatter light inside the telescope and make the diffraction pattern worse.

If you plan to take your telescope on the road, perhaps to enjoy dark skies in the wilderness, you will probably want it to be compact. A popular portable design is called a "catadioptric" telescope, which is a hybrid between a reflector and a refractor. This type of telescope uses both a large mirror and a large lens, called a corrector plate, to focus the light entering the aperture, as illustrated in Figure 33.3. Catadioptric telescopes are generally built using the Cassegrain design (Unit 30.2). This folds the path of the light back on itself, sending the light through a hole in the primary mirror, making the telescope more compact. The idea of a catadioptric telescope is that the corrector lens and mirror are relatively simple shapes to manufacture, keeping the cost lower than a refractor of the same aperture. Similar to a refractor, the telescope body is a solid sealed unit, so dust and misalignment are less problematic than for a reflector. The secondary mirror blocks part of the aperture, but the corrector plate is used to hold it in place, so no additional support structures are needed.

Selecting a telescope with a larger aperture of 20 cm or more will allow you to view faint objects more easily, but at increasing expense and decreasing portability. Using your eyes well can have more effect on what you will be able to see than increasing your telescope's size. Keep your eyes shielded from bright lights for at least half an hour before observing, and use averted vision (looking off to the side) when trying to see dim objects through the telescope. With these steps you will be able to see things through the telescope that would otherwise be invisible to you.

33.4 COMPLETING THE TELESCOPE

Many other aspects than just the telescope design contribute to having a successful experience observing the night sky. The telescope gathers the light and brings it to focus, but viewing the focused image requires additional optics called "eyepieces." The telescope must be held on a steady "mount," but that mount also has to be able to move smoothly and easily so you can follow the motion of the celestial sphere. Choosing the right eyepieces and mount for your purposes is essential.

Notice that we have said nothing about magnification to this point. Some telescopes advertise their "magnifying power," but this is an unimportant number because any telescope can magnify by any amount with different eyepieces. You can determine the magnification of a telescope–eyepiece combination by dividing the **focal length** of the telescope by the focal length of the eyepiece:

$$\text{Magnification} = \frac{\text{Telescope focal length}}{\text{Eyepiece focal length}}.$$

The focal length of a refractor is usually the length of the telescope tube—the distance over which it focuses the light entering the aperture down to a point. With the multiple bounces and shaped mirrors in many reflectors, the focal length may not be obvious. However, you can calculate it by multiplying the telescope's aperture size by its *f-ratio*:

$$\text{Telescope focal length} = (\text{Aperture diameter}) \times (\text{f-ratio}).$$

For example, a 100-mm f/8 telescope has a focal length of 100 mm × 8 = 800 mm.

Eyepieces almost universally have their focal length listed on them in millimeters. Thus, for a telescope with an 800-mm focal length, an 8-mm eyepiece will give a magnification of

$$\text{Magnification} = \frac{800 \text{ mm}}{8 \text{ mm}} = 100.$$

Likewise, a 4-mm eyepiece on this telescope will give a magnification of 200, whereas a 16-mm eyepiece will give a magnification of 50. Distortions caused by the atmosphere often make a magnification of about 100 to 200 the useful limit.

Some telescopes are sold with low-quality eyepieces, and you can improve your viewing substantially by upgrading them. Eyepieces come in a wide range of designs, some working better for people with glasses, others offering a wider field of view, and not all will work well with every telescope. Your best way to find a good match may be to visit an astronomy club convention or "star party," where you may get the chance to try different members' eyepieces with your own telescope.

Finally, it is important to decide how your telescope will be supported. Most important is that the mount be sturdy. At a magnification of a factor of 100, tiny vibrations of the telescope caused by wind or the touch of your hand will make the image jiggle, hopelessly blurring it. There are two main designs to choose between for mounting a telescope.

An **alt–az mount** (Figure 33.4A) is the simplest and least expensive. It allows the telescope to swing up and down in altitude from the horizon to the zenith, and pivot in azimuth to point toward different compass directions (Unit 13). This lets you see all parts of the sky, but unless you are observing from the North or South Pole, it will not be easy to follow objects on the celestial sphere as the Earth rotates. Because of the Earth's rotation, objects will move out of your field of view within a minute or two, or even seconds under high magnification.

An **equatorial mount** (Figure 33.4B) has a pivot axis that can be pointed at the celestial pole to compensate for the Earth's rotation. With an equatorial mount, you can swing the telescope from the polar axis until you match an object's declination, and then pivot the telescope to match its right ascension (Unit 5.5). By rotating the telescope about the pivot axis, you can exactly counteract the effects of the

> A telescope with fine optics under excellent atmospheric conditions may be able to achieve a magnification of up to 2.5 times the aperture in millimeters. Thus, a 200-mm (8-inch) telescope may be able to reach 500 times magnification, but this is rare.

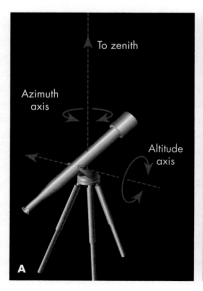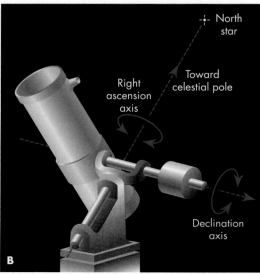

FIGURE 33.4
Telescope mount designs. (A) An alt–az mount allows the telescope to pivot around the horizon (azimuth) and angle up and down (altitude). (B) An equatorial mount pivots around an axis pointing toward the celestial pole (right ascension) and angle north and south (declination).

Binoculars are normally labeled with a pair of numbers such as "8 × 50." These two numbers refer to the magnification (8×) and the diameter of the aperture in millimeters (50 mm). They are often also labeled with a third number, such as 5°, which indicates the angle across the "field of view" you can see within the binoculars.

Earth's rotation and keep the telescope pointed at the same celestial coordinates. Telescopes on equatorial mounts are more difficult to balance and generally must be heavier than alt–az mounts. However, they allow you to follow objects smoothly because they are oriented to rotate parallel to the Earth's axis—and with a motor drive, they can remain pointing at the same object for long periods.

If you exactly align your telescope (no easy task), you can find faint astronomical objects by their right ascension and declination. Generally, though, it is easier to locate an object by pointing to the approximate position relative to bright stars as determined from a sky chart or planetarium software program. Then progressively close in on the position of the object by "star hopping" to fainter stars using the small "finder scope" mounted on the side of the telescope. Some newer telescopes offer computer-guided "GoTo" systems to point the telescope. GoTo systems can also guide alt–az mounts to track celestial objects, and as their price has declined this has become a popular combination.

Before you buy a telescope, you might want to talk with a local amateur astronomer or astronomy instructor. If you can find an astronomy club, some of the members may have secondhand telescopes at reduced prices. For information about new telescopes, browse through magazines like *Astronomy, Sky & Telescope,* or *Astronomy Now* (in Britain)—publications widely read by both amateur and professional astronomers—which contain many advertisements for small telescopes.

33.5 ASTRONOMICAL PHOTOGRAPHY

You do not need a telescope to take astronomical photographs. What you do need is a camera with as much manual control as possible. For example, the images in Figure 33.5 were made with a standard lens, but the camera shutter was kept open for 30 seconds. For astronomical photography, automatic focus and exposure features may frustrate your attempts to use the camera in low-light conditions, but if you can turn these features off or find a camera with manual control, you can take a wide variety of astronomical pictures.

FIGURE 33.5
(A) Early morning image of clouds and stars over Bora Bora. The Large Magellanic Cloud is visible in the right half of the image. This picture was made with a 30-second exposure with the camera sitting on a railing. About 10 attempts yielded one sharp picture. (B) A picture toward the center of the Milky Way. This is another 30-second exposure, with the sensitivity (ISO) setting of the camera set to maximum.

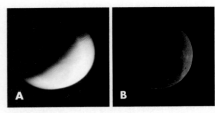

FIGURE 33.6
Two pictures of the Moon made with a handheld digital camera. (A) The automatic features of the camera here overexposed the Moon, and the camera could not be kept steady during the shot. (B) Here the length of the exposure was forced to be just 1/1000 of a second.

For most astronomical photography, the camera focus should be set to infinity and the aperture opened fully (at the lowest f-stop number). For many "SLR" cameras, you can make long exposures by putting the camera on a setting marked "M" or "B" (for "manual" or "bulb"). The shutter release button can then be held down for the desired time. Most modern SLR cameras will also allow you to adjust the sensitivity of the camera by setting the electronics to a high "ISO" number. Start with an exposure of 10 or so seconds, and then try making longer exposures. For such long exposures you should mount the camera on a tripod or other stable platform, then use a remote release to open the shutter without shaking the camera.

Even cell-phone cameras can produce a good image of the Moon with some planning. This is tricky because the bright portions of the Moon are in daylight, so the exposure you want is similar to that for a daytime scene, however, the automatic exposure feature of these cameras will usually lengthen the exposure to try to brighten up the dark sky. The result is that the Moon is overexposed and the camera moves during the long exposure (Figure 33.6). You can obtain a sharp, better-exposed image if you can force the camera to take a short exposure, perhaps by putting other lights in a corner of the image to force a shorter exposure.

If you expose for more than about 30 seconds, the Earth's rotation will smear the star images into streaks. Deliberately allowing the smearing to occur can produce dramatic pictures of what are called "star trails" (see Figure 33.7). To make star-trail pictures, leave the shutter open for 20 minutes or so. Longer exposures are possible, but if the Moon is out or if there is much light pollution (Unit 32), the image may become "foggy."

Most astronomical pictures in books look little like what you see through a telescope, because your eyes cannot store up light as a camera can. Many interstellar clouds and galaxies are large enough that we could easily see them but for the limited sensitivity of our eyes. To photograph them, you do not need a telescope, but you do need to make a long exposure while keeping your camera steady.

To take untrailed long exposures or to use a telephoto lens, you will need a way to compensate for the Earth's rotation. Many telescopes allow you to attach a camera to the body of the telescope. If the telescope has an equatorial mount, you can keep it pointed toward a bright star that is easy to follow while the camera points toward your object of interest. They need not point in the same direction because

FIGURE 33.7
A picture made with a fixed camera showing star trails of the constellation Orion along with the planet Saturn (the bright trail to the left).

> **Concept Question 3**
>
> In the star-trail picture, how long was the exposure? In what direction was the observer looking? Were the stars rising or setting? Which star trails correspond to the three "belt" stars?

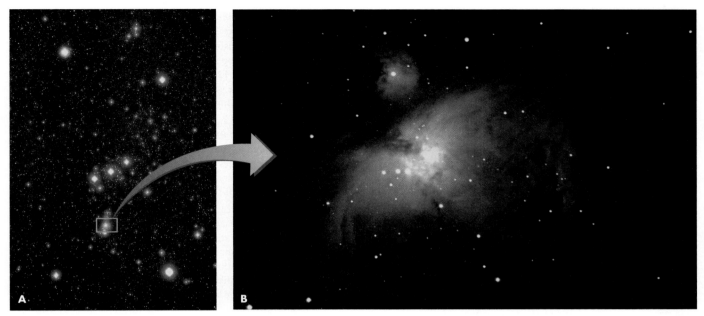

FIGURE 33.8
(A) Photograph of the constellation Orion made with a long exposure using a camera mounted on the body of a telescope tracking the stars.
(B) Picture of the Orion Nebula taken with a small backyard telescope.

Concept Question 4

If you did not have a telescope to compensate for the Earth's rotation, you might be able to attach your camera to a board, itself attached by hinges to a base. How would you need to orient the hinge to compensate for the Earth's rotation? What could you build to slowly tilt the board at the same rate the sky is "turning"?

Visit NASA's Astronomy Picture of the Day (APOD) website for hundreds of examples of amateur photographs in addition to pictures made at major observatories and space observatories.

the equatorial mount is designed to keep the entire telescope (and attachments) in fixed orientation relative to the celestial sphere.

Figure 33.8A shows a picture of the constellation Orion made in this way. You can also attach a camera to a telescope with an alt–az mount, but here you need to point the camera close to the same direction that the telescope is pointing, and even so there will be some twisting of the image in long exposures because the telescope tube rotates relative to the celestial sphere as it tracks the star.

Photography is easier with a motor drive on the telescope, but it is important to keep in mind that the telescope must remain well balanced or you may strain the motor. The weight of the camera and any lenses should be counterbalanced by weights elsewhere on the telescope so there is no net force pulling the telescope to turn in one direction or another. The highest-magnification pictures can, of course, be taken with special adaptors that allow your camera to look through the telescope (Figure 33.8B). You might even have success taking a handheld picture through a telescope eyepiece, although it is tricky to hold the camera lens in just the right spot. It helps if the eyepiece has a rubber eye cup that you can rest the camera against gently.

With the declining price of CCD detectors (Unit 28.2), amateur astronomers are producing spectacular images that were not possible even with professional equipment a decade ago. However, first try taking photos with some of the techniques mentioned earlier—you will be amazed at some of the faint star clusters and nebulae that you can image without investing in expensive equipment.

As you continue your explorations of the night sky, you should look for local astronomy clubs and visit planetariums. There are also organizations such as the American Association of Variable Star Observers (AAVSO), which will help you become involved in international efforts to monitor stars that have exhibited unusual behavior in the past. Through such monitoring projects, amateur astronomers make important contributions to the field of astronomy by alerting the astronomical community to unexpected events, such as outbursts from stars or the appearance of new comets.

KEY POINTS

- The human eye has many features of a telescope: pupil = aperture; retina = detector; cornea = lens.
- The fovea at the center of vision has three types of "cones" to detect different colors, but it is less sensitive to dim light.
- The eye takes 20 minutes or more to reach its peak sensitivity to dim light.
- An equatorial telescope mount rotates on an axis to compensate for Earth's rotation.
- Telescope magnification depends on the ratio of the telescope focal length to the eyepiece focal length.

KEY TERMS

alt–az mount, 236
averted vision, 234
cone, 232
cornea, 232
dark adaption, 233
equatorial mount, 236
focal length, 235
fovea, 233
iris, 232
photopigment, 232
pupil, 232
retina, 232
rod, 232
variable star, 234

CONCEPT QUESTIONS

Concept Questions on the following topics are located in the margins. They invite thinking and discussion beyond the text.

1. The eye's blind spot. (p. 233)
2. Observing stars versus galaxies. (p. 234)
3. Deciphering a star-trail photo. (p. 238)
4. Design mount to compensate for Earth's rotation. (p. 239)

REVIEW QUESTIONS

5. What part of a telescope or observatory corresponds most closely to the parts of your eye: iris, cornea, pupil, lens, retina, rods, fovea, brain?
6. What is meant by *dark adaption*?
7. Why does *averted vision* allow you to see dimmer objects?
8. What determines the magnification of a telescope?
9. What are the advantages of different types of mounts for telescopes?
10. Why does the image of a galaxy look so much dimmer by eye than in a photograph made through the same telescope?

QUANTITATIVE PROBLEMS

11. Suppose you are using a telescope with a 100-mm (4-inch) aperture diameter and a focal length of 1000 mm (40 inches).
    a. What is the f-ratio?
    b. What focal length eyepiece will give you a magnification factor of 100?
12. Suppose you are using a 20-cm (8-inch) f/10 Schmidt-Cassegrain telescope. You have a set of eyepieces with labeled focal lengths of 6, 12, 20, and 40 mm. What magnification will each eyepiece give you? Why will images probably look a bit blurry with the 6-mm eyepiece?
13. Suppose you have a pair of 8×60 binoculars, meaning they have a magnification of 8× and an aperture 60 mm in diameter.
    a. Compared to your eyes with your pupils dilated to 8 mm, how many times more light do the binoculars collect?
    b. If the binoculars have an f-ratio of 4, what is their focal length? What is the focal length of the eyepieces?
14. Similar to an eye, cameras vary f-ratio by changing the aperture. Suppose you have a camera lens labeled as having a focal length of 50 mm with f-ratios varying from f/1.4 to f/30.
    a. What is the range of aperture diameters for this camera?
    b. When the f-ratio is small, the "depth of field" (range of distances over which objects remain in focus) becomes smaller. Why would the focus be likely to remain sharper when the aperture is smaller?
15. If a camera is pointed at a star on the celestial equator and the shutter is left open for 20 minutes, what length of a star trail would show up on the resulting picture? For comparison, the angular size of the Moon is 0.5°. How many moons does the star move?

TEST YOURSELF

16. In very dim light,
    a. your pupils are smallest.
    b. your rods and cones grow much more sensitive to light.
    c. your color vision is at its most sensitive.
    d. you can see things more clearly if you stare straight at them.
    e. All of the above
17. As a star rises and moves across the sky, which of the following change(s)?
    a. Its right ascension
    b. Its declination
    c. Its azimuth
    d. Both (a) and (b)
    e. None of the above.
18. A practical limit to what magnification should be used is often set by the
    a. mount used.
    b. telescope's focal length.
    c. "seeing" conditions.
    d. aperture size.
19. One advantage of an equatorial mount over an alt-az mount is that the equatorial mount
    a. makes it is easy to track the motion of stars.
    b. makes it easier to balance the telescope.
    c. is cheaper to buy.
    d. is more compact and lighter to carry.
20. Suppose you are looking at Jupiter through a 100-mm aperture telescope with an f/10 ratio and a 20-mm eyepiece. if you wanted to make Jupiter look twice as big, you would
    a. change to a 10-mm eyepiece.
    b. change to a 40-mm eyepiece.
    c. reduce the aperture to 50-mm.
    d. increase the aperture to 200-mm.
    e. None of these: you would need to use an f/20 telescope.

PART 3 UNIT 34

The Structure of the Solar System

34.1 Components of the Solar System
34.2 Orbital Patterns in the Solar System
34.3 Compositions in the Solar System
34.4 Density and Composition

Learning Objectives

Upon completing this Unit, you should be able to:
- Recall the main features and locations of the various components of the Solar System.
- Describe the typical orbital and rotation properties of each component.
- Calculate the density of a body and explain what clues this provides to the internal composition of a planetary body.

The **Solar System** contains the Sun, the eight planets, their moons, dwarf planets, and swarms of asteroids and comets. The various objects in the Solar System differ enormously in size, composition, and temperature, but the Solar System as a whole possesses an underlying order. This order is based on physical conditions today as well as when the Solar System formed; and from the patterns they see, astronomers attempt to read the story of how our Solar System came to be. This Unit examines the overall structure of the Solar System and its components, and discusses the patterns in the properties we find. Each of these elements is explored in greater detail in subsequent Units.

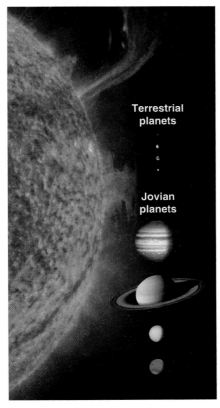

FIGURE 34.1
The planets and Sun to scale.

34.1 COMPONENTS OF THE SOLAR SYSTEM

The Solar System is defined as the Sun and the objects that are held in orbit around it by the Sun's gravity. These other objects are far smaller than the Sun, in total contributing less than one seven-hundredth of the Sun's **mass**—the equivalent in mass to the weight of a car's steering wheel versus the rest of the vehicle. Within the tiny fraction of the Solar System's mass outside the Sun there are probably more than a million celestial objects larger than mountains, with the largest being the planet Jupiter. These objects emit no visible light of their own but shine by reflected sunlight, and astronomers have cataloged only a fraction of them. Even more objects smaller than mountains orbit unseen and uncounted.

On the scale of humans, every one of these objects is big; the fact that altogether they amount to so little compared to the Sun is simply an indication of how enormous the Sun is. The Sun is a star, a huge ball of extremely hot gas. Its enormous gravity crushes and heats matter in its core so intensely that nuclear reactions take place (Unit 51). The Sun's gravity, heat, and light are largely responsible for the nature of everything else within the Solar System. The Sun is the center of the Solar System, but it is very different from the objects that orbit it. A more complete understanding of the Sun is best gained in the context of studying other stars, so Solar System astronomy focuses on the objects in orbit about the Sun.

In order of increasing orbital distance from the Sun, the officially recognized planets are Mercury, Venus, Earth, Mars, Jupiter, Saturn, Uranus, and Neptune. Planets are a big step down in size from stars, as can be seen in Figure 34.1, which shows a portion of the Sun to illustrate how the Sun dwarfs even the large planets. The figure also shows that planets differ greatly in size. Our largest planet Jupiter has only about 1/1000 the mass of the Sun. The Sun contains more than 99% of all the Solar

241

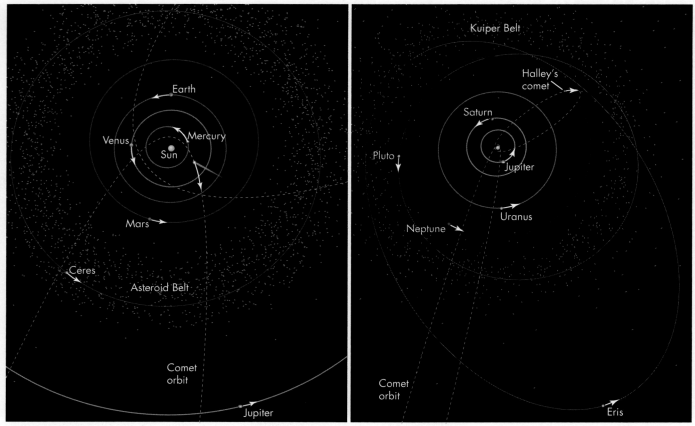

FIGURE 34.2
Diagrams of the inner and outer Solar System seen from above. The planets' orbits are shown in the correct relative scale in the two drawings.

System's mass, and Jupiter itself contains more mass than all the smaller objects in the Solar System combined.

The **inner planets**—Mercury, Venus, Earth, and Mars—are small rocky planets with relatively thin or no atmospheres. This type of planet is often called an Earth-like planet, or a **terrestrial planet**. The Earth is the largest terrestrial planet, although Venus is nearly as big. Mars has about half of Earth's diameter and only about a tenth of Earth's mass. Mercury has 38% of Earth's diameter and only 6% of Earth's mass. From examining the differences and similarities between these planets, we can better understand the processes that shape the Earth.

The planets' orbits are illustrated in Figure 34.2. In this view we are "looking down" on the orbits of the Earth and other planets from far above the Earth's and Sun's north poles. Viewed from this direction, the planets' orbits are roughly circular. Outside of Mars's orbit is a region called the **asteroid belt**. It gets its name because a majority of the small objects known as **asteroids** (Unit 43) orbit there. Asteroids are rocky or metallic bodies with diameters ranging from meters up to hundreds of kilometers. The asteroids are irregular in shape, except for the largest, Ceres, which has a diameter about 7% of the Earth's. Because its gravity has pulled it into a rounded shape, it is termed a **dwarf planet**.

The **outer planets**—Jupiter, Saturn, Uranus, and Neptune (Figure 34.2)—are also called **Jovian** (Jupiter-like) **planets** or **gas giants** because of their thick atmospheres. They are huge compared with the terrestrial planets: 15 to 320 times more massive than Earth, and 4 to 11 times its diameter. The Jovian planets have no distinct surface. Instead, their atmospheres thicken with depth and eventually liquefy under intense pressure. Deeper still, the liquid material gradually compresses to the same density as rock on Earth, but remains gaseous or liquid because of intense heat there. Deep in the center there is probably a core of molten rock and iron. However, we could never

Astronomers sometimes speak of the **superior planets.** This refers to planets having an orbit larger than Earth's and therefore includes Mars.

"land" on Jupiter because we would simply sink ever deeper into its interior until our spaceship was crushed by pressures great enough to compress solid rock.

The outer Solar System has a layout similar to that of the inner Solar System, but on a far grander scale (Figure 34.2). Outside of Neptune's orbit is a region called the **Kuiper** (*KY-per*) **belt** named after a Dutch-born American astronomer. Analogous to the asteroid belt beyond Mars's orbit, the Kuiper belt contains vast numbers of objects that orbit the Sun out to about twice Neptune's orbit. These **trans-Neptunian objects** (or **TNOs**) are composed primarily of ices in this remote, cold region of the Solar System.

Until 2006, one of the TNOs, Pluto, was called a planet. However, over the preceding decade it had become clear that Pluto was just one of the largest among thousands of objects orbiting beyond Neptune. Dozens of the TNOs were found with sizes not much smaller than Pluto. Then in 2005, Eris—a body 25% more massive than Pluto—was discovered. Many questions arose. Was Eris a planet too? What about all of the objects that are just a little smaller than Pluto? And what about potentially huge numbers of large objects that might be discovered in future years as our telescopes and detection techniques improve?

It may seem surprising, but there was no officially accepted definition of "planet" before 2006. The term comes from ancient times and means a "wanderer" because of the changing position of these objects on the celestial sphere (Unit 11). In both popular and astronomical usage, a planet is a large body orbiting the Sun, but how big is big enough? Pluto has only one twenty-fifth the mass of Mercury, the smallest body called a planet since ancient times. Astronomers convened at the *International Astronomical Union* assembly in 2006, and developed a definition with two requirements. A planet must be (1) so massive that its gravity pulls it into a roughly spherical shape and (2) the dominant mass in the neighborhood of its orbit. The second part of this definition means the objects lying in the asteroid belt and Kuiper belt are not planets. However, if they meet the first criterion, they can still be labeled dwarf planets. As a result, Pluto and Eris are both labeled dwarf planets—and several dozen more TNOs may soon be added to that category.

Dwarf planets orbiting beyond Neptune were given the additional designation of "plutoid" in 2008. This now includes two more large objects, Haumea and Makemake (Unit 48).

The outer Solar System continues beyond the Kuiper belt out to perhaps a few thousand times farther than Neptune's orbit—about one-quarter of the distance to the next nearest star. Sunlight is so dim this far from the Sun that little light is reflected back to us, and we know relatively little about this region. Our primary clue about the objects that populate this region comes from **comets** (Unit 49). These are icy bodies that spend most of their time in the outer Solar System; but some have orbits that carry them into the inner Solar System, perhaps no more frequently than once in a million years. A few approach so close to the Sun that their frozen outer layers are vaporized by the Sun's heat, then stream away from the Sun in long "tails" that are sometimes visible to the naked eye. Astronomers hypothesize that a large population of objects orbit in what is called the **Oort cloud,** named after the Dutch astronomer who proposed its existence. The Oort cloud is thought to surround the Solar System (Figure 34.3), with most of its objects remaining there in very large orbits that may never enter the

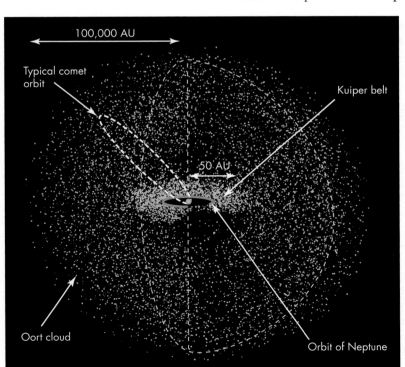

FIGURE 34.3
Sketch of the Oort cloud and the Kuiper belt. The scale shown is only approximate. Orbits and bodies are not to scale.

inner Solar System. Hypotheses about the total mass contained in the Oort cloud vary greatly, from a small fraction of the Earth's mass to as much as Jupiter's mass. Such great uncertainty arises because of the difficulty in determining the likelihood that an Oort cloud object will come close enough to the inner Solar System to be detected by our telescopes. Because we cannot say precisely what percentage of the cloud's population we *do* observe, we cannot determine the total number.

Astronomers suspect that for most of the comets that come into the inner Solar System (and that therefore can be well studied), their orbits have been disturbed by the gravitational effects of a distant star or a close encounter with another object. The comets entering the inner Solar System are generally quite small—diameters of 50 km or less. Until recently it was supposed that all of the Oort cloud objects were this small, but recent discoveries have revealed the presence of objects comparable in size to Pluto beyond the Kuiper belt. One of these, named Sedna, has an orbit that carries it out to 30 times Neptune's distance, but currently it is only about three times farther than Neptune. Sedna would not have been detected with current instrumentation if it had been in a more distant part of its orbit. It seems likely that there are many more even larger worlds awaiting discovery in these most remote regions of the Solar System.

In addition to the objects that orbit the Sun separately, a large number of bodies orbit the planets and other bodies. There are seven **satellites** (moons) of the planets, including our own Moon, that are both larger and more massive than either Pluto or Eris. Many of these are fascinating worlds in their own right. Another dozen satellites are large enough that gravity has pulled them into a round shape, so they would satisfy the definition of dwarf planet if they were not orbiting a planet.

Astronomers are steadily finding more small satellites of the Jovian planets with new satellite missions and improving telescopes (Unit 47). Astronomers have so far identified 67 satellites of Jupiter, 62 of Saturn, 27 of Uranus, and 14 of Neptune. Of the terrestrial planets, Mars has 2 satellites, Earth has 1, while Mercury and Venus have none. Pluto has 5 satellites and Eris has 1, and some smaller TNOs and even some asteroids are known to have small satellites. The satellites generally share some basic characteristics with their host planets, being made of similar materials—rocky materials in the case of our Moon and Mars's two tiny satellites, and ices of some compounds related to the gaseous elements of the Jovian planets.

34.2 ORBITAL PATTERNS IN THE SOLAR SYSTEM

The eight planets move around the Sun in nearly circular orbits, all lying in almost the same plane, as shown in the side view in Figure 34.4. (Figure 34.2 is a view from the "top.") The planets all travel around the Sun in the same direction: counterclockwise as seen from above the Earth's North Pole. Their orbits could be contained in a thin disk, a spinning "pancake," whose thickness relative to its diameter would be about the same as three CDs stacked together.

As the planets orbit the Sun, each also spins on its rotation axis. The spin is again generally in the same direction as the direction in which planets orbit the Sun, and the Sun's own rotation is in the same direction as well—counterclockwise as seen from above the Earth's North Pole. The planets' rotation axes are tilted relative to the plane of planetary orbits, but generally not far from the orientation of the Sun and the planets' orbits. There are exceptions, however. Venus spins slowly but backward (clockwise), while Uranus's axis lies nearly in the **orbital plane** (Figure 34.5) of the planet. The flattened structure of the Solar System and the generally consistent orbital and spin properties of the bodies in it are two of its most fundamental features, and any theory of the Solar System must explain them.

Concept Question 1

There is currently no smallest size in the definition of a satellite. Should any object of any size be labeled a moon? Can you think of any physical basis for setting a smallest size?

The number of known satellites is constantly increasing. The numbers quoted are current as of mid 2013.

The orbits of Pluto and many other TNOs are sufficiently elliptical that they come closer to the Sun than Neptune during part of their orbit. Although these orbits may appear to cross in diagrams, the orbits are generally quite tilted relative to the rest of the Solar System, so collisions are highly improbable.

Unit 34 The Structure of the Solar System   245

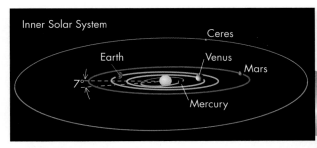
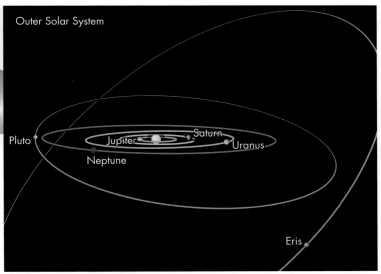

FIGURE 34.4
Planets and their orbits from the side. The left-hand panel illustrates the inner Solar System, including the orbit of the dwarf planet Ceres, which is more tilted than the major planets. The right-hand panel of the the outer Solar System shows the large angle of the orbits of the dwarf planets Eris and Pluto relative to the major planets. The sizes of the Sun and planets are not to the same scale as the orbits.

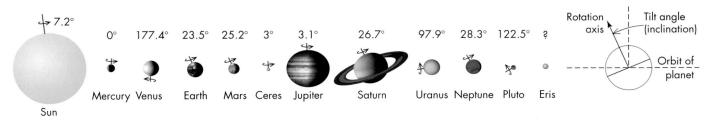

FIGURE 34.5
The tilt of the rotation axes of the Sun, planets, and dwarf planets relative to their orbit around the Sun. Most planets spin around their axis in approximately the same counterclockwise sense as they orbit the Sun; however, Venus spins backward and Uranus's axis almost lies in its orbital plane. The Sun's rotation axis is tilted relative to the average orbital plane of the planets.

Solar System builder

The satellites of planets also generally move in a regular pattern along approximately circular paths that are roughly in the same plane as each planet's equator. Thus, each planet and its moons resemble a miniature Solar System—an important clue to the origin of the satellites. There are a few exceptions among the larger moons, such as Neptune's moon Triton, which orbits in a backward direction—leading astronomers to hypothesize that it was once an independent object in the Kuiper belt that was captured by Neptune.

One feature that is not fully understood about the orbits of the planets is their spacing. The separations between the inner planets are relatively small—between 0.3 and 0.4 AU—while the distances between the orbits of the outer planets are much larger—between 4 and 11 AU. The entire inner Solar System is smaller than the separation between neighboring planets in the outer Solar System. This makes it difficult to show the inner and outer planets' orbits in the same figure when we wish to illustrate their orbits to scale. What is known is that not all configurations of planets are stable for long periods of time because of gravitational interactions between the planets. The Solar System was either formed this way or gradually evolved to a very stable set of orbits. Astronomers are just beginning to uncover the characteristics of other planetary systems. There are not very many where we have detected several planets, but the other systems detected so far look quite different from our own, and it appears that the orbits of some planets become altered over time (Unit 36).

34.3 COMPOSITIONS IN THE SOLAR SYSTEM

By studying the spectrum of light that the Sun emits (Units 24 and 56), we find that the Sun is made mostly of hydrogen (71.1% of its mass) and helium (27.4%). Spectra of the atmospheres of the Jovian planets indicate that they are likewise dominated by the elements hydrogen and helium.

The terrestrial planets have rocky surfaces. Through spectroscopy and direct study, this is found to be composed primarily of silicate rock—compounds of silicon and oxygen (such as SiO_2, SiO_3, SiO_4) with an admixture of other heavy elements such as iron (Fe), magnesium (Mg), calcium (Ca), and aluminum (Al). These are some of the most common elements found in the Sun's atmosphere as well, seen there in vaporized form. The next 14 biggest contributors to the Sun's mass are listed in Table 34.1, making up 99.5% of its remaining mass excluding hydrogen and helium. These same elements also make up over 95% of the mass of the Earth's crust, although in different proportions. Overall, the Earth's crust shows a lack of the elements that are commonly found in gaseous form (listed in blue in the table) versus those that are generally in solid form (listed in black). The exception to this is oxygen, which combines with silicon to make rock.

Spectroscopy of the solid surfaces of the moons and dwarf planets of the outer Solar System show that they are distinct from bodies in the inner Solar System. They are made primarily of *ices* of compounds of hydrogen and other elements common in the Sun's atmosphere, such as oxygen, carbon, and nitrogen, which make water (H_2O), ammonia (NH_3), and methane (CH_4). These are **volatile** substances that would be liquids and gases at typical Earth temperatures, but at the low temperatures found far from the Sun, these hydrogen-rich compounds form solids. Thus, these bodies have a chemical composition more similar to the Jovian planets than to the terrestrial planets.

From spectroscopy and direct sampling we can learn a great deal about the composition of an object's surface or atmosphere, but nothing about a planet's interior. To learn about the overall composition of a planet, astronomers must use alternative methods. One such technique employs the planet's density.

TABLE 34.1 Heavy Elements in the Sun and Earth

Element	Percentage by Mass* Sun	Earth's Crust
Oxygen	44.2%	45%
Carbon	16.6%	0.02%
Iron	9.3%	6%
Neon	8.6%	0.0000005%
Silicon	5.5%	27%
Nitrogen	5.3%	0.002%
Magnesium	4.8%	2.5%
Sulfur	2.8%	0.03%
Argon	0.8%	0.0004%
Nickel	0.5%	0.009%
Calcium	0.5%	5%
Aluminum	0.4%	8%
Sodium	0.3%	2%

*These percentages exclude hydrogen and helium, which make up about 98.5% of the Sun's mass. Estimates for the Earth's crust vary depending on sampling, so these values are approximate. Elements commonly found in gaseous form on Earth are shown in blue.

Concept Question 2

How might you use density estimates to pick or avoid grocery products? What items would you avoid if they were too dense? which if they were too underdense?

34.4 DENSITY AND COMPOSITION

The average **density** of a planet is its mass divided by its volume. Both mass and volume can be measured relatively easily. Newton's law of gravity (Unit 16) allows us to determine a body's mass from its gravitational attraction on a second body orbiting it (Unit 17). From this law, we can calculate a planet's mass by observing the orbital motion of one of its moons or a passing spacecraft. We can determine a planet's volume ($\mathcal{V}$) from the geometric formula for a sphere: $\mathcal{V} = 4\pi R^3/3$, where R is the planet's radius. We can measure R in several ways—for example, from the planet's angular size and distance (Unit 10).

With the planet's mass, M, and volume, $\mathcal{V}$, known, calculating its average density (represented by the Greek letter rho, ρ) is straightforward: We divide M by $\mathcal{V}$ (Figure 34.6). For the Earth, its radius is 6.37×10^6 m and its mass is 5.97×10^{24} kg, so its density is

$$\rho_\oplus = \frac{M_\oplus}{\mathcal{V}_\oplus} = \frac{5.97 \times 10^{24} \text{ kg}}{4\pi(6.37 \times 10^6 \text{ m})^3/3}$$

$$= \frac{5.97 \times 10^{24} \text{ kg}}{1.08 \times 10^{21} \text{ m}^3}$$

$$= 5530 \text{ kg/m}^3 = 5.53 \text{ kg/liter}.$$

Finding the Density of a Planet

Volume

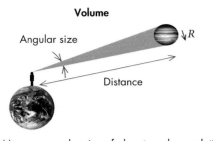

Measure angular size of planet, and use relation between angular size and distance to solve for planet's radius, R. Calculate volume, V, of planet:

$$V = \frac{4\pi R^3}{3}$$

for a spherical body of radius R.

Average Density

Average density, ρ, equals mass, M, divided by volume, V:

$$\rho = \frac{M}{V}$$

Mass

Observe motion of a satellite orbiting planet. Determine satellite's distance, d, from planet and orbital period, P. Use Newton's form of Kepler's third law:

$$M = \frac{4\pi^2 d^3}{GP^2}$$

Insert measured values of d and P, and value of constant G. Solve for M.

FIGURE 34.6
Measuring a planet's mass, radius, and average density. Mass can be determined from the orbit of a satellite (Unit 17), and volume can be measured after we determine the radius of the planet from its angular size (Unit 10).

Mathematical Insights

Densities may be reported in a variety of units. However, many of them are equivalent. For example, 1 kilogram per liter is the same density as 1 gram per cubic centimeter or 1 metric ton per cubic meter (or even 1 pound per pint). To convert from the MKS units of kilograms per cubic meter to one of these more popular sets of units, divide the value in kg/m³ by 1000.

Concept Question 3

How could you rule out a model of Mercury ($\rho = 5.43$ kg/L) with rock in the interior and iron on the exterior to give the correct overall density? What about an iron interior and an ice crust?

The last step in this calculation was made by noting that a cubic meter contains exactly 1000 liters. This density is easy to picture: a liter bottle of water is a familiar volume, and water has a density of 1 kilogram per liter.

Density gives an important clue to an object's composition. For instance, it would be easy to tell the difference if a liter bottle was filled with concrete, water, or styrofoam, even if the exterior were opaque. Understanding how volume, mass, and density are related is the basis of the famous story about the ancient Greek scientist Archimedes leaping from his bath and running down the street shouting "Eureka!" He had been given the task of determining whether the king's crown was solid gold, but he was not allowed to damage it in any way. Sitting in his bath, he realized that his own body displaced a volume of water equal to that of his body. However, a mass of gold with the same mass as his body would have a much smaller volume—because of its high density—and would displace much less water. Thus, if the crown did not displace the same amount of water as an equal weight of gold, it must contain another metal, like lead or silver, which each have lower densities than gold. Likewise, we can use the density of the Earth or any planet to deduce its composition.

From its total mass and volume, the average density of the Earth is 5.53 kilograms per liter. This is substantially more dense than silicate rock, which averages about 3.0 kilograms per liter. This tells us that the interior of the Earth must contain a substance denser than rock, but which one? A number of metals have densities higher than 5.53 kilograms per liter, and might average out with the rocky surface to give the correct overall density. Since it must be a significant portion of the Earth's mass, it is likely to be one of the common elements found in the Sun or the Earth's surface rocks. For example, the only two elements from Table 34.1 with high enough densities are nickel and iron, which both have densities of around 8 kilograms per liter. Other metals such as gold, silver, or lead have even higher densities, but they are extremely rare. Knowing that the average density of the Earth (5.53 kilograms per liter) is intermediate between silicate rock and these metals, it is reasonable to infer that the Earth has an iron and nickel core beneath its rocky, silicate-rich crust, a supposition that has been supported by more complex techniques (see Unit 37).

All the terrestrial planets have an average density similar to the Earth's (3.9 to 5.5 kilograms per liter). On the other hand, all the Jovian planets have a much smaller average density (0.7 to 1.7 kilograms per liter), which is similar to that of ice. There are a number of asteroids for which we have accurate mass and size

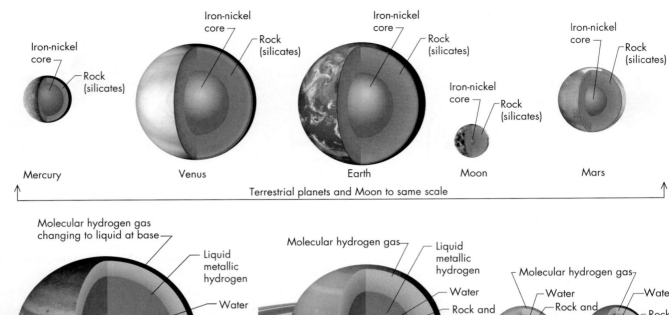

FIGURE 34.7
Sketches of the likely internal structure of the planets and the Earth's Moon.

measurements, and their average densities are in the range of 2.0 to 3.5 kilograms per liter. The solid bodies in the outer Solar System, including Pluto, are generally between 1.0 and 2.0 kilograms per liter. Further analyses indicate that the objects in the inner Solar System are rocky with a decreasing fraction of iron the farther they are from the Sun. In the larger bodies of the inner Solar System, the iron has sunk to the core, as shown in Figure 34.7. In the outer Solar System, the densities of the gas giants and the solid bodies indicate that they must contain mainly low-density materials in their interiors.

Beneath their deep atmospheres the gas giants have molten cores of iron and rock about the size of the terrestrial planets, as illustrated in Figure 34.7. In addition to density determinations, astronomers deduce the existence of these cores in two other ways. First, if the outer planets have the same relative amount of heavy elements as the Sun, they should contain several Earth masses of iron and silicates. Because these substances are much denser than hydrogen, they must sink to the planet's core. Secondly, mathematical analyses of these planets indicate that the distortion to their shape caused by their rotation, along with details of their gravitational fields sensed by spacecraft, imply that they have small, dense cores.

Our discussion of the composition of the planets not only underlines the differences between the inner and outer Solar System but also furnishes another clue to their origin: The planets and Sun are all consistent with having been made from the same material. The giant planets like Jupiter and Saturn have compositions almost identical to that of the Sun, while the terrestrial planets have a similar composition except for the missing volatile compounds, presumably lost because of the higher temperatures of the inner Solar System.

KEY POINTS

- More than 99% of the mass of the Solar System is in the Sun.
- There are four gas-giant planets, four terrestrial planets, and perhaps dozens of dwarf planets and comparable-size satellites.
- There are uncounted millions of smaller bodies orbiting the Sun, primarily in the asteroid belt, Kuiper belt, and Oort cloud.
- Most bodies orbit the Sun in a counterclockwise direction in roughly the same plane, and spin counterclockwise too.
- Bodies in the inner Solar System are primarily made of rock (mostly oxygen and silicon) and iron.
- Bodies in the outer Solar System are primarily composed of gases and ices of hydrogen-rich molecules, with rock and iron cores.
- Density provides an indicator of the composition of planets.

KEY TERMS

asteroid, 242
asteroid belt, 242
comet, 243
density, 246
dwarf planet, 242
gas giant, 242
inner planet, 242
Jovian planet, 242
Kuiper belt, 243
mass, 241

Oort cloud, 243
orbital plane, 244
outer planet, 242
satellite, 244
Solar System, 241
superior planet, 242
terrestrial planet, 242
trans-Neptunian object (TNO), 243
volatile, 246

CONCEPT QUESTIONS

Concept Questions on the following topics are located in the margins. They invite thinking and discussion beyond the text.

1. Setting standards for satellites. (p. 244)
2. Using density to check food quality. (p. 247)
3. Testing different models with matching densities. (p. 247)

REVIEW QUESTIONS

4. What properties distinguish the terrestrial from the Jovian planets?
5. What is a dwarf planet?
6. How are asteroids and TNOs similar? How do they differ?
7. What is meant by a volatile substance, and why would we expect to find objects made out of volatile substances far from the Sun?
8. Why is the density of an object an important clue to its composition?
9. How can we determine the density of a distant planet?
10. What is the Oort cloud? Where is it located, and what kind of objects come from it?

QUANTITATIVE PROBLEMS

11. By what factor would the Sun have to shrink to be the size of a large beach ball, 1 meter in diameter? In Appendix Tables 5 and 6, look up the sizes and distances from the Sun of Jupiter, Neptune, Earth, Mercury, Pluto, and Ceres.
    a. What diameter would these six objects have if scaled down by the same factor?
    b. What would their masses be if their density stayed the same?
12. What would be the distance from the Sun for each of the bodies listed in Problem 11, scaled down by the same factor that would make the Sun 1 meter in diameter?
13. Saturn has a mass of 5.68×10^{26} kg and a radius of 60,300 km. Calculate Saturn's density from these values. According to Archimedes' principle, for something to float, it must have a density less than what it is floating on. Would Saturn float on water, which has a density of 1 kg/L?
14. Use the data in Appendix Table 5, "Physical Properties of the Planets," to list the planets in order according to several different parameters: (i) smallest to largest radius; (ii) smallest to largest mass; (iii) most to least dense. Note that the order of some planets is changed in each case. Can you draw any approximate conclusions from these results? For example, are smaller planets always more dense? What other factors might explain why some planets are out of order with the overall trends you find?
15. You can calculate the acceleration due to gravity at the surface of a planet using the formula: $g = GM/R^2$ where M and R are the mass and radius of the planet, and G is the gravitational constant (see Unit 16). How much do you weigh on Earth? Determine how much you would weigh on Jupiter, Mars, and Ceres. (Hint: Because your mass would not change, you can simply compare the values of g for the other planets with Earth's 9.8 m/sec^2.)

TEST YOURSELF

16. Other than Earth, which three planets are primarily rocky with iron cores?
    a. Venus, Jupiter, Neptune
    b. Mercury, Venus, Pluto
    c. Mercury, Venus, Mars
    d. Jupiter, Uranus, Neptune
    e. Mercury, Saturn, Pluto
17. Why is Pluto not considered a planet?
    a. It orbits inside the Kuiper belt.
    b. It is so far out in the Solar System.
    c. It has less mass than Mercury.
    d. It is smaller than several moons found around other planets.
18. More than 99% of the mass of the Solar System is contained in
    a. the Earth.
    b. the Sun.
    c. Jovian planets.
    d. Jupiter.
    e. asteroids.
19. Terrestrial planets are characterized by what property?
    a. High masses
    b. A lot of satellites
    c. Solid surfaces
    d. Low densities
20. Looking down on the Solar System from far above the Sun's North Pole, ____ planets orbit counterclockwise and ____ planets spin counterclockwise.
    a. all; all
    b. all; most
    c. most; all
    d. all; no
    e. no; all

UNIT 35

The Origin of the Solar System

35.1 The Age of the Solar System
35.2 Birth of the Solar System
35.3 From Dust Grains to Planetesimals
35.4 Formation of the Planets
35.5 Late-Stage Bombardment

Learning Objectives

Upon completing this Unit, you should be able to:
- Explain how the age of the Solar System has been determined.
- Calculate the remaining fraction of a radioactive isotope after a number of half-lives.
- Describe the solar nebula theory and how it explains the Solar System's structure.
- Explain why different elements condensed at different distances from the Sun.
- Discuss how planets form and differentiate, and the sources of atmospheres.

How did the Solar System form? What processes gave it its main features? Because we were not around to witness its birth, our explanation of its origin must be a reconstruction based on observations that we make now, long after the event. The general features of the Solar System discussed in Unit 34 provide powerful clues that have helped astronomers reconstruct how the Solar System must have formed. Any theory of the Solar System's origin must explain why:

- The Solar System is flat, with the planets orbiting in the same direction.
- The composition of the outer planets is similar to the Sun's except the inner planets lack the gases that only condense at low temperatures.
- There are rocky planets near the Sun and massive, gaseous planets farther out.

In addition, there are many other features that must also be explained by a successful theory, such as the properties of asteroids, the number of craters on planetary and satellite surfaces, and the detailed chemical composition of surface rocks and atmospheres. Another essential feature to be explained is *when* the Solar System formed, which we can determine based on one of the properties of atoms.

35.1 THE AGE OF THE SOLAR SYSTEM

Several pieces of information indicate when the Solar System formed. All existing evidence is consistent with the various bodies in the Solar System having formed at the same time—about 4.5 billion years ago. Geological features on Earth point to a history of billions of years for our planet. We find a similar age for the Sun, based on its current brightness and temperature and its rate of nuclear fuel consumption (Unit 62). However, the most precise age information comes from a clock that ticks inside the first rocks that formed.

A number of naturally occurring atoms undergo **radioactive decay,** spontaneously splitting apart into two or more lower-mass atoms. For many of the heaviest elements, all forms of the atom are radioactive, as illustrated by the elements marked in green in the periodic table in the inside back cover of this book. Even the lighter elements have some **isotopes**—nuclei with different numbers of neutrons—that are radioactive.

In the nucleus of an atom that is radioactively unstable, electric repulsion between protons or the weak force (Unit 4) may cause the nucleus to decay. At any given moment there is a small chance of a radioactive atom splitting. The atom may last for billions of years, or it may decay in the next second. This is like a

Unit 35 The Origin of the Solar System

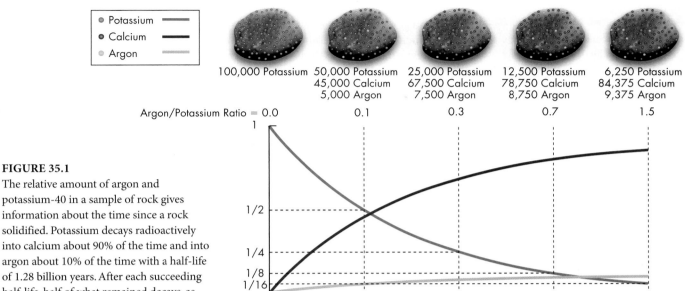

FIGURE 35.1
The relative amount of argon and potassium-40 in a sample of rock gives information about the time since a rock solidified. Potassium decays radioactively into calcium about 90% of the time and into argon about 10% of the time with a half-life of 1.28 billion years. After each succeeding half-life, half of what remained decays, as shown in the graph.

lottery. The atom may "win" (and split apart) on the next tick of the clock or it may "play the lottery" over and over again without winning.

A radioactive atom has a 50-50 chance of decaying in a time called its **half-life**. For example, the isotope potassium-40 has a half-life of 1.28 billion years. If you were to start with 100,000 of these atoms, after 1.28 billion years, 50,000 (on average) of the "parent" atoms would have decayed into "daughter" atoms, which in this case are atoms of calcium and the gas argon (Figure 35.1). After another 1.28 billion years, half of the *remaining* potassium decays, leaving 25,000 of the original potassium atoms. Note that there is no increase in the chances for the remaining potassium to decay—just as your odds for winning the lottery do not improve just because you have lost previously.

After each half-life passes, another half of the remaining atoms decay. So after one half-life, one-half remains; after two half-lives, one-quarter; after three half-lives, one-eighth; and so forth. These fractions can also be written like this:

$$\frac{1}{2} = \frac{1}{2^1} \quad \frac{1}{4} = \frac{1}{2^2} \quad \frac{1}{8} = \frac{1}{2^3}.$$

With the passing of each half-life, the number of remaining parent atoms declines by another power of 2. This can be expressed by a mathematical formula. If the half-life is τ (the Greek letter tau), after a time t the fraction remaining will be

$$\text{Fraction} = \left(\frac{1}{2}\right)^{t/\tau}.$$

t = Time since rock was last molten
τ = Half-life of element

For example, after three half-lives, $t = 3 \times \tau$, so the fraction is $(1/2)^3 = 1/8$, matching what we just found.

We can use this property to determine the age of a mineral that incorporates potassium within its structure. As the potassium-40 decays inside the mineral, an increasing amount of calcium and argon will be bound inside the mineral even though these atoms would not normally be included when the mineral formed. Therefore, the more daughter atoms a rock contains relative to the original radioactive atoms, the older the rock is. The more argon relative to potassium within a rock sample, the older the rock must be, as shown in Figure 35.1.

This method reveals the length of time since the rock was last molten. When the rock is liquid, the decay products (like calcium and argon), which are not normally

Mathematical Insights

This equation also lets you calculate how much of the original isotope remains, for any portion of a half-life; so after 1.5 half-lives ($t = 1.5\tau$), for example, the fraction remaining is $(½)^{1.5} = 0.35$.

> **Concept Question 1**
>
> Why would a smaller body tend to cool more rapidly than a large body? Can you think of any examples of this in your own experience?

part of the chemical structure of the rock, escape. This resets the "clock." By this method geologists find that the oldest rocks on Earth are about 4 billion years old. Such ancient rocks are found in many places—northern Canada, southern Africa, and Australia. Thus, portions of the Earth's crust solidified for a final time 4 billion years ago, and the Earth itself must have formed sometime earlier. Small mineral inclusions in some rocks have been dated to 4.4 billion years. These are thought to be bits of rock that cooled even closer to the beginning of the Earth's history and became imbedded in other rocks without having been melted.

Even older rocks have been found among samples from the Moon and from **meteorites,** which are small pieces of asteroids that have struck the Earth (Unit 50). These have ages up to about 4.56 billion years. Because the Moon and asteroids are small, they are expected to have cooled off rapidly (Unit 37), and hence their ages are probably closer to the age of the Solar System. In addition, a variety of radioactive elements with different half-lives are incorporated in different kinds of minerals. All of them give consistent values for the age of the Solar System.

From such evidence, we assume the Earth also formed about 4.5 billion years ago. Moreover, we can deduce that most of its crust has been molten at one time or another since then. Thus, we can add the following to our clues about the Solar System's formation:

For all bodies in the Solar System whose ages have so far been determined, the evidence indicates that they formed about 4.5 billion years ago.

Four and one-half billion years is an immense age. To illustrate, if those billions of years were compressed into a single year, all of human recorded history would have happened during just the last minute of the year. The brevity of human life compared to the vast age of the Earth prevents us from observing how truly dynamic our planet is. Mountains and seas appear to us to be permanent and unchanging, but in fact they change dramatically over the vast epochs of the Earth's existence. Such changes ultimately are caused by the heat in the Earth's interior, which creates motion in the Earth's interior and crust (Unit 37).

35.2 BIRTH OF THE SOLAR SYSTEM

The modern theory for the origin of the Solar System derives from a hypothesis proposed in the eighteenth century by Immanuel Kant, a German philosopher, and Pierre-Simon Laplace, a French mathematician. Kant and Laplace independently proposed what is now called the **solar nebula theory:** The Solar System originated from a rotating, disk-shaped cloud (*nebula* is Latin for "cloud") of gas and dust, with the outer part of the disk becoming the planets and the center becoming the Sun. This theory offers a natural explanation for the flattened shape of the system and the common direction of motion of the planets around the Sun.

The modern form of the theory proposes that the Solar System was born 4.5 billion years ago from an **interstellar cloud,** an enormous rotating aggregate of gas and dust like the one shown in Figure 35.2. Such clouds are common between the stars in our Galaxy even today, and there is strong evidence today that all stars have formed from them. Thus, although our main concern here is with the birth of the Solar System, we should remember that our theory applies more broadly and implies that most stars could have planets, or at least surrounding disks of dust and gas from which planets might form (Unit 36).

Interstellar clouds are the raw material of the Solar System, so we need to describe them more fully. Although such clouds are found in many shapes and sizes, the one that became our Sun and planets was probably a few light-years in diameter and contained about twice the present mass of the Sun. The cloud

FIGURE 35.2
Hubble Space Telescope image of a dark interstellar cloud seen in silhouette against glowing background gas. This dark cloud may be similar to the one from which the Solar System formed.

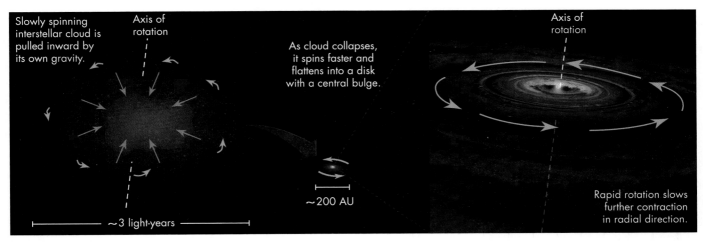

FIGURE 35.3
A sketch illustrating the collapse of an interstellar cloud to form a rapidly spinning disk. Note that the final size of the disk is not shown to scale—in actuality it would be more than 1000 times smaller than the cloud from which it formed!

Solar nebula theory

probably began with a composition similar to what we estimate from spectroscopy in the Sun's atmosphere today (Unit 24). About 71.1% of its mass was hydrogen gas and about 27.4% was helium gas, with the remaining 1.5% included all the other chemical elements. Some elements, such as iron, silicon, and oxygen, form small solid "dust" particles called **interstellar grains.**

Interstellar grains range in size from large molecules to micrometers or larger. They are thought to be made of a mixture of silicates (crystals of silicon and oxygen), iron compounds, carbon compounds, and water frozen into ice. Astronomers deduce the presence of these substances from their spectral lines, which are seen in starlight that has passed through dense dust clouds. Moreover, a few hardy interstellar dust grains, including tiny diamonds, have been found in ancient meteorites. This direct evidence from grains and the data from spectral lines show that these elements occur in proportions similar to those we observe in the Sun. This compositional similarity provides additional support for the view that the Sun and its planets formed from an interstellar cloud.

The cloud began its transformation into the Sun and planets when the gravitational attraction between the matter in the cloud caused it to collapse inward, as illustrated in Figure 35.3. Astronomers hypothesize that the collapse may have been triggered by a star exploding nearby or by a collision with another interstellar cloud. But regardless of its initial cause, the infall was not uniform around the cloud's center. Instead, because the cloud was rotating, it flattened into a spinning disk.

Flattening occurred because rotation retards the inward motion in the plane of rotation, perpendicular to the cloud's rotation axis. This is a result of a basic physical process—the conservation of angular momentum—which causes an object to spin faster if its radius becomes smaller (see Unit 20.2). The flattening caused by spinning happens in a pizza parlor, where the chef flattens the dough by tossing it into the air with a spin. Unlike the pizza dough, however, the inward pull of gravity made the cloud smaller even as it spun faster. It took a few million years for the cloud to collapse and become the **solar nebula.** Most of the matter collected in the center of the nebula, eventually forming the Sun. A fraction of the material was revolving too fast to merge into the center and formed the surrounding disk.

The whole solar nebula was probably about 200 AU in diameter and perhaps 10 AU thick. During the collapse phase, gravitational potential energy was converted into thermal energy, providing the initial heating even before the Sun became a star (see Unit 20.1). The inner parts of the disk were hot, heated by radiation from the young Sun and by the impact of gas falling onto the disk during its collapse. The outer parts of the nebula were cold, both because the radiation from the early Sun was weak and because collisions were slower in this region of weaker gravitational attraction. This set the stage for the formation of the planets.

Technically, *condensation* is the change from gas to liquid, and *deposition* is the change from gas to solid; but it is common practice to use *condensation* to refer to both.

FIGURE 35.4
Water vapor cools as it leaves the kettle. The cooling makes the vapor condense into tiny liquid water droplets, which we see as "steam."

Concept Question 2

Imagine that silicates condensed at 200 K and that the less dense water condensed at 1000 K. How would the structure of the Solar System and composition of the planets differ from what they are today?

35.3 FROM DUST GRAINS TO PLANETESIMALS

Condensation occurs when a gas cools and its molecules stick together to form liquid or solid particles. For condensation to happen, the gas must cool below a critical temperature, the value of which depends on the substance condensing and the surrounding pressure. For example, water condenses at about room temperature, which you can see as steam escapes from a boiling kettle (Figure 35.4). Molecules in the hot water vapor come into contact with the cooler air of the room. As the vaporized water cools, its molecules move more slowly, so that when they collide, electrical forces can bind them together, first into pairs, then into small clumps, and eventually into the tiny droplets that make up the cloud of steam we see above the spout. In the lower-pressure environment of space, this condensation occurs at around 180 K. Some substances condense at much higher temperatures For example, suppose we start with a cloud of vaporized iron at a temperature of 2000 K. If we cool the iron vapor to about 1400 K, the atoms will condense into tiny flakes of solid iron. Silicates condense at a slightly lower temperature, so flakes of rocky material will form if a hot gas containing silicon and oxygen cools to about 1300 K.

An important feature of condensation is that when a mixture of vaporized materials cools, the materials with the highest vaporization temperatures condense first. Thus, as a mixture of gaseous iron, silicate, and water cools, it will make iron grit when its temperature reaches 1400 K; silicate grains form when the temperature drops below 1300 K; and when the temperature drops below about 180 K, water crystals form. This is a bit like putting a bowl of hot chicken soup in the freezer. First the fat condenses and freezes, then the broth.

However, the condensation process will not occur if the temperature never drops sufficiently. Therefore, in a region of the solar nebula where the temperature never cools below 500 K, water will not condense but iron and silicates will. By studying how different atoms and molecules would likely condense in the solar nebula, we can estimate an average condensation temperature for each element. These are given in Table 35.1 for the 10 most common elements in the solar nebula. These condensation temperatures are also indicated on thermometer symbols for each element in the periodic table on the inside back cover of the book.

A sequence of condensation would have occurred in the solar nebula because the young Sun heated the inner part of the disk more than the outer parts (Figure 35.5). Iron and silicates, which have high condensation temperatures, could condense almost everywhere within the disk. However, water could condense only where temperatures were low enough beyond the **frost line,** about 4 AU from the Sun. Water, ammonia, methane, and other easily vaporized substances were present as gases in the inner solar nebula, but they could not form solid particles there. A small portion of these substances could still combine chemically with silicate grains, so that the rocky material from which the inner planets formed contained small quantities of water and other gases.

The tiny particles that condensed from the nebula stuck together into bigger pieces in a process called **accretion.** The process of accretion is a bit like building a snowman. You begin with a handful of loose snowflakes and squeeze them together to make a snowball. Then you add more snow by rolling the ball on the ground. As the ball gets bigger, the larger surface area makes it easier for more snow to stick, and the ball rapidly grows.

Similarly, in the solar nebula tiny grains stuck together and formed bigger grains that grew into clumps. The details of this process are not entirely certain, but dust grains can often become electrically charged and then stick to each other because of electrical forces. This can happen as the grains collide with each other and with

TABLE 35.1 Condensation Temperatures of Major Elements

Element	Percentage by Mass in Sun	Main Form	Condensation Temperature*
Hydrogen	71.1%	water (H_2O)	180 K
Helium	27.4%	He	3 K
Oxygen	0.66%	silicates	1300 K
		water (H_2O)	180 K
Carbon	0.25%	methane (CH_4)	80 K
Iron	0.14%	Fe	1400 K
Neon	0.13%	Ne	9 K
Silicon	0.08%	silicates	1300 K
Nitrogen	0.08%	ammonia (NH_3)	130 K
Magnesium	0.07%	silicates	1300 K
Sulfur	0.04%	iron sulfide (FeS)	700 K

*Condensation temperatures depend on what other elements are present. For example, oxygen will combine with silicon to form silicates that condense at high temperature. However, oxygen is so much more abundant than silicon that a large amount remains after the silicon is all condensed. Similarly hydrogen will condense out in water and other molecules, but most of the hydrogen remains as a gas after it combines with all the other elements that it can.

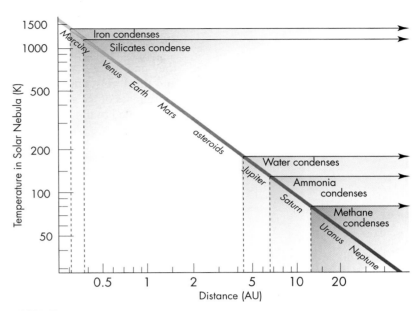

FIGURE 35.5
Temperature decreased with distance from the Sun approximately as shown by the diagonal line. This allowed several common substances to condense beyond the radius indicated in the plot. At smaller distances from the Sun these substances remained gaseous.

gas molecules—this is similar to the buildup of electrical charges on clothes as they tumble in a dryer, making socks stick to shirts. Subsequent collisions, if not too violent, allowed these smaller particles to grow into objects ranging in size from millimeters to kilometers. These larger objects are called **planetesimals**—small, planetlike bodies.

Because the planetesimals near the Sun formed primarily from silicate and iron particles, while those farther out were cold enough that they could incorporate ice and frozen gases, there were two main types of planetesimals: rocky iron ones near the Sun and icy rocky iron ones farther out (Figure 35.6). This, then, explains

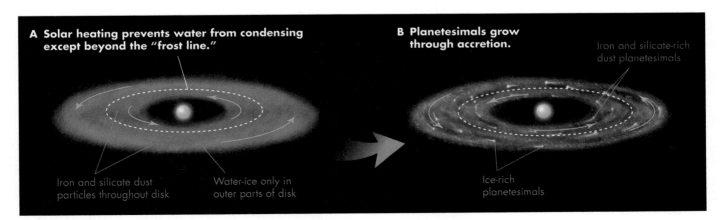

FIGURE 35.6
Depiction of the growth of planetesimals in the solar nebula. (A) Solid particles condensed in the solar nebula where the temperature was low enough, so water and other ices condensed only in outer regions beyond the frost line. (B) The particles stuck together, growing to kilometer-size planetesimals.

the second observation we described at the beginning of this Unit—the change in composition between the inner and outer Solar System.

Planetesimals probably grew much more quickly beyond the frost line because they had more material from which to grow. Hydrogen and oxygen are the first and third most abundant elements, and they can combine to make several times more mass of ice grains than could be made of silicon and iron. This would have contributed to the growth of giant planets in the outer Solar System.

35.4 FORMATION OF THE PLANETS

As planetesimals grew, their masses became large enough that they developed significant gravitational attraction, pulling matter onto them more rapidly and pulling them into collisions with other planetesimals. Computer simulations indicate that some collisions would shatter both bodies, but less violent collisions led to merging into larger bodies.

Merging of the planetesimals increased their mass and thus their gravitational attraction. That, in turn, helped them grow even more massive by drawing planetesimals into clumps or rings around the Sun. Within these clumps, growth proceeded even faster, so that over about 100,000 years, larger and larger objects formed, as depicted in Figure 35.7A. At the same time, their orbits would have steadily grown more nearly circular—averaging out the more eccentric orbits of the individual planetesimals that combined to make them.

Planet growth may have been especially rapid in the outer parts of the solar nebula, where planetesimals could form from abundant ice. Additionally, once a body grew to several times the mass of the Earth, it was able to attract and retain gas by its own gravity. Because hydrogen and helium gas were overwhelmingly the most abundant materials in the solar nebula, planets large enough to tap that reservoir could grow vastly larger than those that formed from solid material only. Thus, Jupiter, Saturn, Uranus, and Neptune may have begun as bodies of ice and rock, but their gravitational attraction resulted in their becoming surrounded by the huge envelopes of hydrogen-rich gases that we see today. The smaller, warmer bodies of the inner Solar System could not capture hydrogen and therefore remained small and lack that gas. This explains the third observation we noted at the beginning of this Unit—that the outer planets have compositions similar to the Sun's.

As planetesimals struck the growing planets, their impact heated them. You can demonstrate how impacts generate heat by hitting a small piece of metal repeatedly with a hammer and then feeling the metal. It will be warmer than it was before you began hitting it. Technically, therefore, impact heating is release of gravitational

Some astronomers argue that planetesimals may not have been necessary for the formation of Jupiter and Saturn. Some computer simulations suggest that such large planets may form directly from slightly denser regions of gas in the solar nebula far from the Sun, where the gas is cold, and gravity can more easily overcome gas pressure.

FIGURE 35.7
Final stages of planet formation. (A) Gravitational attraction between planetesimals causes them to grow in size, although many are left behind in the asteroid belt and Kuiper belt. (B) Planet-sized bodies "sweep up" most of the remaining material orbiting at their distance. The high-speed collisions heat and crater the surface of the body.

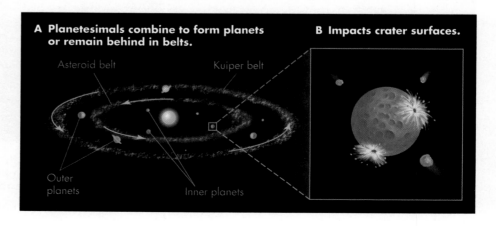

FIGURE 35.8

Differentiation. If a box of mint chocolate chip ice cream warms up, it will "differentiate" as the dense chocolate chips sink to the bottom of the box. Similarly, melting lets much of a young planet's iron sink to its core.

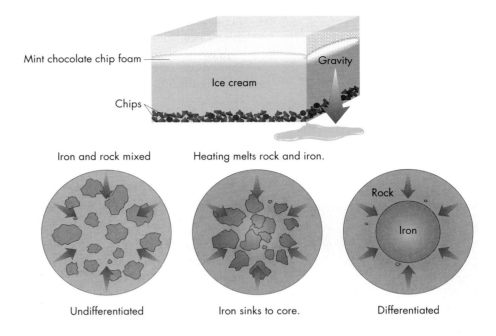

Differentiation

potential energy. Another way to think of this is as an application of the conservation of energy: Energy of motion becomes energy of heat.

For the forming Earth, the impacting objects acted as the hammer, and gravity was the force that drove them into the young Earth (Figure 35.7B). Imagine the vast heating created as mountain-size masses of material slammed into a planet at speeds far greater than the speed at which a bullet is shot from a gun. In addition, many elements in the solar nebula were radioactive and would release heat as they decayed. The heat melted the young planets and allowed matter with high density (such as iron) to sink to their cores, while matter with lower density (such as silicate rock) "floated" to their surfaces in a process called **differentiation** (Figure 35.8). You have seen differentiation at work if you have ever had the misfortune to melt a container of chocolate chip ice cream. When you open the box, you find that all the chips have sunk to the bottom and the air in the ice cream has risen to the top as foam. Besides redistributing material within a planet according to its density, differentiation contributes to the heating. As denser materials are pulled downward through less dense material, heat is generated by their friction.

Although planet building consumed most of the planetesimals, some survive as asteroids and trans-Neptunian objects. It is thought that the asteroid belt consists of planetesimals and their fragments that were unable to assemble into a planet because of constant "stirring" by Jupiter's gravitational force (Unit 43). Jupiter's gravity (and that of the other giant planets) also disturbed the orbits of icy planetesimals in the outer Solar System, tossing some in toward the Sun and others outward in elongated orbits to form the swarm of comet nuclei of the Oort cloud. Those that remain in the disk out to about twice Neptune's orbit form the Kuiper belt (Unit 34).

The moons of the Jovian planets probably were formed from planetesimals orbiting the growing planets. Once the body that would eventually become Jupiter grew massive enough, its gravitational force drew in additional material, and it would have become ringed with debris spinning in a disk in a miniature version of the solar nebula itself. Thus, moon formation was a scaled-down version of planet formation, and so the satellites of the outer planets have the same regularities as the planets around the Sun. All four giant planets have flattened satellite systems in which the satellites (with few exceptions) orbit in the same direction (Unit 47) as the planet spins. This is true even of Uranus, with its oddly tipped axis, so its moons orbit almost perpendicular to the plane of the Solar System.

Many of the satellites of the gas giants are huge themselves. Two exceed Mercury's diameter and would certainly be considered full-fledged planets if they were orbiting the Sun on their own. A few of the larger moons even have atmospheres. However, their small size made their gravity too weak to draw in hydrogen and helium gas, so these moons are composed mainly of rock and ice. This gives them solid surfaces. These distant moons might in the future be ideal bases for studying the planets that have no surface to land on.

35.5 LATE-STAGE BOMBARDMENT

Only a few hundred thousand to a million years were needed to assemble the planets to near their present size from the solar nebula. This is long in the human time frame but only a fraction of a percent of the Solar System's 4.5 billion years. All of the objects within the Solar System are therefore about the same age—the fourth property of the Solar System, mentioned at the beginning of this Unit.

Although the major period of Solar System building was completed relatively quickly, the rain of infalling planetesimals did not cease immediately. It tapered off gradually over hundreds of millions of years, and continues to some degree to this day as comets and asteroids continue to collide with and become part of the planets. Eventually the number of impacts by planetesimals declined enough for the surfaces of the solid bodies to cool and form a crust. Planetesimals continued to collide occasionally, blasting out huge craters, such as those we see on the Moon and on all other bodies in the Solar System with solid surfaces (Figure 35.9). Occasionally an impacting body was so large that it did more than simply leave a crater. For example, one hypothesis proposes that Earth's Moon was created when the young Earth was struck by a Mars-size planetesimal. The extreme tilt of the rotation axis of Uranus may also have arisen from a planetesimal collision. In short, planets and satellites were brutally battered as the large planetesimals collided during the late stages of planet building.

Atmospheres were the last part of the planet-forming process for the terrestrial planets. Unlike the Jovian planets, the inner planets were not massive enough to capture gas from the solar nebula. As a result, they ended up deficient in hydrogen and helium and hydrogen compounds. Moreover, shortly after formation, their surfaces were so hot that they would probably have driven off most gases that might have been present.

Astronomers have proposed several hypotheses to explain how Earth's oceans and atmosphere formed. According to one hypothesis, the gases and liquids on Earth's surface were originally trapped inside the solid material that eventually became the Earth. When that material was heated—in association with volcanic activity (Figure 35.10A) or by the violent impact of asteroids hitting the surface of the young Earth (Figure 35.10B)—the gases escaped and formed our atmosphere.

According to another hypothesis, the gases and liquids were not originally part of the Earth but were brought here by comets after the Earth's surface cooled. Comets are made mostly of a mixture of frozen water and gases. When a comet strikes the Earth, the impact melts the ices and vaporizes the frozen gas. With enough impacts, comets could have delivered sufficient water and gas to form the oceans and atmosphere (Figure 35.10C).

A theory about the origin of atmospheres must also explain why Mercury and the Moon have essentially no atmosphere. Their low masses appear to provide the explanation. First, a small terrestrial planet has a lower escape velocity than a large one (Unit 18), so it cannot hold on to an atmosphere as effectively. Second, smaller

FIGURE 35.9
A portion of the Moon's heavily cratered surface.

Origin of the atmosphere

FIGURE 35.10
Sources of our atmosphere and oceans. (A) Gas from ancient eruptions built some of our atmosphere, and continue to add gases to our atmosphere today. (B) Planetesimals collide with young Earth, melting its surface, which then releases gases. (C) Comets strike young Earth and vaporize. The released gases also contributed to our atmosphere.

planets have less volcanic activity, so there is less volcanic outgassing to start with. If there was no source of replenishment, all terrestrial planets would eventually lose their atmospheres. This is because the molecules in an atmosphere have a wide spread of speeds, and there are always a few moving fast enough in the upper atmosphere to escape. For the Earth, the rate of this loss is so small that it would require far longer than the age of the Earth to lose its atmosphere. For small bodies the loss is much quicker, unless the temperatures are very low, which makes the thermal motions slower. This explains why some moons in the outer Solar System have atmospheres even though their escape velocity is smaller than our Moon's.

One process still had to occur before the Solar System became what we see today: The residual gas and dust between the planets were removed. Just as a finished house is swept clean of the debris of construction, so too was the Solar System. In the sweeping process, the Sun was probably the cosmic broom, with its intense heat driving a flow of tenuous gas outward from its atmosphere—a flow that continues today. As that flow impinged on the remnant gas and dust around the Sun, this debris was pushed away from the Sun to the fringes of the Solar System. Such gas flows are seen in most young stars, and the Sun was probably no exception.

Concept Question 3

How would a planetesimal that survived from the early stages of the Solar System's formation be different today, compared to 4.5 billion years ago? How would it be different from the materials that became planets?

KEY POINTS

- The decay of radioactive materials causes half of the material to break down in each half-life, leaving decay products.
- Radioactive dating and other evidence all indicate that the Solar System formed about 4.5 billion years ago.
- The solar nebula theory explains the main features of the Solar System through the collapse of a rotating interstellar cloud.
- Condensation of gases in the nebula into solid grains only occurred far from the Sun, where the temperature was low enough.
- Only silicates and iron condensed near the Sun, but water ice condensed beyond the "frost line."
- Grains accreted into planetesimals that collided to form planets which reflect the condensation temperature differences.
- Collisions and radioactivity melted the young planets, allowing dense materials like iron to sink to the center (differentiation).
- The forerunners of the gas giants grew large enough to gravitationally capture gas from the solar nebula.
- Atmospheres of the terrestrial planets were probably added later by outgassing and collisions with volatile-rich comets.

KEY TERMS

accretion, 254
condensation, 254
differentiation, 257
frost line, 254
half-life, 251
interstellar cloud, 252
interstellar grain, 253
isotope, 250
meteorite, 252
planetesimal, 255
radioactive decay, 250
solar nebula, 253
solar nebula theory, 252

CONCEPT QUESTIONS

Concept Questions on the following topics are located in the margins. They invite thinking and discussion beyond the text.

1. Cooling of small and large objects. (p. 252)
2. What if silicates condensed at low temperatures. (p. 254)
3. Properties of surviving planetesimals. (p. 259)

REVIEW QUESTIONS

4. How are radioactive elements in rocks used to estimate the age of the Solar System? Why do rocks have different ages?
5. What is the solar nebula? What shape does it have, and why?
6. How do the different types of planets relate to the temperature at which different elements condense?
7. What is the difference between condensation and accretion? What are planetesimals?
8. What is differentiation? Why is it more likely to have occurred in the larger bodies in the Solar System?
9. What hypotheses explain why some terrestrial planets have atmospheres and others do not?

QUANTITATIVE PROBLEMS

10. Use the data from Figure 35.1 to draw a graph showing the ratio of argon atoms to potassium-40 atoms that a rock will contain over a 5.12-billion-year history. Draw a curve through the points. From your curve, estimate the age of a rock that contains half as many daughter argon atoms as the potassium isotope from which they formed. Estimate the age of a rock that contains an equal amount of the two.
11. The common form of uranium (^{238}U) has a half-life of 4.5 billion years. It decays through a sequence of unstable isotopes to lead (^{206}Pb), releasing a number of helium nuclei in the process. (Radioactive decay is the source of most of the helium in Earth's atmosphere today.) Find the ratio of the number of ^{238}U to ^{206}Pb atoms you would expect to find for rocks that last melted 1, 2, 3, and 4 billion years ago.
12. If we estimate that an average planetesimal was 10 km in diameter and had a density of 3 kg/L, approximately how many planetesimals merged together to form the Earth?
13. Silicate molecules can contain different numbers of oxygen atoms per silicon atom. Nearly all of the oxygen in the Earth is locked up in silicates. Based on the data in Table 35.1, calculate how many oxygen atoms there are per silicon atom. (Note that silicon has an atomic mass of 28, while oxygen is 16.)
14. Suppose that the Jovian planets have the same overall composition as the Sun given in Table 35.1, so they are 0.14% iron.
    a. What fraction of their mass would be silicates, assuming that oxygen atoms combined 3-to-1 with silicon?
    b. What would be the mass of rock (silicates) and iron in Jupiter compared to Earth's mass?
    c. What would be the mass of rock and iron in Uranus? What terrestrial planet's mass is this closest to?
15. The conservation of angular momentum (Unit 20) says that a contracting gas cloud must spin faster as it gets smaller.
    a. Suppose the material orbiting today at 1 AU began 1 ly away. By what factor has its orbital radius changed?
    b. Material orbiting at 1 AU today (like the Earth) is moving at 30 km/sec. According to the conservation of angular momentum, how fast was its rotation speed when it was at a radius of 1 ly?
16. A "rule of thumb" in planetary science is that a planet can hold on to a gas for the age of the Solar System if the average speed of molecules in the gas is less than one-sixth the escape velocity of the planet. The average speed of a molecule depends on the gas temperature and the mass of the molecule according to the following equation:

$$V_{gas} = 157 \frac{m}{sec}\sqrt{\frac{temperature}{molecule's\ mass}}.$$

The molecular mass in this formula is found by adding up the atomic masses of its constituents, so hydrogen molecules (H_2) have mass 2, nitrogen molecules (N_2) have mass 28.
    a. What is the average speed of H_2 in Earth's atmosphere if the gas temperature is 300 K? The Earth's escape velocity is 11,200 m/sec. Can the Earth hold on to H_2?
    b. What is the average speed of N_2 in Earth's atmosphere? Can the Earth maintain a nitrogen atmosphere?
    c. What is the average speed of H_2 on Jupiter, where the temperature is 160 K? Jupiter's escape velocity is 59,500 m/sec. Can Jupiter hold on to a hydrogen atmosphere?
    d. If Jupiter were placed at Mercury's distance from the Sun, where the temperature would rise to about 450 K, would it be able to hold on to its hydrogen atmosphere?

TEST YOURSELF

17. One explanation of why the planets near the Sun are composed mainly of rock and iron is that
    a. the Sun's magnetic field attracted all the iron in the young Solar System into the region around the Sun.
    b. the Sun is made mostly of iron, so gas ejected from its surface cooled and condensed to form iron-rich planets.
    c. the high temperatures in the inner part of the Solar System prevented ices and gases from condensing near the Sun.
    d. the Sun's gravity only acts on solid materials, and only iron and rock condensed into a solid.
18. Which of the following features of the Solar System does the solar nebula theory explain?
    a. All the planets orbit the Sun in the same direction.
    b. All the planets move in orbits that lie in nearly the same plane.
    c. The planets nearest the Sun contain only small amounts of substances that condense at low temperatures.
    d. All the planets and the Sun are approximately the same age.
    e. All of the above
19. The numerous craters we see on the solid surfaces of so many Solar System bodies are evidence that
    a. they were hot in their youth with widespread volcanoes.
    b. the young Sun was so hot that it boiled these bodies.
    c. water boiling out of them as the young Sun heated up created hollow pockets in the rock.
    d. they were bombarded in their youth by many solid objects.
    e. all the planets were once part of a single, very large, and volcanically active mass that broke into many smaller pieces.
20. Starting with the Sun and moving outward, what is the first planet that formed beyond the point where water condenses?
    a. Venus
    b. Earth
    c. Mars
    d. Jupiter
    e. Saturn

Other Planetary Systems

36.1 Young Planetary Systems
36.2 Detecting Exoplanets by Doppler Shifts
36.3 Migrating Planets
36.4 New Exoplanet Detection Methods

Learning Objectives

Upon completing this chapter, you should be able to:
- Describe the evidence that other stars form planetary systems as the Sun did.
- Explain the different methods for detecting exoplanets, what they can measure, and the biases of each.
- Summarize properties of exoplanets and how they differ from Solar System planets.
- Describe how planets can migrate and how this may have affected the Solar System.

The solar nebula theory (Unit 35) explains how many of the features of the Solar System probably arose out of its formation process, and the theory predicts that planet formation is a normal part of star formation. Astronomers have long searched for planets orbiting stars other than the Sun. Their interest in planets outside (Greek *exo*) the Solar System, or **exoplanets,** as these distant worlds are called, is motivated not merely by the wish to detect other planets. Astronomers also hope that the study of such systems will help us better understand the Solar System.

Astronomers had long speculated about the likelihood of planets existing around other stars, but it was not until the 1990s that extremely sensitive Doppler shift (Unit 25) measurements confirmed their existence, based on understanding how a planet's gravitational tug affects its star (Unit 17). By 2010 more than 450 exoplanets had been detected. The number is growing rapidly, and new techniques are being developed to detect them. Astronomers are also discovering young star systems that appear to be in their planet-forming phase. Our understanding of the nature of planetary systems is growing rapidly, and although evidence now suggests that planetary systems are common, it may be that ones like our own Solar System are uncommon.

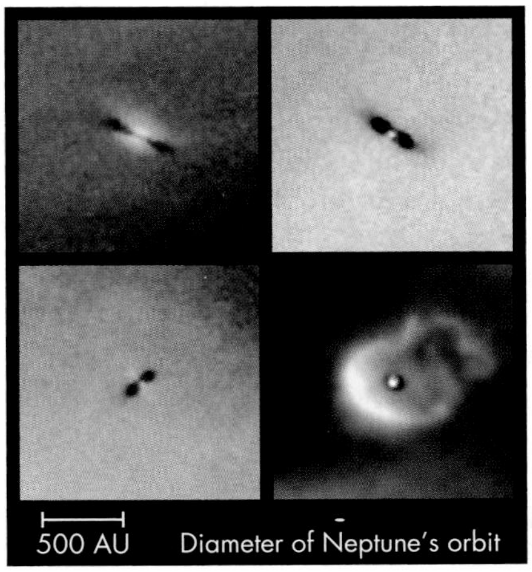

36.1 YOUNG PLANETARY SYSTEMS

If we could find a star just beginning to form, then according to the solar nebula theory it should be surrounded by a disk of dust and gas. The gas and dust would extend out into interstellar space, but in the inner regions—tens of astronomical units across—the disk would be warmed by the young star and heated by collisions.

We can hunt for such systems in interstellar clouds in which we can identify young stars. This has become possible in recent years with high-resolution observations using the Hubble Space Telescope as well as ground-based telescopes employing adaptive optics to correct for the blurring effects of Earth's atmosphere (Unit 32). In addition, a variety of observations at longer wavelengths are able to detect infrared emission from the warm gas and dust. For example, Figure 36.1 shows pictures made with the Hubble Space Telescope of gas and dust disks of young stars forming in the Orion Nebula, one of the nearest regions where stars are currently forming. The pictures show **protoplanetary disks:** disks of dark, dusty material orbiting young stars. The stars at the centers of these disks are probably less than 1 million years old—so young that they have not yet become hot enough to emit much visible light.

FIGURE 36.1
Young stars (the glowing red spot at the center of each image) surrounded by a dark disk of gas and dust. The dust in the disks blocks background light from glowing gas in the Orion Nebula.

LOOKING UP

The Orion Nebula (Messier 42) is easily seen with a small telescope. See Looking Up #6 at the front of the book.

FIGURE 36.2

Disks of dust and perhaps larger debris around three stars. The stars were observed with the Hubble Space Telescope, but the light from the central region was blocked out in order to see the extremely faint light reflected from the disks. (A) HR 4796A is estimated to be less than 10 million years old, and it is surrounded by a ring about the size of the Kuiper belt. (B) Beta Pictoris is somewhat older, perhaps as much as 20 million years old, and is surrounded by a disk seen nearly edge on to us. A secondary disk around Beta Pictoris probably indicates the presence of a planet orbiting at a slight angle to the primary disk. (C) Fomalhaut is much older at more than 400 million years, and it is surrounded by a large ring of dust and debris. Astronomers believed they had detected a planet orbiting just inside the ring, but subsequent observations suggest it is an orbiting cloud of debris.

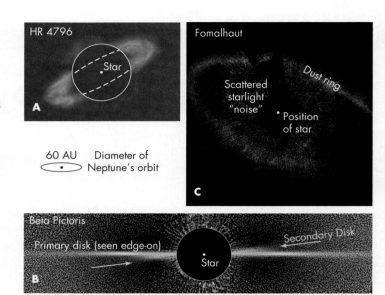

Concept Question 1

If you were observing the Solar System from a distant star, what would the asteroid belt and Kuiper belt look like? What wavelengths would be best for observing them?

The orbiting material has become concentrated enough that it blocks light from the Orion Nebula behind it.

As the gas and dust in a disk continue to condense, they may eventually form planets. Figure 36.2 shows rings of dust and debris orbiting several older stars, ranging from about 10 to about 400 million years old. Material orbiting a star reflects only a tiny fraction of the star's light, so the camera places a disk in front of the star to block out its light. The rings of dust seen in Figure 36.2A and C may be similar to the Kuiper belt in the Solar System, and the lack of dust seen closer to the star may indicate that planets have formed in the inner part of the system and have swept up most of the remaining dust. Figure 36.2B shows another protoplanetary disk or ring around the star Beta Pictoris, which may be as old as 20 million years. This disk is seen almost exactly edge-on, so it is not possible to directly see if there is a hole in its center. However, the Hubble Space Telescope image shows that some of the dust is tilted, making a secondary disk. Models suggest that this tilted disk is probably caused by a planet orbiting at a slight angle to the rest of the disk—much as the planets in the Solar System do not all orbit in precisely the same plane.

These images confirm that processes occur around other stars that are similar to the solar nebula theory; but detecting the planets themselves requires a different approach. Once the scattered material is locked up into a body as concentrated as a planet, it is difficult to image it directly.

36.2 DETECTING EXOPLANETS BY DOPPLER SHIFTS

INTERACTIVE

Exoplanets

Astronomers also refer to exoplanets as extrasolar planets.

Directly observing exoplanets is extremely difficult because the planets are so small compared with the stars they orbit, and they shine primarily by light reflected from their stars. Moreover, they are so far away that from our perspective they are separated from their stars by a very small angle, so they are drowned out by the light of their stars. However, there are other ways of detecting a planet's presence. The first discoveries were made in the 1990s as astronomers developed techniques to detect the gravitational effects of exoplanets on the stars they orbit.

As a planet orbits its star, the planet exerts a gravitational force on the star just as big as the star exerts on the planet as a result of Newton's law of action–reaction

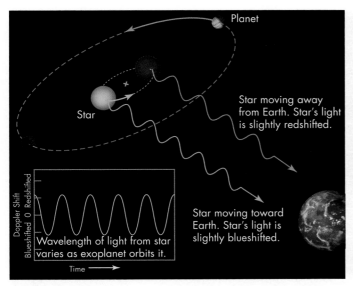

FIGURE 36.3

As an unseen planet orbits a star, the star's position "wobbles." This produces a changing Doppler shift. From the period of the wobble, the planet's distance from the star can be found. From the amplitude of the wobble, the planet's mass can be estimated.

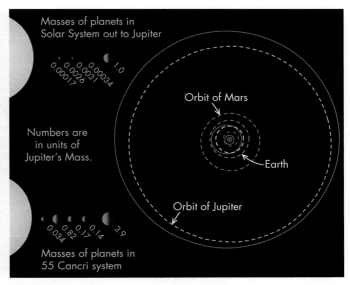

FIGURE 36.4

The 55 Cancri system contains five known planets around a star that is very similar to the Sun. The estimated masses (compared to Jupiter) for these planets and their orbits are compared with the five innermost planets in the Solar System.

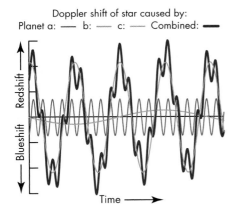

FIGURE 36.5

Astronomers can detect more than one planet by determining the combination of orbits that produce the star's overall pattern of Doppler shifts. In the example illustrated here, planet a is closest and planet c is farthest from the star; a and c have the same mass, and b is 5 times more massive

(Unit 15.3). That force makes the star's position wobble slightly, just as you wobble a little if you swing a heavy weight around you. This allows planets to be detected by the **Doppler shift method,** so named because the wobble creates a changing Doppler shift (Unit 25) in the star's light (Figure 36.3). From that shift, and its change over time, astronomers can deduce the planet's orbital period, mass, and distance from the star. Astronomers have discovered more than 500 exoplanets (as of mid 2013) by this technique.

Figure 36.4 illustrates one of the more complex systems yet found, containing five planets orbiting the star 55 Cancri. As each planet orbits the star, it tugs the star with a regular repeating pattern; after all of the planets have completed at least one orbit, it is possible to reconstruct their masses and distances. An illustration of how the planetary tugs combine for a hypothetical star with three planets is shown in Figure 36.5. In the case of 55 Cancri, despite the star having a mass similar to the Sun's, the planetary system is very unlike our own. Three of the planets—all giants by Solar System standards—orbit closer to their star than Mercury orbits the Sun. The closest takes only 2.8 days to orbit the star! The fourth planet out from the star has an orbit at roughly 1 AU, but its mass is more than twice Neptune's. At this distance, its surface temperature may not be much different from Earth's, but its orbit is so elliptical that its distance from 55 Cancri ranges from about Earth's distance from the Sun to less than Venus's distance from the Sun.

Almost all of the Jovian-mass exoplanets detected by the Doppler technique have orbits closer to their star than Jupiter orbits the Sun, and dozens of them orbit closer to their star than Mercury. Based on our understanding of planet formation in the Solar System (Unit 35), we would not expect gas giants to form so close to a star, so astronomers are examining scenarios that might cause a planet's orbit to migrate inward (Section 36.3). We should also keep in mind that massive planets close to a star are easier to detect by the Doppler method. Figure 36.5 illustrates how a planet orbiting farther out (planet c) produces a weaker and more gradual change in the Doppler shift than the same-mass planet orbiting close to the star (planet a). Thus, the Doppler shift method gives us a biased sample of planets.

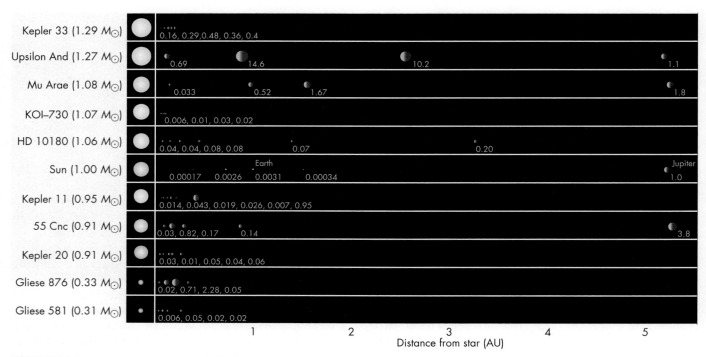

FIGURE 36.6
Comparison of the orbital radii and relative sizes of exoplanetary systems with the Solar System. Systems with four or more confirmed exoplanets are shown, organized according to the mass of the star that they orbit. The sizes shown for the exoplanets indicate their approximate relative size; at this scale, the stars would be about 4 times larger than shown. The numbers indicate the mass of each planet in units of Jupiter's mass.

The first exoplanets discovered were orbiting a pulsar (Unit 68), and several more orbiting pulsars have been found recently. Pulsars generate radiation with a regular "pulse" as they spin. An orbiting exoplanet affects the rate of pulse arrival in the same way as the Doppler shift. These exoplanets probably formed from debris left after the collapse and explosion of the star that became the pulsar.

Concept Question 2

Given their masses and orbital radii, how similar or different do you suppose the exoplanets orbiting Gliese 581 might be in comparison to Solar System planets?

Concept Question 3

How would our Solar System have differed if Jupiter were massive enough to be a brown dwarf?

Astronomers have detected exoplanets orbiting a wide variety of stars, including stars with masses ranging from 0.1 $M_\odot$ to more than 4 $M_\odot$. Planets have even been detected around some "dead stars" called pulsars (Unit 68), and substellar objects known as **brown dwarfs.** A brown dwarf is an object that is intermediate in mass between a star and a planet. Its mass is too low to fuse hydrogen like a true star, but it is so massive that it undergoes nuclear fusion of some isotopes during its formation (see Unit 61). Astronomers are uncertain how low the mass must be to yield a planet instead of a brown dwarf. Computer models suggest that planets have less than about 13 times Jupiter's mass.

Most exoplanet searches have focused on stars with masses similar to the Sun's. Figure 36.6 shows the planetary systems in which four or more exoplanets have been discovered so far. The figure indicates the estimated mass of each of these planets and compares them to planets in the inner part of the Solar System. Most of these planets are probably gas giants. The small dot shown close to the star Gliese 581 in Figure 36.6 has a mass about 2 times the mass of Earth. It orbits only 0.03 AU from the star, taking just 3.1 days to complete an orbit. Gliese 581 is much less luminous than the Sun, but even so, the temperature at 0.03 AU is probably too hot for the planet to have liquid water on its surface.

The least massive exoplanet yet found by the Doppler technique is a planet about 15% more massive than Earth, orbiting in the nearby Alpha Centauri star system. This planet orbits just 0.04 AU from its star, so it too is certainly very hot. This is an intriguing discovery, though, because it demonstrates that planets can form even in multiple-star systems despite the complex gravitational interactions it has had with all the stars.

Most of the exoplanetary systems discovered so far look quite different from the Solar System, but this does not necessarily imply that our model for planet formation is wrong. What is clear, however, is that we need a mechanism to explain how large planets can move close to a star after they form. New hypotheses have been

FIGURE 36.7
Sketch of a large young planet migrating inward due to gravitational torques between it and the protoplanetary disk.

proposed that explain this, and these may clear up some mysteries about the history of our own Solar System.

36.3 MIGRATING PLANETS

Finding gas giant planets so near to stars has led astronomers to reexamine their understanding about how planetary systems form and evolve. Planets interact with each other as well as with the disk of gas and planetesimals around a young star, and those interactions may lead to dramatic changes.

Computer simulations of the early stages of a planetary system show that gravitational interactions between the young planets and leftover material in the protoplanetary disk can shift the planets' orbits either inward or outward. The young planet "eats" into the material like a cosmic Pac-man, being drawn in the direction where more material is present (Figure 36.7). To be more precise, there is an exchange of angular momentum between the planet and disk material as they tug on each other gravitationally.

Other simulations show the importance of gravitational interactions between massive young planets. For example, some simulations suggest that the Solar System initially formed in a more compact configuration with Neptune closer to the Sun than Uranus (Figure 36.8A). Gravitational interactions between the planets gradually shifted their orbits over hundreds of millions of years until a close encounter between Saturn and Neptune threw the planets into highly elliptical orbits (Figure 36.8B). Further interactions with planetesimals in the disk caused the planets' orbits to settle back into their present nearly circular configuration after about 100 million more years (Figure 36.8C). In this process, planetesimals were sent sailing into the outer Solar System, forming the Kuiper belt, and into the inner Solar System, where perhaps they added water and other volatiles to the young terrestrial planets. This may also help to explain the indications that there was a period of intense impact cratering hundreds of millions of years after the Moon formed (Unit 39). Some of the simulations even suggest that there were other Uranus-mass planets ejected from the Solar System by this process.

Migration of planets could have major consequences for smaller planets. For example, if a giant planet migrates inward toward its star, it might swallow smaller, Earth-size planets as it passes them, or send them careening into chaotic orbits. However, other computer simulations suggest that if the giant exoplanet migrates inward early in the planetary system's formation, enough material may remain for terrestrial planets to form after the giant planet completes its migration.

Many of the massive exoplanets detected so far also have very elliptical orbits, rather than the nearly circular ones in our own system. Planets might be left in highly elliptical orbits if too little disk material remained after a major planet–planet interaction to recircularize the orbits. Having a massive planet on a very elliptical orbit would not bode well for the survival of Earth-like planets in these systems. As a massive planet sweeps into the inner portion of a star system, its strongly varying gravitational influence is likely to disturb the orbits of other planets, either ejecting them from the system or causing them to fall into their star. There is possible evidence suggesting that this fate may have befallen some exoplanets. A number of the stars with exoplanets are appreciably richer in iron than our Sun. This could occur if they swallowed Earth-like planets and vaporized them. The iron from the vaporized planet's core then enriches the star's atmosphere.

The higher abundance of metals in stars' atmospheres has an alternative interpretation: it may be that stars that formed out of interstellar gas with a greater fraction of heavy elements are more likely to form planets. Astronomers are finding that the probability of detecting planets around a star directly correlates with the

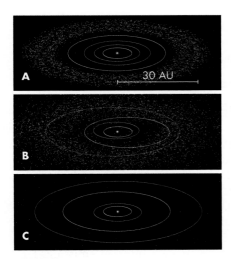

FIGURE 36.8
Illustration of a possible evolution of the young Solar System, based on a simulation carried out by astronomers in Nice, France. (A) Shortly after the major planet formed in a compact configuration, Neptune orbited closer than Uranus. (B) After about 800 million years, the orbits grow quite elliptical. (C) In another 100 million years, the orbits recircularize as they scatter most of the remaining disk material into collisions with the planets or into the far outer reaches of the Solar System.

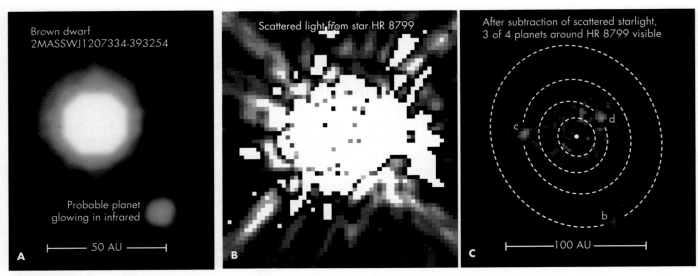

FIGURE 36.9
(A) The first image of an exoplanet was made at infrared wavelengths with the 8-meter Very Large Telescope (VLT) in Chile using adaptive optics. The exoplanet, which is seen glowing red from its infrared emission, is about 50 AU from the star, which is actually a low-mass "brown dwarf." (B) A Hubble Space Telescope infrared image of the star HR 8799 shows the challenge caused by scattered light from the star. (C) Careful analysis of the scattered light allows most of it to be removed, revealing three of the giant planets known to be orbiting the star.

level of iron in the star's atmosphere. Since the amount of metals in stars is generally increasing over time as stars build up heavy elements (Unit 72), planets may be becoming more common as our Galaxy grows older.

36.4 NEW EXOPLANET DETECTION METHODS

Several techniques for detecting exoplanets have been implemented in recent years, and these are helping to broaden our understanding of the planet formation process. Direct imaging techniques have so far succeeded in detecting about two dozen exoplanets. This method works best for exoplanets far from their star. For example, the first found is about 50 AU from its star (Figure 36.9A), detected from its own infrared emission. Figure 36.9B shows how challenging it is to identify planets orbiting closer to their star amid the scattered light from a star. However, by careful analysis of the light pattern, astronomers were able to spot three planets that had been previously detected by the Doppler shift method (Figure 36.9C).

Another technique that has been used to detect exoplanets takes advantage of the ability of gravity to bend the path of light (Unit 27). If we observe a distant star, and a nearer star passes in front of it, the light from the background star can be focused in our direction, making the background star appear to brighten (Figure 36.10). If a planet orbits the nearer star, the planet can produce a smaller brightening event. The gravitational force acts as a lens to

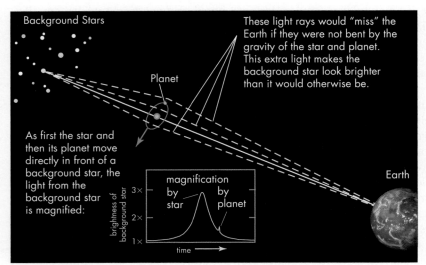

FIGURE 36.10
Detecting an exoplanet by gravitational lensing. When one star passes precisely in front of another, the nearer star can bend the light from the background star. This can focus the light of the background star, making it look brighter. A planet in orbit around the nearer star can cause a smaller brightening event.

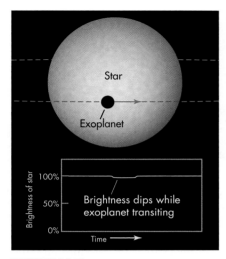

FIGURE 36.11
Detecting an exoplanet by the transit method. If a planet's orbit happens to be at the correct angle, it will pass in front of (transit) its star, diminishing the light we see. A planet with 10% the radius of its star, as shown, blocks 1% of the star's area.

focus the light of the distant star, so this is known as the **gravitational lensing method.** This method has detected fewer than 20 exoplanets so far, but it has the potential for detecting very low mass objects. The method has the disadvantage that detections are likely to be a once-in-a-lifetime chance alignment, so there is no opportunity to observe the exoplanet a second time for confirmation or follow-up.

A future method for detecting exoplanets involves detecting the sideways "wobble" of the central star. This is called the **proper motion method;** it uses the same effect studied with the Doppler shift method, but it observes the sideways shift in the sky. The sideways shift is larger the farther away a planet is from the star, so more-distant planets produce larger shifts. Because orbits this far out take a long time to complete, though, the method will require many years of high-precision position measurements to yield results. Spacecraft missions are currently being planned that could make measurements that might even detect Earth-mass planets at Earth-like distances around nearby stars.

While all of these methods can expand the range of exoplanets detected, by far the most promising development in detecting exoplanets in recent years is the **transit method.** In this method, a planet is detected as it passes in front of its star, temporarily dimming the star's light, as illustrated in Figure 36.11. Like the Doppler shift method, this method also favors the detection of large planets close to the star because small planets block little light, and it is unlikely that a distant planet's orbit will line up precisely to pass in front of the star. The transit method had detected only about 10% as many exoplanets as the Doppler method before the 2009 launch of NASA's *Kepler* spacecraft. This spacecraft measured light from about 150,000 stars, and identified more than 3300 possible transiting planets, about 130 of which have been confirmed by mid-2013.

Transiting exoplanets allow us to determine their radii from the fraction of starlight they block. We can therefore determine the size of the exoplanet relative to its star, and from our knowledge of the type of star we can estimate the actual radius. Sometimes we can combine transits and Doppler shift measurements to build a more detailed picture of the planets. Furthermore, we know that systems that exhibit transits must be oriented so that the plane of the planets' orbits is edge on to our line of sight, and that helps us determine more accurate masses. In the Doppler shift method, we usually do not know the orientation, so the mass estimates are uncertain because the amplitude of the "wobble" we detect only measures the component of motion along our line of sight to the star. As the angle of the orbit becomes more nearly edge on, the Doppler shift wobble becomes larger. We use the size of the wobble to estimate a planet's mass, so the Doppler shift method may underestimate the planet's mass if the system is not edge on.

With estimates of both the radius and mass of a planet, we can obtain a critical piece of information—the exoplanet's density. The current data for exoplanets with both transit and Doppler shift measurements is plotted in Figure 36.12. Lines are also shown corresponding to the densities of the highest and lowest density planets in the Solar System, Earth and Saturn.

For the most part, exoplanets have densities similar to the range of densities in the Solar System, but there are a number of unusual features. For example, the smallest exoplanets with these measurements are still considerably more massive than the Earth. These are probably terrestrial planets, but much larger than the Earth. There are even some the size of Jupiter with densities like the Earth's. Astronomers suspect that these are not terrestrial in character, but so massive that they compress their atmospheres to very high densities. In fact, the most massive exoplanet in the plot has the same radius as Jupiter despite a mass more than 10 times larger. At the opposite

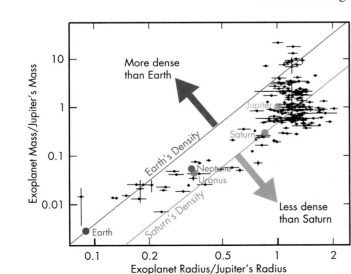

FIGURE 36.12
Plot comparing exoplanet masses to their radii. The values for the five largest planets in the Solar System are shown for comparison.

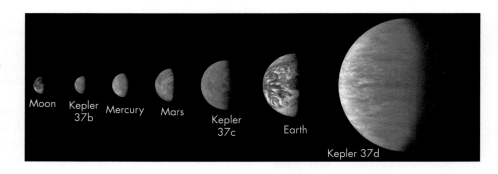

FIGURE 36.13
Artist's impression of the smallest exoplanet found by the *Kepler* spacecraft around a star that also hosts a nearly Earth-sized exoplanet and what appears to be a very large terrestrial planet. Solar System bodies are shown for comparison.

extreme, there are many massive exoplanets with densities even less than Saturn's. These are all orbiting very close to their star, suggesting that the stellar heating has caused their atmospheres to expand to a larger size than they would have if they orbited far from the star. While most exoplanets detected by transits are relatively large, one planet was detected that is not much bigger than the Moon. The same star also has a "supersized" Earth orbiting it as illustrated in Figure 36.13.

When a planet passes in front of its star, astronomers also have an opportunity to study the planet's atmosphere through spectroscopy. Some of the star's light passes through the planet's atmosphere, so that gas in the planet's atmosphere imprints absorption lines on the star's spectrum. For example, a low-density gas giant transiting in front of the star HD 209458 was observed spectroscopically as it transited, then the star's own spectrum was subtracted. This revealed that the element sodium in the exoplanet was at a level consistent with Jupiter's atmospheric composition. Oxygen and water vapor were also detected, substances important for life as we know it. However, this exoplanet has an extremely short orbital period of 3.5 days. Using Newton's version of Kepler's third law (Unit 17), astronomers deduce that the planet orbits a mere 0.045 AU from its star—roughly one-eighth the distance at which Mercury orbits our Sun. Analysis of the spectral lines indicates that the atmosphere is very hot—about 10,000 K (18,000°F). The gases seen are "boiling" off a planet that is almost certainly uninhabitable.

The features so far found for other planetary systems suggest that planetary system formation can take a variety of paths. So far, the Solar System appears to be exceptional in its layout and in the stability of its planets' orbits. However, the bias of our detection methods toward massive planets close to their stars keeps us from developing a more complete picture. There may be many planetary systems more similar to our own, waiting to be discovered.

> **Concept Question 4**
>
> How do you suppose the spectral lines of the gases in a planet might be altered when the gases are at a high temperature?

> **Clarification Point**
>
> Although the majority of planetary systems discovered so far look quite different from the Solar System, we cannot yet conclude that the Solar System is unusual. Most stars like the Sun have not had any exoplanets detected, but even if they had systems like our own, our methods are not yet good enough to detect them.

KEY POINTS

- Astronomers have detected material around some young stars consistent with the early stages of solar nebula theory.
- By measuring the Doppler shift of stars, it is possible to detect a "wobble" due to the gravitational pull of orbiting exoplanets.
- Hundreds of exoplanets have been detected, and several exoplanets have been detected around some stars.
- Most of the planets detected so far are giant planets, orbiting close to their star, unlike in the Solar System, but the Doppler method is biased toward detecting such planets.
- Some exoplanets have been imaged directly, but this is difficult because of their small separation from and dimness relative to the star.
- Astronomers are detecting some exoplanets and measuring their sizes as they transit in front of their star, dimming its light.
- Exoplanets have also been detected by gravitational lensing of background stars.

KEY TERMS

brown dwarf, 264
Doppler shift method, 263
exoplanet, 261
gravitational lensing method, 267
proper motion method, 267
protoplanetary disk, 261
transit method, 267

CONCEPT QUESTIONS

Concept Questions on the following topics are located in the margins. They invite thinking and discussion beyond the text.

1. Appearance of Solar System's belts from afar. (p. 262)
2. Characteristics of exoplanets around a small star. (p. 264)
3. What if Jupiter were a brown dwarf? (p. 264)
4. Effect of high temperatures on an exoplanet's spectrum. (p. 268)

REVIEW QUESTIONS

5. What observations indicate that planets are currently forming around other stars?
6. Why is it so difficult to directly "see" a planet orbiting another star? Can you think of an analogy using objects here on Earth?
7. What different methods have astronomers used to detect planets outside the Solar System?
8. How does Newton's third law apply to our detection of exoplanets? Why is it easier to detect Jupiter-like planets than Earth-like planets?
9. What aspects of the exoplanets discovered so far cause us to question whether our Solar System is "typical" or not?
10. What is planet migration?

QUANTITATIVE PROBLEMS

11. Suppose a Jupiter-size exoplanet (radius 71,500 km) passed in front of a Sun-size star (radius 696,000 km). What percentage of the star's light would be blocked by the exoplanet?
12. The Doppler method for finding planets can be used to see how fast the Sun moves in response to a planet orbiting it.
    a. Using Kepler's third law ($P^2 = a^3$, for P in years and a in AU), find the period of the orbit and its circumference if the semimajor axis is 0.05 AU, 0.5 AU, or 5 AU.
    b. Calculate how fast a planet would be orbiting the Sun in meters per second at each of these three distances.
    c. The Sun also "orbits" the point of balance between the Sun and the planet (although this point may be inside the Sun). According to Newton's third law, the Sun's speed around this point must be smaller than an orbiting planet's in proportion to the ratio of their masses. What is the Sun's speed in each case above if the orbiting planet has Jupiter's mass? the Earth's mass?
13. When planets orbit a star, both orbit their common "center of mass" (Unit 17), which lies at a point between them closer to the center of the more massive object in proportion to the ratio of their masses. (If the planet is 1/10th as massive as the star, the point will be 10× farther from the planet than from the star.)
    a. By how many AU does the Sun shift in each of the cases in Problem 12?
    b. If viewed from 30 light-years away, how large will its angular shift be? (Refer to the angular size formula in Unit 10.)
14. Gliese 581 is a star with a mass 31% of the Sun's mass, and it generates only about 1.3% as much light as the Sun. Interestingly the star may be host to upwards of six exoplanets (although only three have been confirmed and another is probable).
    a. Use Newton's version of Kepler's third law to find the semimajor axis for exoplanets with orbital periods of 3.15, 5.37, 12.9, and 66.6 days.
    b. Compare the brightness of light received from Gliese 581 at the distances found for the four exoplanets to the brightness of light received at Earth from the Sun. (Use the inverse-square law—Unit 21.2.)
15. If Neptune originally formed at 20 AU and then migrated to its present distance of 30 AU, how much was Neptune's orbital speed reduced during the migration?

TEST YOURSELF

16. One of the disadvantages of the transit method for detecting exoplanets is that
    a. it only works for exoplanets larger than Jupiter.
    b. the orbital plane must be viewed nearly edge on to see a transit.
    c. the exoplanet's atmosphere can block the parent star's light.
    d. the method only works using space-based telescopes.
17. Given the characteristics of exoplanets discovered so far, what are the most likely conclusions we might draw about how the solar nebula theory should be modified for the formation of other planetary systems?
    a. Terrestrial planets need not be composed of rock and iron.
    b. Massive planets usually are not composed of light elements.
    c. Not all interstellar clouds spin fast enough to form planets.
    d. Planets' orbits may change size within the nebula.
    e. Small planets do not form in most planetary systems.
18. The Doppler shift method for detecting the presence of exoplanets is best able to detect
    a. massive planets near the star.
    b. massive planets far from the star.
    c. low-mass planets near the star.
    d. low-mass planets far from the star.
19. To date, what is the most successful (in terms of number) method for detecting exoplanets?
    a. Doppler shift method
    b. Proper motion method
    c. Gravitational lensing method
    d. Transit method
20. What kinds of information have been determined by observing planetary transits?
    a. The planet's diameter
    b. The composition of the planet's atmosphere
    c. A more exact mass for the planet
    d. All of the above.

PART 3

UNIT 37

The Earth as a Terrestrial Planet

37.1 Composition of the Earth
37.2 Heating of the Earth's Core
37.3 The Earth's Dynamic Surface
37.4 The Earth's Magnetic Field

Learning Objectives

Upon completing this Unit, you should be able to:
- List the elements and their compounds that make up most of the Earth's mass.
- Explain how seismology is used to investigate the Earth's interior and describe what it shows about the Earth's structure.
- Explain what heats Earth's interior, and why larger bodies keep their heat longer.
- Describe the processes involved in plate tectonics and how the Earth's surface features have changed as a result.
- Explain the source of Earth's magnetic field and the evidence that it reverses.

In the simplest terms, Earth (Figure 37.1) is a huge ball of rock and metal. With a diameter of 12,742 kilometers (7918 miles) and a mass of 5.97×10^{24} kilograms, it is the largest of the "rocky" bodies that inhabit the inner part of the Solar System. It is followed in size by Venus, Mars, Mercury, the Moon, the asteroid Ceres, and then the other asteroids.

Observations of the terrestrial planets and the Moon have been carried out for centuries, revealing many similarities and many differences between terrestrial worlds of different sizes and at different distances from the Sun, as we began to examine in Units 1 and 34. Of course, the Earth is the best-studied planet, so we begin our exploration of the planets with an astronomer's view of our home.

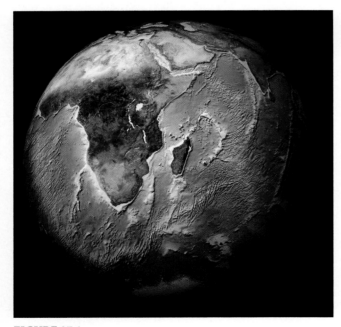

FIGURE 37.1
The solid Earth. A reconstructed image, with the atmosphere and oceans removed.

37.1 COMPOSITION OF THE EARTH

Analysis of the Earth's surface rocks shows that the two most common elements in them are oxygen and silicon, which combine to make **silicates.** For example, ordinary sand (particles of the mineral quartz) is nearly pure silicon dioxide (SiO2). Other types of rocks contain silicon and oxygen in different proportions along with a variety of other elements.

The composition of the Earth's interior is much more difficult to determine. We described in Unit 34 how we can deduce much about the Earth's interior by studying its density. Averaged over the whole planet, the density of the Earth is 5.53 kilograms per liter. This is greater than the density of the crust, which is composed primarily of silicates with an average density of about 3.0 kilograms per liter. We can conclude that the interior parts of the Earth must therefore have a density greater than 5.53 kilograms per liter, which when averaged together with the crust gives the Earth's overall measured density. That by itself does not tell us what lies inside the Earth; but if we examine the common crustal elements, the only one with a high density is iron, which has a density of 7.9 kilograms per liter, making

TABLE 37.1 Composition of the Earth		
	Percentage by Mass	
Element	Whole Earth	Crust
Iron	32%	6%
Oxygen	30%	45%
Silicon	16%	27%
Magnesium	14%	2.5%
Nickel	1.8%	0.008%
Sulfur	1.7%*	0.03%
Calcium	1.6%	5%
Aluminum	1.5%	8%
Chromium	0.5%	0.01%
Potassium	0.2%	2%
Sodium	0.2%	2%

*There is considerable disagreement about the amount of sulfur, with estimates ranging from half to twice this value.

it a likely candidate. Moreover, iron (and the comparably dense element nickel) is underrepresented in the Earth's crust compared to the composition of the Sun (Unit 34), so it makes sense that these dense metals sank to the center of the Earth during its early history. Models based on the suspected composition of the material in the solar nebula from which the Earth formed yield the overall percentages for the Earth listed in Table 37.1. Measured values for the same elements in the crust are shown in the table as well. Obtaining a more detailed picture requires that we examine the Earth's interior more directly.

If we ask how the Earth's core can be studied, your first reaction might be to say, "Why not drill a very deep hole and take a look or pull out samples?" However, the deepest hole yet drilled into the Earth penetrates only 12 kilometers, a mere scratch when compared to the Earth's 6380-kilometer radius. If the Earth were an apple, the deepest drilled holes would not have broken the apple's skin. Thus, to study the Earth's core, we must rely on indirect means such as earthquakes.

When earthquakes shake and shatter rock, they generate **seismic waves** that travel outward from the location of the quake through the body of the Earth. The speed of the waves depends on the properties of the material through which they move. We can measure that speed by carefully timing the arrival of seismic waves at places far from the quake that created them. Analysis of the speed along different paths then reveals to scientists a picture of the Earth's interior. Thus, seismic waves allow us to "see" inside the Earth much as doctors use ultrasound waves to "see" inside our bodies (Figure 37.2). You can use a similar, though obviously much cruder, technique to locate wall studs by thumping areas of a plaster wall with your knuckle and listening for differences in the character of the sound waves.

Seismic waves in the Earth are of two main types: **P waves** and **S waves.** P waves form as matter in one place vibrates against the adjacent matter ahead of it, producing compression and decompression in an oscillating fashion (Figure 37.3A). P waves are like sound waves, and they travel easily through both solids and liquids. By contrast, S waves form as matter is shaken from side to side, like a wriggle in a shaken rope or a long spring (Figure 37.3B). In a liquid, material easily slips past adjacent matter, and S waves do not propagate. Thus, S waves can travel only through solids.

Because of this difference in wave behavior, if a laboratory on the far side of the Earth from an earthquake detects P waves but no S waves, the seismic waves

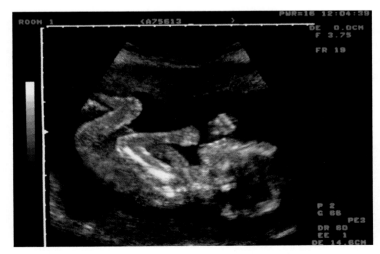

FIGURE 37.2
A sonogram allows a doctor to "see" inside a patient. Here a developing fetus can be "seen" by reconstructing how sound waves reflect off surfaces within the mother's abdomen.

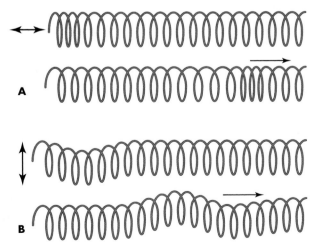

FIGURE 37.3
(A) A P wave can be made when you push and pull a spring, compressing and stretching it. (B) An S wave can be made in a spring when you shake it side to side.

FIGURE 37.4
P and S waves move through the Earth, but the S waves cannot travel through the liquid core.

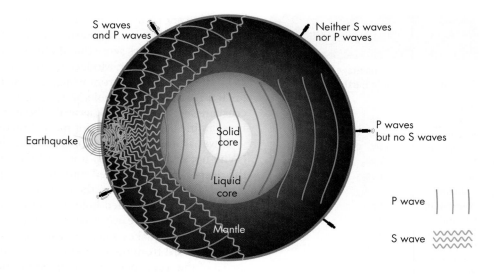

must have encountered a region of liquid on their way from the earthquake to the detecting station (Figure 37.4). Observations show precisely this effect, from which we infer that the Earth has a liquid core. The increasing density toward the interior also causes the waves to curve, much as light waves are refracted when they enter a denser medium (Unit 30), but the core focuses the P waves traveling through it toward the opposite side of the Earth. This leaves a region about two-thirds of the way around the Earth without either P waves or S waves. Finally, the speed of the waves is related to the density and composition of the materials that the waves pass through, allowing seismologists to map out the Earth's internal structure.

Analysis of the seismic waves from many quakes indicates that the Earth's interior has four distinct regions. Figure 37.5 illustrates these different layers and their relative sizes. The surface layer is a solid, low-density **crust** about 20 to 70 kilometers (12 to 43 miles) thick and composed of rocks that are mainly silicates. Beneath the crust is a region of hot, but not quite molten, rock called the **mantle**. This region displays properties indicating that it may be composed largely of the mineral olivine, an iron-magnesium silicate. This mineral gets its name from its olive color. Pieces of it are sometimes carried to the surface in the lava that erupts from volcanoes (Figure 37.6). The mantle extends almost halfway to the Earth's center and, despite not being liquid, is capable of slow motion when stressed, much as a wax candle can be bent by steady pressure when it is warm. Beneath the mantle is a region of dense liquid material, probably a mixture of iron, nickel, and perhaps sulfur, called the liquid or **outer core**. At the very center is a solid or **inner core**, probably also composed of iron and nickel.

P waves are also known as primary or pressure waves. S waves are also known as secondary or shear waves.

Seismic waves

FIGURE 37.5
The Earth's internal structure.

FIGURE 37.6
A fist-sized piece of olivine (the greenish rock) in a lava sample.

You can see from this discussion that the Earth's interior structure is layered so that the densest material (the iron and nickel) is at the center and the lower-density, lighter materials (silicates) are near the surface in the crust and mantle. This separation of materials by density probably occurred while the Earth was forming and still in a molten state—high-density materials sinking while low-density materials floated to the surface—in a process of differentiation (Unit 35.4).

You may be puzzled over why there is a solid core inside a liquid core at the Earth's center. If it is hot enough to melt part of the interior of the Earth, why is the very center not liquid as well? The solid core is not cooler; rather, it has a higher melting temperature because it is under greater pressure. At very high pressures, a liquid material may solidify. You can understand why this happens in the following way. For a solid to form, the atoms composing it must be able to link up to their neighbors to form rigid bonds. Heating makes the atoms move faster, breaking their bonds, so the material becomes liquid. However, if the material is highly compressed, the atoms may be forced so close together that, despite the high temperature, bonds to neighbors may hold and keep the substance solid.

The compression needed to solidify the Earth's inner core comes from the weight of the overlying material. The thousands of kilometers of rock above the Earth's deep interior generate enormous pressure there, which squeezes what would otherwise be molten iron into a solid.

37.2 HEATING OF THE EARTH'S CORE

The Earth's interior is much hotter than its surface (a fact that figures in folklore and theology). Anyone who has seen a volcano erupt can hardly doubt this. As you descend from the surface, the temperature rises about 2 K with every 100 meters in depth. If this temperature increase continued at the same rate all the way to the center, the Earth's core would be 120,000 K! However, by measuring the amount of heat escaping from the deep interior, and from laboratory studies of the properties of heated rock, geologists estimate that the temperature in the core of the Earth is "only" about 6500 K—easily able to melt rock and iron.

What makes our planet's heart so hot? Most scientists think that the Earth formed hot. As we discussed in Unit 35, the Earth formed from many smaller bodies drawn together by their mutual gravity. The impact of these colliding and accreting bodies generated heat. After the bombardment stopped, the Earth's surface cooled, but its interior has remained hot. The Earth has behaved much like a baked potato taken from the oven, cooling on the outside but remaining hot inside because heat leaks only slowly from its interior to its surface. The rate of this heat loss depends on the total surface area exposed to the cold of space. Compared to its total volume of rock, a smaller object has proportionally more surface area through which it can lose heat, as Figure 37.7 illustrates. The result is that larger bodies cool more slowly than smaller bodies. *Thus, we expect that larger objects (such as the Earth) will have a hotter interior than smaller objects (such as the Moon).* Another way of thinking about this is that the core of a smaller object has a thinner outer layer of rock to blanket it from the cold of space, so it cools faster.

The Earth has also been heated since its birth by natural radioactivity in the Earth's interior. Thus, the Earth generates heat much as a nuclear reactor does. All rock contains trace amounts of naturally occurring radioactive elements such as uranium. A **radioactive element** is one that naturally breaks down into lighter elements in a process called **radioactive decay.** For example, the element uranium can spontaneously split apart into thorium and helium, releasing energy that generates heat. The thorium then splits, releasing more heat, and a sequence of such decays continues releasing heat until a stable isotope of lead is

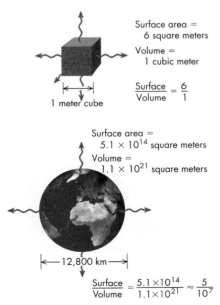

FIGURE 37.7
Heat readily escapes from small rocks but is retained in larger bodies.

reached. The heat so created in a small piece of rock at the Earth's surface escapes into space, and so the rock's temperature does not increase appreciably. However, in the Earth's deep interior, the heat is trapped by the outer rocky layers, slowing its escape. With the heat trapped, the temperature of the Earth's interior gradually rises, and the rock eventually melts. Scientists are unsure whether radioactivity or impacts during formation had the greater effect, but both are thought to have made important contributions to the Earth's internal temperature.

37.3 THE EARTH'S DYNAMIC SURFACE

As heat works its way out from the Earth's core, it drives movement of the rock in the Earth's interior. Heating often causes motion, as you can see by watching a pan of soup on a hot stove. If you look into the pan as it heats, you will see some of the soup, usually right over the burner, slowly rising from the bottom to the top, while some will be sinking again (Figure 37.8A). Such circulating movement of a heated liquid or gas is called **convection.** Convection occurs because heated matter expands and becomes slightly less dense than the cooler material around it. Being less dense ("lighter"), it rises—the principle on which a hot-air balloon operates. As the hotter material flows upward, it carries heat along with it. In this way convection not only causes motion but also carries heat.

Convective motions in a pot of soup are easy to see, but our planet's crust and mantle are not moving so rapidly, and the motions are hidden below many kilometers of rock. Nevertheless, when the rock in the Earth's interior is heated, it too may develop convective motions, but they are very, very slow. It takes hundreds of millions of years for rock to circulate through the mantle, and therefore it appears nearly motionless over the history of human observations.

Deep in the Earth's interior, rock at the base of the mantle is heated and molten. The heat causes the rock to expand, becoming less dense and buoyant. This pushes it upward relative to the surrounding denser rock in wide, slow plumes moving about as fast as your fingernail grows. Near the crust the plume turns and flows parallel to the surface (Figure 37.8B). There, it drags the surface layers, stretching the crust apart in a process called **rifting** (Figure 37.9A). In other places the convection currents force one piece of crust to collide with another, driving one piece beneath the other in a process called **subduction** (Figure 37.9B). These motions resemble the crust that forms when cooking oatmeal. It breaks apart where hot oatmeal bubbles up, and collects where cooler oatmeal sinks.

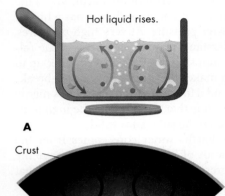

FIGURE 37.8
(A) The soup in a pan on the stove undergoes convection. Hot soup near the burner rises to the surface, where it cools and then sinks. (B) Heat from the Earth's hot core drives slow convection currents in the mantle.

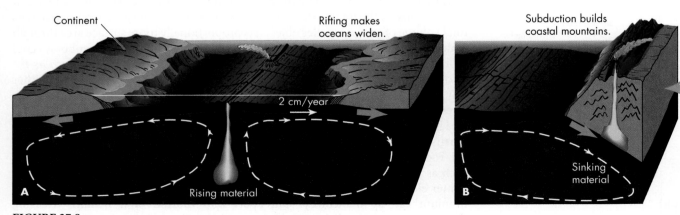

FIGURE 37.9
(A) Rifting occurs where rising material in the mantle pulls crustal plates apart at the planet's surface. (B) Subduction builds mountain chains where plates collide and material sinks back toward the interior of the Earth.

Unit 37 The Earth as a Terrestrial Planet

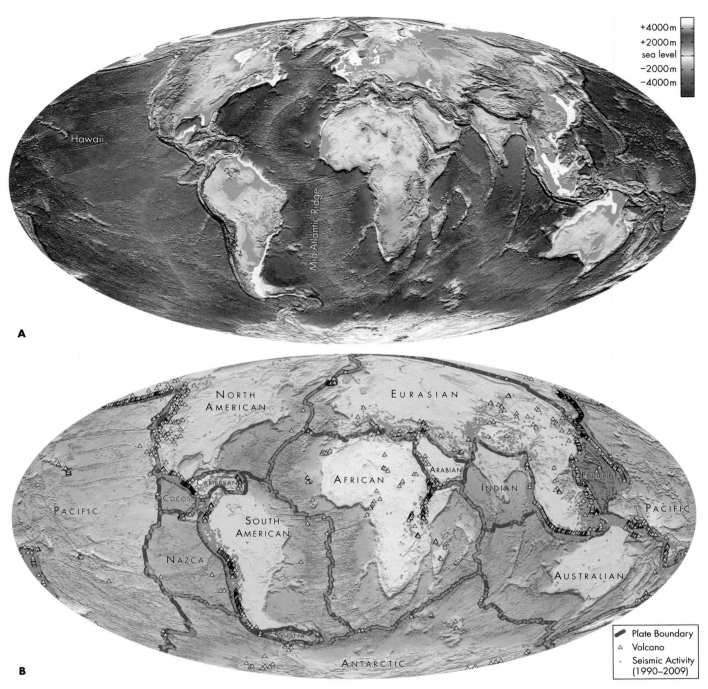

FIGURE 37.10
(A) A topographic map of the Earth's surface, including the ocean floors. Rifting occurs along long ridges on the ocean floor, where plates are moving apart, such as the Mid-Atlantic Ridge. Deep trenches occur where one plate is forced down under another, which can be seen around much of the rim of the Pacific Ocean. (B) Locations of earthquakes and volcanoes identify plate boundaries.

These processes created distinctive surface features. Rifting takes place along long breaks in the crust where convection in the mantle pulls it apart. Molten rock rises into the rifts, producing long ridges on the ocean floors, almost resembling the seams on a baseball (Figure 37.10A). Where sections of the crust are driven into each other, subduction pushes long sections of the crust down into trenches and raises long **mountain ranges** next to them. This does not happen smoothly. The rocks stick, then suddenly break, generating a sudden lurch in the crust—an earthquake. Volcanoes may also form where subducted rock rises back to the surface. A map of the Earth showing the locations of earthquakes and volcanoes (Figure 37.10B) outlines the large stable sections of crust covering Earth's surface.

The shifting of large blocks of the Earth's surface used to be called *continental drift*, but contemporary geologists prefer the term *plate tectonics*.

Plate tectonics

Past plate motions

Concept Question 1

Why do you suppose that such slow motion of the plates can lead to violent earthquakes?

The large stable sections of the crust are known as plates, and the geologic processes shifting them around on the planet's surface is called **plate tectonics**. The term *plate* is used because the pieces of the Earth's crust that move are thin (only about 50 kilometers deep) but thousands of kilometers across. Some plates contain continents, which are typically made of lower density silicates called *granite*. Other plates are entirely ocean floor, which is mostly a higher density silicate called *basalt*.

The motion of a plate can be seen in the geologic record when a plate moves over a narrow plume of heat rising from the Earth's interior. The Hawaiian island chain represents just such a situation. Active volcanoes are seen above the hot plume, but the Pacific plate is moving, so there is a string of successively older extinct volcanic islands carried to the northwest of the hot spot. The oldest of these former volcanoes have subsided back below the surface of the ocean, but they can be seen as a long chain of undersea mountains on the seafloor (Figure 37.10A).

Volcanoes form where hot rock is moving upward because the decreasing pressure on this rock can cause it to melt. This is the principle behind the Earth's core being liquid at shallower depth and solid farther down. A volcano might originate within crustal rock at a depth of about 50 kilometers where the temperature is about 1300 K. As the rock rises, it melts, and since the rock is less dense than its surroundings, it begins flowing upward through fissures in the surrounding rock. If it rises quickly enough that it does not have time to cool and resolidify, it will eventually get to within a few kilometers of the surface, where the pressure becomes so low that gases begin to boil out of the molten rock, forming bubbles within it. This further lowers the molten rock's density and makes it more fluid, helping it to rise even more rapidly, finally erupting through the surface. This last step is a little like releasing the pressure on a can of warm soda—with an ensuing eruption!

The development of plate tectonic theory is an interesting example of how science works. As early as 1596, Abraham Ortelius, a Belgian geographer, noticed that maps depicting the new discoveries of the period showed that the coastlines of South America and Africa approximately matched like two pieces of a giant jigsaw puzzle (Figure 37.11A). In 1858 a French scientist, Antonio Snider-Pellegrini, further found that fossils at matching locales on both sides of the Atlantic were also alike, and conjectured that the continents had broken apart or land bridges were destroyed after the time these fossils were laid down. Again in 1910 the American geologist F. B. Taylor proposed a similar hypothesis, but he too offered only modest supporting evidence.

The scientist who is given most of the credit for developing the modern theory of plate tectonics is the German meteorologist Alfred Wegener. In 1912 he published a paper called "The Origin of the Continents," in which he amassed fossil and geological evidence to support his hypothesis (Figure 37.11A). He made a strong case for the continents' having shifted position, rather than being joined by land bridges that were later destroyed. Wegener proposed that all the continents were originally assembled in a single supercontinent, which he called Pangea

FIGURE 37.11
(A) The idea that continents broke apart and separated was first suggested by how their coastlines fit together. Later, geological features were found that match across the dividing line. (B) Investigations also revealed that some types of fossils lie along bands that fit together only if the continents were once assembled into a single large continent, dubbed Pangea.

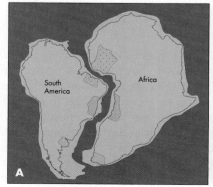
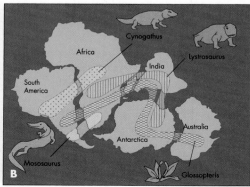

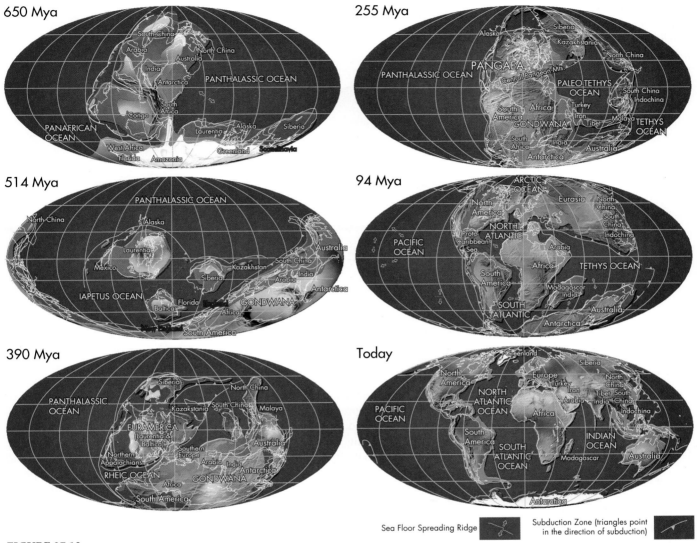

FIGURE 37.12
The appearance of Earth's surface has changed dramatically as the result of plate tectonics. The changes that geologists can trace over the last 650 million years (just 14% of the Earth's history) are illustrated here. Most of the continental material was joined in a single large continent called Pangea several hundred million years ago (Mya). Before that the land areas were split apart in a different configuration.

Concept Question 2

Imagine you are living a few hundred million years from now. How might plate tectonics affect your transcontinental travel plans?

(literally, "all-Earth"). Pangea began to split into smaller plates that became the familiar continents of today (Figure 37.11B), taking about 250 million years for the plates to move into their present locations.

Wegener's ideas were not well received at first, and it was not until the 1960s that the accumulating evidence led to wide acceptance of the idea. In fact, geologists now recognize that there was a long history of plate motion before Pangea. The low-density rock that makes up the continents has broken apart in different ways far back into the remote past. Some of the highly eroded mountain ranges we find in scattered locations today were produced by the collision of ancient plates with completely different configurations of the material that makes up the continents (Figure 37.12).

Today we can directly measure the shifting positions of the continents using global positioning system (GPS) satellites. The continents can be seen to move up to about 10 centimeters per year relative to each other. Our world is literally changing beneath our feet, growing new crust at mid-oceanic ridges and devouring it at subduction zones. This devoured rock is not lost. As it is carried downward, it is heated, and eventually its lower density causes it to rise again toward the surface. These motions within the mantle gradually change the face of the Earth, and the circulation within the core gives rise to another phenomenon—a magnetic field.

FIGURE 37.13
A schematic view of Earth's magnetic field lines and a photograph of iron filings sprinkled around a toy magnet, revealing its magnetic field lines.

37.4 THE EARTH'S MAGNETIC FIELD

Many astronomical bodies have magnetic properties, and the Earth is no exception. Magnetic forces are "carried" by what is called a **magnetic field.** They cause a compass needle to turn so it points in the direction of the field. Magnetism resembles gravity in that it exerts a force on a distant object and requires no direct physical contact. Unlike gravity, magnetism can be either attractive or repulsive—more like electric forces—but a magnet always has one side that is attractive and another side that is repulsive to another magnet.

Magnetic fields are often depicted by a diagram showing magnetic lines of force. Each line represents the direction in which a tiny compass would point in response to the field. The concentration of lines indicates the field's strength, with more lines implying a stronger field. For example, the field lines of an ordinary toy magnet emanate from one end of the magnet, loop out into the space around it, and return to the other end. The Earth's magnetic field has a similar shape, as represented in Figure 37.13. A free-floating compass needle will align itself with the depicted field lines.

An important question about any force is, What generates it? Gravitational fields are generated by masses. Magnetic fields are generated by electric currents, either large-scale currents or currents on the scale of atoms—which is why we speak of magnetism as being part of the electromagnetic force. You can demonstrate that an electric current produces magnetism by wrapping a few coils of insulated wire around an iron nail and attaching the wire ends to a battery. The nail will now act like a magnet: It will be able to pick up small pieces of iron and will deflect a compass needle. Microscopic currents on the atomic scale create the magnetism of permanent magnets.

The magnetic field of the Earth is generated by electric currents flowing in its molten iron core—a process called a **magnetic dynamo.** Scientists are still unsure how the electrical currents originate but hypothesize that complex churning and swirling motions are generated by a combination of the Earth's rotational motion and convection in the molten iron core. Studies of the magnetic fields of other Solar System bodies support this view. For example, bodies with weak or no magnetic fields, such as Mars or Venus, are either too small to have a convecting core or rotate very slowly. On the other hand, bodies with large magnetic fields, such as Jupiter and Saturn, rotate very rapidly and like Earth have liquid metallic regions in their interiors.

The electrical currents that generate a planet's magnetic field need not be aligned with its rotation axis. The orientation of the Earth's magnetic pole has been observed to have shifted by more than 10° in just the last century. The orientation

Clarification Point

There is an erroneous but persistent belief that Earth's magnetic field is caused by magnetic material located underground near the Earth's poles.

The gas giant planets have different internal structures and compositions, but under the high pressure and temperature there, hydrogen can behave like a metal (Unit 45).

of the magnetic field currently places the north magnetic pole about 9° away from the Earth's rotation axis in northern Canada. The south magnetic pole is about 25° from the axis and lies between Antarctica and Australia. Therefore, a compass needle does not in general point to true north but instead points several degrees away to *magnetic north*.

Geologists have discovered that the orientation of the Earth's magnetic pole has changed by much greater amounts in the ancient past. They can deduce this from measurements of the magnetic field preserved in congealed lava. When lava cools, the crystals forming in the rock align with the orientation of the Earth's magnetic field at that particular time. By looking at rocks of various ages, we find that the orientation of the Earth's magnetic field shifts gradually over time, and on occasion it has even completely reversed direction.

One of the best-preserved records of magnetic reversals is found on the ocean floor. As lava wells up in the rift between plates that are moving apart, the fresh lava records the orientation of the magnetic field. This is illustrated in Figure 37.14, which shows that there are regions of normal and reversed magnetic field that are symmetrical around the mid-ocean rift.

From the pattern of the magnetic field recorded in the ocean floor, geologists find that reversals happen, on average, every few hundred thousand years. Reversals follow no regular pattern, however, and there has not been a reversal in nearly 800,000 years. Measurements indicate that the Earth's magnetic field has weakened by about 10% since the 1800s, and this has caused some geologists to speculate whether this might signal the beginning of a field reversal. This could potentially have a serious impact on our environment because the Earth's magnetic field helps prevent energetic particles from the Sun and other astronomical sources from reaching Earth's surface. During a reversal the field may be too weak to protect us from this cosmic radiation.

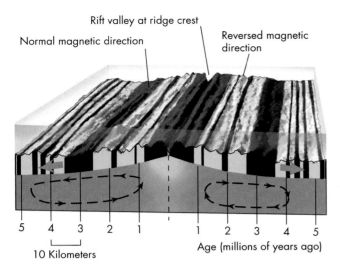

FIGURE 37.14
Magnetic reversals recorded in the ocean floor. As plates spread apart, the upwelling lava cools and records the direction of the magnetic field at that time. Portions of the ocean floor having the same magnetic orientation as we have currently are shown in black, while those with reversed orientation are shown in white.

Concept Question 3

What would be the impacts if the magnetic field of the Earth dropped to zero? This probably happens during the magnetic reversals known to occur every few hundred thousand years.

KEY POINTS

- The structure of Earth's interior can be studied by examining how seismic waves travel through the Earth.
- P (compression) waves travel throughout the planet, but S (side-to-side) waves are blocked by the liquid core.
- Some of Earth's internal heat comes from the initial collapse, some from energy released by radioactive decay.
- The Earth's interior stays hotter than smaller planets' interiors because it has proportionally less surface area to radiate heat.
- Heat from Earth's interior drives convection in the mantle, which moves thin "plates" of crust, causing geological activity and changing the pattern of continents.
- The magnetic field is generated by electrical currents in the Earth's interior, and can reverse direction occasionally.

KEY TERMS

convection, 274
crust, 272
inner core, 272
magnetic dynamo, 278
magnetic field, 278
mantle, 272
mountain range, 275
outer core, 272
plate tectonics, 276

P wave, 271
radioactive decay, 273
radioactive element, 273
rifting, 274
seismic wave, 271
silicate, 270
subduction, 274
S wave, 271

CONCEPT QUESTIONS

Concept Questions on the following topics are located in the margins. They invite thinking and discussion beyond the text.

1. Why slow motions produce violent quakes. (p. 276)
2. The pattern of continents in the distant future. (p. 277)
3. Effects of the loss of magnetic fields. (p. 279)

REVIEW QUESTIONS

4. How do the ways in which P waves and S waves propagate indicate the internal structure of the Earth?
5. What factors cause the interior of the Earth to be hot?
6. How do we know the Earth has an iron core?
7. What is meant by *plate tectonics*?
8. How does plate tectonics cause earthquakes and volcanoes?
9. What generates the magnetic field of the Earth?
10. What is the evidence that the Earth's magnetic field has undergone dramatic changes over time?

QUANTITATIVE PROBLEMS

11. P waves travel at about 6 km/sec.
    a. At this speed, how long would it take P waves to travel to the opposite side of the Earth from the center of an earthquake?
    b. In the iron core, the speed of P waves is approximately two times faster. Sketch a picture of the interior of the Earth and estimate how much sooner the wave would reach the opposite side of the Earth if the wave traveled this much faster while in the core.
12. Draw a scale model of the Earth (radius 6357 km) and the liquid core (radius 3500 km). How large a region is "shadowed" from S waves? Use a protractor to estimate how many degrees away from the site of the earthquake S waves are still detected.
13. S waves travel slower than P waves—about 4 km/sec versus 6 km/sec. Suppose a seismometer detects P waves at 7:00 P.M. and S waves at 7:05 P.M. How far away was the earthquake? At what time did the earthquake occur?
14. Suppose the Earth were made of nothing but iron (at a density of 7.8 kg/L) and silicates (at a density of 3.0 kg/L).
    a. Calculate the percentage (by volume) of iron and silicates that would be required to give a density of 5.5 kg/L. (Hint: you know that the two volume percentages must add up to 100%, so the percentage of silicates equals 100% minus the percentage of iron.)
    b. The outer radius of the core is approximately 3500 km, while the overall radius of the Earth is 6357 km. What percentage of the Earth's volume is occupied by the core?
    c. The percentage of Earth's *mass* that is composed of iron is estimated to be 32% (see Unit 35.3). What might explain why this percentage is smaller than the value in (a) and larger than the percentage in (b)?
15. a. We can get a rough estimate of how fast the rift zone in the Atlantic Ocean is moving the continents of South America and Africa apart simply by assuming that 240 million years ago Natal, Brazil, and Douala, Cameroon, occupied the same neighborhood. Theses cities are presently 5000 km apart. How fast would they have to be moving, on average, to reach their present separation? Express your result in cm/year.
    b. Another way to estimate the same quantity is by analyzing the convection cells in the mantle. If we assume that the convection cells are circular in shape with a diameter of 700 km, and that it takes 100 million years for a piece of rock to complete a trip around a cell, how fast is the rock traveling? How does this value compare to your previous result?
16. In 2005 the Earth's magnetic field was about 10% weaker than it was in 1845 when German mathematician Carl Friedrich Gauss first measured it. If the magnetic field continues to weaken by 10% of its current value every 160 years, approximately when will it be at half of its 1845 value? When will it be only 10% of its 1845 value? (Note that we are supposing, in this question, that after 320 years the strength of the field is 90% of 90%—or 81% of its original value.)

TEST YOURSELF

17. Our knowledge of the composition of the Earth's core comes from
    a. analysis of earthquake waves.
    b. gravitational waves that pass through the Earth.
    c. samples obtained by drilling deep holes.
    d. analysis of the material erupted from volcanoes.
    e. analysis of plate motion.
18. Earth's average density is
    a. slightly greater than the density of water.
    b. the same as the Sun's density.
    c. between the densities of rock and iron.
    d. the same density as rock.
19. What evidence indicates that part of the Earth's interior is liquid?
    a. With sensitive microphones, sloshing sounds can be heard.
    b. We know the core is lead, and we know the core's temperature is far above lead's melting point.
    c. Deep bore holes have brought up liquid from a depth of about 4000 kilometers.
    d. No S-type seismic waves are detectable at some locations after an earthquake.
    e. S-type waves are especially pronounced at all locations around the Earth after an earthquake.
20. The slow shifts of our planet's crust are thought to arise from
    a. the gravitational force of the Moon pulling on the crust.
    b. the gravitational force of the Sun pulling on our planet's crust.
    c. the Earth's magnetic field drawing iron in crustal rocks toward the poles.
    d. heat from the interior causing convective motion, which pushes on the crust.
    e. the great weight of mountain ranges forcing the crust down and outward from their bases.

PART 3
UNIT 38

Earth's Atmosphere and Hydrosphere

38.1 Structure of the Atmosphere
38.2 The Atmosphere, Light, and Global Warming
38.3 The History of the Atmosphere and Oceans
38.4 The Shaping Effects of Water
38.5 Air and Ocean Circulation: The Coriolis Effect

Learning Objectives

Upon completing this Unit, you should be able to:
- List the layers of the atmosphere and explain why atmospheric pressure and temperature vary with elevation.
- Explain atmospheric heating sources and the cause of the greenhouse effect.
- Describe the changes of Earth's temperature over time and discuss global warming.
- Discuss the roles liquid water has played in shaping Earth's surface environment.
- Explain the Coriolis effect and how it affects flows on the Earth's surface.

Surrounding the solid body of the Earth (Unit 37) is a gaseous envelope—the Earth's atmosphere. And most of the Earth's surface is covered with liquid water—the **hydrosphere** (Figure 38.1). This gas and liquid represent a tiny fraction of the Earth's overall mass, but they are major components of the Earth's surface environment, and they play critical roles in shaping the surface and setting its temperature.

Most planets in the Solar System have an atmosphere (Unit 34), but the Earth's has many unique features. Earth also has a unique hydrosphere, and it is the only planet that supports life. After examining the other terrestrial planets in the following Units, we will compare the the planets in Unit 44 to see if we can better understand how the similarities and differences arose.

FIGURE 38.1
The Earth's thin layer of liquid and gas creates our environment.

38.1 STRUCTURE OF THE ATMOSPHERE

When you suck on a straw in a glass of iced tea, you are not "pulling" the liquid up the straw. Actually, you are removing the air from the straw so there is no pressure to counterbalance the pressure pushing down on the surface of the tea in the rest of the glass, so the tea is pushed up the straw by the weight of the atmosphere.

The mass of the atmosphere is about 5.1×10^{18} kilograms—about one one-millionth of the Earth's total mass. The weight of the atmosphere presses down on the ground below. This **atmospheric pressure** at sea level is often described as 1 **atmosphere**, which in metric units is about 1.03 kilograms per square centimeter (about 14.7 pounds per square inch). What this means is that if we could measure a column of air 1 cm × 1 cm all the way up through the atmosphere, its mass would be slightly more than 1 kilogram. Over the roughly 16,000-cm^2 surface area of an adult human's skin, this amounts to about 16 tons of weight! However, because the pressure pushes equally on all sides, the forces balance, and we do not feel any net force. We are so accustomed to a pressure of 1 atmosphere that we often do not notice its effect except in its absence. The total weight of the atmosphere is equivalent to a layer of water about 10 meters deep. This means that the pressure in the ocean increases by about 1 atmosphere with every 10 meters of additional depth—an important consideration for underwater diving.

Our atmosphere extends from the ground and remains detectable to an elevation of hundreds of kilometers; but at the highest altitudes, the air is extremely tenuous. In fact, the density of the atmospheric gases decreases steadily with height. Gases near the ground are compressed by the weight of gases above them. Thus, our atmosphere is a little like a tremendous pile of pillows. The pillow at the bottom is squashed by the weight of all those above (Figure 38.2). Similarly, air near sea level is more compressed and therefore has a greater density than air near the top of the atmosphere. This is why it is so difficult to breathe at 6000 meters (or 20,000 feet), where the air is only about half as dense as at sea level. The atmosphere's density declines to about half its previous value for every 6 kilometers in height from the Earth's surface. Traces of atmosphere can be detected by satellites up to about 800 kilometers above the surface, although the air there is extremely thin and gradually merges with the near vacuum of interplanetary space.

The dense lowest layer of the atmosphere, up to about 12 kilometers (7 miles), is called the **troposphere** (Figure 38.3). In this region of the atmosphere water vapor circulates, clouds form, and the air is very turbulent. Above this is a much more stable layer called the **stratosphere,** which extends up to about 50 kilometers (30 miles) above the surface. Airplanes fly in the stratosphere to avoid the turbulence of the troposphere. More importantly, gases in the stratosphere, such as ozone, shield the Earth's surface from most of the short-wavelength radiation from space. Above about 80 kilometers, the nature of the atmosphere is quite different. Most molecules and atoms there have been broken apart into ions in a region called the **ionosphere.** These

FIGURE 38.2
Like the layers of the atmosphere, a stack of pillows compresses the bottom pillow most and the top one least because of the increasing pressure lower down in the stack.

Concept Question 1

If you weighed yourself on a bathroom spring scale, would you weigh more or less if you weighed yourself in a vacuum chamber with no atmospheric pressure?

FIGURE 38.3
The Earth's atmosphere has a complicated temperature structure, but its density drops steadily lower at higher altitudes. Sunlight heats the surface, which warms the bottom of the troposphere. Sunlight is directly absorbed by low-density gases in the upper atmosphere, breaking apart molecules and ionizing atoms. This leads to warm temperatures in a layer in the stratosphere and in the outer ionosphere. Layers that do not absorb much radiation remain colder.

Concept Question 2

Atmospheric pressure can push water up a tube at most about 10 meters (about 33 feet). How do you suppose trees draw water up into leaves much higher than this?

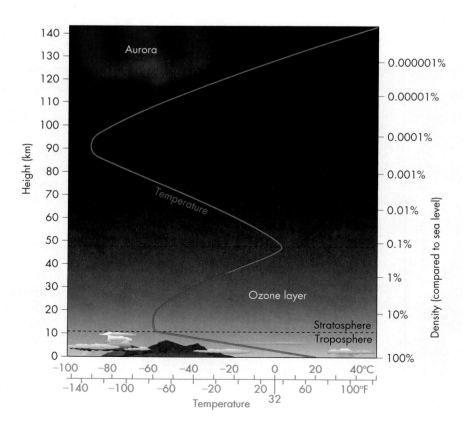

FIGURE 38.4
Electrically charged particles from the Sun spiral in the Earth's magnetic field. Some of these particles become trapped in two regions ranging out to many times the Earth's radius known as the Van Allen radiation belts. "N" and "S" indicate the Earth's North and South Poles.

Astronauts traveling to the Moon had to pass through the Van Allen belts. The radiation doses they received ranged up to the equivalent of about 40 chest X-rays, and it has been suggested as a possible cause of early cataracts in many of the astronauts.

electrically charged particles interact with long-wavelength radio waves, reflecting AM signals back to the surface and blocking our view from the ground of the longest-wavelength electromagnetic radiation from space.

The Earth's magnetic field (Unit 37) also partially screens us from electrically charged particles emitted by the Sun (Unit 53). Such particles are often energetic enough to damage living cells, so they are potentially harmful to us. Because moving charged particles interact with magnetic fields, they are deflected into spiraling motion around the Earth's magnetic field lines (Figure 38.4). Energetic charged particles are trapped by Earth's magnetic field in the **Van Allen radiation belts,** doughnut-shaped regions surrounding the Earth, thousands of kilometers above its surface. Thus, we are spared the full impact of the Sun's charged particles.

As the solar particles stream past the Earth, they apply forces on the magnetic field, and this in turn generates electric currents in the ionosphere. These currents excite the ions to emit the lovely light we see as the **aurora** (Figure 38.5). The exact process by which the aurora forms is still not completely understood, but there is no doubt that its beautiful streamers are shaped by the Earth's magnetic field. The aurora tends to be most visible close to the magnetic poles, where the solar particles can penetrate closest to the Earth's surface as they move along the magnetic field lines.

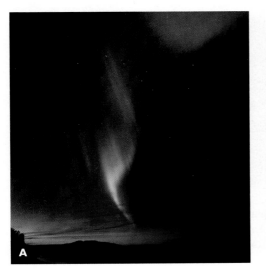

FIGURE 38.5
Photographs of an aurora (A) from Anchorage, Alaska, and (B) from space. The green colors are generally produced by ionized oxygen atoms, while red can be produced by both oxygen and nitrogen.

38.2 THE ATMOSPHERE, LIGHT, AND GLOBAL WARMING

The surface of a planet reaches an **equilibrium temperature** when the thermal radiation it emits (Unit 23) balances its rate of energy absorption. The Sun is the primary source of energy reaching the Earth's surface. The atmosphere affects how much sunlight reaches the surface and how readily thermal energy escapes to space. Earth's atmosphere is primarily a mixture of nitrogen and oxygen (Table 38.1), but some of the rarer molecules have a bigger effect on the Earth's energy balance.

Oxygen blocks most ultraviolet (UV) wavelengths of sunlight from ever reaching the Earth's surface. That shielding is provided by O_2 (the normal form of oxygen) and the molecular form with three oxygen atoms—O_3, or **ozone**. Ozone is a minor constituent in the atmosphere, but it is the main absorber of the Sun's UV light. Most of the ozone is located in the ozone layer at an altitude of about 25 kilometers (80,000 feet) in the stratosphere, and the absorption of UV energy leads to the warm temperature in the stratosphere noted in Section 38.1.

Ozone is formed when solar UV radiation encounters O_2 in the upper atmosphere. If the UV photon's energy is great enough to break the molecular bond in O_2, the molecule flies apart into individual oxygen atoms (Figure 38.6). These atoms then combine with other O_2 molecules to form O_3. Ozone absorbs lower energy, longer wavelength UV radiation than O_2 absorbs, and when an ozone molecule absorbs a UV photon, the energy splits it back into an O_2 molecule and an oxygen atom. The oxygen atom goes on to combine with other oxygen molecules.

Without the ozone layer, life on Earth's surface would suffer radiation injury. The UV photons blocked by ozone can damage the complex, delicate organic molecules of living organisms. This is why scientists are concerned about losses to the ozone layer caused by pollutants. Pollutants that release chlorine atoms in the stratosphere, notably chlorofluorocarbons (or CFCs), are a major source of

> **Clarification Point**
>
> Global warming does *not* cause the ozone hole. This is a common point of confusion, but the destruction of the ozone layer is caused by specific pollutants.

TABLE 38.1 Composition of Earth's Atmosphere

Molecule (Symbol)	Percentage by Volume
Nitrogen (N_2)	78.1%
Oxygen (O_2)	20.9%
Argon (Ar)	0.93%
Water vapor (H_2O)	0 to 5%
Carbon dioxide (CO_2)	0.039%
Neon (Ne)	0.0018%
Helium (He)	0.0005%
Methane (CH_4)	0.0002%
Krypton (Kr)	0.0001%
Sulfur dioxide (SO_2)	0.0001%
Hydrogen (H_2)	0.00005%
Nitrous oxide (N_2O)	0.00003%
Xenon (Xe)	0.000009%
Ozone (O_3)	0.000007%

*Water vapor is not included in overall percentages because of its high variability.

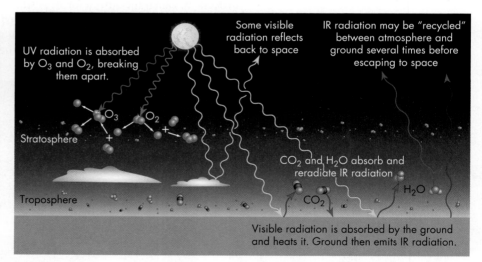

FIGURE 38.6
Schematic diagram of the interaction of light with molecules in the atmosphere. Illustrated on the left, ultraviolet (UV) light from the Sun breaks apart oxygen molecules (O_2) into free oxygen atoms in the stratosphere. The free oxygen atoms then combine with O_2 to form ozone (O_3), which also absorbs UV light. Illustrated on the right, visible light from the Sun heats the ground. The warmed ground emits infrared (IR) radiation, but gases such as carbon dioxide (CO_2) and water vapor (H_2O) can absorb this radiation. They later reemit the energy, but the IR photons may be reabsorbed and reemitted many times before escaping to space.

FIGURE 38.7

The Antarctic ozone hole in September 2012. Colors in the image at bottom indicate the relative amounts of atmospheric ozone. The purple region above Antarctica shows only about one-third of its normal amount of ozone. Chlorine (Cl) atoms released by pollutants appear to be one of the main causes of the ozone hole. As illustrated in the three small panels at top, chlorine can eliminate ozone and oxygen atoms.

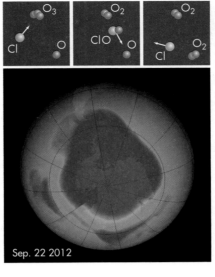

ozone destruction. Losses to the ozone layer were first seen over the South Polar region because chemical pollutants accumulate in the stratosphere during the long polar winter. An "ozone hole" becomes visible each spring when sunlight returns (Figure 38.7), and satellite studies showed that the size and duration of the hole were increasing through the 1980s. When the worst pollutants were identified, a ban was imposed on CFCs, which were widely used in spray cans and refrigeration systems. Recent observations indicate that the growth of the ozone hole has probably halted, and perhaps begun to reverse.

Although atmospheric ozone is responsible for absorbing most UV radiation, this is only about 4% of the total radiation from the Sun. Other molecules and dust particles absorb infrared (IR) radiation from the Sun, and together with ozone this accounts for about 20% of the Sun's energy that reaches Earth. About 30% is reflected back to space, and the remaining 50%, mostly at visible wavelengths, is absorbed by the ground and oceans, warming the planet's surface.

The warmed surface in turn radiates infrared light, but the heat loss is reduced by **greenhouse gases,** which can absorb radiation at infrared wavelengths, as illustrated in the right half of Figure 38.6. Water vapor and carbon dioxide are the two main greenhouse gases, but other trace gases such as methane and nitrous oxide also contribute to the effect. These molecules reemit the infrared radiation, but it then is reabsorbed by the ground or other molecules, or occasionally escapes to space. This makes the Earth's surface warmer than it would be if its infrared energy could escape freely, a phenomenon known as the **greenhouse effect.**

You can get some indication of how effectively water vapor traps heat by noticing how the temperature drops dramatically at night in desert regions or on clear low-humidity nights. All gardeners know that in cold seasons, clear nights (with no clouds and little water vapor) are most likely to produce frost. On cloudy nights, heat is retained.

The greenhouse effect does not generate heat; rather, it slows heat loss to space. The greenhouse effect therefore warms the Earth as a blanket warms you—by slowing the loss of heat. Similarly, water and carbon dioxide slow the loss of heat from the ground by absorbing infrared energy. Eventually they reemit an equal amount of energy, but some is reemitted back toward the ground or is absorbed by other molecules. The extra infrared energy reradiated back to the ground helps keep it warm at night.

Levels of carbon dioxide have risen by about 30% over the last century (Figure 38.8). This increase far exceeds any natural changes over the last 10 million years, so most scientists attribute it to the increased burning of fossil fuels by humans. Human activities such as deforestation have also reduced the places where carbon can be stored. Over the same time period, average ocean and air temperatures have increased by about 0.8°C, a phenomenon referred to as **global warming.** Most climate scientists have concluded that the primary source of global warming is the trapping of heat by the additional carbon dioxide in Earth's atmosphere.

Clarification Point

"Greenhouse effect" is the common term, but some call it the "atmosphere effect." This is a more accurate name since greenhouses retain heat mostly by preventing internal and external air from mixing, rather than by trapping infrared radiation like the atmosphere.

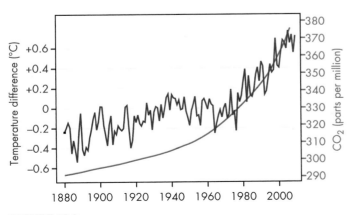

FIGURE 38.8

Carbon dioxide levels (red line) and average global atmospheric temperatures (blue line) since 1880.

38.3 THE HISTORY OF THE ATMOSPHERE AND OCEANS

When the Earth first formed, it was too hot to have liquid water on its surface. As it cooled, gases released from volcanoes and by impacting volatile-rich bodies began to build up an atmosphere (Unit 35). Geological evidence from some of the oldest rock samples suggests that water was present within 100 to 200 million years, but it may have taken several hundred million years before there was an ocean.

Chemical analysis of ancient rocks, particularly those rich in iron compounds that react with oxygen, shows that our atmosphere originally contained much less oxygen than it does today. Originally the atmosphere was probably similar to the current atmospheres of Venus and Mars, containing a large amount of carbon dioxide. The amount of carbon dioxide in the atmosphere has decreased fairly steadily throughout the history of our planet. Part of the reduction has been caused by the presence of liquid water. Carbon dioxide dissolves in water, so as rain falls through the atmosphere, it absorbs carbon dioxide. In fact, the oceans have about 500 times more carbon dioxide dissolved in them than the total contained in the atmosphere. This carbon may then become incorporated into plants and microscopic creatures that settle into sediments after they die. As the layers of sediment are compressed by the kilometers of material deposited on top of them, they can be converted to **sedimentary rock** such as limestone. There the carbon remains locked away until the rock is melted by geological activity, such as a volcanic eruption.

Over the last 65 million years, various tracers in fossils, sedimentary rocks, and geochemical models suggest that the carbon dioxide level has declined by about a factor of 3, as shown in Figure 38.9. The temperature of ocean water can be estimated from oxygen isotope ratios in sediments. The ratio varies because water with heavier isotopes of oxygen requires higher temperatures to evaporate. From sediments deposited at different times, the average ocean temperature can be estimated, as also shown in Figure 38.9. From this graph, it is clear that we live in a relatively cold period in the Earth's history.

Over the longer history of the Earth, the data are less precise, but it appears that the Earth has gone through long warm periods with no ice caps as well as some previous cold periods, some of which may have enveloped nearly the entire Earth in ice. Some of the warm episodes may have been triggered by massive geological activity releasing greenhouse gases, or variations in the Sun's energy output (Unit 53). During some past warm periods, the polar caps have completely melted and sea level has been more than 200 meters higher than it is today.

The chemical compositions of ancient rocks indicate that the amount of oxygen in our atmosphere began to build up about 2.4 billion years ago, when the Earth was about half its current age. Oxygen levels reached modern levels only about half a billion years ago. None of the geological sources of gases can explain the current abundance of oxygen in our atmosphere. Where then did such quantities of this vital ingredient originate? We can thank ancient plants and bacteria. As they consumed and stored away carbon dioxide, they produced oxygen as a waste product. We will examine the evolution of life on Earth further in Unit 85. It appears quite clear that life itself has had a profound effect on the Earth's atmosphere since long before humans entered the picture.

Detailed estimates of temperature and carbon dioxide levels over the past 800,000 years can be made from

> **Concept Question 3**
>
> It has been suggested that we might reduce heating of the Earth by screening sunlight with a cloud of very lightweight materials in orbit between the Earth and Sun. Does this seem like a good idea to you? Can you think of problems this might cause?

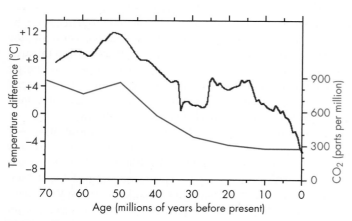

FIGURE 38.9
Carbon dioxide levels (red line) and atmospheric temperatures (blue line) are shown for the past 65,000,000 years. Carbon dioxide levels are estimated from geochemical models. Temperatures are estimated from the relative numbers of oxygen isotopes found in sediments.

FIGURE 38.10

Antarctic ice core data on carbon dioxide content and temperature over the last 800,000 years. Bubbles trapped in the ice give CO_2 levels, while oxygen isotope ratios indicate average ocean temperature. Temperatures more than about 5° below current levels result in widespread glaciation.

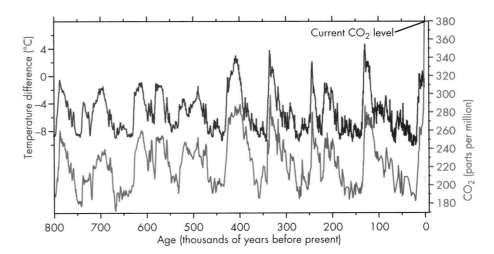

Antarctic ice samples. Layers of ice have accumulated for more than a million years there. Drilling down through the ice layers, scientists can examine bubbles of gas to find ancient CO_2 levels, and examine the percentage of different oxygen isotopes to estimate the temperature. These detailed measurements are shown in Figure 38.10. Within this time range, the Earth's temperature has gone up and down, with **glaciation**—when there is widespread ice cover on the Earth—if the temperature is more than about 5° below the current level. Compared to this more recent record, we are currently in a warm "interglacial" period.

Figure 38.10 shows that there are glacial periods about every 100,000 years. An explanation for this repetitive glaciation was provided by the Serbian mathematician Milutin Milankovitch in the 1920s. He calculated the effect of changes of the Earth's axis and orbit on the heating of the Earth and showed that this could explain the pattern of glacial periods. The Earth's axis is tilted, and its orbit is elliptical, and these vary over time as illustrated in Figure 38.11. We examined precession (Figure 38.11A) on the position of the celestial north pole in Unit 6. Over even longer time scales there are variations in the tilt of the Earth's axis and changes in the ellipticity of Earth's orbit, caused mainly by the gravitational tugs of Jupiter and Saturn. The orbital changes can lead to variations in the overall heating of Earth known as **Milankovitch cycles,** and these agree with the temperature changes seen in Figure 38.10.

Astronomers think that the strong correlation between carbon dioxide level and temperature over the last 800,000 years results from both being driven by the Milankovitch cycles. This could occur naturally because as the oceans are heated, the dissolved carbon dioxide is released into the atmosphere. Furthermore, the addition of carbon dioxide into the atmosphere can amplify slight variations in solar heating into the stronger temperature changes as are seen to occur.

There is no precedent for the current extremely high carbon dioxide level. The concern is that the addition of so much carbon dioxide might push Earth into conditions like the remote past when the polar caps completely melted; sea level would then be so high that coastal cities and environments would become completely submerged. Whatever the ultimate cause of current global warming, it seems likely that reducing greenhouse gases in the atmosphere will help slow the change and reduce the severity of the consequences.

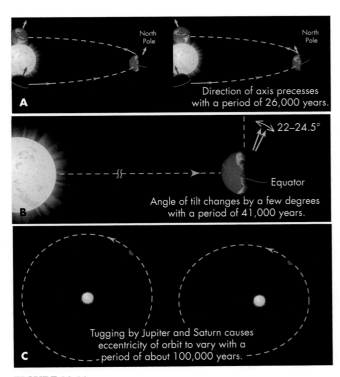

FIGURE 38.11

(A) The Earth's axis precesses every 26,000 years. (B) The tilt of the axis rocks by a few degrees every 41,000 years. (C) The eccentricity of the orbit varies with a period of about 100,000 years.

38.4 THE SHAPING EFFECTS OF WATER

One of the Earth's unique features is the presence of abundant liquid water on its surface. Oceans cover 71% of the Earth's surface, while lakes, rivers, and streams mark the faces of the continents. The mass of all the Earth's water is estimated to be about 1.4×10^{21} kilograms—about 1/4000th of the Earth's total mass. If the rocky surface of Earth were to be ground smooth, the oceans would have an average depth of about 2700 meters (about 1.6 miles) over the planet's surface. Water is vital to the existence of life on our planet, and water also affects the chemistry and geology of the Earth's surface.

Sunlight powers the cycling of water through the atmosphere by evaporating water. The water molecule (H_2O) has a lower mass than oxygen or nitrogen molecules, so air rich in water vapor has a lower density. It therefore rises through the atmosphere, where it cools, condenses, and then rains over the land before flowing back downhill to the oceans (Figure 38.12). The constant flow of water over the Earth's surface wears away rock and other surface materials by **erosion,** which over millions of years can level even high mountains. Erosion carves river channels and grinds rock into sand over much of the surface of the continents (Figure 38.13).

The pulverized rock flows into lake and ocean beds, where it settles into layers that pile up to such depths that they compress into new sedimentary rock. Much of the rock on Earth's surface has experienced erosion multiple times over the Earth's long history. The layers of rock being exposed in the Grand Canyon (Figure 38.13) were themselves laid down hundreds of millions of years ago. In addition, water in the form of glacial ice rips enormous scars into the landscape. If the Earth had no geological activity to rebuild mountains, the action of rain, flowing water, and waves would have gradually ground down the continents over the Earth's history, and the entire surface would be under water.

Different kinds of rock form in the presence of water because of water's chemical interactions. Rocks that form with water melt at a lower temperature and are "runnier" when molten than rocks with little or no water. We can see the effects of dry versus wet lava in Figure 38.14, where we compare Mount Fuji in Japan with the Mauna Loa volcano in Hawaii. Both Fuji and Mauna Loa have been built by erupted lava, but their lavas differ. Mauna Loa's lava is rich in water, so it remains molten at a lower temperature and disperses more evenly and farther from its point of eruption. As a result, Mauna Loa (like other Hawaiian peaks built similarly) has shallow slopes. Mount Fuji's lava, on the other hand, is relatively dry, so it is "stiff," and its slopes are steep.

FIGURE 38.12
Water vapor is less dense than dry air, so it rises until it cools and forms liquid water that falls back down as rain.

Clarification Point

Humid air has a lower density than dry air, other conditions being the same. Humid air may feel heavier because it prevents moisture on your skin from evaporating.

FIGURE 38.13
The Grand Canyon has been carved through rock by the flow of water over millions of years, revealing sedimentary layers laid down over hundreds of millions of years.

FIGURE 38.14
Comparison of the dry-lava Mt. Fuji (top) to the wet-lava Mauna Loa (bottom).

38.5 AIR AND OCEAN CIRCULATION: THE CORIOLIS EFFECT

ANIMATION
Coriolis effect

If you sit with a friend on a rotating merry-go-round and throw a ball back and forth, you will discover that the ball does not travel in the direction in which you aim but instead seems to curve to the side. This is a matter of perception. Someone standing to the side of the merry-go-round and observing the ball's path would see it as having traveled in a straight line once you released it. The curving path you see on the merry-go-round is actually a reflection of your own curving path. A similar effect faces ocean and air currents sweeping across our rotating planet's surface. They are deflected from their original direction of motion in a phenomenon called the **Coriolis effect**.

The Coriolis effect, named for the French engineer who first studied it, alters the paths of objects moving over a rotating body, such as the Earth, other planets, or stars. To understand why the Coriolis effect occurs, imagine standing at the North Pole and throwing a rock toward the equator (Figure 38.15). As the rock arcs through the air, the Earth rotates under it. Thus, if you were aiming at a particular point on the equator, you will miss because the surface has turned beneath the rock's path, making the rock turn to the right from the standpoint of someone in the Northern Hemisphere and to the left in the Southern Hemisphere. From the equator, a rock thrown toward either pole will also be deflected because the conservation of angular momentum (Unit 20) causes objects to move faster around the axis of rotation as they move closer to it. To conserve angular momentum as it moves closer to the Earth's axis, the rock's eastward velocity must grow faster, making it rotate faster than the Earth's surface beneath it. This again turns the path to the right from the standpoint of someone in the Northern Hemisphere and to the left in the Southern Hemisphere.

Air, water, rockets, or anything else moving across the rotating Earth in any direction will always veer to the right in the Northern Hemisphere and to the left in the Southern Hemisphere. This can be seen in the curving direction of ocean currents (Figure 38.15). The Gulf Stream, for example, flows north up the east coast of North America, curves right over to Europe, then flows south past the west coast of Africa before flowing west back across the Atlantic. A similar flow occurs in the north Pacific ocean, but the rotation is in the opposite direction in southern oceans. The Coriolis effect also creates atmospheric currents such as the trade winds. The trade winds form in response to the Sun's heating of equatorial

Clarification Point

It is a myth that the Coriolis effect determines the direction that water spins as it drains down a sink or toilet. Any tiny disturbance in the water will be much stronger than the Coriolis effect unless the distance to the Earth's axis changes significantly as the water flows.

FIGURE 38.15
The Coriolis effect causes the path of material moving over the Earth's surface to curve to the right in the Northern Hemisphere. The initial north-south direction of objects moving away from the pole and equator (left) result in paths that curve across the surface of the rotating Earth (right).

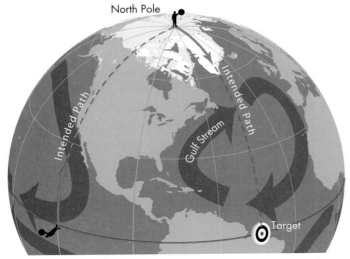
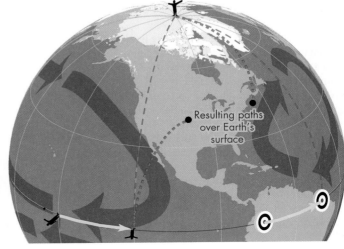

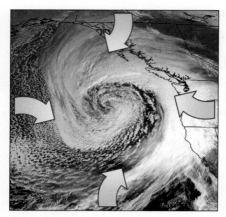

FIGURE 38.16
Weather satellite picture of the spinning air around a storm. Air drawn in toward the low-pressure region is deflected to the right by the Coriolis Effect in this Northern storm, giving it a counterclockwise spin.

air, which rises and expands, flowing away from the equator toward the poles in the upper atmosphere. However, before this air can reach the poles, it cools and sinks toward the surface and flows back toward the equator. As the air approaches the equator, the Coriolis force deflects it to the west (right in the North, left in the South), creating a steady surface wind that traders in sailing ships relied upon.

Likewise, the Coriolis force also establishes the direction of the **jet streams**, which are narrow bands of rapid, high-altitude winds. Jet streams are an important feature of the Earth's weather and are found on other planets as well. On rapidly rotating planets like Jupiter, Saturn, Uranus, and Neptune, the Coriolis effect is much stronger than on the Earth, creating extremely fast jet streams. The striking cloud bands found on Jupiter (Unit 45), for example, are partly caused by this effect.

The Coriolis effect also causes storm systems, hurricanes, and tornadoes to spin in a counterclockwise direction in the Northern Hemisphere, and clockwise in the Southern Hemisphere. This might at first seem to be backward from what you would expect, but these storms arise in low-pressure regions, which draw in air from surrounding regions. As the air is pulled in, it is deflected to the right (in the Northern Hemisphere), giving the air a counterclockwise spin, as illustrated in Figure 38.16, which shows a weather satellite pictures of a large storm system off the west coast of North America. Contrary to urban myth, the Coriolis effect does not determine the direction in which water spirals down a drain or toilet! Air or water must travel over a sizable portion of the Earth's surface to receive any significant spin from the Coriolis effect. In a small bowl, any small disturbance has far more power to set the water spiraling in either direction than the Coriolis effect has.

At one time the Coriolis effect was of special interest because it provided one of the first proofs that the Earth rotates. This was demonstrated with a Foucault pendulum, a massive swinging ball on a long wire such as you may have seen at a science museum. If you watch the pendulum for half an hour or so, you will notice that its swing gradually changes direction. On each swing it is deflected very slightly to the right (in the North), causing its direction to steadily shift. If the Earth did not rotate, the pendulum would always maintain the same direction.

KEY POINTS

- The atmosphere and hydrosphere make up a tiny fraction of the Earth's mass, but are critical in determining surface conditions.
- The pressure and density of the atmosphere decline with height, but the temperature has peaks in several layers.
- The troposphere is heated from below by the ground, which is itself heated by sunlight.
- "Greenhouse" gases in the atmosphere intercept infrared radiation from the warmed ground, trapping additional heat.
- Past changes in temperature correlate closely with the amount of carbon dioxide in the atmosphere.
- Ozone in the stratosphere absorbs ultraviolet photons from the Sun, both warming this layer and preventing these energetic photons from reaching the surface.
- Some pollutants that rise into the stratosphere can deplete ozone.
- The oceans and atmosphere formed more than 100 million years after the Earth, probably from both volcanic gases and comets.
- Atmospheric composition has steadily changed, in large part because of living things.
- Water has altered the surface of the planet through erosion and by changing the characteristics of molten rock.
- The Coriolis effect results from the Earth's rotation, causing material to rotate faster as it nears the axis of rotation.
- The Coriolis effect causes large-scale flows of air or water to turn to the right north of the equator, leading to counterclockwise rotation of storms; the directions are reversed south of the equator.

KEY TERMS

atmosphere, 281	hydrosphere, 281
atmospheric pressure, 281	ionosphere, 282
aurora, 283	jet stream, 290
Coriolis effect, 289	Milankovitch cycles, 287
equilibrium temperature, 284	ozone, 284
erosion, 288	sedimentary rock, 286
glaciation, 287	stratosphere, 282
global warming, 285	troposphere, 282
greenhouse effect, 285	Van Allen radiation belts, 283
greenhouse gas, 285	

CONCEPT QUESTIONS

Concept Questions on the following topics are located in the margins. They invite thinking and discussion beyond the text.

1. Your weight under pressure and without. (p. 282)
2. How trees draw water up more than 10 meters. (p. 282)
3. Advisability of building a sunscreen for Earth. (p. 286)

REVIEW QUESTIONS

4. What produces atmospheric pressure?
5. What are the major layers of the Earth's atmosphere, and what are the major features of each layer?
6. What is the greenhouse effect? How does it relate to global warming?
7. How does the ozone layer protect life on the Earth's surface?
8. What causes the aurora?
9. What are Milankovitch cycles? How do they relate to temperature and carbon dioxide on Earth?
10. What is the Coriolis effect? What motions does it affect? What motions does it not affect?

QUANTITATIVE PROBLEMS

11. Imagine gathering all the 1.4×10^{21} kg of water in the world and forming it into a sphere. What would be the diameter of that sphere and how does it compare to the Moon's diameter of 3400 km? (Hint: the density of water is 1 kg/L.)
12. It is estimated that there are about 3.6×10^{16} kg of carbon (C) dissolved in the oceans. If all that combined with oxygen (O) to form carbon dioxide (CO_2), how much mass would it be? What percentage of the Earth's atmospheric oxygen would that be? (Note: Carbon has an atomic mass of 12, whereas oxygen has an atomic mass of 16.)
13. Geologists estimate that the Grand Canyon was created primarily during a period of about 3 million years. Given its current depth of about 1.6 km (1 mile), on average how many centimeters deeper did the canyon become each year during this time period?
14. The colors in the Aurora Borealis are due to the interactions between electrons and atoms or molecules in the atmosphere. At low densities, collisions with oxygen produce light at 630 nm. At higher densities, collisions with oxygen atoms produce light at 558 nm, and collisions with nitrogen molecules produce light at 428 nm. What colors do we associate with these wavelengths? Examine Figure 38.5A and describe what is producing the aurora seen there.
15. Antarctica holds about 3×10^{19} kg of ice. If all of that ice melted and entered the oceans, approximately how much would sea level rise? Use the fact that the total mass of oceans = 1.4×10^{21} kg, while the average depth of the oceans would be 2.7 km if they covered the entire Earth.
16. As ocean water warms, the water expands in volume by about 0.014% per 1°C rise in temperature. This thermal expansion is small, but it contributes a significant part of the recent rise in sea level. The average depth of the oceans today is 3800 meters. How much has the water risen over the last century due to the 0.6°C rise in ocean temperature? If the temperature rose by 20°C, how much would sea level rise?
17. Air moving along with the surface of the Earth takes 24 hours to circle the Earth's axis. At a latitude of 60°, the air is 3183 km from the axis and its path around the axis is 20,000 km long. Calculate its speed in kph. Air at a latitude of 61° is 3086 km from the axis following a path around the axis that is 19,392 km long. Suppose now that a mass of air at 60° latitude moves to 61° latitude. Use the principle of conservation of angular momentum (Unit 20.3) to estimate how fast air must move around the Earth's axis at the smaller radius there. How fast is this *relative to* stationary air at this location?

TEST YOURSELF

18. Traveling straight up from the surface of the Earth, the temperature in the atmosphere is found
    a. to steadily decrease with altitude all the way to space.
    b. to steadily increase with altitude all the way to space.
    c. to fluctuation randomly with changing weather patterns.
    d. to decrease or increase with altitude depending on the layer of the atmosphere.
19. Auroras are produced when the Earth's atmosphere receives a blast of
    a. energetic electrically charged particles from the Sun.
    b. debris from comets that pass through the inner Solar System.
    c. gamma rays generated by unknown cosmic sources.
    d. ozone-destroying chemicals, mostly generated by refrigerants.
20. The levels of carbon dioxide (CO_2) found in the atmosphere today have not been this high
    a. for over ten years.
    b. for over a century.
    c. for tens of thousands of years.
    d. for millions of years.
    e. ever!
21. Which of the following would contribute to a rise in the surface temperature of Earth if you increased it?
    a. Dissolved carbon dioxide in the ocean
    b. Ozone in the stratosphere
    c. Cloud cover above the troposphere
    d. Humidity in the troposphere
    e. All of the above

Our Moon

- 39.1 The Origin of the Moon
- 39.2 Surface Features
- 39.3 The Moon's Structure and History
- 39.4 The Absence of a Lunar Atmosphere
- 39.5 The Moon's Rotation and Orbit

Learning Objectives

Upon completing this Unit, you should be able to:
- Explain how formation by a major impact can explain the Moon's composition as well as other unusual lunar properties.
- Describe the surface features of the Moon and its internal structure.
- Explain how craters and the maria formed and why the Moon has no atmosphere.
- Describe the Moon's orbit and rotation and what makes the Moon's unusual.

The Moon is a barren ball of rock, possessing no air, water, or life. In the words of lunar astronaut Buzz Aldrin, the Moon is a place of "magnificent desolation." It is our nearest neighbor in space, a natural satellite orbiting the Earth. The Moon is the frontier of direct human exploration, an outpost that we reached in 1969 but from which we have since drawn back. Despite our retreat from its surface, the Moon remains of great interest to astronomers.

Although the Moon is a satellite of the Earth, it is almost six times more massive than Pluto. The Moon is so large that it would be regarded as a planet in its own right if it orbited the Sun separately from the Earth. In this sense the Moon can be regarded as the smallest of the terrestrial planets, making it interesting to compare with the largest of the terrestrial planets, Earth (Unit 37).

FIGURE 39.1
Photograph by Apollo 17 astronaut Gene Cernan taken at the edge of "Shorty Crater." Seen next to the lunar rover, geologist Harrison Schmitt collected samples of unusual soil containing bits of volcanic glass blasted out by the impact that created the crater.

39.1 THE ORIGIN OF THE MOON

Lunar rocks brought back to Earth by the Apollo astronauts (Figure 39.1) led astronomers to radically revise their ideas of how the Moon formed. Before the Apollo program, lunar scientists had three hypotheses of the Moon's origin. In one, the Moon was originally a small planet orbiting the Sun that approached the Earth and was captured by its gravity. In another, the Moon and Earth were "twins," forming side by side from a common cloud of dust and gas. In the third, the young Earth initially spun enormously faster than now and formed a bulge that ripped away from the Earth to become the Moon.

Each of these hypotheses led to different predictions about the composition of the Moon. For example, had the Moon been a captured planet formed elsewhere in the Solar System, its composition might be very unlike the Earth's. If the Earth and Moon had formed as twins, their overall composition should be the same. Finally, if the Moon was once part of the Earth's surface layers, its composition should be nearly identical to the Earth's crust. When the rock samples returned by the Apollo missions were analyzed, astronomers were surprised that for some elements the composition was the same as the Earth, but for others it was very different. For example, Moon rocks have

Birth of the Moon

an abundance of high-melting-point materials such as uranium and a strong deficit of low-melting-point materials such as lead. The Moon also has much less iron than the Earth, which is evident from its low density of just 3.3 kilograms per liter.

Thus, the Moon's composition is quite puzzling. It is rich in high-melting-point materials, as if it formed at a higher temperature than the Earth. However, it has a deficit of iron—the most common of all the high-melting-point elements—as if it formed in a region much cooler than the Earth. None of the three hypotheses predicted this strange dichotomy of characteristics. A new hypothesis for the Moon's origin emerged based on computer simulations. The new hypothesis proposes that the Moon formed from debris blasted out of the Earth by the impact of an object comparable to the size of the planet Mars, as illustrated in Figure 39.2. This event would have had to occur during the Earth's own formation, about 4.5 billion years ago in order to explain the great age of some Moon rocks.

This hypothesis can explain the compositional abnormalities of the Moon if the collision occurred after the proto-Earth had become differentiated. The material blasted out of the Earth's mantle would have relatively little iron because the iron, being dense, had sunk deep into the interior. The colliding body splattered millions of cubic kilometers of the Earth's lower-density mantle and crust, hurling it into space in an incandescent plume. The strong heating of the blasted material would have baked out the more volatile substances, so that what became incorporated in the Moon would have a deficit of them as well. As the debris cooled, its gravity drew it together. This hypothesis was developed to explain the Moon's composition, but the real test is how well it explains other features.

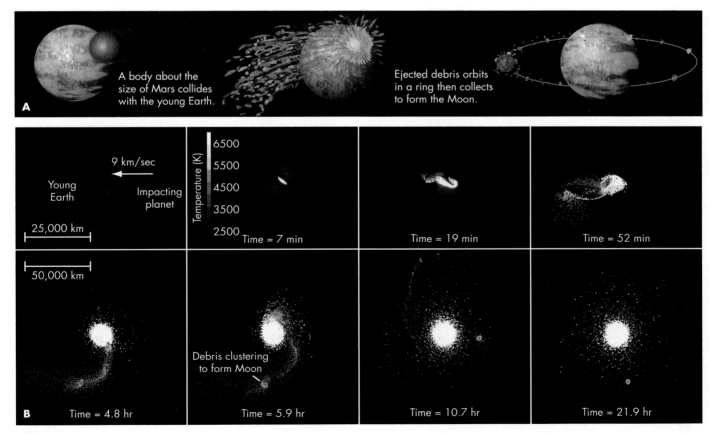

FIGURE 39.2

Origin of the Moon. (A) Sketch of the main stages in the birth of the Moon by a major collision. (B) This computer simulation shows how the Moon might have formed when a Mars-size object hit the young Earth and splashed out debris that later assembled into the Moon.

39.2 SURFACE FEATURES

The Moon's surface reveals a remarkable history. Without wind or water to erode its surface, we can see features that date back to when its crust first cooled. Although the Moon is a world of grays without the vivid colors of Earth's landscape, its surface preserves a record of spectacular events, some dating back to shortly after the Solar System formed (Unit 35).

With just your eyes, you can already see that the Moon's surface has light and dark areas (Figure 39.3). A small telescope reveals that the dark areas are smooth, while the light areas are rough. The dark areas were named **maria** (pronounced *MAHR-ee-ah*), from the Latin word for "seas" by early astronomers. However, these regions, like the rest of the Moon, are totally devoid of water. Early observers gave these dark areas poetic names such as Mare Tranquillitatis (*MAHR-ay tran-KWIL-ih-TAH-tis*—**mare** is the singular of maria) or "Sea of Tranquility," the site of the first Apollo landing.

The bright areas that surround the maria are called **highlands.** The highlands and maria differ in brightness because they are composed of different rock types. The maria are **basalt,** a dark, congealed lava rich in iron, magnesium, and titanium silicates. The highlands, on the other hand, are rich in calcium and aluminum silicates based on rock samples collected by lunar astronauts. Moreover, the samples also show that the highland material is generally less dense than mare rock and considerably older. Highland rocks have been dated radioactively (Unit 35) at about 4.0 to 4.4 billion years, whereas the rocks from the maria range from about 3.1 to 3.9 billion years old.

Not only are the highlands brighter than the maria, they are also more rugged, being pitted with large numbers of circular features called **craters,** as shown in Figure 39.4A. Highland craters are so abundant that they often overlap. The maria also contain craters, but they are well separated (Figure 39.4B). Craters are made by

FIGURE 39.3
Photograph of the Moon showing the smooth dark maria and the heavily cratered highlands. Craters are circular regions caused by high-speed collisions, and some exhibit rays of material splashed out by the impact.

 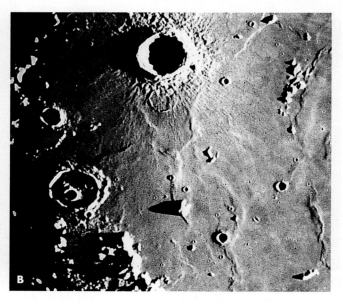

FIGURE 39.4
(A) Overlapping craters in the Moon's highlands. (B) Isolated craters in a smooth mare.

the **impact** of solid bodies striking the Moon's surface. When such an object hits a solid surface at typical orbital speeds of several tens of kilometers per second, it disintegrates in a cloud of vaporized rock and rock fragments. Craters range in size from tiny holes less than a centimeter across to gaping scars in the Moon's crust such as Clavius, about 240 kilometers (150 miles) across. Most lunar craters are named for famous scientists. For example, Cristoph Clavius (1537–1612) was a German astronomer and mathematician.

The kinetic energy released in an impact by an asteroid 1 kilometer in diameter is roughly a million times the energy released by the atomic bomb dropped on Hiroshima. The size of an impact crater reflects the kinetic energy released, and is estimated to be 10 to 100 times the diameter of the impacting object. As the vaporized rock expands from the point of impact, it forces surrounding rock outward, pushing it into a raised rim and making circular craters.

Sometimes a large impact compresses rock at the point of contact so strongly that afterward the ground rebounds upward, creating a central peak, as you can see in Figure 39.5A. Large impacts may also leave behind multiple concentric rings in the impact crater as the crust solidifies after the impact. These phenomena are analogous to how a pond might look if flash frozen a moment after someone dropped a rock into it (Figure 39.5B).

The scarcity of craters in the maria must be related to their younger age. The difference could be explained if most of the impact events ended within several hundred million years of the formation of the Solar System. The maria having formed later would then cover over the record of impacts that preceded them.

Through a small telescope, you can also see long, light streaks called **rays** radiating outward from many craters. A particularly bright set of rays is visible in and around a crater named Tycho close to the southern edge of the Moon (Figure 39.3). The material blasted out of a crater by an impacting body splatters pulverized rock, sometimes melted into small glasslike reflective spheres.

Closer examination of the Moon's surface reveals some less common features. Such features include the lunar canyons known as **rilles,** perhaps carved by ancient lava flows, which wind away from some craters, as shown in Figure 39.6A. Elsewhere, straight rilles gouge the surface, probably the result of crustal cracking (Figure 39.6B). You can observe everyday examples of similar cracks developing

Concept Question 1

Highway surfaces develop "potholes" over time, so you can use the number of potholes as an indication of the "age" of the paving. However, not all roads form potholes at the same rate. What other factors do you think might affect the number of potholes you see?

FIGURE 39.5
(A) Central peak in a crater and slumped inner walls in the crater Eratosthenes. The crater is 58 kilometers (approximately 36 miles) in diameter. (B) A raindrop falling into water exhibits surrounding rings and central rebound.

 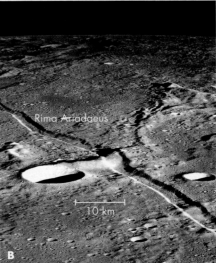

FIGURE 39.6
Photographs of lunar rilles. (A) The sinuous rilles in this region are thought to have formed from lava flows. (B) Linear rilles may have formed from shifting blocks of crust.

FIGURE 39.7
An image made from *Apollo 16* as it orbited the Moon. The left of the image shows part of the side visible from Earth, while the right shows the far side of the Moon.

Concept Question 2

What kind of evidence would remain that could identify an ancient eroded impact crater on Earth?

by monitoring either drying mud or chocolate pudding left uncovered in the refrigerator.

Figure 39.7 shows a view never visible from Earth, revealing part of the far side of the Moon recorded by *Apollo 16* astronauts as they orbited the Moon. The Moon always keeps the same side facing the Earth, as is discussed further in Section 39.4. Spacecraft orbiting the Moon show that the far side looks quite different from the side of the Moon that faces the Earth. The far side consists almost entirely of heavily cratered terrain.

The topographic map in Figure 39.8 helps to highlight the differences between the near and far sides of the Moon. The map is colored according to elevation and shows that the maria are at the lowest elevations on the nearside of the Moon. However, it also shows the surprising result that the lowest region of the Moon's surface is on the far side, near the south pole. This region, called the Aitken Basin, is as much as 13 kilometers below the surrounding terrain and spans about 2500 kilometers (1550 miles). The Aitken Basin appears to be a giant impact crater, the largest impact structure identified in the Solar System. The basin is completely covered with impact craters, so the basin must have formed very early, before the rain of impacting bodies had dwindled.

Presumably Earth also was just as battered by impacts in its youth. Although multitudes of these craters have been wiped out by erosion and by plate tectonics, a few remain, either in ancient rocks or from more recent collisions (Unit 50). The huge collision thought to have led to the Moon's formation is in this context just one of a series of huge collisions—tapering off from enormous globe-shattering events, to a steady rain of mostly smaller objects, and finally to just an occasional violent collision. To more fully understand why the Moon has preserved such clear evidence of a long history of collisions we need to learn more about its internal structure, which we turn to next.

FIGURE 39.8
Topographic map showing the near side (left half) and far side (right half) of the Moon. The elevations were mapped by the *Clementine* satellite and are shown from low to high in colors ranging from violet to red. The maria are generally at relatively low elevations for the near side, but the largest impact feature, the Aitken Basin, is at an even lower elevation on the far side. Several major features and the locations of the six Apollo landing sites are labeled.

39.3 THE MOON'S STRUCTURE AND HISTORY

At the start of this Unit, we noted how the Moon's low overall density (3.3 kg/L) tells us that its interior contains much less iron than the Earth. The Moon also lacks a magnetic field, which is probably a consequence of differences between the internal structure of the Moon and the Earth. as well as the Moon's relatively slow rotation. The Earth's magnetic field requires both a molten, electrically conductive interior and rapid rotation for its generation (Unit 37), To study the Moon's interior in greater detail, the Apollo astronauts placed seismic detectors at their landing sites. The data recorded by that mission were difficult to analyze with computers of the time, but recent reanalysis has yielded a fairly detailed picture.

The Moon's interior is somewhat similar to the Earth's, containing an iron core, rocky mantle, and rigid crust, as shown in Figure 39.9. The Moon's interior is cooler, with the seismic measurements indicating a thin liquid outer core possibly of iron sulfide, which has a lower density and melting temperature than just iron. There also appears to be some partial melting of the inner part of the mantle. The Moon has a cooler internal temperature than the Earth because the Moon's mass is only about one-eighteenth of the Earth's and its radius is only about one-quarter of the Earth's. As a result, it cannot retain its internal heat as effectively, and the Moon has much less seismic activity than the Earth. Some small "Moonquakes" are caused by impacts and the Earth's gravitational tidal pull on the Moon, which varies in strength as the Moon moves closer and farther in its elliptical orbit. This introduces internal stresses in the Moon, and landslides of crater rims occur occasionally.

The Moon's surface layer is shattered rock that forms a **regolith**—meaning "blanket of rock"—tens of meters deep. The regolith is made of rocky chunks and fine powder, the result of successive impacts breaking rock into smaller and smaller pieces as well as impacts of tiny micro-meteorites. This powdery nature can be seen in a photograph of an astronaut's footprint on the Moon (Figure 39.10). Below this surface layer of rubble and dust is the Moon's crust, averaging about 100 kilometers (60 miles) in thickness. Like the Earth's crust, it is composed of silicate rocks relatively rich in aluminum and poor in iron. One puzzle is that the Moon's crust is quite asymmetric. It is about 65 kilometers (40 miles) thick on the side facing the Earth, and about 150 kilometers (90 miles) thick on the far side.

From the great age of the highland rocks (up to nearly 4.5 billion years), astronomers deduce that these rugged uplands formed shortly after the Moon's birth. At

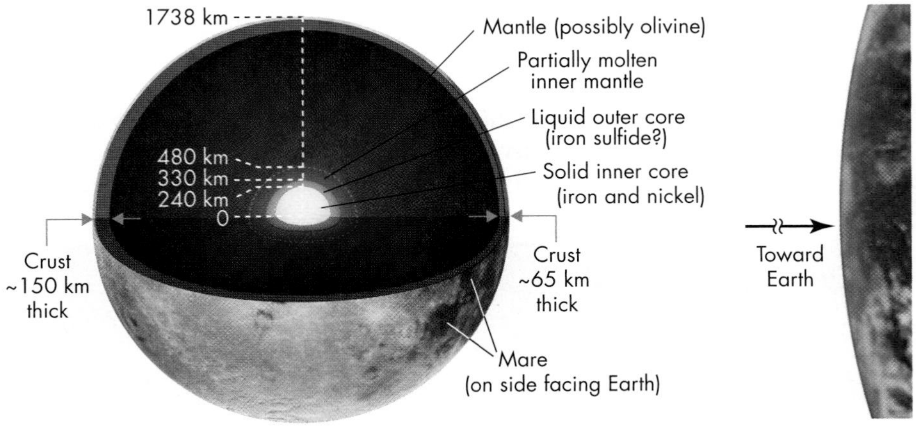

Astronauts generated seismic waves by sending their lunar landers back to crash into the Moon's surface after starting home to Earth.

FIGURE 39.9
Structure of the Moon's interior. Notice the thinner near-side crust, so the Moon's center of mass is displaced by a few kilometers toward the Earth. The Moon's iron core is small, with an outer liquid part, like the Earth's, and the surrounding mantle rock may be partially molten too.

FIGURE 39.10
Footprint of an astronaut on the Moon.

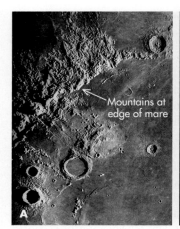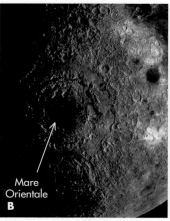

FIGURE 39.11

(A) Mountains along the edge of a mare were probably thrown up by the impact that created the mare. (B) Mare Orientale shows the multiple ring structure from a major impact. The central area of the impact crater was flooded by lava after the impact.

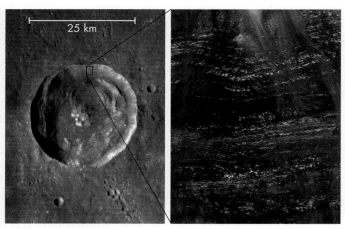

FIGURE 39.12

Euler Crater and a close up view of the wall of the crater made by the *Lunar Reconnaissance Orbiter.* The impact blasted out a section of Mare Imbrium, exposing the layers of lava that built up the mare's floor. Layers are measured to be 3 to 12 meters thick.

that time, much of the Moon's interior was probably molten, allowing denser material to sink to the core while lower-density rock floated to the lunar surface by the process of differentiation (Unit 35). Upon reaching the surface, the less-dense rock cooled and solidified, forming the Moon's crust.

Astronomers think that the maria are also impact features produced by objects over 100 kilometers (60 miles) in diameter. These struck the surface, blasting huge craters and pushing up mountain chains along their edges (Figure 39.11A). The impact that formed Mare Orientale (Figure 39.11B) sent shock waves out into the surrounding crust that formed multiple rings around the mare. These impacts may have produced some impact melting from the energy release, but the radiometric dating of the dark lava that flooded these low areas have ages up to 1.4 billion years after the Moon formed, much later than the period of heavy bombardment. It appears that molten material from within the Moon flooded the vast craters, and this occurred repeatedly, as suggested by the layering seen in the wall of a deep crater in one of the maria (Figure 39.12). The major impacts that formed the maria basins would have thinned and broken up the crust, and combined with the already thinner crust on the side facing the Earth, provided a pathway for lava to flow to the surface at later times as radioactive heating in the upper mantle melted rock there (Figure 39.13). Because the maria formed after most of the impacting bodies were gone, few bodies remained to crater the maria. The maria therefore remain relatively smooth to this day.

In this model of the Moon's formation and later development, it is puzzling that the crust would have become so asymmetric. Since gravitational tidal forces are equally strong on both the near and far sides of a body (Unit 19), a molten young

Concept Question 3

The Earth's distribution of continents is also very asymmetric, with most of the land mass concentrated in the Northern Hemisphere. In what ways is this similar to, and in what ways is it different from, the asymmetry of the Moon's crust?

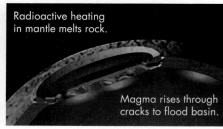

FIGURE 39.13

Large impacts late in the process of the Moon's formation formed huge basins. Lava flooded the basins to make the maria.

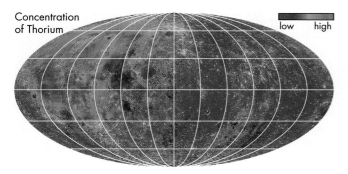

FIGURE 39.14
Concentration of thorium on the lunar surface measured by the gamma ray spectrometer on the *Lunar Prospector* satellite. The map coordinates are the same as in Figure 39.8.

Moon should have formed a symmetric crust. In addition, the composition of the crust shows major asymmetries. Scientists placed the *Lunar Prospector* satellite in orbit around the Moon in 1998. Using a gamma-ray spectrometer, it mapped out the location of radioactive elements such as thorium (Figure 39.14). Rather than being spread throughout the crust, the thorium is concentrated in the region of maria, so radioactive heating must have been most intense in this region, helping explain why the Aitken Basin was not similarly flooded by lava.

A new hypothesis is emerging that the Moon may have undergone a major collision event after it had already begun to cool. This should not be thought of as a separate event from the collision thought to have formed the Moon. Instead, the debris from that collision collected into an object nearly the size of the Moon today, with one or more smaller bodies that all began to cool as they orbited the Earth. After several million years, a smaller moon underwent a relatively slow collision with the already differentiated Moon, depositing materials near the surface that would otherwise have sunk into the Moon's interior. This event, that some lunar geologists are calling the "big splat," could account for the Moon's asymmetries.

39.4 THE ABSENCE OF A LUNAR ATMOSPHERE

The Moon's surface is never hidden by lunar clouds or haze, nor does the spectrum of sunlight reflected from it show obvious signs of gases. The Moon has no atmosphere because the Moon's small mass creates too weak a gravitational force for it to retain the gas. To be more precise, its escape velocity (Unit 18) is so low that gases are able to leak away from the Moon over times that are short compared to its age.

Concept Question 4

If the Moon's mass is only about 1/80 of the Earth's mass, why is its escape velocity one-quarter of the Earth's escape velocity?

If the Moon were currently producing large amounts of gas, it would retain an atmosphere for many millions of years. However, after its formation the interior cooled quite rapidly. Recent high-resolution imaging by the *Lunar Reconnaissance Orbiter* has revealed a few features that appear to be caused by volcanic outgassing, which occurred recently based on the small number of impact craters seen (Figure 39.15). However, the Moon's escape velocity is only about a fourth of the Earth's. Therefore, whatever volcanic gases are produced would escape relatively easily.

The lack of atmosphere has several important consequences for the Moon's physical characteristics. Without an atmosphere's greenhouse effect, the average temperature of the Moon's surface is only about 250 K (about −10°F) versus the Earth's mean temperature of about 290 K (about 60°F). Without an atmosphere to retain heat, there are dramatic temperature changes between day and night—reaching about 390 K (about 240°F), hot enough to boil water, during the two-week-long lunar day, and dropping to about 120 K (about −240°F) during the equally long lunar night.

Near the north and south poles of the Moon, there are craters that are never lit by the Sun, and their temperature probably remains close to about 40 K (about −390°F). At such constant low temperatures, it is possible that water ice survives there, perhaps buried below the surface. NASA space probes have found probable evidence for ice in these regions, based on their reflectivity to radio waves and neutrons emitted when high-energy particles interact with hydrogen atoms within about a meter of the surface, and an impact probe that was sent to crash into one of these dark craters in 2009. The impact sent up a plume of material that showed spectroscopic evidence of water. The presence of water could improve the possibilities for someday building a permanent base for humans on the Moon.

FIGURE 39.15
The peculiar feature Ina. It appears that in the lighter-colored parts of this 3-kilometer-wide D-shaped region, the surface regolith was blown away by venting gas. The lack of cratering within the area suggests this was within the last few million years.

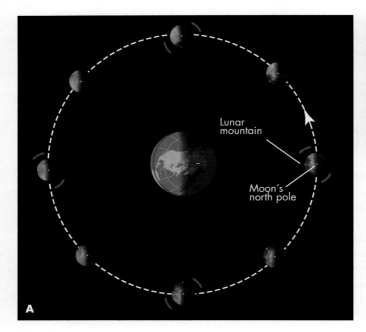
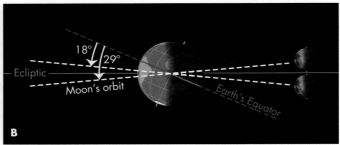

FIGURE 39.16
(A) The Moon rotates once each time it orbits the Earth, as can be seen from the changing position of the exaggerated lunar mountain. Notice that when the Moon is new (left side in the diagram), the lunar peak faces to the right, while when the Moon is full (on right side), the mountain faces to the left. Thus, from the Earth, we always see the same side of the Moon even though it turns on its axis. (B) The Moon's orbit is nearly in the plane of the Earth's orbit (the ecliptic), but is quite tilted with respect to the Earth's equator, which is quite unusual for large satellites of planets. (Separation of Earth and Moon not to scale.)

Rotation of the Moon

Clarification Point

The far side and the dark side of the Moon are often confused, but the half of the Moon that is dark changes throughout the lunar month.

The Moon appears to *librate*, or "wobble," slightly from our perspective on Earth, so that we can see a small percentage of the far side at different times of the month. This effect is mainly due to the Moon's orbital speed varying (because its orbit is elliptical) while its speed of rotation remains almost precisely constant.

Concept Question 5

Would tidal braking in the distant past between the Earth and the Moon have been greater than, equal to, or less than what it is today? Why?

39.5 THE MOON'S ROTATION AND ORBIT

As it orbits, the Moon keeps the same side facing the Earth, as you can see by watching it through a cycle of its phases. You might think from this that the Moon does not rotate. Figure 39.16A shows, however, that the Moon must slowly rotate to keep the same features facing the Earth. Thus, the Moon *does* turn on its axis relative to the stars, but with a rotation period exactly equal to its orbital period, a condition known as **synchronous rotation.** This is a common characteristic of satellites in the Solar System, which almost all keep one face toward their planet.

The Moon may have rotated more rapidly in the distant past than it does now. But the Earth and Moon exert gravitational forces on each other that slow their rotation. The Earth's spin is currently slowing due to tidal forces (Unit 18), and the same may have been true in the Moon's early history. The Earth's tidal force may have slowed the Moon's spin until the Moon now keeps the same side toward the Earth. The asymmetric distribution of matter in the Moon would have settled with one side toward the Earth similar to how a ball with an off-center mass inside settles with its heavier side facing down, However, because tidal forces are equally strong toward and away from the Earth, there would have been an equal chance of the Moon ending up with the opposite face toward the Earth. Tidal braking by the Sun probably explains the slow rotation of Mercury and Venus (Units 40 and 41) as well.

Unlike almost all other large moons, our Moon has an orbit with a large tilt with respect to its planet's equator. The Moon's orbit is tilted by a little more than 5° with respect to the Earth's orbit around the Sun. As a result, its orbit is tilted between 18° and 29° with respect to the Earth's equator, as shown in Figure 39.16B. This is unlike all of the major moons of Jupiter, Saturn, and Uranus, which lie nearly exactly in their planet's equatorial plane. The only large satellite that has a larger discrepancy than the Moon is Neptune's satellite Triton, which may have been an outer Solar System body captured by Neptune, as we will discuss in Unit 48. The tilt of the Moon's orbit is yet another oddity that may be explained by the Moon's formation by a major collision.

KEY POINTS

- The Moon's odd chemical composition suggests that it formed in a collision between a Mars-size object and the just-formed Earth.
- The Moon's gravity is too weak to retain an atmosphere, and the Moon is so small that it cooled rapidly, generating little geological activity.
- The Moon's surface is divided into heavily cratered highlands and smooth maria, flooded with basalt after major early impacts.
- Impacts produce craters with circular rims and sometimes rays of splashed rock, central peaks, or multiple rings, depending on the kinetic energy of the impact.
- The Moon rotates so one side always faces Earth; that side has a thinner crust and almost all of the maria.
- The Moon's largest impact basin, on the far side, is not flooded with basalt, possibly because of the thicker crust there.

KEY TERMS

basalt, 294
crater, 294
highlands, 294
impact, 295
mare (pl. *maria*), 294
rays, 295
regolith, 297
rille, 295
synchronous rotation, 300

CONCEPT QUESTIONS

Concept Questions on the following topics are located in the margins. They invite thinking and discussion beyond the text.

1. Determining the age of a highway from potholes. (p. 295)
2. Identifying eroded impact craters on Earth. (p. 296)
3. Comparing the Moon's and Earth's asymmetries. (p. 298)
4. Relationship of escape velocity to mass. (p. 299)
5. Rate of tidal braking in the past. (p. 300)

REVIEW QUESTIONS

6. Why are craters so much larger than the impacting object?
7. Why are the maria and highlands so different in appearance?
8. Why is the Moon so much less active than the Earth?
9. What is odd about the Moon's rotation? Why does it spin this way?
10. What are the effects of tides on the Earth and on the Moon?
11. How do astronomers now think the Moon formed? What evidence supports this hypothesis?

QUANTITATIVE PROBLEMS

12. Galileo was able to estimate the height of mountains on the Moon by measuring the length of their shadows. Make a drawing of a tall mountain sitting on a flat plane viewed from the side. With a protractor, draw a line with a 10° angle to the plane that touches the mountain peak. This shows how long the shadow is when the Sun is 10° above the horizon.
    a. If a shadow is 5 kilometers long when the Sun is 10° above the horizon, estimate how tall the mountain is.
    b. Estimate at what angle the Sun must be above the horizon for the shadow to be twice as long as the mountain's height.
13. The Moon's density is 3.3 kg/L. Suppose that the Moon is made of just two substances: silicates with a density of 3.0 kg/L and iron with a density of 7.9 kg/L.
    a. What fraction of its mass must be iron?
    b. What fraction of its volume must be iron?
14. To measure the distance to the Moon, astronomers reflect short pulses of light off of retroreflectors placed on the Moon by Apollo astronauts. The round-trip time for these pulses indicate the Moon's current distance.
    a. The Moon's distance varies from 363,000 km to 405,000 km over its elliptical orbit. What is the round-trip time for laser pulse to travel these distances?
    b. The results of the experiment demonstrate that the Moon is moving away from Earth at 38 mm per year. How much longer does it take a light pulse to make a round-trip each year?
15. As the Moon moves away from the Earth, the radius of its orbit grows larger, although its orbital speed grows smaller. The angular momentum of the Moon is equal to the product of its mass and its orbital velocity and radius ($M \times V \times R$). Is the Moon's orbital angular momentum growing larger or smaller? Use Newton's version of Kepler's third law (Unit 17) to determine the Moon's speed as a function of its orbital radius, to show if the angular momentum is increasing or decreasing.
16. Assume that the density of an average asteroid is 3.0 kg/L, and that the craters on the Moon are roughly 20 times the size of the impacting body.
    a. What was the diameter of the asteroid that created the 240-km diameter Clavius crater? Using the density of an average asteroid, calculate the mass of this asteroid.
    b. Calculate the kinetic energy of this impact if the speed of the asteroid was 20 km/sec relative to the Moon.
17. The Apollo 11 command module orbited the Moon at an altitude of about 110 km. Use Newton's modification to Kepler's third law (Unit 17.2) to calculate the module's orbital period.

TEST YOURSELF

18. Compared to the lunar highlands, the lunar maria are
    a. smoother and older.
    b. smoother and younger.
    c. more cratered and older.
    d. more cratered and younger.
19. The lunar maria have a composition most like the
    a. lunar highlands.
    b. Moon's mantle.
    c. Moon's core.
    d. impacting meteor.
20. What evidence supports the idea that the Moon does not have a large iron core?
    a. Most maria face Earth.
    b. The Moon's average density is closer to that of rock than iron.
    c. The Moon lacks an atmosphere.
    d. The Moon rotates once per month.

UNIT 40

Mercury

40.1 Mercury's Surface Features
40.2 Mercury's Interior
40.3 Mercury's Rotation
40.4 Mercury's Temperature and Atmosphere

Learning Objectives

Upon completing this Unit, you should be able to:
- List Mercury's surface features and its temperature extremes.
- Explain how scarps and jumbled terrain are thought to have formed.
- Discuss hypotheses about why Mercury has such a large iron core.
- Describe and explain how Mercury's rotation relates to its orbit.

Mercury is named for the Roman deity who was the speedy messenger of the gods. The name was inspired by the fact that Mercury changes its position in the sky faster than any other planet. Mercury is difficult to observe from Earth because it orbits the Sun at just 0.39 AU and usually lies in the Sun's glare.

Mercury is the smallest of the eight planets, with a radius about one-third of Earth's and a mass 18 times smaller—not very much bigger than the Moon (Figure 40.1). It has been visited by two space probes, *Mariner 10* in 1974–1975 and *Messenger*, which began a long-term rendezvous in 2008. Before *Messenger* arrived, less than half of Mercury's surface had been seen. Images from these spacecraft show that although Mercury's surface resembles that of our Moon (Unit 39) in many ways, there are hints of unique geological processes on this airless planet.

Every 1½ to 2½ months, Mercury reaches its "greatest elongation" (Unit 11) from the Sun, alternately in our evening and morning skies. At these times you may see it at not much more than a hand's span above the horizon.

40.1 MERCURY'S SURFACE FEATURES

At first glance Mercury's landscape is difficult to distinguish from the Moon's; but a careful examination reveals differences (Figure 40.2). First, Mercury's impact craters generally overlap less than the Moon's do in the lunar highlands (Unit 39). In addition, the crater walls tend to be less steep, most likely because Mercury's surface gravity is more than twice as strong as the Moon's, making steep hills less

FIGURE 40.1
Mercury (on the left) and the Moon (on the right) shown with the correct relative size. Note that both these airless objects are heavily cratered, but the Moon has large maria.

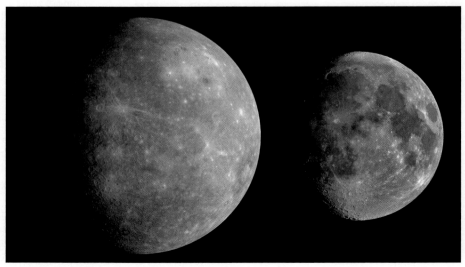

The *Messenger* spacecraft had to make a series of flybys, designed to minimize the need for rocket fuel while it matches Mercury's orbital velocity. *Messenger* went into orbit around Mercury in 2011.

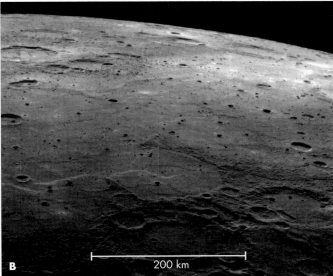

FIGURE 40.2
Two views of Mercury from the *Messenger* spacecraft. (A) Some regions show a long history of overlapping craters. A recent impact produced a bright crater (lower center of image) and splashed material to make rays that cross the old craters. (B) A heavily cratered region in the foreground gives way to a smoother area beyond it flooded by volcanic flows, somewhat like a lunar mare.

stable on Mercury. Congealed lava flows flood not only many of its old craters but much of its surface. On our Moon, these flows are found almost exclusively within the maria. Finally, the lava flows do not appear as dark as the lunar maria, and probably have a different composition from the Moon's basalt.

Mercury's surface contains some features quite different from the Moon's. Enormous **scarps**—cliffs formed where the crust has shifted—cover Mercury's surface (Figure 40.3A). Some run for hundreds of kilometers and range up to 3 kilometers high. Because the scarps cut across many impact craters, they must have formed later in Mercury's history, perhaps as the planet cooled and shrank, wrinkling like a dried apple. Astronomers estimate that Mercury's diameter may have shrunk by about 5 km. In addition to the large lava flows, there are also some indications of volcanic activity dating from after most of the craters formed. In several locations features have been found that appear to be volcanic vents that have expelled material that coats the surrounding area (Figure 40.3B).

> Recent observations have identified small scarps on the Moon as well.

FIGURE 40.3
(A) *Mariner 10* imaged many scarps. This scarp cuts across several older craters. (B) *Messenger* imaged this peculiar light-colored region from two angles. The irregularly shaped depression is thought to be a volcanic vent from which the lighter-colored material erupted.

> Most features on Mercury are named after artists, writers, and other cultural figures.

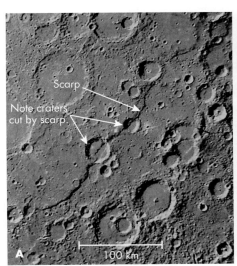
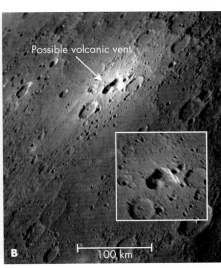

 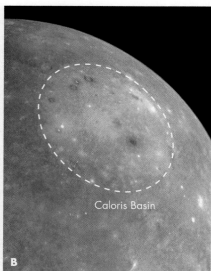 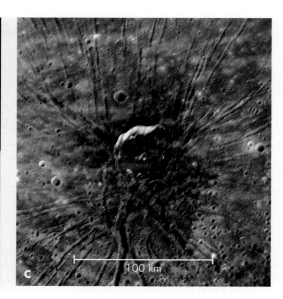

FIGURE 40.4
(A) The edge of the Caloris Basin is seen in the semicircular set of rings surrounding Mercury's largest impact feature. Only this edge of the basin was seen by the passing *Mariner 10* spacecraft in 1974. (B) In 2008 the *Messenger* spacecraft imaged the rest of the basin. The Caloris Basin appears orange in this false-color image, which enhances subtle color differences to show different mineral compositions. (C) The center of the Caloris Basin has a strange spidery pattern of troughs surrounding a 40-kilometer-diameter crater near the center of the basin.

Concept Question 1

If a team of astronauts were sent to visit one small area of Mercury, what region would you select for them to explore?

By far, the largest of the impact features found on Mercury is the vast **Caloris Basin,** only the edge of which was seen by *Mariner 10* (Figure 40.4A). Astronomers waited more than three decades to see the rest of the Caloris Basin, which has now been imaged in detail by *Messenger*. With a diameter of about 1500 kilometers (about 900 miles), this mountain-ringed depression is reminiscent of some of the lunar maria. Moreover, its circular shape and surrounding hills indicate that, like lunar maria, it was formed by a major impact after the young Mercury's surface had cooled, perhaps about 3.8 billion years ago.

Messenger's imaging system detected subtle color differences between different portions of Mercury's surface, which are exaggerated in Figure 40.4B to make them clearer. Lava flows in the Caloris Basin appear orange, old impact craters dark blue, and young ones white. Features identified as possible volcanic vents have a bright orange color around the edge of the basin.

Near the center of the Caloris Basin is an unusual and poorly understood feature, a spider-shaped pattern of troughs radiating away from a small crater (Figure 40.4C). The crater is too small to have been the result of the impact that formed the basin, so it is probably the result of a later collision. The spidery pattern of troughs suggests that this portion of Mercury's surface has been stretched, as if internal activity in Mercury caused a bulge in the surface here, which stretched and split the surface.

The Caloris impact had global effects on Mercury. On the side of the planet opposite the Caloris Basin, the terrain has a hilly, jumbled appearance (Figure 40.5). Astronomers have hypothesized that this was produced by the seismic waves that would have traveled through the planet and along the surface after the Caloris impact event. These waves would have converged from all directions on the far side of the planet almost simultaneously, and might have shaken apart the landscape into its present appearance.

FIGURE 40.5
Mariner 10 image of odd jumbled terrain that lies on the side of Mercury opposite the Caloris Basin.

40.2 MERCURY'S INTERIOR

Mercury is thought to have a large iron-nickel core beneath a relatively thin rocky exterior. Because no spacecraft has landed there to deploy seismic detectors, astronomers base this conclusion primarily on Mercury's density. Although a massive planet's gravity can compress its interior to a higher density than it would normally have, Mercury is too small for this effect to be substantial. Therefore, although its density (5.4 kilograms per liter) is slightly lower than Earth's, models indicate it has an iron core that is a greater proportion of its interior than Earth's. A model of the probable internal structure is shown in Figure 40.6.

There is only a slight difference in the temperatures at which rock and iron condense. Therefore Mercury's richness in iron relative to its rock content is difficult to explain based on where the elements condensed in the solar nebula (Unit 35). One hypothesis proposed to explain this difference is that Mercury once had a thicker rocky crust that was blasted off by the impact of a large planetesimal early in the Solar System's history (Figure 40.7). A collision like this might also explain Mercury's relatively elliptical orbit. Another hypothesis is that after Mercury formed and differentiated, the young Sun became hot enough at some stage to vaporize the planet's surface rock.

Astronomers have bounced radar signals off Mercury, allowing them to measure very precisely the way that Mercury "wobbles" as it rotates. They found that Mercury behaves as though its core is partially molten. The effect they observed is similar to how a raw egg behaves if you spin it, then touch it momentarily to stop it. After letting go, it will continue to spin a little more. Mercury likewise experiences tugs from the Sun that would alter its spin slightly, but it wobbles about twice as much as it would if it were solid all the way through.

Mercury was not expected to have a molten core because of its small mass and radius. Even though it is close to the Sun, this does not provide nearly enough heat

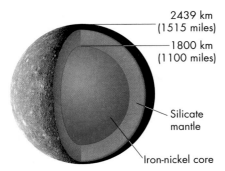

FIGURE 40.6
Artist's depiction of Mercury's interior, showing the radii of the core and surface.

Concept Question 2

If a planet of Mercury's mass had formed at the Earth's distance from the Sun, how might it have differed from Mercury?

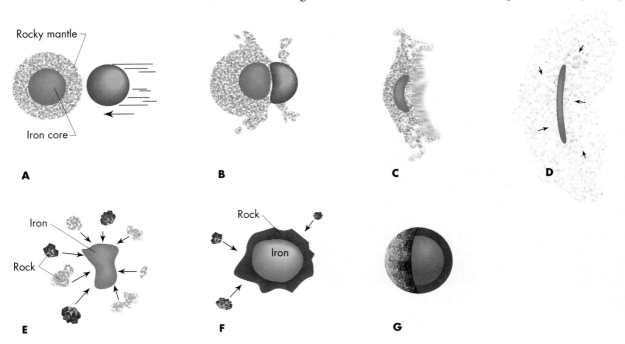

FIGURE 40.7
Diagrams illustrating a computer simulation of a collision between Mercury and a large planetesimal. The impact compresses and flattens the iron core and blasts away most of the outer rocky layers. As the remaining material accumulates back into the planet, its iron core remains highly distorted, surrounded by a thinner crust of rocky debris. Gravity eventually reshapes the planet into a sphere.

to make the core molten. Instead, astronomers now suspect that the outer iron core may be mixed with sulfur, which can lower the melting temperature. This suggests some greater amount of mixing in the solar nebula because iron sulfide should not have condensed so close to the Sun. A partially molten core also helps explain why Mercury has a magnetic field, albeit one that is only about 1% that of the Earth's. Magnetic fields can be generated by circulation of electrically conducting material in a planet's interior (Unit 37).

40.3 MERCURY'S ROTATION

Mercury spins very slowly. Its rotation period relative to the stars is 58.646 Earth days, exactly two-thirds its orbital period around the Sun of 87.969 Earth days. This means that it spins three times for each two trips it makes around the Sun. This is not just coincidence. Mercury probably spun faster when it formed, but the Sun's tidal forces slowed Mercury's rotation. This is similar to how the Earth tidally braked the Moon's rotation until the same side always faced us (Unit 19). However, Mercury did not slow down to the point where the same side always faces the Sun.

Why Mercury did not continue to slow its rotation further is probably a consequence of Mercury's elliptical orbit around the Sun. Because of the ellipticity, Mercury's orbital speed changes in accordance with Kepler's second law of planetary motion. That changing speed means that no fixed rotation rate can possibly keep one side always facing toward the Sun. Mercury would always twist back and forth ("librate") even if its spin rate matched its year. Mercury's spin instead slowed until one side faces toward the Sun on one closest approach and then away from the Sun on the next closest approach (Figure 40.8) when the tidal forces are strongest. This very slow spin achieves a stable rotation rate for a planet in an elliptical orbit.

A secondary effect of this unusual **tidal lock** is that Mercury's day is longer than its year! While Mercury orbits the Sun in 88 Earth days, the time between successive "noons" is two orbital periods, or 176 days. Having a day longer than a year sounds strange, but consider that if one side of Mercury always faced the Sun, the "day" would be perpetual on that side.

As on Earth, Mercury's spin causes the Sun to rise in the east and set in the west. However, the slow rate of Mercury's rotation combined with its large orbital eccentricity gives rise to a peculiar effect. When closest to the Sun, at **perihelion,** Mercury's orbital speed is largest, and for about eight days (of Earth time) the speed of its motion around the Sun exceeds the effect of the planet's slow rotation. As a result, the Sun travels backward across the Mercurian sky, moving from west to east. If you were watching from the Caloris Basin, you would see this backward motion of the Sun around Mercurian "noon." Forty days after sunrise the Sun would stop traveling to the west, reverse direction for eight days, then begin moving again toward the western horizon, setting 40 days later. From some locations on Mercury, the Sun rises slightly above the horizon for several Earth days of time, then dips back below the horizon before rising again to complete the long Mercurian day.

Rotation of Mercury

If Mercury did not rotate with respect to the stars, its orbit would make the Sun rise in the west and set in the east.

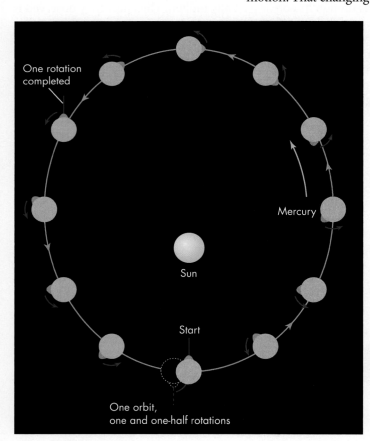

FIGURE 40.8
Mercury's odd rotation. The planet spins three times for each two orbits it makes around the Sun. Only one orbit is shown here. Note that a feature on the surface has completed one and a half rotations after one orbit.

40.4 MERCURY'S TEMPERATURE AND ATMOSPHERE

Mercury's surface is one of the hottest places in the Solar System, and it undergoes some of the most extreme changes of temperature. At its equator, noon temperatures can reach approximately 710 K (about 820°F). On the other hand, nighttime temperatures are among the coldest in the Solar System, dropping to approximately 80 K (about −320°F). These extremes result from Mercury's closeness to the Sun and from its lack of atmosphere. As on the Moon, no atmospheric gas is present that could retain heat during the long Mercurian night.

Mercury has the most elliptical orbit of the eight planets. Its closest approach to the Sun, when Mercury is at its perihelion, is 46 million kilometers (29 million miles). Its orbit carries it about 50% farther away at its most distant, about 70 million kilometers (43 million miles), where the heating it receives from the Sun is substantially less. At perihelion, from Mercury's perspective the Sun looks about 50% bigger, and the planet receives about twice as much solar radiation as at its greatest separation. Thus, noon temperatures on Mercury's equator reach 710 K at perihelion but "only" about 580 K when the separation is largest. The name *Caloris* (from the Greek for "hot") Basin was chosen because it lies near one of two points on Mercury's equator that face the Sun directly when Mercury is at perihelion and therefore reach the highest temperatures.

Mercury lacks an atmosphere for the same reason the Moon does. Although traces of gas, perhaps captured from interplanetary space, have been detected in Mercury's spectrum, Mercury's small mass makes its gravitational attraction too weak to retain any significant amount of gas. Even though Mercury's escape velocity is almost twice as large as the Moon's, Mercury's closeness to the Sun makes its temperature higher. That in turn causes any gas molecules there to move so rapidly that they readily escape into space.

The conditions on Mercury are so hot that astronomers were surprised to find evidence of ice. The initial evidence was based on the reflection of radar waves, which are affected in a characteristic way by ice. Since going into orbit around Mercury, the *Messenger* spacecraft has been able to show that the radar-reflective regions are located in perpetual shadows within craters near the planet's poles. *Messenger* has also detected evidence of excess hydrogen levels in these regions, consistent with what ice would produce. Where would ice have come from? Astronomers are not certain, but perhaps it was delivered by comet impacts.

Concept Question 3

If a comet carrying an ocean of water collided with Mercury, what do you suppose would happen?

KEY POINTS

- Mercury's surface resembles the Moon in being heavily cratered, having large ancient lava flows, and lacking an atmosphere.
- Scarps in the crust suggest that the core has shrunk as it has cooled.
- The Caloris Basin is a huge impact feature, somewhat like a lunar mare; the impact altered terrain on the far side of Mercury.
- Mercury has had more volcanic activity than the Moon; its magnetic field and rotational wobble imply the interior is still molten.
- Mercury has a very large iron core, and some astronomers speculate that part of the rocky crust was lost in its early history.
- Mercury rotates very slowly, spinning three times for every two orbits, making its day longer than its year.
- Mercury's surface experiences extremes of hot and cold because it is so close to the Sun and has such a long night.

KEY TERMS

Caloris Basin, 304
perihelion, 306
scarp, 303
tidal lock, 306

CONCEPT QUESTIONS

Concept Questions on the following topics are located in the margins. They invite thinking and discussion beyond the text.

1. Best site for astronauts to visit on Mercury. (p. 304)
2. Mercury-mass planet at Earth's distance from the Sun. (p. 305)
3. Consequence of a large comet striking Mercury. (p. 307)

REVIEW QUESTIONS

4. How do the craters on Mercury differ from those on the Moon?
5. What are scarps? How might they have formed?
6. What is the evidence for a large iron core in Mercury? What might explain why it is so large?
7. What is the evidence for ice on Mercury? How can there be ice there?
8. How can Mercury's day be longer than its year?
9. Why does Mercury have no atmosphere? How does this affect Mercury's temperature?

QUANTITATIVE PROBLEMS

10. Astronomers think that Mercury's scarps formed as the planet's diameter shrank by about 5 kilometers.
    a. If the 3600-km diameter core shrank by 5 km, by what percentage did its volume decrease?
    b. If the volume of iron changes by about 30 parts per million for each kelvin of temperature change, by how much would the core's temperature have had to change?
11. In this problem we are going to estimate the total number of craters on Mercury that have a diameter greater than 25 km.
    a. Given Mercury's radius of 2440 km, what is the planet's total surface?
    b. Estimate the total surface area shown in Figure 40.3A.
    c. Count the number of craters in Figure 40.3A that appear greater than 25 km in diameter.
    d. Estimate the number of craters on Mercury that are greater than 25 km based on your results in a and b.
12. The highest temperature on Mercury is about 710 K, while the highest temperature on the Moon is about 400 K. The average speed of a molecule in a gas is proportional to the square root of the temperature. The rate of molecules escaping to space depends on the fraction of their average velocity relative to the escape velocity, and Mercury's escape velocity is about twice the Moon's. Given these facts, which would have retained an atmosphere longer, Mercury or the Moon?
13. Mercury's orbit ranges from 46 to 70 million kilometers from the Sun, while Earth orbits at about 150 million kilometers.
    a. The Sun has a 30-arc-minute diameter viewed from Earth; what range of sizes does it have when viewed from Mercury? (Unit 10)
    b. At Mercury's orbital extremes, how many times stronger is the Sun's radiation on Mercury than on Earth? (Unit 21)
14. How many times faster is Mercury orbiting the Sun at perihelion than at aphelion given that Mercury's perihelion and aphelion distances are 46 and 70 million kilometers respectively?
15. The cause of the Sun occasionally appearing to move backward across Mercury's sky is explored in this problem. Begin with the facts that Mercury rotates approximately once every 58.6 Earth days relative to the stars, and takes 88.0 Earth days to revolve about the Sun. (It may also help to consult Figure 40.8 for this problem.)
    a. Estimate how fast the Sun would move across Mercury's sky if Mercury were spinning at its same rate, but not revolving about the Sun. In other words, calculate the angular speed (degrees per Earth day) at which the Sun moves across the sky due to Mercury's rotation but not Mercury's orbital motion.
    b. Calculate the average angular speed (degrees per Earth day) at which the Sun moves across the sky due just to Mercury's orbit of 88 days. (This is slower than the motion of the Sun due to its rotation, which is why the Sun rises in the east and sets in the west on Mercury.)
    c. When Mercury is at perihelion, its orbital speed is about 25% faster than its average orbital speed. In addition, because Mercury is closer to the Sun, the angle the Sun shifts by is about 25% greater too. What is the resulting angular speed when Mercury is at perihelion?

TEST YOURSELF

16. Why does Mercury have so many craters and the Earth so few?
    a. Mercury is far more volcanically active than the Earth.
    b. Mercury is much more dense than the Earth, and therefore its gravity attracts more impacting bodies.
    c. The Sun has heated Mercury's surface to the boiling point of rock, and the resulting bubbles left craters.
    d. Erosion and plate tectonic activity have destroyed most of the craters on the Earth.
    e. Mercury's iron core produces a strong magnetic field that attracts other bodies made of iron.
17. Mercury's average density is about 1.5 times greater than the Moon's, even though the two bodies have similar radii. What does this suggest about Mercury's composition?
    a. Mercury's interior is much richer in iron than the Moon's.
    b. Mercury contains proportionately far more rock than the Moon.
    c. Mercury's greater mass has prevented its gravitational attraction from compressing it as much as the Moon is compressed.
    d. Mercury must have a uranium core.
    e. Mercury must have a liquid water core.
18. The scarps that cut across the surface of Mercury probably were
    a. cut by flowing lava.
    b. produced by impacts pushing portions of the crust outward.
    c. formed when the crust buckled as Mercury cooled.
    d. formed when crustal plates ran together during a plate tectonic phase.
19. Over the course of a Mercurian year, the Sun's motion across the sky is
    a. always from east to west.
    b. always from west to east.
    c. briefly travels east to west but otherwise travels west to east.
    d. briefly travels west to east but otherwise travels east to west.
20. Mercury's core is unusual in that
    a. it is relatively large compared to Mercury's size.
    b. it contains iron.
    c. it is partially molten.
    d. it has a high density.

PART 3
UNIT 41

Venus

41.1 The Venusian Atmosphere
41.2 The Surface and Interior of Venus
41.3 Rotation of Venus

Learning Objectives

Upon completing this Unit, you should be able to:
- Describe Venus's atmosphere and what causes it to be so hot compared to the Earth's.
- Discuss what causes a runaway greenhouse and whether it could happen to Earth.
- Explain how Venus's surface has been studied and describe the features found.
- Discuss hypotheses to explain the differences between Venus's and Earth's geology.
- Explain what is unusual about Venus's rotation and what may have caused it.

Venus is named for the Roman goddess of beauty and love. Perhaps this name was given because of its beautiful appearance as the brightest "star" in the dawn or evening sky. We can only see Venus close to sunrise or sunset because Venus's smaller orbit (at about 0.72 AU) never carries it farther than 48° from the Sun from our perspective.

Of all the planets, Venus (Figure 41.1) is most like the Earth in diameter and mass. Its density is thus also similar to the Earth's density, implying that the deep interior of Venus is like the Earth's—an iron core and a rock mantle (Unit 37). Because Venus is so like the Earth in its overall characteristics, we might expect it to be like the Earth in other ways. However, Venus's atmosphere and surface conditions are radically different from Earth's (Unit 38).

Venus is visible for a period of about 9 to 10 months as a very bright "evening star" that sets after the Sun. Then, after Venus passes between the Earth and the Sun, it reappears as a "morning star" for another 9 to 10 months.

41.1 THE VENUSIAN ATMOSPHERE

The atmosphere of Venus is about 100 times more massive than the Earth's atmosphere. It also differs greatly in composition, consisting mostly (96.5%) of carbon dioxide, whereas Earth's atmosphere consists primarily of nitrogen. Astronomers know the atmospheric composition of Venus from its spectrum and from measurements with space probes. Gases in its atmosphere absorb some of the sunlight falling on the planet and create absorption lines that reveal the composition and density of the gas (see Unit 24). From such observations, we have learned that, in addition to carbon dioxide, Venus's atmosphere contains about 3.5% nitrogen and trace amounts of water vapor and other gases.

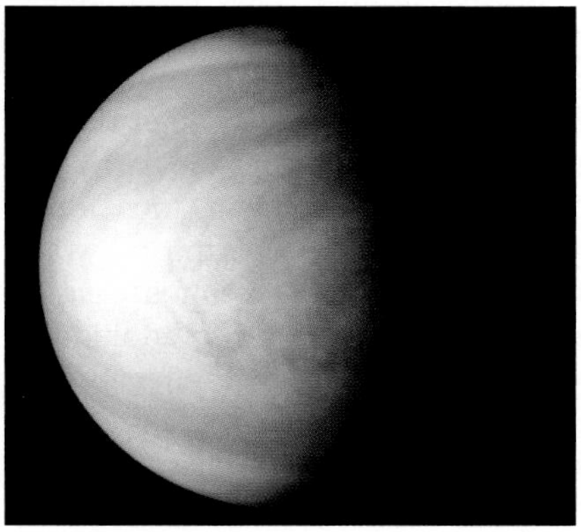

Instruments aboard the European Space Agency's *Venus Express* satellite reveal that Venus has lightning storms but no rain. Spectra reveal that the Venusian clouds are composed of sulfuric acid droplets that form when sulfur compounds—perhaps ejected from volcanoes—combine with the traces of water in the atmosphere. These clouds permanently cover the planet and are very high and thick; in fact, they are so thick that no surface features can be seen through them with ordinary telescopes. However, below the clouds, the Venusian atmosphere is relatively clear, and some sunlight penetrates to the surface. The light is tinged orange, because the blue wavelengths are absorbed in the thick cloud layer.

In Venus's upper atmosphere, wind speeds can exceed 350 kilometers per hour (210 mph), but near the surface the winds move much more slowly, just a few

FIGURE 41.1
Photograph through ultraviolet filter of the clouds of Venus. The picture is artificially colored and enhanced to show the clouds more clearly.

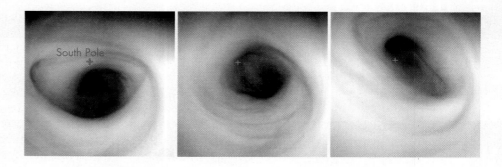

FIGURE 41.2
Series of images of Venus's southern polar vortex made by ESA's *Venus Express* spacecraft at 24-hour intervals. The images are taken in infrared light, and darker parts of the image correspond to where the cloud tops are deeper down. Each image is about 4000 km (about 2500 miles) across.

kilometers per hour. The motion of the atmosphere is driven by the Sun's heating near the equator, which causes the gas to expand most there. Its upper layers then flow toward the cooler polar regions, where they sink and flow back toward the equatorial regions. This produces a huge vortex near each pole, like the water running down a drain. The rapidly changing shape of these vortices has been studied by the European Space Agency's *Venus Express* mission (Figure 41.2).

On Venus's surface, the atmosphere exerts a pressure roughly 90 times that of the Earth's, or "90 atmospheres"—equivalent to the pressure you would feel under 900 meters (about 3000 feet) of water. Its atmosphere is also extremely hot—hot enough that lead would melt there. Observations made with radio telescopes from Earth show that the surface temperature is more than 750 K (about 900°F), a value confirmed by Soviet spacecraft landers in the 1970s. The dense atmosphere allows the surface temperature to cool by only a few degrees at night.

What makes the atmosphere so hot on a planet that is so similar to the Earth in size and only slightly nearer the Sun? Venus's carbon dioxide atmosphere (Table 41.1) creates an extremely strong greenhouse effect. We discussed in Unit 38 how carbon dioxide in Earth's atmosphere is a **greenhouse gas**—that is, it allows sunlight to enter and warm the surface but prevents much of the infrared radiation emitted by the heated surface from escaping to space. On Earth the greenhouse effect is mild. On Venus, however, with about 300,000 times more carbon dioxide than Earth, the greenhouse effect is dramatically stronger. It is so effective at trapping heat that Venus's surface is hotter than Mercury's.

Venus has suffered what scientists term the **runaway greenhouse effect,** in which greenhouse heating has driven the temperature up, and the higher temperature has led to an even stronger greenhouse effect. When it was young, Venus was probably much more similar to Earth. Water would have been delivered by the same mechanisms that gave Earth its oceans, such as comet impacts or volcanic outgassing. This is corroborated by the thermal characteristics of rocks in the higher regions detected by the *Venus Express* satellite. These resemble rock types on Earth that form in the presence of water, suggesting that Venus once had oceans too. Oceans on Venus would have experienced stronger solar heating than on Earth, and would have generated higher levels of water vapor. Water vapor is also a greenhouse gas, so as it evaporated, it would have made Venus's atmosphere even hotter. In addition, as the oceans warmed they would begin releasing the carbon dioxide that water absorbs. With a higher level of greenhouse gases, more heat would be trapped, and the ocean temperature would rise further. The warmer water would put more water vapor and carbon dioxide into the atmosphere, further increasing the heat trapping, spiraling out of control until the oceans were entirely boiled away.

Once water vapor drifts to the upper atmosphere, it can be broken down by **photodissociation,** a process in which ultraviolet photons from the Sun split the molecule into its hydrogen and oxygen atoms. Hydrogen atoms have a low mass, so they travel the fastest in a gas of a particular temperature. Once broken apart from the molecule, they would often have a high enough velocity to escape into space. In

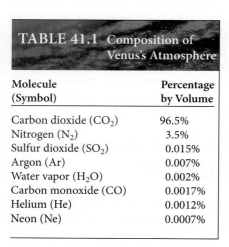

TABLE 41.1 Composition of Venus's Atmosphere

Molecule (Symbol)	Percentage by Volume
Carbon dioxide (CO_2)	96.5%
Nitrogen (N_2)	3.5%
Sulfur dioxide (SO_2)	0.015%
Argon (Ar)	0.007%
Water vapor (H_2O)	0.002%
Carbon monoxide (CO)	0.0017%
Helium (He)	0.0012%
Neon (Ne)	0.0007%

Concept Question 1

Would it be possible to make Venus a habitable place for humans if we could somehow block enough of the Sun's light so that Venus received an amount similar to what Earth receives?

fact, measurements of hydrogen escaping to space made by the *Venus Express* spacecraft are consistent with the idea that Venus used to have much more water, with oceans probably covering much of its surface billions of years ago.

Although Venus today has a much more substantial atmosphere than Earth, this does not consider the total amount of volatiles on Earth. We should compare Venus's atmosphere to the combined content of the Earth's oceans and atmosphere. On Earth, almost all of the carbon dioxide that was originally in the atmosphere has been locked away—dissolved in the oceans and incorporated in rock and soil. If there are processes that could release the carbon back into the atmosphere, Earth could have a carbon dioxide atmosphere even more massive than Venus's. Could a runaway greenhouse effect ever happen to Earth? We do not know if there might be a "tipping point" at which additions of greenhouse gases trigger the runaway effect, but once begun, it might be impossible to stop.

41.2 THE SURFACE AND INTERIOR OF VENUS

Eight Soviet *Venera* landers descended to the Venusian surface between 1970 and 1981, sampling its atmosphere as they parachuted to the surface.

The surface of Venus

The surface of Venus is hidden beneath its thick clouds, but several Soviet *Venera* spacecraft succeeded in transmitting pictures back to Earth from the Venusian surface. The cameras on these spacecraft scanned a narrow strip from horizon to horizon, providing a peculiar perspective. Views from the two cameras were combined in Figure 41.3 to give a better sense of what Venus's surface might look like to someone standing on the surface. The pictures show a barren surface covered with flat, broken rocks and lit by the pale orange glow of sunlight diffused through the deep clouds. Distant hills on the horizon can be seen beneath a yellow sky. The landers generally survived little more than an hour on the planet's surface because of the high temperatures and pressures there, but these robotic spacecraft were able to sample the rocks, showing them to have volcanic origins.

Much of what astronomers know about the geology of the Venusian surface comes from radar maps. Just as radar can penetrate terrestrial clouds to show an

FIGURE 41.3
Two cameras each on the *Venera 13* (left) and *Venera 14* (right) landers scanned narrow strips that overlapped at each end. The images have been pasted together here to give a more natural perspective. Distant hills are visible, as are nearby volcanic rocks. A color panel and portions of the spacecraft are visible in the lower part of the images.

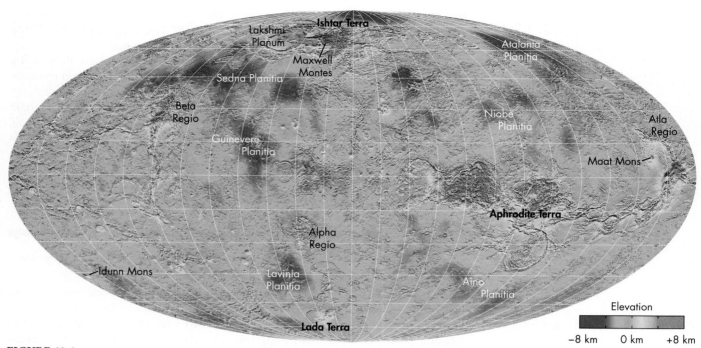

FIGURE 41.4
Global radar map of Venus made by the spacecraft *Magellan* in orbit around Venus. Colors indicate the relative height of surface features. Lowlands are blue; high elevations are red.

By convention, surface features on Venus are named after real or mythological females. The only feature named in honor of a man is Maxwell Montes—named after James Clerk Maxwell, the English physicist who first worked out the mathematical principles of electromagnetism. This mountain was one of the few features identified on Venus from Earth-based radar before more-detailed radar maps were made by orbiting spacecraft.

Concept Question 2

If astronauts were to someday explore the surface of Venus, what kind of protection would they need?

aircraft pilot a runway through fog, so too radar penetrates the Venusian clouds, revealing the planet's surface. Figure 41.4 shows a radar elevation map of Venus's surface made with *Magellan*, a U.S. spacecraft that orbited the planet in 1993–1994. The map is color coded by elevation. Venus is less mountainous and rugged than Earth, with most of its surface being low, gently rolling plains.

Two major highland regions, **Ishtar** and **Aphrodite,** rise above the lowlands to form landmasses similar to terrestrial continents. Ishtar, named for the Babylonian goddess of love, is about the size of Australia and is studded with steep volcanic peaks, the highest of which, **Maxwell Montes,** rises about 11 kilometers (7 miles) above the surrounding plateau. Aphrodite, named for the Greek goddess of love, is about the size of South America and is located near Venus's equator.

Venus is so similar in diameter and mass to the Earth that geologists were surprised by how different Venus's surface is from the Earth's. Together Ishtar and Aphrodite compose only about 8% of Venus's surface—a far smaller fraction than for Earth, where continents and their submerged margins cover about 45% of the planet. Rock on Venus may differ in some characteristics because of the absence of water. We saw that the behavior of rock on Earth is different when water is present (Unit 38). In the absence of water, rock melts at a higher temperature and can form steeper mountains. This would explain some very steep mountains found on Venus, such as Maxwell Montes, and it may help explain the differing behavior of the crustal rock in response to internal heating.

The radar maps show many volcanic landforms on Venus, but these are different from those on Earth. Volcanoes do not lie along lines marking the edges of tectonic plates (Unit 37). Several high-resolution radar images of Venus's surface are shown in Figure 41.5. These include peaks with immense lava flows, "blisters" of lava called **pancake domes,** and grids of long narrow cracks or **faults.**

We can estimate the age of Venus's surface by comparing the number of the largest impact craters found there—those with diameters greater than a few kilometers—with

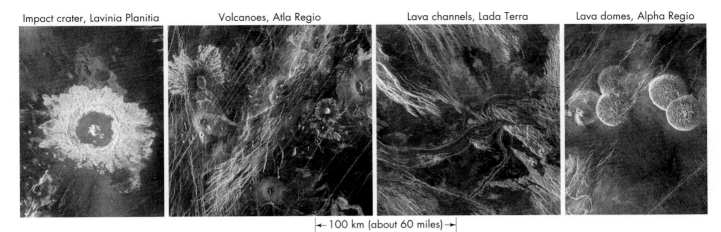

FIGURE 41.5

Magellan radar images of a variety of features on Venus, all shown to the same scale. The views look straight down at the surface, with the radar illumination from the left. In addition to the individually noted features, most of these images show long cracks or faults in the bedrock.

Small craters do not provide a good basis for comparison between planets because smaller impacting bodies may be "burned up" or substantially reduced in size by friction with Venus's dense atmosphere before they strike the surface.

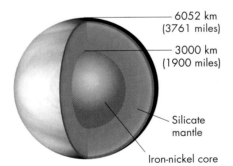

FIGURE 41.6
Sketch of the internal structure of Venus, showing the radius of the core and surface.

the number of those found on the Earth. From the small number of craters, the average age of Venus's surface appears to be roughly 500 million years, with some regions less than 10 million years old, which is similar to Earth. This presents us with a puzzle, though. Earth's young surface is a result of plate tectonics (Unit 37); but on Venus we do not see evidence of the same kind of geological activity. Geologists disagree about the mechanism that caused the renewal of Venus's surface.

The interior of Venus is probably similar to the Earth's, an iron core and rock mantle (Figure 41.6). Planetary geologists have no seismic information, so they must rely on deductions from its density, which is just slightly less than the Earth's, and measurements of the local gravity made by orbiting spacecraft, which indicate that Venus's crust is thicker than Earth's.

Why then are the surface features so different? Some geologists suspect that volcanic activity may be periodic on Venus. Its thick crust may effectively trap the internal heat. As the interior heats up, it might then gradually melt the bottom of the crust, thinning it and allowing it to break up. This may then produce widespread volcanic activity, flooding much of the planetary surface with lava in a brief time. After the internal heat is released, the interior cools, and the crust again thickens. The thicker crust causes heat to be trapped once more, and the process repeats at intervals of hundreds of millions of years.

Another hypothesis is that rock in Venus's mantle does not circulate in the same kind of large convection currents found in the Earth's mantle. If the mantle is not undergoing convection, there would be no plate tectonics. For example, a few locations on Earth, such as Yellowstone Park and the Hawaiian Islands, appear to be heated intensely but locally by "plumes" of rising heat in the mantle. It may be that Venus's internal geology is dominated by such localized plumes. As heat wells upward, it bulges the crust, stretching and cracking it. This might explain the numerous volcanic peaks, domes, and uplifted and fractured surface regions.

According to yet another hypothesis, Venus's mantle is circulating much as the Earth's; but because the crust is thicker and the rock more rigid, it behaves differently. Venus's crust may not so readily break into the crustal plates we find on Earth. At points above where the hot material rises within the mantle, the crust bulges upward, stretches, and weakens; volcanoes form, but without producing the motion of plates seen on Earth. Where the mantle material circulates back down, portions of the crust are pushed together, thickening and crumpling the crust into

FIGURE 41.7
Computer-generated view of Venus's volcano Idunn Mons based on *Magellan* radar data is shown at left. The volcano appears to have relatively recent lava flows surrounding it, based on infrared measurements made by the *Venus Express* satellite shown in image on right.

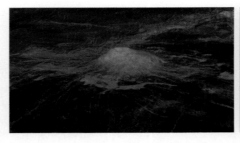
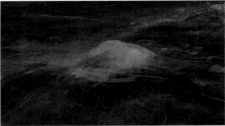

regions like Aphrodite and Ishtar. This might create a planet with few and small continents and whose surface is active but only in isolated spots. The push and pull on the bottom of the crust by convection flows in the mantle may also explain many of the faults seen on the surface of Venus, and may gradually erase surface features such as impact craters.

Eruptions have not been seen directly, but some lava flows appear very fresh. Differences in the infrared radiation from some volcanic peaks (Figure 41.7) seen by *Venus Express* suggest that these are relatively recent lava flows. In addition, electrical discharges, perhaps lightning, have been detected near some of the larger peaks. On Earth, volcanic eruptions frequently generate lightning, and some astronomers think the electrical activity indicates that Venus's volcanoes are still erupting. Such eruptions might also explain brief increases in sulfur content detected in the Venusian atmosphere, changes similar to those produced on Earth by eruptions here.

41.3 ROTATION OF VENUS

Venus spins on its axis more slowly than any other planet in the Solar System, taking 243 days to complete one rotation. Moreover, Venus is unique among the terrestrial planets in that it has a **retrograde spin.** That is, compared with the direction of rotation of the other terrestrial planets, it spins backward. Thus, the Sun rises in the west and sets in the east. Because Venus's rotational period is longer than its orbital period of 225 days, you might at first expect that Venus's day would be longer than its year. That would be true if Venus's spin were in the same direction as its orbit; but by turning in a retrograde direction, a point on the surface turns back to face the Sun more quickly than if it weren't rotating at all—in about half a Venusian year (117 days), as illustrated in Figure 41.8.

A retrograde spin cannot be caused by tidal braking (Unit 39.5), which can only work to slow a planet's spin. However, if Venus had a faster retrograde spin in the distant past, the Sun's tidal effects could have slowed it down to its current rate. Some astronomers hypothesize that Venus was struck late in the formation process by a large planetesimal that collided in a direction that set the planet spinning backward, then solar tides slowed it to its present rate. Venus's slow rotation, however it was caused, also means that despite Venus probably having a molten iron core like the Earth's, the dynamo process for generating a magnetic field does not operate strongly there. In fact, Venus's magnetic field is tens of thousands of times weaker than Earth's.

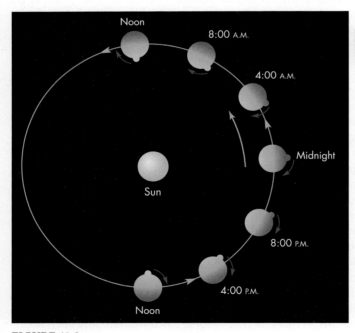

FIGURE 41.8
Venus's rotation is illustrated for a spot on its surface. Venus's slow backward spin results in a day that is about 117 Earth days long—more than half of Venus's year.

Concept Question 3

If the Earth spun as slowly as Venus, what effects might that have on our environment?

KEY POINTS

- Venus is very similar to the Earth in overall size and density, but its dense atmosphere is almost entirely carbon dioxide.
- The CO_2 atmosphere has created conditions for a runaway greenhouse effect, making Venus hotter than Mercury.
- Liquid water may once have been present, but it boiled away, and sunlight broke down the H_2O, allowing the hydrogen to escape.
- The surface has been mapped by radar, showing two small "continents"; smooth low-lying lava plains cover most of the surface.
- Radar images show volcanoes and crustal bulges from internal heating, but it is unclear when volcanic activity last occurred.
- The planet's thick crust may prevent tectonic activity, but geological activity must be occurring to erase impact craters.
- Venus rotates backward slowly, perhaps caused by a major collision when it formed; its day is about half the length of its year.

KEY TERMS

Aphrodite, 312
fault, 312
greenhouse gas, 310
Ishtar, 312
Maxwell Montes, 312
pancake dome, 312
photodissociation, 310
retrograde spin, 314
runaway greenhouse effect, 310

CONCEPT QUESTIONS

Concept Questions on the following topics are located in the margins. They invite thinking and discussion beyond the text.

1. Modifying Venus to be more Earthlike. (p. 310)
2. How astronauts might survive on Venus. (p. 312)
3. Environment on a slowly spinning Earth. (p. 314)

REVIEW QUESTIONS

4. How does Venus compare with the Earth in mass and radius?
5. How does Venus's surface differ from Earth's? How have astronomers determined what the surface of Venus is like?
6. What gas dominates Venus's atmosphere? How is this known?
7. What is the runaway greenhouse effect? Could it happen here?
8. What do crater counts on the Earth and Venus imply about these planets' past geological activity?
9. How is Venus's rotation unusual? What might have caused this?
10. Contrast the magnetic fields of Earth, Mercury, and Venus, and explain why these magnetic fields are so different.

QUANTITATIVE PROBLEMS

11. Venus's atmosphere contains about 4.6×10^{20} kg of carbon dioxide (CO_2). If all of the oxygen atoms in the CO_2 were combined with hydrogen to make water (H_2O) instead of carbon dioxide, what would the total mass of water be? (Note: Hydrogen has an atomic mass of 1; carbon, 12; oxygen, 16.) Compare this to the mass of Earth's oceans, $\sim 1.4 \times 10^{21}$ kg.
12. From your answer to Problem 11, how deep an ocean would this water produce if it were the same depth everywhere on the surface? (Note that 1000 kg of water fills a volume of 1 m^3.)
13. Kepler's third law (Unit 12.2) describes planetary orbital sizes relative to Earth's orbit, which is assigned a size of 1 AU. One method for finding absolute orbital sizes is to use radar ranging between planets. Use Venus's scaled orbital size of 0.723 AU, and the observation that a radar signal takes 276 sec to make a round-trip between Earth and Venus when they are minimally separated, to find how many kilometers are in an AU.
14. Venus and the Earth make their closest approach to each other (inferior conjunction, Unit 11) every 583.9 Earth days. Use a diagram like Figure 41.8 or math to show that nearly the same side of Venus faces Earth at each closest approach.
15. The Earth covers about 1° per day in its orbit about the Sun, and the solar day is slightly longer than the sidereal day (Unit 7). If Earth spun in a retrograde direction like Venus but it still had the same sidereal period (23 hr 56 min), how long would the solar day be? Use a diagram to explain your answer.

TEST YOURSELF

16. Why is Venus's surface hotter than Mercury's?
    a. Venus rotates more slowly, so the Sun "bakes" it more.
    b. Mercury's surface reflects more sunlight preventing it from heating as much as Venus.
    c. Carbon dioxide in Venus's atmosphere traps heat radiating from its surface, thereby making it warmer.
    d. Recent volcanic activity on Venus has heated its atmosphere.
    e. Friction from winds blowing across Venus's surface heat it more than most planets.
17. Water vapor in Venus's atmosphere
    a. was probably destroyed by the Sun's ultraviolet light.
    b. is responsible for the runaway greenhouse effect.
    c. was never present in the planet's history.
    d. is responsible for Venus being covered in clouds.
18. What evidence is there that the surface of Venus was covered by giant flows of lava a few hundred million years ago?
    a. The surface is still hot.
    b. There are relatively few impact craters on Venus.
    c. Radioactive dating was used to determine when it cooled.
    d. The number of cracks in the lava indicate how old it is.
19. Some astronomers think Venus once had oceans. If that is true, where did the water go? The water
    a. flowed down through large cracks in the crust.
    b. was converted into carbon dioxide by solar fusion.
    c. is trapped in volcanic glass "bubbles" in the rock.
    d. boiled, and sunlight broke it into hydrogen and oxygen.
20. What feature of Venus best explains its lack of a magnetic field?
    a. Its dense atmosphere
    b. Its slow rotation rate
    c. Its low core temperature
    d. Its lack of seawater

PART 3 UNIT 42

Mars

- 42.1 The Major Features of Mars
- 42.2 A Blue Mars?
- 42.3 The Martian Atmosphere
- 42.4 The Martian Moons

Learning Objectives

Upon completing this Unit, you should be able to:
- Describe the unique surface features of Mars and how they likely formed.
- Describe missions to Mars, in the past and currently, and their basic findings.
- Discuss the evidence for liquid water on Mars in the past, and where it is now.
- Compare the geology and atmosphere of Mars and Earth.
- Describe Mars's moons and their possible origin.

Mars is named for the Roman god of war, presumably because of its "blood red" color (Figure 42.1). Compared with the harsh environments of Mercury and Venus (Units 37 and 38), Mars seems positively Earth-like. Although its diameter is only about half Earth's and its mass about one-tenth Earth's, its surface and atmosphere are less alien. Mars is colder than the Earth because it orbits 1.52 AU from the Sun; but on a warm day, the temperature at the Martian equator may rise above the freezing temperature of water to about 283 K (about 10°C or 50°F).

Mars rotates with a period similar to Earth's, with a day just 40 minutes longer than our own. Mars has a 25° tilt to its polar axis, similar to the Earth's 23.5° tilt. Thus, Mars experiences seasonal changes like Earth's during its own year of 1.88 Earth years—which takes 669 Martian days. Mars resembles the Earth in so many ways that many searches have been made for life on the planet, but there is no strong evidence that life ever existed there, as is discussed further in Unit 86. Even though life has not been found on Mars, there is growing evidence that a self-sustaining base on the planet is possible, potentially paving the way for future human exploration.

42.1 THE MAJOR FEATURES OF MARS

Telescopic views hinted at the intriguing geology of Mars; but space-based pictures, the legacy of many spacecraft missions, reveal the planet's true marvels. The first successful mission to Mars was a flyby in 1965 by NASA's *Mariner 4*. Over the subsequent decades, orbiters and landers from the U.S., USSR, and Europe have revealed much more. Several spacecraft are currently taking high-resolution pictures and making spectroscopic analyses from orbit around Mars: the European Space Agency's *Mars Express* spacecraft, along with NASA's *Global Surveyor*, *Odyssey*, and *Mars Reconnaissance Orbiter*. NASA also landed rovers, two of which are currently exploring the surface: *Opportunity*, which landed in 2004 and has traveled more than 35 kilometers, and *Curiosity*, which landed in 2012 and is still early in its mission to study Martian geology and habitability. Several more missions are planned before 2020.

There is a variety of indirect evidence that Mars's interior is differentiated, like the Earth's, into a crust, mantle, and iron core. For example, using spacecraft in orbit around the planet, scientists have measured Mars's magnetic field and internal structure from

FIGURE 42.1
Picture of Mars constructed from *Viking Orbiter* images.

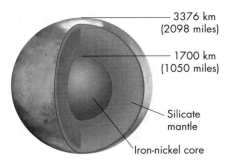

FIGURE 42.2
Sketch of the internal structure of Mars showing the radius of the core and surface.

detailed measurements of its gravitational field. This evidence implies that Mars has a metallic core with about half the overall radius of the planet (Figure 42.2). This is very similar to the fractional size of the Earth's iron core, but in Mars the iron core appears to have a significant fraction of sulfur—which helps explain why its overall density is lower. Recent measurements of the way Mars is distorted by tidal effects of the Sun also indicate that its core is partially molten, but not to the degree of the Earth's core. Spacecraft have also detected the magnetization of ancient volcanic flows, indicating that Mars's core generated a much stronger magnetic field when the planet was young.

The elevation map in Figure 42.3 shows that much of the planet is cratered like the Moon and Mercury. Mars shows a major dichotomy between hemispheres (somewhat similar to the difference to the Moon's near and far sides), with the northern hemisphere several kilometers lower in elevation and the crust perhaps 20 km thinner than in the southern hemisphere. The origin of this difference is uncertain, but it may have resulted from major impacts as Mars formed.

There are also several distinctive features visible on a global scale that indicate Mars has had an active and complex geology. Most of the most remarkable features are concentrated around the highest region on the planet, called the **Tharsis bulge**. This spans the equator and is about the size of North America. It is dotted with volcanic peaks. Three of these can be seen poking through the clouds along the western side of the planet in Figure 42.1.

Another of these remarkable features is a huge chasm called **Valles Marineris** whose western end lies near the center of the Tharsis bulge. This is also visible across the face of Mars in Figure 42.1. Named for the *Mariner* spacecraft, in whose pictures it was first seen, it is 4000 kilometers (2500 miles) long, 100 kilometers (60 miles) wide, and 10 kilometers (6 miles) deep. This chasm would span the continental United States, dwarfing the Grand Canyon, which is less than a tenth as long or a fifth as deep. A view of what Valles Marineris would look like from

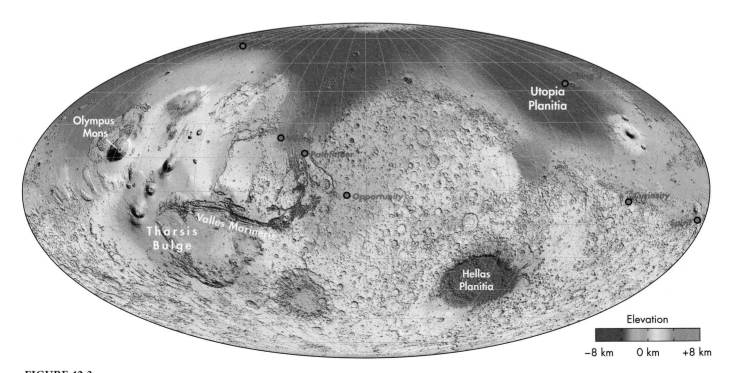

FIGURE 42.3
Topographic map of Mars showing its major features. The map is color coded according to elevations. Several of the major features are named, and the landing sites of the seven craft to successfully reach the surface are also indicated.

FIGURE 42.4

A reconstructed view down Valles Marineris, the Grand Canyon of Mars. This image was constructed from hundreds of thousands of laser altimeter measurements made by the *Mars Odyssey* orbiter. This enormous gash may be a rift that began to split apart the Martian crust but failed to open further. The canyon is about 4000 kilometers (approximately 2500 miles) long. Were it on Earth, it would stretch from California to Florida.

Concept Question 1

Examining the image of Olympus Mons, in what ways can you distinguish between an impact crater and the crater (caldera) of a volcano?

a high-flying airplane is shown in Figure 42.4, based on detailed laser altimetry measurements.

A huge volcano called **Olympus Mons** (Figure 42.5), is at the edge of the Tharsis bulge. It rises to nearly three times the height of Earth's highest peaks, some 26 kilometers (15 miles) above its surroundings, making it the biggest volcano in the Solar System. If interplanetary parks are ever established, Olympus Mons should lead the list!

Mars's immense volcanoes give testimony to an active geological past. However, Mars does not show evidence of large-scale crustal motion, like folded mountain ranges generated by plate tectonics on Earth. Astronomers therefore think that Mars cooled relatively rapidly. As a result, its now weak internal heat sources can no longer break its crust, which is about twice as thick as Earth's, to drive tectonic motion. The thick Martian crust may also explain why Mars has a small number of very large volcanoes. The heat from the interior may have found a few weaker regions to break through, in contrast to the Earth, where a thin, shifting crust allows less dramatic volcanic activity to occur at a large number of sites.

Geologists think that the Tharsis region formed as hot material rose from the deep interior of the planet. As this molten material neared the crust, it forced the surface upward. The hot matter then erupted through the crust to form the volcanoes,

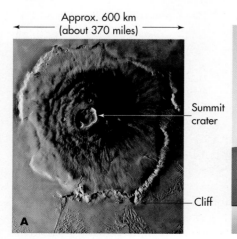

FIGURE 42.5

(A) Image from orbit of Olympus Mons, the largest known volcano in the Solar System. (B) Profile of Olympus Mons; Mount Everest, the highest mountain above sea level on Earth; and the Hawaiian volcano Mauna Kea, rising from the seafloor.

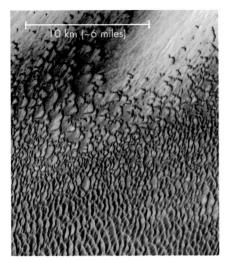

FIGURE 42.6
Mars Odyssey image of dunes around the northern polar cap. Color ranges from blue for colder and yellow for warmer, based on the thermal imaging system.

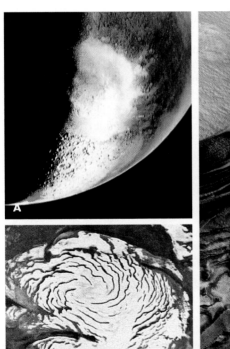
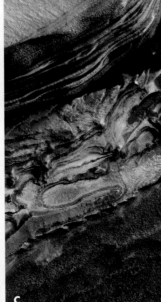

FIGURE 42.7
Pictures of (A) the south Martian polar cap and (B) the north Martian polar cap. (C) An image from the *Mars Reconnaissance Orbiter* shows the wall of a chasm approximately 1 kilometer (0.6 mile) deep in the north polar cap. The image covers a section of the ice cliff about 1.3 kilometers wide. The alternating layers of ice are mixed with more or less dust, laid down during cycles of changing climate conditions.

some of which appear to be relatively young. For example, the small number of impact craters on its slopes implies that Olympus Mons was probably last active less than 250 million years ago. Some planetary geologists think the Tharsis bulge may be linked to the formation of Valles Marineris. As the Tharsis region swelled, it stretched and cracked the crust. Perhaps this vast chasm is something like the early stages of plate tectonic activity on Earth, circulation in the Martian mantle having begun to split the planet's crust. Unlike on Earth, however, the tectonic activity ceased as the planet aged and cooled.

Mars's current low level of tectonic activity is also demonstrated by the many craters that cover most of its terrain—far more than are seen on either Earth or Venus. The number of those craters implies that most of Mars's surface has been geologically quiet for billions of years. Thus, Mars appears to be entering a phase of planetary old age similar to Mercury or the Moon. Unlike them, though, Mars retains an atmosphere, and winds sweep its surface. This produces immense deserts, at midlatitudes, with dunes blown into parallel ridges (Figure 42.6). The dust has been blown into a thin layer all over the planet, and its iron oxide content gives Mars its distinctive rusty red color.

At its poles, Mars has frozen polar caps (Figure 42.7). These frozen regions change in size during the cycle of the Martian seasons. Although the tilt of Mars's rotation axis is similar to Earth's, the Martian seasons are more extreme because the Martian atmosphere is much less dense. The thin atmosphere does not retain heat well, permitting greater temperature extremes. Because Mars's seasonal changes are so extreme, its polar caps vary greatly in size, shrinking during the Martian summer and growing again during the winter.

Martian seasons also show effects due to its elliptical orbit. Mars's distance from the Sun varies by about 45 million kilometers during its orbit. This represents a 20% change in Mars's distance from the Sun and means that the solar heating Mars receives varies substantially during its orbit compared to the Earth, whose distance from the Sun varies by only about 3%. Mars is farthest from the Sun during its southern winter and closest during southern summer. As a result, the seasons in the south are more extreme, and the southern polar cap shows dramatic changes in size. The visible changes in the polar caps appear to be due primarily to frozen carbon dioxide—dry ice. In winter, its frost extends in a thin layer across a region some 5900 kilometers (about 3660 miles) in diameter, from the south pole to latitude 40°, much as snow cover extends to middle latitudes such as New York during our Earth winters. But because the frost is very thin over most of this vast Martian ice cap, it shrinks in the summer to a diameter of about 350 kilometers (approximately 220 miles).

The northern hemisphere seasons are less extreme because the planet is farther from the Sun in summer and closer in winter. The northern cap shrinks less, to a diameter of about 1000 kilometers (about 600 miles). The northern cap consists of numerous separate layers called **laminated terrain,** as can be seen in Figure 42.7C, with each layer representing a separate long-term episode of ice deposition. Planetary scientists think that these strata are evidence for cyclical changes in the Martian climate like Earth's Milankovitch cycles (Unit 38). Such change might occur because Mars's

> The hemisphere that experiences summer when Mars is closest to the Sun alternates as Mars precesses. This produces especially hot summers in that hemisphere that lead to severe dust storms.

polar axis precesses over a period of about 50,000 years. This means that 25,000 years ago Mars's northern hemisphere would have had more extreme seasons and the southern hemisphere less extreme. Patterns of dust storms would vary with these global climatic changes, leading to darker and lighter layers of ice, each laid down over the course of tens of thousands of years.

Although the polar regions have a surface layer of CO_2, the bulk of the frozen material in the caps is ordinary water ice, which remains almost permanently frozen. A ground-penetrating radar instrument aboard the *Mars Express* orbiter has been able to probe through the layers of ice at the poles. It showed that each polar cap contains about the same volume of ice averaging about 2 kilometers (1.2 miles) thick. If both caps melted, there would be enough water to cover the entire surface of Mars to an average depth of 22 m (72 feet).

42.2 A BLUE MARS?

Perhaps the most intriguing features revealed by the two *Viking* orbiters in the 1970s were dry riverbeds, such as those seen in Figure 42.8A. We infer that water once flowed on Mars along these winding channels, which resemble tributaries converging to make a large river channel on Earth. Some of the channels appear to have once had major flows that carved teardrop-shaped islands around crater rims (Figure 42.8B). There are even features that suggest there once were standing bodies of water. Such features include smooth terraces that look like old beaches around the inner edges of craters and basins, as you can see in the "crater lake" in Figure 42.8C. Narrow canyons breach this crater's rim, showing where water flowed in from the south and drained out into lowland areas to the north, with the ancient shoreline still visible set in from the crater rim even though no surface liquid is present now.

The images from the *Viking* orbiters provided strong evidence that liquid water was once present on Mars, but they raised even more questions. How long ago was the water present? Was it ever a long-term feature of Mars, or did it occur in violent

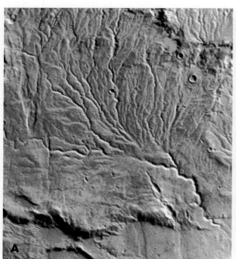 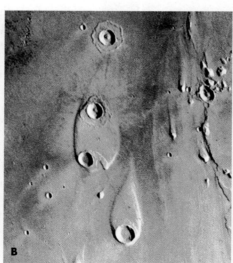 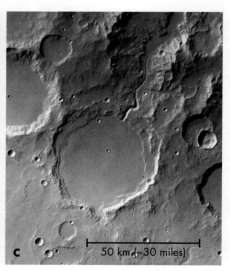

FIGURE 42.8
Images suggesting past water on Mars from the *Viking* orbiter. All three images have the same scale and are oriented with north at top. (A) Picture of channels probably carved by running water on Mars. (B) Teardrop-shaped islands formed as water flowed (from bottom toward top of figure) around the rims of craters. (C) A Martian crater thought to have once been a "crater lake." Note the inflow channel at the south end of the crater, and the outflow channel to the northeast. The smooth floor (apart from a few small craters) suggests that the crater bottom is covered with sediment left behind as the lake dried out

FIGURE 42.9
A panoramic view of the Bonneville Crater, lying in the floor of the much larger Gusev Crater on Mars. This spot is near where the *Spirit* spacecraft landed.

episodes of melting? For example, if there was ice buried under the soil, perhaps it was melted by an impact, with a sudden catastrophic flood. Furthermore, some features, such as the "crater lake" in Figure 42.8C, could also be interpreted as arising from lava flows, so were there ever standing bodies of water on Mars? And does much water remain today, perhaps frozen in the ground, or has it almost all evaporated and dissociated as on Venus, with the hydrogen lost into space?

The *Viking* landers provided our first view of the Martian surface in the 1970s, and in the 1990s the Mars *Pathfinder* mission provided a demonstration that we could operate a roving science vehicle across the surface of Mars. The next two NASA landers, *Opportunity* and *Spirit*, reached Mars in 2004 and carried out a remarkable set of explorations that far exceeded the original plan. These two missions were designed to explore Mars's surface, landing at sites that were chosen because pictures and spectral data taken from orbit suggested that water might have been present there long ago. *Spirit* landed in the center of the 150-km (90-mile) diameter Gusev Crater, a smooth-floored crater at the end of a narrow Martian valley that appeared to have once been flooded. Figure 42.9 shows a panorama from the *Spirit* lander site. *Opportunity* landed on the flat plains along the Martian equator on the opposite side of Mars from *Spirit*. Both craft deployed rovers—small, wheeled vehicles that can move away from the landing site and explore interesting features.

Both rovers were highly successful in their searches. For example, *Opportunity* took the pictures shown in Figure 42.10, which both suggest processes that involved liquid water. Figure 42.10A shows small spheres (dubbed "Martian blueberries") that are made of hematite, which normally occurs from depositions of minerals in water. Figure 42.10B shows a rock outcropping thought to be material deposited in an ancient, now dried-up small sea. Examination of the rocks shows that they contain layers that are typical of sediment that sank to the bottom of a body of water and later was transformed into rock. This image also shows that the layers are wavy, similar to the ripple marks you see at the beach as water washes back and forth across the sand. Moreover, minerals in the rocks at the *Opportunity* site have a chemical makeup consistent with their having been deposited in a salty lake or small ocean. Half a world away *Spirit* found more layered rocks and other minerals that normally form in water. The evidence from the rovers has convinced many scientists that Mars once had large areas under liquid water.

FIGURE 42.10
(A) Small spherical "blueberries" most likely formed from iron depositions in standing water. The lighter circular area was swept off by the *Opportunity* rover to study the rock type. (B) A close-up image of a rock outcropping at the landing site. The rocks show thin layers and contain minerals that suggest that they were formed on the bottom of a salty lake.

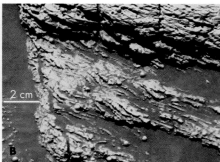

FIGURE 42.11

Opportunity traveled to Victoria Crater, imaged at right by the *Mars Reconnaissance Orbiter*. The interior edge of this crater (visible near the horizon in the image at left from *Opportunity*) exposed layers of subsurface rock when the rim collapsed and slid into the crater.

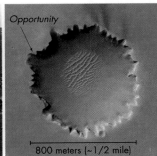

The original mission for the two rovers was planned for only 90 days, but the two rovers operated long past that, surviving cold winters and global dust storms. *Spirit* finally stopped responding in 2010, but *Opportunity* continues traveling over the Martian surface in 2013. *Opportunity* has traveled on to the rim of a large crater about 20 kilometers from its landing site, studying layered rock outcrops in an unusually shaped crater on its way (Figure 42.11).

Another important experiment was carried out by the *Phoenix* lander, which reached the northern polar region of Mars in 2008. *Phoenix* carried out experiments on the soil, showing that it is quite different from Earth's—about as alkaline as baking soda with high levels of oxidizing chemicals. As the lander scooped up soil samples, it exposed ice below the surface (Figure 42.12).

Meanwhile, a wide assortment of satellites have continued to send back detailed pictures of the surface of Mars with high resolution, allowing planetary scientists to build a more detailed understanding of Martian geology. The terraced layers in Figure 42.13A are at the bottom of one of Mars's great canyons. Such features on Earth are usually laid down underwater. A detailed examination of the scouring of features in Figure 42.13B implies that water carved these features with a flow exceeding that of the Mississippi River. Figure 42.13C shows a crater at a latitude of about 70° North that contains a large ice field, perhaps the remains of a lake.

In addition to being able to make highly detailed images, each orbiter includes a variety of other detectors. Some instruments can measure the spectrum of light reflected from the ground, others collect radiation at other wavelengths. From these data, geologists can deduce what minerals are present at different spots on the Martian surface. Matching the composition of those minerals with data on whether water is needed to produce those minerals gives additional evidence that

FIGURE 42.12

The *Phoenix* lander scooped up soil samples, revealing ice under the polar dust.

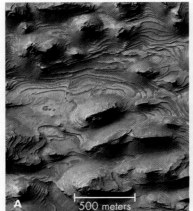

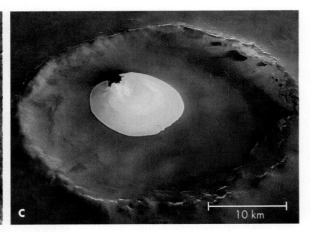

FIGURE 42.13

(A) Image from the *Mars Global Surveyor* of terraced features at the bottom of a Martian canyon. (B) *Mars Odyssey* image of former river bed in which teardrop-shaped islands formed behind craters, and the surface layers were scoured by the stream flow. (C) *Mars Express* image of a crater with a lake of ice in its interior.

FIGURE 42.14
Mars Reconnaissance Orbiter image of fanning outflow in a former crater lake. Spectral analysis of the surface features identified minerals normally found underwater, and different types of materials, such as clays, have been color coded in this image.

Concept Question 2

What materials besides water ice might explain the hydrogen detected in Mars's surface by the *Odyssey* spacecraft? Would these be likely?

FIGURE 42.16
Curiosity self-portrait made from a series of pictures by a camera on a robotic arm.

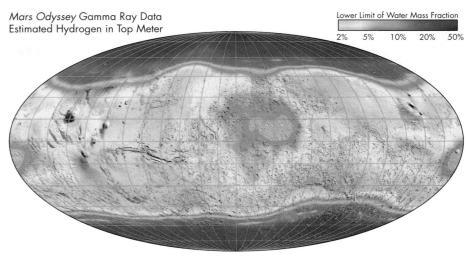

FIGURE 42.15
Global map of the likely percentage of water in Martian surface layers. *Mars Odyssey* measured high-energy radiation that is generated when cosmic rays from space strike the planet's surface. This radiation is reduced where there are hydrogen atoms, permitting scientists to estimate the amount of water that is present.

Mars was once much wetter. For example, spectroscopic analysis of the feature in Figure 42.14 indicates the presence of clays and minerals that form in water over thousands of years. Lava flows can sometimes resemble water features, but this mineralogical analysis provides strong confirmation that the fan of material in the figure was flowing into a standing lake or sea.

An instrument on the *Mars Odyssey* orbiter can detect hydrogen atoms locked up in water in the Martian soil by studying gamma rays. These data have revealed what appears to be huge amounts of water, probably ice, in the upper meter of the Martian surface (Figure 42.15). Based on the accumulating results, most planetary scientists think that Mars had extensive liquid water on its surface when it was young. It is still uncertain how long the water remained, but perhaps a billion years after the planet formed. It seems entirely possible that Mars might have been habitable, and perhaps life formed there.

In order to explore these possibilities, a sophisticated rover called *Curiosity* was sent to Mars in 2012 (Figure 42.16). One of the primary goals of Curiosity is to determine if Mars was indeed once habitable (see also Unit 86). It will climb Mount Sharp, which appears to contain a series of sedimentary deposits (Figure 42.17), and hopefully provide a much more detailed history of Mars's climate.

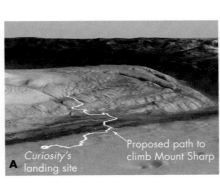 

FIGURE 42.17
(A) The rover *Curiosity* landed in Gale crater, near the base of Mount Sharp, which it will climb and study. (B) View of the layered structure of Mount Sharp as seen by *Curiosity*.

TABLE 42.1 Composition of the Martian Atmosphere	
Molecule (Symbol)	Percentage by Volume
Carbon dioxide (CO_2)	95.32%
Nitrogen (N_2)	2.7%
Argon (Ar)	1.6%
Oxygen (O_2)	0.13%
Carbon monoxide (CO)	0.08%
Water vapor (H_2O)	0.021%
Nitric oxide (NO)	0.010%
Neon (Ne)	0.00025%
Krypton (Kr)	0.00003%
Formaldehyde (CH_2O)	0.000013%
Xenon (Xe)	0.000008%
Ozone (O_3)	0.000003%
Methane (CH_4)	0.000001%

42.3 THE MARTIAN ATMOSPHERE

Clouds and wind-blown dust are visible evidence that Mars has an atmosphere. Spectra and direct sampling by spacecraft landers confirm this and show that the atmosphere is mostly (95%) carbon dioxide with small amounts (3%) of nitrogen and traces of oxygen and water (Table 42.1). From these studies, astronomers can determine the density of Mars's atmosphere, which turns out to be very low—only about 1% the density of Earth's.

This density is so low that, although the Martian atmosphere is mostly carbon dioxide, it creates only a very weak greenhouse effect. The consequent lack of heat trapping and Mars's greater distance from the Sun leaves the planet very cold. Temperatures at noon at the equator may reach a little above the freezing point of water, but at night they plummet to far below freezing. The resulting average temperature is a frigid 218 K (−67°F). Thus, although water exists on Mars, it is frozen solid, locked up either below the surface in the form of permafrost or in the polar caps as solid water ice.

Clouds of dry ice (frozen CO_2) and water-ice crystals (H_2O) drift through the Martian atmosphere carried by the Martian winds. These winds, like the large-scale winds on Earth, arise because air that is warmed near the equator rises and moves toward the poles. This flow from equator to poles, however, is deflected by the Coriolis effect arising from the planet's rotation. The result is winds that blow around the planet approximately parallel to its equator, and small tornadoes or "dust devils" (Figure 42.18A). The Martian winds are generally gentle, but seasonally and near the poles they become gales, which sometimes pick up large amounts of dust from the surface. The resulting vast dust storms occasionally cover the planet completely and turn its sky pink.

No rain falls from the Martian sky, despite its clouds, because the atmosphere is too cold and contains too little water. In fact, there is so little water in the Martian atmosphere that even if all of it were to fall as rain, it would make a layer only about 12 micrometers deep (less than 1/2000 inch). For comparison, Earth's atmosphere holds enough water to make a layer a few centimeters (inches) deep. Despite such dryness, however, fog sometimes forms in Martian valleys (Figure 42.18B), and frost condenses on the ground on cold nights (Figure 42.18C). In addition, during the Martian winter, CO_2 "snow" falls on the Martian poles.

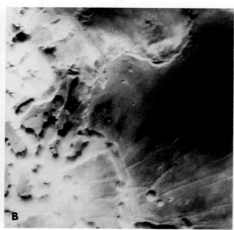

FIGURE 42.18
(A) A "dust devil" (small tornado) spotted by the *Mars Reconnaissance Orbiter*. (B) Fog in Martian valleys seen by the *Viking* orbiter. (C) Frost on surface rocks near the *Viking* lander.

FIGURE 42.19
Sequence of images of the wall of a crater from early to mid summer. As the crater warms, dark streaks extend down the crater wall, resembling wet patches. It is possible that a briny solution, melting and seeping out from underground layers, might produce these features. Pure water would evaporate too quickly to keep the soil moist.

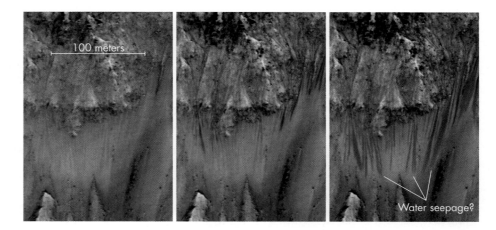

Mars has not always been so dry, as we saw from the numerous channels in its highlands and other evidence from many Mars missions. But for a planet to have liquid water, it must have a temperature above freezing with a high enough pressure to keep the liquid from immediately boiling away. If the pressure on a liquid is very low, molecules can break free from its surface, evaporating easily because no external force restrains them. On the other hand, if the pressure is high, molecules in a liquid must be heated strongly to turn them into gas. For example, at normal atmospheric pressure, water boils at 100° Celsius. If the pressure is reduced, however, the boiling point drops, an effect used by food producers to make "freeze-dried" foods, such as instant coffee. Coffee is brewed normally, then frozen and placed in a chamber from which the air is pumped out. The reduced pressure makes the liquid "boil" without heating, and it evaporates leaving only a powder residue—instant coffee. Similarly, on Mars, any liquid water on its surface today would evaporate.

The existence of channels carved by liquid water on Mars is therefore strong evidence that in its past, Mars was warmer and had a denser atmosphere. However, that milder climate must have ended billions of years ago. How do we know? The large number of impact craters on Mars shows that its surface has not been significantly eroded by rain or flowing water for about 3 billion years. Why did Mars dry out, and where have its water and atmosphere gone?

Some water probably lies buried below the Martian surface as ice, as indicated by measurements from the *Mars Odyssey* orbiting spacecraft. If the Martian climate was once warmer and then cooled drastically, water would condense from its atmosphere and freeze, forming sheets of surface ice. Wind might then bury this ice under protective layers of dust, as happens in polar and high mountain regions of Earth. Figure 42.19 shows possible evidence that this buried ice continues to melt and trickle out of subsurface layers even today. The series of images made by the *Mars Reconnaissance Orbiter* shows dark lines growing down the inner wall of a crater over several months of the Martian summer. Scientists hypothesize that frozen groundwater is melting and seeping down the side of the crater.

If Mars had a denser atmosphere in the past, as deduced from the higher pressure needed to allow liquid water to exist, then the greenhouse effect might have made the planet significantly warmer than it is now. The loss of such an atmosphere would have weakened the greenhouse effect and plunged the planet into a permanent ice age. Such a loss could happen in at least two ways. According to one idea, repeated asteroid impacts on Mars when it was young may have blasted its original atmosphere off into space. Such impacts, although rare now, did occur for hundreds of millions of years after the Solar System began to form. For example, some of the maria on the Moon date to impacts about 3.9 billion years ago.

Concept Question 3

One possible plan for sending a human mission to Mars is a one-way trip. Materials to build a base would be sent, but there would be no launch craft for a return trip. From what you know of Mars, would you consider going on such a mission?

Concept Question 4

The dream of many science fiction stories is that it might be possible to "terraform" Mars (to make it more like Earth) so that its environment would be suitable for humans. How might this be done?

Some astronomers hypothesize that Mars lost its atmosphere when it was struck by a large asteroid, or a series of midsize asteroids. A large enough impact could blast the planet's atmosphere off into space.

A less dramatic explanation for how Mars lost most of its atmosphere is that Mars's low gravity allowed gas molecules to escape over the first 1 to 2 billion years of the planet's history. Regardless of how Mars lost its atmosphere, because the planet has a relatively cool interior and low level of geological activity, Martian volcanoes have not been able to replenish its atmosphere. Without a thicker atmosphere, the planet has only a very weak greenhouse effect, so its atmosphere has grown cool and its water has become locked up as permafrost.

42.4 THE MARTIAN MOONS

Mars has two tiny moons, **Phobos** and **Deimos** (Figure 42.20), which are named for the demigods of Fear and Panic, appropriate companions to the god of war. Phobos and Deimos were discovered in 1877, but centuries earlier Kepler had guessed that Mars had two moons because the Earth has one moon and Jupiter, at least in Kepler's time, was known to have four. He thought that Mars, lying between these two bodies, should therefore have a number of moons lying between one and four, and he chose two as the more likely case. This was just a lucky guess, but it was a popular notion that even appears in Jonathan Swift's book *Gulliver's Travels*, nearly two centuries before the moons were discovered. Today features on the two moons are named after Swift and characters in his book.

These satellites are only about 20 kilometers across and may be captured asteroids (Unit 43). They are far too small for their gravity to have pulled them into spherical shapes. Both moons are cratered, implying bombardment by smaller objects. Both moons orbit very close to Mars. Deimos orbits only 20,000 kilometers above the surface of Mars and takes about 30 hours to circle the planet. This is only about six hours longer than Mars's rotation period, so from the planet's surface, Deimos appears to move very slowly across the sky. In fact, there are about two and a half Martian days between Deimos's rising and setting.

Phobos is in an even lower orbit, only about 6000 kilometers above the surface. It orbits in little more than a quarter of a Martian day. Even though it orbits in the same direction as Mars rotates, because it revolves around the planet faster than Mars rotates, it rises in the west and sets in the east. Phobos is so close to Mars that tidal drag on Phobos will eventually cause it to spiral in and crash into the planet.

These unusual orbits are part of the reason that astronomers suspect Phobos and Deimos are captured asteroids (Unit 43). It is possible, however, that they were assembled in place around Mars, somewhat as the Moon was around Earth. According to this hypothesis, a large collision with Mars scattered a large amount of debris into orbit around Mars, and pieces of this material later reassembled to make the two moons.

FIGURE 42.20
Pictures of Phobos and Deimos, the tiny moons of Mars.

KEY POINTS

- In its surface conditions, rotation period, and axis tilt, Mars is more similar to Earth than any other planet in our Solar System.
- Winds occasionally raise global dust storms that have covered the entire planet with red dust.
- Mars has huge volcanoes and a long chasm that may have opened up as the result of internal tectonic or geological activity.
- The interior of Mars appears to have cooled, but volcanic activity may still have occurred in the last few hundred million years.
- Water ice is present under the surface in many regions and is kilometers thick at the poles.
- Surface features imply that Mars once had bodies of liquid water, flowing streams, and deposited layers of sedimentary rock.
- The atmosphere is now too thin for liquid water to remain present on the surface, but there may be seepage from underground.
- Thinning of the atmosphere several billion years ago allowed the surface to cool to below the freezing temperature of water.
- Mars has two small moons that may be captured asteroids or debris from a collision like the one that created our Moon.

KEY TERMS

Deimos, 326
laminated terrain, 319
Olympus Mons, 318
Phobos, 326
Tharsis bulge, 317
Valles Marineris, 317

CONCEPT QUESTIONS

Concept Questions on the following topics are located in the margins. They invite thinking and discussion beyond the text.

1. Differences between impact craters and volcanic craters. (p. 318)
2. Potential sources of hydrogen in surface. (p. 323)
3. One-way ticket to Mars. (p. 325)
4. How to terraform Mars. (p. 325)

REVIEW QUESTIONS

5. How does Mars compare with the Earth in mass and diameter?
6. What are the major geological features on Mars?
7. What evidence indicates that Mars was once geologically very active but has been geologically inactive for hundreds of millions or billions of years?
8. Valles Marineris on Mars is often compared to the Grand Canyon on Earth. In what ways are these two features similar? In what ways do they differ?
9. What are the Martian polar caps composed of? Why do they show layering?
10. What is the Martian atmosphere like?
11. What is the evidence that Mars once had running water on its surface? Is there evidence for water in any form today?
12. Why is it impossible for Mars to have liquid water on its surface today?

QUANTITATIVE PROBLEMS

13. In Figure 42.3, suppose that the areas with negative elevations were actually oceans. Estimate from the figure the surface area covered by the northern ocean. Multiply this by an average depth of 5 kilometers to get a volume. Compare your result to the 1.4×10^{18} cubic meters of Earth's oceans.
14. Mars's orbit is quite elliptical, so at different times when it is in opposition, its distance from Earth can vary substantially. From the data given in this chapter, estimate the largest and smallest distance to Mars at opposition and when it is on the far side of the Sun. Assume that the Earth's orbit is circular at 1 AU.
15. The volume of a cone is given by $V = \pi R^2 \times h/3$, where h is the height of the cone and R is the radius of its base. Use this formula, and the dimensions shown in Figure 42.5, to estimate the volume of Olympus Mons. Compare its volume to that of Mauna Kea, which has base radius of approximately 60 km.
16. From the surface of Earth under very good conditions, we can resolve an angular size of about 1 arc second. When Mars is 0.5 AU away, calculate the linear size (in kilometers) of a feature on the surface that would appear to be 1 arc second across. The Hubble Space Telescope can resolve about 0.04 arc seconds; what size on Mars does this correspond to?
17. What angular size would Phobos and Deimos have as viewed from the surface of Mars? Could either produce a total solar eclipse?
18. Earth's atmospheric pressure drops by about a factor of 2 every 5 km higher up you go, so it is one-fourth the surface pressure at 10 km, for example. How high up in Earth's atmosphere would you have to go to reach a pressure similar to the 0.6% atmospheric pressure on Mars? (You can do this mathematically with logarithms or plot the changing pressure with elevation and estimate the height where this pressure is reached.)
19. The average surface pressure on Mars is almost exactly equal to that below which water cannot be liquid. Review the topographic map of Mars in Figure 42.3. If you wanted to search an area that had the greatest probability of any location on Mars for liquid water today, where would you go and why?

TEST YOURSELF

20. Why are daily temperature variations on Mars much larger than we experience on Earth?
    a. Mars spins slowly, so its nights are very long.
    b. Mars is much darker than the Earth, so it absorbs more sunlight.
    c. Mars's atmosphere is too thin to insulate the surface.
    d. Mars has smaller internal heat sources than the Earth.
21. Which of the following is not true of the Tharsis Bulge?
    a. The largest Martian volcanoes are located here.
    b. The large chasm, Valles Marineris, is located here.
    c. The region is at a higher elevation than the rest of Mars's surface.
    d. The surface material is relatively older than the rest of Mars's surface.
22. What evidence suggests that liquid water was once present on the Martian surface?
    a. Branching channels in the shape of riverbeds
    b. The wobble in Mars's rotation, implying there is liquid water under an iced-over surface
    c. High levels of humidity measured spectroscopically in the atmosphere
    d. Microscopic fossil creatures found by the Martian rovers
    e. All of the above.
23. In what way are the atmospheres of Mars and Venus similar to each other?
    a. Temperature
    b. Pressure
    c. Chemical composition
    d. Density

UNIT 43

Asteroids

43.1 The Discovery of Asteroids
43.2 Asteroid Sizes and Shapes
43.3 Asteroid Compositions and Origin
43.4 Asteroid Orbits

Learning Objectives

Upon completing this Unit, you should be able to:
- Recall the history of how asteroids were discovered, their approximate sizes and numbers, and the debate about how to classify them.
- Explain why Ceres is now designated a dwarf planet.
- Describe the surface features of asteroids that have detailed observations.
- Explain why there are gaps in the asteroid belt and what causes them.

Asteroids are small, generally rocky bodies that orbit the Sun. More than 90% lie in the asteroid belt, a region between the orbits of Mars and Jupiter stretching from about 2 AU to 4 AU from the Sun, as shown in Figure 43.1. Several have been visited by spacecraft, but the first-discovered and largest of them, Ceres, will not be reached until 2015. Most of the quarter million cataloged asteroids are in stable orbits, and perhaps represent a planet that never formed because of the gravitational influence of Jupiter. If Ceres could somehow sweep them all up, we would be speaking of the fifth planet instead of the asteroid belt, but it would be a planet much smaller than the Moon.

Some asteroids appear to be nearly solid metal, and perhaps someday we will mine their ores. Others appear to be carbon-rich, and these are probably very primitive bodies—planetesimals that were never incorporated into the planets (Unit 35). Actually, the process of planet formation continues to this day, and from time to time an asteroid collides with and becomes part of one of the planets. These collisions release an enormous amount of energy that can create dramatic changes in a planet's environment (Unit 50).

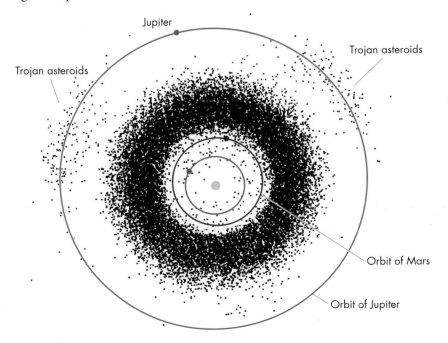

FIGURE 43.1
Diagram showing the distribution of more than 20,000 asteroids—about 1 in every 25 of the total number known. Notice that most lie between Mars and Jupiter, but a small number form two loose clumps—the Trojan asteroids—located along Jupiter's orbit. Making the plotted points large enough to see causes the asteroids to appear far more closely packed than they really are. A spacecraft traveling through the asteroid belt is highly unlikely to encounter an asteroid.

43.1 THE DISCOVERY OF ASTEROIDS

The discovery of the asteroids was based on a prediction that astronomers argue about to this day. A pattern in the distances of the planets from the Sun, pointed out in 1766, is known today as **Bode's rule.** It reproduced the distances of the planets known at the time and suggested that there was a gap between Mars and Jupiter where a planet ought to exist. With what we now know about other planetary systems, most astronomers think Bode's rule does not express a broader universal law; nevertheless, it led to the discovery of the asteroids.

Bode's rule, also known as Bode's law or the Titius-Bode law, was first pointed out by a Prussian astronomer, Johann Titius. Johann Bode confirmed the calculation and published it six years later.

The pattern found among the planets can be expressed by a simple mathematical progression. Bode's rule works as follows: Write down 0, 3, and then successive numbers by doubling the preceding number. That is, 0, 3, 6, 12, 24, and so on. Next add 4 to each and divide the result by 10, as shown in Table 43.1. The resulting numbers are close to the distances of the planets from the Sun in astronomical units, except there was no planet at 2.8 AU and none known beyond Saturn at the time the rule was devised.

TABLE 43.1 Bode's Rule

Bode's Rule	Number	Planet	True Distance
(0 + 4)/10 =	0.4	Mercury	0.39
(3 + 4)/10 =	0.7	Venus	0.72
(6 + 4)/10 =	1.0	Earth	1.00
(12 + 4)/10 =	1.6	Mars	1.52
(24 + 4)/10 =	2.8	Ceres (dwarf)	2.78
(48 + 4)/10 =	5.2	Jupiter	5.20
(96 + 4)/10 =	10.0	Saturn	9.58
(192 + 4)/10 =	19.6	Uranus	19.2
(384 + 4)/10 =	38.8	Neptune	30.1
(768 + 4)/10 =	77.2	Pluto (dwarf)	39.5
(1536 + 4)/10 =	154.0	Eris (dwarf)	67.7

The planet Uranus was discovered in 1781, and when astronomers realized that it fit the rule, they began a systematic search for a planet in the "gap" at 2.8 AU. In 1801 Giuseppe Piazzi, a Sicilian astronomer, discovered the largest of the asteroids, which he named **Ceres** in honor of the patron goddess of Sicily. Ceres fit the rule splendidly (Table 43.1), and it was widely accepted as the "missing" planet.

However, within six years, three more objects were found orbiting in the same general region of the Solar System. Some astronomers called these planets too, but others pointed out that they appeared to be a different class of object. The astronomer William Herschel (who discovered Uranus) noted that unlike the other planets, these objects were so small that they looked like stars—pinpoints of light—through even the best telescopes of the day. That is how they got the name *asteroid*—because they look like stars (*aster* in Latin). They are also sometimes called *minor planets,* reflecting their original status.

Even today, asteroids appear merely as points of light through research telescopes, so their status as asteroids cannot be determined from a single image. Instead, astronomers must track their motion for weeks or months, using their shifting position to determine the size and shape of their orbits. Ceres's and other asteroids' orbits are less circular and tilted to the plane of the major planets' orbits, also indicating their difference from the major planets.

By 1850 more than a dozen asteroids had been found, and the number has climbed steadily since. Once an orbit has been confirmed, an asteroid is given a number that indicates the order in which its discovery was recorded, and the discoverer is allowed to name it. Thus, Ceres is designated "1 Ceres." The asteroids discovered fourth and second, "4 Vesta" and "2 Pallas," are estimated to be the next two most massive.

Initially asteroids were named after goddesses; but as the names ran out, the discoverers were given wide latitude. There are now more than 15,000 named asteroids. Some have been given the names of discoverers' family members, well-known people, or cities. After an asteroid was named for a pet cat, the International Astronomical Union stepped in and set guidelines for their naming.

The introduction of automated search programs in recent years has dramatically increased the count of asteroids, reaching over 600,000 in 2013. For about half of these an orbit has not yet been accurately determined; this sometimes requires several years of observations. Astronomers continue to search for these objects, both because they provide valuable information about the formation of the Solar System and because there is a significant possibility that a few may have orbits that could someday cause them to collide with Earth (Unit 50). Many of the asteroids currently being discovered by automated searches are estimated to be under a kilometer in size. These are rarely named, but just given a catalog number and a numeric designation indicating when and how they were found.

> **Clarification Point**
>
> In movies and cartoons the asteroid belt is often depicted as crowded, but it is a very empty place. The large numbers of asteroids reside in an even larger volume of space. It is so large that if you crossed through the asteroid belt, you would be unlikely to see any asteroids by chance, let alone two at the same time.

The asteroids might have been able to collect together to form a planet at about 2.8 AU, but, probably because of the effect of Jupiter's gravity (see Section 43.4), these planetesimal-size objects were never able to coalesce into a single body. However, even if all the asteroids present today in the asteroid belt were put together, they would form a body with a mass less than 1% of Earth's mass.

Bode's rule, then, did seem to match a pattern in the planets' spacing. Ironically, though the rule inspired searches that ultimately led to the discovery of Neptune, that planet turned out *not* to continue the pattern (Table 43.1). Today astronomers note in their simulations of planetary system formation that similar patterns of planet spacing sometimes arise, with each planet roughly 1.5 to 2 times farther out than the preceding planet. This appears to be due to each planet's influence on the formation of neighboring planets. However, depending on many factors, such as the particular masses of the forming planets, the final pattern may look quite different from what Bode's rule predicts. These differences from Bode's rule are also exhibited by the planetary systems found around other stars (Unit 36).

43.2 ASTEROID SIZES AND SHAPES

No spacecraft missions have yet made a close encounter with the largest asteroid, Ceres, and it is too small to resolve much detail from Earth. However, the Hubble Space Telescope has imaged this tiny body well enough to determine that it is nearly spherical (Figure 43.2) with a diameter of about 930 kilometers. While this is the largest of the asteroids, it is very small compared to any of the terrestrial planets or even the Moon, which is about 80 times more massive.

The round shape of Ceres indicates that it has a gravitational force strong enough to compress itself approximately into a sphere. This is one of the criteria for being labeled a planet. However, Ceres fails to meet the other requirement: It is estimated that the rest of the asteroids in the asteroid belt have a combined mass about twice that of Ceres. Ceres has not "cleared its orbit," in the sense that there is a comparable mass in other asteroids that cross Ceres's orbital path. Ceres is therefore designated a "dwarf planet." The second most massive asteroid, Vesta, was recently orbited by the *Dawn* space probe. Vesta is almost massive enough to have pulled itself into a spherical shape; it appears to be just below the boundary where gravitational forces overcome the inter-molecular forces in a rocky body, although part of its irregular shape appears to stem from a major impact that left huge fractures that completely circle the asteroid (Figure 43.3).

FIGURE 43.2
Hubble Space Telescope image of the largest asteroid Ceres, which is more than 900 km (550 miles) in diameter. The *Dawn* spacecraft is scheduled to reach it in 2015.

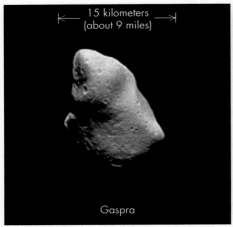 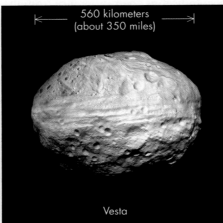 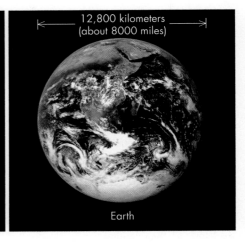

FIGURE 43.3
Earth is round, but the asteroids Gaspra and Vesta are not. Gaspra is too small for its gravity to make it spherical, but Vesta is nearly big enough.

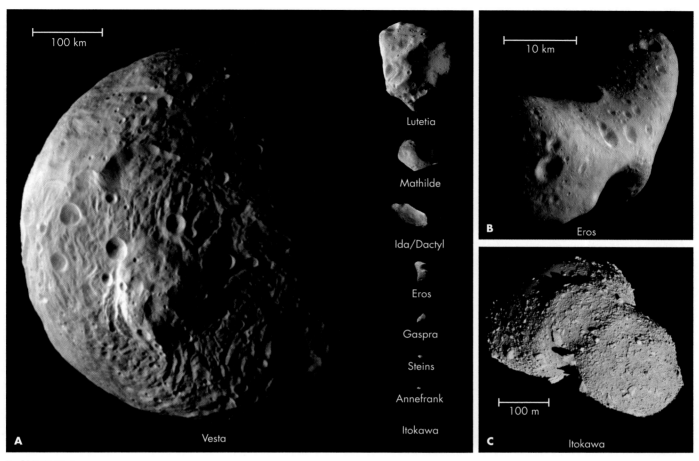

FIGURE 43.4
(A) Images of asteroids that have been visited by spacecraft shown to the same scale. Vesta is the second largest asteroid, about half the diameter of Ceres. (B) Eros is very elongated, shaped like a fat banana. (C) Tiny Itokawa is only a few hundred meters across.

FIGURE 43.5
Six views of the asteroid Eros from the *NEAR* spacecraft.

To date only a few asteroids have been viewed in detail by spacecraft, and these are mostly much smaller objects than Vesta as shown in Figure 43.4A. Two small asteroids have been visited by landers. NASA's *NEAR* (Near Earth Asteroid Rendezvous) spacecraft landed on the asteroid Eros in 2001. Eros (Figure 43.4B) is 33 kilometers (21 miles) long, one of the largest of the asteroids that comes close to Earth's orbit. *NEAR* studied its highly irregular shape (Figure 43.5) and cataloged more than 100,000 craters. The surface of Eros is blanketed with a regolith of rubble and dust like the Moon. The collisions that produced the largest craters must have had nearly enough power to break the asteroid apart. More numerous smaller collisions have left the asteroid pitted and lumpy, and the fragments blasted out by these impacts either have fallen back onto the surface to become part of the regolith or have become other small asteroids. It is likely that Eros originated in the asteroid belt but was shifted to an orbit between Earth and Mars by some of the major collisions that shaped it.

An even more ambitious mission was carried out by the Japanese spacecraft *Hayabusa,* sending a probe to land on the asteroid Itokawa (Figure 43.4C) and returning samples of the asteroid back to Earth. This asteroid is about 500 meters long and has an orbit that actually crosses the Earth's. Close-up pictures of the asteroid show a surface covered with boulders but no obvious impact craters. Scientists hypothesize that it may be the merger of two asteroids, erasing evidence of past impacts and creating a "pile of rubble" that is only relatively loosely held together by gravity.

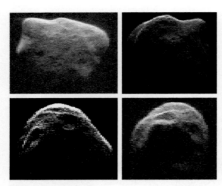

FIGURE 43.6
Radar images of a 3.5 kilometer diameter asteroid that passed close to Earth in 1999.

In addition to the images from spacecraft, a few asteroids have passed close enough to Earth to allow radar techniques to map their structure (Figure 43.6; also see Unit 25). However, the only information we have about most asteroids is the amount of light they reflect. This allows us to learn some things about the asteroid. For example, most asteroids vary in brightness in a cyclical fashion. These variations in brightness probably occur because the asteroid is spinning and either it is irregularly shaped or some portions are darker than others. The variations indicate that asteroids typically rotate with a period of 3 to 10 hours.

We can estimate the size of asteroids from the amount of sunlight reflected, but this can be quite inaccurate because a poorly reflective large object will look as bright as a highly reflective small one. For this reason, the thermal radiation (Unit 23) of asteroids is often used to estimate their size. Asteroids are heated by the Sun, and the amount of emitted infrared radiation is a good indicator of diameter. Bigger bodies emit more radiation than smaller ones of the same temperature.

From such measurements, astronomers have found that asteroids range tremendously in diameter. Based on the reflection measurements, it appears that about 200 of the asteroids have diameters larger than 100 kilometers. Astronomers estimate there are about 700,000 asteroids larger than 1 kilometer in diameter, based on a recent sensitive sky survey, which provided a statistical sample of the population of asteroids. The smallest asteroids are far too tiny to detect unless they come extremely close to Earth. For example, the diameter of the tiny asteroid 2012-DA14, which passed just 28,000 kilometers from Earth (less than 1/8 the distance to the Moon) in February 2013, is about 30 meters (100 feet).

Even very small asteroids can have devastating effects if they strike Earth (Unit 50), but a near miss with a planet can also have a catastrophic effect on an asteroid. Because gravity pulls more strongly on the side of the asteroid closer to the planet, material on that side tries to follow a tighter trajectory around the planet than material on the asteroid's far side. This stretching (tidal force) can rip an asteroid into many pieces.

43.3 ASTEROID COMPOSITIONS AND ORIGIN

When sunlight falls on an asteroid, the minerals in its surface create absorption features in the spectrum of the reflected light, from which we can determine the asteroid's composition. Such spectra indicate that asteroids belong to three main compositional groups: carbonaceous bodies, silicate bodies, and metallic iron-nickel bodies. The carbonaceous material is a carbon-rich, coal-like substance.

The groups are not mixed randomly throughout the asteroid belt: inner-belt asteroids tend to be silicate-rich, and outer-belt asteroids tend to be carbon-rich. Spectra of some carbonaceous asteroids indicate that minerals on the surface are hydrated, implying the presence of water ice.

Masses have also been estimated for about a dozen asteroids based on the gravitational effect they have on each other and on spacecraft. Several asteroids have now been discovered that have small moons orbiting them as well, such as the asteroid Ida and its tiny moon Dactyl (Figure 43.7). An object orbiting an asteroid allows us to estimate the asteroid's mass from Newton's laws (Unit 17). The average densities found are typically between about 1.5 and 3.5 kilograms per liter, suggesting that asteroids are mostly rocky and probably include some lower-density solids—such as the carbonaceous compounds and water ice.

The properties of asteroids give us clues to their origin and support the solar nebula theory for the origin of the Solar System.

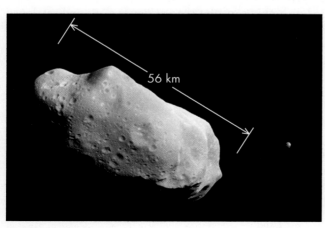

FIGURE 43.7
The asteroid Ida and its moon Dactyl (at right).

FIGURE 43.8
Sketch depicting (A) differentiation in asteroids and (B) their subsequent breakup by collision to form iron and stony bodies.

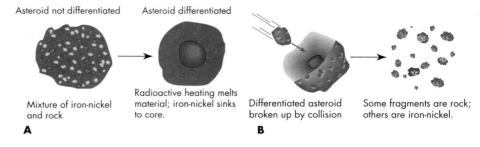

Differentiation and breakup of asteroids

Concept Question 1

What might be some advantages of mining material from asteroids instead of from the Earth or the Moon?

The asteroids are probably fragments of planetesimals, the bodies from which the planets were built. According to the solar nebula theory, because the inner asteroid belt is warmer than the outer belt, bodies that condensed in the inner belt are made primarily of silicate and iron, which condense at high temperatures, whereas the outer belt contains more-volatile substances: carbon-rich materials and water (Unit 35).

The existence of iron asteroids might at first seem to be evidence against the solar nebula theory. However, chemical elements can be separated by gravity if they are in a molten state—in the process called differentiation (Unit 35). When an object that is a mix of rocky and iron-rich material melts (perhaps as a result of naturally occurring radioactive substances heating it), the denser iron will sink to the object's core. We know this has happened to the Earth, and it also appears to have occurred in large asteroids (Figure 43.8A). In fact, some asteroids seem to have been sufficiently heated to have had volcanic eruptions. This activity, now long extinct, is deduced from their spectra, which show the presence of basalt, a volcanic rock.

After differentiation, collisions with neighboring asteroids and close encounters with planets broke up some of these large bodies (Figure 43.8B). The fragments are what we see today: pieces of crust became stony asteroids, whereas pieces of core became iron asteroids.

43.4 ASTEROID ORBITS

The solar nebula theory also offers an explanation for the asteroid belt being located near Jupiter's orbit. Any small planet beginning to form there would have to compete for material with Jupiter. Moreover, Jupiter's gravitational tidal force would disturb the accretion process, tugging apart clumps of matter in the solar nebula before they could grow very sizable.

Even today Jupiter affects the asteroid belt. Figure 43.9 shows a partial census of the number of asteroids found at specific distances from the Sun within the belt. A clear gap can be seen at 2.5 AU. The seemingly empty region in the asteroid belt is called a **Kirkwood gap** after the American astronomer Daniel Kirkwood who discovered it in 1866. The location of the gap corresponds to the distance at which an orbiting asteroid would have an orbital period of 3.95 years—exactly one-third of Jupiter's orbital period. Orbiting with this period, an asteroid would be at the exact same point in its orbit once every three orbits when it makes its closest approach to Jupiter. Jupiter's proximity would tend to pull the asteroid toward it, and by always tugging outward in the same part of the asteroid's orbit, it would cause the orbit to gradually become increasingly elongated toward Jupiter.

Concept Question 2

One early hypothesis was that the asteroids were once part of a planet that later broke up. What observational findings argue against this scenario?

Eventually the orbit would stretch close enough to Jupiter for the asteroid to be flung off in an entirely new direction. Compare this to an orbit that does not have a simple numerical relationship with Jupiter's orbital period. The asteroid is tugged in one direction during one of its closest approaches to Jupiter; but then, in successive orbits, it is pulled in different directions, so that the effects tend to cancel each other out.

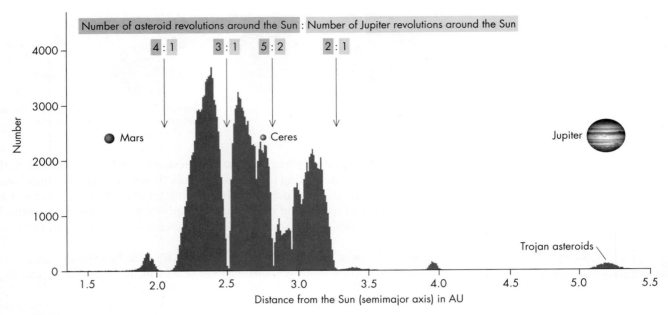

FIGURE 43.9
The number of asteroids at each distance from the Sun within the asteroid belt. Notice the conspicuous gaps where there are very few objects. These empty zones are called *Kirkwood gaps*.

Resonance of orbits

Kirkwood gaps are also found at distances corresponding to ratios of orbital periods of 2:1 and 4:1, as well as at some more complicated combinations such as 5:2—in other words, the asteroid orbits five times while Jupiter orbits twice. These orbits with simple whole-number ratios of their periods have what are called **resonances** with Jupiter's orbit. But resonances with Jupiter's motion not only create gaps; they can also create concentrations of asteroids. For example, a larger-than-average number of asteroids orbit at 4.0 AU. Asteroids orbiting at this distance have a period two-thirds of Jupiter's, and are therefore in a 3:2 resonance.

Not all asteroids are found in the main belt. The *Trojan asteroids* travel along Jupiter's orbit in two loose swarms 60° ahead of and 60° behind that planet, as can be seen in Figure 43.1. These turn out to be spots where gravitational forces from the Sun and Jupiter balance and the asteroids' orbits are stable.

Sometimes asteroids suffer collisions with each other that send fragments into new orbits. By comparing their orbital parameters, astronomers have identified dozens of **asteroid families** that each appear to have originated from a single object that underwent a collision. Supporting this idea, spectroscopic studies of the members of these families show that they have compositions similar to each other.

Many asteroids travel well outside the asteroid belt, possibly because they were disturbed by Jupiter's gravity or a collision with another asteroid, which sent them into highly elliptical orbits. These asteroids can pose a serious threat to Earth. For example, astronomers have identified the Baptistina asteroid family (named after the largest member of the family) as the likely source of the asteroid that killed the dinosaurs (Unit 50). The orbits of the members of this family point to an origin when a large carbonaceous asteroid broke up in a collision about 160 million years ago.

An assortment of nearly 10,000 other asteroids have orbits that cross or come very close to the Earth's orbit. These are known as **near-Earth objects.** The chance of a collision with Earth is slim but, on average, a body 100 meters in diameter hits the Earth about every few thousand years. Collisions of even such a "small" asteroid with a planet can have devastating effects, as we will discuss in Unit 50.

Concept Question 3

Itokawa is a near-Earth object. If it were on a collision path with Earth, what strategies could we use to avoid the collision?

KEY POINTS

- Asteroids are small rocky bodies that mostly orbit between Mars and Jupiter, and their total mass adds up to less than the Moon's.
- The largest asteroid is a dwarf planet because its gravity has pulled it into a round shape, but it has not cleared its orbit.
- Orbits have been determined for hundreds of thousands of asteroids, but far more are too small to have been detected yet.
- Some asteroids were large enough to differentiate and subsequent collisions between asteroids have broken them apart.
- Asteroids have been cleared out from Kirkwood gaps where orbits have a resonance with Jupiter's orbit.
- Thousands of asteroids have orbits that come close to Earth's orbit, and astronomers are searching for any others.

KEY TERMS

asteroid, 328
asteroid family, 334
Bode's rule, 329
Ceres, 329
Kirkwood gap, 333
near-Earth object, 334
resonance, 334

CONCEPT QUESTIONS

Concept Questions on the following topics are located in the margins. They invite thinking and discussion beyond the text.

1. Advantages of mining asteroids. (p. 333)
2. Asteroids probably were not formerly a planet. (p. 333)
3. Avoiding a collision with a small asteroid. (p. 334)

REVIEW QUESTIONS

4. What is Bode's rule, and how did it lead to the discovery of asteroids?
5. Where are asteroids located? Why did they not form a planet?
6. What is different about the shape of Ceres compared to other asteroids? Why is there this difference?
7. How do we know what asteroids are made of?
8. How have differentiation and impacts led to the variety of asteroid types we observe today?
9. What are the Kirkwood gaps? How were they formed?
10. Where are asteroids found, other than in the main belt? What are their orbits like?

QUANTITATIVE PROBLEMS

11. Asteroids are often discovered on long-exposure images because they are moving relative to the stars during the exposure. They therefore leave a short trail on the image while the stars remain stationary. Suppose that an exposure 25 minutes long is made, and that this image has a "plate scale" of 1.1 arc minutes per millimeter.
    a. If the asteroid leaves a trail 3 mm long on the image, how many arc minutes did it move during the exposure?
    b. If the asteroid is about 0.5 AU away, what is its velocity through space?
    c. How would Earth's motion affect this estimate? Draw a diagram to support your explanation.
12. Given Jupiter's orbital period of 11.9 years, use Kepler's third law (Unit 12.2) to calculate the distances from the Sun to each Kirkwood gap.
13. If an asteroid 500 km in diameter is destroyed by an impact, how many 1 km sized asteroids can result from the collision? Assume that the initial and resulting asteroids are spherical in shape, which means their volumes are given by $4\pi R^3/3$.
14. The asteroid Icarus has an elliptical orbit that carries it between 0.19 and 1.97 AU from the Sun. What is its semimajor axis? How often does it cross the Earth's orbital radius?
15. The asteroid 2004 FH was discovered on March 15, 2004, by the NASA-funded LINEAR asteroid survey. It is approximately 30 meters in diameter. Through careful observation, astronomers determined its orbital period to be 270.1 days.
    a. What is its semimajor axis?
    b. If its orbit varies from 1.05 AU to 0.58 AU, could this be a potentially hazardous asteroid? Explain your answer.
    c. If it has a density of 2 kg/L, what is its mass?
    d. If this asteroid were to strike the surface of the Earth at 30 km/sec (Earth's orbital velocity), how much kinetic energy would it have?
16. The *NEAR* spacecraft landed on Eros with a velocity estimated at 1.6 m/sec. Convert this speed to kph. If the spacecraft had a mass of about 800 kg, what was the kinetic energy of its impact? Discuss how much this would have affected the asteroid's orbit.
17. Ceres has a mass of approximately 8.7×10^{20} kg and a diameter of approximately 930 km.
    a. What is the acceleration of gravity on its surface (Unit 16)?
    b. What is the escape velocity from its surface (Unit 18)?

TEST YOURSELF

18. Asteroids
    a. are made of silicates only.
    b. vary widely in size.
    c. are only found between Mars and Jupiter.
    d. are spherical in shape.
19. How do astronomers estimate the sizes of most asteroids?
    a. By observing angular diameters from space telescopes
    b. By timing how long they take to pass in front of stars
    c. By determining their gravitational influence on other asteroids
    d. By measuring the amount of light they reflect
20. Which of the following are you *least* likely to find in an asteroid?
    a. Liquid water
    b. Rocky materials
    c. Metallic materials
    d. Carbon-based materials
21. The asteroids failed to form a planet because of
    a. gravitational interactions with Jupiter.
    b. ongoing collisions amongst the asteroids.
    c. an intense solar heating event in the young Solar System.
    d. the Sun's magnetic field interacting with metallic asteroids.

UNIT 44

Comparative Planetology

44.1 The Role of Mass and Radius
44.2 The Role of Water and Biological Processes
44.3 The Role of Sunlight
44.4 The Outer Versus the Inner Solar System

Learning Objectives

Upon completing this chapter, you should be able to:
- Explain what general processes affect each planet, and why differences arise.
- Discriminate differences in terrestrial planets caused by size, distance from the Sun, water, and life.
- Describe the broad differences and similarities of terrestrial and Jovian planets and the icy worlds orbiting in the outer Solar System.

The study of the similarities and differences between planets—what astronomers call **comparative planetology**—allows us to better understand all the planets, including our own. Many factors affect a planet, such as its size, distance from the Sun, and internal and atmospheric composition. Seeing how physical processes operate under different conditions lets us test our hypotheses and perhaps better predict the consequences of global changes for our own planet's environment. In this comparison, some planets are clearly more similar to the Earth than others, although all have stories to tell that can help us better understand the Earth.

The terrestrial planets (and the Moon) are roughly similar in diameter and distance from the Sun, but their surfaces look very different, as illustrated in Figure 44.1, particularly when examined in detail as we saw in Units 37–42. Yet, as great as the differences between the terrestrial planets appear to be, even our brief look at the worlds of the outer Solar System at the end of this Unit makes it clear just how similar the terrestrial planets are to each other. The worlds populating the outer Solar System are examined further in Units 45–48.

FIGURE 44.1
Pictures of (left to right) Mercury, Venus, Earth, Moon, and Mars, showing representative surface features.

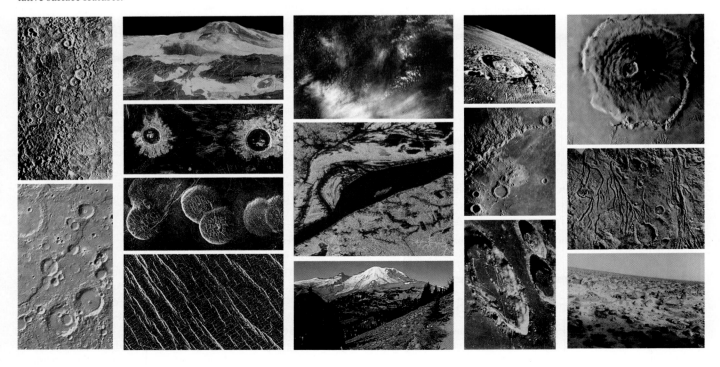

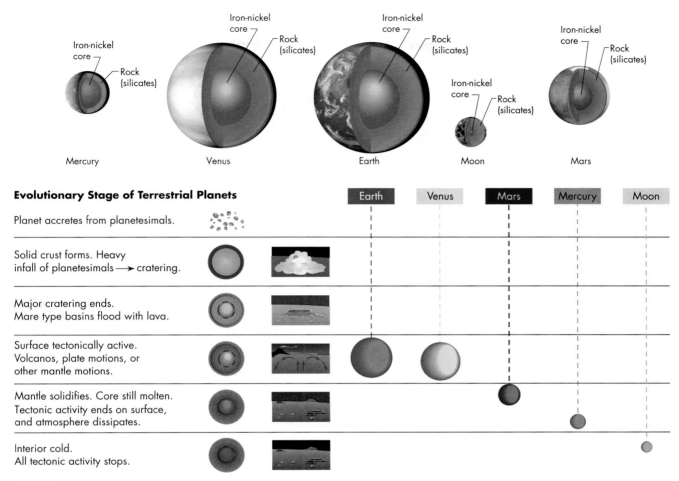

FIGURE 44.2
Gallery comparing the interiors of the terrestrial planets and the Moon, and indicating their relative stage of geological cooling.

Planetary variations

44.1 THE ROLE OF MASS AND RADIUS

The terrestrial planets and the Moon are shown to the same scale in Figure 44.2. Because of their similarity to Earth, the planets Mercury, Venus, and Mars are called *terrestrial planets*. However, Mercury and our Moon are essentially airless. Venus, on the other hand, has an atmosphere 100 times more massive than our own. The surfaces of Mercury and the Moon are covered with craters, but they lack most other landscape features. In contrast, Earth and Mars have huge volcanic peaks and canyons, but Mars has many more impact craters. Some of their basic properties relating to their sizes are summarized in Table 44.1. These properties play a major role in determining the observed characteristics of each planet.

A planet's mass and radius affect its interior temperature and thus its level of tectonic activity. Low-mass, small-radius planets cool inside more rapidly than larger bodies. We see, therefore, a progression of activity from small, relatively inert Moon and Mercury, to slightly larger and formerly active Mars, to the larger and far more active surfaces of Venus and Earth (Figure 44.2). Mercury's surface still bears the craters made as it was assembled from planetesimals. Mars has some craters, but it also has younger surface features such as volcanoes and riverbeds carved by running

TABLE 44.1 Comparison of the Terrestrial Planets and Our Moon—Factors Related to Size

Property	Mercury	Venus	Earth	Moon	Mars
Radius (Earth units)	0.383	0.950	1.000	0.273	0.533
(km)	2439	6052	6378	1738	3398
Mass (Earth units)	0.055	0.815	1.0	0.012	0.107
(kg)	3.30×10^{23}	4.87×10^{24}	5.98×10^{24}	7.35×10^{22}	6.42×10^{23}
Density (kg/L)	5.43	5.25	5.52	3.34	3.94
Uncompressed density (kg/L)	5.4	4.2	4.2	3.3	3.3
Escape velocity (km/sec)	4.3	10.4	11.2	2.4	5.0
Pressure at surface (Earth atmospheres)	0	About 90	1.0	0	About 0.007
Magnetic field (relative to Earth)	0.0006	None	1	Remnant	Remnant

water. In contrast, Earth and Venus retain almost none of their original crust, their surfaces having been modified enormously by interior activity over their lifetimes.

Mars, intermediate in mass and radius between Mercury and Earth, has a surface that bears the traces of an intermediate level of activity. Much of it is cratered, implying that a long time has passed since geological activity ended. Some regions, however, show volcanic uplift like that seen on Venus, implying that Mars's interior was once much hotter than it is now. Studies of the magnetization of surface rocks on both Mars and the Moon show that both of these worlds must have had strong magnetic fields in the distant past, although neither still has a global magnetic field like the Earth's. Moreover, Mars's surface shows an amazing variety of landforms: polar caps of ordinary ice and frozen carbon dioxide, immense deserts with dune fields, and canyons created by both crustal cracking and erosion from running water.

The small number of impact craters on Venus and Earth implies that they have young surfaces, extensively altered by volcanic activity. On Earth, the flow of heat from its core has created plate motion. On Venus, perhaps because of its thicker, more rigid crust, the flow of heat to its surface is less uniform. As a result its surface has isolated regions of intense volcanic uplift.

The extent of the atmosphere of a planet is also affected by the planet's size and mass. All of the terrestrial planets and the Moon probably received some gases in the late stages of formation, delivered to them by comets; but volcanic activity was probably an even larger source of atmospheric gases. Because larger planets maintain their internal heat longer (Unit 37) they also have greater volcanic activity, which continues to this day for the largest terrestrial planets, Venus and Earth.

A small planet has an additional problem: Any atmosphere it forms or captures is likely to leak away faster than it can be replenished by volcanoes or comets. Mercury and the Moon have so little mass that their escape velocity is low and gases rapidly escape to space. (The escape velocity and atmospheric pressure for each planet are listed in Table 44.1.) Mars is an interesting transitional case. It was active once but is now quiescent. To have had liquid water on its surface as the evidence suggests, its atmosphere must once have been quite dense, but today its atmosphere is very thin.

Concept Question 1

Why is it surprising that Mercury has a weak magnetic field whereas Venus has none?

44.2 THE ROLE OF WATER AND BIOLOGICAL PROCESSES

The atmospheres and surface temperatures of the terrestrial planets are compared in Figure 44.3. Venus and Mars both have atmospheres rich in carbon dioxide, although Venus's atmosphere is about 100 times more massive, while Mars's is about 200 times less massive than Earth's. The great mass of the Venusian atmosphere creates a strong greenhouse effect, and high in the Venusian atmosphere thick clouds of sulfuric acid droplets block our view of the Venusian surface.

Astronomers think that the atmospheres on Venus, Earth, and Mars have been appreciably changed in composition over time by chemical processes. The atmospheres of all three bodies were probably originally nearly the same: primarily CO_2 with small amounts of nitrogen and water. But these original atmospheres have been modified by sunlight, tectonic activity, water, and, in the case of the Earth, life.

The great difference in the water content of the atmospheres of Earth and Venus has created a drastic difference between these planets' atmospheres. Ultraviolet light from the Sun is energetic enough to break apart water molecules into oxygen and hydrogen atoms. The hydrogen atoms that are liberated, being very light, move faster than average and are likely to escape into space, while the heavier oxygen atoms remain. Fortunately, other gases in the atmosphere can absorb ultraviolet light and shield water near the surface from destruction. In the warm Venusian atmosphere, water can rise to great heights, so over billions of years water molecules have steadily been dissociated when they reach the upper atmosphere and almost completely lost from the planet's atmosphere. In Earth's cooler atmosphere, however, water has survived, and its transitions between gaseous, liquid, and solid forms are a large part of the complex weather phenomena we experience.

The presence of liquid water permits chemical reactions that can profoundly alter the composition of a planet's atmosphere. For example, CO_2 dissolves in liquid water and creates carbonic acid (which is also present in carbonated drinks). In fact, the bubbles in a carbonated drink are just carbon dioxide that is coming out of solution. As rain falls through our atmosphere, it absorbs CO_2, making the rain slightly acidic even in unpolluted air. When the rain reaches the ground, the carbonic acid reacts chemically with silicate rocks to form carbonates, locking some CO_2 into the rock.

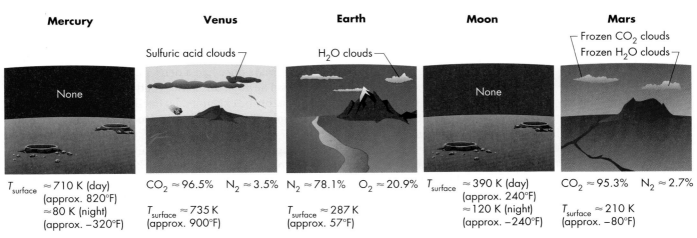

FIGURE 44.3
Comparison of the atmospheres and surface temperatures of the terrestrial planets and the Moon.

Biological processes can also remove CO_2 from the atmosphere. For example, plants use CO_2 to make the large organic molecules, such as cellulose, of which they are composed. This CO_2 is usually stored for years or centuries before decay or burning releases it back into the atmosphere. More permanent removal occurs when sea creatures use dissolved carbon dioxide to make shells of calcium carbonate. As these creatures die, they sink to the bottom, where their shells form sediment that may eventually be compressed into rock. Thus, carbon dioxide is swept from our atmosphere and locked away, both chemically and biologically, in the crust of our planet. With most of the carbon dioxide removed from our atmosphere, the relatively inactive gas nitrogen is left as the primary component. In fact, Earth's atmosphere contains a total amount of nitrogen comparable to that in the atmosphere of Venus. This suggests that these two planets may have begun with similar overall atmospheres—but the carbon dioxide has not been removed from Venus's atmosphere, so it remains the dominant constituent.

Our atmosphere is also rich in oxygen, a gas found nowhere else in the Solar System in such high relative abundance. Our planet's oxygen is the product of plants and single-celled creatures breaking down CO_2 and H_2O molecules during photosynthesis. Thus, much of the cause of the great difference between the atmospheres of Earth and Venus may be that life was able to start on Earth. On our planet, liquid water removed much of the carbon dioxide from our atmosphere and life absorbed most of the rest of it, leaving only the tiny amount (0.03%) we have today. In removing the carbon dioxide, plants also produced the oxygen on which all animal life depends.

What evidence supports this idea that liquid water and living things removed most of the early Earth's carbon dioxide? It is the carbonate rock we mentioned earlier. If the carbon dioxide locked chemically in that rock were released back into our atmosphere and the oxygen were removed to pre-plant-life levels, our planet's atmosphere would closely resemble Venus's.

If water is so effective at removing carbon dioxide from our atmosphere, why does any CO_2 remain? We add small amounts of CO_2 by burning wood and fossil fuels, but the major contribution is from natural processes. Atmospheric chemists hypothesize that tectonic activity gradually releases CO_2 from rock into our atmosphere. At plate boundaries, sedimentary rock is carried downward into the mantle, where it melts. Heating breaks down carbonate rock, releasing its carbon dioxide, which then rises with the heated rock to the surface and reenters the atmosphere.

Carbon dioxide may have been removed from Mars's atmosphere and locked in rocks there during the planet's early history, when it apparently contained much more liquid water. Mars's lower level of tectonic activity, however, prevents its CO_2 from being recycled. With so little of its original carbon dioxide left to provide greenhouse warming, Mars has grown progressively colder. Mars's atmosphere is too thin now for a strong greenhouse effect; but the old river channels, presumably cut by running water, imply that Mars once had a denser and warmer atmosphere, perhaps even resembling Earth's. The great differences between the atmospheres of Venus, Earth, and Mars can probably be explained by the differences in their water content and the fact that life evolved on Earth, leading to plants and their release of oxygen.

Our Earth has retained enough CO_2 in its atmosphere to maintain a moderate greenhouse effect, making our planet habitable. Some geologists hypothesize that living organisms may respond to global changes, such as long-term changes in the Sun's brightness and changes in the Earth's activity, through feedback mechanisms to keep the Earth's climate temperate (Unit 85). Thus, at its location between one planet that is too hot and another that is too cold, Earth has been favored with a relatively stable temperature, one factor in the complex web of the environment to which we owe our existence.

Concept Question 2

Why would the discovery of oxygen in an exoplanet's atmosphere be significant? What opposing hypotheses might explain its presence?

44.3 THE ROLE OF SUNLIGHT

Although the Sun's energy is negligible for a planet's geology, it is the major heating source for the planet's surface and atmosphere. A planet's distance from the Sun determines the amount of sunlight it receives. Thus Venus, which is about twice as far from the Sun as Mercury, receives about a quarter as much energy over each square meter of its surface. The shape of a planet's orbit and the tilt of its rotation axis also affect how sunlight is distributed over the surface and how the amount of sunlight varies during an orbit. A tilted rotation axis, such as Earth's and Mars's, causes seasonal shifts in opposite hemispheres. An elliptical orbit, such as Mercury's and Mars's, adds a significant variation to a planet's global surface heating during each year.

Sunlight affects a planet's atmosphere in several ways. First, of course, it warms a planet by an amount that depends on the planet's distance from the Sun, and so we expect that Venus will be warmer than Earth and Earth will be warmer than Mars. These expected temperature differences are increased, however, by the atmospheres of these bodies. Moreover, even relatively small differences in temperature can cause large variations in physical behavior and chemical reactions within an atmosphere. For example, Earth is cool enough that water vapor condenses to ice at an elevation of about 10 kilometers (about 30,000 feet) and falls back down as precipitation, leaving higher layers of our atmosphere almost totally devoid of water. Venus is just enough nearer the Sun that even without a strong greenhouse effect, water would rise much higher in the atmosphere before condensing. Water molecules reaching the upper atmosphere can be broken apart by the Sun's ultraviolet light, and the hydrogen lost from the planet. Without water, carbon dioxide remains dominant in the atmosphere.

Distance from the Sun does not alone determine a planet's temperature. The role of the atmosphere in retaining the energy from the Sun is important too. We see this in the fact that Mercury has lower temperatures than Venus, despite Mercury's being half as far from the Sun. We also see this in the lower mean temperature of the Moon versus the Earth, even though both are at the same average distance from the Sun.

The distance from the Sun in the early history of the Solar System also affected the basic composition of the planets. This is relatively straightforward to see in the overall density of the planets between the inner and outer Solar System (Unit 34). A more refined examination of the terrestrial planets shows that they, too, exhibit a trend of density (Table 44.2). This is not obvious at first, because the third planet, Earth, has the highest density—slightly higher than Mercury's. However,

TABLE 44.2 Comparison of the Terrestrial Planets and Our Moon—Factors Related to Distance from Sun

Property	Mercury	Venus	Earth	Moon	Mars
Distance from Sun	0.387 AU	0.723 AU	1 AU	1 AU	1.524 AU
Orbital period	87.969 Earth days	224.70 Earth days	365.26 Earth days	—	686.98 Earth days
Density (kg/L)	5.43	5.25	5.52	3.34	3.94
Uncompressed density (kg/L)	5.4	4.2	4.2	3.3	3.3
Surface temperature (K)	440	735	287	250	210
Temperature range	80–710	720–750	185–331	120–390	133–293

the density of a material is affected by the planet's gravitational force. For example, the pressures at the center of a massive planet like Jupiter may squeeze rock, which has a normal density of 3 kilograms per liter, to a density of 7 or 8 kilograms per liter. In fact, the pressures at the center of Jupiter would be sufficient to squeeze the Earth to half its current volume! Thus, if we compare the density of two planets made of the same materials, the more massive one will be more dense.

The Earth is the most massive of the terrestrial planets, so its gravitational force squeezes its interior most. It is estimated that if Earth's gravity were not squeezing the material within it, Earth's **uncompressed density** would be 4.2 kilograms per liter. This affects the smaller planets less. As a result, in Table 44.1 we see the highest uncompressed density for Mercury, which contains the largest proportion of iron (a substance that condenses at high temperatures), and the lowest uncompressed density for Mars.

> **Concept Question 3**
> What features of each planet might be explained by a large collision late in its process of formation?

The Moon has a density even lower than Mars, which is thought to be a consequence of the Moon's formation in a collision that tore out low-density crust and mantle from the young Earth. The density the Earth would have had if this collision had not occurred should be an average of the densities of the Earth and Moon, which would then follow the trend of declining densities.

The low-density of Mars indicates that it contains a smaller proportion of iron and a larger proportion of low-density substances. This trend continues through the asteroid belt, with larger fractions of water ice and carbon rich materials in outer-belt objects (Unit 43)—pointing toward the characteristics of the bodies in the outer Solar System.

44.4 THE OUTER VERSUS THE INNER SOLAR SYSTEM

FIGURE 44.4
The gas giants are dominated by their enormous atmospheres. They are shown here to the same scale and compared to the largest terrestrial planet, Earth.

Beyond Mars and the asteroid belt, the Solar System is a realm of ice and frozen gas. In this frigid zone far from the Sun, where solar heat is only a vestige of what we receive on Earth, the giant planets formed (Figure 44.4). Beyond a distance called the **frost line**, the low temperatures—below 180 kelvin (−140°F)—allowed water (H_2O) to condense. This is critical because hydrogen is the most abundant element, and oxygen the third most abundant, in the universe (Unit 35). Because water remained gaseous inside the frost line, the terrestrial planets could incorporate only the scarcer rock and iron that condensed there. Giant planets could become giants because they could add to the rock and iron much larger quantities of ice.

> The *frost line* is also sometimes called the *water line* or the *snow line*. "Frost" is perhaps the most descriptive because it is at this distance that gaseous water freezes out into solid ice particles.

The low temperatures beyond the frost line not only provided greater amounts of raw materials to accrete a planet; they also slowed the movement of gas molecules, making them more likely to be captured. As a result, these cold planets were able to draw from the vast reservoir of gaseous hydrogen and helium in the solar nebula and became far larger than planets that formed near the Sun. Thus, beyond the asteroid belt we see a sudden jump to planet masses ranging from 15 to 320 times the mass of Earth (Table 44.3).

Planets forming in this way became very different from the terrestrial planets in structure and composition. They consist mostly of hydrogen and its compounds. These gases form a deep atmosphere, often richly colored, that grows denser with depth and eventually becomes liquid. The interiors of Jupiter and Saturn are composed mainly of hydrogen—liquid just below the atmosphere and metallic deeper down. Each has a rock and iron core with a diameter estimated to be similar to the Earth's.

TABLE 44.3 Comparison of the Four Giant Planets with the Earth

Property	Earth	Jupiter	Saturn	Uranus	Neptune
Distance from Sun (AU)	1	5.203	9.539	19.19	30.06
Orbital period (Earth years)	1	11.8622	29.4577	84.014	164.793
Radius (Earth units)	1.000	11.19	9.46	3.98	3.81
(km)	6378	71,492	60,268	25,559	24,764
Mass (Earth units)	1.0	317.9	95.18	14.54	17.13
(kg)	5.98×10^{24}	1.9×10^{27}	5.68×10^{26}	8.68×10^{25}	1.02×10^{26}
Sidereal day (hours)	23.9345 hours	9.925	10.656	17.24	16.11
Axial tilt	23.45°	3.12°	26.73°	97.86°	29.56°
Density (kg/L)	5.52	1.33	0.69	1.32	1.64
Escape velocity (km/sec)	11.2	59.5	35.5	21.3	23.5
Atmospheric composition	N_2: 78.1%	H_2: 89.5%	H_2: 96.3%	H_2: 82.5%	H_2: 80.0%
	O_2: 20.9%	He: 10.2%	He: 3.3%	He: 15.2%	He: 18.5%
	Ar: 0.9%	CH_4: 0.3%	CH_4: 0.4%	CH_4: 2.3%	CH_4: 1.5%
Temperature (K) at one atmosphere of pressure	287	160	130	80	75
Magnetic field (relative to Earth)	1	19,500	578	47	25
Satellites	1	63	61	27	13

Clarification Point

It is often believed that the gas giants have icy interiors, but their cores are even hotter than those of the terrestrial planets.

Concept Question 4

In Table 44.3 the sidereal day indicates how fast the planets are spinning. Why do you suppose the giant planets spin faster than the Earth?

Uranus and Neptune may have some liquid hydrogen below their hydrogen- and methane-rich atmospheres, but they are probably mostly water, ammonia, and methane surrounding smaller rock and iron cores. Because Uranus and Neptune are smaller, their weaker gravity has not allowed them to capture or retain as much hydrogen and helium as Jupiter and Saturn have; but they have retained the heavier gases.

Measurements of the heat rising from the interiors of the gas giants show that they generate more heat than they receive from the Sun. Their cores are even hotter than the Earth's core, and their silicates and iron are probably molten. The hydrogen and hydrogen compounds surrounding the core's molten mass are liquid because they are so strongly compressed by the weight of all the overlying layers, not because they are cold. This white-hot liquid is more dense than rock. Thus, these giant planets have no surfaces that we can compare to the terrestrial planets.

The giant planets are probably heated by continued gravitational contraction and settling of denser matter toward their cores. As the heat flows outward, it generates convective motions that stir the atmosphere. The planets' rotation creates a Coriolis effect on the rising gas, drawing it into the cloud belts (visible in Figure 44.4). This is similar to the effect that produces high-speed jet streams in the Earth's stratosphere and the spinning motions of our storms, but some of the storm systems on these gas giants are larger than the entire Earth!

All of the gas giants spin more rapidly than the Earth, giving each planet an equatorial radius that is distinctly larger than its polar radius. Rotation also affects the shape of Earth and Mars, although to a lesser extent (Figure 44.5). The outer parts of the planet swing around so rapidly that they move outward, as a child's feet swing outward when you hold the child by the hands and spin around. The material at the equator of a spinning planet has inertia that would keep it going in a straight line (Unit 14), but it reaches a balance with the inward force of gravity.

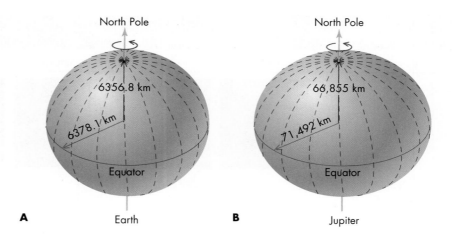

FIGURE 44.5
(A) Rotation makes a planet's equator bulge outward. The Earth's equator bulge is only about 20 kilometers. (B) Jupiter's rapid rotation creates a much greater equatorial bulge that is apparent in most photographs of Jupiter or Saturn (such as Figure 44.4).

Concept Question 5

Jupiter-size exoplanets (Unit 36) have been found close to a number of stars. Can you think of any scenario in which terrestrial planets might be found far from a star?

Concept Question 6

What differences between the planets do not appear to agree with the solar nebula theory?

Each of the four giant planets has a ring system composed of small orbiting particles. The rings are probably debris from small objects that have broken up as the result of either collisions or tidal forces exerted by the planet. Collisions continue to break up small particles, providing fresh ring material. At the same time, some ring materials collide and may fall into the planet's atmosphere. It is possible that the rings come and go, so that if we were to see the Solar System a few billion years from now, Saturn might not have an obvious ring system but maybe Neptune would. Each giant planet also has many orbiting moons. A few of these moons are large, but of the more than 100 cataloged, some are not much more than a kilometer in size. These orbiting systems are similar to mini solar systems (Unit 47).

Although the giant planets have many similarities, closer examination reveals them to be as individual as the terrestrial planets. Some of these differences may have causes parallel to the ones that shape the terrestrial planets. For example, the strong color differences between the giants are related to their distances from the Sun. Ammonia and methane condense at lower temperatures than water, so the chemistry of the outer giants differs from that of the inner giants. The least massive of the giants (Uranus) also seems to generate the least internal heat, again like the terrestrial planets.

One of the interesting aspects of every planet is how it differs from the others. Almost every planet has at least one major feature that makes it appear to be an "oddball" of sorts. Venus and Uranus have peculiar backward rotation directions; Earth and Neptune have unusual moons; Mercury and Earth have thin crusts; Saturn, Mars, and Earth have quite tilted rotation axes; Mars has a smaller atmosphere than might be expected. Every one of these oddities has been explained at one time or another as being caused by a major impact late in the formation of the planet. This may be correct. It is hard to imagine the violence of the final collisions between the last few dozen giant planetesimals shortly before the Solar System settled down to its final configuration. The last big collisions probably involved bodies of comparably large sizes, such as the collision that is thought to have resulted in the Earth's Moon. Jupiter may be the only "normal" planet because it grew so massive that nothing else was big enough to set its spin in an odd direction, stir up its forming satellite system, or strip out much of its material.

Besides the giant planets, the outer Solar System is teeming with smaller bodies. The largest of these worlds orbit the giant planets, but many more orbit beyond Neptune. Pluto and Eris are the largest trans-Neptunian objects discovered so far, but six moons of the gas giants are larger and more massive than either. In fact, the two largest, Ganymede and Titan, are larger than Mercury, although they have less than half as much mass. The reason they can be so big while still having low

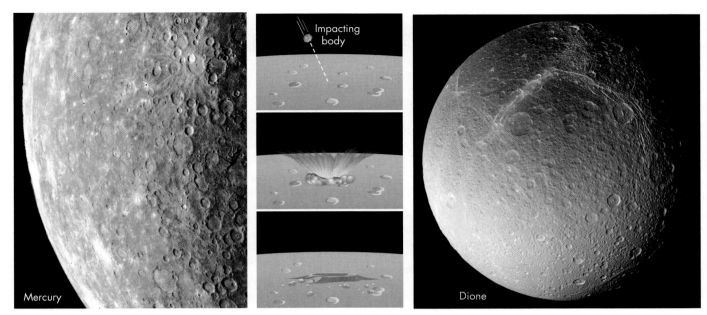

FIGURE 44.6
Pictures taken by spacecraft showing impact craters on Mercury and Saturn's moon Dione, both shown to the same scale.

masses is that they are "ice worlds," containing large quantities of frozen water, ammonia, and methane surrounding small cores of rock and iron.

One common surface feature of almost all of these ice worlds is impact craters that resemble the craters on the terrestrial planets (Figure 44.6). The number of craters is often quite high, suggesting that relatively little geological activity has occurred there to erase them. Given their small masses compared to Mercury, we might not expect to find *any* geological activity, yet some of the larger ice worlds are devoid of impact craters and others have actively erupting volcanoes!

When we study these ice worlds, we need to keep in mind that at their low temperatures, water ice behaves more like the rock here on Earth. We might even think of ice as the solidified lava of these worlds, and with such "lava" we have the possibility of an icy kind of geological activity. With that readjustment in our thinking, we can see similarities between the geological processes that occur on these ice worlds and more familiar processes here on Earth.

Several of these frozen worlds show signs of an icy kind of tectonic activity. For example, some of the larger moons have faults, smooth uncratered regions, and parallel ridges of mountains analogous to the features of plate tectonics on Earth (Unit 37). This suggests that below the crust, circulation is occurring in a mantle of "molten" ice. The heating source for the tectonic activity and volcanic eruptions on many of these worlds appears to be quite different from the terrestrial planets' radioactive heating, caused instead by gravitational tidal forces.

The only ice world that has a substantial atmosphere is Titan, and it is remarkably dense, producing a surface pressure similar to that on Earth. Titan even has lakes and rain—of liquid methane. Based on radar comparisons with Venus, a possible volcano has been identified on Titan (Figure 44.7), but here the "magma" might to us seem like an icy slush. Other ice worlds have thin atmospheres that freeze back onto the surface at night or seasonally. A few of these worlds have surface features unlike anything seen elsewhere in the Solar System (Unit 48). Indeed, astronomers consider these diverse bodies, virtually unknown before the space age, to be some of the most intriguing members of the Sun's family.

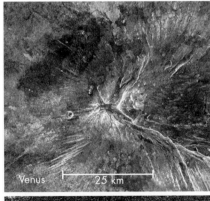
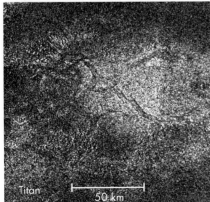

FIGURE 44.7
Radar images show an icy feature on Titan that resembles a volcano on Venus.

KEY POINTS

- A terrestrial planet's mass mainly determines how long its geological activity lasts and how long it holds on to an atmosphere.
- The presence of liquid water cycling through an atmosphere changes its chemical composition, as does life.
- Distance from the Sun determines not only the amount of solar heating, but the starting chemical composition of a planet.
- Conditions in the outer Solar System differ greatly from the inner Solar System because of the presence of large amounts of ice.
- All planets respond to basic physical forces; for example, they bulge at the equator if they spin fast.
- Coriolis effects look somewhat different on gas giants because gas can move vertically, not just along a terrestrial planet's surface.
- Impact cratering is common on all types of solid-surface bodies.

KEY TERMS

comparative planetology, 336   uncompressed density, 342
frost line, 342

CONCEPT QUESTIONS

Concept Questions on the following topics are located in the margins. They invite thinking and discussion beyond the text.

1. Mercury's versus Venus's magnetic field. (p. 338)
2. Oxygen on an exoplanet. (p. 340)
3. Planetary features caused by major collisions. (p. 342)
4. Why giant planets spin faster than the Earth. (p. 343)
5. Terrestrial planets far from a star. (p. 344)
6. Discrepancies with the solar nebula theory. (p. 344)

REVIEW QUESTIONS

7. Why do we expect strong volcanic and tectonic activity on Earth and Venus but little or none on the Moon and Mercury?
8. Qualitatively compare the crater densities for Earth and Venus with those of the Moon and Mercury. What does this information tell us about geological activity on these Solar System bodies?
9. What factors cause the differences in the terrestrial planet atmospheres?
10. How does distance from the Sun affect a planet?
11. What are the different roles that water can play on a planet?
12. What is meant by the "frost line" in the formation of the planets, and what role did it play in planetary sizes and composition?
13. In what ways are the gas giants and ice worlds similar to the terrestrial planets?
14. How do a planet's distance from the Sun and atmosphere combine to determine the temperature of the planet?

QUANTITATIVE PROBLEMS

15. For each terrestrial planet, use the values given in Units 37, 40, 41, and 42 to calculate what percentage of the planet's total volume is comprised of its core. Rank the terrestrial planets according to this percentage.
16. The mass of Uranus is about 14.5 times that of the Earth, yet the escape velocity for Uranus is a little less than twice that of Earth's. Explain the seeming discrepancy.
17. Scientists classify objects based on various characteristics. If we rank the terrestrial planets according to their densities (mass divided by volume), we will get one relationship. If we then rank the terrestrial planets according to their uncompressed densities, we will get a slightly different relationship. Compare the rankings. What do the differences in the rankings tell us about the planets themselves?
18. Examine the data in Tables 44.1, 44.2 and 44.3, and consider that the average speed of a molecule in a gas with temperature T is approximately

$$V_{\text{mol}} = 0.15 \text{ km/sec} \times \sqrt{T/m}$$

where T is the temperature in kelvin and m is the mass of the molecule in atomic mass units; so $m(H_2) = 2$, and $m(CO_2) = 44$. About how many times smaller than the escape velocity does V_{mol} have to be for a planet to retain a gas?

19. Heating by the Sun decreases with distance. For an object absorbing all sunlight, we can predict that the temperature will be $227 \text{ K} \times (1/\sqrt{a})$, where a is the distance from the Sun in AU. Find the expected temperatures at the distances of Venus, Mars, and Jupiter. What factors might cause the temperatures to differ from what this formula yields?

TEST YOURSELF

20. Many of the major unusual features that are unique to a particular planet may be explained by
    a. its surface temperature.
    b. its magnetic field strength.
    c. an active interior.
    d. its distance from the Sun.
    e. a large impact in the late formation stages.
21. Living organisms have transformed the Earth by
    a. removing carbon dioxide from the atmosphere.
    b. speeding up plate tectonic activity.
    c. changing how much sunlight reflects from the planet's surface.
    d. warming the planet enough to maintain liquid water.
22. Suppose a number of planets all have the same mass but different sizes and temperatures. Which of the following planets is most likely to retain a thick atmosphere?
    a. Small, hot   c. Large, hot
    b. Small, cool   d. Large, cool
23. Which of the following planets probably has the coldest core?
    a. Earth   c. Saturn   e. Venus
    b. Jupiter   d. Mars

UNIT 45

Jupiter and Saturn

45.1 The Appearance of Jupiter and Saturn
45.2 The Giants' Interiors
45.3 Stormy Atmospheres
45.4 The Magnetic Fields

Learning Objectives

Upon completing this Unit, you should be able to:
- Describe and name the external atmospheric features and internal layers.
- Explain what physical processes help to generate and maintain the Great Red Spot.
- Explain what makes the interior of these gas giants hot.
- Describe liquid metallic hydrogen and its role in these planets' magnetic fields.

Although we speak of four giant planets in the Solar System, Jupiter is more than twice as massive as all the smaller bodies in the Solar System combined (including Saturn), and Saturn is likewise more than twice as massive as everything smaller. Astronomers have developed models of both planets that indicate these two giants have similar internal structures, and these two largest planets share many other characteristics even though Saturn is less than 1/3 as massive.

Because of Saturn's spectacular ring system (Unit 47), we often overlook the body of the planet. Neither Jupiter nor Saturn has a solid surface, but their massive atmospheres dwarf the Earth (Figure 45.1). These behemoths become progressively hotter and denser deep in their interiors, with cores of molten rock and iron. Jupiter and Saturn also both formed extensive satellite systems with remarkable moons, which have become the focus of recent space missions to these distant systems (Unit 48).

FIGURE 45.1
Jupiter as imaged by the Hubble Space Telescope, and Saturn as it appears without its rings. (The rendering of Saturn with the rings removed was made by Björn Jónsson, based on NASA images.)

45.1 THE APPEARANCE OF JUPITER AND SATURN

Jupiter is at a distance of about 5 AU from the Sun. It has a diameter about 11 times the Earth's and a mass more than 300 times the Earth's. Saturn is almost twice as far from the Sun (9.5 AU) and just slightly smaller than Jupiter, with 9.5 times Earth's diameter and about 100 times its mass.

The ancient Romans did not know that Jupiter was the largest planet in the Solar System, but they nevertheless named it appropriately after the king of the gods. Possibly they chose its name because of its relatively steady brightness and stately motion through the celestial sphere, spending about one year in each constellation of the zodiac before moving to the next. From our Earth-based perspective, only Venus gets brighter than Jupiter at its brightest; but Venus sometimes becomes dimmer than Jupiter's dimmest appearance, and it swings back and forth among the constellations as it orbits the Sun.

Saturn bears the name of an ancient Roman god of agriculture who is identified with the Greek god Cronus (also spelled Kronos). Jupiter banished Saturn and the other Titans to the darkest pit, farthest from heaven; perhaps this explains the naming of Saturn, the dimmest, slowest-moving of the planets known before the invention of the telescope.

Through even a small telescope, Jupiter displays parallel bands of clouds ringing the planet, as shown in Figure 45.2. Dark **belts** alternate with light-colored **zones.** Spectra of the sunlight reflected back through Jupiter's atmosphere show that almost 90% of the molecules are hydrogen (H_2) and 10% are helium (He), and there are only traces of other hydrogen-rich gases such as methane (CH_4), ammonia (NH3), and water (H_2O) as shown in Table 45.1. These gases were also detected directly in December 1995 when a probe from the *Galileo* spacecraft parachuted into Jupiter's atmosphere. The solid particles and droplets that make up the clouds themselves are harder to analyze, but computer calculations of the chemistry of Jupiter's atmosphere suggest they are made of water, ice, and ammonia compounds. The bright colors in Jupiter's belts are probably produced by complex organic molecules.

Saturn also shows bands of color, but the variations are much more subtle (Figure 45.1). Spectral analysis of Saturn's atmosphere shows that it is even more

> **Clarification Point**
>
> Organic molecules are carbon-based; they do not require life or "organisms" to form.

> Atmospheric temperatures generally become lower at higher altitudes. The temperatures given for gaseous planets are usually estimated where the pressure reaches "one atmosphere"—the same as at the Earth's surface.

Rotation of Jupiter

TABLE 45.1 Composition of Jupiter's and Saturn's Atmospheres

Molecule (Symbol)	Jupiter % by Volume	Saturn % by Volume
Hydrogen (H_2)	89.8%	96.3%
Helium (He)	10.2%	3.3%
Methane (CH_4)	0.3%	0.45%
Ammonia (NH_3)	0.026%	0.013%
Hydrogen deuteride (HD)	0.0028%	0.011%
Ethane (C_2H_6)	0.0006%	0.0007%
Water (H_2O)	0.0004%	

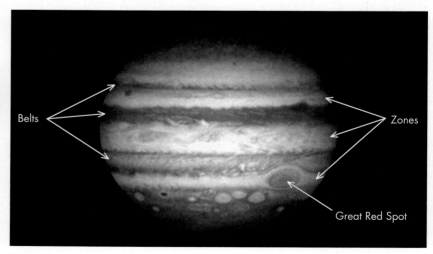

FIGURE 45.2
Jupiter's cloud system has a banded structure of darker belts and light-colored zones, and a "Great Red Spot," which is larger than the Earth.

dominated by hydrogen, which makes up about 96% of the molecules in its atmosphere. Helium again makes up most of the remainder, with small traces of hydrogen-rich compounds.

Part of the difference in appearance between these two gas giants may be caused by their different temperatures. At its distance from the Sun, Jupiter's outer atmosphere is warmed to about 160 K (about −170°F). Saturn's greater distance from the Sun results in a lower temperature of about 130 K (about −230°F). Saturn is cold enough for ammonia gas to condense into cloud particles that veil its atmosphere's deeper layers. This makes markings in lower layers indistinct, like a colorful object seen through fog.

Both planets rotate very rapidly. Jupiter rotates once every 9.9 hours and Saturn once every 10.7 hours. Such rapid rotation results in both planets being visibly flattened. Jupiter's equatorial diameter is about 7% larger than its diameter from pole to pole. Curiously, although Saturn rotates slower, its equatorial bulge is larger, about 10% larger than its polar diameter (see Unit 44.4).

45.2 THE GIANTS' INTERIORS

It is clear from the densities of Jupiter and Saturn that their atmospheres must be extremely deep. A planet's average density is determined by dividing its mass by its volume. The mass can be found by calculating the planet's gravitational attraction on one of its moons (Unit 17), and its radius—and hence its volume—from its angular size and distance (Unit 10). When such calculations are made for Jupiter, we find that its average density is only slightly greater than that of water—1.3 kilograms per liter. For Saturn the density is lower than the density of water—0.7 kilograms per liter. This indicates that the bulk of these planets, not just their surface layers, must be composed mainly of very light elements such as hydrogen and helium.

Astronomers can learn something about the internal density structure of a planet by detailed study of its gravitational pull as space probes pass by. Differences of internal structure would not be detectable if the matter inside were distributed in spherical shells, but when a planet spins, its internal layers bulge out, similar to how the outer regions do at the equator as we saw in the previous section. This produces small variations in the direction and strength of the planet's gravitational pull at

Concept Question 1

In the remote future, the Sun will enter a phase in which it may become thousands of times more luminous than it is now. What will happen to the giant planets when this happens?

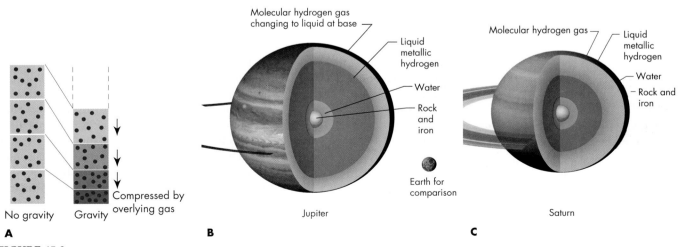

FIGURE 45.3
(A) The density of gas increases with depth because the overlying gas compresses the underlying matter. (B) A sketch of what astronomers think Jupiter's interior is like. (C) The internal structure of Saturn.

different latitudes that depend on the internal density distribution. Astronomers combine theoretical models of different possible internal compositions with the observed flattening to come up with a best fit to the space probe gravity data. The models indicate that both Jupiter and Saturn have similar internal structures.

The immense mass of these planets exerts a tremendous gravitational force on their interiors, compressing the gas to very high densities. Near the cloud tops this compression is only slight, so the gas density there is low. Deep in the interior, however, the weight of thousands of kilometers of gas compresses the matter to a high density. Thus, the density of the gas increases with depth, as illustrated in Figure 45.3A. Everywhere within the planet, the gas is squeezed until it pushes outward with a force that matches the gravitational force—the weight—of the gas above it. If the forces didn't match, the atmosphere would either expand or contract. This same relationship is true for any other atmosphere, including Earth's.

Deep within Jupiter, the compression created by its gravity presses molecules so close together that the gas changes to liquid. Thus, about 10,000 kilometers below the cloud tops—about one-sixth of the way into the planet—Jupiter's interior is a vast sea of liquid hydrogen. The pressures at this depth are about a million times the pressure we experience on the surface of the Earth. Slightly deeper still, the added weight of the overlying layers compresses the hydrogen below them into a state known as **liquid metallic hydrogen,** a form of hydrogen that scientists on Earth have created in tiny high-pressure chambers for fractions of a second. The hydrogen is "metallic" in the sense that the electrons can move about easily, allowing electric currents to travel through it as in a metal.

Close to the center of Jupiter, the pressure has risen to nearly 100 million times Earth's atmospheric pressure, and liquid is compressed to densities greater than that of rock. Assuming that Jupiter formed from gas with the same overall composition as the Sun, it must contain several times more than Earth's own mass in heavy elements such as silicon and iron. Because of their high density, these elements would have sunk to Jupiter's center, forming a core of iron and rocky material (see Figure 45.3B).

Astronomers infer that Saturn's internal structure (depicted in Figure 45.3C) is similar to Jupiter's, with several minor differences. The models suggest that Saturn bulges more because its mass is more concentrated toward its center than is Jupiter's. Saturn's core is also probably rock and iron, with many times the mass of the Earth. The models even suggest that Saturn's rock and iron core may be larger than Jupiter's, but there are large uncertainties in the models.

Measurements of infrared radiation coming from Jupiter and Saturn show that they both emit roughly twice as much energy as they receive from the Sun. As it entered the upper parts of Jupiter's atmosphere, the *Galileo* atmospheric probe measured this heat. The heat output implies that Jupiter's core must be hot, perhaps 30,000 K, which is about five times hotter than the Earth's core. Even 10,000 kilometers below the surface, the temperature is already almost as hot as the Earth's core, about 5000 K. The heat from the interior rises slowly to the planet's surface and escapes into space as low-energy infrared radiation.

Astronomers are not certain what generates Jupiter's and Saturn's heat, although much of it may be left over from the planets' formation. Planet building is a hot process, heat being generated by the kinetic energy of gas and planetesimals colliding with the forming world (Unit 37). Earth also began hot, and would have cooled much more after 4.5 billion years except that energy is being steadily released in its interior as radioactive elements decay. Because Jupiter and Saturn are primarily made of hydrogen and helium, they do not have proportionally as much of these radioactive elements to explain their large heat output.

Additional heat may be generated by slow but steady shrinkage of the planets as the internal material "settles" under its own weight. Thus, giant gas planets such as Jupiter may still be shrinking slightly and therefore heating as the matter is squeezed. You may have experienced this pressure heating if you have pumped up a bicycle tire; the pump grows hot from your compression of the air. Heat is also generated

---

The *Galileo* probe reached about 200 km (about 120 miles) below the cloud tops; it reported a pressure 22 times Earth's atmospheric pressure and a temperature of about 420 K (about 300°F) before the pressure and heat in deeper layers destroyed the craft.

Concept Question 2

How might Jupiter and Saturn change in the remote future after they stop generating heat by differentiation? Which planet is likely to stop first?

The *Cassini* probe has found that the energy output from Saturn is variable, and scientists hypothesize that this may relate to changes in internal circulation and weather on the planet.

by differentiation, as matter denser than hydrogen, such as helium, sinks toward the planet's core. Helium gas may condense into droplets, much as water droplets condense in our atmosphere. As the helium droplets fall toward the core, they generate heat by friction as they move through the hydrogen. Unlike much denser rock, which sank rapidly to their cores when the planets were young, helium sinks more slowly. The settling of this helium toward their cores may be continuing to this day.

In a relative sense, Saturn is generating more heat than Jupiter, given that its mass is less than one-third of Jupiter's. This extra heating may be related to the observation that Saturn has a lower abundance of helium in its upper atmosphere (Table 45.1). Astronomers hypothesize that this low abundance results from the colder temperature on Saturn, which has allowed more of the helium to condense and sink into the interior.

45.3 STORMY ATMOSPHERES

The high temperatures deep inside Jupiter and Saturn drive strong convection currents in the planets' outer layers. The *Galileo* atmospheric probe measured winds faster than 500 kilometers per hour (more than 300 mph). The convection currents carry warm gas upward to the top of the atmosphere. Here the gas radiates heat into space. As the gas cools, it becomes denser and sinks again, as illustrated in Figure 45.4A. Such deep circulation, combined with Jupiter's and Saturn's rapid rotation, makes "weather" on the gas giants very different from that on the Earth.

As material moves on a rotating planet, it is deflected by the Coriolis effect (Unit 38). In particular, winds that move farther away from the planet's axis of rotation (either closer to the equator or upward from the interior) are deflected to the west, and winds that move closer to the axis are deflected to the east, as shown in Figure 45.4B. The rapid rotation of Jupiter and Saturn, along with the deep circulation of gases upward and then back down, creates an extremely powerful Coriolis effect. This deflects rising and sinking atmospheric gases into powerful winds called *jet streams*. The winds in adjacent regions may blow in opposite directions, as shown in Figure 45.4B. We see these winds as the cloud belts, illustrated in Figure 45.4C. Such reversals of wind direction ("wind shear") from place to place also occur on Earth, where equatorial winds generally blow from east to west, while midlatitude winds generally blow from west to east. Time-lapse photographs show that Jupiter's clouds move swiftly, sweeping around the planet in jet streams that are far faster than those of Earth. The dark belts move westward

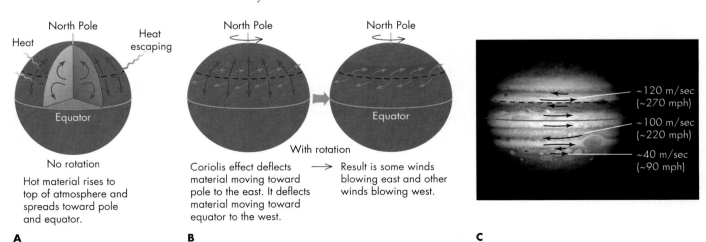

FIGURE 45.4

(A) Rising gas from Jupiter's hot interior cools near the top of the atmosphere and sinks. (B) The Coriolis effect, arising from Jupiter's rotation, deflects the gas, creating winds that blow as narrow jet streams. (C) The wind varies widely in speed and direction from region to region.

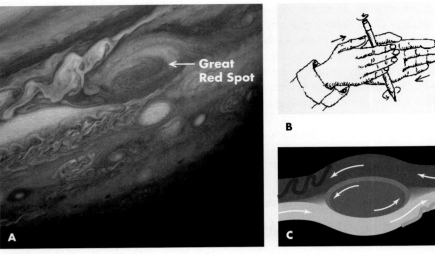

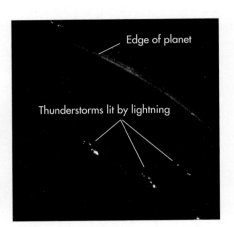

FIGURE 45.5
(A) Vortices form between atmospheric streams of different velocities. Note the Great Red Spot and the white oval below it. (B) The vortex motion is something like the spin of a pencil twirled between two hands. (C) The white zones move around Jupiter faster than the dark belts, creating a relative motion that can produce a vortex.

FIGURE 45.6
Lightning on Jupiter, as observed by the *Voyager 1* spacecraft as it flew over Jupiter's night side.

at speeds hundreds of kilometers per hour relative to the planet, while the white zones move eastward at hundreds of kilometers per hour.

As the various jet streams circle Jupiter, gas between them can be spun into a huge, whirling atmospheric **vortex.** The gas in the vortex is given a spin much as a pencil between your palms twirls as you rub your hands together. Some of these spinning regions are brightly colored, as Figure 45.5 shows. Brown and shades of white dominate, but one exceptionally large vortex—bigger across than the Earth—is nearly brick red. Known as the **Great Red Spot,** this vortex was discovered in the seventeenth century and has remained active for more than 300 years!

The exact chemistry that gives the spots their colors is uncertain, although spectroscopy reveals the presence of methane, ammonia, and more-complex compounds that have not been completely identified. (For complex molecules, spectroscopy can sometimes indicate some of the atomic bonds in the molecules without giving a unique identification.) Infrared imaging indicates that these regions are heated from below.

The Great Red Spot is a high-pressure region that rises to about 50 kilometers above the surrounding regions, and time-lapse images show winds swirling within the spot at about 500 kilometers per hour (300 mph). Over several hundred years of observation, the spot has changed in size and color, but otherwise this giant vortex appears to be very stable. This is unlike hurricanes on Earth, which die out when they are no longer fed by warm water from below. Computer models indicate that as long as Jupiter's interior is able to supply a steady source of heat, the Great Red Spot may persist as a nearly permanent feature of Jupiter's atmospheric circulation.

Figure 45.6 depicts another familiar phenomenon associated with storms occurring in Jupiter's atmosphere: lightning flashes visible on the night side of the planet. These are associated with clusters of Jovian thunderstorms some 1000 km (600 miles) across. Just as on Earth, electrical charges are carried by precipitating water (or perhaps ammonia, on Jupiter). Particles with opposite charges get separated by the rising and sinking motions in the storm cells, which can lead to an electric discharge between clouds.

Saturn's atmosphere looks calmer, but it is much more active than it appears. High haze obscures details of storms deeper in the atmosphere. Figure 45.7 shows

> **Concept Question 3**
>
> What are some examples of a vortex here on Earth?

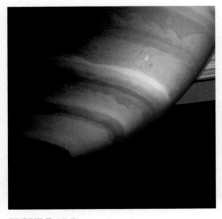

FIGURE 45.7
A high-contrast image of Saturn, made by *Cassini* in 2004, reveals complex cloud belts and storms in Saturn's atmosphere.

FIGURE 45.8
Auroral activity on (A) Jupiter and (B) Saturn. Observed with the Hubble Space telescope at ultraviolet wavelengths, the aurora is shown in violet over visible wavelength images.

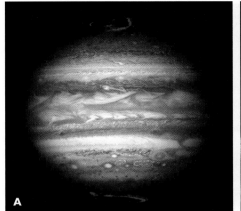
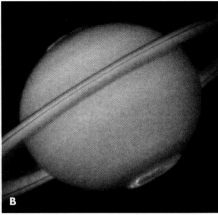

a picture of Saturn's atmosphere made by the *Cassini* spacecraft with the color "stretched" to bring out details of the belts as well as a large storm seen in 2004.

45.4 THE MAGNETIC FIELDS

In the deep interiors of Jupiter and Saturn, convection in the metallic liquid hydrogen combines with the planets' rapid rotation to generate magnetic fields. These magnetic fields are driven by a natural dynamo process similar to that which generates the Earth's magnetic field in its own metallic liquid core. Jupiter's dynamo process creates the strongest magnetic field of any planet in the Solar System, almost 20,000 times stronger than Earth's. Saturn's magnetic field is not as strong, but it is still more than 500 times stronger than Earth's.

The Earth's magnetic field steers incoming energetic particles from the Sun into our upper atmosphere, where the particles trigger the lovely pale glow of the northern and southern lights—the **aurora**. Jupiter's and Saturn's magnetic fields do much the same, but moons and ring particles orbiting close to the planet also contribute charged particles. Charged particles from volcanoes on Jupiter's moon Io (Unit 48) produce an especially strong aurora on Jupiter.

As happens on Earth, particles are trapped magnetically in belts far above the planet, but they spiral through the field toward the magnetic poles, where they can collide with the upper atmosphere and create an aurora. The Hubble Space Telescope photographed this phenomenon on both Jupiter (Figure 45.8A) and Saturn (Figure 45.8B). This also produces strong radio emission as the trapped charged particles that create the aurora accelerate from pole to pole in the planet's magnetic field.

KEY POINTS

- Jupiter and Saturn each outweigh the combined total of everything else in the Solar System that is smaller than them.
- Cloud belts and zones are colored by a complex weather based on hydrogen, methane, ammonia, and organic compounds.
- Saturn's coloration is less distinctive because of a high-altitude haze of frozen ammonia particles.
- The gas giants are made mostly of hydrogen and helium, but with cores of molten rock and iron much more massive than the Earth.
- Gas is compressed to densities exceeding that of rock on Earth.
- At the high pressures in Jupiter and Saturn, hydrogen can be compressed to a metallic, electrically conducting state.
- Because of rapid rotation, there is a strong Coriolis effect creating high-speed "jet streams" moving in opposite directions.
- Jupiter's Great Red Spot is a storm larger than Earth that has persisted for hundreds of years and is driven by rapid, opposing flows of belts and zones.
- Both Jupiter and Saturn have strong magnetic fields, probably produced by electric currents in the metallic hydrogen.
- The magnetic fields funnel electrically charged particles toward the magnetic poles, creating auroras.

KEY TERMS

aurora, 353
belt, 348
Great Red Spot, 352
liquid metallic hydrogen, 350
vortex, 352
zone, 348

CONCEPT QUESTIONS

Concept Questions on the following topics are located in the margins. They invite thinking and discussion beyond the text.

1. Future of giant planets when the Sun grows very luminous. (p. 349)
2. Changes in giant planets when differentiation halts. (p. 350)
3. Examples of vortices. (p. 352)

REVIEW QUESTIONS

4. How do Jupiter's and Saturn's masses and radii compare with the Earth's?
5. How are the interior structures of Jupiter and Saturn similar? How do they differ?
6. What is the source of internal heat for Jupiter and Saturn?
7. What sorts of atmospheric motion and activity are observed in Jupiter and Saturn?
8. What is the Great Red Spot? What allows it to persist for centuries?
9. What are the similarities and differences of the Coriolis effect for Jupiter or Saturn compared to the Earth?

QUANTITATIVE PROBLEMS

10. Given its rate of rotation, at what speed does a point on the equator of Jupiter move? Give your result in km/hr.
11. Given its rate of rotation, at what speed does a point on the equator of Saturn move? Compare this speed to that of Jupiter.
12. a. Jupiter is 318 times more massive and 11.2 times larger than Earth. How much greater is the surface gravity (Unit 16.3) on Jupiter compared to Earth?
    b. Do the same calculation for Saturn, which has a mass 95 times larger than Earth and is 9.5 times bigger.
13. Compare the volume of Jupiter to the volume of Earth. How many bodies the size of the Earth could fit inside Jupiter?
14. If Saturn were compressed until it had the same density as Jupiter, what would Saturn's new radius be? Compare this to Jupiter's radius. If Saturn were compressed until it had the same density as Earth, what would Saturn's new radius be?
15. In the Sun, hydrogen makes up 90% of the atoms, corresponding to 75% of its mass. Iron and other elements with atomic weights of about 56 make up 0.001% of the atoms, while silicon and other elements with atomic weights of about 28 make up 0.01% of the atoms. These represent the percentage of atoms, not the fraction of the mass. If Jupiter has the same proportions of atoms in its interior, what would the total mass of Jupiter's iron- and silicon-like atoms be? Compare the masses you find to the mass of the Earth.
16. Look up information on the speed, structure, and lifetime of a class 5 hurricane on Earth and compare it to the properties of the Great Red Spot. Describe the differences in their causes.

TEST YOURSELF

17. The cores of Jupiter and Saturn are
    a. just high-density gas.
    b. balls of extremely cold liquefied gases.
    c. icy bodies larger than Mars.
    d. hot rock and iron about as large as Earth.
18. The low average density of Saturn suggests that
    a. Saturn is hollow.
    b. Saturn's gravitational attraction has compressed its core into a rare form of iron.
    c. Saturn contains large quantities of light elements, such as hydrogen and helium.
    d. Saturn is very hot.
    e. volcanic eruptions have ejected all the iron that was originally in Saturn's core.
19. Below the thick clouds of Jupiter and Saturn is a layer of liquid hydrogen. Hydrogen is in a liquid form due to
    a. high pressures.
    b. cold temperatures.
    c. convection in the outer clouds.
    d. friction from the planets' rapid rotation.
20. The amount of infrared energy emitted by Jupiter is about twice as great as the amount of sunlight the planet absorbs. What is the significance of this discrepancy?
    a. It implies that the planet is cooler than it should be.
    b. It implies that there are significant energy sources within Jupiter.
    c. It implies that the Sun was once much dimmer than it now is.
    d. It implies that the rotation of Jupiter must be slowing down.
21. What causes Jupiter and Saturn to have substantially larger radii at their equators than at their poles?
    a. They spin so fast that inertia causes the equator to bulge.
    b. High-energy storms near the equator cause the atmosphere to expand there.
    c. Intense magnetic fields pull the atmosphere in at the poles.
    d. These icy planets accumulated most of their planetesimals along their equators as they formed.

UNIT 46

Uranus and Neptune

46.1 Discovery of Two New Planets
46.2 The Atmospheres of Uranus and Neptune
46.3 Oddly Tilted Axes

Learning Objectives

Upon completing this Unit, you should be able to:
- Recall how Uranus and Neptune each were discovered, and what observations are available to study them at present.
- Describe the planets' structure and explain why they have a blue color.
- List differences between the appearance of the two and discuss possible causes.
- Describe the rotation of Uranus, and what its seasons are like.

Uranus and Neptune (Figure 46.1) are the only planets in the Solar System that were unknown in ancient times. They are almost twins. Both are significantly smaller than Jupiter or Saturn, but they are still giant planets with masses of 15 and 17 Earth masses, respectively, and radii about four times the Earth's. Both have been visited by just one spacecraft—the *Voyager 2* probe, which flew by Uranus in 1986 and Neptune in 1989, providing our most detailed images of them. Much of what we know about these two blue worlds came from the brief encounters of that mission. Despite their similarities, Uranus and Neptune exhibit some interesting differences, many of which are not yet fully understood. Both also have complex systems of rings and moons like Jupiter and Saturn (Unit 45).

To distinguish them from the gas giants Jupiter and Saturn, astronomers sometimes call Uranus and Neptune **ice giants**. Ice does not describe their composition since their cores are probably as hot as Earth's, but it may describe a difference in how they formed. While Jupiter and Saturn appear to have gathered a large amount of gas from the solar nebula, Neptune and Uranus contain high percentages of water and other molecules that suggest they were built primarily by accumulating icy planetesimals (Unit 35).

FIGURE 46.1
Uranus and Neptune as imaged from the *Voyager 2* spacecraft, shown to scale with Earth. Because of Uranus's odd tilt, we are viewing it nearly pole-on. While Uranus is almost featureless, Neptune exhibits a variety of atmospheric storm systems.

355

Herschel was a musician by profession, but searched for comets with his sister Caroline in his spare time.

Galileo saw Neptune in 1613 while observing Jupiter's moons. His notes record a dim object whose position changed with respect to the stars, but he failed to appreciate its significance so Neptune eluded discovery for two more centuries.

46.1 DISCOVERY OF TWO NEW PLANETS

Uranus was discovered in 1781 by Sir William Herschel, a German émigré to England. Uranus is sometimes just barely visible to the unaided eye, but it was with his homemade telescope that Herschel first observed this new planet. He saw a pale blue object whose position in the sky changed from night to night. After several months, astronomers determined that the body's orbit was nearly circular and realized that Herschel had found a new planet.

For this discovery, King George III named Herschel his personal astronomer, and to honor the king, Uranus was briefly known as *Georgium Sidus* ("George's Star"). The planet was also sometimes called *Herschel* in honor of its discoverer before astronomers decided that its name should be consistent with those of the planets known since antiquity. The name Uranus was chosen because Uranus was the father of Saturn in Greek mythology, as Saturn was the father of Jupiter.

Neptune was discovered in 1846 from predictions made independently by a young English astronomer, John Couch Adams, and a French astronomer, Urbain Leverrier. Adams and Leverrier both found that Uranus had an **orbital perturbation**—it was not precisely following its expected orbit. They inferred that Uranus's motion was being disturbed by the gravitational force of an unknown planet. From the size and direction of the perturbation, Adams and Leverrier predicted where the unseen body must lie.

Adams completed his calculations in 1845, but his prediction was ignored. In 1846, however, when Leverrier carried out calculations nearly identical to those made by Adams, Neptune was found almost immediately. Assignment of credit for the discovery of the new planet led to a rancorous dispute tinged with national pride that lasted decades. The discovery is now credited equally to Adams and Leverrier. Having a deep blue color, Neptune was named for the Roman god of the sea after several years of dispute over its name.

46.2 THE ATMOSPHERES OF URANUS AND NEPTUNE

Uranus and Neptune have atmospheres rich in hydrogen and helium, like Jupiter and Saturn. Their atmospheres both contain a larger amount of methane (CH_4) than the bigger gas giants, about 2% by volume (Table 46.1). We know this from spectra of their atmospheres, which show strong absorption lines of methane.

It is methane that gives these planets their blue color. As illustrated in Figure 46.2, when sunlight enters the atmosphere, methane absorbs the incoming red and orange light. The remaining light, now mostly blue and green, reflects off clouds

TABLE 46.1 Composition of Uranus's and Neptune's Atmospheres

Molecule (Symbol)	Uranus % by volume	Neptune % by Volume
Hydrogen (H_2)	82.5%	80.0%
Helium (He)	15.2%	19.0%
Methane (CH_4)	2.3%	1.5%
Hydrogen deuteride (HD)	0.0148%	0.0192%

Concept Question 1

Why do Uranus and Neptune have a higher percentage of helium than their more massive neighbors, Jupiter and Saturn?

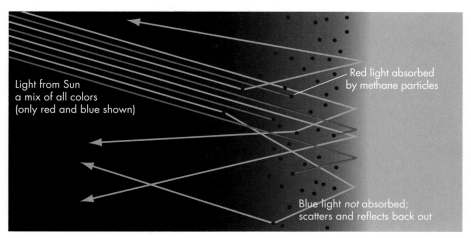
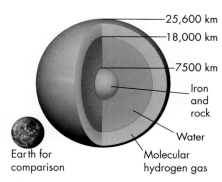

FIGURE 46.2
Sketch illustrating why Uranus and Neptune are blue. Methane absorbs red light, removing the red wavelengths from the sunlight that falls on the planet. The surviving light—now missing its red colors—is therefore predominantly blue. As that light scatters off cloud particles in the planet's atmosphere and returns to space, it gives the planet its blue color.

FIGURE 46.3
Model for the interior of Uranus, which is similar to models of Neptune. The estimated radii of different layers are given. Note how different it is from Jupiter and Saturn, which both contain large regions of liquid and metallic hydrogen.

Concept Question 2

Methane is a very strong greenhouse gas on Earth. How would you expect the methane in Uranus and Neptune to affect their temperatures?

deeper down and is scattered by a haze of frozen methane particles in the upper atmosphere as it travels back into space.

A haze of crystals of frozen methane is present in Uranus's and Neptune's atmospheres because their outer atmospheres are substantially cooler than Saturn's or Jupiter's. Uranus and Neptune are about 19 and 30 AU from the Sun, respectively, about two and three times farther from the Sun than Saturn. The mean temperature is about 80 K (about −320°F) for Uranus, and about 75 K (about −330°F) for Neptune.

Deeper down, the interiors of these two planets are probably similar based on the limited information currently available. To study the interior of these planets, astronomers rely on indirect methods, using their density and shape. Uranus has an average density of about 1.3 kilograms per liter. Its density is nearly twice that of Saturn and almost as high as Jupiter's. Neptune's density is even greater, at about 1.6 kilograms per liter. Given their smaller masses, Uranus and Neptune compress the matter in their interiors to a lesser degree, and astronomers therefore deduce that both must contain proportionally fewer light elements, such as hydrogen, than Jupiter and Saturn. This suggests that they did not accumulate as much gas from the solar nebula when the Solar System was forming (Unit 35).

The masses of Uranus and Neptune do not produce the extremely large pressures needed to liquefy hydrogen as occurs in Jupiter and Saturn. The best model for their density and atmospheric composition is that they contain a relatively high amount of ordinary water mixed with methane and ammonia surrounding a core of rock and iron-rich material, as illustrated in Figure 46.3. Even though this water is very hot, probably several thousand Kelvin, it is not gaseous because of the large pressure exerted by the atmosphere.

Like Jupiter and Saturn (Unit 45), Uranus and Neptune radiate more energy than they receive from the Sun, but unlike the strong output from the other three, Uranus emits only a few percent more than what it receives from the Sun. The limited information about Uranus's gravitational field suggests that its material may not be as differentiated as that of the other giant planets, and may contain relatively less rock or iron. It is not clear what causes Uranus to be dissimilar to the other gas giants in internal composition, but its lower degree of differentiation may explain its lower energy output.

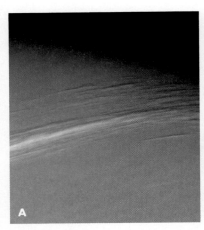 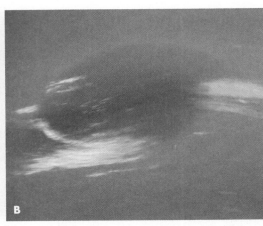 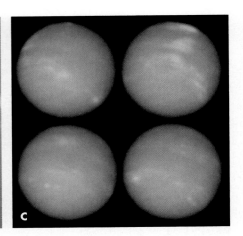

FIGURE 46.4
(A) During its closest approach, *Voyager 2* captured this picture of fast-moving, narrow cloud streams about 50 kilometers above Neptune's cloud deck, which is tinged blue by the methane below the cloud streams. (B) *Voyager 2* imaged a dark vortex on Neptune about half the size of Jupiter's Great Red Spot. (C) Monitoring by the Hubble Space Telescope beginning in 1994 has shown that the dark spot has disappeared, although high-level cloud systems are often seen.

Rotation of Neptune

> **Concept Question 3**
>
> The rotation period of the four gas giants is faster the more massive the planet. Could something about the formation of a larger planet explain this trend?

Uranus and Neptune rotate at about the same rate, with days of length 17 and 16 hours, respectively. Although they do not rotate as fast as Jupiter or Saturn, their rotation produces a strong Coriolis effect, which can lead to high-speed jet streams and vortices in the atmosphere (Unit 45). Even if such motions were present on Uranus, the dense methane haze hid them in images made by the *Voyager 2* spacecraft as it swung by. By contrast, Neptune displayed distinctive cloud belts. Pictures taken of Neptune by *Voyager 2* in 1989 showed markings reminiscent of Jupiter. Cloud bands encircle it (Figure 46.4A), and it even had a **great dark spot** (Figure 46.4B), a huge dark blue atmospheric vortex about 13,000 kilometers across, about half the size of Jupiter's Great Red Spot.

Neptune's winds are extremely fast, reaching some of the highest measured speeds in the Solar System: nearly 2200 kilometers per hour (about 1300 mph). As these gales sweep around Neptune, they can create a large, stable vortex. However, Neptune's giant vortex turned out not to be nearly as stable as Jupiter's. By 1994, when the Hubble Space Telescope studied the planet, the spot had disappeared; but other cloud motions have been tracked (Figure 46.4C).

Why do Uranus and Neptune, which in many ways are so similar, have such different cloud formations? Recall that Neptune, like Jupiter and Saturn, radiates a substantial amount of internal energy. Neptune has both a hotter interior and a cooler surface than Uranus, so it generates much stronger convection currents that rise to its outer atmosphere. There the rising gas is deflected into a system of winds by the Coriolis effect. The resulting winds create cloud bands similar to those seen on Jupiter and Saturn but tinted blue by Neptune's methane-rich atmosphere.

46.3 ODDLY TILTED AXES

One factor that we have not considered in discussing Uranus's atmospheric patterns is its extremely tipped rotation axis. It is tipped 98°. Since it is more than 90°, it is actually rotating slightly backward (like Venus), with its equator nearly perpendicular to its orbit. That is, it spins nearly on its side, as illustrated in Figure 46.5. Moreover, the orbits of Uranus's moons are similarly tilted. They orbit Uranus

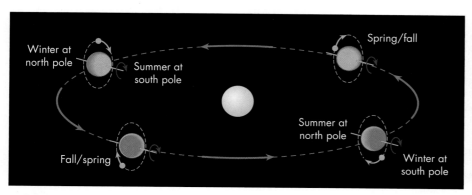

FIGURE 46.5
A sketch of Uranus's odd tilt. Because of this tilt, when Uranus's north pole points toward the Sun (its northern summer), the Sun is above the horizon for many Earth years, while the other pole will be in night. During the Uranian spring and fall, the Sun rises and sets approximately every 17 hours. (Sizes of bodies and orbits are not to scale.)

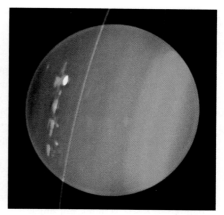

FIGURE 46.6
Uranus observed with the Keck 10-m telescopes in 2004. This false-color infrared image shows banded structure in the atmosphere and a number of storms.

Concept Question 4

Do you think it is possible that there are more planets orbiting the Sun yet to be discovered? Why or why not?

in its equatorial plane; as a result, their orbits are also tilted at the same large angle with respect to the planet's orbit. Some astronomers therefore hypothesize that during its formation Uranus was struck by a large planetesimal and that this impact tilted the planet and splashed out material to create its family of moons.

The strong tilt of its axis gives Uranus an odd pattern of day and night. For part of its 84-year orbit, much of one hemisphere is in "perpetual" day and the other hemisphere is in "perpetual" night. In 1989, at the time of the *Voyager 2* encounter, Uranus's south pole was nearly facing the Sun. With this orientation the polar region receives more heating than the equatorial regions, and the whole northern hemisphere received almost no sunlight. This may have contributed to the lack of cloud bands seen during the flyby.

Uranus's polar axis was perpendicular to the Sun in 2007, so the Sun was shining straight down onto its equator. The solar heating of Uranus in the decade before and after 2007 is more similar to the solar heating received by the other planets—greatest in the equatorial regions and with typical day/night variations. Markings in its atmosphere were seen around this time with the Hubble Space Telescope and ground-based infrared telescopes (Figure 46.6). These markings appear to be large storm systems, so perhaps Uranus is not always as featureless as it looked during the *Voyager 2* flyby in 1986.

The magnetic fields of Uranus and Neptune have peculiar orientations as well. The strengths of the fields are what one might expect, intermediate between those of the larger gas giants and the Earth—about 47 times stronger than Earth's in Uranus, and 25 times stronger in Neptune. However, the magnetic fields of both planets are tilted at a large angle to the rotation axis and offset by a large distance from the planet's center, as illustrated in Figure 46.7. Because these planets do not have liquid metallic hydrogen in their interiors, astronomers suspect that the internal electric currents that generate the magnetic fields occur within the high-temperature water layer. Although the mechanism that produces such highly tilted magnetic fields is uncertain, it may be that the circulation of electric currents is irregular or localized to certain spots in the water layer.

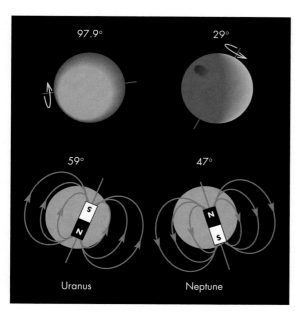

FIGURE 46.7
The magnetic fields of Uranus and Neptune are tilted at a large angle to their rotation axes and are offset from the centers of the planets.

KEY POINTS

- Uranus and Neptune are the only major planets in our Solar System that were discovered since ancient times.
- The two are similar in size, about 4 times Earth's diameter, and in mass, about 16 times Earth's mass.
- They are sometimes called ice giants because they probably formed from icy planetesimals and captured less gas than Jupiter or Saturn.
- These planets are colored blue by methane, but their atmospheres are primarily hydrogen and helium.
- Both planets rotate rapidly and show evidence of atmospheric bands and storm systems, but much detail is hidden beneath the methane haze.
- Uranus's rotation axis is almost "sideways," leading to large seasonal changes in solar heating.
- Both Uranus and Neptune have peculiar magnetic fields that do not align with the rotation axis; they may be generated in the internal water layer.

KEY TERMS

great dark spot, 358 orbital perturbation, 356
ice giant, 355

CONCEPT QUESTIONS

Concept Questions on the following topics are located in the margins. They invite thinking and discussion beyond the text.

1. Cause of higher helium percentage. (p. 356)
2. Methane as a greenhouse gas in the ice giants. (p. 357)
3. Relationship of rotation period to a planet's mass. (p. 358)
4. More planets not yet found in the Solar System. (p. 359)

REVIEW QUESTIONS

5. What are the differences in how Uranus was discovered versus how Neptune was discovered?
6. How do Uranus and Neptune differ from Jupiter internally?
7. Why are Uranus and Neptune blue?
8. What is unusual about Uranus's rotation axis? What might explain this peculiarity?
9. What are seasons like on Uranus?
10. What are the differences between the magnetic fields of Uranus and Neptune and that of Jupiter?

QUANTITATIVE PROBLEMS

11. Find the "surface" gravity on Uranus and Neptune (Unit 16.3). How does this compare with Earth's surface gravity?
12. If Uranus and Neptune had the same density, how many times bigger than Uranus would Neptune be?
13. In order to match the densities of Uranus and Neptune, scientists hypothesize that their interiors are composed of three layers: (i) a dense rock-iron core; (ii) a liquid water-ammonia layer; and (iii) a hydrogen-helium gas layer. Suppose the densities of these layers average 8, 1, and 0.2 kg/liter, respectively.
    a. Find the density of a planet that is 10% (i), 40% (ii), and 50% (iii).
    b. If there were no liquid layer, find a mix of percentages of the other two layers that gives the same density as in a.
    c. What other constraints do astronomers have that help decide on the correct mix?
14. Given Neptune's semi-major axis of 30 AU, use Kepler's third law (Unit 12.2) to calculate its orbital period. How many orbits has Neptune completed since its discovery in 1846.
15. Assume that Earth's axis is tilted 90° with respect to its orbital plane, similar to that of Uranus. How would this affect the length of Earth days? What would happen to the seasons? Describe a complete year for where you live.

TEST YOURSELF

16. It was the winter solstice for the northern hemisphere of Uranus in 1986, when *Voyager 2* passed the planet. Uranus takes 84 years to orbit the Sun, so 21 years later in 2007, what part of the planet would be facing the Sun?
    a. the South Pole
    b. the North Pole
    c. the equator
    d. all parts over the course of the night on Earth
17. What gives Neptune and Uranus their blue color?
    a. Objects with such low temperatures take on a blue color.
    b. Water molecules in the lower clouds reflect blue light.
    c. Methane gas absorbs the longer wavelengths of visible light.
    d. Interplanetary dust absorbs the Sun's red light before it reaches Uranus or Neptune.
    e. The icy blue polar caps of these planets face toward the Earth.
18. Neptune was discovered
    a. by accident while an astronomer was looking for comets.
    b. as it passed in front of a bright star, blocking its light.
    c. by looking at a position predicted by calculations.
    d. in antiquity, but it moves so slowly that people thought it was a star.
19. What is unusual about the magnetic fields of Uranus and Neptune?
    a. There are multiple north and south poles on each planet.
    b. They are weaker than Earth's magnetic field.
    c. They point at a large angle to their rotation axes.
    d. They occasionally disappear for a few years at a time.
20. Uranus and Neptune are sometimes called ice giants instead of gas giants because they are composed largely of
    a. solid ice cores beneath their atmospheres.
    b. blue-colored methane, which has a color resembling ice.
    c. ice crystals that reflect sunlight from their outer atmospheres.
    d. molecules that condense into icy planetesimals at their distance from the Sun.

PART 3 UNIT 47

Satellite Systems and Rings

47.1 Satellite Systems
47.2 Satellite Properties
47.3 Ring Systems
47.4 Origin of Planetary Rings

Learning Objectives

Upon completing this Unit, you should be able to:
- Describe the overall properties of satellite systems, and define an irregular orbit.
- Compare satellite systems and their formation process to the Solar System.
- Explain the differences between larger and smaller satellites.
- Describe ring systems, their origin, and the factors that shape them.

All of the giant planets have large systems of satellites as well as rings of particles too small to be called satellites. Two of the satellites are even larger than Mercury, though less massive, but the vast majority of the satellites are much smaller. More than 150 satellites have been cataloged orbiting the gas giants, and these systems of orbiting bodies bear many resemblances to the Solar System itself. Each gas giant's system has some unique properties, although there are many shared characteristics. The six largest moons are larger than Pluto and Eris and are fascinating worlds in their own right. They are discussed briefly here, but they are explored in greater detail in Unit 48. The general properties of the satellites and ring systems are the focus of this Unit.

47.1 SATELLITE SYSTEMS

When Galileo Galilei first viewed Jupiter with his telescope in 1610, he saw four starlike objects all in a line with the planet (Figure 47.1). As Galileo observed the planet night after night, he discovered that these small objects orbited Jupiter, taking between 2 and 17 days to complete their orbits. He realized that they must be satellites of Jupiter, just as our own Moon is a satellite of Earth (see Unit 12). They are called the **Galilean satellites** in his honor.

Over the past four centuries, astronomers have found many more satellites. More than 60 satellites of Jupiter and Saturn are now known. The total keeps growing as we obtain more-detailed views from spacecraft missions and from large telescopes with improving technologies. Almost all of the recent discoveries are less than a few kilometers across. The distances and sizes of the satellites orbiting each of the gas giants are illustrated in Figure 47.2. The sizes of the satellites are all shown to the same scale along with our Moon for comparison. Note that the spacing between planet and moons in Figure 47.2 is not to a fixed scale but varies to show all the satellites both near and far from each planet.

The Galilean satellites are all large bodies, comparable in size to our Moon. Johannes Kepler suggested naming them after mythological figures who were associated with Jupiter: Io, Europa, Ganymede, and Callisto. This idea took hold, so the satellites of each planet are named after related mythological figures. Today we know that two other satellites, Saturn's moon Titan and Neptune's moon Triton, are also quite large. Together with our Moon, these satellites are the next seven most massive objects in the Solar System after the Sun and eight planets.

The Galilean satellites are bright enough to see without a telescope, but they are lost in Jupiter's glare.

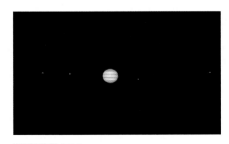

FIGURE 47.1
Jupiter and the Galilean satellites as seen through a small telescope. (This image was made by the Louisiana State Amateur Astronomy Club on March 19, 2004.) The starlike dots are the four large moons of Jupiter. From left to right: Europa, Io, Callisto, and Ganymede.

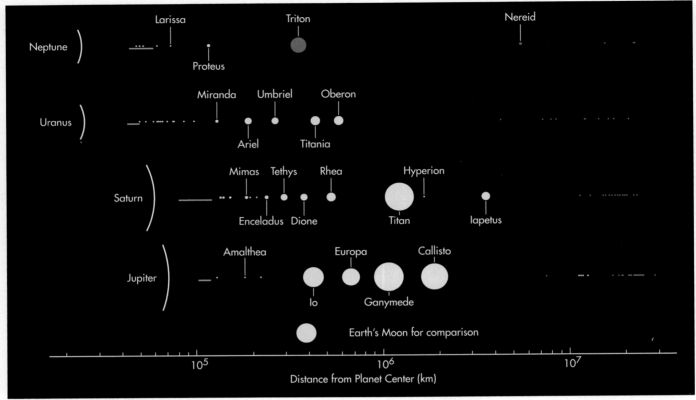

FIGURE 47.2
The satellite systems of the gas giants. The satellites are drawn to the same scale, but their spacing is compressed. The extent of each planet's ring system is shown with a blue line. Satellites with irregular orbits (at a large angle or highly noncircular) are shown in red. Moons with dimensions larger than 200 kilometers (120 miles) are named.

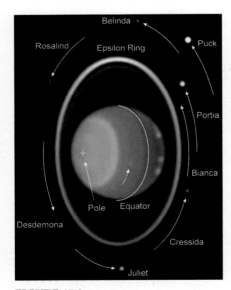

FIGURE 47.3
Hubble Space Telescope infrared image of Uranus. The rings and satellites orbit around Uranus's equator even though the planet is tilted almost perpendicular to the plane of the Solar System.

Figure 47.2 shows the sizes and distances of the satellites around each of the gas giants. Each system is quite different. Jupiter has four large moons, but almost all of the other satellites are tiny. The other planets all have a variety of satellites of intermediate sizes, as well as many small satellites.

Almost all of the satellites orbiting near each planet are in **regular orbits**. Regular orbits are nearly circular, are in the same direction as the planet spins, and are close to the plane of the planet's equator. The satellites with these properties are shown in yellow in Figure 47.2 and include almost all the satellites with radii larger than 100 kilometers. These systems of satellites form a flattened disk like a miniature Solar System. Even the satellites of Uranus orbit around its equator despite its extremely tilted rotation axis (Figure 47.3). This suggests that these satellites formed by a scaled-down version of the process that created the Solar System. That is, they probably aggregated from planetesimals and gas that collected around the gas giants during their formation like a miniature solar nebula (Unit 35).

Many of the smaller satellites discovered in recent years do not have regular orbits. They orbit backward from the planet's spin, or at large angles to the equator, or with highly elliptical orbits. Satellites with these properties are probably captured objects. They may have been asteroids or other objects that passed close enough to the planet to undergo a gravitational encounter. Capturing a satellite is not simple: The same orbit that brings a potential satellite close to a planet will normally carry it away again, following an elliptical trajectory, as predicted by Kepler's second law (Unit 12). For a passing body to end up in orbit around a planet, something must slow it down when it is close to the planet. This may occur if it encounters another satellite.

47.2 SATELLITE PROPERTIES

The six large satellites are shown to scale in Figure 47.4. Each has a variety of unusual features that are discussed in Unit 48. Medium-size satellites, which still have diameters larger than 200 kilometers (120 miles), are also shown in Figure 47.4, but at this scale, it is difficult to see any detail, so they are shown on a larger scale in Figure 47.5. The images were obtained by the *Voyager*, *Galileo*, and *Cassini* spacecraft. Two of these medium-size satellites, along with many smaller satellites, were discovered by the *Voyager 2* spacecraft as it passed Neptune in 1989.

Most of the satellites orbiting the gas giants have densities that imply they are a combination of ices and some rocky material, and their surfaces are peppered with impact craters. Nearly all are "tidally locked," like our own Moon, with the same side always facing the planet they orbit (Unit 39).

Icy bodies with diameters larger than about 400 kilometers settle into spherical shapes. This is perhaps half the diameter needed for a rocky asteroid to be forced into

Concept Question 1

Why do you suppose the giant planets have so many satellites and the terrestrial planets have so few?

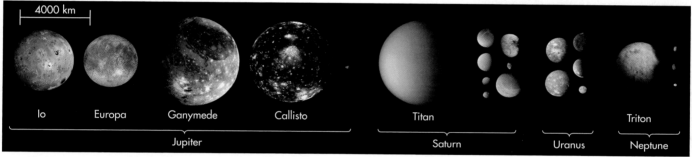

FIGURE 47.4
The six large satellites of the gas giants are named in this figure. The Galilean satellites' images are from the *Galileo* spacecraft, Titan's image is from the *Cassini* spacecraft, and Triton's image is from the *Voyager* flyby. To provide a sense of scale, the next largest moons are shown also to the same scale.

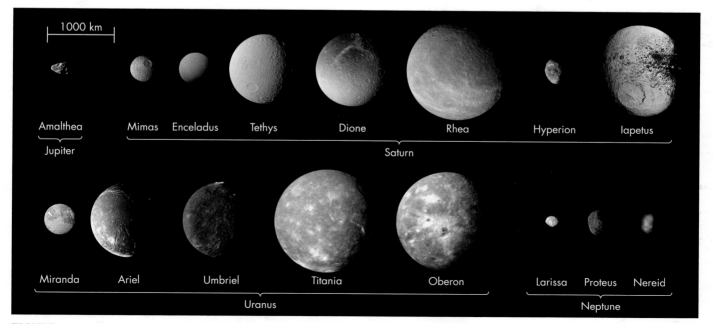

FIGURE 47.5
The medium-size satellites of the gas giants. These are the satellites with sizes between about 200 and 1500 kilometers (120 and 950 miles).

FIGURE 47.6

(A) Two views of Jupiter's fifth largest satellite Amalthea. It is about 250 kilometers (150 miles) in its longest dimension. It is an icy object coated with a layer of sulfur compounds from the moon Io's volcanoes. (B) Saturn's satellite Hyperion, which is 360 kilometers (220 miles) across in its longest dimension. The surface of Hyperion is heavily cratered and it has an extremely low density—below 0.6 kg/liter—suggesting that it may be loosely packed material.

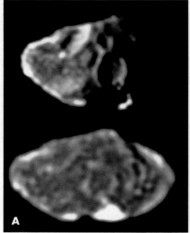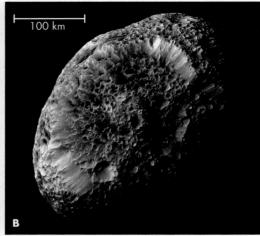

Concept Question 2

Many medium-size satellites are spherical, whereas asteroids of the same diameter are usually irregularly shaped. What might explain this difference?

a spherical shape by gravity (Unit 43). Two of the largest irregularly shaped satellites are shown in Figure 47.6, Jupiter's moon Amalthea (*am-uhl-THEE-uh*) and Saturn's moon Hyperion (*hi-PEER-ee-on*). Neptune's moon Proteus is the largest irregularly shaped satellite, with an average diameter of about 400 km as found by *Voyager 2*.

Amalthea also illustrates another feature of many satellites. Even though the satellites are made mostly of ices, many are dark. The dark colors appear to be thin layers over the surface of the satellite. In Amalthea's case, the dark red coating probably comes from Jupiter's moon Io, which has active volcanoes that loft sulfur compounds into a cloud that surrounds Jupiter (Unit 48). The coating makes Amalthea look rocky, but a small break in its surface at the point of an impact crater reveals a snowy white interior (Figure 47.6A). Measurements of the satellite's mass by the *Galileo* spacecraft indicate that its density is less than 1 kilogram per liter, suggesting that it is made of ices that are held together only loosely. Hyperion has an odd "spongy" appearance and an even lower density of less than 0.6 kilogram per liter

Saturn's moon Iapetus (*eye-YAP-ih-tuhss*) is unusual in that one hemisphere is nearly black while the other is white (Figure 47.7; also see Figure 47.5). Because Iapetus is tidally locked with one side always facing Saturn, there is also one side that always faces in the leading direction of the orbit—this is the black side of the satellite. It is as if the moon encountered a cloud of dark sooty material that has covered the leading half of the moon as it orbited Saturn. Some astronomers have hypothesized that this dark material may be dust from another moon that underwent a large impact. It may have been less concentrated at first but the ices may have sublimated as the darker side absorbed sunlight, and the dust grew more concentrated.

FIGURE 47.7

Three views of Saturn's moon Iapetus photographed by the *Cassini* spacecraft. Half of its surface is snow white as seen in Figure 47.5, while the other half is coated with a layer of dark brown hydrocarbons. The first image is a view of the dark side. Note the strange 13-km-tall ridge along the moon's equator, which is shown in a close-up view in the second image. The third image shows spots of the dark material on the otherwise snow white side of Iapetus.

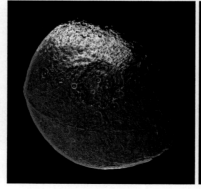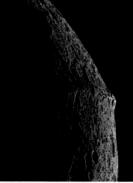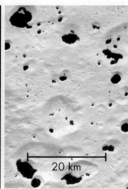

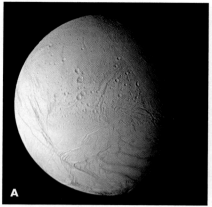  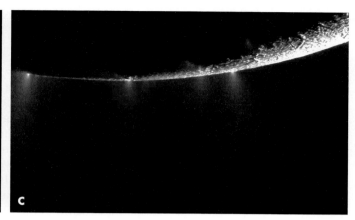

FIGURE 47.8
Saturn's moon Enceladus. (A) A series of long faults are visible near the south pole of Enceladus. Their blue color is enhanced in this image, making them stand out. (B) and (C) Views of the moon lit from behind by the Sun show eruptions of water from the faulted region.

Smooth areas and long cracks or **faults** appear on some of the medium-size moons, such as Saturn's moon Enceladus (*en-SELL-uh-duhss*), shown in Figure 47.8. These features suggest that some internal heat and tectonic activity may have occurred in these bodies. The surface may have melted, erasing older impact craters and leaving a smooth surface. Melting would also allow dusty material to sink and make the surface brighter. There is not as clear a link between size and geological activity as there is for the terrestrial planets. Some of the larger satellites are dark and heavily cratered, suggesting that their surfaces have not been renewed. On the other hand, Enceladus is relatively small, but very active. Enceladus ejects geysers of water into space, as seen in Figure 47.8. The source of the energy giving rise to this activity appears to be tidal gravitational forces (Unit 19) of Saturn on Enceladus that are flexing the moon's crust.

Uranus's larger moons are very dark and heavily cratered (Figure 47.5), but the fifth biggest, Miranda (*mih-RAN-duh*), has a complex faulted surface (Figure 47.9). Miranda's diameter is only about 470 kilometers (290 miles), but its surface is broken into distinct areas that seem to bear no relation to one another. One region shows long parallel grooves, while an adjacent one has small hills and craters, more similar to our Moon. Miranda has a remarkable set of cliffs, shown in the magnified portion of Figure 47.9, that are twice the height of Mount Everest. Miranda's patchwork appearance led some astronomers to hypothesize that it had been shattered by impact with another large body, and that the pieces were subsequently drawn back together by their mutual gravity, giving this peculiar moon its jumbled appearance. The leading hypothesis, however, is that the complex pattern arose as the result of strong tectonic activity that broke the surface into "plates," much as occurred on Earth.

> **Concept Question 3**
>
> Miranda's tortured-looking surface may have resulted from tectonic activity or from a giant impact. If you could send a probe to Miranda, what kind of tests would you make to decide between these two hypotheses?

Miranda

FIGURE 47.9
Miranda, an extremely puzzling moon of Uranus, as observed by the *Voyager 2* spacecraft. The second picture shows a close-up view of enormous cliffs seen at the bottom right of the first picture.

The satellites of Uranus were first named by Herschel's son after spirits in the works of Shakespeare and Alexander Pope, and later other characters in those works.

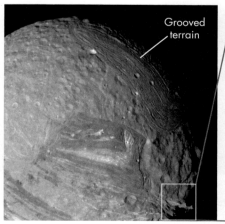 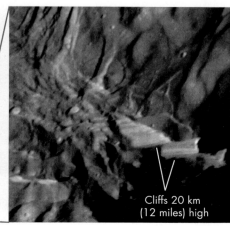

47.3 RING SYSTEMS

Saturn's spectacular rings (Figure 47.10) were first seen by Galileo in 1610, but when he trained his telescope on this planet, Galileo was baffled by what he saw. Through his imperfect lenses, the rings looked like "handles" on each side of the planet, and he thought there must be two large satellites, one on either side of the planet. Two years later Saturn moved into a point of its orbit where the rings were edge-on to Earth and nearly invisible through Galileo's telescope; but then they reappeared several years later, further baffling astronomers of the time.

It was not until 1659 that Christian Huygens, a Dutch scientist, observed that the rings were detached from Saturn and encircled it. Early on, astronomers assumed that the rings were solid; but in 1857 the British physicist James Clerk Maxwell demonstrated mathematically that no material could be strong enough to hold together in a solid sheet of such vast size. He thus deduced that the rings must be swarms of particles. Spectra of the rings support Maxwell's hypothesis: Doppler shift measurements (Unit 25) show that each part of the **ring system** orbits at the velocity appropriate to its distance from the planet—fast near the planet and slower farther away. The rings are composed of myriad individual orbiting bodies.

The ring system is very thin. Saturn's main ring system extends out to a little more than twice the planet's radius (136,000 kilometers, or about 84,000 miles). Some faint inner rings can be seen even closer to Saturn, and faint outer rings extend out to about eight times its radius. Yet despite the rings' immense breadth, they are probably less than a few hundred meters thick—thin enough that we can see stars through them.

Astronomers have determined that the ring particles are relatively small, from less than a centimeter to several meters across. Although these particles are far too small to be seen individually, they reflect radar signals. From the strength of the radar "echo," astronomers can estimate the particle sizes. Cassini also sent radio signals through the rings back to Earth that showed the ring particles are smallest at the inner and outer edges of the bright rings.

> In mythology, Saturn feared that his children would overthrow him, so he ate them when they were born. However, Jupiter's mother hid him from his father and Jupiter eventually overthrew Saturn and forced him to disgorge the other gods. When Saturn's rings were edge-on and disappeared from view, Galileo mused that the planet Saturn had perhaps swallowed and then disgorged its moons like its mythological namesake!

FIGURE 47.10

The rings of Saturn imaged by the *Cassini* spacecraft. At right, images from the Hubble Space Telescope show the changing appearance of Saturn from Earth as it moves through about a quarter of its orbit. Saturn's axis has a tilt similar to Earth's, so the top image is the appearance at the solstice when it is summer in the southern hemisphere, while the bottom image shows the rings seeming to disappear at the equinox.

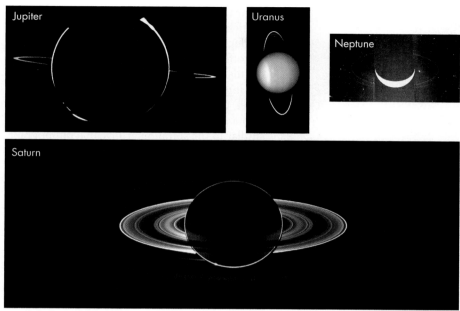

FIGURE 47.12
The rings of Uranus were imaged by *Voyager 2* from behind the planet (left) and in front (right), illustrating that in addition to thin reflective rings, there are many darker rings only visible from behind. The outermost ring does not align in the two views because it has an elliptical orbit.

FIGURE 47.11
Ring systems shown to the same scale. Jupiter's and Neptune's rings were imaged by *Voyager 2* from the night side of the planets. Saturn was also imaged from the night side by *Cassini*, showing a faint outer ring. The Uranus image was made by the Hubble Space Telescope.

FIGURE 47.13
Cassini image detailing the substructure of Saturn's rings. Slight color differences have been greatly exaggerated over half the image to illustrate differences in the ringlets' composition.

Kirkwood also discovered gaps in the distribution of asteroids related to the period of the orbit of Jupiter (Unit 43).

For centuries, astronomers thought that the only Solar System planet with rings was Saturn. But in 1979 the *Voyager 1* spacecraft flew by Jupiter. The photographs it took as it passed the planet clearly showed that Jupiter has a ring, although only a very thin one, as illustrated in Figure 47.11. We now know that all the giant planets have rings surrounding them, but none of the others' rings are nearly as substantial as Saturn's. Uranus and Neptune are each also encircled by a set of narrow rings, as shown in Figure 47.11. Rings made up of dark dust particles do not reflect light back toward Earth very strongly, which is why some ring material shows up more clearly from "behind" the planet (Figure 47.12).

Spectra of sunlight reflected from the rings indicate that Saturn's ring particles are composed primarily of water ice. However, high-contrast spacecraft pictures, such as Figure 47.13, show different coloration at different radii, implying that the composition of the particles varies. Analysis at other wavelengths shows that some particles are made of much darker material, probably consisting of carbon compounds. The Uranian rings are also made of darker carbon compounds, which makes them difficult to see from Earth.

Figure 47.13 also demonstrates that Saturn's rings consist of numerous separate **ringlets.** Large gaps in the rings had been seen from Earth, but the many narrow gaps in the rings came as a surprise. What causes these gaps that create the multitudes of ringlets?

In 1872 the astronomer Daniel Kirkwood noticed that the largest gap in the rings, known as the **Cassini division** (see Figures 47.10 and 47.13), occurs where ring particles orbit Saturn in half the time it takes the moon Mimas to orbit Saturn. This **resonance** (a simple relationship between orbital periods) means that any ring particle orbiting in the gap would undergo a strong and repeated gravitational force from Mimas every other orbit. The cumulative effect of these tugs over long periods would pull particles from the gap and thereby create Cassini's division.

Some gaps apparent in Figure 47.13 have a different cause. They probably arise from a complex interaction between the ring particles and tiny moons that orbit within the rings. As these small moons—some less than 10 kilometers in size—orbit

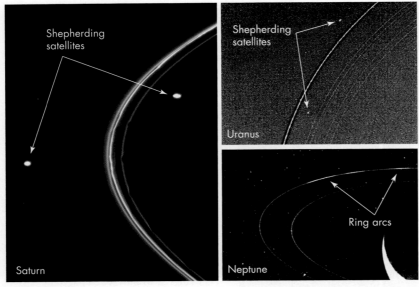

FIGURE 47.14
Small satellites may shape rings by "shepherding" material into narrow rings between the two satellites' orbits. The gravitational influence of unseen satellites is thought to be responsible for dense "arcs" of material in one of Neptune's rings.

Saturn, their gravitational attraction on the ring particles generates waves. These waves spread through the rings much like ripples in a cup of coffee that is lightly tapped. Such ripples are circular in a cup or on a still pond; but because the inner part of the ring orbits faster than the outer part, the spreading waves are stretched into a spiral shape. The tightly wound pattern, similar to the spiraling groove in a phonograph record, is called a **spiral density wave.** The crests of these density waves form the narrow rings.

Thin isolated ringlets can form when two small moons orbit at slightly different radii. The combined gravitational force of these **shepherding satellites** deflects ring particles into a narrow stream between the two moons' orbits (Figure 47.14). Such rings sometimes have complex twisted shapes and are not necessarily circular. The distribution of particles may also be clumpy, as is seen in one of the rings around Neptune, where much of the ring material is collected into arcs along the ring.

47.4 ORIGIN OF PLANETARY RINGS

When planets and their satellite systems first formed, much debris must have been left over, much like the asteroids in the Solar System itself. Most of this material was depleted over time as it collided with either a satellite or the planet. Material on circular orbits in the same plane as the satellite system could maintain long-term stable orbits, although occasional collisions with other debris or gravitational disturbances by the moons would send material into a collision course.

The particles that make up rings are subject not only to gravitational forces but to other forces that may also make their orbits unstable. For example, gas trapped in a planet's magnetic field may exert a frictional force on the ring particles, gradually causing them to slow and spiral into the planet's atmosphere. These nongravitational forces have the strongest effect on small particles. For example, it is estimated that the fine dust that makes up most of Jupiter's thin ring spirals into the planet in just thousands of years.

Many of the narrow rings of dusty material appear to be supplied with particles by one or more satellites. Each of the many impact craters seen on satellites speaks to a collision that would have blasted out material that might become part of a ring. For example, there is a faint ring around Jupiter at Amalthea's orbital radius. It is probably supplied by small impacts that kick dust and debris from the moon's surface. These particles have slightly different orbits from Amalthea, so they spread out around the whole orbital path. We have also seen that geologically active moons such as Enceladus can eject particles, and this is the source of material for the faint outer ring around Saturn seen in Figure 47.11.

Ring systems as big as Saturn's require much larger amounts of material. It is estimated that Saturn's ring system is about as massive as its moon Mimas. It is possible that a satellite of this size broke apart to create Saturn's rings, and no direct collision need have occurred if the satellite or a passing object came too close to Saturn.

FIGURE 47.15
The Roche limit. A planet's gravity pulls more strongly on the near side of a satellite than on its far side, stretching it. At radii smaller than the Roche limit, the tidal force can overcome the satellite's own gravity, pulling loosely connected pieces of the satellite apart.

The Roche limit

Clarification Point

A body will not break up inside the Roche limit if forces other than gravity hold it together. The chemical bonds between neighboring atoms in an unfractured piece of ice or rock can keep an object from falling apart.

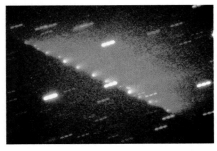

FIGURE 47.16
Comet Shoemaker-Levy 9. This image shows 20 or so of the fragments that the comet was broken into by Jupiter's tidal force.

Concept Question 4

If the space station were left empty for many years, where would tidal forces tend to move any dust or debris left floating in the cabin?

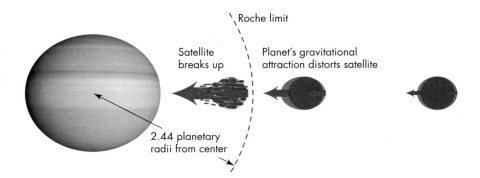

If a moon gets too close to its planet, the planet's gravity can rip the satellite apart. This was shown mathematically in 1849 by the French scientist Edouard Roche (pronounced *Rohsh*). Because gravity weakens with distance, a planet pulls harder on the near side of a satellite than on the far side. This is the effect that gives rise to ocean tides on Earth (Unit 19), but it can have much more extreme results. If the difference in this pull exceeds the moon's own internal gravitational force, the moon will be pulled apart, as illustrated in Figure 47.15. Thus, if a moon—or any body held together by gravity alone—approaches a planet too closely, the planet raises a tide so large it pulls the encroaching object to pieces.

Roche calculated the distance at which the tide becomes fatally large and showed that for a moon and planet of the same density, breakup occurs if the moon comes closer to its planet than 2.44 planetary radii, a distance now called the **Roche limit**. Most planetary rings lie near their planet's Roche limit, suggesting that rings might be created by satellite disintegration. The existence, side by side, of ringlets with different compositions (some rich in ice, others rich in carbon) also suggests that the rings formed from the breakup of many different small objects.

Jupiter's main ring has two small moons orbiting within its Roche limit. These moons orbit faster than Jupiter rotates and are slowly spiraling closer to Jupiter because of tidal braking (Unit 19). Perhaps they were once part of a larger body that has been pulled apart, with the two small moons remaining as the largest structurally stable fragments. Smaller debris left over from the breakup probably continues to collide, adding fresh dust to the ring.

Objects that are not orbiting a planet but happen to pass within the Roche limit can also suffer severe consequences. For example, asteroids or comets occasionally pass too close to a planet. This occurred in 1992 when Comet Shoemaker-Levy 9 (Figure 47.16) passed within about 30,000 kilometers of Jupiter. It was pulled apart into a string of smaller comets, which later collided with the planet.

Astronomers have debated for centuries how old Saturn's ring system might be. Saturn's main rings are highly reflective, and this seemed to point to a young age. Objects tend to grow darker over time as dust particles accumulate on their surfaces. This is apparent on the dark coloration of moons that have not had geological processes refresh their surfaces. This led to some age estimates of no more than a few hundred million years.

A recent model has shown that ring systems as dense as Saturn's need not be so young. Material in the rings may alternately stick together to form larger, more stable clumps, then later break apart to feed material back into the ring. The large clumps are less affected by frictional forces, and in the process of breaking back apart, fresh reflective surfaces are exposed, keeping the rings from growing dark. A "recycling" process like this would explain how Saturn's ring system could survive for billions of years with a similar appearance.

It is also possible that the gravitational interactions between the moons in these complex satellite systems could drive a moon close enough to its planet to break up. In a billion years, the giant planets' ring systems may look very different.

KEY POINTS

- The giant planets all have rings and large satellite systems.
- Most satellites' orbits are "regular": nearly circular, close to the equatorial plane, and in the same direction as the planet spins.
- Regular satellites probably formed with the planet, while irregular satellites (most of which are very small) were probably captured.
- Satellites with diameters larger than about 400 km generally have sufficient gravity to make them round.
- The moons are made mostly of ice, but are often heavily cratered and darkened by dust or other volcanic or carbon-rich material.
- Some moons show geological activity, which may be related to gravitational forces distorting a moon and melting internal ice.
- Rings are made mostly of orbiting bodies a few meters in size or smaller; much of the material is ice, but some is very dark.
- The gravitational pull of satellites can create gaps in wide rings like Saturn's, or "shepherd" narrow rings between two moons.
- Ring material may come from impact debris and satellites that orbit so close to a planet that gravitational tides pull them apart.

KEY TERMS

Cassini division, 367
fault, 365
Galilean satellites, 361
regular orbit, 362
resonance, 367
ringlet, 367
ring system, 366
Roche limit, 369
shepherding satellite, 368
spiral density wave, 368

CONCEPT QUESTIONS

Concept Questions on the following topics are located in the margins. They invite thinking and discussion beyond the text.

1. Why so few terrestrial satellites. (p. 363)
2. Round satellites versus round asteroids. (p. 364)
3. Sending a space probe to study Miranda. (p. 365)
4. Tides and dust in the space station. (p. 369)

REVIEW QUESTIONS

5. What is the range of sizes and distances of the gas giants' moons?
6. What properties of gas-giant satellites tend to be similar?
7. What is a regular orbit for a satellite? Why are satellites that form along with their planet expected to have regular orbits?
8. Why are large moons round and small ones irregularly shaped?
9. What are the different ways in which satellites affect ring systems? How permanent are ring systems?
10. What is the Roche limit? How does it explain the presence of ring systems around the giant planets?

QUANTITATIVE PROBLEMS

11. Astronomers hypothesize that the very diffuse outer rings of Saturn are constantly being replenished by mechanisms such as the ejection of material from Enceladus's geysers. Use the data in Appendix Table 7 to calculate the escape velocity from Enceladus. This is the minimum speed material must be ejected from the moon to replenish Saturn's outer ring.
12. The Earth's Moon experiences tidal forces that are gradually expanding its orbit by a few centimeters per year (Unit 19.4). Explain, using vectors, how a moon orbiting faster than its planet rotates will experience a "tidal drag" that causes its orbit to shrink.
13. The top-right of Figure 47.14 shows a pair of shepherd moons and one of Uranus's rings. Assume that the moons and ring particles are orbiting counterclockwise in the image.
    a. Qualitatively describe the relative speeds of the moons and ring particles as they pass each other.
    b. If a passing moon boosts the speed of a ring particle, what will happen to the size of the ring particle's orbit?
    c. If a ring particle slows when it passes a slower moon, what happens to the size of the ring particle's orbit?
    d. Explain how two shepherding moons might drive ring particles into a narrow stream.
14. Examine Appendix Tables 5 and 7 to determine which moons orbit inside the Roche limit, around each of the giant planets. Make a list of these moons and discuss how it is possible for these moons to orbit there. Would you expect these moons to be spherical or irregular? Why?
15. Suppose a moon that was 100 kilometers in diameter broke up into particles just 1 millimeter in diameter.
    a. What was the surface area of the moon before it broke up?
    b. How many particles 1 millimeter in diameter could be made from the moon?
    c. What is the total surface area of all of the 1-mm particles you found in part b?
    d. What is the ratio of the surface area in part c to the surface area in part a? Explain why this ratio also represents how much sunlight would be reflected by the particles versus the original moon.

TEST YOURSELF

16. Why did Saturn's rings disappear from Galileo's view in 1612?
    a. Planetary rings are only temporary, needing constant replenishment, which stopped at that time.
    b. The very thin rings were edge-on as viewed from Earth.
    c. Only the bottom, dark side of the rings faced toward Earth.
    d. The ring material coalesced into a temporary moon.
17. Particles in Saturn's main rings are made of (or covered with)
    a. water ice.
    b. frozen carbon dioxide.
    c. metallic hydrogen.
    d. a dark tarlike substance.
18. The Roche limit is the
    a. mass a planet must exceed to have satellites.
    b. smallest mass a gas-giant planet can have.
    c. largest stable orbital distance from a planet.
    d. distance a planet's tidal force equals a moon's own gravity.
    e. depth astronomers can see into a planet's clouds.
19. Shepherding satellites are
    a. two satellites on the same orbit.
    b. multiple satellites with similar composition.
    c. two satellites that confine a thin planetary ring.
    d. satellites that are in resonance with a larger satellite.

PART 3 UNIT 48

Ice Worlds, Pluto, and Beyond

48.1 The Galilean Satellites
48.2 Saturn's Moon Titan
48.3 Neptune's Moon Triton
48.4 Pluto
48.5 The Trans-Neptunian Worlds

Learning Objectives

Upon completing this Unit, you should be able to:
- Describe the properties of outer Solar System's "ice worlds."
- Compare geological processes on ice worlds to those on terrestrial planets.
- Explain the evidence and significance of liquids on or in different ice worlds.
- Describe what is known about Pluto and other trans-Neptunian objects and explain why Pluto is now termed a dwarf planet.

Although the outer Solar System is dominated by the four gas giants, there are many icy bodies there that would qualify as planets if they orbited the Sun in isolation. Their icy composition reflects the fact that they formed beyond the frost line, where volatile compounds were able to condense from the solar nebula (Unit 35). Their sizes relative to each other and to the Earth and the Moon are illustrated in Figure 48.1. We begin in this Unit with the nearest of these worlds and work our way out to beyond Pluto. Some of these worlds are listed as satellites and some as dwarf planets, but they might be better classified as "ice worlds," with features found in neither the terrestrial planets nor the gas giants. Their geological processes take place at low temperatures, but some have active volcanoes and atmospheres. At their frigid temperatures, "molten ice" is a major component of their geological activity. By examining them together we can begin to build a better understanding of this kind of world.

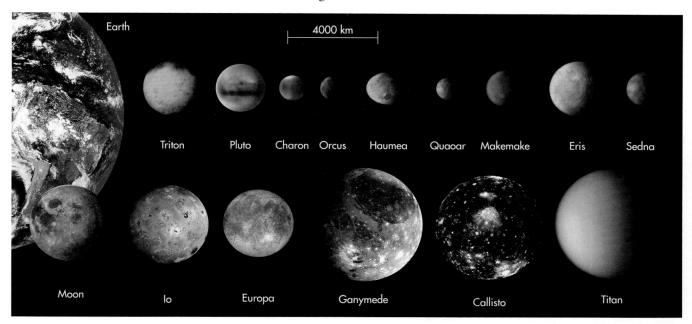

FIGURE 48.1
Ice worlds of the outer Solar System. Images from the *Voyager, Galileo,* and *Cassini* space probes of the large moons of the gas giants are shown to the same scale along with the Earth and Moon for comparison. The images of Pluto and Charon are based on computer models and observational data. Artists' depictions, roughly indicating the size and colors, are shown for several other large trans-Neptunian objects.

48.1 THE GALILEAN SATELLITES

The four large moons of Jupiter—Io, Europa, Ganymede, and Callisto— discovered by Galileo in 1610, are collectively known as the **Galilean satellites.** Each would be considered a planet if it circled the Sun in an independent orbit. Their circular orbits in the plane of Jupiter's equator suggest that they formed along with Jupiter out of the material that made up a disk of gas and planetesimals around the massive planet. At their distance from the Sun, these satellites have surface temperatures of about 110 to 130 K. The density, brightness of the surface, and level of geological activity are all generally higher for the Galilean satellites that orbit closer to Jupiter. This probably reflects Jupiter's gravitational tidal effects more than internal differences when the satellites formed. Over billions of years, Jupiter's gravity has strongly altered the appearance of the inner two moons.

Io (pronounced *EYE-oh* or *EE-oh*) is the Galilean satellite that lies nearest Jupiter. Because Io is so close to Jupiter, it experiences a strong gravitational tidal force from the planet. That tidal force locks Io's spin, so that one side always faces Jupiter, as our Moon's spin is locked with the same side facing Earth. But Io is also in an orbital resonance with Europa, the Galilean satellite next closest to Jupiter, orbiting twice for each of Europa's orbits. This keeps Io from settling into a circular orbit; instead its orbit is quite elliptical, so the strength of the tides varies, and because the orbital speed changes, Io twists from side to side. The result is that Io is subject to strong and changing gravitational forces that distort its shape first in one way then another. This constant flexing heats Io by internal friction, much as bending a paperclip back and forth heats the wire (touch it and it will feel hot).

Over billions of years, the heating has melted not only the ice but also the rocky matter in Io's interior. As molten matter oozes close to the surface, it erupts, creating volcanic plumes that are seen reaching hundreds of kilometers above the surface (Figure 48.2A). Such activity has driven most of Io's water into space, where it is lost because of Io's weak gravitational attraction. As a result, Io's density is 3.5 kilograms per liter, quite similar to the densities of the terrestrial planets.

Sulfur, also common in terrestrial volcanoes, is now the major component of Io's volcanic outpourings, although some eruptions are also rich in ordinary silicate rock. Sulfur compounds give Io its rich red, yellow, and orange colors. A recent lava flow can be seen in Figure 48.2B, where some of the molten material is still glowing. The dark regions in the craters of older volcanoes result when molten sulfur cools.

> The Galilean satellites are named after various characters from Greco-Roman mythology who were pursued by the god Zeus (Jupiter in Roman myths).

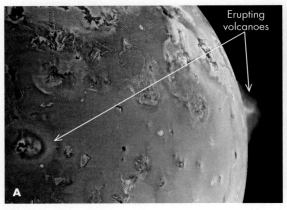 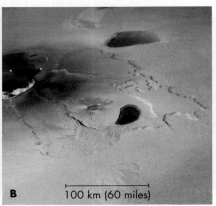 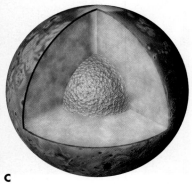

FIGURE 48.2
Pictures of Io. (A) A *Galileo* spacecraft image of two volcanoes erupting on Io. (B) A close-up view of a currently erupting volcano (left side of image) with hot lava seen glowing in the infrared in this false-color image. (C) A model of Io's interior.

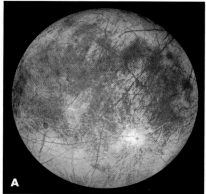
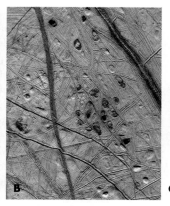
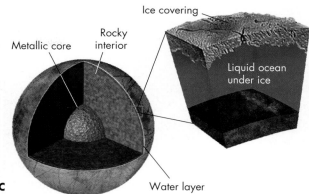

FIGURE 48.3
Pictures of Europa. (A) Europa as imaged by the *Galileo* spacecraft. (B) Close-up picture of Europa's surface showing fractures and "freckles." The dark spots and pits are about 10 kilometers (6 miles) across. (C) A model of Europa's interior. Note the possible liquid water ocean beneath ice.

Europa (*you-ROPE-ah*), the smallest of the Galilean satellites, looks rather like a cracked egg. Long, thin lines score its surface, as shown in Figure 48.3A. The white material is probably a crust of ice, whereas the red material is probably mineral-rich water that has oozed to the surface through the cracks and frozen. From the small number of impact craters, astronomers estimate that Europa's surface is no more than about 30 million years old, suggesting that its surface has been heated and melted recently, obliterating older impact craters.

Close-up pictures taken by the *Galileo* spacecraft show a complex set of interwoven cracks that are reminiscent of ice floes (Figure 48.3B). In addition, small regions known as "freckles" are spots about 10 kilometers (6 miles) across that appear to have melted. In some freckles, material has been pushed up, whereas in others the region has subsided. These may be the result of a volcanic process in which warmer regions of ice rise and melt—somewhat like the silicate rock in lava fields on Earth.

The cracks in the surface ice and the "freckles" suggest that heat is rising from the interior, shifting and melting Europa's ice crust. Such heating probably comes from Jupiter's gravitational force deforming Europa, as occurs with Io. Europa is itself in a 2-to-1 resonance with the next nearest Galilean satellite, Ganymede, orbiting twice for each of Ganymede's orbits. Europa, also like Io, has a relatively high density of nearly 3 kilograms per liter (Table 48.1), similar to our Moon, indicating that it has a relatively large rock and iron core. The tidal heating appears to be sufficient to keep a layer of water melted beneath Europa's crust, forming an ocean (Figure 48.3C) whose total volume of water may exceed the Earth's oceans. The *Galileo* spacecraft also detected a weak magnetic field in Europa, which might be generated within an electrically conductive salty ocean under the surface. Some astronomers even speculate that Europa's ocean might harbor life, and missions are being designed that might send a probe to explore what is beneath the outer ice layer.

Concept Question 1

In what ways are the trends observed among the Galilean satellites similar to those of the planets? In what ways are they different?

TABLE 48.1 Properties of Large Satellites

Name	Diameter	Mass	Density (kg/L)	Distance from Planet	Orbital Period
Moon	3476 km	7.35×10^{22} kg	3.34	384,400 km	27.322 days
Io	3642 km	8.93×10^{22} kg	3.57	421,600 km	1.769 days
Europa	3130 km	4.80×10^{22} kg	2.97	670,900 km	3.551 days
Ganymede	5268 km	14.82×10^{22} kg	1.94	1,070,000 km	7.155 days
Callisto	4806 km	10.76×10^{22} kg	1.86	1,883,000 km	16.689 days
Titan	5150 km	13.46×10^{22} kg	1.88	1,222,000 km	15.945 days
Triton	2705 km	2.14×10^{22} kg	2.06	354,590 km	5.875 days

The outer two Galilean satellites are larger than Io or Europa, but their densities are lower (Table 48.1). They appear to have a large proportion of ice, making them more similar to other ice worlds in the outer Solar System. The surfaces of **Ganymede** (*GAN-ih-meed*) and **Callisto** (*kal-IH-stoh*) look somewhat like our own Moon. They are grayish brown and covered with craters made during the late stages of their formation, much as our own Moon was pockmarked by debris (Figure 48.4). However, spectroscopy indicates that the surfaces are composed mainly of ice. The darkest areas are apparently the oldest, based on the high density of impact craters, whereas recent impact craters that break through the "dirty" surface layers reveal bright white ice beneath.

A sign of geological activity is seen on Ganymede. Large sections are covered with a younger **grooved terrain** (Figure 48.4) that astronomers suspect results from tectonic activity of the icy surface. Long parallel ridges, typically 10 to 20 kilometers apart and up to a

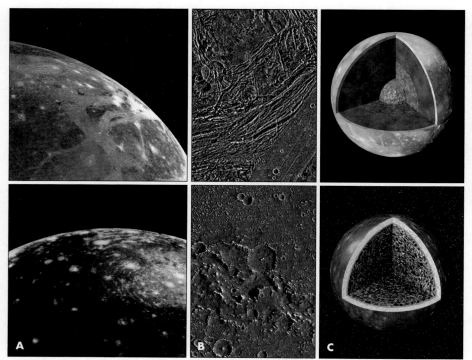

FIGURE 48.4
Images of Ganymede (top) and Callisto (bottom). (A) The surfaces of both satellites are dark, but Ganymede has long linear features. (B) Close-up of regions about 70 kilometers (45 miles) across. (C) Models of the interior of each satellite.

kilometer tall, stretch for hundreds of kilometers over the surface. These are somewhat like the linear features seen on Europa or Miranda (Unit 47). These regions are probably formed by faulting of the icy surface as it is stretched, perhaps driven by convection of "molten slush" layers below.

Although Ganymede and Callisto have similar densities, astronomers think their internal structures are quite different. The distribution of the masses of these satellites was studied by carefully measuring the gravitational effects on the *Galileo* spacecraft during close flybys of each moon. The craft's trajectory was affected by the fine details of the mass distribution; and by analyzing dozens of these flybys, astronomers have determined that Ganymede is composed mainly of ice from its surface to a depth of about 800 kilometers (about 500 miles). This ice appears to surround a rocky mantle and probably an iron core. This core may be hot, and astronomers suspect that Ganymede also has a thick layer of water, as illustrated in Figure 48.4C. A molten core or a thick layer of saltwater can conduct electricity, and this would help explain why this moon has a magnetic field.

Heating of these moons during their formation would probably have allowed differentiation to occur, similar to what occurred in our own planet. Thus, iron sank to their cores, surrounded by less-dense silicates and then thick layers of water and ice nearer the surface. The measurements of Callisto suggest that it may be less differentiated, with rock and ice still intermixed throughout much of its interior. Analysis of the flybys of Io and Europa indicates that they both have large iron and rock cores. Indeed, if their thick icy surface layers were stripped away, Ganymede and Callisto would be similar in density to the inner two Galilean satellites. This may actually be what happened to Io and Europa as a result of the tidal heating they experience—ice and other volatile substances, once vaporized, would escape from the relatively weak gravity of these objects.

Unit 48 Ice Worlds, Pluto, and Beyond 375

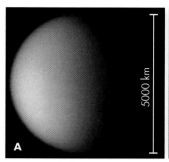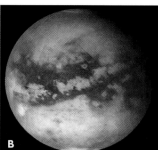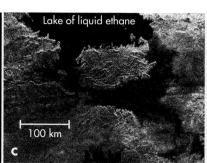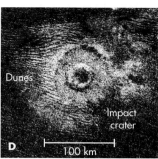

FIGURE 48.5
Titan, as seen from the *Cassini* spacecraft. (A) Visible-light image shows Titan's thick clouds. (B) This false-color infrared image shows some surface features, glimpsed through the clouds. (C) Radar image reveals rivers and a lake. (D) Radar image of impact crater and dunes.

48.2 SATURN'S MOON TITAN

Titan is one of the most intriguing moons in the Solar System. It is by far the largest Saturnian moon, containing more than 20 times the mass of all of Saturn's other satellites combined. Titan has a diameter of about 5000 kilometers (3000 miles), making it slightly bigger in diameter than the planet Mercury and comparable in mass and radius to Jupiter's largest moon, Ganymede. However, it is shrouded by a thick atmosphere and dense clouds that hide its surface (Figure 48.5A). Infrared images made by the *Cassini* spacecraft (Figure 48.5B) reveal complex surface features through the outer cloud layers, and radar imaging reveals the presence of features that appear to be lakes (Figure 48.5C), impact craters, and dunes (Figure 48.5D).

In 2005 a joint mission of the European Space Agency (ESA) and NASA sent the *Huygens* probe parachuting to the surface of Titan. As it descended, the lander sent back a series of remarkable images (Figure 48.6) showing features that resemble rivers and lakes. Astronomers think that these rivers may have been carved by liquid methane cutting through rock-hard water ice. On the basis of chemical models, some astronomers suspect that the hydrocarbons methane (CH_4) or ethane (C_2H_6) may circulate on Titan, evaporating, raining, and flowing in rivers, much as water does on Earth. A picture from the surface (Figure 48.6C) shows a rubble-strewn field. The "rocks" are probably made of ice, and their dark color

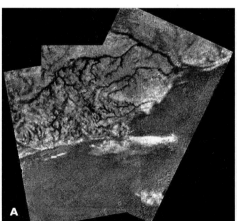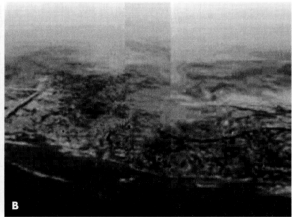

FIGURE 48.6
The landscape of Titan: (A) river networks carved by flowing liquid methane imaged by the *Huygens* lander from a height of 16 kilometers (10 miles); (B) panorama of the horizon made when the probe was about 8 km (5 miles) above the surface; (C) "rocks" of ice on the surface are about 10-15 cm (4-6 inches) across.

> **Concept Question 2**
>
> The topography on Titan resembles river valleys at first glance, but how might you go about testing their similarity to the shapes of river channels on Earth? What specific features do you think might differ?

may indicate that hydrocarbon compounds are mixed in. Moreover, the bases of the "rocks" show signs of erosion, suggesting that a liquid has flowed through the region.

Why is Titan able to retain an atmosphere even though the similarly sized Ganymede cannot? This is likely because Titan is about twice as far from the Sun. As a result, Titan is very cold—about 95 K (−290°F)—and gas molecules move so slowly that they are less likely to escape. Titan's atmosphere is so dense, in fact, that it has a pressure comparable to Earth's. Astronauts visiting there would not need pressure suits, although they would rapidly freeze solid unless they had space suits to keep them warm! Spectra of Titan's atmosphere show that it is mostly nitrogen, like the Earth's, but an astronaut would also need to bring a supply of oxygen.

48.3 NEPTUNE'S MOON TRITON

Neptune's moon **Triton** (*TRY-tuhn*) is larger than Pluto and nearly as large as Europa. Moreover, its orbit is highly unusual for a large moon. Triton orbits backward relative to Neptune's rotation, in an orbit that is highly tilted with respect to Neptune's equator. The small moons orbiting near Neptune are in regular, roughly circular orbits, but the orbit of the medium-size moon **Nereid** (*NEER-ee-id*) is also peculiar, traveling on a highly elliptical orbit.

These orbital peculiarities lead many astronomers to think that Triton was an icy dwarf planet that Neptune captured after the planet and its satellite system formed. When Neptune captured Triton, the encounter would have destroyed or expelled most of the outer moons that Neptune originally possessed—and those interactions also would have slowed Triton enough to leave it in orbit around Neptune. Nereid's elliptical orbit carries it far beyond Neptune's other moons, most likely because the strength of the interaction was just short of forcing it to escape from Neptune altogether. Perhaps the event in which Triton was captured was one of many gravitational encounters that were part of the process of Neptune settling into an orbit outside of Uranus's orbit, hundreds of millions of years after the Solar System formed, as discussed in Unit 36. In this case, Triton may be an example of a body that formed in isolation in the outer Solar System.

Triton is massive enough that its gravity, in combination with its low temperature, allows it to retain gases. Triton is thus one of the few moons in the Solar System with an atmosphere. However, its atmosphere is much less dense than Titan's, and Titan's atmosphere apparently freezes on the night side of the moon, somewhat as water vapor can condense as frost on a cold night on Earth.

Triton's surface has many unusual features, as you can see in Figure 48.7. Wrinkles give much of the surface a texture that looks like a cantaloupe. Craters pock the surface elsewhere, and dark streaks extend from some of them. At least one of these streaks originates from an ice volcano seen erupting by the passing *Voyager 2* spacecraft in 1989. Astronomers think that the matter ejected by the volcanoes is a mixture of nitrogen, ice, and carbon compounds. Sunlight, weak though it is at Triton's immense distance from the Sun, may warm such matter trapped below the surface and make it expand and burst through surface cracks. The erupted material cools and condenses in Triton's cold, thin atmosphere, where winds carry it and deposit it as dark debris on the surface (Figure 48.7).

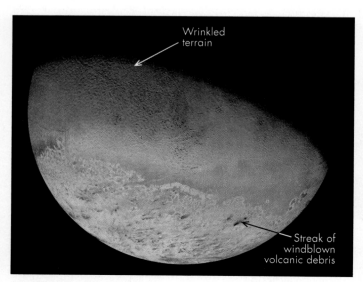

FIGURE 48.7
The surface of Neptune's moon Triton.

48.4 PLUTO

NASA launched the *New Horizons* spacecraft in 2006. It will make a flyby of Pluto in 2015 and possibly continue on to examine other objects in the outer Solar System.

In the 1920s astronomers suspected there was another large planet in the outer Solar System because the orbits of Uranus and Neptune appeared to be disturbed. The predicted body was dubbed "Planet X," and a young American astronomer, Clyde Tombaugh, carefully searched for the missing object near the ecliptic. In 1930, after scanning millions of pairs of star images and searching for objects, he discovered an object whose position changed between the exposures (Figure 48.8A). The motion between images distinguishes an orbiting body in the Solar System from a star, and the rate of motion implied it was orbiting beyond Neptune. The object was named **Pluto**, after the Greco-Roman god of the underworld.

Pluto's great distance from the Sun made it too dim and too small to measure its angular size. For years after its discovery, astronomers had only a rough idea of Pluto's size. They had thought it must be massive, to explain the disturbance of Uranus's and Neptune's orbits, but careful reanalysis showed that actually there were no disturbances. The search for "Planet X" had been based on a mistake! In the decades following Tombaugh's discovery, the size estimates for Pluto were based on the amount of light it reflected from the Sun. This gave very uncertain results, because Pluto could be a small object that reflected most of the sunlight striking it or a large object reflecting only a small percentage of the light.

Concept Question 3

Charon's orbit around Pluto is almost perpendicular to the plane of the Solar System (the ecliptic). Both Charon and Pluto keep the same face toward each other. How would Charon appear to an observer on Pluto in terms of phases and location in the sky?

The mass of Pluto was not determined until 1978 when James Christy of the U.S. Naval Observatory discovered a moon orbiting Pluto (Figure 48.8B). The moon, Charon (*SHARE-uhn* or *CARE-uhn*), is named for the boatman who, in mythology, ferries dead souls across the river Styx to the underworld. The presence of a large satellite helped explain why some observers had thought that Pluto was larger than it ultimately proved to be. Charon takes just 6.4 days to orbit Pluto, and it orbits just 19,600 kilometers away. The entire orbit could fit inside the planet Neptune. From Charon's orbital data we can calculate Pluto's mass using Newton's law of gravity (Unit 17). Pluto's mass is about 0.002 times the Earth's, showing that Pluto is far less massive than any of the eight planets, and even less massive than Earth's Moon or the six largest satellites of the giant planets.

Four more much smaller moons have been discovered since 2005. They orbit Pluto two to three times farther from Pluto than Charon.

Pluto's diameter is a little less than one-fifth the Earth's diameter, and its density is approximately 2.0 kilograms per liter. This density is quite similar to the large satellites of the giant planets, suggesting it has a similar composition of rock and ice. Images from the Hubble Space Telescope show very little detail on Pluto (Figure 48.8C).

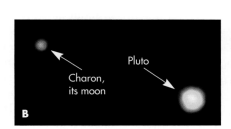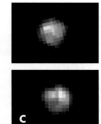

FIGURE 48.8
(A) Pluto looks like merely another dim star in this photograph taken at Lick Observatory in California. Only its change in position from night to night shows it to be orbiting the Sun. (B) A picture of Pluto and its moon Charon obtained with the Hubble Space Telescope—free of the blurring effects of our atmosphere. (C) Two pictures of Pluto made with the Hubble Space Telescope hint at surface features.

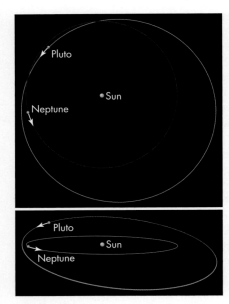

FIGURE 48.9
Pluto's orbit is so eccentric that it comes closer to the Sun than Neptune during part of its orbit. However, Pluto will never collide with Neptune because its orbital plane is so tilted that it is well above Neptune's orbit when it comes closest to the Sun.

Spectra indicate that Pluto has a very thin atmosphere. This is possible despite Pluto's tiny mass because in the bitter cold (40 K, or about −390°F) of this remote part of the Solar System, molecules move slowly enough that they do not escape. Pluto's atmosphere is probably nitrogen and carbon monoxide with traces of methane, and is perhaps similar to Triton's. Bright spots on Pluto's surface may be methane frost on a pale brown surface of other ices.

Pluto's elliptical orbit, which crosses Neptune's orbit, once led some astronomers to hypothesize that it was originally a satellite of Neptune that escaped and now orbits the Sun independently. Today, however, astronomers think almost the reverse—that Neptune has "captured" Pluto into a resonance. Pluto's orbital period is 247.7 years, almost exactly 1.5 times Neptune's. At its orbital radius of 39.5 AU from the Sun, Pluto makes two orbits around the Sun for every three made by Neptune—a 2-to-3 resonance (see, for comparison, Unit 43). Neptune's strong gravitational attraction on Pluto may have tugged it into its current orbit. The arrangement is stable, and Pluto is in no danger of ever coming close to Neptune (Figure 48.9).

Pluto was called a planet for many decades after its discovery, but astronomers grew troubled by this designation. It may seem surprising that there was never a definition for "planet," but it is a term that comes from ancient times, long before formal definitions were developed. When Pluto was first discovered, it was thought to be a large, massive body, but it "shrank" over the decades—it took almost 50 years for a firm measurement of its tiny mass to be made. It became clear finally that Pluto was different from the eight major planets in almost every respect. It was less massive than many moons, its orbit was more elliptical and more tilted than any planet's orbit, and its orbit even crossed the orbit of another planet.

For more than 60 years Pluto remained the only known inhabitant of the deep Solar System other than comets (Unit 49), so astronomers continued to call it a planet despite their misgivings. Pluto's discovery is really a tribute to Clyde Tombaugh's skills and perseverance. It was not until 1992, using new telescopes and techniques, that astronomers began to discover more objects beyond Neptune, even displacing Pluto as the most massive object. This ultimately led to the recognition that Pluto was just one of many icy worlds orbiting in the **Kuiper belt,** an orbiting swarm of objects similar to the asteroid belt. And as happened to the asteroids 150 years earlier, Pluto was demoted from the status of planet to become just one of the largest among the trans-Neptunian worlds.

48.5 THE TRANS-NEPTUNIAN WORLDS

Since 1992, astronomers have detected more than a thousand objects orbiting beyond Neptune, now commonly known as **trans-Neptunian objects** or **TNOs.** The largest of these, named **Eris** (*AIR-iss*) after the Greek goddess of strife and discord, is about 27% more massive than Pluto. The name Eris is particularly appropriate because this object's discovery drove astronomers to "demote" Pluto from the status of planet, causing much discord among astronomers and the public.

With the discovery of so many objects orbiting beyond Neptune, Pluto's position began to resemble that of one of the large asteroids in the asteroid belt. History provided a precedent for this situation. The first four asteroids were discovered between 1801 and 1807 and were called planets in most textbooks for about four decades. Then beginning in 1845, new observations began revealing that there were many more asteroids, so in the 1850s astronomers decided on the label *asteroid* for this population of objects.

In 2006 the International Astronomical Union (IAU) met to decide on a definition of "planet" to settle the debate about Pluto's status. They decided that an object

Concept Question 4

Many people argued that Pluto should continue to be labeled a planet because it had become established as a planet for so many decades. Do you agree with this? When should popular or historical usage take precedence over creating a scientifically consistent nomenclature?

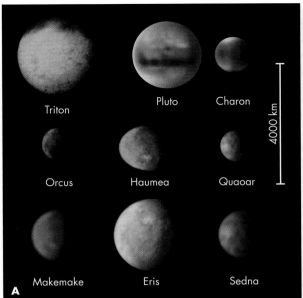
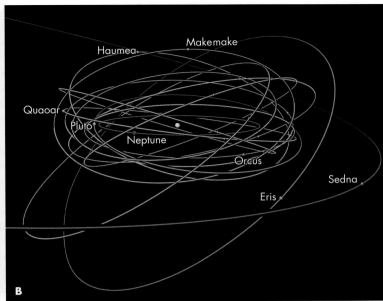

FIGURE 48.10
(A) Renditions of the size, shape, and coloration of several of the largest trans-Neptunian objects, along with a *Voyager 2* image of Neptune's moon Triton, which may be a captured TNO. (B) Orbits of several of the TNOs.

orbiting the Sun (and not itself a satellite) needs to meet two criteria to be classified as a planet. First, it has to be massive enough that its gravitational force can overcome its internal solid-body forces and pull it into a round shape. This criterion is met by even many of the moderate-size moons—those larger than about 400 kilometers (250 miles) in diameter are almost all spherical. Second, to be a planet, an object must have "cleared the neighborhood around its orbit." What this second requirement means is that there is substantially less mass of material sharing a similar orbit. The eight planets meet this criterion very readily, while none of the TNOs, including Pluto, comes even close.

The IAU decided to call objects that do not meet the neighborhood-clearing condition, but that are otherwise orbiting the Sun on their own and have pulled themselves into round shapes **dwarf planets.** Only one asteroid, Ceres (Unit 43), appears to meet this criterion, but it is likely that in addition to Pluto and Eris, dozens of the TNOs will achieve this designation once observations confirm their round shape. The IAU decided in 2008 that TNOs that are dwarf planets will be designated **plutoids.** Table 48.2 shows the properties of several of the largest TNOs, and Figure 48.10 shows the orbits of the several of the largest TNOs besides Eris and Pluto. Demonstrating that these objects have round shapes is difficult because of their great distance, so proving that they are plutoids will be challenging, but the IAU has accepted two more TNOs, Haumea (*how-MAY-uh*) and

TABLE 48.2 Large Trans-Neptunian Objects and Neptune's Moon Triton

Name	Diameter	Mass	Semimajor Axis	Orbital Eccentricity
Triton	3476 km	7.35×10^{22} kg	30.1 AU	Neptune's satellite
Eris	2600 km	1.7×10^{22} kg	67.8 AU	0.44
Pluto	2320 km	1.3×10^{22} kg	39.5 AU	0.25
Makemake	~1600 km	?	45.8 AU	0.16
Sedna	~1500 km	?	526 AU	0.85
Haumea	~1900 × 1000 km	4.2×10^{21} kg	43.3 AU	0.19
Charon	1210 km	1.5×10^{21} kg	39.4 AU	Pluto's satellite
Quaoar	~1000 km	?	43.6 AU	0.04
Orcus	~950 km	7.5×10^{20} kg	39.4 AU	0.22

Makemake (*MAH-kay-MAH-kay*), into the category. Haumea is unusual in that it rotates so fast, in under 4 hours, that it has become very elongated. This does not exclude it from being considered a dwarf planet, because this is a normal shape that a spherical body taks when it is spinning so fast—similar to how the major planets become slightly fatter at their equators because of their rotation.

The TNOs include objects orbiting at a wide range of distances from the Sun. It has become clear that there are classes among them. For example, the members of one category, termed **plutinos,** have orbits similar to Pluto's, with a 2-to-3 resonance with Neptune. There are several hundred plutinos. The largest after Pluto is Orcus (*OR-kuss*), with an estimated diameter of about 1500 kilometers (950 miles), about two-thirds of Pluto's diameter.

Other icy worlds lie farther beyond Neptune. Detecting these more distant objects is difficult because they are so dimly lit. Eris, for example, has an elliptical orbit that stretches from about 38 AU to about 98 AU from the Sun. Another object, named **Sedna,** resides extremely far from the Sun. Its highly elliptical orbit carries it from 76 AU (more than twice Neptune's distance from the Sun) to 928 AU from the Sun. Many of these objects have unusual characteristics that are not yet fully understood. Sedna, for example, has a color nearly as red as Mars.

In recent years astronomers have found that several of these distant objects have companions orbiting around them. Just as for Pluto, this orbital motion permits astronomers to determine the masses of these objects. From the mass, and with an estimate of the radius of the objects, astronomers can calculate their density. Typically the density is about 1 or 2 kilograms per liter, supporting the idea that these objects are similar to Pluto in composition.

It is difficult to estimate how many more large TNOs will be discovered in the outer Solar System as technologies improve. Sedna is currently near its closest approach to the Sun. If it were in the outer part of its orbit, it would not have been detected. We have to wonder how many more objects like Sedna and Eris, or even larger ones, might reside in the outer Solar System. The icy worlds found so far may indeed be just the "tip of the iceberg."

> **Mathematical Insight**
>
> The TNOs become rapidly more difficult to detect the farther they orbit from the Sun. The sunlight a TNO reflects becomes dimmer as $1/d^2$, where d is its distance from the Sun (inverse square law, Unit 21). Then the reflected light must return to Earth nearly the same distance d, so it dims by another factor of $1/d^2$, for a total dimming of $1/d^4$. A Pluto-sized object orbiting twice as far out as Pluto would be only $1/2^4 = 1/16$ as bright!

KEY POINTS

- There is a class of large bodies in the outer Solar System whose composition includes a high percentage of ice.
- The five largest ice worlds formed around Jupiter and Saturn, the sixth (Triton) was probably captured by Neptune after it formed.
- Pluto, Eris, Haumea, and Makemake and probably dozens more TNOs are similar ice worlds, labeled dwarf planets (or plutoids).
- The Galilean satellites are four of the six most massive objects after the planets (Saturn's Titan and our Moon are the other two).
- Io, the Galilean satellite closest to Jupiter, is heated by tidal effects from Jupiter, creating volcanoes.
- Tidal heating of Europa has created an ocean of liquid water under its smooth ice crust, a potential location for life to have formed.
- Io and Europa now have densities like terrestrial planets, but they probably had much more water before tidal heating drove it off.
- Saturn's Titan is cold enough to maintain a substantial atmosphere, with liquid methane forming rivers and lakes.
- After Pluto was discovered, it took more than 60 years to detect other large TNOs, but many are now known.

KEY TERMS

Callisto, 374	Nereid, 376
dwarf planet, 379	plutino, 380
Eris, 378	Pluto, 377
Europa, 373	plutoid, 379
Galilean satellites, 372	Sedna, 380
Ganymede, 374	Titan, 375
grooved terrain, 374	trans-Neptunian object (TNO), 378
Io, 372	Triton, 376
Kuiper belt, 378	

CONCEPT QUESTIONS

Concept Questions on the following topics are located in the margins. They invite thinking and discussion beyond the text.

1. Trends in properties of Galilean satellites. (p. 373)
2. Comparing Titan's rivers to Earth's. (p. 376)
3. Appearance of Charon from Pluto. (p. 377)
4. Recategorizing Pluto as a dwarf planet. (p. 378)

REVIEW QUESTIONS

5. Which ice planets have impact craters on them? Which do not? What are the causes of the differences?
6. What are the satellites of the giant planets thought to be composed of?
7. What sort of activity has been seen on Io? What is Io's heat source thought to be?
8. Describe the evidence of a possible liquid ocean underneath the icy surface of Europa.
9. Jupiter's moon Ganymede and Saturn's moon Titan are similar in size. Why does Titan have a thick atmosphere while Ganymede has none?
10. Why do astronomers think that Triton may be a TNO that was captured by Neptune?
11. Why did the discovery of a moon orbiting Pluto cause astronomers to question its status as a planet?
12. What is a TNO? What is a dwarf planet?

QUANTITATIVE PROBLEMS

13. Io, Europa, and Ganymede orbit at distances from Jupiter's center of 421,800 km, 671,100 km, and 1,070,400 km, respectively. Calculate how many times longer it takes for Europa and Ganymede to orbit Jupiter than for Io to orbit. (Hint: You do not need to know Jupiter's mass to calculate the ratio of times.)
14. Use the modified form of Kepler's third law (discussed in detail in Unit 17) to calculate Pluto's mass from the orbital data for Charon given in the text. Be sure to convert the orbital period to seconds and the orbital radius to meters before putting those numbers into the formula.
15. Calculate the density of Charon, given that its radius is approximately 593 km and its mass is about 1.1×10^{21} kg. (Be sure to convert kilometers to meters, and remember that there are 1000 liters per cubic meter.) Is it likely that Charon has a large iron core? Why?
16. Calculate the angular size of Pluto in arc seconds as seen from Earth. Pluto's diameter is 2320 km and its distance is about 39.5 AU.
17. From conservation of energy considerations it is possible to estimate the ejection speeds from one of Io's volcanoes using the formula $V = \sqrt{2gh}$, were h is the maximum height above the surface the material reaches, and g is the surface gravity (Unit 16.3). If we take the maximum height as 200 km, what is the initial ejection speed? The mass and radius for Io are given in Table 48.1.
18. Triton's orbit is tilted about 157° with respect to Neptune's equator, while Neptune's equator is tilted about 28° with respect to the plane of its orbit around the Sun.
    a. Sketch the Neptune-Triton system, marking the plane of Neptune's orbit and setting up the tilts and inclinations correctly.
    b. Tidal forces cause Earth's Moon to gradually spiral outward. Draw a diagram similar to the one shown in Unit 19, Figure 19.5, for Earth's Moon to explain why Triton will instead spiral inward.
    c. Is it possible that any of the giant planets might have previously "swallowed" a moon? Explain your reasoning.

TEST YOURSELF

19. How was the mass of Pluto determined?
    a. By measuring the effect of its gravity on the terrestrial planets
    b. By observing its effect on the motion of an unmanned flyby of Pluto
    c. By determining the orbit of its satellite, Charon
    d. By measuring how it bends light rays passing near it
20. The small number of craters found on Europa's surface indicates that
    a. the surface is relatively young.
    b. Jupiter's magnetic field shields the moon.
    c. the surface is made of a soft, spongy material.
    d. volcanic activity has resurfaced the moon.
21. In what respect are the atmospheres of Titan and Earth most different?
    a. Chemical composition   c. Density
    b. Presence of clouds   d. Temperature
22. The largest Solar System moon is
    a. Earth's Moon.   b. Triton.   c. Titan.
    d. Ganymede.   e. Rhea.

UNIT 49

Comets

49.1 Comet Structure and Appearance
49.2 Composition of Comets
49.3 Origin of Comets
49.4 Meteor Showers

Learning Objectives

Upon completing this Unit, you should be able to:
- Describe the features and structure of a comet, and the findings of space probes.
- Explain the time sequence of what a comet looks like.
- Describe the origin of comets and explain why astronomers have concluded this.
- Describe a meteor shower and explain why they appear to radiate from a particular constellation.

A bright comet is a stunning sight. Comets have long been feared and revered, and their sudden appearance and rapid disappearance after a few days or weeks of gracing the sky have added to their mystery. Throughout history, comets have been blamed for plagues, or believed to foretell disasters, or taken as a sign to go to war. Today we recognize that comets give us a glimpse of the most remote parts of the Solar System. Many comets appear to originate from distances greater than 10,000 AU from the Sun. Comets are some of the most primitive bodies in the Solar System, probably surviving nearly unchanged from when the Solar System formed (Unit 35).

Discovering comets is one of the areas in which amateur astronomers have made major contributions to astronomy. Finding a comet before it becomes bright requires careful scanning of large amounts of the sky, with the reward that a comet is named for the person who discovers it. Most comets never brighten enough to be seen without a telescope. However, every few years a comet is discovered that grows to spectacular size when it travels so close to the Sun that gas and dust are driven off in large quantities and stretched out by sunlight and the solar wind into a tail millions of miles long, such as Comet McNaught in 2007 (Figure 49.1).

FIGURE 49.1
Photograph of Comet McNaught as seen from Australia in January 2007.

Concept Question 1

The tail of Comet McNaught seen in Figure 49.1 has a bright edge along the bottom and brighter and darker vertical streaks. What might explain this complex structure?

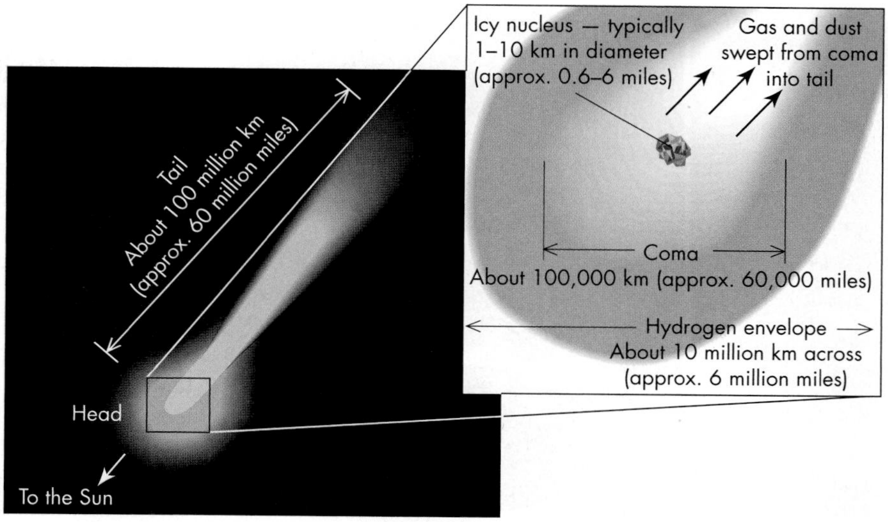

FIGURE 49.2
Artist's depiction of the structure of a comet, showing the tiny nucleus, surrounding coma, and long tail.

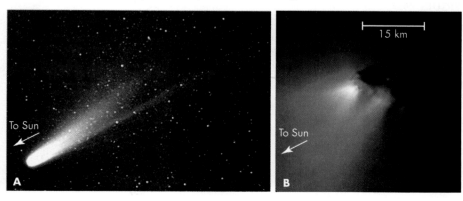

49.1 COMET STRUCTURE AND APPEARANCE

Comets consist of three main parts—the tail, the coma, and the nucleus—as illustrated in Figure 49.2. The largest part is the long tail, a column of dust and gas that may stretch across the inner Solar System. Some comets have had tails as long as 500 million kilometers (about 3 Astronomical Units).

The tail emerges from a cloud of gas called the **coma**, which may be some 100,000 kilometers (60,000 miles) in diameter (roughly 10 times the size of the Earth). However, despite the great volume of the coma and the tail, these parts of the comet contain very little mass. The gas and dust are extremely tenuous, so a liter of the gas contains only a few million atoms and molecules. (In contrast, the air we breathe contains about 10^{22} molecules per liter.) By terrestrial standards, this would be considered a superb vacuum. This extremely rarified gas is matter that the Sun's heat boils off of the heart of the comet, its nucleus.

The comet **nucleus** is a ball of ice and dust whose diameter is typically about 10 kilometers (6 miles). The nucleus of a comet is sometimes described as a giant "dirty snowball," and it contains most of the comet's mass.

One of the best known comets, **Comet Halley** (*HAIL-ee*), reappears about every 76 years (Figure 49.3A), and was visited by the *Giotto* spacecraft during its passage close to the Sun in 1986. The European Space Agency (ESA) sent the *Giotto* spacecraft through Halley's coma, approaching to within 600 kilometers of the nucleus. Images sent back to Earth provided our first look at a comet nucleus. Despite its icy composition, the nucleus is very dark, as you can see in Figure 49.3B, which is one of the

Sir Edmund Halley did not discover the comet bearing his name, but he made the first calculations of the comet's return based on historical records and Newton's laws. The comet returned in 1759 as he predicted, but he did not live to see it.

FIGURE 49.3
Comet Halley in March 1986: (A) photographed from Earth; (B) its nucleus photographed from the *Giotto* spacecraft.

The *Giotto* spacecraft was named in honor of the Italian artist who painted a portrait of a comet, possibly Halley's comet, in 1301.

FIGURE 49.4
Comet Hale-Bopp in 1997, illustrating the two tails, one of dust, one of ions. The ion tail is blown directly away from the Sun by the solar wind. By contrast, dust particles feel the effect of the Sun's gravity as well as its radiation pressure. Responding to the gravity, the dust particles follow orbits around the Sun, but particles farther from the Sun orbit more slowly, so the dust tail arcs behind the comet.

last pictures made by *Giotto* before particles from the comet damaged the camera. Other visible features of the nucleus are its irregular shape and the jets of gas erupting from the frozen surface. The jets form when sunlight heats and vaporizes the icy material. Astronomers think that the dark color comes from dust and carbon-rich material that remains coating the surface of the nucleus after most of the ices have boiled away.

Comet nuclei spend most of their time in a deep freeze. Only when they come close to the Sun do they grow a coma and tail. The Sun's radiation heats the nucleus, and the ices near the surface begin to **sublimate,** passing directly from a solid to a gaseous state in the vacuum of space. At a distance of about 5 AU from the Sun (Jupiter's orbit), the heat is enough to sublimate the ices of water, methane, ammonia, and other volatile compounds, turning them into gases that surround the nucleus to form the coma. The escaping gas carries tiny dust grains that were frozen into the nucleus with it. The "dirty snowball" then appears through a telescope as a dim, fuzzy ball—the initial stage of a comet. As the comet falls ever nearer the Sun, its gases sublimate even faster.

A comet actually develops not one, but two tails, as seen quite clearly in Figure 49.4. Different forces act on dust and gas in the coma that push the material in slightly different directions, creating a **dust tail** and an **ion tail** that are separated from each other.

The dust tail grows as a result of sunlight striking dust grains and imparting a tiny force to them, a force known as **radiation pressure.** Radiation pressure is too weak to have much effect on large bodies, but the microscopic dust grains in a comet's coma do respond to radiation pressure and are pushed away from the Sun, as illustrated in Figure 49.5. The dust particles are massive and remain in orbit around the Sun, but radiation pressure produces an outward force that partially counterbalances gravity. As if orbiting in a weaker gravitational field, the dust particles follow orbits that swing wider around the Sun, sailing farther outward from the Sun than the comet nucleus. Therefore, the dust tail extends in a direction away from the Sun along a curved path.

The ion tail is made of gas and ions from the comet pushed by an outflow of gas streaming from the Sun into space, a flow called the **solar wind** (Unit 51). The solar wind blows away from the Sun at about 400 kilometers per second. It is very tenuous, containing only a few atoms per cubic centimeter. But the material in the comet's coma is also extremely tenuous, so the solar wind is able to push it outward into a long plume. Magnetic fields carried along by the solar wind enhance its

FIGURE 49.5
Sketch of how radiation pressure pushes on dust particles. Photons hit the dust, and their impact drives the dust into a wider orbit around the Sun, forming a dust tail. (Sizes and distances are not to scale.)

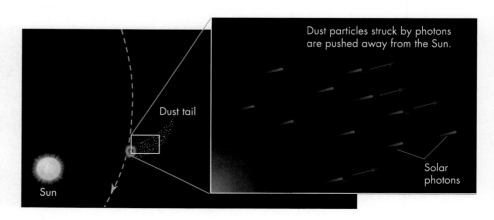

FIGURE 49.6
Sketch illustrating how radiation pressure and the solar wind make a comet's tail always point approximately away from the Sun. (Sizes and distances are not to scale.) Comets may orbit in any direction around the Sun.

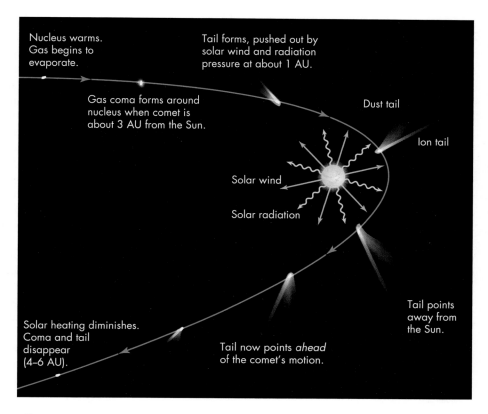

Orientation of comet tails

Concept Question 2

After the volatiles have all sublimated, how would you expect the core of a short-period comet to differ from an asteroid?

effect on electrically charged particles—ions—in the gas from the comet, helping to drag matter out of the coma and channel it, just as magnetic fields in the Earth's atmosphere channel charged particles from the Sun toward the magnetic poles to form the aurora.

Because radiation pressure and the solar wind are both directed away from the Sun, both tails always point away from the Sun (Figure 49.6). Even after a comet passes around the Sun, the tails point out ahead of the comet as it moves away from the Sun. It might help you understand this seemingly odd phenomenon if you think of a runner carrying an Olympic torch. If the air is still, the smoke and flame from the torch will trail behind the runner. However, if a strong wind is blowing from behind the runner, the smoke will be carried along in the direction of the wind regardless of which way the runner moves. Likewise, the high velocity of the solar wind (400 kilometers per second versus about 40 kilometers per second orbital speed for the comet) carries the tail of a comet outward from the Sun regardless of the comet's motion.

49.2 COMPOSITION OF COMETS

The dust and gas that is "boiled" off of a comet offers astronomers a way to probe the comet's composition. The dust particles reflect sunlight, but the gases emit light of their own by a process called **fluorescence.** Fluorescence is produced when light at one wavelength is converted to light at another wavelength. A familiar example is the so-called black light that you may have seen for illuminating posters. Black light is really ultraviolet radiation that we have difficulty seeing because of its short wavelength. When such ultraviolet radiation falls on certain paints or dyes, the chemicals in the pigment absorb the ultraviolet radiation and convert it into visible light.

> Some laundry detergents have "whiteners" in them that fluoresce. These chemicals convert ultraviolet radiation into visible light and thereby make a white shirt or blouse look brighter. You can see this effect strongly if you shine a black light (ultraviolet light) on a freshly washed shirt in a darkened room.

Much of a comet's light is created by fluorescence. A photon of ultraviolet—and thus energetic—radiation from the Sun lifts electrons in the atoms of the comet's gas molecules to an upper, excited level in a single leap. The electron then returns to its original level in two or more steps, emitting a photon each time it drops. The combined energy of these photons equals that of the absorbed ultraviolet photon. Therefore, the energy of each emitted photon must be less than that of the original ultraviolet one. The smaller energy corresponds to a longer wavelength, which may be in the visible part of the spectrum, which we can see with our eyes, or in even longer wavelength bands such as infrared or radio. By observing the precise wavelengths emitted in the spectrum of the fluorescing gas, we can identify what atoms and molecules are present and thereby deduce the comet's composition.

Spectra of gas in the coma and tail show that comets are rich in water (H_2O), carbon dioxide (CO_2), carbon monoxide (CO), methane (CH_3), ammonia (NH_4), and small amounts of other gases that condensed from the primordial solar nebula. Many of these molecules are broken up by solar ultraviolet radiation, so most comets are surrounded by a large cloud of hydrogen and a variety of ions.

Space probes have recently investigated several comet nuclei up close to learn about their composition and structure more directly. Figure 49.7A shows an image from the *Stardust* probe of Comet Wild 2 (pronounced *vilt*) on its mission to collect samples of dust from near the comet. The samples were successfully returned to Earth by parachute in early 2006. The dust particles included small crystals of silicate rock as well as a wide range of organic compounds, and some chemicals that require liquid water in order to form. These materials likely formed at a wide range of distances from the Sun. In addition *Stardust* also found a few particles of interstellar dust that probably date to before the Solar System formed. This suggests a complex "stirring" and a rich chemistry that had not been expected for the early Solar System.

The *EPOXI* spacecraft made even more detailed pictures of Comet Hartley 2 (Figure 49.7B and C), capturing images of chunks of ice ejected from the nucleus by jets of evaporating carbon dioxide. Spectroscopic studies of the gas flowing off Comet Hartley have also confirmed that its isotopic composition closely matches that of Earth's oceans, suggesting that comets could have been a source of the Earth's water.

During these various flyby missions, the gravitational pull of the comet nucleus has been measured, and so it is possible to estimate a mass and density. There are uncertainties in these measurements, but the comet nuclei appear to have a typical density of about 0.6 kilograms per liter—a little over half that of water ice. This suggests an icy composition that is not very tightly packed, truly more like a snowball than a chunk of ice.

FIGURE 49.7
(A) The nucleus of Comet Wild 2. The image was made by NASA's *Stardust* spacecraft, which collected samples of dust from the comet's coma. A longer-exposure image is superimposed, showing faint jets of gas emerging from the comet. (B) The nucleus of Comet Hartley 2. (C) A close-up image of chunks of ice blown off Hartley 2 by jets of evaporating carbon dioxide.

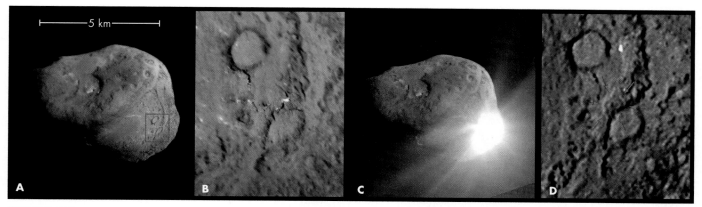

FIGURE 49.8

Comet Tempel 1 before and after NASA's *Deep Impact* mission. (A) The comet nucleus minutes before impact. (B) A composite image made from the impactor probe, showing the collision point. (C) An image made by the main spacecraft about a minute after impact shows a spray of fine particles blasted out by the impact and brightly lit by sunlight. (D) The *Stardust* probe imaged the impact crater about 6 years later.

A more intrusive comet sampling mission was carried out by NASA's *Deep Impact* mission, which in 2005 smashed a 370-kilogram (~800-lb) probe into Comet Tempel 1 at a relative speed greater than 10 kilometers per second (about 23,000 mph). The impact was designed to break through the comet's outer crust and stir up and release dust and gas. The impact event is shown in Figure 49.8. The first two images are from the point of view of the impact probe, which sent pictures up until a few seconds before it smashed into the surface. The third image is from the main spacecraft, showing a cloud of very fine dust blasted out by the impact, which created a crater about 150 meters across. The crater was imaged 6 years later by the *Stardust* spacecraft. The spectra of the material blasted out by the impact showed the presence of water and silicates as well as clays and other water-based crystals with a fine consistency about like that of talcum powder.

49.3 ORIGIN OF COMETS

Comets are perishable objects. Comets that pass by the Sun many times will eventually sublimate all of their ices, and then they will no longer produce a coma or tail. Other comets are lost because they literally fall into the Sun. The *SOHO* and *STEREO* spacecraft (which monitor the Sun) have imaged dozens per year falling into the Sun (Figure 49.9). Since there continue to be comets 4.5 billion years after the Solar System formed, there must be a huge source of comets somewhere far from the Sun.

Based on the portion of comets' orbits that can be observed, astronomers can extrapolate their starting point. The majority of comets follow orbits that indicate their outermost distance from the Sun is more than 50,000 AU. At the same time, only a few appear to have speeds that would carry them completely beyond the Sun's gravitational influence into the vicinity of a neighboring star, so the comet reservoir must be in a very distant part of the Solar System. Comets enter the inner Solar System from all directions without any clear preference, so this great swarm of comets must be a roughly spherical.

This kind of reasoning was used by Dutch astronomer Jan Oort to propose the existence of a vast cloud of trillions of icy bodies orbiting far, far beyond the planets. Known today as the **Oort cloud,** astronomers think it formed from material in the original solar nebula (Unit 35), or perhaps from material orbiting neighboring

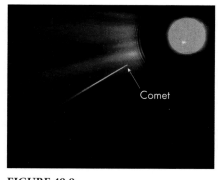

FIGURE 49.9

Comet on collision course with the Sun observed by one of the *STEREO* spacecraft. Regions near the Sun were blocked out to see faint outer features in this image; the image of the Sun used was made at the same time by a different camera.

The Oort cloud and Kuiper belt

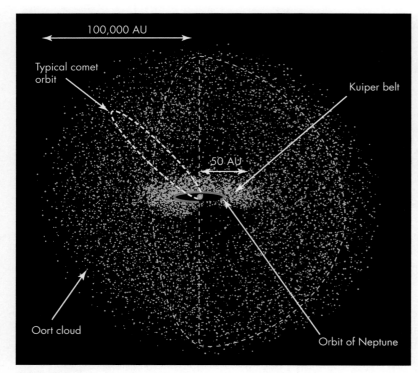

FIGURE 49.10
Schematic drawing of the Oort cloud, a swarm of icy comet nuclei orbiting the Sun out to about 100,000 AU. Also shown is the Kuiper belt, another source of comet nuclei. (The scale is exaggerated for clarity.)

> **Mathematical Insights**
>
> Kepler's third law $P^2 = a^3$ relates the orbital period P and semimajor axis a (see Unit 12). If a comet was on a circular orbit with $a = 100{,}000$ AU, then $P^2 = 100{,}000^3 = 10^{5 \times 3} = 10^{15}$. Therefore, $P = \sqrt{10^{15}} = 3.1 \times 10^7$ years.

> **Mathematical Insights**
>
> To find the semimajor axis of a comet with $P = 200$ years, we apply Kepler's third law: $a^3 = 200^2 = 40{,}000$. Therefore, $a = \sqrt[3]{40{,}000} = 34.2$ AU. Note that if a comet with this semimajor axis approached the Sun to within 1 AU, it would travel out to 33.2 AU.

stars that formed together with the Sun, since we normally see stars forming in clusters of hundreds or thousands (Unit 70). The Oort cloud might be distributed something like the sketch in Figure 49.10. Some comets may be made of material that originally orbited near the giant planets and was tossed into the outer parts of the Solar System by the gravitational force of those planets.

Astronomers suspect that the gravitational influence of stars neighboring the Solar System may alter the orbits of bodies in the Oort cloud occasionally, causing some to drop in toward the inner Solar System. A single disturbance may shift enough orbits to supply comets to the inner Solar System for tens of thousands of years. The discovery of massive objects such as Sedna (Unit 48) suggests that comets may also be disturbed by dwarf planets that orbit in the Oort cloud.

Comet nuclei from the Oort cloud may take millions of years to complete an orbit and are therefore known as **long-period comets.** With orbits so far from the Sun, these icy bodies receive virtually no solar heating, and calculations indicate that their temperature when so far from the Sun is a mere 3 K (−454°F). Thus, their gases and ices remain deeply frozen.

Some comets reappear at intervals of less than 200 years. These **short-period comets** include Halley's, which has a period of 76 years. These comets come mostly from a flatter, less remote region—the **Kuiper** (*KY-per*) **belt,** also shown schematically in Figure 49.10. The Kuiper belt begins at about the orbit of Neptune, 30 AU from the Sun, and seems to thin out near 55 AU from the Sun, but its outer boundary is uncertain. Many much larger bodies, such as Pluto, Eris, and other dwarf planets (Unit 48), also orbit in the Kuiper belt. Their gravitational interactions may be what sends smaller bodies in the Kuiper belt into the inner Solar System to become comets.

The origin of short-period comets is still under study. At one time it was thought that they began in the Oort cloud, and that their orbits were altered as they moved by one of the giant planets. Many astronomers now suspect that most short-period comets came originally from the Kuiper belt, because most of their orbits lie close to the plane of the Solar System and orbit in the same direction as the rest of the Solar System. Comet Wild 2 (see Section 49.2) was probably a Kuiper belt object originally. It was deflected into the inner Solar System in 1974 during a close passage by Jupiter and now has an orbital period of just 6 years.

Comets that reach the inner Solar System are generally smaller than 10 kilometers. Even before astronomers began to detect large numbers of trans-Neptunian objects, they had hints that other, more massive objects were present in the outer Solar System. For example, in 1977 astronomers discovered an object, Chiron (*KY-ruhn*), with an orbit that stretches from just inside Saturn's orbit to almost the orbit of Uranus. This was a surprising discovery because orbits in this region are generally unstable, due to the gravitational perturbations of the giant planets. Astronomers thought that perhaps Chiron should be classified as an asteroid because of its unexpected location and very large size compared with known comets: 180 kilometers (about 110 miles) in diameter versus the more typical 10 kilometers. But Chiron has other properties that make it more similar to comets than to asteroids.

In particular, it changes in brightness, sometimes flaring up and ejecting gas like a comet. It does not display the familiar comet's tail only because it remains too far from the Sun to be strongly affected by its heat and the solar wind.

Astronomers now estimate that the Kuiper belt may contain tens of thousands of objects 100 kilometers or larger in diameter, and millions as large as a typical comet. Chiron may have originated in this region, later to be deflected into its current orbit by a close passage by Neptune. Objects at this distance are still difficult to study because they are so dimly lit by the Sun, but we can extrapolate from the number of large objects to predict the number of smaller ones. The Kuiper belt probably contains a total mass hundreds of times larger than the asteroid belt. It is spread out over a much larger volume of space, so it is even emptier than the asteroid belt. These frozen objects, like the rocky asteroids, are survivors of the Solar System's birth—icy planetesimals still orbiting in the disk.

49.4 METEOR SHOWERS

The Sun steadily whittles away a short-period comet. With each close passage by the Sun, ices sublimate and carry off solid matter (dust and grit), leaving behind a smaller nucleus. This fate is like that of a snowball made from snow scooped up alongside the road, where small amounts of dirt have been packed into it. If such a snowball is brought inside, it melts and evaporates, leaving behind only the grit accidentally incorporated in it. So too, as a comet sublimates, it leaves behind in its orbit grit that can continue to circle the Sun long after the comet has lost so much of its ices that it no longer produces a visible tail.

If you go outside on a clear night after midnight, far from city lights with an unobscured view of the sky, you will see, on average, one "shooting star" or **meteor** every 15 or so minutes. Most of these meteors are tiny stray fragments of asteroids and comets that arrive at the Earth randomly (Unit 50) and then burn up in its atmosphere. However, at certain times of the year, instead of one meteor per quarter hour or so, you may observe one every few minutes. Furthermore, if you watch such meteors carefully, you will see that they appear to come from the same general direction in the sky in what is called a **meteor shower.**

The best-known meteor shower is probably the Perseid meteor shower. It reaches its peak each year between August 10 and 14, when meteors rain into our atmosphere from a direction that lies toward the constellation Perseus (Figure 49.11A). The meteors themselves have no association with Perseus. Rather, they are the debris of a comet (named Swift-Tuttle) and are following an orbit around the Sun that crosses the part of our orbital path that the Earth travels through in mid-August (Figure 49.11B).

This encounter creates an effect similar to what you observe when you drive at night through falling snow: The flakes seem to radiate from a point in front of you, the location of which

Concept Question 3

Given how dust particles are affected by solar radiation pressure, where would you expect to find most of the debris from a comet many orbits after the comet expelled the material?

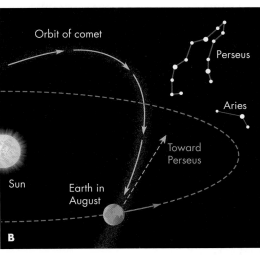

FIGURE 49.11
Sketch of the Perseid meteor shower in mid-August. (A) The meteors appear to radiate away from a point in the constellation Perseus. (B) At the time of the Perseid meteor shower, the Earth, moving along its orbit, crosses the debris strewn along Comet Swift-Tuttle's orbit, the scattered material plunges into our atmosphere with a velocity in the direction from the constellation Perseus. (Bodies and orbits are not to scale.)

FIGURE 49.12
This time-lapse picture shows meteors from the Leonid meteor shower on November 19, 2002.

TABLE 49.1 Major Meteor Showers

Peak Date	Shower Name	Maximum Rate	Constellation	Comet
January 3–4	Quadrantids	30–90 per hour	Bootes	2003 EH1
April 22–23	Lyrids	10–15 per hour	Lyra	Thatcher
May 7–8	Eta Aquarids	10–20 per hour	Aquarius	Halley
July 28–29	South Delta Aquarids	15–20 per hour	Aquarius	Marsden & Kracht
August 12–13	Perseids	40–100 per hour	Perseus	Swift-Tuttle
October 21–22	Orionids	15–20 per hour	Orion	Halley
November 17–19	Leonids	10–? per hour	Leo	Tempel-Tuttle
December 13–14	Geminids	40–90 per hour	Gemini	Phaethon
December 22–23	Ursids	10–15 per hour	Ursa Minor	Tuttle

depends on a combination of the direction and speed of both the falling flakes and your car. Thus, while the Earth crosses the path followed by the cometary debris, the debris seems to diverge from a particular point, called the **radiant**. Meteor showers are generally named for the constellation from which they appear to diverge. Table 49.1 lists several of the brighter and more impressive showers and the dates when they peak. Each shower therefore marks when the Earth crosses the path of the comet listed in the table—sometimes even a "dead" one like Phaethon or 2003 EH1, which were only recently discovered.

On rare occasions the Earth passes through a dense clump of material left by a comet. When that happens, thousands of meteors per hour may spangle the sky. Such a display happened in November 2002 during the Leonid meteor shower (so named because it appeared to originate from the constellation Leo). At various spots around the world, observers reported periods with one meteor every second or two (Figure 49.12). This shower comes from the debris of the short-period comet Tempel-Tuttle, which orbits the Sun approximately every 33 years. Comet Tempel-Tuttle is now difficult to see because it has sublimated most of its volatile material in repeated passages around the Sun. The densest clumps of debris remain close to the comet nucleus; so every 33 years or so the meteor display is spectacular. In 1966 in western North America there were observations of a meteor shower, or storm, containing dozens of meteors *per second!* Beginning around 2030, the chances of meteor storms during the Leonid shower will increase once again.

KEY POINTS

- Comets are leftover material, a mix of ice and dust, from the formation of the Solar System.
- They are basically "dirty snowballs" that spend most of their time far from the Sun at temperatures close to absolute zero.
- When a comet gets too near the Sun, ices sublimate and carry off dust particles, forming a gas cloud (coma) around the nucleus.
- If the comet's orbit brings it to the inner Solar System, the solar wind pulls gas from the coma into an ion tail, and radiation pressure from the Sun pushes the dust particles out into a dust tail.
- A comet's tail always points away from the Sun, so it may be in front of the comet as the comet travels away from the Sun.
- Short-period comets have orbits less than 200 years long and generally originate from the Kuiper belt.
- Long-period comets come from the Oort cloud, which extends roughly 100,000 AU from the Sun.
- It is possible that material in the Oort cloud originated closer in and was sent outward by gravitational interactions with the gas giants.
- Dust lost from a comet continues in orbit and can create meteor showers if the Earth passes through the debris cloud.
- The meteor shower appears to come from the direction of a comet's orbit relative to the Earth's own motion.

KEY TERMS

coma, 383
Comet Halley, 383
dust tail, 384
fluorescence, 385
ion tail, 384
Kuiper belt, 388
long-period comet, 388
meteor, 389

meteor shower, 389
nucleus, 383
Oort cloud, 387
radiant, 390
radiation pressure, 384
short-period comet, 388
solar wind, 384
sublimate, 384

CONCEPT QUESTIONS

Concept Questions on the following topics are located in the margins. They invite thinking and discussion beyond the text.

1. Explanations of complex structure in a comet's tail. (p. 382)
2. How a comet's spent core differs from an asteroid. (p. 385)
3. The orbit of the dust from a comet. (p. 389)

REVIEW QUESTIONS

4. Where do comets originate? What causes them to travel into the inner Solar System?
5. What is sublimation, and how does it differ from evaporation or boiling?
6. What creates a comet's tail? In what direction does the tail point?
7. Some comets have two tails—what is the difference between the tails?
8. What is a meteor shower?
9. Why are meteor showers named for particular constellations?

QUANTITATIVE PROBLEMS

10. Use Kepler's third law to find the semimajor axis of Halley's comet, given that its orbital period is 76 years.
11. Halley's Comet has a mass of 2.2×10^{14} kg and loses about 1/1000 of its mass for every trip it makes through the inner Solar System.
    a. The first recorded observation of Halley's Comet was in 240 BCE. Assuming a 76 year period, how many times has the comet passed through the inner Solar System since it was first observed?
    b. How much mass has Halley's Comet lost since its first sighting? What percentage of the total mass does this represent?
12. Comet McNaught's orbit ranges from 0.17 AU out to 4100 AU. Use Kepler's third law to determine the comet's period.
13. Use data from the text to estimate the total kinetic energy, in joules, of the impact of NASA's *Deep Impact* probe onto comet Tempel 1. If the comet has a mean diameter of about 6 km and a density of about 0.8 kg/L, and is orbiting at about 30 km/sec, what is its kinetic energy?
14. The temperature, T, of a body decreases as the square root of its distance, d, from the Sun increases: $T = \text{constant}/\sqrt{d}$. If the temperature at 1 AU is 250 K, estimate the temperature of a comet nucleus in the Oort cloud.
15. As a comet orbits the Sun, its temperature changes approximately as described in the previous problem.
    a. Calculate the distance from the Sun at which it would reach the sublimation temperature of water (sublimates at $T = 180$ K). Repeat for ammonia (130 K) and carbon dioxide (80 K).
    b. Based on your results in (a), describe how a comet would behave as it orbits the Sun.
    c. What would be the effect on the comet's temperature if it had dark regions on its surface? Qualitatively, how would this affect its behavior as it orbited the Sun? Explain your reasoning.

TEST YOURSELF

16. A comet's ion tail
    a. points directly away from the Sun.
    b. points away from the Sun with a slight curl away from the direction the comet is moving.
    c. is only visible when the comet is far from the Sun.
    d. is what pushes the comet forward in its orbit.
17. Meteor showers such as the Perseids in August are caused by
    a. the breakup of asteroids that hit our atmosphere at predictable times.
    b. the Earth passing through the debris left behind by a comet as it moved through the inner Solar System.
    c. passing asteroids triggering auroral displays.
    d. nuclear reactions in the upper atmosphere triggered by an abnormally large meteoritic particle entering the upper atmosphere.
    e. None of the above.
18. A comet that someone observes traveling through the inner Solar System multiple times in their lifetime probably originated from
    a. the Oort cloud.
    b. the Kuiper Belt.
    c. the asteroid belt.
    d. interstellar space.
19. What does a comet look like when it is 100 AU from the Sun versus 1 AU from the Sun?
    a. Its nucleus, coma, and tail are all about 100 times smaller.
    b. Its tail is much shorter, but its coma and nucleus are the same.
    c. Only its nucleus is present; there is no tail or coma.
    d. In the weaker gravity, the nucleus spreads into the coma.
20. How did astronomers find the Oort cloud?
    a. They deduced its presence from the orbits of comets.
    b. They found such clouds around other stars and realized the Sun must have one too.
    c. They saw many stars winking on and off as distant orbiting bodies blocked their light.
    d. They detected thermal emission from objects in the cloud from space-based infrared telescopes.

PART 3 UNIT 50

Impacts on Earth

50.1 Heating of Meteors
50.2 Meteorites
50.3 The Energy of Impacts
50.4 Impacts with the Earth
50.5 Mass Extinction Events

Learning Objectives

Upon completing this Unit, you should be able to:
- Explain where meteors occur and why they produce light.
- Describe the classes of meteorites and their probable origins.
- Calculate the energy released by an impacting body.
- Describe the evidence for impact events on Earth, and discuss the consequences.

If you have spent even an hour looking at the night sky, you have probably seen a "shooting star," a streak of light that suddenly appears and just as quickly fades away (Figure 50.1). Astronomers call this lovely but brief phenomenon a **meteor.** What we observe as a meteor is actually the glowing trail of hot gas and vaporized debris left by a solid object heated by friction as it moves through the Earth's atmosphere at very high speed. Most meteors are produced by the bits and pieces of material that are released by a comet when it passes the Sun (Unit 49)—small debris that totally burns up about 50 kilometers (30 miles) up in the Earth's atmosphere. However, not all meteors are so small, and some may reach the Earth's surface. The solid body, while in space and before it reaches the atmosphere, is called a **meteoroid.** If it reaches the ground, it is called a **meteorite.**

The idea of stones falling from space seemed so strange that for centuries most scientists found the concept unthinkable, suggesting instead that they must be debris ejected from distant volcanoes. By the early 1800s the evidence that these were objects from space became convincing. Many witnesses reported seeing the dramatic fall of meteors and then collecting the associated meteorites, and many of these meteorites had properties unlike any geological rock specimen—some were giant chunks of pure iron, for example. But this still left the mystery of how and where these bodies formed, and what brought them to Earth. Astronomers think that most meteorites are fragments of asteroids, although a few meteorites are now convincingly identified to be fragments blasted out of the crust of Mars and the Moon.

50.1 HEATING OF METEORS

Meteors heat up on entering the atmosphere for the same reason a reentering spacecraft does. When an object plunges from space into the upper layers of our atmosphere, it collides with atmospheric molecules and atoms. These collisions convert some of the body's energy of motion (kinetic energy) into heat, as illustrated in Figure 50.2. In a matter of seconds, the outer layer of the meteor reaches thousands of degrees Kelvin. The entry speeds are typically from 10 to 60 kilometers per second, depending on the relative directions of the orbits of the Earth and the asteroid, which are each moving at about 30 kilometers per second around the Sun. The collisions with air molecules are extremely violent and tear electrons from the air molecules and vaporize material from the surface of the meteoroid. The hot, ionized gas surrounding the meteoroid produces the glow that we see.

FIGURE 50.1
A time-exposure photograph captures a "shooting star" (meteor) flashing overhead. The curved streaks are star trails.

FIGURE 50.2

Sketch depicting how a meteor is heated by air friction in the upper atmosphere.

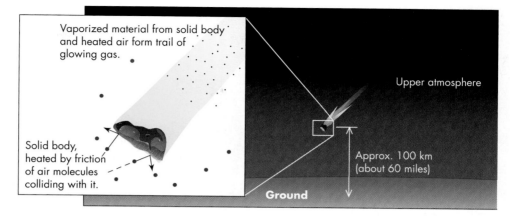

If the meteoroid is larger than a few centimeters, it creates a large ball of incandescent gas around it and may leave a luminous or smoky trail. Such exceptional meteors, sometimes visible in daylight, are called *fireballs*.

Meteoroids bombard the Earth continually—a hail of solid particles that astronomers estimate at hundreds of tons of material each day. More strike between midnight and dawn (actually until noon, but they are not usually visible in the daylight) than in the evening hours, so it is best to watch for meteors before dawn. This difference arises for the same reason that if you run through rain, your front side will get wetter than your back. That is, the dawn side of our planet advances into the meteoritic debris near us in space, while the sunset side moves away from it.

Most meteors that we see last no more than a few seconds and are made by meteoroids the size of a raisin or smaller. These tiny objects are heated so much that they completely vaporize. Larger pieces may survive the ordeal and reach the ground, with their surfaces scorched and partially vaporized, to be found as meteorites.

Clarification Point

It may help you recall the meanings of the meteor-related terms if you remember that *meteoroid* rhymes with *void*, whereas the last two letters of *meteorite* are the first two letters of *terrestrial*.

50.2 METEORITES

Most meteorites are small pieces of asteroids, broken off during collisions. Asteroids generally exhibit many impact craters (Unit 43), and because the gravitational attraction of the asteroids is weak, fragments blasted out by an impact can escape the asteroid and travel along new orbital paths around the Sun. Astronomers classify meteorites into three broad categories based on their composition: *stony* (that is, composed mainly of silicate compounds), *iron*, and *stony-iron*.

About 86% of all meteorites are **chondrites** (*KON-drites*). These are stony meteorites that contain small rounded bits of rocky material stuck together, as shown in Figure 50.3A. These small spheres of rock are called **chondrules** (*KON-drools*), meaning "grain" in Greek. Experiments show that chondrules form when particles of silicate rocks are rapidly heated to form millimeter-size molten droplets that then quickly cool before they grow large. The source of the heating is not known, although there are many hypotheses. Chondrules may have been melted by outbursts from the young forming Sun, collisions between planetesimals, or the explosion of a nearby star—perhaps an explosion that triggered the collapse of the solar nebula. Radioactive ages (Unit 35) of chondrules indicate that they are some of the oldest material in the Solar System—slightly more than 4.5 billion years old.

In ordinary chondrites, the chondrules, along with flakes of metal, are embedded in a matrix of other rocky material that condensed from the solar nebula. This gives us a glimpse of what the inner parts of the solar nebula must have looked like: a disk of hot gas full of these small rocky pebbles and metal flakes. In a small percentage of chondrites, the chondrules are embedded in a black, carbon-rich, coal-like

Concept Question 1

Depending on the different hypotheses about the source of melting of chondrules, would you expect chondrules to be more or less abundant in comets?

FIGURE 50.3

Slices of various types of meteorites. (A) Chondrites are the main kind of meteorite. They contain small rounded chondrules that formed in the solar nebula. Carbonaceous chondrites have a matrix of carbon-rich material and probably condensed in the outer part of the asteroid belt. (B) Meteorites from asteroids that are large enough to have differentiated come in three main types: irons from the core of the asteroid, stony-irons from the core mantle boundary, and achondrites from the rocky surrounding material. The iron slice has been etched with acid to reveal the pattern of crystallization, which only develops when molten iron is allowed to cool gradually over millions of years.

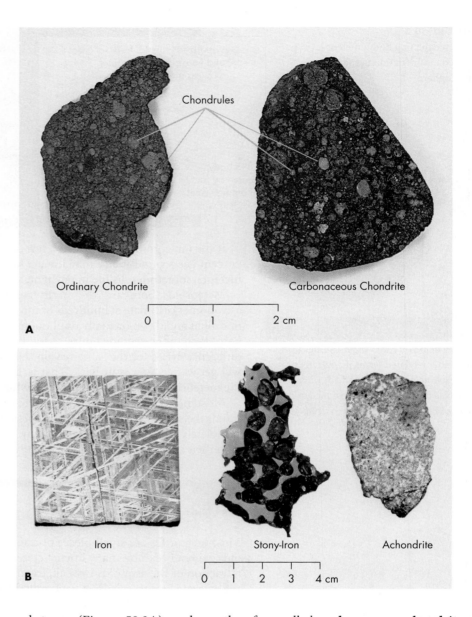

Clarification Point

The term *organic* indicates a complex carbon-based molecule. It does not necessarily indicate that living organisms are involved.

substance (Figure 50.3A), and are therefore called **carbonaceous chondrites.** Spectroscopic comparisons indicate that these came from the carbon-rich asteroids in the outer part of the asteroid belt (Unit 43.3). This carbonaceous matter contains **organic compounds**—carbon-based molecules—as well as water and other volatile chemicals. The presence of these organic molecules demonstrates that natural processes in space produce some of the raw materials essential for life and that those materials would have been raining down on the Earth (in fact, continue to do so) after the young Earth cooled enough to form a solid crust.

Carbonaceous chondrites contain some fairly complex organic compounds, such as amino acids, which are used by living things to construct proteins (Unit 85). We can be fairly certain that the amino acids found in meteorites were not themselves produced through biological processes, however. When living things on Earth build amino acids, the molecules are always put together in what is called a "left-handed" form. Nonbiological processes produce both the left-handed forms and mirror-image "right-handed" forms in about equal numbers, and this 50-50 mix of left and right is what we find in meteorites.

The great majority of meteorites are stony, based on the statistics of meteorites that were collected after being observed falling ("falls"). If they were not observed falling, stony meteorites are not easily distinguished from Earth rocks, so most of the meteorites that are found ("finds") tend to be the more unusual iron meteorites (about 5% of falls) and the very rare stony-iron meteorites (about 1% of falls). Slices of these types of meteorites are shown in Figure 50.3B. Iron meteorites must come from asteroids so large that they melted and differentiated (Unit 35), allowing the dense iron to sink to the core. An asteroid probably has to be at least 20 to 50 kilometers in diameter to form an iron core. Stony-iron meteorites may form at the boundary between the iron core and the rocky mantle, or they may be formed during the collision that smashes the asteroid apart, scattering metal fragments from the core into space.

The final 8% of meteorites are stony, but are **achondrites,** meaning that they have no chondrules (Figure 50.3B). The achondrites are rocky material from the upper layers of differentiated bodies. The metal flakes are gone, having melted and sunk to the center of the asteroid, and the rock shows evidence of melting and undergoing geological processes. Spectroscopic and chemical studies have even been able to link some of the achondrites to particular large asteroids or even Mars and the Moon. Evidently, some impacts are powerful enough to blast out pieces of one planet's crust and send it colliding into another planet.

50.3 THE ENERGY OF IMPACTS

Based on the number of impact craters found on Earth, every few thousand years a large meteoroid, tens of meters or more in size, strikes our planet. The impact velocities of tens of kilometers per second are more than 10 times faster than a rifle bullet. At such high velocities, meteoroids have a large kinetic energy—that is, a large energy of motion.

A meteoroid more than several centimeters in size may not burn up completely during its passage through the atmosphere, so its remaining kinetic energy is released when it hits the ground. The energy released from large meteoroids can be huge. To calculate the energy released by an object striking the Earth, we use the formula for an object's kinetic energy, E_K, which is given by the expression

$$E_K = \tfrac{1}{2} m \times V^2,$$

where m is its mass and V is its velocity (Unit 20).

As an example, consider the meteorite impact depicted in Figure 50.4. About 100,000 kilograms (100 tons) of iron struck the Earth in Russia in 1947. This was not an especially large meteoroid, since 100 tons of iron would fit in a sphere less than 3 meters (10 feet) in diameter. Supposing it struck the Earth at 30 kilometers per second ($= 3 \times 10^4$ meters per second $= 108{,}000$ kilometers per hour, or almost 65,000 miles per hour), the kinetic energy of impact was

$$\tfrac{1}{2} \times 10^5 \text{ kg} \times (3 \times 10^4 \text{ m/sec})^2 = 4.5 \times 10^{13} \text{ joules}.$$

This is about as much energy as is released when 10,000 tons of dynamite are exploded, or approximately the same as the explosive power of the nuclear bomb that destroyed Hiroshima. Events of similar strength are found every few years by Earth monitoring satellites, which detect the atmospheric fireball from these impacts as a brilliant flash of light. Most of these events occur over the ocean

FIGURE 50.4
An iron meteorite from the Ural Mountains in Russia. Approximately 100 tons of iron struck the Earth with an impact energy comparable to the atomic bomb dropped on Hiroshima. A commemorative stamp depicts the fireball and smoke trail from a painting by an eyewitness.

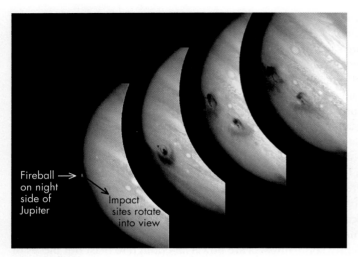

FIGURE 50.5
The impact of Comet Shoemaker-Levy 9 with Jupiter in 1994 was observed with the Hubble Space Telescope. This series of images shows the evolution of the impact of one of the fragments of the comet. The lower-left image was taken a few moments after the impact, which occurred on the night side of Jupiter. A fireball rose over 3000 kilometers (1900 miles) above the cloud tops. When the impact site rotated into view about 1.5 hours later (second image), dark material blasted up from the impact site became visible. The third and fourth pictures show the dispersal of the material over the next several days, along with additional impact sites from other fragments of the comet.

or in unpopulated areas and may not be witnessed from the ground. Were such a body to hit a heavily populated area, it could be lethal.

Astronomers estimate that the meteoroid that produced the 1947 event may have originally had a mass up to 10 times greater than the 100 tons of material that reached the ground. About 90% of the meteoroid was vaporized and most of the meteoroid's kinetic energy was released as it was torn apart and vaporized by its high-speed passage through the air. Meteoroids smaller than about 10 meters in diameter release most of their kinetic energy in the fireball produced during their passage through the atmosphere, but larger bodies retain most of their kinetic energy until they strike the Earth's surface.

An impact of a comet with Jupiter was observed in 1994. Comet Shoemaker-Levy 9 had previously passed close to Jupiter, and tidal forces pulled it apart into a string of smaller fragments (Unit 47), each estimated to be a few kilometers in diameter. The impact of one large fragment, photographed by the Hubble Space Telescope, is shown in Figure 50.5. Dark "sooty" material generated in the blast made large dark areas as big as the entire Earth, traces of which lasted for months after the impacts. The energy released is estimated to have been the equivalent of about 6 million megatons of TNT. This is more than 400 times the total nuclear arsenal amassed by all countries during the Cold War!

Because a similar impact on Earth would have devastating consequences, astronomers have developed an international monitoring system to hunt for asteroids that might impact the Earth in the future. Such bodies might be deflected away from Earth by altering their orbit, but this requires early detection. Relatively little force could deflect the object if this is done early enough, but the force required rapidly grows larger as the collision time grows nearer. Fortunately, we have been spared such disasters recently; but there have been some close calls and some truly horrific impacts in the distant past.

50.4 IMPACTS WITH THE EARTH

Based on the appearance of the Moon, the Earth must also have been struck by objects repeatedly throughout its history, although relatively few obvious traces remain today. High-speed impacts vaporize rock at the impact site and send out a shock wave that peels back the ground around the point of impact, as illustrated in Figure 50.6. The **Barringer crater** (also known as Canyon Diablo) is one of the first documented meteor impact features identified on Earth. It is in northern Arizona, about 70 kilometers (40 miles) east of Flagstaff, and is easily visible

FIGURE 50.6
A computer simulation of an impact crater's formation.

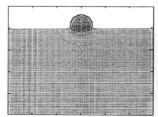
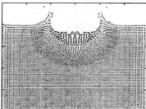
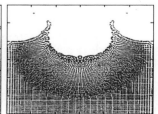

FIGURE 50.7
Photograph of the Arizona meteor crater, which is more than 1 kilometer across. The inset image shows one of the iron meteorites found at the site.

FIGURE 50.8
The site of the Tunguska event photographed by an expedition in 1927.

from the sky. Geological analysis of the fragments of the impacting body recovered there suggests the impact occurred about 50,000 years ago. The meteoroid is estimated to have been some 50 meters in diameter, but it created a crater about 1.2 kilometers across and 200 meters deep (Figure 50.7).

More recently, in 1908 an asteroid broke up in our atmosphere over a mostly uninhabited part of north-central Siberia. This **Tunguska event,** named for the region where it hit, leveled trees radially outward from the blast point to a distance of some 30 kilometers and broke windows up to 600 kilometers away. Eyewitnesses reported seeing a fireball brighter than the Sun traveling through the sky. The blast was followed by clouds of dust that rose to the upper atmosphere. Sunlight reflecting off this dust gave an eerie glow to the night sky for several days. According to some accounts, the blast killed two people. Casualties were few because the area was so remote.

Unfortunately, scientists did not visit the site until about two decades later (or no records of earlier visits survive) because of the political turmoil in Russia at that time. When they did reach the site, they found no crater, just the felled trees. Interestingly, some trees at the center of the damaged area were left standing vertically but with their branches stripped off (Figure 50.8), suggesting that the explosion occurred in the air.

Hypotheses to explain the Tunguska event offer a good example of how science tests and retests ideas. The lack of a big crater and the absence of meteorites at the site at first seemed like evidence against the idea that it was caused by a small asteroid. This led some astronomers to propose that a comet might have caused the blast. Being icy, a comet would break up more easily—perhaps even while still in the atmosphere—so there would be no crater. Moreover, ice fragments would rapidly disappear, explaining the absence of large rock or iron meteorites.

However, computer simulations of how an asteroid travels at high speed through our atmosphere indicated what at first seemed a surprising result: even stony asteroids can create devastation without leaving a crater or telltale fragments. Depending on an asteroid's orbit relative to the Earth's, it will typically approach Earth at speeds between 10 and 40 kilometers per second. When it hits our atmosphere, it compresses and heats the air ahead of it. The hot compressed air obeys Newton's law of action–reaction and exerts a tremendous force back on the asteroid that may shatter it. The resulting fragments plunge deeper into our

Concept Question 2

If an asteroid heading toward Earth could be broken up into many small pieces that would hit the Earth over a larger area, would this be better or worse than a single large impact?

FIGURE 50.9
Pictures from the meteor explosion over Russia in February 2013. From left to right: a "dash-cam" image of the asteroid exploding in the atmosphere, a factory roof and wall collapsed by the shock wave, a hole broken through the ice of a frozen lake by a fragment of the asteroid.

There are false but persistent rumors that radioactivity has been detected at the Tunguska site.

atmosphere, where air resistance heats them further until they vaporize, creating a fireball at a height of 10 kilometers (6 miles) or so above the ground. No traces of the asteroid necessarily survive to form a crater or fragments at the ground. All that remains is the 6000-K incandescent air of the fireball that blasts downward and out, crushing and igniting whatever lies below it, exactly as a nuclear bomb burst would.

Tunguska-like events may not be as rare as once thought. Hundreds of cameras caught a large meteor explosion over Russia in 2013 (Figure 50.9). The shock wave from the blast blew in windows, injuring more than a thousand people in an explosion estimated to have been the equivalent of about 500 kilotons of TNT. Only relatively small amounts of meteorite material were recovered from this event, although it is estimated that the original asteroid was about 17 meters (55 feet) in diameter. Accounts of fireballs and widespread damage in two remote regions of South America in the 1930s also resemble this event. Some tsunamis (tidal waves) may have been caused by impacts in the ocean. One suspected event about 500 years ago left deposits of beach sand more than 200 meters above sea level in New Zealand. This catastrophic event may explain why the Maori people abandoned many coastal settlements at that time, although this is debatable.

More impact scars have been found in many places on our planet. Wolf Creek crater in northwestern Australia is about 0.9 kilometers in diameter and formed in an impact estimated to have occurred about 300,000 years ago. Geological processes and erosion on Earth hide many older craters, although some remain evident. The ring-shaped Manicouagan Lake in Quebec is about 70 kilometers (approximately 43 miles) in diameter (Figure 50.9). It is a meteor crater estimated to be about 210 million years old, produced by a meteoroid approximately 5 kilometers in diameter. Other impact craters that are completely buried have been detected through drilling and seismic studies of underground features. For example, an impact crater from about 35 million years ago has been identified below the mouth of the Chesapeake Bay on the East Coast of the United States. The crater has since been buried by hundreds of meters of sand and silt, but geologists can trace shattered rock and the profile of the buried crater over a region about 90 kilometers (55 miles) in diameter. An even larger impact gave rise to the Sudbury iron mines in southern Canada. They lie in a 1.85-billion-year-old impact crater, now highly eroded, that is about 250 kilometers (150 miles) in diameter.

Many other large arc-shaped features on the surface of the Earth may also have originated as impact features. Two such craters are the vast arc (nearly 500 kilometers across) on the east edge of Hudson Bay and a basin (about 300 kilometers across) in central Europe. Geological activity and erosion so change the Earth's surface that most features become unrecognizable through surface wear, distortion,

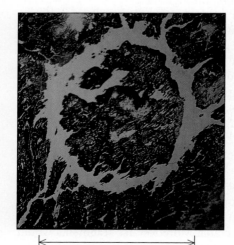

Approx. 70 km
(about 43 miles)

FIGURE 50.10
Picture (from Earth orbit) of the Manicouagan crater in Quebec. This winter view shows the lake that fills the crater frozen and covered with snow.

or burial. The Moon's unchanging surface, however, offers a visual reminder of the crater-forming impacts the Earth also must have suffered during its earliest history.

Astronomers have come to realize that meteor impacts are more common (and more deadly) than once believed. Although there are no recent, clear cases of people being killed by meteorites, there are many documented near misses. Searching old records from Europe and China, astronomers have found past instances of fatalities. One grim report is from a fifteenth-century Chinese chronicle that describes "stones that fell like rain" that killed 10,000 people. Scientists studying the tusks of woolly mammoths from about 35,000 years ago discovered that several were peppered with pockmarks with small bits of iron meteorite embedded in them. It appears that the iron meteoroid burst in the air, sending a spray of possibly lethal shrapnel into a herd of the beasts. If a similar event occurred over a city today, it would have the potential to kill enormous numbers of people.

50.5 MASS EXTINCTION EVENTS

Throughout Earth's history, the geological record shows evidence of mass extinctions of life on Earth. These events are so large that they wipe out entire species, not just a few individuals. Sometimes a relatively small number of species disappear from the geological record, but at other times thousands of species of animals and plants, large and small, are suddenly killed off. For many years the favored explanation for these sudden changes was a geological event, such as massive volcanic eruptions or drastic changes in climate and sea level. Such events may indeed be responsible for some of these episodes of dramatic change on Earth; but in 1980 two scientists from the United States, the father-and-son team of Luis and Walter Alvarez, found compelling evidence that one of the most devastating extinctions was caused by the impact of an asteroid.

About 65 million years ago, some disaster exterminated the dinosaurs and many less-conspicuous but widespread creatures and plants. The sudden disappearance of large numbers of life forms occurred at the end of the **Cretaceous period.** The evidence that an extraterrestrial body caused this cataclysm comes from the relatively high abundance of the otherwise rare element iridium that the Alvarezes discovered in sediments from that time (Figure 50.10). Iridium, a heavy element similar to platinum, is very rare in terrestrial surface rocks. Early in the Earth's history, when our planet differentiated, the high density of iridium caused that element to sink to the Earth's deep interior. However, iridium is relatively abundant in much meteoritic material, which either did not differentiate or includes the dense material from the core. Thus, the presence of so much iridium in a layer of clay laid down 65 million years ago suggests a link to meteoritic material. On the basis of samples gathered worldwide from that layer, scientists calculate that it would have taken the explosion of a meteor 10 kilometers (6 miles) in diameter to disperse that much iridium. Therefore, most astronomers and paleontologists believe the Earth was hit by an asteroid of that size.

A 10-kilometer asteroid hitting the Earth would produce an explosion on impact equivalent to that of several billion nuclear weapons. Not only would the impact make an immense crater, but it would also blast huge amounts of dust and molten rock into the air. The molten rock raining down would ignite global wildfires. The hot fragments and blast would also create nitrogen oxides, which would combine with water to form a rain of highly concentrated nitric acid. This toxic combination of fires, acid rain, and blast damage would then be followed by months of darkness and intense cold caused by the dust shroud blotting out the Sun. It seems likely

FIGURE 50.11
Sedimentary layers deposited 65 million years ago show an abrupt change in character where a layer of dark, iridium-rich clay marks the end of the Cretaceous period on Earth. The layer was probably formed when an asteroid struck the Earth, leading to global devastation and changes to the environment.

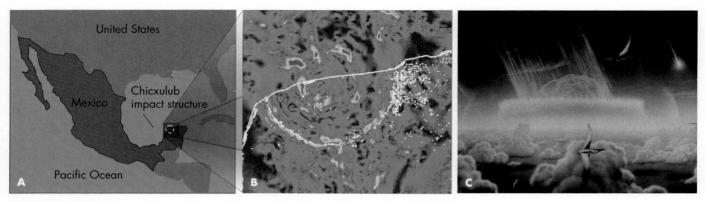

FIGURE 50.12
(A) The Chicxulub impact structure from 65 million years ago is buried under a kilometer of sedimentary rock in the Yucatan, so it is difficult to trace on the surface. (B) However, geological maps of subsurface features clearly trace the edge of a 180-km (110-mile) diameter impact crater. (C) A depiction of the moment of impact by NASA artist Don Davis.

Astronomers have recently identified a family of asteroids that appears to be the probable source of the dinosaur killer (Unit 43.4).

Concept Question 3

Many other major extinction events have been found in much more ancient geological strata. Given Earth's geological activity, how likely is it that the craters from the impacts that might have caused these events can be found?

that the biosphere would be devastated, leading to mass extinctions, just as the fossil record shows.

This frightening picture is supported by the finding that this geological layer contains soot that could have been produced by the blast and by burning, as well as a layer of tiny quartz pellets that could have been created by the melting of surface rocks during a violent impact. Material dating to this time was found at high elevations on islands in the Caribbean, most likely deposited by tsunamis more than a kilometer high, and this focused scientists' search for a large impact in this region. Scientists have now identified a buried crater 180 kilometers (110 miles) wide, dated to 65 million years ago, in the Yucatán region of Mexico near **Chicxulub** (pronounced *cheek-shoo-loob* and meaning "flea of the devil" in Mayan). The Chicxulub crater rim can be clearly seen in a *gravity map* made by an oil company to look for subsurface density variations (Figure 50.11).

The Cretaceous mass extinction may have played an especially important role in our own evolution. Before that event, reptiles were the largest animals on Earth. Subsequently, mammals have assumed that niche. Small mammals may have escaped the fury of the heat and acid rain by remaining in burrows, and they may have survived the subsequent cold by virtue of their fur. You may be running your fingers through your hair, rather than your claws across your scales, because of that impact.

KEY POINTS

- Earth and the other planets continue to be struck by pieces of debris, often produced when asteroids collide and break apart.
- A larger chunk of rock and iron from the asteroid belt may survive its fiery trip to the ground to become a meteorite.
- Some meteorites appear never to have been melted, and contain chondrules that were grains of material in the solar nebula or carbonaceous materials, probably from the outer asteroid belt.
- Some meteorites come from bodies that grew large enough to differentiate, making an iron core and crustal rocks.
- The impact of a body produces a blast that releases the kinetic energy of its speed relative to Earth.
- At orbital speeds, impacts of rocky meteoroids typically produce craters about 10 times larger than the impacting body.
- Several impact events over the last century have exceeded the power of nuclear weapons used in World War II.
- Craters on Earth are often buried or eroded, but many have been identified that indicate major impacts.
- Iridium is so dense that almost all of it differentiated to Earth's center when the Earth formed, but high levels of iridium are found in the sedimentary layer above the last large dinosaur fossils.
- The source of the iridium appears to have been an asteroid about 10 km in diameter that struck Earth about 65 million years ago.
- A major impact event can generate widespread fires from molten debris, cause global cooling from dust blasted into the atmosphere, followed by global warming from the carbon dioxide released.

KEY TERMS

achondrite, 395
Barringer crater, 396
carbonaceous chondrite, 394
Chicxulub, 400
chondrite, 393
chondrule, 393
Cretaceous period, 399
meteor, 392
meteorite, 392
meteoroid, 392
organic compound, 394
Tunguska event, 397

CONCEPT QUESTIONS

Concept Questions on the following topics are located in the margins. They invite thinking and discussion beyond the text.

1. Chondrules in comets. (p. 393)
2. A single large impact versus many smaller ones. (p. 397)
3. Likelihood of finding large impact craters. (p. 400)

REVIEW QUESTIONS

4. What are the differences between a meteor, a meteoroid, and a meteorite?
5. What are chondrules, and what do they indicate about conditions in the solar nebula?
6. What kinds of meteorites come from differentiated asteroids?
7. How do we know that asteroids are similar to some meteorites in composition?
8. What evidence is there that the Earth has been hit by asteroids or comets?
9. Why do some scientists think that asteroids and comets played a role in mass extinctions?
10. How might meteorites be involved in the formation of life on Earth?

QUANTITATIVE PROBLEMS

11. Each day about 100 tons (10^5 kg) of meteors enter the Earth's atmosphere. What fraction of Earth's current mass does this represent?
12. Use the formula for the kinetic energy of a moving body to estimate the energy of impact of a 10^6-kg (1000 metric ton) object hitting the Earth at 30 km/sec. Express your answer in kilotons of TNT, using the conversion that 1 kiloton is about 4×10^{12} joules. *Note:* Be sure to convert kilometers/second to meters/second.
13. Suppose a major impact blasted out rock from a 100-km-diameter crater that was, on average, 1 km deep. If all that material became dust that settled out evenly over the entire surface of the Earth, how thick a layer would it form?
14. There is about a 1 in 45,000 probability that the potentially hazardous asteroid named Apophis will impact Earth in 2036. Based on observations, the asteroid is estimated to have a mass of 2.1×10^{10} kg, a diameter of 0.25 km, and a density of 2.6 kg/L. It has been estimated that its impact velocity would be 12.6 km/sec. If the mass may be uncertain by a factor of 3 and the impact velocity uncertain by a factor of 2, what is the range of predicted impact energy in joules?
15. The "rule of thumb" ratio for crater diameter versus impactor diameter is 10:1—the crater's diameter will be 10 times that of the impactor. If there is an uncertainty in the mass of Apophis of a factor of 3, what is the uncertainty in the asteroid's diameter? What range in crater size does this imply?
16. It is most likely that an asteroid will impact in an ocean rather than on land. Here is a simplified formula that gives the height of the wave (h), in meters, at a distance of 1000 km from the impact site:

$$h = 8 \times \sqrt{\left(\frac{D}{406}\right)^3 \times \left(\frac{V}{20}\right)^2 \times \left(\frac{\rho}{3}\right)},$$

where D is the asteroid diameter in meters, V is the velocity in km/sec, and ρ is the asteroid density in kg/L. Calculate the expected tsunami height in meters at 1000 km from the impact if Apophis happens to land in the middle of an ocean. How would you expect the tsunami height to vary with distance from the point of impact?

TEST YOURSELF

17. The bright streak of light we see as a meteoroid enters our atmosphere is caused by
    a. sunlight reflected from the solid body of the meteoroid.
    b. radioactive decay of material in the meteoroid.
    c. a process similar to the aurora that is triggered by the meteoroid's disturbing the Earth's magnetic field.
    d. frictional heating as the meteoroid speeds through the gases of our atmosphere.
    e. the meteoroid's disturbing the atmosphere so that sunlight is refracted in unusual directions.
18. If the Earth were struck by an asteroid 10 km in diameter, which of the following would have the most serious consequences for life on Earth?
    a. The huge crater produced by the collision
    b. The atmospheric shock wave
    c. The dust cloud raised by the impact
    d. Lava released from the Earth's interior through the crater
    e. Radioactive substances brought by the asteroid
19. Under what circumstances are you most likely to come across a meteoroid?
    a. While traveling on a spaceship
    b. While dog sledding across Antarctica
    c. While submerged in a submarine
    d. While walking through a dense marsh
20. The crater associated with the dinosaur mass extinction at the end of the Cretaceous period is located
    a. in Arizona.
    b. below the mouth of the Chesapeake Bay.
    c. in the Tunguska region of Russia.
    d. off the Yucatan peninsula.

UNIT 51

The Sun, Our Star

51.1 The Surface of the Sun
51.2 Pressure Balance and the Sun's Interior
51.3 Energy Transport
51.4 The Solar Atmosphere
51.5 Solar Seismology

Learning Objectives

Upon completing this Unit, you should be able to:
- Describe the surface features and the properties of the layers within the Sun and in its atmosphere.
- Explain hydrostatic equilibrium and its implications for the internal temperature and density of the Sun.
- Explain how magnetic fields heat the outer atmosphere of the Sun.

The Sun is a star—a dazzling, luminous ball of ionized gas. It is so large that a million Earths would fit inside its volume, and it is about 300,000 times more massive than the Earth. All stars generate power in enormous quantities, lighting up the universe despite their immense distances from one another. For centuries, the source of the Sun's and other stars' power was one of astronomy's greatest mysteries. Solving this mystery requires us to understand almost incomprehensible temperatures and pressures deep in a star's interior.

In its core, the Sun releases the energy of 100 billion atomic bombs every second; yet the Sun does not blow itself apart because this enormous power is counterbalanced by an enormous gravitational pull (Unit 16). Stars pit titanic forces against one another. If these forces somehow became unbalanced, stars would blast themselves apart or collapse in on themselves. Yet the Sun has remained stable for billions of years. Understanding this balance and the imbalances that sometimes occur is one of the main themes of Part Four of this book.

In this Unit we explore the structure of our star and the methods astronomers use to study parts of the Sun that cannot be seen directly. In Units 52 and 53 we will examine the source of energy in the Sun and the activity that occurs on its surface. In later Units we will discover that stars cannot maintain forever their precarious balance between gravity and pressure; and when the balance fails, cataclysmic events result. Along the way, we will learn of the likely fate of our own star.

51.1 THE SURFACE OF THE SUN

Although the Sun exhibits some features in visible light (Figure 51.1A), observations at other wavelengths reveal that the Sun's surface is violently agitated, with enormous storms blasting out fountains of incandescent gas (Figure 51.1B). A single solar storm may release as much energy in a few minutes as the combined energy of all the earthquakes on Earth over the last 10 million years, and such storms are not rare. Yet despite their vast energy, these storms are dwarfed by the Sun's total flow of energy. Every second the Sun radiates a hundred times more energy than a single one of its solar storms.

When we look at the Sun, we see through the low-density, tenuous gases of its outer atmosphere. Our vision is ultimately blocked, however, as we peer deeper into the Sun, because the material there is compressed to higher densities by the weight of the gas above it. In this dense material, the atoms are sufficiently close

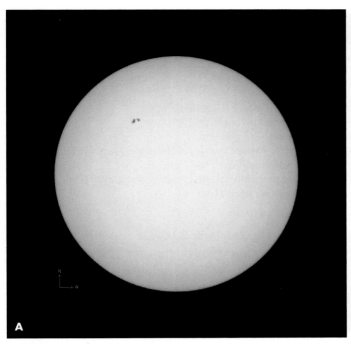
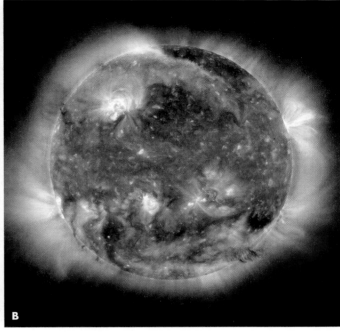

FIGURE 51.1
Images of the Sun made at the same time by the *SOHO* satellite in (A) visible and (B) ultraviolet light. The complex stormy surface of the Sun is revealed in the false-color ultraviolet image. Higher-energy photons are shown in blue, and lower-energy photons are shown in red.

FIGURE 51.2
A cutaway sketch of the Sun.

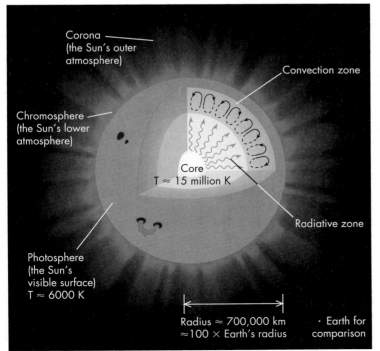

together that they strongly absorb the light from deeper layers, blocking our view of them much as fog obscures our view of what lies beyond a certain distance.

The Sun's "surface" is about 700,000 kilometers (430,000 miles) from its center. This is not a surface in the usual sense of the word. The glowing gas there is thin—only about one thousandth the density of the air we breathe. However, this region is where light can escape into space as the sunlight we see. This layer, where the Sun's gas changes from opaque to transparent—where the photons we see come from—is called the **photosphere.** The photosphere looks like a surface, but it is actually a region about 500 kilometers (300 miles) thick, with gas that grows denser below and thinner above. The location of the photosphere is illustrated in Figure 51.2—this diagram also depicts the other parts of the Sun discussed throughout the rest of this Unit.

From the upper part of the photosphere to its lowest part, the gas density increases by about a factor of 10. If we could see farther into the Sun, we would find that its density rises steadily toward its center, compressing the gas, much as in a pile of laundry the clothes on the bottom are flattened most, while those on top remain puffed up. A similar compression occurs in the atmosphere of the Earth and other planets, but the greater mass of the Sun leads to a vastly greater compression of its gas. Not only does the density of the Sun rise as we plunge deeper into its interior, the temperature rises too. The average temperature of the photosphere is about 5800 K; but at its top, the photosphere's temperature is about 4500 K, while near the bottom, the temperature is about 7500 K.

TABLE 51.1 Properties of the Sun

Property	Value	Method of Determination
Distance	1 AU = 150 million kilometers or 93 million miles	Triangulation or radar (Unit 54)
Radius	7×10^5 km	From angular size and distance (Unit 10)
Mass	2×10^{30} kg	Modified from Kepler's third law (Unit 17)
Average density	1.4 kg/L	From radius and mass
Central density	160 kg/L	Indirectly from need to balance internal pressure and gravity (Unit 51)
Surface temperature	About 5800 K	Color–temperature relation (Wien's law, Unit 23)
Central temperature	15 million K	Indirectly from need to balance internal pressure and gravity (Unit 51)
Composition	71% hydrogen, 27% helium, and 2% vaporized heavier elements such as oxygen, carbon, and iron	Spectra of gases in surface layers (Unit 56)
Luminosity	4×10^{26} watts	From amount of energy reaching Earth and inverse-square law (Unit 55)

Deep in the Sun, the matter is compressed to densities far greater than rock or iron in the Earth. Despite this great density, *the Sun is gaseous throughout* because its high temperature gives the atoms so much energy of motion that they are unable to bond with one another to form a liquid or solid substance. In fact the temperatures are so high that the atoms are ionized, and therefore the matter is properly termed a **plasma.**

Before we discuss how the Sun can remain so hot and dense, we need to understand better some of its overall properties. What forces are at work in its interior? How rapidly does it lose energy? What resources does it have available to supply its energy needs?

Table 51.1 lists a number of the Sun's properties and shows how astronomers measure them. Some of these properties can be calculated in a straightforward way, but others, such as the Sun's internal temperature and density, cannot be observed directly and therefore must be deduced. Astronomers use the known physical properties of gases and computer models to determine these, as we discuss next.

51.2 PRESSURE BALANCE AND THE SUN'S INTERIOR

The Sun dwarfs the Earth and even Jupiter, and its immense mass is what drives it to shine. The gravitational force (Unit 16) of this much concentrated matter compresses the gas in its interior to extremely high temperatures and densities. To balance that crushing force and prevent its own collapse, the Sun must maintain a very high pressure within the gas that pushes outward against gravity's inward pull.

Pressure is the force produced by the particles in a gas as they collide and bounce off each other (or the walls of a container). For example, a balloon is kept inflated by trillions upon trillions of molecules bouncing off the balloon's inside surface, each giving a tiny "punch" outward. Pressure and gravity must be in balance in the Sun, or else it would contract or expand. This balance between pressure and gravity is called **hydrostatic equilibrium.**

In the Sun, as in virtually all stars and planets, the balance of hydrostatic equilibrium requires that at every point, the outward force created by pressure exactly balances the inward force due to the inward gravitational pull—the weight—of material above that point (Figure 51.3). If the forces were unbalanced, the gases would begin flowing, and within minutes the Sun's size would be altered. Newton's law of gravity (Unit 16) determines the gravitational pull. To understand the pressure balance in the Sun, we need to discuss in more detail how its pressure arises.

Gas pressure

Pressure is measured in terms of the force exerted by a gas over an area. This can be expressed, for example, as newtons per square meter or pounds per square inch (see Unit 3).

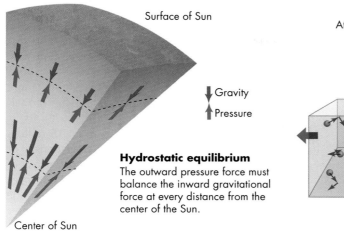

FIGURE 51.3
Hydrostatic equilibrium in the Sun. The pressure (blue arrows) and gravitational force (purple arrows) balance at every radius within the Sun, so there is no net force pushing the gas in or out.

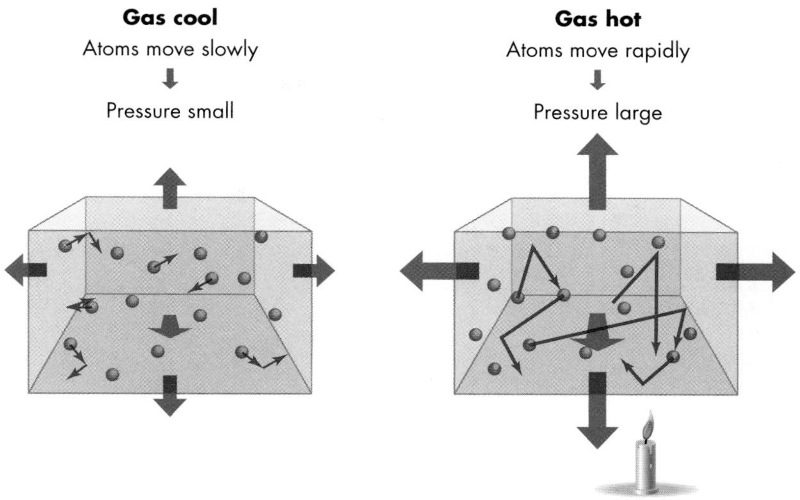

FIGURE 51.4
Sketch illustrating the ideal gas law. Gas atoms move faster at the higher temperature, so they collide both more forcefully and more often than atoms in a cooler gas. Therefore, other things being equal, a hotter gas exerts a greater pressure.

Pressure in a gas comes from collisions among its atoms and molecules. For example, if you squeeze a balloon, the atoms are pushed closer to each other. They collide and rebound more frequently, raising the pressure. When you squeeze a balloon, you experience this as an increasing resistance the harder you squeeze; or perhaps as you squeeze it, a section of the balloon you are not pressing may stretch outward in response to the increased pressure. Temperature also plays an important role. You can raise the pressure in a balloon by heating the air inside it, as in a hot-air balloon. Raising the temperature of the gas speeds up the atoms, making them collide harder and more often (Figure 51.4). Thus, the strength of the pressure is proportional to the density times the temperature of the gas:

$$\text{Pressure} = \text{Constant} \times \text{Temperature} \times \text{Density}.$$

The relationship is known as the **ideal gas law.**

The ideal gas law tells us how a gas responds to a change in any of the three variables in the formula—the other variables must change so that the equation remains true. Some simple examples illustrate the power of this law. For example, if the pressure in a gas is doubled, the product of the density and temperature must also double in response. In some situations, external conditions might hold the pressure steady. In that case if the temperature doubles, the gas must expand until it reaches half its former density. This is how a hot-air balloon works—as it is heated, the gas in the balloon expands until it is in pressure balance with the outside air. The hot air inside the balloon is less dense than the outside air, making the balloon buoyant.

The inward force of gravity grows greater closer to the center of the Sun, so the pressure must increase correspondingly to support the weight of the material above. An enormous pressure is needed at the center of the Sun to support the crushing force of gravity inward. The ideal gas law alone cannot tell us whether it is temperature or density that generates the necessary pressure, but astronomers can also examine other properties to determine this. Astronomers use computers to model not only how gases respond in terms of the ideal gas law, but also to consider how the Sun's internal gravity changes based on the density structure in the interior of the star, and how the temperature structure responds to the way the energy travels out from the center. The computer models divide up the Sun's interior into a series of layers like the layers of an onion. The weight of each layer on the layers below it can be calculated to find the pressure needed to balance that weight.

Mathematical Insight

In chemistry, the ideal gas law is often written $PV = nRT$, where P stands for pressure, V stands for volume, n stands for the number of atoms, R is a constant, and T is the temperature. Because n/V is a measure of the density, the equation can be rearranged as shown in the text.

Concept Question 1

How do you suppose the interior conditions in a very massive star would differ from the conditions in the Sun?

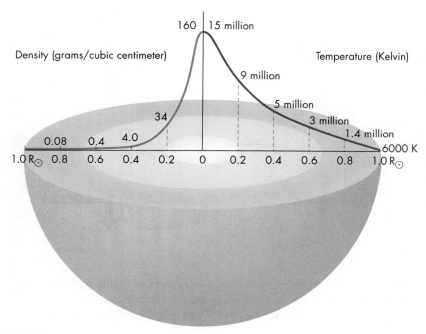

FIGURE 51.5
Plots of how density and temperature change throughout the Sun.

The model is constrained by a variety of requirements. For example, the sum of all the layers must add up to the total mass of the Sun, and the temperatures must be consistent with the flow of energy seen coming out of the Sun.

Below the photosphere, heat is retained more effectively by the denser gas, and the models show that the temperature climbs until at its core the temperature soars to about 15 million K. And while the Sun has an average density of about 1.4 kilograms per liter (close to that of Jupiter), near its core the Sun's density reaches 160 kilograms per liter—about 20 times the density of steel! Figure 51.5, based on computer models, illustrates how the temperature and density change throughout the Sun.

No space probe is ever likely to make direct measurements of the Sun's interior, but we are confident that the computer models are correct. It is true that the same pressure could be produced at the center of the Sun if the temperature were 10 times lower and the density 10 times higher. However, a model that assumes that the Sun is 10 times more dense would also predict a higher mass, and an assumption that the temperature is 10 times lower would predict a lower energy output. Even the composition of the Sun's core used in the computer models will result in slightly different predictions for the radius and surface temperature of the Sun, so astronomers have high confidence in the accuracy of the models.

51.3 ENERGY TRANSPORT

The high temperature in the Sun's core requires a powerful energy source because energy is constantly being lost from the Sun as it radiates energy out into space. We experience the Sun's lost energy as sunshine. Sunshine is essential for life on Earth, but it is a death warrant for the Sun. The Sun is gradually consuming itself to replace the lost energy, and thereby maintain the high temperature and pressure in its core. When its fuel is used up, the Sun must collapse—a fate facing all stars.

Your own experience and experiments in the laboratory show that heat always flows from hot to cold. Applying this principle to the Sun, we can infer that because its core is hotter than its surface, heat will flow outward from the Sun's center. Near the core, the energy is carried by photons through what is called the **radiative zone,** illustrated in Figure 51.2. Because the gas there is so dense, a photon travels less than a centimeter before it is absorbed by an atom and stopped. The photon is quickly reemitted, but it leaves the atom in a random direction—perhaps even back toward the Sun's center—and then is almost immediately reabsorbed by another atom. This constant absorption and reemission slows the rate at which photons escape from the Sun, like people lost in a dense forest walking randomly between trees until eventually they wander close enough to the edge of the forest to see out. Even though photons move at the speed of light between absorptions, it takes them over 10 million years, on average, to travel from the core to the surface. Today's sunshine was born in the Sun's core before the birth of human civilization!

Concept Question 2

In your experience, where else does radiative energy transfer occur? Where else does convection occur?

FIGURE 51.6

Image of a portion of the Sun's surface showing the granulation on the surface. This image was made using a telescope with adaptive optics to capture high-resolution details. The inset black-and-white image shows a region where analysis has found the height of the granulation features. The triangles represent the heights of these features, which are from 200 to 450 kilometers tall.

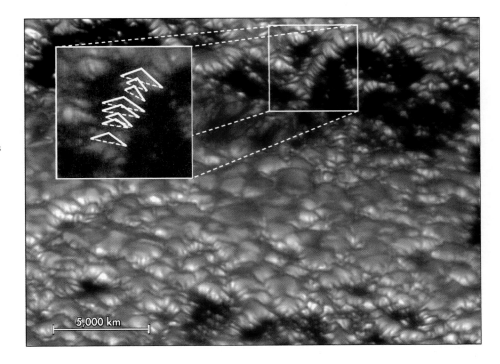

The outward movement of photons is slowed even more in a region that begins about two-thirds of the distance out from the Sun's center. In this region the gas is cooler and the atoms are less completely ionized than in the radiative zone; as a result the atoms are more effective at blocking photons. The energy from the Sun's interior is therefore trapped, and it heats the gas there. Hotter regions in the gas expand, become buoyant, and rise like a hot-air balloon. The gas rises and cools, then sinks back down in a region called the **convection zone** (Figure 51.2).

We can infer the gas's motion in the convection zone by observing the top layer of it—the photosphere. There we see numerous bright regions surrounded by narrow darker zones called **granulation,** as shown in Figure 51.6. The bright areas are bubbles of hot gas, up to about 1000 km (600 miles) across, which rise up through the convection zone. Upon reaching the surface, these hot bubbles radiate their heat to space, causing them to cool. That cooler matter then sinks back toward the hotter interior, where it is reheated and rises again to radiate away more heat. These bubbles of hot gas generally last only 5 to 10 minutes before they sink back into the roiling depths below the visible photosphere.

51.4 THE SOLAR ATMOSPHERE

Astronomers refer to the lower-density gases that lie above the photosphere as the Sun's *atmosphere*. This region marks a gradual change from the relatively dense gas of the photosphere to the extremely low-density gas of interplanetary space. A similar transition occurs in our own atmosphere, where gas density decreases steadily with altitude and eventually merges with the near-vacuum of space.

The Sun's atmosphere consists of two main regions. Immediately above the photosphere lies the **chromosphere,** the Sun's lower atmosphere. It is usually invisible against the glare of the photosphere, but can be seen during a total eclipse of the Sun as a thin red zone around the Sun (Figure 51.7) that is about 2000 kilometers (1200 miles) thick.

FIGURE 51.7
Photograph of a portion of the solar chromosphere during a total solar eclipse.

FIGURE 51.8
Spicules in the chromosphere imaged in the wavelength of hydrogen's H-alpha spectral line. The spicules are the thin, stringy features that look like tufts of grass.

The chromosphere's red color is the source of its name, which literally means "colored sphere." The color comes from the strong red emission line of hydrogen, H-alpha (Unit 24). Telescopic views reveal that the chromosphere contains hundreds of thousands of thin columns or spikes called **spicules** (Figure 51.8). Each spicule is a jet of hot gas that grows during several minutes to be thousands of kilometers tall before it cools and sinks back to the surface.

Astronomers can determine a great deal about a gas by studying the emission lines it produces (Unit 24), and in the chromosphere the emission lines reveal a surprising reversal in the gas temperature. By the top of the photosphere, the temperature has declined to about 4500 K, but in the chromosphere the temperature begins climbing again. At the top of the chromosphere, only about 2000 kilometers above the photosphere, the temperature reaches 50,000 K. Above this, in a thin transition region at the top of the chromosphere, the density drops rapidly and the temperature shoots up to about 1 million Kelvin as we enter the **corona,** the Sun's outer atmosphere.

The corona's extremely hot gas has such low density that under most conditions we look right through it. But like with the chromosphere, we can see it during a total solar eclipse when the Moon covers the Sun's brilliant disk. Then the pale glow of the corona can extend far beyond the Sun's edge to distances several times larger than the Sun's radius (Figure 51.9). Because the corona is so tenuous, it contains little energy despite its high temperature. It is like the sparks from a Fourth of July sparkler: Despite their high temperature, you hardly feel them if they land on your hand because they are so tiny and carry little total heat.

The corona is strongly influenced by the Sun's magnetic field. Scientists do not fully understand how the Sun generates its magnetic field, but magnetic fields

Concept Question 3

What are the speeds of winds in the most severe storms on Earth? How do these compare with the speeds of the motion in spicules?

FIGURE 51.9
Photograph of the corona during a total eclipse of the Sun in 1988.

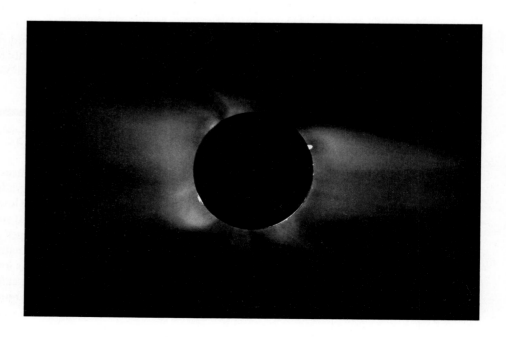

FIGURE 51.10
Magnetic loops in the Sun's lower atmosphere. Gas trapped along the loops is heated by magnetic activity and in turn heats the Sun's corona. This image was made by the *SOHO* satellite at ultraviolet wavelengths.

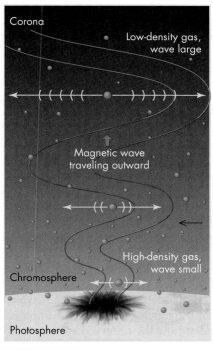

FIGURE 51.11
Diagram illustrating how magnetic waves (blue) heat the Sun's upper atmosphere. As the waves move outward through the Sun's atmosphere, they grow larger, imparting ever more energy to the gas (red dots) through which they move, accelerating and thereby heating it

arise around almost all spinning bodies that have electrically conductive fluids circulating in their interiors. For example, Earth's magnetic field arises in its liquid iron core through the *magnetic dynamo* process (Unit 37.4). The Sun is made of plasma that conducts electricity, and the circulation of this gas inside the Sun generates a magnetic field.

As we will discuss in greater depth in Unit 53, the Sun's magnetic field varies in strength and intensity, and this changes the appearance of the corona. This is because magnetic fields exert a force on charged particles, such as the ions and electrons in a plasma, that tends to make them follow the direction of the field when they move. Pictures of the Sun made at X-ray wavelengths show streamers in the corona that follow the direction of the Sun's magnetic field (Figure 51.10).

Studies of the ionized gas in the chromosphere and corona show that it is stirred by the magnetic fields, and this may cause the heating of these regions. An analogy may help you understand how magnetic waves can heat a gas. When you crack a whip, a motion of its handle travels as a wave along the whip. As the whip tapers, the wave's energy of motion is transferred to an ever smaller piece of material. Having the same amount of energy but with less mass to move, the tip accelerates and eventually breaks the sound barrier. The whip's "crack" is a tiny sonic boom as the tip moves faster than the speed of sound—about 1200 kilometers per hour (about 700 miles per hour).

A similar speedup occurs in the Sun's atmosphere when magnetic waves, formed in the turbulent photosphere, move into the corona along the Sun's magnetic field lines. As the atmospheric gas thins, the wave energy is imparted to an ever smaller number of atoms, making them move faster, as illustrated in Figure 51.11. Solar physicists find that turbulent motions in the photosphere can twist and shake the magnetic field. These disturbances travel up the magnetic field resembling the motion along a whip. In the thin coronal gas the ions are whipped around by the magnetic field producing collisions in the gas that heat it.

The corona contains large cooler regions, called **coronal holes,** where the magnetic field is weak. The corona's high temperature generates pressures high enough to drive gases away from the Sun in the coronal holes because the magnetic field does not trap the gas. The flow of gas away from the Sun is known as the **solar wind,** which results in a gradual loss of mass from the Sun. The expanding gas has a very low density (only a few hundred thousand atoms in a liter, compared to about 10^{22} molecules in a liter of the air we breathe). The amount of material lost from the Sun is "small": about 1.5 million tons each second. This number sounds large, but even after 10 billion years this would amount to only about 0.02% of the Sun's mass.

Near the Sun, the pressure within the corona may be only slightly larger than the Sun's gravitational attraction. The solar wind therefore begins its outward motion slowly. The gravitational pull of the Sun weakens with distance, but the pressure in the corona remains relatively high, so the solar wind gradually accelerates. At the distance of the Earth's orbit, the solar wind speed reaches about 400 kilometers per second (250 miles per second). The flow of the solar wind pushes comets' tails away from the Sun (Unit 49) and interacts with the Earth's magnetic field and ionosphere to create the aurora (Unit 38).

Beyond the Earth, the solar wind coasts at this high speed until it collides with the interstellar gas surrounding the Solar System (Figure 51.12). Here the "bubble" of outflowing solar wind is rapidly slowed at the interface where the gases collide and heat. The *Voyager 1* spacecraft crossed into this boundary region in 2004 and *Voyager 2*, on a slightly slower trajectory, crossed it in 2007, at distances of about 94 and 84 AU from the Sun, respectively—roughly three times Neptune's distance. Rapid changes in cosmic ray levels detected by *Voyager 1* in August 2012 seem to indicate that the spacecraft passed from this boundary region into the denser surrounding interstellar gas when it was about 122 AU from the Sun.

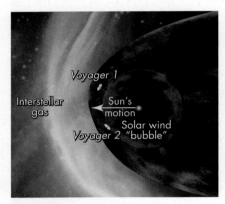

FIGURE 51.12
Diagram illustrating where the solar wind impinges on the interstellar gas and the trajectories of the two *Voyager* spacecraft. The Sun is surrounded by a "bubble" of solar wind gas that pushes into the interstellar gas, producing a bow wave in front of the Sun's direction of motion.

51.5 SOLAR SEISMOLOGY

Just as scientists can study the Earth's interior by analyzing earthquake waves, or waves in a stream reveal the presence of submerged rocks, astronomers can learn about the Sun's interior by analyzing the waves that travel through the Sun. By analogy with the study of such waves in the Earth, astronomers call the study of such waves in the Sun **solar seismology.**

The study of waves in the Sun began in the 1960s, when astronomers noted that the photosphere of the Sun vibrates, oscillating up and down by several meters over periods of several minutes. These oscillations are similar to earthquake waves on the Earth (Unit 37). However, instead of arising from the shifting of rocky material, as in the Earth, waves in the Sun arise from huge convecting masses of material rising and dropping in its outer regions. These motions in turn generate waves that travel through the body of the Sun, causing other portions of the Sun to shake in response.

The rising and falling surface gas makes a regular pattern (Figure 51.13), which can be detected as a Doppler shift (Unit 25) of the moving material. Astronomers next use computer models of the Sun to predict how the observed surface oscillations are affected by conditions in the Sun's deep interior. With this technique, astronomers can measure the density and temperature deep within the Sun. The results provide an independent confirmation of our computer models of the Sun's interior.

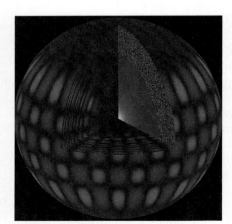

FIGURE 51.13
Computer diagram of solar surface waves.

KEY POINTS

- The Sun is a star, generating enormous amounts of energy in its core. That energy is emitted into space in the form of light.
- The Sun is entirely composed of a gas of ionized atoms, a plasma. Its density at its core is much greater than that of iron, and is a near vacuum in its outermost layers.
- The "surface" of the Sun, the photosphere, is the layer where densities have dropped low enough for visible photons to escape.
- The weight of overlying layers compresses gas in the Sun's interior until its pressure rises enough to counterbalance the weight.
- Pressure in a gas depends on the product of the density and temperature, which both grow higher closer to the Sun's center.
- Energy generated in the Sun's core travels outward, as radiation in the inner parts of the Sun, then by convection nearer the surface.
- Above the photosphere, the chromosphere and corona are hotter but have much lower density and produce little visible light.
- Magnetic fields interact with electrically charged particles, and movements of the magnetic fields appear to heat the outer layers.
- The solar wind is gas driven outward that escapes to deep space.
- The Sun's internal structure can be studied by observing seismic motions on the surface.

KEY TERMS

chromosphere, 407
convection zone, 407
corona, 408
coronal hole, 409
granulation, 407
hydrostatic equilibrium, 404
ideal gas law, 405
photosphere, 403
plasma, 404
pressure, 404
radiative zone, 406
solar seismology, 410
solar wind, 410
spicule, 408

CONCEPT QUESTIONS

Concept Questions on the following topics are located in the margins. They invite thinking and discussion beyond the text.

1. Differences in the interiors of more massive stars. (p. 405)
2. Places with radiative and convective energy transport. (p. 406)
3. Comparing speeds of atmospheric motions on Earth. (p. 408)

REVIEW QUESTIONS

4. How big is the Sun compared to the Earth?
5. What are the photosphere, chromosphere, and corona?
6. How do densities and temperatures vary throughout the Sun? What is hydrostatic equilibrium?
7. What visible evidence do we have that the Sun has a convection zone?
8. Why is the corona hotter than the Sun's surface? Why is it said that the corona contains relatively little heat?
9. Why does light take so long to travel through the Sun's interior?

QUANTITATIVE PROBLEMS

10. The Sun's angular size is 0.5°, and its distance is 1.5×10^8 km. Use this information to calculate the Sun's diameter.
11. The central density of the Sun is about 160 kg/L. Estimate how much a chunk of this material the size of a sugar cube would weigh on Earth.
12. How much mass would the Sun have to lose each second in order to lose 10% of its mass in 10 billion years?
13. At the rate at which the Sun is losing mass in the solar wind, how many years does it take the Sun to lose as much mass as the mass of the entire Earth?
14. Wolf-Rayet stars have very strong stellar winds, losing as much as 10^{-5} $M_\odot$ per year. If a typical Wolf-Rayet star is 20 $M_\odot$, what percentage of its mass might it lose in a million years?
15. Use Wien's law to calculate the wavelength of highest emission for the bottom and top of the photosphere where the temperatures are 7500 K and 4500 K, respectively. What part of the electromagnetic spectrum do these wavelengths correspond to?
16. If a packet of gas rising through the interior of the Sun reduces its temperature from 1.5 million K to 10,000 K and has a density decrease from 120 kg/L to 5.5 kg/L, by what factor has the pressure on it changed?

TEST YOURSELF

17. What region of the Sun is considered its surface?
    a. Radiative zone
    b. Convection zone
    c. Photosphere
    d. Chromosphere
    e. Corona
18. According to the ideal gas law, if the temperature of a gas is made 4 times higher, which of the following is a possible result?
    a. Its pressure increases by 4 times and its density remains the same.
    b. Its density increases by 4 times and its pressure remains the same.
    c. Its pressure and density both double.
    d. Its pressure increases by 4 times while its density decreases by 4 times.
    e. Its pressure and density both decrease by 2 times.
19. Which of the following is not a consideration when modeling the Sun's interior?
    a. The Sun's total mass
    b. The Sun's overall luminosity
    c. The Sun's average density
    d. The Sun's surface temperature
    e. All of the above contribute to modeling the Sun's interior.
20. Hydrostatic equilibrium in the Sun requires that the
    a. amount of hydrogen matches the energy output.
    b. pressure balances the overlying weight at each radius.
    c. temperature is higher where the density is lower.
    d. Sun's average density matches that of liquid water.
    e. flow of gas is limited by magnetic fields inside the Sun.

UNIT 52

The Sun's Source of Power

52.1 The Mystery Behind Sunshine
52.2 The Conversion of Hydrogen into Helium
52.3 Solar Neutrinos
52.4 The Fusion Bottleneck

Learning Objectives

Upon completing this Unit, you should be able to:
- Describe how fusion releases energy in a star.
- List the steps in the proton–proton chain, and the fraction of mass lost.
- Explain the role of neutrinos in fusion, and what observations of them indicate.
- Describe the role of each of the four forces in the production of energy in the Sun.

Energy that leaves the Sun's hot core eventually escapes into space as sunshine. That energy loss in the core must be replaced, or the Sun's internal pressure would drop and the Sun would begin to shrink under the force of its own gravity. The Sun is therefore like an inflatable chair with tiny leaks through which the air escapes. If you sit in the chair, it will gradually collapse under your weight unless you pump air in to replace that which is escaping. What acts as the energy pump for the Sun?

Although the immense bulk of the Sun hides its core from view, we have a number of clues to the possible sources of energy. From a combination of theoretical models and direct observations, we can deduce what the core's temperature and density must be (Unit 51). From the spectra of its atmospheric gases, we know the Sun's composition. From the amount of sunlight we receive, we know how much energy the core must be generating. Moreover, it has become possible in the last few years to measure subatomic particles called **neutrinos** (Unit 4) that the Sun emits. The new field of "neutrino astronomy" has begun to reveal much more about the internal workings of the Sun because it allows us to "view" reactions occurring deep in the Sun's interior more directly than ever before.

These astronomical discoveries, combined with the discovery of nuclear energy in the 1900s, lead us to a picture of almost unbelievable violence in the core of the Sun. Every second, more than 4 million tons of matter are annihilated—not just turned into vapor or ionized or ejected into space, but turned into light—so that the Sun's mass gradually declines. Over 40 million years the Sun obliterates a mass equivalent to the entire Earth! This loss may seem large, but the Sun has such a tremendously large mass that it can afford this extravagant destruction of its mass in its fight to keep from collapsing under its own weight.

52.1 THE MYSTERY BEHIND SUNSHINE

Early astronomers believed that the Sun might burn ordinary fuel such as coal. But even if the Sun were pure coal, it could burn only a few thousand years given its prodigious energy output. In the late 1800s another proposal was that the Sun is not in hydrostatic equilibrium, but that gravity slowly compresses it, making it shrink. In this hypothesis, compression heats the gas and makes the Sun shine, much as the giant planets generate heat in their interiors (Unit 45). However, gravity could power the Sun by this mechanism for only about 10 million years, and the Sun would have to be shrinking by about 10 kilometers each year, which modern observations rule out. Therefore, something else must supply the Sun's energy.

E = energy
m = mass
c = speed of light = 3×10^8 m/sec

Clarification Point

All reactions that release energy result in a reduction in mass, but except for nuclear reactions, the change in mass is a minuscule percentage.

Mathematical Insight

An explosion of 1 kiloton (1000 metric tons) of TNT releases about 4.2×10^{12} joules of energy.

We know the Sun has been shining with approximately the same luminosity for billions of years because we have geological and fossil evidence of water on Earth dating back that far. For Earth to have maintained a climate with liquid water on its surface, the Sun's power output must have remained fairly steady over that whole time. In 1899 T. C. Chamberlin suggested that subatomic energy—energy from the reactions of atomic nuclei—might power stars, but he could offer no explanation of how the energy was liberated. In 1905 Einstein proposed that energy might come from a body's mass. His famous formula,

$$E = m \times c^2,$$

states that a mass, m, can become an amount of energy, E, equal to the mass multiplied by the square of the speed of light, c. It is important to understand that this formula simply states a basic "exchange rate" between energy and mass, similar to the way we might convert between dollars and euros. Einstein's formula indicates how many joules (kg · m²/sec²) of energy (Unit 3) correspond to 1 kilogram of mass. The term c^2 in the equation is the exchange-rate factor.

For example, if you could convert 1 gram (10^{-3} kg) of mass—about the amount of matter in a paperclip—into energy, it would release an energy of

$$E = 0.001 \text{ kg} \times (3 \times 10^8 \text{ m/sec})^2$$
$$= 9 \times 10^{13} \text{ kg} \cdot \text{m}^2/\text{sec}^2$$

or 9×10^{13} joules. This is the equivalent of approximately 20 kilotons of TNT, about as powerful as the atomic bomb that destroyed Hiroshima. You could grind the same paperclip into fine metal powder and burn it—a chemical process. This would release less than one ten-billionth as much energy—about as much as burning a match. With the understanding that mass may be converted into energy, scientists realized that if the Sun can convert even a tiny fraction of its mass into energy, it would have an enormous source of power. Einstein's equation does not say what process might release so much energy, but it provides a clue where to look: reactions that result in a significant change in mass.

In 1919 the English astrophysicist A. S. Eddington, a pioneer in the study of stars, developed a hypothesis for a reaction that might provide a significant source of energy: the conversion of hydrogen into helium. Hydrogen contains one nuclear particle, while helium contains four nuclear particles; however, the mass of helium is significantly less than four times the mass of hydrogen. Eddington recognized that if hydrogen nuclei could be combined to make helium nuclei, it would involve the loss of enough mass to provide the energy necessary to power the Sun. It was not until the late 1930s that the German physicists Hans A. Bethe and Carl F. von Weizsäcker worked out the details of how such a conversion might take place. They showed that the Sun generates its energy by converting hydrogen into helium through a process called **nuclear fusion,** a process that bonds two or more atomic nuclei into a single heavier one.

Under normal conditions hydrogen nuclei repel one another, pushed apart by the electrical charges of their **protons.** Protons are positively charged nuclear particles, and like charges repel each other. This force of repulsion grows greater the closer together the protons get, so they must be moving toward each other at extremely high speeds for the nuclei to come into close contact, about 10^{-15} m apart. When this happens, the nuclear force of attraction—the **strong force** (Unit 4)—overcomes the electrical repulsion between the protons. The strong force can also bind protons together with another kind of nuclear particle, a **neutron,** which has no electrical charge but is otherwise similar to a proton. The strong force is about 100 times stronger than the electrical repulsion between neighboring protons, but it remains strong over only a very short distance, then dies away rapidly at larger distances.

Fusion is possible in the Sun because its interior is so hot. At very high temperatures—above about 5 million Kelvin—atomic nuclei move so fast that they

Nuclei move slowly and electric force repels them and pushes them apart. **No** fusion.

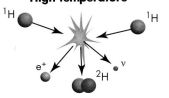

Nuclei move faster and electrical repulsion is overwhelmed. They collide and **fuse.**

FIGURE 52.1
High temperature acts to overcome the electrical repulsion of the nuclei and brings them close enough for the nuclear force to fuse them.

> **Concept Question 1**
>
> How do we know what the Sun is made of?

collide at speeds that bring them close enough together where the strong force overcomes the electrical repulsion and pulls them together (Figure 52.1). The two separate nuclei then "fall" together, somewhat like two large bodies pulled together by their mutual gravitational attraction. They crash into each other and release energy when they collide. Similarly the atomic nuclei merge, or fuse, into a single new nucleus, and the potential energy of the strong force is released in the form of a gamma-ray photon. Because the strong force is 10^{40} (ten thousand trillion, trillion, trillion) times stronger than the gravitational force, the energy released is immensely larger. But because this nuclear fusion process requires such a high temperature, the only place in the Sun hot enough for fusion to occur is its core.

52.2 THE CONVERSION OF HYDROGEN INTO HELIUM

Before we can fully understand how fusion creates energy, we need to look at the structure of hydrogen and helium as they are found in the Sun's core. At the high temperatures there atoms collide so violently that the electrons are stripped away from the nuclei; that is, the atoms are completely ionized. (See also Unit 4.2.) The common form of hydrogen has a nucleus consisting of a single proton, and that of helium consists of two protons and two neutrons (Figure 52.2).

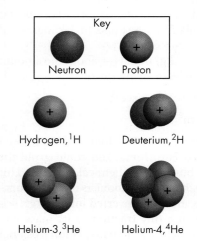

FIGURE 52.2
Schematic diagrams of the nuclei of hydrogen, its isotope deuterium, and two isotopes of helium.

Hydrogen and helium always have one and two protons, respectively, but they have other forms, **isotopes,** with different numbers of neutrons. To identify the isotopes, we write their chemical symbol preceded by a superscript that shows the total number of protons + neutrons. The usual form of hydrogen with one proton and no neutrons is written ^{1}H, whereas the form of hydrogen containing one proton and one neutron is ^{2}H. Most isotopes are not given separate names, but ^{2}H is so common that it has been given its own name: **deuterium.** This isotope of hydrogen is more common than all but about a half dozen elements in the universe.

The most common form of helium, with two protons and two neutrons, is written ^{4}He, whereas helium with two protons but only one neutron is ^{3}He. When speaking, we refer to these as "helium four" and "helium three," respectively. These isotopes of hydrogen and helium play a critical role in the energy supply of the Sun.

Hydrogen fusion in the Sun occurs in three steps called the **proton–proton chain.** In the first step, two ^{1}H nuclei (protons) collide and fuse to form the isotope of hydrogen, ^{2}H or deuterium. During the collision, one proton turns into a neutron through the weak force (Unit 4) and two other particles are ejected: a **positron** (the "antiparticle" of the electron, denoted as e^+) and a neutrino (denoted by ν, the Greek letter "nu"). This step is depicted in Figure 52.3A and can be written symbolically like this:

$$^1H + ^1H \rightarrow {}^2H + e^+ + \nu + \text{Energy.}$$

The terms to the left of the arrow are the normal hydrogen nuclei that start the process. The terms to the right are the deuterium, positron, neutrino, and the potential energy of the nuclear bond that is released.

An indication that energy is released by this reaction is that the mass of ^{2}H (and the by-products of the reaction) is less than the mass of the two ^{1}H that started the process. The amount of energy released can be found from Einstein's formula $E = m \times c^2$, an amount of energy that sends the particles off at high speeds. This sustains the high temperature in the core and ultimately ends up as thousands of visible-wavelength photons in the sunlight we see. The neutrinos produced in this reaction also carry away a small part of the nuclear energy, but they play no further role in the Sun's energy generation. We will encounter them again, however: They

Proton–proton chain

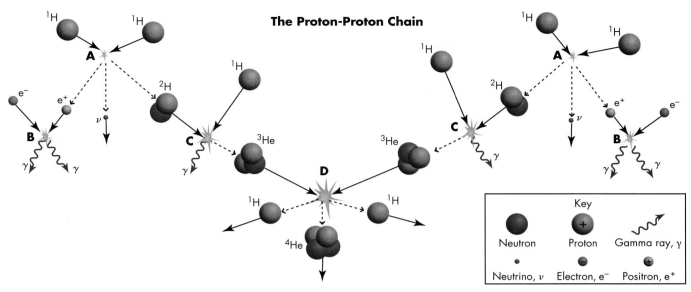

FIGURE 52.3
Diagram of the proton–proton chain. (A) Hydrogen (H) nuclei first combine to make deuterium (^{2}H). (B) The positron produced in the first step annihilates an electron. (C) Deuterium and hydrogen combine to make ^{3}He. (D) ^{3}He nuclei combine to form ^{4}He.

can be detected when they leave the Sun, helping astronomers to learn more about conditions in the Sun's interior.

Each positron produced in the first step of the proton–proton chain almost immediately annihilates one of the many electrons contained in the dense gas in the Sun's core. This creates two energetic photons (gamma rays). This side reaction, shown in Figure 52.3B, contributes even more energy than the fusion of protons into deuterium.

Once deuterium is created, the second step of the proton–proton chain proceeds very rapidly. The ^{2}H nucleus collides with one of the numerous hydrogen nuclei, ^{1}H, to make the isotope of helium containing a single neutron, ^{3}He. This process releases a high-energy gamma ray photon, denoted by the Greek letter gamma, γ. Figure 52.3C shows this step, which can be written as follows:

$$^1H + {^2H} \rightarrow {^3He} + \gamma + \text{Energy}.$$

Here again the resulting particle, ^{3}He, has a smaller mass than the particles from which it was made, and several times more energy is released than from the first reaction. Some of the energy becomes the kinetic energy of the particles, while the rest is carried away by the gamma-ray photon.

The third and final step in the proton–proton chain is the collision and fusion of two ^{3}He nuclei. Here the fusion results not in a single particle, but rather in one ^{4}He and two ^{1}H nuclei. You can think about this reaction as the attempt to form a nucleus with four protons and two neutrons, except that two protons are ejected by their electric repulsion, as shown in Figure 52.3D. This reaction, which releases about twice as much energy as the second reaction, is written as

$$^3He + {^3He} \rightarrow {^4He} + {^1H} + {^1H} + \text{Energy},$$

where again the final mass is less than the initial mass.

Scientists who want to understand the details of fusion in the Sun study each of the steps in the proton–proton chain individually; but for understanding the larger picture of the Sun's energy generation, we are concerned primarily with the overall energy production. We can find the quantity of energy released by comparing the initial and final masses of the reactions and using the mass–energy relationship $E = m \times c^2$. Steps 1 and 2 use three ^{1}H, but the first two steps must occur twice to make the two ^{3}He nuclei needed for the last step. Therefore, six ^{1}H nuclei are used, but two are returned in step 3, and so a total of four ^{1}H nuclei are needed to make each ^{4}He nucleus.

Concept Question 2

The term *solar energy* is also used to refer to the generation of power on Earth from sunlight. This is done using collectors that convert the energy of photons from the Sun into electrical current. What other kinds of power generation ultimately rely on energy that came from the Sun?

The mass of a hydrogen nucleus is 1.673 × 10^{-27} kilograms, whereas the mass of a helium nucleus is 6.645 × 10^{-27} kilograms. Comparing their masses, we find the following:

$$\begin{aligned} 4 \text{ hydrogen} &= 6.693 \times 10^{-27} \text{ kg} \\ -1 \text{ helium} &= -6.645 \times 10^{-27} \text{ kg} \\ \hline \text{Mass lost} &= 0.048 \times 10^{-27} \text{ kg} \end{aligned}$$

Multiplying the mass lost by c^2 gives the energy yield per helium atom produced. That is,

$$\begin{aligned} E &= 0.048 \times 10^{-27} \text{ kg} \times (3.0 \times 10^8 \text{ m/sec})^2 \\ &= 0.048 \times 10^{-27} \times 9.0 \times 10^{16} \text{ kg} \cdot \text{m}^2/\text{sec}^2 \\ &= 4.3 \times 10^{-12} \text{ joules per helium nucleus created.} \end{aligned}$$

To continue to power itself at its current rate, the Sun needs to generate 4×10^{26} joules of energy per second. This means that each second, about 4×10^{38} hydrogen nuclei are converted into 10^{38} helium nuclei, and about 4×10^9 kg—*4 million metric tons*—of mass are converted into energy each second. The total energy released is equivalent to exploding 100 billion-megaton H-bombs per second! Our sunshine has a violent birth.

52.3 SOLAR NEUTRINOS

We saw in the previous section that the Sun makes neutrinos as it converts hydrogen into helium. Neutrinos are unusual subatomic particles. They have no electric charge and only a tiny mass, and they travel at nearly the speed of light. This gives them phenomenal penetrating power. They escape from the Sun's core through its outer 700,000 kilometers, and into space like bullets through a wet tissue. They pass straight through the Earth and anything on the Earth, such as *you*, and keep going. In fact, several trillion neutrinos from the Sun passed through your body in the time it took you to read this sentence.

The number of neutrinos coming from the Sun is a direct indication of how rapidly hydrogen is being converted into helium. This is a more immediate measure of the rate of current nuclear reactions than the light we see, because photons take so long to work their way to space from the Sun's core as they interact with any intervening matter. Thus, if we can measure how many neutrinos come from the Sun, we can directly deduce the conditions in the Sun's core.

Counting neutrinos is extremely difficult. The elusiveness that allows neutrinos to slip so easily through the Sun makes them slip with equal ease through detectors on Earth. However, although they interact weakly, the rate of interaction is not zero. And because so many neutrinos are produced, we need detect only a tiny fraction of them to get useful information. Still, this requires very large detectors.

The detection of solar neutrinos was pioneered by the American physicist Raymond Davis Jr., who advanced the idea of placing tanks of material deep underground that could react with neutrinos from the Sun. Neutrino detectors are buried to shield them from the many other kinds of particles besides neutrinos that constantly bombard the Earth. For example, protons, electrons, and a variety of other subatomic particles shower our planet. These particles, traveling at nearly the speed of light, are called **cosmic rays** and are thought to be particles blasted across space by cataclysmic events, such as when a massive star explodes.

Cosmic rays can penetrate only a short distance into the Earth; so if a detector is located deep underground, nearly all the cosmic rays are filtered out. Neutrinos, on the other hand, are unfazed by a mere mile of solid ground. They have a small but

Raymond Davis and Japanese scientist Masatoshi Koshiba (who headed the first Kamiokande experiment) shared the 2002 Nobel Prize in Physics for their work on solar neutrinos.

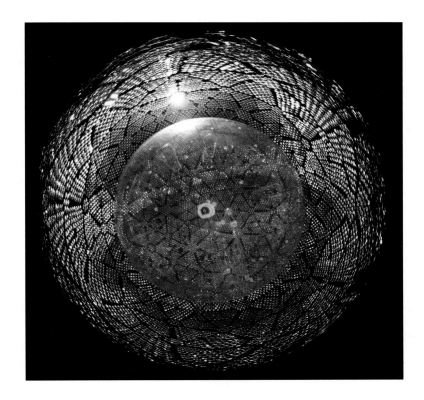

FIGURE 52.4
An inside view of the neutrino detector located in Sudbury, Canada. The sphere is 12 meters in diameter (about 40 feet). It is filled with heavy water, a form of water in which hydrogen atoms contain a neutron as well as the usual proton. When a neutrino strikes one of the neutrons, the neutron may break down into a proton and electron. As the electron streaks off, it produces a tiny flash of light. Ten thousand detectors (which form the grid visible around the sphere) record the emitted light, thereby allowing scientists to detect the neutrino's passage.

FIGURE 52.5
The first neutrino image of the Sun. This image was constructed from 1.3 years' worth of neutrino data collected by the Super-Kamiokande detector. The image covers about one quarter of the sky. Neutrinos come from a small spot at the center of the image, but uncertainties in the measurement of the paths of the neutrinos give rise to the extended shape.

predictable chance of interacting with the matter through which they pass, but that chance is so small that even if they encountered a wall of lead a light-year thick, most of the neutrinos would pass through!

Davis's experiment showed that the count of neutrinos was substantially lower than what physicists had predicted. However, his experiment relied on neutrinos produced in a side process of the proton–proton chain. Newer detectors are capable of detecting the neutrinos produced by the primary proton–proton interaction. Currently the largest neutrino "telescopes" are the Super-Kamiokande detector, located deep in a zinc mine west of Tokyo, and the Sudbury Neutrino Observatory (Figure 52.4) more than a mile underground in a nickel mine in northern Ontario, Canada. The Sudbury detector contains about 1000 tons of *heavy water,* water consisting of molecules in which one of the hydrogen atoms in the molecule is the isotope ^{2}H (deuterium). This heavy water gets its name because the extra neutron in the deuterium nucleus gives the water a slightly higher density.

The principle behind a neutrino telescope is that occasionally a neutrino collides with a neutron in the water, breaking the neutron into a proton and an electron. As the electron shoots off, it emits a tiny flash of light, which is recorded by photodetectors. These experiments detect solar neutrinos, and it is possible to reconstruct the approximate direction the neutrinos came from. This is shown in Figure 52.5—the first neutrino image of the Sun. Although the image is blurry, it confirms that the neutrinos are coming from the Sun. Like Davis's earlier experiment, these telescopes see only one-third the expected number of neutrinos. What does this low number mean? Is the Sun not fusing hydrogen into helium as predicted? Do neutrinos somehow escape detection? The solution to the puzzle offers us another chance to see the scientific process at work.

As a first step toward resolving the solar neutrino discrepancy, astronomers checked that their calculations for the predicted number of neutrinos were correct. All such checks led to roughly the same result, implying that there was no obvious flaw in their understanding of how the Sun works. Was it possible, then, that neutrinos had undiscovered properties that affected their detectability on Earth or their production in the Sun?

The penetrating power of neutrinos is the theme of a poem called "Cosmic Gall" by novelist and poet John Updike:

> Neutrinos, they are very small.
> They have no charge and have no mass
> And do not interact at all.
> The earth is just a silly ball
> To them, through which they simply pass,
> Like dustmaids down a drafty hall
> Or photons through a pane of glass.
> They snub the most exquisite gas,
> Ignore the most substantial wall,
> Cold-shoulder steel and sounding brass,
> Insult the stallion in his stall,
> And scorning barriers of class,
> Infiltrate you and me! Like tall
> And painless guillotines, they fall
> Down through our heads into the grass.
> At night they enter at Nepal
> pierce the lover and his lass
> From underneath the bed—you call
> It wonderful: I call it crass.

As scientists looked more closely at the properties of neutrinos, they realized that two other kinds of neutrinos were associated with different kinds of subatomic reactions (Unit 4). The nuclear fusion reactions in the Sun produce only one type of neutrino (called an *electron neutrino*), the type of neutrino that the early experiments were designed to detect. The standard theory held that the three kinds of neutrinos could not be converted into one another, but a new hypothesis suggested that they could. This hypothesis additionally predicted that the neutrinos have a mass, whereas the established model held that they had no mass. The new hypothesis suggested that the first experiments detected only one-third the expected number because the rest had been converted into the other types of neutrinos.

To test this idea, scientists built a second generation of neutrino detectors—the Sudbury detector is one of these—that could also detect the two other varieties of neutrino. By 2003, results from Sudbury clearly showed that the total number of neutrinos (of all three types) coming from the Sun agreed with the predictions of nuclear fusion rates. This gives astronomers a greater confidence in their understanding of the Sun. But in confirming the nuclear fusion model for energy generation in the Sun, they have helped to confirm a new hypothesis involving neutrinos, which modified the previous Standard Model of subatomic particles. This new work gives us evidence that neutrinos must have a small mass, and that the different varieties can be converted into each other.

52.4 THE FUSION BOTTLENECK

The fusion process in stars depends sensitively on all four of the fundamental forces. We have seen that gravity forces matter together and drives up its temperature to the point where protons can overcome the electromagnetic forces holding them apart. They then come so close to each other that the strong force can take hold. However, there is a problem with this process. Two protons cannot directly combine with each other because there is no stable isotope of helium with just two protons—no "^{2}He."

The first step of the proton–proton chain is actually a highly improbable interaction because in the extremely short moment of time while two protons are colliding and very close to each other, one of them must be transformed into a neutron. This is a process that requires the **weak force** (Unit 4), and such interactions are rare. On average, a particular proton in the Sun is likely to undergo this transformation while colliding with another proton only once in several billion years!

The conversion of a proton into a neutron *requires* energy:

$$^1\text{H} + \text{Energy} \rightarrow n + e^+ + \nu$$

where the "n" denotes a neutron. The amount of energy required for this process is large—more than half as much energy as is released when a proton and neutron combine to make deuterium. This presents a barrier that dramatically slows down the rate of fusion, although it does not prevent the process from occurring, because the overall process still has a net production of energy.

By contrast, the second and third steps of the proton–proton chain do not involve the weak force, and they typically occur just seconds after the first reaction. The weak force introduces a "bottleneck" in the rate of reaction. If the weak force were a bit stronger or weaker, stars would be very different than they are. They would burn at very different rates, or perhaps would not be able to undergo fusion at all. Thus, the weak force—as well as each of the other fundamental forces—is critical for making the Sun release its nuclear energy gradually over billions of years. This long time has allowed life to form and evolve on our planet.

Concept Question 3

Suppose the weak force were stronger so that weak interactions occurred 10 times faster. What consequences can you think of that this might have for the Sun and for the Solar System?

KEY POINTS

- Before nuclear energy was discovered, no source of energy was known that could have kept the Sun shining for billions of years.
- Einstein showed that mass can be converted into energy, producing a huge amount of energy for a small mass.
- The Sun must convert 4 million tons of matter into energy each second to replace the energy lost in emitting light.
- Nuclear fusion of hydrogen results in a reduction of mass of a fraction of a percent, but this is enough to power the Sun.
- Fusion proceeds through a sequence of reactions between protons and other nuclei, resulting in the production of helium.
- Neutrinos produced in nuclear fusion can travel out of the Sun because neutrinos interact so weakly with matter.
- By studying neutrinos from the Sun, astronomers showed that they have properties not previously predicted by particle physics.
- The Sun burns its fuel at a relatively slow pace because the weak interaction keeps nuclear fusion from occurring rapidly.

KEY TERMS

cosmic ray, 416
deuterium, 414
isotope, 414
neutrino, 412
neutron, 413
nuclear fusion, 413
positron, 414
proton, 413
proton–proton chain, 414
strong force, 413
weak force, 418

CONCEPT QUESTIONS

Concept Questions on the following topics are located in the margins. They invite thinking and discussion beyond the text.

1. How we know the Sun's composition. (p. 414)
2. Energy sources that are derived from sunlight. (p. 415)
3. Consequences if weak force were stronger. (p. 418)

REVIEW QUESTIONS

4. Why is it impossible to explain the Sun's power output as coming from a chemical process such as the combustion of hydrogen and oxygen to form water?
5. What is the meaning of Einstein's formula $E = m \times c^2$?
6. How does hydrogen fuse to form helium? Why is a high temperature necessary?
7. What is the solar neutrino discrepancy and how was it resolved?
8. Why are neutrino detectors located deep underground?
9. How does the weak interaction control the rate of fusion?

QUANTITATIVE PROBLEMS

10. In chemical reactions, when energy is released, mass is lost according to Einstein's formula just as it is for nuclear energy. When a kilogram of coal is burned, it releases about 40 million joules of energy.
    a. What amount of mass is lost?
    b. How many carbon atoms (mass = 2.0×10^{-26} kg) would it take to match the amount of mass that is lost?
11. Worldwide, humans use about 4×10^{20} joules of energy each year. A typical power station generates about 10^9 watts of power, while the Sun's luminosity is about 4×10^{26} watts. (Recall that 1 watt = 1 J/sec.)
    a. How many power stations would be needed to produce power equivalent to the Sun's?
    b. How many typical power stations are needed to supply current human energy usage?
    c. If we could store all of the energy that the Sun produces in 1 second, how long would it last humans at current energy use rates?
12. The Sun's luminosity indicates how much energy it radiates each second. Starting from the Sun's luminosity, 3.83×10^{26} watts, calculate how much mass must be converted to energy each second in order to maintain hydrostatic equilibrium. (Recall that 1 W = 1 J/sec.)
13. If the Sun generates 3.83×10^{26} watts throughout its lifetime.
    a. How much mass will our Sun convert to energy over its entire main-sequence lifetime (10^{10} years)?
    b. What fraction is this of the mass of the Sun?
    c. How does this compare to the mass of the Earth?
14. One "kiloton" of explosive energy is 4.2×10^{12} joules. Using Einstein's formula, calculate the mass that is converted into energy in a 1-megaton H-bomb explosion.
15. How long does it take a neutrino produced in the center of the Sun to reach the Sun's surface?

TEST YOURSELF

16. The Sun produces its energy from
    a. fusion of neutrinos into helium.
    b. fusion of positrons into hydrogen.
    c. disintegration of helium into hydrogen.
    d. fusion of hydrogen into helium.
    e. electric currents generated in its core.
17. Nuclear fusion only occurs in the Sun's core because it is where
    a. temperatures are high enough for nuclear fusion to take place.
    b. all the hydrogen is located.
    c. the Sun's magnetic field is generated.
    d. there are free electrons needed for the proton-proton chain.
18. The number of neutrinos measured coming from the Sun permits us to measure
    a. how much deuterium is left in the Sun.
    b. the strength of the strong force within atomic nuclei.
    c. the current age of the Solar System.
    d. the risk of radiation exposure on the Earth.
    e. the rate of hydrogen fusion in the Sun's core.
19. The fusion of hydrogen into helium causes the Sun's total mass to
    a. decrease.
    b. stay the same.
    c. increase.
    d. increase then decrease.

UNIT 53

Solar Activity

53.1 Sunspots
53.2 Prominences and Flares
53.3 The Solar Cycle
53.4 The Solar Cycle and Terrestrial Climate

Learning Objectives

Upon completing this Unit, you should be able to:
- Name solar features related to solar activity and describe their effects on Earth.
- Describe the properties of sunspots, and explain what causes them.
- Describe the solar cycle, and explain its connection the Sun's magnetic field.
- Discuss the connections of solar activity to the Earth's climate.

The Sun is a stormy place. The vast amount of energy pouring out of its interior (Unit 52) heaves upward enormous masses of gas, which then plunge back down with forces far greater than any storm on Earth. Small hot spots send jets of white-hot ionized gas streaming up into the Sun's atmosphere, and occasional eruptions blast billions of tons of matter into space. Astronomers call these various disturbances **solar activity.** Astronomers infer that other stars with convection zones near their surfaces also have activity in their chromospheres and coronas, because they too exhibit phenomena such as flarelike brightenings.

The Sun's storminess exhibits dramatic kinds of activity that can take place in as little as a few minutes or that sometimes persist for months or years. As we look over even longer time scales, we find that the Sun goes through a regular pattern of changes every 11 years, and shows other variations over longer periods still. From our safe distance on Earth, these phenomena are dramatic and lovely. Solar storms often take on forms quite unlike anything we might imagine from our experience on Earth, because the Sun's hot ionized gas interacts with its strong, complex magnetic field. Solar activity can also affect us more directly: It can damage spacecraft, interfere with radio communications, and trigger auroral displays. Long-term changes in the Sun's level of activity also have an impact on the Earth's climate that compete with the greenhouse gases discussed in Unit 38.

53.1 SUNSPOTS

Sunspots are the most easily seen indicator of solar activity. They are large, dark regions (Figure 53.1) ranging in size up to many thousands of kilometers across. Sunspots last from a few days to more than a month. They are darker than the surrounding gas because they are cooler (about 4500 K as opposed to the 5800 K of the normal photosphere), but they actually shine very brightly—they appear dark only by contrast. Sprinkling a few drops of water onto a hot electric stove burner will have a similar effect. Each drop momentarily cools the burner, making a dark spot. Four centuries after Galileo's first telescopic observations of sunspots, solar physicists are just beginning to piece together the complex interactions that drive them.

A major clue to the puzzle of sunspots was the discovery in the early 1900s that sunspots are the location of the Sun's most intense magnetic fields. Astronomers can detect magnetic fields in sunspots and other astronomical bodies by the **Zeeman effect,** a physical process in which the magnetic field causes some of the spectral lines produced by atoms to split into two, three, or more components. The splitting occurs because the magnetic field alters the electron orbitals within an atom, which

FIGURE 53.1
Image of a large group of sunspots. The darker areas are cooler gas, but they are still bright—they are dark in the image just by contrast to the surrounding hotter regions.

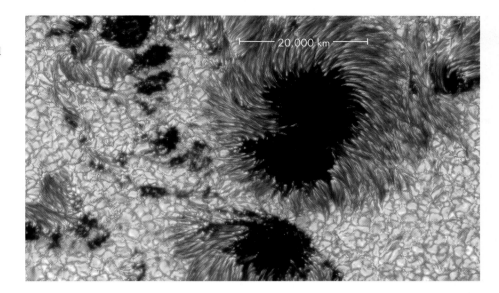

in turn alter the wavelengths of light the electrons can absorb or emit. Figure 53.2A shows the Zeeman effect splitting spectral lines in a sunspot. The spectrum was taken from a region that cuts across a sunspot. The spectral line is single outside the spot but triple within, and the degree of splitting indicates that the magnetic field strength is thousands of times stronger than the average field strength on the Sun's surface.

By measuring the strength of Zeeman splitting across the Sun's face, astronomers can map the Sun's magnetic field, creating a **magnetogram,** as seen in Figure 53.2B. The colors in the magnetogram show the strength as well as the polarity of the magnetic field. The *polarity* indicates whether the magnetic field is "north" or "south"; in other words, it indicates whether a compass needle would point toward or away from the region.

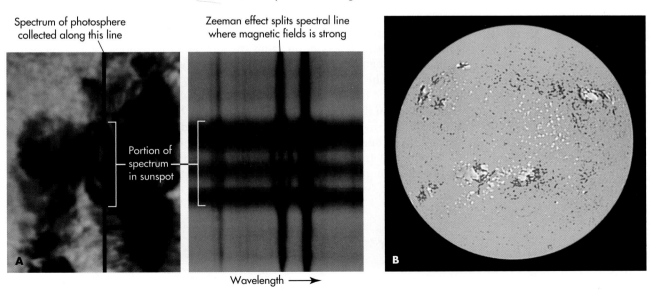

FIGURE 53.2
(A) Spectrum made across a sunspot. The sunspot is shown on the left, and a small part of the spectrum around 525 nm is shown at right. An absorption line produced by iron atoms splits into several slightly different wavelengths in the portion of the spectrum crossing the sunspot. This is because a strong magnetic field slightly alters the atoms' energy levels due to the Zeeman effect.
(B) Magnetogram of the Sun. Yellow indicates regions with north polarity, and dark blue indicates regions with south polarity. Notice that the polarity pattern of spot pairs is reversed between the top and bottom hemispheres of the Sun. That is, in the upper hemisphere, blue tends to be on the left and yellow on the right. In the lower hemisphere, blue tends to be on the right and yellow on the left.

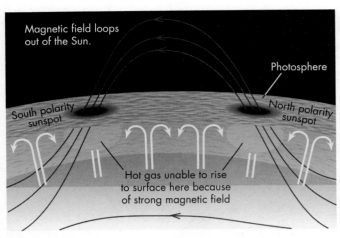

FIGURE 53.3
Sunspots occur where an intense region of the Sun's magnetic field emerges through the surface of the Sun, creating spots of opposite polarity. The strong magnetic field blocks the flow of hot gas upward to the surface making the spot cooler.

Concept Question 1

If you were in a spacecraft that was orbiting just above the Sun's surface, how would a magnetic compass aboard the ship behave?

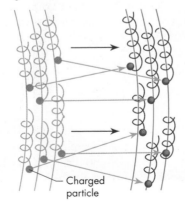

FIGURE 53.4
Charged particles are tied to magnetic field lines, spiraling back and forth along them. If forces move the gas, it can drag the field with it—and vice versa.

Sunspots occur in pairs of opposite polarity, as illustrated in Figure 53.3. A bundle of magnetic field lines (indicating a strong magnetic field) emerges from the surface of the Sun, then returns into the Sun. Curiously, magnetic fields heat the chromosphere and corona (Unit 51.4), but they cool sunspots. This is related to how magnetic fields affect the motion of electrically charged particles. Where magnetic fields are intense, they can block the upward flow of ionized gas, which is composed of charged particles, preventing the heat that the gas carries from reaching the Sun's surface.

Charged particles spiral around a strong magnetic field, tightly tied to it, as illustrated in Figure 53.4. They are thereby forced to follow the direction of the magnetic field as they spiral along it. For example, the Earth's magnetic field deflects particles from striking the Earth in most locations. Instead the particles move along the magnetic fields until they reach low altitudes near the north and south magnetic poles, where they interact with the atmosphere and create the **aurora** (Unit 38.1). Likewise, magnetic fields in the Sun deflect the gas. Regions of strong magnetic fields prevent hot gas in the interior of the Sun from rising to the surface, so that the surface cools and becomes comparatively darker, making a sunspot.

A strong magnetic field cannot be the whole explanation for sunspots, though. A region where the magnetic field is strong will normally push itself apart—just as magnets oriented in the same direction repel each other. Why, then, do sunspots persist for days or weeks, instead of rapidly weakening? This was one of the puzzles explored by a joint NASA/European Space Agency spacecraft called the *Solar and Heliospheric Observatory (SOHO)*. Using detailed Doppler measurements of the flow and wave patterns of the photospheric gas, scientists were able to construct the three-dimensional image of the motions of the gas below the surface shown in Figure 53.5.

The cooling that occurs in the middle of the sunspot causes the gas there to sink, drawing in gas from the surrounding regions, much as a hurricane or tornado on Earth draws in air from the surrounding regions. *SOHO* measured flow speeds in and down toward the sunspot of about 5000 kilometers per hour (3000 mph). The rapidly flowing gas drags along the magnetic field because, just as the magnetic field exerts a force on the charged particles, the charged particles exert a force on the magnetic field, as required by Newton's law of action and reaction (Unit 15.3). As a result, the flow of gas inward keeps the magnetic fields from spreading out and weakening, which keeps the hot gas from rising into the region of the sunspot from the layers below. This feedback between the magnetic field and the flow of gas is what makes sunspots persist on the Sun's surface.

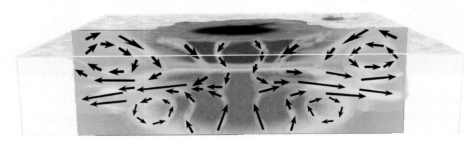

FIGURE 53.5
Flows of gas underneath a sunspot measured by the spacecraft *SOHO* in 2001. The colored diagram at the bottom of the figure shows a slice through the region below the sunspot, indicating where the gas is hottest (red) and coolest (blue). The arrows show the direction of gas flow.

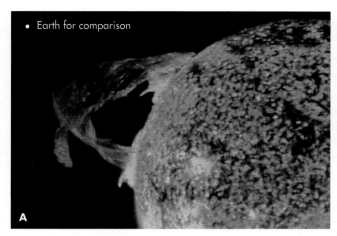 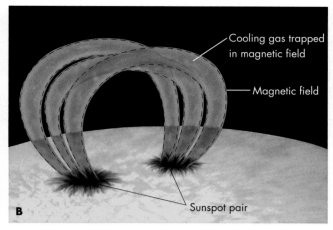

FIGURE 53.6
Solar prominences. (A) An ultraviolet image made from the *SOHO* satellite. The size of the Earth is shown for comparison. (B) Sketch illustrating how magnetic fields between sunspots support a prominence.

Concept Question 2

Prominences have been found to be less hot than the surrounding coronal gas. How then can they appear brighter than the corona in visible-wavelength images?

Solar prominences

53.2 PROMINENCES AND FLARES

The strong magnetic fields associated with sunspots also shape the motions of ionized gas as it flows above the photosphere. **Prominences** are huge plumes of glowing gas that jut from the lower chromosphere into the corona (Figure 53.6A). Time-lapse movies show that gas streams through prominences, sometimes rising into the corona, sometimes raining down onto the photosphere. The flow is channeled by, and supported by, strong magnetic fields that arc between sunspots of opposite polarity (Figure 53.6B). The pressure of the surrounding coronal gas helps confine and support the gas in the prominence. Under favorable conditions, the gas in a prominence may remain trapped in its magnetic prison and glow for weeks.

Sunspots also give birth to solar **flares,** which are brief but bright eruptions of hot gas in the chromosphere. Over a few minutes or hours, gas near a sunspot may brighten, emitting the energy equivalent of millions of atomic bombs. Such eruptions, though violent, are so localized they hardly affect the visible light output of the Sun at all. Generally you need a specialized telescope to detect the visible light of flares, though they can increase the Sun's radio, ultraviolet, and X-ray emission by factors of a thousand in a few seconds (Figure 53.7).

One hypothesis suggests that flares arise when the magnetic field near a sunspot gets twisted by gas motions. Somewhat like winding up a toy powered by a rubber band, the twisting can go only so far before the rubber band breaks. So, too, the magnetic field can be twisted only so far before it suddenly readjusts, whipping the gas in its vicinity into a new configuration. The sudden motion heats the gas, and it expands explosively. Some gas escapes from the Sun and shoots across the inner Solar System to stream down on the Earth.

Closely related to flares, but on a much larger scale, are **coronal mass ejections.** These are enormous bubbles of hot gas, sometimes containing billions of tons of ionized gas, that are blasted from the corona out into space (Figure 53.8). The mechanism for ejecting this gas is so powerful that the ionized gas drags the embedded magnetic field along with it. The cause of these ejections is still not well understood, but they seem to be triggered by flares.

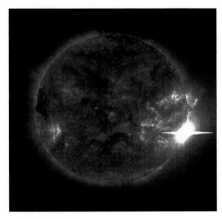

FIGURE 53.7
A solar flare—the bright spot on the right side of the image. This picture was taken through an ultraviolet filter on the *SOHO* satellite.

FIGURE 53.8
A coronal mass ejection recorded at visible wavelengths by the *SOHO* satellite. A disk blocks light from the Sun so the image does not get burned out, but the features continue inward. An ultraviolet image of the Sun's surface made at the same time is superimposed.

Even though the ejected gas moves at a speed of 400 kilometers per second, it takes several days for the bulk of gas to traverse the 150 million kilometers between the Earth and the Sun. However, the most energetic particles from a flare can sometimes reach the Earth in less than an hour. These high-energy "proton storms" are particularly dangerous for astronauts traveling outside of the Earth's magnetic field. Strong flares during trips to the Moon or Mars could expose astronauts to dangerous levels of radiation, as well as disrupting communications and disabling satellites

When gas ejected from the Sun strikes the Earth, it can produce one of the more beautiful astronomical sights, an aurora. The stream of charged particles is mostly guided toward the north and south poles by the Earth's magnetic field, but in a major event, the auroral activity can spread over much more of the Earth, with serious consequences. Electrical power grids can be destabilized as the magnetic field carried by the stream of gas causes electric currents to surge in electrical transmission wires. A solar outburst in 1989 (Figure 53.9) caused blackouts in several locations across North America. An even more remarkable event occurred in 1859, when the aurora was seen even in the tropics, and telegraph equipment was set afire by huge magnetic surges. Today, an array of satellites monitors the Sun and "space weather" to allow sensitive equipment to be placed in a safe mode before the magnetic disturbances reach Earth.

FIGURE 53.9
Photograph of the great aurora of March 1989. The magnetic storm associated with this aurora shut down the power grid in Quebec, Canada.

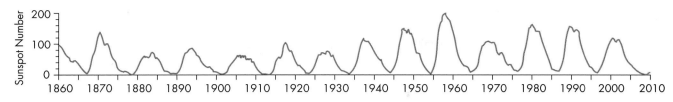

FIGURE 53.10

Plot of sunspot numbers over the last 150 years showing the solar cycle. The number of sunspots grows to a peak about every 11 years.

53.3 THE SOLAR CYCLE

Sunspot and flare activity changes from year to year in what is called the **solar cycle.** This variability can be seen in Figure 53.10, which shows the number of sunspots detected over the last 150 years. The numbers clearly rise and fall approximately every 11 years. For example, the cycle had peaks in 1969, 1980, 1990, and 2001. The interval between peaks varies: It may be as short as 7 or as long as 16 years. The numbers of flares and prominences also follow the solar cycle.

The solar cycle may be caused by the way the Sun rotates. Gas near the Sun's equator circles the Sun faster than gas near its poles; that is, the Sun spins differentially. This is a property common in many spinning gaseous bodies. The Sun's differential rotation is such that its equator rotates in about 25 days and its poles in 30. So a set of points arranged from pole to pole in a straight line would move, over time, into a curve, as shown in Figure 53.11.

Differential rotation should similarly distort the Sun's magnetic field, "winding up" the field below the Sun's surface. Although the exact mechanism is still not well understood, astronomers think this winding of the Sun's magnetic field causes the solar cycle. According to one hypothesis, the Sun's rotation wraps the solar magnetic field into coils below the surface (Figure 53.12), making the field stronger and increasing solar activity: spots, prominences, and flares. The wrapping occurs because the Sun's magnetic field is twisted around by the differential rotation of the ionized gas. Sunspots form where the twisted magnetic field rises to the Sun's surface and breaks through the photosphere.

The 11-year cycle appears to represent the time it takes differential rotation to twist the Sun's magnetic field to the breaking point. When the field becomes too strained, the twisted field lines break and reconnect into a simpler pattern. These events, when the magnetic field suddenly restructures itself, may produce flares and coronal mass ejections. After several years of these violent eruptions the magnetic field has relaxed into an untwisted pattern again. Interestingly, the overall polarity of the Sun's magnetic field reverses after each of these peaks of activity, so that the Sun's overall "north magnetic pole" is in the opposite direction.

> **Concept Question 3**
>
> From time to time claims are made that something like, for example, the stock market correlates with the sunspot cycle. Do you think there *could* be a way for sunspots to cause changes in such things? How could you test your ideas?

Twisting of Sun's magnetic field

FIGURE 53.11

Sketch showing solar differential rotation. Points near the Sun's equator rotate faster than points near the poles.

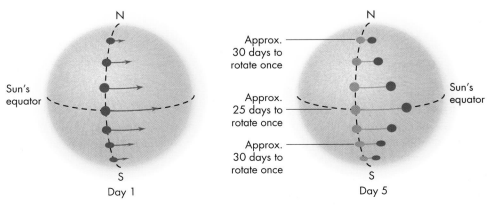

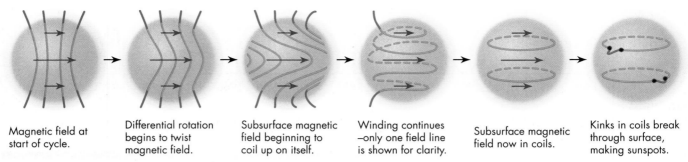

FIGURE 53.12
Sequence showing how differential rotation can wind up the Sun's subsurface magnetic field. The field becomes twisted into coils that can break through the surface, making sunspots.

The twisting of the magnetic fields reverses the internal circulation of the electric currents that produces the Sun's magnetic field.

The mechanism causing these reversals of the magnetic field is not fully understood, but it is probably similar to what makes the Earth's magnetic field also flip direction. On Earth this occurs every 250,000 years, on average, as indicated by geological evidence (Unit 37). Earth's magnetic field reversals are far more sporadic than the Sun's—the last one occurred almost 800,000 years ago, so we are long "overdue" for one. Perhaps through understanding the Sun's magnetic field, we will come to understand the Earth's magnetic field better.

53.4 THE SOLAR CYCLE AND TERRESTRIAL CLIMATE

Climatologists have long wondered how solar activity might affect Earth's climate. Outflows of ionized gas interact with the Earth's magnetic field, but a more important effect may be that as the solar activity varies, the Sun's total power output varies as well. It is counterintuitive, but as the number of sunspots increases, the Sun becomes slightly more luminous by a fraction of a percent. Although sunspots are relatively dim, the surrounding photosphere actually grows slightly brighter, increasing the net output.

The slight changes in the solar luminosity measured during a single 11-year cycle are minor compared to changes that occur over longer time periods. The evidence for such long-term changes is based in part on the work of E. W. Maunder, a British astronomer who studied sunspots. Maunder noted in 1893 that, according

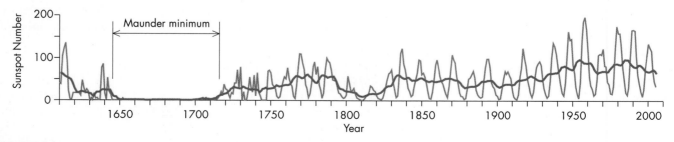

FIGURE 53.13
Plot of the solar cycle extended back to 1610 pieced together from early records. The darker line shows the number of sunspots averaged over 11 years, showing how the Sun's overall level of activity changes with time. During the period of the Maunder minimum, the Sun became almost completely inactive, and Earth experienced a period of cooling.

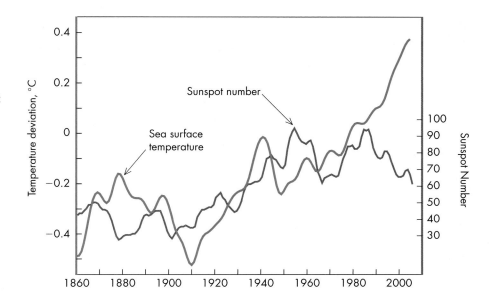

FIGURE 53.14
Curves showing the change in ocean surface temperature on Earth and the change in the average sunspot number since 1860. Notice that the curves change approximately in step until the 1980s. Over the last two decades the oceans have continued to warm while sunspot counts have declined. Astronomers deduce from these curves that solar activity affects our climate, but other factors must explain the recent temperature rise.

Concept Question 4

Graphs like those in Figure 53.14 can be quite controversial. What kinds of changes in measurement techniques have occurred over the last 150 years? How might these changes have affected the values plotted?

to old solar records, almost no sunspots were seen between 1645 and 1715 (Figure 53.13). He concluded that the solar cycle turned off during that period. The period is now called the **Maunder minimum** in honor of his discovery.

The Maunder minimum coincides with an approximately 70-year spell of abnormally cold winters in Europe and North America. Glaciers in the Alps advanced; rivers froze early in the fall and remained frozen late into the spring. The Maunder minimum is in the middle of a period of several centuries when the Earth appears to have been cooler—meteorologists call this time the "little ice age." If only one such episode were known, we might dismiss the sunspot–climate connection as a coincidence; but three other cold periods have also occurred during times of low solar activity. This strengthens the case for solar activity affecting our climate. It is suspected that the Sun's luminosity decreases during these periods of solar activity, but there may be other effects as well. A recent hypothesis suggests that energetic particles from the Sun during periods of solar activity stimulate cloud formation on Earth, thereby slowing the escape of heat from our planet.

Although scientists are still unsure about what creates the link between solar activity and our climate, few now doubt that such a link exists. However, there is substantial debate about whether it has as large an impact on climate as other sources. Figure 53.14 shows how the ocean surface temperature (expressed as deviations from the normal average) changed from 1860 to 2005. The figure also shows how the number of sunspots, averaged over the 11-year cycles, changed over the same time span. Until the mid-1980s, solar activity was rising, and its rise showed some degree of correlation with the rise in sea surface temperature. This led some scientists to propose that solar activity was the main cause of global warming. However, the Sun showed fewer sunspots in the last two solar cycles, while sea temperature has continued to rise. For the great majority of scientists, this suggests that other factors, such as human production of greenhouse gases (Unit 38), is probably the main cause of global warming.

KEY POINTS

- Solar activity describes the "storminess" on the Sun's surface.
- Sunspots are darker regions of the Sun's surface where there are intense magnetic fields.
- The magnetic fields in sunspots prevent hot ionized gas from rising, making the regions cooler than the rest of the photosphere.
- Sunspots occur in pairs where a strong region of the magnetic field emerges from the Sun.
- The strong magnetic fields between sunspots can support prominences of hot gas above the Sun's surface.
- Powerful flares and ejections of large amounts of coronal material appear to occur when the magnetic field becomes twisted.
- The level of activity in the Sun reaches a peak about every 11 years, on average.
- The cycle activity is thought to be caused by winding up of the magnetic field caused by differential rotation in the Sun.
- The overall direction of the Sun's magnetic field reverses north for south after each 11-year cycle.
- The strength of solar activity has an effect on Earth's climate, as shown by cold periods during times of solar inactivity.
- The rise in sea temperature over the last two decades does not correlate with solar activity.

KEY TERMS

aurora, 422
coronal mass ejection, 423
flare, 423
magnetogram, 421
Maunder minimum, 427
prominence, 423
solar activity, 420
solar cycle, 425
sunspot, 420
Zeeman effect, 420

CONCEPT QUESTIONS

Concept Questions on the following topics are located in the margins. They invite thinking and discussion beyond the text.

1. Behavior of magnetic compass orbiting the Sun. (p. 422)
2. Why gas can be brighter although cooler. (p. 423)
3. How to test claims about effects of sunspot cycles. (p. 425)
4. Challenges of graphing historical data. (p. 427)

REVIEW QUESTIONS

5. Why do sunspots appear dark?
6. What roles does magnetic activity play in solar activity?
7. How do we know there are magnetic fields in the Sun?
8. What phenomena are associated with solar activity?
9. What are solar flares? What effects do they have on Earth?
10. What is the solar cycle? What happens to the Sun's magnetic field during the solar cycle?
11. What is the Maunder minimum? Why is it of interest?
12. How might solar activity affect the climate?

QUANTITATIVE PROBLEMS

13. If the surface of the Sun near the equator completes one rotation every 25 days, compared to a 30-day rotation near the poles, how long does it take for the surface near the equator to complete one extra rotation around the Sun compared to the poles? After 11 years, how many extra turns would this result in?
14. What is the speed of material ejected by a solar flare if it takes four days to reach the Earth?
15. How long does it take the material ejected by a solar flare to reach the orbit of Neptune?
16. The amount of light emitted by a hot region is proportional to the temperature to the fourth power ($L \propto T^4$). If a sunspot is at a temperature of 4500 K, how much less luminosity does it generate than if it were at 6000 K?
17. The largest solar flares can release as much as 6×10^{25} joules of energy. How does this energy compare to the total energy output of the Sun each second?

TEST YOURSELF

18. About how many years elapse between times of maximum solar activity?
    a. 3   b. 5   c. 11   d. 33   e. 105
19. Sunspots are dark because
    a. they are cool relative to the gas around them.
    b. they contain 10 times as much iron as surrounding regions.
    c. nuclear reactions occur in them more slowly than in the surrounding gas.
    d. clouds in the cool corona block our view of the hot photosphere.
    e. the gas within them is too hot to emit any light.
20. Which of the following is true about solar prominences but *not* about solar flares?
    a. They can last for many days.
    b. They occur when the Sun is inactive.
    c. They occur where the magnetic field is weakest.
    d. They are related to sunspots.
    e. None of the above because prominence and flare are two names for the same phenomenon.
21. The Maunder minimum represents
    a. the smallest size for a sunspot.
    b. a period in the late 1600s with very little solar activity.
    c. the shortest time between solar cycle peaks.
    d. the shortest time for the Sun's equator to rotate once around.
22. Why do sunspots occur where the magnetic field is strong? The magnetic field
    a. attracts dark iron particles to the area.
    b. shifts the emitted light out of the visible range.
    c. blocks rising hot gas from inside the Sun.
    d. deflects light into different directions.
    e. Does all of the above.

PART 4

UNIT 54

Surveying the Stars

54.1 Triangulation
54.2 Parallax
54.3 Calculating Parallaxes
54.4 Moving Stars
54.5 The Aberration of Starlight

Learning Objectives

Upon completing this Unit, you should be able to:
- Describe how triangulation methods were used to find distances of remote objects.
- Explain the stellar parallax method, and calculate a star's distance from its parallax.
- Define proper motion, and describe how a star's space velocity is found.
- Describe the aberration of starlight and how it provided the first direct proof of Earth's motion.

Determining the distance of a remote object is one of the most fundamental problems that astronomers must tackle. Knowing stars' distances is necessary for determining many of their most basic properties such as their diameters, masses, and the amount of light they emit.

In the last few decades, space exploration has allowed us to measure many distances within the Solar System directly. Spacecraft have traveled to most of the planets and their moons. As a result, we can time how long signals take to travel (at the speed of light) back and forth between the object being visited and the Earth. Ground-based technology has also advanced, so we do not have to rely on space probes to measure distances accurately. Instead, we can bounce radar signals off an asteroid, for example, and time how long the signals take to travel there and back. With this method we can determine a distance accurate to within a few meters (Figure 54.1A).

Beyond the Solar System, however, distances are so vast that such direct techniques are impossible and will remain so for centuries to come. The situation is not hopeless, though. We can take a cue from the ancient Greeks, who were able to estimate the distances to the Moon and Sun by applying geometry (Unit 10) with only the crudest of astronomical instruments. The same geometric methods that a surveyor might use to measure distances on the Earth allow us to measure the distances of nearby stars, and such methods provide a fundamental stepping-stone to learning about the objects that lie far beyond the Solar System.

54.1 TRIANGULATION

Long before radar provided us with high-accuracy distance measurements, and long before space probes ventured out into the Solar System, some planetary distances were very accurately determined by a geometric method, used by surveyors, called **triangulation.** If we can measure the direction to an object from two points that are a known distance apart, we can determine the distance to the object. This is based on a fundamental property of triangles: if you know the length of one side of a triangle and the two angles at either end of that side, the shape of the triangle is determined as illustrated in Figure 54.1B.

As an example, suppose we want to measure the distance across a deep gorge. We can construct an imaginary triangle to a point on the far side of the gorge. Two sides of this triangle span the gorge, while we can choose the length and direction

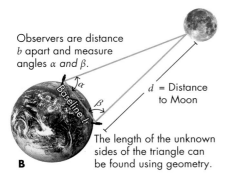

FIGURE 54.1
(A) The distance to the Moon can be measured very precisely by bouncing radar (or laser) signals off the Moon and timing how long they take to return. (B) The Moon's distance can also be found from geometry by two observations at a known separation.

Concept Question 1

To make triangulation measurements more accurate, surveyors usually make the baseline as big as possible. Why would this improve the results?

Mathematical Insight

Using trigonometry, the distance, d, across the gorge equals the baseline length, b, times the "tangent" of the angle α: $d = b \times \tan \alpha$.

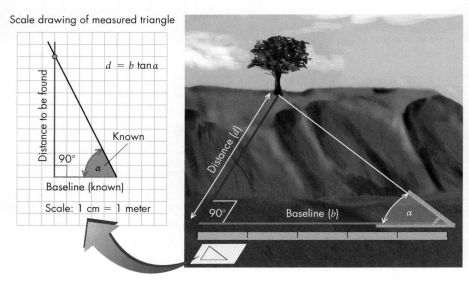

FIGURE 54.2

Sketch illustrating the principle of triangulation. If we know the length of one side of a triangle and the two angles at either end of it, we can determine the length of the other sides.

FIGURE 54.3

Hipparchus used the fact that a partial eclipse in Alexandria was total in a spot farther north. Since he knew the angular diameter of the Sun, he could determine the angle between the two observations.

The exact values of the distances to Venus and Mars at closest approach vary because the orbits are elliptical.

of the third side—the **baseline**—along the edge we are on, as shown in Figure 54.2. It simplifies our calculations if we choose the baseline so it is perpendicular to the line straight across the gorge. Then, by measuring the length of the baseline and the angle α between the baseline and the remaining side (the *hypotenuse*), we can determine the distance across the gorge from a scale drawing of the triangle.

Solving this type of problem is the motivation behind the area of mathematics called *trigonometry* (literally meaning "the measuring of triangles"). Trigonometric formulas allow us to calculate the length of the sides without requiring us to make precision scale drawings of the triangle. In the figure, the baseline is 5 meters, and the angle α is 63.4°. Constructing a right triangle with one of the angles at this value, we find that the unknown distance across the gorge must be two times the baseline, or 10 meters.

As early as about 140 B.C.E. the Greek astronomer **Hipparchus** used triangulation to estimate the distance to the Moon. He compared the shift of the Moon's position with respect to the Sun, as seen from two locations during a solar eclipse (Figure 54.3). Since he knew the angular size of the Sun, this told him the angle between the two sides of the triangle pointing to the Moon. The triangulation is more complicated in this case because of the Earth's curved surface, but Hipparchus was able to estimate that the Moon was about 38 Earth diameters distant—remarkably close to the modern measurement of 30 Earth diameters.

Triangulating distances to objects more distant than the Moon requires greater accuracy. Because of the much larger distances involved, the angles may be only a small fraction of an **arc minute.** An arc minute is 1/60 of a degree, or about 1/30 of the angular size of the Moon. Nonetheless, the ancient Greeks attempted to find the Sun's distance from Earth. Their result was far too small because instruments at the time were simply too crude to make a sufficiently accurate measurement of the angles. Because the Earth–Sun distance—known as the **astronomical unit** or AU—is so fundamental to our knowledge of the scale of the universe, we will discuss it in some detail here.

It turns out that we do not have to measure the distance to the Sun to determine the size of the AU. Johannes Kepler's discoveries in the early 1600s established the *relative* distances of the Sun and planets (Unit 12). For example, Kepler's law relating orbital size and the period of the orbit established that Mars's orbit is, on average, 1.52 times larger than the Earth's—1.52 AU—while Venus's is 0.72 AU. Therefore Mars is about 0.52 AU away from Earth at its closest approach and

Venus is about 0.28 AU. So if we could triangulate the distance to Mars during one of its close approaches, we could determine the size of the AU.

Such a measurement of the distance to Mars was made during one of Mars's close approaches to Earth in 1672 by astronomers in Paris and South America, at sites separated by about 7000 km (about 4000 miles). We know today that the value they obtained was correct to about 10%, but it implied a distance to Mars—and therefore a size for the astronomical unit—that was almost 10 times greater than had been believed previously. This led many to doubt the accuracy of the value obtained. If the new distance was correct, the distance to the Sun that had been used for centuries was far too small. Recall that sizes of objects are measured by the angular size versus distance formula (Unit 10.4). If the distance was actually 10 times larger, then the size of the Sun would have to be 10 times larger as well. Not only was the Sun even bigger than previously thought, the bigger value of the AU also implied that Jupiter and Saturn were much larger than the Earth—a challenging idea to accept in an age when even the concept of the Earth orbiting the Sun was new. Because of the controversial nature of these results, astronomers sought independent measurements to confirm these findings. It took almost a century, however, before the distance of Venus could be determined.

Even though Venus comes closer to Earth than Mars, triangulating its distance is difficult because when it is closest to Earth, it is lost in the glare of the Sun. However, the British astronomer Edmund Halley (of Comet Halley fame) developed a method for determining both Venus's and the Sun's distance. He did this by timing how long Venus takes to move across the face of the Sun, from different locations on Earth, during one of Venus's rare **transits**. Because its orbit is tilted relative to our own, Venus usually lies slightly "north" or "south" of the Sun's disk as it passes between the Earth and Sun. But twice every 120 years or so it crosses directly in front of the Sun. This last happened in 2004 and 2012 (Figure 54.4) and will not occur again until 2117 and 2125.

For the transit of 1761, British astronomers Charles Mason and Jeremiah Dixon sailed to the southern tip of Africa, even braving cannon fire (Britain was at war with France) to get to their destination. The observations of Mason and Dixon provided the long baseline needed for this astronomical surveying project. (Other astronomers were not so lucky, spending years traveling to remote parts of the globe, only to have clouds prevent them from making the critical measurements on the day of the transit.) The results of the Venus transit measurements confirmed the Mars measurements, yielding a size for the astronomical unit to within about 1% of the modern value.

In the end, however, the triangulation method of surveying is limited by the size of the baseline that we can obtain on the Earth. Even the most accurate measurements are nowhere near precise enough to tell us about the vast distances to stars if our baseline is at most the 12,800-km diameter of the Earth. A much bigger baseline is needed.

> **Concept Question 2**
>
> Why do you suppose scientists are usually reluctant to accept new results that are very different from previously accepted results?

> **Mathematical Insight**
>
> Halley's technique is based on the time it takes Venus to cross the face of the Sun seen from different locations north and south on Earth. By comparing the duration, we can calculate how much closer Venus was to crossing near the Sun's equator or pole from each location, giving the precise angular change north or south relative to the Sun.

Mason and Dixon were summoned to the American colonies in 1763 to settle a dispute about the boundary between Pennsylvania and Maryland. This became known as the "Mason–Dixon Line," which figured prominently in the history of the United States in disputes between slave and free states.

FIGURE 54.4
This image shows Venus (the round, dark circle) passing between us and the Sun on June 6, 2012. By precisely timing how long Venus takes to transit the Sun's disk from different locations on Earth, we can find Venus's distance.

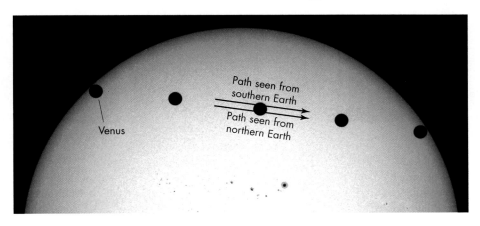

54.2 PARALLAX

There is a baseline available to us that is more than 20,000 times larger than the size of the Earth—the Earth's orbit. As we move around the Sun, we shift our location in space by about 300 million kilometers (185 million miles) from one side of our orbit to the other. Even the ancient Greeks recognized that if the Earth moved around the Sun, stars should show shifts in their positions called **parallax** (see Unit 10).

An easy way to demonstrate the principle of parallax is to hold your thumb pointed up, motionless, at arm's length, and shift your head from side to side. Your thumb seems to move against the background even though in reality it is your head that has changed position. This simple demonstration also illustrates how parallax gives a clue to an object's distance. If you hold your thumb at different distances from your face, you will notice that the apparent shift in your thumb's position—its parallax—is also different. Its parallax is smaller if it is at arm's length, and larger if your thumb is close to your face (Figure 54.5). That is, for a given motion of the observer:

Distance is inversely proportional to parallax.

This law is just as true for stars as it is for your thumb.

Using parallax to find distance may seem unfamiliar, but parallax creates our stereovision—the ability to see things three-dimensionally. When we look around with two eyes, each eye sends a slightly different image to the brain. Your brain processes the two images and determines the distances to various objects in your field of view. The effect of parallax can be seen if you extend your arm in front of you, with your thumb pointed up. Look first through one eye and then the other, while keeping your head stationary. Notice that your thumb seems to shift to the left when looking with the right eye and then to the right when looking with the other eye.

You can demonstrate the importance to your brain of comparing the two images simultaneously by trying some experiments that involve covering one eye. Cover either eye, then drop a coin on a table in front of you. Hold your arm straight out in front of you, and with your first finger pointing down, try to lower it so that the tip of your finger comes down directly on the center of the coin. If you try this again with both eyes open, you'll see this is much easier using two eyes than one. With one eye closed, you may find yourself moving your head side to side (another way to estimate parallax) or looking for other visual cues to compare and adjust your hand's distance from you as you move it downward. Another experiment you

Stellar parallax

Parallax

A Viewed from right eye Viewed from left eye

B Viewed from right eye Viewed from left eye

FIGURE 54.5
(A) When you hold your thumb at arm's length and look through one eye and then the other or shift your head from side to side, you see your thumb shift by a small amount relative to the background. This shift is *parallax*. (B) If you hold your thumb close to your face and do exactly the same thing, your thumb shifts more against the background. This demonstrates that parallax is larger for nearer objects.

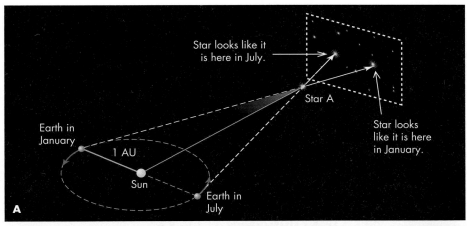
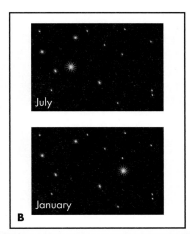
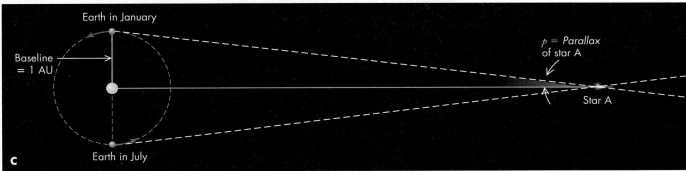

FIGURE 54.6
(A) Triangulation to measure a star's distance. The radius of the Earth's orbit is the baseline. (B) As the Earth moves around the Sun, the star's position changes as seen against background stars. (C) Parallax is defined as one-half the angle by which the star's position shifts. Sizes of bodies and their separation are exaggerated for clarity.

can do begins by standing in front of a hanging cord (from a lightbulb or window shade, for example). Hold one arm out to the side of the cord with a finger pointed toward it. Now bring your pointed finger in from the side, and try to touch the cord with the tip of your finger. Without stereovision, this is surprisingly difficult.

Even after the Copernican revolution convinced most that the Earth moves around the Sun, stellar parallax was not detected for centuries. The failure to detect stellar parallax was one of the reasons some astronomers continued until the 1700s to believe that the Earth did not move through space.

The stellar parallax technique is a little different from the triangulation method. Instead of measuring the angle between two views made along different lines of sight at the same time, we need to compare measurements made six months apart. For even the nearest star, the change in position, or **angular shift,** relative to distant stars is tiny.

To achieve such precision for observations made at different times, astronomers observe a star and carefully measure its position against background stars, as shown in Figure 54.6A. They then wait six months, until the Earth has moved to the other side of its orbit, and make a second measurement. As Figure 54.6B shows, the star will have a different position when compared with the background of stars, as seen from the two different points six months apart. The amount by which the star's apparent position changes depends on its distance from the Earth, just as the angular shift of your thumb depends on how far you hold it away from your eyes.

In the late 1500s, from his observatory at Uraniborg, Tycho Brahe carried out measurements of star positions that could have detected an angular shift of about

1 arc minute. However, he failed to detect a shift, so he concluded that the Earth must be stationary. Astronomers recognized at the time that this failure to detect parallax might have been because the stars are very far away, but the distances required seemed absurdly large to many. In fact, detecting parallax required the invention of the telescope and more than two centuries of improvements in optics and measurement techniques before the parallax shift was finally detected.

The distances of stars are so large that the parallax effect is extremely small—so small that it is measured in the fractions of a degree called **arc seconds**. One arc second is 1/3600 of a degree, or 1/60 of an arc minute. It may help you visualize how tiny an arc second is if you keep in mind that 1 arc second is equivalent to the angular size of a U.S. penny from a distance of 4 kilometers (2.5 miles).

54.3 CALCULATING PARALLAXES

As the Earth orbits the Sun, stars appear to move in the sky in a direction opposite to the Earth's orbit. Each of them moves along a small circle, ellipse, or line, depending on their position relative to the Earth's orbital plane. For all but a few of them, the distances are so large that their positional shifts are virtually undetectable. For some nearer stars, however, we can detect a shift relative to the background stars.

Mathematically, astronomers define a star's parallax, p, as the angle by which a star shifts to each side of its average position (see Figures 54.6C and 54.7). Because the distance d is inversely proportional to parallax, as discussed earlier, we can write

$$d = \frac{\text{constant}}{p},$$

where the constant depends on the size of the baseline (2 AU) and the units in which we measure angles and distances.

To make this equation easier to use, astronomers invented a new distance unit, called the **parsec**, by setting the constant in this equation equal to 1. If we measure p in arc seconds and set the constant to 1, the value of d from this formula comes out as parsecs. That is,

$$d_{\text{pc}} = \frac{1}{p_{\text{arcsec}}}.$$

d_{pc} = distance to star in parsecs
p_{arcsec} = parallax in arc seconds

We have added the subscripts "pc" (for parsecs) and "arcsec" (for arc seconds) as reminders that for this formula to work, the parallax must be measured in arc seconds, which will result in a distance in units of parsecs.

Figure 54.7 shows how we can determine the size of a parsec from the geometry of a right triangle with a baseline of 1 AU and a parallax of 1 arc second. From the way we have defined a parsec, someone one parsec from the Sun would see the Earth move 1 arc second to either side of the Sun as it orbited. Therefore, 1 astronomical unit is to the circumference of a 1-parsec-radius circle as 1 arc second is to 360°, as shown in the figure. Using the known size of an AU then allows us to determine that the size of a parsec is

$$1 \text{ parsec} = 3.09 \times 10^{13} \text{ kilometers}.$$

Comparing this distance to a light-year (9.46×10^{12} kilometers), it turns out that the parsec is just a few times larger:

$$1 \text{ parsec} = 3.26 \text{ light-years}.$$

The word *parsec* comes from a combination of *parallax* and *arc second*. Most astronomers use parsecs when discussing distances—they are easy to calculate

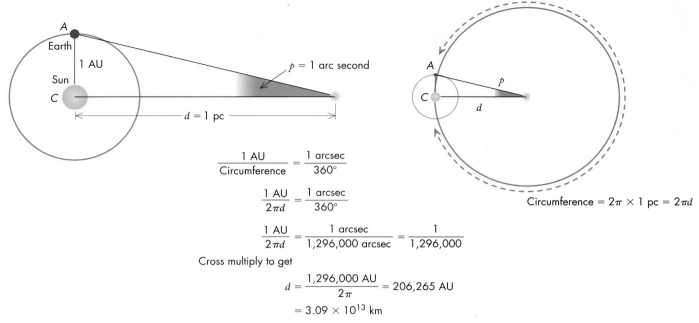

FIGURE 54.7
How to determine the distance of a star with a parallax of 1 arc second.

from parallax measurements. However, light-years are also frequently used, particularly when we discuss the time light takes to travel from an object to Earth—the conversion between travel time and distance is easy to calculate. It is a little like the difference in describing distances in meters versus feet.

To see how we can use this relation, suppose, for example, that we find from the shift in position of a nearby star that its parallax is 0.25 arc seconds. Its distance is then $d = 1/0.25 = 4$ parsecs. Similarly, a star whose parallax is 0.1 arc second is at a distance $d = 1/0.1 = 10$ parsecs from the Sun. Parallax measurements of nearby stars reveal that stars are typically separated by a few parsecs.

The star **Proxima Centauri** has the largest known parallax: 0.772 arc seconds. This indicates that it is at a distance $d = 1/0.772 = 1.30$ parsecs (or 4.22 light-years) from the Sun. Proxima Centauri is a dim companion of two slightly more distant stars, in orbit about each other, that have a parallax of 0.742 arc seconds, indicating a distance of 1.35 parsecs (or 4.40 light-years) from us. The larger of these two companions, Alpha Centauri A, is the third brightest star in the sky and is very similar to the Sun. This system of stars rises above the horizon only from latitudes south of 30° N and is too far south to be seen from most of the Northern Hemisphere.

Recognizing that these three stars are, in fact, close to one another in space already tells us something interesting. Not all stars are as isolated from other stars as is the Sun. Surveying other nearby stars, we see that some are isolated but most are members of small systems of stars.

Although the parallax–distance relation is mathematically a simple formula, obtaining a star's parallax to use in the formula is difficult because the angle by which the star shifts is extremely small. It was not until the 1830s that the first parallax was measured by the German astronomer Friedrich Bessel. Even now, the method fails for most stars farther away than about 100 parsecs. This is because the Earth's atmosphere blurs the tiny angle of their shift, making it very difficult to measure. Astronomers can avoid such blurring effects by observing from above the atmosphere. The European Space Agency launched the satellite *Hipparcos*—a clever acronym standing for HIgh-Precision PARallax COllecting Satellite in honor of the ancient Greek

LOOKING UP

The position of Proxima Centauri is shown in Looking Up #8 at the front of the book, but it is so dim, it is difficult to identify even in this deep photograph.

Concept Question 3

If you lived near Alpha Centauri and could look back and observe the Earth orbiting the Sun, how would the motion of the Earth compare to the parallactic motion of your star as seen from Earth?

FIGURE 54.8
The *Gaia* satellite will measure parallaxes for more than a billion stars.

astronomer Hipparchus—which measured the parallax of almost 120,000 stars from space. With its data, astronomers can accurately measure distances to stars as far away as about 500 parsecs. An even more ambitious mission, launching in 2013, should measure parallaxes for more than a billion stars (Figure 54.8).

The tiny parallaxes of stars tell us that they are so far away that it is hard to even imagine the distances involved. Astronomers throughout history have had as difficult a time understanding the immensity of these distances as have *any* of us, when first hearing of such numbers. Modern measurements tell us that the Sun is about 150 million kilometers (about 93 million miles) away, and that the next nearest star is about 42 trillion kilometers (about 25 trillion miles) away. One way of thinking of these huge distances is in terms of the time light takes to travel that far. Light moves so fast that it could circle the Earth almost eight times in 1 second. Light takes about 1.3 seconds to reach us at this speed from the Moon, and about 8.5 minutes to reach us from the Sun. But even from the very nearest star, Proxima Centauri, light takes 4.22 years to arrive here at Earth, and for most of the stars we can see at night, their light takes hundreds or thousands of years to reach us.

54.4 MOVING STARS

Although it took more than two centuries after Tycho Brahe before telescopes and measurement techniques were accurate enough to detect parallax, astronomers discovered other motions of the stars more than a century earlier. Observed over many human lifetimes, the positions of stars appear to remain fixed on the celestial sphere. This was one of the reasons it was easy for early astronomers to suggest that what we see at night are glowing dots on a celestial sphere that surrounds us. However, if we could watch the stars for millions of years, we would see them moving about like bees in a swarm.

In the short history of human records of the sky, the motion of stars is almost imperceptible. Nonetheless, in 1718, by comparing Tycho Brahe's star positions to those listed in ancient catalogs, Edmund Halley discovered that stars were shifting positions. This effect is called **proper motion**—the term *proper* here indicates a property of the star, as opposed to an *apparent* effect caused by the Earth's own motion.

As a result of proper motion, the pattern of stars we see gradually changes. The configuration of constellations 50,000 years from now will be significantly different from their appearance today. The changing appearance of the Big Dipper, for example, is illustrated in Figure 54.9.

Because they are close to us, some stars move relatively rapidly over time on the celestial sphere. For example, Barnard's Star moves more than 10 arc seconds each year, and other nearby stars, such as Proxima and Alpha Centauri, also move by several arc seconds each year. Most stars, however, shift position much more slowly—not necessarily because their speed through space is slow, but because they are so far away that they must move a large distance for us see a significant angular shift. This is the same effect we see when a nearby car moves past us rapidly, while an airplane overhead seems to move much more gradually, even though it is traveling at a much greater speed than the car.

Concept Question 4

If a star has proper motion, is orbiting another star, and has a measurable parallax, what will its motion through the sky look like to us? How can we distinguish the different motions?

FIGURE 54.9
Proper motion of stars gradually alters their positions on the sky. The three figures show the pattern of the Big Dipper 50,000 years ago (left), today (middle), and 50,000 years in the future (right).

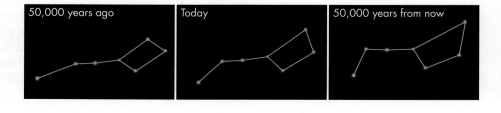

The actual speed of the star across, or *transverse to*, our line of sight is called its **transverse velocity.** The transverse velocity, V_T, can be determined from the proper motion and distance of the star. Astronomers usually use the Greek letter μ (mu) to represent the proper motion in arc seconds per year. In that case the transverse velocity turns out to be

$$V_T = 4.74 \text{ km/sec} \times \mu_{\text{arcsec/yr}} \times d_{\text{pc}},$$

where μ is measured in arc seconds per year and d is measured in parsecs as indicated by the subscripts after each.

By the early 1800s, careful examination of stars' positions also showed that some stars orbited each other (see Unit 57); and by the late 1800s astronomers were also able to detect stars' motions from their spectra, using the Doppler shift (Unit 25). This is the change in wavelength of light from a source as the source moves toward or away from us. If a source moves toward us, its wavelengths are shortened, whereas if it moves away, they are lengthened. The amount of wavelength shift depends on the source's speed along our line of sight—its **radial velocity.** Note that a star will have a radial velocity only if its motion has some component toward or away from us, and it will have a transverse velocity only if it has a motion perpendicular to its direction from us. The overall **space velocity** of the star can be determined by vector addition (see Unit 15) of the star's radial and transverse velocities.

What we learn of stars' motions by measuring Doppler shift and radial velocity can give us clues about an individual star's past. For example, we have found some stars that are moving at many hundreds of kilometers per second relative to all other stars. There is evidence that some of these "runaway" stars were flung out of orbit after a second star they were orbiting exploded!

54.5 THE ABERRATION OF STARLIGHT

The Earth's own space velocity has a surprising effect on the positions of stars. This was discovered in 1729, more than a century before parallax was detected. James Bradley, an English astronomer, discovered a shift in the position of stars during the course of the year. At first he believed that this was the long-sought parallax, but it proved to be a different effect of Earth's motion around the Sun.

Bradley observed that all stars in the sky exhibit a shift in different directions depending on the time of year. The shift reaches a maximum of about 20 arc seconds for all stars at certain times during the year depending on their position on the sky. This is about 25 times larger than the largest parallax shift, and it does not depend on a star's distance. Instead it is equally large for all stars in the same part of the sky. This makes it a difficult effect to observe directly because we do not see it as a shift of one star relative to other stars seen near it.

What Bradley saw is easiest to visualize for a star that passes straight overhead. Imagine a telescope that is cemented into the ground and points straight up to the zenith. We would expect a star that passes straight overhead to pass through the middle of the field of view of the telescope once every rotation of the Earth. An example of the effect we might see instead is that the star was shifted 20 arc seconds to the north when it passed through the field of view in September. Then in December the star crosses through the field of view a little late; in March the star passes 20 arc seconds to the south of the zenith position, and in June it crosses through a little early.

If we marked all stars' positions throughout the year, we would see that they move along little ellipses or circles on the sky. This is similar to the kind of motion

---

V_T = transverse velocity of star
μ = proper motion (arc seconds/yr)
d = distance (parsecs)

Mathematical Insight

Because the radial and transverse velocities are perpendicular to each other, the magnitude of a star's space velocity can be determined by the Pythagorean theorem:
$$V = \sqrt{V_R^2 + V_T^2}.$$

Mathematical Insight

The shift of 20 arc seconds represents the angle of a right triangle with one side represented by the speed of light $c = 300{,}000$ km/sec and another by the speed of the Earth $V_\oplus = 29.8$ km/sec. This angle is $360° \times (V_\oplus/c)/(2\pi) = 0.00569° \approx 20$ arc seconds.

FIGURE 54.10

The "aberration of starlight" is an effect of the Earth's motion through space. If a telescope is pointed toward the actual direction of a star (upper panel) the light won't reach the back of the telescope. We must tilt telescopes into the direction of the Earth's motion (lower panel) so that photons entering the front of the telescope will reach the back of the telescope. Six months later, when the Earth is traveling in the opposite direction, we must tilt the telescope in the opposite direction.

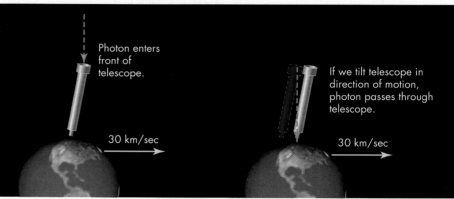

Concept Question 5

Would the aberration of starlight be larger or smaller on Mars than on Earth? Why?

A star will pass through the center of the field of view of a stationary telescope once every sidereal day (Unit 7), equaling 23 hours 56 minutes.

that is caused by parallax, but the stars always shift in a direction *ahead* of the Earth's motion through space as the Earth orbits the Sun.

Today astronomers call this shift in the positions of stars the **aberration of starlight,** and they recognize that it occurs because of the finite speed of light. We must tilt a telescope so it points ahead of the direction to a star because between the time light enters the front end of the telescope and the time it reaches the back end of the telescope, the Earth has shifted position in space (Figure 54.10). This is similar to the reason you should hold an umbrella tilted slightly ahead of you if you are running through vertically falling rain. To keep the lower part of your body dry, you need to block the raindrops that are falling where your body *will* be.

If you ran through the rain around a circular racetrack, you would tilt your umbrella ahead of you in whatever direction you ran, in the same way astronomers must tilt their telescopes to compensate for the Earth's orbital motion. Note that you would have to tilt the umbrella only slightly if you were running slowly, but by a larger amount if you were running faster. Likewise, we tilt the telescope forward by an amount determined by the ratio of the Earth's speed through space and the speed of light.

A century before the aberration of starlight was discovered, Galileo and Kepler had convinced most astronomers that the Earth went around the Sun. However, a few astronomers continued to cling to a belief in a geocentric universe, partly for religious reasons. Tycho Brahe's model in which the planets orbited the Sun while the Sun orbited the Earth (Unit 12) continued to be taught into the early 1700s until Bradley's discovery firmly and finally proved that the Earth circled the Sun.

KEY POINTS

- By measuring the length of one side of a triangle and the angles of the other two sides, it is possible to determine the triangle's size.
- Triangulated distances of Mars and Venus in the seventeenth and eighteenth centuries showed that the AU was much larger than then thought.
- To triangulate stars, astronomers measure parallax: the angle shift of the star from one side of the Earth's orbit to the other.
- The parallax angle grows smaller with distance, making nearer stars easier to measure than distant ones.
- Stars have motions through space relative to the Sun that gradually change the shapes of constellations over thousands of years.
- Earth's motion through space requires us to point a telescope slightly in the direction of Earth's motion to keep the stars in view.
- Detection of stellar shifts proved that Earth orbits the Sun.

KEY TERMS

aberration of starlight, 438
angular shift, 433
arc minute, 430
arc second, 434
astronomical unit, 430
baseline, 430
Hipparchus, 430
parallax, 432
parsec, 434
proper motion, 436
Proxima Centauri, 435
radial velocity, 437
space velocity, 437
transit, 431
transverse velocity, 437
triangulation, 429

CONCEPT QUESTIONS

Concept Questions on the following topics are located in the margins. They invite thinking and discussion beyond the text.

1. Why larger baselines give more accurate results. (p. 430)
2. Scientific reluctance to accept big changes. (p. 431)
3. View of Earth from star where we see parallax. (p. 435)
4. Apparent motions of orbiting star. (p. 436)
5. Aberration of starlight on Mars versus Earth. (p. 438)

REVIEW QUESTIONS

6. How can distances be measured by triangulation?
7. How are stars' distances measured? How do we know that Proxima Centauri is the closest star to Earth?
8. What is the largest distance we are able to measure for stars? Why can't the distances of farther stars be measured?
9. How is a parsec defined? How big is a parsec compared with a light-year?
10. What is proper motion? How is it related to space velocity?
11. Why would a star with a large proper motion be likely to be close to Earth?
12. Why does the speed of the Earth's motion through space make stars' positions appear to shift?

QUANTITATIVE PROBLEMS

13. Using triangulation, make an approximate measurement of the distance from where you live to a neighboring building. Make a scale drawing, perhaps letting 1 cm represent 1 m (or you might let 1 inch represent 10 feet), and choose a baseline. Construct a scale triangle on your drawing and find how far apart the buildings are from one another.
14. Procyon has a parallax of 0.284 arcsec. How far away is it in parsecs and in light-years?
15. A star is believed to be 5,000 pc away. What should its parallax be?
16. If a star's parallax is 0.1 arcsec on Earth, what would it be if measured from an observatory on Mars?
17. Suppose a satellite is developed that can measure the parallax of stars with an uncertainty of 2×10^{-5} arcsec. Suppose a star is measured with a parallax of 3×10^{-5} arcsec with this satellite. What range of distances could the star be at and still fall within the range of uncertainty of the measurement?
18. Barnard's Star has a parallax of 0.545 arcsec, a proper motion of 10.3 arc seconds per year, and a radial velocity of -110 km/sec (the negative sign means it is approaching Earth). What is the star's overall speed through space?
19. How much greater is the angle of aberration of starlight measured from Mercury compared to that measured from the Earth? What would the angle of aberration of starlight be as measured from Saturn?

TEST YOURSELF

20. A star has a parallax of 0.025 arcsec. What is its distance?
    a. 4 pc
    b. 40 pc
    c. 20 pc
    d. 25 pc
    e. 250 pc
21. Star A is 10 times farther than star B. Star A's parallax is therefore _____ star B's.
    a. 10 times larger than
    b. 100 times larger than
    c. 10 times smaller than
    d. 100 times smaller than
    e. the same as
22. The most favorable conditions for observing the angular shift of a star occur when
    a. a large star moves across a field of smaller stars.
    b. a small star moves against a field of larger stars.
    c. Venus transits the Sun.
    d. a nearby star moves against a field of farther stars.
    e. nearby stars move relative to one another.
23. What was the first observation that definitively proved that the Earth orbits the Sun?
    a. Detection of parallax by a few nearby stars
    b. Measurements of Venus transiting the Sun
    c. The aberration of starlight for almost any star
    d. Triangulation of Mars
    e. Radar observations of the Moon

UNIT 55

The Luminosities of Stars

55.1 Luminosity
55.2 Measuring Luminosities Using the Inverse-Square Law
55.3 Distance by the Standard-Candles Method
55.4 The Magnitude System

Learning Objectives

Upon completing this Unit, you should be able to:
- Define luminosity, brightness, and magnitude and give examples of each.
- Explain how brightness changes with distance, and calculate brightness from luminosity and distance.
- Explain and apply the standard-candles method.
- Define absolute magnitude, and relate different magnitudes to each other.

FIGURE 55.1
The brightest star in the image, Alpha Centauri, one of the nearest stars to us, at a distance of about 4.3 ly. Beta Centauri is nearly as bright, but over 500 ly away. Comet Halley looks almost as bright as the stars, but it is less than 1 AU—just light-*minutes*—distant.

It is hard to believe that the stars we see in the night sky as tiny glints of light are in reality huge, dazzling balls of gas similar to our Sun. In fact, many of those gleaming specks are vastly larger and brighter than the Sun. Stars look dim to us only because they are so far away—it is several **light-years** (tens of trillions of kilometers or miles) to even the nearest. Such remoteness creates a challenge for astronomers trying to understand the nature of stars. Stars may look similarly bright in a photograph, yet have vastly different distances and light output (Figure 55.1).

Although we cannot send space probes to study the stars, because of their vast distances, we can learn about many properties of stars from their light, using physical laws and theories to interpret the measurements. This area of astronomy, known as **astrophysics**, has the remarkable ability to give us detailed "pictures" of stars, even though nearly all of them look like points of light under even the highest magnifications of the largest telescopes. We will see in Unit 56 that we can determine stars' temperatures and compositions, and in subsequent Units we will see how we can find their masses and sizes. We begin in this Unit by determining a star's total light output by applying the inverse-square law that we derived in Unit 21.

55.1 LUMINOSITY

One critical property for understanding a star is the total amount of energy it radiates out into space each second: its **luminosity,** L. A star's luminosity tells us how much energy is being generated within it, which is one of the most important differences between star types.

An everyday example of luminosity is the wattage of various lightbulbs. A typical table lamp bulb has a luminosity of 100 watts, whereas a bulb for an outdoor parking lot light may have a luminosity of 1500 watts. By comparison, stars are almost unimaginably more luminous. The Sun, for example, has a luminosity of about 4×10^{26} watts. This "wattage" is fairly typical of stars, although they range from millions of times greater to millions of times smaller.

These numbers are so enormous that it is more convenient to use the Sun's luminosity as a standard unit, where $1\ L_\odot = 4 \times 10^{26}$ watts (Unit 3). When we speak, then, of a star with a luminosity of $2\ L_\odot$ or $100\ L_\odot$, we mean it has a luminosity of 8×10^{26} or 4×10^{28} watts, respectively.

Stars generate their luminosity by fusing the nuclei of atoms together in their cores. A star's luminosity indicates not only the rate at which it is emitting energy into space, but also the rate at which it is fusing atoms in its core. Over thousands of years of observation, the light we see from nearly every star in the night sky has remained steady. From this constancy, we conclude that the energy generation process deep inside stars must be very stable, steadily replacing the energy radiated to space.

The Sun must convert 4 million tons of matter into energy every second to sustain its luminosity (Unit 52). One of the most luminous stars known, discovered by the Hubble Space Telescope in 1997, has 10 million times the Sun's luminosity. At that phenomenal rate of energy release, every few months the star is annihilating as much mass as is contained in the entire Earth.

55.2 MEASURING LUMINOSITIES USING THE INVERSE-SQUARE LAW

When we look at stars at night, some look brighter than others, but this is not necessarily because they are more luminous. We all know that a light looks brighter when we are close to it than when we are far from it. As light travels outward, its energy spreads uniformly in all directions. If you are standing near a light source, the light will have spread out only a little, and so more light enters your eye; if you are farther away, less light enters your eye, which you will perceive as the light source being dimmer.

We examined this dimming with increasing distance in Unit 21. Photons leaving a star or other light source spread out along straight lines in all directions. If you imagine small and large spheres around the light source (Figure 55.2), the same number of photons pass through each sphere in 1 second. However, because the more-distant sphere is larger, the number of photons passing *through a fixed area* on that sphere, such as the aperture of a telescope, is smaller.

The amount of light reaching us from a star is called its **brightness** and is a function of distance. This is different from a star's luminosity, which is the total amount of light emitted. The luminosity is expressed by the number of watts of light energy generated by the star, but we describe brightness as the wattage received *per square meter*. That is, if we had a telescope with a collecting area (Unit 29) of one square meter, and we collected 10 watts of light, we would conclude that the brightness was 10 watts per square meter. The same light source measured with a telescope with a collecting area of just 0.1 square meters would collect only 1 watt of light.

> **Clarification Point**
>
> Some astronomers use the terms *flux* or *apparent luminosity* to describe the brightness. Others call the luminosity the *absolute brightness*.

FIGURE 55.2
Light from a star decreases according to the inverse-square law. Light leaving a star during an instant can be imagined to form a bubble around the star that expands at the speed of light. The fraction of the light that enters a fixed-size telescope aperture decreases as the inverse square of the distance to the star.

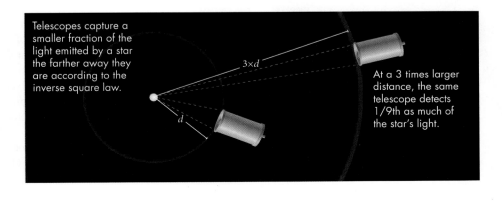

Telescopes capture a smaller fraction of the light emitted by a star the farther away they are according to the inverse square law.

At a 3 times larger distance, the same telescope detects 1/9th as much of the star's light.

B = Brightness of received light (watts/meter^2)

L = Luminosity of star (watts)

d = Distance to star (meters)

Inverse-square law

A standard 40-watt lightbulb produces about 1 watt of luminous power at visible wavelengths. The difference between the electrical power consumption (40 watts) and the light delivered (1 watt) occurs because most of the power is radiated as heat at infrared wavelengths we cannot see. Stars also generate some of their light at infrared and ultraviolet wavelengths, but a star like the Sun emits most of its power at visible wavelengths.

To picture how the brightness of a star depends on its distance from us, imagine all the photons emitted by a star in a single moment. At some later time, those photons will have traveled a distance d, forming a sphere surrounding the star. The star's luminosity L will be spread over the surface of a sphere of radius d, as shown in Figure 55.2. We use d to stand for radius here to emphasize that we are speaking of a distance. The surface area of the sphere is given by the geometric formula $4\pi d^2$, so the brightness we observe, B, is calculated as follows:

$$B = \frac{L}{4\pi d^2}.$$

This relationship is the **inverse-square law** that we derived in Unit 21. It puts into a mathematical form the everyday experience that distant light sources look dimmer.

To get an idea of the brightness levels of stars, you could perform the following experiment. Take a light source that generates 1 watt of luminosity and place it at various distances until it looks similar in brightness to a bright star. It turns out that you will have to place it about 2 kilometers away to look about as bright as the brightest stars in the sky. By applying the brightness formula to this known source, you would find that its brightness is

$$B = \frac{1\text{ watt}}{4\pi(2000\text{ m})^2} = 2 \times 10^{-8}\text{ watt/m}^2.$$

This tells us that the brightest stars in the sky deliver only about 2 hundred-millionths of a watt per square meter. Because the pupil of your eye is at most about 8 millimeters in diameter, your eye is sensing only about a *trillionth* of a watt when you look at a bright star.

The inverse-square law is one of the most powerful tools of astrophysics. We can measure B for a star by using a **photometer,** a device similar to the electronic exposure meter in a camera. A photometer measures the amount of electromagnetic energy that strikes it each second. When calibrated for the size of the telescope, this can be converted into watts per square meter. If we can also determine the star's distance, d, using a technique such as parallax (Unit 54), for example, then we can calculate the star's luminosity, L, by rearranging the inverse-square law:

$$L = B \times 4\pi d^2.$$

Photometric measurements indicate that although many stars have luminosities comparable to the Sun's, some stars are millions of times more or less luminous. Determining what causes this huge range of luminosities is one of our challenges in understanding stars.

55.3 DISTANCE BY THE STANDARD-CANDLES METHOD

We use brightness to estimate distance in many situations. For example, suppose you look at two street lights that have bulbs of the same wattage, but one street light is close and one is far away. From how bright they appear, you can estimate how much farther away the dim one is than the bright one (Figure 55.3). In fact, if you drive at night, your life depends on making such distance estimates when you see traffic lights or oncoming cars. Astronomers use a more refined version of this idea to find the distances to stars and galaxies using a photometer to make precise measurements of sources with known luminosities.

In the early history of photometry, scientists used as their standard of light generation a particular kind of candle that burned wax at a specified rate. These were well-calibrated light sources and were known as "standard candles." Today, if we can

FIGURE 55.3
The brightness of street lights of the same luminosity diminishes with distance.

find any particular type of star whose luminosity is well determined and can therefore serve as a reference for other stars, we call them **standard candles** too. Because stars have such an immense range of luminosities, finding a good standard candle is not easy. However, some stars have peculiar characteristics—such as unusual aspects of their spectra, or curious ways in which their brightness pulsates—that allow them to be identified as unique. Tests have shown that some of these unique classes of stars have well-determined luminosities, making them good standard candles.

The inverse-square law provides a way to determine a star's distance if we know the star's luminosity and can measure how bright it appears. Finding the distances of stars based on their luminosities works only if they are good standard candles—that is, we know the luminosities with good accuracy. For example, if we determine that the properties of some star are almost identical to the Sun's, we might reasonably infer that its luminosity would be the same as the Sun's too. Then we can measure its brightness, B, and solve for its distance using the inverse-square law. Rearranging that equation (see margin note), we get

$$d = \sqrt{\frac{L}{4\pi B}}.$$

> **Mathematical Insight**
>
> Deriving the distance formula:
>
> $$B \times \left(\frac{d^2}{B}\right) = \frac{L}{4\pi d^2} \times \left(\frac{d^2}{B}\right)$$
>
> $$\sqrt{d^2} = \sqrt{\frac{L}{4\pi B}}$$
>
> $$d = \sqrt{\frac{L}{4\pi B}}$$

For example, Alpha Centauri A is a star whose color and spectrum have been found to be very similar to the Sun's. Using a photometer, we find that its brightness is about 3×10^{-8} watts per square meter. If we assume it has the same luminosity as the Sun, we can plug the numbers into the above equation:

$$d = \sqrt{\frac{4 \times 10^{26} \text{ watts}}{4\pi \times 3 \times 10^{-8} \text{ watts/m}^2}} = \sqrt{1 \times 10^{33} \text{ m}^2} = 3 \times 10^{16} \text{ m}.$$

Converting this to parsecs, we find a distance of about 1 parsec, which is slightly less than Alpha Centauri's measured distance of 1.35 parsec. The match is not perfect because Alpha Centauri is slightly more luminous than the Sun, but before astronomers were able to detect parallaxes, this kind of comparison provided the first clue that stars are at such enormous distances.

> **Concept Question 1**
>
> How could you determine the light output of a candle compared to that of a lightbulb?

The Sun is not a particularly good standard candle, because many stars look quite similar to it yet have substantially different luminosities. For the best standard candles, astronomers can use this same method to find distances that may be accurate to better than 10%. This distance-finding scheme is referred to as the *method of standard candles*. This is a powerful method for the following reason: If we identify a set of stars that all have the same luminosity, *we can determine the distances to all of them if we can find the distance to any one of them.*

Care must be taken, though! The accuracy of the standard-candles method relies on the degree to which the stars actually have the same luminosity. If we mistakenly group stars of dissimilar luminosities, we will get incorrect distances. This kind of mistake has happened several times in the history of astronomy, forcing us to revise distance estimates, and even forcing us to revise our notions of the size of the entire universe!

55.4 THE MAGNITUDE SYSTEM

> The magnitude system is complicated to use in calculations. It is not necessary to learn it in detail, but it is handy to know at least a little about the system, because star brightness levels are usually given in magnitudes in articles and on star charts.

About 140 B.C.E., the Greek astronomer Hipparchus measured the apparent brightness of stars using units he called *magnitudes*. He designated the stars that looked brightest "magnitude 1" and the dimmest ones he could just barely see "magnitude 6." For example, Betelgeuse, a bright red star in the constellation Orion, is magnitude 1, and there are only about two dozen first-magnitude stars over the whole celestial sphere. The somewhat dimmer stars in the Big Dipper's handle are approximately magnitude 2, whereas the stars in the bowl of the Little Dipper

FIGURE 55.4
In Ursa Minor, the bowl of the "Little Dipper" is made of stars of magnitudes 2, 3, 4, and 5 at each of its corners. Polaris is also a second-magnitude star.

LOOKING UP

You can use the Big Dipper to help you find the Little Dipper. See Looking Up #2 at the front of the book.

Concept Question 2

Where else do we use systems like the magnitude system, where a step is used to indicate a change by some factor?

Mathematical Insight

The magnitude system is based on logarithms. The difference between the magnitude m of two sources depends on the ratio of their brightnesses as follows:

$$m_1 - m_2 = -2.5 \log(B_1/B_2).$$

The difference between an object's apparent magnitude m and its absolute magnitude M is:

$$m - M = 5 \log(d/10)$$

where d is its distance in parsecs.

range from magnitude 2 to magnitude 5 (Figure 55.4). From a dark clear sky, you may be able to see about 4500 stars, about two-thirds of which are magnitude 6.

Astronomers still use Hipparchus's scheme to measure the brightness of astronomical objects, but they now use the term **apparent magnitude** to emphasize that they are measuring only how bright a star *appears* to an observer. A star's apparent magnitude is a way of indicating its brightness. Astronomers use the magnitude system for many purposes (for example, to indicate the brightness of stars on star charts, like the foldout chart in the back of this book), but it has several confusing properties. First, the scale is "backward" in the sense that bright stars have "smaller" magnitudes, while dim stars have "larger" magnitudes. Moreover, modern measurements show that Hipparchus underestimated the magnitudes of the brightest stars, and so the magnitudes now assigned to the brightest few stars are negative numbers!

Magnitudes are not easy to work with because differences in magnitudes correspond to ratios in brightness. In particular, a *difference* of five magnitudes corresponds to a *ratio* of 100 in brightness. So when we say one star is five magnitudes brighter than another, we mean it is a factor of 100 brighter. That is, if we measure the brightness of a first-magnitude star and a sixth-magnitude star, the first-magnitude star is 100 times brighter than the sixth.

Each magnitude difference corresponds to a factor of about 2.512 (the fifth root of 100) in brightness. A first-magnitude star is 2.512 times brighter than a second-magnitude star and is 2.512 × 2.512, or 6.310, times brighter than a third-magnitude star. It is possible to make finer gradations by measuring fractions of a magnitude. For example, 0.1 magnitudes is a change of about 10% in brightness. Table 55.1 lists the ratios that correspond to various differences in magnitude.

The magnitude system becomes quite confusing when discussing bright objects because they have "negative" magnitudes. For example, at its brightest, Venus is about 100 times brighter than a first-magnitude star. Because a factor of 100 in brightness corresponds to a difference of five magnitudes, the apparent magnitude of Venus is −4 at its brightest. On the other hand, the *faintest* stars seen by the Hubble Space Telescope have magnitudes "greater" than 30.

When using magnitudes, it is important to understand the distinction between what astronomers call **absolute magnitude** and the apparent magnitude we observe. Absolute magnitude is defined as the apparent magnitude a star would have if it were situated at a distance of 10 parsecs from us. Although we cannot, of course, move stars to whatever distance we choose, absolute magnitude serves as a way of indicating a star's luminosity in units that are mathematically compatible with the apparent magnitude system. Table 55.2 illustrates how absolute magnitude is related to a star's luminosity.

TABLE 55.1 Magnitude Differences and Brightness Ratios

Magnitude Difference	Ratio of Brightness		
0	2.512^0	=	1:1
0.1	$2.512^{0.1}$	=	1.10:1
0.5	$2.512^{0.5}$	=	1.58:1
1	2.512^1	=	2.512:1
2	2.512^2	=	6.310:1
3	2.512^3	=	15.85:1
4	2.512^4	=	39.81:1
5	2.512^5	=	100:1
10	2.512^{10}	=	10^4:1
20	2.512^{20}	=	10^8:1

TABLE 55.2 Absolute Magnitude and Luminosity

Absolute Magnitude	Approximate Luminosity in Solar Units	
−10	1,000,000	= 10^6
−5	10,000	= 10^4
0	100	= 10^2
5	1	= 10^0
10	0.01	= 10^{-2}
15	0.0001	= 10^{-4}

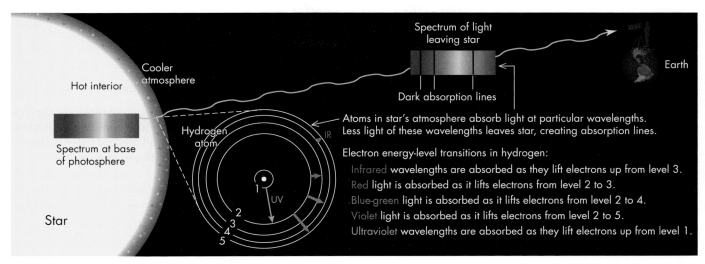

FIGURE 56.1
Formation of stellar absorption lines. Atoms in the cooler atmospheric gas absorb radiation at wavelengths corresponding to jumps between electron orbitals. Absorption lines of hydrogen are illustrated.

Concept Question 1

At the wavelength of an absorption line in a star's spectrum, we can usually still see *some* photons coming from the star. What elevation would you expect these photons to come from, compared to photons that are far from any absorption line?

Not all photons have an equal chance of escaping to space from the photosphere. The photosphere is relatively cool (compared to the interior of the star), so some electrons recombine with nuclei there. Some stars' photospheres are so cool that atoms even stick together to form molecules. These atoms and molecules will absorb photons whose energies match the energies needed to lift electrons into higher energy levels. Because a photon's energy is determined by its wavelength, or color, when we observe the spectrum of light from a star, we find that certain colors are more strongly blocked. This creates dark **absorption lines** (Unit 24) in the star's spectrum, as shown in Figure 56.1.

Each type of atom—hydrogen, helium, calcium, and so on—absorbs photons with a unique set of wavelengths. For example, hydrogen absorbs photons with wavelengths of 656.3, 486.1, and 434.1 nanometers, which are, respectively, in the red, blue-green, and violet parts of the spectrum. Gaseous calcium, on the other hand, absorbs strongly at 393.3 and 396.8 nanometers, producing a dark double line in the violet portion of the spectrum. Because each type of atom absorbs a unique combination of wavelengths of light, the set of dark lines in a star's spectrum—its "fingerprint"—depends on the composition of the gas in the photosphere.

It is possible, therefore, to measure a star's composition by comparing the absorption lines in its spectrum to the wavelengths of lines made by each kind of atom. When we find matching absorption lines, we infer that the element exists in the star. To find the quantity of each atom in the star—each element's abundance—we use the darkness of the absorption line. A darker line generally implies a greater amount of that particular element, although some spectral lines are naturally stronger than others, so laboratory and theoretical work is needed to interpret the line strengths. The temperature of the atmosphere is also important because the hotter the gas, the more frequent and higher the speed of collisions between the atoms. Collisions knock the electrons into different energy levels, changing which spectral lines an atom absorbs. We therefore need to know the temperature in a star's photosphere to interpret its spectrum.

56.2 STELLAR SURFACE TEMPERATURE

To determine how hot a star's surface is, we can use the same method used to judge the temperature of an electric stove burner or a piece of glowing charcoal. A bright orange glow indicates a higher temperature than a dull red glow (Unit 23.2). Even the coolest stars have surface temperatures hotter than a stove burner can reach.

They are about as hot as the 3300-K yellow-white tungsten filament in a bright lightbulb. Many stars are even hotter and glow with a blue or even violet color.

Thus, a star's surface temperature can often be deduced from the color of its emitted light, hotter stars emitting shorter wavelengths more strongly:

Hotter stars emit more blue light, and cooler stars emit more red light.

You can see such color differences if you look carefully at stars in the night sky. For example, Rigel and Betelgeuse, the two brightest stars in the constellation Orion, have distinctly different colors (Figure 56.2). Rigel has a blue tint whereas Betelgeuse is reddish, so even our eyes can tell us that stars differ in temperature.

We can use color in a more precise way to measure a star's temperature with Wien's law (Unit 23.4). Wien's law relates an object's temperature to the wavelength, λ_{max}, at which it radiates most strongly, as shown in Figure 56.2. Wien's law lets us calculate the temperature T from λ_{max} using this formula:

$$T = \frac{2.9 \times 10^6 \text{ K} \cdot \text{nm}}{\lambda_{max}}.$$

T = Temperature of continuum source (kelvin)

λ_{max} = Wavelength of maximum brightness (nanometers)

This law applies specifically to sources of thermal radiation, and the photospheres of stars are fairly good examples of this.

As a demonstration of Wien's law, consider Rigel again. Detailed measurements of its spectrum show that it radiates most strongly (has a wavelength of maximum brightness λ_{max}) at about 240 nanometers, in the ultraviolet range. Its temperature is therefore about

$$T(\text{Rigel}) = \frac{2.9 \times 10^6 \text{ K} \cdot \text{nm}}{240 \text{ nm}} = \frac{2{,}900{,}000}{240} \text{ K} = 12{,}000 \text{ K}.$$

The spectrum of light for 12,000 K thermal radiation is illustrated in Figure 56.2. Notice that even though the spectrum reaches a maximum outside the range of visible light, visible light is produced. The visible light is stronger toward the blue end of the spectrum, so Rigel looks blue to us.

The star Betelgeuse in Orion looks reddish to our eyes (Figure 56.2), but its wavelength of maximum brightness is actually in the infrared, at about 830 nanometers.

FIGURE 56.2
From the photograph at right you can see that Betelgeuse is red compared to Rigel. A plot of each star's relative brightness at different wavelengths shows that they peak at different wavelengths. Note that neither star emits its peak emission within the visible range, but they are brighter at one end or the other of the visible spectrum.

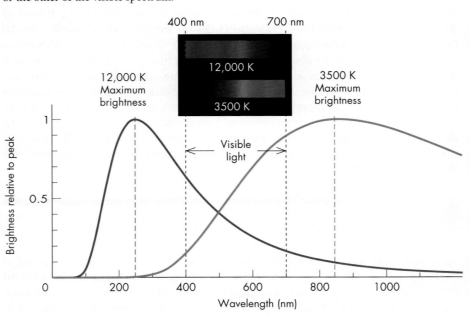

So if we calculate its surface temperature, we find:

$$T(\text{Betelgeuse}) = \frac{2{,}900{,}000}{830}\ \text{K} = 3500\ \text{K}.$$

This is only a little bit hotter than the temperature of the filament inside a standard 100-watt lightbulb. Sometime when you observe stars, compare the color of a distant incandescent bulb with Betelgeuse or another of the reddish stars you see in the sky. That reddish color also helps explain why photographs taken using indoor lighting often look yellow or orange compared with those taken in sunlight. Like Betelgeuse, lightbulbs emit radiation mostly in the infrared. We see their visible light because they also emit electromagnetic radiation at shorter wavelengths than $\lambda_{\max}$.

Wien's law is important for interpreting stellar colors, but for many stars like Rigel and Betelgeuse the maximum brightness lies outside the visible-wavelength range. In addition, light absorption by atoms and molecules in a star's atmosphere may make it difficult to determine the peak wavelength of the thermal radiation. Fortunately, studies of the spectra of stars have revealed another way to determine stars' temperatures.

> **Clarification Point**
>
> Astronomers describe stars as blue or red, but the colors are very pale (or pastel) since they include a mixture of all visible wavelengths, which we see as white.

> **Concept Question 2**
>
> In Unit 23 we saw that cooler sources emit much less light than hot sources. How can Betelgeuse be about as bright as Rigel?

56.3 THE DEVELOPMENT OF SPECTRAL CLASSIFICATION

The classification of stellar spectra is important to astronomers today because it offers a quick way to determine a star's temperature. But this discovery took decades of study. **Stellar spectroscopy**—the study and classification of spectra—originated early in the nineteenth century when the German scientist Joseph Fraunhofer discovered absorption lines in the spectrum of the Sun and later in other stars.

In the late 1800s, Henry Draper, a physician and amateur astronomer, began recording spectra with the then-new technology of photography. By observing a field of stars through a thin prism, he spread each star's light into a spectrum (Figure 56.3). On his death, Draper's widow endowed a project at Harvard to create a compilation of stellar spectra. In this work, stars were assigned to alphabetic types running from A to Q, collecting together similar-looking spectra in each type. These classifications were based mostly on the strength of the spectral lines—how dark they were—particularly the lines of hydrogen that are seen in the visible part of the spectrum. A-type stars have the strongest lines of hydrogen, while O-type stars, for example, have very weak lines.

About 1901, Annie Jump Cannon (Figure 56.4), the astronomer doing most of the classification for the Draper Catalog, determined that the types fell in a more reasonable sequence if she rearranged them by temperature. She eliminated a number of types that had similar temperatures and reordered the rest into the sequence, from hottest to coolest, of O-B-A-F-G-K-M. Her work is the basis for the stellar **spectral types** we use today.

Application of Wien's law and theoretical calculations show that temperatures range from more than 30,000 K for O stars to less than 3500 K for M stars. A-type stars have temperatures ranging from about 7500 to 10,000 K, and G stars, such as our Sun, range between 5000 and 6000 K. Because a star's spectral type is set by its temperature, its type also indicates its color, ranging from violet-blue colors for O and B stars to orange and red colors for K and M stars.

To distinguish finer gradations in temperature, astronomers subdivide each type by adding a numerical suffix—for example, B0, B1, B2, ... , B9—with the smaller numbers indicating higher temperatures. With this system, the temperature of a B0

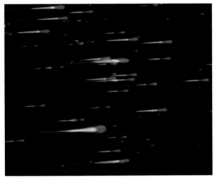

FIGURE 56.3
This is a photograph of the Hyades cluster made through a thin prism. The prism spreads the light of each star into a spectrum.

FIGURE 56.4
Annie Jump Cannon (1862–1941) founded the modern system of classifying stars.

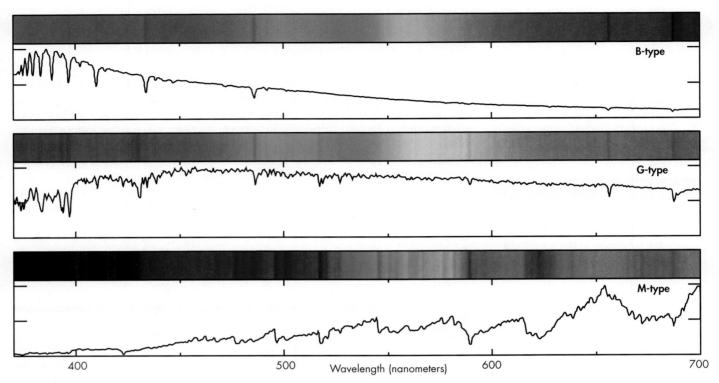

FIGURE 56.5
Spectra of stars of type B, G, and M. The B star is the hottest of the three, and the M star is the coolest. Cooler stars tend to have more absorption lines, but some lines are darker in hotter stars.

Concept Question 3

If the Sun had a temperature like that of Rigel or Betelgeuse, do you think our eyes would have evolved differently? Why or why not?

Stellar spectroscopy

star is almost 30,000 K, whereas that of a B5 star is about 15,000 K. Our Sun is a G2 star with a surface temperature of about 5800 K.

An example of the differences between spectra is given in Figure 56.5, which shows the spectra of stars with three different temperatures. The B-type spectrum is for a hot star like Rigel that produces mostly violet and blue light. The G-type spectrum is similar to the Sun's, with a temperature of about 6000 K and radiation peaks near the middle of the visible-wavelength range. The spectra are shown in Figure 56.5 both as they appear through a prism and as line tracings showing the relative intensity at each wavelength. The M-type spectrum is for a cool star like Betelgeuse. The B-type spectrum has only a few lines, but the lines are strong, and their pattern indicates they come from hydrogen. The M-type spectrum, however, shows a welter of lines with no apparent regularity.

The complicated history of the development of the spectral types resulted in the odd nonalphabetical progression for the spectral types—O-B-A-F-G-K-M. So much effort, however, had been invested in classifying stars using this system that it was easier to keep the types as assigned, with their odd order, than to reclassify them. Cannon, over her lifetime, classified nearly a quarter million stars!

In the late 1990s, astronomers started using sensitive infrared detectors that allow detection of stars even cooler than the M-type stars. Some of these stars are so cool that not only do molecules form in their atmospheres, but solid dust particles condense there as well. To continue the progression of spectral types, astronomers have designated two new types—L and T—to describe these cool objects. The full spectral type sequence is therefore now **O-B-A-F-G-K-M-L-T,** from hottest to coolest.

As a help to remember the peculiar order of spectral types, astronomers and students often use a mnemonic. Before the discovery of the L and T types, the most widely known mnemonic was "Oh, Be A Fine Girl/Guy. Kiss Me," which was printed in textbooks throughout the twentieth century. Perhaps you can develop a better mnemonic for the twenty-first century.

FIGURE 56.6
Cecilia Payne (1900–1979) discovered that hydrogen is the most abundant element in stars. She went on to become the first woman full professor at Harvard.

Balmer lines are named for Johann Balmer, a German scientist who first studied their pattern.

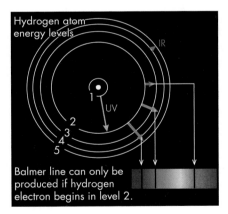

FIGURE 56.7
The visible-wavelength Balmer lines of hydrogen can only be produced if the gas is at the right temperature for electrons to already be in the second energy level.

56.4 HOW TEMPERATURE AFFECTS A STAR'S SPECTRUM

To understand why temperature affects the lines in a star's spectrum, recall that a photon can be absorbed by an atom only if the photon's energy exactly matches the energy difference between two electron orbitals. For an atom to absorb the photon, it must already have an electron in the orbital with the correct lower energy level. An atom may be abundant in a star's atmosphere but create only weak lines at a particular wavelength simply because the gas is so hot or so cold that its electrons are in the "wrong" level to absorb light at that wavelength.

This was first worked out in the 1920s by the U.S. astronomer Cecilia Payne (later Payne-Gaposhkin; Figure 56.6). She pointed out that the strength of the hydrogen lines in a star's spectrum depend strongly on the star's temperature. She recognized that it would be a mistake to assume that a star had little hydrogen if these lines were weak, unless the appropriate corrections for temperature had been made.

The absorption lines of hydrogen that are observable at visible wavelengths are made by electrons orbiting in the hydrogen atom's second level (Figure 56.7). These lines are sometimes called the **Balmer lines** to distinguish them from other hydrogen lines with ultraviolet and infrared wavelengths. The Balmer lines occur at wavelengths where light has exactly the amount of energy needed to lift an electron from a hydrogen atom's second energy level up to the third or higher levels. These lines are no more important than other transitions in hydrogen that occur at infrared or ultraviolet wavelengths, but because their wavelengths are in the visible spectrum they are much more easily observable.

If the hydrogen atoms in a star have very few electrons orbiting in level 2, the Balmer absorption lines will be weak, even if hydrogen is the most abundant element in the star. This may happen in a cool star, because most of the electrons in hydrogen atoms are in level 1, the lowest energy level. In a hot star the atoms move faster, and when they collide, electrons get excited ("knocked") into higher energy orbitals: the hotter the gas, the higher the orbital. In fact, in very hot stars, electrons may be knocked out of the atom entirely, in which case the atom is said to be *ionized*. As a result of this excitation, proportionally more of the electrons in a very hot star will be in level 3 or higher. Hydrogen has the maximum number of its electrons in level 2 when its temperature is about 10,000 K—corresponding to an A0 star. Hence, it is at this temperature that the Balmer lines are strongest. If we are therefore to deduce correctly the abundance of elements in a star, we must correct for such temperature effects.

After accounting for these effects, Payne discovered that virtually all stars are composed mainly of hydrogen. This idea was controversial at the time because most astronomers thought that stars were made primarily of heavy elements, and she was discouraged from publishing her conclusions by her thesis adviser. However, several years after her thesis was published, her ideas gained wide acceptance. Payne's idea led to our current understanding that about 71% of a star's mass is hydrogen, 27% is helium, and the remaining 2% is a mixture of all the other elements known to exist.

56.5 SPECTRAL CLASSIFICATION CRITERIA

Astronomers have put the stellar spectral types in order of temperature by using the pattern of lines in a star's spectrum. The similarity in most stars' compositions means that the main cause of differences in the spectra is the temperature of the photosphere. The effects of temperature on the line patterns are fairly easy to see in

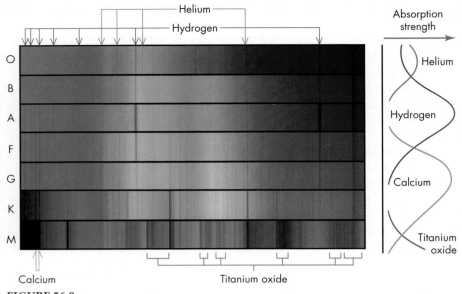

FIGURE 56.8
The stellar spectral types. The types are shown in order from hottest (O) to coolest (M). Several absorption lines are indicated, and their changing strengths are illustrated at right.

Figure 56.8, which shows spectra of the seven main types—O through M.

For example, A-type stars show strong hydrogen Balmer lines because they are near the ideal temperature for hydrogen electrons to be in energy level 2. O stars have weak absorption lines of hydrogen but detectable absorption lines of helium, the second most abundant element, because they are so hot. At their high temperature, O stars' hydrogen atoms collide so violently and are excited so much by the stars' intense radiation that the electrons are stripped from most of the hydrogen, ionizing it. With its electron missing, a hydrogen atom cannot absorb light, making for very weak hydrogen absorption lines. By contrast, helium electrons are more tightly bound, so they produce absorption lines.

Lines of some elements, like ionized calcium, become more prominent at lower temperatures, but in cooler stars the hydrogen Balmer lines grow much weaker. In this case, the lines are weak because the hydrogen's electrons are mostly in level 1 and therefore require much more energetic ultraviolet photons to raise them to a higher energy level. The K and M stars have such cool atmospheres that some molecules, like titanium oxide, are able to form. These molecules often have rich and complex patterns of absorption lines.

The even cooler L stars show strong molecular lines of iron hydride and chromium hydride, while T stars show strong absorption lines of methane. These features, and those found in the other spectral types, are summarized in Table 56.1. Comparing the strength of spectral lines to determine a spectral type is now often done automatically by computers that scan a star's spectrum and match it against standard spectra stored in memory.

Finally, we note that a small number of stars have spectra and temperatures that do not fit into the normal classification sequence. They have compositions or temperatures that are quite different from those of stars in the standard spectral sequence, and they are sometimes assigned their own classification letters. Some of these are now understood to be stars that have undergone dramatic changes late in their lifetimes; others remain mysteries that are still being studied.

Concept Question 4

What might a cool star's spectrum look like if it contained no elements other than hydrogen and helium? What might a hot star's spectrum look like if it contained a large fraction of heavy elements?

Concept Question 5

If a star has faint hydrogen Balmer lines, what are the next steps involved in determining its spectral type?

TABLE 56.1 Summary of Spectral Types

Spectral Type	Temperature Range (K)	Features
O	Hotter than 30,000	Ionized helium, weak hydrogen
B	10,000–30,000	Neutral helium, hydrogen stronger
A	7500–10,000	Hydrogen very strong
F	6000–7500	Hydrogen weaker, metals—especially ionized Ca—moderate
G	5000–6000	Ionized Ca strong, hydrogen weak
K	3500–5000	Metals strong, CH and CN molecules appearing
M	2000–3500	Molecules strong, especially TiO and water
L	1300–2000	TiO disappears. Strong lines of metal hydrides, water, and reactive metals such as potassium and cesium
T	900?–1300?	Strong lines of water and methane

KEY POINTS

- A star's spectrum indicates the temperature and composition of the star's surface regions.
- The surface temperature can be estimated from the wavelength of maximum emission using Wien's law.
- Stars can be divided up into spectral types according to the pattern of dominant absorption lines in their spectra.
- Spectral features depend on the surface temperature of the star, because the atoms' electrons are in different energy levels.
- From hot to cold, the spectral types are O-B-A-F-G-K-M-L-T.
- If one element has stronger spectral lines than another, it does not necessarily indicate a greater abundance, because the strengths depend on characteristics of the atoms and the temperature.
- After correcting for temperature effects, it is typically found that about 71% of a star's mass is hydrogen, 27% helium, and 2% other.
- New low-temperature spectral types L and T have been added to the sequence, based on infrared studies in recent years.

KEY TERMS

absorption line, 447
Balmer lines, 451
O-B-A-F-G-K-M-L-T, 450
photosphere, 446
spectral type, 449
spectrum, 446
stellar spectroscopy, 449

CONCEPT QUESTIONS

Concept Questions on the following topics are located in the margins. They invite thinking and discussion beyond the text.

1. Elevation in star of absorption-line photons. (p. 447)
2. Brightness of cool versus hot stars. (p. 449)
3. Evolution of eyes in response to starlight. (p. 450)
4. Effect on stars' spectra of different compositions. (p. 452)
5. Properties to examine to determine spectral type. (p. 452)

REVIEW QUESTIONS

6. What different ways are there to measure a star's temperature?
7. Why do stars have dark lines in their spectra?
8. Where in the star's atmosphere are the dark lines produced?
9. What are the stellar spectral types? Which are hot and which are cool?
10. What distinguishes the spectral types of stars?
11. If a star has a peak emission in the infrared part of the spectrum, is it hotter or cooler than the Sun? What color would the star be?
12. Why do both O and M stars have weak absorption lines of hydrogen in their spectra?

QUANTITATIVE PROBLEMS

13. Suppose a star radiates most strongly at about 200 nm. How hot is it? What spectral type is it?
14. The bright southern star Alpha Centauri A radiates most strongly at about 500 nm. What is its temperature? How does this compare to the Sun's?
15. Estimate the wavelength of maximum brightness for a star with dark calcium lines.
16. If a T star has a surface temperature of 1000 K, at what wavelength should it be brightest?
17. When our star reaches the end of its main-sequence lifetime, it will become an M-type red giant (Unit 65) with a surface temperature of approximately 3300 K. How will its wavelength of peak radiation differ from the current value?
18. Suppose a star radiates most strongly at 330 nm. Would you expect to see strong absorption lines in its spectrum? Why or why not?

TEST YOURSELF

19. As the surface temperature of a star increases the wavelength of maximum brightness
    a. decreases.
    b. stays the same.
    c. increases.
    d. may change. More information is needed.
20. Which of the following measurements is *not* used by astronomers to estimate how much of an element is present in a star?
    a. The star's density
    b. The star's surface temperature
    c. The strength of the element's spectral lines
    d. The atomic properties of the element
21. Which of the following spectral type of star is hottest?
    a. A   b. B   c. G   d. M   e. O
22. A star has spectral lines of molecules in its atmosphere. Which of the following spectral types is it therefore most likely to belong to?
    a. A   b. B   c. G   d. M   e. O
23. Which type of star has the strongest Balmer lines of hydrogen?
    a. A   b. B   c. G   d. M   e. O

UNIT 57

The Masses of Orbiting Stars

57.1 Types of Binary Stars
57.2 Measuring Stellar Masses with Binary Stars
57.3 The Center of Mass

Learning Objectives

Upon completing this Unit, you should be able to:
- List the classes of binary stars, explaining what distinguishes them.
- Calculate the mass of a binary star system, and explain what observations are necessary to provide the necessary data.
- Define center of mass, and show how individual star's masses are determined.

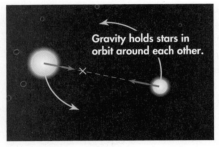

FIGURE 57.1
Two stars orbiting in a binary system, held together as a pair by their mutual gravitational attraction.

Many stars are not alone in space; rather, they have one or more stellar companions held in orbit about each other by their mutual gravitational attraction (Figure 57.1). Astronomers call such stellar pairs **binary stars.** Binary stars are extremely useful, because from their orbital motion astronomers can determine the stars' masses. To understand how, recall that the gravitational force between two bodies depends on their masses. The gravitational force, in turn, determines the stars' orbital motion. Therefore, if we can measure that motion, we can work backward to find the mass, as was shown in Unit 17.

A star's mass is not just another piece of data in our catalog of stellar properties; it is the most critical property for determining the structure and fate of a star. A star's mass tells us how much weight presses down on the core of the star, thereby driving the nuclear reactions that generate the heat and pressure that hold the star up (Unit 51). The mass also tells us how much hydrogen "fuel" is potentially available for the nuclear reactions needed to keep the star from collapsing under its own weight (Units 52 and 62).

Binary star systems are quite common—at least 40% of all stars known have orbiting companions, and the percentage may be much higher. Typically the stars in a binary system are a few AU apart. In many cases, more than two stars are involved. Some stars are found in triplets, others in quadruplets, and at least one six-member system is known. The masses we determine for these stars give us mass estimates for stars of similar type that do not have companions.

57.1 TYPES OF BINARY STARS

A few binary stars are easy to see with a small telescope. Mizar, the middle star in the handle of the Big Dipper, is a good example. When you look at this star with just your eyes, you will also see a dim star, Alcor, near it, but this is not the binary companion. Alcor and Mizar form an **apparent double star:** two stars lying in the same direction but not in orbit around each other. With a telescope, however, we can resolve Mizar to reveal that it consists of two much closer stars in orbit around each other, Mizar A and Mizar B: a **true binary.**

We can directly observe the orbital motion of Mizar A and Mizar B around each other by comparing images made years apart. Such binary stars are called **visual binaries** because we can see two separate stars and their individual motions. In visual binary systems we can usually observe the two stars completing an orbit around each other over several years. Some widely separated binary stars take a

Mizar and Alcor are pictured in Looking Up #2 at the front of the book.

Binary stars

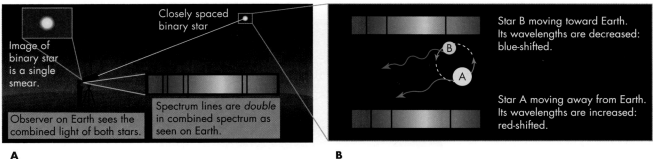

FIGURE 57.2
Spectroscopic binary star. (A) The two stars are generally too close to be seen separately by even the most powerful telescopes. (B) Their orbital motion creates a different Doppler shift for the light from each star. Thus, the spectrum of the stellar pair contains two sets of lines, one from each star, shifted relative to each other by different amounts depending on the stars' movements.

very long time to complete an orbit. For example, binary stars Mizar A and B have completed only a fraction of their orbit since their discovery in 1617; astronomers can trace a partial arc of the orbit, and they estimate that the total length of the orbital period is a few thousand years.

Some binary stars are so close together that their light blurs into a single spot of light that defies separation with even the most powerful telescopes. In such cases, their orbital motion cannot be seen directly, but it may nevertheless be inferred from their combined spectra. As each star moves along its orbit, it alternately moves toward and then away from the Earth. This motion creates a Doppler shift, and so the spectrum of the star pair shows two sets of spectral lines that shift relative to each other. While one star's spectral lines shift to longer wavelengths (because the star is moving away from us), the other's spectral lines shift to shorter wavelengths (because it is approaching us). Then, in a cyclic fashion, half an orbit later, the pattern reverses (Figure 57.2). Astronomers call such star pairs **spectroscopic binaries,** and by observing a full cycle of their spectral shifts, we can determine the orbital motion of the stars. Usually these stars are close together, and the orbital periods are short. For example, Mizar A is itself a spectroscopic binary with another companion orbiting it in a period of just 20 days.

Eclipsing binaries are pairs of stars whose orbits are oriented exactly edge-on to our line of sight; so the stars eclipse each other sequentially. From the duration of the eclipses, it is possible to determine the sizes of the stars (Unit 58). When an eclipsing binary is also detected as a spectroscopic binary, we can determine the parameters of the system better than for other spectroscopic binaries. This is because we need to know the orientation of the orbits to know whether the Doppler shift reflects the full speed of orbital motion or just the part of the motion that happens to be along our line of sight toward the star. In an eclipsing system we know that the orbit is edge-on to us.

57.2 MEASURING STELLAR MASSES WITH BINARY STARS

From their knowledge of orbital motion as just described, astronomers can find the mass of a stellar pair using a modified form of Kepler's third law (Unit 17). Kepler demonstrated that the time required for a planet to orbit the Sun is related to its distance from the Sun. If P is the orbital period and a is the semimajor axis (half the long dimension) of the planet's orbit, then $P^2 = a^3$, a relation called Kepler's third law.

Spectroscopic binary

Concept Question 1

When we observe spectroscopic binaries, we often do not know how the plane of the orbit is oriented to our line of sight. How will the Doppler shift change if the same orbit is oriented edge-on, face-on, or somewhere in between?

Concept Question 2

When Kepler devised his third law, he did not include a term describing mass. Why was he able to omit mass for the Solar System?

FIGURE 57.3
Measuring the combined mass of two stars in a binary system. Data is shown for Alpha Centauri, the combined mass calculated using the modified form of Kepler's third law is a little less than twice the mass of the Sun. The orbit of the less-massive companion, Alpha Centauri B (orange), is shown relative to Alpha Centauri A (yellow).

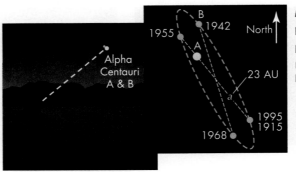

Measuring Mass of Alpha Centauri
Plot of star positions: $P = 80$ years
Find semi-major axis: $a = 23$ AU
Using the modified form of Kepler's third law then gives:

$$M_A + M_B = \frac{a_{AU}^3}{P_{yr}^2}$$

$$= \frac{23^3}{80^2}$$

$$= 1.9 \, M_\odot$$

M_A = Mass of star A in solar masses
M_B = Mass of star B in solar masses
P_{yr} = Period of orbit in years
a_{AU} = Semimajor axis of orbit in AU

LOOKING UP

Alpha Centauri can be seen in Looking Up #8 at the front of the book.

Newton discovered that Kepler's third law could be generalized to apply to *any* two bodies in orbit around each other. As shown in Unit 17, the sum of their masses can be found from the period of their orbit and their separation:

$$M_A + M_B = \frac{a_{AU}^3}{P_{yr}^2}$$

where a_{AU} is expressed in astronomical units, P_{yr} in years, and M_A and M_B are in solar masses. This relationship is our basic tool for measuring stellar masses.

To find the mass of the stars in a visual binary, astronomers first plot their orbital motion, as depicted in Figure 57.3. It may take many years to observe the entire orbit, but eventually we can determine P, the time required for the stars to complete an orbit. From the plot of the orbit, and with knowledge of the stars' distance (to convert angular separations to physical separations), astronomers next measure the semimajor axis, a, of the orbit of one star about the other.

Consider, for example, the orbit of the two stars that compose the visual binary star Alpha Centauri. The two components have an orbital period of 80 years and a semimajor axis of 23 AU. As shown in Figure 57.3, the modified form of Kepler's third law then gives their combined mass as $1.9 \, M_\odot$. That is, the masses of Alpha Centauri A and B add up to 1.9 times the Sun's mass.

57.3 THE CENTER OF MASS

Kepler's law allows us to find the combined mass of two stars that orbit each other. Additional analysis of stars' orbits allows us to find their individual masses. We can tell their masses relative to each other by the amount each star moves. If one star is much more massive than the other, then the massive one will hardly move while the less massive one exhibits nearly all of the motion—like a planet orbiting the Sun. If the two stars are equal in mass, they will move the same amount.

Even in the Solar System, where we usually speak of the planets "orbiting the Sun," we should properly say that the Sun and planets orbit their common center of mass. The Sun is so much more massive than the planets that the center of mass turns out to be inside the Sun. The Sun "wobbles" because of the planets, primarily Jupiter. This kind of wobble seen for other stars is what allows us to detect planets orbiting them (Unit 36).

Pairs of stars do not have such disparities in mass, though, so they orbit a point more nearly equidistant between them called the **center of mass,** as depicted in Figure 57.4. This is located along a line joining the two stars at a position that depends on the stars' relative masses. It is like the point of balance on a children's

FIGURE 57.4
The location of the center of mass of two bodies depends on their relative mass. (A) If the masses are equal, the center of mass is halfway between. (B) If one body is more massive than the other, the center of mass is closer to the more massive body.

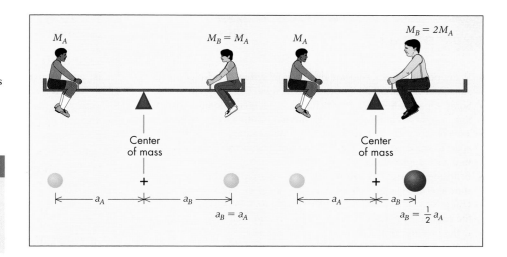

Concept Question 3

The Greek scientist Archimedes is supposed to have said, "Give me a lever and a place to stand and I will move the Earth." Might this be possible? How?

Mathematical Insight

Note that $a_A + a_B$ equals the semimajor axis a used in Kepler's formula.

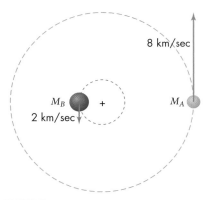

FIGURE 57.5
As two stars orbit their center of mass, the more massive star completes a smaller orbit about the center of mass and therefore travels at a slower speed than the less massive star. The Doppler shift of the more massive star is also smaller in proportion to the stars' relative masses.

playground "seesaw" or "teeter-totter." If one star is three times more massive than the other, the point of balance will be three times closer to the more massive star.

We can express this idea mathematically as follows. If one star has mass $\mathcal{M}_A$ and is orbiting at a distance a_A from the center of mass, and the other has mass $\mathcal{M}_B$ and is orbiting at a distance a_B (Figure 57.4), then

$$\mathcal{M}_A \times a_A = \mathcal{M}_B \times a_B.$$

The *larger* mass has the *smaller* distance from the center of mass. If two stars are equal in mass ($\mathcal{M}_A = \mathcal{M}_B$), then $a_A = a_B$, and they will orbit a point exactly halfway between them. On the other hand, if star B is two times less massive than star A ($\mathcal{M}_B = \frac{1}{2} \times \mathcal{M}_A$), then star B will orbit two times farther from the center of mass than its orbital companion ($a_B = 2 \times a_A$). Similarly, an adult who weighs twice as much as a child will overbalance the child on a seesaw unless he sits half as far from the pivot point as the child.

For spectroscopic binaries, we can compare the relative speed of each star. Each orbits the center of mass in the same period of time, but the more massive one has less distance to move in its orbit. So, for example, we might detect star A moving toward us at 2 kilometers per second while star B moves away from us at 8 kilometers per second (Figure 57.5). Half an orbit later we see star A moving away from us at 2 kilometers per second, while star B moves toward us at 8 kilometers per second. This shows us that star B is moving 4 ($= 8/2$) times faster and farther than star A. Therefore, star B is four times less massive than star A.

For Alpha Centauri, measurements show that star A orbits about 0.7 times as far from the center of mass as star B. So $a_A = 0.7 \times a_B$. We therefore know that $\mathcal{M}_B = 0.7 \times \mathcal{M}_A$. From the previous section we also know that their combined mass equals 1.9 solar masses, so we can write the following equation:

$$\mathcal{M}_A + \mathcal{M}_B = \mathcal{M}_A + 0.7 \times \mathcal{M}_A = 1.7 \times \mathcal{M}_A = 1.9 \, M_\odot.$$

Solving for $\mathcal{M}_A$, we find:

$$\mathcal{M}_A = \frac{1.9}{1.7} M_\odot \approx 1.1 \, M_\odot.$$

So Alpha Centauri A is about 1.1 solar masses, while Alpha Centauri B is about 0.8 solar masses.

From analyzing many star pairs, astronomers have determined that most stars have masses from about 0.1 to 30 $M_\odot$. A few rare stars are even more massive, ranging up to more than 100 $M_\odot$. Our star, the Sun, is fairly average in mass, as it is in many other ways.

KEY POINTS

- Many, possibly most, stars are members of systems of two or more stars in orbit around each other.
- Some apparent double stars are just coincidentally along the same line of sight from our perspective, but not orbiting each other.
- True binaries can be traced visually if the two stars are far enough apart to see separately, or spectroscopically by the regular variation in their Doppler shifts as they orbit.
- Eclipsing binaries have orbits seen edge-on from Earth, so that the stars alternately pass in front of each other as they orbit.
- Using Newton's version of Kepler's third law, it is possible to find the sum of the masses of stars in orbit around each other.
- If the position of the two stars can be determined relative to the center of mass of the orbit, their masses can be determined.
- If one star in a binary is X times more massive than the other, it will orbit X times slower and X times closer to the center of mass.

KEY TERMS

apparent double star, 454
binary star, 454
center of mass, 456
eclipsing binary, 455
spectroscopic binary, 455
true binary, 454
visual binary, 454

CONCEPT QUESTIONS

Concept Questions on the following topics are located in the margins. They invite thinking and discussion beyond the text.

1. Effect of orientation on spectroscopy of binaries. (p. 455)
2. Why Kepler's third law does not depend on mass. (p. 455)
3. Moving a planet with a lever. (p. 457)

REVIEW QUESTIONS

4. What is a binary star?
5. How do visual and spectroscopic binaries differ?
6. What are eclipsing binaries? Under what circumstances do binary stars appear to us to be eclipsing binaries?
7. What properties of stars can astronomers determine from binary stars?
8. What is the center of mass?
9. Applying Kepler's third law to binary stars gives us only the sum of the stars' masses. How do we find the individual masses?

QUANTITATIVE PROBLEMS

10. How far from the center of the Earth is the center of mass between the Earth ($M_\oplus = 5.97 \times 10^{24}$ kg) and the Moon ($M_M = 7.34 \times 10^{22}$ kg)? Is this point outside the Earth? (Hint: The Moon-Earth separation is 3.84×10^5 km.)

11. Two stars are in a binary system. Find their combined mass if they have
    a. an orbital period, P, of 5 years and an orbital separation, a, of 10 AU.
    b. an orbital period, P, of 2 years and an orbital separation, a, of 4 AU.
12. Two stars are orbiting each other, and the sum of their masses is 6 $M_\odot$. It is found that star A of the pair is orbiting 3 arcsec from the center of mass, while star B is orbiting 6 arcsec from the center of mass. What are the two stars' masses?
13. Find the distance from the center of the Sun to the center of mass between the Sun ($M_\odot = 1.99 \times 10^{30}$ kg) and the following planets:
    a. Earth ($M = 5.97 \times 10^{24}$ kg; $a = 1.50 \times 10^8$ km)
    b. Jupiter ($M = 1.90 \times 10^{27}$ kg; $a = 7.78 \times 10^8$ km)
    c. Saturn ($M = 5.68 \times 10^{26}$ kg; $a = 1.43 \times 10^9$ km)
14. When Jupiter and Saturn are on the same side of the Sun, is the center of mass inside or outside the Sun's radius ($R_\odot = 6.96 \times 10^5$ km)?
15. Two stars are orbiting each other, both 4.2 arcsec from their center of mass. Their orbital period is 420.3 years and their distance from the Earth is 104.1 ly. Find their masses.
16. Using Kepler's third law, we find that the combined mass of a binary system is 55.5 $M_\odot$. The stars have separations of 6.3 arcsec and 2.5 arcsec from the center of mass, and the system is at a distance of 64 ly from the Earth. Find their distances from the center of mass of the system, and their individual masses.

TEST YOURSELF

17. It is difficult to get masses for binary stars from spectroscopic information alone because
    a. the stars are constantly eclipsing each other.
    b. the spectral lines are constantly shifting between blueshifts and redshifts.
    c. not enough is known about the orientation of the orbits.
    d. the orbital periods tend to be short.
18. Other stars have masses _____ the Sun's mass.
    a. that range from the Sun's mass to much smaller than
    b. that range from 10 times smaller to more than 10 times larger than
    c. that range from the Sun's mass to much larger than
    d. within about 10% of
19. Which of the following is not a class of orbiting star systems?
    a. A sequential binary
    b. An eclipsing binary
    c. A spectroscopic binary
    d. A true binary
    e. A visual binary
20. Suppose two stars are measured to be orbiting each other, star A at 8 km/sec and star B at 4 km/sec relative to the center of mass. What is the ratio of the mass of star A to that of star B?
    a. 10 to 1
    b. 1 to 10
    c. 5 to 1
    d. 1 to 5
    e. 2 to 1
    f. 1 to 2

PART 4
UNIT 58
The Sizes of Stars

58.1 The Angular Sizes of Stars

58.2 Using Eclipsing Binaries to Measure Stellar Diameters

58.3 Using the Stefan-Boltzmann Law to Calculate Stellar Radii

Learning Objectives

Upon completing this Unit, you should be able to:
- Explain why stellar radii are difficult to measure directly, and describe some of the techniques used.
- Show how eclipsing binary light curves can be used to give both stars' radii.
- Apply the Stefan-Boltzmann law to calculate a star's radius.

A property of stars as basic as diameter seems like it should be easy to measure. However, stars are at such great distances from Earth that this is one of the most challenging stellar properties to determine through direct observation. Astronomers have devised a variety of techniques for determining stars' sizes from observations, but measuring stellar sizes remains near the limits of our present technologies.

Astrophysics offers another solution. A great success of astrophysics in the late 1800s was the discovery of a physical law that describes the amount of light generated by hot objects. This is called the *Stefan-Boltzmann law* (Unit 23), and it can tell us the radii of stars if we know their luminosities (Unit 55) and surface temperatures (Unit 56). We begin this Unit by reviewing techniques for making direct measurements of stars' sizes. We conclude by showing how we can use physical laws determined in Earth's laboratories to find the sizes of stars so distant that no telescope will ever be able to resolve their sizes directly.

58.1 THE ANGULAR SIZES OF STARS

Clarification Point

The "size" of a star can mean different things in different contexts, such as angular size or diameter. Size is not normally used to indicate mass except colloquially.

In principle, we can measure a star's radius from its angular size and distance, as astronomers did for the Sun hundreds of years ago (Unit 10). Unfortunately, the angular sizes of all stars other than the Sun are almost impossibly tiny because they are so far from the Earth. Even under the highest magnification in the largest telescopes, stars look like smeary spots of light. The size of the smeared spot has nothing to do with the star's size, but is caused by the blurring effects of our atmosphere and by a physical limitation of telescopes called *diffraction* (Unit 31).

Atmospheric blurring can be avoided altogether with telescopes that are placed into space (Unit 32). However, diffraction still presents a fundamental limit. Bending of light at the edge of a telescope's aperture smears the light by an amount that depends inversely on the telescope's diameter. That is, diffraction effects are less severe in bigger telescopes.

Figure 58.1A shows a Hubble Space Telescope image of the star Betelgeuse, a star with one of the largest angular sizes other than the Sun. The surface of the star is barely resolved, because the diffraction limit for the telescope's 2.4-meter diameter is about a quarter of the width of the image. To achieve better resolution requires larger telescope apertures, which are only available on the ground, so observations must compensate for the blurring caused by Earth's atmosphere.

Several techniques have been developed for examining stars with ground-based telescopes at higher resolutions. In Unit 32 we described adaptive optics techniques, which rapidly adjust the position and focus of an image to compensate for changes in the Earth's atmosphere. Using some of the largest telescopes, these have achieved

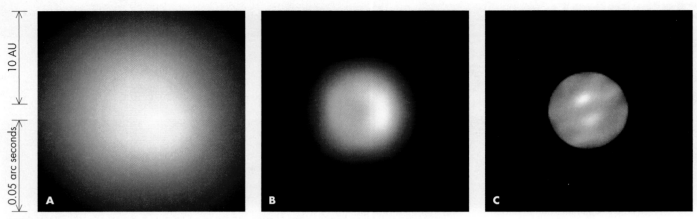

FIGURE 58.1
The star Betelgeuse. (A) Hubble Space Telescope image made at ultraviolet wavelengths. (B) Image made by speckle imaging using a 4-meter ground-based telescope. (C) An image made at infrared wavelengths with the IOTA interferometer (see Unit 31). The angular size scale shown on the left is the same for all three images. The 10-AU length scale assumes a distance of 197 pc, based on recent parallax measurements.

> **Concept Question 1**
>
> The Hubble Space Telescope image in Figure 58.1A was made at ultraviolet wavelengths. How would this affect what parts of the atmosphere are seen versus the optical image in B?

LOOKING UP

Betelgeuse is a bright star in Orion. See Looking Up #6 at the front of the book.

> Speckle imaging techniques have led to a revolution in imaging sources such as planets from the ground. Images are now regularly made by amateur astronomers that show detail only visible with the Hubble Space Telescope or planetary probes in the past.

resolutions better than that of the Hubble Space Telescope. Before adaptive optics, astronomers developed **speckle imaging** techniques that continue to be used for individual bright sources such as stars. Speckle imaging is based on the fact that the blurriness we see on the ground is caused by the image of the star rapidly jumping around as the light passes through moving pockets of air in the Earth's atmosphere. Images captured in a fraction of a second, however, look like tiny "speckles"—each speckle is actually an image of the star formed by momentary focusing of light in the turbulent atmosphere of the Earth. By collecting hundreds or thousands of brief images (as in a video) and realigning them, it is possible to reconstruct an image at nearly the diffraction limit of the telescope. Astronomers have developed a variety of speckle techniques to reconstruct images, which yield images like that of Betelgeuse in Figure 58.1B.

To achieve even higher resolutions, astronomers can combine the light from two (or more) smaller telescopes separated by distances larger than any existing telescope. This method, called **interferometry** (Unit 31), allows astronomers to measure angular sizes with a precision almost equal to that of a single telescope whose diameter is equal to the distance that separates the two smaller ones. Two smaller telescopes separated by 20 meters can measure details almost equivalent to those from a single 20-meter-diameter telescope, although the time required for observation is much greater and generally requires a bright source. A computer can combine the information from the two telescopes to determine the angular size of the star, although it does not provide a detailed image. Interferometers using several telescopes have succeeded in detecting some variations in the brightness of the surface of Betelgeuse (Figure 58.1C).

With speckle and interferometry techniques, astronomers have measured the radii of hundreds of stars, but something is quite unusual about the stars that have the largest angular sizes as seen from Earth. Almost all are *much* larger than the Sun. For example, Betelgeuse's diameter can be determined using the angular size formula (Unit 10) and its distance of about 197 parsecs, found from parallax measurements. The result is a diameter of more than 8 AU—bigger than the orbit of Mars! If the Sun were that big, all of the inner planets would be swallowed up and destroyed.

Smaller stars, those more similar to the Sun, are much more common in the universe, but they are almost all too small and too dim for interferometers to measure their sizes. Only in the last few years have interferometers, using some of the largest optical telescopes in the world, succeeded in determining the sizes of our nearest neighbors, the stars of the Alpha Centauri star system. These are stars similar to the Sun, but at their distance of about 1.3 parsecs, they look a quarter million times smaller, less than one one-hundredth of an arc second in diameter.

58.2 USING ECLIPSING BINARIES TO MEASURE STELLAR DIAMETERS

Eclipsing binary

Perseus and Algol are visible in Looking Up #3 at the front of the book.

Directly measuring the angular sizes of most stars would require telescopes that are almost unimaginably large, but astronomers have developed another technique to measure stellar radii. Many stars are in orbit around a companion star, much as the Earth orbits the Sun. Such gravitationally bound star pairs are called **binary stars,** and astronomers can measure the stars' masses by observing their orbits (Unit 57). Some binary star pairs orbit almost exactly edge-on to our perspective from the Earth. As the stars orbit, one will eclipse the other as it passes between its companion and the Earth. Such systems are called **eclipsing binary stars,** and if we watch such a system, the light from it will periodically dim. The star Algol (Arabic for "Demon Star") in the constellation Perseus is the brightest example of this phenomenon. Normally Algol is fairly bright—it is the second brightest star in Perseus—but once every 68 hours it dims to one-third its usual brightness for 10 hours, becoming the seventh brightest star in that constellation.

During most of the orbit of an eclipsing binary system, we see the light from both stars; but at the times of eclipses, the brightness of the system decreases as one star covers part or all of the other. This produces a cycle of variation in light intensity over time called a **light curve,** as shown by the graph in Figure 58.2.

The duration of the eclipses depends on the diameters of the stars. An eclipse begins as the edge of one star first lines up with the edge of the other, as shown at time 2 in Figure 58.2. Dimming increases until the smaller star is completely in front of the larger star (time 3); the light level remains low until the smaller star reaches the other edge of the large star. Then the process reverses until we see all the light from both stars again. A dip in brightness is also seen when the smaller star passes behind the larger star. The relative size of these dips depends on which star has the hotter surface temperature—not necessarily on which star is bigger. For example, when a large cool star is eclipsed by a smaller, hotter one, the reduction of light may be quite small because the amount of light emitted per square meter depends on the temperature, as is discussed further in Section 58.3. The

Concept Question 2

How might stars in eclipsing binary systems differ from individual stars like the Sun?

FIGURE 58.2
An eclipsing binary and its light curve. From the durations of the eclipses, the diameter of each star may be found. In this illustration, the yellow star is hotter than the red star.

If we imagine star B at rest, star A moves a distance equal to its own diameter between the times 2 and 3. That time interval, when multiplied by star A's orbital speed, gives its diameter. Similarly, the time between points 5 and 6 multiplied by star A's orbital speed will give star B's diameter.

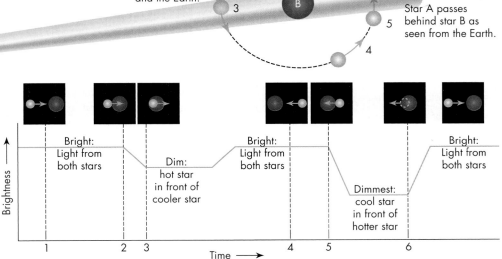

shape of the light curve also depends on whether the stars pass completely behind each other, as we have assumed here, or if they only partially eclipse each other.

Doppler measurements (Unit 25) can tell us how fast the stars are orbiting each other, allowing us to estimate the size of each star as follows: Suppose the stars are moving around each other at a speed of 100 kilometers per second and the light takes 10,000 seconds (about 3 hours) to dim from its full brightness at the start of the eclipse (time 2 in Figure 58.2) to its dim state at full eclipse (time 3). This is how long it takes for the smaller star's disk to move sideways by its own diameter across the edge of the larger star. The star's diameter is therefore equal to the product of its speed and the time for the dimming. In this example the distance is 100 km/sec × 10,000 sec = 1,000,000 km. The time duration from the start of an eclipse (for example, time 5 in the figure) to the beginning of the emergence of the smaller star from behind the larger one (time 6) corresponds to the diameter of the larger star. Suppose this takes 100,000 seconds (about 28 hours); then the larger star is 100 km/sec × 100,000 sec = 10,000,000 km in diameter.

This method of using eclipses lets us estimate the sizes of stars that are too small to detect with an interferometer. It has drawbacks, though. For example, it applies only to the small subset of stars that happen to be in binary systems with orbits oriented in just the right way. Moreover, the method has uncertainties because the orbit may not be exactly edge-on, in which case the duration of the eclipse does not reflect the total diameter of the star.

> For a circular orbit, when the stars are aligned for an eclipse, we will not see any speed difference. We must measure the speed of the stars a quarter orbit later when the stars appear farthest apart.

58.3 USING THE STEFAN-BOLTZMANN LAW TO CALCULATE STELLAR RADII

In the late 1800s, scientists discovered how the luminosity of a hot object depends on its temperature. They found that each square meter of a body radiates a luminosity equal to σT^4, where σ is a constant and T is the temperature, as illustrated in Figure 58.3. This relationship, named for its discoverers, is called the **Stefan-Boltzmann law** (Unit 23.5). It expresses the fact that most matter—such as the relatively dense gas of a star's photosphere—emits an amount of electromagnetic radiation that increases rapidly as the temperature climbs.

The Stefan-Boltzmann law tells us how much light each square meter of a star must radiate if it has a surface temperature T. We can find the total energy the star radiates per second—its luminosity, L—by multiplying the energy radiated from one square meter (σT^4) by the number of square meters of its surface area (Figure 58.3). If we assume that the star is a sphere, its surface area is $4\pi R^2$, where R is its radius. Its luminosity, L, is therefore

$$L = 4\pi R^2 \times \sigma T^4.$$

That is, a star's luminosity equals its surface area times σT^4. The relationship between L, R, and T may at first appear complex, but the equation's meaning is as follows: Increasing either the temperature or the radius of a star makes it more luminous. Making T larger makes each square meter of the star brighter. Making R larger increases the number of square meters.

The Stefan-Boltzmann equation has three "variables"—the quantities luminosity, radius, and temperature—that may vary from star to star or even from one time to another in a particular star. Whenever we have an equation with three variables, if we can measure two of them, we can find the third one from the equation. This means that we can find a star's radius from the Stefan-Boltzmann equation because we have ways of determining a star's luminosity (from its brightness and distance)

> L = Luminosity of star (watts)
> R = Radius of star (meters)
> σ = Constant = 5.67×10^{-8} watts per meter^2 per Kelvin^4
> T = Surface temperature of star (Kelvin)

$$L = \sigma T^4 \times 4\pi R^2$$

Luminosity—total energy radiated per second by the star | Energy emitted per second by 1 square meter | Number of square meters in surface area of the star

FIGURE 58.3
The Stefan-Boltzmann law relates a star's radius to its luminosity and temperature. Each part of the star's surface radiates σT^4. Multiplying σT^4 by the star's surface area ($4\pi R^2$) gives its total power output—its luminosity, $L = 4\pi R^2 \sigma T^4$.

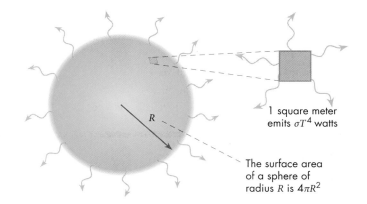

1 square meter emits σT^4 watts

The surface area of a sphere of radius R is $4\pi R^2$

Mathematical Insight

We can solve for a star's radius by rearranging the equation as follows:
$$L \div (4\pi\sigma T^4) = 4\pi R^2 \times \sigma T^4 \div (4\pi\sigma T^4)$$
$$\frac{L}{4\pi\sigma T^4} = R^2.$$
Reverse and take the square root for
$$R = \frac{\sqrt{L}}{\sqrt{4\pi\sigma} \times T^2}.$$

Mathematical Insight

A convenient form of the Stefan-Boltzmann equation is to solve for the radius in terms of the Sun's measured properties:
$$\frac{R}{R_\odot} = \sqrt{\frac{L}{L_\odot}} \times \left(\frac{T_\odot}{T}\right)^2.$$

Concept Question 3

What other equations have you used that have three variables? Can you always solve for one variable if you know the other two?

and temperature (from its color or spectrum). Mathematically, this permits us to solve the equation for R if we know T and L.

We can rearrange the Stefan-Boltzmann equation algebraically (see margin) and solve for a star's radius:
$$R = \frac{\sqrt{L}}{\sqrt{4\pi\sigma} \times T^2}.$$

We can test this formula on the Sun. The Sun's luminosity is measured to be $L_\odot = 3.83 \times 10^{26}$ watts, while its surface temperature is measured to be $T_\odot = 5800$ K (Unit 51). We calculate then that

$$R_\odot = \frac{\sqrt{L_\odot}}{\sqrt{4\pi\sigma} \times T_\odot^2} = \frac{\sqrt{3.83 \times 10^{26}}}{\sqrt{7.13 \times 10^{-7}} \times 5800^2} = \frac{1.96 \times 10^{13}}{2.84 \times 10^4} = 6.9 \times 10^8 \text{ m},$$

or 690,000 kilometers. This is quite close to the Sun's measured radius of 696,000 kilometers, demonstrating the power of the Stefan-Boltzmann law.

To gain a feel for this equation, think about two different cases: (1) when the temperature remains the same or (2) when the luminosity remains the same. If two stars have the same temperature, the more luminous one has the larger radius. On the other hand, for two stars that have the same luminosity, the hotter one must be smaller (because T is in the denominator). We can use this form of the Stefan-Boltzmann equation to solve for the radius of any star whose distance, brightness, and temperature are known—which is many hundreds of thousands of stars.

The Stefan-Boltzmann law is very useful, but it relies on some assumptions about how stars generate light. Eclipsing binary and interferometry measurements provide an important check on the technique. Observations by all three methods show that stars differ enormously in radius. Although many stars have radii similar to the Sun's, some are hundreds of times larger; astronomers call them *giants*. Some are hundreds of times smaller and are called *dwarfs*. We will discover in subsequent Units that the Sun will in future stages turn into both!

KEY POINTS

- A star's diameter can be determined from its angular size and distance, but most stars are so distant that the angular size is too small to measure.
- The smallest object that can be resolved depends on the diffraction limit of the telescope and other blurring effects.
- Using speckle imaging and interferometry, astronomers have measured the angular sizes of several hundred stars.
- Most of the measurable stars turn out to be far larger than the Sun; most normal stars are too small to measure.
- Some binary stars eclipse each other as they orbit, and the duration of the eclipses can give both stars' diameters.
- The Stefan-Boltzmann law provides a relationship between a star's luminosity, surface temperature, and radius.
- If the luminosity can be found (from brightness and distance), and surface temperature measured, the radius can be estimated.

KEY TERMS

binary star, 461
eclipsing binary star, 461
interferometry, 460
light curve, 461
speckle imaging, 460
Stefan-Boltzmann law, 462

CONCEPT QUESTIONS

Concept Questions on the following topics are located in the margins. They invite thinking and discussion beyond the text.

1. Ultraviolet versus visible appearance of a star. (p. 460)
2. Differences between individual and binary stars. (p. 461)
3. Equations with three variables. (p. 463)

REVIEW QUESTIONS

4. How does interferometry work to measure stars' diameters? What sort of stars does this method work for?
5. What is speckle imaging?
6. What is a light curve?
7. What are eclipsing binaries? How can they be used to measure stellar sizes?
8. The Stefan-Boltzmann law is based on what assumptions?
9. How does the Stefan-Boltzmann law allow us to measure stars' sizes?
10. For two stars of the same surface temperature, how does their relative luminosity depend on their radii?

QUANTITATIVE PROBLEMS

11. Alpha Centauri is actually two stars that orbit each other. Recent measurements show that Alpha Centauri A has an angular diameter of 0.0085 arcsec, while Alpha Centauri B has an angular diameter of 0.0060 arcsec. The pair are at a distance of 4.4 ly. What are their diameters, compared to the Sun's?
12. A star with the same color as the Sun is found to produce a luminosity 81 times larger. What is its radius, compared to the Sun's?
13. A star with the same radius as the Sun is found to produce a luminosity 81 times larger. What is its surface temperature, compared to the Sun's?
14. A certain class of stars, known as hypergiants, have extremely large radii. One particular hypergiant, VY Canis Majoris, has an estimated luminosity 270,000 times greater than the Sun's but a surface temperature of only 0.6 times that of the Sun. What is the radius of this star? If it was located in our Solar System, which planets would be inside the star?
15. If a star's surface temperature is 3000 K, how much power does a square meter of its surface radiate?
16. Betelgeuse is roughly the same temperature as Proxima Centauri, the closest star to the Earth aside from the Sun. However, Betelgeuse is 10^9 times brighter. How much larger in radius is Betelgeuse than Proxima Centauri?
17. A star is five times as luminous as the Sun and has a surface temperature of 9000 K. What is its radius, compared to that of the Sun?
18. Using the information that Betelgeuse has an angular size 0.05 arcsec and is 197 pc from the Earth, show how you can calculate its diameter in kilometers.

TEST YOURSELF

19. Only a handful of individual stars have been resolved primarily because
    a. stars are so far away.
    b. most stars are much smaller than the Sun.
    c. most stars are too red.
    d. stars twinkle too much.
20. A star that is very hot but not very luminous must
    a. have a large size.
    b. have a small size.
    c. have a small mass.
    d. be far away.
21. If the radius of a star doubles, but its surface temperature remains the same, its new luminosity is _____ its old luminosity.
    a. 16 times smaller than
    b. 4 times smaller than
    c. the same as
    d. 4 times larger than
    e. 16 times larger than
22. If the surface temperature of a star doubles, but its radius remains the same, its new luminosity is _____ its old luminosity.
    a. 16 times smaller than
    b. 4 times smaller than
    c. the same as
    d. 4 times larger than
    e. 16 times larger than

PART 4
UNIT 59

The H-R Diagram

59.1 Analyzing the H-R Diagram
59.2 The Mass–Luminosity Relation
59.3 Luminosity Classes

Learning Objectives

Upon completing this Unit, you should be able to:
- Explain how an H-R diagram is made, and identify the kinds of stars in each region.
- Describe trends in the mass and other properties of stars along the main sequence.
- Explain how astronomers can estimate a star's luminosity without distance information, and indicate where the luminosity classes are located in the H-R diagram.

By the early 1900s astronomers had learned how to measure stellar luminosities (Unit 55), temperatures and compositions (Unit 56), masses (Unit 57), and radii (Unit 58), but they understood little about how stars worked. We have now reached a similar point in our study of stars. We know that stars exhibit a wide range of properties. But why does such variety occur?

An important method in science for clarifying a wide array of data is to look for patterns and relationships among the different variables. A basic tool for doing this is to plot one variable against another in an X–Y diagram. Astronomers in the early 1900s discovered that a plot of temperature versus luminosity gave new insights into the nature of stars. That plot is now known as the **H-R diagram**—the letters H and R standing for the initials of the Danish astronomer Ejnar Hertzsprung and the U.S. astronomer Henry Norris Russell, who independently discovered its usefulness.

To construct an H-R diagram of a group of stars, astronomers plot each star according to its temperature on the X axis and its luminosity on the Y axis, as shown in Figure 59.1. The H-R diagram shows that stars do not have just any combination of temperature and luminosity. Instead, there are patterns in this diagram, suggesting relationships between these quantities for certain classes of stars.

The H-R diagram is like a crime detective's bulletin board. We pin up the various pieces of information we have about stars and see what they suggest about the stars. It takes many billions of years for a star like the Sun to be born and to die, so we cannot follow a single star over its lifetime. But if we look at all the stars with masses similar to the Sun's, we can begin to piece together clues about how a star with the Sun's mass might appear at different stages of its life. In this Unit, we will begin the job of sorting out the properties of stars on this bulletin board. In subsequent Units we will develop a detailed understanding of stars' lives.

Concept Question 1

Think of two properties that describe humans. If you made a plot of these two properties, what would the diagram look like? What could you learn about people from it?

FIGURE 59.1
Constructing an H-R diagram from the spectra and luminosities of a set of stars.

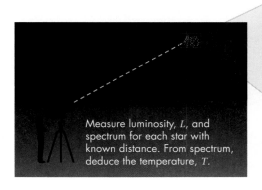

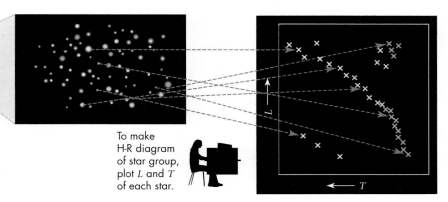

59.1 ANALYZING THE H-R DIAGRAM

Figure 59.2 shows an H-R diagram based on modern measurements of parallax, brightness, and color from the *Hipparcos* satellite. These data have been converted into luminosities and temperatures, and one × is plotted for each star. It is important to keep in mind that the H-R diagram does *not* depict the position of stars at some location in space. It shows how stellar properties correlate, much as a height-weight table does for people, where we can see the trend that in general taller people are heavier than shorter people.

By tradition, the H-R diagram is plotted with the most luminous stars at the top and the least at the bottom—O stars (hot, blue) on the left and M stars (cool, red) on the right. Notice that temperature therefore increases to the left, which is the opposite of the usual convention in graphs. The majority of the stars lie along a diagonal band in the diagram that runs from upper left (hot and luminous) to lower right (cool and dim). This line is called the **main sequence,** and stars within this region of the diagram follow a trend that you might expect: The hotter stars are more luminous than the cooler stars.

Although about 90% of the stars lie along the main sequence, some others are in the upper right part of the diagram, where stars are cool but very luminous, and some are below and to the left of the main sequence, where stars are hot but dim. To understand what makes stars in these two regions different, we turn to

H-R diagram

Clarification Point

The position of a star in the H-R diagram is unrelated to its position in space. Astronomers may speak of a star moving in the H-R diagram if its temperature or luminosity changes, but this is unrelated to spatial motion.

FIGURE 59.2
Hertzsprung-Russell diagram of bright and nearby stars with accurately measured distances. Each "×" is plotted according to one star's estimated temperature and luminosity. The symbols are colored according to the approximate color of the star. Notice that bright stars are at the top of the diagram and dim stars are at the bottom. Also notice that hot (blue) stars are on the left and cool (red) stars are on the right.

The H-R diagrams in this Unit display both the brightest stars in the sky and dim, nearby stars to illustrate the full range of stellar properties. The brightest stars are all about the Sun's luminosity or greater, while the nearest stars are almost all less luminous than the Sun (see Unit 70).

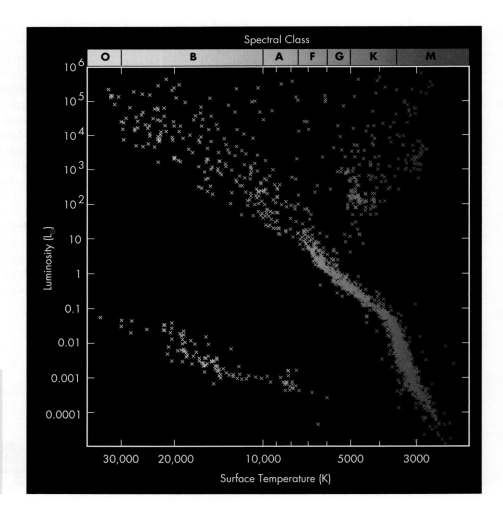

FIGURE 59.3

Regions of the H-R diagram. Most stars lie along a diagonal band from upper left to lower right, known as the main sequence. Giant stars lie above the main sequence and are sometimes labeled according to their color. White dwarf stars lie along a narrow band near the bottom of the diagram. Some well-known stars are marked on the diagram.

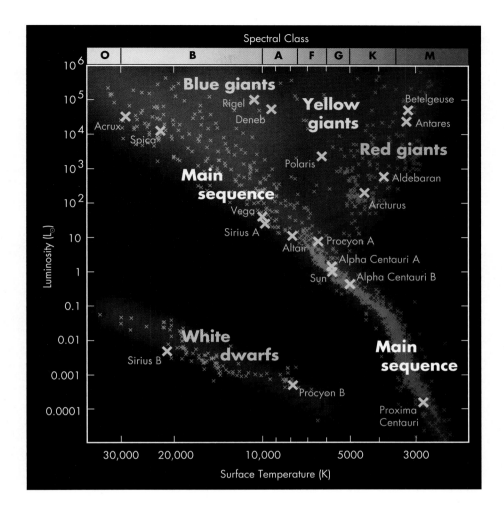

LOOKING UP

Many of the stars named in this H-R diagram can be found in various Looking Up photographs at the front of the book.

Mathematical Insight

If two stars have the same surface temperature, but one is 100 million (10^8) times more luminous than the other, then the surface area must be 10^8 times larger as well. Surface area is proportional to R^2, so the radius R is proportional to the square root of the surface area = $\sqrt{100,000,000}$ = 10,000 times larger.

the Stefan-Boltzmann law (Unit 58). This equation shows that if two stars have the same surface temperature but differ in their size, the larger one emits more energy—that is, it is more luminous—than the smaller one. For stars of the same temperature, more luminous implies a larger radius.

We can apply this reasoning to the H-R diagram. In Figure 59.3 we have labeled a number of the better-known stars. Consider first the stars on the right (cool) side of the diagram. For example, Proxima Centauri and Antares lie along an approximately vertical line because they have the same surface temperature. However, Antares, being nearer the top of the diagram, is much more luminous than Proxima. Because these two stars have the same surface temperature, each square meter of their surfaces emits the same amount of light. Therefore, to emit more light, Antares must have a larger size. Because Antares is large and red, it is called a **red giant**.

We can go one step further, however. By comparing how much more luminous one of these stars is, we can calculate how much larger its radius is. For example, we see that Antares and Proxima Centauri have about the same surface temperature, so the fact that Antares is about 100 million times more luminous than Proxima must be because the surface area that is 100 million times larger. Since the surface area depends on the square of the radius, Antares must have a radius 10,000 times larger than Proxima's. If the Sun were as big as Antares, it would swallow the Earth.

Not all giant stars are red. They span a range of temperatures and colors. For example, the North Star, Polaris, is a giant. However, it has about the same temperature and color as the Sun—yellow—but it is about 50 times larger. To distinguish

FIGURE 59.4

Sizes of stars in the H-R diagram. Green lines show where stars of a given radius will lie. The sizes and colors of some well-known stars are suggested in the figure. However, the size range has been greatly compressed because of the enormous difference between the largest and smallest stars (about a factor of 100,000).

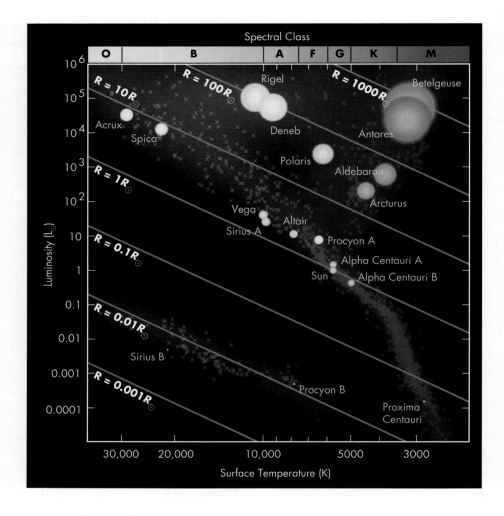

Clarification Point

The colors used in the names of dwarfs and giants are not always good descriptions. Most white dwarfs are bluish in color, and many astronomers refer to all giants above the main sequence as red giants, although their colors span the whole range of colors seen in stars.

Concept Question 2

Would you expect most of the stars visible to you in the night sky to be giants or main-sequence stars? What factors determine which is more common?

between different classes of giants, we sometimes refer to them by their actual color—a blue or yellow giant, for example.

A similar analysis shows that the stars lying below the main sequence must have small radii if they are both hot and dim. Because these stars are so hot compared to main-sequence stars of the same luminosity, they are much bluer. Of the first of these stars to be discovered, many were very hot, glowing with a white heat, and these stars were therefore called **white dwarfs.** With modern observations we can detect some that have surface temperatures cooler than the Sun; but nonetheless, stars falling in this region below the main sequence are all called white dwarfs (Figure 59.3). The nearest known white dwarf is Sirius B, a dim companion of the bright star Sirius 8.6 ly from us. Its radius is about 0.008 the Sun's—a little smaller than the Earth's radius—and its surface temperature is more than 20,000 Kelvin.

We can illustrate the range of star diameters by drawing lines in the H-R diagram that represent a star's radius. That is, we use the Stefan-Boltzmann equation ($L = 4\pi R^2 \times \sigma T^4$) and set R constant. As we make T larger, L must get larger. The result is a line for each value of R that slopes from upper left to lower right as shown in Figure 59.4. Any stars lying along such a line have the same radius. If we move up or to the right in the H-R diagram, we find stars of progressively larger radii. Thus the largest stars lie in the upper right of the diagram, while the smallest stars lie in the lower left. We can see that the giants span a wide range of radii. By contrast, the white dwarfs are all about the same size, about 1/100th of the Sun—about the size of the Earth. The main-sequence stars increase gradually from about under 1/10th the Sun's radius at the cool end to nearly 10 times larger at the hot end.

59.2 THE MASS–LUMINOSITY RELATION

Some of the stars we have plotted in the H-R diagram are in binary systems, allowing us to determine their masses (Unit 57). When we label the stars in the H-R diagram by their masses, we notice that the masses are not randomly scattered around the diagram. As on a detective's bulletin board, we see certain patterns emerging. Stars near the same position on the main sequence all have approximately the same mass. Thus, main-sequence stars of the same spectral type as the Sun all have approximately the same mass.

Farther up and left along the main sequence we find stars of steadily larger masses, while farther down and right on the main sequence we find that the stars have smaller masses. From top to bottom along the main sequence, the star masses decrease from about 20 $M_\odot$ to 0.2 $M_\odot$, as shown in Figure 59.5A. If we expand our sample of stars, we find that we can extrapolate this trend to even more luminous stars that lie along the line of the main sequence to masses of more than 50 $M_\odot$, while there are extremely dim stars extending down from the main sequence that have masses less than 0.1 $M_\odot$.

This trend does not hold for stars in the rest of the H-R diagram. The giants have a wide range of masses, similar to that of the stars on the main sequence but in no particular order. White dwarfs, on the other hand, tend to all be around one solar mass or less, but again in no organized sequence. We will note these facts about the giants and white dwarfs, but set these types aside for the moment and concentrate on the main-sequence stars.

The trend of masses among main-sequence stars is illustrated in Figure 59.5B, where the stars' luminosities are plotted against their masses. This pattern was first recognized in 1924 by the English astrophysicist A. S. Eddington. The relationship

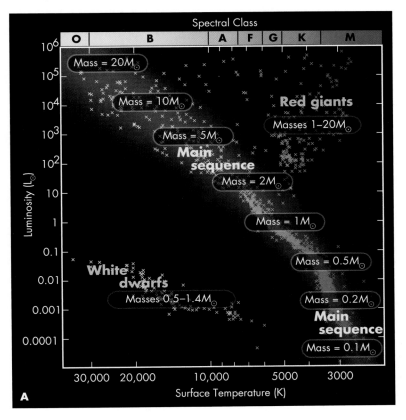
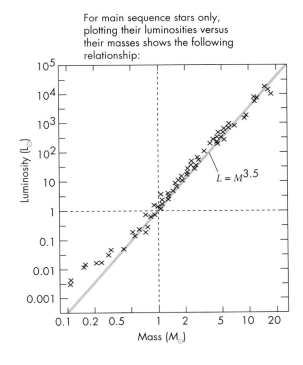

FIGURE 59.5
(A) On the H-R diagram, high-mass stars lie higher on the main sequence than low-mass stars. The giants and white dwarfs exhibit a range of masses in no particularly obvious order. (B) The mass–luminosity relation shows that along the main sequence more-massive stars are more luminous. The relationship does not hold for red giants or white dwarfs.

$\mathcal{L}$ = Luminosity in solar luminosities

$\mathcal{M}$ = Mass in solar masses

Mathematical Insight

Taking $\mathcal{M}$ to the power of 3.5 is the same as multiplying $\mathcal{M} \times \mathcal{M} \times \mathcal{M} \times \sqrt{\mathcal{M}}$. Many calculators have a key labeled $\boxed{x^y}$. To find $10^{3.5}$ you would type this into the calculator:

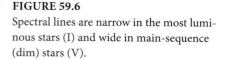

between mass and luminosity can be expressed by a formula. If we let $\mathcal{M}$ and $\mathcal{L}$ represent the mass and luminosity in solar units, then for main-sequence stars

$$\mathcal{L} \approx \mathcal{M}^{3.5}.$$

This **mass–luminosity relation** is approximate and varies between $\mathcal{L} \approx \mathcal{M}^3$ and $\mathcal{L} \approx \mathcal{M}^4$ over different portions of the main sequence, but it gives a fairly accurate description of how the luminosity of a main-sequence star is related to its mass. If we apply this relation to the Sun, whose mass and luminosity are each 1 in these units, the relation is obeyed because 1 raised to any power still equals 1.

If we consider a more massive star—for example, one 10 times as massive as the Sun—then the mass–luminosity relationship tells us that $\mathcal{L} \approx 10^{3.5} \approx 3 \times 10^3$. That is, its luminosity is about 3000 $L_\odot$. Alternatively, a star with a mass of just 1/10th the Sun's mass would have a luminosity $\mathcal{L} \approx 0.1^{3.5} \approx 3 \times 10^{-4}$. That is, its luminosity is about 0.0003 $L_\odot$.

We can understand in general terms how such a mass–luminosity relationship arises. For increasingly larger stellar masses, the gravitational pull among all the parts of a star squeezes the interior more strongly so that the gas is driven to a higher temperature. A higher temperature generates more of the energetic collisions between hydrogen nuclei that can result in fusion, so more massive stars are more luminous.

59.3 LUMINOSITY CLASSES

We cannot calculate a star's luminosity from its brightness without knowing its distance. We might know from a star's color that it is a G-type star, but without a luminosity, it might be a main-sequence star similar to the Sun, a red giant, or even a white dwarf. Luminosity is so important in characterizing a star that astronomers have developed other ways to estimate it. Fortunately, a method was already developed in the late 1800s by Antonia Maury, who discovered that the more luminous the star, the narrower the absorption lines in its spectrum, as illustrated in Figure 59.6. Maury, an American astronomer, together with Annie Jump Cannon (Unit 56), developed the basic scheme for stellar classification still used today.

The reason the width of spectral lines reflects the luminosity of a star is that the lines are affected by the density of the gas producing them. In a dense gas the atoms collide more frequently, "jostling" the electrons' orbitals, altering their energy levels. This in turn makes the atoms able to absorb photons with slightly different energies. The net effect in the entire star's atmosphere is to spread the range of photon energies that are absorbed in the photosphere and to broaden the absorption lines. By contrast, when the atoms are spread farther apart, they collide less frequently and tend to absorb only the exact energy required for an electron to jump to a higher energy orbital, leading to narrow absorption lines.

Giant stars have narrower lines because the atmospheres are less dense. We can see why this must be true for giant stars. Recall that the average density of a body is its mass divided by its volume. Giant stars have masses similar to those of main-sequence stars. So with similar masses but much larger volumes, the giant stars are much less dense. For example, the average density of main-sequence stars

FIGURE 59.6

Spectral lines are narrow in the most luminous stars (I) and wide in main-sequence (dim) stars (V).

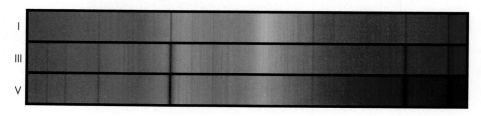

TABLE 59.1 Stellar Luminosity Classes		
Class	Description	Examples
Ia	Bright supergiants	Betelgeuse, Rigel (brightest stars in Orion)
Ib	Supergiants	Antares (brightest star in Scorpius)
II	Bright giants	Polaris (the North Star)
III	Ordinary giants	Arcturus (brightest star in Boötes)
IV	Subgiants	Procyon A (brightest star in Canis Minor)
V	Main sequence	The Sun, Sirius A (brightest star in the sky)

Some astronomers call the very rarest and most luminous stars "hypergiants," and designate them as luminosity class 0.

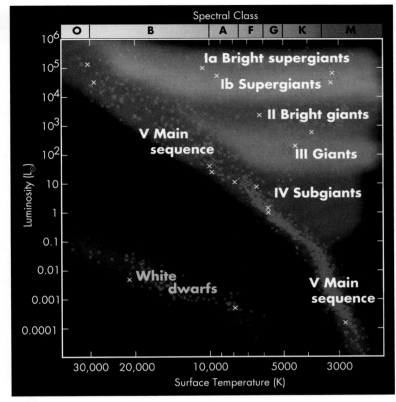

FIGURE 59.7
Stellar luminosity classes in the H-R diagram.

Concept Question 3

If two stars have atmospheres of the same density, but one of the stars has a higher temperature, how would this affect the width of the absorption lines of the hotter star? Why?

is roughly 1 kilogram per liter, whereas the density of a typical giant star is about 10^{-6} kilogram per liter—about a million times less dense.

Using the relationship between spectral line width and luminosity just described, astronomers divide stars into five **luminosity classes,** denoted by the Roman numerals I through V. Class V stars are the densest and dimmest (for a given surface temperature), and class I stars are the largest and most luminous. Class I stars—**supergiants**—are split into two classes, Ia and Ib. Figure 59.7 and Table 59.1 show the correspondence between class and luminosity. Although the scheme is not precise, it allows astronomers to get an indication of a star's luminosity from its spectrum. A star's luminosity class is often added to its spectral type to give a more complete description of its light. For example, our Sun is a G2V star, whereas the blue supergiant Rigel is a B8Ia star.

The high luminosity of the giants is a surprise, given their low densities. After all, in discussing the mass–luminosity relation we argued that the more-luminous main-sequence stars became that way because of the greater compression of their cores. The giants seem at first to defy this relationship—growing more and more luminous at lower and lower densities. The key to this puzzle is that the density of a star's outer atmosphere does not necessarily tell us about the density deep in a star's core. Giant stars differ in structure significantly from the main-sequence stars, the giants having far denser cores. As we will explore in subsequent Units, this difference is a result of how stars age. We will track these changes employing the H-R diagram to help us understand how the different classes of stars arise during stars' long lifetimes.

KEY POINTS

- Graphs of different properties of stars (or other objects) in an X–Y plot can provide clues about their interrelationships.
- The Hertzsprung-Russel (H-R) diagram plots stars' luminosities against their spectral types or temperatures.
- In the H-R diagram, most stars lie along a diagonal line (from hot and luminous to cool and dim) known as the main sequence.
- Stars lying above and to the right of the main sequence must be large and are called giants.
- Stars below and to the left of the main sequence are known as white dwarfs.
- The Sun is a main-sequence star.
- Main-sequence stars exhibit a relationship in which luminosity rises as mass to the power 3.5.
- Stars can be classified into luminosity classes by their spectra, because the density of gas is lower in giant stars, leading to fewer collisions and narrower spectral lines.

KEY TERMS

H-R diagram, 465
luminosity class, 471
main sequence, 466
mass–luminosity relation, 470
red giant, 467
supergiant, 471
white dwarf, 468

CONCEPT QUESTIONS

Concept Questions on the following topics are located in the margins. They invite thinking and discussion beyond the text.

1. H-R diagram for human populations. (p. 465)
2. Giants or main-sequence stars more easily seen. (p. 468)
3. Effect of temperature on spectral line width. (p. 471)

REVIEW QUESTIONS

4. What is an H-R diagram? What are its axes?
5. Where would a very hot and very luminous star be located on the H-R diagram?
6. What is the main sequence?
7. How do we know that giant stars are big and dwarfs are small?
8. How does mass vary along the main sequence?
9. What is the mass–luminosity relation?
10. How do spectral lines vary between main-sequence stars and giant stars of the same temperature? Why?

QUANTITATIVE PROBLEMS

11. Estimate the luminosity and temperature of the star Acrux by examining its vertical and horizontal position in the H-R diagram (Figure 59.3). (You should make an estimate of these values by interpolating between the values marked on each axis.) Using the values you estimated, determine the size of Acrux as compared to the Sun, using the Stefan-Boltzmann law (Unit 58).
12. Alnitak is the brightest O-class star in the night sky, located at the eastern end of Orion's belt. Proxima Centauri is the nearest M-class main-sequence star. Estimate the luminosities and temperatures of these two stars from their positions on the H-R diagram (Figure 59.3) as in Problem 11. Calculate the ratio of their sizes using the Stefan-Boltzmann law.
13. The red supergiant Antares has a mass of 12 $M_\odot$ and a radius of 880 $R_\odot$. Barnard's Star is a main sequence star with a mass of 0.14 $M_\odot$ and a radius of 0.20 $R_\odot$. Which star has a greater average density and by what factor?
14. If a main-sequence star has a luminosity of 3000 $L_\odot$, what is its mass in relation to the Sun's?
15. Estimate the luminosity of a main-sequence star that has a mass twice the Sun's. By what factor is such a star brighter than the Sun?
16. How massive would a main-sequence star need to be, to be 100 times more luminous than the Sun?
17. Use the mass–luminosity relationship and the Stefan-Boltzmann equation from Unit 58 to work out the surface temperature of a star with half the radius of the Sun and one-eighth the mass.

TEST YOURSELF

18. Suppose Star X is straight above Star Y in the H-R diagram. Which of the following is true of the two stars?
    a. Star Y is more luminous than Star X.
    b. Star Y is hotter than Star X.
    c. Star Y is smaller than Star X.
    d. Star Y is more massive than Star X.
    e. All of the above.
19. A star that is cool and very luminous must have
    a. a very large radius.
    b. a very small radius.
    c. a very small mass.
    d. a very great distance.
    e. a very low velocity.
20. Stars that appear red but are not very luminous are located in what part of the H-R diagram?
    a. Upper left
    b. Lower left
    c. Upper right
    d. Lower right
    e. Near the Sun
21. The variation in stellar masses is _____ the variation in luminosities.
    a. about the same as
    b. much smaller than
    c. much larger than
22. If Stars X and Y have the same spectral type, but the absorption lines for Star X are wider, then compared to Star Y, Star X must have
    a. a dusty envelope blocking some of its light.
    b. the dense atmosphere of a smaller star.
    c. a lower surface temperature.
    d. a higher percentage of heavy elements.
    e. one or more orbiting companions.

PART 4

UNIT 60

Overview of Stellar Evolution

- **60.1** Stellar Evolution: Models and Observations
- **60.2** The Evolution of a Star
- **60.3** The Stellar Evolution Cycle
- **60.4** Tracking Changes with the H-R Diagram

Learning Objectives

Upon completing this Unit, you should be able to:
- Describe the phases stars pass through as found by theory and observation.
- Explain the recycling of matter that occurs as stars are born and die.
- Sketch how stars of different masses evolve through the H-R diagram.

To us, the stars appear permanent and unchanging; but they, like us, are born, grow old, and die. Over their lifetimes they slowly change, transforming themselves in many basic ways but taking millions or billions of years to do so. Over the time span of one human life, it is rare to see change in any one star. Even during recorded human history, few stars have been seen to change significantly. We would need to go back to times before humans evolved to notice major changes in the properties of the stars we see today. Astronomers refer to the changes that occur while stars age as **stellar evolution.**

The English astronomer Sir William Herschel offered an analogy to the effort of humans seeking to understand the evolution of stars, and in particular our own Sun. He suggested it is like being allowed to spend a brief time in a forest, and then trying to deduce the life story of trees from the information you've collected. One minute in the forest compared to the lifetime of a tree would be about the equivalent of the entire human record of astronomical observations compared with the lifetime of the Sun.

Figuring out a tree's life story from a minute's observation might sound like an impossible task, but think of all we could photograph in a forest. We could match the leaves and bark to identify and track the life stages of a particular type of tree. We might see some trees of that type as saplings, others that were fully grown, and yet others as rotting stumps. We might find seeds on the tree and on the ground, and perhaps even some seedlings. We might see a leaf fall. If we piece together our snippets of information correctly, we could develop a description of a tree's life without ever having witnessed an individual tree change at all.

This, then, is the challenge of studying stellar evolution—to piece together the story of stars from the vast array of data we have gathered as described in Unit 59. In this Unit, we summarize the story of stellar evolution; the individual pieces of the story are explored in more detail in the Units that follow.

Concept Question 1

Visit a spot where trees grow. List the different things you find there. What evidence can you put together about their relationship in the life cycle of trees?

Clarification Point

Astronomers and biologists use the term *evolution* differently. In astronomy, *evolution* refers to the gradual changes occurring during a single object's lifetime, whereas in biology *evolution* refers to the gradual changes that occur between generations as a result of natural selection.

60.1 STELLAR EVOLUTION: MODELS AND OBSERVATIONS

From observations, astronomers can deduce some features of how a star must evolve. For example, the existence of main-sequence stars, red giants, and white dwarfs suggests to astronomers a picture of how stars age, much as a snapshot of a baby with its parents and grandparents depicts human aging. However, we need not wait vast spans of time to watch a star age. We can "see" a star's aging with computer calculations that

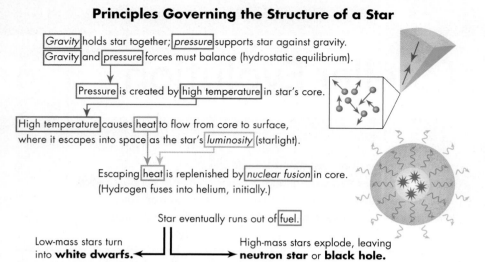

FIGURE 60.1
Outline of the processes that govern the structure of stars.

solve the equations that govern the star's physics. **Stellar models,** as such calculations are called, allow us to trace a star's life from birth to death.

A stellar model is based on the principles that govern the structure of a star (Figure 60.1). Such models suggest that the entire life story of a star is largely determined at its birth. How a star evolves and then dies depends primarily on its mass, because the mass determines the force with which the matter in a star is squeezed, and it also indicates the total amount of fuel that is available for the star to consume in its lifetime. A star evolves in response to the gravity that is constantly squeezing it. In order for stars to shine steadily, at each moment they must supply pressure forces to balance the gravitational compression. For the Sun, the stellar models indicate the matter at its core is squeezed to a density 20 times greater than that of steel (Unit 51), and there are stars that reach much higher densities.

In the competition between gravity and pressure, the force of gravity never runs down or gets used up. In fact, it grows stronger if the star shrinks at all. Meanwhile, a star's heat energy steadily escapes into space (mostly as starlight), so the gas pressure will drop unless the star can replenish its escaping energy.

Technically, when we refer to an object as a "star," we are speaking of an object like the Sun that uses nuclear fusion to generate the heat that maintains the pressure resisting gravity (Unit 62). Nuclear energy is a finite resource, and over its lifetime a star will exhaust all possible sources of nuclear fuel (Units 63 and 67). As a star is forming, when it is a *protostar* or *pre-main-sequence star,* the heat is generated from the process of contraction itself (Unit 61). At the end of its lifetime, a star becomes a **stellar remnant** that is supported by forces between the atoms and particles that it is composed of (Units 65 and 68)—more like the forces between atoms that keep the Earth from collapsing under its own weight.

The stellar models show that any change in the source of energy or in the way a star produces its internal pressure will result in a different internal structure of the star. Sometimes during these changes, the forces can become unbalanced, and observations show that this can lead to violent collapses and explosions.

60.2 THE EVOLUTION OF A STAR

Figure 60.2 summarizes what we might see if we could watch the changes that occur in a star like the Sun over many billions of years. This figure is based on the findings of observations and stellar models over the last century.

The formation of a star begins when a region of interstellar gas—an **interstellar cloud**—starts to contract. Initially this contraction is very slow, and other events may disrupt it. Eventually, though, if the knot of material grows dense enough, the gravitational force increases to a point where the material falls together, collapsing in on itself as the gravitational force grows steadily stronger. The gas heats up as it falls together and collides, converting gravitational potential energy (Unit 20)

Evolution of a low-mass star

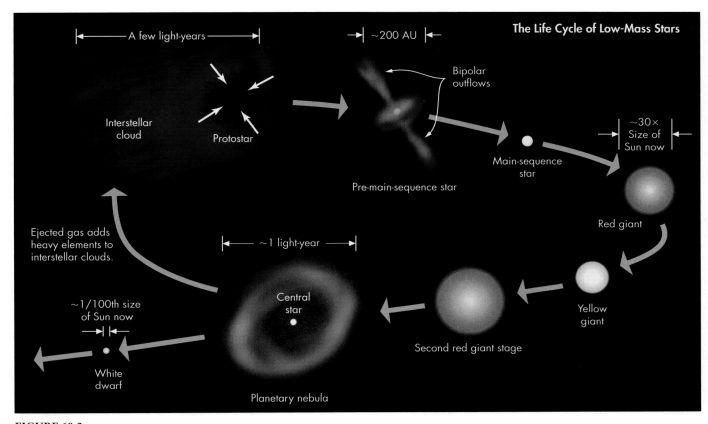

FIGURE 60.2
Evolution of a low-mass star such as the Sun, based on observations and computer calculations. Stars with masses below 8 solar masses have a similar evolution.

into thermal energy or heat, much as water at the bottom of a waterfall rises in temperature as the water crashes down. The cloud at this point has become a **protostar.** During this stage, inflowing matter hides the protostar from view. As the matter is pulled in by gravity, it begins spinning, like water down a drain, forming a rotating disk of gas and dust around the protostar. The gas eventually grows hot enough that the gas pressure slows down the collapse to a more gradual contraction. At this point the surrounding cocoon of gas and dust often clears away enough to reveal the hot pre-main-sequence star at the middle. Some of the gas is heated and driven back outward in **bipolar outflows,** driven off in opposite directions. The formation process is described further in Unit 61.

A ball of contracting gas becomes a **main-sequence star** when its core grows hot enough to start fusing hydrogen into helium (Unit 62). The amount of energy available from fusion reactions is enormous, so the star becomes stable for a long time. However, eventually it will consume all the hydrogen in its core and, because gravity's pull remains, it resumes the process of compressing its core. Compressing a gas causes it to heat up, and this heating raises the temperature in regions surrounding the core so that they now become hot enough to fuse hydrogen. The core may grow hot enough for helium fusion to begin there as well. This yields even more energy than before, which makes the star brighter yet. As the energy floods through the star's outer layers, they swell, and the outermost layers are pushed so far from the core that they become cooler. The star becomes a **red giant** (Unit 63).

Stars differ slightly in many details of their evolution, depending on their exact mass, but the differences grow greater late in the stars' lifetimes. In particular, the ultimate fate of stars is much different for stars whose mass is more than eight times greater than the Sun's. We will therefore divide stars into two groups—**low-mass stars,** such as the Sun, and **high-mass stars,** whose mass is greater than about eight times the Sun's mass.

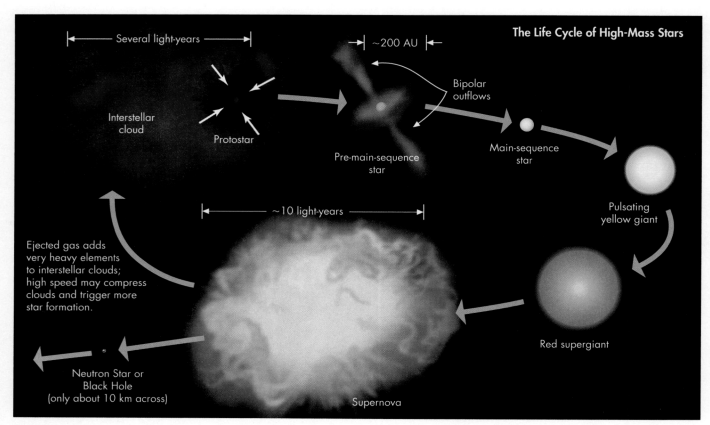

FIGURE 60.3
Evolution of a high-mass star, more than about 8 solar masses, based on observations and computer calculations.

Clarification Point

The term *planetary nebula* can be easily misinterpreted. The name arose because to astronomers with early telescopes these objects looked somewhat like the planets, but they have no relation otherwise to planets.

High-mass stars are blue during their main-sequence stage because their surfaces are hotter than those of low-mass stars, but the more important differences lie deep inside. Whereas a star like the Sun consumes its core's hydrogen in about 10 billion years, high-mass stars may use up their hydrogen in only a few million years (Unit 62). High-mass stars fuse their fuel quickly to supply the energy they need to support their great weight. The sequence of stages that low-mass stars experience as they evolve beyond the main-sequence stage (Figure 60.2) also differs from that of high-mass stars (Figure 60.3). High-mass stars evolve through these stages very rapidly.

In both high- and low-mass stars, the core eventually compresses and heats to the point where the helium can begin fusing, although this happens in somewhat different sequences in the two kinds of stars. In low-mass stars, the core takes a long time to reach helium fusion temperatures. During this time the star grows into a red giant. But when helium fusion begins, the energy released expands the core, and the outer layers shrink, making the star's surface hotter and hence more yellow than red. The cores of massive stars are able to reach helium fusion temperatures fairly rapidly, allowing them to expand directly into *yellow giants*. The surfaces of these yellow giants are often seen to pulsate in and out, and their total light output varies, making them **variable stars** (Unit 64).

When a star exhausts its helium, its core again shrinks, and the star grows into a red giant—for the second time, in low-mass stars. These red giants are so luminous now that their radiation drives off the outer layers of gas in a **stellar wind**. In low-mass stars, the stellar wind carries away so much material that the intensely hot core of the star is laid bare. The core's energetic photons may ionize the inner parts of this wind of gas flowing away from the star. This faintly luminous cloud of gas surrounding the dying star is called a **planetary nebula**. The core remains as a small, dense stellar remnant: a **white dwarf** (Unit 65), which gradually cools and dims.

High-mass stars fuse a sequence of successively heavier elements—such as carbon, oxygen, and silicon. They fuse these fuels rapidly to supply the energy they need to support their great weight. They soon exhaust their fuel and die (Unit 67). Therefore, the stars that begin their lives on the blue end of the main sequence burn out quickly and die "young" by the standards of other stars. They also die more violently because of the tremendous forces at work inside them. At the end of its life, a high-mass star undergoes a dramatic implosion in its core that leads to a **supernova** explosion. The stellar remnants left behind, a **neutron star** or **black hole,** are some of the most bizarre objects in the universe (Units 68 and 69).

There is a potential alternative fate for a low-mass stellar remnant resembling the fate of a high-mass star. Some are brought back to life in explosive fashion when a companion star deposits material on them. If enough material is deposited, a white dwarf may collapse and undergo a supernova explosion (Unit 66).

60.3 THE STELLAR EVOLUTION CYCLE

In a larger context, the process of star formation and death can be seen as a process of cycling matter between interstellar clouds to stars and back again, but with a steady change in composition. A star forms from gas in an interstellar cloud that is drawn together by gravity. This process heats the gas to the point where it can fuse hydrogen into helium. Low-mass stars fuse the helium into carbon and sometimes oxygen before dying, but high-mass stars can fuse a series of heavier elements.

Stellar winds from low-mass stars during their red giant phase (Unit 65) return some of these newly created elements to become part of the interstellar gas. The internal structure of many stars does not allow much mixing, so most of the material returned to interstellar space by the winds is the outer atmospheric gases—which are changed little from when the star formed. However, the more massive of the low-mass stars eject more than 80% of their mass before they shut down, and this may include a significant amount of carbon and oxygen. Some low-mass stars blast all the heavy elements they formed into space when a companion star dumps material onto their surface, inducing them to undergo a supernova explosion.

High-mass stars build in their cores a series of layers of heavy elements: carbon, oxygen, neon, silicon, and eventually iron. Once an **iron core** forms in a high-mass star, the star must soon die, because iron cannot release energy through fusion to continue supporting the star. When the iron core collapses under its own weight, it triggers a supernova explosion. The heavy elements produced during the high-mass star's life are blasted into space. Some heavy elements, such as gold, are formed during the explosion itself. Supernovae therefore enrich the gas in interstellar clouds with heavy elements.

In the end, some of the mass that initially formed stars is locked away in stellar remnants—the white dwarfs, neutron stars, and black holes—that remain after the stars die. However, most of the material is returned to interstellar space, enriching the gas there with heavy elements formed through nuclear fusion, as illustrated in Figures 60.2 and 60.3.

The material flowing outward from these dying stars may also run into interstellar clouds, compressing them and triggering new episodes of star formation. In this way, dying stars may initiate the birth of new generations of stars as well as provide elements vital to human life, such as carbon, oxygen, and iron. Each new generation of stars incorporates a greater amount of these heavy elements, so that stars in later generations, such as the Sun, are able to have planets made of heavy elements, such as the Earth, and carbon-based life forms, such as us. We therefore owe our existence to stellar evolution.

60.4 TRACKING CHANGES WITH THE H-R DIAGRAM

A good way of tracking the changes a star undergoes as it evolves is to plot its "path" in an H-R diagram (Unit 59). A star's position in the diagram changes whenever its surface temperature or luminosity changes, which it usually does in response to any changes in its interior.

In thinking about evolution in terms of an H-R diagram, it may be helpful to consider a similar sort of diagram for a population of people. Suppose we plotted a "height–strength diagram" for a random sample of people. On the X axis we plot height from tall to short (in the backward tradition of the H-R diagram), while on the Y axis we plot the amount of weight the person is able to lift. The analogy to our traditional H-R diagram would be to find a large sample of people, measure their heights and the amount of weight they can lift, and plot one point for each person. A hypothetical diagram like this is shown in Figure 60.4.

If we took a random sample of people, everyone from infants to the elderly, we would generally find a trend that taller people can lift more weight, something like the main sequence in the H-R diagram. Some especially muscular people can lift a much larger weight than we would predict from their heights, while others are particularly weak for their height—analogous to the giant and dwarf regions of the H-R diagram.

If you traced various individuals' heights and strengths throughout their lifetimes, you would see that they "move" through this diagram. Perhaps they follow a relatively normal pattern like that illustrated by the red line in Figure 60.4, growing up and gaining strength as they grow taller, but losing strength (and maybe even a little height) in old age. Some individuals might have jobs that require a lot of lifting, and therefore they grow stronger after reaching adulthood, moving above the "main sequence" into the upper part of the diagram. Perhaps they suffer debilitating injuries, which reduce their strength drastically so that they drop below the main sequence. Every person's track through this diagram might be a little different, although the common features of their "evolution" will create certain patterns in the diagram.

> **Concept Question 2**
>
> Suppose you made a height-strength diagram for all types of animals instead of just humans. How would it differ in appearance from diagram for humans alone? Would this be more analogous to the actual H-R diagram for stars?

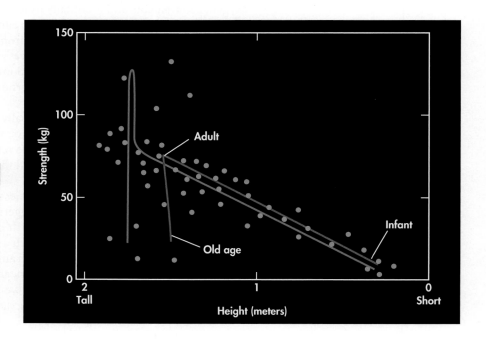

FIGURE 60.4
A hypothetical diagram of the heights of people versus their strength, as measured by the amount of mass they can lift. Each person in a sample is marked by a green dot. Paths of two individuals throughout their lifetimes are also shown.

> **Concept Question 3**
>
> Try drawing what you would expect a height–strength diagram to look like for different populations (such as preschoolers, college students, retirees). What characteristics would allow you to identify the age of the people plotted in such a diagram if you were not told their ages in advance?

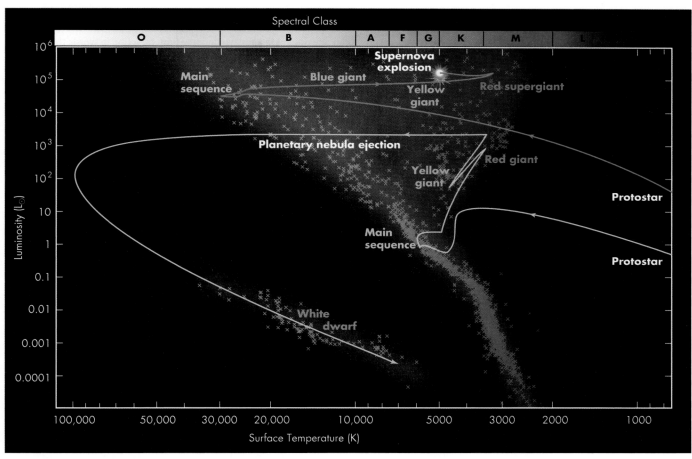

FIGURE 60.5
Schematic diagram of the evolutionary tracks of a low-mass star like the Sun (yellow) and a high-mass star (blue) through the H-R diagram.

Concept Question 4

If stars move through almost all parts of the H-R diagram during their lifetimes, why do we see most of them concentrated in isolated regions of the diagram?

Similarly, an individual star's path through the H-R diagram can be quite complex. The paths of a low-mass star like the Sun and a high-mass star through the H-R diagram are illustrated in Figure 60.5. Compare these paths to Figures 60.2 and 60.3. Each is a way of illustrating a star's lifetime.

Finally, we can also find clues to the changes that result with aging if we examine groups of stars or people who are all of a similar age. If we draw samples of humans from a nursery school, a college, and a retirement community, we find very different distributions of height and strength in each group. Similarly, we can test our ideas about stellar evolution if we can identify groups of stars that are all of similar ages, where physical traits will also differ within the group. We will return to the idea of testing sets of stars of the same age in Unit 70.

KEY POINTS

- The subject of stellar evolution describes how stars change over their long lifetimes and is based on both observations of stars at different stages and computer models.
- The mass of a star is the main factor deciding its fate, because the mass determines the gravitational force and the amount of fuel.
- All types of stars initially form from interstellar clouds, whose gravity causes them to contract and heat up.
- When the center of the contracting object reaches fusion temperatures, a main-sequence star begins its lifetime.
- When the hydrogen in a star's core is used up, the core continues contracting, heating the surrounding gas, forming a red giant.
- Stars below about 8 solar masses expel most of their atmosphere while giants, producing a planetary nebula that expands into space, and leaving behind a white dwarf, the star's former core.
- Massive stars consume their fuel rapidly, then collapse, triggering an explosion when no fuel remains.
- Most of the gas that forms a star is returned to interstellar space at the end of a star's life, enriched with elements made by fusion.
- Most of the matter that makes up Earth and our bodies was formed in the cores of earlier generations of stars.
- The evolution of a star can be traced by its path through the H-R diagram.

KEY TERMS

bipolar outflow, 475
black hole, 477
high-mass star, 475
interstellar cloud, 474
iron core, 477
low-mass star, 475
main-sequence star, 475
neutron star, 477
planetary nebula, 476
protostar, 475
red giant, 475
stellar evolution, 473
stellar model, 474
stellar remnant, 474
stellar wind, 476
supernova, 477
variable star, 476
white dwarf, 476

CONCEPT QUESTIONS

Concept Questions on the following topics are located in the margins. They invite thinking and discussion beyond the text.

1. Gathering evidence of stages of a tree's life. (p. 473)
2. "H-R diagram" for all animals versus just humans. (p. 478)
3. Human "H-R diagram" for different populations. (p. 478)
4. Patterns of stars in H-R diagram despite "movement." (p. 479)

REVIEW QUESTIONS

5. What is a stellar model?
6. What makes a star's mass the most important parameter for predicting its future evolution?
7. What two forces need to be balanced inside a star to keep it at a constant size?
8. What determines when a star becomes a main-sequence star?
9. What is a protostar? What is a planetary nebula?
10. How can we track stellar evolution in an H-R diagram?
11. What kinds of stars end up as neutron stars or black holes?

QUANTITATIVE PROBLEMS

12. Examine the evolutionary path of the Sun-like star in Figure 60.5. Compare its luminosity on the main sequence to its luminosity as a red giant and as a white dwarf. How many times brighter and fainter will the Sun become than it is currently? What effects will this have on the Earth?
13. Examine the evolutionary path of the Sun-like star in Figure 60.5. Estimate the size of the Sun when it becomes a red giant and white dwarf by referring to Unit 58. If the Sun is currently 0.5° across, how big will it appear to be when it is in its later stages of evolution?
14. Imagine that a large collection of stars with a wide assortment of masses forms in a group. Draw sketches of the Hertzsprung-Russell diagram at three times: (a) soon after they formed; (b) at the end of the main-sequence life of a star similar in mass to the Sun; (c) at the end of the main-sequence life of a star much less massive than the Sun. Describe in words the differences between the three sketches.
15. The solid line that illustrates the evolutionary path of a star like the Sun in Figure 60.5 can lead to some confusion. It may give the impression that we are as likely to see stars like the Sun in their protostar phase as in their main-sequence phase. However, a star moves through some phases rapidly and others much more slowly. If instead of a line you put down one point for every 10 million years, you would get a better idea of where you would be likely to see stars like the Sun in the H-R diagram. Lightly sketch or trace the Sun's evolutionary path in an H-R diagram and then use the following times to estimate how many dots to put down in each portion of the Sun's evolutionary path: protostar and pre-main-sequence phases ~20 million years (Myr); main-sequence phase ~10,000 Myr; first red giant phase ~1000 Myr; yellow giant phase ~100 Myr; second red giant phase ~10 Myr; planetary nebula phase 0.01 Myr; white dwarf phase ~3000 Myr. (The star remains a white dwarf beyond this time, but fades from visibility.)

TEST YOURSELF

16. Suppose we could measure one property of a protostar. Which property would tell us most about its future evolution?
    a. Its temperature
    b. Its radius
    c. Its color
    d. Its luminosity
    e. Its mass
17. Which of the following is not a possible stellar remnant?
    a. Brown dwarf
    b. White dwarf
    c. Neutron star
    d. Black hole
18. Massive stars have shorter lifetimes than low mass stars because massive stars
    a. have low core temperatures which means the average particle speeds are slower.
    b. have less hydrogen in their cores to fuse into helium.
    c. fuse hydrogen more rapidly in order to supply the energy needed to support their masses.
    d. have lower luminosities, meaning they produce less energy.
19. After its main sequence phase, the Sun's luminosity will
    a. decrease.
    b. stay approximately the same.
    c. increase.
    d. steadily fluctuate about its main sequence luminosity.
20. Suppose a thousand solar masses of interstellar gas forms into a group of stars of all masses. If you came back a billion years later, where would that gas now be?
    a. Much still in the same low-mass stars, but some back as interstellar gas or later generations of stars.
    b. Almost all in white dwarfs, black holes, and neutron stars.
    c. Most of it collected into the heaviest stars that formed.
    d. Nearly all of it would have turned back into interstellar gas except for a small amount in black holes.

UNIT 61

Star Formation

61.1 The Origin of Stars in Interstellar Clouds
61.2 Proto- and Pre-Main-Sequence Stars
61.3 Star Formation in the H-R Diagram
61.4 Stellar Mass Limits

Learning Objectives

Upon completing this Unit, you should be able to:
- List the stages of star formation from interstellar cloud to main-sequence star.
- Explain the physical changes that occur internally, approximate time scales, and how the outward appearance changes during the stages of a star's formation.
- Draw the path of a forming star through the H-R diagram.
- Explain why there are upper and lower limits to the mass of a star.

We described in Unit 35 the idea that the Solar System formed from a large cloud of gas, contracting because of its gravity. We saw how it spun faster as it contracted, much like water going down a drain, but with the bits of material that did not fall into the star becoming the planetary system. The idea that the Solar System formed from a gas cloud was proposed in the 1700s, long before astronomers had any evidence of gas between the stars. But the idea made sense because it successfully explained so many features of the Solar System.

Today we can study interstellar gas clouds directly and hunt for clouds that may be forming stars. The idea that a collapsing cloud of gas can produce a star is convincingly borne out by our current observations. These data confirm the early stages of star formation sketched in Unit 60, and they link the evolution of stars to the overall evolution of interstellar gas in our Galaxy, discussed further in Unit 73.

FIGURE 61.1
Radio image of a star-forming molecular cloud in the constellation Taurus. The Taurus molecular cloud is about 450 light-years distant, and the region shown in the figure covers an area of about 60 × 100 light-years. The brightest regions occur in the densest gas, where stars are beginning to form.

61.1 THE ORIGIN OF STARS IN INTERSTELLAR CLOUDS

Stars—whether they are of high or low mass—form from **molecular clouds.** These clouds are the densest and coldest kinds of interstellar clouds, in which atoms combine to form molecules and heavier elements combine to form small solid particles called **interstellar dust.** Molecular clouds can be hundreds of light-years across and can contain anywhere up to 10,000 solar masses of cold gas and dust. Some of the most active star formation occurs in **giant molecular clouds,** which can have masses up to about 1,000,000 $M_\odot$.

Molecular clouds orbit along with the stars inside our Galaxy (Unit 73). The gas is mostly hydrogen (71%) and helium (27%); the dust is composed of solid, microscopic particles of silicates, carbon, and iron compounds with a coating of ices. The presence of dust in these clouds blocks most light from entering the clouds, while the molecules can radiate energy even at low temperatures (Figure 61.1). The combination of these effects allows the gas to cool to low temperatures, which is essential for star formation.

FIGURE 61.2
Bok globules seen in silhouette against the background glow of a hot gas cloud. The image shows two large overlapping globules, each over 1 light-year across and containing about 15 solar masses of gas. The smaller dark clouds in the image may eventually dissipate or perhaps form objects too small to become stars.

A Bok globule near the south celestial pole is shown in Looking Up #9 at the front of the book.

Molecular clouds cool down to about 10 Kelvin—not far above absolute zero. At such low temperatures, atoms and molecules in the gas cloud move too slowly to generate much pressure. As a result, the pressure may be barely enough to support the gas cloud against its own gravity, so if the gas cloud grows even slightly more dense, it can collapse. Such a collapse might be triggered by a collision with a neighboring cloud, strong stellar winds from a nearby star, or even the explosion of such a star. Whatever disturbance triggers the collapse, the cloud does not collapse as a whole. Instead, denser regions in the cloud will begin to contract on their own, and these may fragment even further because the rate of collapse is faster wherever the density is higher.

At any given time, many clumps exist within a molecular cloud. Astronomers map the structure of these clumps at radio wavelengths by observing spectral lines from molecules in the cold gas. The radiation coming from these molecules allows us to trace where the gas is densest, as indicated by the brightest regions in Figure 61.1. The emission of this radiation is also significant because it is one of the main ways the clouds lose energy, so they can continue to collapse.

Gravitational energy is released whenever something falls. For example, if a cinder block falls onto a box of tennis balls, the impact scatters the balls in all directions, giving them kinetic energy—energy of motion. As an interstellar cloud contracts, the atoms and molecules collide and scatter like the tennis balls. These randomly directed motions are what we mean by thermal energy or heat. This type of compression heating is something you can observe for yourself with a bicycle pump—the pump cylinder grows warmer as you compress the gas to high pressure to fill a tire. Gravity squeezes the star, much as you use your weight and muscles to squeeze the gas inside the pump cylinder, and the gas responds by heating up.

If a cloud had no way to radiate the heat produced by this gravitational collapse, the cloud would grow hotter until its pressure was high enough to stop it from contracting any further. The emission of radiation allows the cloud to continue contracting. Because it radiates away energy, this is sometimes called "cooling" radiation. However, in allowing the cloud to contract, the gravitational force grows stronger so the collisions grow more violent, and the cloud grows hotter.

As the cloud contracts, the dust becomes so concentrated that virtually no visible light can pass through. The resulting dense, dark blobs are called **Bok globules** in honor of Bart Bok, the Dutch-born U.S. astronomer who first studied them in the 1940s. Because these small dark clouds emit virtually no visible light, they are most easily seen in silhouette against the background glow of hot gas, as in Figure 61.2.

A large, cold molecular cloud may break up into thousands of separate clumps, each clump collapsing on its own. The time that it takes a clump to collapse to a star depends on its mass. In this case bigger is faster, because the gravitational pull is larger. The times range from hundreds of thousands of years for massive stars to hundreds of millions of years for the lowest-mass stars. All of these times are relatively short by astronomical standards. The result is that protostars generally form in groups, not in isolation, and all the stars within a group form at approximately the same time. These groups of stars often persist for millions of years, or longer, after the stars form, and they can tell us a great deal about star formation (Unit 70).

61.2 PROTO- AND PRE-MAIN-SEQUENCE STARS

In the constellation Orion, one "star" visible to your unaided eye is no star at all. The middle object of Orion's sword is actually a cloud of gas in the process of forming several thousand stars. Because stars do not form in isolation and are not all born simultaneously, star-forming regions are complex in appearance. The glowing core of the Orion Nebula, heated by new massive stars there, is still surrounded by the original dark cloud. This can be seen in Figure 61.3, where the

The Orion nebula is shown in Looking Up #6 at the front of the book.

firstborn stars have begun heating up and ionizing gas in the cloud even while other clumps of gas are still in the early stages of contraction.

The transformation from molecular cloud to star proceeds in several distinct stages, illustrated in Figure 61.4. Initially a dense clump within a much larger molecular cloud begins to collapse, as described in Section 61.1. The clump of gas grows hotter as a result of collisions from matter pulled inward by the growing gravitational pull. During this *accretion phase* the object is called a **protostar.** The protostar is surrounded and hidden by the cloud of infalling material.

When the cloud of infalling material is mostly used up, the protostar becomes visible for the first time, in what astronomers refer to as the **pre-main-sequence phase.** Although they do not yet fuse hydrogen into helium, the surfaces of these pre-main-sequence stars are as hot as many main-sequence stars and generate even higher luminosities. Pre-main-sequence stars are often surrounded by a disk of gas and dust, as illustrated in Figure 61.4 and seen in the inset image of a disk surrounding a forming star in the Orion Nebula. Pre-main-sequence stars can be highly variable in brightness, and high-speed jets of gas are sometimes seen flowing from them.

The turbulent life of these young stellar objects is powered by the energy released as a result of its gravitational contraction. In the accretion phase, gravitational compression causes the protostar to grow from a gas clump, just slightly warmer than the interstellar cloud, to a dense ball with a temperature of about 3000 K. The thermal radiation of the growing object is substantial. An accreting protostar that

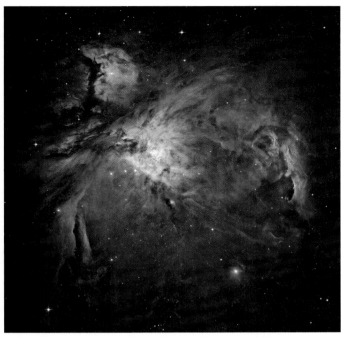

FIGURE 61.3
The Orion Nebula (M42) as imaged with the Hubble Space Telescope. Dark, dusty gas still surrounds the central glowing region where young massive stars have "hollowed out" the center of the cloud. The image shows a region about 11 light-years across.

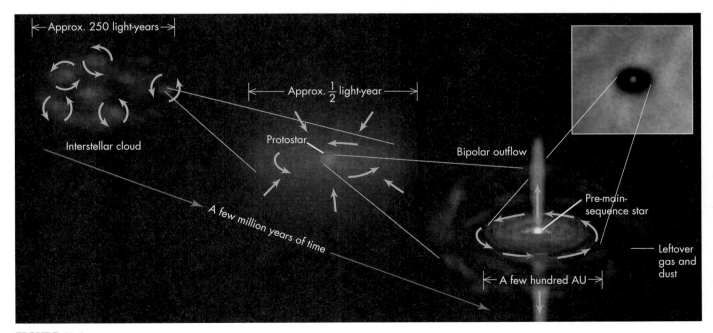

FIGURE 61.4
An artist's depiction of the birth of a star. A Hubble Space Telescope image of a protostar with a dusty disk of gas surrounding it is inset at right. The disk is seen in silhouette against glowing gas in the Orion Nebula.

will evolve into a star like the Sun may have a luminosity about 10 times greater than the Sun's. Throughout most of the accretion phase, the radiation from the protostar is absorbed by the surrounding cloud of infalling material and reradiated at infrared and radio wavelengths. Computer models suggest that most of the surrounding material accretes during several hundred thousand years. The protostar at the end of the accretion stage is several times larger than its eventual size on the main sequence, but otherwise it does not look much different from an ordinary star.

With most of the infalling gas consumed, the pre-main-sequence star becomes visible. When it has contracted to the point where its surface temperature reaches about 4000 K, the gas becomes ionized, which allows it to absorb the radiation generated in its hot interior. With the radiation prevented from escaping, the star contracts more slowly. The pre-main-sequence star keeps a nearly constant surface temperature as it shrinks, descending vertically in the H-R diagram. The contraction phase for a star like the Sun is estimated to take tens of millions of years.

The Hubble Space Telescope photograph of the Eagle Nebula in Figure 61.5A shows a star-forming region that is estimated to be about 2 million years old. A few young stars are visible, but many more remain hidden in the dark pillars of gas and dust in the image. Stars forming inside these dark clouds are not completely hidden, however. Astronomers can detect infrared emission from them, as seen in Figure 61.5B. This image was made with one of the 8-meter-diameter telescopes of the VLT. Many of the stars in this false-color image lie behind the Eagle Nebula. But comparison of the visual and infrared images reveals a number of protostars (appearing bright orange in this false-color image) that are hidden by the dark veil of dust in the visible-wavelength image.

Throughout the accretion process of the protostar, as the cloud shrinks, any small amount of rotation in the initial gas cloud becomes magnified because of the conservation of angular momentum (Unit 20.3). This makes the rotating cloud flatten into a disk as it shrinks, with a rapidly spinning protostar at its center. The disk material is in orbit, perhaps eventually to turn into planets (Unit 35); this remaining material can fall onto the star only if friction or collisions slow it

> **Concept Question 1**
>
> If parallaxes are not available, how might it be possible to decide whether stars seen at infrared wavelengths are inside a molecular cloud instead of behind it?

FIGURE 61.5
The Eagle Nebula (M16), a region where stars are forming from molecular clouds. (A) Dust hides many protostars in dense portions of the cloud in this visible-wavelength image made with the Hubble Space Telescope. (B) A false-color infrared image of the Eagle Nebula made with the VLT (Very Large Telescope) 8-m telescope in Chile. At infrared wavelengths, some stars hidden within the dust become visible.

down enough. In fact, the rotation of all the material that comes together to make the protostar poses a problem for star formation. The protostar cannot continue to contract, because the outer layers are held up by the rapid spin. It must lose some of its angular momentum to continue to contract. This is difficult for astrophysicists to model precisely, but it appears that magnetic fields play a critical role. The rapidly rotating pre-main-sequence star drags its magnetic field through the slower-rotating disk, and this produces a reaction force on the star that slows its spin down.

When the temperature in a pre-main-sequence star's core reaches about 1 million Kelvin, some nuclear fusion begins. This temperature is not sufficient to begin the proton–proton chain (Unit 52), but fusion of naturally occurring deuterium (the isotope of hydrogen with one proton and one neutron) begins. Because naturally occurring deuterium is less than one ten-thousandth ($<10^{-4}$) as abundant as hydrogen, it is not a long-lived energy source. Nevertheless, this fusion sustains the temperature and pressure in the core, slowing further contraction for a while.

The star formation process is not purely about infall. These objects sometimes eject gas in opposite directions called a **bipolar outflow,** as shown in Figure 61.6. In these images, protostars still hidden in dark clouds are ejecting gas that is traveling outward at high speed in opposite directions, carving out a path through the surrounding interstellar gas. Doppler measurements of bipolar outflows show that the gas is shooting outward at speeds of several hundred kilometers per second. Where the jets of material run into surrounding interstellar clouds, they heat and ionize the gas, producing bright glowing regions. The cause of these outflows and jets is not yet fully understood. They are probably powered by energy that is released as matter falls from the inner parts of the disk surrounding the star, crashing into the protostar's surface layers and becoming very hot. The jets' narrow shape may be caused by the star's magnetic field, which confines ionized gas to flow out along the directions of the magnetic poles, much as such fields confine prominences on the Sun (Unit 53). The disk and infalling matter surrounding a protostar may also funnel the hot gas into the direction of least resistance, perpendicular to the disk.

Another phenomenon observed among some forming stars is a very strong wind. This outward flow of gas is similar to the solar wind, but far more intense, and it seems to occur only episodically during the formation process. This outflow may be powered by the spinning star's magnetic field. In the same process that slows the star down, some of the hot ionized gas surrounding the star is accelerated to such a high speed that it is flung outward—carrying away some of the protostar's angular momentum.

Finally, pre-main-sequence stars are also often seen to vary erratically in brightness. Such variable stars are called **T Tauri stars** after a star in the constellation Taurus, which is the first pre-main-sequence star seen to exhibit this behavior. Some of this variability is probably caused by magnetic activity, much like that observed on our Sun. In addition, it is suspected that the inflow of material from the disk onto the star may occur irregularly—the gas in the disk may be "lumpy"—and this could contribute to the variability as well.

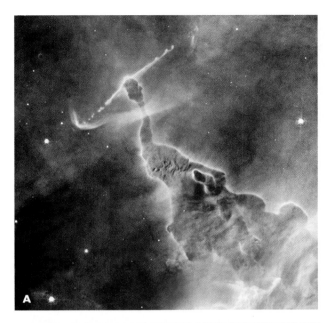

FIGURE 61.6
Bipolar outflows. (A) A protostar lies hidden at the tip of a "pillar" of gas and dust in the Carina Nebula in this Hubble Space Telescope image. The protostar is generating jets of gas that extend more than one-fourth of a light-year in each direction. At the end of the leftward jet, the outflow strikes interstellar gas, which is heated and glows. (B) A protostar lies hidden inside this Bok globule, one of the jets of glowing hot gas (of its bipolar outflow) can be seen emerging from the center of the dark cloud in this VLT image. The globule is about 1 light-year across, and infrared observations reveal several more protostars forming in the cloud.

61.3 STAR FORMATION IN THE H-R DIAGRAM

When pre-main-sequence stars are plotted in an H-R diagram (Unit 59), they lie a little above the main sequence because they are still shrinking and are therefore larger and brighter than when they become main-sequence stars. Figure 61.7 shows the H-R diagram for a young group of stars that contains a number of T Tauri stars and Bok globules, and is still surrounded by gas and dust left over from the period of star formation. The H-R diagram for this and other young star-forming regions shows that the massive stars reach the main sequence before the lower-mass stars.

Computer models indicate that the mass of the forming star is the most important variable in determining how long the protostar stage lasts. An example of the predicted evolutionary tracks for a range of star masses is shown in Figure 61.8. The protostars begin their contraction at low temperatures far off to the right of the diagram. The dashed lines show the late stages of accretion as the protostars approach their final masses. The protostars are buried so deep inside their dusty interstellar "cocoon" that they cannot be directly observed. As they begin their pre-main-sequence phase, illustrated by the solid red lines, lower-mass protostars like the Sun finish accretion then drop in luminosity at a relatively constant temperature. The star settles down to a configuration where it is hot enough in its core to stably fuse hydrogen to helium. The curve of stellar

The nearly vertical paths in the late stages of formation of stars similar to the Sun are known as *Hayashi tracks*, named after the Japanese astrophysicist who predicted them.

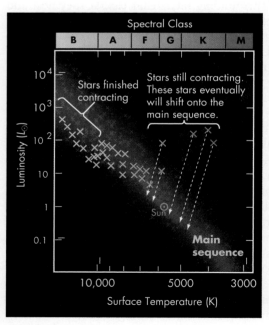

FIGURE 61.7
H-R diagram plotting pre-main-sequence stars in the young star cluster NGC 2264. Surface temperatures and luminosities of stars in the cluster are shown with large X symbols. The cluster is estimated to be about 10 million years old, so some lower-mass pre-main-sequence stars are still contracting. Stellar spectral types (Unit 56) are shown along the top.

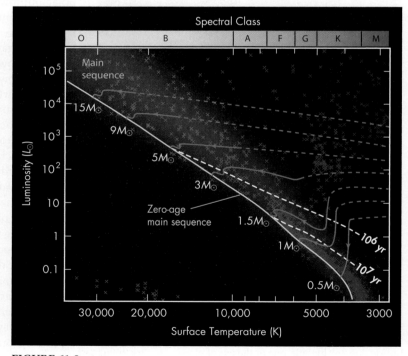

FIGURE 61.8
Theoretical evolutionary tracks for young stellar objects in the H-R diagram. The dashed red lines show predicted properties for protostars hidden inside dark star-forming clouds as they heat up and grow more luminous. The solid red lines are models for pre-main-sequence stars, showing the changing state of the surface temperature and luminosity for stars of various masses during their final stage of contraction. The white dashed lines show where a star can be found along these tracks 1 million and 10 million years after the cloud begins to collapse.

temperatures and luminosities where fusion begins is known as the **zero-age main sequence.**

For a star like the Sun, the whole pre-main-sequence phase lasts approximately 20 million years. For high-mass stars it may be less than a tenth as long, whereas for very low-mass stars it may be more than 10 times longer. The difference in the duration of the protostar stage explains why the higher-mass stars have already reached the main sequence in Figure 61.7, while lower-mass stars in that star-forming region are still in their protostar stage.

The varying timescales for formation imply that the more massive stars may also influence the formation of lower-mass stars. In most star-forming regions, the high-mass, high-luminosity stars form rapidly and begin to ionize and heat up the surrounding gas. The radiation and stellar winds from these massive stars may stop the growth of the slower-forming low-mass stars.

61.4 STELLAR MASS LIMITS

Observationally, we find few stars outside the range of about 0.2 $M_\odot$ to 20 $M_\odot$. Lower-mass stars are dim and hard to detect, while higher-mass stars are rare. However, it is not just observational limitations that determine this range. Theoretical studies of the physical conditions necessary to allow a star to form indicate that stars must have masses between 0.08 $M_\odot$ and 150 $M_\odot$.

Stars below 0.08 $M_\odot$ do not occur because this is too little mass to compress the gas in the core enough to reach hydrogen fusion temperatures. Although lower-mass objects form, they never establish a stable balance between gravity and fusion's internal heat production. Some unstable fusion of hydrogen may occur in slightly less massive objects, and fusion of deuterium may occur in objects down to about 0.016 $M_\odot$ (about 17 times Jupiter's mass). However, naturally occurring deuterium is too scarce for this to be a stable state. Objects with masses between 0.016 $M_\odot$ and 0.08 $M_\odot$ fall between our definitions of planet and star and have been dubbed **brown dwarfs.** These are dim, cool objects that infrared technologies began detecting only in the mid-1990s. Most of the objects in the L and T spectral types (Unit 56) appear to fall in this mass range. These objects are probably very numerous, but astronomers are still in the early stages of understanding this population of "failed stars" because they are so difficult to detect.

By contrast, the upper limit to star masses might be described as too much of a good thing! A star more massive than 150 $M_\odot$ never stabilizes for the following reasons: If a giant cloud of gas begins to contract, the intense gravitational compression drives up its temperature and luminosity very rapidly. The center of a contracting clump of gas more massive than 150 $M_\odot$ emits such intense radiation that it heats the surrounding gas to an extremely high temperature. This raises its pressure so much that no additional material can fall in, setting a limit on the amount of material the largest stars can accumulate. Furthermore, once formed, high-mass stars become so luminous (recall the mass–luminosity relation) that their radiation rapidly drives gas from their outer layers into space in a strong stellar wind. The most massive stars—or would-be stars—essentially "shine" themselves apart.

In reality, conditions governing the protostar stage are rarely conducive to the formation of stars much larger than 30 solar masses. Moreover, such massive stars burn themselves out extremely rapidly (Unit 62). Nevertheless, some isolated examples have been found, like the star Eta Carina, which is estimated to have a mass 100 to 150 times the Sun's mass.

> **Concept Question 2**
> If pre-main-sequence stars lie above the main sequence in H-R diagrams, how can they be distinguished from giant stars?

> **Concept Question 3**
> How would young brown dwarfs differ in appearance from pre-main-sequence stars with masses greater than 0.08 $M_\odot$?

The star Eta Carina is shown in Looking Up #8 at the front of the book.

KEY POINTS

- Stars form from clumps in large interstellar clouds that contain dust and molecules and have very low temperatures.
- Bok globules are isolated dense clumps of gas and dust where individual stars may be forming.
- As clumps within these clouds contract, the molecules radiate away the heat that is generated, allowing further contraction.
- The initial stage of collapse of a clump of gas is called a protostar, which accretes material, pulled in by its growing gravity.
- As the accreting material is used up, a pre-main-sequence star is exposed, which continues heating by gravitational contraction.
- Material surrounding the protostar forms a disk that spins due to the conservation of angular momentum of the infalling material.
- Bipolar jets may form during star formation, expelling material and angular momentum, possibly through the spinning star's magnetic field interacting with the surrounding gas.
- Pre-main-sequence stars tend to have variable brightness, possibly varying as accreting material falls onto the star.
- High-mass stars contract more rapidly, reaching the main-sequence phase faster than low-mass stars.
- Contracting clumps with masses below 0.08 solar masses do not have a strong enough gravity to heat to fusion temperatures.
- Clumps more massive than about 150 solar masses produce so much light that they probably tear themselves apart.

KEY TERMS

bipolar outflow, 485
Bok globule, 482
brown dwarf, 487
giant molecular cloud, 481
interstellar dust, 481
molecular cloud, 481
pre-main-sequence phase, 483
protostar, 483
T Tauri star, 485
zero-age main sequence, 487

CONCEPT QUESTIONS

Concept Questions on the following topics are located in the margins. They invite thinking and discussion beyond the text.

1. Determining whether stars are within a dark cloud. (p. 484)
2. Distinguishing pre-main-sequence from giant stars. (p. 487)
3. Distinguishing young stars from brown dwarfs. (p. 487)

REVIEW QUESTIONS

4. What are molecular clouds, and how are they related to star formation?
5. What are Bok globules, and how are they relevant to stars?
6. At what wavelengths is it best to observe a protostar when it is still accreting gas? Why?
7. What generates the heat in a protostar?
8. What is a T Tauri star? What are bipolar outflows?
9. What determines the upper and lower mass limits of stars?
10. What is a brown dwarf?

QUANTITATIVE PROBLEMS

11. Use Wien's law to find the wavelength of maximum radiation from a 2000-K protostar. To what part of the electromagnetic spectrum does this wavelength belong?
12. Use Figure 61.8 to estimate by what factor a pre-main-sequence star shrinks as it forms a one solar mass star.
13. The jets from a T Tauri star are measured to have speeds of about 300 km/sec. At this speed, how long would it take the ejected material to travel 1 ly away from the protostar?
14. Suppose a molecular cloud begins with a mass of 1 $M_\odot$ and has a radius of 1 ly. Imagine a dust grain at the outer edge of the cloud pulled straight inward, accelerating the entire way. When it reaches a radius of 1 million km from the center of the forming star, how fast will it be moving? (*Hint:* This problem is much easier to answer through the conservation of energy.)
15. Compare the average density of a newly formed star of mass 20 $M_\odot$ (radius $\approx$ 10 $R_\odot$) and 0.1 $M_\odot$ (radius $\approx$ 0.1 $R_\odot$) to the Sun's density. You should find that the higher-mass star has a lower density. Relate these relative densities to the formation processes that determine the upper and lower mass limits for stars.
16. a. Assuming that a 0.08-$M_\odot$ brown dwarf has the same density as Jupiter (see Appendix Table 5 for data), what is its radius?
    b. The coldest brown dwarfs have surface temperatures near 600 K. For this brown dwarf, what is its luminosity?

TEST YOURSELF

17. A molecular cloud that will eventually form a star is _____ the final star system size.
    a. many thousands of times bigger than
    b. a few hundred times bigger than
    c. roughly ten times bigger than
    d. about the same size as
18. Which kinds of stars reach the main sequence the quickest?
    a. Massive stars
    b. Low-mass stars
    c. All masses about the same
    d. The rate of formation differs widely, independent of mass.
19. The high luminosities associated with pre-main-sequence stars arise from
    a. the fusion of hydrogen into helium.
    b. heat generated from the contracting material.
    c. chemical reactions.
    d. radioactive decay.
20. Why are there no stars less massive than 0.08 $M_\odot$?
    a. With such a low mass they take longer than the age of the universe to contract.
    b. Their mass is so small that deuterium fusion blasts them apart.
    c. They exist, but they are so dim that we cannot detect any.
    d. They cannot compress their cores to hydrogen fusion temperatures.
    e. All of the above apply.

UNIT 62

Main-Sequence Stars

62.1 Mass and Core Temperature
62.2 Structure of High-Mass and Low-Mass Stars
62.3 Main-Sequence Lifetime of a Star
62.4 Changes During the Main-Sequence Phase

Learning Objectives

Upon completing this Unit, you should be able to:
- Explain how conditions in a star's core are determined by hydrostatic equilibrium.
- Compare the proton–proton chain and CNO cycle.
- Calculate the main-sequence lifetime of a star, and explain the calculation's basis.
- Explain how stars are affected by convection and how they change as they age.

A collapsing clump of interstellar gas like those described in Unit 61 completes its conversion into a **main-sequence star** when it begins fusing hydrogen into helium in its core. Fusion generates the heat needed to maintain pressure in the core, which stops the steady contraction caused by gravity. At this stage of its life, the star's interior structure is much like the Sun's (Unit 51): the star contains a core, where hydrogen is fusing into helium, and an envelope of gas around the core, where energy is transported to the star's surface.

In this Unit we will continue our examination of how stars evolve, as outlined in Unit 60. Although all main-sequence stars fuse hydrogen into helium, the properties of the core and envelope depend greatly on the star's mass. We will also see that main-sequence stars are stable only in a relative sense. We expect the Sun's luminosity to change steadily over the Sun's lifetime, which raises questions about the past and future of life on our planet.

62.1 MASS AND CORE TEMPERATURE

In examining the physical properties of stars along the main sequence, we saw in Unit 59 that the higher the mass of a star, the hotter and more luminous it is. This relation between mass, temperature, and luminosity results from the fact that for a star to be stable, the gravity and pressure inside it must be in balance, a condition called **hydrostatic equilibrium.** With computer modeling of stars, we can determine how these processes balance with one another and can learn many details about a star's internal structure—its density, temperature, pressure, and energy generation (Unit 60).

The lowest-mass stars take the longest time to complete their journey from protostar to main-sequence. Their weaker gravity causes them to contract slowly, so their core temperature rises very gradually, and the heat generated by compression has time to leak away into space as radiation. As a consequence, a $0.1\text{-}M_\odot$ star must compress the gas in the core to a very high density—about 500 kilograms per liter—just to reach a temperature where hydrogen fusion can begin. By contrast, massive protostars contract rapidly so heat does not leak away, and they reach high temperatures in their cores without compressing the gas to such a high density (Figure 62.1). A $10\text{-}M_\odot$ star's core reaches a density of only a few kilograms per liter when fusion begins.

The higher density of low-mass stars may seem a little surprising: It would seem that high-mass stars would compress matter more. Later in their lifetimes that

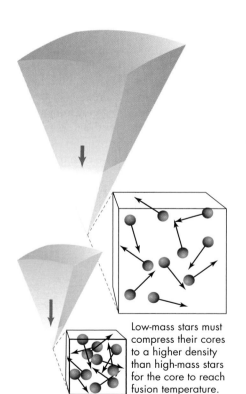

FIGURE 62.1
Schematic cutaway view of the interiors of low- and high-mass stars, illustrating the higher density and pressure, but lower temperature, of a low-mass star's core when its core reaches fusion temperature.

does eventually prove true, but during the main-sequence phase, the highest-mass stars are lowest in density. They achieve hydrostatic equilibrium despite their lower density precisely because they reach hydrogen fusion temperatures while the core is still at a low density. Recall that the gravitational pull rises as the inverse square of the separation between objects (Unit 16), so the gravitational force inward is not (yet) as intense near the center of the massive, but low-density star.

At the low-mass end of the main sequence, the cores reach only 5 million Kelvin after compressing the gas to extremely high density. At the upper end, the core temperature is about 10 times hotter. Thus, high-mass stars have hotter, lower-density cores than do low-mass stars. The rate of nuclear fusion rises rapidly with temperature, and additional fusion pathways become possible at high temperatures as well. The high temperatures in high-mass stars thus allow nuclear reactions to generate the *very* large luminosities observed in massive stars.

62.2 STRUCTURE OF HIGH-MASS AND LOW-MASS STARS

In the Sun and other low-mass stars, hydrogen is converted to helium through the **proton–proton chain** (Unit 52). In the proton–proton chain, hydrogen nuclei, or protons, fuse to form successively more massive isotopes, finally producing helium ^{4}He, along with radiation, neutrinos, and positrons (Figure 62.2). A positron is the anti-particle of the electron, so when it encounters one of the many electrons, they annihilate each other, releasing additional energy.

At the higher temperatures in the cores of stars more massive than about 2 $M_\odot$, a faster set of fusion reactions converts hydrogen to helium. In these larger stars, where core temperatures rise to more than about 20 million Kelvin, hydrogen fusion takes place primarily by means of the **CNO cycle.** In this cycle, illustrated in Figure 62.3, carbon atoms already present in the star's core act as catalysts to aid the reaction through a series of steps that build nitrogen and oxygen nuclei. No carbon, nitrogen, or oxygen is created or destroyed in the end by the CNO cycle, and the net energy release is essentially the same per created helium atom as in the proton–proton chain. However, at high temperatures the CNO cycle is much faster than the proton–proton chain reaction.

The high rate of energy production at these high temperatures makes the internal structure of high-mass stars different from that of low-mass stars. The photons

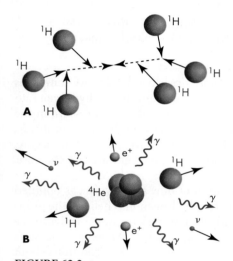

FIGURE 62.2
In the proton–proton chain, (A) six protons react in sequence to produce (B) a helium nucleus, 2 neutrinos, 2 positrons, and radiation. Two protons are also returned at the end of the process.

FIGURE 62.3
The nuclear reactions of the CNO cycle. Four hydrogen nuclei (^{1}H) combine with carbon, nitrogen, and oxygen (C, N, and O) in a cycle whose net result is to produce one helium (^{4}He) nucleus.

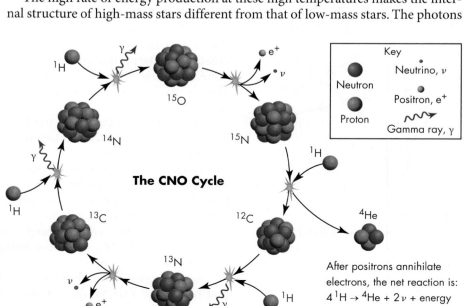

After positrons annihilate electrons, the net reaction is:
$4\,^1H \rightarrow\,^4He + 2\nu + $ energy

Concept Question 1

What differences might there be in a massive star if there were no heavy elements for the CNO cycle?

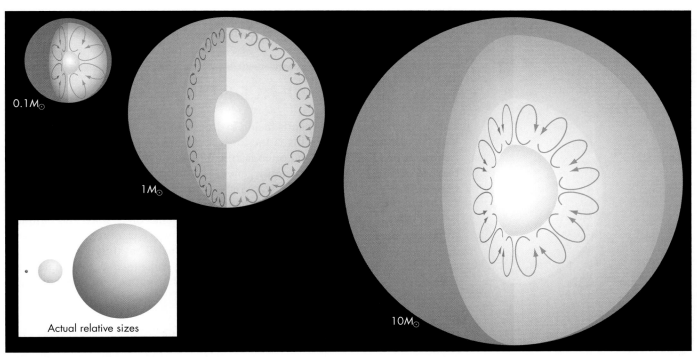

FIGURE 62.4
The comparative structure of main-sequence stars of different masses. A massive star undergoes convection in the neighborhood of its core, while low-mass stars undergo convection in their surface layers. For very low-mass stars (less than about 0.4 $M_\odot$) the convection reaches all the way from the surface to the core of the star. The inset box shows the stars' true relative sizes.

(gamma rays) generated by the fusion reactions can travel only in short hops, only a centimeter or so at a time, before interacting with another atom. Because the photons cannot carry the energy away quickly, clumps of gas in the core region become superheated. These clumps rise toward the star's surface, traveling a sizable fraction of the star's radius, and then sink back down after releasing their heat and cooling—the process of **convection.**

Convection occurs wherever the rate of energy flow by photons becomes too slow. Convective motions occur just below the surface in stars like the Sun, where the cooler surface gas strongly absorbs the photons and impedes the flow of energy by radiation (Unit 51). In stars less massive than the Sun, the surfaces are even cooler, and the convection zone reaches even deeper into the star. These different internal structures are illustrated in Figure 62.4, and they have consequences for stellar lifetimes because they affect the amount of fuel in a star's core.

A star's core is the region where temperatures are high enough for hydrogen to fuse into helium. The core typically contains only about one-tenth of the star's total mass. For a star like the Sun, the convection zone does not carry gas all the way from the surface into the core. So as the hydrogen fuel is gradually consumed in the core, the Sun's convection is unable to reach deep enough to mix unfused hydrogen from the outer parts into the core. Therefore, the core will run out of hydrogen fuel, even though there is a large amount in surrounding regions.

Computer models indicate that the highest- and lowest-mass stars can replenish their cores with hydrogen to varying degrees. Convection within the inner regions of massive stars can mix some of the unburned hydrogen from the surrounding layers into the core and thereby increase, by up to several tenths of the star's mass, the amount of hydrogen that is available for fusion. In stars less massive than about 0.4 $M_\odot$, the gas in the outer layers mixes down into the core completely. Such stars are said to be **fully convective.** By contrast, the outer layers of more-massive stars do not undergo any mixing, so they retain their original composition throughout the star's lifetime. Because the surface layers do not change, the spectra of these more-massive stars offer no clue about the changing composition in the star's interior.

62.3 MAIN-SEQUENCE LIFETIME OF A STAR

The length of time a star spends fusing hydrogen to helium in its core is called its **main-sequence lifetime**. This is the longest stage of a star's life, typically about 90% of its overall lifetime, and often lasting billions of years. Because of the stable balance between gravity and energy generation during the main-sequence phase, the external properties of these stars remain nearly constant, and we find most stars in the main-sequence region of the H-R diagram.

Any individual star's main-sequence lifetime depends on its mass and luminosity. To see why, we can use a simple analogy. Suppose we want to know how long a camping lantern will run on a tank of propane fuel. That time clearly depends on how much fuel it has and how rapidly the fuel is consumed. For example, if the tank contains 2 liters of propane, and at maximum brightness the fuel burns at a rate of 0.2 liters per hour, the lamp will last for

$$t = \frac{\text{Amount of fuel}}{\text{Rate of burning}} = \frac{2 \text{ liters}}{0.2 \text{ liters/hour}} = \frac{2}{0.2} \text{ hours} = 10 \text{ hours}$$

before it runs out. Clearly we can increase the amount of time the lamp will run if we increase the amount of fuel or dim the light so it burns the fuel more slowly. We can apply the same formula to a star to determine its main-sequence lifetime if we know how much fuel it has and how rapidly the fuel is being consumed.

The amount of fuel a star has is set by its mass, or to be a little more precise, the amount of a star's mass that is hydrogen—about 71% of the mass of most stars. This represents the total amount of potential fuel. But only within the core—approximately the central one-tenth of the star's overall mass, for most stars—are temperatures high enough for fusion to occur. Thus, only one-tenth of 71%, or about 7%, of a star's mass is available as fuel. Some stars increase this amount by mixing gas from the outer parts of the star into the core, but this results in significant adjustments only for very high- and very low-mass stars.

The rate at which a star consumes its fuel is directly indicated by its luminosity. More-luminous stars fuse their fuel faster. This makes sense because the energy that a star puts out is supplied by its fuel. Again, it is like a lantern. If it is brighter, it is burning the fuel faster; if it is dimmer, it is burning the fuel more slowly.

Thus, a star's lifetime, t, is proportional to its mass divided by its luminosity, M/L. To work out the actual times involved, we have to calculate what mass of hydrogen is needed to produce a given quantity of light. This is determined by the amount of nuclear energy released in the fusion process. For the Sun, astronomers have determined that 360 million tons of hydrogen are being fused each second to generate its observed luminosity (Unit 52). This creates 356 million tons of helium and "4 million tons of light"—that is, according to Einstein's law, $E = m \times c^2$, 4 million tons of mass are turned into energy each second.

Taking the amount of hydrogen in the Sun's core (7% of its total mass of 2×10^{30} kg) and dividing it by the amount of fuel consumed each second (360 million tons per second), we find that the total duration of the Sun's main-sequence lifetime is

$$t_\odot \approx \frac{0.07 \times 2 \times 10^{30} \text{ kg}}{360 \times 10^9 \text{ kg/sec}} = 3.9 \times 10^{17} \text{ sec} = 1.2 \times 10^{10} \text{ yr}.$$

This lifetime of 12 billion years is a bit longer than found by more-detailed stellar evolution models, which indicate that slightly less hydrogen will be fused in the core before the Sun leaves the main-sequence phase. Given the uncertainties,

Concept Question 2

Suppose in the remote future we wanted to extend the Sun's lifetime. A government official proposes collecting interstellar hydrogen and dumping it on the Sun. If this were possible, would it be a good idea? Why or why not?

the Sun's main-sequence lifetime is usually rounded off to 10 billion (10^{10}) years. Given its present age, the Sun is about halfway through its main-sequence lifetime.

We can estimate the lifetimes of other stars by comparing them with the Sun. If $\mathcal{M}$ is its mass and $\mathcal{L}$ its luminosity in solar units, a star's main-sequence lifetime is

$$t = \frac{\mathcal{M}}{\mathcal{L}} \times 10^{10} \text{ years.}$$

t = Main-sequence lifetime
$\mathcal{M}$ = Star's mass in solar masses
$\mathcal{L}$ = Star's luminosity in solar luminosities

This assumes that all stars fuse the same fraction of their mass as the Sun does, so it will underestimate somewhat the lifetimes of the highest- and lowest-mass stars; but it gives a good lifetime estimate for most stars.

We might have expected massive stars to live longer than other stars because they have more fuel to fuse. However, the luminosities of massive stars go up by even more than their masses. For example, the star Sirius has an estimated mass of about 2 $M_\odot$, but its luminosity is about 20 $L_\odot$. Therefore, it has 2 times more fuel than the Sun, but it is fusing it 20 times faster, so its lifetime will be 2/20 as long as the Sun's. In other words, Sirius will have a main-sequence lifetime, t, of

$$t(\text{Sirius}) = \frac{2}{20} \times 10^{10} \text{ years} = 10^9 \text{ years.}$$

Mathematical Insights

The mass–luminosity relation (Unit 59) states that $\mathcal{L} \approx \mathcal{M}^{3.5}$ for a main-sequence star whose mass and luminosity are measured in solar units. Using this relationship to substitute for the luminosity, we find that a star's lifetime is approximately

$$t \approx \frac{\mathcal{M}}{\mathcal{M}^{3.5}} \times 10^{10} \text{ years}$$

$$\approx \frac{1}{\mathcal{M}^{2.5}} \times 10^{10} \text{ years.}$$

So Sirius is expected to live on the main sequence only about a tenth as long as the Sun. For a star whose mass is 10 $M_\odot$ and whose luminosity is 10^4 $L_\odot$, the main-sequence lifetime is a mere 10^7 years.

These results demonstrate that massive stars have much shorter lifetimes than the Sun. Despite having more fuel, they fuse it much faster to supply their greater luminosity. In other words, massive stars are like gas-guzzling cars that, despite having large fuel tanks, run out of fuel sooner than fuel-efficient cars with smaller tanks. This brevity of massive stars' lives implies that those we see must be relatively "young" by the standard of the Sun. Because massive stars are blue, we can conclude that, in general, blue stars have formed recently. Because of their recent formation, we can understand why this type of star is often seen associated with clouds of interstellar gas. They die so quickly that there has been little time for the interstellar gas to disperse in response to stellar winds, or for the star to drift away from the cloud if it has any relative motion. Therefore, blue stars are frequently embedded in matter left over from their formation, as you can see in Figure 62.5. In these recently formed groups of stars, the hot blue stars produce ultraviolet light that ionizes the interstellar hydrogen, producing the pink glow of interstellar hydrogen seen in the image.

FIGURE 62.5
Photograph of the young star cluster NGC 2264. The massive, short-lived blue stars are still surrounded by the interstellar gas from which they formed. NGC 2264 is also called the "cone nebula" for the shape of the dark column of gas and dust (on left side of image), the tip of which is ionized by the brightest blue star near the center of the image.

Concept Question 3

If a star is red, what can you say about its age?

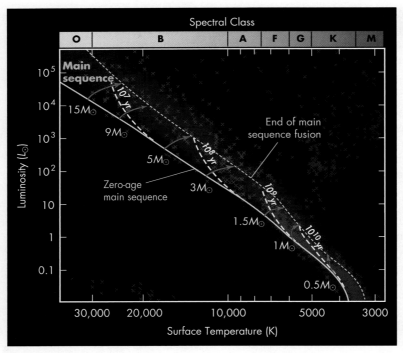

FIGURE 62.6

Stellar evolution while stars are on the main sequence. Luminosities typically increase by about a factor of 3 during this phase. Tracks of stars of different masses are shown in red. The white dashed lines indicate where the stars will be along the tracks at different times after they begin main-sequence fusion.

Concept Question 4

The star Alpha Centauri A has the same spectral type as the Sun, but it is about 20% more luminous. Based on this, how would you expect its mass and age to compare to the Sun's?

Concept Question 5

As the Sun's power output continues to increase, what else can humankind do to compensate for the Sun's increasing luminosity?

62.4 CHANGES DURING THE MAIN-SEQUENCE PHASE

A star fusing hydrogen to helium in its core maintains a relatively constant structure, temperature, and luminosity. Therefore, it remains in almost the same position in the H-R diagram. Since stars spend a large percentage of their lives in the main sequence, their luminosities and temperatures as a function of their masses establishes the diagonal band seen in the H-R diagram.

While stars change relatively little during their main-sequence phase, they do change. Stellar structure models tell us that as a star depletes the hydrogen in its core, the hydrogen nuclei become more spread out and less likely to collide. The decline in energy production causes the star to contract slightly, compressing and heating the core. This drives up the temperature until the overall rate of reactions once again generates enough pressure to support the star. Over the long time a star spends in its main-sequence phase, it gradually increases in luminosity by a factor of about 3, as illustrated in Figure 62.6. The stellar models show that the star's atmosphere expands in response to the more luminous core. Tripling the luminosity is not a minor change, but on the H-R diagram (Unit 59), which displays a factor of over a million in luminosity, the change leaves the stars in a narrow band.

It is possible to estimate how much a star has aged by its shift from the expected initial characteristics when it began fusing hydrogen—also known as its **zero-age main sequence** (or **ZAMS**) position in the H-R diagram. This requires such precise measurements of a star's mass, luminosity, temperature, and chemical composition that we have accurate age estimates for relatively few stars.

Astronomers estimate that the Sun is about 1.4 times more luminous today than when it began on the ZAMS. Its luminosity will continue to increase, approximately doubling from its current value over the remaining 5 or 6 billion years of its main-sequence lifetime. As the Sun's core temperature climbs, the CNO cycle will become the dominant source of its energy production.

Although these changes are small compared to the range of other stars' luminosities, they are very significant for the environment of the Earth and other planets. Remarkably, through most of the Earth's history the temperature has remained stable, apparently due to the steady removal of "greenhouse gases" by microscopic life forms (Units 38 and 85). The consequences for the Earth of the future increase in the Sun's luminosity are not clear. Just a few percentage points of increase in the Sun's luminosity would drive the Earth's temperature up by about 2°C (about 4°F), enough to cause polar ice cap melting, which would dramatically raise ocean levels. Unless compensating effects occur, the Earth's climate and habitability will change in the distant future.

KEY POINTS

- Contracting massive stars reach and exceed the minimum fusion temperature relatively rapidly while the core's density is still low.
- Low-mass stars contract for far longer and compress their cores to a much higher density before they reach fusion temperature.
- Instead of the proton–proton chain, at the high temperatures in massive stars, hydrogen fusion occurs through the CNO cycle, which uses carbon as a catalyst to fuse hydrogen at a faster pace.
- The high heat in massive stars causes convection around the core.
- The low-temperature gas at the surface of low-mass stars blocks light, causing surface convection.
- The surface convection reaches down to the core in very low-mass stars, which are the only stars that exchange core and surface gas.
- A star's main-sequence lifetime is proportional to the amount of fuel (the mass) divided by the rate of energy loss (the luminosity).
- The Sun's main-sequence lifetime is about 10 billion years.
- More-massive stars are so much more luminous that they have much shorter lifetimes; less-massive stars live much longer.
- As stars exhaust their hydrogen during the main-sequence phase, their luminosity steadily grows to become several times higher.
- Millions of years from now, the increasing luminosity of the Sun will pose a serious challenge to the Earth's habitability.

KEY TERMS

CNO cycle, 490
convection, 491
fully convective, 491
hydrostatic equilibrium, 489
main-sequence lifetime, 492
main-sequence star, 489
proton–proton chain, 490
zero-age main sequence (ZAMS), 494

CONCEPT QUESTIONS

Concept Questions on the following topics are located in the margins. They invite thinking and discussion beyond the text.

1. Massive star with no carbon content. (p. 490)
2. Extending the Sun's lifetime by adding hydrogen. (p. 492)
3. Age of red stars. (p. 493)
4. Comparing Alpha Centauri to the Sun. (p. 494)
5. Coping with Sun's increasing luminosity. (p. 494)

REVIEW QUESTIONS

6. What determines when a star becomes a main-sequence star? What is the ZAMS?
7. What factors determine the density and temperatures of the cores of main-sequence stars?
8. Why do high-mass stars last a shorter time on the main sequence than low-mass stars?
9. How does convection differ in very low-mass stars, low-mass stars, and high-mass stars?
10. How does a star's luminosity change during its main-sequence lifetime? What consequences does this have for Earth?

QUANTITATIVE PROBLEMS

11. Calculate the luminosity of the 2.0-$M_\odot$ star Sirius predicted by the mass–luminosity equation (Unit 59). Its observed luminosity is about 20 $L_\odot$. How might you explain the difference?
12. Calculate the main-sequence lifetime of a 25-$M_\odot$ star using the mass–luminosity relationship (Unit 59) to estimate its luminosity.
13. Given that the Universe is 13.8 billion years old, if a 0.1 $M_\odot$ star with a luminosity of 10^{-4} $L_\odot$ was born in the early days of the Universe would it have died out by now? If so, how long ago did it die out? If not, how long until it does die out?
14. Use the main-sequence lifetime equation and the mass–luminosity relationship to estimate the mass of a star that would have a main-sequence lifetime of 500 million years.
15. According to a recent computer model of the Sun's evolution, when the Sun was young its surface temperature was 130 K cooler (than its current value of 5800 K) and its luminosity was 0.71 $L_\odot$. When the Sun is approaching the end of its main-sequence lifetime 5.5 billion years from now, its surface temperature will be 650 K cooler than it is currently, and its luminosity will be 2.02 $L_\odot$. Use the Stefan-Boltzmann law (Unit 58) to estimate the Sun's diameter at these past and future times.
16. Suppose the Sun had a luminosity of 0.71 $L_\odot$ when it was young. Compare the brightness of sunlight (Unit 55) on Venus (at 0.72 AU) when the Sun was young to the brightness of sunlight on Earth today.
17. Estimate the size of the core for a 10 $M_\odot$ star. Assume the values discussed in the text: a star's core is roughly 10% of the star's overall mass and the core has a density of roughly 3 kg/L. Express your result in terms of solar radii.

TEST YOURSELF

18. A star whose mass is 2 times larger than the Sun's has a main-sequence lifetime _____ than the Sun's lifetime.
    a. many times longer
    b. a few percent longer
    c. a few percent shorter
    d. many times shorter
19. Blue main-sequence stars have
    a. a large mass and a short lifetime.
    b. a large mass and a long lifetime.
    c. a low mass and a short lifetime.
    d. a low mass and a long lifetime.
    e. You can't say, because colors of main-sequence stars are unrelated to their masses or lifetimes.
20. A star is considered to be in its main sequence phase when it
    a. first becomes luminous enough to be observed.
    b. uses convection to transport hydrogen to the core.
    c. uses the CNO cycle to maintain hydrostatic equilibrium.
    d. is fusing hydrogen into helium in its core.
21. On an H-R diagram the main sequence is not a narrow line because main sequence stars have
    a. large uncertainties in measuring their stellar properties.
    b. luminosities that fluctuate randomly.
    c. luminosities and temperatures that change as they age.
    d. become steadily dimmer as they age.

PART 4 · UNIT 63

Giant Stars

63.1 Restructuring Following the Main Sequence
63.2 Helium Fusion
63.3 Electron Degeneracy and the Helium Flash in Low-Mass Stars
63.4 Helium Fusion in the H-R Diagram

Learning Objectives

Upon completing this Unit, you should be able to:
- Describe the structural changes that take place as a star leaves the main sequence.
- Explain why and how a star transitions to helium fusion with hydrogen shell fusion.
- Describe a degenerate gas, and explain which stars reach and leave this condition.
- Explain how giant stars move through the H-R diagram as they evolve.

The main-sequence stage of a star's life is a time of relative stability (Unit 62), but the next stage of its life is one of constant change. After hydrogen runs out in its core, a star is constantly readjusting its structure to different sources of fuel that run out much more rapidly than hydrogen. The star's core resumes the gravitational contraction that was taking place during the pre-main-sequence phase (Unit 61), but an unexpected thing happens.

Even though the core of the star begins contracting and heating again, the surface of the star expands and cools, and the star grows into a **giant.** Stellar models indicate that a star's lifetime as a giant is about 10% to 20% as long as its main-sequence lifetime. In this Unit we will explore why stars grow into giants and examine some of the unusual aspects of this phase of a star's overall evolution (Unit 60).

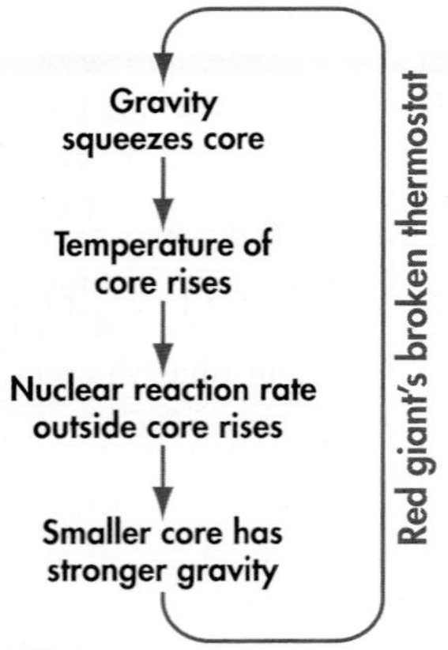

63.1 RESTRUCTURING FOLLOWING THE MAIN SEQUENCE

Stars are stable during their main-sequence phase because nuclear fusion in the core acts as a thermostat. For example, if a star contracts slightly, the compression of the gas in its core heats it, driving up the rate of nuclear reactions. Consequently, the core produces more heat and pressure, reversing the contraction. A slight expansion of a star reduces nuclear reactions, leading to less pressure, which reverses the expansion. A main-sequence star is therefore quite stable. However, once nuclear reactions cease in the core, the thermostat is broken, and this leads to unusual effects, which are the hallmarks of the next phase of a star's life.

When hydrogen is used up in the core of a star, the star's weighty outer layers still press downward. The core is squeezed smaller. This compresses the gas in the core and therefore heats it (as discussed also in Units 51 and 61). The result for the star is that its core temperature rises. However, without hydrogen in the core, the star has no way to generate energy to counteract this compression. The smaller core has an even stronger gravitational pull because gravity grows stronger as separations become smaller. The stronger gravity squeezes the core even harder, leading to further compression in a runaway cycle of heating (Figure 63.1).

Although the core itself has no hydrogen to fuse, regions outside the core are still rich in hydrogen because they never reached temperatures high enough for hydrogen to fuse during the star's main-sequence lifetime. Now, however, as the interior of the star shrinks and is compressed, the regions immediately

FIGURE 63.1
A red giant does not have the thermostatic effect of gravity and pressure in the core counteracting each other if they rise. This leads to runaway heating of the region surrounding a red giant's core.

Gravity squeezes core → Temperature of core rises → Nuclear reaction rate outside core rises → Smaller core has stronger gravity
Red giant's broken thermostat

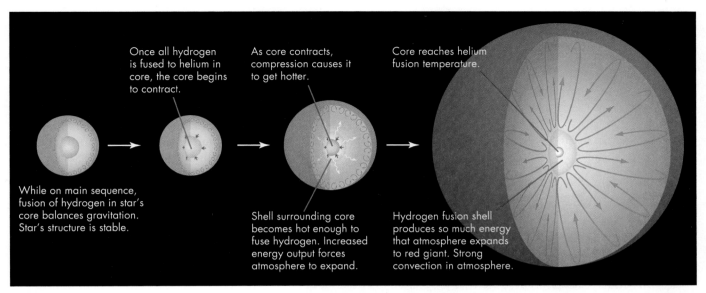

FIGURE 63.2
(left) A star's core begins to shrink as the star uses up the hydrogen in its core. This compresses and heats the core. (middle) The hotter core ignites the surrounding gas and the outer layers of the star expand, turning it into a red giant. (right) When the core temperature rises to about 100,000,000 K, helium fusion begins.

Evolution to red giant phase

Some energy is also released from gravitational potential energy as the star's core contracts.

surrounding the core are also compressed and heated to fusion temperatures. Hydrogen fusion begins in a layer outside the core in a process called hydrogen **shell fusion** (Figure 63.2). Although shell fusion generates heat and pressure, it is outside the core, and therefore it does not provide the internal pressure needed to counteract the core's continuing contraction.

This situation is almost the opposite of the thermostat effect that makes a main-sequence star so stable. The core continues contracting and growing hotter, causing more and more hydrogen to fuse in regions surrounding the core. The star grows steadily more luminous, but the rise in energy production now does nothing to reverse the contractions. For low-mass stars like the Sun, the luminosity may increase by a factor of 1000. In other words, it will be fusing 1000 times more hydrogen each second in its shell region during its giant phase than it fused each second in its core during its main-sequence phase.

By contrast, when high-mass stars complete hydrogen fusion in their core, they change relatively little in luminosity. They have already consumed more of their hydrogen outside the core because of greater convection activity during their main-sequence lifetimes, so the transition to shell fusion is more gradual. The shell fusion during this phase does not represent a substantial increase over their vigorous main-sequence rate of fusion. Indeed, even while on the main sequence, these massive stars already have the luminosities of supergiants (Unit 59).

Shell fusion pushes the surface layers outward for both low- and high-mass stars, and this is reflected by a significant drop in surface temperature. This drop in surface temperature is perhaps the most counterintuitive aspect of star evolution after the main sequence. The core shrinks and heats up, but the outer layers of the star expand and cool.

We can explain this unusual behavior as follows. Shell fusion raises the temperature and pressure of the gas in the shell. The heating and higher pressure in the shell pushes the surrounding gas outward, causing the star's surface to expand. The star may grow in radius by a factor of anywhere from five to several hundred, depending on the star's mass. This expansion leaves the outer layers much farther from the source of the star's luminosity, the fusion in and around its core.

At this larger distance from the core, even if the star has grown more luminous, the intensity of radiation is reduced. This is a consequence of the inverse-square law

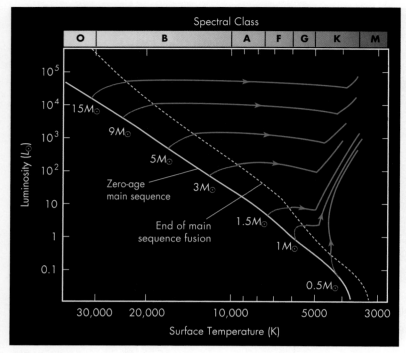

FIGURE 63.3
Post-main-sequence evolutionary tracks followed by stars with masses from 0.5 to 15 times the Sun's mass in the H-R diagram. The tracks show how stars' outward characteristics change from the end of hydrogen fusion in the core to the start of helium fusion. Very low-mass stars, 0.5 $M_\odot$ and lower, never achieve helium fusion before their cores become too dense and stop contracting.

($B = L/4\pi d^2$, Unit 55.2): the brightness of light is inversely proportional to the square of the distance from the source of light. The luminosity coming from the interior of the star is thus spread out over a much larger area because the distance (d) from the core has grown much larger. The decreased intensity of radiation results in surface layers that are cooler, and because cooler bodies radiate more strongly at longer wavelengths, the star becomes redder. Thus, the main-sequence star has evolved into a red giant or supergiant.

In the H-R diagram (Figure 63.3), low-mass stars move up and to the right of the main-sequence region along a part of its evolutionary track called the **red giant branch.** A star like the Sun will grow steadily larger and more luminous over nearly a billion years as it searches for a new equilibrium, its core gradually shrinking and heating throughout this time. High-mass stars rapidly drop in surface temperature at a relatively steady luminosity, so their evolutionary tracks traverse the H-R diagram along a nearly horizontal path.

A red giant has a very different structure from that of a main-sequence star. Its enormous size is deceptive. Most of the star's volume is taken up by its very low-density atmosphere, with a tiny, compressed core in the central 1% of its radius. Hydrogen shell fusion supplies most of the energy during the beginning of the giant phase, but if the star's core becomes hot enough, it will begin to generate energy by the fusion of helium, leading to a new stage of the giant star's evolution.

63.2 HELIUM FUSION

Helium nuclei can fuse to form heavier elements, but the process requires a much higher temperature than hydrogen fusion does. Two nuclei fuse when they are brought close enough together for the strong force (Unit 4) to bind them. That force, which is what holds the protons and neutrons together in a nucleus, operates only over very short distances. If nuclei are more than a few diameters apart, the nuclear strong force is too weak to bond them. In fact, they will be repelled by the electrical force between the similarly charged protons, as illustrated in Figure 63.4. The electric repulsion grows as the number of protons increases. The repulsion is smaller for hydrogen with its single proton, but larger for heavier elements with more protons. It is more difficult, therefore, for helium nuclei, with their two protons, to fuse than for hydrogen atoms to do so.

Helium fusion occurs when three ^{4}He nuclei combine to make a carbon nucleus, ^{12}C. This is called the **triple alpha process** because helium nuclei were given the name *alpha particles* in some of the first nuclear radioactivity experiments around 1900, before their identity was known. For the triple alpha process to become a steady source of energy, a temperature in excess of 100 million K is required. However, despite operating at such a high temperature, the triple alpha process releases much less energy than hydrogen fusion—only about one-tenth the amount of energy per kilogram of fuel as hydrogen fusion yields. Thus, although the initiation of helium

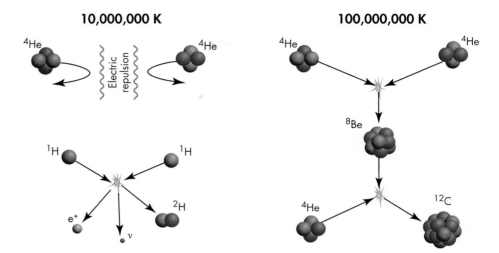

FIGURE 63.4
At 10 million K, hydrogen nuclei can fuse, but the larger electric repulsion between helium nuclei keeps them from fusing. At 100 million K, helium nuclei can fuse, making first beryllium and then carbon.

Concept Question 1

The triple alpha process requires two steps in rapid succession. First, two helium nuclei fuse to make a beryllium nucleus, ^{8}Be. However, this nucleus is unstable and breaks back apart into the helium nuclei in only about 2 millionths of a second, so another helium nucleus must collide with it in that short interval. If ^{8}Be were stable, how might this change stars' giant phase?

fusion stabilizes the core against further contraction, the triple alpha process does not produce the majority of a giant star's luminosity.

Calculations indicate that the compression that occurs in stars more massive than about 0.5 $M_\odot$ is sufficient to reach temperatures that allow helium fusion, but the process begins very differently in low- and high-mass stars. For example, a 10-$M_\odot$ star needs to compress its core to only about 100 times its main-sequence density before helium begins to fuse, because its core is already extremely hot. A low-mass star like the Sun, however, must compress its core by about a factor of 10,000 before it becomes hot enough for helium fusion. This requires the Sun's core to reach a density of about 1 million kilograms per liter. A teaspoonful of such matter would weigh as much as a truck! As you might suppose, such a high density strongly affects the properties of a gas. To see how, we need to look more closely at the nature of extremely dense gases.

63.3 ELECTRON DEGENERACY AND THE HELIUM FLASH IN LOW-MASS STARS

To see why gases behave differently when they are densely packed, imagine a few dozen tennis balls inside a large box. If you shake the box, the tennis balls bounce about inside the box. This is similar to the behavior of the particles in a normal gas. You can make the box smaller, and the tennis balls will still bounce—until you make the box so small that the balls are packed tightly against one another. At that point the balls no longer bounce, and they prevent you from making the box any smaller.

Gases display a similar behavior—they can be compressed until the particles are so closely packed that they fill the volume. At that point the gas becomes almost rigid and extremely difficult to compress. The particles that resist being packed so closely together are the electrons in a hot ionized gas. The behavior of subatomic particles like electrons is different from how solid balls interact. The electrons obey what is called an **exclusion principle:** No two electrons can occupy the same space if they have the same energy. Note that this rule *does* allow electrons to occupy the same volume of space if they have *different* energies.

If a gas is compressed to the point where electrons of the same energy are trying to occupy the same space, they behave like the tennis balls, resisting coming any closer together. A gas in which its electrons are packed together like this is called a **degenerate gas.** This packing limit makes the gas as "stiff" as a solid. As a result, a degenerate gas does not contract when it cools, any more than a cooled

Electrons have an additional property called spin, which can be in opposite directions, so actually the exclusion principle allows two electrons of the same energy to occupy the same space if their spins are opposite.

brick shrinks. This stiffness has important consequences for nuclear fusion in a degenerate gas.

For very low-mass stars of 0.5 $M_\odot$ or less, the compression makes the core degenerate before it grows hot enough to start helium fusion. Hydrogen fusion in the shell continues briefly after degeneracy is reached, but without the rising temperature coming from the continued contraction of the core, there is nothing to keep the surrounding gas hot enough for fusion to continue in the shell. As the core cools, hydrogen fusion slows in the shell and then halts. The star then eventually cools and dims.

The cores of stars with masses from about 0.5 $M_\odot$ to 2 $M_\odot$ get compressed to the degeneracy limit when they are close to the helium ignition temperature, and their fate is quite different. Normally the thermostatic feedback behavior of nuclear fusion in a gas prevents the nuclear reactions from becoming too violent. If the gas becomes hot, the rate of nuclear fusion increases, heating the gas and raising its pressure, so the gas expands. The expansion cools the gas, and the rate of nuclear fusion goes back down. This is not true when electron degeneracy pressure supports the core: In that case, the pressure resisting contraction no longer rises if the temperature rises.

When the core becomes degenerate, it cannot be compressed because the electrons are occupying all the available space. However, compression is not completely impossible. Initially, the electrons in a degenerate gas are all in the lowest energy level possible. However, if some electrons absorb energy, they have a different energy from their neighbors, so according to the exclusion principle they can be packed even more tightly. As the core is heated by gravitational contraction and the radiation from the surrounding fusion shell, the electrons absorb some of this energy. This allows the electrons to pack together more tightly, the core shrinks, and the gravity grows more intense. The greater compression on the hydrogen fusion shell makes it hotter so it generates more energy, which continues the process—although the compression proceeds more slowly than before the electrons became degenerate.

Under these circumstances, the initiation of helium fusion can occur explosively. When the helium fusion temperature is reached, the energy released by the fusion reactions is soaked up by the electrons. This relieves some of the degeneracy, so the core shrinks. But this drives the temperature up, causing more-rapid helium fusion, which gives the electrons more energy, allowing more compression and heating. This runaway process accelerates the star's energy production explosively in what is called a **helium flash.**

During the helium flash, a star's energy production increases by many thousand times in just a few minutes. The outburst is totally hidden from our view by the star's outer layers, just as a firecracker set off under a mattress creates little visible disturbance. When the released energy heats the core enough that the gas is no longer degenerate, it stabilizes. With continued energy generation from helium fusion, the gas in the core can adjust to maintain a stable pressure balance with the gravitational weight of the layers pressing down on it.

63.4 HELIUM FUSION IN THE H-R DIAGRAM

The changes taking place deep inside a giant star after it begins to fuse helium are reflected in the star's outward appearance as the outer layers adjust to the new source of power. The star becomes stable again because a "thermostat" is once more established in the core (Section 63.1). As happened during hydrogen fusion in the main-sequence phase, compressing the helium-fusing core heats it up, which drives up energy production so rapidly that the pressure expands the core back out again. This results in a sort of *helium-fusing main sequence* region in the H-R diagram,

---

LOOKING UP

Aldebaran, a red giant shown in Looking Up #5 at the front of the book, has a mass estimated to be less than 2 $M_\odot$. It is probably in the stage of compressing its core toward helium-fusion temperature.

Concept Question 2

In a very low-mass star, what would you observe from the outside after it fails to ignite helium fusion? Would the changes be fast or slow?

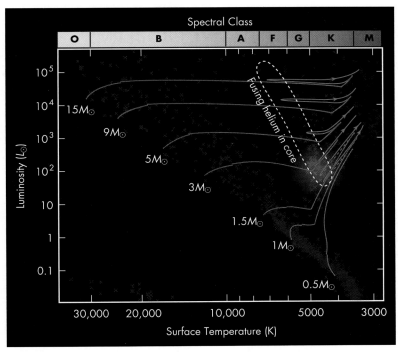

FIGURE 63.5

Evolutionary tracks (in red) of stars in the H-R diagram after the start of helium fusion. When helium fusion begins in the core, stars rapidly readjust and stabilize in the region outlined by the dashed white line. After helium is used up in the core, helium begins fusing in the region surrounding the core, and stars begin climbing back up to the upper right part of the H-R diagram. Many red giants are found in the region highlighted by a red cloud, because this is where the common class of stars with masses similar to the Sun are positioned while fusing helium.

Betelgeuse, shown in Looking Up #6 at the front of the book, is a red supergiant. Astronomers are uncertain whether it has completed its helium fusion or not.

Concept Question 3

When the Sun becomes a red giant, it will be about a thousand times more luminous than it is now. What effect will this have on the outer planets?

as illustrated in Figure 63.5. After having grown to an extreme size and luminosity as its core was shrinking and heating, the star now stabilizes at a lower luminosity. With helium fusion in the core, the star's surface settles to a smaller size, although it grows hotter and therefore changes color from red to yellow.

In the H-R diagram, low-mass stars undergoing a helium flash shift abruptly down and to the left (dashed lines) as they adjust to helium fusion in the core. Then for about 100 million years, these stars remain relatively stable as they fuse helium in their cores. During this phase they have similar luminosities of about 100 $L_\odot$ and are similar in temperature to each other, so they show up as a "clump" or "branch" in the H-R diagram.

Like lower-mass stars, when stars more massive than 2 $M_\odot$ exhaust the hydrogen in their cores, they leave the main sequence and shell fusion begins. Meanwhile, their surfaces expand and cool, just like in the low-mass stars. However, the cores are already so hot that they ignite helium with relatively little compression, and so they do not become degenerate at this stage of their lives. After helium fusion begins, these stars also readjust, typically shrinking in size and luminosity, but they do this in a controlled way rather than an explosive helium flash.

The time it takes massive stars to adjust to begin fusing helium is only about 1% of their main-sequence lifetime. As a result, their positions in the H-R diagram shift rapidly from the main-sequence region over to the low-temperature side of the plot. After their cores are compressed sufficiently to initiate helium fusion, they generally stabilize at a luminosity just slightly higher than when the star was on the main sequence, and this helium-fusion phase may last 10% to 20% as long as the main-sequence phase. Even after helium fusion begins in the core, a substantial portion of the light generated by a star is produced by hydrogen shell fusion.

When the helium in the core is exhausted, the core begins contracting and heating again, just as after the main sequence phase. Unconsumed helium surrounding the core begins fusing while hydrogen fusion continues in a shell around that. The luminosities climb and temperatures decline, and the stars travel back up in the H-R diagram, growing to be even more luminous red giants than before. The large compressive forces raise the temperature in the core high enough, to about 200 million K, for heavy elements to fuse with helium (often called *alpha capture*), which becomes the dominant source of energy production within the giant. For example, a ^{4}He nucleus can fuse with ^{12}C to form oxygen, ^{16}O, releasing energy. Convection inside these stars can drag some of this material to the surface, where it may be expelled in winds (Unit 65). This is one of the major sources of oxygen and carbon—elements essential to life.

The evolutionary paths of low- and high-mass stars diverge at this point. In low-mass stars, the cores quickly become degenerate and halt further contraction. This final bright stage marks the end of fusion within stars like the Sun, and is accompanied by the loss of atmosphere discussed further in Unit 65. Massive stars may continue to fuse heavier elements, leading to a catastrophic death as described in Unit 67.

KEY POINTS

- When fusion is completed in a star's core, the core begins compressing and heating as the star enters its red giant phase.
- Hydrogen surrounding the core rises to fusion temperature leading to shell fusion and driving the atmosphere to expand.
- The outer parts of the atmosphere expand in response to the hotter core and shell fusion; the surface becomes cooler and redder.
- The core continues contracting and heating until it reaches helium fusion temperature, unless it becomes degenerate first.
- Helium fusion requires a temperature about 10 times higher than hydrogen fusion, and leads to the production of carbon.
- A degenerate gas prevents further compression because the electrons cannot occupy the same space if they have the same energy.
- In stars smaller than about 0.5 solar mass, contraction of the core is halted by degeneracy before the core reaches helium fusion.
- The cores of stars with about 0.5 to 2 solar masses become degenerate, but gradually heat enough to ignite helium in a "flash."
- Stars more massive than 2 solar masses begin fusing helium more smoothly.
- Shell fusion produces much of a giant's overall luminosity.
- Helium core fusion lasts only a fraction as long as the main sequence phase, but the overall luminosity ranges from 10 to 1000 times greater than during the main sequence.

KEY TERMS

degenerate gas, 499
exclusion principle, 499
giant, 496
helium flash, 500
red giant branch, 498
shell fusion, 497
triple alpha process, 498

CONCEPT QUESTIONS

Concept Questions on the following topics are located in the margins. They invite thinking and discussion beyond the text.

1. Effects on stars' fusion if beryllium-8 were stable. (p. 499)
2. End stages of very low-mass stars. (p. 500)
3. Outer planets when the Sun is a red giant. (p. 501)

REVIEW QUESTIONS

4. What makes a star move off the main sequence?
5. Where do main-sequence stars end up as they evolve?
6. Why do high- and low-mass stars evolve differently as they become giants?
7. What are the ingredients and the products of the triple alpha process?
8. What is the exclusion principle for electrons, and why is it important for the evolution of giant stars?
9. What is the helium flash triggered by?

QUANTITATIVE PROBLEMS

10. Using the plots in Figures 63.3 and 63.5, estimate the luminosity and temperature of a 5-solar-mass star (a) at the start of the main sequence phase, (b) at the end, (c) while it fuses helium in its core, and (d) at its most luminous final stage.
11. When the Sun expands to be a red giant, its radius may be 100 $R_\odot$. What will the average density of the Sun be at that time?
12. When the Sun becomes a red giant, its core region (containing about 20% of its mass) may contract to 0.01 $R_\odot$. What will the density of the core be at that time?
13. By how many times its current radius would the Sun have to expand before its outer atmosphere reached the Earth?
14. Suppose a Sun-like star is just barely visible to the naked eye. If an identical star has already become a red giant, with a radius 100 times larger and a surface temperature half as large, how many times farther away must it be to look equally bright?
15. We observe a red giant with a luminosity 127 times the Sun's and a wavelength of maximum radiation of 675 nm. What is its radius? (Use Wien's law, Unit 56.2, to find its temperature.)
16. Estimate the radius of the 15-$M_\odot$ star shown in the H-R diagrams in this Unit from when it is initially a blue supergiant to when it becomes a red supergiant. (You will need to estimate luminosities and temperatures from the H-R diagram and then use the Stefan-Boltzmann equation from Unit 58 to solve for radius.)

TEST YOURSELF

17. As a star like the Sun evolves into a red giant, its core
    a. expands and cools.
    b. contracts and heats.
    c. turns into iron.
    d. expands and heats.
    e. turns into uranium.
18. A star leaves the main sequence when
    a. it has converted all of its hydrogen into helium throughout the star.
    b. it has converted all of its hydrogen into helium in its core.
    c. it has converted all of its helium into carbon in its core.
    d. it begins to fuse hydrogen into helium in its core.
19. As low mass stars evolve off the main sequence they become more luminous because their
    a. surface temperatures increase.
    b. surface temperatures decrease.
    c. sizes increase.
    d. sizes decrease.
    e. surface temperatures and sizes both increase.
20. Approximately how long will the Sun's red giant phase last, compared to its main-sequence phase?
    a. Much less than 1% (just a few hundred years)
    b. About 10% as long
    c. Approximately the same length of time
    d. About twice as long
    e. About 100 times longer

UNIT 64

Variable Stars

- 64.1 Classes of Variable Stars
- 64.2 Yellow Giants and Pulsating Stars
- 64.3 The Period–Luminosity Relation

Learning Objectives

Upon completing this Unit, you should be able to:
- List the types of variable stars, describing the characteristics of their behavior.
- Describe the mechanism that causes some giant stars to undergo regular pulsations.
- Explain how a distance to a star can be calculated from the period of the pulsation for some classes of pulsating variable.

Many stars pass through a stage sometime during their lives in which their luminosity varies. For many stars this stage occurs after they enter the red giant phase. Astronomers call stars whose luminosities change **variable stars.** All stars vary slightly in brightness—even the Sun varies by a few percentage points because of its magnetic activity cycle (Unit 53)—but this kind of variation is detectable only with careful photometric measurements. All stars also experience gradual changes in their luminosity and temperature while they are on the main sequence, and faster changes while they are giants.

Although virtually all stars vary to some degree, traditionally the stars cataloged as variables change in brightness over a decade or less. Furthermore, these changes can be seen just by looking through a telescope and comparing the variable star with neighboring stars. An example of a variable star is shown in Figure 64.1—the star Mira (*MY-rah*) in the constellation Cetus. For hundreds of years Mira has been known to regularly brighten and fade in a little less than a year. These changes often signal large fluctuations in the radius of the stars, as can be deduced from the Stefan-Boltzmann law (Unit 58). These variations are more than just curiosities. As we shall see in this Unit, they signal interesting happenings in a star, and sometimes they prove to be excellent distance indicators.

FIGURE 64.1
These two photographs show the star Mira in the constellation Cetus. Mira varies over a period of 332 days from being visible to the unaided eye to being about 100 times fainter.

64.1 CLASSES OF VARIABLE STARS

The first variable stars identified were stars that brightened dramatically, often from previous invisibility, before fading again. A "new star" suddenly appearing this way in 1572 was called a *nova stella* by Tycho Brahe (Unit 12). Today this is shortened to **nova.** These are associated with explosive events sometimes marking the final stages of a star's lifetime, and they may remain visible for weeks or months before fading away. Actually the nova seen in 1572 was an especially luminous type that today is called a **supernova** (Unit 66). Tycho's discovery was revolutionary at the time, because it proved that even the highest celestial realm was changeable, running counter to beliefs of the day.

To characterize a star's variability, astronomers measure its brightness at frequent intervals and plot these against time. This results in a graph called a **light curve.** Over a hundred types of variable-star patterns have been identified, but these can be divided into a few classes. Stars like novae belong to the class of **irregular variables,** which undergo unpredictable changes that do not follow a repeating pattern. Another example of an irregular variable is a T Tauri star

FIGURE 64.2

Schematic light curves of (A) an RR Lyrae variable; (B) a Cepheid variable; (C) a Mira variable; and (D) a T Tauri irregular variable star. The timing and unique pattern of variability allow astronomers to identify these stars. The scale of each graph suggests the range of variability. Thus, Mira variables can vary by more than a factor of 100 from dimmest to brightest, but the other types mostly vary by, at most, a factor of about 2.

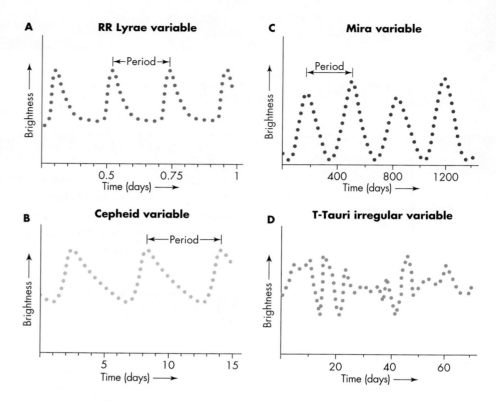

(Unit 61), which is a stage that pre-main-sequence stars pass through if they are less than about 2 solar masses. These objects probably brighten as material falls sporadically onto the forming star. Dozens of patterns have been identified in the irregular class, and these are often associated with very young or very old stars.

Some stars also vary because of periodic eclipses by other stars in orbit about them (Unit 58), but the majority of cataloged variable stars fall in the class of **pulsating variable.** These stars change in luminosity in a rhythmic pattern. Based on the shapes of their light curves, astronomers have identified more than a dozen types of regularly pulsating variable stars. For example, some stars pulsate with a period of about half a day, while others take more than a year to complete a cycle of pulsation. The interval of time it takes the pattern of brightness to repeat is called the **period,** as shown in Figure 64.2.

When we analyze the temperatures and luminosities of pulsating variable stars, we find that these stars also vary in size, rhythmically swelling and shrinking. These changes in diameter are caused by the pressure rising below a star's surface, which makes the surface expand outward. Some classes of these stars may more than double in size over their period, but the size variations are more commonly less than 20%. Such variations in size also cause a star's luminosity to change. If the star expands, its surface area increases so that it has more area to radiate. This would make the star more luminous. However, as the gas expands, its temperature usually drops, sometimes to half its former value. This would make the star dimmer. The result of both changes is that, as the star pulsates, its luminosity changes in a more complex way that can be determined using the Stefan-Boltzmann law (Unit 58). Stars like Mira can vary by more than a factor of 100 in luminosity, although the majority of variables exhibit a factor of two or less.

Amateur astronomers play a critical role in the study of variable stars (Unit 33). Many variables can be monitored with a relatively small telescope, and groups such as the American Association of Variable Star Observers have kept track of stars for the last century. They are often the first to discover that a star is behaving in some unexpected way and to alert astronomers around the world to the changes.

Concept Question 1

If the Sun maintained the same average luminosity, but began varying by a factor of 2 following a pattern like a Cepheid variable, what would it be like on Earth? Could life survive on the planet?

Information about joining the American Association of Variable Star Observers can be found at www.aavso.org.

64.2 YELLOW GIANTS AND PULSATING STARS

Pulsating variable stars are important to astronomers because they can serve as "standard candles" (Unit 55) and thereby offer a way to measure distances within the Milky Way and even to other galaxies (Unit 75). When such stars are plotted on the H-R diagram according to their average luminosities and temperatures, many of them lie within a narrow region called the **instability strip** (Figure 64.3). Because many of these stars have temperatures of about 5000 to 6000 K, they are yellow. Most pulsating variables are giants (Unit 63) and are sometimes called **yellow giants** or **yellow supergiants**, according to their color and luminosity.

One type of pulsating star useful for finding distances is the **RR Lyrae** (pronounced *LIE-ree*) **variable**. RR Lyrae variables have a mass comparable to the Sun's and are yellow giants with about 40 times the Sun's luminosity. Their pulsation cycle lasts about half a day. They are named for RR Lyrae, a star in the constellation Lyra, the harp, which was the first star of this type to be identified. From their known luminosity, we can apply the standard-candles method (see Unit 55) to RR Lyrae variables. Knowing that their luminosity is $40 \times L_\odot = 1.6 \times 10^{28}$ watts, we can plug this value into the equation for distance ($d = \sqrt{L/4\pi B}$), and if we measure the brightness B we can find the distance (see sample calculation in margin).

Other pulsating stars important for finding distances are the **Cepheid** (pronounced *SEF-ee-id*) **variables.** Cepheid variables are yellow supergiants that are more massive than the Sun, typically with about 10,000 times its luminosity. They are named for the star Delta Cephei, and their periods range from about 1 to 70 days.

Mathematical Insights

Suppose an RR Lyrae variable has a brightness of 10^{-11} watts/m². If its distance is d, then its luminosity L is spread over a sphere with a surface area of $4\pi d^2$, so we can calculate
$d = \sqrt{L/4\pi B} = \sqrt{1.6 \times 10^{28}/10^{-11}}$
$= \sqrt{16 \times 10^{38}} = 4 \times 10^{19}$ m,
or about 1300 parsecs.

FIGURE 64.3
The locations of several types of variable stars in the H-R diagram. Some classes of variables pulsate along the narrow region known as the "instability strip" in the diagram.

Delta Cephei, the prototype of Cepheid variables, is visible in Looking Up #1 at the front of the book.

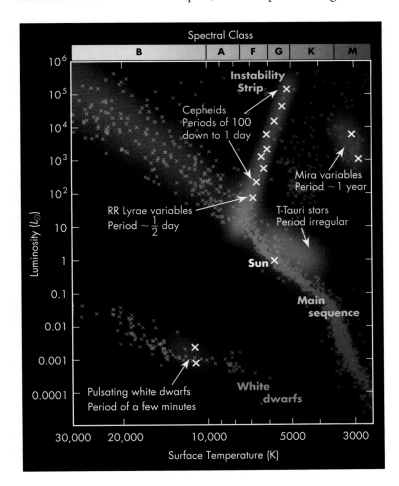

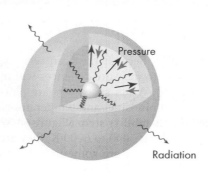 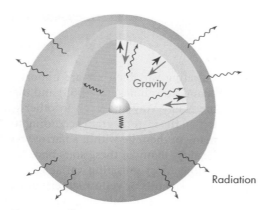 

Radiation (purple wiggles) partially trapped. Pressure (blue arrows) increases, overcomes star's gravity (red arrows), and makes it expand.

Expansion allows trapped radiation to escape. Star cools and pressure decreases. Gravity now exceeds pressure, so star contracts.

Pressure rises inside star as gas is compressed. After contraction halts, the cycle begins again.

FIGURE 64.4
Schematic view of a pulsating star.

> **Concept Question 2**
>
> If you put a heavy lid on a pan of boiling water instead of a light lid, how would this change the frequency of the steam escaping? What does this suggest about the variability of a star with a greater depth of atmosphere that must be adjusted?

Why these stars make such good standard candles has to do with the physical mechanisms that regulate their pulsations. Giant stars pulsate because their atmospheres trap some of their radiated energy. This heats their outer layers, raising the pressure and making the layers expand. The expanded gas cools and the pressure drops, so gravity pulls the layers downward and recompresses them. The recompressed gas begins once more to absorb energy, leading to a new expansion. These stars alternately trap and release energy, continuing to swell and shrink in a regular way that depends on their composition and gravity (Figure 64.4).

A covered pot of boiling water behaves similarly. The lid will trap the steam so that pressure inside rises. Eventually the pressure becomes strong enough to tip the lid, and steam escapes. The pressure decreases, and the lid falls back. It again traps the steam, the pressure builds up, and the cycle is repeated. In pulsating stars, the role of steam in this process is played by the star's radiation, and the role of the lid is played by partially ionized helium gas in the star's atmosphere. This is sometimes called the **valve mechanism,** because it is like a valve that opens to relieve excessive pressure.

Stars in the instability strip have surface temperatures of about 5000 K. The reason these stars are susceptible to pulsation is that stars of this temperature have an unstable region, not far below their surface, where partially ionized helium is able to absorb and then release the energy flowing out from the center of the star. This unstable region can alternately push the outer layers of the star upward and release the pressure so they fall back. The instability strip shifts to lower temperatures for higher-luminosity, larger-radius giants because their surface gravity is weaker, so the partially ionized helium layer can be deeper in the star and still provide enough pressure to push the surface layers upward.

After stars complete their main-sequence phase, they move along evolutionary paths through the giant region of the H-R diagram (Unit 63). When a star crosses through the instability strip, it begins to pulsate, and it continues to pulsate until its surface temperature changes enough to remove it from the instability strip. When a low-mass star crosses the instability strip, it becomes an RR Lyrae variable. When a high-mass star crosses, being more luminous, it lies above the RR Lyrae variables in the instability strip and instead becomes a Cepheid variable.

The amount of time a given star spends in the instability strip depends on its mass. Cepheids evolve across the strip in less than 1 million years. RR Lyrae variables spend more time in the strip, perhaps a few million years. In either case, stars pulsate for only a brief portion of their lives.

Astronomers have identified several other types of pulsating variables besides Cepheids and RR Lyrae stars. For example, there are pulsating white dwarfs with

Concept Question 3

At optical wavelengths, Miras sometimes vary in brightness by a factor of 10,000. At infrared wavelengths the variation is much smaller. What might explain this difference?

periods as short as a few minutes. Another kind of variable star, Mira variables, have pulsation periods of about a year. These are bright red giants that lie in the upper right portion of the H-R diagram. It is thought that the Sun will become a Mira variable close to the end of its lifetime. Mira variables are stars that have already completed both hydrogen and helium fusion in their cores, and now have shell fusion of both. It appears that the pulsation of Mira variables is due in part to light being blocked by several molecules that form in the cool atmosphere.

64.3 THE PERIOD–LUMINOSITY RELATION

Cepheids and some other classes of pulsating variable stars obey a law that relates their luminosity to their period—the time it takes them to complete a pulsation. Observations indicate that the more luminous a Cepheid is, the slower it pulsates. This **period–luminosity relation** is plotted in Figure 64.5.

Why does a more luminous star pulsate more slowly? We can trace the reason for this back to the gravitational force for stars of different sizes. Because Cepheids all lie in the instability strip and have similar surface temperatures, a higher luminosity implies a larger radius according to the Stefan-Boltzmann law (Unit 58). Because an object's gravity weakens as the distance from the center increases, a large-radius star has weaker surface gravity than a small-radius star. (Differences in mass along the instability strip are relatively small compared to the differences in radii.) So if the layers of a star with a large radius are pushed outward, the feeble gravity pulls them inward more slowly than in a small-radius star. Therefore, the pulsations in a big star take longer than in small, less luminous stars.

The period–luminosity law is one of the astronomer's most powerful tools for determining distances. A Cepheid with a period of a few months has a luminosity of tens of thousands of solar luminosities. Such a luminous star can be seen even in galaxies millions of light-years distant, and it can be identified by its characteristic light curve. Once its period is determined, its distance can be determined by using the standard-candle formula (Unit 55.3).

Some other classes of variables, such as Miras, also obey their own version of a period–luminosity relationship. On the other hand, RR Lyrae stars have a nearly fixed luminosity, independent of their pulsation period. In short, if we can identify the class of a variable star from its light curve, we can determine the star's luminosity from the period–luminosity law appropriate to it. If the variable star is associated with another object or other stars, we can learn the distance to them as well. This makes variable stars one of the astronomer's best tools for finding distances to objects too far away to measure by parallax.

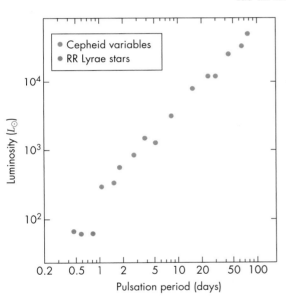

FIGURE 64.5
The period–luminosity relation. More-luminous Cepheid variables tend to pulsate more slowly. RR Lyrae variables all have about the same relatively low luminosity, and they pulsate faster than any of the Cepheids.

KEY POINTS

- All stars vary in brightness to some degree, but variable stars are generally those that vary obviously during a decade or less.
- There are many types of variable stars, some varying irregularly and some varying regularly.
- The pattern of variability often allows a determination of the type of star and sometimes its luminosity.
- Irregular variables are frequently stars undergoing rapid changes during their formation or death.
- Pulsating variables occur because of unstable physical conditions in the star's atmosphere that cause cyclical changes.
- Ionized helium can block radiation in a way that increases with temperature, in a way that leads to pulsations of Cepheids and RR Lyrae stars in the "instability strip" in the H-R diagram.
- Cepheids with longer pulsation periods have higher luminosities, allowing their luminosities to be determined even when they are very far away, making them excellent "standard candles."

KEY TERMS

Cepheid variable, 505
instability strip, 505
irregular variable, 503
light curve, 503
nova, 503
period, 504
period–luminosity relation, 507
pulsating variable, 504
RR Lyrae variable, 505
supernova, 503
valve mechanism, 506
variable star, 503
yellow giant, 504
yellow supergiant, 505

TABLE 64.1 Temperature and Luminosity of a Variable Star

Phase	Luminosity	Temperature
0.00	10,000 $L_\odot$	6700 K
0.25	7,800 $L_\odot$	5900 K
0.50	6,100 $L_\odot$	5400 K
0.75	5,000 $L_\odot$	5500 K
1.00	10,000 $L_\odot$	6700 K

CONCEPT QUESTIONS

Concept Questions on the following topics are located in the margins. They invite thinking and discussion beyond the text.

1. Living near a variable star. (p. 504)
2. Steaming pot analogy for Cepheid variability. (p. 506)
3. Differences in optical and infrared variability. (p. 507)

REVIEW QUESTIONS

4. What is a variable star?
5. How is a pulsating variable different from other types of variables?
6. What is meant by the "period" of a variable star?
7. What is the "instability strip"?
8. What is the difference between an RR Lyrae star and a Cepheid variable?
9. What makes the period–luminosity relationship so valuable to astronomers?

QUANTITATIVE PROBLEMS

10. An interferometer measured the change in radius of one Cepheid as about 10% during its 10-day period. The star was measured to have an average radius of 40 million km. How fast was the surface of the star moving outward, on average, if it took 5 days to expand by 10%?
11. The same interferometer as in Problem 10 measured the change in radius of an RR Lyrae star to be 10% during its 10-hour period. If the star was measured to have a radius of 10 million km and took half a period to expand by this amount, what would be the speed of the surface of the star as it expands?
12. A variable star varies in luminosity and temperature throughout its cycle according to the values listed in Table 64.1. The *phase* indicates the fraction of time between one maximum and the next. Graph the luminosity and temperature change, then plot the positions of the star in the H-R diagram. Connect the sequential points with a line to show the pattern of variation.
13. Using the information from Table 64.1, calculate the average luminosity of the variable star at the four distinct phases listed (the phase listed as 1.00 is back around to the same phase as 0.00). Using your average luminosity, estimate the period this variable should have if it is a Cepheid from Figure 64.5.
14. Use the information from Table 64.1, along with the Stefan-Boltzmann law, to calculate the radius of the variable star at each phase. When is the star largest? When is it smallest?
15. Suppose a Cepheid has a period of 5 days. Estimate its luminosity from Figure 64.5, then determine its distance if its measured brightness is 4×10^{-12} watts/m^2.
16. Use Figure 64.3 to estimate the range of sizes for Cepheid variables. Note that, these sizes only represent averages since Cepheid variables are constantly changing in size.

TEST YOURSELF

17. Cepheid variables are important to astronomers because
    a. they are one of the only types of stars whose radii are known.
    b. they are the only type of stars that varies in brightness at regular intervals.
    c. the precise rate of their pulsation allows them to be used as time standards.
    d. they can be used as standard candles.
    e. All of the above.
18. Which of the following is an irregular variable?
    a. T Tauri star
    b. Cepheid
    c. RR Lyrae star
    d. Mira variable
19. The period–luminosity relation indicates that
    a. a dimmer Cepheid pulsates more slowly.
    b. a brighter Cepheid rotates more slowly.
    c. a dimmer Cepheid rotates more slowly.
    d. a brighter Cepheid pulsates more slowly.
20. Variable stars change in
    a. luminosity.
    b. temperature.
    c. size.
    d. density.
    e. All of the above.

Mass Loss and Death of Low-Mass Stars

PART 4 · UNIT 65

- 65.1 The Fate of Stars Like the Sun
- 65.2 Ejection of a Star's Outer Layers
- 65.3 Planetary Nebulae
- 65.4 White Dwarfs

Learning Objectives

Upon completing this Unit, you should be able to:
- Describe the end-stages that a low-mass star passes through when fusion ends.
- Explain the process by which a red giant drives off much of its atmosphere.
- Explain how planetary nebulae form, how they evolve, and their variety of shapes.
- Describe the properties of a white dwarf and how it evolves in the H-R diagram.

FIGURE 65.1
Matter ejected from a dying star can produce complexly shaped planetary nebulae.

The ultimate fate of a star depends on its original mass, as outlined in Unit 60. As stars consume the hydrogen in their cores, their evolution is fairly similar into the red giant phase (Unit 63). At this point the story of stars' evolution splits into quite different fates. In this Unit we examine the fate of stars that begin their lives with less than about 8 $M_\odot$. Toward the end of the red giant phase, the cores of these stars can no longer contract. This prevents the core temperature from rising further, and as a result the star's nuclear fusion ends. The end does not come quietly, however.

In its terminal stages such a star ejects great quantities of matter, forming dramatically shaped and brightly colored objects (Figure 65.1) known as **planetary nebulae.** How a star ejects its outer layers is still not fully understood, but in this Unit we discuss some of the ideas proposed by astronomers to explain what happens when a low-mass star approaches the end of its life. The planetary nebula signals the death of a star. The remnant it leaves behind, a **white dwarf,** is initially hot but does not generate power, so it gradually cools and fades from view.

65.1 THE FATE OF STARS LIKE THE SUN

When a star like the Sun begins to fuse helium in its core, it is nearing the end of its life. As death approaches for the Sun, its evolution will speed up. This is shown in Figure 65.2, which illustrates the Sun's evolutionary track through the H-R diagram. A similar sequence is followed by other low-mass stars, although, as for all other stages of stellar evolution, the time taken to pass through the phases depends on mass—taking several times longer for lower-mass stars than for higher-mass stars.

We can summarize the Sun's life as follows. The Sun will spend approximately 10 billion years in its main-sequence phase, consuming the hydrogen in its core. It will then spend about one-tenth as long—about a billion years—as a red giant. The Sun will spend most of its red giant phase fusing hydrogen into helium in a shell surrounding the core (Unit 63). However, when its core grows hot enough to fuse helium, the Sun's structure will rapidly change from that of a red giant to that of a smaller yellow giant. The evolution during that phase will be even faster because helium yields less energy than hydrogen when it is fused, so the helium must be fused faster to produce enough energy to support the star. Helium fusion will occur in the core for about the last 100 million years of the Sun's giant phase.

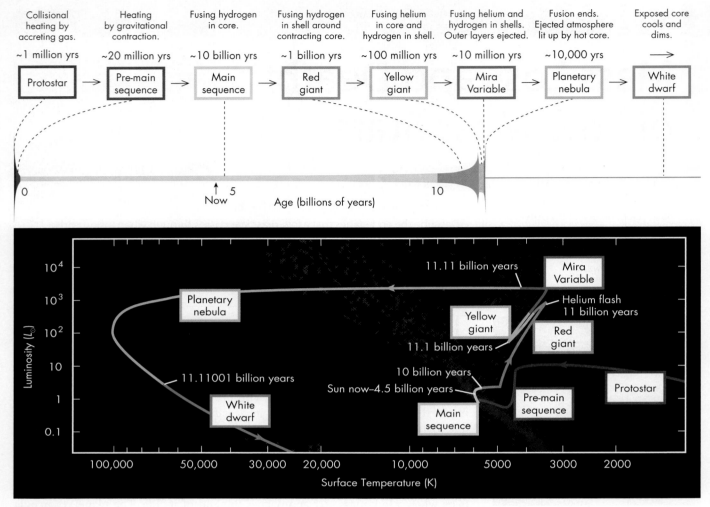

FIGURE 65.2

The evolution of a low-mass star such as the Sun plotted in an H-R diagram. The time line along the top illustrates the relative times spent in each stage and provides a sense of the changing size of the star (not to scale). The fuel powering the star in each stage is described above the time line.

When the helium in the Sun's core is consumed, the core, now primarily made of carbon, will contract and heat. A surrounding shell of helium will begin fusing, and a hydrogen shell around that will continue fusing. The Sun's atmosphere will expand to an even greater diameter than it had when it first entered the red giant stage—160 times its current diameter, according to one calculation. It will be classified as a bright giant (Unit 59) with a luminosity about 2000 times greater than it has now. However, the core's contraction will not raise temperatures enough to begin fusing carbon. Contraction will cause the core's density to climb to more than a million kilograms per liter—approximately 16 tons per cubic inch! This phase has an even shorter duration of about 10 million years. By the end of this phase, electron degeneracy (Unit 63) will halt the compression, and the Sun's core will begin to cool.

During this final phase of intense fusion of hydrogen and helium in shells, the Sun will become a Mira variable (Unit 64). These luminous red giants drive away most of the star's atmosphere, as explored in Section 65.2. As the Sun drives off its atmosphere, it will shift rapidly all the way across the H-R diagram, horizontally from right to left, before dropping into the white dwarf region (Figure 65.2). Stars maintain their luminosity during the brief transition to the planetary nebula phase, but the observed color and surface temperature of the central star will

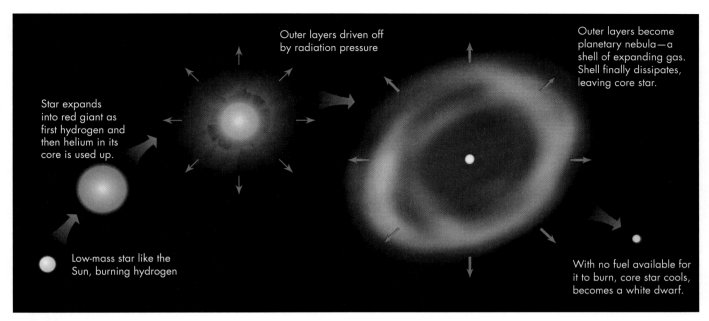

FIGURE 65.3
Final stages of the evolution of a low-mass star. After helium is used up in its core, the star expands to a very luminous red giant, gradually driving off its outer atmosphere. The ejected gas forms a shell around the star, lit up by the still-hot core, creating a planetary nebula. The remaining core of the former star becomes a white dwarf.

rapidly climb as deeper layers are exposed. Once exposed, though, the core—now a white dwarf—steadily cools, dropping in luminosity and surface temperature.

Although the Sun may not expand quite enough to swallow the Earth, the intense luminosity will heat the Earth's surface to be nearly molten. More and more of the Sun's outer layers will flow into space, thinning out the remaining atmosphere and exposing deeper levels of what remains of the Sun. Ultraviolet photons from the very hot interior of the Sun will now be able to travel out large distances, ionizing some of the Sun's former atmosphere—which by that time will have expanded far beyond Pluto's orbit. This gas will glow briefly as a planetary nebula (Section 65.3). Finally the remainder of the core will be exposed and will gradually cool to become a white dwarf (Section 65.4). These final stages are illustrated in Figure 65.3.

65.2 EJECTION OF A STAR'S OUTER LAYERS

When a star reaches the Mira variable stage, it is so large that its surface gravity is quite weak and the escape velocity is low. This allows the development of a strong **stellar wind** of gas flowing away from the star. A red giant's stellar wind is similar to the solar wind (Unit 51), but different mechanisms are at work for red giants. As the star swells, its outer layers cool to about 2500 K—so cool that carbon and silicon atoms condense and form **grains,** much as water in our atmosphere forms snowflakes when it cools. These are not like compact grains of sand, however, but rather are expected to be like flakes of carbon and "rock," very loosely assembled as their atoms and molecules stick together.

The grains do not fall into the star. As illustrated in Figure 65.4A, they are pushed outward by the flood of photons pouring from the star's luminous core, just as the dust in a comet is blown by the Sun's radiation into a tail that streams away from the Sun (Unit 49). The rising grains in a red giant drag gas with them, driving it into space. The mass loss is also driven by the pulsations occurring in the Mira variable stage. Like Cepheid variables (Unit 64), the atmosphere cyclically becomes opaque, is pushed outward, grows transparent again, and sinks back down. The pulsations in Mira variables are so forceful that they may drive some of the gas to expand outward faster than the escape velocity.

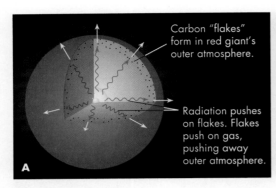 

FIGURE 65.4
(A) Diagram illustrating atmosphere loss by red giant. Solid carbon flakes form in the cool gas of the red giant's outer atmosphere. Radiation pressure pushes on the particles and creates a stellar wind. (B) Ultraviolet image of the variable red giant Mira made by the *Galex* space telescope. The star is located in the clump of gas at the right of the image. The former atmosphere of Mira has been ejected into space, and as Mira moves through interstellar gas clouds, the ejected gas trails behind it. The portion of the tail seen in this image is about 4 light-years long.

Concept Question 1

It is estimated that the Sun will have lost about half of its mass by the time it becomes a white dwarf. How will this mass loss affect the orbits of the planets that survive its red giant phases?

The combination of radiation pressure and pulsations produces a strong stellar wind, in which as much as about 10^{-4} solar masses of material flow each year into space, typically at tens of kilometers per second. At this rate, a star can lose an entire solar mass in as little as 10,000 years. The stellar winds eventually decline as the star's core is exposed and cools, leaving what remains of the star's inner regions in the middle of an expanding cloud of gas that used to be the star's atmosphere.

The results of this process can be seen around the red giant Mira (Figure 65.4B). Using the ultraviolet space telescope *Galex*, astronomers discovered a long tail of gas trailing behind Mira. The star is moving through interstellar gas, and the atmosphere it has expelled is left behind like the smoke from a train. The "lumpiness" of the tail probably indicates how the outflow of atmosphere from Mira has varied over many thousands of years.

The gas expelled from stars like Mira is rich in carbon and oxygen. These two elements are produced by fusion reactions in and around the core, and strong convection during the red giant phases carries them into the star's outer atmosphere (Unit 63). Stellar winds from Mira-type variables then carry these outward and are probably one of the main sources of carbon and oxygen in interstellar clouds.

65.3 PLANETARY NEBULAE

Clarification Point

A planetary nebula has nothing to do with planets. The name was developed when astronomers had only poor telescopes, through which planetary nebulae looked like small disks, similar to planets.

As the atmosphere of a star flows away into space, the weight compressing the core of the star decreases. The core of the star is still furiously hot, and fusion of hydrogen and helium is taking place in shells around the core. However, with the declining compression, these inner regions begin to expand and cool enough that the rate of fusion drops. At the same time, the expanding atmosphere thins and becomes progressively more transparent, which allows high-energy photons from the hot core to travel farther within the remaining atmosphere. These energetic photons drive what is called a *fast wind*, pushing much of the remaining atmosphere of the star outward at high speed. The fast wind pushes into the slower-moving gas of the star's earlier stellar wind, clearing out the region surrounding the star and exposing the star's core.

Because the core is so hot, its radiation is rich in ultraviolet light. Photons at these energetic wavelengths heat and ionize the inner portion of the expanding cloud of gas around it, causing it to glow. Astronomers have observed many such glowing shells around dying stars (Figure 65.5) and call them *planetary nebulae*. The unusual colors of planetary nebulae come from a mix of emission lines (Unit 24) of a number of elements, but particularly oxygen (green) and nitrogen (red). These spectral lines occur in addition to the more common emission lines of hydrogen that give rise to the pink-colored glow from many ionized gas clouds (the Balmer lines, Unit 24). Initially astronomers had difficulty identifying many of the spectral lines in planetary nebulae. The oxygen and nitrogen lines are sometimes called *forbidden lines* because they do not normally occur under laboratory conditions. Even low-density gases studied in a lab are much denser than the gases in a planetary nebula. When the gas is very diffuse, electrons may undergo uncommon energy transitions during the long time between atoms colliding with each other.

For many years, astronomers thought that the typical planetary nebula was ejected uniformly in all directions, so that it formed a huge "bubble" around the star's core, as you might deduce from Figure 65.5A. Astronomers knew of oddly shaped planetary nebulae but thought them to be the exception. This view has changed markedly over the last decade, especially now that more detailed images (such as Figure 65.5C and D, made with the Hubble Space Telescope) are available. Most of the planetary nebulae revealed by these photos show that the shell is not spherical, but that instead the gas has been ejected primarily in two directions. Such ejection creates not a bubble but two oppositely directed cones, as sketched in Figure 65.5E. Why this happens is not entirely clear, although a ring of dust and gas in orbit around the star—perhaps part of a planetary system—may physically block the ejection in those directions. Magnetic fields in the stars probably also play a role, much as they are thought to affect the bipolar flow seen ejected from some protostars (Unit 61). Some of the more complex geometries may be produced when the stellar wind interacts with a companion star. Spherically shaped planetary nebula, perhaps the result of isolation from other bodies, may be the exception, because we suspect that few stars are truly isolated.

> **Concept Question 2**
>
> Take a clear, cylindrical glass and shine a light through it so it casts a shadow. What kinds of shapes can you make with the shadow?

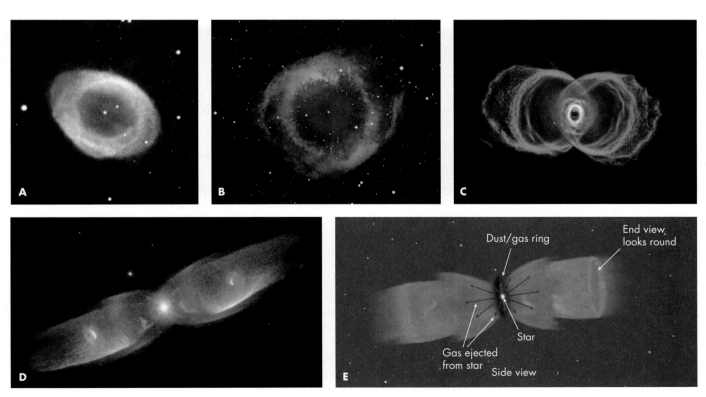

FIGURE 65.5
Pictures of several planetary nebulae. (A) The Ring Nebula. (B) The Helix Nebula. (C) The Hourglass Nebula. (D) The Butterfly Nebula. Notice the central star in each. Other stars that look as if they are inside the shell are foreground or background stars. (E) Sketch of a planetary nebula.

Many details of the final stages of a star depend critically on details of mass loss and the timing of when the core of a star is exposed. Some models suggest that not all low-mass stars produce a planetary nebula on their path to becoming a white dwarf.

Planetary nebulae eventually grow so big and diffuse that light from the central star no longer ionizes the gas much, and the nebula fades to invisibility. This occurs on a very short time scale by the standards of stellar evolution. In a mere 15,000 years, gas traveling at 10 kilometers per second in the expanding envelope around the star will have expanded to a light-year in diameter—about the largest size seen among planetary nebulae. We can directly detect this expansion in less than a human lifetime by careful comparison of pictures made several decades apart. The expanding gas mingles with the general interstellar gas, and the cycle of a star's life is complete: gas to star and back to gas. Only the hot core of the star remains: a tiny, glowing ball that gradually cools.

65.4 WHITE DWARFS

The hot core of the former star at the center of a planetary nebula becomes a white dwarf. A white dwarf's dim light does not come from fusion, for it has exhausted all its fuel supplies. Rather, the light is produced by the residual heat left from when the star was fusing hydrogen and helium. This light very slowly drains away the remaining heat from the star's interior.

White dwarfs are tiny but hot objects that lie in the lower part of the H-R diagram, as illustrated in Figure 65.2. A white dwarf has the same composition as the core of the star from which it evolved, mainly carbon and oxygen, the end products of the parent star's nuclear fusion. The white dwarf's surface is a very hot, thin layer of hydrogen and helium, but its carbon-oxygen core has too little mass, and therefore too little gravity, to contract and heat itself to the ignition temperature of carbon. So, unable to fuse any further, the star's "corpse" gradually cools and grows dim.

The stars that reach this final stage may begin with masses up to 8 $M_\odot$, but with the mass lost from stellar winds during the red giant phase, the remaining mass is found to be less than 1.4 $M_\odot$. The diameter of a white dwarf is about a hundredth of the Sun's—roughly the same as the Earth's. Because all white dwarfs have a similar size, they lie along a line of fixed radius in the H-R diagram (Unit 59). Their tiny size gives white dwarfs so small a surface area that they are very dim despite having hot surface temperatures up to about 100,000 K.

Stars with masses less than a third the Sun's are fully convective (Unit 62), so their evolution to a white dwarf is likely to be quite different. They may fuse most of the hydrogen throughout the star because they mix core and atmosphere material during their main-sequence lifetime. As a result, they will not produce the same kind of hydrogen shell fusion when the core stops fusing hydrogen, and may not have the large mass loss or planetary nebula production. They will probably contract and grow hotter for more than a billion years after hydrogen fusion ceases, until degeneracy pressure halts their contraction, leaving them as white dwarfs made almost entirely of helium. Because very low-mass stars evolve so slowly, however, they require longer than the current age of the universe to reach this point in their evolution, so the models will not be tested for billions of years.

Concept Question 3

If you could actually retrieve a sample of white dwarf matter, would it be stable if you had it back in a laboratory on Earth?

The enormous timescales over which white dwarfs cool provide an intriguing test for when the first stars formed in our Galaxy. Although a white dwarf is initially very hot, it has no fuel, so it cools like a dying ember. Astronomers calculate that it takes about 10 million years for the surface temperature of a white dwarf to drop to 20,000 K. Subsequent cooling becomes progressively slower, requiring billions of years to drop to the current surface temperature of the Sun. The coolest white dwarfs astronomers have yet found have a surface temperature of about 3900 K, which is estimated to have been cooling for about 9 billion years. In the remote future, our Galaxy may entirely consist of hundreds of billions of these former stars, darkly orbiting one another.

KEY POINTS

- Stars born with masses less than 8 $M_\odot$ end up as white dwarfs.
- During their final stage as a red giant, low-mass stars are highly luminous pulsating Mira variables.
- The atmosphere is so cool that flakes of solid matter form and radiation pressure can push these flakes away from the star.
- Pulsations and a strong stellar wind drive off most of the atmosphere, leaving behind a core of at most about 1.4 $M_\odot$.
- The hot core can ionize the escaping atmosphere, creating a glowing planetary nebula up to about a light-year across.
- The star's former core is about the size of the Earth, but extremely dense and hot—a white dwarf.
- The white dwarf cools slowly, taking billions of years to drop to a temperature similar to the Sun's.
- The coolest and therefore oldest white dwarfs have been cooling for about 9 billion years.

KEY TERMS

grains, 511
planetary nebula, 509
stellar wind, 511
white dwarf, 509

CONCEPT QUESTIONS

Concept Questions on the following topics are located in the margins. They invite thinking and discussion beyond the text.

1. Orbits of planets around eventual white dwarf. (p. 512)
2. Modeling nebula shapes with a glass cylinder. (p. 512)
3. Stability of white dwarf matter. (p. 514)

REVIEW QUESTIONS

4. What phases will the Sun pass through in the future?
5. Why do red giants have strong winds?
6. What is a planetary nebula?
7. How does a planetary nebula change over time?
8. How are the different shapes of planetary nebulae explained?
9. Why do white dwarfs take so long to cool?
10. What is the energy source for a white dwarf?

QUANTITATIVE PROBLEMS

11. As a dying star moves across the H-R diagram during its planetary nebula ejection phase, it maintains an almost constant luminosity as successively deeper and hotter layers inside the star are exposed. If it begins in the upper right part of the H-R diagram as a 3000-K, 100-$R_\odot$ red giant, use the Stefan-Boltzmann law (Unit 58.3) to calculate the central star's radius when its surface temperature is
    a. 6000 K.   b. 30,000 K.   c. 100,000 K.

12. How long will it take a planetary nebula shell moving at 20 km/sec to expand to a radius of one-fourth of a light-year?
13. The tail of gas behind the star Mira spans about 13 ly in total. The star's proper motion indicates that it is moving at 130 km/sec. How long ago was Mira at the position of the start of the tail? Give your answer in years.
14. How does the luminosity of a white dwarf at a temperature of 20,000 K compare to its luminosity when it was first exposed and had a temperature of 100,000 K? What would be the wavelength of peak radiation of the white dwarf at these temperatures?
15. a. Estimate the average density (in kg/L) of a planetary nebula around a solar-type star, assuming that a star like the Sun loses half its mass to a spherical nebula that expands to a light-year in diameter.
    b. For the same system, estimate the average density of the resulting white dwarf assuming a radius a hundredth of the Sun's.
16. Using the result of Problem 15a and the fact that the mass of a hydrogen atom is 1.7×10^{-27} kg, calculate the average number of hydrogen atoms per liter in a planetary nebula. What radius would the planetary nebula have to be so that its density matched the interstellar medium—about 1000 atoms per liter?

TEST YOURSELF

17. Which of the following sequences is in the correct order for the evolution of the Sun from birth to death?
    a. White dwarf, red giant, main-sequence, protostar
    b. Red giant, main-sequence, white dwarf, protostar
    c. Protostar, red giant, main-sequence, white dwarf
    d. Protostar, main-sequence, white dwarf, red giant
    e. Protostar, main-sequence, red giant, white dwarf
18. Planetary nebulae glow because
    a. the outward expanding gas collides with the surrounding interstellar gas.
    b. radiation emitted from the core stimulates emission in the nebula gases.
    c. nuclear fusion is still occurring in the expanding nebula.
    d. the gas retains residual heat from when the material was still part of the star.
19. How does a 4 solar mass star end up as a 1 solar mass white dwarf?
    a. A stellar wind carries away most of the star's mass when it is a red giant.
    b. Helium fusion turns most of the star's mass into energy.
    c. When an object contracts, its mass turns into density instead.
    d. Neighboring stars and planets pull the star apart when it is a giant.
    e. It spins so much faster as it contracts that much of the gas is thrown off.
20. After a white dwarf forms it will _____ on the H-R diagram.
    a. stay forever at one spot
    b. move horizontally, toward the right
    c. move vertically downward
    d. move down and to the right

UNIT 66

Exploding White Dwarfs

66.1 Novae
66.2 The Chandrasekhar Limit
66.3 Supernovae of Type Ia

Learning Objectives

Upon completing this Unit, you should be able to:
- Describe the processes that cause a white dwarf to undergo a nova or supernova.
- Explain why there is an upper limit to the mass of a white dwarf and what happens when that mass is exceeded.
- Describe the properties and timing of a supernova of Type Ia.

After a low-mass star has run out of nuclear fuel, the stellar remnant left behind is a **white dwarf**. This fate awaits the Sun about 6 billion years from now (Unit 65). Gravity will squeeze its remains into an object about 100 times smaller than the present Sun, or roughly the size of the Earth. Initially hot, the white dwarf will gradually cool and fade. In squeezing a white dwarf to its small dimensions, gravity squeezes the matter to densities far higher than anything we can create in Earth laboratories. For example, a piece of white dwarf material the size of an ice cube would weigh about 20 tons.

Because most stars eventually reach the white dwarf stage, our Galaxy is littered with billions of these tiny dead stars. They shine only with heat left over from their previous phase and are extremely dim. None are visible to the naked eye, and most are challenging to detect even telescopically. However, as we will explore in this Unit, some white dwarfs have an opportunity for a new life—and a spectacular death—if they accrete fresh matter.

66.1 NOVAE

A white dwarf can accrete matter when it has a nearby companion star. This is a possibility for a fairly large number of white dwarfs because many stars begin life in multiple-star systems (Unit 57). For example, the nearest white dwarf, Sirius B, orbits the brightest star in our sky, Sirius A. Material in the stellar winds from Sirius A can fall onto the white dwarf in a process astronomers call **mass transfer.**

Mass transfer can grow strong if the white dwarf's companion enters the red giant phase, as illustrated in Figure 66.1. As the atmosphere expands, it can eventually expand out to a distance where the gravitational pull from the white dwarf exceeds the pull of the red giant itself. The red giant's atmosphere can grow to fill a volume called the **Roche** (pronounced *rohsh*) **lobe** before it starts "spilling" onto the companion star. This teardrop-shaped volume, illustrated in

Mass transfer

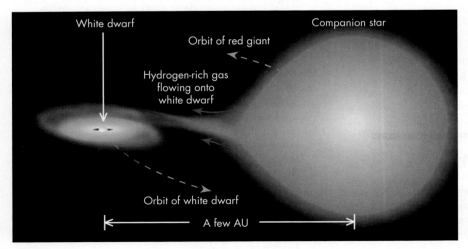

FIGURE 66.1
A white dwarf accreting (pulling in) mass from a binary star companion, which in this case is shown as a red giant. The typical separation of the stars is a few AU.

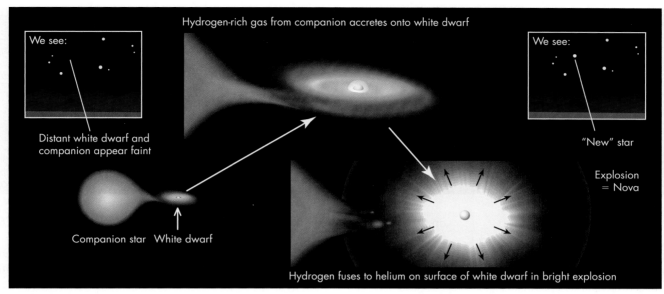

FIGURE 66.2

A nova explosion. The hydrogen that falls onto the surface of a white dwarf from its companion may suddenly fuse into helium. For several weeks the explosion typically generates a luminosity hundreds of thousands times greater than that of the Sun.

Concept Question 1

In what ways are the disks around forming stars similar to and different from white dwarf accretion disks?

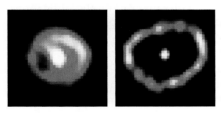

FIGURE 66.3

Hubble Space Telescope images made about 1.3 and 1.9 years after a nova exploded in Cygnus. They show a shell of gas expanding away from a white dwarf.

The Latin heritage of *nova* is reflected by plural form *novae,* although *novas* is also acceptable. The same applies to other Latin-derived words, such as *supernovae versus supernovas.*

Figure 66.1, is named after the French astronomer Edouard Roche, who calculated the gravitational effects in a rotating binary system. The process of sweeping up additional matter tends to add a drag on the white dwarf's orbit, gradually drawing the two stars closer together and further increasing the rate of transfer.

Coming from the companion's outer layers, the gas deposited on the white dwarf is generally rich in hydrogen. This gas forms a layer on the white dwarf's surface, where gravity compresses and thereby heats it. Additional heating occurs as more matter falls onto the star's surface. Such matter gains a great deal of kinetic energy as it falls and releases that energy as heat when it strikes the star's surface (Unit 20).

Most of the gas falling toward a white dwarf from a companion star spirals around the white dwarf before it reaches the white dwarf's surface. Astronomers call this spiraling gas an **accretion disk.** Gas in the accretion disk orbits the white dwarf rapidly, and differences in orbital speed at different radii lead to collisions and strong frictional heating of the gas as it spirals in toward the star's surface.

Gas accumulates on the white dwarf's surface, and it may eventually grow hot and dense enough to begin fusing hydrogen into helium. Even if the fusion begins in a small region initially, the heat released will trigger fusion in surrounding areas. As a result, more and more of the hydrogen ignites in a **thermonuclear runaway,** a detonation that blasts some of the gas into space. This forms an expanding cloud of hot gas, as Figure 66.2 illustrates. This gas radiates far more energy than the white dwarf itself. In fact, the luminosity may grow to 100,000 times its previous level, so that the explosions on distant stars may become visible to the naked eye.

When early astronomers saw such an event, they called it a *nova stella,* from the Latin for "new star," because the explosion would make a bright point of light appear in the sky where no star was previously visible (see also Unit 64 on variable stars). Today we shorten *nova stella* to **nova.** Modern observations allow us to observe the nova material expanding outward, forming a shell of hot gas (Figure 66.3).

Novae (plural form, pronounced *NO-vee*) may occur repeatedly for the same white dwarf, depending on the rate at which material is transferred from the companion to the white dwarf. A **recurrent nova** is a white dwarf that explodes repeatedly, and it is possible that many novae that have been recorded only once will repeat in the future. The timing depends on how rapidly material accumulates

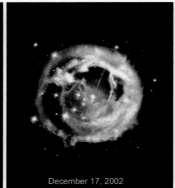

May 20, 2002

September 2, 2002

December 17, 2002

February 8, 2004

FIGURE 66.4
Hubble Space Telescope images of the nova V838 in Monoceros. This sequence of images does *not* show an expanding gas cloud. Instead, we are seeing the expanding "light echo" off of material surrounding a star that underwent an unusual type of nova. As the flash of light from the nova travels outward, it illuminates more-distant parts of the cloud of gas and dust that surrounds the star.

on the surface, whether it has time to cool, and how much the rate of mass transfer fluctuates—in other words, many factors that are difficult to predict.

Not all novae necessarily explode by this same mechanism. The exact cause of a nova in the constellation Monoceros (the unicorn) first seen in 2002 is still debated by astronomers. It initially appeared to be a fairly ordinary nova, but exceptionally luminous—becoming about a million times more luminous than the Sun. The initial burst of light was then followed by two more brightenings that did not resemble a standard nova. Astronomers have offered a variety of hypotheses: a series of planets swallowed by an expanding red giant; a collision and merger of two stars; or perhaps a more standard nova that triggered some further explosions in the star or its companion. One of the most remarkable aspects of this nova is that the intense flash of light is illuminating the surroundings of the star through a "light echo." As the flash of light travels outward, it lights up more and more distant regions surrounding the star (Figure 66.4), providing a detailed glimpse of the gas and dust that may have been ejected during the star's red giant phase.

66.2 THE CHANDRASEKHAR LIMIT

Despite the ability of a nova to blast away some of the material accumulated from a companion star, a white dwarf gradually increases its mass through mass transfer. This gradual accumulation of mass may have dire consequences for the white dwarf.

Because white dwarfs are very compact and have no fuel supply, their structure differs significantly from that of ordinary stars. Although they are in hydrostatic equilibrium, with pressure balancing gravity, their pressure arises from **electron degeneracy,** the peculiar interaction that prevents electrons of the same energy from occupying the same volume (Unit 63.3). This gives white dwarfs an odd property: Adding material to them makes them shrink. Even more crucial, however, the mass of white dwarfs must remain below a critical limit, or they collapse. To see why, we must consider conditions in a white dwarf's interior.

Concept Question 2

Some astronomers have suggested that cooled white dwarfs are made of diamond. If we could travel to such a white dwarf, why might it be impractical to mine it?

White dwarfs are very dense, having formed from the dense cores of their parent stars. If we divide the mass of a white dwarf by its volume, we find that its density is about 10^6 kilograms per liter—one ton per cubic centimeter (about 16 tons per cubic inch). This high density packs the star's particles so closely that they are separated by less than the normal radius of an electron orbit. Even as the white dwarf cools, then, the electrons cannot drop into lower-energy orbits around their atomic nuclei, as we would normally expect when matter cools.

If mass is added to a white dwarf, the extra mass increases the star's gravity. This increases the white dwarf's gravitational potential energy and drives some electrons into higher energy levels. The electrons that shift into higher energy levels can now overlap with neighboring electrons, allowing the white dwarf to compress into a slightly smaller volume. When the white dwarf's mass grows large

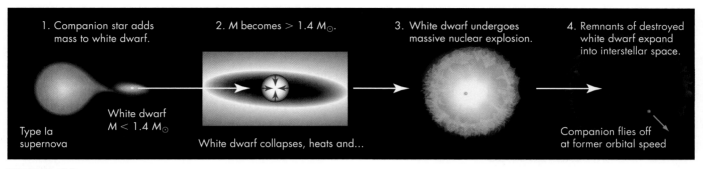

FIGURE 66.5
A white dwarf exploding as a Type Ia supernova. If too much hydrogen from a companion accumulates on the white dwarf, it may raise its mass to a value above the Chandrasekhar limit (about 1.4 $M_\odot$). A white dwarf more massive than this limit collapses, rapidly heats up from the resulting compression, and explodes.

enough, however, any further additional mass will cause the white dwarf's gravity to increase so much that it will contract more and more in a runaway process. The mass at which this occurs was determined by theoretical calculations made in 1931 by the young Indian astrophysicist Subrahmanyan Chandrasekhar (pronounced *Chahn-drah-say-car*). Chandrasekhar won the Nobel Prize for physics in 1983 for these calculations and related work. The limiting mass of a white dwarf is now called the **Chandrasekhar limit** in his honor.

For stars made of a helium-carbon mix—the expected composition of many white dwarfs—the Chandrasekhar limiting mass is about 1.4 $M_\odot$. If a white dwarf's mass exceeds this limit, it begins to contract uncontrollably—until other forces come into play. Chandrasekhar's work has been confirmed observationally by the fact that all white dwarfs for which we can measure the mass (those in binary systems) always have masses below the limit he determined theoretically.

66.3 SUPERNOVAE OF TYPE IA

As a white dwarf accumulates mass from a companion star, it may eventually approach the Chandrasekhar limit. When the white dwarf exceeds the limit, the white dwarf collapses in on itself. The resulting compression of the matter drives the temperature up, rapidly climbing to several billion Kelvin.

A star becomes a white dwarf because it did not originally have enough mass to compress and heat its core to temperatures that would allow carbon or heavier elements to fuse. It is therefore full of fuel. Carbon and oxygen in the collapsing white dwarf rapidly reach their ignition temperature and begin nuclear fusion. Carbon and oxygen can fuse to form silicon either directly ($^{12}C + ^{16}O \rightarrow ^{28}Si$) or through a chain of reactions (as in the core of a high-mass star; see Unit 67.1). Much of the silicon in turn fuses into nickel ($^{28}Si + ^{28}Si \rightarrow ^{56}Ni$). The energy released in these fusion reactions is enough to blow the entire white dwarf apart, resulting in a **Type Ia supernova** (Figure 66.5). The isotope of nickel, sprayed into space by the explosion, is highly radioactive and rapidly decays into cobalt (^{56}Co), which then decays to iron (^{56}Fe), each decay adding additional energy to the outburst.

A Type Ia supernova explosion can produce nearly 10^{10} times the Sun's luminosity at its brightest. This makes the supernova nearly as bright as an entire galaxy of stars. Figure 66.6 is a photograph of a distant galaxy with a Type Ia supernova: The exploding star gleams like a beacon. Supernovae can also occur when the core of a massive star is no longer generating heat and pressure from fusion in its core. When the core becomes massive enough, it too will collapse (Unit 67.3). This also produces supernovae that are called by a variety of names: Type Ib, Type Ic, and Type II supernovae (Unit 67).

FIGURE 66.6
A Type Ia supernova in a distant galaxy. The brilliant white supernova seen in 2002 shines as brightly as its galaxy of approximately a hundred billion stars.

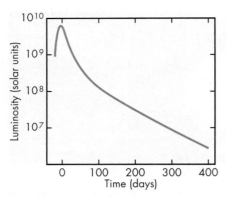

FIGURE 66.7
The light curve of a Type Ia supernova. The exploding white dwarf initially shines many billion times brighter than the Sun, then dims steadily following a curve characteristic of the decay of the radioactive elements produced in the explosion.

At their brightest, Type Ia supernovae are about two times brighter than Type II supernovae in visible light; however, the total power output is greater in Type II, in which much more of the power goes into the energy of expansion and neutrinos.

Concept Question 3

Are there any circumstances under which the Sun could become a nova or supernova?

Type Ia supernovae have several features that indicate their origin. First, the spectra of Type Ia supernovae show no signs of hydrogen because they are the remnants of stars that have lost their outer atmospheres of hydrogen in the stellar wind and planetary nebula phases. Second, the spectra contain strong lines of silicon, which forms from the rapid fusion of carbon and oxygen during the fatal last moments. Finally, the characteristics of the Type Ia **light curve** (Figure 66.7)—the way the luminosity changes over time (Unit 64)—closely match theoretical predictions for the thermonuclear runaway of a white dwarf of this mass.

These characteristics not only confirm the white dwarf origin of the Type Ia supernovae, but allow astronomers to identify them when they occur. Studies of other galaxies indicate that Type Ia supernovae are the most common. No supernova has been directly observed in our own Galaxy for more than 400 years, but the last two were Type Ia—one in 1572 known as Tycho's supernova and another in 1604 known as Kepler's supernova. Both occurred before the invention of the telescope.

Type Ia supernovae appear to be excellent "standard candles" (Unit 55.3), all having almost identical maximum luminosities. This is because the pattern of their collapse and explosion occurs the same way each time. The mass gradually increases to the Chandrasekhar limit, and then the collapse occurs. In other words, the Chandrasekhar limit is a leveling factor, which causes each Type Ia supernova to ignite in the same way, with roughly the same raw materials at its disposal.

Type Ia supernovae leave no star behind except the former companion of the white dwarf. The matter from the white dwarf becomes a rapidly expanding cloud of hot gas, known as a **supernova remnant**. X-ray images of Tycho's supernova remnant as well as another that exploded in 1006 are shown in Figure 66.8. Astronomers have detected a star moving away from the position of Tycho's supernova at more than 100 kilometers per second. This appears to be the former companion of the destroyed white dwarf, flung out of orbit when its partner exploded. Relatively few former companions have been identified despite careful searches. This has led to a hypothesis that some Type Ia supernovae are produced when binary white dwarfs spiral together, merge, and explode, destroying both.

The supernova remnant is rich in leftover carbon and oxygen, plus the silicon, iron, and other elements made during the supernova explosion in the final frenzy of nuclear fusion. Magnetic fields in the rapidly expanding remnant can accelerate some atomic nuclei to close to the speed of light. These high-energy particles, or **cosmic rays,** travel vast distances throughout space. Currently cosmic rays from millions of past supernovae are detected reaching the Earth's surface, and they pose a serious radiation risk to astronauts in space, who are not shielded by the Earth's own magnetic field. Although supernova explosions pose some radiation risks, they are also essential for our planet and life. The silicon in a chunk of rock, and the iron in your blood, were probably made when a white dwarf met its doom.

FIGURE 66.8
X-ray false-color images of two Type Ia supernova remnants. (A) Tycho's supernova was seen exploding in 1572 and is now about 20 light-years across. (B) A supernova seen exploding in the year 1006 is now about 60 light-years across. The images are colored blue for the highest energy X-rays, yellow for intermediate energies, and red for the lowest energies.

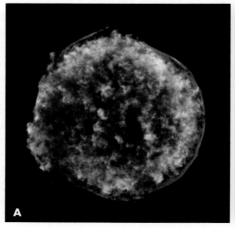
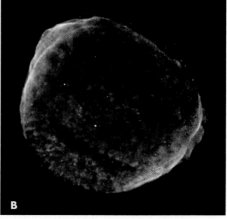

KEY POINTS

- Many white dwarfs are in binary systems with red giant companions that expel material that drops onto the white dwarf's surface.
- The atmosphere of a red giant in orbit around another star will expand to fill a volume known as a Roche lobe, then additional gas expanding further tends to flow onto its companion.
- Hydrogen deposited onto the surface of a white dwarf may ignite in a thermonuclear explosion that produces a bright nova.
- Novae may occur repeatedly, and mass transfer may gradually increase the mass of the white dwarf.
- A white dwarf grows smaller as mass is added to it, and above the Chandrasekhar limit of 1.4 $M_\odot$ the white dwarf will implode.
- A collapsing white dwarf heats to fusion temperatures extremely rapidly, leading to the ignition of almost all unfused matter.
- The explosion of a white dwarf is known as a Type Ia supernova, and may briefly shine as bright as ten billion Suns.

KEY TERMS

accretion disk, 517
Chandrasekhar limit, 519
cosmic rays, 520
electron degeneracy, 518
light curve, 520
mass transfer, 516
nova, 517
recurrent nova, 517
Roche lobe, 516
supernova remnant, 520
thermonuclear runaway, 517
Type Ia supernova, 519
white dwarf, 516

CONCEPT QUESTIONS

Concept Questions on the following topics are located in the margins. They invite thinking and discussion beyond the text.

1. Disks around protostars and white dwarfs. (p. 517)
2. Mining white dwarf diamonds. (p. 518)
3. Could the Sun become a nova or supernova. (p. 520)

REVIEW QUESTIONS

4. Where did the term *nova* originate?
5. What causes a nova explosion? What is a Roche lobe?
6. How can a nova be recurrent?
7. What is meant by *electron degeneracy*? What is the Chandrasekhar limit?
8. What is a supernova? Why are Type Ia supernovae "standard candles"?
9. What is distinctive about the appearance of a Type Ia supernova—an exploding white dwarf—that we can use to distinguish it from other supernovae?
10. How bright is a Type Ia supernova?

QUANTITATIVE PROBLEMS

11. How fast does an accretion disk particle orbit a 1 $M_\odot$ white dwarf when the particle is at a radius of 10^4 km? How does this value compare to the speed of light?
12. How does the acceleration of gravity (Unit 16) on the surface of a 1-$M_\odot$ white dwarf, which has a radius of 10^4 km, compare to g, the acceleration due to gravity on the surface of the Earth?
13. What is the escape velocity (Unit 18) from the surface of a 1-$M_\odot$ white dwarf that has a radius of 10^4 km?
14. The brightest star in the night sky, Sirius A, is located 8.60 ly from Earth and has a luminosity of 25.4 $L_\odot$. What is the maximum distance a Type Ia supernova could occur such that it would have the same brightness as Sirius A?
15. If the maximum luminosity of a Type Ia supernova is 10^{10} $L_\odot$, and the supernova remains at this brightness for 15 days, estimate how long our Sun would take to emit the same amount of energy.
16. IK Pegasi, a white dwarf/A-type main-sequence binary system 150 ly away, is a candidate to become a Type Ia supernova in the future. Assuming that it will reach a luminosity of 10^{10} $L_\odot$, how bright will it appear from Earth? How many times brighter than an average star in our sky would it appear?

TEST YOURSELF

17. If mass is added to a white dwarf, which of the following does *not* occur?
    a. Its radius increases.
    b. Its mass increases.
    c. Its density increases.
    d. It may exceed the Chandrasekhar limit and collapse.
    e. Some of its electrons gain higher energies.
18. Why do Type Ia supernovae have no hydrogen in their spectra?
    a. They have converted all of their hydrogen into iron.
    b. They have lost their hydrogen atmospheres during their planetary nebula phase.
    c. They do not have enough power to excite the hydrogen lines.
    d. They are so powerful that they fuse all of the hydrogen in the explosion.
    e. They become black holes, which swallow up the hydrogen.
19. For a nova to occur the white dwarf must
    a. have some nuclear fusion still occurring in its core.
    b. be a member of a binary system.
    c. have a large magnetic field.
    d. be spinning rapidly.
20. The afterglow of a Type Ia supernova typically lasts
    a. less than an hour.
    b. a couple of days.
    c. a few months.
    d. over a year.

UNIT 67
Old Age and Death of Massive Stars

67.1 The Fate of Massive Stars
67.2 The Formation of Heavy Elements
67.3 Core Collapse of Massive Stars
67.4 Supernova Remnants
67.5 Gamma-Ray Bursts and Hypernovae

Learning Objectives

Upon completing this Unit, you should be able to:
- Describe the end stages of massive stars' evolution, sketching this in an H-R diagram.
- List the sequence of elements fused in a star's core, and explain the endpoint.
- Describe the causes and events of a star's core collapse, and how this may produce a gamma ray burst in the most massive stars.
- Describe a supernova remnant, indicating which elements are uniquely found in it.

High-mass stars (masses greater than about 8 $M_\odot$) have violent ends. As outlined in Unit 60, high- and low-mass stars shed mass as they age and the temperature in their cores grows hotter. As we will explore in this Unit, high-mass stars have cores that grow so hot they can fuse much heavier elements than low-mass stars can. This prolongs their short lives—for a while. Less energy is available from these fuels, and the demands of the star grow steadily larger.

Once a massive star forms iron in its core, the end is near, because elements heavier than iron do not release energy when they fuse. The inability of an iron core to provide energy triggers the collapse and explosion of these stars. When massive stars explode, they spew into space the products of their nuclear fusion, adding heavy elements to the interstellar medium. They also leave behind some of the densest objects in the universe—neutron stars and black holes—which are explored in Units 68 and 69.

67.1 THE FATE OF MASSIVE STARS

The life of a high-mass star is plotted in the H-R diagram in Figure 67.1 from its birth on the main sequence up to its spectacular death in a supernova. Similar tracks, with minor variations, are followed by all stars more massive than about 8 $M_\odot$.

The massive star begins its life on the upper main sequence as a luminous blue star. When its core hydrogen is consumed, it leaves the main sequence. The core contracts, hydrogen shell fusion begins, and the surface swells. This is similar to the pattern followed by lower-mass stars, but everything happens much faster because fuels are consumed so quickly. This is shown in Table 67.1, which lists the various

FIGURE 67.1

A massive star's evolution through the H-R diagram. Massive stars usually begin fusing helium in their cores while they are still moving across the H-R diagram, but depending on their mass and heavy-element composition, this may occur when the star is in its blue, yellow, or red supergiant phase.

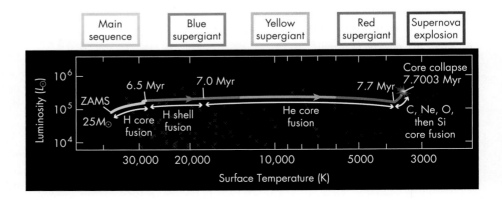

TABLE 67.1 Major Fusion Reactions in Stars

Reaction	Primary Fusion Products	Minimum Temperature	Energy Released (% of Mass)	Duration of Fusion in 25 $M_\odot$ Star
Hydrogen fusion	$4\,{}^1\mathrm{H} \to {}^4\mathrm{He}$	5,000,000 K	0.71%	7,000,000 yr
Helium fusion	$3\,{}^4\mathrm{He} \to {}^{12}\mathrm{C}$	100,000,000 K	0.065%	⎫
Alpha capture	${}^{12}\mathrm{C} + {}^4\mathrm{He} \to {}^{16}\mathrm{O}$	200,000,000 K	0.016%	⎬ 700,000 yr
Carbon fusion	${}^{12}\mathrm{C} + {}^{12}\mathrm{C} \to {}^{20}\mathrm{Ne} + {}^4\mathrm{He}$	600,000,000 K	0.021%	300 yr
Neon fusion	${}^{20}\mathrm{Ne} + {}^{20}\mathrm{Ne} \to {}^{24}\mathrm{Mg} + {}^{16}\mathrm{O}$	1,500,000,000 K	0.012%	8 months
Oxygen fusion	${}^{16}\mathrm{O} + {}^{16}\mathrm{O} \to {}^{28}\mathrm{Si} + {}^4\mathrm{He}$	2,000,000,000 K	0.032%	3 months
Silicon fusion	${}^{28}\mathrm{Si} + {}^{28}\mathrm{Si} \to {}^{56}\mathrm{Fe}$	2,500,000,000 K	0.034%	1 day

Key: H = hydrogen; He = helium; C = carbon; O = oxygen; Ne = neon; Mg = magnesium; Si = silicon; Fe = iron.

nuclear fusion reactions that take place in a 25-$M_\odot$ star. Main-sequence fusion of hydrogen to helium lasts only about 7 million years in such a star. By comparison, the Sun's main-sequence lifetime is more than a thousand times longer.

After the massive star leaves the main sequence, its surface is still so hot that it has a blue color, and it is so luminous that it is labeled a blue supergiant (Unit 59). The contracting core reaches helium fusion temperature relatively quickly because its core reached such a high temperature during its main-sequence phase. This may begin even while the atmosphere is just beginning to swell and cool. The star continues to expand, and on the H-R diagram it crosses from the blue through the yellow to the red supergiant regions. By the time it is a red supergiant, the star may have a diameter more than a thousand times larger than the Sun—so much larger that in the Sun's place it would swallow Jupiter. When the helium in the core is consumed, the core contracts and heats again. When it reaches 600 million K, it can begin to fuse carbon and, after the carbon is consumed, a series of heavier elements (Table 67.1). Now the changes in the core happen so fast that the bloated surface of the star has too little time to react to the series of new fuels being ignited and exhausted.

Throughout all of these changes, the star's luminosity remains approximately constant, but the star changes in both size and surface temperature. Computer models yield a variety of results for these stages, for several reasons: The amount of mass lost during the supergiant stage is uncertain, and it appears that factors like the star's rotation rate and initial composition may have major effects. For some mass ranges and chemical compositions, computer models indicate that a massive star may loop back toward the blue side of the H-R diagram as different nuclear fuels are fused. As a result, a massive star may follow a path back and forth in the H-R diagram, similar to the path of a low-mass star in its red giant phase but at a higher luminosity.

As the star consumes one nuclear fuel after another, the "ashes" of one set of nuclear reactions become the fuel for the next set. Oxygen, neon, magnesium, and silicon are formed; but because progressively higher temperatures are needed for each new fusion process, each new fuel is confined to a smaller and hotter region in the star's core. The star develops a layered structure, as illustrated in Figure 67.2. Hydrogen remains plentiful in the star's atmosphere, because the

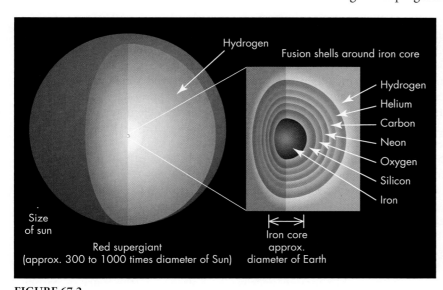

FIGURE 67.2
The structure of a massive star at the end of its life. Shells of lighter elements forming heavier elements surround an inert iron core.

temperature has remained too low there for nuclear reactions to have occurred. At the base of the hydrogen atmosphere, the temperature is high enough for hydrogen fusion, and the hydrogen fusion shell "eats" its way out into the surrounding atmosphere, depositing the helium it produces. At the base of the helium layer, helium fusion produces carbon and some oxygen (through the alpha capture process listed in Table 67.1). At the base of that layer, carbon fusion produces a neon layer, and so on—a series of nested shells, each made of an element heavier than the one surrounding it.

By the time the star is fusing silicon into iron, its core has been compressed to a size smaller than the Earth, to drive the temperature up to about 2.5 billion K. At such temperatures, the silicon is being consumed at a fantastic pace. To appreciate how rapidly the star consumes its silicon, consider the following: A massive star of 25 $M_\odot$ takes almost 8 million years from its birth to form an iron core. The main-sequence phase lasts about 7 million years; the helium fusion phase lasts only about 700,000 years; the carbon fusion phase lasts just a few hundred years; and the silicon fusion phase lasts only about one day!

The time decreases for each phase for several reasons. First, the star must produce ever larger amounts of heat and energy to support the core against the greater intensity of gravity as the core compresses. Second, in most of the heavy-element fusion reactions, less energy is released per kilogram of material fused than in hydrogen or helium fusion. Third, much of the energy released in these heavy-element reactions is carried away by neutrinos, which travel out of the star, reacting very little with its gases. The neutrinos therefore offer little help in either heating the gas or providing the pressure needed to support the star.

When silicon fuses to form an iron core, the death of a star is very near. It is the end of the line because iron cannot fuse to provide the energy needed to keep the star going. To see why, we need to examine how atomic nuclei interact.

67.2 THE FORMATION OF HEAVY ELEMENTS

To understand massive stars, we need to understand atomic nuclei. The fusion of lighter nuclei into heavier nuclei is what powers stars. Astronomers call the formation of heavy elements by such nuclear fusion processes **nucleosynthesis**. The details of these reactions were first worked out in the 1950s by a team of astronomers—E. Margaret Burbidge, Geoffrey R. Burbidge, William A. Fowler, and Fred Hoyle. They hypothesized that all of the chemical elements in the universe heavier than helium were made by nuclear fusion in stars. Detailed calculations and measurements since then confirm that the observed abundances of the elements in the universe agree extremely well with this hypothesis. The fact that oxygen and carbon are the third and fourth most common elements in the universe, and that the Earth is made largely of elements such as iron and silicon, are direct consequences of nucleosynthesis in stars.

The process of nucleosynthesis depends on a balance of repulsive and attractive forces between nuclei. Because protons are positively charged, atomic nuclei repel each other. However, if nuclei can come close enough to each other, the short-range *strong force* (Unit 4) takes over and the nuclei can fuse. The fusion of larger nuclei requires successively higher speeds. To see why, consider the electrical repulsion between two carbon nuclei: With 6 protons each, the repulsion is $6 \times 6 = 36$ times stronger than the repulsion between two hydrogen nuclei, which have only 1 proton each. The collision speed needed to overcome this repulsion can be achieved if the gas is hot enough, because the hotter a gas, the faster its particles move. To fuse progressively heavier nuclei, the star's core must grow progressively hotter. For example, fusing carbon requires a temperature of about 600 million K, whereas fusing heavier elements may require temperatures greater than a billion K.

LOOKING UP

The star Eta Carina, shown in Looking Up #8 at the front of the book, is thought to be more than 100 times more massive than the Sun and a likely candidate to undergo a supernova explosion "soon," but this could be up to a million years.

Concept Question 1

The even-numbered elements are generally more common throughout the universe than the odd-numbered elements. How might this be understood from the fusion reactions we have examined?

Mathematical Insights

The superscript number listed before an element symbol equals the number of protons + neutrons. The element's name depends on the number of protons (the atomic number in the periodic table). So ^{13}C, or "carbon-13," is element #6 and must therefore have 6 protons and $13 - 6 = 7$ neutrons.

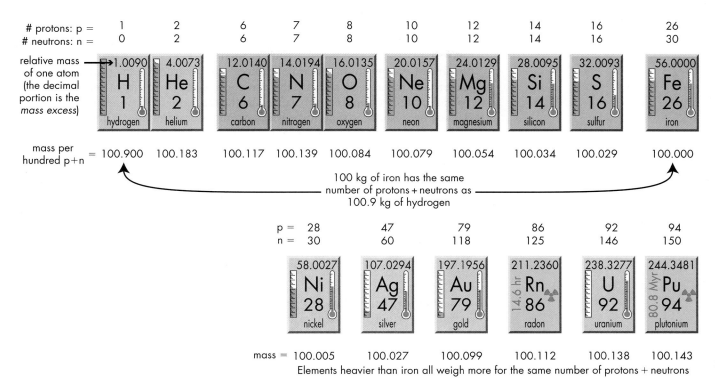

FIGURE 67.3
Some of the most common elements in the universe (top row) and an assortment of elements that are heavier than iron. The boxes are taken from the periodic table, with color coding and other symbols that indicate a number of element properties—see the periodic table on inside back cover of this book. The figure shows how the element masses reflect the mass excess of each element and how this can be used to calculate the energy released by a nuclear reaction.

Mathematical Insight

Sample calculation of energy release from conversion of mass excess to energy:

100.117 kg of carbon all fusing into 100.054 kg of magnesium would release 0.063 kg of energy.

This amount of energy can be converted into joules using Einstein's formula:

$E = m \times c^2$
$= 0.063 \text{ kg} \times (2.998 \times 10^8 \text{ m/sec})^2$
$= 5.66 \times 10^{15}$ joules
$= 1350$ kilotons of explosive

Concept Question 2

Why do you suppose that heavier atoms' nuclei tend to have a larger proportion of neutrons than is found in lighter atoms?

When nuclei fuse, the sum of the number of protons and neutrons always remains the same. However, the *weak force* (Unit 4) allows neutrons and protons to be converted into each other. For example, suppose two ^{28}Si nuclei fuse. Silicon has 14 protons and 14 neutrons, so the fusion reaction begins with 28 of each nuclear particle. Their fusion ultimately results in the production of ^{56}Fe, which contains 2 fewer protons, but 2 more neutrons. The number of protons and neutrons in the main isotopes involved in stellar fusion reactions are shown in Figure 67.3, which also shows the properties of an assortment of other elements in the periodic table.

The most common nuclear fusion reactions occurring in stars are listed in Table 67.1, along with the approximate temperatures required for these fusion processes to take place. Besides the reactions listed in the table, a wide variety of less common interactions create other combinations of nuclei. For example, as you can see in the table, oxygen fusion produces the element silicon, but it may also produce sulfur, although normally in smaller quantities. Note that helium nuclei (alpha particles) are created in a number of these fusion reactions, and these are usually quickly "captured" in further fusion interactions with the heavier elements. By tracing through all of the nuclear interactions, it is possible to predict the amount of each element that is produced, and this agrees remarkably well with the mix of elements we observe in the universe.

Each nuclear reaction converts some of the nuclei's mass into energy, in accordance with Einstein's law $E = m \times c^2$. The percentage of mass turned into energy is listed in the second to last column of Table 67.1. Nearly 0.7% of the fusing nuclei's mass is converted into energy during hydrogen fusion. However, the sum of all of the other reactions converts only about 0.2% of the matter's mass into energy.

The declining amount of energy available from heavier elements is reflected in the masses of the elements. This is indicated in Figure 67.3, by the **mass excess** for each of the elements, which is the amount of mass—or equivalently, energy, according to Einstein's formula—potentially available from an element through nuclear reactions. The mass of each element is given relative to ^{56}Fe, which has the lowest-energy configuration of protons and neutrons of any nucleus. In nuclei larger than the iron-56 nucleus, the attraction of the strong force between nuclear particles begins to be counterbalanced by the electrical repulsion between protons. This is because the strong force weakens much more rapidly with distance than does the electrical force. As a result, attempting to fuse additional nuclei onto an iron nucleus weakens the bonds and absorbs energy, rather than releasing it.

The formation of an iron core signals the end of a massive star's life. Iron cannot fuse and release energy. Thus, a star with an iron core is out of fuel. Nucleosynthesis does not quite end here, though. In one final burst, these stars create the rest of the elements beyond iron during their final collapse.

67.3 CORE COLLAPSE OF MASSIVE STARS

When an iron core forms in a star, the reaction of the star at first looks the same as when previous fuels were exhausted. The star's core shrinks and heats, and just as for low-mass stars, the shrinkage presses the matter tighter and tighter until electron degeneracy (Unit 63.3) finally halts the contraction. However, the pressure is so enormous, and the most energetic electrons have so much energy, that a new reaction can occur. Protons and electrons may themselves combine to become neutrons by this reaction:

$$p + e^- + \text{Energy} \rightarrow n + \nu.$$

Besides converting a proton (p) and electron (e$^-$) into a neutron (n) and neutrino (ν), this reaction *absorbs* energy, triggering a catastrophe for the star.

To understand why a reaction that absorbs energy can have such dire consequences, consider the following: Each layer of the star is supported by the one below it, with all of them finally supported by electron degeneracy at the very center, where the density reaches more than 1 million kilograms per liter. When the electrons are swallowed up by protons to become neutrons, the pressure provided by electron degeneracy disappears. The center of the star begins to implode.

Neutrons also have a degeneracy limit, but the volume each degenerate neutron occupies is about a billion times smaller than the volume the electrons filled. This effectively creates a hole in the middle of the star. In a fraction of a second, the core is thus transformed from a sphere of iron, with a radius of 10,000 kilometers, into a sphere of neutrons with a radius of just 10 kilometers held up by neutron degeneracy pressure. This is something like pulling out the bottom brick from an enormous stack of loose bricks. Nothing remains to support the star, and so it begins falling inward.

The plunging outer layers of the star strike the neutron core, reaching speeds greater than 100,000 kilometers per second—about one-third of the speed of light. Some of the material that smashes into the extremely dense core "bounces" back outward at nearly the same speed, colliding with other infalling matter. The impact heats the infalling gas to billions of degrees. Several solar masses of matter are almost instantaneously raised to fusion temperature and detonate all at once. Meanwhile an enormous quantity of neutrinos is pouring out from the core—one neutrino for each of the protons that turned into a neutron. Normally neutrinos interact so weakly that they escape freely from a star; but at the enormous densities of the collapsing material, many of the neutrinos flooding out from the core collide with the infalling matter and give up their energy to it. This causes nuclear reactions, further heats the infalling matter, and pushes matter back out.

Concept Question 3

Hold a small rubber ball on top of a basketball, and drop them together toward the floor. What happens to the small ball? How is this similar to a Type II supernova?

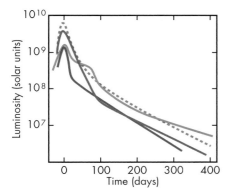

FIGURE 67.4
Some examples of the light curves of supernovae of stars undergoing core collapse. The dashed red line shows the light curve of a Type Ia supernova for comparison.

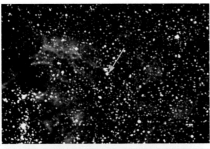

FIGURE 67.5
Photograph of a portion of the Large Magellanic Cloud (a small, nearby galaxy) before Supernova 1987A exploded, and a second photograph taken during the event.

All of these processes generate more energy in a few seconds than the Sun generates in its entire lifetime. A powerful blast wave travels outward from the core into the atmosphere of the star. The blast wave tears through the surface of the supergiant in an eruption that sends most of the star's atmosphere flying outward into space. This catastrophe marks the death of a high-mass star in a tremendous supernova explosion.

We saw in Unit 66 that the white dwarfs left behind by lower-mass stars can also undergo supernova explosions if a companion star deposits enough matter on them. This process is similar in the sense that the mass of the object grows too large for degenerate electrons to support the weight. The collapse of a white dwarf is called a Type Ia supernova, whereas a massive star undergoes a **Type II supernova.** The types of supernovae were defined before the mechanisms were known, and astronomers classified these explosions according to the spectrum of the light they generated. Type II supernovae show spectral lines of hydrogen, produced by the superheated gas of the star's former atmosphere, whereas Type I do not exhibit hydrogen lines.

The lack of hydrogen in a Type Ia supernova is a result of that element's having been completely consumed in the core of the star that ended up as a white dwarf. However, some massive stars explode without hydrogen lines visible in their spectra. These supernovae, which are classified as Type Ib and Type Ic, appear to be massive stars undergoing core collapse that have lost their hydrogen atmospheres to a companion star or stellar winds. Astronomers can distinguish these from Type Ia supernovae because they do not exhibit some of the secondary characteristics of Type Ia, such as silicon or helium lines. Supernovae of types Ib, Ic, and II are sometimes collectively referred to as **core-collapse supernovae.**

Core-collapse supernovae come from stars that have a wide range of masses and histories. A star of 20 $M_\odot$ will build up the critical mass for the iron core to collapse much more quickly than a star of 10 $M_\odot$. When the iron core in each of these stars is about to collapse, their atmospheres are structured in significantly different ways in response to their many layers of shell fusion and the amount of atmosphere lost in stellar winds. Core-collapse supernova blasts may therefore look quite different for stars of different masses. These differences are illustrated by a wide variety of light curves—the amount of light emitted as a function of time—as illustrated in Figure 67.4.

In death, a massive star releases more energy in a few minutes than it generated by nuclear fusion during its entire existence. Core-collapse supernovae are generally slightly less luminous at visible wavelengths than the Type Ia supernovae, but they are more energetic. Most of their energy is carried by the burst of neutrinos. By some estimates, almost 99% of the energy of a Type II supernova is carried away by the neutrinos produced when the protons and electrons combine. Several neutrino detectors around the world detected just such a burst in February 1987 when a supernova—SN 1987A—blew up. Several hours later, light from the exploding star became visible as the blast wave emerged through the surface of the supergiant star.

Images of this star had been obtained before it exploded, making it the first supernova for which we had "before" and "after" photographs (Figure 67.5). A surprise for astronomers was that the star that exploded was a blue supergiant. Before that time it had been widely expected that supernovae would occur in red supergiants. It is now understood that some massive stars may "loop back" to bluer colors in the H-R diagram after their red supergiant phase (Figure 67.1), and can have almost any color when the iron core develops.

Supernova 1987A did not even occur in our own Galaxy, but in the Large Magellanic Cloud, a small galaxy neighboring our own (Unit 77). Nonetheless, even though the blast took place about 160,000 light-years away—hence 160,000 years ago—trillions of the neutrinos from that explosion passed through every person on Earth!

67.4 SUPERNOVA REMNANTS

Gas ejected by a supernova blast plows into the surrounding interstellar space, sweeping up and compressing other gas that it encounters. Because massive stars have such short lives, they are often surrounded by large amounts of the gas and dust in the interstellar cloud from which they formed (Unit 61). The **supernova remnant**—a huge, glowing cloud of stellar debris and interstellar gas—may expand to a diameter of several light-years within a century, but its expansion slows as it runs into more surrounding gas. Figure 67.6 shows photographs of three Type II supernova remnants of different ages. Notice how ragged the remnants in Figure 67.6 look compared with the smoothly ejected bubbles of planetary nebulae (Unit 65), which reflects the more violent death of massive stars.

The explosion of a massive star mixes the elements synthesized by nuclear fusion during the star's evolution with the star's outer layers, and blasts them into space. This incandescent spray expands away from the star's collapsed core at more than 10,000 kilometers per second. Depending on the star's mass, many solar masses of matter may be flung outward, ultimately mixing with surrounding material and, in time, forming new generations of stars. Interstellar gas is thereby enriched in heavy elements, the atoms needed to build the rock of planets and the bones of living creatures here on Earth.

The supernova outburst itself generates even heavier elements than were originally created in the star's core. Elements such as silver, gold, and uranium are created in a supernova explosion from the abundant energy of its blast. This occurs because the explosion creates free neutrons, which rapidly combine with atoms from the star to build up elements heavier than iron.

One such violent outburst was seen almost a thousand years ago in 1054 by astronomers in China and elsewhere in the Far East (Unit 28.4). This is perhaps the most famous of all supernova remnants, now known as the Crab Nebula (Figure 67.6B). With even a small telescope, you can still see the glowing gases ejected from that dying star. Another intriguing supernova shown in Figure 67.6A, Cassiopeia A, was first seen as a radio source—the brightest radio source in the sky outside the Solar System. Astronomers have determined from its expansion rate that it exploded about 300 years ago, making it one of the most recent known supernovae in our Galaxy. However, there are no clear records that its explosion

Some of the light seen coming from supernovae in the weeks and months following the explosion has been shown to originate from heavy radioactive elements. These must have been produced in the explosion because they decay rapidly.

The Crab Nebula is shown in Looking Up #5 at the front of the book.

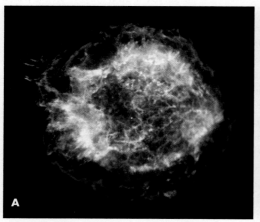
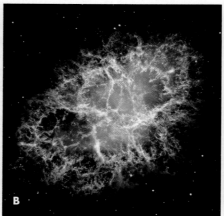
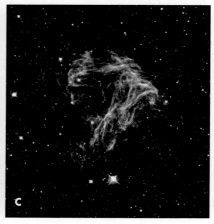

FIGURE 67.6

Several supernova remnants arranged according to their age: (A) Chandra X-ray image of Cassiopeia A supernova remnant, which is about 300 years old. In this image, the red on the left is radiation from hot iron atoms and the bright green is radiation from silicon and sulfur atoms. (B) Hubble Space Telescope image of the Crab Nebula, which is about 1000 years old. (C) Hubble Space Telescope image of a supernova remnant that is several thousand years old

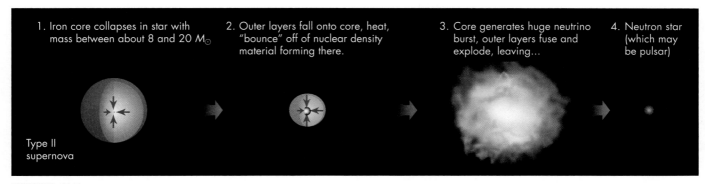

FIGURE 67.7
Formation of a supernova explosion and a neutron star by the collapse of the iron core of a star with a mass between about 8 $M_\odot$ and 20 $M_\odot$.

was seen. The explosion appears to have been hidden from view by dense dust surrounding the young massive star.

A supernova explosion marks the death of a massive star, but a part of the star—the collapsed core of neutrons that initiated the explosion (Figure 67.7)—may survive. Such *neutron stars* (Unit 68) have been detected, for example, in the middle of the expanding debris of the three supernova remnants in Figure 67.6. In some other supernova remnants, no remaining core is detected. It is possible that these stars destroyed themselves entirely, like the Type Ia supernovae; or perhaps they remain but are hidden from sight in the form of an even more compressed body—a strange object known as a black hole (Unit 69).

67.5 GAMMA-RAY BURSTS AND HYPERNOVAE

Astronomers have made many of their most exciting discoveries when new telescopes have allowed them to observe the sky at wavelengths not previously detectable. Gamma-ray astronomy began in 1965 with a small and (by modern standards) primitive satellite designed to detect cosmic gamma rays. Gamma rays are the electromagnetic waves with the highest of energies, so naturally they are associated with the most energetic events in the cosmos. By the 1970s astronomers had discovered that many energetic sources, such as the center of our Galaxy and other galaxies, the Crab Nebula, and the remnants of other exploded stars, emitted gamma rays. Ironically, and unknown to astronomers, the most intriguing gamma-ray sources had already been discovered accidentally in the 1960s.

In 1967 the United States placed several military surveillance satellites in orbit to watch for the gamma rays produced when a nuclear bomb explodes. The satellites were designed to monitor the United States–Soviet Union ban on nuclear bomb tests in the atmosphere. On a number of occasions, the satellites detected brief, but intense, flashes of this extremely short wavelength radiation. Curiously, these **gamma-ray bursts** came not from the Earth but from space. This caused great concern among political and military leaders, but fortunately some scientists involved had enough background to recognize these as astronomical sources and avert a crisis. Unfortunately for astronomers, the discovery of the bursts was top secret at the time and was not made public until 1973.

Strangely, gamma-ray bursts did not appear to be associated with any identifiable object. They would appear suddenly in otherwise blank areas of the sky, flare in intensity for a few seconds, and then fade to invisibility. Despite intense study, it was not until 30 years after the first gamma-ray burst was observed that astronomers could even say whether the bursts came from nearby or far away. Without a distance, it was impossible to know how powerful these bursts were.

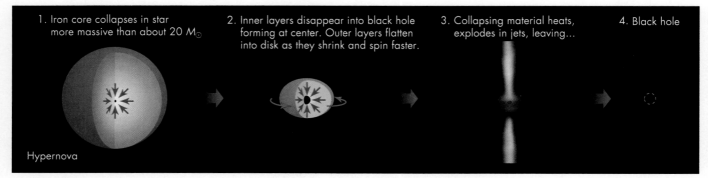

FIGURE 67.8

Formation of a hypernova blast and a black hole by a star more massive than about 20 $M_\odot$.

The breakthrough came in 1997, when astronomers detected a gamma-ray burst that they were able to show came from a distant galaxy. At the time, gamma-ray telescopes had very poor resolution, but by rapidly pointing an X-ray telescope to the direction of the sky where the burst was detected, astronomers caught the fading afterglow of the burst and were able to pinpoint its location. Over the following decade, astronomers succeeded in making hundreds more identifications with improving detail, even observing the afterglow of bursts at visible and radio wavelengths. A number of these events are associated with supernovae explosions.

Based on their brightnesses and distances, gamma-ray bursts herald the most luminous explosions in the universe, even brighter than normal supernovae. A hypothesis is emerging that explains most of their features in terms of an explosion called a **hypernova.** A hypernova is an intense burst generated during the core collapse of a star so massive that it collapses not to a neutron star but to a black hole—an object with such intense gravity that nothing, not even light, can escape (Unit 69).

Instead of forming a dense core of neutrons, the core and surrounding material collapse inward and disappear into a growing black hole. Models suggest that a black hole is likely to form upon the death of stars that began their life with more mass than about 20 $M_\odot$. If a black hole forms, there can be no core bounce nor the huge outpouring of neutrinos that characterize the supernova we described earlier.

If all of the matter fell into the black hole, we would see absolutely nothing at all, so it may seem odd that these hypernovae are so bright. However, not all of the matter can fall into the black hole. The black hole is only about 20 kilometers across, and material falling in from thousands or even millions of kilometers away is very unlikely to fall straight in. This is illustrated in Figure 67.8. Because of the conservation of angular momentum (Unit 20), as the outer parts of the star collapse inward toward the black hole, they will spin faster and flatten out. A large percentage of the material will end up in a flattened disk of material, similar in appearance to the disk around a protostar (Unit 61). The material collapsing toward the new black hole will crash together at near the speed of light and rise to extremely high temperatures. A huge thermonuclear explosion occurs, with jets of gamma rays escaping as the explosion bursts through the infalling matter along the direction of the poles of the former star.

Because stars more massive than 20 $M_\odot$ are rare, hypernovae are also rare. Furthermore, the burst of gamma rays can be seen only by observers who happen to lie close to the direction of the star's polar axis. If we do not lie along the axis, we would not see the gamma-ray burst, but we might see something more similar to other types of supernova explosions. Astronomers have identified some peculiar supernovae that they suspect may be such events. It is a good thing that we are unlikely to see the gamma rays from these bursts. Estimates suggest that one of these gamma-ray jets could cause mass extinction of life on the Earth if it occurred within 1000 light-years of us.

> Not all gamma-ray bursts are produced by hypernovae. Other proposed events that might produce such bursts are collisions between neutron stars or the collapse of a neutron star to a black hole when too much material falls on it from a companion star (analogous to a white dwarf supernova).

KEY POINTS

- Stars that begin their lives more massive than about 8 $M_\odot$ fuse heavy elements in their cores at the end of their red giant phase.
- Elements with masses below iron (atomic number 26) can release energy when they fuse to make heavier elements.
- The release of energy in a nuclear reaction is in proportion to the mass decrease in the reaction products according to $E = m \times c^2$.
- Fusion of heavier elements yields small amounts of energy compared to hydrogen fusion, so the fusion stages are brief.
- In its red supergiant phase, a massive star develops a series of shells fusing heavier elements closer to the star's core.
- When an iron core develops, it cannot generate energy to counteract the gravity, so it is supported by electron degeneracy.
- Eventually the weight of the growing iron core drives electrons to merge with protons, so the core collapses.
- Electrons combining with protons make neutrons; these collapse inward and emit neutrinos that drive some matter outward.
- Nuclear fusion and a "bounce" of some of the infalling neutrons add to the neutrinos to make a core-collapse supernova explosion.
- Stars that began with 8 to 20 $M_\odot$ are thought to leave behind neutron stars; more-massive stars probably become black holes.
- Before a black hole forms, from which nothing can escape, collapsing material may produce intense bursts of gamma radiation.

KEY TERMS

core-collapse supernova, 527
gamma-ray burst, 529
hypernova, 530
mass excess, 525
nucleosynthesis, 524
supernova remnant, 528
Type II supernova, 527

CONCEPT QUESTIONS

Concept Questions on the following topics are located in the margins. They invite thinking and discussion beyond the text.

1. Abundances of even and odd elements. (p. 524)
2. Increasing fraction of neutrons in heavy elements. (p. 525)
3. Modeling core bounce with rubber balls. (p. 526)

REVIEW QUESTIONS

4. What is nucleosynthesis?
5. Why can't a star fuse iron in its core?
6. What happens that causes a high-mass star's core to collapse?
7. What is a core-collapse supernova?
8. What is the observational difference between Type I and Type II supernovae?
9. How are Type Ib and Type Ic supernovae thought to be related to Type II supernovae?
10. What role do neutrinos play in a core-collapse supernova?
11. What is a hypernova? How does it relate to gamma-ray bursts?

QUANTITATIVE PROBLEMS

12. If every proton in the core of a massive star turns into a neutron and releases one neutrino, how many neutrinos are produced? Assume the core contains 2 $M_\odot$, and half the mass is made of protons ($M_{proton} = 1.67 \times 10^{-27}$ kg).
13. Assume that 10^{12} neutrinos passed through every square meter of surface perpendicular to the radiation from Supernova 1987A (which was 160,000 ly away). How many neutrinos were produced by the explosion?
14. How long does it take the blast wave from the center of a star undergoing a supernova explosion to reach the star's surface if the blast wave is traveling at 10,000 km/sec and the star is about 100 times the size of the Sun?
15. Assuming an average collapse speed of 50,000 km/sec, how long does a star take to collapse during the supernova stage if it starts off with a surface radius of 7 AU and ends up at the surface of the neutron star core, 10 km from the center of the star?
16. The Crab Nebula is today measured to be about 10 ly across. If it exploded in 1054, what has its average speed been in kilometers per second?
17. Betelgeuse is a red supergiant about 640 ly from Earth. Astronomers expect that within the next million years Betelgeuse will explode as a Type II supernova. Use Figure 67.4 to estimate how bright Betelgeuse will appear. Will it outshine the full Moon, which has a brightness of 3×10^{-3} watts/m^2?
18. It is estimated that in the final stages of fusion, a giant star produces a core of iron more massive than the Sun. From the energy release given in Table 67.1,
    a. how much energy is this (in joules)?
    b. how long would it take the Sun to emit this much energy at its rate of 4×10^{26} joules/sec?

TEST YOURSELF

19. Stars like the Sun do not form iron cores during their evolution because
    a. all the iron is ejected when they become planetary nebulae.
    b. their cores never get hot enough to fuse silicon into iron.
    c. the iron they make is all fused into uranium.
    d. their strong magnetic fields keep iron out of the core.
    e. only higher-mass stars can attract iron when they form.
20. What is the heaviest element that fusion can produce in the core of a massive star?
    a. Carbon b. Gold c. Iron d. Neon e. Silicon
21. What one characteristic is the most influential in determining a star's evolution?
    a. If it is a member of a binary or not
    b. The number of nearby stars
    c. Its mass
    d. Its initial composition
    e. The strength of its magnetic field
22. The study of gamma ray bursts progressed slowly because
    a. nobody could confidently say if gamma bursts were a local phenomena or originated in a distant location.
    b. it was difficult to pinpoint the location of gamma ray bursts.
    c. they are extremely brief.
    d. All of the above.

PART 4

UNIT 68

Neutron Stars

68.1 Pulsars and the Discovery of Neutron Stars
68.2 Emission from Neutron Stars
68.3 High-Energy Pulsars

Learning Objectives

Upon completing this Unit, you should be able to:
- Recall how pulsars were discovered, and why they are thought to be neutron stars.
- Explain why neutron stars spin so fast, and why they produce pulses of radio waves.
- Describe how millisecond and high-energy pulsars arise.
- Explain what we know about the internal structure of neutron stars.

We saw in Unit 67 that when a massive star forms an iron core late in its lifetime, the pressures at the center of the star can force electrons to merge with protons to make neutrons, and the whole star collapses to an extraordinarily small size. The resulting star resembles a giant atomic nucleus, with a density a billion times greater than a white dwarf—about a trillion tons per liter!

In 1934 Walter Baade (pronounced *BAH-deh*) and Fritz Zwicky, astrophysicists at Mount Wilson Observatory and the neighboring California Institute of Technology, respectively, suggested that when a massive star reaches the end of its life, its gravity will crush its core and make the star collapse. They predicted that the collapse of the core would trigger a titanic explosion, which they named a *supernova* (Unit 67). Almost as an afterthought, they speculated that the collapsed core might be so dense that the star's protons and electrons would be driven together and merge into neutrons, forming a **neutron star**.

Astronomers accepted Baade and Zwicky's idea that a massive star dies as a supernova, because such violent explosions had been seen (Unit 67); but they paid little attention to their suggestion that a neutron star might be born in the blast. Such stars would have radii of only about 10 kilometers (about 6 miles), incredibly tiny compared with even white dwarfs, and probably unobservable if they existed at all. The idea of these tiny, collapsed stars lay dormant for more than 30 years.

68.1 PULSARS AND THE DISCOVERY OF NEUTRON STARS

In 1967 a group of English astronomers led by Anthony Hewish was observing fluctuating radio signals from distant, peculiar galaxies. Jocelyn Bell, a graduate student working with the group, noticed an odd radio signal with a rapid and astonishingly precise pulse rate of one burst every 1.33 seconds. The precision of the pulses led some astronomers to wonder whimsically if they had perhaps stumbled onto signals from another civilization, and informally the signal became known as LGM-1, for "little green men #1." Over the next few months Hewish's group found several more pulsating radio sources, which came to be called **pulsars** for their rapid and precisely spaced bursts of radiation (Figure 68.1).

The ensuing discovery of many more pulsars, and the unchanging repetitiveness of their signals, convinced astronomers that they were observing a natural phenomenon. But it was difficult to come up with a plausible hypothesis for an astronomical phenomenon that could be so regular and rapid.

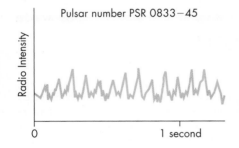

FIGURE 68.1
Pulsar signals recorded by a radio telescope.

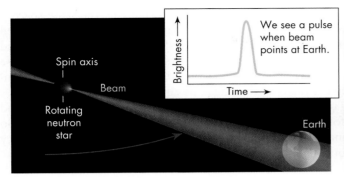
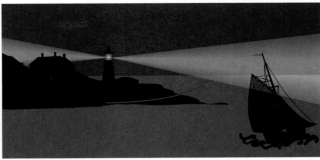

FIGURE 68.2
A pulsar's pulses are like the flashes of a lighthouse as its lamp rotates.

Concept Question 1

What do you think an artificial signal from an extraterrestrial life form would look like? Do you think it would be easy for the extraterrestrial to tell that radio signals from Earth were artificial?

Pulsar model

The solution to this puzzle came from the work of the Italian astronomer Franco Pacini and the Austrian-born British astronomer Thomas Gold. Their work led to the idea that a pulsar is a rapidly spinning neutron star. This idea soon gained support from the discovery of a pulsar at the center of the Crab Nebula, a supernova remnant.

The pulses from a spinning neutron star are produced by its magnetic field. Collapsed to such a small radius, the star's magnetic field is squeezed into a far smaller volume, amplifying the field's strength to more than a trillion times that of the Earth's magnetic field. A neutron star is thus an extremely powerful magnet. Its intense magnetic field causes it to emit radiation in two narrow beams along its north and south magnetic poles (Section 68.2). Like the Earth's and most other astronomical bodies' magnetic fields, a pulsar's magnetic field is not aligned with the rotation axis; so as the star spins, its beams sweep across space. We see a burst of radiation at radio wavelengths when a beam points at the Earth. A pulsar shines like a cosmic lighthouse whose beam swings around as its lamp rotates, as illustrated in Figure 68.2.

But what could make a neutron star spin so fast? For example, the pulsar at the center of the Crab Nebula rotates 30 times per second. By comparison, the Sun and most other stars take weeks to make a single rotation. However, when a massive star collapses to a tiny radius, the **conservation of angular momentum** (Unit 20) requires it to spin faster. This is the same effect that allows ice skaters to spin rapidly by pulling their arms and legs close to their axis of rotation.

An object's angular momentum is approximately given by the body's mass, M, times its equatorial rotation velocity, V, times its radius, R:

$$\text{Angular momentum} \approx M \times V \times R.$$

The principle of conservation of angular momentum states that the product of these three quantities must remain constant unless some force brakes or accelerates the spin. Therefore, in the absence of external forces, if a fixed mass contracts, it must increase its rotational speed by the same factor by which its radius decreases—if R shrinks, V must increase to keep the angular momentum constant, as illustrated in Figure 68.3. Therefore, the contracting core of a star must spin faster and faster.

A neutron star is supported by neutron degeneracy much as the core of a red giant is supported by electron degeneracy (Unit 63). This is because neutrons

FIGURE 68.3
Conservation of angular momentum causes a collapsing star to spin faster. If its radius R shrinks, its rotation velocity V will increase to maintain a constant angular momentum.

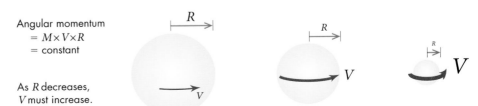

Mathematical Insights

To convert days to seconds, multiply by the number of hours per day, the number of minutes per hour, and the number of seconds per minute:
25 days = 25 × 24 × 60 × 60 seconds
= 2,160,000 seconds.

Neutron stars

obey a rule like electrons that prevents them from occupying the same space if they have the same energy. This allows a solar mass of matter to be compacted into a sphere with a radius of about 7 kilometers! Imagine the Sun contracted from its current radius of about 700,000 kilometers to just 7 kilometers. Because its radius would be 100,000 times smaller, conservation of angular momentum requires that the speed of material rotating at its equator would be 100,000 times greater. The period—the time it takes a point on the equator to come back around again—would therefore be $(100,000)^2 = 10,000,000,000 = 10^{10}$ times faster. So instead of rotating once every 25 days, the collapsed Sun would rotate once every 0.0002 seconds—about 5000 times per second!

Such a rapid spin is what is expected when the iron core of a massive star collapses. Observed pulsars have slower periods than this, presumably because they gradually slow down. The Crab Nebula pulsar period, for example, is currently observed to be slowing down by about 10^{-5} seconds each year, so when the nebula formed nearly a thousand years ago, it may have been spinning as fast as our calculation here suggests.

68.2 EMISSION FROM NEUTRON STARS

When a magnetic field moves, it creates an electric field—a principle we use here on Earth by spinning magnets in dynamos to generate electricity. Similarly, the rapid spin of a neutron star and its intense magnetic field generates powerful electric fields. These fields rip positively and negatively charged particles off the star's surface and accelerate them to nearly the speed of light. The electrically charged particles are channeled by the pulsar's intense magnetic field to travel along the magnetic field lines. This is similar to how the Earth's magnetic field directs solar wind particles, except in the Earth's case they are directed *inward,* creating the aurora near the magnetic poles (Unit 38). The magnetic field of a spinning pulsar, in a like manner, generates two narrow beams of charged particles flowing outward at the magnetic poles.

As the charged particles move along the pulsar's magnetic field, they generate radio waves along their direction of motion. This is somewhat like a radio broadcast antenna on Earth, where electric currents are pulsed through the antenna, accelerating electrons in it; the accelerated electrons in turn produce the radio waves we detect. In a pulsar, too, the charges radiate as they accelerate, in this case along the magnetic field, spiraling as they go (Figure 68.4). The pulses of radiation we see are the collective emission from myriad charges pouring off the neutron

FIGURE 68.4

The strong magnetic field of a spinning pulsar generates a powerful electric field, which pulls electrically charged particles from its surface. The charged particles spiral in the star's magnetic field and emit radiation along their direction of motion. Because the particles stream in a narrow beam confined by the magnetic field, their radiation is also in a narrow beam at each magnetic pole of the star.

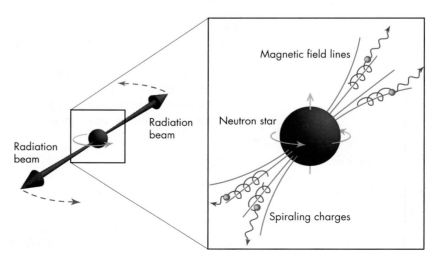

FIGURE 68.5
The rapid rotation of the pulsar in the Crab Nebula makes the star appear to turn on and off 30 times per second as its beams of radiation sweep across the Earth. The top portion of the figure shows video images, illustrating that the pulsar blinks on and off at visible wavelengths as well as radio wavelengths. The lower portion of the figure shows a light curve as the pulsar completes one revolution.

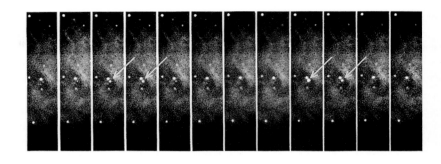
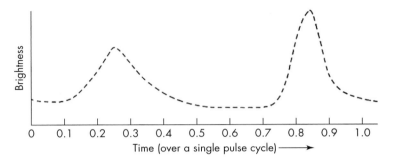

star's surface. This radiation is beamed because the charges are traveling along the field lines that emerge from the star's surface at each magnetic pole. Because of this beaming, we see only neutron stars that have a magnetic pole that points toward Earth at some time as they rotate. Pulsars radiating in other directions are invisible to us, so many more neutron stars must exist than the approximately 1500 we have detected to date.

The emission created by the accelerating charges is called **synchrotron radiation,** named for a kind of particle accelerator used in physics experiments that accelerates charged particles using a magnetic field. This kind of radiation produces electromagnetic waves across a broad and continuous range of wavelengths. Synchrotron radiation is different from thermal radiation, which also produces radiation over a broad range of wavelengths (Unit 23), because its properties depend on the charged particles' acceleration and on the strength of the magnetic field, rather than on the temperature of a heated gas.

Most pulsars generate synchrotron radiation that is detectable only at radio wavelengths. However, some young pulsars, like the Crab Nebula, generate synchrotron radiation across the entire electromagnetic spectrum, including visible light and gamma rays. A high-speed camera has captured the flashes of visible light from the Crab Nebula's pulsar as it spins (Figure 68.5).

In everyday life, spinning objects slow down. Pulsars are no exception. As a pulsar spins, it drags its magnetic field through the particles that boil off its surface into the surrounding space. That drag slows down the pulsar. Astronomers measure this **spindown** of a pulsar by precisely timing the interval between pulses. Such measurements indicate that the time interval—the period of a spinning pulsar—gradually lengthens. The slowing rotation also reduces the energy of the radiation that the pulsar emits. Thus, young, rapidly spinning pulsars emit electromagnetic waves from visible light to radio waves, whereas old, slowly spinning ones generate only radio waves.

Although pulsars gradually slow down overall, occasionally they suddenly speed up by a small amount, as illustrated by the graph in Figure 68.6. Such jumps in rotation speed are called **glitches,** and they can tell us about the interior structure of the neutron star. The sudden speedup occurs because the neutron star's

Concept Question 2

Astronomers suspect that the slowdown rate of pulsars is probably greater when they are young. Why might this be?

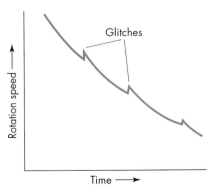

FIGURE 68.6
A pulsar's rotation speed gradually slows, but occasionally it suddenly speeds up during a glitch. This probably occurs as the crust readjusts its shape.

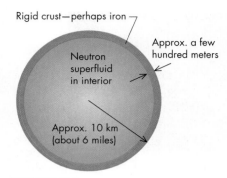

FIGURE 68.7
Schematic structure of the interior of a neutron star.

Concept Question 3

If the Earth's polar caps melt, will the Earth's rotation speed change? How does this relate to the changes of rotation speed of a pulsar?

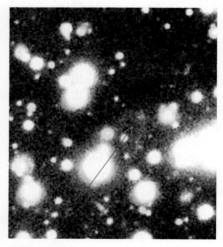

FIGURE 68.8
One of the nearest known neutron stars to the Solar System, RX J1856.5–3754, is an extremely faint object shown in this image made with the VLT. An arrow points to the neutron star and shows its direction of motion through space. As the star plows through interstellar gas at high speed, it creates a "bow wave," seen here in pink from hydrogen emission.

shape changes during the star's overall slowdown. When it is spinning rapidly, the pulsar is larger at its equator because of the centripetal force outward. The Earth, Sun, and other spinning planets all bulge outward at their equators because of this same effect (Unit 44), but at the high spin rate of a pulsar, the distortion is much larger. As the spin slows, the neutron star settles toward a more spherical shape, and therefore its equatorial radius shrinks slightly. Because the matter moves closer to the axis of rotation, conservation of angular momentum causes the star to speed up a small amount before continuing its gradual slowdown.

Pulsar glitches reveal that the crust of the neutron star is rigid, because it does not deform gradually. Computer models suggest it is probably made of iron, a few hundred meters thick, as illustrated in Figure 68.7. The material inside appears to be a "fluid" of neutrons. The models suggest the neutron star also has a gaseous atmosphere—about 1 millimeter thick.

As the pulsar's rotation slows, its radiation weakens, and ultimately the star becomes undetectable. Thus, a pulsar "dies" by becoming invisible. But just as a white dwarf can be resurrected with infalling gases from a nearby companion star, so too can a pulsar in a binary system be reawakened. As material from the companion spirals into a pulsar, it adds angular momentum to the pulsar, causing it to spin faster and faster. The fastest pulsars known are called **millisecond pulsars** because their periods are just a few milliseconds long. They all appear to have been "revived" by this process.

Millisecond pulsars tend to be the exception, however, because pulsars are not frequently found in binary star systems. The supernova explosion that creates a pulsar usually expels so much mass that the gravitational attraction between the stars is no longer great enough to hold them together at their existing orbital speed. The sudden loss of mass can be compared to a ball spun around at the end of a string: If the string breaks, the ball flies off with the speed at which it was revolving. After the mass loss of a supernova explosion, the pulsar may fly off at its orbital speed—perhaps several hundred kilometers per second. It eventually escapes the debris of the explosion that spawned it. Such speeding pulsars have been detected moving through our Galaxy at hundreds of kilometers per second.

One of the nearest known neutron stars to the Solar System with the catalog name RX J1856.5–3754 is shown in Figure 68.8. This object has a rapid motion seen from its shifting position in Hubble Space Telescope measurements made several years apart. It probably escaped from a binary star system after it underwent a supernova explosion. Its distance is estimated to be about 120 parsecs (about 400 light-years). RX J1856.5–3754 is not a pulsar, but faint thermal emission from its hot surface has been detected at visible and X-ray wavelengths. Its surface temperature is estimated to be about 700,000 K. Even though it is so hot, it is thought to be 1 million years old, based on its trajectory and distance from the nearest star-forming region. The image in Figure 68.8 was made with the VLT 8-meter telescope in Chile, showing that the neutron star is creating a "bow wave" like a boat does as it plows through interstellar gas.

Observations of the emission from the surface of a neutron star are essential for understanding the neutron star's internal structure. For example, astronomers initially estimated that the temperature of RX J1856.5–3754 was higher; this in turn implied that its size was so small that the neutrons should be broken down into their constituent quarks in the center of the neutron star. However, when it was found that the neutron star is farther away than initially thought, astronomers revised the estimates of size and temperature downward to values that appear to be consistent with the interior being made of neutrons down to the core. Studies of the sizes of other neutron stars, and a determination of the highest possible mass for a neutron star, will allow astronomers to better understand whether some neutron stars reach such high densities that their cores form a new state of matter before they collapse in on themselves.

68.3 HIGH-ENERGY PULSARS

Astronomers have identified some additional classes of neutron stars that produce very high-energy pulsed radiation at X-ray and gamma-ray wavelengths. These appear to be powered by different sources than normal pulsars are—by either accreting matter or extremely strong magnetic fields.

X-ray pulsars generate pulses of X-ray radiation, as their name implies. However, besides the higher energy of the radiation, their pulsations are more erratic. The strength of the pulses is sometimes fairly steady and sometimes arrives in strong bursts, and the rate of pulsation is seen to speed up sometimes and to slow down at others. These appear to be neutron stars that remain in a binary star system after the supernova explosion.

An X-ray pulsar generates its emission from small "hot spots" near its magnetic poles. Gas from a companion star flows in along magnetic field lines. As the gas falls toward the neutron star, the star's intense gravity accelerates the gas. As it crashes onto the neutron star at the magnetic poles (Figure 68.9), the gas is compressed and heated intensely, causing it to emit X-rays. As the neutron star spins, the hot regions near each pole move into and then out of our field of view, creating the X-ray pulses that we observe. These X-ray pulsars are often measured to be speeding up—and perhaps someday will be millisecond pulsars.

Another type of pulsar is known as a **magnetar.** These are neutron stars with extremely intense magnetic fields. The magnetic field of a magnetar may be a thousand times stronger than the already intense magnetic fields of a normal pulsar, and more than a trillion times stronger than the Sun's magnetic field. Magnetars have been identified from extremely intense bursts of X-ray and gamma-ray radiation. These bursts last for minutes, pulsing during the burst at the rate at which the neutron star is known to be spinning. During a burst, the magnetar can generate as much energy as the Sun produces in hundreds of thousands of years. A magnetar outburst in December 2004 was so powerful that it briefly shut down a number of satellites and ionized gas in the Earth's upper atmosphere. These effects occurred despite the huge distance to the magnetar—about 50,000 light-years.

Only a few magnetars have been identified to date, perhaps because their strong magnetic fields cause them to spin down much faster than normal pulsars. The leading hypothesis for explaining the enormous outbursts is that these neutron stars undergo massive glitches as they slow down, and these create huge shifts and rearrangements of the magnetic field. These might be analogous in some ways to the flare activity seen on the Sun, but vastly stronger because of the enormous strength of the magnetic field.

> **Concept Question 4**
>
> In a science fiction novel by Robert Forward, he imagines life developing on a neutron star based on nuclear chemistry. What do you think such life might be like, how it might evolve, how it might be similar to us?

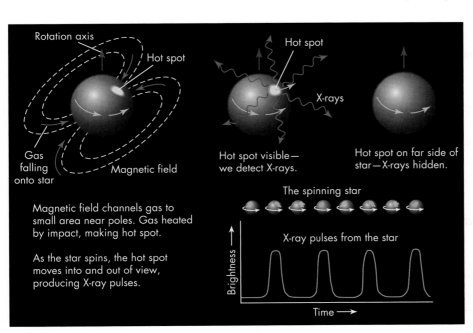

FIGURE 68.9
Gas falling onto a neutron star follows the magnetic field lines and makes a hot spot on the star's surface, creating X-rays. As the star rotates, we observe X-ray pulses.

KEY POINTS

- Pulsars are rapidly rotating neutron stars that send out beams of electromagnetic energy like a lighthouse.
- When a star's core collapses down to the size supported by neutron degeneracy, it is only about 20 km in diameter.
- Conservation of angular momentum causes a star rotating in weeks to rotate in just milliseconds when shrunk to neutron star sizes, and the magnetic field intensifies by a similar factor.
- Young pulsars like the Crab Nebula pulsar are found in the remnants of their supernova explosions and still spin rapidly.
- Slowing of a pulsar's rotation results from the energy it expends accelerating particles with its magnetic field as it spins.
- Glitches (sudden small speedups) of the otherwise slowing spin of a pulsar are probably caused by a hard crust cracking and readjusting as the pulsar slows down.
- Some old neutron stars can be spun up to high speed by material falling onto them from a companion star.
- Some pulsars emit X-rays and gamma rays, probably due to extremely hot spots on the surface of the neutron star or extremely intense magnetic fields.

KEY TERMS

conservation of angular momentum, 533
glitch, 535
magnetar, 537
millisecond pulsar, 536
neutron star, 532
pulsar, 532
spindown, 535
synchrotron radiation, 535
X-ray pulsar, 537

CONCEPT QUESTIONS

Concept Questions on the following topics are located in the margins. They invite thinking and discussion beyond the text.

1. How to identify artificial radio signals. (p. 533)
2. Rate of pulsar slowdown when young versus old. (p. 535)
3. Effect on Earth's spin if polar caps melt. (p. 536)
4. Imagining life on a neutron star. (p. 537)

REVIEW QUESTIONS

5. What are the mass and radius of a typical neutron star, compared with those of the Sun?
6. What is a pulsar? In what sense does it pulsate?
7. How does a pulsar generate the pulses we see?
8. When in its lifetime does a neutron star produce pulses? What determines how rapid the pulses are?
9. What is a pulsar "glitch"? How does it affect the pulsar?
10. What is a magnetar? How can we tell magnetars from normal pulsars?

QUANTITATIVE PROBLEMS

11. Calculate the escape velocity (Unit 18) from a white dwarf and a neutron star. Assume that each has $1.4\,M_\odot$. Let the white dwarf's radius be 10^4 km and the neutron star's radius be 10 km.
12. Suppose a neutron star with a radius of 10^4 km is spinning with a period of 0.001 sec. How fast is its equator moving? Compare this speed to the neutron star's escape velocity calculated in the previous problem.
13. The mass of a neutron is 1.67×10^{-27} kg. How many neutrons are in a neutron star that has $1.4\,M_\odot$?
14. The volume of a neutron is about 10^{-45} m^3. Suppose you packed the number of neutrons you found in the previous problem into a sphere of radius R. The volume of a sphere is $4/3\,\pi R^3$. What is R in kilometers?
15. What is the luminosity of neutron star that has a surface temperature of 700,000 K and a radius of 10 km? Where on the H-R diagram should neutron stars be placed?
16. What is the orbital separation for a binary composed of two $1.4\,M_\odot$ neutron stars orbiting each other once every 10 seconds? Such theoretical systems are targets for current gravitational wave detectors (Unit 27.4).

TEST YOURSELF

17. When the first pulsars were first detected, some astronomers thought they might be produced by an extraterrestrial civilization because the pulses
    a. had an extraordinarily regular and rapid period.
    b. were produced by stars that had earth like planets.
    c. all had precisely the same strength.
    d. produced a sound resembling classical music.
    e. were at a wavelength where no natural emissions occur.
18. Which of the following has a radius (linear size) closest to that of a neutron star?
    a. The Sun c. A basketball e. A gymnasium
    b. The Earth d. A small city
19. What causes the radio pulses of a pulsar?
    a. The star vibrates like the quartz crystal in a watch, which sends electromagnetic waves through space.
    b. As the star spins, beams of radio radiation from it sweep over the Earth once each revolution, producing pulses.
    c. The star undergoes a series of nuclear explosions that generate radio emission.
    d. The star's dark orbiting companion periodically eclipses the radio waves emitted by the main star.
    e. Convection inside the pulsar produces twisted magnetic fields much like what occurs during the Sun's sunspot cycle, but at a much faster rate.
20. Which of the following pulsar properties is not considered extreme in comparison to most stellar objects?
    a. Density
    b. Mass
    c. Magnetic field strength
    d. Rotation rate

PART 4
UNIT 69
Black Holes

69.1 The Escape Velocity Limit
69.2 Curving Space Out of Sight
69.3 Observing Black Holes
69.4 Hawking Radiation
69.5 Small and Large Black Holes

Learning Objectives

Upon completing this Unit, you should be able to:
- Define a black hole, and calculate its radius from its mass.
- Explain how a Newtonian description of a black hole is inconsistent with relativity.
- Describe the effects on time and space approaching the event horizon.
- Explain what observations astronomers use to infer the presence of black holes.
- Show the range of properties black holes may have at low and high masses.

The iron core of a massive star is supported by pressure between the electrons in its interior. If enough mass is added to the core, the gravitational force grows so large that the star implodes into a neutron star (Unit 68). What happens if we gradually add mass to a neutron star? Will it collapse to a body of yet a smaller size? The answer appears to be that something fundamentally stranger happens. The nature of space and time around the object changes in such a way that the collapsing object disappears from view. It becomes an object that astronomers call a **black hole,** from which nothing can escape, not even light.

When gravity grows large enough, it changes the nature of space in a way that prevents any other possible force from supporting the star. The effect that mass has on space was one of the extraordinary discoveries made by Albert Einstein in the early 1900s. He developed the theory of **general relativity** (Unit 27) to describe these effects. Initial ideas about black holes date back to the 1700s, but general relativity is required to fully understand how space and time are altered by matter. As we shall see in this Unit, collisions of matter *outside* the black hole may be extremely violent, so the region immediately outside of black holes is full of radiation. However, general relativity makes it clear that no radiation, neutrinos, or any other particles can come out once they get closer than a certain distance from the center of the black hole. Anything that gets within that distance is lost from the universe as we know it.

69.1 THE ESCAPE VELOCITY LIMIT

To understand black holes, we need to review the concept of escape velocity (Unit 18). **Escape velocity** is the speed a mass must acquire to avoid being drawn back by another object's gravity. As illustrated in Figure 69.1, the speed, V_{esc}, needed to escape from an object of mass M starting at radius R is

$$V_{esc} = \sqrt{\frac{2GM}{R}}.$$

Here G is Newton's gravitational constant, which has a value of 6.67×10^{-11} in MKS units, so if R is in meters and M is in kilograms, V_{esc} is in meters per second.

For the Earth, the escape velocity from its surface is about 11,000 meters per second; for the Sun, it is about 600,000 meters per second. Note that the escape velocity depends on the distance of the escaping body from the center of the body it is trying to escape. For a planet or star we usually select its radius to find the

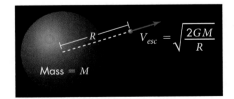

FIGURE 69.1
Escape velocity. The small brown ball in the figure will escape from the gravity of the large mass M only if it has at least the escape velocity for its distance from the center of the object R.

speed needed to escape from the surface. However, we can just as well calculate the speed needed to escape starting at an arbitrary radius.

Since R is in the denominator, the escape velocity of an object grows larger at smaller radii, reaching a maximum at the surface. The escape velocity from the surface grows larger if the radius of the object shrinks. In other words, the same mass packed into a smaller radius creates a larger force of gravity at its surface.

The escape velocity from the surface of the Sun is 600 kilometers per second. However, if you packed the Sun's mass into a neutron star, the escape velocity from the surface of the neutron star would be much larger, because the neutron star's radius is so small. In fact, a neutron star with the mass of the Sun would have a radius about 10^5 times smaller than the Sun's, so its escape velocity is about $\sqrt{10^5} \approx 300$ times larger than the Sun's escape velocity, or about 180,000 kilometers per second. This is over half the speed of light! Hence, if a neutron star is compressed just a little, its escape velocity could exceed the speed of light. Such an object becomes a black hole.

As long ago as 1783, the English cleric John Michell discussed the possibility that objects with escape velocities exceeding the speed of light might exist. A little more than a decade later, the French mathematician and physicist Pierre Simon Laplace entertained the same idea. Following their logic, we find then that a mass M becomes a black hole if it is compressed to a radius

$$R_S = \frac{2 \times G \times M}{c^2}.$$

The radius of the black hole is found by equating the escape velocity to the speed of light (see derivation in margin at left). We label this radius R_S and call it the **Schwarzschild radius** in honor of the German astrophysicist Karl Schwarzschild, who studied how to modify Newton's theory of gravity to account for effects near the speed of light and when gravity is extremely strong. His results, published in 1915, showed that the simple approach we have adopted here provides the correct result.

In principle, a body of any mass can be turned into a black hole if it is compressed to a small enough radius. For example, suppose we wanted to find out how small we would need to squeeze the Sun to make it a black hole. The Sun has a mass of 1.99×10^{30} kilograms, so if we plug in this mass and the speed of light (3.00×10^8 m/sec), we find that the Schwarzschild radius for the Sun is

$$R_S(\text{Sun}) = \frac{2 \times G \times M_\odot}{c^2}$$

$$= \frac{2 \times 6.67 \times 10^{-11}\,\text{m}^3\,\text{kg}^{-1}\,\text{sec}^{-2} \times 1.99 \times 10^{30}\,\text{kg}}{(3.00 \times 10^8\,\text{m/sec})^2}$$

$$= 2950\,\text{m} = 2.95\,\text{km}.$$

So if the Sun could be compressed to a radius slightly smaller than 3 kilometers (1.9 miles), it would become a black hole. Similarly, a 2-$M_\odot$ would become a black hole if it were compressed to a radius of about 6 kilometers; a 3-$M_\odot$ object would become a black hole if it were compressed to a radius of 9 kilometers; and so on. So the radius of a black hole approximately equals 3 kilometers times its mass in solar masses.

When the iron in the core of a high-mass star collapses into neutrons, it may become a black hole instead of a neutron star. Suppose the iron core grows to a mass of 3-$M_\odot$. If it collapses to neutron density, it will have a radius of about 9 kilometers—very close to the Schwarzschild radius. Any additional mass added to this ball of neutrons will increase the mass, and probably decrease the radius, just as white dwarfs decrease in size as mass is added to them.

At that point, an odd thing happens. When we think of particles like neutrons pushing against each other, the "push" is a force between the neutrons. These forces

---

R_S = black hole's Schwarzschild radius
M = mass of black hole
G = Gravitational constant
c = speed of light

Mathematical Insights

If we set the escape velocity to the speed of light, $c = \sqrt{2GM/R_S}$, then we can solve for the Schwarzschild radius, first by squaring both sides of the equation:

$$c^2 = \frac{2GM}{R_S}$$

$$c^2 \times \frac{R_S}{c^2} = \frac{2GM}{R_S} \times \frac{R_S}{c^2}$$

$$R_S = \frac{2GM}{c^2}.$$

must themselves travel from one point to another, and they can travel no faster than the speed of light. Anywhere inside the Schwarzschild radius, where the escape velocity exceeds the speed of light, none of the neutrons can communicate its "push" to the particle above it. No force can be communicated outward, so the matter collapses inward, disappearing like a boat sinking below the surface of water. What happens next is unobservable because it occurs inside the black hole, but it is thought that the collapse continues until the matter collapses to a single point—what astronomers call a **singularity.** Black holes are remarkable objects, and they reveal that our understanding of gravity based on Newton's law (Unit 16) is incomplete. To better understand how matter behaves in gravitational fields, we need to look in a little more detail at the relation between mass, gravity, and space.

69.2 CURVING SPACE OUT OF SIGHT

We described in Unit 27 how Albert Einstein's theory of general relativity provides a new mathematical and physical understanding of gravity. General relativity describes how mass alters the nature of space, creating what astronomers call **curvature of space.** A black hole is a place where the curvature of space has become so extreme that the matter in the black hole disappears from view.

An analogy may help illustrate how gravity and curvature of space are related. Imagine a water bed on which you have placed a baseball (Figure 69.2). The baseball makes a small depression in the otherwise flat surface of the bed. If a marble is now placed near the baseball, it will roll along the curved surface into the depression. The bending of its environment made by the baseball therefore creates an "attraction" between the baseball and the marble. Now suppose we replace the baseball with a bowling ball. It will make a bigger depression, and the marble will roll in farther and be moving faster as it hits the bottom. We therefore infer from the analogy that the strength of the attraction between the bodies depends on the amount by which the surface is curved. Gravity also behaves this way, according to the general theory of relativity. According to that theory, mass creates a curvature of space, and gravitational motion occurs as bodies move along the curvature.

We can extend our analogy to black holes by supposing that we remove the bowling ball and put on the water bed a large boulder. The boulder presses down so hard on the vinyl that it tears a hole in it. A marble placed on the water bed will roll into the depression made by the boulder and disappear into the hole it made.

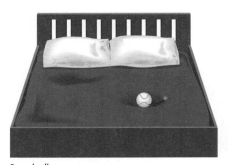
Baseball:
Marble rolls into depression

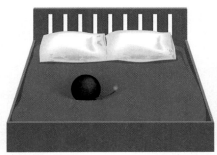
Bowling ball:
Marble rolls in faster

Big rock:
Marble disappears into hole

FIGURE 69.2
Objects on a water bed make depressions analogous to the curvature of space created by a mass. According to the general theory of relativity, the curvature of space produces the effect of gravity. We can see the similarity by placing a marble at the edge of the depression and watching it roll inward, as if attracted to the body that makes the depression. Bigger bodies make bigger depressions, so a marble rolls in faster. However, a very big body may tear the water bed, creating an analog of a black hole.

So too, a black hole creates a "rip" in space where the curvature has become so strong that the structure of space is disrupted.

General relativity is a major departure from Newton's ideas about gravity. For example, in the old idea of a black hole, like that envisioned by Michell and Laplace, photons trying to leave the black hole should behave like objects thrown upward at less than the escape velocity. That is, they would reach some maximum height and then fall back down again. If this were the correct description of how a black hole works, a black hole would be surrounded by photons that had traveled out and were momentarily stopped before they fell back in again. This clearly contradicts the observation that photons always travel at the speed of light (Unit 26).

Problems such as these are resolved by general relativity. As we discussed in Unit 27, another way of describing the curvature of space is that space has a motion caused by massive objects, like the flow of water in a lake with a drain hole at its middle. If you were in a motorboat on such a lake, you would find the boat being pulled in toward the middle of the lake. At the edges of the lake this pull would be gentle, but if you moved closer to the position of the drain, the current would grow faster. If the flow were powerful enough, you might even find that at a certain distance from the drain your boat was being pulled in by a flow of water so rapid that, even when operating at its top speed, the boat could make no progress away from the drain. This flow or curvature can be pictured as a curved surface, as illustrated in Figure 69.3.

Clarification Point

The motion of space into a black hole does not "remove space" from the universe like water flowing down a drain. Another analogy for the motion is an escalator that carries passengers downward unless they run upward at a greater speed than the escalator is moving down.

FIGURE 69.3
Models of curved space around the Sun and around a black hole of one solar mass. The curvature in the regions outside the Sun's radius is the same for both. At very small radii, the black hole produces such a strong inward motion of space at the event horizon that a photon moving outward at this radius cannot make any progress away from the black hole, and all photons and matter inside this radius must flow inward. The motion of space toward the massive object is illustrated by blue arrows. Far outside the event horizon, space moves slowly, so photons move at near-normal speed relative to a distant observer.

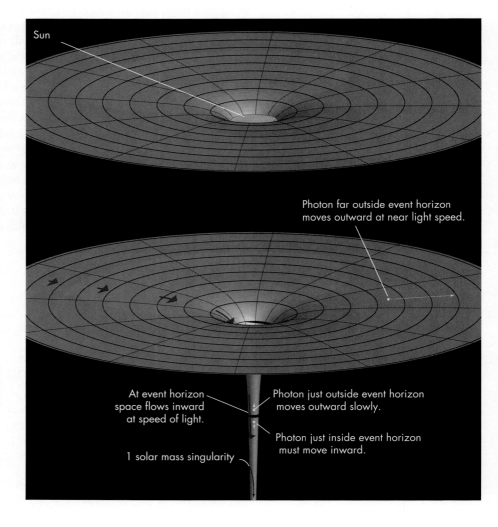

Concept Question 1

Suppose you were traveling in a spaceship and calculated that your trajectory might carry you within a million kilometers of the center of a 5-$M_\odot$ black hole, or the same distance from the center of a 5-$M_\odot$ main-sequence star. Why would the main-sequence star present the bigger danger of your crashing?

Clarification Point

Matter that falls into a black hole cannot come back out, but black holes do not "suck matter into them like a vacuum cleaner" as is sometimes said. They pose no more threat (and perhaps less) than any other object with the same mass because they are so much smaller, you are less likely to encounter them.

γ_G = gravitational time dilation
R_S = Schwarzschild radius
R = distance from singularity

Concept Question 2

If you were orbiting just outside a black hole, there might be differences in how fast time ran in parts of the ship closer to and farther from the event horizon. Could your feet age slower than your head?

At the Schwarzschild radius of a black hole, space is flowing inward at the speed of light, just as the water in the imaginary lake is flowing toward the drain. A photon inside the Schwarzschild radius finds itself, like the boat, fighting a current moving faster than it can travel. Even though the photon is moving at the speed of light through space, the space through which it is traveling is "falling" into the hole at the speed of light, and thus the photon makes no progress outward and cannot escape. Such extreme curvature of space in a black hole, which prevents light from escaping, creates a kind of boundary that astronomers call the **event horizon** (Figure 69.3). Just as the curvature of the Earth's surface blocks our view of what lies beyond the horizon, so too the curvature of space at the Schwarzschild radius prevents our seeing beyond the event horizon into the interior of the black hole.

All that happens within the black hole is hidden forever from our view. No radiation of any sort, nor any material body—rocket, spacecraft, or other object—can break free of the black hole's gravity if it travels past the event horizon. Because we cannot observe the interior of a black hole even in principle, there are only a limited number of physical properties we can ascribe to it. For example, it is meaningless to ask what a black hole is made of. It could be made of neutrons or cornflakes—only the amount of mass is important, not what it is composed of.

We can measure the mass of a black hole because its mass generates a gravitational field. In fact, at large distances the gravitational field generated by a black hole is no different from that generated by any other body of the same mass. For example, if the Sun were suddenly to become a black hole with the same mass it has now, the Earth would continue to orbit it without any change: It would *not* be pulled in. What makes a black hole so special is that it is so small for its mass. The small size means that it is possible for other objects to get within an extremely small distance of the center. And when the separation between objects becomes small, gravity's force grows extremely strong because the gravitational force depends inversely on the separation.

We saw in Unit 27 that curvature of space causes time to slow down. We showed that there is a **gravitational time dilation** γ_G that tells us how much slower clocks run in a gravitational field based on the escape velocity. Using the relationships derived in Section 69.1, we can rewrite the equation for time dilation in terms of the Schwarzschild radius R_S as follows:

$$\gamma_G = \frac{1}{\sqrt{1 - R_S/R}}$$

where R is the distance from the singularity. This equation shows that time runs slower the closer you get to the Schwarzschild radius because the denominator gets smaller as R approaches R_S.

At the Schwarzschild radius, where $R = R_S$, the denominator $\sqrt{1 - R_S/R}$ goes to zero, and time runs infinitely slowly. For someone remaining stationary just outside the event horizon, time can be slowed by any factor. In principle, then, if you could spend some time just outside the Schwarzschild radius of a black hole and then travel away from the black hole's vicinity, you might find that hundreds or even millions of years passed in the rest of the universe!

The time dilation formula given above seems to suggest some kind of impossible time flow inside the Schwarzschild radius, where the formula contains the square root of a negative number. However, the formula cannot be applied, because general relativity does not allow a stationary observer inside the black hole. Space is flowing inward at greater than the speed of light inside the event horizon, so everything is carried down to the singularity. Calculations show that for the unfortunate observer falling into a black hole, time would continue to run all too rapidly, and the observer would be crushed into the singularity in a finite time.

Besides having a gravitational field detectable beyond the Schwarzschild radius, black holes may also have an electric charge if, for example, an excess of positive charges falls into them. They may also have a spin—which will change the shape of the event horizon to be fatter at the equator than at the poles and alters the way light flows around the black hole.

X-ray binary system

69.3 OBSERVING BLACK HOLES

An object that emits no light or other electromagnetic radiation is not easy to observe. But just as you can "see" the wind by its effect on leaves and dust, so too astronomers can see black holes by their effects on their surroundings.

Suppose a massive star in a binary system undergoes core collapse, directly forming a 10-$M_\odot$ black hole. Gas from the companion star may be drawn toward the hole by its gravity, just as we know happens for neutron stars and white dwarfs that have companions. The infalling matter swirls around the black hole and forms an **accretion disk** whose inner edge lies just outside the Schwarzschild radius, as depicted in Figure 69.4A. Near the Schwarzschild radius, where the disk orbits at nearly the speed of light, turbulence and friction heat the swirling gas to a furious 10 million K, causing it to emit X-rays and gamma rays.

As the black hole's companion star orbits it, the X-ray-emitting gas may disappear from our view as it is eclipsed by the companion star. An X-ray telescope trained on such a star system will show a steady X-ray signal that disappears at each eclipse, as shown in Figure 69.4B. Such a signal might be the sign of a black hole, and at least three cosmic X-ray sources bear that signature.

But how do we know an X-ray source is not just a neutron star? The answer involves the escape velocities we calculated earlier. If X-ray-emitting gas surrounds

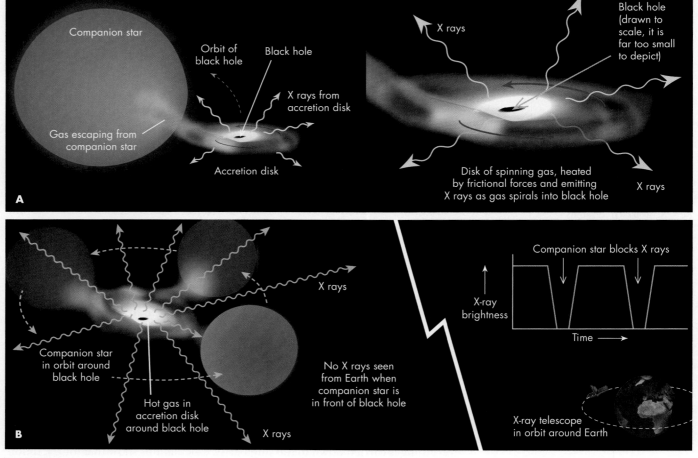

FIGURE 69.4

(A) Black holes may reveal themselves by the X rays emitted by gas orbiting them in an accretion disk. (B) The X-ray emission from gas around a black hole is sometimes eclipsed by its companion star. Sizes and separations are not to scale.

a body we cannot see, but whose mass exceeds 3 $M_\odot$, we can be reasonably confident that the invisible object is a black hole. This is because if more than 3 $M_\odot$ of neutron-density matter is packed together, its radius will be smaller than the Schwarzschild radius and it therefore must become a black hole. No special transition is necessary for the neutron matter to become a black hole, simply enough of it has to be collected in one place.

Astronomers have good evidence for several such black holes. For example, Cygnus X-1, the first X-ray source detected in the constellation Cygnus, consists of a B supergiant star and an invisible companion whose mass—based on an application of the modified form of Kepler's third law—is estimated to be between 10 and 15 $M_\odot$. Nothing we know of but a black hole can be so massive and yet invisible. Astronomers have detected more than a dozen other X-ray sources that appear to be black holes in binary systems.

The location of Cygnus X-1 is shown in Looking Up #4 at the front of the book.

69.4 HAWKING RADIATION

Up to this point, we have described black holes as intrinsically black. Although no radiation can leave the interior of a black hole, the English mathematical physicist and cosmologist Stephen Hawking showed in 1974 that black holes must nevertheless produce radiation. As so often happens in science, others were working on similar ideas. For example, in the previous year, the physicist Jacob Bekenstein, who was studying the relation between gravity and thermodynamics, noted that black holes can be assigned a temperature. But Hawking went further and showed that a black hole emits blackbody radiation, such as we discussed in Unit 23. In fact, we can use Wien's law, which relates a body's temperature to the wavelength at which it radiates most strongly, to estimate the temperature of a black hole.

Hawking and Bekenstein calculated that the wavelength of maximum emission is approximately 16 times the black hole's Schwarzschild radius (R_S). Wien's law states that an object's temperature is inversely proportional to the wavelength of the radiation it emits, and so the temperature of a black hole is equal to a constant divided by R_S. For a solar-mass black hole, this turns out to be about 6×10^{-8} K. This is very cold, but it is not absolute zero. Accordingly, a black hole, like any other body whose temperature is not absolute zero, emits energy in the form of electromagnetic waves, energy now known as **Hawking radiation**.

Hawking radiation is created by quantum physical processes near the event horizon of a black hole. Quantum processes allow tiny fluctuations in energy, everywhere in space, in the form of pairs of photons, or a particle and its antiparticle (such as an electron and a positron—see Unit 4). These pairs are called **virtual particles** because they exist as a tiny energy fluctuation for only a minuscule fraction of a second. Conservation of energy requires them to rapidly recombine, annihilating each other and returning the energy they have momentarily "borrowed." However, if one of these fluctuations occurs near an event horizon, one of the pair may cross the event horizon, leaving the other free to travel away, carrying away energy (Figure 69.5). Because the only source of energy available to a black hole is its mass, as a black hole "shines," its mass must decrease according to $m = E/c^2$. In other words, black holes must eventually "evaporate." However, the time it takes for a solar-mass black hole to disappear by "shining itself away" is very long: approximately 10^{67} years! This is immensely long—vastly longer than the current age of the Universe—but the implications are important: even black holes evolve and die. For black holes of stellar mass, the long-wavelength Hawking radiation they emit is far too weak to be detected by current telescopes, but if much smaller black holes exist, as we consider in the next section, they would self-destruct in a blaze of high-energy radiation that might be detectable.

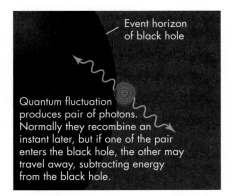

FIGURE 69.5

Quantum fluctuations near the event horizon of a black hole can create pairs of virtual particles that gradually subtract energy (and therefore mass) from a black hole.

69.5 SMALL AND LARGE BLACK HOLES

We have seen that massive stars can become black holes as a natural part of their evolution. This happens when matter of several solar masses is compressed to a volume with a radius of less than 10 kilometers.

Other objects can, in principle, also become black holes if compressed to a small enough size. The Earth ($M_\oplus = 5.97 \times 10^{24}$ kg) could be turned into a black hole if it could be compressed to a radius

$$R_S(\text{Earth}) = \frac{2 \times 6.67 \times 10^{-11} \times 5.97 \times 10^{24}}{(3.00 \times 10^8)^2} = 0.0088 \text{ m} = 8.8 \text{ mm}.$$

Thus, an Earth-mass black hole would have a radius of just under 1 centimeter—about the size of a marble! It would require forces and circumstances far beyond anything we think exists in the universe to turn the Earth into a black hole. Nonetheless, some astronomers hypothesize that early in the history of the universe forces might have existed that could have formed tiny black holes that survive to the present. Some of these might even be detectable from their Hawking radiation. At the other extreme, there is strong evidence that galaxies harbor giant black holes with masses up to billions of times greater than the Sun's, as we will explore in Unit 78.

If we calculate the density of an Earth-mass black hole, it is fantastically high, more than 10^{27} kilograms per liter. This is about one hundred billion times more dense than the density of a stellar-mass black hole. On the other hand, the density of a black hole with a mass of about 100 million solar masses is close to just 1 kilogram per liter—about the density of water. This illustrates that black holes are not always dense. If we could somehow gather enough water from interstellar space to fill the inner part of the Solar System out to Mars's orbit, we would create a giant black hole. The enormous mass of water would collapse in on itself without requiring any special events to squeeze it to higher density.

Another difference between low- and high-mass black holes is the difference in the tidal forces (Unit 19) near their event horizons. The strength of tidal forces depends on the fractional differences in distance to an object, because gravity grows stronger as separations become smaller. This is why the nearby Moon has a larger tidal effect on Earth than the more distant Sun, despite the Sun's larger gravitational pull on the Earth. Likewise, the tidal effects across 1 meter of a spaceship near a small black hole will be larger than the tidal effects outside a giant black hole. Thus, a space probe (or an astronaut) passing near an Earth-mass black hole would experience so much larger a gravitational force on the side nearer the black hole than on the far side that it would likely be torn apart. On the other hand, the tidal forces near a giant black hole at the center of a galaxy would be relatively gentle across the ship. In fact, a probe could probably fall into a giant black hole without being destroyed until it approached the singularity deep within the event horizon. However, it still could never report back to the rest of the universe what it saw.

Concept Question 3

The density of a black hole with the mass of the visible universe is just a few hydrogen atoms per cubic meter of space. Could we be living in a universe that is a giant black hole?

KEY POINTS

- Black holes are objects that have an escape velocity greater than the speed of light, at the Schwarzschild radius.
- General relativity shows that the curvature at the Schwarzschild radius prevents any motion outward, even that of light.
- Because space has an inward motion greater than the speed of light inside the event horizon, everything must move inward to the singularity.
- Gravity far from a black hole is no different from that of any other body of the same mass.
- The gravitational time dilation for a stationary object grows infinite at the Schwarzschild radius.
- High-mass stars are thought to end as neutron stars of several solar masses, which are smaller than their Schwarzschild radius.
- Matter falling into a black hole may collide violently outside it, generating high-energy radiation.
- Several binary star systems contain stars with probable black hole companions detected by X-ray radiation from accreting material.
- Black holes can break up pairs of virtual particles, generating radiation that causes a black hole to gradually lose mass.
- Small and large black holes are possibilities, with the density and tidal effects growing smaller the more massive the black hole.

KEY TERMS

accretion disk, 544
black hole, 539
curvature of space, 541
escape velocity, 539
event horizon, 543
general relativity, 539

gravitational time dilation, 543
Hawking radiation, 545
Schwarzschild radius, 540
singularity, 541
virtual particle, 545

CONCEPT QUESTIONS

Concept Questions on the following topics are located in the margins. They invite thinking and discussion beyond the text.

1. Risk of colliding with a black hole versus a star. (p. 543)
2. Time rate differences outside a black hole. (p. 543)
3. Could the universe be a black hole. (p. 546)

REVIEW QUESTIONS

4. What is a black hole? Why is Newton's law of gravity insufficient to explain black holes?
5. What is the Schwarzschild radius? Why is it also called the "event horizon"?
6. Stars of what mass are expected to become black holes? What mass of neutron stars must turn into black holes?
7. What would you observe if you watched a spaceship travel into a black hole?
8. What happens to time near the event horizon of a black hole?
9. What observational evidence is there of black holes?
10. What is Hawking radiation? What are virtual particles?

QUANTITATIVE PROBLEMS

11. Calculate the Schwarzschild radius of a 3-$M_\odot$ object.
    a. The orbital period and separation of the Cygnus X-1 binary system are 5.6 days and 0.2 AU respectably. If the visible B supergiant has a mass of 19 $M_\odot$, what is the mass of the unseen companion?
    b. What is the Schwarzschild radius of the invisible companion?
12. Calculate *your* Schwarzschild radius. What in nature has a comparable size?
13. Calculate the radius of a billion-$M_\odot$ black hole at the center of a galaxy. Calculate its density in kilograms per liter. (*Reminder:* There are 1000 liters in a cubic meter.) Use $3/4\, \pi R_S^3$ as the volume of the black hole.
14. If a neutron star is 10% bigger than its Schwarzschild radius, how slowly would a clock run on its surface?
15. Suppose you observe two friends. One takes her spaceship to an orbit around a black hole a distance of 1.1 R_S from the singularity (i.e., 10% farther away from the singularity than the event horizon is). The other moves in a straight line at 0.1 c. Which friend has the greatest time-stretching factor (as observed by you)? See Unit 26 for the special relativity Lorentz factor.
16. The evaporation time for a black hole can be calculated from $\tau_{ev} = 2.67 \times 10^{-24}\, M^3$, where M is the mass of the black hole expressed in kilograms and the resulting time is in years. Calculate the evaporation time for black holes with masses equal to (a) the Sun, (b) Earth, and (c) a medium size asteroid, 10^{15} kg. Which of these black holes have an evaporation time shorter than the current age of the Universe (13.8 billion years)?
17. Neutron-density matter is close to being incompressible. Assuming that it is, we can calculate how big a mass of neutron-density matter turns into a black hole.
    a. Calculate the density ρ of a 1.4-$M_\odot$ neutron star, which has a radius of 7 km.
    b. The mass of a neutron star is equal to its volume times its density. Using the density you found in (a), multiply it by the volume $(3/4\, \pi R^3)$ and use the Schwarzschild radius formula to determine the radius when $R = R_S$.
    c. Use the radius you found in (b) to calculate the mass of this neutron star.

TEST YOURSELF

18. What evidence leads astronomers to believe that they have detected black holes? They have detected
    a. eclipses of X-ray accretion disks in massive binary systems.
    b. bursts of radiation from the merger of two black holes.
    c. ultraviolet pulses coming from dark regions of space.
    d. stars suddenly disappear when swallowed by a black hole.
    e. Hawking radiation from nearby black hole candidates.
19. The Schwarzschild radius of a body is the distance from its center at which
    a. nuclear fusion ceases in the core of a star.
    b. an orbiting companion will be broken apart.
    c. a white dwarf grows before it collapses.
    d. a neutron star begins ejecting material in a supernova.
    e. its escape velocity equals the speed of light.
20. Which of the following properties must have a large value for all black holes?
    a. Density
    b. Mass
    c. Escape velocity
    d. All of these.
21. If by some unknown process the Sun suddenly collapsed in on itself and became a black hole tomorrow, the planets would
    a. fly off into space in the direction they were already moving.
    b. be dragged inward and sucked into the black hole.
    c. be blasted to pieces by neutrinos emitted by the black hole.
    d. continue orbiting the black hole in the same orbits.
    e. also collapse into black holes because of the strong gravity.

UNIT 70
Star Clusters

70.1 Types of Star Clusters
70.2 Testing Stellar Evolution Theory
70.3 The Initial Mass Function

Learning Objectives

Upon completing this Unit, you should be able to:
- Define and contrast the different classes of star clusters and associations.
- Explain how and why the H-R diagram for different star clusters will differ in appearance depending on the age of the cluster; define the turnoff point.
- Describe the initial mass function, and explain why the Sun is less luminous than nearly all stars we see at night, yet is more luminous than the vast majority of stars.

Groupings of stars called **star clusters** provide a unique environment to test the ideas about stellar evolution outlined in Unit 60. They also provide us with an important transition toward understanding the behavior and dynamics of the much-larger collections of stars known as galaxies, which we examine beginning in Unit 71.

Stars usually form in clusters out of huge clouds of interstellar gas (Unit 61). When a cloud containing thousands or even millions of solar masses of matter begins to contract under its own gravity, smaller clumps begin to form. Each clump forms a star or system of stars, and possibly planets as well. Some clumps are disrupted by the firstborn stars before they too can form a star, but typically hundreds or thousands of stars are born almost simultaneously out of nearly identical material, and we can trace their evolution using an H-R diagram (Unit 59).

Each star in a cluster moves along its own orbit about the center of mass of the cluster, held to the cluster by the gravitational attraction of all the other stars. Many clusters are loosely bound, and after the remnants of the original gas cloud are driven away—by stellar winds, radiation pressure, and exploding stars—the stars may drift apart. Because the stars within a cluster form at approximately the same time, they are all nearly the same age. This allows us to determine what kinds of stars die more quickly than others, and we can learn about the relative numbers of stars of each type that normally form.

70.1 TYPES OF STAR CLUSTERS

Star clusters range from loose associations of tens of stars to dense concentrations of millions of stars. The spacing between the stars in a cluster varies enormously. In some clusters, the stars are loosely packed, so the cluster is only slightly denser than its surroundings. In other clusters, however, the stars are so closely spaced that their separation may be as little as a tenth of a light-year. Yet even in such dense star clusters, the spacing between stars is still vast compared with the sizes of stars, similar to grains of salt separated by football fields. Although the stars in a cluster sometimes look quite crowded in a photograph, the likelihood of a physical collision between them is extremely low.

The star cluster nearest to us is about 75 light-years away and is about 30 light-years across. It is so close that it covers a large portion of the sky. Called the **Ursa Major group,** it includes most of the stars of the Big Dipper and a number of

The Big Dipper is shown in Looking Up #2 at the front of the book.

FIGURE 70.1
Open clusters shown in order of age. NGC 3293 in Carina is estimated to be about 10 million years old. The Pleiades in the constellation Taurus are about 100 million years old. M67 in Cancer is roughly 4 billion years old.

LOOKING UP

The Pleiades and Hyades are visible in Looking Up #5 at the front of the book. Other open clusters are shown in Looking Up #1, #3, and #8.

Clarification Point

Note that the brightest stars in the Pleiades form almost a miniature "dipper" shape. They are sometimes confused with the "Little Dipper" of the constellation Ursa Minor.

stars from neighboring constellations. About a hundred stars have been identified as members of this group by their common location and motion through space, including the brightest star in the constellation Corona Borealis and the second brightest in Auriga.

With your unaided eyes you can see several clusters that are richer than the Ursa Major group. One of the most obvious is the **Pleiades** (Figure 70.1 middle), or the "Seven Sisters," named for the daughters of the giant Atlas, who in Greek mythology carried the world on his shoulders. The Pleiades (pronounced *PLEE-ah-deez*) are visible as a tiny group of stars north of the V in the constellation Taurus. The V in Taurus is another cluster of stars called the Hyades (pronounced *HI-ah-deez*), named for the half-sisters of the Pleiades in mythology. The two clusters both happen to lie in the same constellation, but they are not otherwise associated. The Hyades are much closer to us—about 150 light-years away—while the Pleiades are 440 light-years distant. These clusters are highest in the evening sky in December and January. With binoculars you can see that the Pleiades and Hyades each contain hundreds of stars—far more than the half dozen or so stars visible to the unaided eye.

More than a thousand other **open clusters** are cataloged in our Galaxy; they contain up to a few thousand stars in a volume with a radius of typically 7 to 20 light-years. They are called *open* because their stars are scattered loosely, as you can see in the sampling of other open clusters shown in Figure 70.1. Astronomers think that open clusters form when giant, cold interstellar gas clouds are compressed and collapse, breaking up into hundreds of stars whose mutual gravity binds them into the cluster. Once formed, the stars of an open cluster continue to move through space together; but over hundreds of millions of years the stars gradually drift off on their own, so the cluster eventually dissolves. The high number of blue stars in the Pleiades suggests that this cluster is fairly young, probably about 100 million years old. The Hyades and the Ursa Major group, by contrast, are both estimated to be many hundreds of millions of years old, and they appear to be less tightly bound than the Pleiades. Our own Sun was probably a member of such a star group, but its companion stars have long since scattered across space.

Other stars sometimes occur in loose groups called **stellar associations** that are a few hundred light-years across. Associations typically spread out from a single large open cluster near their center and may contain other, smaller star groupings. Moreover, the stars in associations are usually still mingled with the massive clouds of dust and gas from which they formed. The stars in associations probably form

FIGURE 70.2
Examples of globular clusters. Omega Centauri in the constellation Centaurus and M13 in Hercules are each about a million solar masses.

Clarification Point

Although stars appear very close together in photographs of globular clusters, the spacing between stars is still millions of times larger than their diameters.

LOOKING UP

Omega Centauri is visible in Looking Up #8 at the front of the book. Another globular cluster is shown in Looking Up #7.

at about the same time, perhaps from the same triggering event. They have at most a very weak gravitational link to one another and often have different motions, so they disperse rapidly.

A far denser type of cluster, called a **globular cluster,** is much more strongly bound together, and is larger in size and stellar content than an open cluster. These clusters contain from a few hundred thousand to several million stars and have radii of about 40 to 160 light-years. The stronger gravity in globular clusters pulls their stars into a dense ball, as you can see in Figure 70.2. About 150 globular clusters have been cataloged. Although they are part of our Galaxy, most of them are so distant that despite the many stars they contain, they are not easily seen without a telescope. A notable exception to this is the globular cluster Omega Centauri in the southern constellation Centaurus (Figure 70.2, left). Cataloged as the 24th brightest "star" in the constellation, it is clearly fuzzy even to the naked eye. In northern skies in the constellation Hercules, a globular cluster cataloged as M13 can sometimes be seen by eye as a faint fuzzy patch (Figure 70.2, right). A small telescope shows both of these clusters to be a large swarm of stars.

Table 70.1 summarizes some properties of open clusters, globular clusters, and stellar associations.

Concept Question 1

What would the sky look like if you lived on a planet orbiting a star within a globular cluster?

TABLE 70.1 Properties of Clusters and Associations

Type	Number of Stars	Radius*
Open cluster	Tens to a few thousand	7–20 ly (2–6 pc)
Globular cluster	10^5–10^6	40–160 ly (12–50 pc)
Associations[†]	5–70 O or B stars	65–325 ly (20–100 pc)

*Because star clusters do not have sharp edges but instead gradually thin out from their center, quoted dimensions differ substantially.
[†]Astronomers identify several other types of associations as well. For example, T associations are regions with above-average numbers of T Tauri stars.

70.2 TESTING STELLAR EVOLUTION THEORY

Because the stars in a given cluster are approximately the same age, a cluster's H-R diagram shows a snapshot of the state of evolution of its stars. For example, in a young cluster, we expect that more-massive stars will have completed their contraction stage from protostars and will lie on the main sequence, and that no stars will have had time to become red giants and shift off the main sequence. If we look at an older cluster, however, some massive stars will have consumed their core hydrogen and evolved off the main sequence. We can use such differences between cluster H-R diagrams to check our theory of stellar evolution. We can do this by calculating evolutionary tracks (such as those in Figures 65.2 and 67.1) for every star on the main sequence to show where each star will be at 10 million years, 100 million years, and so on. The resulting curves show us what the H-R diagram of the entire cluster should look like at each of these times after its birth.

Conversely, astronomers can deduce the age of a star cluster from the pattern of its H-R diagram. Recall that the main sequence is determined by the location of stars fusing hydrogen in their cores. All the stars in a newly formed cluster lie on or near the main sequence. However, massive stars use up their fuel more rapidly (to maintain their high luminosity) than low-mass stars do. With their hydrogen used up, high-mass stars leave the main sequence and turn into red supergiants. The low-mass stars, on the other hand, still have hydrogen to fuse, so they remain on the main sequence longer.

Because the high-mass stars in an older cluster have moved off the main sequence, a line in the H-R diagram connecting the position of the cluster's stars bends away to the right, as shown in Figure 70.3. The point where that line bends away from the main sequence is called the **turnoff point.** A star just below the turnoff point is about to run out of hydrogen. But all stars in the cluster are the same age—namely, the age of the cluster. To get the cluster's age, we therefore determine how long a star at the turnoff point can live by calculating its main-sequence lifetime from its mass or luminosity, as derived in Unit 62. The answer we get is the age of the star cluster.

FIGURE 70.3

The pattern of stars in the H-R diagram of a star cluster indicates its age. Low-mass stars can take many millions of years to reach the main sequence, while the most massive stars begin evolving off the main sequence after some tens of millions of years.

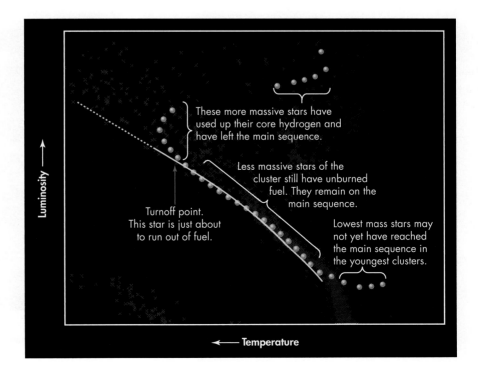

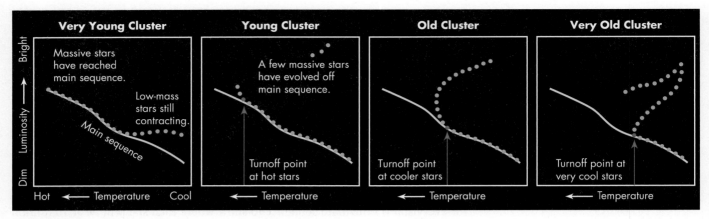

FIGURE 70.4
Schematic H-R diagrams of clusters of different ages illustrating the turnoff point and how it shifts down and to the right for older clusters.

Figure 70.4 shows a series of schematic H-R diagrams for clusters ranging from very young to very old (a few million years for the youngest to more than 10 billion years for the oldest). Notice that old clusters have few, if any, stars on the upper part of the main sequence. On the other hand, short-lived stars are still present on the upper main sequence of young clusters.

When we compare the curves generated by computer models with the H-R diagrams of actual star clusters, as in Figure 70.5, the match is excellent. The examples in the figure are very similar to the first and last panels of Figure 70.4, giving estimated ages of 10 million and 7 billion years for the open cluster and globular cluster, respectively. If our theory of stellar evolution were wrong, the shapes would be unlikely to agree so well. The models not only offer a natural explanation of properties of main-sequence, red giant, and white dwarf stars; they are sophisticated enough to explain how pulsating variables, planetary nebulae, and supernovae fit into the overall scheme of stellar evolution. This success in interpreting such stellar diversity is evidence that our theory of stellar evolution is essentially correct. This success gives astronomers confidence in the estimated ages of a star cluster.

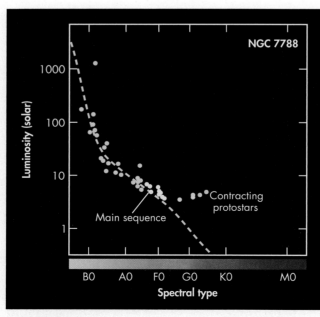
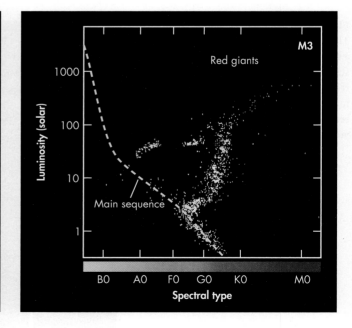

FIGURE 70.5
The H-R diagrams of two clusters. NGC 7788 is an open cluster in Cassiopeia with an estimated age of about 10 million years. M3 is a globular cluster in Canes Venatici with an estimated age of 7 billion years.

70.3 THE INITIAL MASS FUNCTION

If you studied animal life in a forest, your first impression might be that the main forms of life there are birds, deer, and other large animals. Closer examination would reveal smaller animals in burrows. A careful tally, however, would probably reveal at least 10 times more mass in insects than in larger animals. And you would probably find an even larger mass of microscopic "animals" than all the rest combined.

The stars we see at night turn out to be as unrepresentative of the general stellar population as large animals are to the overall animal population. When we look at the night sky, we see the stars whose brightness is great enough for us to detect their light. We primarily see the stars that are unusually luminous and that, therefore, stand out relative to the others. As a result, when we look at the night sky, the stars we see tend to be giants. This is demonstrated in Figure 70.6, which shows an H-R diagram of the stars bright enough to be seen without the aid of a telescope.

Giant stars are actually quite rare; but because we can see them out to very large distances, we see greater numbers of them than the far more numerous stars that are too dim to be visible. How bright a star looks to us is determined by its luminosity and distance according to the inverse-square law, $B = L/4\pi d^2$ (Unit 55). From this law we can see that if star A is a million times more luminous than star B, but star B is a thousand times closer than star A, they will appear to have the same brightness because $1000^2 = 1,000,000$ (see mathematical derivation in margin at left). So according to the inverse-square law, star A will appear just as bright

Mathematical Insights

If star A has a luminosity a million times greater than star B, then $L_A = 10^6 \times L_B$. If star B is a thousand times closer than star A, then $d_A = 10^3 \times d_B$. Comparing these, we find

$$B_A = \frac{L_A}{4\pi d_A^2}$$

$$= \frac{10^6 \times L_B}{4\pi(10^3 \times d_B)^2}$$

$$= \frac{10^6 \times L_B}{10^6 \times 4\pi d_B^2}$$

$$= \frac{L_B}{4\pi d_B^2} = B_B.$$

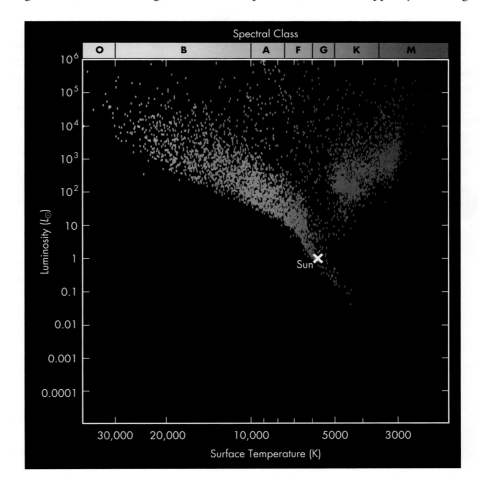

FIGURE 70.6
The H-R diagram of the stars that are visible to the unaided eye in the night sky. Nearly all are more luminous than the Sun.

> **Concept Question 2**
>
> Can you think of everyday situations in which choosing people based on one set of criteria might give you an erroneous sense of the characteristics of the whole population?

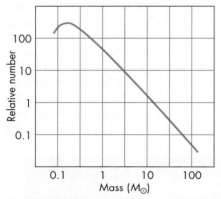

FIGURE 70.7
The initial mass function of stars indicates the relative number of stars of each mass. Studies of star clusters indicate that there are *many* more low-mass stars than high-mass stars, as indicated by the curve.

Proxima Centauri can be found in Looking Up #8 at the front of the book.

to us even if it is a thousand times farther away than star B. In other words, great luminosity can compensate for sizable distance.

Many giant stars are *more* than a million times more luminous than the lowest-luminosity stars. As a result, when we see a set of stars that looks bright to us, we are looking at the most luminous stars out to distances more than a thousand times greater than those of the low-luminosity stars. This biases what stars we are able to see heavily in favor of luminous stars. Astronomers call this problem of sample biasing a **selection effect.** This is illustrated in Figure 70.6, which shows an H-R diagram of all the stars in the sky that are visible to the unaided eye. Notice that the Sun is one of the least luminous among all of these stars.

When astronomers observe a star cluster, they avoid this problem because all of the stars are at (nearly) the same distance. If we can count all the stars within the cluster, we have an unbiased sample. The relative numbers of stars of different luminosities in the sample will then indicate the true proportion of stars of each luminosity. If we use the mass–luminosity relation (Unit 59) to deduce the masses of the main-sequence stars from their luminosities, we can estimate how many stars of each mass were born when the cluster formed. Astronomers call such a census of a cluster's stars its **initial mass function.**

Studies of star clusters show that stars similar to or smaller than the Sun vastly outnumber more massive stars, as shown by the initial mass function plotted in Figure 70.7. The most numerous stars turn out to be dim, cool, red stars. These stars lie on the main sequence but have a mass that is less than 0.5 $M_\odot$. Astronomers do not yet understand what determines the distribution of stellar masses, but it appears to be similar in most clusters, so it must be a property of how a collapsing interstellar cloud breaks up into stars.

The initial mass function found from these studies indicates that the number of stars having a certain mass increases sharply as mass decreases. Typically there are about 20 times more stars with masses between 1 $M_\odot$ and 2 $M_\odot$ than between 10 $M_\odot$ and 20 $M_\odot$. The relative numbers of stars of different masses are also suggested by Figure 70.8, which shows an H-R diagram for all the stars currently known within 10 parsecs (32.6 light-years) of the Sun.

Despite their great numbers, low-mass stars are hard to see because they are so intrinsically dim. Not a single main-sequence M-type star is visible to the unaided eye, even though these are by far the most common type of star. Even the star nearest to the Solar System, the M-type star Proxima Centauri, is more than 10,000 times too dim to be seen with the unaided eye. The initial mass function suggests that M-type stars (masses ranging from 0.08 $M_\odot$ to 0.48 $M_\odot$) outnumber all other types of stars combined by a factor of 10. In fact, the combined mass of all stars of this dim spectral type is approximately half the combined mass of all types of stars.

Objects less massive than about 0.08 $M_\odot$ were once also believed to be rare, but astronomers now think they may be common and were simply missed in previous stellar surveys. How could astronomers miss these objects if they are so abundant? Recall that low-mass stars tend to be cool. In fact, if a contracting mass of gas is less massive than about 0.08 $M_\odot$, it is so cool that it is unable to fuse hydrogen into helium and is termed a *brown dwarf* (Unit 61). These are in some ways more like huge planets than tiny stars, and their extreme dimness and small size make them difficult to detect. However, searching the sky at infrared wavelengths at which brown dwarfs are brightest greatly increases the chance of finding them. Studies of the Pleiades and other young open clusters at infrared wavelengths have revealed many probable brown dwarfs. Astronomers now think brown dwarfs may outnumber ordinary stars, making them one of the most common astronomical objects in our Galaxy, although it appears that they do not contain as much mass in total as is contained in ordinary stars.

FIGURE 70.8
The H-R diagram of the known stars within 10 pc of the Sun. Few of the nearest stars are more luminous than the Sun.

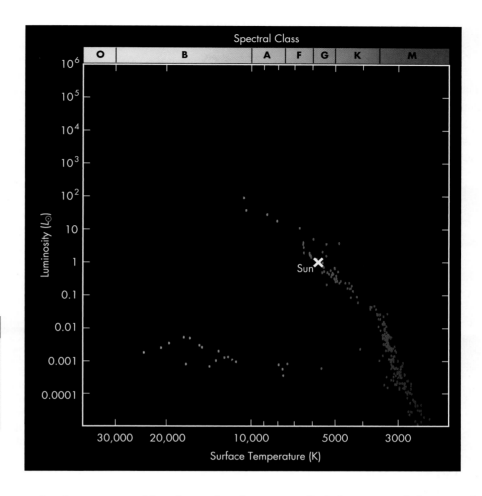

Concept Question 3

What does the number of white dwarfs in Figure 70.8 versus the number of red giants suggest about the relative duration of these stages of stellar evolution?

In almost any unbiased sample of stars, we find that most of the mass is contained in the lowest-mass objects, yet most of the light comes from the highest-mass objects. In this wide range of properties, the Sun is sometimes called a "typical" star. It is true that the Sun sits somewhere in the middle of the range of stellar properties, yet it is typical neither in mass nor in luminosity. The Sun is more massive than about 97% of all stars, yet more than 99% of the total luminosity in a typical collection of stars is produced by stars more massive than the Sun. However, when we add up the total mass and the total luminosity of stars in the neighborhood of the Sun, it happens that the ratio of mass to luminosity is similar to that of the Sun. In other words, if a thousand solar masses are contained in the collection of stars, then this star collection will generate about a thousand solar luminosities of light. Hence, the mass and luminosity of the stars in our neighborhood balance in a way that makes the Sun seem fairly average.

KEY POINTS

- Star clusters range from associations of a few dozen stars, to open clusters of thousands of stars, to globular clusters of millions.
- Globular clusters have the highest density of stars, but the spacing between stars remains very large compared to the size of stars.
- The stars in clusters are thought to have formed nearly simultaneously as a cloud of gas collapsed and fragmented into clumps.
- By examining the pattern of stars in an H-R diagram, it is possible to estimate the cluster age, particularly from the most massive star still remaining on the main sequence.
- The most luminous stars can be seen to far greater distances than dim stars, so star samples tend to be biased toward including the more-luminous varieties and overlooking numerous dim ones.
- Low-mass stars are much more commonly formed than high-mass stars, and provide most of the mass of stellar populations.
- The Sun is one of least luminous stars that we can see with the unaided eye, but it is one of the most luminous among nearby stars.

KEY TERMS

globular cluster, 550
initial mass function, 554
open cluster, 549
Pleiades, 549
selection effect, 554
star cluster, 548
stellar association, 549
turnoff point, 551
Ursa Major group, 548

CONCEPT QUESTIONS

Concept Questions on the following topics are located in the margins. They invite thinking and discussion beyond the text.

1. Appearance of sky from within a globular cluster. (p. 550)
2. Biased samples of a human population. (p. 554)
3. Interpreting populated areas of an H-R diagram. (p. 555)

REVIEW QUESTIONS

4. How do star clusters form?
5. What are the different types of star clusters?
6. How are star clusters useful for testing stellar evolutionary theory?
7. What kind of cluster did our Sun form in? Do we know what other stars were born in this cluster?
8. Why does the H-R diagram of a cluster have a "turnoff point"? Why don't we see a turnoff point when we make an H-R diagram of all the stars we see in the night sky?
9. Comparing H-R diagrams for clusters that are 10^8, 10^9, and 10^{10} years old, what parts of the diagrams look similar? What parts differ, and what causes the differences?
10. Is the Sun a typical star in terms of mass? in terms of luminosity? Why is it called a typical star?

QUANTITATIVE PROBLEMS

11. The globular cluster Omega Centauri looks like a "fuzzy star" to the unaided eye, and is about 100 times less bright than the brightest star in the sky. It is approximately 16,000 light-years distant. At what distance would it be as bright as the brightest star in the sky? At what distance would it appear as bright as the full moon, which is about 4 million times brighter than Omega Centauri at its present distance? What problems are there with this result?

12. We can see associations of O and B stars in the Andromeda Galaxy, 2 million ly away. What would be the brightness of an association of 70 O stars at this distance?
13. The Hyades is 150 ly away from Earth, and is 75 ly in diameter. The globular cluster Omega Centauri is 18,000 ly away from Earth, and is 200 ly in diameter. What are the angular diameters of these two objects as seen from Earth?
14. A globular cluster contains no main-sequence stars with surface temperatures higher than 5000 K and luminosities greater than 0.80 $L_\odot$. How old is the cluster?
15. If a globular cluster contains 1 million stars in a volume with a radius of 100 ly, what is the average separation between stars? Calculate this by assuming that each star fills one millionth of the total volume of the cluster. The diameter of the sphere it occupies is then the average separation. (The volume of a sphere is $3/4 \pi R^3$.)
16. Estimate the escape velocity for an open cluster that contains 1,000 stars and is 6 pc in radius. For this calculation, make the assumption that each star is like our Sun. How does this value compare to Earth's escape velocity (11.2 km/sec)?

TEST YOURSELF

17. The most numerous type of star has
    a. less than half the Sun's mass.
    b. about the Sun's mass.
    c. about twice the Sun's mass.
    d. more than 20 times the Sun's mass.
    e. none of the above. The stars are fairly evenly distributed as far as mass goes.
18. When looking at the night sky with our unaided eyes, most of the stars we see are
    a. brown dwarfs.
    b. lower in mass than the Sun.
    c. about equal in mass to the Sun.
    d. more luminous than the Sun.
    e. young, hot O stars.
19. The turnoff point for a star cluster is used to estimate
    a. the initial mass of the interstellar material that formed the cluster.
    b. the age of the cluster.
    c. the direction the cluster is moving through space.
    d. the most massive star that could form in the cluster.
20. Astronomers study star clusters in order to understand how stars evolve because the stars all have similar
    a. ages and distances.
    b. masses and ages.
    c. compositions and masses.
    d. distances and temperatures.
    e. temperatures and compositions.

UNIT 71

Discovering the Milky Way

71.1 The Shape of the Milky Way
71.2 Star Counts and the Size of the Galaxy
71.3 Globular Clusters and the Size of the Galaxy
71.4 Galactic Structure and Contents

Learning Objectives

Upon completing this Unit, you should be able to:
- Recount the history of how we came to determine our Galaxy's shape and size.
- Describe the appearance of the Milky Way on the sky, its structure as presently known, and how the two relate.
- Explain how globular clusters were used to find the distance to the Galaxy's center, and why this value disagreed with star-count estimates.

On a clear, moonless night, far from city lights, you can see a pale band of light spangled with stars stretching across the sky (Figure 71.1). The ancient Hindus thought this shimmering river of light in the heavens was the source of the sacred river Ganges. To the ancient Greeks, this dim celestial glow looked like milk spilled across the night sky, so they called it the **Milky Way.** Astronomers also call this our **Galaxy,** from the Greek word for "milk" (*galactos*).

A view of the Milky Way on a clear, dark night is one of nature's finest spectacles. The band stretches in a full circle around us on the celestial sphere, but it is at a an angle of about 60° with respect to both the Earth's equator (the celestial equator) and to the plane of the Solar System (the ecliptic). As a result, when you observe the Milky Way it crosses the sky at different angles, depending on when or where you see it, and the Sun, Moon, and planets cross through it twice during each of their cycles around the sky.

Superimposed on the dim background glow are most of the bright stars and star clusters that we can see, which all belong to our Galaxy. Here and there dark blotches interrupt the glowing backdrop of stars, as you can see in Figure 71.1. The Incas of ancient Peru, who observed the Milky Way from their temple observatories in the Andes, gave these dark areas names, just as peoples of the classical world named the star groups. Today we know the dark regions are clouds of dust and gas that give birth to new stars.

Our Galaxy is an enormous system of stars—home to the Sun and hundreds of billions of other stars. It contains huge interstellar clouds, stars that are forming and stars that are dying. Yet the Galaxy is more than just a collection of all these other things. It has its own structure on a scale far larger than stars or solar systems. Understanding that structure provides clues to why and where stars form and the Sun's relationship to other stars.

FIGURE 71.1
Long-exposure photograph of the Milky Way. A portion of the Milky Way is visible above the horizon. The dark lane within the Milky Way is caused by dark dusty interstellar clouds.

FIGURE 71.2
Through a telescope the Milky Way is resolved into millions of stars. Several clusters of stars, some pink from the emission lines of hydrogen, are seen scattered across this visible-light image.

71.1 THE SHAPE OF THE MILKY WAY

When Galileo Galilei pointed a telescope at the Milky Way in 1609, he discovered that it contains millions of stars too dim to be seen as individual points of light with the unaided eye (Figure 71.2). This had been hypothesized 2000 years earlier by the Greek philosopher Democritus, who also proposed the idea that matter was made of atoms. In the twenty-first century we know that these stars, along with our Sun, form a huge, slowly revolving disk—our Galaxy.

Our understanding of the Milky Way as a star system developed further in the mid-1700s when Thomas Wright, an English astronomer, and Immanuel Kant, a German philosopher, independently suggested that the Milky Way is a flattened swarm of stars. They argued that if the Solar System were near the center of a spherical cloud of stars, we would see roughly the same number of stars in all directions. However, from inside a disk-shaped system, we would see vastly more stars in directions toward the outer edge of the disk than in directions perpendicular to it, as illustrated in Figure 71.3. This explains why the stars of the Milky Way appear to stretch around us in a great circle.

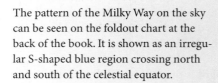

The pattern of the Milky Way on the sky can be seen on the foldout chart at the back of the book. It is shown as an irregular S-shaped blue region crossing north and south of the celestial equator.

An analogy may help you better understand this argument. Imagine making lemon JeLL-O® in a big, round flat dish. Imagine that you have spread blueberries throughout the JeLL-O®. If an ant fell into the gelatind and was clinging to a blueberry in it, the ant would see few berries if it looked up through the top of the gelatin or down through the bottom. However, it would see lots of blueberries if it looked in directions that lie in the plane of the dish. So too is it with us and stars in the disk of the Milky Way.

The Milky Way

FIGURE 71.3
If the Milky Way had a spherical shape, we would see about the same number of stars in every direction. However, because it is a disk, we see stars concentrated into a band around the sky.

71.2 STAR COUNTS AND THE SIZE OF THE GALAXY

The first quantitative attempts to determine the Milky Way's size and shape were made in the 1780s. William and Caroline Herschel seem an unlikely pair to have revolutionized astronomy, but this brother and sister team of musicians made a hobby of studying the skies. William (1738–1822) and Caroline (1750–1848) spent their free nights scanning the stars with high-quality telescopes built by William, discovering comets and the planet Uranus (Unit 46). Their growing fame won them support from the king of England, and they became full-time astronomers.

William decided to attempt to measure the shape of the Milky Way by observing hundreds of areas over the sky and counting the stars. He reasoned that the number of stars that he could see within his telescope's field of view would tell him the extent of the star system in that direction. When Caroline cataloged and indexed the results of William's counts, there were typically several hundred stars in each field throughout the band of the Milky Way, but on average only a few stars when looking at directions 90° away from it. William assumed that he could see the stars all the way to the edge of the Milky Way and that the stars were equally spaced on average. In that case, the number of stars should be larger in proportion to the volume encompassed by his telescope's field of view out to the edge of the Milky Way. Based on this, he produced the cross-sectional diagram shown in Figure 71.4. The overall size of this star system could be estimated by assuming that the other stars were like the Sun, and knowing how light dims with distance (Unit 55.2). This suggested that the Milky Way was a disk about 2500 parsecs in diameter, with the Sun near the center, and about one-fifth as thick as it was wide.

The left side of Herschel's diagram illustrates a problem with this method. The stars extend out a large distance, except in the middle, where their distribution looks like a pair of open alligator jaws. This odd shape is caused by a dusty interstellar cloud that blocks light in that direction—however, astronomers did not learn about interstellar clouds until 150 years later. William recognized another problem after he had built larger telescopes. He could see even more stars, so he realized his earlier observations had not seen to the edge of the Milky Way. Nevertheless, for more than 100 years this remained the best model of the star system in which we live.

It was not until the early 1900s that much more extensive studies gave us a clearer idea of the size of the Milky Way. By this time it was clear that stars did not all shine with the same luminosity as the Sun, and that stars had to be studied to much fainter limits to detect the "edge" of the Galaxy. The Dutch astronomer Jacobus C. Kapteyn (pronounced *CAP-tine*) carried out an extensive study along the lines of the Herschels', but with modern instruments, photography, and knowledge of the different types of stars. Because most of the stars were too distant to make direct parallax estimates, Kapteyn made his distance estimates for various types of stars by determining how much they appeared to shift their position on

Concept Question 1

If the Earth's axis pointed into the Milky Way, how would the Milky Way appear to change position as the Earth rotates? It the Earth's axis were tilted 90° with respect to the disk of the Milky Way, how would the Milky Way appear to move in our sky?

LOOKING UP

Large sections of the Milky Way are pictured in Looking Up #4, #7, and #8 at the front of the book. Regions obscured by dark clouds can be seen in each.

FIGURE 71.4
A copy of William Herschel's cross-sectional diagram of the Milky Way made in 1785. This was based on the number of stars Herschel could see in different directions. From this picture, he correctly deduced that the Milky Way is wider than it is thick. However, Herschel *incorrectly* concluded that the Sun (the orange dot) is near the Milky Way's center. He was led astray by his lack of knowledge that dust clouds blot out distant stars and prevent us from seeing our Galaxy's true extent.

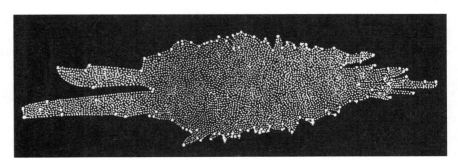

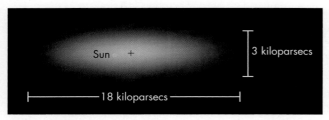

FIGURE 71.5
"Kapteyn's Universe," a model of the Milky Way as a roughly disk-shaped system of stars about 3 kpc thick, with the Sun near its center.

the sky over the course of many years (their *proper motion*—Unit 54.4). If two stars are moving through space at the same speed relative to the Sun, the one that is farther away will appear to have a smaller motion across the sky, just as a ball tossed past your head moves across your field of view more rapidly than one thrown at the far end of a playing field. This method avoids the problem of assuming that stars all have the same luminosity.

Kapteyn's resulting model (Figure 71.5) proposed a greatly revised size of the Milky Way, suggesting that it was 18,000 parsecs in diameter, again with the Sun fairly near the center. This model became known as **Kapteyn's Universe,** because at the time it was not known that there were other galaxies beyond the Milky Way. Galaxy dimensions are so huge that even parsecs are inconvenient for measuring their size, so astronomers often use **kiloparsecs** (kpc) for that purpose. One kiloparsec = 1000 parsecs = 3300 light-years. Thus, the Herschel model of the Milky Way was just 2.5 kpc in diameter, while the Kapteyn Universe was 18 kpc in diameter.

In both the Kapteyn model and the Herschel model, our Galaxy is depicted essentially as a disk of stars. Midway between the two faces of this disk is what astronomers call the **Galactic plane.** Within the disk, different kinds of stars are concentrated more or less tightly around the Galactic plane. Young stars, gas, and dust are found close to the plane, on average, but older stars span a much larger range of distances from the plane, extending "above" or "below" to several kiloparsecs.

71.3 GLOBULAR CLUSTERS AND THE SIZE OF THE GALAXY

At about the same time as Kapteyn was working, the U.S. astronomer Harlow Shapley argued that the Milky Way was far larger—about 100 kiloparsecs across—and that the Sun was not near the center but rather was about two-thirds of the way out in the disk. He based his claim on new ideas and a different set of objects.

To determine the Milky Way's size, Shapley used an approach entirely different from the methods used by the Herschels or Kapteyn. He studied the locations of **globular clusters** (Unit 70)—these are dense groupings of up to a million stars (an example of a globular cluster is shown in Figure 71.6). Because these clusters contain so many stars, they are very luminous and can be seen at large distances—across the Galaxy and beyond. Moreover, many of them have orbits that carry them far outside the Galactic plane, so we have a clear view of them above or below the dense disk of the Milky Way. Shapley argued that these massive star clusters must orbit the center of the Milky Way.

Shapley noticed that the globular clusters are *not* scattered uniformly across the whole sky but are concentrated in the direction where the Milky Way looks brightest to us, toward the constellation Sagittarius. He hypothesized that the middle of the Galaxy lay somewhere in the direction of Sagittarius, and the system of globular clusters was centered on it. By mapping where the clusters lay, he could deduce our distance from the center and therefore the size of the Milky Way.

To map the positions of the clusters, Shapley needed to estimate their distances from us. He did this by observing the variable stars in them (Unit 64). Certain kinds of variable stars have predictable luminosities, and globular clusters often contain stars of a class known as RR Lyrae variables. From the luminosity and apparent brightness of these stars, he could calculate their distance using the inverse-square law. The distance to the stars in a cluster gave the cluster's distance. With good estimates of the clusters' distances, Shapley plotted where they lay in the Milky Way. Figure 71.6 shows his results. The clusters fill a roughly elliptically shaped region,

LOOKING UP

The center of the Milky Way is in the direction of Sagittarius, shown in Looking Up #7 at the front of the book. Some objects that look like stars in this image are actually globular clusters.

For northern observers, Sagittarius is most easily seen above the southern horizon during summer evenings.

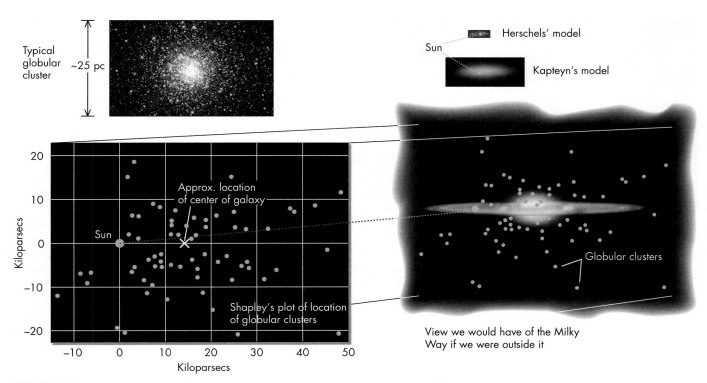

FIGURE 71.6
Schematic version of Shapley's plot of globular clusters, from which he inferred the size of the Milky Way and the Sun's location in it. Notice that the clusters fill a roughly elliptical region and that the Sun is *not* at the center of their distribution. An image of a typical globular cluster is shown at top, and Herschel's and Kapteyn's models are shown to the same scale as Shapley's model.

and the Sun lies *not* at the middle of the region, but about two-thirds of the way from its center. Shapley therefore concluded that our Galaxy is roughly 100 kiloparsecs in diameter and that the Sun is nearer to its edge than to its center.

The great difference between Shapley's and Kapteyn's findings about the Milky Way created a major controversy among astronomers. As in so many other controversies, both sides were partially right and partially wrong. However, Shapley's results about the Milky Way were closer to the truth. The main reason both results differed from what we know today is that neither recognized the effects of **interstellar dust**—small particles of solid matter that are found in deep space (Unit 73). Close to the Galactic plane, the dust is so dense that it blocks light from distant stars, which limited Kapteyn's observations to the portion of the Milky Way's disk that surrounds the Sun, making it appear that the Sun is near the center. Most globular clusters, on the other hand, are located far "above" or "below" the Galactic plane where there is little dust, so their light is dimmed only slightly. Still, this dimming makes them look a little more distant than they really are, so Shapley overestimated the distance to the globular clusters and deduced that our Galaxy is bigger than it really is. Several decades later, astronomers also discovered that variable stars of the class Shapley used to estimate the cluster distances were less luminous than he had assumed, which caused him to further overestimate the distances.

Modern estimates place the center of the Galaxy about 8 kiloparsecs from us in the direction of the constellation Sagittarius. Observations also indicate that the outer part of the Galaxy has no distinct edge—the density of stars steadily declines, becoming so sparse that it is difficult to detect stars at about twice the Sun's distance from the center. Overall, though, the Milky Way's disk has a diameter of about 30 to 40 kiloparsecs, or roughly 100,000 light-years.

Concept Question 2

What would the Milky Way look like in the night sky of an observer who lived in a globular cluster? What would it look like to an observer at the Galactic nucleus?

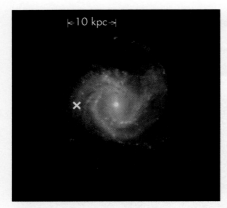

FIGURE 71.7
Visible-light image of the spiral galaxy NGC 4303 (also known as M61), illustrating its spiral arms. This galaxy is thought to be fairly similar in size and shape to the Milky Way. The X marks a position about 8 kpc from the galaxy's center where the Sun would be found in the Milky Way.

Concept Question 3

How might the presence of dusty clouds inside the Milky Way bias our understanding of what lies outside the Milky Way?

FIGURE 71.8
Artist's sketch of the Milky Way, showing top and side views, illustrating the disk, bulge, and halo. Notice how thin the disk is and how the halo surrounds the disk much as a bun surrounds a hamburger. Because the Milky Way's stars gradually thin out and do not just stop at some distance, its size is labeled only approximately.

71.4 GALACTIC STRUCTURE AND CONTENTS

Today our understanding of the structure of the Milky Way is aided greatly by studies of other galaxies. In the early 1900s, it was not widely accepted that there *were* other galaxies; this is explored further in Unit 75. Because we live inside the Milky Way, we cannot observe our Galaxy in its entirety—it is difficult "to see the forest for the trees." Having other examples of galaxies gives us a better idea of the kinds of structures we might find in the Milky Way.

Many external galaxies are flat disks with conspicuous **spiral arms** (Unit 76), as shown in Figure 71.7. Such arms are hard to detect in our own Galaxy because from our location in the Milky Way we cannot get an overview of its disk. Nevertheless, given that the Milky Way is a flat disk and that other disk galaxies have spiral arms, it is reasonable to infer that ours does too, and observations of the locations of interstellar gas, dust, and bright stars are consistent with this picture. By combining observations of our own Galaxy with the general features known to occur in other galaxies, astronomers can assemble a more detailed model for the Milky Way, as described next.

The Milky Way consists of three main parts, illustrated in Figure 71.8: a **disk** about 30 to 40 kiloparsecs in diameter, a more spherically distributed component called the **halo**, and a flattened, somewhat elongated **bulge** of stars at its center. Within the disk, numerous bright young stars collect into spiral arms that wind outward from near the center. Our Solar System lies about 8 kiloparsecs from the center in a region between spiral arms. You may find the following analogy helpful in visualizing the scale of our Galaxy: If the Milky Way were the size of the Earth, the Solar System would be the size of a large cookie.

Mingled with the stars of the disk are huge clouds of gas and dust that amount to about 15% of the disk's mass. We can see some of these clouds by the visible light they emit, whereas others reveal themselves because their dust blocks the light of background stars, as can be seen in Figure 71.2. In fact, dust scattered throughout the Galaxy prevents us from seeing farther than about 3 kiloparsecs away from the Sun in almost any direction close to the Galactic plane. For example, we cannot see our Galaxy's **nucleus**—its core—at visible wavelengths. However, radio, infrared, and X-ray telescopes can "see" through the dust and reveal that the core of our Galaxy contains a dense swarm of stars and gas as well as a supermassive black hole (Unit 74).

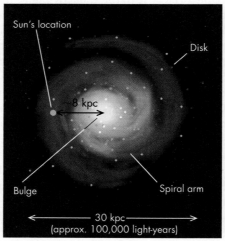

Top view

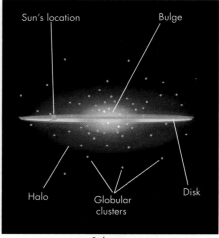

Side view

KEY POINTS

- The Milky Way appears from Earth as a faint band of light that wraps around the entire sky.
- The Milky Way's light is produced by billions of stars, most too distant to see individually.
- The stars lie in a thin disk, so our view from within the disk looks like a narrow band of stars.
- Interstellar clouds that contain dust block light from portions of the Milky Way.
- Because dust blocks light from distant stars, astronomers initially thought the Milky Way was smaller than we now know it to be.
- Tracing the positions of globular clusters revealed that the Milky Way's center is about 25,000 light-years distant.
- The Sun is located in the outer part of a disk of stars that form a spiral pattern like a pinwheel around the Milky Way's center.
- At the center of the Milky Way is a dense bulge of stars, and a spherical halo distribution of stars extends out 50,000 light-years.

KEY TERMS

bulge, 562
disk, 562
Galactic plane, 560
Galaxy, 557
globular cluster, 560
halo, 562
interstellar dust, 561
Kapteyn's Universe, 560
kiloparsec, 560
Milky Way, 557
nucleus, 562
spiral arm, 562

CONCEPT QUESTIONS

Concept Questions on the following topics are located in the margins. They invite thinking and discussion beyond the text.

1. Motions of the Milky Way if Earth's axis were different. (p. 559)
2. Appearance of the Milky Way from different locations. (p. 561)
3. Effect of dust on observations beyond the Milky Way. (p. 562)

REVIEW QUESTIONS

4. What does the Milky Way look like in the night sky?
5. How do we know that the Sun is not located at the edge of the Milky Way?
6. How do we know our Galaxy is a flat disk?
7. How did star counts determine the shape of the Milky Way?
8. What are globular clusters, and how do we determine their distances from the Sun?
9. How did Shapley deduce the Milky Way's size and the Sun's position in the Milky Way?
10. What effects did dust have on the determination of the size of the Milky Way?
11. What are the major components of the Milky Way?

QUANTITATIVE PROBLEMS

12. If the total mass of the Milky Way is $10^{12}\ M_\odot$, and 10% of that is in the disk, and 15% of the disk is dust and gas, how many solar masses of dust and gas are in the Galaxy?
13. If a star located at the center of the Galaxy goes supernova, how long does the supernova's light take to reach the Solar System?
14. Traveling at 60,000 kph, how long would it take us to reach the center of the Milky Way?
15. Suppose a galaxy like the Milky Way is about 2 million ly away (the distance to the nearest galaxy similar in size to the Milky Way). What would its angular size be, given that its disk is about 100,000 ly across?
16. If we made a model of the Milky Way that had the diameter of the Earth, how big would the Earth itself be in this shrunken model? (Assume that the Milky Way's diameter is 35 kpc for this problem.)
17. Kapteyn used observations of the smaller shift in the position of stars that are farther away to estimate their distances. This is essentially a consequence of the angular size versus distance relationship discussed in Unit 10, as illustrated by the following: (a) If stars move at a typical speed of 20 km/sec, how far will they move in 10 years? (b) If a star is 10 pc distant, what will its angular shift be over these 10 years? (c) If a star is 100 pc distant, what will its angular shift be?
18. Interstellar dust limits our view into the Galactic disk to only the nearest 3 kpc. If the entire galactic disk has a radius of 15 kpc, what fraction of the disk is visible from Earth? Recall that the area of a disc is given by $\mathcal{A} = \pi R^2$.

TEST YOURSELF

19. Roughly how many stars are in the Milky way?
    a. Dozens
    b. Hundreds
    c. Tens of thousands
    d. Millions
    e. Hundreds of billions
20. By studying the positions of globular clusters astronomers deduced
    a. that the Solar System is not located near the center of our Galaxy.
    b. that the Galaxy has three main components: a disk, a halo, and a central bulge.
    c. the presence of spiral arms in the Galaxy.
    d. the existence of interstellar dust.
21. Our Solar System is located in
    a. a globular cluster.
    b. the Galactic bulge.
    c. the Galactic halo.
    d. the Galactic disk.
22. One of the reasons Kapteyn underestimated the size of the Milky Way and Shapley overestimated it is that they did not recognize the
    a. effects of the motion of the Sun from the center.
    b. dimming effect of interstellar dust.
    c. age of the globular clusters.
    d. existence of dark matter.

UNIT 72

Stars of the Milky Way

72.1 Stellar Populations
72.2 Formation of Our Galaxy
72.3 Evolution Through Mergers
72.4 The Future of the Milky Way

Learning Objectives

Upon completing this Unit, you should be able to:
- List and contrast the characteristics of Population I and II stars.
- Explain how distinct stellar populations may result from the Galaxy's formation.
- Describe evidence for mergers and cannibalism in our Galaxy.
- Explain what the very first stars would be like, and how the Galaxy's stars age.

The Milky Way (Unit 71) contains many types of stars: giants and dwarfs, hot and cool, young and old, stable and exploding. All of these star types combine to define the overall population of stars in our Galaxy. Such studies of stellar populations reveal that despite the wide range of star types and the enormous range of luminosities, the typical star in the Milky Way is rather small, dim, and cool.

Astronomers can also make a "stellar census" in different regions of the Milky Way by counting the relative numbers of each type of star. The types of stars and their relative numbers reveal similarities and dissimilarities in when and how many stars formed in these regions. In this Unit, we will put this information together to explore the history of our Galaxy, much as archaeologists can learn about life in an ancient city from the locations and kinds of buildings it contained.

72.1 STELLAR POPULATIONS

Hidden in the great diversity of star types is an underlying simplicity that astronomers first noted in the 1940s. At that time, the 100-inch reflector at Mount Wilson Observatory, located outside of Los Angeles, California, was the largest telescope in the world and the best for observing galaxies. The glow from city lights made it hard to see faint galaxies, but blackouts during World War II darkened the night sky, and Walter Baade (*BAH-deh*), an astronomer at Mount Wilson, took advantage of the darkness to make a series of photographs of neighboring galaxies. He noticed that stars in these nearby galaxies were segregated by color. Red stars were concentrated in the bulges and halos of the galaxies, whereas blue stars were concentrated in their disks and especially in their spiral arms. To distinguish these groups, Baade called the blue stars in the disk **Population I,** and the red stars of the bulge and halo **Population II** (Pop I and Pop II, for short).

While pursuing his studies of the stellar populations of other galaxies, Baade realized that Milky Way stars showed the same division. Furthermore, when he examined the properties of Population I and II stars in our Galaxy, he found further differences between the two populations. They differ not only in color and location in the Galaxy, but in their age, motion, and composition. Pop I stars are young, typically less than a few billion years old, and many of them are blue. They lie in the plane of the Galaxy's disk and follow approximately circular orbits, as shown by the blue orbits in Figure 72.1. Their atmospheric composition is like the Sun's: mostly hydrogen and helium, with a few percent of their mass consisting of **metals,** which for an astronomer means any element heavier than helium. Thus, Pop I stars are relatively rich in metals.

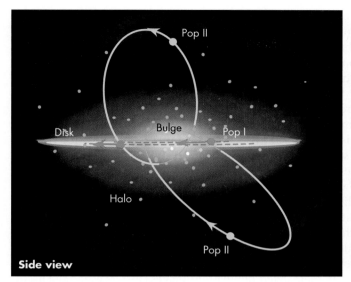
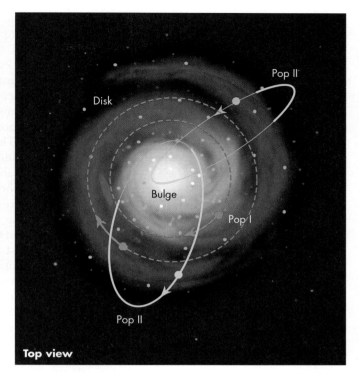

FIGURE 72.1
Stellar orbits in the Milky Way. Population I stars orbit in the disk (blue lines). Population II stars orbit in the halo (yellow lines). Notice the elongation and inclination of their orbits.

Pop I and II

Concept Question 1

Suppose the Sun had an orbit similar to that of a Pop II star in the halo of the Milky Way. In what ways would that make studying the Milky Way easier? Is it likely there are Earth-like planets orbiting halo stars?

Population II stars are generally red and more than about 10 billion years old. They lie in the bulge and halo of the Galaxy, moving along highly elliptical orbits that are often tilted strongly with respect to the Galactic disk, as illustrated by the yellow orbits in Figure 72.1. Pop II stars are almost entirely hydrogen and helium, with only a few hundredths percent of their mass composed of heavy elements—roughly a hundred times less than stars like the Sun. Table 72.1 summarizes some of these properties.

Division of all stars into two broad categories is an oversimplification. For example, the Sun's age does not fit precisely into either category, but the Sun is considered a Pop I star because of its heavy element content and approximately circular orbit in the disk. To avoid forcing stars into population categories they do not fit, astronomers often refer instead to disk, bulge, and halo stars, recognizing that the great majority of disk stars are Pop I, while most of the bulge and halo stars are Pop II.

The fairly sharp distinction in properties between the two populations appears to relate to major changes in the structure of the Milky Way during its history. Distinct populations suggest that star formation has not occurred continuously in the Milky Way. Population II stars probably formed in a major burst at the Galaxy's birth during its initial collapse, whereas Population I stars began forming later and continue forming today. This hypothesis explains the differences between the two populations, as we will discuss in greater detail in the next section.

TABLE 72.1 Properties of Population I and II Stars		
Property	**Pop I**	**Pop II**
Location	Disk and concentrated in arms	Halo and bulge
Age	Young (< few billion years)	Old (> 10 billion years)
Color	Blue (overall)	Red
Orbit	Approximately circular in disk	Plunging through disk on approximately elliptical orbit
Heavy element (metals) content	High (a few percent—similar to Sun's)	Low (10^{-2} to 10^{-3} times Sun's)

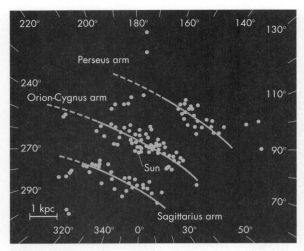

FIGURE 72.2
Young star clusters show parts of the spiral arms in the Sun's neighborhood. The arms are named after the constellations in which they appear bright from our vantage point in the Solar System.

One of the first uses astronomers made of differences between stellar populations was to map the Milky Way's spiral arms. As Baade noted, the spiral arms of other galaxies gleam with the blue light of their Pop I stars. The blue color of Pop I is produced primarily by the luminous stars with spectral types O and B (Unit 56) that use up their fuel rapidly, "burning out" in just tens of millions of years. Therefore, by measuring the location of O and B stars near the Sun, astronomers can make a picture of the spiral arms in the Milky Way's disk. This method is limited, however, because dust in space prevents us from seeing even the most brilliant O and B stars if they are farther from the Sun than about 3 kiloparsecs. Nevertheless, maps such as that shown in Figure 72.2 provided some of the first direct evidence that we live in a spiral galaxy.

The stars in open clusters and globular clusters (Unit 70) also exhibit the differences between Populations I and II. Open clusters are almost all Pop I, located within the Milky Way's disk, whereas globular clusters are always Pop II and orbit in the Milky Way's halo. This difference in their stellar populations makes clusters especially useful to astronomers for studying the structure of the Milky Way: Globular clusters outline the halo and bulge, while young, open clusters trace the Galaxy's arms.

72.2 FORMATION OF OUR GALAXY

One of the major research topics in astronomy today is how galaxies form and evolve. The process is presumably a large-scale version of star formation (Unit 61)—that is, a gas cloud collapses under the influence of gravity and breaks up into stars. But this does not explain why the Galaxy contains two main categories of stars, Population I and II, that differ so greatly in their properties. Based on an analysis of the orbits of Pop I and Pop II stars, the British astronomer Donald Lynden-Bell and the U.S. astronomers Olin J. Eggen and Allan R. Sandage proposed in 1962 a two-stage collapse model to explain the birth of the Milky Way.

According to the two-stage collapse model, our Galaxy began as a vast, slowly rotating gas cloud, a few million light-years (a million parsecs) in diameter, containing several hundred billion solar masses of gas. The cloud was composed of almost pure hydrogen and helium. As gravity began shrinking this immense cloud, clumps of gas within it grew in density, forming stars, as illustrated in Figure 72.3A.

These first stars would have had a composition unlike stars today. The heavy elements are produced by nuclear fusion inside stars, so the first stars would have been composed of essentially pure hydrogen and helium. The most massive of these stars would have evolved quickly and blown up as supernovae (Units 66 and 67), adding the first heavy elements to the gas cloud even before it had finished collapsing. It may seem odd that stars could form and die before the cloud's collapse was complete. But the collapse of such an immense cloud takes hundreds of millions of years, whereas massive stars can evolve and die in less than a few tens of millions of years.

As the cloud's collapse continued, the next generation of stars now contained at least a few heavy elements. Formed from gas falling in from all directions, the resulting inward trajectories of these bodies would have given them highly elliptical orbits. Hence, their orbits continue to carry them in and out of the inner part of the Galaxy, distributed all around the Galaxy in the halo. These stars, many of which formed in globular clusters, are the Population II stars that we see today.

> The structure of the Solar System also provides clues to its origin. Its disklike shape, the common age of the planets, and the existence of different families of planets allowed astronomers to develop the solar nebula theory (Unit 35).

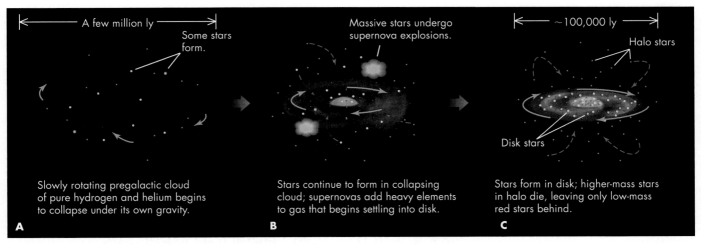

FIGURE 72.3
(A) Birth of the Milky Way as a gas cloud collapses. A first generation of stars (Population II) forms. (B) Collapse continues with more Population II stars forming and some exploding as supernovae. Heavy elements created by the stars exploding as supernovae are mixed into the gas. Gas not used in making stars settles into a rotating disk. (C) Population I stars form in the disk.

We can use the globular clusters to estimate the Milky Way's age. Using the techniques for measuring stellar ages in star clusters (Unit 70), astronomers calculate that our Galaxy's most ancient globular clusters are about 13 billion years old, a number we will take as the approximate age of our Galaxy.

Star formation during this first stage of mostly radial collapse did not turn all of the gas into stars. The remaining clouds of gas continued to fall inward. Clouds and stars behave very differently under these circumstances. Stars are so small, relative to the vast space between them, that it is extremely improbable that they will collide, and so nearly all continued past each other along their original trajectories. On the other hand, clouds fill a large fraction of the space they occupy, and as they fell toward the center of the growing Galaxy, they crowded closer together. In the second stage of the collapse, these clouds collided with each other, losing orbital energy in the process, but conserving angular momentum (Unit 20.2), and settling into a smaller rotating disk of gas, as sketched in Figure 72.3B. This is similar to how drops of water might slide down a shower floor toward the drain in the middle, where they merge into a small spinning pool.

The infalling gas that formed the disk was enriched with heavy elements from the death of additional stars formed during the collapse. Therefore, even the first Population I stars born in the disk had a higher level of heavy elements. Continuing the motion of their parent gas, they circled the center of the young Galaxy in the disk (Figure 72.3C).

The two-stage collapse model explains the difference in the properties of Pop I and Pop II stars quite well, but it also suggests that there was a third, even earlier population of stars: the very first stars that formed from pure hydrogen and helium. These first, or **Population III,** stars would presumably also have orbits much like Pop II stars, but they would be older and would have no metals at all.

Astronomers have been searching for "Pop III" stars for decades, but the closest they have come are two stars discovered in the last decade that have less than 1/100,000 the heavy element abundance of the Sun. They have merely 1/1000 the typical metal content of Pop II stars, but they do still contain heavy elements.

It is possible that no Population III stars survive today. The only ones that might remain today would have to be less massive than about $0.9\ M_\odot$. Only such low-mass stars fuse their hydrogen slowly enough to survive for 13 billion years (Unit 62). One hypothesis is that conditions in the initial cloud of gas may have prevented

Concept Question 2

If you randomly selected two Population I stars, one red and one blue, which would you expect to have higher metallicity?

> **Concept Question 3**
>
> How might our Galaxy have been different if stars had formed at a significantly faster rate? at a significantly slower rate?

low-mass stars from forming; for example, the gas may have been too hot or turbulent, or a cloud of pure hydrogen and helium may collapse differently from the clouds of gas we observe today. According to this idea, only massive stars formed from the pure hydrogen and helium gas, and all of those stars have already died.

Another possibility is that Population III stars may exist but are masquerading as Population II stars. For example, the two stars with extremely low metal content just noted are estimated to be about 0.7 $M_\odot$ to 0.8 $M_\odot$ in mass and 13 billion years old. Such low-mass stars take more than 100 million years to contract and form—longer than the entire lifetime of massive stars forming out of the same collapsing cloud (Unit 60). Their outer layers may have been contaminated by gas ejected at the death of their more massive brethren. Stars like these may, then, be the only survivors from the initial collapse of the Milky Way.

72.3 EVOLUTION THROUGH MERGERS

Some features of the Milky Way call for refinements to the two-stage collapse model we have just described. First, according to the model, all Population II stars should be about the same age, having formed during the relatively brief period of the Galaxy's initial collapse. But observations of Pop II stars show that they formed over a significantly longer time span than the model predicts. Second, studies of the ages and distribution of stars in the disk and halo show some complex patterns that do not obviously result from the collapse model. For example, there appear to have been episodes of intense star formation, and there are streams of stars that follow unusual orbits within the Galaxy.

One hypothesis for explaining some of these anomalies is that the Milky Way has collided with other galaxies throughout its history. As we saw regarding the formation of our Galaxy, galaxy collisions are unlike the collision of two solid bodies. Because of the large spaces between stars, when two galaxies collide, their individual stars would almost never strike each other, although the gas and dust between the stars *would* collide. The compression of the gas clouds in both galaxies could lead to huge bursts of star formation. Computer modeling of such collisions suggests that two colliding galaxies often merge into a single larger system, a process that astronomers call **galactic cannibalism**.

We see evidence that some small galaxies are merging with the Milky Way even now. For example, astronomers recently detected a small "dwarf" galaxy on the far side of the Milky Way. The dwarf galaxy is elongated, probably because the Milky Way's gravity has stretched it out, and a long stream of stars within the Milky Way's halo appears to have been pulled from the smaller galaxy (Figure 72.4). Other streams of stars seen in the Milky Way may be all that remains of other smaller galaxies that have been pulled apart and merged with our Galaxy. Astronomers conclude that large galaxies like the Milky Way may have begun life with the collapse of a single gas cloud, but they continue to grow by swallowing smaller galaxies.

The interpretation of these observations is greatly aided by computer simulations. Computers can track the simultaneous effects that large numbers of stars and gas clouds have on each other. Modeling gravitational interactions with computers is, in principle, straightforward. The computer simply needs to calculate the forces and reactions between every pair of objects, using Newton's laws. The computer then projects where these objects will have moved in response to these forces after a short

FIGURE 72.4
Diagram illustrating a stream of stars (shown in red) pulled out of a dwarf galaxy orbiting the Milky Way (depicted in blue). Based on estimated distances to the stars in the stream, the figure shows what the stream and Milky Way might look like from a distant galaxy.

period of time (for example, one year) and then, for the new configuration, carries out the calculation again. Of course, with hundreds of billions of stars in the Milky Way, there are a lot of calculations to carry out! Computers cannot yet model every star in the Milky Way, but they have become fast enough to carry out such calculations for millions of stars at a time, which appears to be enough to sample many of the effects that occur.

Even the most advanced supercomputers, though, do not have the capacity to simulate all of the details of collisions between interstellar clouds. Such collisions generate rapid pressure changes, shock waves, heating, and turbulence, as well as a complex string of subsequent events. As the gas cools, it might become gravitationally unstable and form new stars. By treating a cloud as millions of small interacting "blobs" of gas, modern simulations are able to reproduce many known results. In fact, modern simulations attempt to add in the effects of stellar evolution, supernova explosions, and metal enrichment in the interstellar clouds. The great power of computer simulations is that they allow us to observe the consequences of large numbers of interactions and to predict the changes that will result over very long periods of time.

Computer simulations support the two-stage collapse model and merger model for the birth and evolution of the Milky Way. For example, Figure 72.5 shows how an unstructured region of gas, of the appropriate size and rotation speed, collapses. You can see that it forms a thin disk with spiral arms, surrounded by a halo. Simulations are an important part of astronomy today, but it should always be kept in mind that the simulations only reflect the aspects of the physical interactions that were designed into them. If, for example, magnetic fields are important in these collisions, a new set of computer models would need to be developed to incorporate the effects of magnetic interactions.

FIGURE 72.5
Early computer simulation of the birth of a galaxy similar to the Milky Way. The top three frames (A), (B), and (C) show the initial state of the gas and the development of a clump that becomes the galaxy, illustrated in the bottom two frames (D) and (E). Notice the thin disk and spiral arms.

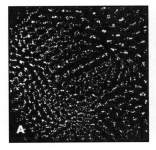
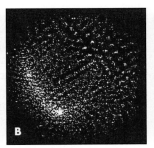
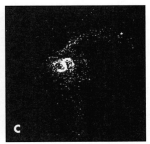
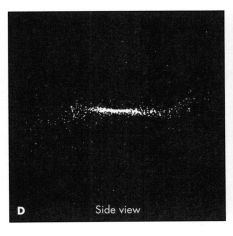

72.4 THE FUTURE OF THE MILKY WAY

Throughout the Milky Way's history, as stars form, the amount of gas remaining is depleted. With a smaller reservoir of gas, the rate of star formation must also steadily decline, except for the occasional influx of new material caused by a merger event. Dying stars return a large fraction of their material to space as planetary nebula shells, supernova remnants, and stellar winds; but with each generation of stars, more and more of the Galaxy's mass is locked away from further use, trapped in stellar remnants such as white dwarfs, neutron stars, and black holes (Unit 60). An even larger fraction remains tied up in the very low-mass stars that can live even longer than the current age of the Galaxy.

Throughout this process, the interstellar gas becomes steadily more enriched with heavy elements, so that later generations of stars have a greater percentage of metals (elements heavier than helium). At the time of our Galaxy's birth, the birthrate of stars was probably very high and has since declined steadily. The enrichment of metals in the Sun suggests that it is about a "fifth-generation" star, in the sense that its enrichment of metals would require that the gas had cycled through about five stars before forming the Sun and Solar System.

By dividing the number of stars in the Milky Way by its age, astronomers estimate that, on average, about 10 stars have been born each year. A current star formation rate of a few stars per year is consistent with studies of star-forming clouds today. This seemingly small number of new star births will be sufficient to keep our Galaxy shining for many billions of years into the future. At current rates, the Milky Way may run out of material for making new stars about 10 billion years from now. The lowest-mass stars will continue to shine for several hundred billion years, and the Galaxy will grow dimmer and redder. In about a trillion years, even these low-mass stars will have died, and all that will remain will be the dark remnants of former stars. The Milky Way will fade, slowly spinning in space, a dark disk of stellar cinders.

The future will not be entirely quiet. Large merger events can be anticipated. Using computer simulations, astronomers predict that in about a billion years, two fairly large satellite galaxies—the Large and Small Magellanic Clouds—will spiral into our Galaxy and eventually merge with it. When they do, they will bring fresh gas and will probably initiate a period of strong star formation. In about 5 billion years, the Milky Way will probably collide with a nearby galaxy larger than our own, the Andromeda Galaxy (Unit 77), and eventually merge with it. And so this cannibalism will continue, the galaxies consuming one another even as their stars fade and die.

Concept Question 4

When the Milky Way is very old, the interstellar gas may contain little hydrogen. In what ways might this affect the evolution of stars?

KEY POINTS

- The stars in the Milky Way mostly fall into two populations, with different ages, compositions, and locations.
- Pop I stars are younger, contain more heavy elements, and are located in the disk of the Milky Way.
- Pop II stars appear to have formed early in the Milky Way's history before many heavy elements formed.
- The location of Pop II stars in the halo suggests they may have formed as a vast cloud collapsed to form the Galaxy.
- There may have been an even earlier "Pop III" generation of stars, with no metals, that produced the first heavy elements.
- Star streams and differences in stars' ages indicate the Galaxy has accreted material and small galaxies throughout its lifetime.
- The Galaxy is growing steadily more metal-rich and dimmer and redder as gas is recycled through stars, with a fraction locked away in stellar corpses.

KEY TERMS

galactic cannibalism, 568
metals, 564
Population I (Pop I), 564
Population II (Pop II), 564
Population III (Pop III), 567

CONCEPT QUESTIONS

Concept Questions on the following topics are located in the margins. They invite thinking and discussion beyond the text.

1. Observing the Milky Way from a Pop II star. (p. 565)
2. Color and metallicity in Pop I stars. (p. 567)
3. Effects on Galaxy of different star formation rates. (p. 568)
4. Star formation in the remote future. (p. 570)

REVIEW QUESTIONS

5. What are some differences between Pop I and Pop II stars?
6. How do the orbits of Pop I stars differ from the orbits of Pop II stars? Why?
7. Why must O and B stars be Pop I?
8. Why are spiral arms outlined by Pop I stars?
9. Describe one model for the origin of the Milky Way. How does this model explain the differences between Pop I and Pop II stars?
10. What are Pop III stars?
11. How are elements heavier than helium produced?
12. What is the eventual fate of the Milky Way?

QUANTITATIVE PROBLEMS

13. A Pop II star has a highly eccentric orbit 12 kpc in diameter. What is its orbital period?
14. Conservation of angular momentum (Unit 20) requires that the product of the speed at which an object is orbiting times the distance from the point around which it is orbiting remain constant. If the Sun is currently orbiting at 220 km/sec, 8 kpc from the center of the Milky Way, how fast did the material that formed the Sun rotate if the gas began 400 kpc distant (about halfway to the next large galaxy)?
15. For a Pop III star to survive the 13 billion years since the formation of the Galaxy, it can only have a maximum mass of about 0.9 $M_\odot$. Estimate the corresponding luminosity for this mass using the mass-luminosity relationship. Why do you think it may be difficult to find Pop III stars today?
16. Suppose that after x billion years, the Milky Way converted half of its gas into stars. In the next x billion years, it converted half of the remaining gas into stars—and so on, each x billion years. The Milky Way is estimated to be about 13 billion years old, and only about 15% of its disk remains today in the form of gas.
    a. What value of x would give the Milky Way's current gas fraction? (*Hint:* You can calculate this mathematically, or you can draw an approximate graph of the declining gas fraction and estimate x.)
    b. In how many billion years will the gas fraction be just 1%?
17. Suppose we found a nearby Pop II star that is on a highly elliptical orbit, but it is currently near its most distant point from the center of the Milky Way in its orbit. To stars like the Sun in orbit in the disk, what would be the approximate relative speed of this Pop II star? In what direction?

TEST YOURSELF

18. A dim, red star moving on a highly elliptical and inclined orbit to the galactic disk is most likely
    a. a Pop I star.
    b. a Pop II star.
    c. a Pop III star.
    d. orbiting near the Earth.
    e. None of the above.
19. Modeling galaxy evolution using computer simulations is largely limited by
    a. the inability of current computers to monitor the motion of every star that makes up a galaxy.
    b. a poor understanding of how to model the gravitational interactions between stars.
    c. the impossibility of modeling collisions between interstellar clouds.
    d. the difficulty of including complex processes such as star formation and supernovae.
20. The greater number of heavy elements seen in the spectra of Pop I stars relative to Pop II stars is explained by the fact that Pop I stars
    a. are older, and therefore have fused more hydrogen into heavy elements.
    b. formed more recently, and therefore have been made from enriched interstellar gas.
    c. have planetary systems, and debris from these fall onto the star's surface.
    d. are colder and therefore exhibit strong "metal" lines.
21. Several billion years from now, the Milky Way will probably
    a. be redder.
    b. have more Pop II stars.
    c. have a lower mass.
    d. have lost most of its stars.
    e. experience all of the above.

UNIT 73
Gas and Dust in the Milky Way

73.1 The Interstellar Medium
73.2 Interstellar Dust: Dimming and Reddening
73.3 Radio Waves from Cold Interstellar Gas
73.4 Heating and Cooling in the ISM

Learning Objectives

Upon completing this Unit, you should be able to:
- Describe the components of the interstellar medium and how each is observed.
- Explain how dust particles interact with electromagnetic radiation to produce reflection nebulae, dark nebulae, reddening, and infrared emission.
- Explain how atoms and molecules are detected at radio wavelengths.
- Describe the processes that cause an interstellar cloud to heat up or cool down.

The space between stars is not empty but contains gas and dust particles that compose what astronomers call the **interstellar medium,** or **ISM.** By terrestrial standards, this space is almost a perfect vacuum. On average, each liter contains only a few thousand atoms of gas. For comparison, the air we breathe contains about 10^{22} atoms per liter. The density of gas in interstellar space compared to the Earth's air is like having one marble in a box 8 kilometers (5 miles) on a side compared to the same box being solidly filled with marbles.

Despite the interstellar medium's very low density, small particles of dust block our view of distant parts of our Galaxy, as we saw in Unit 71. The ISM plays a vital role as the reservoir of material from which new stars form, and into which dying stars deposit the ashes of their fusion processes, as we explored in Unit 61. In this Unit, we will study the physical and chemical properties of interstellar clouds themselves, bringing to bear the techniques of spectroscopy developed in Unit 24.

73.1 THE INTERSTELLAR MEDIUM

Interstellar matter is not spread smoothly throughout the Milky Way. Gravity pulls most of it into a thin layer in the disk, as can be seen by the dark band of dust in the disks of other galaxies (Figure 73.1). Interstellar matter behaves quite differently from stars as it orbits within the Galaxy. Because the gas and dust particles fill space and

Stars get scattered into new orbits when they make gravitational encounters with other massive objects orbiting in the galaxy, so their orbits spread out rather than settling into the plane like the gas.

FIGURE 73.1
Interstellar matter, outlined by dark dust clouds, is concentrated in the plane of a galaxy, forming the dark band. This galaxy is popularly called the "Sombrero Galaxy." Its official designation is either NGC 4595 or M104.

  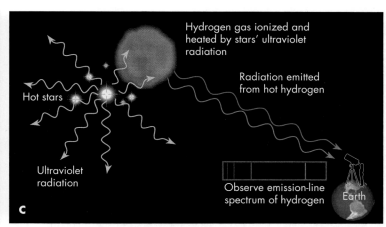

FIGURE 73.2
Emission nebulae. (A, B) Gas near hot, luminous stars often glows pink because of hydrogen emission lines. (C) Stars that emit ultraviolet radiation ionize nearby gas. Electrons dropping from hydrogen's third to second energy level generate the pink light.

Pink dots in the photographs of other galaxies are usually emission nebulae as well.

The Orion Nebula is shown in Looking Up #6 at the front of the book. Other emission nebulae are shown in Looking Up #7 and #9.

collide with each other, they lose energy and settle down into the common orbital plane. Stars, by contrast, are so dense relative to the gas that they can pass through the gas near the Galactic plane, which offers too little "friction" to slow them down significantly. Within this layer, gravity and gas pressure cause the gas and dust to clump into clouds, which are embedded in a very low-density background gas.

Sometimes the gas in an interstellar cloud is hot and emits visible light, as you can see in Figure 73.2A and B. Such a glowing cloud is called an **emission nebula** (*nebula* is the Latin word for "cloud"). The power source causing the gas to light up is usually young, luminous O and B stars. They produce ultraviolet radiation that can ionize the hydrogen and other atoms in these clouds. An example of such a gas cloud is the Orion Nebula, which is easy to see with a small telescope or even a pair of binoculars. If you look at the middle star in Orion's sword, you will see a pale, fuzzy glow created by luminous interstellar gas surrounding a dense star cluster.

When the interstellar gas has been heated, it produces emission lines (Figure 73.2C). On the other hand, radiation is absorbed by cool interstellar gas, allowing astronomers to detect cool clouds that lie in front of stars. Atoms and molecules absorb the specific wavelengths of light that correspond to energy-level changes of their electrons (Unit 24). Therefore, when astronomers observe the spectrum of a star behind an interstellar cloud, they see the dark lines of the interstellar gas's absorption superimposed on the star's own spectral signature (Figure 73.3).

By studying the spectral emission and absorption lines, astronomers can deduce the composition, temperature, density, and motion of the cloud. Such spectra show that interstellar gas has a composition similar to that of the Sun and other stars—about 71% hydrogen and 27% helium, with the remainder made up of heavier elements.

Dust particles also emit and absorb light. But because dust particles are solids, they do not generate such clear spectral signatures as isolated atoms and molecules do. Still, astronomers can deduce dust particles' compositions from the elements that are "missing" from the gases that are detected, and they can also detect some spectral signatures that are produced by, for example, carbon bonds in dust grains. Based on piecing together a variety of information like this, it is

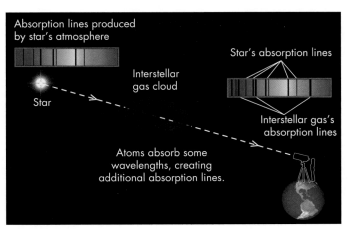

FIGURE 73.3
Gas atoms in an interstellar cloud between us and a distant star absorb some of the star's light, adding their own absorption lines to those in the star's spectrum.

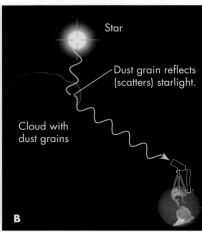

FIGURE 73.4
Reflection nebulae. (A) The Pleiades star cluster exhibits bluish reflection of the light from many of the brighter stars. (B) Light from a star near a dust cloud is reflected by dust particles, making the cloud visible.

deduced that the dust particles are made up primarily of silicates and carbon compounds, coated with a thin layer of ices made of molecules such as water, carbon monoxide, and methanol. Based on the way they interact with light, these particles must be extremely tiny, ranging from about a micrometer to a nanometer or so in diameter (not much larger than some large molecules).

Dust particles are sometimes detected when they reflect light from a nearby star, producing what is called a **reflection nebula** (Figure 73.4). The reflection effect is something like the beams of light you see coming from the headlights of a car as it drives through fog. When the photon strikes a dust particle, it is reflected in a fairly random direction in a process called **scattering.**

Although interstellar clouds are similar in overall composition, they differ greatly in size, ranging from a fraction of a light-year to a few hundred light-years in diameter. They also range widely in mass from about one to millions of solar masses. Finally, they also have a wide range of density and temperature. For example, some regions have temperatures of millions of kelvins, while others are only a few kelvins above absolute zero.

The complexity of the interstellar medium is a little like "weather." If we could watch a high-speed movie of interstellar clouds, we would see that they constantly form and dissolve, driven by powerful physical phenomena such as stellar winds and supernova explosions. Although matter in interstellar space undergoes huge changes in temperature and density, this occurs on timescales of millions of years. Our observations of interstellar clouds reveal a momentary snapshot of a dynamic and turbulent medium.

73.2 INTERSTELLAR DUST: DIMMING AND REDDENING

Only about 1% of interstellar matter is in the form of dust grains, but the effects of that dust on light are strong. Not only does dust dim the light of distant stars, it also alters the light's color, much as haze in our atmosphere dims and reddens the setting Sun. These effects occur because visible light interacts strongly with dust.

The dust in a cloud may block the light of background stars and appear as a **dark nebula** against the starry background (Figure 73.5A and B). Clouds like these are too cold to emit visible light, but they emit lower-energy radiation at infrared or radio wavelengths, as discussed in Section 73.3. Some dark nebulae are also visible against emission nebulae, as can be seen in Figure 73.2A. Dark nebulae are the starting point for the formation of the next generation of stars.

Even when dust grains do not absorb a star's light, the scattering they cause dims the star because some of the radiation heading toward us is sent off in random directions and is lost to our sight. That same scattered light will appear as a reflection nebula to an observer in a different direction. However, if a dust cloud is very thick or dense, the photons are scattered repeatedly and their energy is eventually absorbed by the dust grains. Astronomers call the overall dimming **extinction.** If the cloud is thick enough, the extinction may be so strong that little or no light of any wavelength will pass through it, and the cloud will be seen as a dark nebula.

Concept Question 1

How many different kinds of nebulae could an interstellar cloud appear to be to different observers?

FIGURE 73.5
Dark nebulae. (A) Dust in Barnard 86 blocks our view of the background stars. The stars that appear to be within the outline of the dark nebula lie between us and the nebula. (B) A string of dark clouds called the "Snake Nebula." Short-wavelength light is blocked more strongly than long wavelengths, so where the clouds are thin enough for light to pass through, the background stars are reddened. (C) Dust in an interstellar cloud scatters the blue light from a distant star, removing it from the radiation reaching Earth and making the star look redder than it really is.

Clarification Point

The sky is blue because short wavelength light from the Sun is scattered in Earth's atmosphere.

Concept Question 2

Suppose we observe a main-sequence star behind a thin dust cloud, so its light is slightly reddened and dimmed. If we used its color to predict its luminosity, would we tend to overestimate or underestimate the luminosity? What if we then used its brightness to estimate its distance?

A dust particle has the strongest scattering effect on electromagnetic waves whose wavelength is close to or slightly smaller than the size of the particle. If the dust particle is much smaller than the wavelength of radiation, it is much less effective at scattering the radiation. An analogy to this is that a boat will block short-wavelength ocean waves, which "slap" against the side of the boat, while it does little to affect long-wavelength waves, which raise and lower the boat and continue past it.

A sunny day shows the effect that small particles in our atmosphere have on sunlight. Here oxygen and nitrogen molecules scatter short-wavelength blue photons of sunlight much more strongly than longer-wavelength red photons. The sky is not blue because the molecules in the atmosphere are glowing with this color. Instead, the blue photons we see are part of the mix of all wavelengths coming from the Sun. Because of the size of Earth's atmospheric molecules, the Sun's blue wavelengths are scattered more strongly than longer, redder wavelengths that pass through to the surface. This process is similar to the effect of small dust grains in interstellar space, and this is why reflection nebulae usually have a bluish color (Figure 73.4).

Because shorter-wavelength photons are scattered in random directions, the balance of the remaining light contains a relatively larger number of longer-wavelength photons. This effect is called **reddening** because the color balance shifts toward the red end of the visible spectrum. It might more accurately be called "de-bluing" because no red color is added to the light, but instead the bluer colors are subtracted by scattering. The effect of reddening by dust in the Earth's atmosphere can make the Sun look yellow, orange, or even red as more of the short-wavelength photons are scattered. The effect is strongest at sunset when the sunlight reaching us has traveled a long path through our atmosphere, so that more of the short-wavelength photons are scattered away from their path toward us.

Small interstellar dust grains are very effective at scattering short-wavelength blue and ultraviolet light (Figure 73.5C). As a result, when we look at a star through a cloud containing dust, the light appears reddened. Reddening of background stars can be seen at the edges of the dark nebulae in Figure 73.5.

The extinction effects of dust on visible wavelengths of light make distant parts of our Galaxy hard to study. However, the longer wavelengths of infrared and radio waves are affected much less because they are much larger than the dust particles. As a result, infrared pictures reveal the structure of the Milky Way far better than visible-light pictures can, as shown in Figure 73.6. The disk and central bulge of our Galaxy are easy to see in this wide-angle computer image of stars mapped by infrared cameras.

FIGURE 73.6
An all-sky view of our Galaxy, as we see it from Earth. The image was made by plotting on a map of the whole sky the millions of stars observed in the infrared by the 2-Micron All Sky Survey (2MASS). It is therefore like a map of the Earth flattened out to show all continents. The map clearly reveals the flatness of our Galaxy and the bulge of its central regions. The reddening along the plane is caused by dust clouds, which allow more of the longer-wavelength infrared light to reach us.

73.3 RADIO WAVES FROM COLD INTERSTELLAR GAS

Unless an interstellar cloud is within a few parsecs of a hot star, it may cool down to just a few kelvins. Such cold material emits no visible light because it has too little energy to generate visible photons. However, even at such low temperatures it may still emit the low-energy, long-wavelength radiation that we can detect with radio telescopes.

One of the most important sources of radio emission is cold hydrogen atoms, which radiate at a wavelength of 21 centimeters. This radiation is called **H I** (pronounced *H-one*) **emission** because a cold hydrogen atom with electron attached is called H I, while ionized hydrogen is called H II (*H-two*).

H I emission arises because subatomic particles such as protons and electrons possess a property, known as *spin,* that makes them behave like tiny magnets. The energy of the magnetic interaction between the proton and electron is a tiny bit higher in the hydrogen atom when the proton and electron spin in the same direction. If the electron flips its spin direction, as shown in Figure 73.7, the energy is slightly lower. The atom emits the energy difference as a radio wave with a wavelength of 21 centimeters.

This **21-centimeter radiation** has proved extremely valuable for studying the Milky Way and other galaxies. First, this type of radiation is not absorbed by interstellar dust. Second, hydrogen is abundant in space, so the signal of 21-centimeter radiation is strong. From the strength of this signal, astronomers can deduce the amount of hydrogen in a region. If the gas is moving, the wavelength will be Doppler-shifted (Unit 25), and the shift will enable us to determine the gas's motion toward or away from us. Radio observations thereby allow astronomers to map not only where the gas is concentrated in the Milky Way but also how it is moving. Astronomers still face complications in interpreting these measurements, because the 21-centimeter radiation coming from a particular direction in the Milky Way is a blend of the signals from nearby and more distant gas along our line of sight. However, by modeling known motions in the Galaxy, astronomers can estimate the distances and produce maps of the entire Milky Way at the 21-centimeter wavelength. Such maps show that H I gas is confined to a thin disk in the inner regions, but the gas flares out in the outer parts of the Galaxy.

In addition to 21-centimeter radiation, many other wavelengths of radio radiation are emitted by interstellar gas. In the densest and coldest clouds, molecules form. These regions are known as **molecular clouds,** and molecules within them

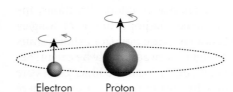

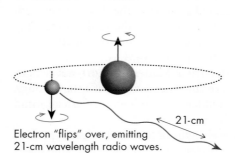

FIGURE 73.7
Radio radiation at 21 centimeters is emitted by cold hydrogen as an electron "flips" and spins in the opposite direction.

TABLE 73.1	Some Interstellar Molecules
Molecule	Chemical Formula
Molecular hydrogen	H_2
Hydroxyl radical	OH
Cyanogen radical	CN
Carbon monoxide	CO
Water	H_2O
Hydrogen cyanide	HCN
Ammonia	NH_3
Formaldehyde	HCOH
Acetylene	HCCH
Formic acid	HCOOH
Methanol	CH_3OH
Ethanol	CH_3CH_2OH

Concept Question 3

Why are H, C, N, and O the most common constituents of interstellar molecules?

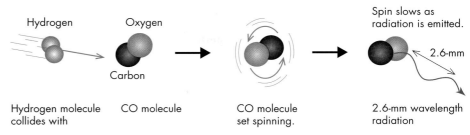

FIGURE 73.8

Emission of radio waves by spinning molecules in cold interstellar clouds. Collisions set the molecules spinning. When their rotation speed changes, they emit radio waves. The wavelength emitted depends on the kind of molecule and by how much its rotation speed changes.

emit primarily at radio wavelengths. More than 100 interstellar molecules have been identified so far, including common compounds such as molecular hydrogen (H_2), carbon monoxide (CO), formaldehyde (HCHO), and ethanol (CH_3CH_2OH), along with even more complex molecules. Table 73.1 lists a number of the more common molecules detected.

Collisions between molecules may make them spin or vibrate. As they slow down, the decrease in energy of spinning or vibration is emitted as radio-wavelength radiation. For example, carbon monoxide is a common interstellar molecule, and it can emit at wavelengths of 2.6 millimeters, 1.3 millimeters, and a series of shorter wavelengths as its spin slows (Figure 73.8). Astronomers can analyze these waves, just as they do 21-centimeter radiation, to learn about gas in cold dense clouds. In fact, these clouds emit little 21-centimeter radiation because their atomic hydrogen has combined to form molecular hydrogen. The observations of molecules therefore provide information about the densest interstellar environments, which complements the 21-centimeter observations.

73.4 HEATING AND COOLING IN THE ISM

The temperature of any object is set by a balance between the energy it receives (heating) and the energy it loses (cooling), and interstellar clouds are no exception. Gas can be heated in several ways. Two especially important heat sources are radiation and winds from hot stars, and blasts and radiation from supernovae. These processes add energy to the cloud's gases. Atoms and molecules are physically set in motion, and the high-energy photons from hot stars or explosions give atoms so much energy that they are ionized and eject electrons at high speed. If we could watch microscopically, we would see this energy translated into the random motions of thermal energy as electrons, atoms, and molecules collide with one another and rebound in different directions.

Heating of interstellar gas is strongly affected by the distance from the source and by shielding. Greater distances spread out the impact of photons, winds, and blast waves—the energy being spread over a larger and larger surface area with distance. **Shielding** occurs wherever material near a star "casts a shadow" that blocks energy from reaching interstellar clouds that are farther away. Dust is a particularly important source of shielding. Just as dust blocks the light of some stars from us, it can shield material behind it.

Cooling occurs when the gas emits radiation of its own. Radiation often results from collisions between gas atoms or molecules. To see how the thermal energy generated by a collision is turned into radiation energy, we look again at the interactions microscopically. If two molecules collide, part of the kinetic energy of their collision may be absorbed by electrons being knocked into higher energy levels. Energy is conserved because the two molecules rebound from each other with slightly lower speeds than when they collided, reducing the thermal energy. The electrons later drop down to lower energy levels, emitting one or more photons.

The net effect of collisions between molecules is to transfer thermal energy into radiation energy. Indeed, the radiation that allows us to see a cloud is also the means by which the cloud cools. Because collisions are more frequent where the atoms or molecules are closer together—that is, where interstellar matter is denser—high-density regions tend to cool more rapidly. This is especially interesting because cool high-density regions are also where gravitational forces are largest, so these regions tend to cool even more rapidly as they grow denser.

Dust grains also heat up when they absorb radiation or undergo collisions. The grains cool by emitting thermal radiation, which is generally at infrared wavelengths for the typical temperatures inside interstellar clouds (Unit 23). In dense clouds containing lots of dust, the combination of high density and shielding allows the gas and dust deep in the cloud to cool down to only a few kelvins (about −270°C or −450°F).

Applying these ideas, we find that near hot, blue stars (spectral type O or B, Unit 56), gas and dust will itself be hot, typically about 10,000 K. Such temperatures may even be hot enough to vaporize the dust grains or boil off some of their surface molecules. O and B stars are especially effective at heating nearby material because they emit large amounts of ultraviolet photons, which have shorter wavelengths and much more energy than visible photons. Ultraviolet light with a wavelength less than 91.2 nanometers is so energetic that when it is absorbed by hydrogen, it tears the electron free of the nucleus, ionizing the gas. The free electrons collide with atoms of oxygen, nitrogen, and other elements and excite them, making them emit light of their own. Eventually the electrons recombine with the ionized hydrogen atoms and emit more radiation as they drop from higher-energy orbitals down to lower-energy orbitals. Visible-wavelength photons are produced when the electrons drop down to the second energy level from higher levels, producing the pink light of hydrogen Balmer lines (Unit 56.4). This gives these nebulae their characteristic pink color, as seen in Figure 73.2. Because hydrogen in these hot gas clouds is ionized, these clouds are called **H II regions**.

The free electrons in H II regions are themselves a powerful source of radio emission. Astronomers can detect that emission and thus see, in radio wavelengths, H II regions that are otherwise hidden from us by dust in the interstellar medium. Because these H II regions are generally located in spiral arms, radio maps of their location help to reveal the spiral structure of our Galaxy. In fact, maps made by combining radio and optical observations of H II regions provide some of the best maps we have of the Milky Way's spiral structure (Figure 73.9).

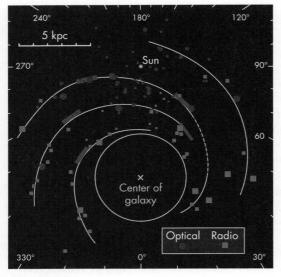

FIGURE 73.9
Map of the Milky Way made by combining radio observations (radio can penetrate dust and thereby allow us to see all the way across the Milky Way's disk) and optical observations of the position of H II regions. Such regions outline the spiral arms of other galaxies. In our Galaxy, too, they form a spiral pattern.

KEY POINTS

- The interstellar medium fills the space between stars with very sparse gas and dust of varying density and temperature.
- Several kinds of clouds or nebulae can be identified based on their gas and/or dust content.
- Emission nebulae generally are heated and ionized by O and B stars and emit visible spectral lines of hydrogen and other atoms.
- Gas in cool clouds between us and stars produces absorption lines in the star's spectrum in addition to the star's own.
- Dust particles in an interstellar cloud scatter and absorb starlight, dimming the light from background stars.
- Dust scatters short wavelengths more than long wavelengths, so reflected light is bluish, while light passing through is reddened.
- Hydrogen atoms and molecules radiate radio-wave photons from low-energy transitions.

- More than 100 molecules, some quite complex, have been identified in colder interstellar clouds.
- The absorption of energy from stars heats the ISM, while emission of radiation lets clouds cool; dust can sometimes shield a cloud from radiation.
- The final temperature of a cloud depends on the balance between the rates of absorption and emission of energy.

KEY TERMS

21-centimeter radiation, 576
dark nebula, 574
emission nebula, 573
extinction, 574
H I emission, 576
H II region, 578
interstellar medium (ISM), 572
molecular cloud, 576
reddening, 575
reflection nebula, 574
scattering, 574
shielding, 577

CONCEPT QUESTIONS

Concept Questions on the following topics are located in the margins. They invite thinking and discussion beyond the text.

1. Different nebulae for different observers. (p. 574)
2. Effect of reddening on distance estimates. (p. 575)
3. Why only a few elements are seen in molecules. (p. 577)

REVIEW QUESTIONS

4. How do we know interstellar matter exists?
5. What are the differences between emission, dark, and reflection nebula?
6. How do we know that some interstellar matter is dust?
7. How does the gas of the ISM cool?
8. How does interstellar dust affect our observations of stars and the Milky Way?
9. What type of radio radiation does the ISM emit? How?
10. What evidence makes astronomers believe the Milky Way has spiral arms?

QUANTITATIVE PROBLEMS

11. What is the density of a gas cloud that is 10 ly in diameter and contains 100 $M_\odot$ of material?
12. Given that a single hydrogen atom has a mass of 1.67×10^{-27} kg, how large would a spherically shaped interstellar cloud, with a density of 10^6 atoms/liter, need to be to contain a solar mass of material?
13. Convert the following radio wavelengths to frequencies (see Unit 22):
    a. 21-cm line of atomic hydrogen
    b. 2.6-mm line of carbon monoxide
14. How do the energies of photons with the wavelengths given in Problem 13 compare to the wavelength of a visible photon at 500 nm?
15. Find the ratio of the energy of a 21-cm photon to the energy of an electron shifting from the second energy level of hydrogen to the ground state, which produces a photon of 122-nm wavelength.
16. On average, a hydrogen atom emits a 21-cm photon once every 15 million years. How many 21-cm photons does a 100-$M_\odot$ gas cloud emit per second?
17. The dust in an interstellar cloud blocks blue light in the following way: For every 1 pc light travels through the cloud, only 90% of the light continues. Thus, after 2 pc, 90% of the remaining 90%, or 81% ($0.9 \times 0.9 \times 0.81$), remains.
    a. You might initially expect that after 10 pc, the ten 10% reductions would have removed all of the blue light. How much blue light actually remains?
    b. The same cloud removes about 7% of the red light every parsec, so after 1 pc, the ratio of blue to red has dropped to $90/93 \approx 97\%$ of its unreddened value. What is the ratio of blue to red after 5 pc? after 10 pc?

TEST YOURSELF

18. Which of the following is *not* part of the interstellar medium?
    a. Atomic gases
    b. Molecular gases
    c. Dust
    d. Stellar remnants
19. What is the most common source of energy that makes emission nebulae glow?
    a. Heat from hot white dwarfs within the nebula
    b. Nearby hot O and B type stars
    c. Nearby black holes
    d. Intense radio waves
    e. Gravitational compression
20. The reddening of starlight that passes through interstellar matter is caused by
    a. interstellar gases absorbing the blue light and thus making the sky blue.
    b. the addition of red light emitted by the intervening interstellar matter.
    c. the blue light being randomly scattered out by the interstellar dust.
    d. the Doppler effect from the interstellar matter moving toward Earth.
21. What kind of interstellar cloud generally has a bluish color?
    a. H I cloud
    b. H II region
    c. Emission nebula
    d. Reflection nebula
    e. Dark nebula

UNIT 74

Mass and Motions in the Milky Way

74.1 The Mass of the Milky Way and the Number of Its Stars

74.2 The Galactic Center and Edge

74.3 Density Waves and Spiral Arms

Learning Objectives

Upon completing this Unit, you should be able to:
- Define a rotation curve and explain what is unusual about the Galaxy's being flat.
- Explain what mass the rotation speed can be used to measure at a given radius.
- Describe the evidence for a massive black hole at the center of the Galaxy.
- Describe the physical mechanisms that may produce spiral arms in our Galaxy.

It is a challenge for us to discern the structure of the Milky Way, because we reside inside it (Unit 71). But we can determine much about its structure by applying our understanding of motions. Astronomers' measurements have shown that our Sun and neighboring stars move around the center of the Milky Way at a speed of about 220 kilometers per second. The Sun's large orbital speed (more than seven times the speed of the Earth around the Sun) may make you think the Milky Way spins rapidly; but because our Galaxy is so huge, the Sun takes approximately 220 million years to complete one trip around it. Since the extinction of the dinosaurs about 65 million years ago, our Galaxy has made less than one-third of a revolution.

Were it not for the collective gravity of all the components of the Milky Way, the motions of its stars and gas would cause the Galaxy to fly apart and disperse within a few billion years. At the same time, were it not for the motion, gravity would draw all the stars and gas inward, causing the Galaxy to collapse. The motion and gravity are in balance. As we saw in Unit 17, astronomers can use this balance to determine the mass of the body being orbited. As we will see in this Unit, the motions of the stars within our Galaxy do not give a simple result for the mass of the Milky Way. Rather, they hint at the presence of something dark residing in our Galaxy.

74.1 THE MASS OF THE MILKY WAY AND THE NUMBER OF ITS STARS

The rotation of stars in a flattened disk around the Galaxy's center is somewhat similar to the orbital motion of the planets around the Sun. Unlike a spinning solid disk, such as a wheel or a Frisbee, different parts of the Milky Way complete their orbits in different amounts of time. This is similar to how planets near the Sun complete their orbits faster than planets farther out—so too do stars near the center of the Galaxy complete their orbits faster than stars at the edge of the Galaxy. The phenomenon of the inner and outer parts of the Galaxy taking different times to complete an orbit is called **differential rotation.**

We can describe the unique properties of the Milky Way's rotation by observing stars at a variety of distances from the Galactic center to obtain a relation between orbital velocity and radius. Astronomers call this relation a **rotation curve.** Rotation curves for a solid disk and the Solar System are illustrated in Figure 74.1A and B. When we measure the speeds of stars rotating around the center of the Milky Way, though, we see neither a steadily rising rotation speed nor a steadily declining one. The stars orbit at about the same speed at most radii (Figure 74.1C).

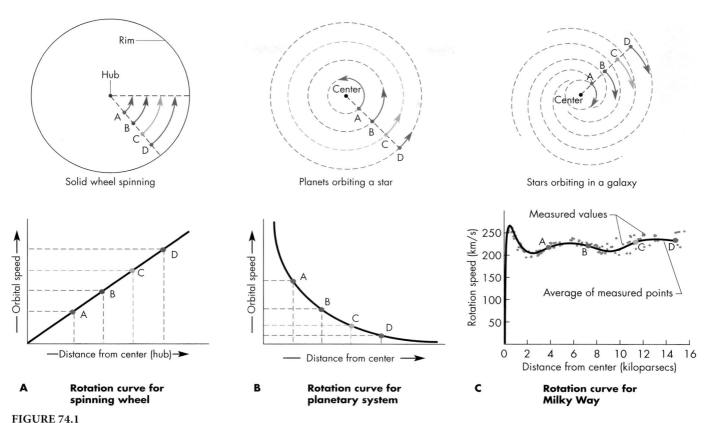

FIGURE 74.1
(A) When a wheel rotates, the outer parts move faster, creating a rising rotation curve. (B) Planets orbiting a star have slower speeds, the farther they are from the star. (C) The Milky Way's rotation speed remains nearly constant at all radii, as illustrated by actual data measured within our Galaxy.

The Milky Way's rotation curve reflects the distribution of mass in our Galaxy. It is very different from the distribution of mass in the Solar System, where more than 99% of the mass is concentrated in one central object (the Sun). A planet orbiting the Sun is subject to the gravitational pull essentially of just that one object. In the Milky Way, on the other hand, an orbiting star is held in its orbit by the collective gravitational pull of all the stars and other matter lying inside its orbit. For example, there might be a total of 1 billion solar masses out to a radius of 1 kiloparsec, but 2 billion solar masses within a radius of 2 kiloparsecs of the center, so a star orbiting at 2 kiloparsecs will be feeling the pull of much more mass. As a reminder that the amount of mass depends on the radius out to which we measure it, we will write $M_{<R}$. The subscript "<R" indicates "within a radius smaller than R." This peculiar notation will be helpful when we discuss the Milky Way and other galaxies, because modern observations suggest that galaxies have no clear edge and that their masses increase as we measure them out to larger radii.

The mass outside of a star's orbital radius has a gravitational pull, but it pulls outward in a variety of directions, canceling the net force.

Astronomers can calculate the Milky Way's mass from the gravitational attraction needed to hold the Sun and other stars in orbit (Figure 74.2). This is yet another application of Newton's law of gravity (Unit 17). From the speed V of a star orbiting at radius R from the center of the Galaxy, we can calculate the mass interior to its orbit as follows:

$$M_{<R} = \frac{V^2 \times R}{G},$$

$M_{<R}$ = mass within radius R (kilograms)
R = radius (meters)
V = velocity of rotation (meters per second)
G = 6.67 × 10^{-11} m³/kg·sec²

where G is Newton's gravitational constant. By looking at the rotation speeds of other stars at different distances from the center, we can build up a picture of the mass structure of the Galaxy.

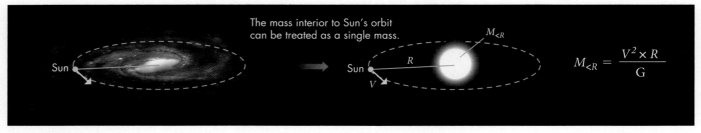

FIGURE 74.2
The Sun orbits all of the mass interior to its orbit $M_{<R}$. We can represent all of this mass by a single mass (the white ball in the figure) and apply Newton's law to find the mass within the Sun's orbit.

If we put in the rotation speed of the Sun (220 km/sec = 2.2×10^5 m/sec) and distance from the Galaxy's center (8 kpc = 2.5×10^{20} m), we find that the Milky Way's mass out to 8 kiloparsecs is

$$M_{<8\text{ kpc}} = \frac{(2.2 \times 10^5 \text{ m/sec})^2 \times 2.5 \times 10^{20} \text{ m}}{6.67 \times 10^{-11} \text{ m}^3/\text{kg·sec}^2}$$

$$= 1.8 \times 10^{41} \text{ kg.}$$

This is 9×10^{10} times the Sun's mass ($M_\odot = 1.99 \times 10^{30}$ kg). Thus, there are 90 billion solar masses of material in the Milky Way out to the Sun's orbit—the total of stars, gas, and all other matter.

This enormous total mass of material within the Sun's orbit is somewhat higher than estimates of the number of stars and other known material in the inner part of the Milky Way. The level of disagreement is not considered significant, given the uncertainties in our observations because of problems such as interstellar extinction. What is peculiar, though, is that outside the Sun's orbit, the rotation speed remains high even though the number of stars declines rapidly toward our Galaxy's outer reaches. At twice the Sun's distance from the center (16 kpc), the measured rotation speed remains at about 220 kilometers per second. So in the calculation of mass interior to 16 kiloparsecs, everything remains the same except that we multiply by a radius that is two times larger. The mass within 16 kiloparsecs is therefore twice the mass within 8 kiloparsecs, even though there appears to be only a fraction as many stars in this outer region.

We can even estimate the Milky Way's mass out to about 100 kiloparsecs from its center—beyond where we detect any stars—by examining the speeds of small satellite galaxies orbiting the Milky Way. We do not have very complete information about their orbits, but the Milky Way's gravitational effect on them implies that the Milky Way contains approximately 2 trillion solar masses ($2 \times 10^{12}\ M_\odot$). This is about 20 times more mass than we just found within the Sun's orbit, which is difficult to understand because we see almost nothing at these large distances that could account for so much mass. This discrepancy leads astronomers to conclude that the Milky Way's halo contains much more matter than the disk, and that most of this material emits no as-yet detectable light. They therefore describe this unseen mass as **dark matter.**

Measurements of the speed of rotation at different radii suggest that dark matter is present everywhere in the Milky Way. Its density is highest near the Galaxy's center, but in those regions the density of normal matter is much greater than the dark matter density. Based on rotation curve measurements, it appears that the overall fraction of the mass that consists of dark matter is greater in the outer parts of the Galaxy, eventually constituting the majority of mass. The dark matter apparently continues well beyond the area where we detect the outermost stars and gas.

In the vicinity of the Sun, astronomers have another clue about the nature of dark matter, from the way stars move within the Galactic disk. As the Sun and

Mathematical Insights

Because 16 kpc = 5.0×10^{20} m, we find that

$M_{<16\text{ kpc}}$

$= \dfrac{(2.2 \times 10^5 \text{ m/sec})^2 \times 5.0 \times 10^{20} \text{ m}}{6.67 \times 10^{-11} \text{ m}^3/\text{kg·sec}^2}$

$= 3.6 \times 10^{41}$ kg

$= 2 \times M_{<8\text{ kpc}}.$

Concept Question 1

If there were no mass beyond the Sun's orbit around the center of the Milky Way, what would the rotation curve look like?

other stars orbit the center of the Milky Way, their orbits do not remain precisely in the midplane of the disk. Instead small "vertical" motions through the disk carry them up and down through the plane, similar to how horses on a merry-go-round go up and down as they all circle the center. This occurs because as a star travels away from the midplane, the concentration of mass there pulls on it until it falls back. It then crosses through to the other side of the plane, and the process repeats, producing an oscillating motion. The Sun, for example, crosses through the midplane about once every 30 million years.

In the 1930s the speed of these oscillations was used to calculate the mass of material in the disk. The calculations showed that there is about twice as much mass in the disk as is evident based on the known stars, gas, and dust. This factor of 2 discrepancy is not as big as in the outer parts of the Milky Way, but it is nearby where telescopes can detect such things as dim, cold stars, old white dwarfs, or neutron stars. After decades of searching, the failure to identify the source of all this mass even near to the Solar System indicates that the dark matter must be something quite unusual.

Some of the most powerful evidence for the nature of dark matter comes from studying other galaxies and the beginnings of the universe. These studies suggest that dark matter is unlike the matter we are familiar with—and that the Milky Way lies at the center of a vast cloud of dark matter particles, which make up most of the mass of our Galaxy. We return to the question of dark matter in Unit 79.

Despite the uncertainty about the Milky Way's total mass, in the inner portions of the disk it appears that stars dominate the overall mass, so we can still estimate the number of stars from the rotation speed. From the rapid decline in stars seen at radii larger than the Sun's, we can deduce that the great majority of the Milky Way's stars are contained within about twice the Sun's distance from our Galaxy's center. In total, the disk contains about 100 billion stars, along with a comparable amount of mass consisting of interstellar matter, extremely dim stars or brown dwarfs, and stellar corpses. Each object moves along its own orbit around the center of the Galaxy, held in its path by the collective gravity of all the other objects.

74.2 THE GALACTIC CENTER AND EDGE

Within this spinning cloud of stars, the Sun is nearly lost, like a speck of sand on a beach. But unlike grains of sand that touch others, stars are widely separated. For example, in the vicinity of the Sun, stars are typically more than a parsec apart. The Sun's nearest neighbor is 1.3 parsecs away, a separation akin to sand grains spaced 10 kilometers (6 miles) apart. Near the core of the Milky Way, stars are packed far more densely, with a separation roughly 100 times smaller than near the Sun. Here the typical separation can be represented by sand grains placed at opposite ends of a football field.

The density of stars at the Sun's orbital radius can be measured in terms of the number of stars within a cube 1 parsec on a side (stars/pc^3). This is found to be about 0.1 stars/pc^3 or equivalently one star every 10 cubic parsecs. At twice the Sun's orbital radius (about 16 kpc) from the center, stars are spread even more thinly—less than 0.01 stars/pc^3—much as our atmosphere thins out as it merges into space. Thus, the Milky Way has no sharply defined outer edge.

Toward the center of the Galaxy, the density in the disk rises to about 1 star/pc^3 just outside the bulge—about 2 kpc from the center. Within the bulge, the density is about 10 times higher still. At the center of the Galaxy the density rises sharply, exceeding 100,000 stars/pc^3 in the central few parsecs, and the stars there orbit around the center quite rapidly.

What lies at the very center of the Milky Way Galaxy? The core is difficult to observe because interstellar dust clouds almost completely block the visible light emitted by objects there. However, it is possible to examine the Galactic center in

Concept Question 2

Would it be difficult to detect planets elsewhere in the Galaxy? Are they likely to make up much of the dark matter? Why or why not?

LOOKING UP

A visible-light image toward the center of the Milky Way is shown in Looking Up #7 at the front of the book.

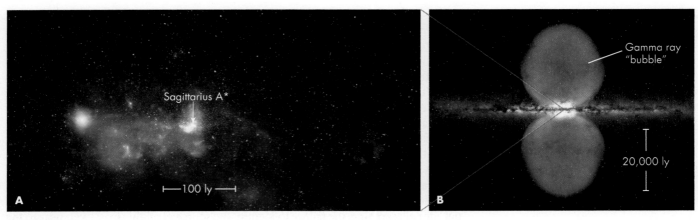

FIGURE 74.3
(A) The central region of the Milky Way as observed at X-ray wavelengths by the orbiting Chandra X-ray telescope. In this false-color image, the colors represent X-rays of different energies. The small bright spots are white dwarfs, neutron stars, and probably gas surrounding a few black holes. The diffuse glow is from extremely hot gas surrounding the Galaxy's core. (B) Illustration of giant "bubbles" of faint gamma-ray radiation discovered with NASA's Fermi telescope. The gamma-ray bubbles appear to be the result of energetic particles emitted from the Galaxy's center.

X rays as seen in Figure 74.3A. This image reveals a region emitting intense X-rays right at the center of the Galaxy surrounded by swirling clouds of hot gas. A recent gamma-ray study also reveals immense "bubbles" of energetic radiation that originate from our Galaxy's center and span almost half the sky (Figure 74.3B).

Within the X-ray bright region is an intense source of radio waves known as **Sagittarius A***, which is abbreviated "Sgr A*" (but usually pronounced *sadj ay star* by astronomers). The position of this mysterious object is shown in Figure 74.4A, an infrared image of the stars within the central 2 light-years of the Galaxy. Sgr A* appears to mark the very center of the Milky Way. Gas and stars orbit it, reaching

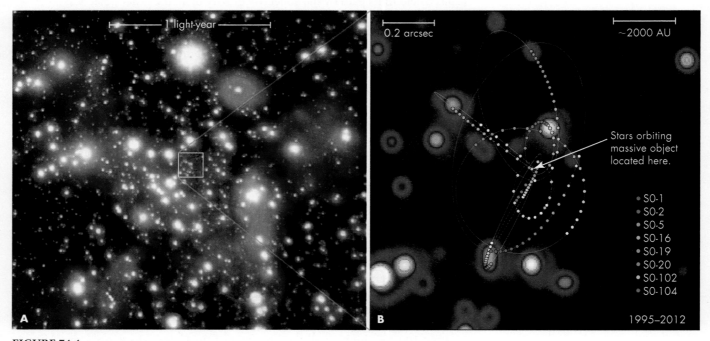

FIGURE 74.4
(A) Infrared image shows the inner 2 light-years of the core of the Milky Way. (B) The position of Sagittarius A* has been determined by tracking the orbits of stars as they orbit the massive black hole at the center of the Galaxy. The positions of eight stars tracked by the Keck/UCLA Galactic Center Group since 1995 are shown by dots of different colors, which are given a stronger color in each successive year along with an ellipse that extrapolates the rest of the orbit.

speeds of thousands of kilometers per second. Many of those orbits have been tracked with infrared cameras over the last decade (Figure 74.4B). We can use these stars' orbital motions to estimate the mass of Sgr A* using Newton's law of gravitation, just as we did to find the mass of the Galaxy. For example, the object SO-2 in Figure 74.4B was measured to be traveling at 5000 kilometers per second (5×10^6 m/sec) when it was about 20 billion kilometers (2×10^{13} m) from Sgr A*. The mass that SO-2 is orbiting, lying within this distance, should thus be

$$M_{\text{Sgr A*}} = \frac{(5 \times 10^6 \text{ m/sec})^2 \times 2 \times 10^{13} \text{ m}}{6.67 \times 10^{-11} \text{ m}^3/\text{kg·sec}^2}$$

$$\approx 8 \times 10^{36} \text{ kg}.$$

This is about 4 million solar masses. Yet no object is visible there, only a strong source of radio emission. That emission comes from a very small object, less than 10 AU in diameter (about the size of Jupiter's orbit). The almost inescapable conclusion is that a huge black hole lies at the center of the Milky Way and holds SO-2 and the other rapidly orbiting stars in its grip. In fact, these observations are perhaps the best evidence yet that black holes exist.

How might such a supermassive black hole form? When a massive star reaches the end of its life, it may collapse, leaving a black hole as a remnant. Such black holes contain only 10 or so solar masses; but in the central regions of a galaxy, where there is a greater density of objects, ordinary black holes may grow vastly more massive by drawing in nearby gas with their gravitational attraction. In normal star clusters, stars virtually never collide. But stars are packed more tightly in the center of the Milky Way than within clusters, and they may have been even more tightly packed at its center when the Milky Way formed. In such a crowded environment, stars may have collided, tearing away one another's gas, which then fell into the growing black hole. Once a black hole starts growing, it will not stop until it runs out of available matter, and some computer simulations suggest that a 10-$M_\odot$ black hole could grow to 10^6 $M_\odot$ in less than 1 billion years.

Astronomers are constructing the Event Horizon Telescope, an array of millimeter wave telescopes, to study the giant black hole at the center of the Milky Way. At millimeter wavelengths it should be possible to achieve high enough resolution to see light-bending effects at the black hole's Schwarzschild radius (Unit 69).

74.3 DENSITY WAVES AND SPIRAL ARMS

The motions of stars and gas give rise to the structure of the Milky Way. Although the stars in the Milky Way's disk orbit its center approximately in circles, we also know that the Milky Way displays concentrations of stars in its spiral arms. Spiral arms appear to be a common feature of galaxies, so some process must drive stars to collect together in this pattern.

The presence of a feature like spiral arms is puzzling, because as we noted earlier, the stars in the Milky Way exhibit differential rotation—stars take different periods of time to orbit the Galaxy's center. The Sun takes about 220 million years, but a star at a radius of 4 kiloparsecs takes about 110 million years, and a star at 12 kiloparsecs takes about 330 million years. Yet a spiral arm may stretch from 4 kiloparsecs to 12 kiloparsecs or even farther—so within a few hundred million years it should be torn apart by the different rotation speeds. Astronomers call this the **winding problem,** because it seems that differential rotation should cause spiral arms to become tightly "wound up" as the inner parts wrap around the Galaxy several times for every single orbit of the outer part of the arm (Figure 74.5).

We know that gravity holds a galaxy together, so we might guess that the gravity of the stars in a spiral arm might hold the arm together. Gravity does in fact help hold a spiral arm together, but not as a fixed rotating mass of material. Instead the stars and matter in different regions become alternately more or less tightly packed together due to gravity's effects, creating a phenomenon known as a **density wave.**

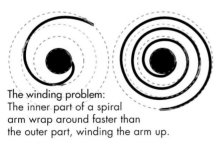

The winding problem: The inner part of a spiral arm wrap around faster than the outer part, winding the arm up.

FIGURE 74.5

Stars in the inner part of the Galaxy complete their orbits much more quickly than stars in the outer part, so if the same stars always remained in an arm, it should grow more tightly wound up over time.

The winding dilemma

A sound wave passing through the air is another kind of density wave—molecules in the air are alternately pressed closer together and then spread farther apart.

In the density-wave model, waves of stars and gas sweep around the Galactic disk. The waves are not an up-and-down motion like an ocean wave; instead, they are places where the density of stars and gas is large compared to the surroundings. The densest region is analogous to the wave crest, with the concentration of stars in the wave crest resulting in what we observe as a spiral arm. Spiral waves are not unusual in rotating matter; for example, liquid in a food blender often forms a spiral wave with the liquid moving through it.

As stars circle the Galaxy, a group of closely spaced stars at a given location will make the local gravity slightly stronger than elsewhere in the disk. That excess gravity draws stars toward the region so that the clump grows. Now imagine that clump stretched out into a spiral arm. Stars cannot remain in the arm continuously because their orbits move around the center of the Galaxy at speeds different from the **pattern speed** at which the arm goes around the center. However, the gravitational pull of the arm first hurries stars toward the arm as they approach it and then slows them as they depart it. Because all of the stars linger a little longer in that portion of their orbits where they are closest to the arm, the higher density of the arm is sustained.

When interstellar clouds move into an arm, they become compressed and more crowded, raising the density of the clouds and increasing the probability that they will collide with one another. This may in turn trigger the collapse of a cloud so that it forms a cluster of stars as illustrated in Figure 74.6. The brilliant blue O and B stars that form illuminate the arms and make them stand out against the more ordinary stars of the disk. These massive stars remain concentrated in the arms because they die before they have a chance to move past the area of clumping. Longer-lived stars are born there too, but they survive long enough to continue orbiting within the disk, so they are not so obviously concentrated in the arms.

Concept Question 3

Where else do we see spiral patterns? What causes them? Are there any connections to the spiral pattern in the Milky Way?

To better understand how arms form in the Milky Way, consider the following analogy. As cars move along a freeway, most move at nearly the same speed and keep approximately the same separation. If one car moves slightly slower than the others, however, traffic will begin to bunch up behind it. Cars can pass the slower vehicle but must change lanes to do so. The result is a clump of cars behind the slow one. But that clump is not composed permanently of the same cars. Cars join the clump from behind, pass the slower car, and then leave the clump at its front. Of course, in the spiral arms no stars move more slowly than all the rest; rather the *concentration* of stars in the arm is moving slower than the stars themselves.

The density-wave model explains several features of spiral arms. For example, in the galaxy shown in Figure 74.7A, the gas and stars are orbiting faster than the pattern speed, so they enter the arms on their inner edge. The raised density of the clouds makes them more opaque to starlight, producing a dark edge to that side

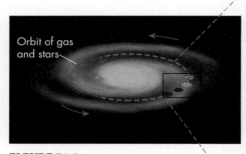

FIGURE 74.6
Sketch of a density wave in a galaxy, showing the progression from dust and gas to stars across the arm.

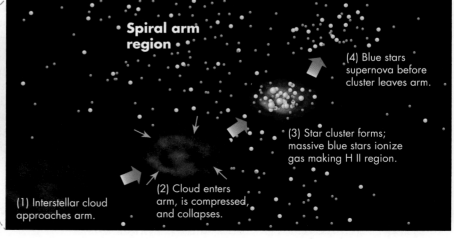

FIGURE 74.7
(A) The galaxy M81 has smooth spiral arms, but (B) M63 has ragged arms whose spiral pattern is difficult to trace. M63's arms may be the result of self-propagating star formation.

of the arm. Star formation makes the region just behind this dark edge bright, and then the arm gradually fades because the bright, short-lived stars die out as they cross the arm. Radio wavelength studies of the Doppler shifts of the gas also clearly show the gas moving into the arm on one side and leaving on the other.

Although many spiral galaxies exhibit a distinct spiral pattern, others have much more ragged arms, such as the galaxy shown in Figure 74.7B. An alternate hypothesis has been proposed to explain these ragged arms called the **self-propagating star formation** model.

In this second model, star formation starts at some random point in the disk of a galaxy when a gas cloud collapses and turns into stars. As the stars drive winds into the gas around them and explode as supernovae, the pressure they generate makes the surrounding gas clouds collapse and turn into stars. The original stars die, but the new stars—formed as the old ones evolve and explode—trigger more gas clouds to collapse and form additional stars. In this fashion, the region of star formation spreads across the galaxy's disk much as a forest fire burns outward in a circle through a forest. But unlike a forest, where the trees stand still, stars orbit in a galaxy, and those near the center orbit in less time than those farther out. The zone in which star formation occurs is drawn into a spiral shape by the difference in rotation rates between the inner and outer parts of the disk. The result is a spiral pattern without clearly delineated arms, as illustrated in Figure 74.8.

Based on what we can determine about the Milky Way's structure, astronomers suspect that both density waves and self-propagating star formation are important, but in many spirals just one of these processes may be dominant. As we shall see in Unit 76, there are yet other types of galaxies that have no spiral structure at all, indicating that star formation can occur in a variety of ways. Our understanding of how galaxies form and evolve is an area of active research and current debate.

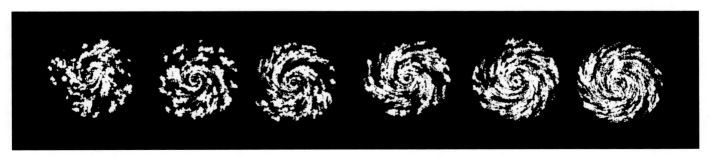

FIGURE 74.8
An early computer model of self-propagating star formation illustrates how this process can create the appearance of spiral structure despite the lack of clearly defined arms.

KEY POINTS

- Motions of stars within the Milky Way reveal how much mass is concentrated in each region.
- A rotation curve shows the average rotation speed at each radius, and indicates the amount of mass interior to that radius.
- In the Milky Way, the rotation speed is nearly constant outside the bulge, implying that the mass steadily increases with radius.
- The rotation speed remains high far outside the Sun's orbit, although little is visible there, implying the presence of dark matter.
- At the center of the Milky Way, individual stars can be seen orbiting an object of about 4 million solar masses.
- No visible object is present at the Galactic center, so the 4-million-solar-mass object is thought to be a supermassive black hole.
- Spiral arms are denser regions than average within the Galaxy's disk, and their gravity affects stars' orbits in a density wave.
- Compression of gas clouds as their orbits carry them through spiral arms may trigger star formation.
- A fragmentary spiral pattern can be produced by differential rotation stretching star-forming regions into short arcs.

KEY TERMS

dark matter, 582
density wave, 585
differential rotation, 580
pattern speed, 586
rotation curve, 580
Sagittarius A* (Sgr A*), 584
self-propagating star formation, 587
winding problem, 585

CONCEPT QUESTIONS

Concept Questions on the following topics are located in the margins. They invite thinking and discussion beyond the text.

1. Rotation curve if Sun is at edge of Milky Way. (p. 582)
2. Could planets be made of dark matter? (p. 583)
3. Spiral patterns in nature. (p. 586)

REVIEW QUESTIONS

4. How can we determine the Milky Way's mass?
5. What does differential rotation tell us about the mass distribution in the Milky Way?
6. What is dark matter? Why do astronomers conclude that the Milky Way may contain such unobserved material?
7. What types of material and densities are found close to the Galactic center?
8. What is the evidence for a black hole at the center of the Milky Way?
9. What are possible causes of the Milky Way's spiral arms?
10. How does a density wave induce star formation?
11. How do supernovae induce star formation?

QUANTITATIVE PROBLEMS

12. Given that the Sun moves in a circular orbit of radius 8 kpc around the center of the Milky Way, and its orbital speed is 220 km/sec, work out how long it takes the Sun to complete one orbit of the Galaxy. How many orbits has the Sun completed in the 4.5 billion years since it formed?
13. If the Milky Way were to double in mass, what would the Sun's velocity be?
14. Near the center of the Milky Way, the rotation speed climbs in proportion to the radius; so at 400 pc from the center, the speed is twice as fast as at 200 pc from the center.
    a. What does this imply about the mass within 400 pc versus the mass within 200 pc?
    b. Is there any differential rotation in this region? Explain.
15. Suppose there were no additional mass in the Milky Way beyond the distance of the Sun's orbit. What would the rotation speed be at a radius of 16 kpc (twice as far from the center as the Sun)?
16. What is the Schwarzschild radius of a black hole with the mass of the black hole thought to be at the center of the Milky Way? What is the angular size of the black hole as viewed from Earth?
17. If spiral arms are about 2 kpc in width, how long does the Sun, at a distance of 8 kpc from the Galactic center and an orbital speed of 220 km/sec, spend inside a spiral arm on its motion around the Milky Way?

TEST YOURSELF

18. What kinds of objects are least likely to be found outside a spiral arm?
    a. Interstellar gas
    b. The most massive stars
    c. Stars like the Sun
    d. Globular clusters
    e. Brown dwarfs
19. Evidence for the existence of dark matter comes from
    a. the inferred existence of a supermassive black hole at the center of the Galaxy.
    b. dark nebulae blocking the starlight from distant stars.
    c. the frictional slowing of the Sun's motion about the Galaxy.
    d. stars in the outer regions of the Galaxy orbiting faster than expected.
20. The solution to the winding problem is that
    a. spiral arms are young and have not had time to wind up yet.
    b. the rotational speed of the spiral arms is different than the orbital speed of stars.
    c. the spiral structure observed in the Galaxy is the result of random, unrelated star formation regions.
    d. stars travel up and down instead of in a flat plane.
21. Astronomers have determined the mass of the supermassive black hole at the center of our Galaxy by
    a. observing orbits of stars going around it.
    b. measuring the size of its Schwarzschild radius against background stars.
    c. using X-rays to see inside the hidden black hole.
    d. calculating from the speed of the Sun as it orbits the Galaxy.
    e. Doing all of the above.

UNIT 75

A Universe of Galaxies

75.1 Early Observations of Galaxies
75.2 The Distances of Galaxies
75.3 The Redshift and Hubble's Law

Learning Objectives

Upon completing this Unit, you should be able to:
- Relate the history of how we learned that the Milky Way was just one of many galaxies, and explain why interstellar dust caused so much confusion.
- Explain how astronomers find the distances of galaxies independent of redshifts.
- Define Hubble's law, and calculate the distance of a galaxy from its redshift.
- Explain where Hubble's law can be applied and why it is not accurate everywhere.

Beyond the edge of the Milky Way and filling the depths of space are billions of other star systems similar to our own. These remote, immense star clouds are called "external galaxies" or simply galaxies. When they were first discovered they were called "spiral nebulae" because of their unusual appearance, but it was not yet clear that they were objects like the Milky Way. Indeed, for centuries the Milky Way, described in Unit 71, was thought to be the entire universe, the only galaxy. When it was realized in the 1920s that the spiral nebulae were in fact huge star systems, one name that was suggested was "island universes," which was an apt description because each is an island of stars in a vast dark space.

Today we have the technology to detect more galaxies than the total number of stars in the Milky Way, and every one of these galaxies itself contains enormous numbers of stars. It is interesting to realize that when Einstein made his great discoveries about space and time in the early 1900s (Units 26 and 27), he thought that the universe consisted of a flattened collection of just a billion stars surrounded by a vast emptiness. It was not until a few years later that the picture of the universe as we understand it today began to take shape. As we shall see in this Unit, each **galaxy** is an immense cloud of hundreds of millions to trillions of stars, with each star moving along its own orbit, held within its galaxy by the combined gravitational force of all the other stars and matter. Thus, each galaxy is an independent, isolated star system.

FIGURE 75.1
M31, the Andromeda Galaxy, the nearest large galaxy to us.

75.1 EARLY OBSERVATIONS OF GALAXIES

All other galaxies are extremely distant from us, generally millions or billions of light-years away. Their great distance makes galaxies look dim to us, and only a few can be seen with the unaided eye. For example, on clear, dark nights in autumn and winter in the Northern Hemisphere, with your unaided eyes you can see the Andromeda Galaxy (Figure 75.1), which is more than 2 million light-years away. It looks like a pale elliptical smudge on the sky. Early observers such as the tenth-century Persian astronomer Al-Sufi noted this object, and it appeared on early star charts as the "Little Cloud." From the Southern Hemisphere, you can easily see by eye the Large and Small Magellanic Clouds, two small satellite galaxies of the Milky Way that are "only" 150,000 light-years away from us (Figure 75.2).

FIGURE 75.2
Photograph of the Large and Small Magellanic Clouds, two small galaxies that orbit the Milky Way.

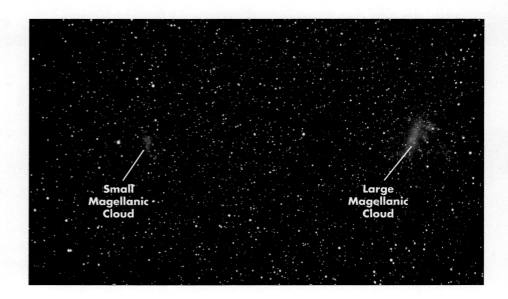

LOOKING UP

M31 is shown in Looking Up #3 at the front of the book. The Magellanic Clouds are both visible in Looking Up #9.

Clarification Point

Astronomers normally capitalize *Galaxy* when referring to the Milky Way and refer to other galaxies with the lower-cased *galaxy*, much as we capitalize *our Moon* but not *moons*.

One thing you can learn when looking at these galaxies by eye or through a telescope is that galaxies, even though magnified, still shine with only a pale glow—similar to the pale light from the Milky Way itself. Even if they contain billions of stars, galaxies do not appear bright. The reason for this is that the large distances between stars in any galaxy mean that the light is always spread out. The galaxies also look pale through a large telescope, because although the starlight is magnified, so are the vast spaces between the stars. Astronomers refer to this property as the **surface brightness** of galaxies (also see Unit 29.4). The surface brightness does not change with distance or magnification because the tiny fraction of a galaxy's total area that is covered by the area of stars remains the same. To study details of the low surface brightness features of galaxies requires the use of photography or modern imaging systems to collect light over a long period of time.

The study of external galaxies began in the eighteenth century when the French astronomer Charles Messier (pronounced *MESS-yay*) accidentally discovered many galaxies during his searches for new comets. He realized that some of the faint, diffuse patches of light never moved; so to avoid confusing them with comets, he assigned them numbers and made a catalog of their positions. Although many of Messier's objects have since been identified as star clusters or glowing gas clouds in the Milky Way, several dozen are galaxies. These and the other objects in the Messier catalog are still known by their Messier number, such as M31, the Andromeda Galaxy mentioned earlier.

In the early part of the nineteenth century other astronomers, such as William and Caroline Herschel, began to map faint objects in the heavens systematically, finding and cataloging numerous astronomical objects, including galaxies. The Herschels' work was continued by William's son John, and in the late 1800s was revised and added to by John Dreyer to include many thousands of galaxies and additional nebulae. This compilation is now known as the New General Catalog (or NGC for short), and it gives many galaxies their name, such as NGC 1275. Many galaxies appear in more than one catalog and so bear several names. For example, Messier's M31 is the same galaxy as NGC 224.

For these early observers, the nature of the galaxies was a mystery. They looked different from other types of cataloged objects—planetary nebulae, globular clusters, emission nebulae, and so on—and because individual stars could not be resolved, many believed they were clouds of glowing gas rather than the merged light of billions of stars. One notable aspect of galaxies was that they were found only in portions of the sky away from the plane of the Milky Way. Even a modern

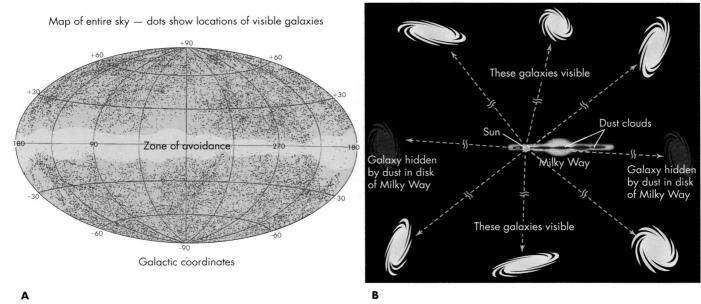

FIGURE 75.3

Dust limits our view in the disk of our Galaxy and creates the zone of avoidance. We see no galaxies in the plane of our own Galaxy because dust in the disk blocks our view. (A) A plot of where galaxies appear to be distributed on the sky. Notice how few lie along the central line (the galactic equator). This blank region is the zone of avoidance. (B) A sketch illustrating why we see so few galaxies along the plane of the Milky Way. Dust clouds in the disk of the Milky Way block our view toward directions that lie close to the Milky Way's plane. (Note that the galaxies are in reality about the same size as the Milky Way and are much farther away than can properly be shown in the figure.)

catalog of deep-sky galaxy observations contains very few galaxies near the galactic plane, as shown in Figure 75.3A. Many astronomers once interpreted this **zone of avoidance** around the Milky Way as evidence that the galaxies must somehow be interacting with (and therefore be near) the Milky Way. They asked, If these objects were far away, how could they "know" to avoid this direction? What astronomers did not realize at the time was that the zone of avoidance is simply an effect of the dust in our own Galaxy blocking the light from distant galaxies (Figure 75.3B). This is the same effect that blocks our view of the center of the Milky Way.

As early as 1755, the German philosopher Immanuel Kant, whose views on the origin of the Solar System were far ahead of their time, suggested that galaxies might be remote star systems—island universes—similar to the Milky Way. However, until the 1920s there was little observational evidence to clearly show whether galaxies were huge distant systems of stars like the Milky Way or were small nearby objects associated with our own Galaxy, and astronomical opinion was strongly divided.

This controversy was aired in detail during a series of debates organized by the National Academy of Sciences in which two U.S. astronomers, Harlow Shapley (Unit 71) and Heber Curtis, took opposing sides. Shapley argued that the Milky Way was huge and that other galaxies (then called spiral nebulae) were merely small, nearby companions. Curtis, on the other hand, argued that the Milky Way was smaller than Shapley claimed, that the nebulae were star systems like our own, and that they were immensely distant from us.

The debate reflects the challenges of an observational science like astronomy. For example, Curtis argued that objects such as M31 must be far away based on a supernova that had been observed in 1885. Assuming this was as luminous as the supernovae that had been observed in our own Galaxy hundreds of years earlier (Unit 66), he found that M31 must be very distant; and if it was distant, then M31 must be huge. Shapley countered that the object seen in 1885 was not a supernova but a nova—a much less luminous explosion—and therefore M31 was nearby,

Adriaan van Maanen reported that he detected rotational shifts in the positions of features in spiral galaxies in the early 1900s. These shifts were later shown to be incorrect, possibly biased by van Maanen's belief that galaxies were nearby objects.

perhaps at a distance more like he had found for the globular clusters orbiting the center of the Milky Way. Both arguments were reasonable based on the evidence available at the time, and both fit into the model each side had developed.

Shapley based part of his argument on a claimed detection of motions inside galaxies—observations that were later found to be incorrect. If such motions had been detectable, they would have ruled out the possibility that the galaxies were far away. Curtis correctly doubted these reported motions, even though the report was made by a well-respected astronomer of the time. As in any mystery story, not all clues are useful or correct.

Historians differ on who "won" the debate. Shapley was more nearly correct about the Milky Way's size and structure (also see Unit 71). Curtis was right about the nebulae being distant systems similar in structure to the Milky Way.

One of the biggest problems for both sides in the debate was the lack of understanding about the dimming caused by dust in the Milky Way. Curtis adopted *Kapteyn's Universe,* a model of a small Milky Way (Unit 71), whereas Shapley interpreted the zone of avoidance as evidence that the galaxies were nearby. The other major problem was the lack of a definitive method for measuring the galaxies' distances. Breakthroughs in the 1920s solved both of these problems. Astronomers discovered how starlight was dimmed by dust in the Milky Way, and they were also able to find *standard candles* (Unit 55.3) in external galaxies. With these discoveries, the true nature of galaxies as enormous systems containing billions of stars was finally confirmed.

75.2 THE DISTANCES OF GALAXIES

Galaxies' distances are so enormous that we cannot measure them using the same methods that we use to find distances to stars. Parallax cannot be used to measure such vast distances because the angle by which the galaxy's position changes as we move around the Sun is too tiny to detect. Individual stars within an external galaxy are extremely difficult to detect because the angular separations are so small that atmospheric blurring (Unit 32.2) blends their light together. To apply the method of standard candles, then, astronomers must find stars of known luminosity that are extremely bright, so that their light can be picked out from the sea of surrounding stars.

Concept Question 1

If the intergalactic space between the Milky Way and M31 contained dust, how would that have affected Curtis's measurement of M31's distance?

In using the standard-candle method to find a galaxy's distance, astronomers use the principle that the farther from us an object is, the dimmer it looks. More precisely, if we know the luminosity of a light source—our standard candle—and how bright it looks to us, we can find its distance by using the inverse-square law (Unit 55).

Astronomers use many different astronomical objects as standard candles, but one of the most reliable is the class of luminous, pulsating variable stars known as Cepheids (Unit 64). Cepheids have luminosities more than 10,000 times the luminosity of the Sun, and because of that brilliance it is possible to detect them in nearby galaxies. Furthermore, because their brightness changes as they pulsate, their light can be distinguished from other stars. In the early 1900s, the U.S. astronomer Henrietta Leavitt discovered the period–luminosity relation (Unit 64.3) by careful examination of the light from Cepheids in the Magellanic Clouds. She showed that a Cepheid's period (the time it takes to go from bright to dim to bright again) can be used to determine its luminosity.

Once astronomers know the variable star's luminosity, they can find its distance from a measurement of how bright it appears. For example, suppose we observe a Cepheid in a distant galaxy, as illustrated in Figure 75.4A. From observations of Cepheids in our own Galaxy, we know what luminosity a Cepheid has based on its period, which calibrates the period–luminosity relationship. Therefore, we can

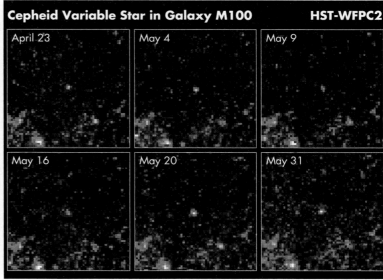

FIGURE 75.4
(A) Cepheids in the Milky Way have known distances, d_*, so we can determine their luminosities. Knowing those luminosities, we can find the distance to a galaxy, d_g, if we can measure the brightness of a Cepheid in it. (B) Hubble Space Telescope observations of a Cepheid variable in the galaxy M100.

d_g = Distance to Cepheid and galaxy
L_c = Luminosity of Cepheid from period–luminosity relationship
B_c = Measured brightness of Cepheid

Concept Question 2

Why do multiple measurements give a more accurate value?

Concept Question 3

When Edwin Hubble originally measured the distance to M31, the values of the luminosity in the period–luminosity relationship he was using were four times lower than they are known to be today. How would this have affected his distance measurement?

determine the luminosity L from the period of variation, and we can measure the brightness B directly. With these values we can determine the distance to the Cepheid using the standard-candle method:

$$d_g = \sqrt{\frac{L_c}{4\pi B_c}}.$$

Astronomers generally measure the distances of many Cepheids within the same galaxy to get an accurate average for the galaxy's distance.

Cepheids are very luminous, but few galaxies are close enough that we can make successful ground-based measurements of their light. Leavitt's discovery allowed the U.S. astronomer Edwin Hubble to determine a distance to M31 in 1923, finally proving that it was millions of light-years away and even larger than the Milky Way. Over the subsequent half century, Cepheids were observed in only a handful more galaxies. It required the Hubble Space Telescope to carry out similar observations for dozens of more-distant galaxies, such as that shown in Figure 75.4B. The Cepheid in the figure is in the galaxy M100 and was measured to have a period of 51 days. The period–luminosity relationship indicates that this Cepheid has a luminosity of $3 \times 10^4\ L_\odot$ (or 1.2×10^{31} watts). The apparent brightness measured with the Hubble Space Telescope was 3.4×10^{-18} watts per square meter. Therefore, the distance to the Cepheid (and its galaxy) is

$$d_g = \sqrt{\frac{1.2 \times 10^{31}\ \text{watts}}{4\pi \times 3.4 \times 10^{-18}\ \text{watts/m}^2}} = 5.3 \times 10^{23}\ \text{m} = 1.7 \times 10^7\ \text{parsecs}.$$

Thus, the galaxy M100 is about 17 million parsecs (56 million light-years) distant. Even for the Hubble Space Telescope, though, Cepheids are undetectable in any but the nearest few hundred galaxies.

To measure greater distances, astronomers must use even more-luminous standard candles, such as supernovae. However, we cannot predict when a supernova of known luminosity might occur in a galaxy of interest. This might have been an insurmountable problem but for a very strange discovery.

75.3 THE REDSHIFT AND HUBBLE'S LAW

In the early 1900s the U.S. astronomer Vesto Slipher discovered that the spectral lines of nearly all of the galaxies he observed were strongly shifted toward longer wavelengths—to the red end of the visible spectrum. This so-called *redshift* is seen in all galaxies, except for a few nearby ones whose motion is influenced by the gravitational attraction of the Milky Way and its neighbors.

Redshift is defined as the fraction by which the light's wavelength changes—that is, the change in wavelength divided by the original wavelength. For example, suppose we observed a galaxy's spectrum and saw that the pattern of spectral lines was shifted to longer wavelengths (Figure 75.5A). Among these we might identify an absorption line that normally occurs at 500 nanometers, but is shifted to 505 nanometers in this galaxy. The line has thus been shifted by 5 nanometers, 1% of its normal wavelength of 500 nanometers. This galaxy's redshift is 1%, or 0.01.

Astronomers use the letter z to stand for redshift, and mathematically we can define the redshift as follows:

$$z = \frac{\Delta\lambda}{\lambda}.$$

z = Redshift
$\Delta\lambda$ = Change in wavelength from its original value
λ = Original value of the wavelength

In this formula, $\Delta\lambda$ stands for the change in wavelength, and λ is the original, unshifted wavelength. If we observed a second galaxy for which the wavelength of the 500-nanometer line had shifted to 515 nanometers, we would say $\Delta\lambda = (515 \text{ nm} - 500 \text{ nm}) = 15 \text{ nm}$, and $\Delta\lambda = 500 \text{ nm}$, so

$$z = \frac{15 \text{ nm}}{500 \text{ nm}} = 0.03.$$

This galaxy's redshift is 0.03, or 3%. Incidentally, we chose a 500-nanometer line for this example for numerical simplicity. If we looked at any other spectral line, we would find that they have all been shifted by 3% relative to their normal value, whether they were generated at radio wavelengths or X-rays.

The discovery of galaxy redshifts proved to be even stranger when Edwin Hubble and others found that the redshift was larger for more-distant galaxies.

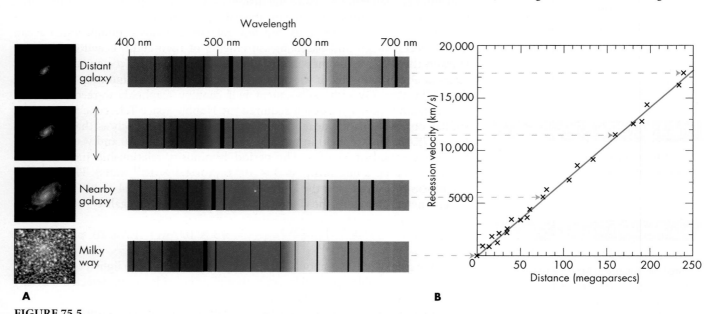

FIGURE 75.5

(A) Galaxies show a larger redshift of their spectral lines the more distant they are. The spectrum of the integrated light from the Milky Way (which has zero redshift) is shown for comparison. (B) Hubble's law (shown by the red line) indicates that the recession velocity increases in proportion to the redshift.

They interpreted the redshift as a Doppler shift (Unit 25), and the increasing redshifts then implied that the speed with which a galaxy moves away from us—its **recession velocity**—increases with distance, as Figure 75.5 shows. That is,

The faster a galaxy recedes, the farther away it is.

Assuming the Doppler effect, Hubble calculated the recession velocity $V = c \times z$. A 1% redshift implies a recession velocity of 1% of the speed of light, or about 3000 kilometers per second. A 3% redshift implies a recession velocity of about 9000 kilometers per second.

Galaxy redshifts look like Doppler shifts, but today we understand from Einstein's theory of general relativity (Unit 27) that the redshifts are caused by a different effect: the expansion of the space between the galaxies as the light travels through it. Does this sound like nothing more than a semantic difference? It turns out to be fundamentally important for understanding the universe as a whole. We will use the notation and language of velocities here that were adopted by Hubble and are still used by astronomers today, but we must keep in mind that astronomers today interpret them quite differently. In particular, astronomers now view redshift as arising from the expansion of space. We will return to the implications of the redshift for understanding the universe in Unit 80.

In 1920 Edwin Hubble discovered that a simple formula relates the recession velocity, V, and the distance, d. When he graphed the recession velocities and distances of a number of galaxies (Figure 75.5B), he found that they lay along a straight line. The line can be described by the equation

$$V = H \times d$$

where H is a constant. Because of his discovery, this relation is called **Hubble's law,** and H is called **Hubble's constant.**

The value of the Hubble constant depends on the units used to measure V and d. Astronomers generally measure V in kilometers per second and d in **megaparsecs** (abbreviated **Mpc**), where 1 megaparsec is a million parsecs (= 3.26 million light-years). In these units, H is about 70 km/sec per Mpc.

As recently as the 1990s, astronomers could not agree on the value of H to within a factor of 2. The reason for this controversy was that to determine H, we must know both the distance and the velocity of at least a few galaxies to calibrate the law. H is simply the average value of each calibration galaxy's velocity divided by its distance. However, as we have just seen, accurately measuring the distance to even a nearby galaxy is difficult. Based on galaxies for which Cepheid variable distances have been calculated as well as other techniques, the value of Hubble's constant is approximately $H = 70$ km/sec per Mpc, with an uncertainty possibly as large as 5 km/sec per Mpc.

Once a value for H is known, we can turn the method around and use the Hubble law to find the distance to a galaxy. The method is as follows: Take a spectrum of the galaxy whose distance you want to know. From the spectrum, measure the shift of the spectral lines. From the Doppler-shift formula, calculate the recession velocity. Finally, use that velocity in the Hubble law to find the distance. Mathematically, we divide both sides of the Hubble law by H, obtaining

$$d = V/H.$$

We insert the measured value of V and our choice of H and solve for the distance. The power of this formula is also why getting H right is so important—if H is incorrect, our whole scale of the universe will be off.

As an example of applying Hubble's law, suppose we find from a galaxy's spectrum that its recession velocity is 49,000 kilometers per second. We find

---

Mathematical Insight

If $V = c \times z$, then for $z = 3\%$, we find
$$\begin{aligned} V &= 3.00 \times 10^8 \text{ m/sec} \times 0.03 \\ &= 9 \times 10^6 \text{ m/sec} = 9000 \text{ km/sec.} \end{aligned}$$

V = Recession velocity of galaxy (km/sec)
H = Hubble's constant (km/sec per Mpc)
d = Distance to galaxy (Mpc)

Hubble's constant is often written H_0 to indicate its value as measured today at redshift $z = 0$. The Planck spacecraft (Unit 84) has recently measured $H_0 = 67.1$ km/sec per Mpc, but the most recent and accurate Cepheid measurements find $H_0 = 74.2$ km/sec per Mpc. We adopt $H_0 = 70$ km/sec per Mpc for simplicity.

its distance by inserting this value for V in the previous equation and then divide by H:

$$d = \frac{49{,}000 \text{ km/sec}}{H}.$$

For $H = 70$ km/sec per Mpc, the value of d is

$$d = \frac{49{,}000 \text{ km/sec}}{(70 \text{ km/sec})/\text{Mpc}} = 700 \text{ Mpc}.$$

The galaxy's recession velocity implies a distance of about 700 Mpc.

In applying this method, we must be careful about units. When we divide V in kilometers per second by H in km/sec per Mpc, the answer we get will be in megaparsecs. The 700 we just found is not in parsecs or light-years but in megaparsecs. To get the distance in light-years, we multiply by 3.26 light-years per parsec to get about 2300 million light-years, or 2.3 *billion* light-years. Detecting a galaxy this far away was time-consuming until the 1990s, but newer technologies can measure thousands of galaxies at these distances in a single night of observation.

The Hubble law does not work for nearby objects. For example, M31, the spiral galaxy nearest to us, is moving *toward* the Milky Way at about 90 kilometers per second, so it has a "negative redshift." We cannot apply Hubble's law to it, or we would get the absurd result that M31 has a negative distance. Similarly, we cannot apply Hubble's law to a nearby star that is moving away from the Sun, even though its spectrum is redshifted. Its redshift is caused by orbital motion inside our Galaxy, and therefore, again, Hubble's law does not apply.

Hubble's law applies only to redshifts caused by the general motion of galaxies away from each other—what astronomers term the *overall expansion of the universe*. When we are observing distant galaxies, the redshift we measure is caused primarily by the expansion of the universe, but in part by the galaxy's orbital motion around another galaxy or system of galaxies. To make the distinction clear, astronomers refer to the first as the **expansion redshift,** and to the second simply as the *Doppler shift*, as illustrated in Figure 75.6. If we assume the redshift we measure is entirely due to expansion, the distance we calculate from Hubble's law might be too high or too low by an amount depending on the size of the Doppler shift. For an isolated galaxy, the Doppler shift motion is usually under 100 kilometers per second. So if we calculate a velocity of 700 kilometers per second for such a galaxy from its redshift, the true recession velocity might be anywhere in the range of 600 km/sec to 800 km/sec. By using 700 km/sec in the Hubble law, we will find this distance:

$$d = \frac{V}{H} = \frac{700 \pm 100 \text{ km/sec}}{70 \text{ km/sec per Mpc}} = 10 \pm 1.4 \text{ Mpc}.$$

The uncertainty in the expansion redshift produces an uncertainty in the distance equal to the orbital speed divided by H. This uncertainty may cause significant errors in the distances of galaxies orbiting within a large system of galaxies where orbital speeds in excess of 1000 km/sec have been observed. If we assumed these galaxies' redshifts were caused entirely by expansion, we could make an error of up to $(1000/70) \approx 14$ Mpc in calculating their distances. This is one reason it was so difficult to determine Hubble's constant accurately: the nearby galaxies with good distance measurements all have uncertainties in their expansion redshifts.

With Hubble's constant now determined to within about 5%, we can use the redshift to find the distance to a galaxy a hundred megaparsecs away with an uncertainty of only a few megaparsecs if the galaxy is isolated. No other method for finding the distances of remote galaxies is as easy to use or as accurate—provided that the rate of expansion has remained constant.

> **Clarification Point**
>
> The Hubble law will not give an accurate distance for nearby objects where the orbital motions exceed or are comparable to the recession velocity caused by the expansion of the universe.

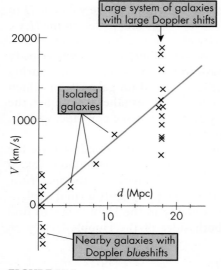

FIGURE 75.6
The Hubble law works best for isolated galaxies. Orbital motions in groups of galaxies give Doppler shifts that add or subtract from the expansion redshift, even causing some nearby galaxies to have blueshifts.

KEY POINTS

- The existence of galaxies outside the Milky Way was not established until the 1920s.
- Although galaxies may contain hundreds of billions of stars, they appear dim because of their huge distance from us.
- Galaxies appear to avoid the plane of the Milky Way because they are hidden from view by dust in our Galaxy.
- The remoteness of galaxies was finally established with the detection of Cepheids in them, which could be used to find distances.
- The light from almost all galaxies is redshifted—with all wavelengths lengthened by the same percentage.
- A galaxy's redshift is often described as a recession velocity, as if we are stationary and the galaxy is Doppler shifted as it moves away from us, but the redshift is actually due to the overall expansion of the universe.
- Hubble discovered that galaxies that are more distant have a larger redshift in proportion to their distance.
- Hubble's law can be inverted to estimate a galaxy's distance from its redshift or recession velocity.
- Galaxies may have a Doppler shift in addition to their redshift, so there is an uncertainty in distances found using Hubble's law.

KEY TERMS

expansion redshift, 596
galaxy, 589
Hubble's constant, 595
Hubble's law, 595
megaparsec (Mpc), 595
recession velocity, 595
redshift, 594
surface brightness, 590
zone of avoidance, 591

CONCEPT QUESTIONS

Concept Questions on the following topics are located in the margins. They invite thinking and discussion beyond the text.

1. Effect of intergalactic dust on distance estimates. (p. 592)
2. Why multiple measurements produce better results. (p. 593)
3. Effect of miscalibration on distance estimates. (p. 593)

REVIEW QUESTIONS

4. How do astronomers measure the distances to nearby galaxies?
5. What is Hubble's law?
6. Why are astronomers uncertain about the precise value of Hubble's constant?
7. Why is a galaxy difficult to see with the naked eye even though its apparent brightness may be greater than that of stars we can see?
8. What is the zone of avoidance?
9. How do astronomers measure the distance to nearby galaxies?
10. Why is parallax not used to determine the distances to galaxies?
11. What is an expansion redshift?

QUANTITATIVE PROBLEMS

12. A Cepheid variable with a period of 20 days and an apparent brightness of 5.28×10^{-16} watt/m^2 is observed in a nearby galaxy. Estimate the Cepheid's luminosity from the period-luminosity relation (Figure 64.5) and convert this to watts. How far away is the host galaxy?
13. Suppose that a galaxy's 21-cm line of atomic hydrogen is shifted to 22 cm.
    a. What is the redshift of the galaxy?
    b. What is its recession velocity?
14. Two identical galaxies have redshifts of 0.036 and 0.072. Which has the larger angular size? What is the ratio of their apparent sizes?
15. For most of the second half of the twentieth century, measured values of Hubble's constant ranged from about 50 to 90 km/sec per Mpc. For a galaxy with a recessional velocity of 28,000 km/sec, what range of distances would Hubble's law return?
16. What is the approximate recession velocity of a galaxy that we are seeing as it looked 1 billion years ago?

TEST YOURSELF

17. A galaxy has a recession velocity of 7000 km/sec. If the Hubble constant is 70 km/sec per Mpc, how far away from Earth is the galaxy?
    a. 10^6 Mpc
    b. 100 Mpc
    c. 0.01 Mpc
    d. 10^8 Mpc
    e. 10^{-6} Mpc
18. In order to apply Hubble's law to find the distance to a galaxy, what quantity is actually measured?
    a. Redshift
    b. Rotational velocity
    c. Hubble's constant
    d. Surface brightness
19. The zone of avoidance is
    a. an unusually large dark nebula.
    b. the portion of a galaxy for which individual stars cannot be resolved.
    c. the region on the sky blocked by the plane of the Milky Way Galaxy.
    d. the nearby region to the Milky Way Galaxy in which Hubble's law does not apply.
20. Why does a galaxy like M31 have a blueshift instead of a redshift?
    a. All nearby galaxies have a blue color.
    b. Local orbital motions are bigger than the expansion redshift.
    c. M31 contains no Cepheid variables.
    d. Young galaxies have more high-mass blue stars.
    e. M31 is actually behind us, but seen reflected by dust in the Milky Way.

UNIT 76

Types of Galaxies

76.1 Galaxy Classification
76.2 Differences in Star and Gas Content
76.3 The Evolution of Galaxies
76.4 Galaxy Mergers and Changing Types

Learning Objectives

Upon completing this Unit, you should be able to:
- Describe our galaxy classification system, and identify a galaxy's type from an image.
- Explain how galaxies' physical properties and composition relate to their type.
- Describe how galaxies have changed over the history of the universe, and discuss how mergers may have caused galaxies to grow and evolve between types.

In the vast universe of galaxies we began to explore in Unit 75, many galaxies look similar to the Milky Way in size and shape (Unit 71). Other galaxies are distinctly different. For example, some do not have a disk and spiral arms. Rather, they are shaped more like a rugby ball or a distorted sphere, with their stars distributed in a vast, smooth elliptical cloud. Others are neither disks nor smooth elliptical shapes but are completely irregular in appearance.

Galaxies differ in more than shape; they also differ in their content. The Milky Way has populations of young and old stars (Unit 72), but some galaxies contain almost exclusively one or the other. Some are rich in interstellar matter, whereas others have hardly any at all. Some galaxies—despite having as much total mass as the Milky Way—have formed far fewer stars, leaving them dim and scarcely visible to us. Others have a small but tremendously powerful energy source at their core that emits as much energy as the entire Milky Way but from a region about 0.001% its size, less than a few light-years in diameter. Many of these *active galactic nuclei* eject gas in narrow jets at nearly the speed of light (Unit 78).

Astronomers can see all this diversity in galaxies, but we are only beginning to understand how it arises. Unlike stars, whose structure and evolution are well understood, many aspects of galaxies remain a mystery. Astronomers do not all agree, for instance, on what causes some galaxies to be disk-shaped while others are more nearly egg-shaped. Although this lack of certainty can be confusing, it offers us a chance to see how scientists work. In particular, we will learn how astronomers begin with basic observations and construct hypotheses to explain what they observe. For example, many galaxies appear to be undergoing collisions with their neighbors. Could past interactions of this kind account for the different galaxy shapes?

76.1 GALAXY CLASSIFICATION

The first objects recognized as potentially being other star systems similar to the Milky Way were the "spiral nebulae," but by the early twentieth century it became clear that galaxies had a variety of other shapes. Edwin Hubble, working first as a graduate student at the University of Chicago and later at Mount Wilson in California in the 1920s, developed one of the first schemes for classifying galaxies on the basis of their shape. He sorted them into three main types: spirals, ellipticals, and irregulars.

The first type has two or more arms winding out from the center. Astronomers call these **spiral galaxies,** a term often abbreviated to simply **S.** The Milky Way is of this type, and Figure 76.1 shows photographs of several others, illustrating how

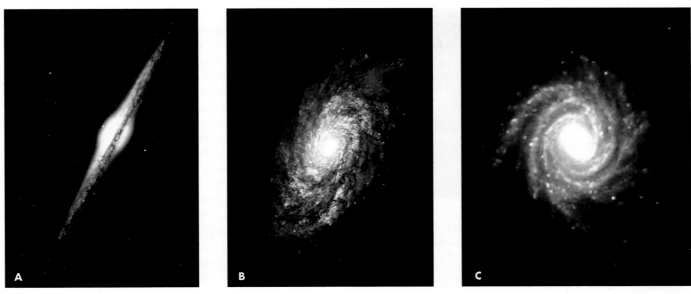

FIGURE 76.1
Photographs of typical spiral galaxies. (A) NGC 4565; (B) NGC 4414; and (C) NGC 1288. These examples show spiral galaxies with their disks oriented from nearly edge-on (A) to nearly face-on (C).

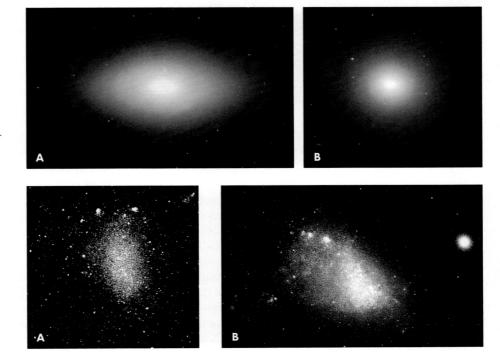

FIGURE 76.2
Hubble Space Telescope images of the elliptical galaxies (A) NGC 4660 and (B) NGC 4458. The images show how elliptical galaxies can sometimes be very elongated and other times almost spherical.

FIGURE 76.3
Photographs of two irregular galaxies (A) NGC 6822, and (B) the Small Magellanic Cloud.

their appearance differs depending on whether the disk is seen more nearly edge-on or more nearly face-on.

The second galaxy type shows no signs of spiral structure or a disk. These galaxies have a smooth, featureless appearance and an overall elliptical shape, sometimes highly elongated, and sometimes nearly circular, as can be seen in Figure 76.2. Accordingly, astronomers call them **elliptical galaxies,** abbreviated as **E.**

Galaxies of the third major type show neither arms nor a smooth uniform appearance. In fact, they generally have stars and gas clouds scattered in random patches. For this reason, they are called **irregular galaxies** (Figure 76.3), abbreviated as **Irr.**

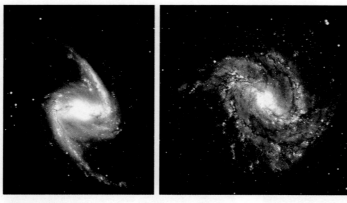

FIGURE 76.4
Barred spirals. NGC 1365 (left) has a prominent bar, while NGC 5236 or M83 (right) has a weaker bar—more like the one thought to be at the center of the Milky Way.

FIGURE 76.5
Two S0 galaxies. Although resembling ellipticals in having no spiral arms, S0 galaxies are different because their distribution of stars shows evidence of a disk.

The Milky Way is estimated to be an SBb or SBc galaxy.

Concept Question 1

Surface brightness is discussed in Unit 29. Unlike the apparent brightness of a galaxy, why does the surface brightness not change with distance?

In addition to these three main types, Hubble recognized two additional subtypes of the spiral systems. The first of these is called a **barred spiral galaxy,** abbreviated as **SB**. These have arms that emerge from the ends of an elongated central region, or bar, rather than from the core of the galaxy (Figure 76.4).

Hubble thought that bars might be permanent features of these galaxies, but astronomers now think that bars are just temporary features that can arise in any spiral galaxy. Computer models of galaxies show that gravitational interactions can cause the stars to form a bar-shaped pattern that may last for hundreds of millions of years. This happens particularly when a galaxy undergoes a gravitational disturbance as the result of a close encounter with a neighboring galaxy. Infrared observations of the stars in the center of the Milky Way show that it currently has a weak bar, and this might be due to its gravitational interactions with small companion galaxies.

Hubble also identified another subtype of galaxy similar to spirals, but that oddly show no spiral structure. These disk systems with no evidence of arms (Figure 76.5) are called **S0** ("*S-zero*") **galaxies.** Astronomers think S0 galaxies are spiral galaxies whose gas has been removed. For example, if a spiral galaxy collides with intergalactic gas at high speed, gas and dust can be swept out of the galaxy, much as a light-weight objects might be blown out of a speeding car with its windows down. Without gas, the galaxy makes no more young blue stars, and the star orbits spread out until there is no obvious spiral structure.

Hubble developed a sequence of gradations among his major types, seeking a pattern that might explain why galaxies exhibit such diversity in their appearance. He subdivided the ellipticals into classes E0 to E7 according to how circular or elongated they appeared. Differences among spirals are indicated by lowercase letters, ranging from Sa for spirals with a large bulge and tightly wound arms, to Sd for a spiral with a small bulge and loosely wrapped arms.

Hubble organized the galaxy types in a sequence, starting with spherical E type galaxies, then moving to flatter E galaxies, to S0 galaxies, then to the sequence of spirals, and finishing with irregulars. Hubble proposed that as a galaxy ages, its type might evolve through this sequence. This hypothesis seems plausible if you look at the "tuning fork" diagram that Hubble proposed (Figure 76.6), but astronomers now think it is incorrect. Nevertheless, the diagram is still widely used today because it offers a convenient way to organize the galaxy types and subtypes.

Although Hubble's classification system categorizes the shapes of most galaxies, it does not describe their size or brightness. Photographic technology in Hubble's time allowed him to study primarily galaxies that are large and had formed large numbers of stars. However, modern observing instruments reveal that there are many **dwarf galaxies** (small galaxies that may be the building blocks of ordinary galaxies) and **low surface brightness galaxies** (dim, often large-diameter systems with far less star formation than ordinary galaxies). Examples are shown in Figure 76.7. These types of galaxies are very numerous, but their properties make them harder to detect, so few were known before modern observing technologies developed. Many of these galaxies can still be categorized according to Hubble's scheme according to the pattern of stars within them, but a dwarf elliptical may contain a million times fewer stars than a typical giant elliptical studied by Hubble.

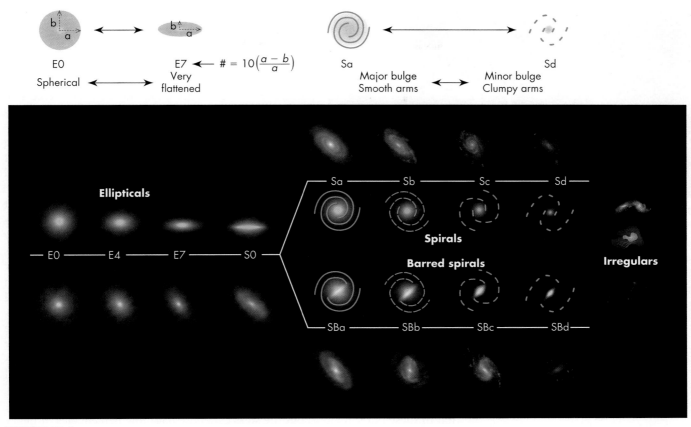

FIGURE 76.6
The Hubble tuning-fork diagram. Elliptical galaxies are subdivided according to how "flattened" they look. Regular and barred spirals are subdivided into subclasses according to how large their bulges are and how tightly wound their arms are. The images are drawn from the Sloan Digital Sky Survey with foreground stars removed to make the galaxies' structures clearer. The spiral galaxies all have flat circular disks, but galaxies were selected that are tilted at similar angles to make comparisons easier.

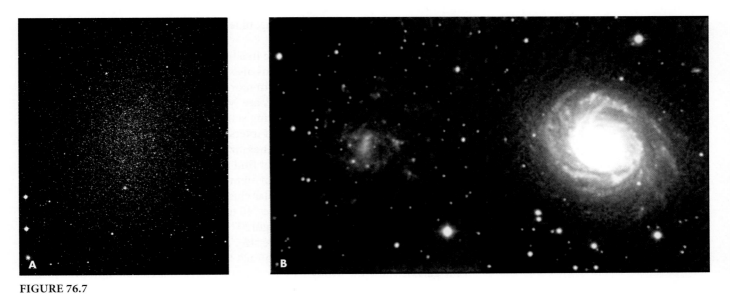

FIGURE 76.7
(A) Dwarf galaxy, Leo I, a nearby galaxy found in 1977. (B) Comparison between two distant galaxies of matching size. The one on the left is a low surface brightness spiral of the type often missed in visible-wavelength surveys.

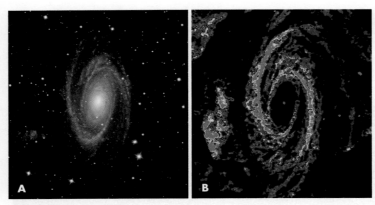

FIGURE 76.8
The galaxy M81. (A) Visible-light image of M81. Note also its small companion to its lower left. (B) False color map of the 21-centimeter radio emission from hydrogen. Hydrogen is also visible in a faint irregular galaxy left of M81. Note the concentration of hydrogen in M81's spiral arms, and the lack in the bulge, where the stars are yellow.

Concept Question 2

Are there circumstances in which you might expect to find young stars in elliptical galaxies? Are there likely to be many planetary systems in elliptical galaxies?

76.2 DIFFERENCES IN STAR AND GAS CONTENT

Since Hubble's time, astronomers have discovered that spiral, elliptical, and irregular galaxies differ not only in their shape but also in the types of stars they contain. For example, spiral galaxies contain a mix of young and old stars (Population I and II—see Unit 72.1); but elliptical galaxies contain mostly old (Pop II) stars, and irregulars contain mostly young (Pop I) stars.

This difference in the kind of stars in the different galaxy types is understandable in terms of their different gas content. To make young (Pop I) stars, a galaxy must contain dense clouds of gas and dust. Spiral systems typically have at least 10% of the mass of their disk in the form of such interstellar clouds, and so they can make the short-lived blue stars that light up their spiral arms (Figure 76.8).

The central bulge of spiral galaxies typically contains very little gas, as can be seen in Figure 76.8B. As a result, there is little star formation and a lack of any young blue stars. The remaining population of stars in the bulge has an overall yellow color, as seen in Figure 76.8A. By contrast, the hydrogen gas often extends well outside the visible disk of the spiral, where it has too low a density to form stars. Deep optical photographs show the presence of a small amount of star formation in these outer regions, but the stars generally have low percentages of heavy atomic elements or "metals" because so little previous star formation (and death) have occurred in these regions.

Ellipticals generally contain small amounts of interstellar gas, and in many ways they resemble the bulge of a spiral. In fact, astronomers used to think that the E galaxies contained essentially no interstellar gas or dust. However, more recent observations made with X-ray telescopes show that many E galaxies contain hot (10-million K) low-density gas. Such gas contains few or no high-density cold clumps that might collapse gravitationally to form stars, so it is no surprise that elliptical galaxies rarely contain young, blue stars like the ones we see in the disks of spiral systems.

Although young, blue stars are rare in elliptical systems, such stars are common in irregular galaxies. These Irr systems also often contain large amounts of interstellar matter, amounting sometimes to more than 50% of their mass. Accordingly, astronomers think that these galaxies are "young" in the sense that they have not yet used up much of their gas in making stars.

Apart from their different star and interstellar matter content, S, E, and Irr galaxies have few other differences to generalize about. For example, if we look at the mass and radius of galaxies, we do not find that all ellipticals are huge and all spirals are small. Instead, we discover that elliptical galaxies range enormously in size. Some are not much larger than globular clusters and contain only 10^7 solar masses of material; others are monster galaxies, 10 to 100 times the mass of the Milky Way. This spread in sizes can be seen in Figure 76.9, which shows an assortment of galaxies all scaled to their correct relative size.

What fraction of galaxies falls into each type classification? If we take a census of the galaxies near the Milky Way, we find that most galaxies are dim dwarf E and Irr systems, sparsely populated with stars. These dim galaxies are undetectable at greater distances, where we can see only the more luminous galaxies. If we take a

FIGURE 76.9

An assortment of galaxies showing their actual relative sizes and providing examples of the range of appearances that a galaxy may have and still be classified as the same type. All types of galaxies are found in a wide range of sizes, containing anywhere from fewer than 10^8 stars to more than 10^{12}. There are many times more of the smallest galaxies than the larger galaxies—just a few are shown for size comparison. The images are drawn from the Sloan Digital Sky Survey. Foreground stars have been digitally removed to make the galaxies' structures clearer.

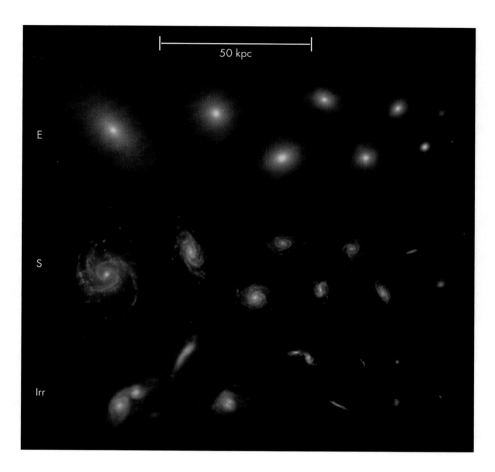

census of those luminous galaxies, we find that the majority of galaxies are spirals, but the percentage depends on whether we look at galaxies that occur in large groups or in more sparsely populated regions. In the largest groups of galaxies, fewer than 50% of the members are spirals or S0 galaxies, whereas in the sparsely populated regions, the proportion of spirals is about 80%. The diversity of galaxy form and content requires an explanation, and a major question in extragalactic research today is why galaxies come in such distinct and different forms.

76.3 THE EVOLUTION OF GALAXIES

Galaxies exhibit a wide variety of shapes, but is this a case of nature or nurture? In other words, were elliptical galaxies born as ellipticals, spirals as spirals, and irregulars as irregulars? Or do all galaxies begin similar to each other and later change form as result of environmental effects?

The Hubble Space Telescope provides a way of studying galaxies when they were very young, allowing us to see how galaxies have changed throughout the history of the universe. Because light takes a finite time to reach us, when we observe a very distant galaxy, we are seeing it as it was when the light left it. This was the idea behind the Hubble Deep Field project—a very long observation of a small "blank" patch of sky (Figure 76.10). By collecting light for more than 100 hours, it was possible to detect extremely distant and therefore very young galaxies.

Concept Question 3

Hubble proposed that his tuning-fork diagram might represent evolutionary changes, with E galaxies gradually changing to S galaxies. What evidence, based on the amount of interstellar matter and the kinds of stars in E and S galaxies, indicates that this is unlikely?

FIGURE 76.10
A section of the Hubble Deep Field, a small region observed for more than 100 hours to detect extremely distant galaxies. The galaxies in this image are not close together in space, but are spread out over more than 10 billion light-years along our line of sight.

LOOKING UP

The position of the Hubble Deep Field is shown in Looking Up #2 at the front of the book. A grain of sand held at arm's length could completely cover the area observed.

The time it takes light to reach us is the distance it travels divided by the speed of light: $t = d/c$. Solving for the time is particularly easy if we write the distance in light-years, because the distance in light-years also tells us how long ago the light left the galaxy. For example, suppose we observe a galaxy that is about 3000 megaparsecs (3 billion parsecs) distant. Multiplying by 3.26 light-years per parsec, we find that the galaxy is about 10 billion light-years away from us. Thus, the light we see today left the galaxy 10 billion years ago, and we are seeing the galaxy as it looked at that time. At such a distance a galaxy has a very small angular size; but with the Hubble Space Telescope, which is unencumbered by the blurring effects of our atmosphere, astronomers can see details in such a very distant—and thus very young—galaxy.

The earliest stages of galaxy formation have not been detected yet, but the Hubble Space Telescope has allowed astronomers to observe galaxies less than a billion years after they formed. Many of the most distant and therefore youngest galaxies are much smaller than systems like the Milky Way. Some examples of distant galaxies detected with the Hubble Space Telescope are shown in Figure 76.11A. Not only are these young galaxies small, they are more numerous than ordinary galaxies today. Thus, these young galaxies must somehow disappear. How can a galaxy disappear? From what is currently known, astronomers tend to think that they collide and merge into bigger systems. This would account both for the formation of large systems like the Milky Way and for the smaller number of galaxies we see today.

The importance of interactions, or what some astronomers call "harassment," is strikingly confirmed in the images of galaxies at later stages, seen at lower redshifts (Figure 76.11B, C). Collisions continue to occur at present between nearby galaxies, as shown in Figure 76.12. Some of the earliest computer simulations of galaxies were able to model these irregularly shaped galaxies. Computers can calculate

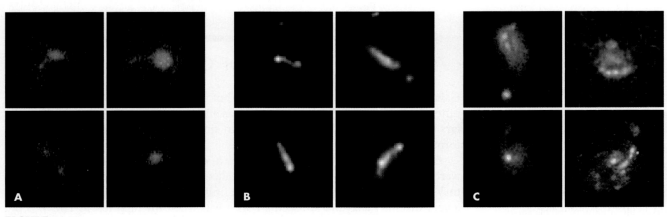

FIGURE 76.11
Hubble Space Telescope images of distant galaxies. Each small square shows an area approximately 25 kiloparsecs on a side, about the size of the Milky Way's disk today. The panels show galaxies (A) less than 1 billion years old, (B) a few billion years old, and (C) about 6 billion years old—half the current age of the Milky Way. ((A) and (B) are from the Hubble Ultra Deep Field, (C) is from the Medium Deep Survey.)

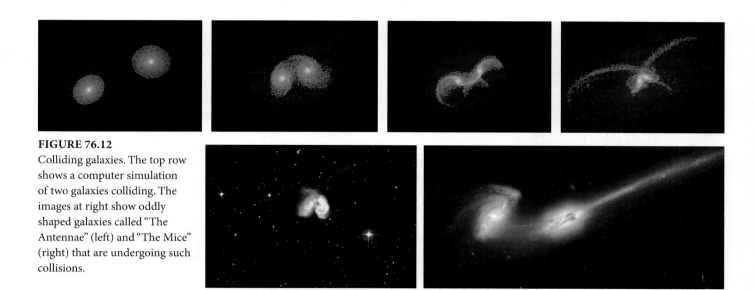

FIGURE 76.12
Colliding galaxies. The top row shows a computer simulation of two galaxies colliding. The images at right show oddly shaped galaxies called "The Antennae" (left) and "The Mice" (right) that are undergoing such collisions.

the gravitational interactions between millions of stars in two galaxies as they encounter each other. Collisions between galaxies exhibit irregular shapes, long tails, and twisted disks. These features appear to be more frequent among distant (younger) galaxies.

Although interactions of this type may radically alter the appearance of galaxies, they do not destroy the individual stars. Stars are so far apart within a galaxy that almost all of them move harmlessly past one another during a "collision." Galaxy interactions are much like tossing two handfuls of sand at each other: Most of the sand particles simply pass by one another without colliding. Dust and gas clouds in the galaxies are not so lucky. Filling each galaxy's space much more completely than stars, clouds do collide, and their impact may compress them enough to trigger a burst of star formation. Such **starburst galaxies,** as these collisionally stimulated galaxies are called, are among the most luminous galaxies known (Figure 76.13A).

Galaxy interactions can create other bizarre forms. Figure 76.13B shows a picture of what can happen when a small galaxy plows directly into a large spiral galaxy. The small galaxy has punched a hole through the larger one, creating what astronomers call a **ring galaxy.** The hole did not result from the smaller galaxy destroying the stars in the other galaxy's disk. Rather, it made the hole by shifting the orbits of stars that were near the center of the larger galaxy into wider orbits farther from the center—like the motion of waves that travel out from the spot where a rock strikes a pond.

> The distances between stars are so large that if stars were represented by grains of sand, the spacing between the grains would be tens of kilometers.

FIGURE 76.13
(A) A starburst in the central portion of the galaxy M82 may be the result of a galaxy collision. The red in this image shows glowing filaments of hydrogen gas. (B) The "Cartwheel Galaxy" has a rim-and-spokes shape that was probably produced when another galaxy passed through it a few hundred million years ago. The likely trajectory is shown, and several candidate galaxies can be found in the direction of the arrow.

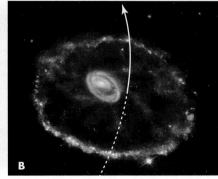

76.4 GALAXY MERGERS AND CHANGING TYPES

We saw in Unit 72 that there are streams of stars in our Galaxy's halo, which are suspected to be debris from small galaxies that fell into the Milky Way and were torn apart by its gravity. Streams of stars are seen around other galaxies as well, as shown in Figure 76.14. Our neighboring galaxy, M31, also appears to have recently cannibalized a small galaxy. Astronomers have detected a small bright clump of stars near the center of M31 that may be the remains of a galaxy it has absorbed. Over time, these streams of stars spread out and become part of the overall population of stars orbiting in the larger galaxy. Astronomers call this process of a large galaxy consuming a small galaxy **galactic cannibalism.**

Repeated cannibalism by a galaxy may turn it into a giant. This might explain why some galaxy groups have an abnormally large elliptical galaxy at their center. This model has the added attraction of explaining a puzzling feature of our Galaxy. Astronomers at one time thought that the stars in our Galaxy's bulge formed in one single, massive event at the Galaxy's birth. However, stars in the bulge have a range of ages, suggesting that the bulge grew over time. This growth probably occurred not just by one galaxy consuming a series of small neighbors, but by a process of **mergers,** in which galaxies of similar size collided and settled into a single larger galaxy. In the process of merging, some of the gas in the galaxies sinks to the center of the new galaxy and forms new stars. Successive mergers lead to successive generations of stars, producing the observed wide range of ages.

Cannibalism and mergers add gas and stars to a galaxy, but does this cause one galaxy to evolve into a spiral galaxy, while another becomes an elliptical galaxy? Astronomers have proposed a model for how this might happen, which is illustrated in Figure 76.15. As we saw in the previous section, most galaxies start out as small, gas-rich systems, which actively form stars and are therefore blue in color. Subsequently, when these galaxies collide and grow into larger systems, they undergo bursts of star formation, which can consume much of their

FIGURE 76.14
Streams of stars around the galaxy NGC 5907 are the remnants of a small galaxy that has been pulled apart by the larger galaxy's gravity.

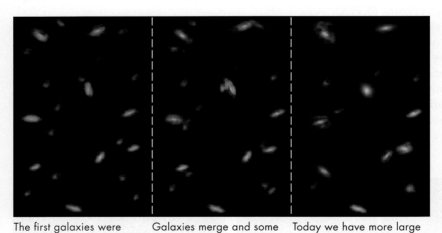

The first galaxies were mostly small, gas-rich, and blue from star-formation.

Galaxies merge and some use up or lose their gas, so star-formation ceases.

Today we have more large and gas-poor galaxies, but some small ones remain.

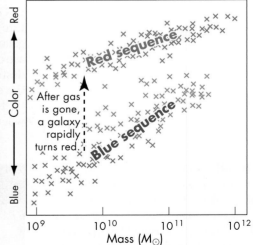

FIGURE 76.15
A possible scenario of how galaxies evolve. The first galaxies were mostly small and gas-rich (left panel). Larger galaxies formed through mergers, as illustrated by the next two panels. Some of these galaxies consumed their gas or lost it during collisions, and ceased forming stars, so they grew redder in color. Others merged to form large spiral galaxies that continue to form stars and remain blue in color. When we plot the colors and magnitudes of galaxies today (right panel), some galaxies are distributed along a "blue sequence." These are probably galaxies that grew through mergers without losing their gas. After losing their gas, they shift to the "red sequence."

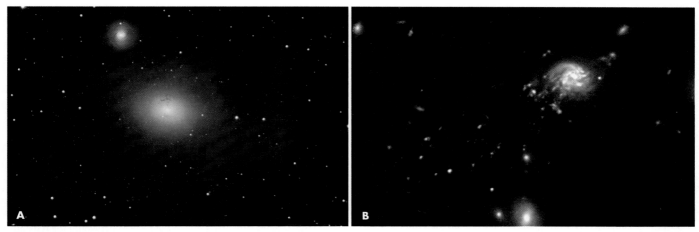

FIGURE 76.16
Galaxies changing character. (A) A merger in which two spirals are colliding and forming an elliptical galaxy. (B) As the spiral galaxy above and to the right of the middle of this image orbits within its galaxy cluster, gas is being stripped from the galaxy. The blue blobs below and to the left of the galaxy are regions where stars are forming in the stripped gas.

Galaxies with tails

Concept Question 4

A number of galaxies can be found in Looking Up #1, #2, and #8 at the front of the book. What type of galaxy does each appear to be? Do you see any evidence of merging?

gas. Computer models suggest that if enough stars form at once, the supernova explosions of many massive stars may drive gas out of the galaxy. If the collision/merger leaves little gas remaining in the system, an elliptical galaxy results. If the collision/merger leaves moderate to large amounts of gas in the system, the gas will settle into a disk, and a spiral results. Such interactions potentially allow some ellipticals to turn into spirals. This can happen if an elliptical captures a gas-rich system. The gas then becomes the disk of a "new" spiral, while the elliptical becomes its bulge.

This hypothesis of the origin of spiral and elliptical systems may explain why galaxies exhibit a division into red and blue "sequences" when their colors are plotted against their masses as shown in the graph in Figure 76.15. As galaxies merge, they grow more massive, and as their stellar population ages they grow redder. Galaxies that retain gas continue to form young stars that are so luminous that they make the galaxy relatively blue overall. Once its gas is exhausted, however, a galaxy rapidly shifts to the red sequence because massive blue stars have such short lives.

Astronomers have identified some galaxies that appear to be undergoing the conversion from a spiral to an elliptical form as they merge. For example, the galaxies in Figure 76.16A are colliding. Both galaxies show evidence of clouds of dust and gas, like spirals, but the collision has disrupted any clear disklike pattern of circular rotation in the larger galaxy. Stars have been sent into new orbits oriented in various directions, and the resulting distribution of stars resembles an elliptical galaxy.

S0 galaxies are particularly interesting because they appear to be transitional between ellipticals and spirals. They have almost no gas, which is thought to be necessary to produce a thin disk, yet they still have stars orbiting in a flat disk. As discussed in Section 76.1, this may result when S0 galaxies have gas and dust swept from them as they collide with intergalactic gas. Possible evidence of gas that has been swept from a galaxy can be seen below and to the left of the blue spiral galaxy shown in Figure 76.16B. The blue blobs in the swept-out gas are regions of forming stars. Another hypothesis for the lack of gas in S0 galaxies is that gas originally in a disk galaxy might have almost all turned into stars early in the galaxy's youth, perhaps because of an interaction with a neighboring galaxy that compressed the gas and made it collapse into stars in one large burst. Subsequent supernovae might then sweep away any remaining gas. In either case, without gas to settle into a thin disk, the stars' orbits gradually spread out into a smooth distribution.

KEY POINTS

- Most galaxies are classified as spirals, ellipticals, or irregulars.
- Spirals, like the Milky Way, have a disk and bulge, and some contain a linear "bar"-shaped feature in their center, which is probably a temporary gravitational alignment of stars.
- Spirals can be further subdivided as types Sa–Sd. Sa galaxies have smoother arms, brighter bulges, and a smaller percentage of gas.
- Ellipticals have a roughly spherical distribution of stars and almost no gas or young blue stars.
- Irregulars are often rich in gas and young blue stars; Some have peculiar shapes because they are interacting with other galaxies.
- S0 galaxies have a disk of stars but little or no interstellar medium.
- The light from high-redshift galaxies takes so long to reach us that we see the galaxies as they looked when they were young.
- Early galaxies were small and numerous, then merged to form larger galaxies; merging galaxies generally have irregular shapes.
- Merging galaxies can cause bursts of star formation that may produce enough supernovae to drive out the galaxies' interstellar gas.
- Merging can produce elliptical galaxies by scrambling stellar orbits and loss of interstellar gas.
- An elliptical capturing a gas-rich galaxy could become a spiral; stripping a spiral of its gas may turn it into an S0.

KEY TERMS

barred spiral (SB) galaxy, 600
dwarf galaxy, 600
elliptical (E) galaxy, 599
galactic cannibalism, 606
irregular (Irr) galaxy, 599
low surface brightness galaxy, 600
merger, 606
ring galaxy, 605
S0 galaxy, 600
spiral (S) galaxy, 598
starburst galaxy, 605

CONCEPT QUESTIONS

Concept Questions on the following topics are located in the margins. They invite thinking and discussion beyond the text.

1. Surface brightness independent of distance. (p. 600)
2. Young stars in elliptical galaxies. (p. 602)
3. Problems with Hubble's evolution ideas. (p. 603)
4. Classifying galaxies from images. (p. 607)

REVIEW QUESTIONS

5. What are the three main types of galaxies?
6. How do the basic galaxy types differ in shape, stellar content, and interstellar matter?
7. What does it mean for a galaxy to have a bar?
8. What criteria distinguish between types of spiral galaxies?
9. What happens to their stars when two galaxies collide?
10. Why do bursts of star formation occur during galaxy interactions and mergers?

QUANTITATIVE PROBLEMS

11. For each of the following distances, if a galaxy the size of the Milky Way (about 30 kpc in diameter) were at that distance from the Milky Way, what would its angular size (Unit 10.4) appear to be? Express your answers in arc minutes.
    a. 10 Mpc
    b. 1000 Mpc
12. If the separation between galaxies is 1 million pc (1 Mpc), how many galaxies are found in a volume of radius 1,000,000 Mpc?
13. For the majority of galaxies astronomers observe, the light travel time is less than a billion years. This is only a fraction of the Sun's age, and therefore we do not expect to find many differences in the galaxies' properties due to their age. What would the redshift be for a galaxy that we are seeing as it was 1 billion years ago?
14. The Hubble Deep Field covers an area of about 5 square arc minutes and contains about 3000 galaxies. If we observed the entire sky to this depth, about how many galaxies would we find? If we assume each galaxy has 100 billion stars (the same as the Milky Way), how many stars are there?
15. The radial velocity of M31 toward the Milky Way is about 100 km/sec. If it maintains that speed, how long will it take to travel the 700-kpc distance between the two galaxies?
16. Assume that the disk of the Milky Way has a diameter of 40 kpc and has a surface area R^2. Likewise, each star covers an area determined by R^2.
    a. If the Milky Way contains 100 billion stars each the size of the Sun, what fraction of the galaxy's disk is actually "covered" by a star?
    b. The fraction found in (a) expresses the approximate chance that when a single star passes through the Milky Way's disk, it will strike a star. If another galaxy with the same number of stars passed through the Milky Way, how many stars would you estimate would undergo direct collisions?

TEST YOURSELF

17. What type of galaxy would not be a good place to observe Type II supernova?
    a. Irr b. S c. SB d. E
18. A galaxy has a bright linear feature in its central region. What kind of galaxy is it most likely to be?
    a. Irr b. S c. SB d. S0 e. E
19. During galaxy collisions the stars that make up the galaxies are most likely to
    a. collide and merge, destroying the original stars.
    b. form binary systems with small separations between the stars.
    c. get ejected from both galaxies.
    d. have their orbits disturbed but otherwise remain intact.
20. Which of the following would be a probable result of a collision between two galaxies?
    a. Complete destruction of all the stars
    b. Formation of a giant spiral from two elliptical ones
    c. Formation of a single open star cluster
    d. Formation of a ring galaxy
    e. All of the above.

UNIT 77
Galaxy Clustering

77.1 The Local Group
77.2 Rich and Poor Galaxy Clusters
77.3 Superclusters
77.4 Large-Scale Structure
77.5 Probing Intergalactic Space

Learning Objectives

Upon completing this Unit, you should be able to:
- Describe how galaxies collect into larger structures, naming the nearest of these.
- Explain how and why galaxy types differ in rich and poor galaxy clusters.
- Describe the large-scale structure of galaxies and the features they form.
- Explain how we observe intergalactic gas, and how it is distributed in space.

Galaxies are not spread uniformly across the sky. Just as stars often lie in clusters, so too galaxies often lie in **galaxy clusters**, and galaxy clusters themselves lie in clusters of galaxy clusters called **superclusters**. Galaxy clusters and superclusters are held together by the mutual gravity of the galaxies and other matter within them. Each galaxy within a cluster moves along its own orbit, just as within a galaxy each star moves on its own orbit. Furthermore, the environment in galaxy clusters may influence the types of galaxies (Unit 76) that form there.

The clustering of galaxies can be seen when we plot the brightest galaxies on the sky. Each dot in Figure 77.1 represents one galaxy. The clusters seen in this figure are typically 10 million light-years across and may contain anywhere from a handful to a few thousand member galaxies. These immense collections of galaxies are not themselves stationary. They are separating from one another, moving apart in the overall expansion of the universe discovered by Edwin Hubble in 1920 (Unit 75).

In this Unit we take the next steps in the revolution that began when Copernicus recognized that the Earth is just one of the planets. In the early 1900s, we did not know about the universe beyond the Milky Way. We find now that our Galaxy is not even the largest galaxy in its own rather small galaxy cluster.

FIGURE 77.1
An all-sky plot of the positions of more than a million galaxies detected by 2MASS, an infrared survey of the sky. Each white dot represents a galaxy; in clusters and superclusters of galaxies, they merge to form brighter areas. The blue shaded portion of the diagram shows where the plane of the Milky Way crosses through this map.

FIGURE 77.2

Images of the galaxies in the Local Group show what the Local Group might look like from outside. The Local Group's largest member is M31. The Milky Way is intermediate in size between M31 and M33. An image of galaxy M67 is used to represent the Milky Way because they are thought to be similar in appearance. The relative positions and sizes of several dozen other dwarf galaxies in the group are also illustrated.

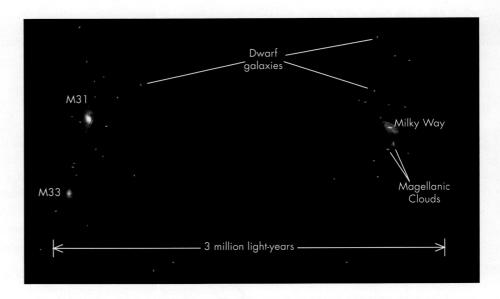

77.1 THE LOCAL GROUP

On the small end of galaxy clustering, astronomers usually refer to collections of galaxies as **galaxy groups.** Our own Galaxy, the Milky Way, belongs to the (unimaginatively named) **Local Group.** The Local Group contains more than 40 known members, as illustrated in Figure 77.2. Its three largest members are spiral galaxies: M31, the Milky Way, and M33. The smaller members include the satellite galaxies orbiting M31 and the Milky Way, and dozens of small, low-luminosity **dwarf galaxies** scattered about the group.

The most famous of the dwarf galaxies are the **Magellanic Clouds,** which appear to be close to the south celestial pole from Earth (see Looking Up #9 at the front of the book). They are about one-tenth the size of the Milky Way. They appear to be irregular galaxies, but infrared photographs of the large cloud show that it has a faint barlike structure. Both galaxies are approaching the Milky Way in highly elliptical orbits, and the shapes of these two galaxies are being distorted by the Milky Way's gravity.

The Magellanic Clouds are visible in Looking Up #9 at the front of the book.

The nearest galaxy to the Milky Way was discovered in 2003 (Figure 77.3). Known as the Canis Major dwarf galaxy, it orbits only about 13 kiloparsecs (42,000 light-years) from the center of the Milky Way. It was not detected earlier because it is merging with the disk of the Milky Way and therefore is largely hidden by dust in our Galaxy. At least eight more small galaxies have been detected in recent sensitive imaging surveys, and several of these appear to be in the process of merging with the Milky Way.

The Local Group is particularly interesting to astronomers, not only because it is our home galaxy cluster, but also because it allows us to study the demographics of galaxies. That is, we can see the true population of galaxies in this region of space, close to home. For example, most of the galaxies in

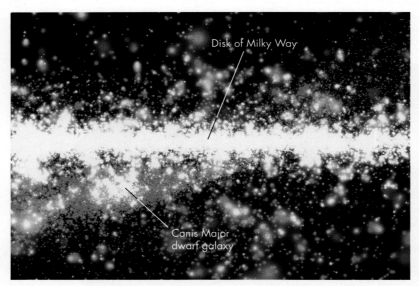

FIGURE 77.3

The nearest galaxy to the Milky Way is called the Canis Major dwarf galaxy. It is hidden from view at visual wavelengths because it is so close to the galactic plane that it is obscured by dust. The image here is based on infrared data from the 2MASS survey. This galaxy appears to be merging with the Milky Way.

Concept Question 1

How could we determine whether most of the mass of the Local Group is in the three spiral galaxies or in the several dozen smaller galaxies?

the Local Group are faint dwarf galaxies. Even with modern telescopes, such galaxies are so dim that they are difficult to detect beyond the Local Group.

Without the Local Group to let us observe the abundance of these small systems, we might grossly underestimate their true numbers in other galaxy clusters. When we examine larger clusters, we may find that they contain hundreds or in some cases thousands of galaxies that are as big as the three largest in the Local Group. Based on what we see in the Local Group, it is likely that there are at least 10 times more uncounted dwarf galaxies.

77.2 RICH AND POOR GALAXY CLUSTERS

The largest groups of galaxies are called **rich clusters** because they contain hundreds to thousands of member galaxies (Figure 77.4). The great mass of such a galaxy cluster draws its members into an approximately spherical cloud, with the most massive galaxies near the center.

In contrast, galaxy clusters that contain relatively few members, such as the Local Group, are called **poor clusters** or just groups. Poor clusters have so little mass that their gravitational attraction is relatively weak, and their member galaxies are not held in so tight a grouping. These clusters generally have a ragged, irregular appearance, with no central concentration.

Not only do rich and poor clusters differ in their numbers of galaxies, but they also tend to contain different types of galaxies. Rich clusters usually contain mainly elliptical and S0 galaxies. Moreover, the few spiral galaxies that they contain tend to be found in the outer parts of the cluster. On the other hand, poor clusters tend to have a high proportion of spiral and irregular galaxies. Spirals are rare in the inner regions of rich clusters because of galaxy collisions. In the core of a rich cluster, galaxies are close together, so collisions between them are frequent and their interstellar medium may be swept out of them, as discussed in Unit 76. Such collisions and removal of interstellar gas may change spiral galaxies into ellipticals and S0s. Thus, although rich clusters may have once contained many spirals, today they contain few, and those are the "lucky" ones that escaped running into a neighbor. The many collisions in the crowded inner parts of these clusters also create **giant elliptical galaxies** by cannibalism as small galaxies merge with the larger galaxies they are

FIGURE 77.4
The Hercules Cluster, a rich cluster of galaxies.

orbiting. A giant elliptical galaxy is shown in Figure 77.5. It is at the center of the rich cluster nearest to us, the Virgo Cluster at a distance of about 15 megaparsecs (50 million light-years).

Rich and poor clusters differ in yet another way. Observations with X-ray telescopes indicate that rich clusters often contain a large amount (10^{12} to 10^{14} $M_\odot$) of extremely hot intergalactic gas that emits X-rays (Figure 77.6A). In fact, rich clusters typically contain 5 to 10 times more mass in hot gas than they do in the stars of all the galaxies within the cluster. Astronomers are not certain where that gas originates. Some of it is probably primordial, part of the original material from which the cluster formed. The surprisingly high percentage of heavy elements in clusters' intergalactic gas indicates that much of it must have been processed inside of stars. For example, when a star explodes as a supernova, some of the gas it ejects may leave the star's galaxy and end up in the intergalactic space within the cluster, carrying heavy elements with it. The intergalactic gas also includes interstellar gas stripped from galaxies when they collide.

Galaxy clusters form in regions that are dense relative to the overall average in the universe. Hubble's law (Unit 75) indicates that galaxies' motion away from each other is small if they originated close to each other. At the same time, gravity grows weaker with distance. Therefore, if two galaxies form near each other, they will be moving away from each other slowly but will feel each other's gravitational pull strongly. This combination of effects tends to make the universe very clumpy, forming groups and small clusters early in its history.

Over time, these groups pull in more galaxies and merge with other groups to make larger clusters, eventually growing into the rich clusters we see today. Much of the evidence for this process comes from our ability to "see back in time" by looking out in space. In particular, galaxies in remote (younger) clusters generally have a clumpier distribution than in nearby (older) clusters, as if they have recently been pulled in and have not had time to spread themselves smoothly. Moreover, X-ray observations of these young clusters (Figure 77.6B) show clumps of hot gas that appear to be falling toward one another to form the cluster.

FIGURE 77.5
The central 1-Mpc region of the Virgo Cluster, centered around the giant elliptical galaxy M87. Giant ellipticals at the centers of galaxy clusters grow through mergers of galaxies within the cluster. The other galaxies visible in this picture will eventually merge with M87.

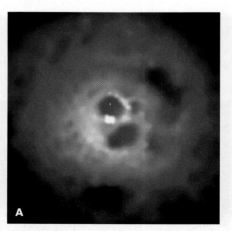 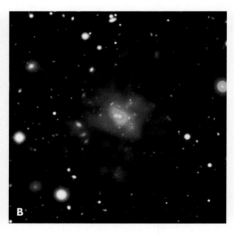

FIGURE 77.6
(A) A false-color X-ray image of the Perseus galaxy cluster. The violet glow is hot gas in the cluster. The picture shows a region about 300,000 light-years across. Darker spots within the hot gas show areas where the hot gas has been pushed out by the outflows from an active galactic nucleus (Unit 78). (B) A combined false-color image of a distant cluster in the constellation Hydra. X-ray emission is shown in violet, while the optical emission from galaxies is shown in other colors. The picture spans a region about 3 million light-years across. This cluster is more than 8 billion light-years away, so it shows the appearance of the cluster when it was less than half its current age.

77.3 SUPERCLUSTERS

The great mass of galaxy clusters creates gravitational interactions between them and holds them together into clusters of clusters, or *superclusters.* These large structures may contain many dozen galaxy clusters spread throughout a region of space, ranging from tens to hundreds of millions of light-years across. For example, the Local Group is part of the **Local Supercluster,** which contains the Virgo galaxy cluster and more than a dozen other small clusters spanning nearly 30 megaparsecs. It is sketched in Figure 77.7.

The presence of large concentrations of galaxies can also be deduced from the motions of galaxies. In general, galaxies are moving away from one another as part of the overall expansion of the universe, and their distances follow Hubble's law (Unit 75.3). However, in regions where there is a large concentration of mass, we see deviations from the velocities predicted by Hubble's law. For example, from our location in the Local Group, we measure a recession velocity relative to the Virgo Cluster of about 1000 kilometers per second, yet based on its distance we would expect the recession velocity to be about 1200 kilometers per second. The motion of the Local Group away from the Virgo Cluster has been slowed by the gravitational pull of the mass concentration there. The Local Group might eventually be pulled into the cluster, but not for hundreds of billions of years, and possibly never.

The gravitational effect of clusters and superclusters on their member galaxies is quite different. In clusters, the gravitational pull and small separations combine to pull the galaxies into collisions with each other. By contrast, the gravitational pull within superclusters has had only enough time to slow down the motion of clusters away from each other. This makes the region of a supercluster more dense than its surroundings, but it will probably be a very long time before the outermost members of the supercluster reverse direction and start moving toward each other.

The technique of studying deviations from Hubble's law has revealed that there are probably even larger structures in our vicinity. In the 1990s astronomers discovered a deviation in the motion of the Virgo Cluster, the Local Group, and other galaxies and clusters out to a distance of about 50 megaparsecs from the Milky Way. These peculiar motions suggest the presence of a huge concentration of mass

> **Mathematical Insights**
>
> The masses of galaxy clusters can be estimated from the typical speed at which galaxies move and the radius of the cluster using the mass formula derived in Unit 17. For the Virgo Cluster, which has a radius of about 5 million light-years ($\sim 5 \times 10^{22}$ m) and galaxy speeds averaging about 700 km/sec (7×10^5 m/sec), its mass is about
>
> $$M \approx \frac{R \times V^2}{G}$$
> $$= \frac{5 \times 10^{22}\,\text{m} \times (7 \times 10^5\,\text{m/sec})^2}{6.67 \times 10^{-11}\,\text{m}^3\,\text{kg}^{-1}\,\text{sec}^{-2}}$$
> $$\approx 4 \times 10^{44}\,\text{kg}.$$
>
> This is approximately 2×10^{14} solar masses.

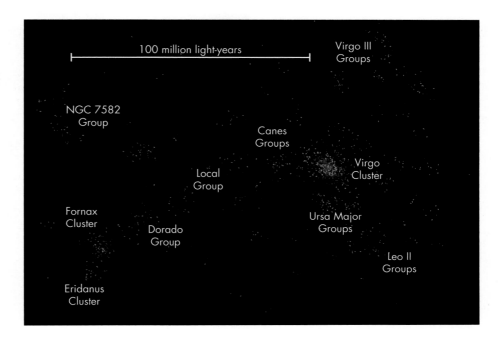

FIGURE 77.7

A depiction of the structure of the Local Supercluster. A number of groups and moderately rich clusters surround the rich Virgo Cluster, which contains most of the supercluster's mass.

If the universe is filled with dark energy, as is suggested by recent observations, the Local Group will probably be kept from falling into the Virgo Cluster (Unit 84).

in the direction of the constellation Centaurus. The whole Local Supercluster has a deviation of more than 500 kilometers per second in that direction. To cause an entire supercluster to move at such a speed would require an enormous gravitational pull. The source of this enormous pull has been named the **Great Attractor** by astronomers. The direction of our motion is toward a position near the center of the all-sky map shown in Figure 77.1. There is a fairly large concentration of galaxies in this direction, although many galaxies are hidden from our view in this direction because of obscuration by the plane of the Milky Way.

The Great Attractor is probably not a single large mass, but an assemblage of many superclusters covering a large region beyond 50 megaparsecs (160 million light-years) away from us. It is estimated to contain about $3 \times 10^{16}\ M_\odot$—the equivalent of 10,000 or more galaxies like the Milky Way—in excess of the mass in comparably sized regions in other directions. The Great Attractor may hint at an even larger kind of structure than a supercluster—a cluster of superclusters! It is not expected that such structures will ever pull their farthest-flung members back toward the center of the attractor region, but they alter the motion of galaxies from that predicted by Hubble's law.

77.4 LARGE-SCALE STRUCTURE

Outside of our local region, astronomers have traced a variety of immense structures out to the limits of where we have surveyed galaxies. Figure 77.8 shows a map of the distribution of galaxies out to a distance of a few billion light-years.

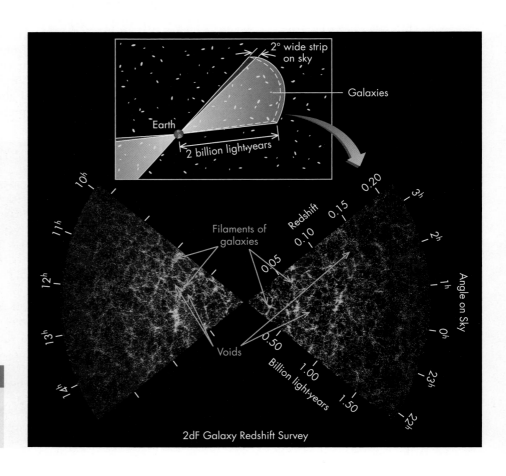

FIGURE 77.8
A map of how galaxies are distributed in space. The shape of the region shown in the figure is like two slices of pizza, tip-to-tip, as illustrated in the upper sketch. Each dot in the lower figure represents one of some 100,000 galaxies that lie in the wedge-shaped regions that extend out to about 2 billion light-years from Earth. Note the nearly empty regions (voids), the stringy structures, and the overall appearance sometimes described as the "cosmic web." The distances to the galaxies were measured from their redshifts. (The angles refer to Right Ascension coordinates on the sky.)

Concept Question 2

What is the difference between the information shown in a map like that in Figure 77.8 and that in Figure 77.1?

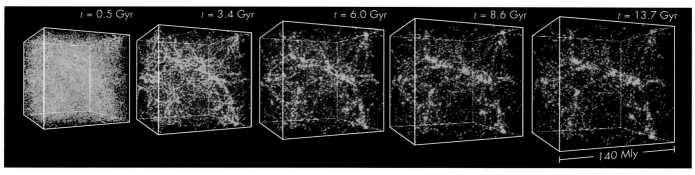

FIGURE 77.9
A computer simulation of the growth of structure as the universe expands. You can see how an initially smooth distribution of matter becomes stringy, much like the features in Figure 77.8. Each box spans about 140 million light-years. The first box on the left represents a time when the universe was roughly one-tenth its present age. The last box on the right shows that same piece of the model of the universe as it would be today.

On this map each galaxy is represented by a single dot. Each galaxy's distance from us is based on its recession velocity (Unit 75). This map represents a thin slice of the sky in which the redshifts of about 100,000 galaxies have been observed by a large team of astronomers. It gives us a cross-sectional look at the distribution of galaxies out to a redshift (z) greater than 0.2—a distance of about 900 megaparsecs.

On this scale, individual galaxies and clusters are no longer obvious, but we *can* see that many appear to be arranged into long filaments or shells surrounding regions nearly empty of galaxies. These latter spaces are called **voids** and, although they are not totally empty, the few galaxies they contain are generally small and dim. Some of these voids are immense. One found in 2007 is estimated to be 300 megaparsecs (about a billion light-years) across—a vast region of space nearly empty of galaxies.

Large-scale structures such as superclusters, voids, and filaments are at the frontier of our knowledge of the distribution of galaxies in space. The patterns of large-scale structure we see at this scale are probably the consequence of the initial conditions in the universe, and astronomers have proposed a variety of hypotheses to explain the features we see. For example, when the huge voids were first found, some astronomers suggested that they might be evidence that even before galaxies began forming, huge explosions pushed the gas out of these regions. Other astronomers have hypothesized that these features were probably produced by gravitational interactions of dark matter (Unit 79).

To attempt to understand large-scale structure, astronomers have turned to computer simulations. In such simulations, astronomers write out the equations that govern the motion of matter (such as Newton's laws) and then solve the equations, assuming that the material filling space is expanding so as to mimic the general expansion of the universe. They then watch what happens as the gravitational force of the material acts within the simulated volume. Figure 77.9 shows a series of images from such a simulation. You can see that matter that initially had a relatively smooth distribution is drawn by gravity into a network of strings. Astronomers identify the places where strings meet as locations where galaxies form. They adjust the equations by varying the amount of ordinary and dark matter and initial lumpiness, thus experimenting to discover what mix best creates structures similar to what is actually observed. The outcome of such simulations by many teams of astronomers confirms that no explosions are needed; but without the gravitational pull of dark matter, the distribution of galaxies in the simulations fails to match the observed distribution of galaxies.

Concept Question 3

If the universe was filled with gas of a uniform density everywhere except for a small region containing a bubble of vacuum, would the bubble expand or contract?

77.5 PROBING INTERGALACTIC SPACE

The large-scale structure of galaxies indicates where galaxies are concentrated, but astronomers are finding evidence that a large amount of matter lies spread out thinly between the galaxies in intergalactic space. Observations using a variety of techniques are beginning to uncover large amounts of matter that never formed into the collections of stars we call galaxies.

We can see the hot intergalactic gas inside rich clusters of galaxies from the X-rays hot gas emits. Outside these regions, intergalactic gas is cooler and more difficult to detect. On the other hand, gas that is cool enough can be detected spectroscopically by its absorption against a bright background source. **Quasars** are bright sources of light that are extremely distant and therefore are useful as probes of intergalactic space along our line of sight toward them. Quasars are intriguing objects in their own right (Unit 78), but for absorption studies we are interested only in the fact that they are extremely distant light sources.

Because quasars are so distant, their light passes through many other galaxies and intergalactic gas clouds as it travels to Earth. Even if an intervening galaxy or cloud is too dim to see, its matter will imprint a signature on the quasar's spectrum, allowing us to detect these otherwise invisible objects. Because these absorbing objects all have different redshifts, each produces a set of absorption lines shifted to different wavelengths. This results in a whole "forest" of absorption lines, each one corresponding to a cloud of gas along the line of sight, as illustrated in Figure 77.10.

Most of these absorption lines come from clouds that have not yet formed into galaxies themselves but are condensing and cooling. Studies of these clouds suggest that they follow the large-scale structure seen in the pattern of galaxies. The intergalactic gas is located in huge sheets, filaments, and voids out to the very largest distances we can see—and therefore the earliest times we can see in the universe's history. The amount of this gas appears to decline with time—in other words, there is less of it in the regions nearer to us. Simulations suggest that over time intergalactic gas accretes onto galaxies, providing a source of additional gas for galaxies and perhaps affecting their evolution. These simulations also suggest that such accretion may continue even to the present, and astronomers are searching for evidence of this very tenuous gas and its possible effects on the Milky Way and other nearby galaxies.

Concept Question 4

How might the absorption lines produced by a spiral galaxy in front of a quasar look different from those produced by a small cloud of gas?

FIGURE 77.10
The light from a distant quasar produces a uniform light source against which we can see hundreds of absorption lines produced by hydrogen clouds along the line of sight to the quasar.

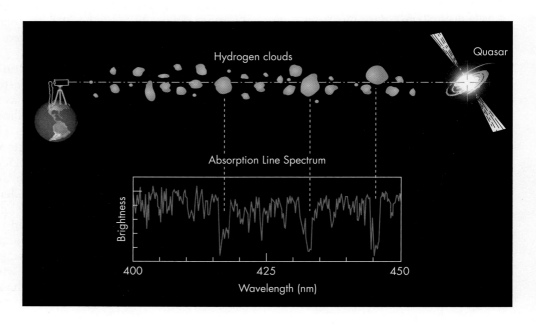

KEY POINTS

- Galaxies are attracted by their gravity into groups and clusters, and even galaxy clusters pull together to form superclusters.
- The Milky Way is part of a group or poor cluster of several dozen mostly dwarf galaxies, in which M31 is the largest galaxy.
- Rich clusters of galaxies usually contain a high proportion of elliptical galaxies, presumably formed by collisions and stripping of interstellar gas from member galaxies.
- Clusters, particular rich clusters, tend to contain a large amount of intergalactic gas, exceeding in mass all the mass contained in stars within the member galaxies.
- Measurements of the motions of local galaxies suggest that there is a concentration of mass, which astronomers call the Great Attractor, attracting all nearby galaxies toward it.
- Other large-scale structures can be mapped using redshift surveys of large numbers of galaxies.
- Absorption-line studies against distant bright sources of light show the presence of intergalactic gas, which is clumpy in a similar way as the distribution of galaxies.

KEY TERMS

dwarf galaxy, 610
galaxy cluster, 609
galaxy group, 610
giant elliptical galaxy, 611
Great Attractor, 614
large-scale structure, 615
Local Group, 610

Local Supercluster, 613
Magellanic Clouds, 610
poor cluster, 611
quasar, 616
rich cluster, 611
supercluster, 609
void, 615

CONCEPT QUESTIONS

Concept Questions on the following topics are located in the margins. They invite thinking and discussion beyond the text.

1. How to estimate masses of galaxies. (p. 611)
2. Comparing position and redshift data. (p. 614)
3. Does a bubble of vacuum expand or contract? (p. 615)
4. Comparing absorption lines of galaxies and clouds. (p. 616)

REVIEW QUESTIONS

5. What is the Local Group?
6. What kind of galaxies are the neighbors to the Milky Way?
7. What are the differences between rich and poor clusters?
8. Over what distance scales do galaxies cluster together?
9. What is the source of the X-ray emission from clusters of galaxies?
10. What is the Local Supercluster?
11. What is large-scale structure?
12. What produces quasar absorption lines?

QUANTITATIVE PROBLEMS

13. Use Wien's Law (Unit 23) to estimate the temperature of the intergalactic medium if the peak emission wavelength is 0.1 nm. How does this temperature compare to the Sun's core temperature of 15 million Kelvin?
14. The Great Attractor, located some 50 Mpc away, is suspected to have caused the Local Group to accelerate to a speed of 500 km/sec over the last 13 billion years.
    a. What is the average acceleration over that time, expressed in m/sec^2?
    b. Using Newton's law of gravitation, how big a mass would be needed at that distance to produce the observed acceleration?
15. Compare the gravitational attraction between the Milky Way ($M = 10^{12}\ M_\odot$) and the Andromeda galaxy ($M = 10^{12}\ M_\odot$, $d = 2.5$ Mly), to the gravitational attraction between the Local Group ($M = 5 \times 10^{12}\ M_\odot$) and the Virgo Cluster ($M = 10^{15}\ M_\odot$, $d = 54$ Mly).
16. Suppose that the typical galaxy has a mass of $10^{11}\ M_\odot$. If the total mass of a cluster of 100 galaxies is $10^{15}\ M_\odot$, how much dark matter is contained in the cluster by the percentage of the total mass (ignoring the hot X-ray gas)?
17. Same as Problem 16, but now the hot X-ray gas has a mass 10 times greater than the galaxies. How much dark matter is contained in the cluster by the percentage of the total mass?
18. A spectral line of hydrogen has an absorption laboratory wavelength of 122 nm. It is observed in the spectrum of a quasar at 130 nm, 180 nm, and 330 nm. What are the redshifts (z) of these clouds? What would be their distances, assuming Hubble's law (and using $V = c \times z$)?

TEST YOURSELF

19. The Local Group
    a. contains about 40 member galaxies.
    b. is a poor cluster.
    c. is the galaxy cluster to which the Milky Way belongs.
    d. mostly contains galaxies much smaller than the Milky Way.
    e. Is all of the above.
20. You write your home address in the order of street, town, state, and so on. Suppose you were writing your cosmic address in a similar manner. Which of the following is the correct order?
    a. Local Supercluster, Local Group, Milky Way, Solar System
    b. Solar System, Local Group, Milky Way, Local Supercluster
    c. Solar System, Milky Way, Local Group, Local Supercluster
    d. Local Supercluster, Local Group, Solar System, Milky Way
    e. Local Group, Solar System, Milky Way, Local Supercluster
21. The majority of galaxies in rich clusters are _____, whereas poor clusters contain high proportions of _____ galaxies.
    a. elliptical and S0 type; spiral and irregular
    b. elliptical and spiral; irregular
    c. spiral; irregular
    d. elliptical and irregular; spiral and S0 type
    e. spiral and irregular; elliptical and S0 type

UNIT 78
Active Galactic Nuclei

78.1 Active Galaxies
78.2 Quasars
78.3 Cause of Activity in Galaxies
78.4 Black Hole/Galaxy Interactions

Learning Objectives

Upon completing this Unit, you should be able to:
- Describe the characteristics of activity in galactic nuclei.
- Explain how variations in brightness set limits on the size of active regions in nuclei.
- Describe how the types of galaxy activity may be explained by viewing angle.
- Discuss the relationship between the size of galaxies and their central black holes.

If our eyes could see radio wavelengths instead of visible light, the night sky would look completely different to us. Very few of the brightest sources of visible light produce much radio emission, and many of the brightest sources at radio wavelengths vary in brightness over time, so that if we worked out a pattern of constellations, in just a few years' time some "stars" would have changed brightness so much that the constellations would no longer look the same. None of these bright radio sources is associated with the bright stars we see at night, but some are related to galaxies, nebulae, and the galactic center.

These oddities of the radio sky were discovered as radio telescopes were being developed during the 1950s and 1960s. It was found that some of the brightest radio sources were identified with very dim starlike objects, and astronomers soon determined that these had some of the highest known redshifts and therefore must be some of the most distant objects in the universe. Meanwhile astronomers working at other wavelengths were discovering a wide variety of galaxies with strange properties in their central regions.

Today we recognize these luminous sources as **active galactic nuclei,** or **AGNs.** They are tiny regions in the centers (nuclei) of galaxies that emit abnormally large amounts of energy. Not only is the emitted radiation intense, but it usually fluctuates in intensity as well. At least 10% of all known galaxies have active nuclei, and many of these galaxies exhibit intense radio emission and other activity outside their central region in addition to their AGNs.

FIGURE 78.1
The Seyfert galaxy NGC 7742, imaged with the Hubble Space Telescope, is a spiral galaxy with an extremely luminous nuclear region.

78.1 ACTIVE GALAXIES

Over the past 60 years different classes of galaxies with unusually vigorous energy output have been found. Their discovery occurred in a variety of ways at different wavelengths, and at first they were thought to be independent classes of objects. Although these galaxies vary in the amount of activity they exhibit, it now appears that most of this activity shares a common explanation. Here we describe two major classes of activity discovered in nearby galaxies.

A **Seyfert galaxy** is a spiral galaxy whose nucleus is abnormally luminous (Figure 78.1). These unusual galaxies are named for the U.S. astronomer Carl Seyfert, who first drew attention to their peculiarities in the 1940s. The core luminosity of a typical Seyfert galaxy is immense, amounting to the entire radiation output of the Milky Way but coming from a region less than a parsec across. Moreover, the radiation is at many wavelengths: optical, infrared, ultraviolet, and X-ray. Despite

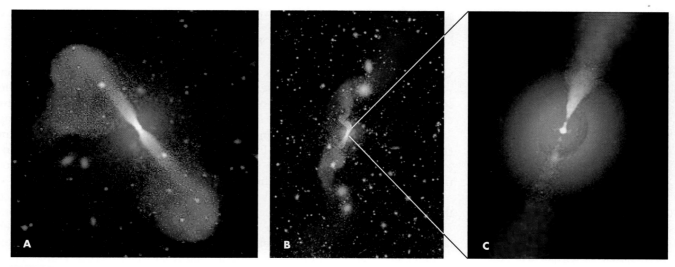

FIGURE 78.2
These images show two radio galaxies, NGC 5532 (A) and NGC 383 (B and C). The figures show the radio emission in red—the brightness indicating its intensity—superimposed on an optical image, shown in blue. The extended red regions in A and B are called radio lobes. Image (C) is a close-up of the central region of the radio emission in NGC 383, showing the bright nuclear source and jets of hot plasma being shot out of the center of the galaxy.

its immense luminosity, the radiation from a Seyfert galaxy's core fluctuates rapidly in intensity, sometimes changing appreciably in a few minutes.

In spectra of the intense radiation in these galaxies' nuclei, Seyfert found very broad emission lines, implying that they contain gas clouds moving at speeds of up to about 10,000 kilometers per second near their cores. A second type of Seyfert galaxy is similarly luminous in the core, but does not show the same evidence of high-speed gas. Subsequent observations suggest that these are actually the same type of active nucleus, but observed at an angle where dust in a surrounding disk has blocked our view of the very central region (see Section 78.3).

Another type of activity is seen in **radio galaxies.** As their name indicates, they emit large amounts of energy in the radio part of the spectrum. Some emit millions of times more energy at radio wavelengths than do ordinary galaxies. This energy is generated primarily in two types of regions: the galaxy's nucleus and enormous regions outside the galaxy and on opposite sides of it, as you can see in Figure 78.2. This figure shows false-color images in which the intensity of radio emission is shown in red while the visible emission (showing light from stars in the galaxy) is shown in blue. The **radio lobes,** as the regions of strong emission outside the galaxy are called, may span hundreds of thousands of parsecs (about a million light-years). On the other hand, the core source is typically less than one-tenth of a parsec across.

What causes the intense emission of radio energy from these galaxies? Astronomers can tell from its spectrum that the emission is **synchrotron radiation,** generated by electrons traveling at nearly the speed of light and spiraling around magnetic field lines (Figure 78.3). The electrons are part of a hot ionized gas that is shot out of the core in narrow **jets** (such as the one shown in Figure 78.2C) and eventually collides with intergalactic gas in the vicinity. There the jets of ionized gas spread out to form the lobes that emit radio waves.

Clearly there are some similarities between radio galaxies and Seyfert galaxies. They both involve rapidly moving gas and a small, bright nuclear region. Furthermore, some Seyfert galaxies have been found that have jets emanating from the nucleus, though smaller than those seen in radio galaxies. Some further clues about such strange behaviors of galactic nuclei come from some of the most distant objects we know of in the universe.

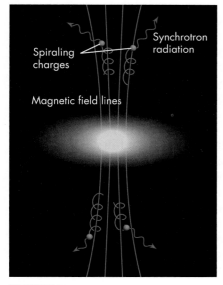

FIGURE 78.3
Charged particles spiraling around magnetic field lines at close to the speed of light generate synchrotron radiation.

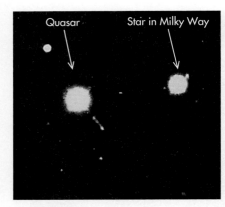

FIGURE 78.4
The quasar 3C273 from a ground-based telescope image. It is the brightest visually of any quasar, but is a dim "star." Even though it is one of the brightest radio sources in the sky, it is a thousand times fainter than the dimmest star visible to the unaided eye. "3C" stands for the Third Cambridge Catalog of Radio Sources.

Concept Question 1

If a quasar is seen as it looked 13 billion years ago, was it 13 billion light-years distant when the light was emitted? Is it this far away now?

FIGURE 78.5
The quasar PG 0052 251 imaged with the Hubble Space Telescope. The image reveals that the quasar lies in the nucleus of a spiral galaxy. ("PG" stands for Palomar Green survey.)

78.2 QUASARS

Quasars are extremely luminous, extremely distant, active galactic nuclei. They were originally identified because many of them are among the brightest radio sources found in early surveys of the "radio sky." Quasars get their name by contraction of the term *quasi-stellar radio source,* where *quasi-stellar* (meaning "almost starlike") refers to their appearance in the first photographs taken of them (Figure 78.4), in which they look like stars—that is, like points of light. Early photographs also revealed that some quasars have small wispy clouds near them or tiny jets of hot gas coming from their cores, like the one seen in Figure 78.4.

Pictures of quasars lack detail because these objects are so far away. We deduce their immense distance from the huge redshift of their spectra. (Recall from Unit 75.3 that a galaxy's distance can be found from its redshift using Hubble's law.) In fact, the redshifts of quasars are so large that their visible spectra are often shifted completely to infrared wavelengths, and what we observe at visible wavelengths was originally emitted by the quasar in the ultraviolet. The visible-light spectra observed were therefore unfamiliar to astronomers. In 1963 the Dutch-American astronomer Maarten Schmidt realized that they were highly redshifted and therefore extraordinarily distant objects.

On the basis of its redshift, the most distant quasar yet found emitted the light we see today nearly 13 billion years ago, less than 1 billion years after the universe began (Unit 80). To be visible at such an immense distance, an object must be immensely luminous, and quasars turn out to be about 1000 to 100,000 times more luminous than the Milky Way. In fact, the brightest quasars have a power output equivalent to a supernova explosion occurring every hour!

During the first decades after quasars were discovered, their extraordinary properties caused some astronomers to question the large distances that had been deduced for them. Alternative hypotheses were offered, suggesting that quasars were nearby stars and that their observed redshifts had a different cause—such as a large gravitational redshift (Unit 27). However, the Hubble Space Telescope was able to settle the question in the 1990s by making high-resolution images of quasars' surroundings (Figure 78.5). These images revealed that quasars lie at the centers of galaxies, like much of the activity in Seyfert and radio galaxies, although quasars are generally much more luminous. Quasars resemble radio galaxies in other ways too, producing jets of material and radio lobes.

Despite their huge power output, quasars appear to be very small. Similar to Seyfert galaxies, quasars fluctuate in brightness (Figure 78.6). Slow changes occur over months, but short-term flickering also occurs in periods as brief as hours. Rapid changes of this sort give clues to the size of the emitting object. For example, suppose an object is one "light-hour" in diameter (about 7 AU). Light from its near side therefore reaches us one hour earlier than the light from its distant side. Even if every part of the object lit up simultaneously, after light from the near side reached us, light crossing from the distant portions of the object would not reach us for an additional hour (Figure 78.7). And if it dimmed all at once, it would again take an hour for the object to drop back down to its original brightness.

It is hard to imagine a physical situation in which simultaneous brightening might occur, so fluctuations in the light output should take longer than it takes light to travel across it. An object's light cannot fully appear or disappear in less the light-crossing time. We can write this as a formula:

$$\text{Diameter} < c \times \Delta t,$$

where c is the speed of light and Δt (pronounced "delta t") is the time interval over which the quasar's brightness changes substantially. We write that the diameter is

FIGURE 78.6
Plot of light variation of a quasar.

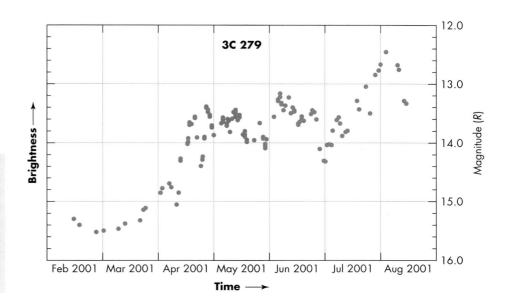

Sound waves are affected by the size of their source much like light waves. Imagine, for example, everyone in a football stadium snapping their fingers simultaneously. The sound you would hear outside the stadium would not be a single loud snap; it would be spread out by the time sound takes to cross the stadium.

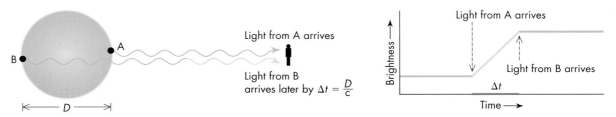

FIGURE 78.7
The effect of size on light arrival times. If a large source lights up instantly, we see the light from its near side before we see the light from its far side. For the objects that we encounter in everyday life, the delay is so short we do not notice it. However, for astronomical bodies, the effect can create delays ranging from days to months, from which we can deduce that the emitting body is light-days or light-months across.

less than $c \times \Delta t$ because the actual diameter can be smaller than this limit if the luminosity changes are not instantaneous.

The time for light variations imply that many quasars are smaller than our Solar System. If a quasar shows substantial variations in $\Delta t = 10$ hours (36,000 seconds), its diameter is smaller than

$$c \times 36{,}000 \text{ sec} = 3.0 \times 10^8 \text{ m/sec} \times 3.6 \times 10^4 \text{ sec} = 1.1 \times 10^{13} \text{ m}.$$

An object of this diameter would fit within the orbit of Pluto. Astronomers have confirmed the remarkably small size of a number of quasars using radio interferometer telescopes. The incredible power output of a quasar comes from a very small object.

Today it is thought that the whole range of activity found in radio galaxies, Seyfert galaxies, and quasars are all powered by this very small engine. What, though, could possibly explain such extraordinary luminosity from such a small object? Many ideas have been proposed. For example, at one time astronomers thought the luminosity might be produced by a single star of 10^8 solar masses that had formed in the center of the galaxy. Astronomers also speculated about more exotic ideas such as a proposed new class of objects called "white holes." These would be the opposite of black holes in the sense that they would spew material out into outer space rather than draw it in. This is an intriguing idea, but no other evidence for white holes has been found, and a much more compelling explanation has been developed.

Concept Question 2

Why are explanations that require heretofore unknown kinds of objects not favored for scientific hypotheses?

78.3 CAUSE OF ACTIVITY IN GALAXIES

Because radio galaxies, Seyfert galaxies, and quasars share many common features, astronomers have sought a unified model for their activity that could explain all three. Any successful model for active galactic nuclei must explain how such a small central region can emit so much energy over such a broad range of wavelengths. For example, at least one quasar is a powerful source of both radio waves and gamma rays. Astronomers therefore hypothesize that the core must contain something very unusual. No ordinary single star could be so luminous. No ordinary group of stars could be packed into so small a region.

The need for an extreme object in the cores of these objects is demonstrated by the very high speeds of jets measured in a number of quasars and at least one Seyfert galaxy. Clumps of gas shoot out from the AGN at speeds clocked at greater than 90% of the speed of light! Whatever is in the core must be able to accelerate matter to almost the speed of light.

Based on the accumulated evidence—the huge energy output, the small size, the enormous speeds—astronomers have ruled out all "ordinary" objects. The best remaining hypothesis is that active galactic nuclei contain **supermassive black holes.** These black holes are more than a million times the mass of the Sun, and around them swirls a huge **accretion disk** that powers the activity we see. Most matter falling toward a black hole does not fall straight in, but ends up orbiting just outside the black hole in the accretion disk where it collides with other matter and emits intense X-ray radiation, as described for stellar-mass black holes in Unit 69.

A 50-million-solar-mass black hole would have a radius about the size of the Earth's orbit around the Sun, while the accretion disk would have a radius of tens to hundreds of AU. Gas within the disk spirals toward its center as it loses gravitational energy through collisions, and is ultimately swallowed. Drawn in by the black hole's fatal attraction, the gas orbits faster and faster, growing hotter and hotter as it is pulled deeper into the gravitational field of the black hole—converting the gravitational energy into kinetic and thermal energy. The gas is heated to millions of kelvins by frictional forces in the disk.

Detailed observations support this model. Figure 78.8 shows a disk of gas and dust in the central regions of a radio galaxy. This disk feeds the much smaller accretion disk around the black hole. Although most of the matter ultimately falls into the black hole, some material is ejected out into space. The accretion disk surrounding the black hole channels the ejected gas into outward flows from the top and bottom surfaces of the disk, producing the observed jets.

A recent observation made with the Chandra X-ray space telescope has provided even more direct evidence of the supermassive black hole at the center of one galaxy. A dark cloud passed in front of and eclipsed the accretion disk around the central black hole (Figure 78.9). From the timing of the eclipse, astronomers determined that the diameter of the accretion disk is only 7 AU. This is about 10 times larger than the event horizon of the supermassive black hole estimated to be in the center of this galaxy, which is about the size astronomers expected for the black hole's accretion disk.

> **Clarification Point**
>
> Quasars often have redshifts greater than 1. This does not mean that they are traveling at speeds greater than the speed of light; instead it indicates that the universe has more than doubled in size since light left the quasar (see Unit 79).

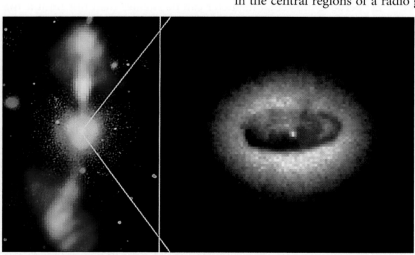

FIGURE 78.8
An image (at right) made with the Hubble Space Telescope of material orbiting and feeding the accretion disk in the center of the radio galaxy NGC 4261. The image on the left shows a wider view of this galaxy's central region and shows the jets of hot gas that were ejected from the nuclear region.

FIGURE 78.9

In 2007 the Chandra X-ray space telescope observed a dark cloud crossing in front of the accretion disk around the black hole in the galaxy NGC 1365. From the speed of the dark cloud and the time it took to block the X-rays, astronomers were able to determine that the accretion disk is less than 7 AU in diameter.

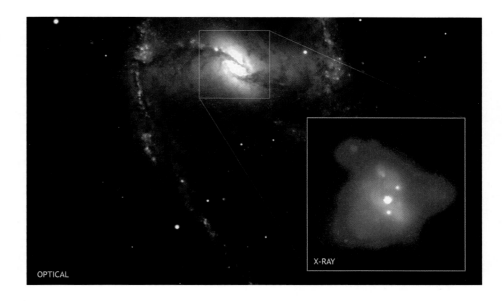

Because blazar jets point nearly straight at the Earth, the high-speed motion can sometimes make clumps of gas appear to be moving faster than the speed of light. This is an illusion of timing because the gas is moving toward us, so light emitted from it at later times has less distance to travel before reaching us.

When such a system is viewed from different directions, it may appear to be different kinds of active galaxies, as illustrated in Figure 78.10A. Rapid motions near the black hole give rise to the broad spectral lines seen in some Seyferts. A much larger dusty outer disk, several parsecs across, can block our view of the fast-moving gas, in which case it would appear to be a narrow-line Seyfert. This model also explains another kind of active galaxy called a **blazar.** These objects were known for rapid and extreme variability. The properties of blazars match what we would expect if we observed an AGN straight down one of the jets. The relativistic speed of the material in the jets moving straight toward us can explain the variability.

Although this model explains the overall features of active galactic nuclei, it cannot fully explain why the jets remain in such narrow beams. Astronomers have proposed a model in which the accretion disk has a strong magnetic field that is twisted as the disk spins (Figure 78.10B). The magnetic field traps charged particles that exist in the hot ionized gas, in much the same way as the Sun's magnetic

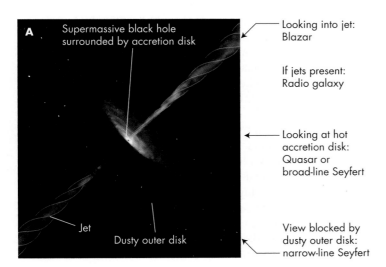
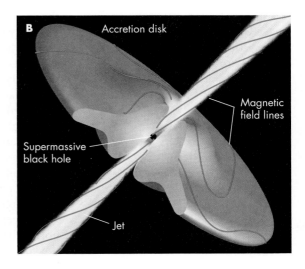

FIGURE 78.10

(A) Sketch of the nuclear region of an AGN. Depending on the angle from which it is viewed, it may appear to be many different kinds of active galaxies. (B) Close-up diagram of the accretion disk around the black hole. A thick accretion disk helps funnel gas into jets emitted perpendicular to the disk, and magnetic fields, twisted up in the hot plasma, may help keep the jet of hot gas tightly confined.

field traps the hot gas in prominences (Unit 53). The spinning and twisting of the magnetic field allows it to capture hot gas from the accretion disk and confine its outward motion along the axis of rotation. This model also provides a powerful supply of energy for the jet—the rotational energy of the spinning black hole—and it is estimated that only a few percent of the infalling matter need be ejected to power the jets.

Over time, if no new sources of matter are added, the accretion disk gradually drains into the black hole. Once this supply is exhausted, the activity ceases. The disk, however, may be replenished with matter from at least two sources. Sometimes stars within the galaxy may approach too closely to the black hole and be torn apart by its gravitational tidal forces. The debris from the star then forms a new accretion disk, providing a new source of energy. Alternatively, collisions with neighboring galaxies may occasionally send gas falling into the galaxy's central regions. Each such event can give the central black hole a new source of matter for its accretion disk, causing the active galactic nucleus to "reawaken." Thus, the same supermassive black hole that powered a quasar may later produce a radio or Seyfert galaxy when some fresh food has been thrown in its direction.

78.4 BLACK HOLE/GALAXY INTERACTIONS

Our own Galaxy has a black hole of several million solar masses at its center (Unit 74.2). According to one hypothesis, the Milky Way's central black hole formed in several steps. First, a massive star in the central region of the Galaxy reached the end of its life, exploded as a supernova, and formed a black hole perhaps as small as 5 $M_\odot$. Normally a black hole like this has little opportunity to grow—it is only 30 kilometers in diameter, so the odds of anything hitting it are small. However, at the center of the Galaxy, matter is packed densely and was falling toward this region during the early stages of the galaxy's formation. As a result, the initially small black hole at our Galaxy's center had a special opportunity to grow by swallowing infalling matter. Bit by bit, the mass of the black hole (and therefore its radius) increased, making it easier for the hole to attract and swallow yet more material. Eventually the hole became large enough to swallow entire stars, at which point it grew rapidly to a million or more solar masses. During its growth the hole was surrounded by an accretion disk, so our Milky Way had an active galactic nucleus. As the Milky Way aged, its black hole consumed the gas around it, and the activity subsided to the low level we see today.

Some galaxies have central black holes that are far larger—more than a billion times larger than the mass of the Sun. The evidence for these large masses is similar to that for the black hole at the center of the Milky Way. Astronomers measure the speed at which stars and gas orbit the cores of the galaxies. Material moving at such high speeds could only be kept in orbit by an enormous mass. Yet this huge mass seems to emit no starlight, and black holes are the only known objects that are both massive and dark enough. While the steady accumulation of mass into an initially small black hole might explain our Galaxy's central black hole, such a gradual process is incapable of producing black holes with masses thousands of times larger within the lifetime of these galaxies.

A recent discovery provides a clue to how the growth of supermassive black holes is linked to the overall growth of their host galaxies. Observations made with the Hubble Space Telescope show that where central black holes can be measured, they generally have a mass of a few tenths of a percent of the mass of the host galaxy's central stellar bulge (Figure 78.11). Why does such a correlation exist? It is difficult to see how black holes that form in a region less than 1 parsec across are

---

Concept Question 3

Suppose you currently lived in one of the galaxies that we are presently observing as a quasar. What would your galactic center look like to you today? What would the Milky Way look like to you?

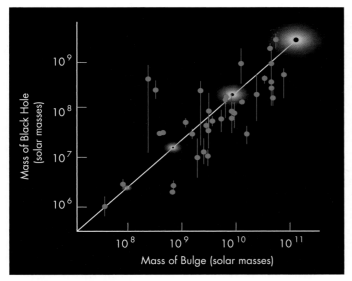

FIGURE 78.11
The correlation between the mass of a central black hole and a galaxy's bulge. The green dots indicate the mass of the central black hole and the bulge in more than 30 galaxies studied with the Hubble Space Telescope. The vertical green bars indicate the uncertainty in the measurements. The size of the central black hole grows with galaxy mass, as illustrated by the background diagrams of galaxy bulges.

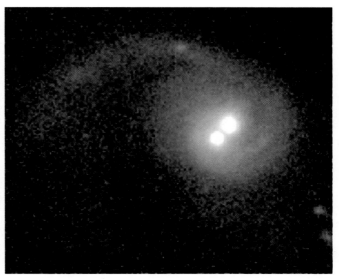

FIGURE 78.12
Hubble Space Telescope image of a galaxy in which two active nuclei appear to be orbiting each other. The long stream of stars at the top of the image implies that this galaxy is the product of a recent merger, and the central black holes may eventually spiral into each other to form a single larger black hole.

"aware" of the total mass of stars extending outward many kiloparsecs. Astronomers are not yet certain, but the correlation may be explained if central black holes grow at the same time galaxies grow through mergers. When a galaxy cannibalizes a small neighbor, some of the inflow of extra mass could reach the central black hole in the larger galaxy, allowing it to grow. In addition, the massive black holes of two galaxies may sink toward the center of merged galaxy and then themselves merge. Optical and X-ray observations support this model, showing a number of galaxies that appear to contain a pair of massive orbiting black holes in the process of merging (Figure 78.12). When such massive black holes merge, they will produce intense gravitational waves (Unit 27), which astronomers hope to detect in the near future.

Even though only about 10% of galaxies show signs of activity, astronomers suspect that all large galaxies have a supermassive black hole at their cores, following the relation found in Figure 78.11. The central black hole is active mainly when the galaxy is young and has substantial matter accreting onto it. Thus, quasars are mostly found at relatively large redshifts, where we are seeing back to a time when galaxies were young. After a few billion years, any material that has an orbit carrying it near the black hole will have been swallowed by the black hole or ejected by the process that produces the AGN jets. With no more matter feeding the black hole, it grows quiescent. However, each time the galaxy undergoes a collision with another galaxy, additional matter may end up in orbits that lead it close to the central black hole, reawakening the activity as a Seyfert or radio galaxy as the black hole grows in size.

Supermassive black holes at the centers of galaxies are not only a product of the way galaxies form, they also generate *feedback* that affects how galaxies evolve. Recent observations and computer simulations of AGN jets suggest that the furious winds they drive into a galaxy's interstellar medium may impede star formation. It is beginning to appear that there is a coevolution linking the growth of each galaxy's tiny black heart to its grander structure.

KEY POINTS

- Large energy outputs are seen from active galactic nuclei (AGNs), although the form of this activity takes on different appearances.
- Seyfert galaxies exhibit a very bright nuclear region; in some the spectral emission lines are very broad, implying rapid rotation.
- Radio galaxies have jets of hot gas emitted in opposite directions from the AGN; the synchrotron radiation from the jets implies magnetic fields, which help confine the jets into narrow beams.
- Quasars, or quasi-stellar radio sources, appear to be similar to other AGNs, but more distant and more luminous.
- The high redshifts and small sizes of quasars were mysterious for many decades, but it is now possible to image the host galaxy.
- AGN emission can vary strongly over very short times, implying that emission comes from regions as small as about 1 AU.
- The source of an AGN's power is thought to be a supermassive black hole surrounded by a hot accretion disk.
- A surrounding dusty disk may block our view of the hot accretion disk when seen edge-on.
- In blazars we are looking straight into one of the jets.
- All large galaxies probably contain supermassive black holes, usually inactive because of the lack of material to feed an accretion disk.
- The mass of a galaxy's bulge is correlated with the mass of its central supermassive black hole, implying that they form in unison.

KEY TERMS

accretion disk, 622
active galactic nucleus (AGN), 618
blazar, 623
jet, 619
quasar, 620
radio galaxy, 619
radio lobe, 619
Seyfert galaxy, 618
supermassive black hole, 622
synchrotron radiation, 619

CONCEPT QUESTIONS

Concept Questions on the following topics are located in the margins. They invite thinking and discussion beyond the text.

1. Distance to quasar when light emitted. (p. 620)
2. Inventing new objects in scientific hypotheses. (p. 621)
3. Appearance of Milky Way from present-day quasar. (p. 624)

REVIEW QUESTIONS

4. How are the types of active galaxies similar and different?
5. In what wavelength ranges do active galaxies emit their energy?
6. What is the source of radio emission from active galaxies?
7. What is a quasar?
8. Why do astronomers think that quasars are very distant?
9. How is it known that active galaxies have small core regions?
10. What mechanism has been suggested to power active galaxies?
11. How does a galaxy and its central black hole interact?

QUANTITATIVE PROBLEMS

12. If the radio galaxy NGC 383 has lobes that are 300 kpc in length, how long does it take material traveling at 90% of the speed of light to travel from the base to the tip of a single lobe?
13. The radius of a black hole is approximately 3 km times the mass of the black hole in solar masses (Unit 69).
    a. Find the diameter in AU of a black hole with a mass of 1 billion $M_\odot$.
    b. For an accretion disk 10 times the radius of the black hole in (a), how many light-hours across would it be?
14. Hydrogen has a spectral line at 122 nm in the ultraviolet portion of the spectrum. What are the largest and smallest redshifts a quasar could have for this line to be seen in the visible portion of the spectrum (400 to 700 nm)?
15. An AGN is observed to double in brightness over seven days and then to grow dim again in the next seven days. What is an upper limit to the probable size of the AGN? Give your result in AU.
16. Spectral broadening observed in a particular Seyfert galaxy suggests that the material is orbiting at 1200 km/sec at a radius of 0.5 parsecs from its center. Calculate the amount of mass within this radius.
17. A hydrogen line of gas orbiting in the nucleus of an active galaxy is Doppler shifted on the right side of the AGN to 123 nm, and on the left side to 121 nm. How fast is the gas rotating?

TEST YOURSELF

18. A recently discovered galaxy shows variations in brightness that last only a few hours. This implies that the source of the variation is
    a. very small.
    b. extremely hot.
    c. rotating rapidly.
    d. highly magnetized.
    e. about to explode.
19. What produces the synchrotron radiation seen in radio galaxies?
    a. High-speed electrons spiraling around magnetic field lines
    b. Supernova explosions
    c. Visible starlight that has been heavily redshifted
    d. Hydrogen molecules
    e. Radiation from a black hole at the center of the galaxy
20. Because the large redshifts of quasars arise from the expansion of the universe, we can conclude that
    a. quasars must be very small.
    b. quasars must be within the Local Group.
    c. quasars must be single stars with extremely large masses.
    d. quasars must be moving toward Earth at a large velocity.
    e. quasars must be very luminous.
21. The correlation between the mass of a central massive black hole and the galactic bulge may imply that the
    a. bulge is transferring its material to the black hole.
    b. black hole and bulge grow simultaneously during galaxy mergers.
    c. central black hole grows in mass independent of the bulge.
    d. black hole formed first and then attracted material to the center of the galaxy.

PART 5 UNIT 79

Dark Matter

79.1 Measuring the Mass of a Galaxy
79.2 Dark Matter in Clusters of Galaxies
79.3 Gravitational Lenses
79.4 What Is Dark Matter?

Learning Objectives

Upon completing this Unit, you should be able to:
- Describe the evidence that dark matter exists in galaxies and larger scales.
- Explain how gravitational lensing is used to estimate the masses of galaxy clusters.
- Describe the known characteristics of dark matter determined from observations.
- Discuss current hypotheses for the nature of dark matter.

When we map the positions of galaxies throughout the universe, it is something like studying the lights seen from an airplane flying over a city at night. The pattern of lights suggests the general shape of things, but there is a whole landscape that can only be guessed at. Astronomers have begun to realize that galaxies hint at a very strange landscape, in which most of the matter in the universe is of a type unlike anything we know.

We can deduce the existence of this other kind of matter because it exerts a gravitational force on the things we see, such as stars and galaxies, using Newton's gravitational law (Unit 17). In studies of the Milky Way (Unit 74), astronomers labeled this **dark matter** to indicate that they knew mass was present but that it emitted no light. But a growing body of evidence says that it is a kind of matter that in fact *cannot* be seen. It may be made of particles that do not interact with photons, and therefore it may be all around us without being detected.

It is interesting to consider how far we have moved from our Earth-centered view of the universe in our exploration of galaxies. We have learned that the Sun lies in the outskirts of a galaxy, which is not a particularly significant galaxy, which is in a minor cluster of galaxies. And now we are realizing that the kind of matter that makes up everything we know is just a minor kind of matter in the universe. This is the Copernican revolution taken to extremes! Such a revolutionary idea requires exceptionally strong evidence, and we will see in this Unit why the evidence is strong enough that it is now widely accepted among astronomers.

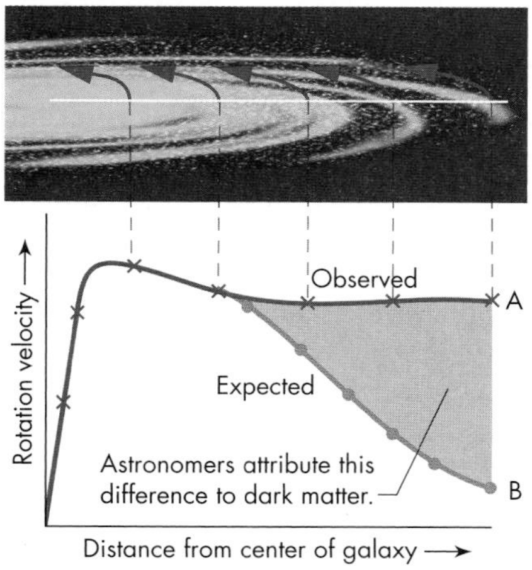

79.1 MEASURING THE MASS OF A GALAXY

We saw in Unit 74 that when astronomers use orbital motion to measure the mass of the Milky Way, they find a puzzling discrepancy between the mass they calculate and the mass they can account for among the stars and other material they detect. The discrepancy in the Milky Way is difficult to quantify in some ways, however. Because of our location within the disk, dust blocks our view of many stars and distances can be difficult to determine.

Studies of other galaxies reveal that the discrepancy found in the Milky Way is a universal phenomenon. Almost without exception, the mass calculated from Newton's law of gravity is much larger than the mass of stars and interstellar matter detectable by telescopes at all wavelengths. This discrepancy is very large, typically amounting to 10 times a galaxy's visible mass. This means that when we look at the starlight from a galaxy, we are observing only about one-tenth of its matter.

FIGURE 79.1
A schematic galaxy rotation curve. The line with dots represents the curve expected if the galaxy's mass comes only from its luminous stars. The line with crosses represents the observed curve, implying that the galaxy contains dark matter.

Dark matter

$M_{<R}$ = Mass within radius R (kilograms)
R = Radius (meters)
V = Velocity of rotation (meters per second)
$G = 6.67 \times 10^{-11}$ m³/kg·sec²

Concept Question 1

If there is the same amount of dark matter between radii of 10 and 20 kpc as between 40 and 50 kpc, what does that imply about the *density* of the dark matter at different radii?

The most direct evidence for dark matter comes from **rotation curves** of spiral galaxies, as we saw for the Milky Way in Unit 74. A galaxy's rotation curve is a plot of the orbital velocity of the stars and gas moving around it at each distance from its center, as shown by curve A in Figure 79.1. A rotation curve reveals how a galaxy's mass is distributed, because the speed at each radius from the galaxy's center reveals how much mass must lie interior to the orbit at that distance. We can understand how that dependence arises by thinking about what keeps a star in orbit.

According to Newton's first law, if an object does not move in a straight line (for example, if a star moves in a circular orbit around a galaxy), a force must be acting on it. We know that for a star orbiting in a galaxy, that force is gravity. We also know that the strength of the gravitational force depends on the mass of the attracting body (in this case, the galaxy). It turns out, however, that for a good approximation, only the material inside of the star's orbit contributes to the net gravitational force. If the speed of rotation is V at radius R, we can write this as an equation for the mass interior to that radius as:

$$M_{<R} = \frac{V^2 \times R}{G}$$

where the subscript "<R" is used as a reminder that the measurement represents the mass interior to this radius, as explained in Unit 74.

Stars near the center of a galaxy will therefore have only a small net force acting on them, because there is little mass between them and the center. Because the force acting on them is small, such stars do not orbit very rapidly. On the other hand, stars orbiting very far from the center should also feel a small gravitational force. They feel the effect of the galaxy's full mass, but the gravitational force on them should nevertheless be weak because the force of gravity grows rapidly weaker at greater distances from a mass. Therefore, we would predict that the outermost stars should also be orbiting slowly to be in balance with the weak gravitational pull. Detailed mathematical models bear out this analysis and indicate that a galaxy's rotation curve should start at small velocities near the core. The curve should then rise to higher velocities at middle distances and finally drop to lower velocities again far from the center, as indicated in curve B of Figure 79.1.

However, when astronomers measure the rotation curves of galaxies, they do not find this behavior. In work pioneered by the U.S. astronomer Vera Rubin, it was discovered that the rotation curves do rise near the center, but they then flatten out and almost never drop off to low velocities again. These **flat rotation curves** are found in nearly all spiral galaxies studied, as shown by a number of examples in Figure 79.2. Rotation curves have been studied for many galaxies, to distances far beyond those we have been able to study in the Milky Way, and those galaxies maintain flat rotation curves out to radii exceeding 50 kiloparsecs.

A constant rotation speed in the mass formula implies that the mass climbs steadily at larger radii. This is because with a constant value of V, the mass formula indicates that the interior mass rises in proportion to the radius R. For example, the amount of matter measured out to 10 kiloparsecs from a galaxy's center might be found to be 10^{11} $M_\odot$. Then out to 20 kiloparsecs the constant velocity implies there is 2×10^{11} $M_\odot$; out to 30 kiloparsecs there is 3×10^{11} $M_\odot$; and so forth.

This can be contrasted with a galaxy's light curve. In a typical spiral galaxy, about 90% of the light comes from the inner 10 kiloparsecs, while 99% is contained within about 20 kiloparsecs. The number of stars in the outer regions declines very rapidly, so stars cannot contribute enough additional mass at larger radii to explain the flat rotation curve. Astronomers have hunted for other components, such as interstellar gas, which often extends out farther than the stars, but this also provides too little mass. The predicted rotation speeds based on stars and gas alone are illustrated in Figure 79.3.

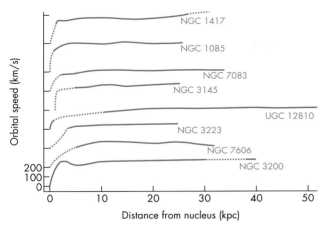

FIGURE 79.2
The rotation curves of an assortment of spiral galaxies. The curves shown are offset vertically so they do not overlap. Each rises to a rotation velocity of about 200 km/sec and remains there to the limits of the observations.

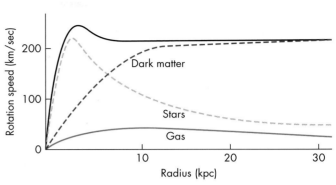

FIGURE 79.3
The rotation curve of a typical spiral galaxy. The overall rotation speed is shown in black. The amounts of rotation produced by the detected stars and gas are shown. Dark matter becomes dominant beyond several kiloparsecs from the center of the galaxy.

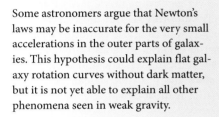

Some astronomers argue that Newton's laws may be inaccurate for the very small accelerations in the outer parts of galaxies. This hypothesis could explain flat galaxy rotation curves without dark matter, but it is not yet able to explain all other phenomena seen in weak gravity.

Analyzing elliptical and irregular galaxies is more complex because the stars and gas do not orbit in circles. However, a similar mass formula with modifications for random orientations of the velocities can be applied. Again, the results indicate that stars alone cannot account, by a factor of about 10, for the orbital speeds observed.

What can explain such a major discrepancy between theory and observation? If our laws of physics are correct, only one thing can explain such behavior: Galaxies must contain large amounts of unseen mass. This unseen mass appears to be present throughout a galaxy, but its presence is particularly evident in the outermost parts of the galaxy where its huge cumulative mass exerts a gravitational force that holds the stars in orbit despite their large velocity. We deduce that what we describe as a galaxy is a small pocket of "normal matter" embedded in a large **dark matter halo** about 10 times more massive than the visible galaxy.

The term *halo* can be a little confusing because it sometimes makes people think of a ring. The dark matter is actually distributed throughout a galaxy. The rotation curve studies imply that the dark matter density is highest at the center of the galaxy. Nonetheless, it represents only a fraction of the mass in the inner parts of a galaxy where the stars and gas there are packed tightly. At larger radii, however, the mass of stars and gas drops off rapidly, while the dark matter content declines only gradually. In the outermost parts of the galaxy, dark matter is left as the primary component.

79.2 DARK MATTER IN CLUSTERS OF GALAXIES

Galaxies in clusters (Unit 77) often have speeds of more than 1000 kilometers per second with respect to one another. These great speeds reflect the enormous mass present in the cluster, pulling whole galaxies into high-speed orbits. If we add up the masses of all the galaxies within a cluster, we come up with a total mass far short of what is needed to hold the cluster together. Just as the stars within a galaxy orbit too rapidly for the gravitational force that can be attributed to its luminous mass, so too the galaxies in a cluster have speeds too fast to be held in orbit by the observed mass.

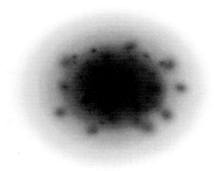

FIGURE 79.4
Artist's depiction of what the dark matter distribution might look like in a galaxy cluster. Individual galaxies reside within each smaller dark matter "halo," but dark matter is also spread throughout the cluster.

Gravitational lensing

This stark discrepancy was discovered in the 1930s by the Swiss astronomer Fritz Zwicky. Once again, applying the mass formula we can estimate the mass interior to any radius in the cluster by measuring the speeds of the galaxies at that radius. The discrepancy is immense—a typical cluster must have a mass more than 100 times larger than would be predicted from its combined starlight. Galaxy clusters, like galaxies themselves, must contain huge amounts of dark matter.

For decades astronomers studied clusters at other wavelengths to search for matter that might explain the discrepancy. Hot intergalactic gas was one possibility, and X-ray telescopes do reveal the presence of a lot of such gas in clusters (Unit 77). The amount of mass estimated to be within this hot gas is actually larger than that contained within the galaxies' stars, but it is still only about a tenth of the amount of mass predicted from the galaxies' orbital speeds.

Hot gas actually provides additional evidence for the dark matter in galaxy clusters. To hold the hot gas within the cluster and prevent its expansion, the cluster must exert an inward gravitational pull on the gas that matches the pressure within the gas. (This is the same condition of *hydrostatic equilibrium* that pertains to our Sun and other stars—Unit 52.) The gravity needed to confine the gas depends on the cluster's total mass; and the mass needed is consistent with the high mass calculated from the motions of the galaxies.

Combining the evidence for dark matter in galaxies and galaxy clusters, it appears that these systems contain approximately 10 times more dark matter than any kind of normal matter that is detectable via electromagnetic emission. The dark matter appears to be distributed relatively smoothly within these systems; it is most dense toward the center of each galaxy as well as at the center of a galaxy cluster, but it extends out much farther than the galaxy's stars and other matter. The picture that emerges is that if we could see dark matter, it would look something like the depiction in Figure 79.4. The stars and gas that we can detect are concentrated in the regions where the dark matter is most dense.

79.3 GRAVITATIONAL LENSES

Dark matter may not be observable by electromagnetic radiation, but it can be detected in a different way. It not only affects star and galaxy orbits, but it can change the path of light through space. This gravitational bending of light is one of the predictions of *general relativity*, Albert Einstein's theory of gravity (Unit 27).

The bending of light was detected soon after Einstein developed his theory of gravity, and in 1937 Zwicky suggested the possibility of a **gravitational lens.** A gravitational lens would focus the light from an object behind it, making background objects look brighter. A gravitational lens does not require dark matter in principle, but generally the masses of galaxies and galaxy clusters without it were known to be too small to produce detectable lensing effects. As a result, the idea was mostly forgotten for several decades.

Then, in the late 1970s, a quasar was discovered that seemed to have a nearby companion quasar with an essentially identical spectrum but with a slightly different brightness and shape. The existence of two so nearly identical quasars so close together is extremely improbable. Today several dozen such quasar pairs and even "quintuplets" are known (Figure 79.5A). They are not actually companions. Instead, they are images of a single quasar created by a gravitational lens.

An ordinary lens forms an image because light bends as it passes through the lens's curved glass. A gravitational lens forms an image because light bends as it passes through the curved space around a massive object such as a galaxy. The galaxy's gravitational force bends the space around it so that light rays that would otherwise travel off in other directions and never reach the Earth are bent so that they do reach us, as depicted in Figure 79.5B.

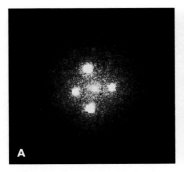 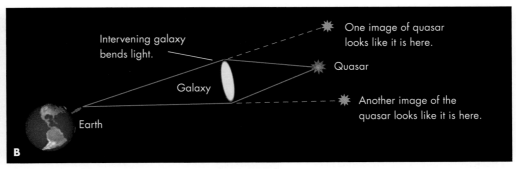

FIGURE 79.5
(A) An Einstein cross, a complex image of a quasar created by a gravitational lens. (B) Sketch of how a gravitational lens forms an image.

Because the matter is distributed unevenly, a gravitational lens is similar to a magnifying glass made of wavy glass. It produces magnified but distorted images of background objects. For example, Figure 79.6 shows the arcs created when light from a distant galaxy was bent into elongated shapes as it passed through a galaxy cluster. The shape of these arcs depends on the mass of material within the cluster, allowing astronomers to measure the cluster's mass. The strong bending of light that is seen requires large amounts of dark matter, confirming the high mass implied by the high speeds of the galaxies.

Measurements of gravitational lensing provide strong confirmation of the dark matter hypothesis. It also appears to rule out some other possibilities that astronomers had previously considered. For example, one hypothesis argued that the rapid motions in clusters did not indicate the amount of mass present because the clusters were not held together by their gravity, but were instead expanding. Another hypothesis suggested that the laws of motion developed by Newton behaved differently at the large distances in galaxy clusters, so the formula we used for finding the mass would not apply. Neither of these hypotheses predicted the large bending of light seen. The agreement of the light-bending measurements with other results shows that there is in fact an immense amount of unseen mass within the clusters.

Recent observations of a pair of galaxy clusters that have collided with each other provide further insights into the characteristics of dark matter. When clusters

FIGURE 79.6
The arc-shaped features in this Hubble Space Telescope picture of the galaxy cluster Abell 2218 are the images of distant galaxies distorted by the gravitational lens effect created by the galaxy cluster. Most of the galaxies in the cluster appear yellow, while the arcs have a variety of colors depending on their type and redshift.

FIGURE 79.7

Dark matter (the blue blobs) mapped out after the collision of two galaxy clusters. The image has been made by superimposing three images: a picture made at optical wavelengths (showing the galaxies in the two clusters); an image made at X-ray wavelengths (showing in red the hot gas in each cluster); and a map of the dark matter as deduced from the gravitational lensing of the background galaxies (the blue blobs). In the collision, the dark-matter blobs have passed through each other. The gas in the two clusters has collided and clumped up between the clusters. The merged cluster is officially known as 1E 0657–56, but more popularly as the *Bullet Cluster*.

collide, the gas from the two clusters slams together and heats up to millions of kelvins and emits X-rays. Astronomers observed this hot gas with the Chandra X-ray telescope in the "Bullet Cluster" in 2006, allowing them to trace where the gas was located after the clusters collided. This is shown in red in Figure 79.7.

The dark matter is, of course, invisible, but its location can be mapped by its gravitational lensing effect on the light of background galaxies. The derived distribution is shown in blue. Unlike the gas, the separate masses of dark matter in the two clusters simply passed through each other as the clusters collided. This is similar to the way that stars pass by each other when galaxies collide (Unit 76.4), yet the bulk of the known matter in clusters is in the hot gas. Thus, dark matter must be something that produces no light and that does not interact with other matter in collisions. Astronomers have described the dark matter as behaving like a "collisionless gas," although it is challenging to determine whether it is a "gas" made up of particles as massive as stars, or as small as subatomic particles.

79.4 WHAT IS DARK MATTER?

What could this dark "collisionless gas" be? Astronomers have ruled out normal gas because it does collide, and searches with radio telescopes would have revealed cold gas, and optical, radio, or X-ray telescopes would have revealed hot gas.

We know that stars in galaxies are so widely separated that they can travel through a galaxy with little likelihood of colliding. We might speculate, then, that dark matter is composed of vast numbers of "dark stars" that add up to all of the mass that we do not see. However, astronomers can rule out ordinary dim stars and even brown dwarfs (Unit 70) because such objects would emit at least some detectable infrared radiation.

It is possible to imagine other objects whose radiation would be much more difficult to detect, such as planets, very old and cold white dwarfs, old neutron stars, or black holes. This idea is called the **MACHO (massive compact halo object)** hypothesis. This would require some peculiar sort of star formation scenario in which the formation of almost all normal long-lived stars was suppressed when galaxies were young, leaving behind an enormous population of dark objects.

Astronomers are currently searching for MACHOs within the Milky Way's halo using the gravitational lensing technique. We saw in Unit 36 that astronomers

Concept Question 2

If dark matter were made of old white dwarfs and neutron stars, what would this predict about the amount and kinds of stars that formed early in the Milky Way's history?

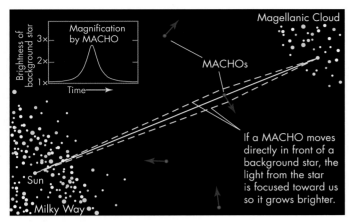

FIGURE 79.8
If a massive object in the Milky Way's halo passes precisely in front of a star in one of the Magellanic Clouds, gravitational focusing of the light can make the star temporarily appear brighter.

Concept Question 3

Does any evidence rule out the possibility that dark matter interacts with itself to form "dark stars" and "dark planets"?

FIGURE 79.9
The Alpha Magnetic Spectrometer on the International Space Station has detected cosmic rays that might have been produced by dark matter annihilation events.

have developed techniques sensitive enough to detect lensing caused by a planet. If a dark MACHO in the Milky Way's halo passes in front of a background star as seen from Earth, the gravity of the MACHO will bend the light of the background star and focus it, brightening the light we see (Figure 79.8).

The probability of one star aligning closely enough with another to produce a measurable brightening has a low probability, so astronomers have set up telescopes to monitor the light from millions of stars within the Magellanic Clouds, the small companion galaxies of the Milky Way (Unit 77.1). If a MACHO passes in front of one of those stars, it focuses the star's light for several days until it moves out of alignment. The brightening caused by lensing follows a known pattern, and dozens of halo objects have been detected in this way.

Even though the foreground objects are themselves too dim to see, astronomers can estimate their mass from the lensing effect they create. Moreover, from a statistical analysis of the number of such events, astronomers can deduce the number of objects creating the lens effect. From the evidence so far available, the objects creating the lens effect are primarily dim, low-mass stars rather than a new population of dark objects. It does not appear that MACHOs can account for more than about 10% of the halo mass.

In the novels of Sherlock Holmes, the fictional detective often states that "when you have excluded the impossible, whatever remains, however improbable, must be the truth." The nature of dark matter appears to be one of these improbable truths. Having excluded objects made of normal kinds of matter, the prevailing hypothesis today is that dark matter is entirely unlike the matter familiar to us.

Consider, for example, how difficult it is to detect neutrinos (Unit 4). Neutrinos can pass through the Earth, almost as if it was totally transparent. Evidence from studies of solar neutrinos (Unit 52) indicates that neutrinos have mass, but this mass is so tiny that neutrinos cannot themselves account for the dark matter. However, some physicists think there may be more massive particles, similar to the neutrino, that also interact very weakly with normal matter. This has been dubbed the **WIMP (weakly interacting massive particles)** hypothesis. Such particles are proposed in various models of particle physics, and physicists around the world have designed experiments to detect them.

Several experiments have reported possible detections, although these are inconclusive. A detector aboard the International Space Station (Figure 79.9) has detected cosmic rays that might be produced if dark matter can annihilate itself in some interactions. Other particle detectors in deep mines, where they are shielded from other cosmic rays, have also recorded events that might have arisen from rare interactions with dark matter.

If dark matter is eventually shown to be made of WIMPs, then it would be more accurate to describe a galaxy as a dense region of these particles. After all, the rotation curve measurements indicate that the dark matter represents about 80–90% of a galaxy's mass. The normal matter—the stuff that stars, planets, and we are made of—was probably attracted by the gravity of this dark matter and fell into the region, collecting at the center. Normal matter passes through this sea of WIMPs, registering its presence only by the effect its mass has on the matter we can detect. Hence, WIMP particles would be present within the Solar System, but their density is so low as to be almost undetectable with current instruments.

WIMPs, or some other as yet undetected particles, are the leading hypotheses for explaining dark matter. However, the case is not yet settled, and the nature of dark matter remains one of the central mysteries in astronomy today.

KEY POINTS

- The mass of a galaxy as a function of its radius can be determined from the speed of rotation at each radius—its rotation curve.
- The distribution of light in a galaxy suggests that the rotation curve should decline outside the bulge if stars provide the mass.
- Most galaxies have flat rotation curves, implying that their mass grows steadily larger with radius, with approximately 10 times more dark matter than can be observed in stars and other matter.
- In galaxy clusters, the high speeds of galaxy orbits also imply much more mass than is contained in the galaxies or cluster gas.
- Gravitational lensing provides a second way of testing cluster masses, and confirms the large values found from galaxy orbits.
- The gravitational distortion of shapes of background galaxies provides an estimate of the distribution of mass in a cluster.
- Evidence from one pair of colliding galaxy clusters shows that the dark matter passes through as cluster gas collides.
- The search for gravitational lensing has revealed too few dim or dark stars in our Galaxy's halo for those to be the dark matter.
- Dark matter may be an unknown kind of subatomic particle that does not interact through the electromagnetic force.

KEY TERMS

dark matter, 627
dark matter halo, 629
flat rotation curve, 628
gravitational lens, 630
massive compact halo object (MACHO), 632
rotation curve, 628
weakly interacting massive particle (WIMP), 633

CONCEPT QUESTIONS

Concept Questions on the following topics are located in the margins. They invite thinking and discussion beyond the text.

1. Density of dark matter at different radii. (p. 628)
2. Star formation consistent with MACHO hypothesis. (p. 632)
3. Possibility of "dark stars." (p. 633)

REVIEW QUESTIONS

4. How do we use gravity to measure the mass of a galaxy?
5. What is a galaxy rotation curve?
6. What is meant by dark matter?
7. How do we measure the mass of clusters of galaxies?
8. What is a gravitational lens?
9. How do we know that dark matter is not made of stars or gas?
10. What are MACHOs and WIMPs?

QUANTITATIVE PROBLEMS

11. Use figure 79.3 to calculate the mass within 30 kpc for the overall rotation curve and separately for the contribution just from stars. By what factor is the implied mass from the total rotation curve greater than the curve from the luminous stars?

12. What is the mass-to-light ratio of a galaxy with 10^{11} $M_\odot$ in stars and gas and 5×10^{11} $M_\odot$ of dark matter?
13. A spiral galaxy is observed in which the rotation speed remains 250 km/sec from 1 kpc out to the largest distances at which we can detect anything. How much mass exists out to 10 kpc? 11 kpc? 12 kpc? Can you generalize your results as to how much additional matter there is for each kiloparsec farther out you look?
14. In the Virgo Cluster there are galaxies measured to be traveling at about 1000 km/sec at a distance of 500 kpc from the center of the cluster. If you apply the mass formula to this velocity and radius, what interior mass do you find?
15. Suppose the Milky Way contains 10 billion $M_\odot$ of dark matter at radii between 8 and 9 kpc from its center.
    a. What is the density of matter in the "shell" between the radii of 8 and 9 kpc? (*Hint:* Subtract the volume of a sphere of radius 8 kpc from the volume of a sphere of radius 9 kpc.)
    b. How much mass of dark matter would there be in a spherical volume with a radius equal to the Earth's radius? Compare your result to the mass of the Sun.
16. Most spiral galaxies show a decline in the amount of light they produce that drops by a factor of 2 every 2 kpc farther out. For example, 3×10^{10} $L_\odot$ are produced within 2 kpc of the center; 1.5×10^{10} $L_\odot$ are produced between 2 and 4 kpc of the center; half as much again between 4 and 6 kpc; and so on. Compare the amount of luminous matter found this way to the total amount of matter contained within these same regions if the galaxy has a flat rotation curve with a rotation speed of 250 km/sec. Find the ratio of dark matter to luminous matter within 2 kpc of the center; between 2 and 4 kpc from the center; between 10 and 12 kpc from the center.

TEST YOURSELF

17. Which of the following is not considered observational evidence for dark matter?
    a. Dark nebulae observed in our Galaxy
    b. Gravitational lensing by galaxy clusters
    c. Rapid orbital motions of stars in the outer regions of spiral galaxies
    d. Hydrostatic equilibrium of hot gas in galaxy clusters
    e. Large galaxy velocities within galaxy clusters
18. The rotation curves out to distances that comprise the bulges of typical spiral galaxies are dominated by
    a. gas. c. dark matter.
    b. stars. d. All of the above contribute equally.
19. Gravitational bending of light
    a. shows that space can be curved.
    b. gives the illusion that some quasars have companions.
    c. allows astronomers to detect MACHOs.
    d. provides a means to measure the mass of galaxy clusters.
    e. All of the above are correct.
20. A WIMP has properties most similar to a(n) _____, but with a much larger mass.
    a. proton c. neutrino e. neutron
    b. electron d. positron

UNIT 80

Cosmology

80.1 Evolving Concepts of the Universe
80.2 The Recession of Galaxies
80.3 The Meaning of Redshift
80.4 The Age of the Universe

Learning Objectives

Upon completing this Unit, you should be able to:
- Recall models of the universe that have been proposed throughout history.
- Define the cosmologic principle and explain why Hubble's law applies everywhere.
- Calculate and explain what causes the redshift of galaxies in an expanding universe.
- Explain how it is possible to estimate the age of the universe from Hubble's constant.

Cosmology is the study of the structure and evolution of the universe as a whole. Cosmologists ask: Is the universe infinite? Does it have an edge? Has it existed forever, or does it have a definite age? How did it form? What will happen to it in the future? Given our insignificant size in the cosmos, such questions may seem futile or even arrogant, but most cultures have tried to answer them. Many of the attempted answers have become part of humanity's religious heritage. In cosmology we address some of these same questions based on astronomical evidence and the scientific method. It is important to set aside personal beliefs when examining what the weight of scientific evidence indicates.

We have made many surprising discoveries. Current evidence indicates that the universe was born about 13.8 billion years ago out of a hot, dense, violent state of matter and energy called the **Big Bang.** It has been expanding ever since, and it is filled with radiation from the early stages of the explosion. That radiation carries an imprint of information about the earliest stages of structure formation that we will describe in Unit 81. Within the last few decades, cosmologists have begun to extend our knowledge of the universe to the very beginning of time. They have discovered that the Big Bang may have been born out of even more turbulent events known as the *inflation* (Unit 83), when the entire universe that we see today may have fit in a volume smaller than a proton! Our cosmological models even allow us to predict the ultimate fate of the universe (Unit 84). In this Unit, we will examine some early cosmological ideas and then reexamine Hubble's discovery of the galaxy redshift law (Unit 75) and what it implies about the nature of the universe.

80.1 EVOLVING CONCEPTS OF THE UNIVERSE

Over the centuries, our understanding of the universe has steadily changed. For millennia the idea of a central, stationary Earth seemed natural—a moving Earth seemed absurd, both to scientists and to others just expressing their common sense. As concepts such as inertia (Unit 14) developed, the idea of a moving Earth no longer seemed physically impossible, and by the early 1600s the idea of planets orbiting the Sun seemed plausible. Such a model explained the planet's motions and other observed phenomena more simply, although it required a break with common beliefs.

At the time this idea was not yet supported by strong evidence—such as the predicted parallax effect—and therefore it was not accepted by all. Some astronomers continued to work with a geocentric model, which was still taught in many textbooks until the early 1700s. However, additional evidence accumulated, such as the

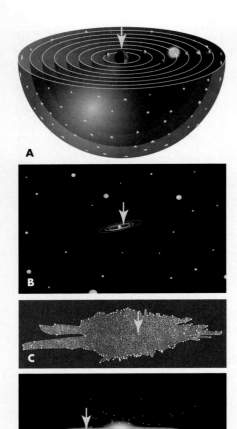

aberration of starlight and parallax of stars (Unit 54), that so strongly contradicted the old Earth-centered model that the model was finally abandoned. In the following several centuries, our ideas about the nature and extent of the universe have repeatedly changed, as illustrated by selected examples in Figure 80.1.

Even as the debate over Earth-centered and Sun-centered models continued, others began proposing even more complex cosmologies that recognized the Sun as one of many stars. One of the most commonly accepted cosmologies was that the universe consisted of the stars of the Milky Way with the Sun near its center, and that this island of stars was surrounded by an empty void. An alternative idea was proposed by the German philosopher Immanuel Kant in the mid-1700s, anticipating some of our current ideas. He suggested that the Milky Way was merely one of a multitude of galaxies. This idea was not widely accepted until the 1920s, when astronomers could demonstrate that galaxies were millions of light-years from the Milky Way and comparable to it in size (Unit 75).

The pace of changes in our models of the universe has grown more rapid in the last century. Even as we were just learning to accept the idea that the universe contains billions of galaxies, the Milky Way being just one of them, observations revealed a strange phenomenon—that the other galaxies are moving away from us at high speeds (Unit 75.3). Edwin Hubble first demonstrated this in the 1920s, and it meshed with ideas about the nature of space and gravity developed by Albert Einstein around the same time. This new picture, of an expanding universe, had implications for how the universe must have begun.

The extraordinary conclusions we have reached about the universe today are rooted in observations that began as we asked simple questions about how the things we see came to be, followed a train of reasoning, and then pursued observations that led to the present model. Given the history of changing ideas about the universe, it needs to be stressed that we are speaking of models of the universe, rather than certain knowledge about a clearly defined "Universe," which we might write with an uppercase "U." This is *not* to say that the scientific evidence supporting modern cosmology is weak, but rather that our understanding of what the entirety of the universe comprises continues to evolve and be refined.

The scientific approach requires that we weigh the evidence and work with the best current model, *and* that we accept that the model may change as new evidence accumulates. In addressing the fundamental questions about the universe, it would be unscientific to work toward "proving" one model of the universe. This is a subtle point: Individual astronomers do attempt to argue the merits of particular models, but as scientists they accept that their models may be inaccurate or incomplete, and that a model should not be selected because of personal beliefs.

80.2 THE RECESSION OF GALAXIES

Nearly all galaxies have a redshift of their spectral lines. Astronomers in the early 1900s interpreted this as meaning that most of them are moving away from us (Unit 75). Hubble showed that the redshift of a galaxy increases with distance, d, in such a way that the recession velocity, V, of a galaxy obeys Hubble's law:

$$V = H \times d$$

where H is Hubble's constant.

When examined carefully, this equation is startling. It seems to say that all the galaxies in the universe are flying away from us, making it appear as if the Milky Way is at the center of the universe—that a vast explosion (the Big Bang) has sent galaxies flying away from us in all directions. But this is incorrect.

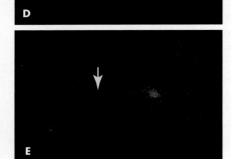

FIGURE 80.1
Changing models of the universe. (A) A geocentric model surrounded by crystalline spheres. (B) The Earth as one planet orbiting the Sun. (C) The Sun as one star near the center of a cloud of stars (the Milky Way) surrounded by an empty void. (D) The Sun orbiting far from the center of our Galaxy, which contains hundreds of billions of stars. (E) Our Galaxy just one among hundreds of billions of clustered galaxies in an expanding universe. Our position in each of these universes is indicated by a yellow arrow.

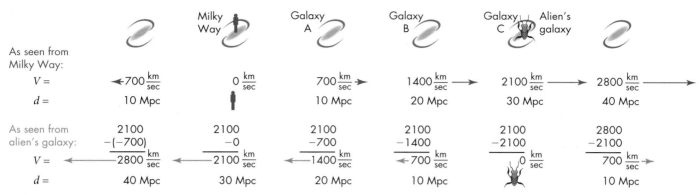

FIGURE 80.2
A line of galaxies illustrating that in a universe that obeys a Hubble law of expansion ($V = H \times d$), an observer in any galaxy sees the same law.

A simple example illustrates why the Milky Way's position is not special. Imagine a string of galaxies distributed evenly along a line, as shown in Figure 80.2. Each galaxy is separated from its nearest neighbor by 10 megaparsecs. From the galaxy marked as the Milky Way in the figure, we would observe that each galaxy has a recession velocity given by Hubble's law: $V = $ (70 km/sec per Mpc) $\times$ (distance). That is, galaxy A recedes from us at 700, B at 1400, and C at 2100 km/sec. Next, suppose we could communicate with an alien in galaxy C and ask what it sees. It would see us receding at 2100 km/sec, galaxy A receding at $2100 - 700 = 1400$ km/sec, and galaxy B receding at $2100 - 1400 = 700$ km/sec. In fact, the alien would see galaxies receding from it in exactly the same way as we see galaxies receding from us. Furthermore, the galaxies seen by the alien will obey precisely the same form of Hubble's law.

A similar argument can be made for an observer in any of the galaxies along this line. Each will see all of the other galaxies flying away from it. This argument can be extended to two or three dimensions, and in each case for whatever galaxy we imagine being "home," we will observe all the other galaxies "flying away" from us with speeds that agree with Hubble's law. A universe obeying Hubble's law has no preferred "center" of expansion. Astronomers describe this lack of a preferred location as the **cosmological principle.** It is a statement of cosmic modesty—an extension of the Copernican revolution (Unit 11.4): There are no special positions in the universe. *We* are not at the center of the universe, nor is anyone else.

If there is no central location from which galaxies are moving away, then it does not make sense to think of the universe's expansion as being like the explosion of a bomb, sending fragments in all directions from a central blast point. But how then *do* we explain this motion? Einstein's theory of **general relativity** offers an answer.

Einstein developed general relativity in 1916, just a few years before the nature and motion of galaxies were discovered. He was trying to solve other puzzles having to do, for example, with gravity and the way it affects light (see Unit 27). The resolution of these problems led him to develop the idea that space can itself have motion.

The idea that space can move is alien to us, and it may even seem a bit silly. How can "nothing" be moving? But this idea is needed to explain many observed phenomena. We can detect changes in the arrival of light signals from galaxies that seem to imply that the light has changed speed as it moved through space. Yet we can measure the speed of light extremely accurately, and every experiment shows that it moves through space at a constant speed. The explanation for the changes in arrival time is that the light traveled through a region where the space itself was moving relative to us. Its arrival time is advanced or delayed much as an airplane's arrival time may be affected if it is traveling through tailwinds or

> **Concept Question 1**
>
> If astronomers calculated Hubble's constant when the universe was half its current age, would the value they found for H be the same?

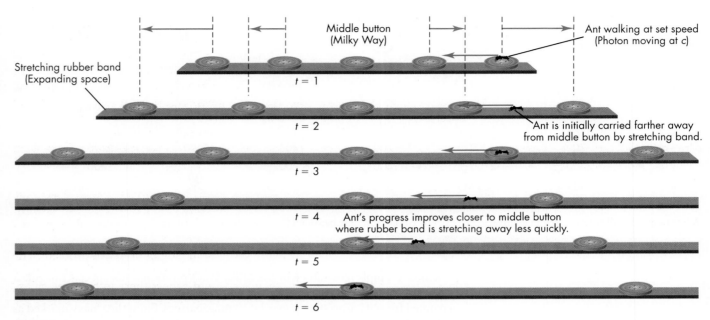

FIGURE 80.3
Galaxies in expanding space are like buttons on a stretching rubber band. An ant walking at a steady speed on the rubber band is like a photon traveling through expanding space: It moves at a speed relative to a distant button that is a combination of its own speed and the speed of the space it is traveling through at that moment.

Mathematical Insights

One effect of the stretching of space is that distant galaxies look much bigger than expected. A galaxy whose light left it 12 billion years ago has an angular size that we would have expected if it was 4 billion light-years away, and looks bigger than some nearer galaxies of the same physical size.

headwinds—even though the airplane maintained a steady speed relative to the air throughout its trip.

As remarkable as it may seem, the more distant galaxies we can see are moving away from us at greater than the speed of light! More properly we should say it is the region of space that those galaxies are in that is moving so fast. For example, the most distant galaxies we have detected are currently moving away from us at about three times the speed of light, and they were moving away even faster at the time the light was emitted. At the time they emitted the light we are currently receiving, they were less than 3 billion light-years from the Milky Way, but because of the motion of space, the light has taken over 13 billion years to finally reach us.

How can we understand these bizarre findings? Returning to the example of the galaxies located along a line (Figure 80.2), one way we might picture the idea of space having motion is to think of the galaxies as buttons glued at intervals along a rubber band, as shown in Figure 80.3. If the rubber band is steadily stretched, each button remains stationary relative to the rubber band, but the space *between* the buttons expands. Buttons that are farther apart on the rubber band move away from each other faster because there is more of the stretching rubber between them. The relative speed increases with distance just as for the Hubble law.

The effect of expanding space on a photon moving through space can be illustrated by an ant walking along the rubber band in our analogy. The ant starts walking from a button on one end of the rubber band with its ultimate destination being the middle button, but what was initially a small distance grows steadily larger. Because the rubber band is stretching faster than the ant can walk, the ant is at first carried farther from its goal by the stretching band. As it keeps walking, though, the ant reaches parts of the rubber band that are not moving away from its target so quickly, and then it makes more rapid progress toward its goal.

A photon traveling through expanding space between galaxies is like the ant. Even though the photon is traveling steadily at a constant speed c through space,

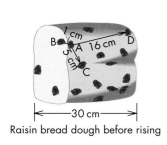 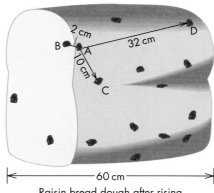

FIGURE 80.4
As a loaf of raisin bread dough rises, the raisins move farther apart from each other at a rate proportional to their separation.

it makes slower progress than we would expect if space were motionless. In motionless space, a photon traveling from one galaxy to another would need to traverse only the original distance separating the two galaxies. If the rubber band had not been stretching, the ant might have made the trip by $t = 3$ instead of $t = 5$ in the figure.

The rubber-band analogy can be extended to two dimensions by imagining a stretching rubber sheet with the buttons glued to it. A three-dimensional analogy would be the relative motion of raisins in the dough of rising raisin bread (Figure 80.4). The separations between buttons on the rubber sheet or raisins in the dough grow because of the expansion of the medium in which they are set.

However, these analogies present a problem. The rubber band, rubber sheet, and the dough all have edges. Based on these analogies, then, we would expect there to be some galaxies close to the edge of the universe; if we lived in such a galaxy we would see galaxies in one direction but not the opposite. Because we do not see such an imbalance from the Milky Way, does this imply that our location is close to the center of the universe after all? We can avoid a conflict with the cosmological principle that this might imply by hypothesizing that the universe is infinite, or at least so large that none of the edges is anywhere close to being visible to us. (We will examine some alternative possibilities in Unit 82 by using the idea that space can be curved.)

In a similar fashion as these analogies, general relativity predicts that space itself is expanding, and galaxies are carried apart by that expansion, not by their own motion *through* space. However, although space expands, matter is not "glued" in place like the buttons or raisins in our analogies. Galaxies, stars, planets, and other bodies can move within the expanding space. In fact, the force of gravity that attracts them toward other nearby objects is often strong enough to overcome the expansion of space that would otherwise carry them apart. As a result, these objects can move toward each other despite the expansion of space, gathering into clusters, galaxies, and solar systems. Finally, there is no contradiction between general relativity and special relativity because although a region of space may be moving away from us at greater than the speed of light, no object travels through space at a speed greater than the speed of light c.

80.3 THE MEANING OF REDSHIFT

The expansion of space affects more than just motion. It also produces an observable effect on the wavelengths of light. As light waves travel through expanding space, they are stretched out by its expansion. The redshift we see from distant galaxies is caused by this stretching. Hubble interpreted this redshift as the recession velocity V used in Hubble's law (Unit 75); but as we discussed in the previous section, this interpretation does not explain the light travel time.

The stretching of wavelengths is the same effect you would find if you drew a wiggly line on the rubber band in the earlier analogy. As the rubber band is stretched, the crests and troughs of the wiggles become more widely spaced. The ant walking on the stretching rubber band would similarly feel its feet being spread apart as it walked (although an ant would pull them together after each step). The

Expansion and redshift

FIGURE 80.5
As space expands, it stretches radiation moving through it, making the wavelengths longer. Because longer wavelengths have lower energy and are associated with cooler objects, the stretching of the radiation has the effect of cooling it.

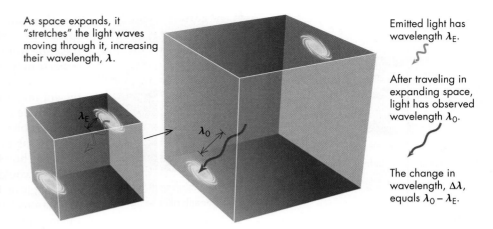

As space expands, it "stretches" the light waves moving through it, increasing their wavelength, λ.

Emitted light has wavelength λ_E.

After traveling in expanding space, light has observed wavelength λ_0.

The change in wavelength, $\Delta\lambda$, equals $\lambda_0 - \lambda_E$.

Concept Question 2

The Solar System formed about 4.5 billion years ago, when the universe was about 70% of its current size. What else would have been smaller back then—stars? the Milky Way? the Local Group? the Virgo Cluster? Or were they all just as big as now?

Mathematical Insights

Hubble interpreted the redshift as a Doppler shift and would have said $z = 0.1$ implies that the galaxy was traveling away from us at 10% of the speed of light, or about 30,000 km/sec.

Mathematical Insights

If the size of the universe today is R_0 and the size of the universe when light was emitted from a galaxy was R, then

$$z = \frac{R_0 - R}{R} = \frac{R_0}{R} - 1.$$

Adding 1 to both sides of the equation, we find that

$$\frac{R_0}{R} = z + 1.$$

effect on a light wave traveling between galaxies is illustrated in Figure 80.5—the wavelength of the light is stretched by the same amount as the space itself.

Understanding that the redshift we observe is caused by the stretching of wavelengths in expanding space gives us a better way to interpret redshift. The redshift indicates how much the universe has expanded since light left the object we are observing. Astronomers determine the redshift z from the fractional change in the wavelength λ of light emitted by a galaxy. We can calculate the redshift from the following formula:

$$z = \frac{\text{Change in wavelength}}{\text{Emitted wavelength}} = \frac{\Delta\lambda}{\lambda}$$
$$= \frac{\text{Change in size of universe}}{\text{Size of universe when light left galaxy}}.$$

Because the universe has no measurable edge, for *size* here we might substitute "the average distance between galaxies." For example, if we detect a spectral line that was emitted at 500 nanometers but is redshifted to 550 nanometers from a distant galaxy, the change in wavelength $\Delta\lambda$ is 50 nanometers. Therefore, the redshift is

$$z = \frac{550\,\text{nm} - 500\,\text{nm}}{500\,\text{nm}} = \frac{50}{500} = 0.1,$$

and we can say that the average distance between galaxies (or the size of the universe) has expanded by 10% since the light left that galaxy.

Astronomers find it helpful to use z when they describe the expansion of the universe, because the redshift is related to the changing separations and "size" of the universe. In particular, z compares the size of the universe when we receive the light from a galaxy to the size of the universe when the light was emitted by the galaxy. That relation is

$$\frac{\text{Size of universe today}}{\text{Size of universe when light left galaxy}} = z + 1.$$

In the preceding example, with $z = 0.1$, we can say that the universe is 1.1 times bigger today than when light left the galaxy.

We observe some extremely distant galaxies and quasars for which $z > 1$. This does *not* indicate that they are traveling faster than the speed of light. Rather, the result is telling us that $z + 1 > 2$—that is, space has expanded by more than a factor of 2 since the light we are now seeing left those objects. If a galaxy has a redshift $z = 1$, then separations and the size of the universe have exactly doubled since the light left the galaxy. For $z = 2$, space has expanded by $(z + 1) = 3$ times since the light left the galaxy, or we might equivalently say that the universe was three times smaller when the light left that galaxy.

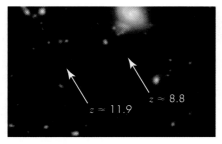

FIGURE 80.6
Infrared measurements made by the Hubble Space Telescope suggest that the very faint red fuzzy spots at the tips of the two arrows in this image are very high-redshift galaxies.

The current record holder for redshift is a galaxy with $z = 8.55$. For this redshift, the universe today is 9.55 times larger than when the light was emitted. This means that the matter in the universe was packed much more tightly together when the light we see left that quasar. The average separation between galaxies was just $1/9.55 = 0.105$ times the current separation. The smaller separation would have been in every direction, so the *volume* of space (length × width × height) would be just $0.105 \times 0.105 \times 0.105 = 0.0012$ times what it is today.

Some galaxies have been identified that, based on their infrared colors, may have redshifts as high as about $z = 12$ (Figure 80.6). However, direct determinations of the redshift using spectral lines has not yet confirmed these estimates. Galaxies at such high redshifts look very different from nearby galaxies, in part because the matter was just in the beginning stages of gathering together to form a galaxy (Unit 76.4). If we look out farther and farther, to earlier and earlier times in the age of the universe, ultimately we should see a time when the whole universe was packed together at extremely high density.

80.4 THE AGE OF THE UNIVERSE

The model of the universe we have just discussed, as an expanding space peppered with galaxies, allows cosmologists to estimate its age and to calculate conditions near the time of its birth. Alexander Friedmann, a Russian scientist, made such calculations in the 1920s, but his work attracted little attention. Not until 1927, when Abbé Georges Lemaître, a Belgian cosmologist and priest, independently made similar calculations, did astronomers appreciate that conditions at the birth of the universe could be deduced from what is observed today.

Lemaître pointed out that because the separations of galaxies are growing today, the galaxies must in the past have been closer together, as we saw in the last section. In fact, if you simply follow the expansion implied by Hubble's law back in time, there must have been a period long ago when all of the galaxies in the universe were crowded together. Galaxies and stars as we know them today could not have existed in such a dense environment. With so much mass in such a relatively small volume, the matter that now composes the far-flung galaxies and their stars must have been packed into an extremely dense ball that Lemaître called the *Primeval Atom*. From this dense state, the universe must have expanded at a tremendous speed—the Big Bang.

It is straightforward to estimate the time since the Big Bang as follows. We know the rate at which space is expanding between any pair of galaxies—how fast they are moving away from each other. We solve for how long it would take them to get this far apart moving at their current speed, just as we might solve for how long a car takes to travel 200 kilometers at a speed of 50 kilometers per hour.

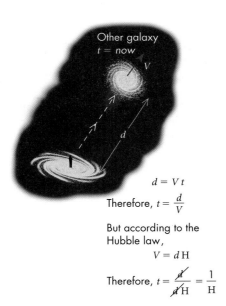

FIGURE 80.7
Estimating the age of the universe by the recession of one galaxy from another galaxy.

Mathematical Insights

Because Speed = Distance/Time, or $V = d/t$, multiply both sides of the equation by t/V to get $t = d/V$.

Consider two representative galaxies separated by a distance d and separating with a velocity V (Figure 80.7). We will assume that V has remained constant and call the time we calculate that it took the galaxies to reach this separation the **Hubble time**, t_H. We can calculate the Hubble time as follows: The time it takes to travel a particular distance, d, is the distance divided by the speed. (For example, 200 km divided by 50 km per hour gives 4 hours.) Therefore, the Hubble time is

$$t_H = d/V.$$

Hubble's law, however, tells us that a galaxy's velocity grows with distance according to the formula $V = H \times d$. Substituting this into our equation, we find

$$t_H = \frac{d}{H \times d} = \frac{1}{H}.$$

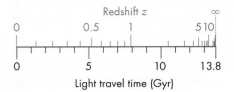

FIGURE 80.8
Comparison of the redshift scale to length of time since the light was emitted—the light travel time. The scale runs from the universe today at left to its beginning at right.

TABLE 80.1 Relationships Between Redshift, Distance, and Time

Redshift (z)	Size of Universe Compared to Now	Distance When Light Was Emitted	Length of Time Since Emitted (Light Travel Time)	Time After Big Bang
0	1	0	Today	13.8 billion yr
0.1	0.91	1.3 billion ly	1.3 billion yr	12.5 billion yr
0.5	0.67	4.2 billion ly	5.2 billion yr	8.6 billion yr
1.0	0.50	5.5 billion ly	7.9 billion yr	5.9 billion yr
2.0	0.33	5.8 billion ly	10.5 billion yr	3.3 billion yr
3.0	0.25	5.3 billion ly	11.6 billion yr	2.2 billion yr
5.0	0.17	4.3 billion ly	12.6 billion yr	1.2 billion yr
7.0	0.13	3.6 billion ly	13.0 billion yr	800 million yr
10	0.09	2.9 billion ly	13.3 billion yr	500 million yr
1000	0.001	45 million ly	13.8 billion yr	400,000 yr

The distance d cancels out, so it doesn't matter which two galaxies we choose—they began moving apart a time equal to $1/H$ ago. In other words, the inverse of Hubble's constant indicates a length of time that approximately measures the age of the universe.

To get a value for t_H in years, we have to carry out some unit conversions because H is expressed in units of kilometers/second per megaparsec. One megaparsec is 3.09×10^{19} kilometers, so if Hubble's constant is 70 km/sec/Mpc, this gives

$$t_H = \frac{1 \text{ Mpc}}{70 \text{ km/sec}} = \frac{3.09 \times 10^{19} \text{ km}}{70 \text{ km}} \text{sec} = 4.41 \times 10^{17} \text{ sec}.$$

Because there are 3.16×10^7 seconds per year, $1/H = 1.40 \times 10^{10}$ years. Therefore, we find that the age of the universe is about 14 billion years. In this simple calculation we assumed that the expansion of space has remained constant. Gravity can slow the expansion, but other effects on space can cause acceleration (Unit 82.4). Detailed calculations that take into account these different effects give an answer that is very close to the age we have calculated: 13.8 billion years.

Table 80.1 shows how much the universe has expanded for different values of the redshift according to the best current model. There are some surprising consequences of the higher speed of expansion when the universe was younger. A galaxy at redshift 7 was actually closer to us at the time when the light was emitted than a galaxy at redshift 0.5. We cannot see where that galaxy is *today*, but if we could, we would see that the redshift 7 galaxy is farther away than any galaxy at lower redshift. This can be confirmed in reference to Table 80.1, by multiplying the distance when the light was emitted by $(z + 1)$, the factor by which space has expanded. This means, for example, that the most distant material that telescopes can detect, radiation from a redshift $z = 1000$ (Unit 81.2), is currently at a distance of about 45 billion light-years.

The complications of an expanding universe create an ambiguity for what we mean by the "distance" of a distant galaxy. Its distance when the light was emitted? Its distance when the light was received? Some other distance? To avoid confusion, astronomers often use the **light travel time distance** when speaking of the distance to remote galaxies. The light travel time distance measures how far the light would have traveled in the time since it was emitted if the universe were not expanding. With that definition of distance, we can then say that a galaxy that we observe today and whose light was emitted 10 billion years ago, is 10 billion light-years away from us. The light travel time is compared to the redshift in Figure 80.8 and Table 80.1.

Mathematical Insights

From Table 80.1, an object at $z = 1000$ emitted the light when it was only 45 million light-years from our region of space. Because of the expansion of space by a factor of $z + 1 = 1001$, that region is now 1001 times farther away than it was then.

Concept Question 3

What were your beliefs about the nature of the universe prior to reading about modern cosmology? Are you willing to change any of them if the scientific evidence disagrees with them?

KEY POINTS

- Religions express beliefs about the size, shape, and age of the universe, but astronomers are finding evidence about these things.
- From all regions of the universe observers see the same motions because all portions move away from each other equally fast.
- Distant portions of the universe can move away from us faster than the speed of light, but light still reaches us eventually.
- The redshift of light is caused by the stretching of light's wavelengths as it travels through expanding space.
- The value of redshift is determined by the ratio of the current size of the universe divided by the size when the light was emitted.
- The increasing speed of galaxies with distance implies that galaxies were all packed together a Hubble time ago.
- Galaxies have been detected up to a redshift of about 10, when everything in the universe was 11 times closer together.
- Because of expansion, the distance when a galaxy emitted its light is smaller than when the light arrives; distances are usually reported as the light travel time, which lies between the two.

KEY TERMS

Big Bang, 635
cosmological principle, 637
cosmology, 635
general relativity, 637
Hubble time, 641
light travel time distance, 642

CONCEPT QUESTIONS

Concept Questions on the following topics are located in the margins. They invite thinking and discussion beyond the text.

1. Hubble's constant at different times. (p. 637)
2. Effect of expansion on sizes of objects. (p. 641)
3. Changing beliefs in light of scientific evidence. (p. 642)

REVIEW QUESTIONS

4. What does cosmology study?
5. Why do astronomers think the universe is expanding?
6. What is the cosmological principle?
7. Why is there no place we can call the center of the universe?
8. What causes redshift?
9. How do we measure the size of the universe?
10. How old is the universe? How is its age found?

QUANTITATIVE PROBLEMS

11. When Hubble first estimated the Hubble constant, galaxy distances were still very uncertain, and he got a value for H of about 600 km/sec per Mpc. What would this have implied about the age of the universe? What problems would this have presented for cosmologists?
12. A galaxy at a distance of 25 Mpc has a recession velocity of 1875 km/sec. What is Hubble's constant based on this one galaxy? Why is this not a good way to determine Hubble's constant?
13. For many years opposing opinions on the value of Hubble's constant were debated. Some astronomers argued that $H =$ 50 km/sec per Mpc while others claimed a value close to 90 km/sec per Mpc. Calculate t_H for these two values of H.
14. What wavelength would visible light (400 to 700 nm) have if it were seen from a galaxy formed just half a billion years after the Big Bang? What if it came from matter when the universe was just 400,000 years old? (Refer to Table 80.1.)
15. Plot the light travel time (Y axis) versus the redshift (X axis) based on the data from Table 80.1. Estimate the redshift for galaxies that are seen as they were 1, 2, and 3 billion years ago.
16. Plot the distance when light was emitted (Y axis) versus the redshift (X axis) based on the data from Table 80.1. Estimate the redshift at which a galaxy of a fixed physical size would have the smallest angular size. Estimate the distance at which the galaxy would look the same size as it does when at a redshift of 0.1.

TEST YOURSELF

17. From what evidence do astronomers deduce that the universe is expanding?
    a. They can see the disks of galaxies getting smaller over time.
    b. They see a redshift in the spectral lines of distant galaxies.
    c. They see the edge of the universe moving away from us.
    d. They can see distant galaxies dissolve, pulled apart by the expansion of space.
    e. All of the above.
18. If we discovered tomorrow that the distances we measured to galaxies were incorrect, and that Hubble's constant is 35 km/sec per Mpc instead of 70 km/sec per Mpc, we would conclude that the universe's age is _____ previously thought.
    a. 4 times older than
    b. 2 times older than
    c. the same as
    d. 2 times younger than
    e. 4 times younger than
19. If a galaxy has a redshift $z = 4$, what fraction of the universe's current size was the universe when the light was emitted from that galaxy?
    a. One-half
    b. One-third
    c. One-fourth
    d. One-fifth
    e. One-sixteenth
20. If a distant galaxy has a redshift of 3, it was 5.3 billion light-years distant when the light was emitted. How many billions of light-years distant from us is it today?
    a. 2.3
    b. 5.3
    c. 10.6
    d. 21.2
21. If you lived in a galaxy far away from the Milky Way, your redshift measurements would show that galaxies move
    a. toward you out to the Milky Way; away at larger distances.
    b. away from the Milky Way but toward you.
    c. away from you in the same way as Hubble found.
    d. toward you in half the sky, away in the other half.
    e. away from the point where the Big Bang occurred.

PART 5

UNIT 81

The Edges of the Universe

81.1 Olbers' Paradox
81.2 The Cosmic Microwave Background
81.3 The Era of Galaxy Formation

Learning Objectives

Upon completing this Unit, you should be able to:
- Describe Olbers' paradox and explain how it is resolved.
- Explain how the cosmic microwave background arose and its current appearance.
- Describe how galaxies first began to form and explain how fluctuations in the cosmic microwave background trace their earliest stages of formation.

Some simple questions can have profound answers. One such question puzzled astronomers for centuries: Why is it dark at night? At first glance this question seems to have an obvious answer. However, answering it fully requires a surprising amount of cosmology (Unit 80), involving the expansion and curvature of space, the speed of light, the cosmic density of matter, and the age of the universe.

To understand why the sky is dark at night, we need to know what conditions were like when all of the matter in the universe was packed together just after the expansion of the universe began. Extrapolating to what the physical conditions must have been like at these times, cosmologists have been able to make several predictions about characteristics of the universe. The confirmation of these predictions has provided strong support for the Big Bang theory of the universe's origin.

In this Unit, we explore the most distant light we can see: light that has been traveling through the universe since shortly after the Big Bang, when all of the matter in the universe was merged into a great ocean of hot gas. We begin, though, by trying to understand why a dark sky at night is so surprising.

81.1 OLBERS' PARADOX

Our discussion of why the sky is dark at night requires that we look first at the distribution of the objects that generate visible light. We show a map of galaxies on the largest scales in Figure 81.1. This wide-angle view shows the distribution of nearly 1 million galaxies as seen looking up out of the disk of our own Galaxy, the Milky Way. Around the edge of the picture, there are fewer galaxies because dust in the Milky Way hides the galaxies in that part of the sky. However, the unobscured portion of the picture provides an important clue about the structure of the universe.

No matter what direction you look (ignoring the dust layer of the Milky Way), you see approximately the same number of galaxies. That is, averaged over a very large area, the universe is more or less the same in all directions, and galaxies are spread throughout the universe much as raisins are spread throughout raisin bread. They may clump a bit in one region or another, but overall, they are fairly evenly distributed. If we look over large enough distances, even the superclusters and voids of large-scale structure (Unit 77.4) average out, and the matter in the universe appears to be approximately smoothly distributed, or **homogeneous,** out to the largest distances we can measure. This homogeneity presents a problem, though. If there are galaxies in every direction, and the universe is infinite, then our sky should be completely covered by them.

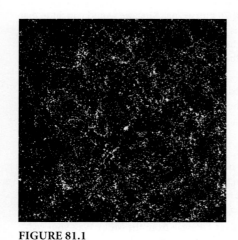

FIGURE 81.1
Positions on the sky of nearly a million galaxies from a visible-wavelength survey. This is the view looking up and out of the Milky Way. Note the relatively uniform distribution of the galaxies.

We can understand the issue more clearly by dividing up the universe into shells of equal thickness surrounding us as in Figure 81.2. The galaxies in each shell are at a distance equal to the shell's radius. Because the galaxies are uniformly distributed, the number of galaxies in each shell will be proportional to the volume of the shell. The volume of each shell is equal to its surface area times its thickness, but the thickness is constant. Therefore, the number of galaxies depends just on the shell's surface area, which is proportional to the distance squared (d^2). However, the brightness of each galaxy declines with distance inversely proportional to the square of the distance ($1/d^2$).

The declining brightness of galaxies is exactly canceled by the increasing number of galaxies with distance. As a result, we should receive equal amounts of light from each shell of galaxies around us. And if we add up every shell surrounding us out to infinite distance, we will find an infinite amount of light. The night sky should be blazingly bright!

This is known as **Olbers' paradox** after the German astronomer Heinrich Olbers, who in 1823 popularized the problem. He phrased the problem in terms of stars (galaxies were not yet known), but the result is the same. The paradox arises from premises that are reasonable (or so it seemed at that time) but lead to a conclusion clearly at odds with the simple fact that the sky is dark at night.

Actually, it is not correct to say that the amount of light would be infinite. It is more accurate to say that if the universe has no edge, then in every direction we look, our line of sight should eventually encounter the surface of a star. So the sky should be as bright as the surface of a star everywhere—still blazingly bright. This is easiest to understand from a simple analogy. Suppose you stand in a small grove of trees and look out between the trees to the surrounding landscape. If the grove is small, your line of sight will be blocked by trees in a few directions, but it will pass between the trees in most directions (Figure 81.3). If the grove is larger, more-distant trees in it will block your view in what previously were clear gaps. In that case, no matter where you look, your line of sight will be intercepted by a tree.

Now suppose that rather than looking between trees, you look out into space through the galaxies and stars that compose the universe. If space extends sufficiently

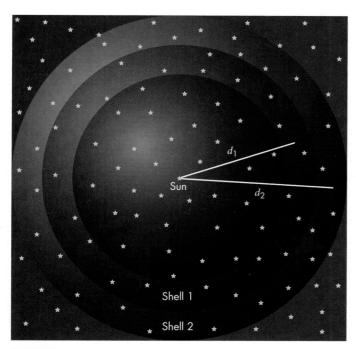

FIGURE 81.2

Olbers' paradox. If we add up the light from all the galaxies (yellow star symbols in the illustration) within a series of shells that we draw around us, we will find that the number of galaxies increases with distance at the same rate as the brightness of each galaxy declines.

Although the paradox bears Olbers' name, it was discussed by various astronomers over the two previous centuries. Kepler pointed out that in an infinite universe, the night sky should be bright, and he therefore concluded the universe was finite.

In a small grove of trees, only a few block your view. Lots of space to view between trees.

In a larger forest, more distant trees block your view. No open space visible between trees.

FIGURE 81.3

(A) An observer in a small grove of trees can see out through gaps between the trees to the surrounding countryside. (B) In a larger forest, distant trees block the gaps so that no matter where you look, your line of sight ends on a tree.

Concept Question 1

How might dust in galaxies affect Olbers' paradox? What would be its effect at infrared wavelengths?

far and is populated with galaxies homogeneously, then no matter what direction you look, every line of sight will ultimately intercept a star—just as in a sufficiently large forest every line of sight ultimately hits a tree. Even though the stars in distant galaxies are faint to us, there are so many of them that their collective brightness should be large. Therefore, the sky should be covered with starlight, glowing brilliantly with no dark spaces in between. The night sky should not be dark!

This argument must contain a false assumption, because the most obvious observation one can make about the sky is that it is dark at night. That is the paradox: Logical arguments imply that the night sky should be bright, but it is not. Where, therefore, is the error in our reasoning?

The first way the paradox can be avoided is if there are no stars beyond some distance. That is, just as we can see out of the woods if the forest is small, so too we will see a dark night sky if we run out of stars and galaxies beyond some distance. However, we do not see any evidence of an edge to the universe.

There can be an edge in a second sense—an edge in time. Even if the universe is infinite in extent, if it has a finite *age*, then we can see only the stars whose light has had time to reach us within the time limit set by the age of the universe. So if the universe is 13.8 billion years old, we can see light from a star that is 10 billion light-years away, but not one that is 20 billion light-years away. The maximum distance that light can travel within the age of the universe defines the **cosmic horizon**. Astronomers call the volume of space within the cosmic horizon the **visible universe** (Figure 81.4). As the universe ages, the visible universe grows larger. At present, we don't know how large the entire universe might be, but it is suspected to be far, far larger than the visible universe. This would suggest that if we could wait long enough, in the future the night sky *might* become bright.

However, realizing that we are looking back in time presents another problem. In every direction we look, we should eventually see back to early times when the universe was filled with hot gas from the Big Bang, and this should be just as bright as the surface of a star. Olbers' paradox predated the discovery that the universe expanded from a fiery hot state, but we need to understand the effects of that expansion to complete our explanation of why the sky is dark at night.

FIGURE 81.4
We cannot, even in theory, see beyond the cosmic horizon. Light from there takes more time than the age of the universe to reach us.

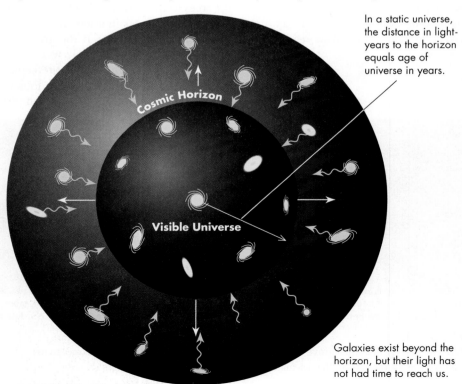

In a static universe, the distance in light-years to the horizon equals age of universe in years.

Galaxies exist beyond the horizon, but their light has not had time to reach us.

Mathematical Insight

The distance to the cosmic horizon can be measured in different ways. For example, some of the most distant regions whose light we can see today emitted that light at about redshift $z = 1000$ when they were only about 45 million light-years away. Those regions are today at about 45 *billion* light-years $-(z + 1)$ times farther away.

81.2 THE COSMIC MICROWAVE BACKGROUND

The idea that we should see light from the Big Bang was first suggested by George Gamow (pronounced *GAM-off*), a Russian astrophysicist who fled to the United States before World War II. Gamow built upon Lemaître's idea that the young universe was extremely dense (Unit 80.4), and he went on to conclude that it must also have been extremely hot. Gamow's argument was based on the observation that compression heats a gas, and so the enormous compression of the early universe must have heated the matter to very high temperatures at the earliest times.

At large distances from Earth we see the universe as it was when it was young, and we should see evidence of the initially high temperatures. However, the radiation from this hot gas has been diluted and redshifted to long wavelengths by the expansion of the universe. Looking outward from Earth, every line of sight *does* encounter a hot glowing surface, and the sky *is* bright, but at radio wavelengths. Thus, the darkness of the night sky results from the limitations of our senses. If we look at radio wavelengths, we should see the afterglow of the Big Bang.

This idea was developed in 1948 by two of Gamow's collaborators, the U.S. physicists Ralph Alpher and Robert Hermann. They showed that this radiation would come not directly from the first moments of the Big Bang but rather from a later time, after the universe had been expanding for several hundred thousand years.

Just minutes after the Big Bang, the temperature of the universe would have been several billion kelvins. After a few hours it cooled to 100 million K; after about 10 years, about 100,000 K; and after 400,000 years, about 3000 K. This cooling is straightforward to predict from laboratory observations of expanding gases. During this whole time, the high temperature of the gas kept it ionized. In this ionized gas, photons interact strongly with the free electrons and, as a result, can travel only a short distance (Figure 81.5A). Because light could not travel far through the gas, the gas was essentially opaque.

Throughout the first few hundred thousand years of the universe's existence, the temperatures and densities of the original gas were similar to those in a star's core. As time progressed, the gas temperature and density fell, making the conditions more similar to those farther from the core of a star. The conditions at 400,000 years finally reached temperatures like those at the surface of a star—low enough that the ionized hydrogen and helium combined with the free electrons, and the universe became transparent (Figure 81.5B). This period is called the **last scattering epoch** because the photons had their final interaction with the expanding gas of the Big Bang.

The last scattering epoch is also sometimes called the *recombination epoch* in reference to the electrons and ions combining to form neutral atoms. This name is misleading, though, because this was actually the first time these particles ever combined.

FIGURE 81.5
(A) In the young universe, free electrons in the hot ionized gas constantly interacted with photons, making the universe opaque. (B) When the universe became cool enough, electrons combined with the nuclei to make neutral atoms (mostly hydrogen). Because the photons did not have enough energy to ionize the atoms, they no longer interacted with the matter, and the universe became transparent.

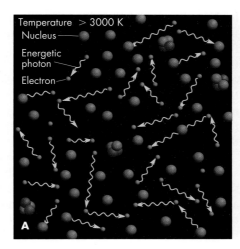
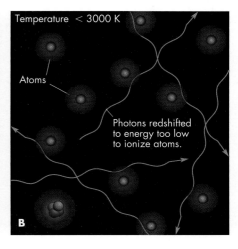

Even though we cannot see to times before 400,000 years, a variety of clues have survived, and these help take us back to less than 1 second after the Big Bang occurred. These clues are explored in Unit 83.

Thus, after 400,000 years the photons from the Big Bang were able to move freely through space—unless they hit something, like a telescope on Earth! These photons are the oldest radiation we can see. Just as we cannot see photons directly from the core of the Sun, we cannot see the photons from the Big Bang itself. In both cases, the photons travel through a hot ionized gas that constantly interacts with the photons for hundreds of thousands of years before the photons can move freely through space. Through these interactions the photons exchange energy with the gas particles and reach the same temperature as those particles. Eventually the photons reach a place (in the Sun) or a time (in the history of the universe) where the gas they encounter has thinned out, and they are then free to travel through space. The photons travel away, carrying with them energies set by the temperature of the last atoms they interacted with.

So what should this radiation look like? According to Wien's law, the thermal radiation from the photosphere of a star or the last scattering epoch should have a spectrum that is most intense at a wavelength determined by the material's temperature (Unit 24). In particular, hot matter radiates most strongly at short wavelengths. Just after the last scattering epoch, the radiation would have been at a temperature of about 3000 K and therefore looked much like the surface of an M-type star such as Betelgeuse (Unit 56). If we lived then, we would have experienced Olbers' picture of a sky everywhere as bright as the surface of a red star. The radiation at that time would have peaked at a wavelength of about 1000 nanometers.

The expansion of space modifies this radiation. The redshift of the wavelengths makes the radiation appear cooler. The universe has expanded by a factor of about 1000 since the last scattering epoch, so the wavelength of the radiation has also stretched by a factor of 1000 (Unit 80). According to Wien's law, the temperature of thermal radiation drops in proportion as the wavelength grows longer, so the apparent temperature of this radiation would drop by a factor of 1000—from 3000 K down to about 3 K today.

If the universe began with a Big Bang, it should today be filled with long-wavelength radiation created when the universe was young. This prediction was confirmed in 1965, when Arno Penzias and Robert Wilson of the Bell Telephone Laboratories accidentally discovered cosmic microwave radiation having just that property. Penzias and Wilson were trying to identify sources of background noise on telephone satellite links. They detected a radio signal with the unusual property that its strength was constant no matter in what direction they pointed their detector. They soon demonstrated that the signal came not from isolated objects such as stars or galaxies but rather from all of space. Initially mystified by the signal, they learned that scientists at Princeton University had repeated the calculations of Alpher, Hermann, and Gamow and were searching for the predicted radiation. Once aware of that work, Penzias and Wilson realized that the background interference they had found was radiation created in the young universe. For their discovery of the **cosmic microwave background**, or **CMB** for short, they won the 1978 Nobel Prize in Physics.

Figure 81.6 shows measurements of the intensity of the CMB and demonstrates that it has exactly the thermal spectrum predicted by theory. Moreover, the radiation is most intense at about 1 millimeter (10^6 nanometers), in the microwave part of the radio spectrum. Knowing this wavelength and using Wien's law, we find that the temperature describing the radiation is about 3 K, close to the value predicted nearly 20 years earlier. A more precise measurement shows that the temperature that describes the CMB is 2.725 K. This is only a little warmer than absolute zero, the lowest temperature anything can have.

Today the cosmic microwave background results provide one of the cornerstones that support the Big Bang theory. In earlier years, though, Gamow's predictions were considered fantastic. The high temperature and rapid expansion from a hot, dense state led the English astronomer Sir Fred Hoyle (a proponent of a competing theory called *steady state cosmology*) to jokingly refer to this theory of the birth of the universe as the *Big Bang*, a name that has stuck despite his sarcasm.

Concept Question 2

If the universe expanded more slowly so the last scattering was at a redshift of 3, what would the sky look like?

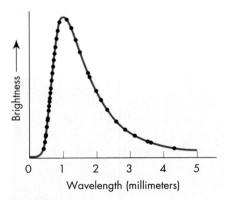

FIGURE 81.6
Spectrum of the cosmic microwave background. The shape of the spectrum matches perfectly a thermal spectrum for a temperature of 2.725 K. Notice the wavelength at which it peaks. Black dots show data points.

81.3 THE ERA OF GALAXY FORMATION

After the cosmic microwave background last scattered off the rapidly thinning and cooling gas, the universe entered what some astronomers have dubbed the "dark ages." The radiation from the microwave background was rapidly growing more and more redshifted, and no matter had yet condensed into stars, so the universe grew darker and darker. After a few hundred million years, the first stars and galaxies formed, and their radiation began to re-ionize much of the gas distributed throughout intergalactic space—lighting up the universe again. How this process occurred is one of the current areas of research in astronomy.

On the basis of the ages of stars within galaxies, astronomers deduce that the Milky Way and other galaxies are at least 10 to 12 billion years old. We can observe galaxies when they were very young in deep images such as the Hubble Deep Field (Unit 76) and Ultra Deep Field (Figure 81.7), where we find some extremely faint galaxies at redshifts that imply they had formed less than 1 billion years after the Big Bang. Galaxies must therefore have formed fairly rapidly once the universe had cooled enough. To understand how the universe evolved from a sea of hot gas so quickly into galaxies, astronomers are carrying out simulations using the most advanced supercomputers available.

Presumably gravity pulled gas clouds into protogalaxies, much as gravity pulls interstellar clouds into protostars; but there appear to be important differences. In particular, the amount of matter observed in the universe exerts too feeble a gravitational attraction for this process to have been able to form galaxies within the age of the universe, let alone within the first billion years. Most astronomers therefore conclude that additional matter must be present to provide a strong enough gravitational pull to speed up galaxy formation. This need for unseen matter to aid galaxy formation strengthens astronomers' conclusion that dark matter (Unit 79) plays a major role in the structure of the universe.

> During the "dark ages" stars had not yet formed, but there was hydrogen throughout the universe. Astronomers are developing a giant radiotelescope called the Square Kilometer Array to try to study this ancient era.

FIGURE 81.7
Portion of the Hubble Ultra Deep Field. This image shows 1/8 of the area observed in the constellation Fornax that you could cover with the period at the end of this sentence held at arm's length. Nearly every dot in the image is a distant galaxy. Some are so far away that we are seeing them the way they appeared nearly 13 billion years ago, when the universe was only about one-twentieth of its present age.

FIGURE 81.8
Schematic illustration of peaks in the density of dark and normal matter at the time of last scattering. Dark matter has pulled itself into dense clumps. Normal matter is attracted by the gravity of the dark matter clumps, but the normal matter cannot settle there as it is constantly struck by high-energy photons.

Concept Question 3

If we could have been present at the location where the Milky Way was going to form, not long after the last scattering epoch, what might we have observed?

Dark matter could begin clumping even before the last scattering epoch, precisely because it does not interact with electromagnetic radiation as ordinary matter does. Just as the gravity of gas in an interstellar cloud begins pulling itself into dense clumps that eventually become stars, regions of slightly higher than average dark matter density began growing shortly after the Big Bang. By 400,000 years after the Big Bang, at the last scattering epoch, the seeds of large-scale structure were already well established, as illustrated in Figure 81.8.

The gravitational pull of dark matter clumps pulled normal matter toward them, but high-energy photons buffeted the normal matter, keeping it from settling into the clumps. This is a little like how high heat makes water boil out of a pan. The gas would have been slightly more compressed in the regions where the dark matter clumped, making it slightly hotter and brighter. This produced small fluctuations in the temperature of the CMB, which astronomers can detect with very precise measurements. Figure 81.9 shows the CMB over a region a few degrees across, mapped by a microwave telescope located at the South Pole. The warmer regions are just a few hundred micro-kelvins brighter than the average temperature of 2.7260 K. These are regions of slightly higher density where clusters of galaxies are going to form.

The strongest variations in the CMB (between highest and lowest density regions) occur at a separation determined by the size of the cosmic horizon at the last scattering epoch. The cosmic horizon had a radius of about 400,000 light-years at that time, because that was as long as the universe had lived. This means that regions of gas and dark matter could have interacted with each other—sharing their heat and experiencing each other's gravitational pull—only if they were both within a radius of 400,000 light-years of a common point. After the last scattering epoch, the high-density regions of dark matter began pulling in normal matter more rapidly, jump-starting galaxy formation.

The clumping of the dark matter itself arose from variations in density that were in place during even earlier moments immediately after the Big Bang (see Unit 83). In the density variations visible in the CMB, we are seeing the distribution of matter in distant regions of space as they were before any galaxies formed. It is possible that those regions today, almost 13.8 billion years later, contain galaxies in which astronomers are looking in our direction and seeing a bright region where the Milky Way and thousands of other galaxies are going to form.

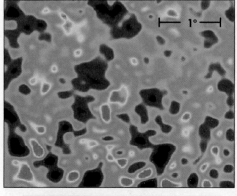
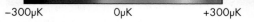
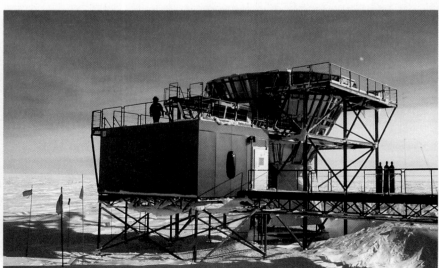

FIGURE 81.9
A portion of the cosmic microwave background radiation (left) mapped by a microwave telescope at the South Pole (right). The red bumps are slightly hotter and denser regions where galaxies will likely form. At this redshift, 1° is about 800,000 light-years.

KEY POINTS

- Over very large scales, the universe appears to be homogeneous.
- If the universe were infinite and homogeneous, every line of sight would eventually encounter a star, so it would be bright at night.
- The fact that the night sky is dark implies an "edge" to the universe, though the edge may be only apparent, caused by finite age.
- The size of the visible universe grows as the universe ages, giving light more time to reach us from more remote regions.
- We are looking back to earlier times at larger distances because the light took longer to reach us.
- Because the universe is expanding and cooling, we can look back to a time when the universe was hot and small in every direction.
- Light did not begin traveling freely until the temperature of the universe dropped below about 3000 K, when electrons and atomic nuclei combined.
- The cosmic microwave background (CMB) is today redshifted to 2.725 K and is almost perfectly uniform across the sky.
- Small fluctuations in the brightness of the CMB indicate where normal matter began to clump in the young universe.
- Dark matter was able to begin gravitational contraction even before light moved freely because it does not interact with light.

KEY TERMS

cosmic horizon, 646
cosmic microwave background (CMB), 648
homogeneous, 644
last scattering epoch, 647
Olbers' paradox, 645
visible universe, 646

CONCEPT QUESTIONS

Concept Questions on the following topics are located in the margins. They invite thinking and discussion beyond the text.

1. Interstellar dust and Olbers' paradox. (p. 646)
2. Appearance of the sky in a slowly expanding universe. (p. 648)
3. The Milky Way just after the last scattering epoch. (p. 650)

REVIEW QUESTIONS

4. What is meant by a cosmic horizon?
5. What is Olbers' paradox?
6. What is the cosmic microwave background? What is its origin?
7. How was the cosmic microwave background discovered?
8. What is special about the thermal spectrum of the CMB?
9. What is the significance of fluctuations in the cosmic microwave background?
10. What effects did dark matter have on the universe when it was young?

QUANTITATIVE PROBLEMS

11. The temperature of the universe needs to be less than 100 K before stars can form. At what redshift does this occur?
12. When our Solar System formed about 4.5 billion years ago, approximately what would the temperature of the CMB radiation have been? Would the structure in the CMB have looked the same as today? Explain your answer.
13. Imagine that there were so many galaxies that 30-kelvin dust inside their interstellar clouds blocked our view in every direction. What would their redshift have to be for the dust to have a temperature like the CMB? What problems are there with this model for explaining the observed properties of the CMB?
14. At the time of the last scattering epoch (at a redshift $z \approx 1000$),
    a. what was the average density of normal matter, if today it is about 3×10^{-31} kg/L?
    b. how large a sphere would have contained 10^{11} $M_\odot$, about as much normal matter as in the Milky Way Galaxy? Express your answer in parsecs.
15. Table 80.1, in Unit 80, shows that material we see at a redshift of 1000 was emitted when the material was 45 million ly away from us. This distance determines the angular size of regions at this redshift, as given by the angular size formula in Unit 10.4.
    a. Show that a region that is 1° across at this redshift has a diameter of approximately 800,000 ly.
    b. How big is that region today?

TEST YOURSELF

16. The cosmic horizon is
    a. at the radius of the entire universe.
    b. at the distance that light travels in the age of the universe.
    c. a shell centered on the Earth and size given by the farthest visible galaxy.
    d. at the redshift where visible light shifts to the infrared.
    e. an alternative name for the surface of last scattering.
17. The cosmic background radiation comes from a time in the evolution of the universe
    a. when protons and neutrons were first formed.
    b. when the Big Bang first began to expand.
    c. when the first quasars began to shine.
    d. when X-rays had enough energy to penetrate matter throughout the universe.
    e. when electrons began to combine with nuclei to form atoms.
18. If our eyes could see at microwave wavelengths the sky would appear
    a. black except for a few bright spots due to isolated objects.
    b. black except for a bright spot in the direction where the Big Bang occurred.
    c. to have a nearly uniform glow in all directions.
    d. bright in the plane of the Solar System, but dark in the other directions.
19. Dark matter may have been essential for galaxy formation because dark matter
    a. acted as a seed for normal matter to clump together.
    b. shielded normal matter from the background radiation.
    c. aided in the cooling of the universe allowing intergalactic clouds to collapse..
    d. first clumped together and then transformed into normal matter.

UNIT 82

The Curvature and Expansion of Universes

82.1 Quantifying Curvature
82.2 Curvature and Expansion
82.3 The Density of the Universe
82.4 A Cosmological Constant

Learning Objectives

Upon completing this Unit, you should be able to:
- List the differences between spaces that have positive, negative, or no curvature.
- Describe the connection between curvature and expansion and explain the significance of the critical density and the Omega parameter.
- Discuss the effect of a cosmological constant on the expansion of the universe.

We know from Einstein's theory of general relativity (Unit 27) that mass curves space and gives it motion, but our ability to perceive curvature within our universe is a little like the problem faced by an ant that crawls into a rubber hose lying on the ground. If the hose is straight, the ant walking in a straight line will reach the other end of the hose and emerge. However, suppose the hose is bent and twisted in large loops. If the curves are gradual, the ant may still feel like it is traveling in a straight line, even though its direction is changing. And if the other end of the hose was brought around so that the ends met, the ant could crawl forever in a seemingly straight line and would never reach the hose's end.

In this Unit, we will examine some of the ways that matter and energy can alter the curvature and the way space flows in a universe. This will prepare us for Units 83 and 84, where we will examine current evidence about our universe, including such strange possibilities as the existence of regions of the cosmos with different curvatures that are forever separated from each other—effectively creating separate universes.

82.1 QUANTIFYING CURVATURE

Einstein's general theory of relativity shows that mass and energy "curve" space and time (Unit 27). It is easier to picture the bending of space and time by an object like a black hole (Unit 69) or a cluster of galaxies (Unit 79), because the strong bending is localized to one region. But the enormous mass and energy of the entire universe can alter how space connects to itself, bending and shaping the universe.

We experience a kind of curved space on the surface of the Earth. Athletic fields and parking lots appear flat, and, if we ignore hills, the surface of the Earth from horizon to horizon certainly looks flat. We know, however, that if we could travel in a "straight" line along this "flat" surface, we would return eventually to our starting point. The expanding universe could be similarly curved.

Consider the following analogy: galaxies in an expanding curved universe are like buttons glued to a balloon (Figure 82.1). As the balloon inflates, the space between buttons expands, and the buttons separate at a rate proportional to their distance from each other, as in Hubble's law. Moreover, no button is near an edge, even though space is finite. Could our universe be curved like this, only appearing to extend off to distances beyond what we can see?

In the balloon analogy, only two of the dimensions of space can be represented by the balloon's surface. A photon traveling from one button to the next travels

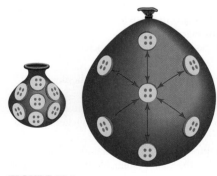

FIGURE 82.1
Representation of a curved but expanding universe. Curved space can expand between galaxies, similar to how the space expands between buttons on a balloon as the balloon is inflated. The space between buttons grows proportionally to distance, just as Hubble's law indicates.

along the curved surface of the balloon. General relativity predicts that the three dimensions of space can also curve back around on themselves as in the balloon analogy. Such a space looks as if it extends forever in all directions; but if we could travel in a spaceship along an apparently straight line, we would eventually return to our starting position. We could move through this space forever without coming to an end point. A universe with this property is said to have **positive curvature** and is called a **closed universe** because it closes back on itself. The Earth's surface is also positively curved—if we travel in a straight line on it, we return to our starting point. Positively curved space has other properties as well: Parallel lines meet when extended, and the sum of the interior angles of a triangle drawn on a positively curved surface is greater than 180°, as illustrated in Figure 82.2A.

General relativity holds open the possibility that the universe might have other types of curvature. It might be **flat** (that is, have no curvature) or have **negative curvature**. Flat space is what most people picture when they think of space. In it, parallel lines do not meet, and a triangle's interior angles add up to exactly 180°, as illustrated in Figure 82.2B. Negatively curved space is harder to visualize, but you can think of it as being bent into a saddle shape, as shown in Figure 82.2C. Such curvature corresponds to bending space one way along one line and the opposite way along another line. In negatively curved space, the sum of the interior angles of a triangle is less than 180°.

Note that the paths of light through curved space can have surprising properties. If the universe had a very strong positive curvature, we might be able to see all the way around the universe to see ourselves—or perhaps our own Galaxy as it looked long ago. Positive curvature is like the curvature of light around a black hole or galaxy cluster. It acts as a lens (Unit 79.3), making distant objects look larger, as illustrated in Figure 82.2A. On the other hand, if space is negatively curved, it will make distant objects look smaller than they would look in flat space. Perhaps like a rearview mirror, a negatively curved universe should have a label saying "Galaxies in a negatively curved universe are closer than they appear"!

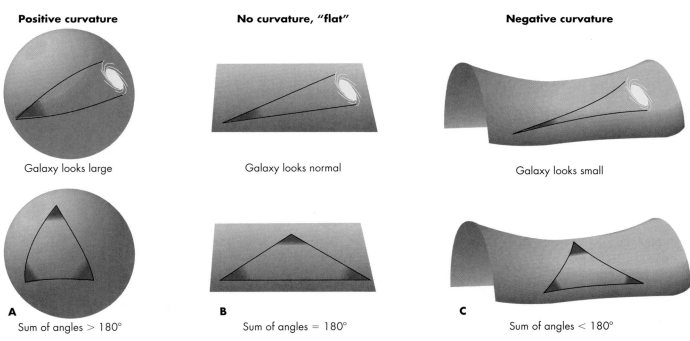

FIGURE 82.2
Figures illustrating how the shape of a surface determines the geometry of a triangle on it. (A) For a positive curvature, distant objects appear larger, and the interior angles of a triangle on the surface of a sphere add up to more than 180°. (B) When the curvature is zero (flat geometry), objects have their normal size, and the interior angles of a triangle on a flat surface add up to 180°. (C) For a negative curvature, objects appear smaller, and the interior angles of a triangle on a saddle-shaped surface add up to less than 180°.

82.2 CURVATURE AND EXPANSION

The curvature of the universe is related to the amount of mass it contains, which also determines the strength of the gravitational pull between parts of the universe. The universe is currently expanding, but could gravity eventually cause it to collapse back on itself? This is similar to the question of whether a stone thrown upward will fall back down. Normally, the Earth's gravity slows then stops it, and the stone falls back to Earth. However, if you could fire the stone at a speed greater than the Earth's escape velocity, it would have enough energy to travel away from the Earth forever. The stone's behavior depends on the strength of gravity and the upward impulse given to it. So too with the expansion of the universe.

It is thought that the initial expansion of the universe was caused by a process called **inflation** (Unit 83). Inflation is an exponential expansion in which space keeps redoubling in size in fixed time intervals. Without the accelerating effects of inflation, the gravitational forces between galaxies and all the other contents of the universe work to slow its expansion. If the universe's gravitational force is weak enough, or if the initial impetus of expansion is strong enough, we would expect the universe to expand forever. On the other hand, if the mass is large enough (so the gravitational force is strong) or the initial impetus is weak, the expansion will stop and the universe will begin contracting, ultimately collapsing together in a **Big Crunch.**

To measure the relative strength of these effects for the universe, we can compare the gravitational energy holding the universe together to the energy of its expansion. If gravity dominates, the expansion will stop. If the expansion energy dominates, the expansion will continue forever (Figure 82.3). Therefore, we can learn (at least in principle) whether the universe will expand forever or collapse back on itself by seeing if the universe has a gravitational pull strong enough to stop its expansion.

To see how strong the effect of gravity is, astronomers use Newton's law of gravity. Essentially, gravity's effect depends on the universe's mass. It turns out, however, that it is far easier to work with the universe's density (the amount of mass contained within a given volume). The reason is simple: Astronomers can measure the density of the universe but not its mass. To measure its mass, we would need to observe all of space. To measure its density, we need only measure the mass within some large, representative volume of the cosmos.

> **Concept Question 1**
>
> Do you have a personal belief about the eventual fate of the universe? What did your parents or religion indicate about this?

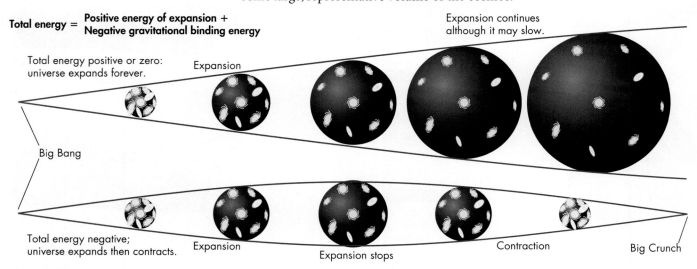

FIGURE 82.3

A sketch of how gravity and the energy of expansion determine the behavior of the universe. The spheres in the figure portray representative volumes of the matter in the universe. If the energy of expansion is small compared to the gravitational pull, the universe eventually collapses back on itself. If the energy is large enough, however, the universe expands forever.

82.3 THE DENSITY OF THE UNIVERSE

To find the density of the universe, astronomers choose a large representative volume of space and attempt to find all of the galaxies within it. Next we measure the mass of each galaxy, add the masses, and divide by the volume (Figure 82.4). For example, to measure the local density, astronomers might choose the Local Group, which contains three large galaxies and a few dozen smaller ones. The total mass of gas and stars in the Local Group is estimated to be about 10^{12} solar masses, which we can convert to kilograms if we multiply by the Sun's mass, 2×10^{30} kilograms. This gives a mass for the group of about 2×10^{42} kilograms.

Next that mass is divided by the Local Group's volume, assumed to be a sphere whose radius is the distance from the center of the Local Group to the next nearest galaxy group, which is about 3 Mpc (about 9×10^{22} meters) away. Using the formula for the volume of a sphere ($\frac{4}{3}\pi R^3$) gives a volume for the Local Group of about 3×10^{69} cubic meters. Dividing this volume into the mass gives a density of about 7×10^{-28} kilograms per cubic meter, or 7×10^{-31} kilograms per liter.

The vicinity of the Local Group is somewhat richer than the average environment, however. To get a more representative sample, we must look at a much larger region that includes both clusters and voids, and we must include intergalactic gas, particularly in clusters (Unit 77). A similar calculation for these much-larger volumes of the universe gives a slightly smaller value of about 3×10^{-31} kilograms per liter, or, on average, roughly 2 hydrogen atoms per 10 cubic meters. This indicates that, on average, the universe today has a fantastically low density.

To determine whether the universe will expand forever or recollapse, astronomers compare the observed density to a theoretically calculated **critical density,** which is written with the Greek letter rho as ρ_c. If the actual density is greater than the critical density, the universe will recollapse; if it is less, the universe will expand forever. We can calculate the critical density by comparing the kinetic energy and gravitational potential energy for a galaxy at the edge of our representative volume (Figure 82.5). It turns out that mathematically the critical density ρ_c is

$$\rho_c = \frac{3H^2}{8\pi G},$$

where H is the Hubble constant and G is Newton's gravitational constant. Without carrying out the full derivation, we can see why the equation has this form. The gravitational potential energy depends on the density and Newton's gravitational constant, whereas the kinetic energy depends on the square of the expansion speed, which can be related to Hubble's constant. Thus, we are equating values proportional to $G \times \rho$ and H^2 and then solving for ρ. The first steps of the derivation are shown in Figure 82.5.

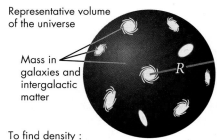

Representative volume of the universe

Mass in galaxies and intergalactic matter

To find density:
 Measure radius R of volume
 Measure mass M of all matter in volume
 Divide M by volume $= \frac{4}{3}\pi R^3$

FIGURE 82.4
The density of the universe can be found from the mass and radius of a representative volume of space.

ρ_c = Critical density
H = Hubble's constant
G = Gravitational constant

FIGURE 82.5
The energy balance of the universe. The kinetic energy of a galaxy at the edge of a representative volume can be compared to its (negative) potential energy (see Unit 20). The net energy will be zero if the volume has the critical density.

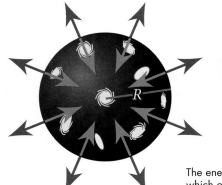

By Hubble's law, at edge of volume, recession velocity $V = H \times R$

Galaxy at edge of volume with mass m

Inward gravitational pull by matter with total mass $M = \rho \times \frac{4}{3}\pi R^3$

Net energy of galaxy = Kinetic energy + Potential energy
$= \frac{1}{2} m \times V^2 - G\frac{m \times M}{R}$
$= \frac{1}{2} m \times H^2 \times R^2 - G\frac{m \times \rho \times 4\pi R^3}{3R}$

The energy equals zero when $\rho = \rho_c$, the critical density, which occurs when the two terms in the equation are equal.

Astronomers use a quantity called **Omega** (Ω) to indicate how close the observed density is to the critical density. Omega is used because, as the last letter of the Greek alphabet, it suggests the "end of things," and the overall density is important for predicting the fate of the universe. For the amount of matter in the universe, then, cosmologists define Ω_M as the value of the actual density of matter divided by the critical density:

$$\Omega_M = \rho/\rho_c.$$

With this notation:

If $\Omega_M > 1$, the universe will eventually collapse.
If $\Omega_M < 1$, the universe will expand forever.
If $\Omega_M = 1$, the universe is exactly at the critical density.

We use the subscript M to indicate that this is the value of Ω for *matter* because a variety of quantities can contribute to the overall value of Ω, and not all of them have the same effect on the universe's expansion, as we will see in Section 82.4.

The changing "size" of the universe is illustrated for various values of Ω_M in Figure 82.6. What is meant by the size of the universe in this diagram is actually the overall relative size, which might be represented by the radius of our representative volume or the distance between two widely separated galaxies. Models like these must match the present size and rate of expansion of the universe. The rate of expansion is determined by Hubble's constant, and it is indicated by the slope of the curve today—steeper slopes indicate faster expansion.

If Ω_M is larger than 1, the universe expands rapidly after the Big Bang, but the large density of matter causes the universe to slow its expansion quickly. In this case the expansion halts sometime in the future, and the universe then collapses back in on itself, ending in a "Big Crunch."

If Ω_M is 0, the expansion of the universe is not slowed at all, and the rate of expansion remains constant. The time since the Big Bang in this type of universe is equal to the Hubble time $1/H \approx 14$ billion years (Unit 80.4), because the speed of expansion is the same today as it was in the past. In models of the universe with values of Ω_M greater than 0, the time since the Big Bang is smaller than $1/H$.

The $\Omega_M = 1$ case is interesting because it implies that the universe's gravitational potential energy exactly cancels its kinetic energy—a net zero energy. This condition makes the most sense, according to some theories of the beginning of the universe. In the $\Omega_M = 1$ cosmology, the universe's expansion slows, approaching zero speed in the infinite future but never quite stopping.

The future of expansion

Cosmology

FIGURE 82.6
The size of the universe in the past and future as calculated for different densities. Curves are shown for three cases, in which: there is no matter ($\Omega_M = 0$), so the universe expands without deceleration; the pull of gravity and the expansion rate balance ($\Omega_M = 1$), so the universe constantly slows down but never recollapses; and the pull of gravity greatly exceeds the energy of expansion ($\Omega_M = 10$), so the universe recollapses in about 12 billion more years. All curves have the same slope today, which is determined by the Hubble constant. Curves lying in the blue region of the diagram have enough energy to expand forever, while those in the yellow region will eventually collapse back on themselves.

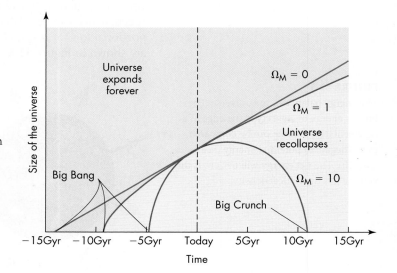

Concept Question 2

If the universe contained nothing but the visible matter we have observed, about how old would it be, based on Figure 82.6?

When we compare the observed and critical densities, what do we find? If we use the value for H of 70 kilometers/second per megaparsec, the critical density is 9.4×10^{-30} kilograms per liter. Remarkably, this tiny density—less than a billionth-billionth of the density of the best vacuum ever created in a laboratory on Earth—would be sufficient to halt the universe's expansion. As small as this density is, it is about 25 times larger than the estimated density of visible stars and gas throughout the universe. This low value of the density of normal matter is also consistent with the estimates determined from the amount of fusion that occurred during the first few minutes of the universe (Unit 83.1). Therefore, the gravitational pull of all the normal matter in the universe is too weak to stop the expansion. This can be written as $\Omega_{VM} = 0.04$, where VM stands for "visible matter."

However, this result does not take into account dark matter. Observations of galaxies and clusters imply large quantities of unseen matter, many times more than is directly observed (Unit 79). To explain the rapid rotation of the outer parts of spiral galaxies and motions within galaxy clusters, astronomers estimated that there might be anywhere from 5 to 20 times more dark matter than normal matter, so $\Omega_{DM} = 0.2$ to 0.8. Combined with the estimates for visible matter, and given the uncertainties in the measurements, the majority of astronomers in the 1990s suspected that the total amount of matter, $\Omega_M = \Omega_{VM} + \Omega_{DM}$, was equal to 1 or less.

This was the state of cosmology at the end of the twentieth century. Data on the density of matter in our vicinity, combined with the measured expansion rate of the universe, indicated that the galaxies should continue sailing apart forever. The fate of this universe would be determined by the total amount of dark matter. Even though most estimates of the amount of dark matter in galaxies and clusters of galaxies suggested that Ω was still below 1, many cosmologists predicted that more dark matter would be found, pushing Ω_M to a value of 1. One of the cosmologists' arguments for a value of 1 was that the inflation that "launches" the Big Bang should make the universe almost exactly flat, which can be shown to imply that $\Omega = 1$ (Unit 83.4).

There were problems with models of the universe with so much dark matter, however. As you can see in Figure 82.6, a value of $\Omega_M = 1$ implies that the Big Bang occurred less than 10 billion years ago. This is because a decelerating universe used to be expanding faster, so it would have reached a large size more quickly than a universe that had a slower fixed rate of expansion. An age for the universe of less than 10 billion years was a major problem, though, because there are globular clusters that are estimated to be 12 to 13 billion years old—and nothing in the universe can be older than the universe itself.

Things were not adding up right, so astronomers thought there might be problems with the value of Hubble's constant, or perhaps the universe was not flat, but instead had a smaller value of Omega. We will return to these issues in Unit 84, but first we need to examine another aspect of Einstein's theory of general relativity and how it might affect the expansion of a universe.

82.4 A COSMOLOGICAL CONSTANT

Einstein developed the theory of general relativity (Unit 27) in 1916, before Hubble had discovered the expansion of the universe. At that time, the universe was believed to remain static, rather than expanding or contracting, but the theory of general relativity predicted that the universe should be in motion.

In 1917 Einstein found a way to reconcile the equations of general relativity with the belief that the universe was static. The energy equations he had put together to describe gravity allowed for the possibility of an extra term, a constant value that could be added to the energy. This extra term did not depend on position and remained the same throughout time. He labeled this extra term with the Greek letter **Lambda (Λ)** and called it the **cosmological constant.**

The cosmological constant would imply space is filled with an energy everywhere, and the equations of general relativity show that the presence of this energy will tend to make space expand. If the value of the cosmological constant is just right, it balances gravity's attraction, allowing the universe to be static. The Lambda term represents a property of space. Unlike other kinds of energy, it cannot be collected or transferred to another object, and it remains the same everywhere and at all times. If the universe expands, the Lambda term implies that new energy would be created out of nothing. The creation of energy out of nothing seems counterintuitive, but the Lambda term was formally permitted by the equations, and it could potentially explain a static universe.

Astronomers quantify the contribution of the Lambda term as Ω_Λ, to make it simpler to compare the density of the energy implied by this term to the density of matter in the universe. Einstein's static universe requires that $\Omega_\Lambda = \Omega_M$, as illustrated in Figure 82.7. The curvature of such a universe can vary.

If $\Omega_\Lambda + \Omega_M > 1$, the curvature of the universe is positive.
If $\Omega_\Lambda + \Omega_M < 1$, the curvature of the universe is negative.
If $\Omega_\Lambda + \Omega_M = 1$, the universe is "flat."

Depending on the combination and size of the terms, the universe may expand or contract in a wide variety of ways.

If the cosmological constant acts alone, or if Ω_Λ is much larger than Ω_M, the expansion of the universe accelerates over time, resulting in the process astronomers call inflation (Section 82.2), which is illustrated by the red curve in Figure 82.7. Einstein's static universe, where Ω_Λ and Ω_M are equal, is in an unstable balance. If the universe expands just slightly, Ω_M decreases and the cosmological constant term then exceeds it, so the expansion accelerates. However, if the universe contracts slightly, the density increases, and collapse ensues.

Other scenarios also are possible. For example, a universe might initially appear to be closed based on a high density of matter ($\Omega_M > 1$), but the cosmological constant can reverse its fate. This scenario was popularized in the 1920s to explain how the universe might have a finite age, yet look static today. This could happen if the density of matter in the expanding universe dropped to a value equal to the cosmological constant just as the universe approached its maximum size. At that point $\Omega_\Lambda \approx \Omega_M$, so the universe would hesitate at a nearly static size, but if the density dropped to just slightly below the cosmological constant term, the universe would eventually begin expanding again, as illustrated by the blue curve in Figure 82.7.

These various universes were fascinating to contemplate, but by the late 1920s it was established that the universe is indeed expanding. Einstein concluded he should never have introduced the cosmological constant into his equations. He called this his "greatest blunder," because he might have predicted the expansion of the universe in addition to his many other remarkable discoveries. However, Einstein's "blunder" now appears to have just been ahead of its time, as we shall see in Units 83 and 84.

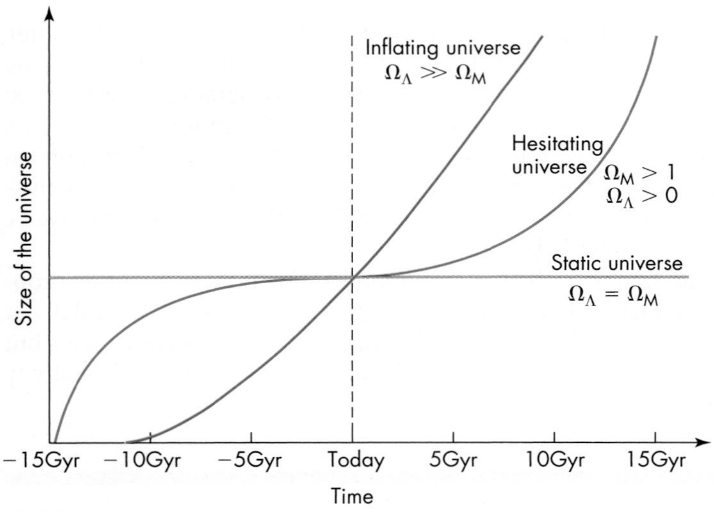

Mathematical Insights

If you have taken a calculus course, you may recognize the cosmological constant as a constant of integration. In calculus, when you take the integral of a function, the result is the antiderivative of the function plus a constant.

Concept Question 3

Comparing Figures 82.6 and 82.7, how would the Hubble constant today compare in the different universes with a cosmological constant?

FIGURE 82.7

Universes with a cosmological constant. A few of the many possible behaviors are shown if $\Omega_\Lambda > 0$. If the cosmological constant balances the gravitational potential energy, it is possible for the universe to be static (yellow line). If the cosmological constant is large, the universe inflates at a rate that grows exponentially faster with time (red curve). A universe that expands just to the point where Ω_M is slightly smaller than Ω_Λ can "hesitate" at a near-constant size before beginning to inflate (blue curve).

KEY POINTS

- The space making up the universe can be curved into different forms that might cause the universe to connect back on itself.
- Positively curved space curves back toward itself like the surface of a sphere; it may have no edge yet be finite.
- Negatively curved space curves away from itself like a saddle; distant objects are farther apart than they appear to us to be.
- Flat space has no curvature, and in that space angles and sizes follow the rules of standard geometry.
- An expanding universe may be open, expanding forever, or closed, collapsing back on itself; this is possible for all curvatures.
- The density of matter and energy that produces a flat universe is the critical density marking the boundary between open and closed universes (if matter alone affects expansion).
- The equations of general relativity permit a constant term that has a repulsive effect, almost like "antigravity."
- Einstein used the cosmological constant term to explain why the universe was static (as was mistakenly believed at the time).
- The cosmological constant can produce a variety of different forms of expansion of the universe.

KEY TERMS

Big Crunch, 654
closed universe, 653
cosmological constant, 658
critical density, 655
flat universe, 653
inflation, 654
Lambda (Λ), 658
negative curvature, 653
Omega (Ω), 656
positive curvature, 653

CONCEPT QUESTIONS

Concept Questions on the following topics are located in the margins. They invite thinking and discussion beyond the text.

1. Beliefs about the fate of the universe. (p. 654)
2. Pattern of expansion if only visible matter exists. (p. 657)
3. Hubble constant for different cosmological constants. (p. 658)

REVIEW QUESTIONS

4. How is the curvature of the universe defined? What observations might we make to determine the curvature?
5. What are the possible fates for the universe?
6. Why do astronomers think the universe will expand forever?
7. How is the density of the universe determined?
8. What is the critical density and how does it relate to the expansion of the universe?
9. Why did Einstein invent the cosmological constant?
10. How does the cosmological constant affect the expansion of the universe?

QUANTITATIVE PROBLEMS

11. On a flat surface, the angles of a triangle always add up to 180°, but on a spherical surface they may add up to a larger number.
    a. Find a triangle on a spherical surface that contains three 90° angles. Illustrate your example.
    b. What is the largest sum of the angles you can find for a triangle on a spherical surface?
12. For a surface with negative curvature, what is the smallest sum of angles you can find for a triangle? Illustrate your example.
13. If the average density of matter in the universe is 7×10^{-31} kg/L, what is the radius of the volume that would form a sphere containing the equivalent mass of the Sun?
14. If our Galaxy contains 10^{11} $M_\odot$ within a radius of 100,000 parsecs, find its density. How does this density compare to the critical density?
15. The critical density depends on the value of Hubble's constant, H. The value of H is still not precisely known. If H turns out to be 73 km/sec per Mpc, calculate how much the critical density changes compared to $H = 70$ km/sec per Mpc.
16. With no cosmological constant, there are basically three distinct patterns of expansion possible: open and positively curved; open and flat; and closed and negatively curved. How many different kinds of universes can you identify if a cosmological constant is permitted? Sketch what the expansion diagram looks like for each.

TEST YOURSELF

17. If visible matter was spread evenly throughout the universe, the spacing between atoms would be roughly a few
    a. nanometers.   c. centimeters.   e. kilometers.
    b. micrometers.   d. meters.
18. Which of the following is true of the expansion of the universe if there is no dark energy?
    a. The expansion rate will increase.
    b. The expansion rate will decrease.
    c. The expansion rate will remain constant.
    d. The expansion rate will first increase and then decrease.
19. Which of the following set of parameters would describe a universe that has matter in it but also is expanding at an accelerating rate?
    a. $\Omega_M = 1, \Omega_\Lambda = 0$
    b. $\Omega_M > 1, \Omega_\Lambda = 0$
    c. $\Omega_M = 0, \Omega_\Lambda = 0$
    d. $\Omega_M > 1, \Omega_\Lambda > 0$
    e. $\Omega_M = 0, \Omega_\Lambda > 0$
20. Suppose you first estimate the age of the universe by assuming it has always been expanding at the same rate as it expands currently. If you discover that the universe is decelerating, what should its true age be, compared to the first estimate?
    a. Longer   b. Shorter   c. Unchanged
21. A universe with no cosmological constant and a density of matter greater than the critical density must
    a. expand forever.   c. form galaxies.
    b. collapse back on itself.   d. have a negative curvature.

UNIT 83

The Beginnings of the Universe

83.1 The Eras of the Universe
83.2 The Origin of Helium
83.3 Radiation, Matter, and Antimatter
83.4 The Epoch of Inflation
83.5 Cosmological Problems Solved by Inflation

Learning Objectives

Upon completing this Unit, you should be able to:
- List and describe the stages of the universe as it expanded from the Big Bang.
- Explain the amounts of helium and other elements produced in the first minutes.
- Explain why it is puzzling that there is any matter in the universe.
- Describe inflation and explain the aspects of the universe that it helps explain.

The early universe was hot and dense. The first 400,000 years after the Big Bang are hidden from our direct view because the universe was full of radiation that interacted constantly with the matter. As we discussed in Unit 81, the universe became transparent only after the *last scattering epoch,* when the density and temperature had dropped low enough that electrons combined with protons to form neutral hydrogen atoms. However, it is possible to reconstruct the history at these early times from a variety of evidence.

In this Unit, we work our way toward the very first moments that physics allows us to understand—just a tiny fraction of a second after time began. Some of our basic understandings about the nature of matter and forces, as described in Unit 4, begin to break down when looking at the earliest times because the universe was far hotter and more dense than we have been able to achieve in laboratories. Despite these challenges to our understanding, scientists have been able to extrapolate back to a time when the universe was just a fraction of a second old, and everything in the visible universe today was far smaller than an atom.

83.1 THE ERAS OF THE UNIVERSE

Figure 83.1 summarizes our current understanding of the history of the universe from its beginning, 13.8 billion years ago. If we could journey back to the times before the last scattering epoch, we would find that the radiation from the Big Bang actually "weighed" more than ordinary matter! Energy generates a gravitational influence just as mass does—the relationship between the two is based on Einstein's formula, $E = m \times c^2$.

Today, when we add up all the energy carried by all of the photons in the cosmic microwave background, it amounts to about one 1/1000 of the mass in ordinary matter. At earlier times, though, the balance was shifted. As the universe expanded, the photons and matter spread out equally, but the photons lost energy as their redshift stretched them out to longer wavelengths. At a redshift of 1000 at about the same time as the last scattering, the photons' equivalent mass was 1000 times greater than today, so the photons matched the mass in ordinary matter. When the universe was younger, the importance of radiation was even greater. The first 50,000 years after the Big Bang was a time that astronomers call the **radiation era,** when radiation generated even more gravity than dark matter.

During the radiation era, there were none of the large gravitational structures we see today. The universe was a sea of subatomic particles and photons at extremely

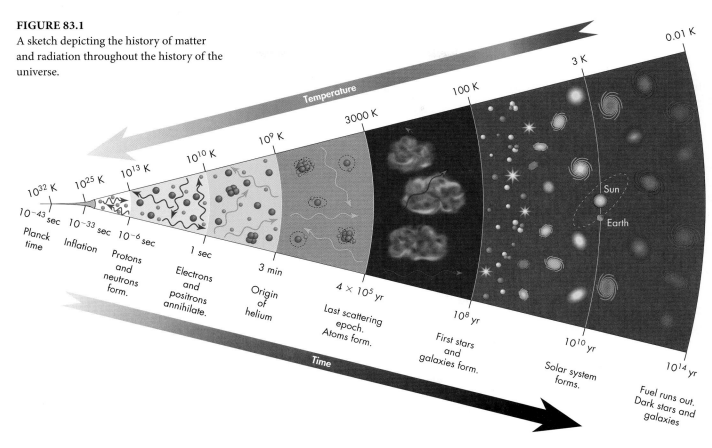

FIGURE 83.1
A sketch depicting the history of matter and radiation throughout the history of the universe.

high temperatures. The behavior of matter at these high temperatures has been explored by physicists with particle colliders. These are machines that accelerate subatomic particles to nearly the speed of light, then crash them together, and for an instant create conditions similar to those in the universe just after the Big Bang. This allows us to hypothesize what happened back then and to search for confirming evidence.

83.2 THE ORIGIN OF HELIUM

When astronomers extrapolate back to times before the last scattering epoch, their first major hypothesis is that a significant fraction of the hydrogen in the newborn universe would have fused to form helium. We can see why from the following argument: If we take all the matter and radiation in the universe today and calculate how compressed it was a few minutes after the Big Bang, the temperature and density throughout the universe would have been similar to that in the core of a massive star. Therefore, fusion should have occurred, just as it does in the cores of stars.

Such reasoning led George Gamow to hypothesize in 1948 that helium and heavier elements should have been created in the young universe—if the Big Bang model was correct. Gamow's hypothesis is supported by more recent detailed simulations as well as by observations.

We now estimate that about 1 second after the universe began, the temperature of the universe would have been 10 billion (10^{10}) kelvins, and its matter would have had a density of about 0.1 kilogram per liter. This may not sound very dense, but it is roughly a thousand-trillion-trillion times denser than the universe is at present.

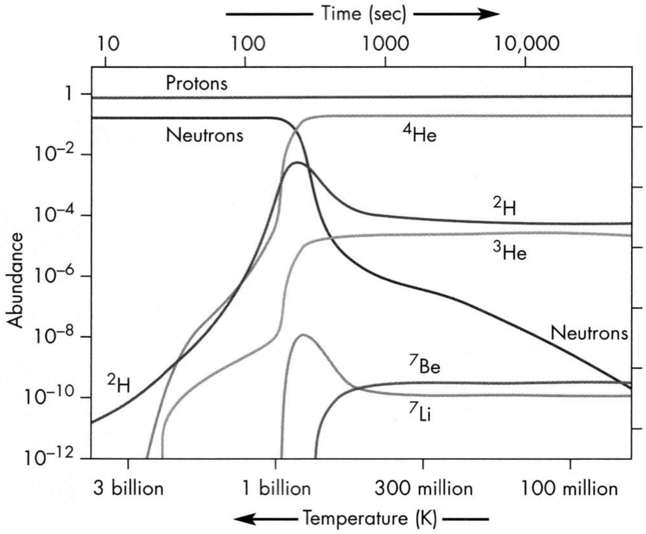

FIGURE 83.2
During the first few minutes after the Big Bang, the universe was hot and dense enough to fuse hydrogen into heavier elements. Only protons and neutrons existed at the start of this period. The production of different elements as they combined took place over the next few minutes. Neutrons and ^{7}Be are both unstable, and they decay after this period of nuclear reactions ends. The calculated amounts of the elements closely match what we observe today.

In stars there are no free neutrons, so fusion begins with the rarer nuclear reaction during which a proton is converted to a neutron through the weak reaction as the protons collide.

T = temperature of matter and radiation during the early history of the universe

t_{sec} = time after the Big Bang in seconds

At these densities and temperatures, all of the matter in the universe would have broken apart into its constituent particles: electrons, protons, and neutrons. In gas this hot, particles might collide and stick together momentarily, but the intense radiation would have broken them apart as fast as they could form.

As the universe expanded, it steadily cooled, and the radiation became weaker. Protons and neutrons could fuse, forming hydrogen's isotope deuterium (^{2}H), so deuterium nuclei could begin building up after 1 second, as shown in Figure 83.2. Creating deuterium is also the first step in the process of hydrogen fusion in the Sun and other stars (Unit 52.2).

Throughout the radiation era, the temperature dropped as the inverse square-root of the time, so if we measure the time in seconds, the temperature is

$$T \approx \frac{10^{10} \text{ K}}{\sqrt{t_{\text{sec}}}}.$$

After 10 seconds the temperature of the universe had therefore cooled to about 3 billion ($10^{10}/\sqrt{10} \approx 3 \times 10^9$) kelvins, which is cool enough for deuterium nuclei to begin fusing to form helium. By the time the universe was 100 seconds old, expansion had cooled the temperature to about 1 billion kelvins, and particles of deuterium were rapidly fusing into helium-3 (^{3}He) and helium-4 (^{4}He), as illustrated in Figure 83.3. The universe was not only cooling, it was also rapidly growing less dense. By 100 seconds, the density of matter was only about 10^{-4} kilograms/liter, so the rate of collisions became infrequent, and fusion slowed and then stopped after a few more minutes.

The total amount of helium that formed was determined by how hot and dense the universe was and how long those conditions lasted. These conditions were in turn determined by how rapidly the universe was expanding. Big Bang models predict that about 24% of the matter should have been fused into helium. Stars also fuse hydrogen into helium, but in regions where there has been little star formation the percentage of helium is always found to be very close to 24%.

Trace amounts of other light elements would also have been left over from the fusion during these first few minutes after the Big Bang. The amount left of the different elements depends in detail on the density of matter in the universe. For example, at lower densities, collisions leading to fusion would have been less likely. In this circumstance, slightly less ^{4}He would have had a chance to form, leaving substantially more unfused deuterium and ^{3}He. Higher densities would have allowed more complete fusion, resulting in much lower amounts of ^{3}He and deuterium, and providing an opportunity for more of the heavier elements such as

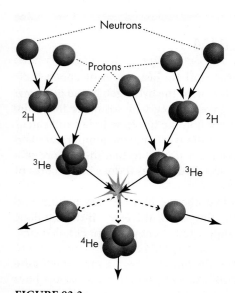

FIGURE 83.3
Nucleosynthesis in the early universe.

FIGURE 83.4
The amount of different elements expected to be produced by nuclear fusion in the young universe. The curves show how the amount of each element varies depending on the density of matter in the universe. The values along the bottom axis show the density of normal matter in the universe today, while the vertical axis shows the amount of each element relative to the amount of hydrogen that would be produced for that density. The red horizontal bars indicate the best estimates of the amount of each element measured in primitive systems, while the vertical blue bar shows the range of densities consistent with these measurements.

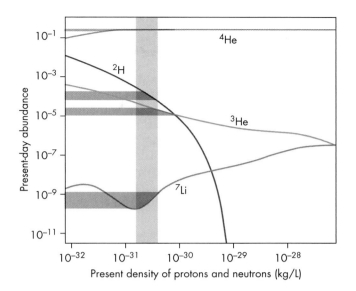

Concept Question 1

If there had not been any fusion in the early universe, how might the evolution of stars have been altered?

lithium to form. The consequences of these different densities of matter (expressed in terms of the density we would find today) are shown in Figure 83.4.

Measuring the levels of these elements as they were after the Big Bang is difficult, because stars also fuse elements in their cores and then "pollute" interstellar space when they die and expel their atmospheres. On the other hand, the amount of deuterium seen today tends to be lower because processes in stars destroy deuterium. Deuterium survives from the Big Bang only because the universe expanded and cooled too quickly for the normal completion of the fusion process. Astronomers have been able to measure the amounts of deuterium in intergalactic clouds by observing quasar absorption lines (Unit 77.5). This measured amount of deuterium—about 1 deuterium for every 30,000 hydrogen nuclei—is our most sensitive measure of the density of matter in the Big Bang.

Astronomers have also been able to measure the level of ^{3}He in gas clouds where little star formation appears to have occurred. In addition, measurements of the amounts of ^{4}He and the element lithium (^{7}Li) have been made by examining the atmospheres of very old stars, whose composition reflects conditions relatively early in the history of the universe. Figure 83.4 shows that the amount of each of these elements is consistent with the density of the universe estimated from the deuterium measurement.

The amounts of these products of the Big Bang suggest that normal matter represents only about 4% of the critical density (Unit 82). When these measurements were first made, the density of matter they implied was puzzling. It was much smaller than most astronomers had expected, given the amount of matter that appeared to be present from the strength of gravitational forces between galaxies. It is now understood that this seeming discrepancy was yet another consequence of the prevalence of dark matter in the universe. The dark matter makes up most of a galaxy's mass, but because it is not composed of protons and neutrons, it could not participate in the fusion that took place shortly after the Big Bang.

The excellent consistency between predictions about Big Bang helium production and the percentages of all the light elements in the oldest matter in the universe provides further strong support for the Big Bang theory. It is interesting that in more than 13 billion years since this early epoch, stars have succeeded in fusing only about 3% more of the matter in the universe into helium. All of the fusion in all of the stars in the whole history of the universe has produced less helium than just a few minutes of the Big Bang!

83.3 RADIATION, MATTER, AND ANTIMATTER

Next let us take another step closer to the beginning of the universe. During the first second after the Big Bang, the universe was so hot that matter and radiation mingled without the sharp distinction we see between these two entities today. Even the weakly interacting neutrinos were constantly colliding with other particles during this early time. Astronomers call this period the **early universe.**

The distinction between matter and radiation can become blurred because of the equivalence between an amount of energy, E, and an amount of mass, m, as given by Einstein's equation $E = m \times c^2$. (The term c^2 is the conversion factor between mass and energy—the speed of light, c, squared.) This equation lets us calculate, for example, the amount of energy that is released when hydrogen fuses into a slightly smaller mass of helium in the core of a star (Unit 52). The reverse reaction is also possible: If an amount of energy E is concentrated in one place, it may be turned into a particle whose mass equals E/c^2. Scientists observe matter created in this way from high-energy gamma rays in laboratory experiments.

In the early universe, extremely high-energy radiation filled all of space. Creation of matter from energy would have occurred everywhere. Such transformations of energy into mass, however, cannot occur in just any way. Laboratory experiments show that the conversion of energy into matter always creates a pair of particles, as illustrated in Figure 83.5A. Moreover, the particle pair has two special properties: The particles must have opposite charge (or no charge), and one of the pair must be made of ordinary matter, whereas the other must be made of what is called **antimatter.** For example, the electron has an antimatter particle, the **positron,** with identical mass but opposite electric charge (see Unit 4).

Antimatter has the important property that a particle and its antiparticle are annihilated on contact, leaving only high-energy radiation, as depicted in Figure 83.5B. Thus, radiation (electromagnetic energy) can become matter in the form of a particle–antiparticle pair, and matter can become radiation. In the early universe, this shifting between mass and energy happened continually as particles formed and were annihilated.

Energetic radiation cannot turn into just any pair of particles: The pair's combined mass multiplied by c^2 must be less than the radiation's energy. For example, to become an electron–positron pair, the radiation needs an energy of at least 9.1×10^{-31} kg $\times c^2$ (a wavelength of 0.0012 nanometers). Any excess energy of the photon becomes the kinetic energy of the electron and positron, which race away from each other until they encounter and annihilate another of their antiparticle partners. Radiation with such large energies occurs only at extremely high temperatures. To have enough energy to make an electron–positron pair, the temperature must be at least about 6×10^9 kelvins, which was the universe's temperature after about one second of expansion.

More massive particle–antiparticle pairs were created at earlier times. Protons, for example, have a corresponding antiparticle called the antiproton. To become a proton–antiproton pair, the radiation needs roughly 1800 times more energy than to create the electron–positron pair. This requires a temperature 1800 times higher, or about 10^{13} kelvins. Such high temperatures do not occur in ordinary stars, but they were reached about 1 microsecond (10^{-6} sec) after the Big Bang. At that time, the universe we can presently see was packed into a volume about 10 times bigger than the Solar System. At earlier times, the universe was hot enough for its radiation to be converted not only into protons and antiprotons but even into **quarks** and antiquarks, the particles that make up protons and neutrons.

The early universe can be traced back to about 10^{-33} seconds after the Big Bang. At the high temperatures found at these early times, even more massive fundamental particles such as tauons and top quarks (and their antiparticles) were present

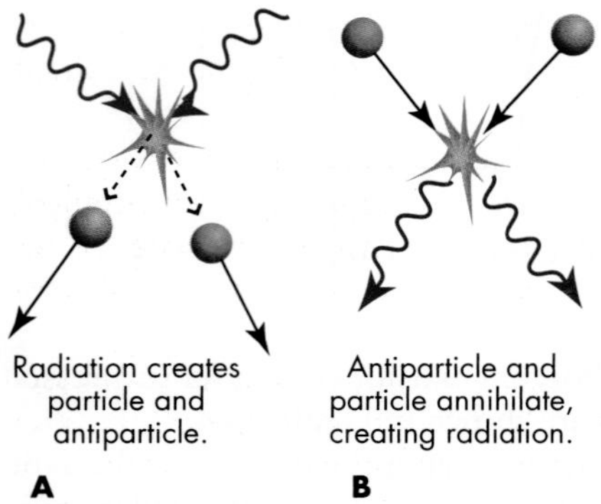

FIGURE 83.5
(A) The energy of electromagnetic radiation can be converted into mass with the formation of a particle and its antiparticle. (B) The collision of a particle and its antiparticle leads to their annihilation and the conversion of their mass back into energy.

A Radiation creates particle and antiparticle.
B Antiparticle and particle annihilate, creating radiation.

Concept Question 2

Physicists can generate antimatter in particle colliders. How do you suppose they can keep it from immediately destroying their experimental apparatus?

Concept Question 3

If some regions of the universe had been left with a surplus of antimatter instead of matter, atoms would form out of antiprotons, antineutrons, and antielectrons. How could we tell whether a distant galaxy was made of antimatter instead of matter?

FIGURE 83.6
Particle accelerator. This view is of a small portion of the 27-km-long tunnel of the Large Hadron Collider. Subatomic particles are accelerated to near the speed of light in these evacuated tubes before they collide at speeds like those in the early universe.

along with extremely intense high-energy radiation. At 10^{-33} seconds, everything in the universe that we can see today was jammed into a volume less than a meter across, expanding at enormous speeds.

We can picture how this progressed as time moved forward. At 10^{-33} seconds the universe was a dazzling sea of particles, antiparticles, and radiation constantly interacting. As time progressed, the universe expanded and cooled. Soon the radiation did not have enough energy to create the most massive particle–antiparticle pairs. Those that existed when the universe had cooled to this point continued to travel until they encountered one of their antiparticle partners and were annihilated. This happened for successively lower mass particles at later times until, at 10^{-6} seconds, the radiation was no longer energetic enough to create the lowest-mass quark–antiquark pairs (the so-called "up" and "down" quarks—see Unit 4). Thereafter, quarks annihilated antiquarks, and the quarks that remained combined to make protons and neutrons.

It is a puzzle why any quarks remained at all, because quarks and antiquarks should have been created from energy in equal numbers. For some reason, though, an imbalance left an excess of quarks over antiquarks. This is fortunate for us, because otherwise there would have been no matter in the universe at all. There would have been nothing in the universe but radiation.

Physicists have been using high-energy collider experiments to try to understand the source of this imbalance (Figure 83.6). They have discovered that the weak force (Unit 4) has what they describe as an **asymmetry.** An asymmetry can allow a pathway for an antiquark to decay, for example, but no equivalent pathway for its partner quark to decay. The result is that at the end of the first microsecond of the universe's history there was a tiny excess of quarks and electrons—one in a billion surviving with no antiquark or positron to annihilate.

The ordinary matter and radiation continued to expand, and for the first second the universe remained hot enough to create electron–positron pairs. At the end of that brief interval—about the time for one heartbeat—the universe cooled below 10 billion kelvins, and the positrons annihilated the electrons, generating more photons. These photons, along with all of the others generated by these final annihilations, eventually became part of the cosmic microwave background, about a billion photons for every proton or neutron that survived.

The annihilations during the first second of the universe also would have produced a similar number of neutrinos. These neutrinos are today redshifted to energies that are currently too low to detect, only about 2 kelvins above absolute zero. Someday, new detectors might allow us to observe these neutrinos, giving us a direct glimpse of the first moments of the universe. If eventually we can detect these primordial neutrinos, it will help us to better understand the early universe, much as the detection of solar neutrinos has proved to be helpful for understanding the inner workings of the Sun (Unit 52.3).

83.4 THE EPOCH OF INFLATION

The time before 10^{-33} seconds remains at the limits of our understanding. Particle colliders cannot reach the energies of such early times, but a variety of evidence suggests what may have happened in the first instants of the universe.

The very earliest time to which we can extrapolate our current understanding of the forces of nature is known as the **Planck time.** The Planck time is a moment about 10^{-43} seconds after the universe began, when the entire universe we can see today would have been packed into a volume far smaller than a proton, and the temperature would have been greater than 10^{33} kelvins. At the Planck time we encounter an interface between our understanding of the subatomic nature of matter and the way space is curved by gravity. Today physicists studying the

FIGURE 83.7

Changes in the size of the universe through time. The size of the total portion of the universe visible to us today is indicated by the blue curve. During the inflation epoch, from about 10^{-35} to 10^{-33} second, the universe increased in size by about a factor of 10^{30}.

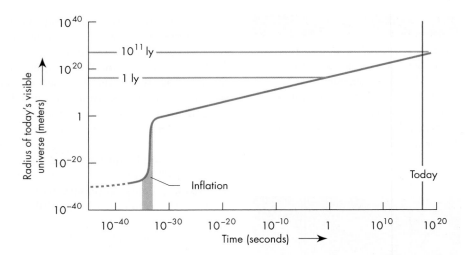

Note that the universe expanded by about the same factor during inflation as in the rest of the history of the universe.

subatomic behavior of matter observe **quantum fluctuations,** where the existence of particles and their properties can be described only according to probabilities rather than certainties. With the huge density of energy at the Planck time, extremely dense bits of matter would have fluctuated in and out of existence, causing space to curve in an unpredictable way that seems almost impossible to reconcile with our current understanding of the nature of gravity. Physicists are attempting to develop a "theory of everything" to reconcile these ideas.

Experiments that physicists have conducted with particle colliders give us a number of clues about conditions after the Planck time. The universe consisted of a boiling "soup" of energy so intense that all types of subatomic particles flashed in and out of existence—their masses almost nothing in comparison to the enormous concentration of energy present. It appears possible that the fundamental forces of nature (Unit 4) interacted at these extremely high energies quite differently from the way they do today. The transition of the strong, weak, and electromagnetic forces into their current form may have provided the mechanism that "launched" the rapid expansion of the universe in a process that cosmologists call **inflation.** Physicists hypothesize that during this brief time the size of the universe visible to us today expanded by a factor of about 10^{30} times its previous size—increasing in size far faster than the expansion before or after, as illustrated in Figure 83.7.

Inflationary models are based on hypotheses that attempt to unify the basic forces of nature. For example, in the 1970s it was discovered that the electromagnetic force and the weak force, which is responsible for radioactive decay, are really two aspects of the same force. The primary difference between these forces is in the mass of the carriers of the force. Photons, which carry the electromagnetic force, have no mass and can therefore be generated easily and travel large distances. W and Z particles, which are corresponding carriers of the weak force, have a very large mass. They therefore require high energies to be created and rarely travel more than a subatomic distance before decaying and giving up their energy. However, in the early universe, when there was so much free energy, W and Z particles could be created virtually as easily as photons. Physicists have found that the carriers of the electromagnetic and weak force form a closely linked family that becomes almost indistinguishable hotter than about 10^{15} kelvins. Today physicists generally refer to the two forces in combination as the *electroweak force* to emphasize the concept of unification.

Temperatures in the inflation epoch were far higher than can be reached with current experiments, but cosmologists have developed hypotheses for how the strong force unifies with the electroweak force at the temperatures found shortly after the Planck time. For an analogy to how these forces change at different energy levels, imagine a BB hitting a pane of glass. If you flick the BB at the glass with your

FIGURE 83.8
Schematic diagram of the relative strength of the forces through time. Physicists suspect that the electroweak and strong forces may merge at the energies present at the time of inflation into a "Grand Unified Theory." To understand extremely early times, they hope to develop a theory of quantum gravity.

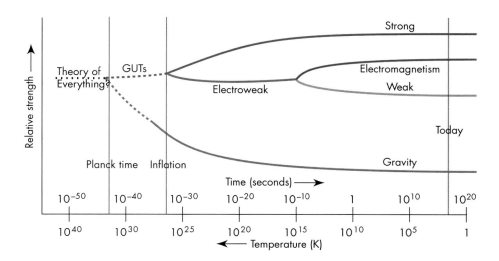

finger, the level of energy is so low that the BB bounces off the glass as if repelled. However, if the BB is shot from a pellet gun, the energy level is high enough that the BB pierces the glass—the repelling effect vanishes at high energy. So, too, in the high-energy environment of the early universe, forces may have interacted and behaved differently than they do today in our low-energy world.

Currently there are several competing hypotheses of unification, called **grand unified theories**, or **GUTs**. According to most GUT models, the strong and electroweak forces would have remained unified until the universe's temperature dropped to about 10^{27} kelvins, as illustrated in Figure 83.8. This would have happened about 10^{-35} seconds after the Big Bang. At this temperature the strong force began to "freeze out," separating from the other forces in the unified mix. The breakdown of unification had extremely important consequences. GUT models suggest that the separation released a vast amount of energy.

The energy release that occurs when the strong force freezes out can be compared to the heat released when a liquid crystallizes. A crystalline solid must absorb energy to melt, and it releases the same amount of energy when it crystallizes. You may have experienced this phenomenon if you have ever used a "heat pack"—these are plastic, liquid-filled pouches that solidify and heat up when used. The liquid in the pack would normally be crystalline at room temperature, but it has been gently cooled and kept liquid. It requires just a small trigger to cause the liquid to crystallize and release the stored-up energy.

As the universe began the transition to a separate strong force, it was caught briefly in a peculiar state similar to the heat pack. The strong force remained part of the mix of forces after the universe had cooled to temperatures below which it would normally freeze out. With the transition to a separate strong force, the creation of energy everywhere created a condition similar to the presence of Einstein's cosmological constant (Unit 82). This led to an accelerating expansion of the universe. In each 10^{-35} second, the universe approximately doubled in size, growing exponentially. A region far smaller than a proton at first grew slowly, but after doubling perhaps a hundred times over, by 10^{-33} second it had grown to the size of a beach ball.

Regions of space at the end of this inflationary expansion were moving away from one another at speeds far exceeding the speed of light. This might sound like an impossibility, but it is not. Motions of different regions of space relative to one another can have any speed, even greater than the speed of light. No light or matter can travel *through* any part of space at greater than light speed, and no observers will ever measure a photon traveling *by* them through space at any speed other than *c*. But different portions of the universe can travel extremely fast *relative* to one another, carrying along the matter within them.

83.5 COSMOLOGICAL PROBLEMS SOLVED BY INFLATION

The rapid acceleration generated during the inflation epoch offers an explanation of how the Big Bang might have been launched. It also explains several features of our universe that had previously seemed like mysterious coincidences.

The first of these features is the uniformity of the cosmic microwave background, as depicted in Figure 83.9A. In every direction we look, the radiation we observe has the same temperature, 2.725 kelvins to within about 1 part in 10,000. This is much smoother than the basic Big Bang model predicts because at the time of the last scattering epoch (when the CMB radiation last interacted with matter) regions we see separated by more than 1° never had an opportunity for their radiation and matter to mix and reach a common temperature. Astronomers call this the **horizon problem** because regions outside each other's cosmic horizon (Unit 81.1) are totally independent and there is no obvious reason they should have the same temperature.

> The horizon problem is sometimes called the *uniformity problem*.

With the inclusion of an inflation epoch, however, the visible universe we see today was originally one-thousandth-trillionth the size of a proton and was not expanding rapidly at first. Even though only 10^{-35} second had gone by in the history of the universe, light had traveled back and forth millions of times across the region that became our present visible universe. This allowed the temperature to even out. That smoothness was preserved during inflation and is retained to this day in the high degree of uniformity of the cosmic microwave background. The faster-than-light expansion speed of inflation carried these regions so far apart in a billionth-trillionth-trillionth of a second that the regions remained out of sight of each other for billions of years. Cosmic microwave background photons (Unit 81) are only now reaching us from regions that, before inflation began, had been closer to us than the distance between two quarks in a proton.

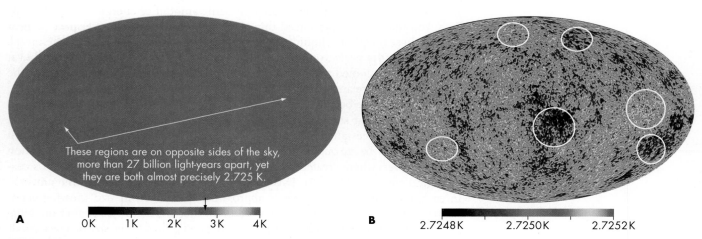

FIGURE 83.9
Maps of the cosmic microwave background, made by the Wilkinson Microwave Anisotropy Probe satellite (WMAP). The maps show the brightness distribution of the CMB across the entire sky, much as a map of the Earth depicts its features on a flat chart. (A) This map depicts the uniformity of the radiation seen—equal to 2.725 K everywhere to within about one-hundredth of 1 percent. (B) The small deviations from uniformity are brought out in this map. Red regions are warmer (by about 200 microkelvins) and blue regions are cooler than average. They show irregularities in the temperature of the universe at the time of recombination, when the universe was about 400,000 years old. Note that some warmer and cooler regions, such as those that are circled, are so large they must have been present before inflation. The satellite was named for David T. Wilkinson, a pioneer in the study of the CMB. The term *anisotropy* refers to the variations in temperature of the CMB.

FIGURE 83.10
An initially small region with obvious curvature looks flatter and flatter if it is inflated by a large factor. At the same time, inflation carries regions of space far away from each other very rapidly.

Although inflation makes the universe uniform on large scales, it also leads to the development of smaller irregularities. The cosmic microwave background is the same in every direction except for tiny variations that amount to about 1 part in 100,000. Just as water will spread and level out to a uniform height (sea level), it still contains small fluctuations (waves) that deviate from this uniform level. The map shown in Figure 83.9B shows the whole sky, spanning 360° across its center. Several of the large warm and cool regions (circled in the map) are far bigger than the 1° limit for communication—the size of the cosmic horizon—at the time when light left the CMB. These large-scale fluctuations must have been produced before inflation began.

The inflation model explains the *origin of large-scale structure* in the universe. This structure grew out of microscopic quantum fluctuations from the time before 10^{-35} second, which inflated tremendously, growing to the size of superclusters of galaxies today (Unit 77.3) or even larger. Based on the inflationary model, cosmologists have predicted a range and distribution in the sizes of denser regions generated by quantum fluctuations and then preserved in the patterns of large-scale structure much later. This closely matches what is seen imprinted on the CMB. The slightly higher-density regions acted as the sites where dark matter began to congregate, and where normal matter collected later on. Thus, the first steps toward the formation of large-scale structure in the universe (Unit 77.4) appear to have been taken when the universe was smaller than an atom.

Inflation also helps to explain another problem, known as the **flatness problem.** Einstein's general theory of relativity predicts that space may be curved in different ways by gravity (Unit 82.1), depending on the overall density of matter and energy in the universe. A universe arising from a Big Bang might equally well have any positive or negative curvature, yet measurements show that the universe is almost precisely "flat"—having zero curvature. Inflation actually drives the universe toward flatness because the huge expansion removes whatever curvature the universe initially had. Inflation removes this curvature much as inflating a balloon makes its surface less curved. While the balloon is small, the curvature of its surface is obvious. However, if the balloon could be inflated to the size of the Earth, its surface would appear nearly flat, just as the surface of the Earth appears to a local observer as nearly flat (Figure 83.10). The rapid inflation of the early universe does the same to space, taking any initial curvature and expanding it until it appears flat over the size of our visible universe.

Inflationary models of the first moments of the universe mark the frontier of our understanding of the cosmos. They offer tentative explanations of some of its features, and they offer some intriguing possibilities about the nature of the universe beyond what we can see. As we shall discuss in Unit 84, the universe may be entering a new period of inflation, and some cosmologists speculate that inflation may lead to such seeming impossibilities as multiple universes.

KEY POINTS

- The temperature of the universe has steadily declined since the Big Bang; before an age of 50,000 years, the universe was so hot that its mass was dominated by the energy of radiation.
- Matter was mostly broken down into basic particles, whose properties have been studied with high-energy physics experiments.
- For the first 10 minutes, fusion converted about 24% of the protons and neutrons into helium, with traces of other isotopes.
- Measurements of isotopes formed in the Big Bang show that normal matter comprises only about 4% of the critical density.
- If photons' energies exceed the mass of a particle, the photons can collide and produce a particle–antiparticle pair; when matter and antimatter collide, they annihilate, producing photons.
- Before about 1 millionth of a second, an asymmetry between matter and antimatter led to the creation of a slight excess of matter.
- At around 10^{-35} second, the strong and electroweak forces may have been unified, trapping the universe in a high-energy state, causing a period of exponential expansion or inflation.
- Rapid inflation helps explain why the universe is so uniform and has a curvature so close to zero.

KEY TERMS

antimatter, 664
asymmetry, 665
early universe, 664
flatness problem, 669
grand unified theories (GUTs), 667
horizon problem, 668
inflation, 666
Planck time, 665
positron, 664
quantum fluctuation, 666
quark, 664
radiation era, 660

CONCEPT QUESTIONS

Concept Questions on the following topics are located in the margins. They invite thinking and discussion beyond the text.

1. Differences in the universe with no early fusion. (p. 663)
2. Identifying a galaxy made of antimatter. (p. 664)
3. How antimatter can be kept in a lab apparatus. (p. 665)

REVIEW QUESTIONS

4. What leads astronomers to think that the early universe was hot and dense?
5. Why do astronomers call the times before the universe was about 50,000 years old the radiation era?
6. How was helium formed from hydrogen in the early universe?
7. What is antimatter?
8. What is the importance of asymmetry in the production of matter?
9. What is required for matter to be created from energy?
10. What is "unified" in grand unified theories?
11. What problems of the Big Bang theory are resolved by inflationary models?

QUANTITATIVE PROBLEMS

12. One second after the Big Bang, the density of the universe was about 0.1 kg/L. What was the radius of a spherical volume of the universe that contained as much mass as the Milky Way Galaxy?
13. What wavelength of radiation is required to produce a proton-antiproton pair? Take the mass of proton to be 1.67×10^{-27} kg.
14. As inflation began, the size of the currently visible universe expanded from about 10^{-30} m to twice that size in 10^{-35} sec. Calculate Hubble's constant at that time, converting it to the same units used today. In the next 10^{-35} sec the size doubled again. Show that Hubble's constant remains the same.
15. The map in Figure 83.9B is 360° wide and 180° tall. The radiation was produced at a redshift of about 1000. An angle of 1° on this map corresponds to about 250 kpc.
    a. Approximately how big, in degrees, are the largest high-density regions you see? How large is this in kiloparsecs?
    b. How large a region would that be today? (Hint: Use the redshift.)
    c. Relate this size to components of the large-scale structure we see today (Unit 77).
16. If the entire universe was 1 meter in diameter before inflation began, and if our visible universe was located at a random place in this volume, what is the probability we would be close enough to the "edge" of the universe to see it?

TEST YOURSELF

17. What is meant by *inflation* in the early universe?
    a. The force of gravity suddenly grew stronger in the past.
    b. Protons expanded to enormous sizes, making new stars.
    c. The universe expanded enormously in a very brief time.
    d. The number of galaxies rises dramatically with distance.
    e. Distant galaxies are much larger than nearby galaxies.
18. Which of the following statements about the first few minutes of the Big Bang is true?
    a. The universe was transparent.
    b. The universe was very cool.
    c. Hydrogen fused into helium.
    d. Galaxies began to form.
    e. Stellar cores began to form.
19. The temperature variations seen across large areas in the cosmic microwave background are caused by
    a. the recent formation of galaxy clusters.
    b. superclusters located between Earth and the background.
    c. microscopic quantum fluctuations in the early universe that grew during inflation.
    d. the collision of large clumps of dark matter after the recombination epoch.
20. Inflationary theory does not explain
    a. why the cosmic microwave background has a nearly uniform temperature in all directions.
    b. why the curvature of the universe appears to be nearly flat.
    c. the sizes of small temperature differences observed in the cosmic microwave background.
    d. what happened before the Big Bang.

PART 5

UNIT 84

Dark Energy and the Fate of the Universe

84.1 The Type Ia Supernova Test
84.2 An Accelerating Universe
84.3 Dark Matter, Dark Energy, and the Cosmic Microwave Background
84.4 A Runaway Universe?
84.5 Other Universes?

Learning Objectives

Upon completing this Unit, you should be able to:
- Explain the Type Ia supernova results and why they imply an accelerating universe.
- Describe the nature of dark matter and its consequences for future expansion.
- Describe how CMB fluctuation studies can measure the components of the universe.
- Discuss scenarios for the universe's future and the possibility of a multiverse.

The universe is currently expanding, but we saw in Unit 82 that the curvature and expansion of the universe may take many different forms. Some initial conditions lead to a universe that expands forever. Others result in a universe that collapses back on itself. How can we determine the fate that awaits us?

This Unit reviews recent discoveries that may reveal the fate of the universe. Over the last century, astronomers have offered dozens of speculations about the ultimate fate of the universe, so it requires a lot of hubris to suggest that we now really know the actual fate. Even so, recent measurements have convinced many astronomers of a new scenario that would have seemed unthinkable as recently as 1998. We will examine the growing evidence for a universe that tears itself apart in the future, and measurements that give precise new information about the overall composition of the universe. We will conclude with some speculations that ask whether the universe we know is just one part of a far larger and stranger "multiverse."

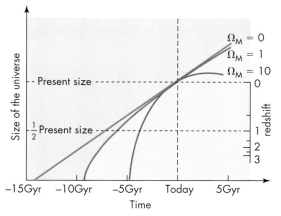

FIGURE 84.1

By studying the rate of expansion at large redshifts, such as at $z = 1$ (when the universe was half its current size), we can determine which cosmological model applies. For the $\Omega_M = 10$ model, for example, light from an object at $z = 1$ will not have had to travel as far as it would have in the $\Omega_M = 0$ model.

84.1 THE TYPE IA SUPERNOVA TEST

To make predictions about the eventual fate of the universe, we can examine the history of past expansion to see if we can determine the energy balance of the universe. Figure 84.1 illustrates three possible histories of expansion that might occur if the universe was launched by the Big Bang at a particular expansion rate and has been coasting ever since while gravity's pull gradually slows it. All three models have the same size and expansion rate at the present time, but each predicts something different about how distances depend on redshift.

When we observe a galaxy at a redshift of $z = 1$, the wavelength of the light coming from it has been stretched by a factor of 2 (Unit 80.2). This means the universe was only half its current size when the light left the galaxy. The horizontal dotted line in Figure 84.1 indicates when the universe was half as big as it is now.

In the $\Omega_M = 1$ model, the rate of expansion does not slow; so when the universe was half as big, it was half as old as it is now, and the light took 7 billion years to reach us. In the high-Ω_M model, the universe contains so much mass that the universe slows down quickly. In this model, the light from an object at $z = 1$ travels for less than 5 billion years.

FIGURE 84.2
A Type Ia supernova in a distant galaxy. This brightest kind of supernova can be seen even against the light of billions of stars in a remote galaxy.

In other words, the distance to a galaxy with a particular redshift will be measured to be smaller if Ω_M is larger. In principle, then, we can determine the value of Ω_M if we can measure both the redshifts and distances of remote galaxies. This test is straightforward, but doing it well enough to decide between expansion models requires very precise distance measurements to faraway galaxies.

In Unit 66 we examined Type Ia supernovae. These occur when a white dwarf is gradually loaded with extra matter by a binary companion until the white dwarf's mass reaches the Chandrasekhar limit. At that point, the white dwarf collapses inward and rapidly heats up its remaining matter to fusion temperatures. This leads to an explosion in which the white dwarf literally blows itself apart.

Type Ia supernovae have been found to be excellent standard candles, meaning that they all have almost precisely the same luminosity. It would be surprising if they were not nearly identical, because the physical conditions leading up to implosion are the same, and the implosion always occurs at the same mass, which is determined by fundamental properties of electrons. Because these supernovae have such a well-determined luminosity, we can find accurate distances by measuring their brightness. These supernovae are bright compared to the light from an entire galaxy (Figure 84.2), so astronomers can spot them even in very remote galaxies.

Type Ia supernovae were observed in samples of remote galaxies in the late 1990s with the Hubble Space Telescope. The result is shown in Figure 84.3, and it delivered quite a shock! The lines drawn in the figure show the predicted supernova brightnesses based on the same three models displayed in Figure 84.1. Most astronomers expected that the measured distances would all lie close to the green $\Omega_M = 1$ line, which would imply a flat, decelerating universe. Instead, the measurements lie to the right even of the $\Omega_M = 0$ line, as if there is less than zero matter in the universe. How can this be?

84.2 AN ACCELERATING UNIVERSE

The results plotted in Figure 84.3 imply that the universe is not decelerating, but just the opposite. For decades cosmologists had assumed that the expansion of the universe is affected only by the gravity of the matter within it. However, gravity's pull can only slow down the expansion, but the expansion is accelerating.

General relativity allows for an acceleration in the form of a constant added to the energy and density equations, which is called the cosmological constant

FIGURE 84.3
A graph of the redshift and brightness of Type Ia supernovae in distant galaxies. The brightness is labeled according to how many times fainter the supernovae appeared compared to how bright they would have been if they were in a galaxy 1 Mpc away. The curves show the predicted brightness if the universe has expanded according to models of the universe in which $\Omega_M = 0$, 1, or 10 (as in Figure 84.1). The horizontal lines through each data point show the uncertainty for each supernova distance measurement. The graph is based on observations of S. Perlmutter and collaborators. Note that the distant galaxies tend to lie below any of the models—indicating that the light is dimmer than any model in which the universe has undergone only deceleration.

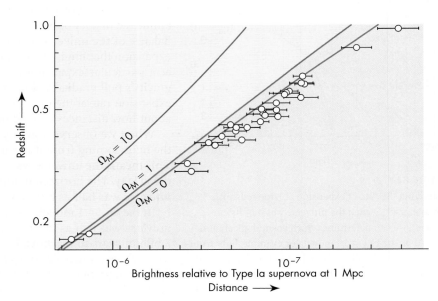

FIGURE 84.4

The data from Type Ia supernovae is shown in a graph of the expansion of the universe. At times in the past, the data lie above curves based on expansion with no dark energy. The best-fit curve, in which the universe contains nearly 3 times as much dark energy as matter, is shown in blue. A universe with dark energy will accelerate more and more rapidly in the future.

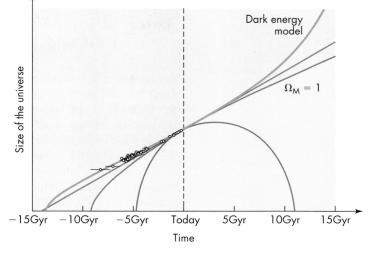

Concept Question 1

How would the data points in Figures 84.3 and 84.4 shift if Type Ia supernovae were twice as bright in the past as is currently thought? If they were half as bright as currently thought? Does such a difference seem likely?

Clarification Point

The term *dark energy* may create the impression that this is a form of energy like others. However, it is a property of the fabric of space, the same everywhere. It cannot be collected and cannot be converted to other types of energy.

The inflation of the universe in the first tiny fraction of a second was another episode of exponentially increasing expansion (Unit 83.4).

(Unit 82.4). In 1918 Einstein noted that the cosmological constant could explain how the universe might be static, as it was believed to be at that time. The cosmological constant seemed somewhat unnatural because it implied a kind of uniform energy field everywhere in space. And if the energy field was larger than the matter density, it would cause a runaway acceleration. When the expansion of the universe was discovered by Hubble, Einstein called the cosmological constant his "greatest blunder," and most cosmologists dropped it from further consideration.

However, a cosmological constant can explain the Type Ia supernova data very well, as shown by the blue curve in Figure 84.4. In this model, the universe initially expanded rapidly, and the gravitational pull of its matter began to slow it down. As the universe expanded, the matter was thinning out, and its gravitational pull was weakening. Meanwhile, though, the cosmological constant maintained its value. Based on current measurements, about 4 billion years ago, the matter density dropped so low that the effect of the cosmological constant began to take over, accelerating the expansion of the universe.

The cosmological constant is a property of space, and it does not grow more dilute as space expands. By contrast, matter and electromagnetic energy spread out and become more dilute as the universe expands. A constant level of energy everywhere in space drives space to expand ever faster, almost like a force of repulsion. Astronomers have given the cosmological constant the descriptive name **dark energy** because it is a kind of counterpart to dark matter.

With acceleration caused by dark energy, our estimates of the age of the universe must be adjusted. The result is an age that agrees well with other estimates, such as the ages of the oldest known stars. The current best estimate of the universe's age, accounting for both gravity and dark energy, is 13.8 billion years. This model uses a value for Hubble's constant of 67.1 kilometers/second per megaparsec, and it assumes that the density of dark energy in the universe is about 2.1 times that of the total of normal and dark matter at present. The comparison of energy and matter in this way is based on Einstein's formula $E = m \times c^2$, so matter (both normal and dark) represents about 3.0×10^{-30} kilograms per liter, while dark energy is about $6.4 \times 10^{-30} \, c^2$ kilograms per liter $\approx 5.8 \times 10^{-13}$ joules per liter.

Over the last decade, astronomers have measured many more high-redshift supernovae. They have tested whether there might be some systematic problem with supernova measurements at high redshift, or whether Type Ia supernovae might have behaved differently when the universe was younger. All tests to date confirm the results of the supernova studies, and in the last several years an independent approach has confirmed the existence of dark energy as well as dark matter.

84.3 DARK MATTER, DARK ENERGY, AND THE COSMIC MICROWAVE BACKGROUND

The cosmic microwave background (Unit 81) gives us a snapshot of the universe about 400,000 years after the Big Bang. We noted in Units 81 and 83 that detailed observations of the background reveal tiny variations in brightness from place to place. Those tiny variations come from slight "lumps" in the universe's density shortly after its birth. The differences probably originated in the initial moments of the universe, and formed the seeds from which large-scale structure and galaxies later formed. Astronomers can measure the relative strength of different-size fluctuations in the CMB, as plotted in Figure 84.5.

The sizes of the fluctuations in the CMB depend on a number of cosmological parameters, like the Hubble constant, the curvature of the universe, and the amounts of normal and dark matter. Dark matter does not interact with radiation in the young universe, but normal matter does, and this influences the pattern of sizes of the clumps that appear in the CMB.

Figure 84.5 shows that features about 1° across occur with the greatest frequency. This was because of the size of the cosmic horizon 400,000 years into the universe's history: Gravity could not act on regions beyond the cosmic horizon, so clumps tended to grow to this size and no larger. The dark matter pulled itself into these structures even before the last scattering epoch, while the normal matter was attracted by the gravity of these clumps (Unit 81.3). The high-energy radiation constantly striking the normal matter prevents it from settling into these regions. The normal and dark matter "sloshed around" in these regions of strong gravitational pull, creating a pattern of smaller-size clumps that astronomers can model.

> **Concept Question 2**
>
> The ripples in the CMB are a little like waves on a lake or an ocean, which often originate from unseen causes. What differences would you predict for waves caused by a distant storm, an underwater earthquake, a distant ship, or a whale just beneath the surface?

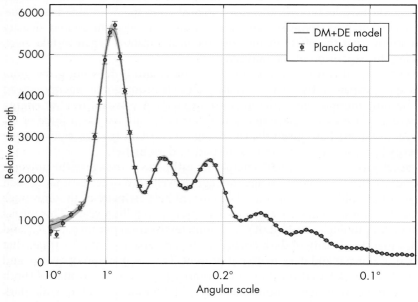
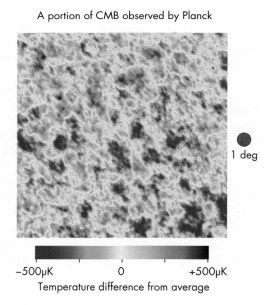

FIGURE 84.5
Plot showing the strength of fluctuations of different angular sizes measured in the cosmic microwave background (CMB). The fluctuations in brightness are caused by the clumping of matter at the last scattering epoch. The position of the first, strongest peak at about an angular size of 1° implies that the geometry of the universe must be very nearly flat. Other peaks in the graph depend sensitively on the amount of dark matter and cosmological parameters. The green line shows a plot of the predicted size of the fluctuations for the best-matching model containing dark matter and dark energy. The map at right shows the pattern of bumps in the CMB as found by the *Planck* satellite in 2013.

FIGURE 84.6
The makeup of the universe as deduced from observations of the brightness variations in the cosmic microwave background and other data. The percentages have been rounded, so they do not add up to exactly 100%.

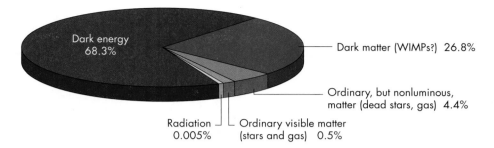

The green curve in Figure 84.5 shows a theoretical model, which almost perfectly matches the distribution of sizes of the observed clumps. To obtain such a good fit requires selecting the correct cosmological parameters, generally to within a few percent. The relative amplitudes of the bumps on the curve imply that there is about 5 times more dark matter in the universe than normal matter. In addition, the position of the first strong peak indicates that the universe is nearly flat (zero curvature), based on the size and distance to these regions, so the overall density of the universe must equal the critical density (Unit 82), or $\Omega = 1$. However, the models indicate that only about 32% of the density is in normal and dark matter combined ($\Omega_M = 0.317$). The remainder must be in dark energy: or $\Omega_\Lambda = 0.683$. The breakdown of components is illustrated in Figure 84.6.

84.4 A RUNAWAY UNIVERSE?

The Type Ia supernova results and the CMB model imply that the universe is now dominated by dark energy. This has not been true for the entire history of the universe. After inflation ended, the density of matter was far greater than the dark energy level. Now, though, the universe is picking up speed, and if we extrapolate into the future, the universe should expand faster and faster.

For many billions of years to come, the acceleration will have little visible effect. The Sun will remain on the main sequence, and the Solar System will orbit the center of the Milky Way. As discussed in Unit 76, the Milky Way will merge with M31 in several billion years, and over the next trillion (10^{12}) years the most distant small galaxies and clouds of gas in the Local Group will probably all coalesce into a single galaxy. Eventually, star formation will consume the remaining hydrogen clouds and probably leave behind a large elliptical galaxy in place of the Local Group. With no new fuel, no new stars will be born to replace the ones that reach the end of their lifetime. Our Galaxy will grow steadily dimmer and redder, the last dim red stars fading away in 100 trillion (10^{14}) years, based on projections of stellar lifetimes and the amount of hydrogen fuel that galaxies have available.

In this remote future, intelligent life could still survive. Pockets of hydrogen fuel might remain. We might even imagine future technologies that permitted us to nudge asteroids or planets toward black holes, where their matter would shine brightly for a while in a hot accretion disk of infalling matter around the black hole. Eventually, though, the available fuel will be consumed, converted into heat from which no further energy can be extracted. This is described as the **heat death** of the universe—not because it is hot, but because what energy there is has all been converted to the same low-grade disordered motion from which no further power can usefully be extracted.

Dark energy at the level currently observed will cause a curious thing to occur as this remote future unfolds. The expansion rate of the universe will continue to accelerate because of dark energy, as illustrated in Figure 84.7. Without dark energy, the Local Group would probably plunge into the Virgo Cluster (Unit 77) in a few hundred billion years. Instead, dark energy will cause the expansion of space

Concept Question 3

The distant future of the universe will be long after the Sun has gone through its red giant phase and Earth has been destroyed or is no longer habitable. Do you think it is likely that intelligent life will find a way to overcome such enormous obstacles?

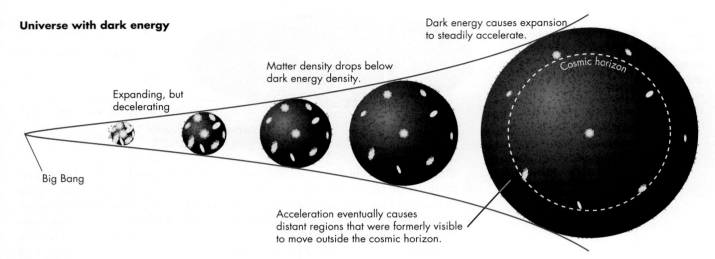

FIGURE 84.7
An expanding universe containing dark energy will initially slow down until the matter density drops below the level of the dark energy. Dark energy then causes accelerating expansion, which can move remote regions of space away so fast that light can never reach us from them. This moves them beyond our cosmic horizon.

to overcome the Local Group's gravitational attraction toward the Virgo Cluster, and then the separation will grow exponentially faster. The exponential expansion of space will move distant parts of space away from us at speeds increasing to far beyond the speed of light. All galaxies beyond the gravity of our Local Group will seem to disappear beyond our cosmic horizon. Eventually, our Galaxy will be essentially alone in the universe, an island of stars in a vast empty universe.

In even more remote times, orbits themselves must gradually shrink because they lose energy through gravitational radiation (Unit 27). After about 10^{20} years, the orbits of any remaining binary or planetary systems will shrink until the objects collide and coalesce. Much of the matter in our Galaxy will end up in black holes as orbits decay or chance collisions collect matter in these gravitational traps.

Matter that does not end up in a black hole faces another insidious problem. Physicists suspect that matter itself may not be stable. One of the consequences of grand unified theories—the same theories that predicted inflationary expansion in the very early universe (Unit 83.3)—is that they predict **proton decay**. Protons appear to be stable particles, but if the forces are unified, there should be pathways by which a proton can break down into lighter particles. Physicists have conducted many experiments hunting for this phenomenon without success, but this may be because of the extremely improbable nature of such decays. It is still highly uncertain, but protons may live 10^{40} years before disintegrating.

Even black holes will not last forever. As Hawking showed, they can radiate away energy through quantum effects (Unit 69). However, the bigger the black hole, the slower this process. Evaporation will take in excess of 10^{60} years for stellar-mass black holes and 10^{100} years for the supermassive black hole at the center of our Galaxy. In the end, then, the universe might contain nothing but low-energy highly redshifted radiation.

84.5 OTHER UNIVERSES?

We end our discussion of cosmology with some highly speculative ideas. Scientists have successfully explained an amazing array of observations using current theories of matter and forces. We are able to piece together the history of the universe from a fraction of a second after the Big Bang to times trillions of years into the future. Yet there are still pieces of the puzzle that do not fit together.

Physicists have not been able to conceive of a way to fit a dark energy field into the standard model of particles and forces that is otherwise so successful. If a

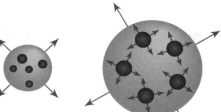

"Bubbles" grow.   Expansion cools universe.   "Bubbles" become separate universes?

WFIGURE 84.8
A sketch depicting bubbles that might become separate universes forming in an inflating universe. Each inflating bubble is completely outside of the region that will ever be visible within other universes.

In his 1920 poem *Fire and Ice,* Robert Frost contemplates possible fates for the world similar to those we have discussed for the universe:

Some say the world will end in fire,
Some say in ice.
From what I've tasted of desire
I hold with those who favor fire.
But if it had to perish twice,
I think I know enough of hate
To say that for destruction ice
Is also great
And would suffice.

dark-energy universe expands exponentially into the infinite future, vast quantities of energy will be created everywhere as part of the fabric of the ever-expanding space. It is strange to think of a cold, dying universe that is generating vast quantities of limitless but unusable dark energy.

Many cosmologists think that we do not yet have a complete description of dark energy. One suggestion is that dark energy is not a constant, but will undergo changes in the future, much as inflation ended after a period of time. This opens up many possibilities. Dark energy may grow with time, not just gradually pulling galaxies away from each other but eventually tearing apart our Galaxy, then even our Solar System, and eventually even smaller structures. In this alternative fate, called the **Big Rip,** space on smaller and smaller scales begins to expand rapidly as during inflation, maybe even launching new universes. Or else dark energy may change in other ways, even leading to contraction and a Big Crunch (Unit 82).

Some physicists think that the accelerating expansion of the universe hints at additional dimensions of space beyond the three we know. In fact, according to *string theories,* space has as many as 10 or 11 dimensions, and perhaps only our familiar dimensions expanded during inflation. In these scenarios, our universe may undergo gravitational interactions with other universes, giving rise to effects like the acceleration attributed to dark energy.

Some inflation models also suggest the existence of multiple universes. Multiple universes could have arisen if inflation did not occur everywhere at the same time. After the Planck time, different regions of space may have inflated independently, and today they may be completely separated from one another with no possible contact between them. Such isolated regions might well be considered separate universes, independent of ours and completely unobservable by us. Furthermore, according to some inflationary theories, these other universes may be either expanding at different speeds or even have expanded a while and then contracted back on themselves, or even have different physical laws in operation. Such a collection of independent universes is sometimes called the **multiverse.**

Inflation may create such sections of space much as bubbles form in a pot of boiling water. As water begins to boil, regions of the liquid suddenly change from liquid to gas (steam) and begin expanding. We see that region of steam as a bubble. Some bubbles expand, rise to the surface, and burst, whereas others collapse. So, too, in inflationary theories, entire universes may form and dissolve, entirely unknowable to us (Figure 84.8). It would be amazing indeed if all the wonderful intricacy and beauty of our universe is merely a single bubble in an even vaster sea of space, a space that we can never see.

All of these different ideas illustrate that astronomers are in the early stages of studying dark energy and the nature of the universe. They are still developing tests to understand the cosmos. The science of cosmology is beginning, however, to sketch the overall structure, history, and future of the universe. It is a universe in which the stars and galaxies—and all of the kinds of matter and energy that we are familiar with—are just minor players. We are along for the ride in a universe driven by far larger forces generated by dark matter and dark energy.

KEY POINTS

- For decades it was thought that after the Big Bang the universe could only slow down because gravity was the only force at work.
- Distances for Type Ia supernovae have enough accuracy to show that the universe is accelerating, which an energy field can cause.
- The cosmological constant is like an energy field associated with all space; it acts to drive the expansion at a growing rate.
- Measurements of the frequency of different-size clumps in the CMB confirm the presence of dark energy and dark matter.
- The CMB measurements indicate that there is about 2.5 times more dark energy than dark matter, and about 5 times more dark matter than normal matter.
- The high level of dark energy found should cause the universe to begin expanding exponentially faster in the future.
- In exponential expansion, distant regions of space eventually move away so fast that they pass outside our cosmic horizon.
- It is thought that matter and black holes will eventually decay, leaving nothing but weak radiation.
- If dark energy does not remain constant, the universe may have many alternate possible fates.

KEY TERMS

Big Rip, 677
dark energy, 673
heat death, 675
multiverse, 677
proton decay, 676

CONCEPT QUESTIONS

Concept Questions on the following topics are located in the margins. They invite thinking and discussion beyond the text.

1. Possibility that supernovae were different in the past. (p. 673)
2. Wave patterns produced by different sources. (p. 674)
3. The long-term survival of intelligent life. (p. 675)

REVIEW QUESTIONS

4. Why are Type Ia supernovae useful for cosmology?
5. What evidence indicates the existence of dark energy?
6. How is dark energy related to Einstein's cosmological constant? How does it differ from other kinds of energy?
7. How is the CMB used to confirm the existence of dark matter?
8. What effects will dark energy have on the future expansion of the universe? on the cosmic horizon?
9. What is meant by heat death of the universe?
10. How might inflation lead to multiple universes?

QUANTITATIVE PROBLEMS

11. In Figure 84.4, the straight-line extrapolation of the current expansion rate (red line) gives the same age for the universe as the much more complex dark energy model (blue curve). This is why the simple calculation of the age of the universe in Unit 80 gave nearly the correct value. If we lived 10 billion years in the future or in the past along the dark energy curve in that figure, what would a straight-line extrapolation have indicated the age was, compared to the actual age at those times?

12. The following calculations can help us begin to understand why the Local Group is predicted to survive the effects of dark energy, but more distant galaxies are expected to be pulled away:
    a. Using the values of mass and size in the text, what is Ω_M for the Local Group? Note that in the text, this is calculated out to the distance of neighboring groups.
    b. Instead calculate the density inside of the central 1-Mpc radius region of the Local Group (where nearly all of the group's mass resides). How does this density compare to critical density?

13. Dark energy retains a constant density even as the universe expands. Today it has a value of $\Omega_\Lambda = 0.7$. Matter, by contrast, grows more dilute as space expands, so its density would have been larger in the past by a factor that depends on the redshift, as $(z + 1)^3$. At what redshift would dark energy and matter (currently $\Omega_M = 0.3$) have had the same density?

14. The Super-Kamiokande experiment searches for proton decays by monitoring 20 million kilograms of water for decay events. Given that a single proton has a mass of 1.67×10^{-27} kg, how many decays would you expect on average in one year of observations if it takes 10^{40} years for a single proton to decay?

15. If the universe now begins doubling in size every 10 billion years, regions that are currently moving away from us at half the speed of light will start receding at c in 10 billion years. Using Hubble's law, how far away from us are such regions currently? About how many doublings and how many years will it take a region that is currently receding at 3000 km/sec to be receding at the speed of light?

TEST YOURSELF

16. Dark energy
    a. is like a cosmological constant.
    b. is repulsive.
    c. is distributed evenly through space, never dissipating.
    d. causes space to expand ever faster.
    e. exhibits all of the above.

17. If the universe ends with a heat death it will be filled with
    a. low-energy highly redshifted radiation.
    b. only black holes, neutron stars, and white dwarfs.
    c. gas clouds too hot to form new stars.
    d. high-energy particles perpetually colliding with one another.

18. Fluctuations of a size of 1° in the CMB are so strong because
    a. there is about 5 times more dark matter than normal matter.
    b. the universe has a curvature that is nearly flat.
    c. the current temperature of the radiation is 2.7 K.
    d. the universe has begun to experience a Big Rip.

19. The possibility that protons decay is a prediction of
    a. general relativity.
    b. the dark energy model.
    c. Type Ia supernovae.
    d. grand unified theories.
    e. the Big Bang theory.

PART 5 UNIT 85

Astrobiology

- 85.1 The History of Life on Earth
- 85.2 The Chemistry of Life
- 85.3 The Origin of Life
- 85.4 Life, Planets, and the Universe

Learning Objectives

Upon completing this Unit, you should be able to:
- Describe the fossil evidence for life on Earth and its overall time line.
- Describe the chemical and biological similarities found across all forms of life.
- Discuss how life may have begun on Earth and the major steps in its evolution.
- Explain the ideas behind the Gaia hypothesis and the anthropic principle.

The evolution of the universe from Big Bang to galaxies, stars, and planets has along the way created us. The "quark soup" of the early universe evolved into stars, which created heavy elements that gathered into simple creatures, who evolved into beings who seek to understand how they fit into this grand scheme. Our own bodies bear clues about our origins, providing evidence of our close relation to simpler creatures on our planet. These observations provide us with some perspective on what life might be like elsewhere in the universe.

We begin this Unit by examining the long history of life on Earth, looking at the factors thought to play a role in its development here. We conclude with some speculations about the role of life in the history of our planet—the so-called Gaia hypothesis—and whether our existence may even say something about the nature of the universe itself.

85.1 THE HISTORY OF LIFE ON EARTH

Life has existed on Earth for most of the history of our planet. The first indications of life are found in the composition of particles of carbon in rocks as old as 3.8 billion years. These particles show an unusual mix of carbon isotopes that is found to be particular to living organisms today. Carbon has two stable isotopes, ^{12}C and ^{13}C, which have 6 and 7 neutrons in their nuclei, respectively. In photosynthesis and other biological activities, microscopic organisms prefer the lighter isotope ^{12}C. The remains of living organisms therefore show a deficit of ^{13}C relative to its naturally occurring abundance.

Few ancient rocks survive that have not undergone major modification over the last several billions of years, but some probable microscopic fossils have been identified in sediments laid down more than 3.4 billion years ago in Western Australia. These are suggestive of single-cell organisms (Figure 85.1), but appearance alone is not sufficient to confirm a microfossil because some physical processes can generate similar-looking features. To help confirm the biological origin of these features, scientists measured their chemical makeup, which shows carbon and sulfur compositions that appear similar to sulfur-metabolizing microorganisms today.

Some microscopic creatures, such as modern day blue-green bacteria or **cyanobacteria,** produce much larger structures. These organisms grow in dome-shaped colonies called **stromatolites.** Stromatolites form as the bacteria trap sediments and then successive generation form new layers of bacteria on the surface. Stromatolites are found today in shallow, warm, salty bays such as those along the coast of

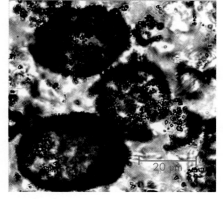

FIGURE 85.1
Microscopic image of a probable microfossil of cells in rock 3.4 billion years old.

FIGURE 85.2
Stromatolites on the western coast of Australia, seen at low tide, are formed by colonies of a cyanobacteria. They trap sediments, forming a layered structure in this characteristic dome shape. The inset shows a 3.4-billion-year-old stromatolite fossil, sliced to show the layered growth pattern.

western Australia, as shown in Figure 85.2. Fossils with very similar dome-shaped structures have been found with ages up to about 3.4 billion years, as shown in the inset in Figure 85.2.

Figure 85.3 shows a time line for the history of life on Earth. The Earth formed a little more than 4.5 billion years ago, but very few rocks older than 4 billion years survive today. This is probably because the Earth was undergoing heavy bombardment and large parts of its surface were molten for several hundred million years, to perhaps as recently as 3.9 billion years ago. Liquid water may have first appeared a few hundred million years after the Earth formed, and life not long thereafter. The diagram illustrates some of these ancient life forms and shows the epoch in which they first appear in the fossil record. The ages have been measured by radioactive dating of their associated rocks (see Unit 35.1).

Scientists conclude from the available evidence that only extremely simple life forms, such as cyanobacteria and other single-celled creatures, were present for more than three-fourths of the Earth's history. The only nonmicroscopic indications of life would have been stromatolites until about 1 billion years ago, when we find the first fossils of multicellular algae. Stromatolites can be traced throughout the geological record, and it is thought that they produced much of the oxygen in Earth's atmosphere as a waste product of photosynthesis.

The rise of oxygen in the atmosphere allowed the evolution of more complex organisms, and ironically these fed on stromatolites and reduced their numbers to the low levels we see today. By about 500 million years ago, only a little more than one-tenth of the Earth's lifetime, shells and crustaceans appeared, followed by the rapid appearance of many more new forms of life, including the first on land. This marks the beginning of the Cambrian geological period, and the sudden appearance of such a diversity of life is called the **Cambrian explosion.**

Mammals and dinosaurs both appeared roughly 250 million years ago, but the latter were wiped out about 65 million years ago in the great Cretaceous extinction, caused, we think, by a totally chance event—an asteroid's collision with the Earth (Unit 50). The fossil record is broken by many other cataclysms as life developed.

Hominids, our immediate ancestors, appeared roughly 5.5 million years ago; but our own species, *Homo sapiens,* evolved only about 500,000 years ago. Earth has therefore existed as a planet roughly 10,000 times longer than we have as a species. If all of Earth's history were compressed into a year, humans would appear only in the last hour, and what we call civilization only in the last few minutes.

Life on Earth shows a dazzling variety of complex forms, from butterflies to whales, from mushrooms to trees, and so on (Figure 85.4). Life also builds upon life.

FIGURE 85.3
A time line illustrating the history of the Earth and life on it.

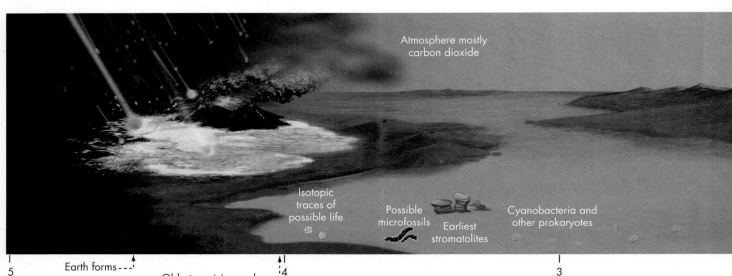

FIGURE 85.4
A few of Earth's many complex life forms.

To survive, complex organisms depend on **symbiosis,** mutually beneficial relationships with other organisms. These can become highly specialized as species evolve together. For example, the human body contains hundreds of kinds of bacteria, many of which are essential to our survival. Bacteria that break down food in our intestines enable us to extract nutrients essential to our own cells. Without those nutrients we would starve. Indeed, there are approximately 10 times more bacteria in our bodies than all of our other cells combined. From the bacterial point of view, a human body is a bag of oceanlike water that travels about in order to feed them!

85.2 THE CHEMISTRY OF LIFE

Despite their variety, all living beings on Earth have an amazing underlying unity. All living organisms use primarily the same kinds of atoms for their structure and function: hydrogen, oxygen, carbon, and nitrogen. These atoms are not only those most commonly used by Earth's life forms, but they are also some of the most abundant throughout the universe. Even widely differing organisms use the same chemical substances. Life tends to draw on the materials with which our planet is most richly endowed; we are made primarily of the same substances that make up our ocean and atmosphere, the same elements that are the primary products of nucleosynthesis in stars.

The chemical elements that compose living things are linked into long-chain molecules for their structure and function. Many of these chain molecules are built up from various arrangements of some 20 amino acids. **Amino acids** are organic molecules containing atoms of hydrogen, carbon, nitrogen, oxygen, and sometimes

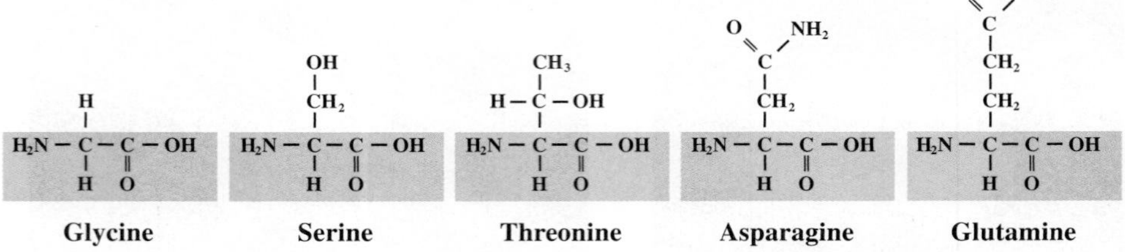

FIGURE 85.5
Diagrams illustrating the structure of several amino acids. H, C, N, and O represent atoms of hydrogen, carbon, nitrogen, and oxygen. The basic building block (shown in yellow) of amino acids is the same.

Glycine   Serine   Threonine   Asparagine   Glutamine

sulfur. Their structure is the same except for a "side chain" that makes each one slightly different (Figure 85.5).

Moving up to the next level of complexity, we find that all living things use amino acids as the structural units of more complicated molecules called proteins. Protein molecules give living things their structure. For example, the scales of reptiles and fish are composed of the protein keratin, and the stiff cartilage of our own bodies is composed of the protein collagen. Proteins not only give cells their structure but also supply their energy needs. That energy supply reveals yet another underlying unity among living things: All cells use the same kind of molecule, adenosine triphosphate (abbreviated as ATP), to supply energy for action and growth.

This unity of structure and energy supply is echoed in the reproduction of life forms. All single-celled organisms divide in two to create new cells; all multicellular organisms produce egg cells, which, once fertilized, divide to create new cells. Moreover, in all cases, parents pass on genetic information to their offspring by the same molecule, **DNA,** which provides the instructions for building proteins.

DNA stands for *deoxyribonucleic acid.* It is a massive molecule consisting of two long chains of molecules wound around each other and linked by molecules called base pairs. DNA has the appearance of a twisted rope ladder with the role of the connecting rungs played by the base pairs (Figure 85.6). When a cell reproduces, the DNA divides in half along its length into two chains, each carrying along half of each base pair. The chains then chemically attract the missing half of each base pair at each point along their length, reproducing the original pattern. At the end of this process, two new DNA molecules exist that are identical to the original one. Nonetheless, copying errors—*mutations*—sometimes occur. Some can be harmful and kill the organism, but the nonlethal ones make evolution possible. That all terrestrial organisms use DNA for their reproduction gives life a truly remarkable unity.

Unity can also be seen in how more-complex organisms function, such as the chemical processes that move biologically important molecules into and out of cells. Complex organisms also use similar molecules for different purposes. For example, plants derive their food from carbohydrates they manufacture during photosynthesis. The chlorophyll they use in this process is structurally similar to the hemoglobin that transports oxygen in the blood of animals. These two molecules differ primarily in that chlorophyll is built around a magnesium atom, whereas hemoglobin is built around an iron atom. Similarly, cellulose, the structural material of plants, has a molecular structure very similar to chitin, the structural material in insect bodies and crustacean shells.

What conclusions can we draw from the chemical and biological similarities of living beings and their ancient history as deduced from the time line? First, such chemical similarity almost inescapably suggests that all life on Earth had a common origin. Second, the antiquity of life suggests that under suitable conditions, life develops rapidly from complex molecules. Therefore, we might infer that wherever such conditions occur, so too will life.

We should treat these conclusions with caution, however, because they are based on only a single case: life on Earth. For example, if the first time you played a lottery you won the grand prize, that would not mean that you would be likely to do so again. Likewise, the fact that life exists on Earth may say next to nothing about its chances elsewhere. Supposing, however, that we do choose to speculate, what *can* we say about the origin of life and its likelihood elsewhere?

Clarification Point

An organic molecule is one that contains carbon atoms. Organic molecules do not require biological organisms for their creation.

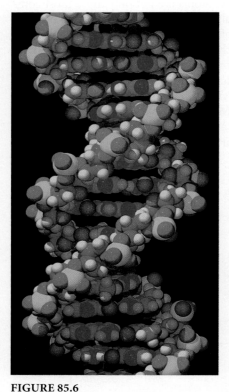

FIGURE 85.6
A diagram illustrating the structure of the DNA molecule. The blue horizontal structures in this model are the base pairs.

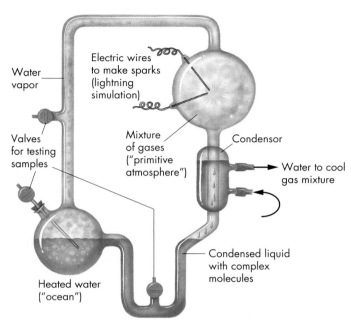

FIGURE 85.7
The Miller-Urey experiment attempted to simulate conditions on the primitive Earth. A mixture of gases received sparks (simulating lightning) and were repeatedly condensed and evaporated as they might have been in the young oceans. A variety of complex organic molecules were generated.

FIGURE 85.8
Spheres about 1 to 2 microns in diameter form when proteinoids are heated and cooled. Similar proteinoids have been discovered around undersea vents of Hawaiian volcanoes.

85.3 THE ORIGIN OF LIFE

Most scientists today think that terrestrial life originated from chemical reactions among complex molecules present on the young Earth. This idea dates back to at least the time of Charles Darwin, who speculated that life arose "in some warm little pond." The idea was strongly bolstered in 1953 by an experiment performed by Stanley Miller, then a graduate student at the University of Chicago, and his professor, Harold Urey. They filled a sterile glass flask with water, hydrogen, methane, and ammonia—gases thought to be present in the Earth's early atmosphere. They then passed an electric spark through this mixture, as illustrated in Figure 85.7. The spark generated both visible and ultraviolet radiation, which triggered reactions in the gas and water mixture. At the end of a week the mixture contained a variety of organic molecules as well as five of the amino acids used to make proteins. They concluded that some of the ingredients necessary for life could have been generated spontaneously on the early Earth. The success of the **Miller-Urey experiment** motivated other researchers to perform similar but more realistic experiments, and these have demonstrated that a variety of conditions can generate amino acids.

The U.S. chemist Sydney Fox took complex organic molecules, such as those made by Miller and Urey, and by repeatedly heating and cooling them in water was able to create short strands of proteins called **proteinoids.** Intriguingly, Fox discovered that these proteinoids spontaneously formed small spheres reminiscent of cells (Figure 85.8). These spheres have no power to reproduce, nor do they show any of the complex structure commonly found in living cells; but they demonstrate that organic material can spontaneously give itself structure and wall itself off from its environment, a step toward the first living things.

Even the energy needs of living things can be produced in Miller-Urey types of experiments. For example, the Sri Lankan–American biochemist Cyril Ponnamperuma and the U.S. astronomer Carl Sagan showed that ATP can be synthesized from compounds similar to those generated by Miller and Urey. ATP is used by cells to store and transport energy, demonstrating one additional step in the creation of life from nonliving material.

While the Miller-Urey and subsequent experiments show that organic material could have developed on the young Earth, it is possible that this material was present on our planet right from the start. Interstellar clouds contain a rich mix of organic molecules (Unit 73.4), including several that are precursors to amino acids. Because the Solar System and Earth formed from such clouds, complex organic molecules must have been present on or near the Earth from its birth. Even if these molecules were destroyed during the Earth's formation, fresh molecules may have been carried to its surface by planetesimals at the end of the period of heavy bombardment (Unit 35.5). In fact, carbonaceous chondrite meteorites—probably surviving fragments of these ancient planetesimals—are rich in organic material, including amino acids (Unit 50.2). Finally, some scientists have demonstrated experimentally that the compression and heating of gases resulting from planetesimal bombardment may be as fruitful as the electric discharges used by Miller and Urey for forming complex organic materials. Given these findings, astronomers are confident that organic molecules were common on the young Earth.

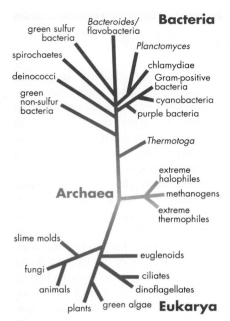

FIGURE 85.9
Evolutionary tree showing the relationships between different species of bacteria based on their genetic differences.

Regardless of the origin of organic matter on the early Earth, whether created by lightning, ultraviolet radiation, or planetesimal bombardment, many steps are needed to progress from droplets filled with amino acids to living cells. Cells in Earth's present life forms replicate using the double-chained molecule DNA. Some researchers have hypothesized, however, that the precursors of current forms of life used the simpler molecule called *ribonucleic acid* (RNA). Biochemists have demonstrated that in an early **RNA world,** if certain enzymes were present, RNA could have replicated in the absence of living cells and could have replaced some of the functions that proteins and DNA provide in modern cells. Indeed, RNA continues to play critical low-level roles within modern cells, which may be "fossil evidence" of this earlier stage in the evolution of life.

Biochemists have also discovered several other molecules far simpler in structure than RNA that can replicate in a mechanical sense. That is, if such molecules are put in a solution containing the necessary raw materials, the ensuing chemical interactions allow the molecules to spontaneously produce copies of themselves. One particular mix of molecules not only replicates but also can "mutate" into a slightly better replicator. Thus, there may have been several pathways toward self-replication long before the complex biochemistry of modern life evolved.

Life also could only have developed when the atmosphere was quite different from today's atmosphere. Ironically, the most essential component of Earth's atmosphere from the human perspective today, oxygen, would have prevented life from forming if it had always been present. Oxygen is highly corrosive to unprotected organic molecules, breaking them down, so the absence of oxygen was crucial for the development of life. Analysis of ancient rocks indicates that Earth's atmosphere did in fact lack oxygen, making it more similar to the other terrestrial planets.

The first cellular life forms on Earth had no internal structures, such as a nucleus to contain the DNA, making them similar to modern bacteria. **Prokaryotes,** as these cells without nuclei are known, appear to have arisen relatively rapidly, and then remained the only form of life until about 2 billion years ago. The prokaryotes are divided into two kingdoms, the bacteria and the **archaea** (ar-KEE-ah), which are some of the most primitive single-celled organisms. The relationships between these species are illustrated in Figure 85.9.

FIGURE 85.10
(A) Undersea volcanic vents expel hot sulfur-rich minerals, which provide the nutrient source for very primitive types of bacteria.
(B) These bacteria in turn are at the bottom of the food chain for colonies of unusual undersea creatures at such depths that sunlight never reaches them.

Today some archaea are found in extreme conditions. One of the most intriguing of these locations is around sea floor vents deep in the ocean. These environments are rich in sulfur minerals, where they support a variety of life forms, as shown in Figure 85.10. The archaea that feed off these sulfur minerals are simpler than cyanobacteria, because they did not need to develop photosynthesis to supply their energy. It is not difficult to imagine that this kind of organism could have thrived on the young Earth. After a few hundred million years, the surface was cool enough for liquid water, at which time the oceans formed. The mineral-rich regions around vents would have been conducive to the production of amino acids, proteinoids, and the cell-like spheres found in Fox's experiments. In analyses of the genetic differences between living organisms, the archaea also reside closest to the "center" of the evolutionary tree, as shown in Figure 85.9. They are probably closest in form to the first life on Earth, although it must be remembered that modern archaea have evolved for billions of years.

More-complex prokaryotes, able to tap into the energy in sunlight through photosynthesis, probably developed within the first billion years of Earth's history. These ancestors to cyanobacteria probably account for the fossil record of

> **Concept Question 1**
>
> One objection to the Miller-Urey type of proposal for the origin of life is that chemicals in the ocean are far too dilute to undergo long series of interactions. In what kinds of environments on the young Earth do you imagine chemicals might have become concentrated?

stromatolites. Stromatolites grew steadily more common until about 3 billion years into the Earth's history, and their success radically altered Earth's environment. One of the waste products of photosynthesis is oxygen. At first this oxygen was absorbed by the Earth, oxidizing minerals such as iron. By about 2.5 billion years (2 billion years ago), oxygen levels had risen to about 1% of present levels. Not until about 500 million years ago was there enough oxygen that we might be able breathe.

The rise of oxygen also permitted the development of an ozone layer, shielding the Earth's surface from the Sun's ultraviolet light. The new atmosphere was caustic to the bacteria that created it, but the presence of oxygen offered opportunities for more complex organisms to develop and eventually spread to the land surfaces of Earth. Cells today take advantage of oxygen to enhance their metabolism, but they needed to develop more-complex structures to handle this caustic chemical. Cells in plants and animals today contain a *nucleus* in which genetic information is stored in long strands of DNA, as well as *mitochondria,* which are tiny bodies that convert food into energy that the cell uses. They also have a complex *membrane* that holds them together and allows food to enter and waste products to leave. The development of more complicated cells may have begun when one prokaryote merged with another, producing a symbiosis that benefited both cells by allowing them to grow and reproduce more rapidly. For example, a cell good at storing energy (a mitochondrion) might have combined with a cell good at reproducing. From such simple beginnings, **eukaryotes** (cells with nuclei) probably evolved. In this scenario mitochondria carry their own DNA because they were originally separate life forms.

Microscopic fossils of eukaryotes, mostly algae, have been found that are up to 2.1 billion years old (Figure 85.11). Gauging by the amount of time it took for prokaryotes to appear on Earth (probably much less than 1 billion years), compared to the 1.5 to 2 billion years for complex cells to appear, this step must have been a difficult one. From the timing of their appearance, it seems that the development of eukaryotes was spurred by the development of an oxygen-rich atmosphere. The eukaryotes proved much more adaptable than the prokaryotes, and by about 1 billion years ago, stromatolites nearly disappear from the fossil record as they were replaced by and preyed upon by eukaryotes.

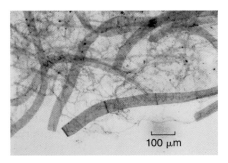

FIGURE 85.11
A 1-billion-year-old fossil of eukaryote algae. The earliest eukaryotes show up in fossils from about 2.1 billion years ago.

The final major step in the evolution toward the complex beings of today required individual cells to work cooperatively as a single organism. Fossils of what appear to be complex bacterial colonies appear at about the same time as the first eukaryotes, about 2.1 billion years ago, but complex multicellular animals began to appear only about 600 million years ago, seen today in fossil impressions of soft-bodied creatures that resemble sea worms and jellyfish (Figure 85.12). The development of ways in which cells could cooperate to form complex multicellular organisms was apparently another difficult step in evolution, given that it took more than a billion years. However, once achieved, tremendous varieties of plants and animals arose during the Cambrian explosion, in an extraordinarily brief time, geologically speaking. The distance between bacteria and sea worms is apparently far larger than the distance between sea worms and humans!

The greater complexity of multicellular organisms also allows for a large amount of variability between individual organisms, and these can provide new opportunities for exploiting resources in different environments. This idea was developed in 1859 by Charles Darwin in his towering work *On the Origin of Species by Means of Natural Selection*. For example, an organism that can move a little faster from a nutrient-poor to nutrient-rich environment will be naturally selected to reproduce more prolifically. The inheritance of successful traits and occasional random mutations can alter characteristics over many generations, eventually yielding new species. The evidence indicates that, over billions of years, natural selection combined with changing environments has yielded the incredible diversity of life that we see around us today.

FIGURE 85.12
A fossil impression of a multicellular animal from about 600 million years ago.

85.4 LIFE, PLANETS, AND THE UNIVERSE

Life on our planet can be viewed in broader contexts. Life has played a critical role in shaping the evolution of our planet. In 1974 James Lovelock, a British chemist, and the U.S. microbiologist Lynn Margulis suggested that life creates a single "larger entity" with a planet, a symbiosis of life and planet that they called the **Gaia (*GUY-uh*) hypothesis** after the Greek goddess of the Earth.

According to the Gaia hypothesis, life does not merely respond to its environment but actually alters its planet's atmosphere and temperature to make it more hospitable. Plant life has almost certainly done just that on Earth. For example, by photosynthesis, plants have created an oxygen-rich atmosphere on Earth that shields them from dangerous ultraviolet radiation. Similarly, photosynthesizing organisms have altered our planet's temperature by removing CO_2—a greenhouse gas—and helping to lock it away in soil and ultimately rock. This means that plants can modify the Earth's temperature by adjusting the greenhouse effect (Unit 38). If the planet is too cold, plant metabolism is slowed and less CO_2 is converted to oxygen. The greater abundance of CO_2 warms the planet and enhances plant growth. Conversely, if the planet gets warmer, plants grow faster and produce lots of oxygen, reducing CO_2 abundance and thereby also reducing greenhouse warming.

Changing levels of CO_2 on Earth have prevented a catastrophe for life on our planet. It is estimated that over the last 4 billion years, the Sun's luminosity has increased by about 40% (Unit 62.4). If greenhouse gases had not been removed steadily from Earth's atmosphere throughout this time, the planet might have suffered a runaway greenhouse effect such as that on Venus (Unit 41). By contrast, the Gaia hypothesis would suggest that Mars never had life, because if life had ever taken hold, conditions conducive to life would probably have been maintained. Other scientists disagree, arguing that such extreme changes as Mars has suffered could have killed off life on that planet.

The existence of intelligent life perhaps implies something significant about the universe itself. In 1961 Robert Dicke, an American physicist, noted some interesting cosmological coincidences. Dicke noted, for example, that the age of the universe is not much different from the lifetime of stars like the Sun, and he went on to argue that this "coincidence" is in fact not remarkable at all. Rather, it is a *necessity*, if cosmologists are to exist to note it.

Dicke pointed out that life requires elements such as carbon, silicon, and iron that are made in massive stars. So for life to form, enough time must pass for massive stars to evolve and make the heavy elements and then eject them into space with a supernova explosion. Moreover, additional time is needed for the ejected material to be incorporated back into interstellar clouds and to form new stars. Only with this second generation of stars does life become possible, so intelligent life cannot evolve in a universe only a few billion years old. He argued that life requires stars, and so it can exist only in a universe young enough that stars have not all died. Therefore, the existence of an intelligent observer in the universe requires that the age of the universe fall within certain limits.

In 1974 the English physicist Brandon Carter took Dicke's idea further and proposed what he called the **anthropic principle.** According to the anthropic principle, "what we can expect to observe must be restricted by the conditions necessary for our presence as observers." Since Carter's work, many scientists have shown how "fine-tuned" the universe must be for life to exist. Thus, no matter how unlikely some aspects of the universe may appear to us, if those aspects are necessary for life, then that is what we will observe. For example, we might argue how truly marvelous it is that conditions on Earth are just right for life. According to the anthropic principle, the rude reply is, "Of course conditions are just right for life. If they were not, there would be no life here to do the marveling."

Concept Question 2

If the runaway greenhouse effect were to begin to take hold, what biological processes might naturally bring it into check?

While the Gaia hypothesis has led to important understandings of the connection between life and evolution of Earth's atmosphere, many scientists consider other Gaia claims, such as effects of life on plate tectonics, to be untestable and therefore unscientific.

Concept Question 3

It has been suggested that if physical constants, such as the gravitational constant G, or properties of forces and particles were different, life in the universe might have been impossible. What possible changes do you imagine could have this effect?

The anthropic principle fits well with the idea of a multiverse (Unit 84). Only those universes well-suited to forming intelligent life evolve philosophers to comment on the fortunate conditions in that universe!

KEY POINTS

- Life probably arose less than 1 billion years after the Earth formed, possibly a few hundred million years after oceans formed.
- Some of the earliest fossils appear to be cyanobacteria and their colonies, although carbon isotope evidence is present earlier.
- Complex organisms did not arise for about 4 billion years, but they then rapidly spread and diversified.
- Life on Earth shares many chemical properties, beginning with the elements most frequently produced in the cores of stars.
- All forms of life on the planet use the same amino acids to build proteins and the DNA molecule to provide the instructions.
- The Miller-Urey experiment shows that simple molecules may form complex organic molecules in young-Earth conditions.
- The steps to the origin of life are unknown, but the first organisms may have used simpler RNA at first.
- Primitive archaea may have formed around places like deep-sea vents, with complex processes like photosynthesis coming later.
- Eukaryotic cells, with a nucleus, probably did not arise until about 2 billion years ago, and multicellular organisms evolved later.
- In some ways the whole Earth functions like an organism (Gaia hypothesis), which helps regulate the planet's environment.
- The existence of intelligent life requires many special conditions that appear coincidental, but are necessary for intelligent life to arise.

KEY TERMS

amino acid, 681
anthropic principle, 686
archaea, 684
Cambrian explosion, 680
cyanobacteria, 679
DNA, 682
eukaryote, 685
Gaia hypothesis, 686
Miller-Urey experiment, 683
prokaryote, 684
proteinoid, 683
RNA world, 684
stromatolite, 679
symbiosis, 681

CONCEPT QUESTIONS

Concept Questions on the following topics are located in the margins. They invite thinking and discussion beyond the text.

1. Origin of life despite dilute chemicals in ocean. (p. 685)
2. Response of life to runaway greenhouse effect. (p. 686)
3. Effect of physical constants on existence of life. (p. 686)

REVIEW QUESTIONS

4. What is the earliest evidence for life on Earth?
5. For how long has complex life existed on Earth?
6. What features of life suggest a common origin?
7. What is the importance of the Miller-Urey experiment?
8. How do prokaryotes and eukaryotes fit in the evolution of life?
9. What is the Gaia hypothesis?
10. What is the anthropic principle?

QUANTITATIVE PROBLEMS

11. For what fraction of Earth's lifetime has *Homo sapiens* existed?
12. A bacterium has a mass of about 10^{-15} kg. It can replicate itself in one hour. If it had an unlimited food supply, how long would it take the bacterium, doubling in number every hour, to match the mass of the Earth? (Hint: Solve for the number of hours N where 2^N gives the number of bacteria you estimate.)
13. If a DNA chain is ionized it can lead to mutations in the cell's structure, which can ultimately lead to cancers in humans. The ionization energy for DNA is roughly 8.0×10^{-19} J. What wavelength of light does this correspond to? Where in the electromagnetic spectrum does this radiation fall?
14. From 1950 to 2000, the world population increased from 2.5 billion to 6 billion. If this trend continues, what will be the world population in 2050? in 2100?

TEST YOURSELF

15. Proteins are made of chains of
    a. cyanobacteria.   c. eukaryotes.   e. amino acids.
    b. DNA.   d. archaea.
16. How did early life forms alter our planet, making it suitable for complex animals to later arise?
    a. Their mounded dead bodies gave rise to dry land.
    b. They produced oxygen, which blocked ultraviolet light.
    c. They killed off even earlier poisonous species on the planet.
    d. Their weight altered the Earth's spin axis to make seasons.
    e. They removed the high acidity of the oceans.
17. If you could travel back in time and visit Earth a billion years ago you would find
    a. no life at all.
    b. only single cell organisms.
    c. single cell and some multicellular organisms.
    d. large, complex organisms confined to the oceans.
    e. life spread throughout the oceans and across the land.
18. The Miller-Urey experiment demonstrated that
    a. very simple bacteria can be created from the chemicals present in the atmosphere of the ancient Earth.
    b. Earth's early atmosphere was too harsh for life to have formed until recently.
    c. conditions on the early Earth were suitable for the creation of many of the complex organic molecules found in living things.
    d. if simple life forms landed on the early Earth, they could have survived.
    e. early life forms were silicon-based.
19. Multiple generation of stars are needed before life can form because
    a. the early universe was too hot to support life.
    b. the first generation stars were too luminous to allow life to form.
    c. stars produce heavy elements, such as carbon and nitrogen, that are required by life.
    d. collisions between early star systems prevented stars from surviving the billions of years needed for evolution.

UNIT 86

The Search for Life Elsewhere

86.1 The Search for Life on Mars
86.2 Life on Other Planets?
86.3 Are We Alone?
86.4 SETI

Learning Objectives

Upon completing this Unit, you should be able to:
- Describe the searches for life on Mars, and other likely places for life to have formed.
- Explain the Drake equation, and discuss the chances that there are other communicating civilizations in our Galaxy.
- Define SETI, and explain how searches have been conducted.

Humans have speculated for thousands of years about life elsewhere in the universe. For example, about 300 B.C.E., the ancient Greek philosopher Epicurus wrote in his "Letter to Herodotus" that "there are infinite worlds both like and unlike this world of ours." Epicurus went on to add that "in all worlds there are living creatures and plants." In a similar vein, the Roman scholar Lucretius wrote about 50 B.C.E. that "it is in the highest degree unlikely that this Earth and sky is the only one to have been created." Such views were not universal, however. For example, Plato and Aristotle argued that the Earth was the only abode for life, a view that prevailed through most of the Middle Ages.

Today, however, influenced by science fiction movies, television, and books, many people have no difficulty at all in believing that extraterrestrial life exists. In fact, supermarket tabloids make silly claims almost weekly about encounters with aliens. If we look at the facts, however, there is not yet a shred of direct evidence that life exists elsewhere. The absence of hard evidence may seem discouraging; but as the British astronomer Sir Martin Rees has said, "Absence of evidence is not the same as evidence of absence." However, lacking direct evidence, what can we say meaningfully about life elsewhere in the universe? What can we say about the existence of other intelligent life? In this Unit, we will attempt to answer these questions, relying heavily on what we have learned in Unit 85 about the one planet we know where life has formed: the Earth.

86.1 THE SEARCH FOR LIFE ON MARS

Mars became the focus of searches for life after it was discovered in the 1700s that it has polar caps, an atmosphere, and a rotation period about the same as the Earth's (Unit 42). Belief in life on Mars was so widespread that scientists searched for ways to communicate across interplanetary space. In 1820 the German mathematician Karl Friedrich Gauss even proposed sending a mathematical message to the Martians by carving out a huge right triangle in the Siberian pine forests, planting wheat fields inside. He calculated that the triangle would be visible to the Martians and would show them that intelligent life was present on Earth.

The belief in Martian life took on a new twist based on a mistranslation. In 1877, the Italian astronomer Giovanni Schiaparelli recorded features on Mars that he called *canali*, by which he meant "channels," but this was translated as *canals*, with the implication that intelligent beings built them. Perhaps inspired by the power of suggestion, some astronomers thought they could see a complex system of interconnected canals (Figure 86.1) just at the limits of telescope resolution.

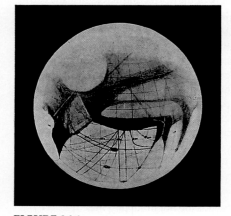

FIGURE 86.1
Drawing of Mars made by Percival Lowell around 1900. Lowell thought he could see straight-line features that he believed were canals for irrigation or travel.

Most astronomers could see no trace of the alleged canals, but they did note seasonal changes in the shape of dark regions—changes that some interpreted as the spread of plant life in the Martian spring.

Popular belief continued to fuel speculation about life on Mars into the mid-1960s, when the first of a series of *Mariner* spacecraft sent back pictures of a far more desolate world than had been previously imagined. The images of Mars made by the early *Mariner* spacecraft (Unit 42) demonstrated that the straight "canal" features were not real except for the giant Valles Marineras canyon, and that the changes in darkness were caused by widespread dust storms. When the United States landed two *Viking* spacecraft on the planet in 1976, all tests for signs of carbon chemistry in the soil or metabolic activity in soil samples were negative or ambiguous.

In 1996 a group of American and English scientists reported possible signs of fossil life in a rock from Mars. The rock was a meteorite found in Antarctica. It arrived on Earth after being blasted off the surface of Mars by the impact of a small asteroid. Such impacts are not uncommon, but most fragments are scattered in space or fall back to Mars. Detailed analysis of the composition of this and several other meteorites shows a close match with Martian rock, and they are unlike any other terrestrial or asteroidal source of rock.

Microscopic examination of samples from the interior of one such meteorite revealed tiny, rod-shaped structures (Figure 86.2). These look like ancient terrestrial bacteria but much smaller. To some scientists they look like fossilized primitive life, but others have shown that ordinary chemical weathering can form similar structures. As a result, most scientists today doubt that any meteorite yet studied shows evidence of Martian life.

The *Spirit* and *Opportunity* rovers and *Phoenix* lander have carried out a variety of measurements that have clearly established that liquid water was once present on Mars, showing some soils contained nutrients suitable for certain forms of microbial life. Often, though, the conditions appear inhospitable for life, with high concentrations of oxidizing chemicals, acids, or salts. If any life exists or once existed on Mars, it probably lies buried well below the hostile surface of the planet.

The *Curiosity* rover is currently studying a region of Mars that appears promising for establishing whether Mars was ever habitable. While it climbs through a landscape laid down over billions of years of Martian history, it is making a detailed analysis of the rocks that should help establish how the Martian environment changed. By early 2013, *Curiosity* had analyzed rock that was part of a lakebed with a chemistry that should have been favorable for life (Figure 86.3).

As *Curiosity* continues its search, the meteorite discovery raises an interesting issue. If evidence of life is ever found on Mars, we will have to consider the possibility that a bacteria-laden rock might have been blasted off Earth's surface in the remote past and reached Mars when it was more habitable. Or perhaps if life formed on Mars first, our own ancestors might have been Martians!

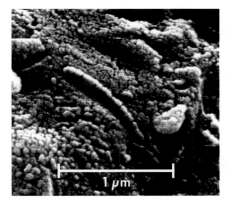

FIGURE 86.2
Fossil of ancient Martian life? The tiny elongated structure in this microscopic image of a Martian rock is about 1 μm long. It is much smaller than most bacteria, although similar to some bacterial traces found in rocks on Earth. However, most scientists now think these structures formed chemically.

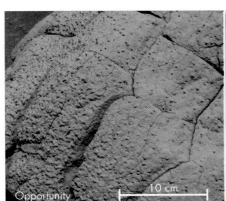 

FIGURE 86.3
Two rocks formed in water on different parts of Mars. Chemical studies of the rock found by the *Opportunity* rover (left) show it formed in highly acidic water. However, the rock found by *Curiosity* (right) suggests a much more habitable environment. The water was neutral in acidity, and the rock contains a variety of chemicals that scientists think would have been conducive to microbes.

> **Concept Question 1**
>
> Based on their chemical and geological composition, many meteorites have been traced back to several of the asteroids, the Moon, Mars, and possibly Mercury. Why do you suppose none have been identified as coming from Venus? Should other planets have Earth meteorites?

The idea that terrestrial life descended from organisms created elsewhere in the universe has been discussed for thousands of years and is known as the **panspermia** hypothesis. According to this hypothesis, simple life forms (perhaps bacteria) from some other location drifted from their place of origin across space to Earth. However, panspermia does not really simplify the problem of the origin of life: It just shifts it elsewhere. Moreover, it further requires getting the life to Earth, a perilous voyage even for a bacterium.

86.2 LIFE ON OTHER PLANETS?

Is there any evidence for life elsewhere in the universe? No, but we have so far only analyzed rock and soil specimens from a tiny portion of the universe: some half dozen spots on the Moon, six on Mars (by robot space landers), and some asteroid fragments picked up on Earth. The Moon's lack of atmosphere and water makes it so inhospitable that astronomers did not expect to find life there, nor did they. Mars seemed more promising because its environment, though harsh, is more like the Earth's than that of any other planet, and laboratory experiments indicated that some terrestrial organisms (bacteria and lichens) might survive Martian conditions, even though they did not reproduce or grow.

The limited sample of planets available to explore in the Solar System has led scientists to consider other ways to search for signs of life on other planets. One interesting method involves looking for unexpected combinations of gases in the atmosphere of a planet—gases that might be produced by living things, as is suggested by the Gaia hypothesis (Unit 85.4). For example, Earth's atmosphere contains both oxygen and methane, which would normally not be expected in combination because of their strong interaction.

Recent discoveries of Earth-sized **exoplanets** (a planet orbiting another star—Unit 36) might permit us to study their atmospheres, but this is extremely challenging for such small planets. A number of larger exoplanets with orbits that carry them in front of their star have been studied spectroscopically in search of absorption lines produced by the exoplanet's atmosphere when it is blocking part of the star's light. For example, astronomers have detected oxygen and carbon in the atmosphere in HD 209548b, a Jupiter-mass exoplanet in a tight orbit around its star. This finding is quite unlikely to signify life, however. The planet orbits so close to its star that the planet's atmosphere is boiling away. The oxygen and carbon were probably the by-products of other molecules being broken down when they were heated by the star and driven away from the planet. What we probably detected was a large cloud of gas streaming off the planet, similar to a comet's tail. Nonetheless, this first indication of oxygen in the atmosphere of another planet gives us encouragement that the next generation of space telescopes may eventually lead to the discovery of more hospitable planets.

Based on the example of the Earth, it seems that liquid water and organic molecules are likely starting points for life to arise. There are environments very different from Earth's where these ingredients are present. Jupiter's moon Europa (Unit 48) and several other satellites of the giant planets probably have oceans underneath an icy shell, possibly with as much water as the Earth's oceans (Figure 86.4). The interiors of these moons are heated in large part by tidal effects produced by the gravity of the giant planet, and this may create conditions not all that different from undersea volcanic vents on Earth (Unit 85). Astronomers envision probes that could melt their way through the outer crust by heating the ice around them and sinking in. A long wire would trail behind the probe, connected to a transmitter on the surface to send back data. Similar techniques have been used in Antarctica to place probes deep inside the polar ice cap.

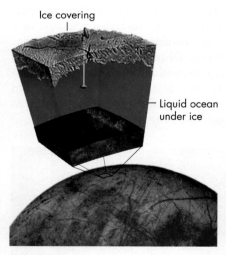

FIGURE 86.4
Beneath the icy crust of Jupiter's moon Europa, there may be an ocean of water harboring life. Astronomers are designing probes that could tunnel down to the water layer.

Perhaps there are forms of life that use a different chemistry than on Earth. It seems likely that some liquid solvent is essential to allow molecules to move into contact with each other to permit complex reactions, but this might be liquid ammonia, which is stable in cold environments, or liquid methane in even colder environments. The *Huygens* probe revealed that Saturn's moon Titan has a surprising resemblance to Earth, with river valleys and lakes produced by liquid methane, and its atmosphere contains complex organic molecules. Could life have formed in such a cold environment?

Some have even questioned whether completely different chemistries might permit life to form. Organic (carbon-based) chemistry is extremely rich because of the flexibility of the carbon atom to form bonds with one, two, three, or four neighboring atoms. The silicon atom is similarly flexible, and it too is produced in large quantities by stellar evolution. Is silicon-based life possible? So far we can only guess.

In searching for life in alien environments, we need a very general definition of what life is. This proves to be more difficult than it might seem at first. There are mechanical and natural processes that share some functional similarities to living organisms, and some aspects of organisms on Earth may not be essential for life to exist. One suggestion is that life is characterized by metabolism (collecting energy from the environment and building the molecules and structures necessary for an organism to exist) and a reproductive process, but even that may not be sufficient to identify all kinds of life.

Concept Question 2

Are viruses alive? They require other organisms' metabolism for their reproduction. Are mules alive? They are unable to reproduce.

86.3 ARE WE ALONE?

Beyond the existence of life elsewhere in the universe, what is the possibility of other intelligent life forms? Hundreds of light-years from Earth, is there perhaps a "student alien" who is right now reading a book discussing the possibility of life on planets such as Gzbhλx? Scientists do not know, and in fact they are strongly divided into two groups. One group of scientists (let us call them **many-worlders**) thinks that millions of planets with life exist in the Milky Way, and that many of these may have advanced civilizations. The other group (let us call them **loners**) argues that we are the only intelligent life in the Galaxy. We will discuss below how these two radically different points of view arise.

Many-worlders argue as follows: Earth-like planets are common. Life has formed on Earth. Therefore, it would be surprising if life did not exist elsewhere. In fact, even if only a small fraction of such planets have life, many of them have probably developed advanced civilizations. This may even vastly underestimate the number of planets inhabited by intelligent beings, because life may be able to arise in environments intolerable to terrestrial organisms.

To make this argument more quantitative, the U.S. astronomer Frank Drake showed how we can estimate the number of intelligent civilizations by multiplying together the probability of each condition necessary for them to exist. He developed a mathematical estimate of the number of such civilizations known as the **Drake equation.** We will follow a modified form of Drake's method here. We can estimate the number of planets in our Galaxy that contain intelligent life, N_I, by taking the number of suitable stars, N_*, and multiplying that by the fraction of them that have a habitable planet, f_P, and the probability of life forming on such a planet, f_L, and the probability of intelligent life forming given that life had formed, f_I. This can be written in equation form as follows:

$$N_I = N_* \times f_P \times f_L \times f_I.$$

Drake originally wrote his equation in terms of the rate of star formation in our Galaxy and estimated the number of civilizations that were currently sending out communications that we might detect.

Unfortunately, it is difficult to establish accurate values for each of the probabilities that go into this equation, but this approach helps focus our discussion. To see how

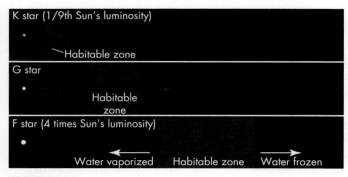

FIGURE 86.5
The habitable zone around different type stars. In the habitable zone, the star's luminosity is high enough to permit liquid water without vaporizing it.

> It is possible that around some stars there are satellites of giant planets in the life zone that are large enough to sustain an atmosphere and oceans.

such numbers are deduced, let us begin by estimating the number of Earth-like planets in the Milky Way.

We start by asking how many Sun-like stars are in the Milky Way. We limit ourselves to F-, G-, and K-type stars (Unit 56), because these are most similar to the Sun. More-luminous types live for less than a few billion years, which is unlikely to be enough time for complex life to form, based on Earth's example. On the other hand, dim red M-type stars are so cool that their planets would have to be very close to them to be warm enough for a terrestrial type of life. At such small distances, planets will almost certainly be tidally locked, with one face very hot facing the star and the other very cold. This may or may not spell trouble for a planet to maintain liquid water on the surface, but to be conservative, we will exclude them. Given this restriction on star types, we turn to census counts of stars, which reveal that these types of stars make up about 20% of our Galaxy's hundred billion stars. If we assume that half of these are in binary star systems (Unit 57) and do not have stable conditions for orbiting planets, there are still about ten billion candidate stars: $N_* = 10^{10}$.

Of these 10^{10} stars, we next ask, How many of them have planets with Earth-like environments? There is a relatively narrow range of distances, sometimes called the "habitable zone," where the luminosity of the star will permit liquid water to persist on a planet's surface (Figure 86.5). In the Solar System, the Earth is in the habitable zone, and it appears Mars is close to it. As illustrated in Figure 86.5, if each planet is about twice as far out as the planet interior to it, which is roughly true of the Solar System, there is almost always one planet in the habitable zone. We saw in Unit 36 that exoplanets are fairly common, although the planetary systems detected so far are mostly made of giant planets. These are often much closer to their star than the giant planets in the Solar System. Techniques are not yet sensitive enough to detect Earth-mass and smaller planets, so there remain considerable uncertainties, but it seems conservative to estimate that 1 in 10 stars has a habitable planet orbiting them, so $f_P = 0.1$.

How many of these habitable planets might actually have life on them? That depends on how easy it is for life to form, a probability that the Miller-Urey experiment (Unit 85) suggests may be high. The difficulty in assessing the probability of life forming is that we have only the single case of life here on Earth as our basis. Many-worlders point to the rapidity with which life developed on Earth. They also argue that the success of the Miller-Urey experiment in making molecular precursors to life in just a few days indicates that the chance of life starting within billions of years is fairly high: 1 in 100, say, so that $f_L = 0.01$.

We now go to the next step and ask, On how many of these life-bearing planets do intelligent beings arise? There is no certain way to know whether the probability of more-complex life evolving is high or low. Moreover, even if life succeeds in starting, perhaps it will be annihilated if its star dies before intelligence develops. One intriguing idea is that natural disasters, such as asteroid impacts, provide opportunities for new species to develop *until* intelligence arises. After all, if an asteroid had not killed the dinosaurs (Unit 50), we might not have gotten a chance to evolve. Now, thanks to human intelligence, we are developing technologies that could protect us from an asteroid collision, and we may potentially be able to colonize other planets in the future. Thus, global disasters might provide a means of natural selection for intelligence. Suppose we are conservative and say that if life forms, it has a 1 in 1000 chance of developing an intelligent, technological civilization. This sets our final factor, the fraction of life-bearing planets where intelligence arises, to $f_I = 0.001$.

The many-worlder carries out this calculation and estimates that the number of intelligent civilizations in our Galaxy is approximately

$$N_\text{I} = 10^{10} \times 0.1 \times 0.01 \times 0.001 = 10{,}000.$$

Therefore, with conservative estimates (from a many-worlder's perspective), there should be many intelligent civilizations within our Galaxy with whom we might communicate.

By contrast, the loners suggest we are the only advanced life in the Milky Way. The loner argument was first proposed by the Italian physicist Enrico Fermi and is sometimes called the **Fermi paradox.** Fermi argued that, based on the rapidity with which our species has developed technology, it is just a matter of time before we spread across the Galaxy, or at least fill it with the radio waves of our own transmissions. Even if it takes a technological civilization thousands of years to travel to other stars, once they have mastered space flight they will need at most a few million years to colonize the entire Galaxy. A few million years is a brief time compared to the billions of years since such civilizations could have arisen, so evidence of their presence should be easy to find if they existed.

Accordingly, loners argue that because no other civilization has been seen, no other technological civilization has yet arisen. Therefore, they argue, we are probably alone in the Milky Way, or nearly alone with at most only a handful of other civilizations. However, this argument, like that of the many-worlders, rests on a number of assumptions that are difficult or impossible to substantiate. For example, it assumes that (1) civilizations are driven to colonize, (2) civilizations seek rather than avoid contact, and (3) it is possible to master interstellar travel.

> **Concept Question 3**
>
> Suppose astronomers found clear evidence of civilization on another planet. What effect do you think this would have on our society? What effect would it have on you?

The loners' argument is not based solely on our failure to have already found such a civilization, though. They point out that the various probabilities for life forming may be significantly lower than the many-worlders suggest. The British astrophysicist Fred Hoyle compared the likelihood of life arising spontaneously to the likelihood of a box containing several hundred tons of aluminum being shaken and by chance assembling itself into a 747 jet. Although many scientists find such a comparison silly, it illustrates the divergence of opinion about the likelihood of life forming in the universe.

It is possible that the Earth and Solar System have some unusual attributes that have permitted life here, but not elsewhere. Perhaps a large moon is critical for stabilizing a planet's axis. Or maybe a migrating outer planet is necessary to perturb the orbits of water-bearing comets, thereby delivering a shower of water after planets have cooled. In addition, among the exoplanetary systems so far detected (Unit 36), it is clear that the nearly circular, regularly spaced orbits of the planets in the Solar System are unusual. What if such orbits are critical for creating stable conditions for the formation of life? It may be that the fraction of planets that are suitable for life f_P is much smaller than the many-worlders suggest.

FIGURE 86.6
A possible Galactic habitable zone. Tightly crowded stars in the inner part of the Galaxy may cause catastrophic events that prevent life from forming or cause planetary disasters—radiation from supernovae, gravitational encounters, impacts with comets, and the like. In the outer Galaxy, there may be too few heavy elements from earlier generations of stars to form planets and life.

Perhaps a star's location in the Milky Way is critical for the survival of life, so there might be a Galactic habitable zone (Figure 86.6). This is based on the idea that because the Sun is fairly far from the Galactic center, where most of the stars are concentrated, the Earth has not been exposed to as much radiation from nearby nova and supernova explosions. When stars are crowded closer to each other, there might be

> **Concept Question 4**
>
> What if a civilization exhausted its supply of fossil fuels and had to live on limited energy supplies? How important has an abundant energy resource been for human development of technologies?

more frequent disturbances in the outer Solar System, perhaps causing many catastrophic impacts on the planets. On the other hand, the amount of star formation in the outer Galaxy is much lower, so there may be too few heavy elements for terrestrial planets and life to form. Some have argued that the Sun might be in a relatively limited part of the Galaxy where life could form, but others dispute this.

Earth also has a much stronger magnetic field than either Mars or Venus, and maybe that also has been critical for protecting life from high-energy particles. Maybe the great majority of exoplanets are sterile because of the radiation they receive, and therefore the fraction where life forms f_L is smaller too.

All in all, there could be enough factors such as these to greatly reduce the probability of life forming. Supposing f_P and f_L were each just 1% of what the many-worlders estimate, we would find

$$N_I = 10^{10} \times 0.001 \times 0.0001 \times 0.001 = 1.$$

By the loners' calculation, there is just one planet with intelligent life in the Galaxy—the Earth.

86.4 SETI

On the chance that there may be technologically capable life elsewhere in our Galaxy, some astronomers are searching for radio signals from other civilizations. Such signals might be either deliberate broadcasts sent to us or communications directed elsewhere that we could simply overhear. On Earth, radio transmissions have been generated for about a century, so our signals might be picked up by an alien civilization on another planet if it is within 100 light-years of us.

We can estimate how far away the next nearest civilization might be by assuming that the N_I civilizations we predict to be present are spread evenly throughout the disk of the Milky Way, as depicted in Figure 86.7. We can calculate the average distance, d, separating them as follows: Imagine drawing a sphere of radius $d/2$ around each civilized world. Then we ask what the size of the average sphere would have to be for N_I of them to cover the Milky Way's disk. If we make the approximation that the Milky Way's disk is a circle of radius R_{MW}, then its area is given by the formula $\pi(R_{MW})^2$. So we set the area covered by the N_I spheres (each of radius $d/2$) equal to the area of the Galaxy's disk:

$$N_I \times \pi \left(\frac{d}{2}\right)^2 = \pi(R_{MW})^2.$$

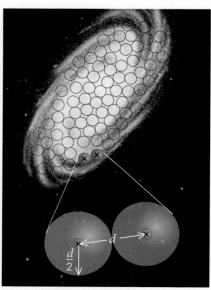

FIGURE 86.7
A sketch illustrating how to estimate the distance between potential civilizations in the Milky Way, if there are N_I civilizations spread evenly within our Galaxy.

We can cancel π on both sides of the equation, multiply both sides by $(2^2/N_I)$, and take the square root of both sides to give

$$d = \frac{2R_{MW}}{\sqrt{N_I}}.$$

The Milky Way's radius R_{MW} is approximately 40,000 light-years, and let us suppose that the number of intelligent civilizations N_I is 10,000, the value deduced from the many-worlder argument. These choices give a value for d of approximately 800 light-years. Therefore, even if 10,000 civilizations exist in our Galaxy, they are all likely to be very far from us, and none has likely yet seen any of the transmissions we have generated.

Other civilizations may have begun transmitting a much longer time ago than we did, so there may be detectable signals passing by Earth right now. The electromagnetic spectrum is very wide, and potential communications signals very narrow, so to have any hope of success, astronomers need to choose the most likely wavelengths to monitor in the *Search for Extra-Terrestrial Intelligence* (**SETI**). For

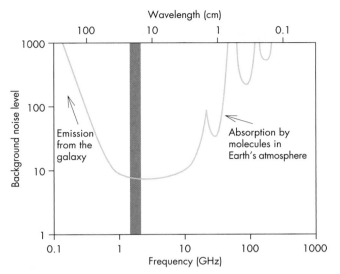

FIGURE 86.8
The *waterhole* is a range of wavelengths between prominent spectral lines of H and OH at 21 and 18 centimeters. It falls in a clear region of the radio spectrum that is free of interference from natural sources of radio emission and is not blocked by absorption in an Earth-like atmosphere.

FIGURE 86.9
When completed, the Allen Telescope Array, being constructed in California, will be able to simultaneously carry out astronomical surveys while it conducts a deep, targeted search for signals from other civilizations. The array will be made up of hundreds of small radio telescopes, using interferometry to simultaneously examine large areas of the sky over a wide part of the radio spectrum.

> Project Ozma was named for the fictional princess in the series of stories of the *Land of Oz*. The princess supposedly sent radio reports about Oz to the author.

example, at very long wavelengths, interstellar gas in the Galaxy emits strongly, overwhelming all but the most powerful signals. On the other hand, at very short wavelengths the molecules in our atmosphere block such signals.

It turns out that the optimum wavelengths for detection are between about 3 and 30 centimeters (Figure 86.8). This clear window for radio signals spans a range that includes the 21-centimeter line emitted by neutral hydrogen and the 18-centimeter line emitted by the hydroxyl (OH) molecule in the cold gas clouds of our Galaxy. These two wavelengths are distinctive, and any other technologically capable civilization would probably have discovered their significance, so they would have equipment that could receive or broadcast at these wavelengths. Therefore, many searches have focused on signals in the range of 18 to 21 centimeters. This wavelength range is sometimes called the **waterhole** because it is bracketed by the spectral lines of H and OH, which combine to make H_2O, an important molecule for life.

Astronomers began listening for extraterrestrial radio signals in 1960 with Project Ozma, which monitored several nearby star systems at radio wavelengths, hoping to detect signals other than those produced naturally. The search continued in more recent years with some targeted searches of nearby stars and other projects designed to "ride piggyback" on the major radio telescopes around the world—collecting radio signals from whatever direction is being studied by astronomers for other experiments. A major new dedicated facility (Figure 86.9) is presently being built, mostly with private contributions. When completed, this array of radio telescopes will expand the SETI search capabilities dramatically.

Searching for extraterrestrial signals is not easy, because even if they exist, they are probably very weak and therefore buried in cosmic static. Improving radio technologies now allow SETI projects to monitor billions of narrow radio wavelength ranges, covering most of the wavelength range where signals from other civilizations might be detected. If any transmissions are received, they will almost certainly be so weak that they will require detailed analysis to be identified. The SETI projects therefore require huge amounts of computer time to search within

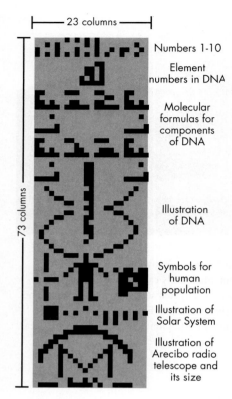

FIGURE 86.10
Message sent from Arecibo in 1974. The message communicates several fundamental ideas through this simple image sent as a series of pulses at different frequencies to represent the "pixels" of the image. To help an alien decipher the message, it was broadcast with 23 columns and 73 rows—both prime numbers.

Concept Question 5

What message would *you* send to an alien civilization? What dangers might sending a message pose for us on Earth?

the collected data for potential signals. To achieve the computer processing power needed to analyze these data, one project, called SETI@home (http://setiathome.berkeley.edu), has developed a system to transmit data from the central search telescope to millions of volunteers who allow their home computers to process the information when the computers are not otherwise in use.

Despite SETI's many clever techniques to utilize existing telescopes and computers, no signals have yet been detected. SETI researchers point out, however, that the search to date has been limited to relatively nearby stars. With more sensitive telescopes under construction, in the future it will be possible to search out to greater distances, increasing the chances of detecting other civilizations.

Even supposing many other civilizations have arisen in our Galaxy, there are other possible reasons why SETI has not yet detected any signals. The other civilizations may not be transmitting radio signals if they never developed radio communications technology, or they may rely on more carefully targeted transmission systems, such as fiber optics or laser communications. If other civilizations are like ours, however, they would inadvertently broadcast a wide range of signals. Our own radio and television stations and airport radars transmit extremely powerful signals that would be detectable, with sensitive equipment, over most of our Galaxy.

There is a more disturbing possibility. The number of worlds with the ability to communicate might be reduced if intelligence does not have the survival value that we would like to believe it has. Technology also gives rise to a host of problems that are potentially dangerous to the species that invents them. Pollution, nuclear war, or other self-inflicted disasters may cause civilizations to destroy themselves.

If technologically capable civilizations do not survive, we need to multiply N_I by another factor. We might call this the survival factor, f_S, which is the fraction of the star's lifetime that a communicating civilization lasts. After all, we have survived as an intelligent species capable of communicating with other intelligent species in our Galaxy for only about a century, that is only 0.000002% (2×10^{-8}) of the life of the Sun so far. If our descendants survive for the rest of the life of the Sun, then this fraction will rise to more than 50%. However, if most civilizations' ability to communicate typically ends after 10,000 years, then f_S might be as little as 10^{-6}. Such a small survival factor would imply that more than a million civilizations would have to have arisen for there to be much chance that another is capable of communicating at the same time as us. For example, if a civilization existed in the nearby Alpha Centauri system, but it somehow destroyed itself even as recently as 100 years ago, we could not have communicated with them by radio, and the chance has now been lost forever.

The immense distances that almost certainly separate us from any other civilizations also make it unlikely that we will be able to exchange information with them. To communicate with another civilization within a human life span, we would need to be within about 30 light-years from them. At that distance, if you send a message when you are 10 years old, and they reply immediately upon receiving it, you will get a response when you are 70! For it to be likely that another civilization is that close, our earlier calculation would have to have given us the result that there are millions of other civilizations in our Galaxy.

Humans *have* sent at least one intentional message to any aliens who might be listening. During a ceremony in 1974 celebrating the renovation of the huge Arecibo radio telescope in Puerto Rico (Unit 29), astronomers broadcast a signal toward the globular cluster M13 (Figure 86.10). This cluster is more than 20,000 light-years distant, so the message of our existence is still just beginning its voyage to the stars.

KEY POINTS

- Belief in life on Mars has led some to think they found signs of it, but the evidence has so far proved unfounded.
- Conditions on the young Mars were similar enough to Earth, but no microscopic life has been found in tests of surface soils.
- It seems possible that bacteria could be transferred from planet to planet by meteorites, but probably not more complex organisms.
- An inhabited planet may be identifiable if a spectrum shows combinations of gases that would be unstable without life.
- Life may have formed in very different environments, like water oceans beneath the icy crust of the moons of giant planets.
- The Drake equation attempts to estimate whether intelligent life exists by determining the probability of each step it requires.
- Because of the uncertainty in the probability for some of the steps toward intelligent life, conclusions diverge enormously.
- SETI searches for radio signals like those we generate through our technologies, but from distant stars the signals will be very weak.

KEY TERMS

Drake equation, 691
exoplanet, 690
Fermi paradox, 693
loner, 691
many-worlder, 691
panspermia, 690
SETI, 694
waterhole, 695

CONCEPT QUESTIONS

Concept Questions on the following topics are located in the margins. They invite thinking and discussion beyond the text.

1. Lack of meteorites from Venus. (p. 690)
2. Viruses, mules, and definitions of life. (p. 691)
3. Effect of discovery of other civilizations. (p. 693)
4. Role of carbon fuels in technology. (p. 694)
5. What message to send to aliens. (p. 696)

REVIEW QUESTIONS

6. What is the panspermia theory?
7. What is the significance of the meteorites from Mars?
8. How might we detect a life-sustaining planet?
9. What is the Drake equation?
10. What is Fermi's paradox?
11. What is SETI?
12. Why are radio wavelengths near 20 cm considered a likely place to find interstellar radio communications?

QUANTITATIVE PROBLEMS

13. How long would it take to get to the nearest star at the space shuttle's speed of 30,000 kph?
14. Use a method similar to the Drake equation to estimate the number of people getting a haircut at some given moment. That is, identify the different factors involved in deciding to get a haircut, estimate the probabilities of each, then multiply the probabilities times the number of people in, for example, the United States (approximately 260 million).
15. Appendix Table 10 lists the nearest stars to Earth. What fraction of these stars have a spectral type that may be suitable for life based on the criterion stated in the text? If we use this fraction instead of the 20% used in the example application of the Drake equation, how would the estimated number of intelligent civilizations change?
16. Make your own estimate of the number of habitable planets in the Milky Way. What is the average distance between them?
17. *Voyager 1* is currently leaving the Solar System at a speed of 17 km/sec. Assuming a constant speed, how long would it take *Voyager 1* to travel 4.24 ly, the distance to the nearest star?
18. Suppose that we are the only civilization in the Galaxy, and all the other civilizations have died out. For a possible 10,000 civilizations calculated by the Drake equation and a Galaxy with an age of 10 billion years, what is the typical life span of a civilization? How old is our civilization?
19. One solution suggested for our growing population in the far future is to construct a sphere surrounding the Sun with a radius of 1 AU (Earth's orbital distance). The amount of light received per square meter on the inside of this sphere would be similar to what we receive on Earth now. What is the surface area of such a sphere compared to the surface of the Earth?

TEST YOURSELF

20. The Drake equation attempts to determine the
    a. conditions under which life originated on Earth.
    b. best wavelength for communicating with extraterrestrials.
    c. age of life on Earth.
    d. number of other technically advanced civilizations.
    e. lifetime of our own civilization.
21. It may be difficult for life to form on planets around low mass stars because
    a. the planets would be too cold to have liquid water on their surfaces.
    b. the habitable zone is so close to the star that habitable planets would be tidally locked.
    c. the gravitational pull from such a low mass star is not strong enough to hold a planet in orbit.
    d. these stars do not radiate enough ultraviolet light needed for breaking down organic molecules.
22. Current searches in the "waterhole" for signals from intelligent life elsewhere in the Galaxy are made in which electromagnetic band?
    a. X-rays c. Visible e. Infrared
    b. Ultraviolet d. Radio
23. The panspermia hypothesis proposes that life formed
    a. elsewhere in the universe and then traveled to Earth.
    b. almost inevitably because of the early conditions on Earth.
    c. only on Earth because the necessary conditions are so rare.
    d. in very many locations in the Galaxy.

Appendix

SCIENTIFIC NOTATION

Scientific notation is a shorthand method for expressing and working with very large or very small numbers. We described this method in Unit 3.2 and discuss it a little further here. We can express any number as a few digits times 10 to a power, or exponent. The power indicates the number of times that 10 is multiplied by itself.

For example, $100 = 10 \times 10 = 10^2$. Similarly, $1{,}000{,}000 = 10 \times 10 \times 10 \times 10 \times 10 \times 10 = 10^6$. Note that we do not always need to write out $10 \times \ldots$. Instead we can simply count the zeros. Thus 10,000 is 1 followed by four zeros, so it is 10^4. To express the numbers 1 and 10 in scientific notation: $1 = 10^0$ and $10 = 10^1$.

To write a number like 300, we break it into two parts: $3 \times 100 = 3 \times 10^2$. Similarly, we can write $352 = 3.52 \times 100 = 3.52 \times 10^2$. Any number can be expressed in scientific notation as a value between 1 and 10 multiplied by 10 to a power.

We can also write very small numbers (numbers less than 1) using powers of 10. For example, $0.01 = 1/100 = 1 \times 10^{-2}$. We can make this even more concise, however, by writing 1×10^{-2} as 10^{-2}. Similarly, $0.0001 = 10^{-4}$. Note that for numbers less than 1, the power is 1 more than the number of zeros after the decimal point.

We can write a number like 0.00052 as $5.2 \times 0.0001 = 5.2 \times 10^{-4}$.

Suppose we want to multiply numbers expressed in powers of 10. The rule is simple: We add the powers. Thus $10^3 \times 10^2 = 10^{3+2} = 10^5$. Similarly, $2 \times 10^8 \times 3 \times 10^7 = 2 \times 3 \times 10^8 \times 10^7 = 6 \times 10^{15}$. In general, $10^x \times 10^y = 10^{x+y}$.

Division works similarly, except that we subtract the exponents. Thus $10^5/10^3 = 10^{5-3} = 10^2$. In general, $10^x/10^y = 10^{x-y}$.

The last operations we need to consider are raising a number to a power and taking a root. In raising a power-of-ten number to a power, we multiply the powers. Thus "one thousand to the fourth power" is $(10^3)^4 = 10^{3 \times 4} = 10^{12}$. Care must be used if we have a number like $(2 \times 10^4)^3$. Both the 2 and the 10^4 are raised to the third power, so the result is $2^3 \times (10^4)^3 = 8 \times 10^{4 \times 3} = 8 \times 10^{12}$.

Taking a root is equivalent to raising a number to a fractional power. Thus the square root of a number is the number to the ½ power. The cube root is the number to the ⅓ power, and so forth. For example, $\sqrt{100} = 100^{1/2} = (10^2)^{1/2} = 10^1 = 10$.

SOLVING DISTANCE, VELOCITY, TIME (d, V, t) PROBLEMS

Many problems in this book (and in science in general) involve the motion of something. In such problems, we often know two of the three quantities distance, velocity, and time (d, V, t), and we want to know the third. For example, we have something moving at a speed V and want to know how far it will travel in a time t. Or we know that something travels with a speed V and want to find out how long it takes for the object to travel a distance d. We can usually solve such problems in our heads if the motion involves automobiles. For example, if it is 160 kilometers to a city and we travel at 80 kilometers per hour, how long does it take to get there? Or how far can we drive in 2 hours if we are traveling at 60 miles per hour? Because we solve such problems routinely, you might find it easier to think of astronomical (d, V, t) problems in terms of cars.

Regardless of your approach, the method of solution is simple. Begin by making a sketch of what is happening. Draw an arrow to indicate the motion. Label the known quantities and put question marks beside the things you want to find. Then write out the basic relation $d = V \times t$. If you want to find d and know V and t, just multiply them for the answer. If you want to find the time and are given V and d, solve for t by dividing both sides by V to get $t = d/V$. If you want the velocity, divide d by t: $V = d/t$.

In some problems the motion may be in a circle of radius r. In that case the distance traveled will be related to the circumference of the circle, $2\pi r$. For such cases you may need to use the expression $V = 2\pi r/t$.

In most problems you will find that it is helpful to write the units of the quantities in the equation. For example, suppose you are asked how long it takes to travel 1500 km at a velocity of 30 kilometers per second. Insert the quantities so that

$$t = \frac{d}{V} = \frac{1500 \text{ km}}{30 \text{ km/sec}} = \frac{1500 \text{ km}}{30 \text{ km}} \text{ sec} = 50 \text{ sec}.$$

Note that the units of kilometers cancel out and leave us with units of seconds, as the problem requires.

APPENDIX TABLE 1 Physical and Astronomical Constants

Physical Constants

Velocity of light (c)	$= 2.99792458 \times 10^8$ m/sec
Gravitational constant (G)	$= 6.67259 \times 10^{-11}$ m$^3 \cdot$kg$^{-1} \cdot$s^{-2}
Planck's constant (h)	$= 6.62608 \times 10^{-34}$ joule $\cdot$ sec
Mass of hydrogen atom (M_H)	$= 1.6735 \times 10^{-27}$ kg
Mass of electron (M_e)	$= 9.1094 \times 10^{-31}$ kg
Stefan-Boltzmann constant (σ)	$= 5.6705 \times 10^{-8}$ watts$\cdot$m$^{-2}\cdot$deg^{-4}
Wien's law constant ($T \times \lambda_{max}$)	$= 2.90 \times 10^6$ K$\cdot$nm

Astronomical Constants

Astronomical unit (AU)	$= 1.495978706 \times 10^{11}$ m
Light-year (ly)	$= 9.4605 \times 10^{15}$ m $= 63,240$ AU
Parsec (pc) $= 3.26$ ly	$= 3.085678 \times 10^{16}$ m $= 206,265$ AU
Year (synodic or "tropical")	$= 365.2422$ days $= 3.1557 \times 10^7$ sec
Mass of Sun ($M_\odot$)	$= 1.989 \times 10^{30}$ kg
Radius of Sun ($R_\odot$)	$= 6.96 \times 10^8$ m
Luminosity of Sun ($L_\odot$)	$= 3.83 \times 10^{26}$ watts
Hubble's constant (H)	≈ 70 km/sec per Mpc

APPENDIX TABLE 2 Metric Prefixes

nano (n)	$= 10^{-9}$	$= 1$ billionth
micro (μ)	$= 10^{-6}$	$= 1$ millionth
milli (m)	$= 10^{-3}$	$= 1$ thousandth
centi (c)	$= 10^{-2}$	$= 1$ hundredth
kilo (k)	$= 10^3$	$= 1$ thousand
mega (M)	$= 10^6$	$= 1$ million
giga (G)	$= 10^9$	$= 1$ billion

APPENDIX TABLE 3 Conversion Between English and Metric Units

Length

1 km	$= 1$ kilometer	$= 1000$ meters	$= 0.6214$ mile
1 m	$= 1$ meter	$= 1.094$ yards	$= 39.37$ inches
1 cm	$= 1$ centimeter	$= 0.01$ meter	$= 0.3937$ inch
1 μm	$= 1$ micrometer	$= 10^{-6}$ meter	$= 3.93 \times 10^{-5}$ inch
1 nm	$= 1$ nanometer	$= 10^{-9}$ meter	$= 3.93 \times 10^{-8}$ inch
1 mile	$= 1.6093$ km	1 inch	$= 2.5400$ cm

Mass

1 metric ton	$= 10^6$ grams	$= 1000$ kg	$= 2.2046 \times 10^3$ lb
1 kg	$= 1000$ grams	$= 2.2046$ lb	
1 g	$= 1$ gram	$= 0.0022046$ lb	$= 0.0353$ oz
1 lb	$= 0.4536$ kg	1 oz	$= 28.3495$ g

APPENDIX TABLE 4 Some Useful Formulas

Geometry

Circumference of circle $= 2\pi R$

Area of circle $= \pi R^2$

Surface area of sphere $= 4\pi R^2$

Volume of sphere $= \frac{4}{3}\pi R^3$

Distance Relationships

Distance–Velocity–Time: $d = V \times t$

Linear size–Angular size: $\ell = d \times \alpha \cdot 57.3°$

Distance from parallax: d(in parsecs) $= 1 \cdot p$(in arcsec)

Hubble's law (for distant galaxies): $d = V \cdot H$

Gravity

Kepler's 3rd law—orbits around Sun with semimajor axis a (in AU) and period P (in years): $P^2 = a^3$

Gravitational force between masses M and m: $F_G = G\frac{M \times m}{d^2}$

Gravitational potential energy: $E_G = -G\frac{M \times m}{d}$

Newton's modified form of Kepler's 3rd law for the total mass of two orbiting bodies: $M = \frac{4\pi^2}{G} \times \frac{d^3}{P^2}$

Mass of object for orbital speed V at distance d: $M = \frac{d \times V^2}{G}$

Escape velocity from a mass M at radius R: $V_{esc} = \sqrt{\frac{2GM}{R}}$

Light

Frequency (ν)–Wavelength (λ) relation: $\lambda \times \nu = c$

Energy of a photon: $E = h \times \nu = \frac{h \times c}{\lambda}$

Stefan-Boltzmann law—luminosity L of thermal source at temperature T: $L = \sigma T^4 \times$ (Surface area)

Wien's law—temperature of thermal source from wavelength of maximum emission: $T = \frac{2.9 \times 10^6 \text{ K} \cdot \text{nm}}{\lambda_{max}}$

Brightness (B)–Luminosity (L) relation: $B = \frac{L}{4\pi d^2}$

Doppler effect: Radial velocity $= V_R = c \times \frac{\Delta\lambda}{\lambda}$

Other Physical Relationships

Density $= \frac{\text{Mass}}{\text{Volume}}$: $\rho = \frac{M}{V}$

Newton's 2nd law—acceleration a produced by force F on mass m: $a = F \cdot m$

Kinetic energy: $E_K = \frac{1}{2} m \times V^2$

Conservation of angular momentum:
(Mass) $\times$ (Circular velocity) $\times$ (Radius) $=$ Constant

Lorentz factor for special relativistic contraction at speed V:
$\gamma = \frac{1}{\sqrt{1 - V^2/c^2}}$

APPENDIX TABLE 5 Physical Properties of the Planets

Name	Equatorial Radius (Earth units)	Equatorial Radius (km)	Mass (Earth units)	Mass (kg)	Average Density (kg/L or g/cm^3)
Mercury	0.383	2440	0.055	3.30×10^{23}	5.43
Venus	0.949	6052	0.815	4.87×10^{24}	5.24
Earth	1.00	6378	1.00	5.97×10^{24}	5.52
Mars	0.532	3396	0.107	6.42×10^{23}	3.94
Ceres (dwarf planet)	0.076	487	0.00016	9.43×10^{20}	2.08
Jupiter	11.21	71,492	317.9	1.90×10^{27}	1.33
Saturn	9.45	60,268	95.16	5.68×10^{26}	0.69
Uranus	4.01	25,559	14.54	8.68×10^{25}	1.27
Neptune	3.88	24,764	17.15	1.02×10^{26}	1.64
Pluto (dwarf planet)	0.181	1160	0.0021	1.31×10^{22}	2.04
Haumea (dwarf planet)	0.11	~700*	0.00067	4.01×10^{21}	~2.8
Makemake (dwarf planet)	0.12	~750	0.0005	$\sim 3 \times 10^{21}$	~2
Eris (dwarf planet)	0.182	1163	0.0028	1.67×10^{22}	2.52

* Haumea is highly elongated, with its long axis about twice its shortest axis.

APPENDIX TABLE 6 Orbital Properties of the Planets

Name	Distance from Sun* (AU)	Distance from Sun* (10^6 km)	Period (Years)	Period (Days)	Inclination of Orbit†	Eccentricity of Orbit‡
Mercury	0.387	57.9	0.2409	87.97	7.00	0.206
Venus	0.723	108.2	0.6152	224.70	3.39	0.007
Earth	1.00	149.6	1.0	365.26	0.00	0.017
Mars	1.524	227.9	1.8809	686.98	1.85	0.093
Ceres (dwarf planet)	2.766	414.7	4.5990	1679.82	10.59	0.080
Jupiter	5.204	778.6	11.8622	4332.59	1.31	0.048
Saturn	9.582	1433.5	29.4577	10,759.22	2.49	0.056
Uranus	19.20	2872.5	84.011	30,685.4	0.77	0.046
Neptune	30.05	4495.1	164.79	60,189	1.77	0.010
Pluto (dwarf planet)	39.48	5906	247.68	90,465	17.15	0.248
Haumea (dwarf planet)	43.1	7710	283.3	103,500	28.2	0.195
Makemake (dwarf planet)	53.1	7940	309.9	113,200	29.0	0.159
Eris (dwarf planet)	67.7	10,120	557.4	203,600	44.2	0.442

* Semimajor axis of the orbit.
† With respect to the ecliptic.
‡ Eccentricity ranges from 0 for a circular orbit to a maximum of 1 for the most extremely elongated orbit.

APPENDIX TABLE 7 Larger Satellites of the Planets*

Primary / Satellite	Radius[†] (km)	Distance from Planet (10^3 km)	Orbital Period[‡] (days)	Mass[§] (10^{15} kg)	Density (g/cm^3)
Earth					
Moon	1738	384.4	27.322	73,490,000	3.34
Mars					
Phobos	13 × 11 × 9	9.38	0.319	11	2.2
Deimos	8 × 6 × 5	23.5	1.263	1.8	1.7
Jupiter					
Metis	30 × 20 × 17	127.69	0.295	~55	
Adrastea	13 × 10 × 8	128.98	0.298	~6	
Amalthea	131 × 73 × 67	181	0.498	2080	0.86
Thebe	58 × 49 × 42	222	0.6745	~650	
Io	1822	422	1.769	89,320,000	3.53
Europa	1561	671	3.551	48,000,000	3.01
Ganymede	2631	1070	7.155	148,200,000	1.94
Callisto	2410	1883	16.689	107,600,000	1.83
Leda	5	11,188	241.8	~1	
Himalia	85	11,452	250.4	~3300	
Lysithea	12	11,740	259.9	~9	
Elara	40	11,778	261.1	~350	
Ananke	10	21,450	642R	~5	
Carme	15	23,200	692R	~18	
Pasiphae	18	23,600	742R	~32	
Sinope	14	24,100	762R	~15	
Saturn					
Pan	17 × 16 × 10	133.58	0.574	5	0.42
Atlas	20 × 18 × 9	137.67	0.602	7	0.46
Prometheus	68 × 40 × 30	139.35	0.613	160	0.48
Pandora	52 × 41 × 32	141.7	0.629	140	0.49
Epimetheus	65 × 57 × 53	151.42	0.694	530	0.64
Janus	102 × 93 × 76	151.47	0.695	1900	0.63
Mimas	199	185.52	0.942	37,900	1.15
Enceladus	252	238.02	1.37	108,000	1.61
Tethys	531	294.66	1.888	618,000	0.99
Calypso	15 × 12 × 7	294.66	1.888	4	1
Telesto	16 × 12 × 10	294.66	1.888	7	1
Dione	561	377.4	2.737	1,100,000	1.48
Helene	22 × 19 × 13	377.4	2.737	30	1.3
Rhea	764	527.04	4.518	2,306,500	1.24
Titan	2575	1221.8	15.945	134,550,000	1.88
Hyperion	180 × 133 × 102	1481.1	21.277	5600	0.55
Iapetus	734	3560.8	79.322	1,810,000	1.09
Kiviuq	~7	11,294	448.2	~2	
Ijiraq	~5	11,355	451.8	~1	1.64
Phoebe	107	12,870	545.1R	8300	
Paaliaq	~10	15,103	693	~5	
Albiorix	~13	16,227	774.6	~12	
Siarnaq	~16	17,777	884.9	~22	
Tarvos	~7	18,563	944.2	~2	
Ymir	~9	22,430	1254R	~4	
Uranus					
Cordelia	~20	49.75	0.336	~44	
Ophelia	~21	53.77	0.377	~51	
Bianca	~26	59.16	0.435	~96	
Cressida	40	61.77	0.465	~350	
Desdemona	~32	62.65	0.476	~180	
Juliet	~47	64.63	0.494	~570	

(continued)

APPENDIX TABLE 7 (continued)

Primary / Satellite	Radius[†] (km)	Distance from Planet (10^3 km)	Orbital Period[‡] (days)	Mass[§] (10^{15} kg)	Density (g/cm³)
Portia	68	66.1	0.515	~1700	
Rosalind	~36	69.93	0.56	~250	
Cupid	~9	74.8	0.618	~4	
Belinda	40	75.25	0.624	~350	
Perdita	~10	76.42	0.638	~5	
Puck	81	86.01	0.764	~2900	
Mab	~12	97.73	0.923	~9	
Miranda	236	129.4	1.413	66,000	1.2
Ariel	579	191.2	2.520	1,350,000	1.7
Umbriel	585	266.3	4.144	1,170,000	1.4
Titania	789	435.8	8.706	3,520,000	1.7
Oberon	761	582.6	13.463	3,010,000	1.6
Francisco	~11	48.2	0.296	~7	
Caliban	~36	7231	579.7R	~250	
Stephano	~16	20,901	2805.5R	~22	
Trinculo	~9	20,901	2805.5R	~4	
Sycorax	~75	12,179	1288.3R	~2300	
Margaret	~10	14,345	1687	~5	
Prospero	~25	16,256	1978.3R	~85	
Setebos	~24	17,418	2225.2R	~75	
Ferdinand	~10	20,901	2805.5R	~5	
Neptune					
Naiad	48 × 30 × 26	48.23	0.294	~200	
Thalassa	54 × 50 × 26	50.08	0.311	~380	
Despina	90 × 74 × 64	52.53	0.335	~2300	
Galatea	102 × 92 × 72	61.95	0.429	~3700	
Larissa	104 × 102 × 84	73.55	0.555	~4900	
S/2004N1	~9	105,283	0.94	~4	
Proteus	220 × 208 × 202	117.65	1.122	~50,000	
Triton	1353.4	354.76	5.877R	21,400,000	2.1
Nereid	170	5513.4	360.14	~27,000	
Halimede	~30	15,730	1879.7R	~150	
Sao	~20	22,420	2914.1	~40	
Laomedeia	~20	23,570	3167.9	~40	
Psamathe	~30	46,700	9115.9R	~150	
Neso	~20	48,390	9374.0R	~40	
Pluto (dwarf planet)					
Charon	603	17.53	6.38718	1,520,000	1.83
Styx	~10	42	20.2	~5	
Nix	~46	48.71	24.9	~500	
Kerberos	~13	59	32.1	~10	
Hydra	~61	64.75	38	~1000	
Haumea (dwarf planet)					
Namaka	~80	25.657	18.278	~1800	
Hi'iaka	~150	49.88	49.462	17,900	
Eris (dwarf planet)					
Dysnomia	~150	37.4	15.77	18,400	

* Only satellites with a mass of at least 10^{15} kg or a radius of at least 5 km are included. Smaller objects are difficult to detect, but are steadily being discovered. A relatively up-to-date list is available at the website solarsystem.nasa.gov/planets. Click on the planet of interest and then click on the tab labeled "moons."
† Satellite radii marked with an approximate symbol (~) are estimated from their brightness. Some satellites have detailed measurements that show them to be irregular in shape. It the difference between their longest and smallest axes is larger than 10%, radial dimensions of the form a × b × c are given.
‡ Satellites with retrograde orbits have an "R" after the orbital period.
§ Satellite masses marked with an approximate symbol (~) are estimated by assuming they have a density of 1.3 g/cm³.

APPENDIX TABLE 8 Properties of Main-Sequence Stars*

Spectral Type	Luminosity ($L_\odot$)	Temperature (K)	Mass ($M_\odot$)	Radius ($R_\odot$)
O5	790,000	44,500	60	12.0
B0	52,000	30,000	17.5	7.4
B5	830	15,400	5.9	3.9
A0	54	9520	2.9	2.4
A5	14	8200	2.0	1.7
F0	6.5	7200	1.6	1.5
F5	3.2	6440	1.3	1.3
G0	1.5	6030	1.05	1.1
G5	0.79	5770	0.92	0.92
K0	0.42	5250	0.79	0.85
K5	0.15	4350	0.67	0.72
M0	0.08	3850	0.51	0.6
M5	0.01	3240	0.21	0.27
M8	0.001	2640	0.06	0.1

*Authorities differ substantially on many of the above values, especially at the upper- and lower-mass values. Note also that the values are generally not consistent with the Stefan-Boltzmann law.

APPENDIX TABLE 9 The Brightest Stars*

Name	Apparent Visual Magnitude	Distance (ly)	Designation	Spectral Type	Luminosity Class	Luminosity ($L_\odot$)
Sun	−26.72	0.0000158		G2	V	1.00
Sirius	−1.46	8.58	α CMa A	A1	V	26
Canopus	−0.74	310	α Car	F0	Ib–II	14,000
Rigel Kentaurus	−0.27	4.36	α Cen {A / B}	G2 / K0	V / V	1.5 / 0.5
Arcturus	−0.05	37	α Boo	K2	III	220
Vega	0.03	25	α Lyr	A0	V	59
Capella	0.08	42	α Aur {Aa / Ab}	G8 / G1	III / III	79 / 78
Rigel	0.10	860	β Ori	B8	Ia	110,000
Procyon	0.37	11.5	α CMi A	F5	IV–V	7
Betelgeuse	0.42	640	α Ori	M1.5	Iab	120,000
Achernar	0.45	140	α Eri	B3	V	3200
Hadar	0.61	350	β Cen {A / B}	B1 / B1	III / III	40,000 / 40,000
Altair	0.77	16.7	α Aql	A7	V	11
Acrux	0.78	320	α Cru {Aa / Ab / B}	B0.5 / B3 / B1	IV / V / V	25,000 / 7000 / 20,000
Aldebaran	0.85	65	α Tau A	K5	III	400
Antares	0.96	550	α Sco A	M1.5	Ib	60,000
Spica	1.04	250	α Vir {A / B}	B1 / B2	III–IV / V	12,000 / 1500
Pollux	1.14	34	β Gem	K0	III	46
Fomalhaut	1.16	25	α PsA	A3	V	17
Deneb	1.25	1400	α Cyg	A2	Ia	60,000
Mimosa	1.25	280	β Cru {A / B}	B0.5 / B2	IV / V	20,000 / 5000
Regulus	1.35	79	α Leo A	B7	V	360
Adhara	1.50	405	ε CMa	B2	II	22,300
Castor	1.58	51	α Gem {Aa / Ba}	A1 / A5	V / V	37 / 13

*Many of these stars are actually multiple-star systems. Data is provided for companion stars that contribute a significant fraction of the visible light we see.

APPENDIX TABLE 10 The Nearest Stars

Common Name	Distance (ly)	Spectral Type and Luminosity Class	Apparent Visual Magnitude	Absolute Visual Magnitude	Estimated Mass ($M_\odot$)
Sun	0.0000158	G2 V	−26.72	4.83	1.00
Proxima Centauri	4.24	M5.5 V	11.09	15.48	0.11
Alpha Centauri	4.36	G2 V	0.01	4.38	1.14
		K0 V	1.34	5.71	0.92
Barnard's Star	5.98	M3.5 V	9.57	13.25	0.16
Wolf 359	7.78	M5.5 V	13.53	16.64	0.09
Lalande 21185	8.29	M2 V	7.47	10.44	0.46
Sirius	8.58	A1 V	−1.46	1.44	2.02
		DA2 (white dwarf)	8.44	11.34	0.98
BL Ceti	8.73	M5.5 V	12.61	15.47	0.11
		M6 V	13.06	15.93	0.10
Ross 154	9.67	M3.5 V	10.44	13.08	0.17
Ross 248	10.31	M5.5V	12.29	14.79	0.12
Epsilon Eridani	10.48	K2 V	3.73	6.20	0.85
Lacaille 9352	10.69	M1 V	7.34	9.76	0.53
Ross 128	10.94	M4 V	11.16	13.53	0.16
EZ Aquarii	11.27	M5 V	13.03	15.33	0.16
		M V	13.27	15.58	0.11
		M V	15.07	17.37	0.11
61 Cygni	11.40	K5 V	5.20	7.48	0.7
		K7 V	6.03	8.31	0.63
Procyon	11.44	F5 IV–V	0.37	2.65	1.42
		DA (white dwarf)	10.70	12.98	0.60
Struve 2398	11.49	M3 V	8.90	11.17	0.35
		M3.5 V	9.69	11.96	0.26
Groombridge 34	11.65	M1.5 V	8.08	10.31	0.49
		M3.5 V	11.06	13.30	0.16
Epsilon Indi	11.81	K4 V	4.68	6.98	0.77
		T1 V (brown dwarf)	24.12	26.42	0.03
		T6 V (brown dwarf)	~26.6	~28.9	0.03
DX Cancri	11.83	M6 V	14.81	17.01	0.09
Tau Ceti	11.90	G8.5 V	3.49	5.68	0.92
Gliese-Jahreiss 1061	11.99	M5 V	13.09	15.26	0.11
YZ Ceti	12.12	M4 V	12.1	14.25	0.13
Luyten's Star	12.25	M3.5 V	9.85	11.98	0.26

Data based mainly on RECONS (Research Consortium on Nearby Stars) website.

APPENDIX TABLE 11 Known and Suspected Members of the Local Group of Galaxies*

Name of Galaxy	Right Ascension (hours and minutes)	Declination (degrees and minutes)	Galaxy Type	Distance (Mpc)	Diameter (kpc)	Visual Apparent Magnitude	Approximate Luminosity ($10^6 L_\odot$)
WLM	0 02	−15 28	Irr	1.0	3	11	50
IC10	0 20	+59 18	Irr	0.7	1	11	300
Cetus dw	0 26	−11 02	E	0.8	—	14	1
NGC 147	0 33	+48 30	E5	0.7	2	10	100
Andromeda III	0 35	+36 30	E5	0.8	1	15	1
NGC 185	0 39	+48 20	E3	0.7	2	9	150
NGC 205 (M110)	0 40	+41 41	E5	0.8	5	8	300
Andromeda VIII	0 42	+40 37	E	0.8	14	9	150
M32	0 43	+40 52	E2	0.8	2	8	300
M31	0 43	+41 16	Sb	0.8	40	3	25,000
Andromeda I	0 46	+38 02	E3	0.8	0.5	14	4
SMC	0 53	−72 50	Irr	0.06	5	2	600
Sculptor dw	1 00	−33 42	E3	0.09	1	9	1
LGC 3	1 04	+21 53	Irr	0.8	0.5	18	1
IC 1613	1 05	+02 07	Irr	0.7	3	9	100
Andromeda V	1 10	+47 38	E	0.8	—	15	1
Andromeda II	1 16	+33 25	E2	0.7	0.7	13	4
M33	1 34	+30 40	Sc	0.8	16	6	3000
Phoenix dw	1 51	−44 27	Irr	0.4	0.6	12	1
Fornax dw	2 40	−34 27	E3	0.14	0.7	9	15
UGCA 92	4 32	+63 36	Irr	1.44	0.6	14	5
LMC	5 23	−69 45	Irr	0.05	0.6	1	2000
Carina dw	6 42	−50 58	E4	0.10	0.7	16	1
Canis Major dw	7 35	−28 00	Irr	0.08	220	—	—
Leo T	9 35	+17 03	Irr	0.420	0.34	16	0.04
Leo A	9 59	+30 45	Irr	0.7	1	12	3
Sextans B	10 00	+05 20	Irr	1.41	1.6	12	30
NGC 3109	10 03	−26 09	Irr	1.38	5	10	100
Antlia dw	10 04	−27 20	E3	1.41	0.6	15	2
Leo I	10 08	+12 18	E3	0.2	0.7	10	4
Sextans A	10 11	−04 43	Irr	1.60	1.8	12	35
Sextans dw	10 13	−01 37	E	0.09	1	12	1
Leo II	11 13	+22 09	E0	0.2	0.7	12	1
GR 8	12 59	+14 13	Irr	2.4	0.4	14	8
Ursa Minor dw	15 09	+67 13	E5	0.06	0.5	11	1
Draco dw	17 20	+57 55	E3	0.08	0.8	10	1
Milky Way	17 46	−29 00	SBc	0.01	40	—	20,000
SagDEG	18 55	−30 29	Irr	0.03	—	10	30
SagDIG	19 30	−17 41	Irr	1.2	1.5	15	5
NGC 6822	19 45	−14 48	Irr	0.5	2	9	200
Aquarius dw	20 47	−12 51	Irr	1.0	0.6	14	2
Tucanae dw	22 42	−64 25	E4	0.9	0.5	15	1
UGCA 438	23 26	−32 33	Irr	1.4	0.5	14	5
Andromeda VII	23 27	+50 42	E	0.7	0.5	13	5
Pegasus dw	23 29	+14 45	Irr	0.8	1	13	7
Andromeda VI	23 52	+24 35	E	0.8	0.9	14	3

* Listing based largely on the SEDS Messier compilation of Local Group data (http://messier.seds.org/more/local.html).

APPENDIX TABLE 12 The Brightest Galaxies Beyond the Local Group*

Name of Galaxy	Right Ascension (hours and minutes)	Declination (degrees and minutes)	Galaxy Type	Distance (Mpc)	Diameter (kpc)	Visual Apparent Magnitude	Approximate Luminosity ($10^6 L_\odot$)
NGC0253	00 48	−25 17	Sc	2.6	21	7.3	17,000
NGC0300	00 55	−37 41	Scd	2.2	13	8.2	2700
NGC1068 (M77)	02 43	−00 01	Sb	13.8	28	8.7	60,000
NGC1291	03 17	−41 06	S0-a	9.2	27	8.7	25,000
IC 342	03 47	+68 06	SBc	2.5	15	8.5	51,000
NGC2403	07 37	+65 36	Sc	3.4	23	8.4	6800
NGC3031 (M81)	09 56	+69 04	Sab	3.7	23	6.9	33,000
NGC3034 (M82)	09 56	+69 41	Irr	3.7	10	8.3	15,000
NGC4258 (M106)	12 19	+47 18	Sb	7.4	41	8.4	34,000
NGC4472 (M49)	12 30	+08 00	E	16.3	43	8.3	110,000
NGC4486 (M87)	12 31	+12 24	E	15.6	32	8.6	81,000
NGC4594 (M104)	12 40	−11 37	Sa	9.2	24	8.1	58,000
NGC4736 (M94)	12 51	+41 07	Sab	5.2	11	8.1	14,000
NGC4826 (M64)	12 57	+21 41	Sab	7.4	17	8.4	30,000
NGC4945	13 05	−49 28	SBc	4.6	29	8.6	37,000
NGC5055 (M63)	13 16	+42 02	Sbc	7.7	21	8.6	25,000
NGC5128 (Cen A)	13 25	−43 01	S0	3.7	32	6.7	32,000
NGC5194 (M51)	13 30	+47 12	Sbc	8.0	26	8.3	33,000
NGC5236 (M83)	13 37	−29 52	Sc	4.6	15	7.3	28,000
NGC5457 (M101)	14 03	+54 21	Sc	7.4	47	7.8	34,000
NGC6744	19 10	−63 51	Sb	10.7	65	8.4	62,000

* Galaxies in this table were selected from the HyperLeda database (http://leda.univ-lyon1.fr) based on their total visual apparent magnitude.

APPENDIX TABLE 13 The Messier Catalog*

Messier Number	Popular Name	Object Type	Right Ascension (hours and minutes)	Declination (degrees and minutes)	Visual Apparent Magnitude	Angular Size (arc minutes)	Distance (kpc)
1	Crab Nebula	Supernova remnant	05 35	+22 01	8.4	6 × 4	1.9
2		Globular cluster	21 34	−00 49	6.5	12.9	11
3		Globular cluster	13 42	+28 23	6.2	16.2	10
4		Globular cluster	16 24	−26 32	5.6	26.3	2.2
5		Globular cluster	15 19	+02 05	5.6	17.4	7.5
6	Butterfly Cluster	Open cluster	17 40	−32 13	4.2	25	0.4
7	Ptolemy's Cluster	Open cluster	17 54	−34 49	4.1	80	0.2
8	Lagoon Nebula	Diffuse nebula	18 04	−24 23	6	90 × 40	1.5
9		Globular cluster	17 19	−18 31	7.7	9.3	8.1
10		Globular cluster	16 57	−04 06	6.6	15.1	4.4
11	Wild Duck Cluster	Open cluster	18 51	−06 16	6.3	14	1.8
12		Globular cluster	16 47	−01 57	6.7	14.5	4.9
13	Hercules Cluster	Globular cluster	16 42	+36 28	5.8	16.6	7.6
14		Globular cluster	17 38	−03 15	7.6	11.7	8.8
15		Globular cluster	21 30	+12 10	6.2	12.3	10
16	Eagle Nebula	Open cluster	18 19	−13 47	6.4	7	2.1
17	Omega or Swan Nebula	Diffuse nebula	18 21	−16 11	7	11	1.5
18		Open cluster	18 20	−17 08	7.5	9	1.5
19		Globular cluster	17 03	−26 16	6.8	13.5	8.7
20	Trifid Nebula	Diffuse nebula	18 03	−23 02	9	28	1.5

(continued)

APPENDIX TABLE 13 (continued)

Messier Number	Popular Name	Object Type	Right Ascension (hours and minutes)	Declination (degrees and minutes)	Visual Apparent Magnitude	Angular Size (arc minutes)	Distance (kpc)
21		Open cluster	18 05	−22 30	6.5	13	1.3
22		Globular cluster	18 36	−23 54	5.1	24	3.1
23		Open cluster	17 57	−19 01	6.9	27	0.6
24	Sagittarius Star Cloud	Milky Way patch	18 17	−18 29	4.6	90	3.0
25		Open cluster	18 32	−19 15	4.6	32	0.6
26		Open cluster	18 45	−09 24	8	15	1.5
27	Dumbbell Nebula	Planetary nebula	19 60	+22 43	7.4	8.0 × 5.7	0.3
28		Globular cluster	18 25	−24 52	6.8	11.2	5.7
29		Open cluster	20 24	+38 32	7.1	7	1.2
30		Globular cluster	21 40	−23 11	7.2	11	8
31	Andromeda Galaxy	Spiral galaxy	00 43	+41 16	3.4	178 × 63	800
32		Elliptical galaxy	00 43	+40 52	8.1	8 × 6	800
33	Triangulum Galaxy	Spiral galaxy	01 34	+30 39	5.7	73 × 45	860
34		Open cluster	02 42	+42 47	5.5	35	0.4
35		Open cluster	06 09	+24 20	5.3	28	0.8
36		Open cluster	05 36	+34 08	6.3	12	1.2
37		Open cluster	05 52	+32 33	6.2	24	1.3
38		Open cluster	05 28	+35 50	7.4	21	1.2
39		Open cluster	21 32	+48 26	4.6	32	0.2
40		Binary star	12 22	+58 05	8.4	0.8	0.1
41		Open cluster	06 46	−20 44	4.6	38	0.7
42	The Orion Nebula	Diffuse nebula	05 35	−05 27	4	85 × 60	0.4
43		Diffuse nebula	05 36	−05 16	9	20 × 15	0.4
44	Praesepe or Beehive Nebula	Open cluster	08 40	+19 59	3.7	95	0.1
45	Pleiades or Seven Sisters	Open cluster	03 47	+24 07	1.6	110	0.1
46		Open cluster	07 42	−14 49	6	27	1.6
47		Open cluster	07 37	−14 30	5.2	30	0.4
48		Open cluster	08 14	−05 48	5.5	54	0.4
49		Elliptical galaxy	12 30	+08 00	8.3	9 × 7.5	16,000
50		Open cluster	07 03	−08 20	6.3	16	0.9
51	Whirlpool Galaxy	Spiral galaxy	13 30	+47 12	8.3	11 × 7	8000
52		Open cluster	23 24	+61 35	7.3	13	1.5
53		Globular cluster	13 13	+18 10	7.6	12.6	18
54		Globular cluster	18 55	−30 29	7.6	9.1	27
55		Globular cluster	19 40	−30 58	6.3	19	5.3
56		Globular cluster	19 17	+30 11	8.3	7.1	10
57	Ring Nebula	Planetary nebula	18 54	+33 02	8.8	1.4 × 1.0	0.7
58		Spiral galaxy	12 38	+11 49	9.7	5.5 × 4.5	17,000
59		Elliptical galaxy	12 42	+11 39	9.6	5 × 3.5	16,000
60		Elliptical galaxy	12 44	+11 33	8.8	7 × 6	16,000
61		Spiral galaxy	12 22	+04 28	9.7	6 × 5.5	15,000
62		Globular cluster	17 01	−30 07	6.5	14.1	6.9
63	Sunflower Galaxy	Spiral galaxy	13 16	+42 02	8.6	10 × 6	7700
64	Blackeye Galaxy	Spiral galaxy	12 57	+21 41	8.4	9.3 × 5.4	7400
65		Spiral galaxy	11 19	+13 05	9.3	8 × 1.5	8000
66		Spiral galaxy	11 20	+12 59	8.9	8 × 2.5	8000
67		Open cluster	08 50	+11 49	6.1	30	0.8
68		Globular cluster	12 40	−26 45	7.8	12	10
69		Globular cluster	18 31	−32 21	7.6	7.1	8.5
70		Globular cluster	18 43	−32 18	7.9	7.8	9

(continued)

APPENDIX TABLE 13 (continued)

Messier Number	Popular Name	Object Type	Right Ascension (hours and minutes)	Declination (degrees and minutes)	Visual Apparent Magnitude	Angular Size (arc minutes)	Distance (kpc)
71		Globular cluster	19 54	+18 47	8.2	7.2	3.8
72		Globular cluster	20 54	−12 32	9.3	5.9	16
73		System of four stars	20 59	−12 38	9	2.8	0.6
74		Spiral galaxy	01 37	+15 47	9.4	10.2 × 9.5	7400
75		Globular cluster	20 06	−21 55	8.5	6	18
76	Little Dumbbell Nebula	Planetary nebula	01 42	+51 34	10.1	2.7 × 1.8	1.0
77		Spiral galaxy	02 43	−00 01	8.7	7 × 6	14,000
78		Diffuse nebula	05 47	+00 03	8.3	8 × 6	0.4
79		Globular cluster	05 25	−24 33	7.7	8.7	12
80		Globular cluster	16 17	−22 59	7.3	8.9	10
81	Bode's Galaxy	Spiral galaxy	09 56	+69 04	6.9	21 × 10	3700
82	Cigar Galaxy	Irregular galaxy	09 56	+69 41	8.3	9 × 4	3700
83	Southern Pinwheel Galaxy	Spiral galaxy	13 37	−29 52	7.3	11 × 10	4600
84		Elliptical galaxy	12 25	+12 53	9.1	5	17,000
85		S0 galaxy	12 25	+18 11	9.1	7.1 × 5.2	16,000
86		Elliptical galaxy	12 26	+12 57	8.9	7.5 × 5.5	17,000
87	Virgo A	Elliptical galaxy	12 31	+12 24	8.6	7	16,000
88		Spiral galaxy	12 32	+14 25	9.6	7 × 4	17,000
89		Elliptical galaxy	12 36	+12 33	9.8	4	16,000
90		Spiral galaxy	12 37	+13 10	9.5	9.5 × 4.5	17,000
91		Spiral galaxy	12 35	+14 30	10.2	5.4 × 4.4	16,000
92		Globular cluster	17 17	+43 08	6.4	11.2	8.1
93		Open cluster	07 45	−23 52	6	22	1.1
94		Spiral galaxy	12 51	+41 07	8.1	7 × 3	5200
95		Spiral galaxy	10 44	+11 42	9.7	4.4 × 3.3	10,000
96		Spiral galaxy	10 47	+11 49	9.2	6 × 4	10,000
97	Owl Nebula	Planetary nebula	11 15	+55 01	9.9	3.4 × 3.3	0.7
98		Spiral galaxy	12 14	+14 54	10.1	9.5 × 3.2	17,000
99		Spiral galaxy	12 19	+14 25	9.9	5.4 × 4.8	17,000
100		Spiral galaxy	12 23	+15 49	9.3	7 × 6	16,000
101	Pinwheel Galaxy	Spiral galaxy	14 03	+54 21	7.8	22	7400
102	Spindle Galaxy	S0 galaxy	15 07	+55 46	9.9	5.2 × 2.3	15,000
103		Open cluster	01 33	+60 42	7.4	6	2.6
104	Sombrero Galaxy	Spiral galaxy	12 40	−11 37	8.1	9 × 4	9200
105		Elliptical galaxy	10 48	+12 35	9.3	2	11,000
106		Spiral galaxy	12 19	+47 18	8.4	19 × 8	7400
107		Globular cluster	16 33	−13 03	7.9	10	6.4
108		Spiral galaxy	11 12	+55 40	10	8 × 1	14,000
109		Spiral galaxy	11 58	+53 23	9.8	7 × 4	17,000
110		Elliptical galaxy	00 40	+41 41	8.5	17 × 10	830

* Data in Table 13 drawn primarily from the SEDS Messier Catalog Web pages. By Hartmut Frommert, Christine Kronberg, and Guy McArthur. SEDS, University of Arizona Chapter, Tucson, Arizona, 1994–2005. http://www.seds.org/messier/.

APPENDIX TABLE 14 A Cosmic Periodic Table of the Elements

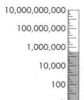

Cosmic Abundance Scale
The blue scale shows the relative amount of each element within the Solar System. Each full step on the scale corresponds to a factor of 10 greater abundance. Note that the scale goes two orders of magnitude higher for the two most abundant elements, hydrogen and helium.

Radioactive Lifetimes
For elements with no stable isotope, the half-life of the most stable isotope is indicated. Rhenium, bismuth, uranium, and thorium are also radioactive, but their half-lives are at least several billion years, and they are found in nature.

Condensation Temperature
The "thermometer" shows the temperature at which about half of each element condenses (often after first forming molecules with other elements) from a low density gas cloud such as the presolar nebula. Elements with lower condensation temperatures condense farther from the proto-Sun.

Atomic Masses
The masses listed in the table are based on setting the mass of iron to 56.0. Iron has 26 protons and 30 neutrons, and has the least energetic nucleus of any element. On this scale, the integer part of each element's mass indicates the total number of protons+neutrons, while the decimal portion shows the mass excess that can potentially be released as energy by fission or fusion (according to $E=mc^2$).

$$4.0073 \\ \| \\ (p+n).(energy)$$

Source: Condensation temperatures and abundance data drawn from K. Lodders (2003). *The Astrophysical Journal*, vol. 591, pp. 1220–1247.

Answers to Questions

Part 1

Unit 1: 16-e; 17-d; 18-b; 19-e; 20-b
Unit 2: 16-c; 17-e; 18-b; 19-d; 20-c
Unit 3: 18-b; 19-d; 20-b
Unit 4: 15-d; 16-d; 17-d; 18-c; 19-a; 20-c
Unit 5: 16-b; 17-a; 18-d; 19-a
Unit 6: 17-b; 18-a; 19-a; 20-b; 21-c
Unit 7: 16-b; 17-d; 18-a; 19-c; 20-c
Unit 8: 18-c; 19-b; 20-d; 21-c
Unit 9: 16-b; 17-b; 18-a; 19-a; 20-a
Unit 10: 17-a; 18-d; 19-c; 20-a
Unit 11: 16-c; 17-b; 18-a; 19-a; 20-c; 21-c
Unit 12: 16-e; 17-a; 18-b; 19-a; 20-b; 21-c
Unit 13: 16-a; 17-d; 18-e; 19-b

Part 2

Unit 14: 16-e; 17-a; 18-d; 19-b; 20-c
Unit 15: 16-c; 17-d; 18-b; 19-b
Unit 16: 16-c; 17-d; 18-d; 19-c; 20-a
Unit 17: 17-e; 18-d; 19-d; 20-d
Unit 18: 17-b; 18-c; 19-c; 20-a
Unit 19: 17-a; 18-e; 19-b; 20-a; 21-d
Unit 20: 17-c; 18-a; 19-d; 20-a
Unit 21: 16-b; 17-c; 18-b; 19-e; 20-a
Unit 22: 17-e; 18-e; 19-a; 20-c; 21-b
Unit 23: 17-c; 18-b; 19-e; 20-c; 21-c
Unit 24: 17-d; 18-d; 19-e; 20-e
Unit 25: 17-b; 18-e; 19-b; 20-b
Unit 26: 18-e; 19-c; 20-b; 21-b
Unit 27: 19-c; 20-e; 21-d; 22-a
Unit 28: 16-d; 17-b; 18-b; 19-a
Unit 29: 18-c; 19-a; 20-c; 21-c
Unit 30: 16-c; 17-b; 18-a; 19-c
Unit 31: 17-c; 18-d; 19-a; 20-a
Unit 32: 17-d; 18-c; 19-d; 20-d; 21-c
Unit 33: 16-b; 17-c; 18-c; 19-a; 20-a

Part 3

Unit 34: 16-c; 17-a; 18-b; 19-c; 20-b
Unit 35: 17-c; 18-e; 19-d; 20-d
Unit 36: 16-b; 17-d; 18-a; 19-d; 20-d
Unit 37: 17-a; 18-c; 19-d; 20-d
Unit 38: 18-d; 19-a; 20-d; 21-d
Unit 39: 18-b; 19-b; 20-b
Unit 40: 16-d; 17-a; 18-c; 19-d; 20-a
Unit 41: 16-c; 17-a; 18-b; 19-d; 20-b
Unit 42: 20-c; 21-d; 22-a; 23-c
Unit 43: 18-b; 19-d; 20-a; 21-a
Unit 44: 20-e; 21-a; 22-b; 23-d
Unit 45: 17-d; 18-c; 19-a; 20-b; 21-a
Unit 46: 16-c; 17-c; 18-c; 19-c; 20-d
Unit 47: 16-b; 17-a; 18-d; 19-c
Unit 48: 19-c; 20-a; 21-d; 22-d
Unit 49: 16-a; 17-b; 18-b; 19-c; 20-a
Unit 50: 17-d; 18-c; 19-a; 20-d

Part 4

Unit 51: 17-c; 18-a; 19-e; 20-b
Unit 52: 16-d; 17-a; 18-e; 19-a
Unit 53: 18-c; 19-a; 20-a; 21-b; 22-c
Unit 54: 20-b; 21-c; 22-d; 23-c
Unit 55: 18-a; 19-e; 20-e; 21-b
Unit 56: 19-a; 20-a; 21-e; 22-d; 23-a
Unit 57: 17-c; 18-b; 19-a; 20-f
Unit 58: 19-a; 20-b; 21-d; 22-e
Unit 59: 18-c; 19-a; 20-d; 21-b; 22-b
Unit 60: 16-e; 17-a; 18-c; 19-c; 20-a
Unit 61: 17-a; 18-a; 19-b; 20-d
Unit 62: 18-d; 19-a; 20-d; 21-c
Unit 63: 17-b; 18-b; 19-c; 20-b
Unit 64: 17-d; 18-a; 19-d; 20-e
Unit 65: 17-e; 18-b; 19-a; 20-b
Unit 66: 17-a; 18-b; 19-b; 20-d

Unit 67: 19-b; 20-c; 21-c; 22-d
Unit 68: 17-a; 18-d; 19-b; 20-b
Unit 69: 18-a; 19-e; 20-c; 21-d
Unit 70: 17-a; 18-d; 19-b; 20-a

Part 5

Unit 71: 19-e; 20-a; 21-d; 22-b
Unit 72: 18-b; 19-d; 20-b; 21-a
Unit 73: 18-d; 19-b; 20-c; 21-d
Unit 74: 18-d; 19-d; 20-b; 21-a
Unit 75: 17-b; 18-a; 19-c; 20-b
Unit 76: 17-d; 18-c; 19-d; 20-d
Unit 77: 19-e; 20-c; 21-a
Unit 78: 18-a; 19-a; 20-e; 21-b
Unit 79: 17-a; 18-b; 19-e; 20-c
Unit 80: 17-b; 18-b; 19-d; 20-d; 21-c
Unit 81: 16-b; 17-e; 18-c; 19-a
Unit 82: 17-d; 18-b; 19-d; 20-b; 21-b
Unit 83: 17-c; 18-c; 19-c; 20-d
Unit 84: 16-e; 17-a; 18-b; 19-d
Unit 85: 16-e; 17-b; 18-c; 19-c; 20-c
Unit 86: 20-d; 21-b; 22-d; 23-a

Glossary

This glossary includes definitions of the key terms and other important terms relevant to the study of astronomy. Section numbers (in parentheses) indicate where terms are first presented in the text. (Section "0" refers to the introductory section of a Unit.)

SYMBOLS

- a – acceleration
- a – semimajor axis of an orbit
- A – area, such as surface area of sphere
- b – baseline length, for triangulation
- B – brightness, of observed light
- c – speed of light
- d – distance
- D – diameter
- e – electron
- E – energy
- f – fraction or percentage
- F – force
- g – acceleration due to gravity
- G – Newton's gravitational constant
- h – Planck's constant
- H – Hubble's constant
- ℓ – linear size
- L – luminosity, or total light output
- $\mathcal{L}$ – luminosity of star in solar luminosities
- m – mass
- M – mass, generally of a larger object
- $\mathcal{M}$ – mass of star in solar masses
- n – neutron
- n – energy level, as for hydrogen
- N – number, such as number of stars
- p – proton
- p – parallax
- P – period, of rotation or of an orbit
- R – radius
- t – time
- T – temperature
- V – velocity
- $\mathcal{V}$ – volume, such as volume of sphere
- z – redshift, or relative shift of wavelength

GREEK SYMBOLS

- a – acceleration
- α (alpha) – angle or angular size
- β (beta) – angle
- γ (gamma) – Lorentz factor
- γ (gamma) – gamma ray or photon
- Δ (delta) – change in…
- λ (lambda) – wavelength, crest-to-crest
- Λ (lambda) – the cosmological constant
- μ (mu) – proper or angular motion
- ν (nu) – frequency, of wave oscillation
- ν (nu) – neutrino
- ρ (rho) – density, or mass per volume
- σ (sigma) – Stefan-Boltzmann constant
- τ (tau) – half-life, of a radioactive substance
- Ω (omega) – density parameter of universe

A

aberration of starlight the angular shift in the apparent direction of a star caused by the orbital speed of the Earth. (54.5)

absolute magnitude the apparent magnitude a star would have if it were at a distance of 10 parsecs. (55.4)

absolute zero the lowest possible temperature, at which a material contains no extractable heat energy. Zero on the kelvin temperature scale. (23.3)

absorption the process in which an atom or molecule intercepts light or other electromagnetic radiation and stores its energy, often by shifting an electron to a higher energy orbital. (21.4)

absorption lines dark lines superimposed on a continuous spectrum, produced when a gas absorbs specific wavelengths of light from a background source of light. (56.1)

absorption-line spectrum a spectrum in which certain wavelengths are darker than adjacent wavelengths. The missing light is absorbed by atoms or molecules between the source and the observer. (24.2)

acceleration the rate of change in an object's velocity (either its speed or its direction). This can include speeding up as well as slowing down (deceleration). (15.1)

accretion the addition of matter to a body. Examples are gas falling onto a star and asteroids colliding and sticking together. (35.3)

accretion disk a nearly flat disk of gas or other material held in orbit around a body by its gravity before it falls onto the body. (66.1, 69.3, 78.3)

achondrite a stony meteorite of rocky material from the upper layers of a differentiated body. Achondrites have no **chondrules.** (50.2)

active galactic nucleus (AGN) a small central region in a galaxy that emits abnormally large amounts of electromagnetic radiation and sometimes ejects jets of matter. Examples are radio galaxies, Seyfert galaxies, and quasars. (78.0)

A.D. abbreviation of Latin *anno domini*, "in the year of the Lord" in the Christian calendar. Equivalent to the nondenominational designation **C.E.** or "common era." (9.5)

adaptive optics a technique for adjusting a telescope's mirror or other optical parts to compensate for atmospheric distortions, such as **seeing,** thereby giving a sharper image. (32.2)

æther a substance that was once hypothesized to fill space and through which light waves propagated. Einstein's theory of special relativity overturned this idea. (26.1)

alpha particle a helium nucleus: two protons plus (usually) two neutrons.

alt–az mount a support structure for a telescope that allows it to rotate up and down (in altitude) and parallel to the horizon (in azimuth). (33.4)

altitude an object's angular distance above the horizon. (13.1)

amino acid a carbon-based molecule used by living organisms to build protein molecules. (85.2)

analemma the figure-8 pattern the Sun traces on the sky if observed at the same clock time each day during a year. (13.5)

angstrom unit a unit of length used in describing wavelengths of radiation and the sizes of atoms and molecules. One angstrom is 10^{-10} meters.

angular momentum a measure of an object's tendency to keep rotating and to maintain its orientation. Mathematically, it depends on the object's mass, M, radius, R, and rotational velocity, V, and is proportional to the product $M \times V \times R$. (20.3)

angular shift the change in the apparent position of an object. For example, an object will show an angular shift relative to background objects when viewed from different places. (54.2)

angular size a measure of how large an object looks to an observer. It is defined as the angle between lines drawn from the observer to opposite sides of the object. For example, the angular size of the Moon is about $1/2°$ as seen by an observer on Earth. (10.4)

annular eclipse an eclipse in which the Moon passes directly in front of the Sun but does not completely cover the Sun. A bright ring of the Sun's photosphere remains visible around the dark disk of the Moon. We therefore see a ring (annulus) of light around the Moon. (8.2)

Antarctic Circle the latitude line of 66.5° south, marking the position farthest from the South Pole where the Sun does not rise on the date of the summer solstice (the start of winter in the Southern Hemisphere). (6.5)

ante meridian (A.M.) before the Sun has crossed the **meridian**; before noon. (7.1)

anthropic principle the principle that the properties of the universe are limited to those that permit our existence—otherwise we could not be present to observe the universe. (85.4)

antimatter a type of matter that, if brought into contact with ordinary matter, annihilates it, leaving nothing but energy. The positron is the antimatter partner of the electron. The antiproton is the antimatter partner of the proton. Antimatter is observed in cosmic rays and can be created from energy in the laboratory. (83.3)

aperture the opening in a telescope or other optical instrument that determines how much light it collects. (28.1, 29.2)

aphelion the point in an orbit where a body is farthest from the Sun.

Aphrodite a continent-like highland region on the planet Venus. (41.2)

apparent double star two stars that lie close to each other in the sky but that are not orbiting each other. (57.1)

apparent magnitude a unit for measuring the brightness of a celestial body. Also simply called *magnitude*; the word *apparent* helps to distinguish it from **absolute magnitude**. The smaller the magnitude, the brighter the star. Bright stars have magnitude 1, whereas the faintest stars visible to the unaided eye have magnitude 6. The largest telescopes can detect stars of about magnitude 30; bright objects like the Sun, Moon, and planets can have "negative" magnitudes. (55.4)

apparent noon the time at which the Sun crosses the meridian as seen from a particular location. This usually differs from noon on a clock, which represents an average within a time zone and over the year. (7.1)

archaea a major category of simple single-celled organisms that have no nucleus, like bacteria, but genetically are more similar to eukaryotes. (85.3)

arc minute (abbreviation: arcmin) an angular measure equal to 1/60 of 1 degree. (12.1, 54.1)

arc second (abbreviation: arcsec) an angular measure equal to 1/60 of 1 arc minute, or 1/3600 of 1 degree. (54.2)

Arctic Circle the latitude line of 66.5° north, marking the position farthest from the North Pole where the Sun does not rise on the date of the winter solstice (the start of winter in the Northern Hemisphere). (6.5, 7.2)

Aristarchus (about 310–230 B.C.E.) Greek astronomer who first estimated the sizes and distances of the Moon and Sun. (10.2)

Aristotle (384–322 B.C.E.) Greek philosopher who developed many early ideas of physics and astronomy. (10.1)

association see **stellar association**.

asterism an easily identified grouping of stars, sometimes a part of a larger constellation (such as the Big Dipper) or extending across several constellations (such as the Summer Triangle). (5.2)

asteroid a small, generally rocky, solid body orbiting the Sun and ranging in diameter from a few meters to hundreds of kilometers. (34.1, 43.0)

asteroid belt a region between the orbits of Mars and Jupiter in which most of the Solar System's asteroids are located. (34.1)

asteroid family a group of asteroids that appear to have originated from a single object that underwent a collision. The asteroids all have orbital parameters and spectroscopic properties that indicate they have similar compositions. (43.4)

astronomical unit (AU) a distance unit based on the average distance of the Earth from the Sun. (1.6, 54.1)

astrophysics the branch of science in which physical laws are applied to interpret astronomical observations and derive chemical and physical properties of remote objects. (55.0)

asymmetry a variation in properties where none is expected. For example, the weak force shows an asymmetry in its behavior with some particles and their antiparticles. (83.3)

atmosphere layer of gas held close to a planet or star by the force of gravity. Also a unit of pressure in which 1 atmosphere is the average pressure of Earth's atmosphere at sea level. (38.1)

atmospheric pressure the pressure produced by the weight of overlying atmosphere. (38.1)

atmospheric window a wavelength range in which our atmosphere absorbs little radiation. For example, our atmosphere allows a range of wavelengths at about 300 to 700 nm to pass through. This is our "visible" atmospheric window. The atmosphere also has a wide window at radio wavelengths and narrow windows at infrared wavelengths. (32.1)

atom a submicroscopic particle consisting of a nucleus and orbiting electrons. The smallest unit of a chemical element. (4.2)

atomic mass the mass of an atom expressed in units approximately equal to the mass of a proton or neutron. The isotope of carbon with 6 protons and 6 neutrons has an atomic mass of 12.0. (4.2)

atomic number the number of protons in the nucleus of an atom. Unless the atom is ionized, the atomic number is also the number of electrons orbiting the nucleus of the atom. (4.2, 21.3)

aurora the light emitted by atoms and molecules in the upper atmosphere. This light is a result of magnetic disturbances caused by the solar wind, and appears to us as the Northern or Southern Lights. (38.1, 45.4, 53.1)

autumnal equinox the beginning of fall in the Northern Hemisphere, on or near September 22 when the Sun crosses the celestial equator. (6.5)

averted vision looking slightly to one side of a dim object so that its light strikes slightly off-center in your field of vision. This allows you to more easily discern faint objects, although at a sacrifice of sharpness. (33.2)

azimuth a coordinate for indicating the direction of objects on the sky. The azimuth direction is measured as the angle eastward from north to the point on the horizon below the object. A star that is directly above the point due east is at azimuth 90°; south is at 180°; west at 270°; and north at 360° or 0°. (13.1)

B

Balmer lines a series of absorption or emission lines of hydrogen seen at visible wavelengths. (24.2, 56.4)

barred spiral (SB) galaxy a galaxy in which the spiral arms wind out from the ends of a central bar rather than from the nucleus. (76.1)

Barringer crater an impact crater more than 1 km in diameter located in Arizona, probably produced about 50,000 years ago. (50.4)

basalt a dense, dark rock of volcanic origin. (39.2)

baseline the distance between two observing locations used for the purposes of triangulation measurements. The larger the baseline, the better the resolution attainable. (54.1)

B.C. an abbreviation used in the Christian calendar to indicate dates before the birth of Jesus ("before Christ"). Biblical scholars estimate that the actual birth year was between 4 and 8 years earlier than the date indicated by this calendar system. (9.5)

B.C.E. (before the Common Era) a nondenominational designation for referring to B.C. dates. (9.5)

belt a dark, low-pressure region in the atmosphere of Jovian planet, where gas flows downward. (45.1)

Big Bang the event that began the universe according to modern cosmological theories. It occurred about 13.7 billion years ago and generated the expanding motion that we observe today. (2.5, 80.0)

Big Crunch point of final collapse of a bound universe. (82.2)

Big Rip a hypothesis about the future of the universe suggesting that dark energy will grow with time, ultimately leading to such rapid expansion on all scales that matter will be torn apart, even down to subatomic scales. (84.5)

binary star two stars in orbit around each other, held together by their mutual gravity. (57.0, 58.2)

bipolar outflow the narrow columns of high-speed gas ejected by a protostar in two opposite directions. (60.2, 61.2)

blackbody an object that is an ideal radiator when hot and a perfect absorber when cool. It absorbs all radiation that falls upon it, reflecting no light; hence it appears black. Stars are approximately blackbodies. The radiation emitted by blackbodies obeys Wien's law and the Stefan-Boltzmann law. (23.1)

black hole an object whose gravitational attraction is so strong that its escape velocity equals the speed of light, preventing light or any radiation or material from leaving its "surface." (60.2, 69.0)

blazar a type of active galactic nucleus that is often highly variable and is thought to have its jet pointing straight toward us. (78.3)

blueshift a shift in the wavelength of electromagnetic radiation to a shorter wavelength. For visible light, this implies a shift toward the blue end of the spectrum. The shift can be caused by the motion of a source of radiation toward the observer or by the motion of an observer toward the source. For example, the spectrum lines of a star moving toward the Earth exhibit a blueshift. See also **Doppler shift**. (25.1)

Bode's rule a numerical formula for predicting the approximate distances of most of the planets from the Sun. (43.1)

Bohr model first theory of the hydrogen atom to explain the observed spectral lines. In this model the electron can orbit the nucleus only at discrete radii, each orbit having a particular energy. An electron can jump between orbits only if it absorbs or emits the exact energy difference between the orbits. (24.1)

Bok globule a small, dark, interstellar cloud, often approximately spherical. Many globules are the early stages of protostars. (61.1)

B.P. a designation used primarily by archaeologists and geologists to indicate the number of years "before the present" era. (9.5)

Brahe, Tycho (1546–1601) Danish astronomer who made the finest pretelescopic measurements. His observations disproved the idea that comets and novae were atmospheric phenomena and provided the basis for Kepler's laws about the motions of the planets. (12.1)

brightness a measure of the amount of light received from a body, often measured in watts per square meter detected by a telescope or other detector. (21.2, 55.2)

brown dwarf a body intermediate between a star and a planet. A brown dwarf has a mass between about 0.017 and 0.08 solar masses—too low to fuse hydrogen in its core, but massive enough to fuse deuterium as it contracts. (36.2, 61.4)

bulge the dense, central region of a spiral galaxy. (71.4)

C

Callisto the second largest of Jupiter's satellites. (48.1)

Caloris Basin the largest impact crater region yet seen on Mercury. (40.1)

Cambrian explosion the rapid appearance of a wide diversity of animals about 540 million years ago. (85.1)

canali straight-line features on Mars observed in 1877 by Italian astronomer Giovanni Schiaparelli. He called them *canali*, by which he meant channels, but some English-speakers interpreted *canali* as canals, with the implication of intelligent beings to build them. (86.1)

carbonaceous chondrite a type of meteorite containing many tiny spheres (chondrules) of rocky or metallic material embedded in carbon-rich material. (50.2)

Cassegrain telescope a type of reflecting telescope in which incoming light hits the primary mirror and is then reflected upward toward the prime focus, where a secondary mirror reflects the light back down through a small hole in the main mirror into a detector or eyepiece. (30.2)

Cassini division a conspicuous 1800-kilometer-wide gap between the outermost rings of Saturn. (47.3)

catadioptric telescope a hybrid telescope between a reflector and a refractor, using both a large mirror and a large lens, called a corrector plate, to focus the light entering the aperture. (33.3)

CCD charged-coupled device: an electronic device that records the intensity of light falling on it. CCDs have replaced film in most astronomical applications. (28.2)

C.E. (Common Era) a designation for the year-numbering system of the most widely used calendar system. A nondenominational equivalent of A.D. (9.5)

celestial equator an imaginary line on the celestial sphere lying exactly above the Earth's equator. It divides the celestial sphere into northern and southern hemispheres. (5.3)

celestial pole an imaginary point on the sky directly above the Earth's North or South Pole. (5.3)

celestial sphere an imaginary sphere surrounding the Earth. Ancient astronomers pictured the stars as all being attached to it, all at the same distance from the Earth. (5.1)

center of mass the "average" position of a collection of massive bodies weighted by their masses. (16.2, 57.3)

centripetal force a force that causes the path of a moving object to curve. (17.1)

Cepheid variables a class of yellow giant pulsating stars. Their pulsation periods range from about 1 day to about 70 days. Cepheids can be used to determine distances. See also **standard candle**. (64.2)

Ceres the first-discovered and largest of the asteroids, over 900 km in diameter. Often called a planet during the early 1800s. (43.1)

Chandrasekhar limit the maximum mass of a white dwarf stellar remnant. Approximately 1.4 solar masses—above this limit the remnant will collapse. Named for the astronomer who first calculated that such a limit exists. (66.2)

chemical energy the energy stored or released when bonds form or break between atoms and molecules. (20.1)

Chicxulub the site of a major impact crater that is the most likely explanation for the extinction of the dinosaurs about 65 million years ago. (50.5)

chondrite a meteorite containing small spherical granules called chondrules. (50.2)

chondrule a small spherical silicate body embedded in many meteorites. (50.2)

chromatic aberration a distortion of an optical system that results in different colors not focusing in the same way, often producing color fringes or distorted shapes. (30.4)

chromosphere the lower part of the Sun's outer atmosphere that lies directly above the Sun's visible surface (photosphere). (51.4)

circumpolar close enough to a celestial pole that always remains above the horizon. Circumpolar stars are closer to the celestial pole than the latitude of the observer. (5.3)

closed universe geometry that the universe as a whole would have if its gravity were great enough to curve around back on itself (positive curvature). A closed universe is finite in extent but has no edge, like the surface of a sphere. (82.1)

cluster a group of objects (for example, stars or galaxies) held together by their mutual gravitational attraction.

CNO cycle/process a reaction involving carbon, nitrogen, and oxygen (C, N, and O) that fuses hydrogen into helium and releases energy. The process begins with a hydrogen nucleus fusing with a carbon nucleus. Subsequent steps involve nitrogen and oxygen. The carbon, nitrogen, and oxygen act as catalysts and are released at the end of the process to start the cycle again. The CNO cycle is the dominant process for generating energy in main-sequence stars that are hotter and more massive than the Sun. (62.2)

coherent a relationship indicating that electromagnetic waves are in synchronization with each other. Waves that are coherent exhibit persistent interference effects. (29.1)

collecting area the total aperture over which a telescope can gather light. (29.2)

coma the gaseous atmosphere surrounding the head of a comet. (49.1)

comet a small body in orbit around the Sun, consisting of a tiny, icy core and a tail of gas and dust. The tail forms only when the comet is near the Sun. (34.1)

compact stars very dense stars whose radii are much smaller than the Sun's. These stars include **white dwarfs, neutron stars,** and **black holes.**

comparative planetology the study of bodies in the Solar System, examining and understanding the similarities and differences among worlds. (44.0)

condensation conversion of free gas atoms or molecules into a liquid or solid. A snowflake forms in our atmosphere when water vapor condenses into ice. (35.3)

cone light-sensing cell in the eye that gives us our color vision. (33.1)

conjunction the appearance of two astronomical objects in approximately the same direction on the sky. For example, if Mars and Jupiter happen to appear near each other on the sky, they are said to be in conjunction. If a planet is in conjunction with the Sun, sometimes it is simply said to be in conjunction without specifying the Sun. (11.3)

conservation law a theory describing how certain properties, such as electric charge, energy and angular momentum, remain constant despite interactions. (20.0)

conservation of angular momentum a principle of physics stating that the angular momentum of a rotating body remains constant unless forces act to speed it up or slow it down. Mathematically, conservation of angular momentum states that $M \times V \times R$ is a constant, where M is the mass of a body moving with a velocity V in a circle of radius R. One extremely important consequence of this principle is that if a rotating body shrinks, its rotational velocity must increase. (68.1)

constellation an officially recognized grouping of stars on the night sky. Astronomers divide the sky into 88 constellations. (5.2)

continuous spectrum a spectrum with neither dark absorption nor bright emission lines. The intensity of the radiation in such a spectrum changes smoothly from one wavelength to the next. (24.3)

convection the rising and sinking motions in a liquid or gas that carry heat upward through the material. Convection is easily seen in a pan of heated soup on a stove. (37.3. 62.2)

convection zone the region immediately below the Sun's visible surface in which its heat is carried upward by convection. (51.3)

Copernicus, Nicolaus (1473–1543) Polish astronomer who proposed that the Earth was just one of the planets orbiting the Sun. (11.3)

core the innermost, usually hot, region of the interior of the Earth or another planet. (34.1)

core-collapse supernova a supernova of type Ib, Ic, or II. A star that begins with more than about 8 $M_\odot$, will fuse elements all the way up to iron in its core, which then collapses catastrophically. (67.3)

Coriolis effect the deflection of an object moving within or on the surface of a rotating body. It is caused by the conservation of angular momentum when the moving object gets closer to or farther from the rotation axis of the rotating body. The Coriolis effect makes storms on Earth spin, generates large-scale wind systems, and creates cloud belts on many planets. (38.5)

cornea the curved transparent layer covering the front part of the eye that does most of the focusing of light that we see. (33.1)

corona the outer, hottest part of the Sun's atmosphere. (51.4)

coronal hole a low-density region in the Sun's corona, probably related to the structure of the Sun's magnetic field. The solar wind may originate in these regions. (51.4)

coronal mass ejections a blast of gas moving outward through the Sun's corona and into interplanetary space following the eruption of a prominence. (53.2)

cosmic horizon the greatest distance it is possible to see out into the universe given its age and rate of expansion. The horizon lies at a **light travel time distance** in light-years equal to the age of the universe in years. (81.1)

cosmic microwave background (CMB) the radiation that was created during the Big Bang and that permeates all space. At this time, the temperature of this radiation is 2.73 K. (81.2)

cosmic rays extremely energetic subatomic particles (such as protons and electrons) traveling at nearly the speed of light. Some rays are emitted by the Sun, but most come from more distant sources, such as exploding supernovae. (52.3, 66.3)

cosmological constant a term in the equations that Einstein developed to describe the effects of gravity on the entire universe. The term has the effect of a repulsive "force" opposing gravity. (82.4)

cosmological principle the hypothesis that on average, the universe looks the same to all observers, no matter where they are located in it. (80.2)

cosmology the study of the evolution and structure of the universe. (80.0)

Crab Nebula a supernova remnant in the constellation Taurus. Astronomers in ancient China and the Far East saw the supernova explode in A.D. 1054. Astronomers discovered a **pulsar** in the middle of the nebula in the 1960s. (28.4)

crater a circular pit, generally with a raised rim and sometimes with a central peak. A crater can be formed by the impact of a solid body, such as an asteroid, with the surface of a planet, moon, or another asteroid. Crater diameters on the Moon range from centimeters to several hundred kilometers. (39.2)

crescent moon a lunar phase during which the Moon appears less than half full. (8.1)

Cretaceous period a period of time in the geologic timescale ranging from about 146 million to 65 million years **b.p.** (50.5)

critical density the density necessary for a closed universe. If the density of the universe exceeds the critical density, the universe will stop expanding and collapse. If the density is less than the critical density, the universe will expand forever. (82.3)

crust the solid surface of a planet, moon, or other solid body. (37.1)

curvature of space the "bending" of space or motion given to space by gravity, as described by Einstein's general theory of relativity. For example, stars curve the space around them so that the path of light traveling by them is bent from a straight line. The universe too may be curved in such a way as to make its volume finite. (27.2, 69.2)

cyanobacteria a kind of photosynthesizing bacteria, formerly known as blue-green algae. Colonies of them form **stromatolites**. Thought to have been found in some microfossils as old as 3.5 billion years, but not confirmed. (85.1)

D

dark adaption the process by which the eye changes, through chemical changes in the retina and by expanding the pupil, to become more sensitive to dim light. (33.2)

dark energy a form of energy detected by its effect on the expansion of the universe, also known as the **cosmological constant.** It causes the expansion to speed up. The nature and properties of dark energy are unknown. (2.5, 84.2)

dark-line spectrum see **absorption-line spectrum.**

dark matter matter that emits no detectable radiation but whose presence can be deduced by its gravitational attraction on other bodies. (2.5, 74.1, 79.0)

dark matter halo the extended "cloud" of dark matter in which galaxies lie. The dark matter comprises most of the mass of a galaxy, but is only detected from the rapid speed of rotation of the outer parts of the galaxy. (79.1)

dark nebula a dense cloud of dust and gas in interstellar space that blocks the light from background stars. (73.1)

daylight saving time the time kept during summer months by setting the clock ahead one hour. This gives more hours of daylight after the workday. (7.4)

declination (dec) one part of a coordinate system (with **right ascension**) for locating objects in the sky. The declination indicates an object's distance north or south of the celestial equator. Declination is analogous to latitude on the Earth's surface. (5.5)

degenerate gas an extremely dense gas in which the electrons and nuclei are tightly packed. Unlike a normal gas, the pressure of a degenerate gas is not affected by its temperature. (63.3)

Deimos one of the two small moons of Mars, probably a captured asteroid. (42.4)

density the mass of a body or region divided by its volume. (3.1, 34.4)

density wave a spiral-shaped compression of the gas and dust in a spiral galaxy kept from dispersing by the gravitational attraction between all the matter in the wave. Density waves may account for the spiral arms of galaxies. According to the theory, stars and gas in one region of a galaxy may become packed more closely together (reaching a higher density). As the galaxy spins, these dense regions are pulled apart, but their gravitational influence on neighboring regions may cause them to compact in succession. This can produce "waves" of higher density that travel through the disk of a galaxy, pulling stars and interstellar gas into a spiral pattern. (74.3)

deuterium an isotope of hydrogen containing one proton and one neutron. (52.2)

differential gravitational force the difference between the gravitational forces exerted on an object at two different points. The effect of this force is to stretch the object. Such forces create **tides** and, if strong enough, may break up an astronomical object. See also **Roche limit.**

differential rotation rotation in which the rotation period of a body varies with latitude or radius. Differential rotation occurs for gaseous bodies like the Sun and Jovian planets as well as for the disks of galaxies. (74.1)

differentiation the separation of previously mixed materials inside a planet or other object. An example of differentiation is the separation that occurs when a dense material, such as iron, settles to the planet's core, leaving lighter material on the surface. (35.4)

diffraction bending of the path of light or other electromagnetic waves as they pass through an opening or around an obstacle. Diffraction limits the ability of a telescope to distinguish fine details. (31.1)

diffraction pattern a pattern that a telescope imparts to light from even something as simple as a single point of light. This is caused by diffraction around various structural elements of a telescope—such as the spikes seen surrounding the image of a star, which are caused by the support structure for secondary mirrors. (31.1)

disk the flat, round portion of a galaxy. The Sun lies in the disk of the Milky Way. (71.4)

dispersion the spreading of light or other electromagnetic radiation into a spectrum. A rainbow is an example of the dispersion of light caused by raindrops. (30.4)

DNA deoxyribonucleic acid. The complex molecule that encodes genetic information in all organisms here on Earth. (85.2)

Doppler broadening the widening of a spectral line caused by motions of the material producing the line both toward and away from the observer. (25.2)

Doppler shift the change in the observed wavelength of radiation caused by the motion of the emitting body or of the observer. The shift is an increase in the wavelength if the source and observer move apart and a decrease in the wavelength if the source and observer approach. See also **redshift** and **blueshift.** (25.0)

Doppler shift method a means of detecting planets orbiting other stars by detecting the Doppler shift caused by the star's own reflex motion in response to the orbiting planet. (36.2)

Drake equation expression that gives an estimate of the probability that intelligence exists elsewhere in the galaxy, based on a number of conditions thought to be essential for intelligent life to develop. (86.3)

dust tail a comet tail containing dust that reflects sunlight. The dust in a comet tail is expelled from the nucleus of the comet. (49.1)

dwarf a small dim star.

dwarf galaxy small galaxy containing a few million stars. (76.1, 77.1)

dwarf planet an object orbiting the Sun that is so massive that its gravity pulls it into a roughly spherical shape but, since it is not the dominant mass in the neighborhood of its orbit, it cannot be called a planet. Pluto, Eris, and Ceres are considered dwarf planets. Several dozen trans-Neptunian objects (TNOs) may soon be added to that category. (34.1, 48.5)

dynamo see **magnetic dynamo.**

E

early universe the first second or so of the universe's history after the Big Bang when the universe was composed of nothing but radiation and elementary particles. (83.3)

eccentricity how round or "stretched out" an orbital ellipse is. A circular orbit has zero eccentricity, while extremely elongated orbits have eccentricities close to one. (12.2)

eclipse the blockage of light from one astronomical body caused by the passage of another between it and the observer. The shadow of one astronomical body falling on another. For example, the passage of the Moon between the Earth and Sun can block the Sun's light and cause a solar eclipse. (8.2)

eclipsing binary a binary star pair in which one star periodically passes in front of the other, totally or partially blocking the background star from view as seen from Earth. (57.1, 58.2)

ecliptic the path followed by the Sun around the celestial sphere. The path gets its name because eclipses can occur only when the Moon crosses the ecliptic. (6.2)

Einstein, Albert (1879–1955) a German-born theoretical physicist, probably best known for his theories of special and general relativity, which describe the effects of motion and mass on space and time. (26.0)

electric charge the electrical property of objects that causes them to attract or repel one another. A charge may be either positive or negative. Unlike charges (+ and −) attract each other, but like charges (+ and + or − and −) repel each other. (4.2)

electromagnetic force the force arising between electrically charged particles, between charges and magnetic fields, and between magnets. This force holds electrons to the nuclei of atoms, makes moving charges spiral around magnetic field lines, and deflects a compass needle. (4.3)

electromagnetic spectrum the assemblage of all wavelengths of electromagnetic radiation. The spectrum includes the following wavelengths, from long to short: radio, infrared, visible light, ultraviolet, X-rays, and gamma rays. (22.0)

electromagnetic wave a wave consisting of alternating electric and magnetic energy. Ordinary visible light is an electromagnetic wave, and its wavelength determines the light's color. (21.1)

electron a low-mass, negatively charged subatomic particle. In an atom, electrons orbit the nucleus, but may at times be torn free if **ionized.** (4.2)

electron degeneracy a condition resisted by electrons because they cannot occupy the same space if they have the same energy. (66.2)

electronic transition a shift in energy level that an atom undergoes when an electron shifts from one energy level to another. (21.4)

electroweak force a combined form (or unification) of the **electromagnetic force** and **weak force** that occurs at high energies.

element a fundamental substance, such as hydrogen, carbon, or oxygen, that cannot be broken down into a simpler chemical substance. Approximately 100 elements occur in nature. (21.3)

ellipse a geometric figure related to a circle but elongated along one axis. (12.2)

elliptical (E) galaxy a galaxy in which the stars smoothly fill an ellipsoidal volume. Abbreviated as E galaxy. The stars in such systems are generally old (Pop II). (76.1)

emission the production of light or other electromagnetic radiation, by an atom or molecule when the atom drops to a lower energy level. (21.4)

emission-line spectrum a spectrum consisting of bright lines at certain wavelengths separated by darker regions in which little or no radiation is emitted. (24.2)

emission nebula a hot gas cloud in interstellar space that emits light. (73.1)

energy a quantity that measures the ability of a system to do work or cause motion. (20.1)

energy level any of the numerous levels that an electron can occupy in an atom, roughly corresponding to an electron orbital. (24.1)

epicycle a fictitious, small, circular orbit superimposed on another circular orbit. Epicycles were proposed by early astronomers to explain the retrograde motion of the planets and to make fine adjustments to the predictions of planets' positions. (11.2)

equation of time a relationship describing the offset of time measured by a sundial from the average 24-hour period from noon to noon. The offsets vary through the year because of the Earth's elliptical orbit and the tilt of its axis. (13.5)

equator the imaginary line that divides the Earth (or any other body) symmetrically into its Northern and Southern Hemispheres. The equator is perpendicular to a body's rotation axis.

equatorial mount a support structure for a telescope that allows it to rotate along celestial coordinates: north and south (in declination) and around the celestial poles (in right ascension). (33.4)

equilibrium temperature the temperature reached when heating and cooling effects are in balance. A planet generally reaches an equilibrium temperature when the rate of heating caused by the absorption of sunlight matches the rate of cooling caused by the emission of thermal radiation from the warmed surface. (38.2)

equinox the time of year when the Sun crosses the celestial equator. The number of hours of daylight and night are approximately equal, and the two dates mark the beginning of the spring and fall seasons. (6.5)

Eratosthenes (276–195 B.C.E.) Greek astronomer who first estimated the diameter of the Earth. (10.3)

Eris the largest known **trans-Neptunian object.** Eris is larger and about 27% more massive than Pluto. (48.5)

erosion the wearing down of geological features by wind, water, ice, and other phenomena of planetary weather. (38.4)

escape velocity the speed an object needs to move away from another body in order not to be pulled back by its gravitational attraction. Mathematically, the escape velocity, V, is defined as $\sqrt{2GM/R}$, where M is the body's mass, R is its radius, and G is the gravitational constant. (18.2, 69.1)

Europa the smallest of the four Galilean satellites of Jupiter. (48.1)

eukaryotes cells with nuclei that evolved after prokaryotes. (85.3)

Evening Star any bright planet, but most often Venus, seen low in the western sky after sunset. (13.4)

event horizon the location of the "boundary" of a black hole. An outside observer cannot see in past the event horizon. (69.2)

excited state the condition in which the electrons of an atom are not in their lowest energy level (lowest orbital). (21.4)

exclusion principle the condition that no two electrons may have the same state in an atom. Because this limitation forces electrons in extremely dense materials to occupy high energy levels, it leads to a **degenerate gas**. (63.3)

exoplanet a planet orbiting a star other than our Sun. (36.0, 86.2)

expansion redshift a lengthening of the wavelength of electromagnetic radiation caused when the waves travel through expanding space. The redshift of distant galaxies is an example of this phenomenon. (75.3)

extinction the dimming of starlight due to absorption and scattering by interstellar dust particles. (73.2)

extrasolar planet see **exoplanet**.

F

false-color image a depiction of an astronomical object in which the colors are not the object's real colors. Instead, they are colors arbitrarily chosen to represent other properties of the body, such as the intensity of radiation at other than visible wavelengths. (28.3)

fault in geology, a crack or break in the crust of a planet along which slippage or movement can take place. (41.2, 47.2)

Fermi paradox the question why, if there are many advanced civilizations in the galaxy, none of them are visiting us or have left indications of their existence. (86.3)

filter (for light) a material that transmits only particular wavelengths of light. (29.3)

first-quarter moon a phase of the waxing Moon when Earth-based observers see half of the Moon's illuminated hemisphere. (8.1)

fission the splitting of an atom into two or more smaller atoms.

flare an outburst of energy on the Sun. (53.2)

flat rotation curve a plot showing that the orbital speed in a galaxy remains nearly constant out to large radii. (79.1)

flat universe a universe that extends forever with no curvature. (82.1)

flatness problem astronomers' measurements show the universe is almost precisely flat, with zero curvature, although Einstein's theory of general relativity predicts that space is likely to be curved by gravity. This mystery is solved by the theory of inflation, which drives the universe toward flatness. (83.5)

fluorescence the conversion of ultraviolet light (or other short-wavelength radiation) into visible light. This occurs, for example, when an atom is **excited** into a high energy level by an ultraviolet photon, and then descends to the ground state in a series of steps, emitting lower-energy photons. (49.2)

focal length the distance between a mirror or lens and the point at which the lens or mirror brings a distant source of light into focus. (33.4)

focal plane the surface where the lenses and/or mirrors of a telescope form an image of a distant object. (30.1)

focus (1) one of two points within an ellipse used to generate the elliptical shape. The Sun lies at the focus of each planet's elliptical orbit. (12.2) (2) A point in an optical system in which light rays are brought together; the location where an image forms in such systems. (28.1, 30.0)

force a push or pull; anything that can cause a body to change velocity. (14.3)

fovea a central part of the retina, where most of our color vision is concentrated. This region is not very sensitive at low light levels. (33.1)

free fall the state of motion of a body that is responding to the force of gravity alone. (27.1)

frequency the number of times per second that a wave vibrates. (22.1)

frost line the distance from the Sun in the solar nebula beyond which it was cold enough for water to condense and therefore become a major component of planetesimals forming there. (35.3, 44.4)

full moon phase of the Moon in which it appears as a completely illuminated circular disk in the sky. (8.1)

fully convective a condition seen in some stars (such as very low-mass main-sequence stars) in which convection circulates matter between the core and the surface. (62.2)

G

g the acceleration due to gravity on the surface of the Earth—about 9.8 meters/second^2. (16.3)

G Newton's universal gravitational constant, which allows us to determine the force between objects if we know their masses and separation. (16.2)

Gaia hypothesis the hypothesis that life does not merely respond to its environment but actually alters its planet's atmosphere and temperature to make the planet more hospitable. For example, by photosynthesis, plants have created an oxygen-rich atmosphere on Earth, which shields the plants from dangerous ultraviolet radiation. Gaia is pronounced *GUY-uh* in this context. (85.4)

galactic cannibalism the capture and merging of one galaxy into another. (72.3, 76.4)

Galactic plane the flat region defined by the disk of the Milky Way. (71.2)

galaxy a massive system of stars, gas, and dark matter held together by their mutual gravity. Typical galaxies have a mass between about 10^7 and 10^{13} solar masses. The Milky Way is our Galaxy, which is usually capitalized. (2.4, 71.0, 75.0)

galaxy cluster a set of hundreds or thousands of galaxies held together by their mutual gravitational attraction. The Milky Way lies on the outskirts of the Virgo Cluster. (2.4, 77.0)

galaxy group a system of from two to several dozen galaxies held together by their mutual gravitational attraction. The Milky Way belongs to the Local Group galaxy cluster. (2.4, 77.1)

Galilean relativity a method for determining the relative speeds of motion seen by observers who are moving with respect to each other. This method successfully describes motions at slow speeds, but not when speeds grow to an appreciable fraction of the speed of light. (26.1)

Galilean satellites the four moons of Jupiter discovered by Galileo: Io, Europa, Ganymede, and Callisto. (12.3, 47.1, 48.1)

Galilei, Galileo (1564–1642) Italian physicist and astronomer who studied inertia and published the first studies of the sky with a telescope. (12.3)

gamma ray electromagnetic waves that have the shortest wavelength. (22.4)

gamma-ray burst a brief flash of very high-energy radiation from what initially appeared to be random positions on the sky. Recent observations have connected many of these bursts with supernovae in extremely distant galaxies. (67.5)

Ganymede the largest moon of Jupiter and the largest satellite in the Solar System—even larger than Mercury. (48.1)

gas giant a Jovian planet. (34.1)

general relativity Einstein's theory relating how mass and energy can change the structure of space and time, "curving" them. (27.0, 69.0, 80.2)

geocentric model a hypothesis that held that the Earth is at the center of the universe and all other bodies are in orbit around it. Early astronomers thought that the Solar System was geocentric. (11.2)

giant a star of large radius and large luminosity. (63.0)

giant elliptical galaxy a large galaxy found in the crowded inner regions of rich clusters. These probably form by galactic cannibalism. (77.2)

giant molecular clouds interstellar clouds, in which atoms combine to form molecules and heavier elements combine to form interstellar dust, with masses up to about a million solar masses. (61.1)

gibbous appearance of the Moon (or a planet) when more than half (but not all) of the sunlit hemisphere is visible from Earth. (8.1)

glaciation the condition of becoming covered with glaciers and ice. Earth has undergone periods of glaciation about every 100,000 years during the last million years. (38.3)

glitch abrupt change in the pulsation period of a pulsar, perhaps the result of adjustments of its crust. (68.2)

global warming a phenomenon in which average ocean and air temperatures on Earth have increased by about 1.6 degrees Celsius over the last century. Most scientists attribute global warming to **greenhouse gases** produced by human activities such as burning of fossil fuels and deforestation. (38.2)

globular cluster a dense grouping of old stars, containing generally about 10^5 to 10^6 members. They are often found in the halos of galaxies. (70.1, 71.3)

globule see **Bok globule**.

gluon a particle that carries (or exerts) the strong force between quarks. (4.3)

grains small solid particles that condense from the heavy elements in a gas when it cools. (65.2)

grand unified theory (GUT) a theory that describes and explains the four fundamental forces as different aspects of a single force. (83.4)

granulation texture seen in the Sun's photosphere. Granulation is created by clumps of hot gas that rise to the Sun's surface. (51.3)

grating a piece of material with many fine, closely spaced parallel lines used to create a spectrum. Light may be reflected off a grating or diffracted as it passes through a transparent grating. (24.0)

gravitational force force exerted on one body by another due to the effect of gravity. The force is directly proportional to the masses of both bodies involved and is inversely proportional to the square of the distance between them. (4.2)

gravitational lens an object whose gravity curves space in a way that can focus the light from an object behind it. The light from the more distant object may be magnified, distorted, or turned into multiple images by the lens. See also **curvature of space**. (79.3)

gravitational lensing method a technique for detecting a planet orbiting a star by observing the gravitational force of the planet acting as a lens to focus the light of the star. (27.2, 36.4)

gravitational potential energy the energy stored in a body subject to the gravitational attraction of another body. As the body falls, its gravitational potential energy decreases and is converted into kinetic energy. (20.1)

gravitational redshift the shift in wavelength of electromagnetic radiation (light) caused by a body's gravitational field as the radiation moves away from the body. Only extremely dense objects, such as white dwarfs or neutron stars, produce a significant gravitational redshift of their radiation. (27.3)

gravitational time dilation the slowing of the passage of time caused wherever a gravitational field is stronger. (27.3, 69.2)

gravitational waves a wavelike disturbance in the curvature of space that is generated by the acceleration of massive bodies. (27.4)

graviton a particle hypothesized to carry the gravitational force. (4.3)

gravity the force of attraction between two bodies that is generated by their masses. (2.0)

Great Attractor a large concentration of mass toward which everything in our part of the universe apparently is being pulled. (77.3)

great dark spot a large atmospheric vortex discovered on Neptune, similar to Jupiter's Great Red Spot. The spot was discovered by the *Voyager II* spacecraft in 1989, but has since disappeared. (46.2)

greatest elongation the position of an inner planet (Mercury or Venus) when it lies farthest from the Sun on the sky. Mercury and Venus are particularly easy to see when they are at greatest elongation. Objects may be at greatest eastern or western elongation according to whether they lie east or west of the Sun. (11.3)

Great Red Spot a reddish elliptical spot about 40,000 km by 15,000 km in size in the southern hemisphere of the atmosphere of Jupiter. The Red Spot has existed for over three centuries. (45.3)

greenhouse effect the trapping of heat by a planet's atmosphere, making the planet warmer than would otherwise be expected. Generally, the greenhouse effect operates if visible sunlight passes freely through a planet's atmosphere, but the infrared radiation produced by the warm surface cannot escape readily into space. (38.2)

greenhouse gas a molecule (such as carbon dioxide, methane, or water vapor) that efficiently absorbs infrared radiation. (41.1)

Gregorian calendar the calendar devised at the request of Pope Gregory XIII in 1582. It is the common civil calendar used throughout the world today. It improved upon previous calendars by omitting the leap year for century years not divisible evenly by 400. (9.4)

grooved terrain regions of the surface of Ganymede consisting of parallel grooves. Believed to have formed by repeated fracture of the icy crust. (48.1)

ground state the lowest energy level of an atom, generally when the electrons are in the smallest possible orbitals. (21.3)

H

half-life the time required for half of the atoms of a radioactive substance to disintegrate. (35.1)

Halley's comet a comet that reappears about every 76 years, famous because it was the first comet whose return was predicted. (49.1)

halo the approximately spherical region surrounding spiral galaxies. The halo contains mainly old stars, as are found, for example, in globular clusters. The halo also appears to contain large amounts of dark matter. (71.4)

Hawking radiation radiation that black holes are hypothesized to emit as a result of quantum effects. This radiation leads to the extremely slow loss of mass from black holes. (69.4)

heat death a scenario in which the universe dies as all available fuel is consumed, converted into heat from which no further energy can be extracted. (84.4)

heliocentric model a model of the Solar System in which Earth and the other planets orbit the Sun. (11.3)

helium flash the beginning of helium fusion in a low-mass star. The fusion begins explosively and causes a major readjustment of the star's structure. (63.3)

hertz the MKS unit of frequency: one wave or cycle per second. Named for Heinrich Hertz, who first produced radio radiation. (22.1)

H I emission 21-cm wavelength radio waves produced by atoms of un-ionized hydrogen. (73.3)

H II region a region of **ionized** hydrogen. H II regions generally have a luminous pink/red glow and often surround luminous, hot, young stars. (73.4)

highlands the older, most heavily cratered regions on the Moon. (39.2)

high-mass star a star born with a mass above about $8 M_\odot$. High-mass stars end their lives by fusing elements up to iron in their cores and then explode as core-collapse supernovae. (60.2)

Hipparchus (about 190–120 B.C.E.) Greek astronomer who made one of the first detailed charts of the stars. (54.1)

homogeneous having a consistent and even distribution of matter that is the same everywhere. (81.1)

horizon the line separating the sky from the ground. See also **cosmic horizon** and **event horizon**. (5.1)

horizon problem the **cosmic microwave background** has almost precisely the same temperature in every direction, yet regions more than one degree apart are outside each other's **cosmic horizon**, are totally independent, and there is no obvious reason they should have the same temperature. This uniformity is explained with the theory of an inflation epoch. (83.5)

H-R diagram (or Hertzsprung-Russell diagram) a graph on which stars are located according to their temperature and luminosity. Most stars on such a plot lie along a diagonal line, called the **main sequence**, which runs from cool, dim stars in the lower right to hot, luminous stars in the upper left. (59.0)

Hubble's constant the multiplying constant H in **Hubble's law**: $V = H \times d$. The reciprocal of Hubble's constant gives the approximate age of the universe. (75.3)

Hubble's law a relation between a galaxy's distance, d, and its recession velocity, V, which states that more distant galaxies recede faster than nearby ones. Mathematically, $V = H \times d$, where H is Hubble's constant. (75.3)

Hubble time an estimate of the age of the universe obtained by taking the inverse of Hubble's constant. The estimate assumes that the universe has expanded at a constant rate. (80.4)

hydrosphere refers to the "layer" of water on the Earth consisting of oceans, lakes, rivers, ice caps, and other liquid water and ice. (38.0)

hydrostatic equilibrium the condition in which pressure and gravitational forces in a star or planet are in balance. Without such balance, bodies will either collapse or expand. (51.2, 62.1)

hyperbola the flattened curve of a trajectory at speeds higher than the **escape velocity**. (18.3)

hypernova a term used to describe a hypothesized kind of supernova explosion of a star so massive that its core collapses directly into a black hole. (67.5)

hypothesis an explanation proposed to account for some set of observations or facts. (4.1)

I

ice giants a term sometimes used to distinguish Uranus and Neptune from the gas giants Jupiter and Saturn. Despite the name, these planets have interiors hotter than Earth's, but the name is descriptive of their likely origin from the accretion of icy planetesimals. (46.0)

ideal gas law a law relating the pressure, density, and temperature of a gas. This law states that the pressure is proportional to the density times the temperature. (51.2)

impact the collision of a small body (such as an asteroid or a comet) with a larger object (such as a planet or moon). (39.2)

inclination the angle by which an astronomical object or its orbit is tilted.

inertia the tendency of an object at rest to remain at rest and of a body in motion to continue in motion in a straight line at a constant speed. See also **mass**. (14.1)

inferior conjunction see **conjunction**.

inferior planet a planet whose orbit lies between the Earth's orbit and the Sun. Mercury and Venus are inferior planets.

inflation the enormously rapid expansion of the early universe. Between about 10^{-35} and 10^{-33} seconds after the Big Bang, the universe expanded by a factor of perhaps 10^{30}. (82.4, 83.4)

infrared a wavelength of electromagnetic radiation longer than visible light but shorter than radio waves. We cannot see these wavelengths with our eyes, but we can feel many of them as heat. The infrared wavelength region runs from about 700 nm to 1 mm. (22.3)

initial mass function the relative number of stars born onto the main sequence at each mass. (70.3)

inner core the inner portion of Earth's iron–nickel core. Despite its high temperature, the inner part of the core is solid because it is under great pressure. (37.1)

inner planet a planet orbiting in the inner part of the Solar System. Usually taken to mean Mercury, Venus, Earth, or Mars. (34.1)

instability strip a region in the H-R diagram containing stars that pulsate. These stars have layers below the surface at a temperature that makes them unstable, so the outer layers expand and contract rhythmically. (64.2)

instrumentation a branch of astronomy dealing with the development of new kinds of detectors. (28.1)

interfere, interference a phenomenon in which electromagnetic waves mix together such that their crests and troughs can alternately reinforce and cancel one another. (21.1, 31.1)

interferometer a device consisting of two or more telescopes connected together to work as a single instrument. Used to obtain a high resolution, with the ability to see small-scale features. Most interferometers currently operate at radio, infrared, or visible wavelengths. (31.3)

interferometry the use of an **interferometer** to measure small angular-size features. (58.1)

international date line an imaginary line extending from the Earth's North Pole to the South Pole, running approximately down the middle of the Pacific Ocean. It marks the location on Earth at which the date changes to the next day as one travels across it from east to west. (7.3)

interstellar cloud a cloud of gas and dust in between the stars. Such clouds may be many light-years in diameter. (35.2, 60.2)

interstellar dust tiny solid grains in interstellar space, thought to consist of a core of rocklike material (silicates) or graphite surrounded by a mantle of ices. Water, methane, and ammonia are probably the most abundant ices. (61.1, 71.3)

interstellar grain microscopic solid dust particles in interstellar space. These grains absorb starlight, making distant stars appear dimmer than they truly are. (35.2)

interstellar medium (ISM) gas and dust between the stars in a galaxy. (73.0)

inverse-square law (1) any law in which some property varies inversely as the square of the distance, d (mathematically, as $1/d^2$). (2) The law stating that the apparent brightness of a body decreases inversely as the square of its distance. (21.2, 55.2)

Io the innermost of the Galilean satellites of Jupiter, noted especially for its active volcanoes. (48.1)

ionization energy the energy needed to remove an electron from an atom. (24.4)

ionize remove one or more electrons from an atom, leaving the atom with a positive electric charge. Under some circumstances, an extra electron may be attached to an atom, in which case the atom is described as negatively ionized. (21.4)

ionosphere the upper region of the Earth's atmosphere in which many of the atoms are ionized. (38.1)

ion tail a stream of ionized particles evaporated from a comet and then swept away from the Sun by the solar wind. (49.1)

iris the region of the eye surrounding the pupil that can control the amount of light entering the eye. (33.1)

iron core the endpoint of fusion in a massive star. When the iron core grows massive enough, the star undergoes a supernova explosion. (60.3)

irregular (Irr) galaxy a galaxy lacking a symmetrical structure. (76.1)

irregular variables a class of variable star that changes in brightness erratically. (64.1)

Ishtar a continent-like highland region on Venus. (41.2)

isotope an atom that has the same number of protons but a different number of neutrons in its nucleus. (35.1, 52.2)

J

jet narrow stream of gas ejected from any of several types of astronomical objects. Jets are seen extending from protostars and active galactic nuclei. They may be funneled into their direction of flow by magnetic fields or accretion disks surrounding the central object. (78.1)

jet stream a narrow stream of high-speed wind that blows in the atmosphere of a planet, usually at high altitudes. (38.5)

joule the MKS unit of energy. One joule per second equals one watt. (20.1)

Jovian planet a giant gaseous planet such as Jupiter, Saturn, Uranus, or Neptune. The term *Jovian* means Jupiter-like and is based on "Jove," an alternative name for Jupiter. (34.1)

Julian calendar a 12-month calendar devised under the direction of Julius Caesar. It includes a leap year every four years. (9.3)

K

Kapteyn's Universe a model of the known universe in wide usage in the early 1900s. Based on observations of the time, it was limited to stars within the disk of the Milky Way surrounding the Sun. (71.2)

kelvin the most commonly used temperature scale in science, defined such that absolute zero is 0 K and water freezes at 273.15 K. (23.3)

Kepler, Johannes (1571–1630) German astronomer who first discovered the elliptical orbit of Mars. (12.2)

Kepler's three laws mathematical descriptions of the motion of planets around the Sun. The first law states that planets move in elliptical orbits with the Sun off-center at a focus of the ellipse. The second law states that a line joining the planet and the Sun sweeps over equal areas in equal time intervals. The third law relates a planet's orbital period, P, to the semimajor axis of its elliptical orbit, a. Mathematically, the law states that $P^2 = a^3$, if P is measured in years and a in astronomical units. (12.2)

kiloparsec a unit of distance, equal to 1000 parsecs (pc), often used to describe distances within the Milky Way or the Local Group of galaxies. (71.2)

kinetic energy energy of motion. Kinetic energy is calculated from half the product of a body's mass and the square of its speed ($½mV^2$). (20.1)

Kirchhoff's laws a set of three "rules" that describe how continuous, bright line, and dark line spectra are produced. (24.3)

Kirkwood gaps regions in the asteroid belt with a lower-than-average number of asteroids. Some of the gaps result from the gravitational force of Jupiter removing asteroids from orbits within the gaps. (43.4)

Kuiper belt a region containing many large icy bodies and from which some comets come. The region appears to extend from the orbit of Neptune at about 30 AU, past Pluto, out to approximately 55 AU. (34.1, 48.4, 49.3)

L

Lambda (λ) a quantity in the equations of general relativity indicating the value of the cosmological constant. (82.4)

laminated terrain alternating layers of ice and dust seen in the ice caps of Mars. The layers appear to reflect climate changes caused by long-term orbital variations. (42.1)

large-scale structure the structure of the universe on size scales larger than that of clusters of galaxies. (77.4)

last scattering epoch the period of time a few hundred thousand years after the Big Bang when the density and temperature of gas in the universe dropped sufficiently that photons could begin to travel freely through space. The time of the formation of the **cosmic microwave background**. (81.2)

latitude the angular distance of a point north or south of the equator of a body. The equator has a latitude of 0° while the poles are at latitudes of 90° north and south. (5.4)

law in science, generally a theory that can be expressed in a mathematical form. (4.1)

law of gravity a description of the gravitational force exerted by one body on another. The gravitational force is proportional to the product of their masses and the inverse square of the distance between them. If the masses are M and m and their separation is d, the force between them, F, is $F = G\,Mm/d^2$, where G is a physical constant. (16.2)

leap second a time adjustment added every few years to clocks around the world to adjust them to account for Earth's gradually slowing spin. (7.5)

leap year a year in which there are 366 days implemented under the Julian and Gregorian calendar systems. (9.4)

lens an optical instrument, made of glass or some other transparent material, shaped so that parallel rays of light passing through it are bent to arrive at a single focus. (30.1)

light electromagnetic energy. (21.1)

light curve a plot of the brightness of a body versus time. (58.2, 64.1, 66.3)

light pollution the illumination of the night sky by waste light from cities and outdoor lighting. Light pollution makes it difficult to observe faint objects. (32.1)

light travel time distance a measure of how far light would have traveled in the time since it was emitted if the universe were not expanding. (80.4)

light-year a unit of distance equal to the distance light travels in one year. A light-year is roughly 10^{13} km, or about 6 trillion miles. (2.2, 3.3, 55.0)

liquid core see **outer core**.

liquid metallic hydrogen a form of hot, highly compressed hydrogen that is a good electrical conductor, found in the interiors of Jupiter and Saturn. (45.2)

lobe see **radio lobe**.

Local Group the small group of three spiral galaxies and several dozen small galaxies to which the **Milky Way** belongs. (2.4, 77.1)

Local Supercluster the cluster of galaxy clusters in which the **Milky Way, Local Group**, and **Virgo Cluster** are located. (77.3)

loners those who argue that humans are probably the only intelligent beings within our Galaxy. (86.3)

longitude a coordinate indicating the east–west location of a point on the Earth's surface. The line of 0° longitude runs north–south through an observatory in Greenwich, England. (5.4)

long-period comet a comet whose orbital period is longer than 200 years. These comets may originate from the Oort cloud. (49.3)

Lorentz factor a term that describes by how much time, space, and mass are altered as a result of their motion. The Lorentz factor is very close to 1 (indicating very little change) except at speeds approaching the speed of light. (26.2)

low-mass star a star born with a mass less than about 8 times that of the Sun; progenitor of a white dwarf. (60.2)

low surface brightness galaxy a galaxy with a low density of stars, making it difficult to detect in visible-wavelength surveys. (76.1)

luminosity the amount of energy radiated per second by a body. For example, the wattage of a lightbulb defines its luminosity. Stellar luminosity is usually measured in units of the Sun's luminosity (approximately 4×10^{26} watts). (23.2, 55.1)

luminosity class a set of categories describing stars' luminosities relative to their luminosity as a main-sequence star. The most luminous is class Ia, then Ib, II, etc.; main-sequence stars are defined as class V. (59.3)

lunar eclipse the passage of the Earth between the Sun and the Moon so that the Earth's shadow falls on the Moon. (8.2)

lunar month the time period of approximately $29\frac{1}{2}$ days during which the Moon cycles from the new phase through all of its phases back to new. (8.1)

lunisolar calendar a calendar system that maintains links to both the lunar month and the solar year, such as the Chinese or Jewish calendars. Because the lunar month does not divide evenly into the year, lunisolar calendar systems must insert an extra month every few years. (9.2)

M

Magellanic Clouds two small companion galaxies of the **Milky Way**. (77.1)

magnetar a highly magnetized neutron star that emits bursts of gamma rays. (68.3)

magnetic dynamo a process whereby magnetic fields are generated by electric currents circulating inside an astronomical body. The process is thought to arise when there is electrically-conductive material flowing inside a rotating body. (37.4)

magnetic field a representation of the means by which magnetic forces are transmitted from one body to another. (37.4)

magnetogram an image of the Sun coded in a way to show the strength and direction of the magnetic field. (53.1)

magnitude see **apparent magnitude**.

main sequence the region in the H-R diagram in which most stars, including the Sun, are located. The main sequence runs diagonally across the H-R diagram from cool, dim stars to hot, luminous ones. See also **H-R diagram**. (59.1)

main-sequence lifetime the time a star remains a main-sequence star, fusing hydrogen into helium in its core. (62.3)

main-sequence star a star, fusing hydrogen to helium in its core, whose surface temperature and luminosity place it on the **main sequence** on the Hertzsprung-Russell diagram. (60.2, 62.0)

mantle the outer part of a rocky planet immediately below the crust and outside the metallic core. (37.1)

many-worlders those who argue that the development of life and intelligence are probably common occurrences within our galaxy. (86.3)

mare a vast, smooth, dark area on the Moon with a roughly circular shape, composed of basalt, a dark, congealed lava. (39.2)

maria plural of **mare**.

maser an intense radio source created when excited gas amplifies some background radiation. *Maser* stands for "microwave amplification by stimulated emission of radiation."

mass a measure of the amount of material an object contains. A quantity measuring a body's **inertia**. (14.1, 34.1)

mass excess the amount of mass—or equivalently, energy, according to Einstein's formula $E = m \times c^2$—potentially available from an element through nuclear reactions. (67.2)

massive compact halo object (MACHO) collective name for bodies of planetary or stellar mass that are hypothesized to contribute a significant amount of mass to our Galaxy's halo, but that generate relatively little detectable light. (79.4)

mass–luminosity relation a relation between the mass and luminosity of stars. Higher-mass stars have higher luminosity, so that $L \approx M^{3.5}$ where L and M are measured in solar units. (59.2)

mass transfer process by which one star in a binary system transfers matter onto the other. (66.1)

matter era the times after the end of the **radiation era**, when radiation from the Big Bang cooled so that matter became the dominant constituent of the universe.

Maunder minimum the period from about A.D. 1645 to 1715 during which the Sun was relatively inactive. Few sunspots were observed during this period, and Earth's temperature was cooler than normal. (53.4)

Maxwell Montes a set of mountains on Venus, including the tallest volcano; one of the first features detected by radar imaging. (41.2)

mean solar day the standard 24-hour day. The mean solar day is based on the average day's length over a year. Using a mean is necessary because the time interval from noon to noon, or sunrise to sunrise, varies slightly throughout the year. (7.1, 13.5)

megaparsec a distance unit equal to 1 million **parsecs** and abbreviated Mpc. (75.3)

merger the process in which galaxies collide and form a single larger system. (76.4)

meridian the line passing north–south from horizon to horizon and passing through an observer's zenith. The meridian is the dividing line between the eastern and western halves of the sky. (7.1)

metals astronomically, any chemical element more massive than helium. Thus, carbon, oxygen, iron, and so forth are metals. (72.1)

meteor the bright trail of light created by small solid particles entering the Earth's atmosphere and burning up; a "shooting star." (49.4, 50.0)

meteor shower an event in which many meteors occur in a short space of time, all from the same general direction in the sky, typically when the Earth's orbit intersects debris left by a comet. The most famous shower is the Perseids in mid-August. (49.4)

meteorite the solid remains of a meteor that falls to the Earth. (35.1, 50.0)

meteoroid small, solid bodies moving within the Solar System. When a meteoroid enters our atmosphere and heats up, the trail of luminous gas it leaves is called a **meteor**. When the body lands on the ground, it is called a **meteorite**. ("A meteoroid is in the void. A meteor above you soars. A meteorite is in your sight.") (50.0)

method of standard candles see **standard candle**.

metric prefix a term that can be attached to the front of a measurement unit to indicate a power of 10 times the unit. For example, the prefix *kilo-* means a thousand, so a kilometer is a thousand meters. (3.1)

metric system   a set of units developed in the late 1800s to replace the confusing systems then used. See also **MKS system**. (1.1, 3.1)

microwaves   a range of electromagnetic radiation lying between radio and infrared wavelengths, typically with wavelengths of from about 1 to 100 millimeters. (22.4)

mid-ocean ridge   an underwater mountain range on Earth created by plate tectonic motion.

Milankovitch cycles   changes to Earth's climate occurring over tens to hundreds of thousands of years as a result of cyclical changes in Earth's orbital eccentricity and in the direction and tilt of its axis. (38.3)

Milky Way   the Galaxy to which the Sun belongs. Seen from Earth, our Galaxy is a pale, milky band in the night sky. (2.3, 71.0)

Miller–Urey experiment   an experimental attempt to simulate the conditions under which life might have developed on Earth. Miller and Urey discovered that amino acids and other complex organic compounds could form from the gases that are thought to have been present in the Earth's early atmosphere, if the gases are subjected to an electric spark or ultraviolet radiation. (85.3)

millisecond pulsar   a pulsar whose rotation period is about a millisecond. (68.2)

MKS system   the form of the **metric system** using meters, kilograms, and seconds to measure length, mass, and time, respectively. (3.1)

model   a theoretical representation of some object or system. (4.1)

molecular cloud   relatively dense, cool interstellar cloud in which molecules are common. (61.1, 73.3)

molecule   two or more atoms bonded into a single particle, such as water (H_2O—two hydrogen atoms bonded to one oxygen) or carbon dioxide (CO_2—one carbon atom bonded to two oxygen atoms).

month   an interval of time loosely resembling the lunar month in some calendar systems, but closely tied to the cycle of moon phases in others. (8.0)

Moon illusion   the optical illusion that the Moon appears larger when nearer the horizon than when seen high in the sky. (10.5)

Morning Star   the planet Venus when it is seen in the eastern sky. (13.4)

mountain ranges   long parallel ridges of mountains formed by the buckling surface of colliding plates. See also **plate tectonics**. (37.3)

multiverse   the set of all "universes" that make up the whole of all physical reality on a vastly larger scale than our own universe. This construct was developed from models of the nature of space, time, and the formation of our universe, which suggest that universes may form that are completely separate from our universe and perhaps have physical properties and dimensions that are entirely different from those of our universe. (84.5)

N

nadir   the point on the celestial sphere directly below the observer and opposite the zenith. (5.4)

nanometer   a unit of length equal to 1 billionth of a meter (10^{-9} meters) and abbreviated nm. Wavelengths of visible light are several hundred nanometers. The diameter of a hydrogen atom is roughly 0.1 nm.

neap tide   a weak tide occurring when the Moon is at first or third quarter, so the Sun's and Moon's gravitational effects on the ocean partially offset each other. (19.3)

near-Earth object   an asteroid with an orbit that crosses or comes close to the Earth's orbit. (43.4)

nebula   a cloud in interstellar space, usually consisting of gas and dust.

negative curvature   a form of curved space sometimes described as being "open" because it has no boundary. Negative curvature is analogous to a saddle shape. (82.1)

Nereid   a moderately large satellite of Neptune with an irregular orbit. (48.3)

net force   the final resulting force on an object from the combination of all the individual forces acting upon it. (14.3)

neutrino   tiny neutral particle with little or no mass and immense penetrating power. These particles are produced by stars when they fuse hydrogen into helium. They are also created in the early universe and during a supernova when a star's core collapses to form a neutron star. (4.4, 52.0)

neutron   a subatomic particle of nearly the same mass as the proton but with no electric charge. Neutrons and protons comprise the nucleus of the atom. (4.2, 52.1)

neutron star   a very dense, compact star composed primarily of neutrons. Protons and electrons combine to form neutrons because of the enormous gravitational pressure squeezing the iron core of a massive star late in its evolution. (60.2, 68.0)

new moon   phase of the Moon during which none of the sunlit side is visible so it appears completely dark. (8.1)

newton   the standard unit of force in the metric (MKS) system. (15.2)

Newtonian telescope   a reflecting telescope designed so that the focused light is reflected by a small secondary mirror out to the side of the telescope, where it can be viewed. (30.2)

Newton, Isaac   (1642–1722) English physicist, mathematician, and astronomer who pioneered the modern studies of motion, optics, gravity, and calculus, and showed how to apply them to the universe. (14.1)

Newton's first law of motion   a body continues in a state of rest or in uniform motion in a straight line unless acted upon by an external force. See also **inertia**. (14.2)

Newton's second law of motion $F = ma$. The amount of acceleration, a, that a force, F, produces depends inversely on the mass, m, of the object being accelerated. (15.2)

Newton's third law of motion   when two bodies interact, they exert equal and opposite forces on each other. (15.3)

Newton's version of Kepler's third law   a generalized version of the law ($P^2 = a^3$) relating the periods and semimajor axes of orbits for systems in which the mass is not dominated by the Sun. The modified form can be expressed as $(M_1 + M_2) P^2 = a^3$ where the sum of the masses is measured in solar masses. See also **Kepler's three laws.** (17.2)

nonthermal radiation radiation emitted by charged particles moving at high speed in a magnetic field. The radio emission from pulsars and radio galaxies is nonthermal emission. More generally, *nonthermal* means "not due to high temperature."

north celestial pole the point on the **celestial sphere** directly above the Earth's North Pole. The star Polaris happens to lie close to this point, and other objects in the northern sky appear to circle around this point.

North Star any star that happens to lie very close to the north celestial pole. Polaris has been the North Star for about 1000 years, and it will continue as such for about another 1000 years, at which time a star in Cepheus will be nearer the north celestial pole. (6.6)

nova an explosion of a surface layer of hydrogen that builds up on a white dwarf and then fuses rapidly into helium, causing the white dwarf to shine brightly for several weeks. Nova explosions may be recurrent. (64.1, 66.1)

nuclear fusion the binding of two lighter nuclei to form a heavier nucleus, with some nuclear mass converted to energy—for example, the fusion of hydrogen into helium. This process supplies the energy of most stars and is commonly called "burning," but burning is a chemical process that does not alter atoms' nuclei. (52.1)

nucleosynthesis the formation of elements, generally by the fusion of lighter elements into heavier ones—for example, the formation of carbon by the fusion of three helium nuclei. (67.2)

nucleus the core of an atom around which its electrons orbit. The nucleus has a positive electric charge and comprises most of an atom's mass. (4.2)

nucleus (of a comet) the core, typically a few kilometers across, of frozen gases and dust that make up the solid part of a comet. (49.1)

nucleus (of a galaxy) the central region of a galaxy where there is a supermassive black hole and energetic activity. (71.4)

O

O-B-A-F-G-K-M-L-T the sequence of stellar-spectral classifications from hottest to coolest stars. (56.3)

Occam's razor the principle of choosing the simplest scientific hypothesis that correctly explains any phenomenon. (11.2)

Olbers' paradox an argument that the sky should be bright at night because of the combined light from all the distant stars and galaxies. (81.1)

Olympus Mons a volcano on Mars, the largest volcanic peak in the Solar System. (42.1)

Omega (Ω) the ratio of the average density of matter or energy in the universe to the **critical density.** In a **flat universe,** $\Omega = 1$. (82.3)

Oort cloud a vast region in which comet nuclei orbit. This cloud lies far beyond the orbit of Pluto and may extend halfway to the next nearest star. (34.1, 49.3)

opacity the blockage of light or other electromagnetic radiation by matter.

open cluster a loose cluster of stars, generally containing a few hundred members. (70.1)

opposition the configuration of a planet when it is opposite the Sun in the sky. If a planet is in opposition, it rises when the Sun sets and sets when the Sun rises. (11.3)

orbit the path in space traveled by a celestial body.

orbital a three-dimensional electron wave in an atom. The closest analog at larger scales is an orbit, however because of the wave nature of an electron at microscopic scales it does not orbit the atoms nucleus like a planet. Each orbital has a specific **energy level.** (21.3)

orbital perturbation a deviation from the normal elliptical orbit of one body around another, generally caused by the gravitational pull of a third body. (46.1)

orbital plane the flat plane defined by the path of an orbiting body. In the Solar System, most of the bodies orbiting the Sun have similar orbital planes. (34.2)

orbital velocity the speed needed to maintain a (generally) circular orbit. (18.1)

organic compound a compound containing carbon, especially a complex carbon compound. Organic compounds can be produced by chemical processes, so they do not necessarily indicate the presence of living organisms. (50.2)

outer core the molten portion of the Earth's iron-nickel core surrounding the solid inner core. (37.1)

outer planet a planet whose orbit lies in the outer part of the Solar System. Jupiter, Saturn, Uranus, and Neptune are outer planets. (34.1)

ozone a molecule consisting of three oxygen atoms bonded together. Its chemical symbol is O_3. Because it absorbs ultraviolet radiation, ozone in our planet's stratosphere shields us from the Sun's harmful ultraviolet radiation. Too much ozone in the lower atmosphere, however, is harmful to human respiration. (38.2)

P

pancake dome an unusual surface feature on Venus, they appear as "blisters" of uplifted rock that may be produced by volcanic activity. (41.2)

panspermia a theory that life originated elsewhere than on Earth and came here across interstellar space either accidentally or deliberately. (86.1)

parabola the mathematical curve of a trajectory at **escape velocity.** (18.3)

parallax the shift in an object's perceived position caused by the observer's motion. Also a method for finding distance based on that shift. (10.2, 54.2)

parsec a unit of distance equal to about 3.26 light-years (3.09×10^{13} km). It is the distance at which an object would have a parallax of one arc second. (3.3, 54.3)

partial eclipse an alignment of two celestial bodies during which only a part of the light from one body is blocked by the other. (8.2)

pattern speed the speed at which a spiral arm orbits the center of a galaxy, which differs from the speed of any individual star. (74.3)

penumbra the outer part of the shadow of a body where sunlight is only partially blocked by the body. (8.2)

penumbral eclipse a lunar eclipse in which the Moon passes only through the Earth's **penumbra**. (8.2)

perfect gas law see **ideal gas law**.

perihelion the point in a planet's orbit where it is closest to the Sun. (40.3)

period the time required for a repetitive process to repeat. For example, *orbital period* is the time it takes a planet or star to complete an orbit. *Pulsation period* is the time it takes a star to expand and then contract back to its original radius. (12.2, 64.1)

periodic table a table displaying the properties of each type of atom, organized in order of the number of protons, or the atomic number. (4.2)

period–luminosity relation the relationship between the period of brightness variation and the luminosity of a variable star. Cepheid variables with longer periods are more luminous. (64.3)

phase the changing illumination of the Moon or other body that causes its apparent shape to change. The following is the cycle of lunar phases: new, crescent, first quarter, gibbous, full, gibbous, third quarter, crescent, new. (8.1)

Phobos one of the two small moons of Mars, thought to be a captured asteroid. Its close orbit will cause it eventually to spiral into the planet. (42.4)

photodissociation the breaking apart of a molecule by intense radiation. (41.1)

photoelectric effect emission of electrons from a material when light of a high-enough frequency strikes it. Regardless of the brightness of the light, no electrons are emitted unless the photons' energy is greater than a value that depends on the material. (28.2)

photometer a device that measures the total amount of light received from a celestial object over a chosen range of wavelengths. (55.2)

photon a particle of visible light or other electromagnetic radiation. (4.2, 21.1)

photopigment a chemical that undergoes a chemical or physical change when light shines on it. Vision is possible because of cells in the eye that have photopigments, and the chlorophyll in plants is a photopigment. (33.1)

photosphere the layer in the Sun or any other star where photons can escape into space. This appears to be a surface when we look at the Sun, but there is higher-density gas below and lower-density gas above the photosphere. (51.1, 56.1)

pixel a "picture element," consisting of an individual detector in an array of detectors used to collect light to construct an image. (28.2)

Planck time the brief interval of time, about 10^{-43} second immediately after the Big Bang, when quantum fluctuations are so large that current theories of gravity can no longer describe space and time. (83.4)

planet a body in orbit around a star. (1.1, 11.0)

planetary nebula a shell of gas ejected by a low-mass star late in its evolutionary lifetime. Planetary nebulae typically appear as a glowing gas ring around a central star. (60.2, 65.0)

planetesimal one of the numerous small, solid bodies that, when gathered together by gravity, form a planet. (35.3)

plasma a fully or partially ionized gas. (51.1)

plate tectonics a model of the movement of the Earth's crust, caused by slow circulation in the mantle. The Earth's surface is divided into large regions (plates) that move very slowly over the planet's surface. Interaction between plates at their boundaries creates mountains and activity such as volcanoes and earthquakes. (37.3)

Pleiades a nearby star cluster in the constellation Taurus. (70.1)

plutino a trans-Neptunian object whose orbital period (like that of Pluto) is in a 2:3 resonance with the orbit of Neptune. (48.5)

Pluto a small icy "planet" discovered in 1930. Recent discoveries of similar and larger bodies in the outer Solar System have resulted in Pluto being recategorized as a **dwarf planet** or **plutoid**. (48.4)

plutoid a trans-Neptunian object (TNO) that is also a **dwarf planet**. (48.5)

Polaris a moderately bright star in the constellation Ursa Minor. Polaris is also known as the "North Star" because it currently lies near the north celestial pole, but this steadily changes because of the **precession** of Earth's axis. (5.3)

poor cluster a galaxy cluster with few members. The galaxy cluster to which the Milky Way belongs, the Local Group, is a poor cluster. (77.2)

Population (Pop) I the younger stars, some of which are blue, that populate a galaxy's disk, especially its spiral arms. (72.1)

Population (Pop) II the older, redder stars that populate a galaxy's halo and bulge. (72.1)

Population (Pop) III a hypothetical stellar population consisting of the first stars that formed, composed of only hydrogen and helium. (72.2)

positive curvature bending of space leading to a finite volume. A **closed universe** with positive curvature is analogous to the surface of a sphere. (82.1)

positron a subatomic particle of antimatter with the same mass as an electron but a positive electric charge. It is the electron's antiparticle. (52.2, 83.3)

post meridian (P.M.) after the Sun has crossed the **meridian**; after noon. (7.1)

potential energy the energy stored in an object as a result of its position or arrangement. For example, a mass lifted to a greater height gains gravitational potential energy. (20.1)

powers of 10 see **scientific notation**.

precession the slow change in the direction of the pole (rotation axis) of a spinning body. (6.6)

pre-main-sequence phase the stage of a star after it is a protostar and before it begins fusing hydrogen in its core. In this stage heat is generated by gradual gravitational contraction. (61.2)

pressure the force exerted by a substance such as a gas on an area, divided by that area. That is, pressure = force/area. (51.2)

primary mirror the large, concave, light-gathering mirror in a reflecting telescope. (30.2)

prime meridian the position of 0° longitude, passing through Greenwich, England, and marking the line between Eastern and Western Hemispheres. (5.4)

principle of equivalence the idea, central to general relativity, that a gravitational field is equivalent to an accelerated frame of reference. (27.1)

prism a wedge-shaped piece of glass that is used to disperse white light into a spectrum. (30.4)

prokaryotes cells without nuclei. The first life forms on Earth were probably prokaryotes. (85.3)

prominence a cloud of hot gas in the Sun's outer atmosphere. This cloud is often shaped like an arch and is supported by the Sun's magnetic field. (53.2)

proper motion the position shift of a star on the celestial sphere caused by the star's own motion relative to the Sun. (54.4)

proper motion method a technique for detecting a planet orbiting another star by observing the side-to-side "wobble" of the star on the sky caused by the pull of the orbiting planet. (36.4)

proteinoids strings of amino acids that can form through chemical and physical interactions. A possible precursor to the origin of life. (85.3)

proton a positively charged subatomic particle. The constituent of an atom's nucleus that determines the type of the atom—any atom with six protons is a carbon atom, for example. (4.2, 52.1)

proton decay a prediction by **grand unified theories** that protons, long thought to be stable particles, might have pathways to break down into lighter particles. (84.4)

proton–proton chain the nuclear fusion process that converts hydrogen into helium in stars like the Sun and thereby generates their energy. This is the dominant energy generation mechanism in main-sequence stars with masses smaller than the Sun's. (52.2, 62.2)

protoplanetary disk a disk of material encircling a protostar or a newborn star. (36.1)

protostar a star still in its early formation stage when matter from the original interstellar cloud is still falling onto it. (60.2, 61.2)

Proxima Centauri the nearest star to the Solar System at a distance of 1.30 parsecs. Despite its proximity, this M-type main-sequence star is too dim to be seen without a small telescope. (54.3)

Ptolemy, Claudius (about 100–170 C.E.) Roman astronomer who developed a geocentric model of the Solar System that could predict most of the observable motions of the planets. (11.2)

pulsar a spinning **neutron star** that emits beams of radiation each time the star spins. These beams flow out along the directions of the poles of the strong magnetic fields of these stars. If the beam happens to point in the Earth's direction, we observe the radiation as regularly spaced pulses as the star spins. (68.1)

pulsating variable a variable star that expands and contracts in size, producing a repetitive pattern of variations in its luminosity and surface temperature. (64.1)

pupil the aperture of the eye, which can be adjusted in size to allow more or less light into the eye. (33.1)

P wave a "primary" or "pressure" wave. P waves form as matter in one place vibrates against the adjacent matter ahead of it, producing compression and decompression in an oscillating fashion. Unlike S waves, P waves can travel through liquids. (37.1)

Pythagoras (about 560–480 B.C.E.) ancient Greek mathematician who developed early ideas about the nature of the Earth and sky and geometrical techniques for studying them. (10.1)

Q

quadrature the points in the orbit of an outer planet when it appears (from Earth) to be at a 90° angle with respect to the Sun. From the outer planet at the same times, the Earth would appear to be at **greatest elongation**. (11.3)

quantized the property of a system that allows it to have only discrete values. Subatomic matter is generally quantized in nearly all of its properties. (21.1)

quantum fluctuation a temporary random change in the amount of energy at a point in space. These tiny subatomic variations are constantly occurring everywhere in space. (83.4)

quark a fundamental particle of matter that interacts via the strong force; basic constituent of protons and neutrons. (4.4, 83.3)

quarter moon a phase of the Moon when it is located 90° from the Sun in the sky. (8.1)

quasar a highly energetic source in the nucleus of a galaxy generally seen at a large redshift. Quasars are among the most luminous and most distant objects known to astronomers. (77.5, 78.2)

R

radial velocity the velocity of a body along one's line of sight. That is, the part of a body's motion directly toward or away from the observer. (25.1, 54.4)

radiant the point in the sky from which meteors in showers appear to originate. See also **meteor shower**. (49.4)

radiation era the period of time up until about 50,000 years after the Big Bang, when radiation rather than matter was the dominant constituent of the universe. (83.1)

radiation pressure the force exerted by photons when they strike matter. (49.1)

radiative zone the region inside a star where its energy is carried outward by radiation (that is, photons). In the Sun, this region extends about two-thirds of the way out from the core. (51.3)

radioactive decay the spontaneous breakdown of an atomic nucleus into smaller fragments, often involving the emission of other subatomic particles as well. (4.3, 35.1, 37.2)

radioactive element an element that undergoes **radioactive decay** and spontaneously breaks down into lighter elements. (37.2)

radio galaxy a galaxy, generally an elliptical system, that emits abnormally large amounts of radio energy. (78.1)

radio lobe a region lying outside the body of a radio galaxy where much of its radio emission comes from. Radio lobes are usually located on opposite sides of the galaxy and contain hot gas ejected from an **active galactic nucleus** into intergalactic space through **jets**. (78.1)

radio waves long-wavelength electromagnetic radiation. (22.4)

rays long, narrow, light-colored markings on the Moon or other bodies that radiate from young craters. Rays are debris "splashed" out of the crater by the impact that formed it. (39.2)

recession velocity the apparent speed of a galaxy (or other distant object) moving away from us, caused by the expansion of the universe. (75.3)

recurrent nova a white dwarf in a binary system which undergoes repeated **nova** outbursts. (66.1)

reddening the alteration in a star's color seen from Earth as the star's light passes through an intervening interstellar dust cloud. The dust preferentially scatters the star's blue light, leaving the remaining light redder. (73.2)

red giant a cool, luminous star with a radius much larger than the Sun's. Red giants are found in the upper right portion of the H-R diagram. (59.1, 60.2)

red giant branch the section of the evolutionary track of a star corresponding to intense hydrogen shell burning, which drives a steady expansion and cooling of the outer envelope of the star. As the star gets larger in radius and its surface temperature cools, it becomes a red giant. (63.1)

redshift a shift in the wavelength of electromagnetic radiation to a longer wavelength. For visible light, this implies a shift toward the red end of the spectrum. The shift can be caused by a source of radiation moving away from the observer or by the observer moving away from the source (see **Doppler shift**). Light traveling through expanding space undergoes an **expansion redshift**, and light traveling away from a strong gravitational source undergoes a **gravitational redshift**. (25.1, 75.3)

reflection nebula an interstellar cloud in which the dust particles reflect starlight, making the cloud visible. (73.1)

reflector a telescope that uses a mirror to collect and focus light. (30.2)

refraction the bending of light when it passes from one substance and enters another, generally with a different density. (30.1)

refractor a telescope that uses a lens to collect and focus light. (30.1)

regolith the surface rubble of broken rock on the Moon or other solid body. (39.3)

regular orbit an orbit of a satellite that is nearly circular, in the same direction as the planet spins, and close to the plane of the planet's equator. (47.1)

relativistic mass the larger mass an object appears to have when it is moving. This differs significantly from the "rest mass" only when an object moves at a sizable fraction of the speed of light. (26.3)

resolution the ability of a telescope or other optical instrument to discern fine details of an image. (28.1, 31.0)

resonance a condition in which the repetitive motion of one body interacts with the repetitive motion of another to reinforce the motion. For example, planets or satellites orbiting with orbital periods that have a simple fractional ratio (2:1, 3:2, etc.) are in resonance and can have a strong influence on each other's orbit. (43.4, 47.3)

rest frame a system of coordinates that appear to be at rest with respect to the observer. (26.1)

retina the back interior surface of the eye onto which light is focused. (33.1)

retrograde motion the westward shift of a planet against the background stars. Planets usually shift eastward because of their orbital motion, but they appear to reverse direction from our perspective when the Earth overtakes and passes them (or when an inferior planet overtakes and passes the Earth). (11.1)

retrograde spin a spin backward from the usual orbital direction. For example, seen from above their north poles, most of the planets orbit and spin in a counterclockwise direction; however, a few have a retrograde spin. (41.3)

revolve to orbit around another body. (6.1)

rich cluster a galaxy cluster containing hundreds to thousands of member galaxies. (77.2)

rifting the pulling apart of a geological plate by currents in the mantle beneath it. (37.3)

right ascension (RA) a coordinate for locating objects on the sky, analogous to longitude on the Earth's surface. Usually measured in hours and minutes of time. Together with **declination,** this defines the position of an object on the celestial sphere. (5.5)

rille narrow canyon on the Moon or other body. (39.2)

ring galaxy a galaxy in which the central region has an abnormally small number of stars, causing the galaxy to look like a ring. Caused by the collision of two galaxies. (76.3)

ringlet any one of numerous, closely spaced, thin bands of particles in Saturn's ring system. (47.3)

ring system particles organized into thin, flat rings encircling a giant planet, such as Saturn. (47.3)

RNA world hypothesized early stage in the formation of life with organisms using RNA instead of DNA and proteins. (85.3)

Roche limit the distance from an astronomical body at which its gravitational force breaks up another astronomical body. (47.4)

Roche lobe the region around a star in a binary system in which the gravity of that star dominates. Matter pushed outside the Roche lobe (which may occur during a star's red giant phase) may be captured by the companion star. (66.1)

rod light-sensing cell in the eye that gives us our black-and-white vision. (33.1)

rotate to spin on an axis. The distinction between revolving and rotating can be confusing—a galaxy may rotate, but the stars in it revolve around the center. (6.3)

rotation axis an imaginary line through the center of a body about which the body spins. Analogous to the handle and point of a spinning top. (6.3)

rotation curve a plot of the orbital velocity of the stars or gas in a galaxy at different distances from its center. (74.1, 79.1)

round off to express a number with fewer digits by replacing the trailing digits with the nearest decimal value. For example, 12.34567 can be rounded off to 12.346, to 12.3, or to 12 depending on the precision needed. (3.4)

RR Lyrae variables a type of pulsating variable star with a period of about one day or less. They are named for their prototype star in the constellation Lyra, RR Lyrae. (64.2)

runaway greenhouse effect an uncontrollable process in which the heating of a planet leads to an increase in its atmospheric greenhouse effect and thus to further heating. The process significantly alters the composition of the planet's atmosphere and the temperature of its surface. (41.1)

S

S0 galaxy galaxy that shows evidence of a disk and a bulge, but that has no spiral arms and contains little or no gas. (76.1)

Sagittarius A* (Sgr A*) the powerful radio source located at the core of the Milky Way Galaxy. (74.2)

satellite a body orbiting a planet. (1.2, 34.1)

scarp a cliff produced by vertical movement of a section of the crust of a planet or satellite. (40.1)

scattering the random redirection of a light wave or photon as it interacts with atoms or dust particles in its path. (73.1)

Schwarzschild radius the distance from the center of a black hole to its **event horizon.** No light (or anything else) can escape from within the Schwarzschild radius. (69.1)

scientific method the process of observing phenomena, proposing hypotheses to explain them, and testing the hypotheses. Scientifically valid hypothesis must offer a means of being tested and rejected. (4.1)

scientific notation a shorthand way to write numbers using 10 to a power. For example, $1{,}000{,}000 = 10^6$. Numbers that are not simple powers of 10 are written as a number between 1 and 10 and 10 to a power—for example, $123.45 = 1.2345 \times 10^2$. (3.2)

scintillation the twinkling of stars, caused as their light passes through moving regions of different density in the Earth's atmosphere. (32.2)

secondary mirror in a reflecting telescope, a mirror that directs the light from the primary mirror to a focal position. (30.2)

sedimentary rock a type of rock formed from the compression of layers of particles, often the product of erosion of other rocks. (38.3)

Sedna a large body (about two-thirds the diameter of Pluto) orbiting far beyond Pluto's orbit. (48.5)

seeing a measure of the steadiness of the atmosphere during astronomical observations. Bad seeing results from **scintillation** and makes fine details difficult to observe. (32.2)

seismic waves waves generated in the Earth's interior by earthquakes. Similar waves occur in other bodies. Two of the more important varieties are S and P waves. S waves can travel only through solid material; P waves can travel through either solid or liquid material. (37.1)

selection effect an unintentional selection process that omits some set of the objects being studied and leads to invalid conclusions about the objects. For example, selecting the brightest stars in the night sky does not provide a representative sample of all star types. (70.3)

self-propagating star formation a model that explains a galaxy's spiral arms as arising from the explosion of massive stars, triggering the birth of other stars around them. The resulting pattern is then drawn out into a spiral by the galaxy's differential rotation. (74.3)

semimajor axis half the long dimension of an ellipse. (12.2)

SETI Search for Extra Terrestrial Intelligence. Some SETI searches involve automated "listening" to millions of radio frequencies for signals that might be from other civilizations. (86.4)

Seyfert galaxy a type of galaxy with a small, abnormally bright **active galactic nucleus.** Named for the astronomer Carl Seyfert, who first drew attention to these objects. (78.1)

shell fusion nuclear energy generation in a region surrounding the core of a star rather than in the core itself. This occurs during the red giant phase of stellar evolution. (63.1)

shepherding satellite a satellite that by its gravitational attraction prevents particles in a planet's rings from spreading out and dispersing. Saturn's F-ring is held together by shepherding satellites. (47.3)

shielding the process in which dust particles in an interstellar cloud block light from entering the interior of the cloud, thereby allowing it to cool. (73.4)

short-period comet a comet whose orbital period is shorter than 200 years. For example, Halley's comet has a period of 76 years. (49.3)

sidereal day the length of time from the rising of a star until it next rises. The length of the Earth's sidereal day is 23 hours 56 minutes. (7.1)

sidereal month the length of time (about 27.3 days) required for the Moon to return to the same apparent position among the stars. Compare with **lunar month.** (8.1)

sidereal period the time it takes a body to turn once on its rotation axis or to revolve once around a central body, as measured with respect to the stars.

sidereal time a system of time measurement referenced to the motion of stars across the sky rather than of the Sun. (7.1)

significant digits all the leftmost digits in a number (excluding any leading zeros) whose values are well determined by measurement or calculation. For example, someone's height might be measured as 1.93524 meters, but if the accuracy of the measurement is only about a millimeter (0.001 meters), the significant digits are 1.935. (3.4)

silicate mineral composed primarily of silicon and oxygen. Most ordinary rocks are silicates. For example, quartz is silicon dioxide, and a wide variety of minerals are made of silicates containing an additional element. (37.1)

singularity a theoretical point of zero volume and infinite density to which any object that becomes a black hole must collapse, according to the **general theory of relativity.** (69.1)

solar activity phenomena that cause changes in the appearance or energy output of the solar atmosphere, such as sunspots and flares. (53.0)

solar cycle the cyclic change in solar activity, such as that of sunspots and of solar flares, that peaks once every 11 years, approximately. (53.3)

solar day the time interval from one sunrise to the next sunrise or from one noon to the next noon. That interval is not always exactly 24 hours but varies throughout the year. For that reason, we use the **mean solar day** (which, by definition, is 24 hours) to keep time. (7.1)

solar eclipse the passage of the Moon between the Earth and the Sun so that our view of the Sun's photosphere is partially or totally blocked. During a **total eclipse** of the Sun we can see the Sun's chromosphere and corona. (8.2)

solar nebula the rotating disk of gas and dust from which the Sun and planets formed. (35.2)

solar nebula theory the theory that the Sun and planets formed all at approximately the same time from a rotating cloud of gas and dust. (35.2)

solar seismology the study of pulsations or oscillations of the Sun to determine the characteristics of the solar interior. (51.5)

Solar System the Sun, planets, their moons, and other bodies, such as meteors and comets, that orbit the Sun. (1.5, 34.0)

solar wind the outflow of low-density, hot gas from the Sun's upper atmosphere. It is partially this wind that creates the tail of a comet by pushing a comet's gases away from the Sun. (49.1, 51.4)

solid core see **inner core.**

solstice (winter and summer) the beginning of winter and summer. The solstice occurs when the Sun is at its greatest distance north (about June 21) or south (about December 21) of the celestial equator. (6.5)

south celestial pole the imaginary point on the **celestial sphere** directly over the Earth's South Pole. Objects on the sky appear to circle around this point as viewed from Earth's Southern Hemisphere.

space velocity the overall velocity of a star, determined by vector addition of the star's **radial velocity** and **transverse velocity.** (54.4)

special relativity a theory developed by Einstein to explain why light is always measured to travel through space at the same speed—regardless of the motion of the source of light or the observer. This very well-established theory shows that the measurements of lengths, times, masses, and other quantities change depending on the speed of the observer relative to the object being measured. (26.0)

speckle imaging a technique to overcome the limitations imposed by the effects of scintillation in our atmosphere by analyzing a set of images collected over very short time intervals. (58.1)

spectral type an indicator of a star's temperature. A star's spectral type is based on the appearance of its spectral lines, with different strengths and weaknesses indicating stellar composition and temperature. The fundamental types are, from hot to cool, O, B, A, F, G, K, M, L, and T. (56.3)

spectrograph a device for making a spectrum by spreading light out into its component wavelengths.

spectroscopic binary a type of binary star in which the spectral lines exhibit alternating red and blue Doppler shifts. This is caused by the orbital motion of one star around the other, causing the stars to alternately move away from us and back toward us. (57.1)

spectroscopy the study and analysis of spectra. (24.0)

spectrum electromagnetic radiation (for example, visible light) spread into its component wavelengths or colors. A rainbow is a spectrum produced naturally by water droplets in our atmosphere. (56.0)

spicule a thin column of hot gas in the Sun's chromosphere. (51.4)

spindown the slowing down of a body's rotation, often referring to pulsars. (68.2)

spiral arm a long, narrow region containing young stars and interstellar matter that winds outward in the disk of spiral galaxies. (71.4)

spiral density wave a mechanism for the generation of spiral structure in galaxies by the gravitational influence of a denser concentration of material in a spiral arm to influence the orbits of other material in the galaxy to reinforce the spiral structure. A density wave can interact with interstellar matter and trigger the formation of stars. Spiral density waves are also seen in the rings of Saturn. (47.3)

spiral (S) galaxy a galaxy with a disk in which its brighter stars form a spiral pattern of two or more **spiral arms.** (76.1)

spring tide a strong tide that occurs at new and full moon, when the Moon's and Sun's tidal forces work in the same direction to make the tide more extreme. (19.3)

standard candle a type of star or other astronomical light source in which the luminosity has a known value, allowing its distance to be determined by measuring its apparent brightness and applying the **inverse-square law.** Good examples include Cepheid variable stars and Type Ia supernovae. (55.3)

Standard Model the current theoretical model that describes the fundamental particles and forces in nature. (4.4)

standard time a common time kept within a given region so that all clocks in that region agree. (7.3)

star a massive, gaseous body held together by gravity and generally emitting light. Stars generate energy by nuclear reactions in their interiors. (1.4)

starburst galaxy a galaxy in which a very large number of stars have formed recently, making them very luminous. (76.3)

star cluster a group of stars numbering from hundreds to millions held together by their mutual gravity. (70.0)

steady-state theory a model of the universe in which the average properties of the universe do not change with time. An early competitor of the **Big Bang** model.

Stefan-Boltzmann constant constant that appears in the laws of thermal radiation. (23.5)

Stefan-Boltzmann law the amount of energy radiated by one square meter in one second by a hot, dense material depends on the temperature T raised to the fourth power (σT^4). (23.5, 58.3)

stellar association an extended, loose grouping of young stars and interstellar matter. (70.1)

stellar evolution the gravity-driven changes in stars as they are born, age, and finally run out of fuel. (2.1, 60.0)

stellar model the result of a theoretical calculation of the physical conditions in the different layers of a star's interior. (60.1)

stellar remnant the body remaining after a star ceases supporting itself with the heat and pressure generated by nuclear fusion in its interior, such as a white dwarf, neutron star, or black hole. (60.1)

stellar spectroscopy the study of the properties of stars based on information that can be learned from their spectra. (56.3)

stellar wind an outflow of gas from a star, sometimes at speeds reaching hundreds of kilometers per second. (60.2, 65.2)

stratosphere the region of the Earth's atmosphere extending from about 12 to 50 km above the surface. A protective layer of **ozone** is located in the stratosphere. (38.1)

stromatolites layered fossil formations caused by ancient mats of algae or bacteria, which build up mineral deposits season after season. (85.1)

strong force the force that binds quarks together and holds protons and neutrons in an atomic nucleus. Sometimes called the *strong nuclear force*. (4.3, 52.1)

subatomic particles particles making up an atom, such as electrons, neutrons, and protons, as well as other particles of similar small size.

subduction the sinking of one crustal plate under another where they are driven together by plate tectonics. (37.3)

sublimate to change directly from a solid into a gas without passing through a liquid phase. (49.1)

sunspot a dark, cooler region on the Sun's visible surface created by intense magnetic fields. (12.3, 53.1)

supercluster a cluster of galaxy groups and clusters. Our Milky Way belongs to the Local Group, which is but one of many galaxy groups making up the Local Supercluster. (2.4, 77.0)

supergiant a very large-diameter and luminous star, typically at least 10,000 times the Sun's luminosity. (59.3)

superior conjunction see **conjunction**.

superior planet a planet orbiting farther from the Sun than the Earth. Mars, Jupiter, Saturn, Uranus, and Neptune are superior planets. (34.1)

supermassive black hole a huge black hole with a mass millions to billions times larger the mass of the Sun. Observations suggest that they may be present in the nuclei of all large galaxies. (78.3)

supernova an explosion marking the destruction of some white dwarfs that reside in binary systems (**Type Ia supernova**), as well as the end of most massive stars' evolution (**core-collapse supernova**). (60.2, 64.1)

supernova remnant the debris ejected from a star when it explodes as a supernova. Typically this material is hot gas, expanding away from the explosion at thousands of kilometers per second. (66.3, 67.4)

surface brightness the amount of light emitted by a source divided by the area from which the light is emitted. For example, a galaxy with a high density of stars will have a high surface brightness, but one with the same number of stars more widely spread out will have a lower surface brightness. (29.4, 75.1)

surface gravity the acceleration an object will experience near the surface of a planet (or other body) because of the planet's gravitational pull. (16.3)

S wave a "secondary" or "shear" wave. A type of seismic wave produced by side-to-side motion. S waves can travel only through solid material and move more slowly than **P waves**. (37.1)

symbiosis the mutually beneficial relationships of organisms with other organisms. For example, the human body needs the bacteria in its intestines to break down its food, and the bacteria are also fed in the human body. Neither would survive without the other. (85.1)

synchronous rotation the condition that a body's rotation period is the same as its orbital period. The Moon rotates synchronously. (39.5)

synchrotron radiation a form of nonthermal radiation emitted by charged particles spiraling at nearly the speed of light in a magnetic field. Pulsars and radio galaxies emit synchrotron radiation. The radiation gets its name because it was first seen in synchrotrons, a type of atomic accelerator. (68.2, 78.1)

synodic period the time between successive configurations of a planet or moon. For example, the time between oppositions of a planet or between full moons. (13.4)

T

tail the plumes of gas and dust from a comet. These are produced by the solar wind and radiation pressure acting on gas and dust evaporated from the comet's nucleus. The **dust tail** and **ion tail** point away from the Sun and get longer as the comet approaches perihelion.

telescope an instrument for gathering and focusing light and magnifying the resulting image of remote objects. (28.1)

terrestrial planet a rocky planet similar to the Earth in size and structure. The terrestrial planets are Mercury, Venus, Earth, and Mars. (34.1)

Tharsis bulge a large volcanic region on Mars rising about 10 km above surrounding regions. (42.1)

theory a hypothesis or set of hypotheses that have become well-established through repeated and diverse testing. (4.1)

thermal energy the kinetic energy associated with the motions of the molecules or atoms in a substance. (20.1, 23.3)

thermal radiation the continuous spectrum of electromagnetic radiation emitted as a result of the thermal energy in any relatively dense material. Also see **blackbody.** (23.0)

thermonuclear runaway a condition in which a nuclear reaction generates enough heat to cause an increasing rate of reaction, rapidly accelerating into an explosion. (66.1)

third-quarter moon a phase of the waning Moon when Earth-based observers see half of the Moon's illuminated hemisphere. This occurs three-quarters of the way through the lunar month. (8.1)

tidal braking the slowing of one body's rotation as a result of gravitational forces exerted on it by another body. (7.5, 19.4)

tidal bulge a bulge on one body created by the gravitational attraction on it by another. Two tidal bulges form, one on the side near the attracting body and one on the opposite side. (19.1)

tidal force relative forces arising between the parts of a body as a result of differences in the strength of the gravitational attraction by another body. (19.1)

tidal lock circumstance in which tidal forces have caused a body to rotate at a rate closely tied to its orbital period. For example, Mercury rotates 3 times in each 2 revolutions around the Sun. The Moon always keeps the same face turned toward the Earth, in **synchronous rotation,** where it rotates once for each revolution. (40.3)

tides the rise and fall of the Earth's oceans created by the gravitational attraction of the Moon. Tides also occur in the solid crust of a body and its atmosphere. (19.0)

time zone one of 24 divisions of the globe at every 15 degrees of longitude. In each zone, a single standard time is kept. Most zones have irregular boundaries, usually because they follow geopolitical borders. See also **universal time.** (7.3)

Titan the largest moon of Saturn, even larger than Mercury. Titan has a substantial atmosphere and "weather," but its surface temperature is far colder than the terrestrial planets. (48.2)

torque a twisting force that can change an object's angular momentum. (20.3)

total eclipse an eclipse in which the eclipsing body completely covers the other body. (8.2)

transit the passage of a planet directly between the observer and the Sun. At a transit, we see the planet as a dark spot against the Sun's bright disk. From Earth, only Mercury and Venus can transit the Sun. (54.1)

transit method a method for detecting planets orbiting other stars by detecting the slight dimming if the planet's orbit causes it to cross in front of the star. (36.4)

trans-Neptunian object (TNO) an object orbiting in the **Kuiper belt** or outer Solar System with a semimajor axis larger than Neptune's. These include Pluto and many other objects ranging up to sizes even larger than Pluto. (34.1, 48.5)

transverse velocity the actual speed of a star across, or transverse to, our line of sight, determined from the **proper motion** and distance of the star. (54.4)

triangulation a method for measuring distances. This method is based on constructing a triangle, one side of which is the distance to be determined. That side is then calculated by measuring another side (the baseline) and the two angles at either end of the baseline. (54.1)

triple alpha process the fusion of three helium nuclei (**alpha particles**) into a carbon nucleus. This process is sometimes called *helium burning,* and it occurs in many old stars. (63.2)

Triton the largest moon of Neptune, somewhat larger than and probably fairly similar to Pluto, possibly a **TNO** captured by Neptune. Triton shows evidence of a thin atmosphere and some volcanic activity. (48.3)

Tropic of Cancer the latitude line of 23.5° north, marking the distance farthest north where the Sun can pass directly overhead (on the summer solstice). (6.5)

Tropic of Capricorn the latitude line of 23.5° south, marking the distance farthest south where the Sun can pass directly overhead (on the winter solstice). (6.5)

troposphere the lowest layer of the Earth's atmosphere, extending up to an elevation of about 12 km, within which convection produces weather. (38.1)

true binary a pair of stars that actually orbit each other as opposed to just appearing to be close to each other on the sky (see **apparent double star**). (57.1)

T Tauri star a type of extremely young star or protostar that varies erratically in its light output as it settles toward the main-sequence phase. (61.2)

Tunguska event a large aerial explosion that occurred over Siberia in 1908, probably when a large meteor entered the atmosphere and heated up and exploded before reaching the ground. (50.4)

turnoff point the location on the main sequence where a star's evolution causes it to move away from the main sequence toward the red giant region. The location of the turnoff point can be used to deduce the age of a star cluster. (70.2)

21-centimeter radiation radio emission from a hydrogen atom caused by the flip of the electron's spin orientation. (73.4)

twin paradox a supposed paradox in special relativity arising from the differences in time measurement for two observers moving at speeds relative to each other. The paradox is usually phrased in terms of the relative aging of one twin taking a trip in a spacecraft while the other remains on Earth. Special relativity (and experiment) shows that there is no paradox—the twin on the spacecraft ages less. (26.5)

Type Ia supernova an extremely energetic explosion produced by the abrupt fusion of carbon and oxygen in the interior of a collapsing white dwarf star. See **supernova.** (66.3)

Type II supernova A type of supernova explosion which shows spectral lines of hydrogen, as opposed to type I supernovae, which do not exhibit hydrogen lines. Also see **core-collapse supernova.** (67.3)

U

ultraviolet a portion of the electromagnetic spectrum with wavelengths shorter than those of visible light but longer than those of X-rays. By convention, the ultraviolet region extends from about 10 to 300 nm. (22.3)

umbra The inner portion of the shadow of a body, within which sunlight is completely blocked. (8.2)

uncompressed density the density a planet would have if its gravity did not compress it. (44.3)

unit a quantity used for reporting measurements. (1.6, 3.0)

universality the assumption that the physical laws observed on Earth apply everywhere in the universe. (4.2)

universal time (UT) the time kept at Greenwich, England. Universal time is the same as Greenwich mean time. This is the starting point from which the Earth's 24 **time zones** are calculated. (7.3)

universe the largest astronomical structure we know of. The universe contains all matter and radiation and encompasses all known space. (2.4)

Ursa Major group the nearest star cluster to the Solar System, including the stars of the Big Dipper and many other bright stars over a large part of our sky. (70.1)

V

Valles Marineris a huge canyon feature on Mars stretching thousands of kilometers. (42.1)

valve mechanism a process causing some kinds of stars to pulsate. Radiation becomes trapped (like a closed valve), making the atmosphere heat and expand. The radiation is no longer trapped when the atmosphere is expanded enough (open valve), and the star shrinks back down in size. (64.2)

Van Allen radiation belts doughnut-shaped regions surrounding the Earth containing charged particles trapped by the Earth's magnetic field. (38.1)

variable star a star whose luminosity changes over time. (33.2, 60.2, 64.0)

vector a quantity, such as a force or a velocity, that has both a magnitude (size) and a direction. (15.1)

velocity a physical quantity that indicates both the speed of a body and the direction in which it is moving. (14.2)

vernal equinox or spring **equinox**. Spring in the Northern Hemisphere begins on the vernal equinox, which is on or near March 20. (6.5)

Virgo Cluster the nearest large galaxy cluster. The gravity of the thousands of galaxies in this cluster is predicted to eventually pull in the Milky Way and Local Group. (2.4)

virtual particles a particle and its antiparticle, which are created simultaneously in pairs and which quickly disappear. Virtual particles are created from **quantum fluctuations**. (69.4)

visible spectrum the part of the electromagnetic spectrum that we can see with our eyes. It consists of the familiar colors red, orange, yellow, green, blue, and violet. (22.1)

visible universe the portion of the universe in which light has had time to reach us within the history of the universe. This region is limited, therefore, to distances for which the travel time of the light is less than about 13.7 billion years. (81.1)

visual binary a pair of stars, held together by their mutual gravity and in orbit about each other, that can be seen with a telescope as separate objects. (57.1)

void a region between clusters and superclusters of galaxies that is relatively empty of galaxies. (77.4)

volatile element or compound that vaporizes at low temperature. Water and carbon dioxide are examples of volatile substances. (34.3)

vortex a strong spinning flow within a gas or liquid, such as the Great Red Spot on Jupiter. (45.3)

W

wane to gradually decrease, as in the "waning crescent Moon" or the "waning gibbous Moon." (8.1)

waterhole the interval of the radio spectrum between the 21-cm hydrogen radiation and the 18-cm OH radiation. Proposed as likely wavelengths to use in the search for extraterrestrial life. (86.4)

watt the MKS unit of power. It represents a rate of energy generation or consumption of one joule per second.

wavelength the distance between wave crests. Wavelength determines the color of visible light and is generally denoted by the Greek letter lambda (λ). (22.1)

wave–particle duality the theory that electromagnetic radiation may be treated as either a particle (photon) or an electromagnetic wave. All subatomic particles appear to have both wavelike and particle-like properties. (21.1)

wax to gradually increase, as in the "waxing crescent Moon" or the "waxing gibbous Moon." (8.1)

weak force the force responsible for radioactive decay of atoms. Now known to be linked to electric and magnetic forces in what is called the **electroweak force**. (4.3, 52.4)

weakly interacting massive particle (WIMP) a hypothetical class of subatomic particles that interacts only through the weak force, making it difficult to detect except by gravitational interactions. A possible dark matter candidate. (79.4)

weight the gravitational force exerted on a body by the Earth (or another astronomical object). Sometimes more broadly used to indicate the net force on a body after other forces (centripetal forces, buoyancy, and so on) and effects of the motion of the surrounding rest frame are accounted for. Thus, astronauts in an orbiting spacecraft are "weightless" because their spacecraft is in orbit; however the gravitational force on them is only slightly less than on the Earth's surface. (14.3)

white dwarf a dense star whose radius is approximately the same as the Earth's but whose mass is comparable to the Sun's. White dwarfs burn no nuclear fuel and shine by residual heat. They are the end stage of stellar evolution for low-mass stars like the Sun. (59.1, 60.2, 65.0, 66.0)

white light visible light exhibiting no color of its own but composed of a mix of all colors. Our eyes adapt to color differences, making many artificial light sources appear "white." (22.3)

Wien's law a relation between a body's temperature and the wavelength at which it emits radiation most intensely. Hotter bodies radiate more intensely at shorter wavelengths. Mathematically, the law states that $\lambda_{max} = 2.9 \times 10^6/T$, where λ_{max} is the wavelength of maximum emission in nanometers, and T is the body's temperature on the kelvin scale. (23.2)

winding problem because of a galaxy's differential rotation, its spiral arms should become tightly "wound up" as the inner parts wrap around the galaxy several times for every single orbit of the outer part of the arm. This seeming problem is explained by **spiral density waves.** (74.3)

X

X-ray the part of the electromagnetic spectrum with wavelengths longer than gamma rays but shorter than ultraviolet. (22.4)

X-ray pulsar a neutron star from which periodic bursts of X-rays are observed. These are thought to consist of a neutron star and a normal star in a close binary system. Mass from the normal star spills onto the neutron star, where it slowly accumulates and then undergoes a burst of fusion. The hot region of the neutron star where the burst occurred spins causing more fusion and leading to an explosion that we observe as the X-ray burst. (68.3)

Y

year the time that it takes the Earth to complete its orbit around the Sun; that is, the **period** of the Earth's orbit. (6.1)

yellow giant a phase stars pass through after moving off the main sequence, and sometimes again after starting helium fusion in their cores. Many yellow giants become unstable and pulsate as they cross the **instability strip in the H-R diagram.** (64.2)

yellow supergiant a yellow giant star with a luminosity greater than about 10,000 solar luminosities. (64.2)

Z

Zeeman effect the splitting of a single spectral line into two or three lines by a magnetic field. The Zeeman effect allows astronomers to detect magnetic fields in objects from the appearance of their spectral lines. (53.1)

zenith the point on the celestial sphere that lies directly above the observer's location. (5.4)

zero-age main sequence (ZAMS) main sequence on the H-R diagram for a system of stars that have completed their contraction from interstellar matter and are just beginning to derive all their energy from hydrogen-to-helium fusion in their core. (61.3, 62.4)

zodiac a set of 12 constellations along the ecliptic, in a band around the celestial sphere. The Sun, Moon, and planets move through the constellations of the zodiac from our vantage point on Earth. (6.2)

zone a bright, high-pressure region in the atmosphere of a Jovian planet, where gas flows upward. (45.1)

zone of avoidance a band running around the sky in which few galaxies are visible. It coincides with the plane of the Milky Way and is caused by dust that is within our Galaxy. This dust blocks the light from distant galaxies. (75.1)

Credits

Text Credits

Unit 24 Figure 24.10a: Based on data courtesy of Doug McGonagle, FCRAO; 24.10b Based on data courtesy of P.F. Winkler, Middlebury College.

Unit 40 Figure 40.7: Based on a computer simulation, courtesy of W. Benz, University of Arizona; W. Slattery, Los Alamos; A.G.W. Cameron, CFA.

Unit 43 Figure 43.1: Based on data courtesy of E.L.G. Bowell, Lowell Observatory.

Unit 73 Figure 73.9: Courtesy of Y.M. and Y.P. Georgelin.

Unit 74 Figure 74.1: Based on diagram by Dan P. Clemens, Ap. J. 295, 422; 74.8: This sequence was generated by P. Seiden (deceased) and H. Gerola, Ap. J. 233, 56, 1979. Reproduced by permission of AAS.

Photo Credits

Image Research by Mary Reeg

Page i: ©Akira Fujii/DMI; p. xi: ©StockTrek/Getty Images RF.

Looking Up #1 Northern Circumpolar Constellations
Background: ©Akira Fujii/DMI; M52: Hartmut Frommelt; M81 and M82: ©Robert Gendler; M101: Adam Block/NOAO/AURA/NSF.

Looking Up #2 Ursa Major
Background: ©Akira Fujii/DMI; M97: Gary White and Verlenne Monroe/Adam Block/NOAO/AURA/NSF; Mizar and Alcor: Courtesy of DSS/Processing by Coelum (www.coelum.com); M51: ©Tony and Daphne Hallas.

Looking UP #3 M31 & Perseus
All photos: © Akira Fujii/DMI.

Looking Up #4 Summer Triangle
Background: ©Akira Fujii/DMI; M57: Courtesy of H. Bond et al., Hubble Heritage Team (STScI/AURA), NASA; M27: ©IAC/RGO/Malin; Albireo: Courtesy of Randy Brewer.

Looking Up #5 Taurus
Background: ©Akira Fujii/DMI; M1: Courtesy of R. Wainscoat; M45: Courtesy of Anglo-Australian Observatory, photographs by David Malin; Hyades: ©Akira Fujii/DMI.

Looking Up #6 Orion
Background: ©Akira Fujii/DMI; Horsehead Nebula: Courtesy of Anglo-Australian Observatory, photograph by David Malin; Close-up of Horsehead Nebula: Courtesy NOAO/AURA/NSF; Betelgeuse: Courtesy of Dupree (CFA), NASA, ESA; M42-Orion Nebula: Courtesy of Carol B. Ivers and Gary Oleski; Protoplanetary Disk: C.R. O'Dell/Rice University, NASA.

Looking Up #7 Sagittarius
Background: Courtesy of Till Credner, AlltheSky.com; M16 Close-up: Courtesy of NASA, HST, J. Hester & P. Scowen (ASU); M16 with Gas Cloud: Courtesy of Bill Schoening/NOAO/AURA/NSF; M20: Courtesy of Jason Ware; M22: Courtesy of N.A. Sharp, REU program/NOAO/AURA/NSF.

Looking Up #8 Centaurus and Crux, The Southern Cross
Background: Anglo-Australian Observatory/David Malin Images; Centaurus A: Courtesy Peter Ward, 2004; Omega Centauri: Al Kelly; Eta Carinae: J. Hester/Arizona State University, NASA; The Jewel Box: Research School of Astronomy and Astrophysics, the Australian National University.

Looking Up #9 Southern Circumpolar Constellations
Background: Christopher J. Picking; Hourglass Nebula: Raghvendra Sahai and John Trauger (JPL), the WFPC2 science team, and NASA; Thumbprint Nebula: Courtesy of STScI; Tarantula Nebula: WFI/2.2-m/ESO.

Part 1: ©Dynamic Graphics/JupiterImages RF.

Unit 1 Figure 1.1: NASA; 1.2a: ©Vol. 74 PhotoDisc/Getty Images RF; 1.2b: ©Dr. F.A. Ringwald; 1.2c: NASA; 1.3(Mercury): NASA/John Hopkins University Applied Physics Laboratory/Carnegie Institution of Washington; 1.3(Venus): Courtesy of NASA/JPL; 1.3(Earth): ©Lunar and Planetary Institute/NASA; 1.3(Mars): NASA/JPL/MSSS; 1.3(Earth): ©Lunar and Planetary Institute/NASA; 1.3(Jupiter): NASA/JPL/Cassini; 1.3(Saturn, Uranus, Neptune): Courtesy of NASA/JPL; 1.4: Courtesy of SOHO, ESA & NASA.

Unit 2 Figure 2.1: Courtesy of Australian Astronomical Observatory, photograph by David Malin; 2.2: NASA, ESA, and M. Livio and the Hubble 20th Anniversary Team (STScI); 2.3a: Courtesy of 2MASS/UMass/IPAC-Caltech/NASA/NSF; 2.3b: NASA/JPL-Caltech; 2.5: ©Robert Gendler; 2.6(Milky Way): NASA/JPL; 2.6(Earth): NASA.

Unit 3 Figure 3.1: ©The Bridgeman Art Library/Getty Images; 3.2: Courtesy of NASA/JPL; 3.3: Photograph courtesy of the BIPM.

Unit 5 Figure 5.3a(left): ©Roger Ressmeyer/Corbis; 5.3a(right): ©Roger Ressmeyer/Corbis, digitally enhanced by Jon Alpert; 5.3b(both): Courtesy of Eugene Lauria.

Unit 6 Figure 6.1: ©JupiterImages RF; 6.9a: ©English Heritage Library; 6.9b: ©Andrew Ward/Life File/Getty Images RF; 6.10b: Courtesy Mag. Iva'n Ghezzi.

Unit 7 Figure 7.3: ©Arnulf Husmo/Stone/Getty Images.

Unit 8 Figure 8.1: NASA/Goddard Space Flight Center Scientific Visualization Studio; 8.5: ©Roger Ressmeyer/Corbis; 8.6: John Walker; 8.7b: ©CNES/Jean-Pierre Haignere', 1999; 8.8b: ©JAXA/NHK; 8.9a-e: ©Stephen E. Schneider; 8.11: ©Roger Ressmeyer/Corbis.

Unit 9 Figure 9.1: ©Araldo de Luca/Corbis; 9.2: ©Stephen E. Schneider; 9.3: Chinese Rare Book Collection, Asian Division, Library of Congress; 9.4: ©Stephen E. Schneider; 9.5: ©Bettmann/Corbis.

Unit 10 Figure 10.1a: ©2009 Anthony Ayiomamitis; 10.1c: ©Stephen E. Schneider.

Unit 11 Figure 11.5: ©Art Resource/Erich Lessing; 11.10(both): Smithsonian Institution Libraries, Library of Congress, 46031925.

Unit 12 Figure 12.1: ©Bettmann/Corbis; 12.2: ©Art Resource/Erich Lessing; 12.5: Courtesy of the Bridgeman Art Library/Getty Images; 12.6a,b: Rare Book and Special Collections Division, Library of Congress; 12.8: Statis Kalyvas and ESO VT-2004 programme.

Unit 13 Figure 13.6: Courtesy of Bausch and Lomb Optical Co.; 13.7(both): ©Stephen E. Schneider; 13.8: Laurent Laveder-PixHeven.net; 13.9a: ©2008 Anthony Ayiomamitis.

Part 2: NASA.

Unit 14 Figure 14.1: ©Bettmann/Corbis; 14.3: ©Iconotec/Alamy RF.

Unit 15 Figure 15.5b: ©Stone/Getty Images.

Unit 16 Figure 16.4: NASA/Apollo Lunar Surface Journal; 16.6: NASA John Space Center (NASA-JSC).

Unit 19 Figure 19.1b: NASA, ESA, and The Hubble Heritage Team (STScI/AURA); 19.2(all): Tom Arny.

Unit 20 Figure 20.2: ©Don Davis; 20.3: ©Alamy Images RF.

Unit 21 Figure 21.7a,b: ©McGraw-Hill Education/Joe Franek.

Unit 22 Figure 22.1: ©Ewing Galloway, Inc.; 22.3b: ©Stephen E. Schneider; 22.4(Pulsar): Courtesy of NASA/CXC/SAO; 22.4(Sun): SOHO (ESA & NASA); 22.4(Interstellar cloud): Courtesy of Australian Astronomical Observatory, photograph by David Malin; 22.4(Cosmic microwave background): Courtesy of NASA/WMAP; 22.4(Active galaxy): NRAO/AUI/NSF; 22.5(X-ray): ROSAT, MPE, NASA; 22.5(Ultraviolet): NASA/Swift/Stefan Immler (GSFC) and Erin Grand (UMCP); 22.5(Visible): NOAO/AURA/NSF; 22.5(Infrared): ESA/Herschel/PACS & SPIRE Consortium, O. Krause, HSC, H. Linz; 22.5(Radio): Max-Planck-Institut für Radioastronomie (Rainer Beck & Philipp Hoernes).

Unit 23 Figure 23.1: ©Science Source; 23.6(both): Spitzer Science Center and Infrared Processing and Analysis Center; 23.8: Courtesy of SOHO/MDI consortium. SOHO is a project of international cooperation between ESA and NASA.

Unit 24 Figure 29.4a: ©Courtesy of Mees Solar Observatory, University of Hawaii.

Unit 25 Figure 25.4(both): NASA.

Unit 26 Figure 26.4 & 26.6: ©Hulton-Deutsch Collection/Corbis.

Unit 28 Figure 28.1: ©Roger Ressmeyer/Corbis; 28.4a: Courtesy of NRAO/AUI; 28.4b: Courtesy of NASA/CXC/SAO/Rutgers/J. Hughes; 28.5a: Courtesy of NRAO/AUI; 28.5b: Courtesy of Richard Wainscoat; 28.5c: Courtesy of NRAO; 28.5d: NASA/CXC/ASU/J. Hester et al.

Unit 29 Figure 29.2: Herbert Raab; 29.3: Photo courtesy of John Hill and the Large Binocular Telescope Observatory; 29.4: ESO/F. Kamphues; 29.5: ESO; 29.6: Courtesy of the NAIC-Arecibo Observatory, a facility of the NSF; 29.7: The Large Millimeter Telescope; 29.10: Courtesy of Australian Astronomical Observatory, photograph by David Malin; 29.11(both): Courtesy of George Greaney.

Unit 30 Figure 30.2: ©Stephen E. Schneider; 30.4: ©Special Collections, Jean and Alexander Heard Library, Vanderbilt University; 30.8: Image courtesy of NRAO/AUI; 30.9a: ©Roger Ressmeyer/Corbis; 30.9b: US Gemini Project/AURA/NOAO/NSF; 30.10b: ©Stephen E. Schneider.

Unit 31 Figure 31.1: ©2011 Richard Megna, Fundamental Photographs, NYC; 31.2a: R. Thompson, (University of Arizona) and NASA; 31.3: ©UC Regents/Lick Observatory; 31.4: California Institute of Technology; 31.5: Courtesy of Steve Criswell; 31.6a,b: Courtesy of Andrea Ghez (UCLA).

Unit 32 Figure 32.2a: Courtesy of Kitt Peak National Observatory; 32.2b: Courtesy Astronomical Society of the Pacific; 32.2c: W.T. Sullivan; 32.4a: Courtesy of USAF; 32.4b: Canada France Hawaii Telescope; 32.6: Pekka Parviainen/Polar Image; 32.7(HST): Courtesy of NASA/JPL; 32.7(EUVE): Courtesy of NASA, ESA, and Max Planck Institute for Extraterrestrial Physics; 32.7(Spitzer Infrared Space Telescope): Courtesy NASA/JPL-Caltech; 32.7(Chandra X-ray Telescope Satellite): CXC/TRW; 32.8(Sombrero Galaxy): Courtesy of NASA and The Hubble Heritage Team; 32.8(Hourglass Nebula): Courtesy of NASA and The Hubble Heritage Team.

Unit 33 Figure 33.3: ©Dave King/Getty Images RF; 33.5a-33.6b: ©Stephen E. Schneider; 33.7: ©Joe Orman; 33.8a: ©Roger Ressmeyer/Corbis; 33.8b: ©Carol B. Ivers and Gary Oleski.

Part 3: NASA /JPL-Caltech/T. Pyle (SSC).

Unit 34 Figure 34.1(Sun): Courtesy of SOHO/ESA/NASA; 34.1(planets): Courtesy of NASA/JPL.

Unit 35 Figure 35.2: NASA, ESA, and The Hubble Heritage Team (STScI/AURA); 35.4: ©Comstock Images/PictureQuest RF; 35.9: ©UC Regents/Lick Observatory.

Unit 36 Figure 36.1: NASA, ESA and L. Ricci (ESO); 36.2a: Smith (University of Hawaii), G. Schneider (University of Arizona), E. Becklin and A. Weinberger (UCLA) and NASA; 36.2b: NASA, ESA, D. Golimowski (Johns Hopkins University), D. Ardila (IPAC), J. Krist (JPL), M. Clampin (GSFC), H. Ford (JHU), and G. Illingworth (UCO/Lick) and the ACS Science Team; 36.2c: NASA, ESA, P. Kalas, J. Graham, E. Chiange, E. Kite (University of California, Berkeley), M. Clampin (NASA Goddard Space Flight Center), M. Fitzgerald (Lawrence Livermore National Laboratory), and K. Stapelfeldt and J. Krist (NASA Jet Propulsion Laboratory); 36.9a: ESO; 36.9b,c: NASA, ESA, and R. Soummer (STScI); 36.13: NASA/Ames/JPL-Caltech.

Unit 37 Figure 37.1: ©2005 Planetary Visions Limited; 37.2: ©Taxi/Getty Images; 37.6: ©Stephen E. Schneider; 37.13: Tom Arny.

Unit 38 Figure 38.1: NASA; 38.2: ©Stephen E. Schneider; 38.5a: Courtesy Paul Schneider; 38.5b: NASA; 38.7: SVS/TOMS/NASA; 38.12: ©Stockbyte/Getty Images RF; 38.13: ©Sylvain Grandadam/Science Source; 38.14(top): Pete Turner/Getty Images; 38.14(bottom): ©Chris Butcher, by permission; 38.16: Courtesy of NOAA.

Unit 39 Figure 39.1: NASA; 39.3: NASA/JPL/USGS; 39.4a,b: ©UC Regents/Lick Observatory; 39.5a: Courtesy of NASA; 39.5b: ©John Gillmoure/Corbis; 39.6a: Courtesy of NASA, KeithLaney.com; 39.6b: ©Lunar and Planetary Institute/NASA; 39.7-39.11a: NASA; 39.11b: ©UC Regents/Lick Observatory; 39.12(both) & 39.15: NASA/GSFC/Arizona State University.

Unit 40 Figure 40.1-40.2b: NASA/Johns Hopkins University Applied Physics Laboratory/Carnegie Institution of Washington; 40.3a: NASA; 40.3b: NASA/Johns Hopkins University Applied Physics Laboratory/Carnegie Institution of Washington; 40.4a: NASA: 40.4b,c: NASA/Johns Hopkins University Applied Physics Laboratory/Carnegie Institution of Washington; 40.5: NASA.

Unit 41 Figure 41.1: NASA; 41.2(all): ESA/VIRTIS/INAF-IASF/Obs. de Paris-LESIA/University of Oxford; 41.3(both): ©Ted Stryk; 41.5(all): NASA/JPL; 41.7(both): NASA/JPL-Caltech/ESA.

Unit 42 Figure 42.1: NASA; 42.4: NASA/JPL/Arizona State University; 42.5: Courtesy of A.S. McEwen, USGS; 42.6: NASA/JPL-Caltech/ASU; 42.7a: Courtesy of NASA/JPL; 42.7b: Courtesy of A.S. McEwen, USGS; 42.7c: NASA/JPL-Caltech/University of Arizona; 42.8a-c: Data from NASA/JPL, processed by the authors; 42.9: Courtesy of Mars Exploration River Mission, JPL, NASA; 42.10a: NASA/JPL/Cornell; 42.10b: Courtesy of NASA/JPL/Malin Space Science Systems; 42.11(left): NASA/JPL-Caltech/Cornell University; 42.11(right): NASA/JPL-Caltech/University of Arizona/Cornell/Ohio State University; 42.12: NASA/JPL-Caltech/University of Arizona/Texas A&M University; 42.13a: NASA /JPL/Malin Space Science Systems; 42.13b: NASA/JPL-Caltech/ASU; 42.13c: ESA/SLR/FU Berlin (G. Neukum); 42.14: NASA/JPL/JHUAPL/MSSS/Brown University; 42.16: NASA/JPL-Caltech/MSSS; 42.17a: NASA/JPL-Caltech/Cornell University; 42.17b: NASA/JPL-Caltech/University of Arizona/Cornell/Ohio State University; 42.18a: NASA/JPL-Caltech/University of Arizona; 42.18b,c: Courtesy of NASA/JPL; 42.19(all): NASA/JPL-Caltech/University of Arizona; 42.20: Courtesy of NASA/JPL.

Unit 43 Figure 43.2: NASA, ESA, J. Parker (Southwest Research Institute), P. Thomas (Cornell University), L. McFadden (University of Maryland, College Park), and M. Mutchler and Z. Levay (STScI); 43.3(Gaspra): NASA/JPL; 43.3(Vesta): NASA/JPL-Caltech/UCLA/MPS/DLR/IDA; 43.3(Earth): NASA; 43.4a: NASA/JPL-Caltech/JAXA/ESA; 43.4b: NASA Jet Propulsion Laboratory (NASA-JPL); 43.4c: ISAS/JAXA; 43.5: NASA/JPL/JHUAPL; 43.6: NASA/JPL-Caltech; 43.7: Courtesy of NASA/JPL.

Unit 44 Figure 44.1(swirling clouds): NASA Johnson Space Center; 44.1(all others): Courtesy of NASA/JPL/USGS; Table 44.1 & Table 44.2(Mercury): NASA/John Hopkins University Applied Physics Laboratory/Carnegie Institution of Washington; Table 44.2(Venus): Courtesy of NASA/JPL/USGS; Table 44.2(Earth, Moon): NASA; Table 44.2(Mars): NASA/JPL/MSSS; 44.4: Courtesy of NASA/JPL; Table 44.3(all): Courtesy of NASA/JPL; 44.6(Mercury): Courtesy NASA's Planetary Geology and Geophysics Program/Arizona State University; 44.6(Dione): NASA; 44.7(both): NASA/JPL-Caltech/ASI.

Unit 45 Figure 45.1(Jupiter): NASA/JPL/Cassini; 45.1(Earth): ©Lunar and Planetary Institute/NASA; 45.1(Saturn): Courtesy of Cassini Imaging Team/SSI/JPL/ESA/NASA; 45.2 & 45.4c: NASA; 45.5a & 45.6: Courtesy of NASA/JPL; 45.7: NASA/JPL/Space Science Institute; 45.8a: NASA/ESA, The Hubble Heritage Team. Acknowledgment: H. Weaver (JHU/APL) and A. Simon-Miller (NASA/GSFC); 45.8b: J.T. Trauger (Jet Propulsion Laboratory) and NASA.

Unit 46 Figure 46.1(Uranus): NASA/JPL; 46.1(Earth): ©Lunar and Planetary Institute/NASA; 46.1(Neptune)-46.4b: NASA/JPL; 46.4c: NASA, ESA, and the Hubble Heritage Team (STScI/AURA); 46.6: Lawrence Sromovsky, University of Wisconsin-Madison/W.M. Keck Observatory.

Unit 47 Figure 47.1: Courtesy Don Wheeler and Brian Dunn/Louisiana Delta Community College; 47.3: Erich Karkoschka (University of Arizona), and NASA; 47.4(Io): NASA Jet Propulsion Laboratory (NASA-JPL); 47.4(Europa): NASA; 47.4(Ganymede): NASA Jet Propulsion Laboratory (NASA-JPL); 47.4(Collisto, Titan): NASA Jet Propulsion Laboratory (NASA-JPL); 47.4(Triton): National Space Science Data Center; 47.5(Amalthea): NASA/JPL/Cornell University; 47.5(Mimas): NASA/JPL/Space Science Institute; 47.5(Enceladus): NASA/JPL; 47.5(Tethys-Hyperion): NASA Jet Propulsion Laboratory (NASA-JPL); 47.5(Lapetus): NASA/JPL/Space Science Institute; 47.5(Miranda-Oberon): NASA Jet Propulsion Laboratory (NASA-JPL); 47.5(Larissa, Proteus): NASA Goddard Space Flight Center; 47.5(Nereid): NASA Jet Propulsion Laboratory (NASA-JPL); 47.6a: NASA/JPL/Cornell University; 47.6b & 47.8c: NASA/JPL/Space Science Institute; 47.9(both): NASA/JPL; 47.10: NASA/JPL/Space Science Institute; 47.11(Jupiter): NASA/JPL; 47.11(Uranus): NASA, ESA, and M. Showalter (SETI Institute); 47.11(Neptune, Saturn): NASA/JPL; 47.13 & 47.14(Saturn): Courtesy of NASA/JPL/Space Science Institute; 47.14(Uranus, Neptune): Courtesy NASA/JPL; 47.15: Bjorn Jonsson; 47.16: Courtesy of STScI/HST.

Unit 48 Figure 48.1(Earth): ©Stocktrek/Getty Images RF; 48.1(Triton): National Space Science Data Center; 48.1(Pluto, Charon): Images courtesy of Marc W. Buie/Lowell Observatory; 48.1(Orcus,

Haumea): NASA Jet Propulsion Laboratory (NASA-JPL); 48.1(Quaoar, Makemade, Eris, Sedna): NASA; 48.1(Moon): ©Digital Vision RF/PunchStock; 48.1(Io, Europa, Ganymede, Callisto, Titan): NASA Jet Propulsion Laboratory (NASA-JPL); 48.2a–48.4a(top): NASA Jet Propulsion Laboratory (NASA-JPL); 48.4a(bottom): NASA Goddard Space Flight Center; 48.4b–48.5a: NASA Jet Propulsion Laboratory (NASA-JPL); 48.5b: Courtesy of NASA/JPL/University of Arizona; 48.5c,d: NASA/JPL-Caltech/ASI; 48.6a–48.7: NASA/JPL; 48.8a,b: ©UC Regents, UCO/Lick Observatory Image; 48.8c(both): Courtesy of STScI; 48.10a(Triton): National Space Science Data Center; 48.10a(Pluton, Charon): Images courtesy of Marc W. Buie/Lowell Observatory; 48.10a(Orcus, Haumea): NASA Jet Propulsion Laboratory (NASA-JPL); 48.10a(Quasar, Makemake, Eris, Sedna): NASA.

Unit 49 Figure 49.1: ©Robert McNaught/Science Source; 49.3a: National Space Science Data Center Photo Gallery, NASA; 49.3b: ©1986, Max Planck Institute for Astronomy; courtesy of Harold Reitsema, Ball Aerospace; 49.4: Courtesy of Mike Skrutskie, University of Virginia; 49.7a: Courtesy NASA/JPL-Caltech; 49.7b–49.8a: NASA/JPL-Caltech/UMD; 49.8b: NASA/JPL-Caltech/University of Maryland/Cornell and NASA/JPL-Caltech/UMD; 49.8c: NASA/JPL-Caltech/UMD; 49.8d: NASA/JPL-Caltech/Cornell; 49.9: STEREO (NASA); 49.12: Courtesy Juan Carlos Casado and Isabe Graboleda.

Unit 50 Figure 50.1: Courtesy of Ronald A. Oriti, Santa Rosa Junior College, Santa Rosa, Calif.; 50.3a–50.4: ©Stephen E. Schneider; 50.5: NASA Jet Propulsion Laboratory (NASA-JPL); 50.7(background): Courtesy of David Roddy Meteor Crater, Northern Arizona, USA; 50.7(inset): Photo by Tom Arny; 50.9(left): ©Associated Press; 50.9(middle): AFP/Getty Images; 50.9(right): ©Kyodo/Associated Press; 50.10: Courtesy of Landsat/EOS; 50.11: NASA Jet Propulsion Laboratory (NASA-JPL); 50.12b: A. Hildebrand, M. Pilkington, and M. Connors; 50.12c: NASA (Artist: Don Davis).

Part 4: ©Stock Trek/Getty Images.

Unit 51 Figure 51.1a: SOHO/MDI project; 51.1b: Courtesy of SOHO/Extreme Ultraviolet Imaging Telescope (EIT) consortium; 51.6: Prof. Goran Scharmer/Dr. Mats G. Lofdahl/Institute for Solar Physics of the Royal Swedish Academy of Sciences. Courtesy of Jacques Guertin, Ph.D.; 51.7: Courtesy of Jacques Guertin, Ph.D.; 51.8: Courtesy NOAO/AURA/NSF; 51.9: 1988 eclipse image courtesy of High Altitude Observatory (HAO), University Corporation for Atmospheric Research (UCAR), Boulder, Colorado. UCAR is sponsored by the National Science Foundation; 51.10: Courtesy of LMSAL and NASA; 51.12: NASA/Walt Feimer; 51.13: Courtesy of NOAO/AURA/NSF.

Unit 52 Figure 52.4: Courtesy of Ernest Orlando, Lawrence Berkeley National Laboratory; 52.5: R. Svoboda and K. Gordan (LSU).

Unit 53 Figure 53.1: Courtesy of Royal Swedish Academy of Sciences; 53.2a–53.6a: Courtesy of NOAO/AURA/NSF; 53.7: Courtesy of SOHO-EIT Consortium, ESA, NASA; 53.8: Courtesy of NOAO/AURA/NSF; 53.9: Courtesy of Eugene Lauria.

Unit 54 Figure 54.4: NASA/Goddard Space Flight Center Scientific Visualization Studio; 54.8: ESA-C. Carreau.

Unit 55 Figure 55.1: ©Edward P. Flaspoehler, Jr.; 55.3: ©Goodshoot/PunchStock RF; 55.4: ©2001 Bert Katzung.

Unit 56 Figure 56.2: ©Akira Fujii/DMI; 56.3: Courtesy of Dean Ketelsen; 56.4: Courtesy of Harvard College Observatory; 56.5: NOAO/AURA/NSF; 56.6: Courtesy of Katherine Haramundanis; 56.8: NOAO/AURA/NSF.

Unit 58 Figure 58.1a: Andrea Dupree (Harvard-Smithsonian CfA), Ronald Gilliland; 58.1b: Steve Schneider based on Richard Wilson's data; 58.1c: Xavier Haubois (Observatoire de Paris) et al.

Unit 59 Figure 59.6: NOAO/AURA/NSF.

Unit 61 Figure 61.1: Courtesy of Karen Strom, Mark Heyer, Ron Snell, FCAD, and FCRAO; 61.2: NASA and The Hubble Heritage Team (STScI/AURA); 61.3: NASA, ESA, M. Robberto (Space Telescope Science Institute/ESA) and the Hubble Space Telescope Orion Treasury Project Team; 61.4: ©European South Observatory; 61.5a: Courtesy of STScI; 61.5b: ©European Southern Observatory; 61.6a: NASA, ESA, N. Smith (University of California Berkeley), and The Hubble Heritage Team (STScI/AURA); 61.6b: J. Alves (ESO), E. Tolstoy (Groingen), R. Robury (ST-ECF), & R. Hook (ST-ECF), VLT.

Unit 62 Figure 62.5: Courtesy of Australian Astronomical Observatory, photograph by David Malin.

Unit 64 Figure 64.1(both): ©Akira Fujii/DMI.

Unit 65 Figure 65.1: Courtesy of NASA, ESA, and Valentin Burjarrabal-Observatorio Astronomico National, Spain; 65.4b: NASA/JPL-Caltech; 65.5a: Courtesy of Hubble Heritage Team (AURA,STScI/NASA); 65.5b: Courtesy of Australian Astronomical Observatory, photograph by David Malin; 65.5c: Raghvendra Sahai and John Trauger (JPL), the WFPC2 Science Team, and NASA; 65.5d: Bruce Balock, University of Washington; Vincent Icke, Leiden University, Netherlands; Garrelt Mellema, Stockholm University/NASA.

Unit 66 Figure 66.3: Courtesy of Paresce, R. Jedrezejewski (STScI), NASA, ESA; 66.4(all): NASA, ESA, and Z. Levay (STScI); 66.6: Pablo Candia; 66.8a,b: NASA/CXC/Rutgers/J. Warren & J. Hughes et al.

Unit 67 Figure 67.5(both): Courtesy of Australian Astronomical Observatory, photograph by David Malin; 67.6a: NASA/CXC/MIT/UMass Amherst/M.D. Stage et al; 67.6b: NASA, ESA, and J. Hester (Arizona State University); 67.6c: NASA and The Hubble Heritage Team (STScI/AURA).

Unit 68 Figure 68.5: Photo courtesy of NOAO; 68.8: ©European Southern Observatory.

Unit 70 Figure 70.1(NGC 3293): James McHugh; 70.1(Pleiades): NASA, ESA and AURA/Caltech; 70.1(M67): Courtesy of Australian Astronomical Observatory, photograph by David Malin; 70.2(Omega Centauri): Max Kilmister; 70.2(M13): Robert Lupton and the Sloan Digital Sky Survey Consortium.

Part 5: ©Stock Trek/Getty Images.

Unit 71 Figure 71.1: Kerry-Ann Lecky Hepburn (Weather and Sky Photography); 71.2: Courtesy of Australian Astronomical Observatory, photograph by David Malin; 71.6: Courtesy of Hubble Heritage

Team (AURA/STScI/NASA); 71.7: Courtesy Sloan Digital Sky Survey Collaboration, www.sdss.org.

Unit 72 Figure 72.4: David Law/University of Virginia; 72.5a-e: Courtesy of Neal Katz, University of Massachusetts, and James E. Gunn, Princeton University.

Unit 73 Figure 73.1: Courtesy of NASA and The Hubble Heritage Team (STScI/AURA); 73.2a-73.5a: Courtesy of Australian Astronomical Observatory, photograph by David Malin; 73.5b: Canada-France-Hawaii Telescope/J.C. Cuillandre/Coelum; 73.6: Atlas Image courtesy of 2MASS/UMASS/IPAC-Caltech/NASA/NSF.

Unit 74 Figure 74.3: Courtesy of NASA, UMass, D. Wang et al.; 74.4a: Courtesy of Rainer Scholdel (MPE) et al., NAOS-CONICA, ESO; 74.4b: Courtesy of Andrea Ghez (UCLA); 74.7a: Stefan Seip/Adam Block/NOAO/AURA/NSF; 74.7b: Bruce Hugo and Leslie Gaul/Adam Block/NOAO/AURA/NSF.

Unit 75 Figure 75.1: Courtesy of George Greaney; 75.2: Courtesy of William C. Keel; 75.4b, 75.5b: Dr. Wendy L. Freedman, Observatories of the Carnegie Institution of Washington, and NASA.

Unit 76 Figure 76.1a: Courtesy of Bruce Hugo and Leslie Gaul, Adam Block, NOAO, AURA, ASF; 76.1b: Courtesy of Hubble Heritage Team, AURA/STScI/NASA; 76.1c: ©European Southern Observatory; 76.2a,b: ESA, NASA and E. Peng (Peking University, Beijing); 76.3a,b & 76.4(both): Courtesy of Australian Astronomical Observatory, photograph by David Malin; 76.5(both) & 76.6: Courtesy Sloan Digital Sky Survey Collaboration, www.sdss.org; 76.7a: Courtesy of Australian Astronomical Observatory, photograph by David Malin; 76.7b(both): ©Stephen E. Schneider; 76.8a: R. Jay GaBany-Collaboration: A. Sollima (IAC), A. Gil de Paz (U. Complutense Madrid), D. Martínez-Delgado (IAC, MPIA), J.J. Gallego-Laborda (Fosca Nit Obs.), T. Hallas (Hallas Obs.); 76.8b: Courtesy of NRAO/AUI; 76.9: Courtesy Sloan Digital Sky Survey Collaboration, www.sdss.org; 76.10: Roff/Lowell Observatory; 76.11a,b: Courtesy of NASA and The Hubble Heritage Team; 76.11c: Richard Griffiths (JHU), STScI, NASA; 76.12(all): Courtesy of Joshua Barnes, University of Hawaii; 76.13a: NASA, ESA, and The Hubble Heritage Team (STScI/AURA); 76.13b: Courtesy STScI and NOAO; 76.14: R. Jay GaBany (Blackbird Observatory)-collaboration: D. Martinez Delgado (IAC, MPHIA), J. Penarrubia (U. Victoria), I. Trujillo (IAC), S. Majewski (U. Virginia), M. Pohlen (Cardiff); 76.16a: Martin Pugh; 76.16b: NASA, ESA and J.P. Kneib (Laboratoire d' Astrophysique de Marseille).

Unit 77 Figure 77.1: Two Miron All Sky Survey, a joint project of the University of MA and the Infrared Processing and Analysis Center/CA Institute of technology, founded by NASA/National Science Foundation/T.H. Jarrett, J. Carpenter, & R. Hurt. Perha; 77.3: Martin & Rodrigo Ibata, Observatoire de Strasbourg, 2003; 77.4: Bob Franke/Focal Pointe Observatory; 77.5: ©Robert Gendler; 77.6a: Courtesy of A. Fabian (IoA Cambridge) et al., NASA; 77.6b: NASA/CXC/ESO/P. Rosati; 77.7: ©Stephen E. Schneider; 77.9: Numerical simulations performed at the National Center for Supercomputing Applications, NCSA, Urbana-Champaign, Illinois; by Andrew Kravtsov (The University of Chicago) and Anatoly Klypin, New Mexico State University.

Unit 78 Figure 78.1: AURA/STScI/NASA; 78.2a-c: Images courtesy of NRAO/AUI; 78.4: Courtesy of NOAO; 78.5: Courtesy of John Bahcall, Institute for Advanced Study; M. Disney, University of Wales and NASA; 78.8(background): Courtesy of W. Jaffe, Leiden Observatory, and H. Forde, Johns Hopkins University, Space Telescope Science Institute; and NASA; 78.8(inset): Courtesy of NRAO and California Institute of Technology; 78.9(background): NASA/CXC/CfA/INAF/Risaliti; 78.9(inset): Optical: ESO/VLT; 78.10a: ©Stephen E. Schneider; 78.12: Julie Comerford; Cosmic Evolution Survey; Hubble Space Telescope.

Unit 79 Figure 79.5a: NASA; 79.6: NASA, ESA, and Johan Richard (Caltech, USA) Acknowledgement: Davide de Martin & James Long (ESA/Hubble); 79.7: NASA/STScI; 79.9: NASA.

Unit 80 Figure 80.6: NASA, ESA, R. Ellis (Caltech), and the HUDF 2012 Team.

Unit 81 Figure 81.1: Courtesy of Seldner, M., Siebers; 81.7: Courtesy S. Beckwith and the HUDF Working Group (STScI), HST, ESA, NASA; 81.9(both): Arcminute Cosmology Bolometer Array Receiver.

Unit 83 Figure 83.6: ©CERN; 83.9b: Courtesy NASA/WMAP Science Team.

Unit 84 Figure 84.2: Pablo Candia; 84.5(both): based on Planck 2013 results. I. Overview: Fig. 19, ESA/Planck, and the Planck Collaboration.

Unit 85 Figure 85.1: ©David Wacey; 85.2(background): ©Paul Hoffman; 85.2(inset): Courtesy of Donald R. Lowe; 85.4(mushroom): Gallo Images-Nigel Dennis/Digital Vision/Getty Images RF; 85.4(butterfly): ©Corbis RF; 85.4(giraffes): ©DV/Getty Images RF; 85.8: ©Science VU/Sidney Fox/Visuals Unlimited, Inc.; 85.10a: ©Dudley Foster/Woods Hole Oceanographic Institution; 85.10b: ©Tom Pfeiffer/Volcano Discovery; 85.11 & 85.12: Courtesy of Andrew H. Knoll.

Unit 86 Figure 86.2: NASA; 86.3: NASA/JPL-Caltech/Cornell/MSSS; 86.6: Courtesy Sloan Digital Sky Survey Collaboration, www.sdss.org; 86.9: ©Seth Shostak; 86.10: National Astronomy & Ionosphere Center (NAIC).

Index

A

Abell 2218 galaxy cluster, 631
Aberration of starlight, 437–38
Absolute magnitude, 444
Absolute zero, 164
Absorption lines, 447
Absorption-line spectra. *See also* Spectra
    basic principles, 171–72, 173, 447
    of quasar emissions, 616
    of Sun, 174
Absorption of electromagnetic radiation, 152, 224–26
Accelerating universe, 672–73
Acceleration, 22, 114–18, 122–23
Accretion, 254, 616
Accretion disks, 517, 544, 622–24
Accretion phase of protostars, 483–84
Achondrites, 394, 395
Achromatic lenses, 235
Action-reaction law, 117–18
Active galactic nuclei, 598
Active galaxies, 618–25
Adams, John Couch, 356
Adaptation, 233–34
Adaptive optics, 227, 459–60
Adenosine triphosphate, 682, 683
Aether, 183, 184
Age of universe, 13–14, 641–42, 673
Agriculture, 41
Aitken Basin, 296, 299
Albireo, xxviii
Alcor, xxvi, 454
Aldebaran, xxix
Algae, 685
Algol, 461
Allen Telescope Array, 695
ALMA, 221
Alpha capture fusion reactions, 501
Alpha Centauri
    location in night sky, xxxii
    mass, 456, 457
    nearness to Earth, 436, 440
Alpha Centauri A, 435, 443, 457
Alpha Centauri B, 457
Alpha Magnetic Spectrometer, 633
Alpha particles, 498
Alpha Regio highlands, 313
Alpher, Ralph, 647
Al-Sufi, 589
Alt–az mounts, 236
Altair, xxviii, 102
Altitude of stars, 101
Alvarez, Luis and Walter, 399
Amalthea, 364, 368
Amateur astronomy, 232–39, 504
American Association of Variable Star Observers, 239, 504
Amino acids, 394, 681–82, 683
Ammonia, 348, 386
AM radio waves, 160
Analemmas, 106
Ancient astronomy
    eclipse forecasting, 64
    observations of Earth, Sun, and Moon, 76–80, 430
    origins of constellations, 34–35
    prehistoric observatories, 40, 46–47
Andromeda constellation, xxvii, 102
Andromeda Galaxy (M31)
    celestial coordinates, 38
    details in photos, 208
    future of, 570, 675
    Hubble's measurement to, 593
    in Local Group, 12–13, 15, 610
    location in night sky, xxvii
    mergers, 606
    negative redshift, 596
    views of in Shapley/Curtis debate, 591–92
    visibility of, 589, 590
Angular momentum, conservation of, 145, 253, 533–34
Angular shift, 433
Angular size, 80–81, 100, 459–60
Anisotropy, 668
Annefrank, 331
Anno Domini, 73–74
Annual motions of Sun, 41–49
Annular eclipses, 65
Antarctic Circle, 48
Antarctic observatories, 204, 650
Antarctic ozone hole, 285
Antares, 467
Ante meridian, 51
The Antennae, 605
Anthropic principle, 686
Antimatter, 664
Antiprotons, 664
Antiquarks, 664, 665
Apertures of telescopes
    discontinuous, 220–21
    f-ratio, 236
    relation to collecting power, 197, 206, 218
    relation to resolution, 214–15
Aphrodite, 312
Apochromatic lenses, 235
Apollo spacecraft, 123, 292, 296
Apparent double stars, 454
Apparent luminosity, 441
Apparent magnitude, 444
Apparent motion of celestial sphere, 35–36
Apparent noon, 51
Approximation, 17, 23
Apus constellation, xxxiii

Aquarius constellation, 42, 102
Aquila constellation, 102
Arabic scholarship, 87
Archaea, 684
Archimedes, 247
Arcminute Cosmology Bolometer Array Receiver, 650
Arc minutes, 93, 430
Arc seconds, 220, 434
Arctic Circle, 48, 52, 53
Arcturus, 101
Arecibo radio telescope, 205, 220, 696
Ariel, 363
Aries constellation, 42
Aristarchus, 77–79, 80, 88
Aristotle, 76–77, 91, 688
Arizona meteor crater, 396–97
Arrays, 220–22, 695
Associations (stellar), 549–50
A stars, 449, 452
Asterisms, xxvi, 34, 100–101, 102
Asteroid belt, 5, 242, 257, 328
Asteroid families, 334
Asteroids
    age, 252
    basic features, 328
    collisions with Earth, 334, 397–400
    composition and origins, 332–33
    crater formation from impacts, 295
    defined, 242
    discovery, 329–30, 378
    Doppler imaging, 179
    forms of energy in, 143
    meteorites from, 393, 395, 396
    as moons, 326
    orbits, 333–34
    sizes and shapes, 330–32
Astrobiology, 679–86
Astronomical units
    Copernicus's use, 90
    determining size, 6, 430–31
    in Kepler's laws, 95
    scale of practical use, 10
Astronomy, defined, 1
Astrophysics, 440, 459
Asymmetry of Moon's crust, 298–99
Asymmetry of weak force, 665
Atacama Large Millimeter Array, 221
Atla Regio, 313
Atmosphere (Earth)
    Coriolis effect, 289–90
    for early life forms, 684
    effects on telescopes, 197, 224–28, 459
    global warming role, 285
    history, 286–87
    meteor heating by, 392–93
    protective role, 3, 225–26, 284
    structure, 281–83

Atmospheres
    absence from Moon, 131–32, 299
    escape velocity and, 131–32, 299
    of exoplanets, 268, 690
    formation in Solar System, 258–59
    Jupiter and Saturn, 242–43, 342, 348–49, 351–53
    Mars, 3, 319, 324–26, 338, 339, 340
    Mercury's lack, 307
    Pluto, 378
    Sun, 407–10
    terrestrial planets compared, 338, 339
    Titan, 345, 375, 376
    Triton, 376
    Uranus and Neptune, 242–43, 343, 356–58
    Venus, 3, 309–11, 339, 340
Atmospheres (metric units), 281
Atmospheric pressure, 281–82, 310
Atmospheric windows, 224–25
Atomic clocks, 56, 186
Atomic energy. *See* Nuclear fusion
Atomic mass, 28
Atomic number, 28, 150
Atoms
    basic features, 27–28
    blackbody properties and, 163
    common in living organisms, 681
    development of knowledge about, 144, 150–51
    elementary particles, 30–31
    forces acting on, 29
    identifying from light emissions, 170–73
    interactions with light, 151–53
    radioactive decay, 250–51
    role of structure in properties of matter, 151
ATP (adenosine triphosphate), 682, 683
Auriga, xxvii
Auroras, 283, 353, 422, 424
Automated search programs, 329
Autumnal equinox, 46
Average density, 247–48. *See also* Density
Averted vision, 234
Azimuth, 101

B

Baade, Walter, 532, 564
Babylonian astronomy, 64
Bacteria
    fossil evidence, 679–80, 689
    impact on Earth's development, 286
    importance to humans, 681
    species, 684
Balance scales, 110

I-1

Balmer lines, 170–71, 451, 452, 578
Baptistina asteroid family, 334
Barnard 86, 575
Barnard's star, 436
Barred spiral galaxies, 600, 601
Barringer crater, 396–97
Basalt, 276, 294, 333
Baseline (parallax), 432
Baseline (triangulation), 430, 431
Base pairs, 682
Before Christ, 73–74
Before the present era, 74
Bekenstein, Jacob, 545
Bell, Jocelyn, 532
Belts of Jupiter, 348
Beryllium, 499
Bessel, Friedrich, 435
Beta Centauri, 440
Beta Pictoris, 262
Betelgeuse
  images compared, 460
  location in night sky, xxx
  magnitude, 443
  as red supergiant, 501
  spectral class, 450
  temperature, 448–49
Bethe, Hans A., 413
Bevis, John, 200
Big Bang
  estimating time since, 641
  evidence for, 14, 635, 647–48
  galaxy formation following, 649–50
  helium fusion following, 661–63
  inflationary period following, 635, 654, 665–69
  origin of matter from, 661–65
  origins of name, 648
  radiation era following, 660–61
Big Crunch, 654, 656
Big Dipper
  as asterism, 34
  binary stars in, 454
  identifying, 100
  locating other objects with, 101
  location in night sky, xxv, xxvi
  lore of, 102
  magnitude of stars in, 443
  proper motions, 436
  star cluster containing, 548–49
Big Rip, 677
Binary stars
  with black holes, 544
  eclipsing, 455, 461–62
  gravitational wave evidence from, 194
  mass transfer between, 516–18
  measuring stellar diameter, 461–62
  measuring stellar masses with, 455–57
  pulsars' rarity among, 536
  types, 454–55
Binoculars, 237
Biological processes. See Life
Bipolar outflows, 475, 485
Blackbodies, 162–63, 545
Black holes
  at center of galaxies, 622–25
  at center of Milky Way, 562, 584, 585, 624

curved space and, 541–43
escape velocity and, 132, 193, 539–41
evolution of, 477
formation, 530, 539
general relativity and, 193
near end of universe, 675, 676
observing, 199, 544–45
small and large, 546
Black light, 385
Blazars, 623
Blind spot, 233
Blue moon, 67
Blue sequence of galaxies, 606, 607
Blueshifts, 178
Blue stars
  comparative age, 493
  effects on interstellar medium, 578
  galaxy types favoring, 602
  high-mass stars as, 522–23
  Population I stars as, 564
  supernovae from, 527
Bode, Johann, 329
Bode's rule, 329–30
Bohr, Niels, 169
Bohr model, 169
Bok, Bart, 482
Bok globules, 482, 485, 486
Bonneville Crater, 321
Boötes, 101
Bottom quarks, 30
Bow waves of neutron stars, 536
Bradley, James, 437
Brahe, Tycho
  data compared to ancient records, 436
  inability to detect parallax, 433–34
  overview of contributions, 93–94
  supernova observations, 503
Bright giants, 510
Brightness of astronomical objects, 208, 553–54. See also Luminosity
Brightness of galaxies, 590
Brightness of light. See also Luminosity
  inverse-square law, 149, 441–42, 443, 553–54
  magnitude system, 443–44
  relation to temperature, 164, 166
Brown dwarfs
  defined, 264, 487
  detecting, 554
  exoplanets with, 264, 266
Bruno, Giordano, 91
B stars, 449–50, 566, 578, 586
Bulge of Milky Way, 562, 606
Bulges (galactic), 600, 602, 606, 624, 625
Bulge stars, 565
Bullet Cluster, 632
Burbidge, E. Margaret, 524
Burbidge, Geoffrey R., 524
Butterfly Nebula, 513

C

3C273 quasar, 620
Calcium absorption spectra, 447
Calculus, 109, 125, 126

Calendars, 69–74
California Nebula, xxvii
Callisto
  comparative size, 361, 362, 363, 371
  major features, 374
Caloris Basin, 304, 307
Cambrian explosion, 680
Cameras, using with telescopes, 237–39
Canali, 688
Canals on Mars, 688–89
Cancer (constellation), 41, 42
55 Cancri, 263
Canis Major constellation, 35
Canis Major dwarf galaxy, 610
Canis Minor constellation, 35
Cannon, Annie Jump, 449, 450, 470
Canopus, 77
Capella, xxvii
Capricornus constellation, 42, 102
Captured objects, 326, 362, 376
Carbohydrates, 682
Carbon
  in CNO cycle, 490
  ejection from red giants, 511, 512
  fusion in stars, 501, 519, 523
  helium fusion into, 498–99
  in interstellar dust, 574
  isotopes, 679
  molecular bonding, 691
Carbonaceous asteroids, 332
Carbonaceous chondrites, 394, 683
Carbon dioxide
  in Earth's atmosphere, 285, 286–87, 339–40
  in Earth's oceans, 286, 310
  infrared absorption, 225
  in Mars atmosphere, 286, 324
  in Mars ice, 319, 324
  removal from Earth's atmosphere, 286, 339–40, 686
  in rock, 286, 340
  in Venus's atmosphere, 286, 309, 310
Carbonic acid, 339
Carbon monoxide, 386, 577
Carina Nebula, 485
Carrier particles, 30–31
Carter, Brandon, 686
"Cartwheel" galaxy, 605
Cas A, 199
Cassegrain telescopes, 213, 235
Cassini division, 366, 367
Cassini spacecraft, 352, 363, 364, 366, 375
Cassiopeia A supernova, 528–29
Cassiopeia constellation, xxv, 102
Catadioptric telescopes, 235
Catholic Church, 98
CCDs, 198, 208, 239
Celestial coordinates, 38, 49
Celestial equator, 36, 37, 45, 46
Celestial poles, 36, 37
Celestial sphere, 33–38, 78. See also Night sky
Cells, 682, 684, 685
Cellulose, 682
Celsius scale, 164

Centaurus A galaxy, xxxii
Centaurus constellation, xxxii
Centers of mass, 121–22, 456–57
Central Standard Time, 54
Centripetal force, 125–26
Centuries, 73–74
Cepheid variables
  formation, 506
  light curve, 504
  period-luminosity relation, 507, 592–93
  as standard candles, 505–6, 592–93
Cepheus constellation, xxv, 102
Ceres
  basic features, 330
  orbit, 245, 329
  reclassification, 5
  shape, 379
  size compared to Earth, 242
Cernan, Gene, 292
Cetus constellation, 102, 503
Chamaeleon constellation, xxxiii
Chamberlin, T. C., 413
Chandrasekhar, Subrahmanyan, 519
Chandrasekhar limit, 519, 520
Chandra X-ray telescope, 229–30, 584, 622, 623
Chankillo ancient observatory, 47
Channels on Mars, 320, 325, 688–89
Charge-coupled devices, 198, 208, 239
Charged particles from Sun, 283
Charges, with strong and weak forces, 29
Charm quarks, 30
Charon, 371, 377
Chemical effects of water, 286
Chemical elements, 150–51. See also Elements
Chemical energy, 143
Chesapeake Bay impact crater, 398
Chichén Itzá, 40
Chicxulub crater, 400
Chinese astronomy, 64, 200
Chinese calendar, 71, 74
Chiron, 388–89
Chitin, 682
Chlorofluorocarbons, 284–85
Chlorophyll, 682
Chondrites, 393–95
Chondrules, 393, 394
Christy, James, 377
Chromatic aberration, 216
Chromosphere of Sun, 403, 407–8, 423
Circular orbits, 129–30
Circumference of astronomical objects, 81
Circumference of Earth, 2, 79–80
Circumpolar constellations, xxv, xxxiii
Circumpolar stars, 36, 49
Clavius, Christoph, 295
Clavius crater, 295
Clementine spacecraft, 296
Climate, 49, 426–27, 494
Closed systems, 141–42
Closed universe, 653

Clouds. *See also* Interstellar clouds
  on Jupiter, 348
  on Mars, 324
  on Titan, 375
  on Uranus and Neptune, 356–57
  on Venus, 309
Clumping of dark matter, 649–50, 674–75
CNO cycle, 490
Coal Sack, xxxii
Coherence of light waves, 203, 213
Collagen, 682
Collecting power, 206
Collisions. *See also* Impacts
  of asteroids, 334
  as cause of planets' dissimilarities, 344
  of galaxies, 568–69, 570, 604–7, 611–12, 625
  of galaxy clusters, 631–32
  of interstellar clouds, 567, 569, 586
  ring material from, 368
  satellite origins from, 293
  suggested as cause of Uranus's tilt, 359
  within interstellar clouds, 577–78
Color charge, 29
Colors
  dispersion, 216
  from emission wavelengths, 153
  false-color images, 199
  filtering, 207
  human perception, 157–58, 165, 232–33, 234
  relation to temperature, 163–64, 165, 447–49
  visible spectrum, 155–56, 157–58
  in white light, 157–58
Comas, 383
Comet Hale-Bopp, 384
Comet Halley, 5, 96, 383–84, 440
Comet Hartley 2, 386
Comet McNaught, 382
Comets
  Brahe's observations, 93
  impacts by, 369
  oceans and atmospheres from, 258, 259
  origins, 243–44, 387–89
  relation to meteor showers, 389
  relation to Solar System, 5
  spectra, 175
  structure and composition, 383–87
Comet Shoemaker-Levy 9, 369, 396
Comet Swift-Tuttle, 389
Comet Tempel 1, 387
Comet Tempel-Tuttle, 390
Comet Wild 2, 386, 388
Common era, 74
Companion stars. *See also* Binary stars
  accretion of matter from, 516–18, 536, 544
  of Sirius, 219
Comparative planetology. *See also* Planets; Solar System
  defined, 336
  outer planets compared, 342–45
  terrestrial planets, 336–42

Compression heating, 474, 482, 483–84
Compression of gases, 403–6
Computer models. *See* Models
Computer use with telescopes, 197
Condensation of solar nebula, 254–56
Cone nebula, 493
Cones of retina, 232–33, 234
Conjunctions, 88, 89
Conservation laws, 141–45, 533–34
Constellations
  annual motions, 41–42
  basic features, 34–35
  learning to recognize, 100–102
  zodiacal, 42
Constellations (by name)
  Andromeda, xxvii
  Apus, xxxiii
  Aquarius, 42
  Aries, 42
  Auriga, xxvii
  Boötes, 101
  Cancer, 41, 42
  Capricornus, 42
  Cassiopeia, xxv
  Centaurus, xxxii
  Cepheus, xxv
  Cetus, 503
  Chamaeleon, xxxiii
  Crux, xxxii, xxxiii
  Cygnus, xxviii, 34–35
  Draco, xxv
  Fornax, 649
  Gemini, 41, 42
  Hydrus, xxxiii
  Leo, 34, 35, 41, 42
  Libra, 42
  Lyra, xxviii
  Mensa, xxxiii
  Monoceros, 518
  Musca, xxxiii
  Octans, xxxiii
  Ophiuchus, 42
  Orion, xxx, 41, 238, 239, 448, 482–83, 573
  Perseus, xxvii, 461
  Pisces, 42
  Sagittarius, xxxi, 42, 560
  Scorpius, 41, 42
  Southern Cross, xxxii
  Taurus, xxix, 41, 42
  Ursa Major, xxvi, 34, 35
  Ursa Minor, 35
  Virgo, 42
  Volans, xxxiii
Continuous spectra, 173
Convection
  in Earth's interior, 274
  in giant planets, 351, 358
  in stars, 407, 491
Convection zone of Sun, 403, 407
Coordinates, 37, 38
Copernicus, Nicolas, 88–91
Core collapse, 526–27
Core-collapse supernovae, 527
Cores
  active galaxies, 618–25
  asteroids, 333

  Earth, 270–71, 272, 273–74
  Galilean satellites, 373, 374
  gas giants, 248, 342–43, 350, 357
  Mars, 316–17
  Mercury, 305–6
  Milky Way, 562, 584–85, 624
  Moon, 297
  planets in Solar System, 242, 247
  stars, 476, 477, 490–91, 496–97, 500–501, 522
  Sun, 406
  Venus, 313
Coriolis effect
  on Earth, 289–90
  on gas giants, 343, 351, 358
  on Mars, 324
Cornea of eye, 232
Corona Borealis, 102
Coronal holes, 409–10
Coronal mass ejections, 423–24, 425
Corona of Sun, 63, 403, 408–10, 423–24
"Cosmic Gall," 418
Cosmic horizon, 646, 650
Cosmic microwave background
  dark matter clumping, 650, 674–75
  as evidence for inflation, 668–69
  principles and discovery, 647–48
  wavelength, 648
Cosmic rays, 416, 520
Cosmological constant, 657–58, 672–73
Cosmological principle, 637
Cosmology, 635–42. *See also* Universe
Crab Nebula
  history of observation, 200–201, 528
  location in night sky, xxix
  pulsar in, 533, 534, 535
Craters
  on asteroids, 331, 393
  on comet nuclei, 387
  formation in Solar System, 256, 258
  of Mars, 319, 320, 321, 322, 325
  of Mercury, 302–4
  of Moon, 2, 292, 294–96, 298
  on outer planets' moons, 345, 363, 365, 368, 374
  from past Earth impacts, 396–99, 400
  of Venus, 312–13
Crescent moon, 58, 59
Cretaceous mass extinction, 399–400, 680
Critical density, 655–57, 663
Cronus, 348
Crust
  of Earth, 270–71, 272, 274–77, 338
  of Mars, 318
  of Moon, 297–99
  of neutron stars, 536
  of Venus, 313–14, 338
Crux, xxxii, xxxiii
Crystallization, 667
Curiosity rover, 316, 323
Curtis, Heber, 591–92

Curvature of space, 191–92, 541–43, 652–54, 669
Cyanobacteria, 679–80, 684
Cygnus constellation, xxviii, 34–35, 102
Cygnus X-1, 545

D

2012-DA14, 332
Dactyl, 332
Daily motion of celestial sphere, 35–36
Dalton, James, 144
Dark adaptation, 233–34
Dark energy, 15, 673–77
Dark matter
  discrepancies with Standard Model, 31
  effects on large-scale structure of universe, 615
  evidence for, 14–15, 627–30, 674–75
  gravitational lenses and, 630–32
  in Milky Way, 582–83
  potential impact on Omega quantity, 657
  role in young universe, 649–50
  suggested nature of, 627, 632–33
Dark matter halos, 629
Dark nebulae, 574, 575
Darkness of night sky, 644–46
Darwin, Charles, 683, 685
Daughter atoms, 251
Davis, Raymond Jr., 416–17
Daylight hours, 52–53, 228
Daylight saving time, 55–56
Days
  on Earth, 51–56, 106–7
  on Mars, 316
  on Mercury, 306
  names of, 69–70
  outer planets compared, 343
  on Uranus, 359
  on Venus, 314
De Broglie, Louis, 151
Deceleration, 123
Declination, 38, 45, 47, 236–37
Deep Impact probe, 387
Deflection of light, 191–92
Degenerate gases, 499–500
Deimos, 326
Delphinus constellation, 102
Delta Cephei, xxv, 505
Democritus, 150, 558
Deneb, xxviii, 102
Denebola, 35
Density
  black holes, 546
  comet nuclei, 386
  Earth, 270
  Earth's atmosphere, 281–82
  estimating for universe, 655–57
  exoplanets, 267–68
  Galilean satellites, 372, 373
  interstellar medium, 572
  Jupiter and Saturn, 349–50
  low-mass versus high-mass stars, 489–90

Density—*cont.*
  Mercury, 305
  metric measure, 19
  in Milky Way, 582, 583
  Moon, 293, 297
  outer planets compared, 343
  planets in Solar System, 246–48
  Pluto, 377
  relation to luminosity of stars, 470–71
  Sun, 404, 406
  terrestrial planets compared, 338, 341–42
  Uranus and Neptune, 357
  white dwarfs, 516, 518
  young universe, 661, 662–63
Density waves, 585–87
Deoxyribonucleic acid, 682, 684, 685
De revolutionibus orbium coelestium (Copernicus), 90
Deserts on Mars, 319
Deuterium
  formation at universe's beginnings, 662, 663
  fusion in stars, 485, 487
  in heavy water, 417
  in proton-proton chain, 414, 415
Diameter
  ancient calculations, 77–78, 80
  basic measurement methods, 80–81
  measuring for stars, 460, 461–62
  of quasars, 620–21
Dicke, Robert, 686
Differential rotation of Milky Way, 580, 585
Differential rotation of Sun, 425–26
Differentiation, 257, 273, 333
Diffraction, 218–19, 459
Digges, Thomas, 91
Dimensions, 677
Dinosaur extinction, 399–400, 680
Dione, 362, 363
Direct imaging of exoplanets, 266
Direct proportionality, 122
Disk of Milky Way, 562
Disks of protoplanets, 261–62
Disk stars, 565
Dispersion of light, 216
Distance
  effects on gravity, 121
  effects on light, 149, 441–42, 443
  effects on orbital velocity, 130
Distance measures
  ancient astronomy, 77–79
  angular size and, 80–81
  astronomical units, 6, 10
  Copernicus's calculations, 90
  Doppler shift, 437
  to galaxies, 592–96
  inverse-square law and, 441–42, 443
  light-years, 10–11
  magnitude system, 443–44
  parallax, 432–36
  scales compared, 10, 14
  standard-candles method, 442–43, 505, 592–93
  triangulation, 429–31

Dixon, Jeremiah, 431
DNA, 682, 684, 685
Doppler broadening, 179
Doppler shift
  from binary stars, 455
  calculating, 177–79
  exoplanet detection by, 262–65, 267
  galaxy redshift versus, 595, 596
  of stars, 182, 437
Double Cluster, xxvii
Double stars. *See* Binary stars
Down quarks, 30, 31
Draco constellation, xxv
Drake, Frank, 691
Drake equation, 691
Draper, Henry, 449
Dreyer, John, 590
Dry ice, 319, 324
Dumbbell Nebula, xxviii
Duncan, John, 200
Dunes, on Mars, 319
Dust (interstellar). *See also* Interstellar clouds
  accretion in solar nebula, 254–55
  basic features, 572–74
  composition, 252, 481, 573–74
  detectors used to observe, 198
  dimming and reddening, 574–75
  formation, 477, 481
  heating and cooling, 578
  as obstacle to Milky Way observations, 557, 559, 561, 562, 591
  protoplanetary disks, 261–62
  sweeping from Solar System, 259
Dust in comets, 384, 387
Dust storms on Mars, 324
Dust tails, 384
Dwarf galaxies, 568, 600, 601, 602, 610
Dwarf planets
  Ceres as, 330
  creation of category, 5, 243, 379
  orbits, 5, 242, 245
Dynamos, 278, 353, 409

E

1E 0657-56 cluster, 632
Eagle Nebula, xxxi, 484
Early universe, 664
Earth. *See also* Atmosphere (Earth)
  age, 252
  ancient ideas about, 76–77
  ancient measurement, 79–80
  annual motions around Sun, 41–48
  atmosphere, 281–88
  compared to other terrestrial planets, 336–42
  compared to universe, 14
  composition, 248, 270–73
  days, 51–56
  density, 246, 247, 270, 338, 341
  development of life on, 679–86
  distance from Sun, 90, 95, 341
  escape velocity, 131, 338
  evolutionary stage, 337

gravitational force on Sun and Moon, 117–18
impact heating, 257, 273
interior structure, 337
internal motions, 274–77
magnetic field, 278–79, 283, 426
mass, 2, 338
orbit, 6, 41–48, 106–7
orbital period, 95, 341
orbital velocity, 22, 126
radius, 1–2, 3, 338
as starting point for astronomy, 1–2, 27
Earthquakes. *See* Quakes
Easter, 73
Eastern Standard Time, 54
Eccentricity, 95
Eclipses
  ancient discoveries from, 76–77
  causes, 61–66
  detecting Sun's corona during, 408
  testing general relativity with, 191
Eclipse seasons, 65–66
Eclipsing binaries, 455, 461–62
Ecliptic
  basic features, 42
  eclipses and, 65, 66
  planetary orbits near, 84, 85, 105
  tilt of, 45, 106
Eddington, A. S., 413, 469
Eggen, Olin J., 566
18-centimeter radiation, 695
Einstein, Albert. *See also* General relativity; Special relativity
  cosmological constant, 657–58, 672–73
  dark energy prediction, 15
  discovery of special relativity, 181, 184–85
  general relativity theory, 14, 189, 190, 193, 539
  mass-energy equation, 144, 413, 414, 664
  photoelectric effect explanation, 198
  revisions to Newton's laws, 27
  at Yerkes Observatory, 212
Einstein cross, 631
Electric charge, 28, 29, 413
Electric currents, 148, 278–79, 283
Electric fields, 534
Electromagnetic force, 29, 31, 666
Electromagnetic radiation. *See also* Spectra
  atmospheric absorption, 224–26
  atmospheric refraction, 228
  atomic spectra, 170–73
  detectors, 196–201, 203–8
  forces creating, 153
  light as, 148–49
  scattering effects on, 575
  spectrum, 155–60, 168
  thermal, 162–66
Electromagnetic spectrum, 155–60, 168. *See also* Spectra
Electromagnetic waves, light as, 148–49
Electron degeneracy, 499–500, 518, 526
Electronic transitions, 153

Electron neutrinos, 418
Electrons
  antiparticles, 664, 665
  in atomic structure, 27, 28, 30, 31, 150
  blackbody properties and, 163
  emissions determined by, 170–73
  interactions with light, 152–53
  nature of, 150–51
  photoelectric effect, 198
  role in chemical properties, 151
Electroweak force, 666–67
Elementary particles, 30–31
Elements
  atomic structure, 150–51
  in composition of typical star, 174, 175
  condensation temperatures, 254, 255
  in Earth's crust, 270–71
  formation in stars, 477, 524
  formation moments after Big Bang, 662–63
  of life, 682, 686
  light-emitting and absorbing properties, 151–53
  periodic table, 28
  radioactive, 273–74
  from supernovae, 528
Elliptical galaxies
  basic features, 599
  clusters containing, 611
  formation, 607
  subdivisions, 600, 601
  types of stars in, 602
Elliptical orbits
  escape velocity and, 132–33
  of exoplanets, 263, 265
  impact on day length, 106–7
  Kepler's discovery, 94–95
  Mars, 319
  Mercury, 306, 307
  of trans-Neptunian objects, 244
Emission-line spectra. *See also* Spectra
  basic principles, 170–71, 173
  from planetary nebulae, 512–13
  selected wavelengths illustrated, 175
  from Sun, 408
Emission nebulae, 573
Emissions of electromagnetic radiation, 153, 165
Empirical findings, 96
Enceladus, 363, 365
Energy
  conservation of, 141–44
  dark energy, 15, 673–77
  defined, 141
  electromagnetic spectrum, 155–60, 168
  from impacts, 395–96
  inverse-square law, 149
  ionization threshold, 174–75
  for living organisms, 682
  relation to surface area, 43–44
  relation to wavelength, 160, 162
  special measurement units, 22
  Sun's production, 4, 406–7, 412–16

Energy levels, 168–69
English system, 18, 19
Epicurus, 688
Epicycles, 87, 90, 91
Epoch of inflation, 665–69
EPOXI spacecraft, 386
Epsilon Lyrae, xxviii
Equation of time, 106
Equations. *See* Formulas
Equator, celestial, 36, 37, 45, 46
Equatorial bulges, 343, 344, 349, 536
Equatorial mounts, 236–37, 238
Equilibrium temperature, 284
Equinoxes
   calendars and, 73
   day length near, 53, 107, 228
   Sun's motions near, 46, 47, 104
   varying times of occurrence, 107
Equivalence principle, 189–90
Eratosthenes, 79–80
Eratosthenes crater, 295
Eridanus Cluster, 15
Eris
   comparative size, 378
   discovery, 5, 243
   orbit, 245, 379, 380
   satellites, 244
Eros, 331
Erosion, 288
Escape velocity
   atmospheric retention and, 131–32, 299
   black holes and, 539–41
   calculating, 130–32
   gravitational time dilation and, 192
   as kinetic energy, 143–44
   of Mercury, 307
   of Moon, 131, 299
   outer planets compared, 343
   terrestrial planets compared, 338
Eta Carinae, xxxii, 487
Ethanol, 577
Eudoxus, 86–87
Eukaryotes, 685
Euler crater, 298
Europa
   comparative size, 361, 362, 363, 371
   major features, 373
   water on, 373, 690
European Southern Observatory, 203, 205, 215
European Space Agency, 310, 316, 383
Evening Star, 105, 309
Event horizons, 543, 545
Event Horizon Telescope, 585
Evolutionary stages of terrestrial planets, 337
Evolution of life on Earth, 683–85
Excited state, 153
Exclusion principle, 499
Exiguus, Dionysius, 73
Exoplanets, 261–68, 690, 692
Expansion of space. *See also* Big Bang
   current measurements, 671–73
   light speed and, 667
   potential collapse versus, 654–58
   recession of galaxies, 594–96, 636–39

Expansion redshift, 596
Experiments, 25–26
Exploding white dwarfs, 517–20, 672
Exponents, 18, 20
Extinction of light by dust, 574
Extraterrestrial life, 688–96
Extremely Large Telescope, 205
Extreme Ultraviolet Explorer, 229
Eye, abilities of, 232–34. *See also* Vision
Eyepieces, telescope, 235, 236

F

Fahrenheit temperatures, 164
Failed stars, 10, 487
False-color images, 199
Far side of Moon, 296
Fast winds, 512
Faults, 312, 313, 314, 365
February, 72, 73
Feedback from supermassive black holes, 625
Fermi, Enrico, 693
Fermi paradox, 693
Filtering light, 207
Fire and Ice (R. Frost), 677
Fireballs, 393, 396, 398
First-quarter moon, 59
Fitzgerald, George, 184
Flares, solar, 423, 425
Flatness problem, 669
Flat rotation curves, 628
Flat space, 653, 657
Fluctuations, quantum, 545, 666, 669
Fluorescence, 385–86
Flux, 441
FM radio waves, 160
Focal length, 235–36
Focal planes, 210
Foci of ellipses, 94–95
Focusing light, 197, 210–16
Fog on Mars, 324
Folklore, 66–67, 70, 102
Fomalhaut, 262
Forbidden lines, 513
Forces. *See also* Energy; Gravity
   Newton's laws, 111, 116–18, 121–23
   subatomic, 29
   torque versus, 145
Formaldehyde, 577
Formulas
   acceleration, 115
   angular momentum, 533
   angular size, 81
   average density, 246, 247
   brightness, 149, 442
   centripetal force, 126
   collecting area, 206
   critical density, 655
   Doppler shift, 178
   Drake equation, 691
   energy level, 169
   escape velocity, 130, 539
   Galactic mass, 581
   galactic mass, 628

   gravitational potential energy, 142
   gravitational time dilation, 192, 543
   Hubble's law, 595, 636
   Hubble time, 641
   ideal gas law, 405
   Kepler's third law, 95, 96
   Kepler's third law modified, 127
   kinetic energy, 142, 395
   linear size, 81
   Lorentz factor, 184
   luminosity, 166, 442
   main-sequence lifetime, 492, 493
   mass-energy equation, 413
   mass-luminosity relation, 470
   mass of stars, 126
   Newton's second law, 116, 117
   orbital velocity, 129
   parallax-distance relation, 434
   parsecs, 434
   photon energy, 157
   radial velocity, 178
   radioactive decay, 251
   redshifts, 594, 640
   resolving power, 220
   Schwarzschild radius, 540
   standard-candles method, 443, 593
   Stefan-Boltzmann law, 166, 462, 463
   surface gravity, 123
   telescope magnification, 235, 236
   tidal forces, 136–37
   transverse velocity, 437
   universal law of gravity, 122
   Wien's law, 165, 448
Fornax cluster, 15
Fornax constellation, 649
Fossil bacteria, 679–80, 685, 689
Fossil plants and animals, 685
Fovea, 233, 234
Fowler, William A., 524
Fox, Sydney, 683
Frame dragging, 191
F-ratio, 236
Fraunhofer, Joseph, 449
Freckles, 373
Free fall, 190
French Revolution, 69
Frequency, 156, 157, 178. *See also* Wavelength
Freya, 70
Friedmann, Alexander, 641
Frost, Robert, 677
Frost line, 254, 255, 342
Frost on Mars, 324
F stars, 452
Full moon, 59, 66–67
Fully convective stars, 491
Fusion. *See* Nuclear fusion

G

Gaia hypothesis, 686
Gaia satellite, 436
Galactic cannibalism, 568, 606, 625
Galactic habitable zones, 693–94
Galactic plane, 560

Galaxies. *See also* Galaxy clusters
   active, 618–25
   basic features, 11, 589
   classification, 598–600, 601
   collisions and mergers, 568–69, 570, 604–7, 611–12, 625
   distances to, 592–96
   diversity, 598, 602–3
   early observations, 589–92
   evolution, 603–5, 606
   formation in young universe, 649–50
   large-scale distribution, 614–15
   mass measurement, 627–29
   motions, 179
   recession of, 636–39
   spiral arms, 585–87
   telescopic viewing, 208
   tidal forces, 135
   wavelengths emitted, 160
Galaxies (by name). *See also* Andromeda Galaxy (M31); Milky Way Galaxy
   Canis Major dwarf galaxy, 610
   "Cartwheel," 605
   Centaurus A, xxxii
   Large Magellanic Cloud, 527, 589, 590
   Leo I, 601
   M31, xxvii, 12–13, 15, 38, 589, 590, 591–92, 593, 596
   M33, 15
   M51, xxvi
   M63, 587
   M67, 610
   M81, xxv, 587, 602
   M82, xxv, 605
   M87, 612
   M100, 593
   M101, xxv
   NGC 1288, 599
   NGC 1365, 600, 623
   NGC 383, 619
   NGC 4261, 622
   NGC 4414, 599
   NGC 4458, 599
   NGC 4565, 599
   NGC 4660, 599
   NGC 5236, 600
   NGC 5532, 619
   NGC 5907, 606
   NGC 6822, 599
   NGC 7742, 618
   Small Magellanic Cloud, 589, 590, 599
   Sombrero, 230, 572
Galaxy clusters
   dark matter in, 629–30
   large-scale distribution, 614–15
   Local Group, 12–13, 14, 610–11, 655, 675–76
   overview, 12–14, 609
   rich versus poor, 611–12
   superclusters, 609, 613–14
Galaxy groups, 12–14, 610
Galex space telescope, 512
Galilean relativity, 182
Galilean satellites, 97, 361, 362, 372–74

Galileo
  discovery of Jupiter's satellites, 361
  Milky Way observations, 558
  motion studies, 97, 109, 110, 143
  Neptune observation, 356
  observation of Saturn's rings, 366
  overview of contributions, 97–98
  use of telescope, 210
Galileo spacecraft, 348, 350, 351, 363, 372, 373, 374
Gamma rays, 160, 225, 415, 529–30
Gamma-ray telescopes, 199, 529, 530
Gamow, George, 647, 648, 661
Ganymede
  comparative size, 344, 361, 362, 363, 371
  major features, 374
Gases
  from comets, 383, 386
  compression, 403–6
  condensation, 254–56
  degeneracy in, 499–500
  detectors used to observe, 198, 199
  ideal gas law, 405
  lack of thermal emissions, 163
  spectra formation, 170–73, 179, 447
  supporting life, 690
  sweeping from Solar System, 259
Gases (interstellar). *See also* Interstellar clouds
  absorption and emission of radiation, 573
  continuing discoveries, 616
  dark matter evidence from, 630
  in different galaxy types, 602
  following Big Bang, 647–48
  in galaxy clusters, 612
  heating and cooling, 577–78
  in Milky Way, 11
  properties of, 572–74
  radio waves from, 576–77
  star formation in, 10
Gas giants. *See also* Outer planets
  basic features, 242–43, 248
  compared, 342–45
  cores, 248
  formation of, 256
  in other systems, 263, 267–68
  rings, 366–69
  satellites, 361–65, 372–76
Gaspra, 331
Gauss, Karl Friedrich, 688
Gemini constellation, 41, 42
Gemini telescopes, 215
General relativity
  acceptance of, 27
  Big Bang and, 14
  cosmological constant and, 657–58, 672–73
  curved space and, 652–54
  explanation for universe expansion, 637–38
  gravitational lensing explanation, 630
  stretching of space and time, 191–93, 541–43
Generations of matter, 30
Geocentric models

  abandonment, 635–36
  Galileo's challenges, 97–98
  of Greek astronomers, 78–79, 86–87
  of Tycho, 94
Geological processes on Earth, 2. *See also* Quakes; Volcanoes
Geometry, 429–31
Germanic gods, 70
Giant elliptical galaxies, 612
Giant Magellan Telescope, 205
Giant molecular clouds, 481
Giant planets. *See* Gas giants
Giant stars. *See also* High-mass stars
  density, 470–71
  evolution, 476, 496–501
  placement on H-R diagram, 467–68
  as pulsating variables, 505–6, 507
  range of sizes, 468, 469
  stellar wind, 477, 511–12
  visibility in night sky, 553–54
Gibbous moon, 59
Giotto spacecraft, 383–84
Glaciation, 287
Gliese 581, 264
Glitches, 535–36
Global positioning system, 56, 277
Global warming, 285, 287, 427
Globular clusters, 550, 560–61, 566, 567
Gluons, 29, 30
Gold, Thomas, 533
GoTo systems, 237
Grains, 511. *See also* Dust (interstellar)
Grand Canyon, 288
Grand unified theories, 667, 676
Granite, 276
Gran Telescopio Canarias, 204
Granulation of Sun, 407
Gravitational force. *See* Gravity
Gravitational lensing, 192, 266–67, 630–32
Gravitational potential energy, 142, 143–44
Gravitational redshifts, 193. *See also* Redshifts
Gravitational time dilation, 192–93, 543
Gravitational waves, 194
Gravitons, 29
Gravity. *See also* Escape velocity; Mass
  balance with gas pressure, 404–6
  basic effects on matter, 9, 112
  curved space and, 191–92, 541–43
  dark matter evidence from, 14–15
  distortion of time and space by, 191–93, 539
  effects on asteroid belt, 333–34
  escape velocity and, 130–32, 539–41
  heat from, 474, 483–84
  lack of perception of, 189–90
  in molecular clouds, 482
  Newton's action-reaction law, 117–18

  Newton's universal law, 121–23
  potential energy of, 142, 143–44
  role in orbital motions, 112, 120–21
  role in stellar evolution, 9–10, 474
  within stars, 9, 404
  subatomic forces versus, 27–28, 29
  tidal forces, 135–39, 369
  weight and, 111–12
Gravity maps, 400
Great Attractor, 614
Great dark spot, 358
Greatest elongation, 89, 105, 302
Great Red Spot, 348, 352
Greek science
  concept of Earth as sphere, 27
  early concepts of atoms, 150
  geocentric models, 86–87
  measurement of Earth, Sun, and Moon, 76–80
Green Bank Telescope, 214
Greenhouse effect. *See also* Infrared radiation
  on Earth, 285, 340, 686
  on Mars, 324, 325
  on Venus, 310, 339, 686
Greenhouse gases, 285, 310
Greenwich observatory, 37
Gregorian calendar, 73
Grooved terrain of Ganymede, 374
Ground state, 151, 169
G stars, 449, 450, 452
Guest star, 200
Gulf Stream, 289
Gulliver's Travels (Swift), 326
Gusev Crater, 321

H

HI emission, 576
HII regions, 578
Habitable zones, 692, 693–94
Half-life, 250–51
Halley, Edmund, 383, 431, 436
Halley's Comet. *See* Comet Halley
Halo of Milky Way, 562
Halos of dark matter, 629
Halo stars, 565
Hands, measuring sky with, 100, 101
Harassment, 604
Harriott, Thomas, 210
Harvest moon, 67
Haumea, 243, 371, 379–80
Hawaiian Islands, 276
Hawking, Stephen, 545
Hawking radiation, 545
Hayabusa spacecraft, 331
Hayashi tracks, 486
HD 209458, 268
HD 209548b, 690
Heat. *See also* Infrared radiation; Temperatures
  compression, 474, 482
  in Earth's interior, 273–74
  effects of orientation to Sun, 43–44
  as energy form, 142
  within Galilean satellites, 373

  within gas giants, 343, 350–51, 357, 358
  greenhouse effect, 285
  in interstellar medium, 577–78
  lacking in Moon's interior, 3
  low-mass versus high-mass stars, 490
  from magnetic waves, 409
  within Mars, 318–19
  of meteors, 392–93
  of planetesimal impacts, 256–57
  relation to light, 158
  sources in planets, 2
  thermal radiation, 162–66
  within Venus, 313–14
Heat death of universe, 675
Heat packs, 667
Heavy elements
  buildup in galaxies, 570
  fusion reactions creating, 524–26
  needed for life, 686
  origins in supernovae, 477, 528
  in rich galaxy clusters, 612
  in young galaxies, 566, 567
Heavy water, 417
Height-strength diagrams, 478
Heliocentric models, 78, 88–91, 97–98
Helium
  attraction to developing planets, 256
  discovery, 171
  formation at universe's beginnings, 661–63
  fusion reactions in stars, 475, 476, 477, 498–501
  fusion reactions in Sun, 413, 414–16, 509
  within Jupiter and Saturn, 348, 349, 351
  in molecular clouds, 481
  as percentage of interstellar clouds, 252
  as percentage of stars' mass, 451
  percentage of Sun, 246
  spectral signature, 171
  within Uranus and Neptune, 356
Helium flash, 500, 501
Helix Nebula, 513
Hematite, 321
Hemispheres of Earth, 43–44
Hemoglobin, 682
Hercules Cluster, 611
Hercules constellation, 102
Hermann, Robert, 647
Herschel, Caroline, 559, 590
Herschel, John, 590
Herschel, William
  asteroid observations, 329
  discovery of infrared, 158
  discovery of Uranus, 356
  mapping by, 590
  Milky Way observations, 559
  reflector design, 214
  on stellar evolution, 473
Hertz, Heinrich, 156, 160
Hertz (unit), 156
Hertzsprung, Ejnar, 465
Hewish, Anthony, 532

Higgs particle, 31
High-energy pulsars, 537
High-energy radiation. *See also* Big Bang
 atmospheric absorption, 225
 in early universe, 660–61, 664–65
 gamma-ray bursts, 529–30
 from pulsars, 537
 wavelengths, 160
Highlands, lunar, 294, 297–98
High-mass black holes, 546
High-mass stars
 black holes from, 540–41
 death of, 522–30
 density, 489–90
 evolution of, 476–77, 479, 522–30
 helium fusion in, 499, 501
 main-sequence lifetime, 493, 551
 neutron stars from, 532
 shell fusion in, 497, 498
 size limits, 487
 in spiral arms of galaxies, 586
 structure, 490–92
High tides, 136
Hipparchus, 430, 443, 444
Hipparcos satellite, 435–36, 466
Hominids, 680
Homogeneity of space, 644
Homo sapiens, 680
Horizon, 33, 82, 228
Horizon problem, 668
Horsehead Nebula, xxx
Horsepower, 19
Hourglass Nebula, xxxiii, 230, 513
Hoyle, Fred, 524, 648, 693
HR 4796A, 262
HR 8799, 266
H-R diagrams
 analyzing, 466–71
 high-mass star lifetime, 522–23
 instability strip, 505, 506
 origins, 465
 red giant branch, 498
 of star clusters, 551–52
 tracking stellar evolution on, 478–79, 486–87
 white dwarfs on, 514
Hubble, Edwin
 Crab Nebula observations, 200
 demonstration of universe expansion, 636
 distance measurements to galaxies, 593, 594–95
 galaxy classifications, 598–600, 601
Hubble Deep Field project, xxvi, 603, 604, 649
Hubble's constant, 595, 641–42, 657, 673
Hubble's law
 applying, 595–96
 deviations from, 613–14
 graph of, 596
 measuring universe age with, 641–42
 measuring universe expansion with, 636–37
Hubble Space Telescope
 active galaxy studies, 620, 622

asteroid images, 330
aurora images on Jupiter and Saturn, 353
Betelgeuse image, 459, 460
black hole evidence from, 624, 625
capabilities, 230
collecting area, 206
Comet Shoemaker-Levy 9 impact images, 396
early problems with, 213
galaxy studies with, 135, 593, 603–4
illustrated, 229
Jupiter image, 347
molecular cloud images, 10, 484, 485
Neptune images, 358
nova images, 517, 518
orbit of, 229
Orion Nebula image, 483
planetary nebula images, 513
Pluto images, 377
protoplanetary disk images, 261–62
Saturn ring images, 366
supernova observations, 672
Uranus studies, 359, 362
Hubble time, 641–42, 656
Hubble Ultra Deep Field, 649
Hudson Bay, 398
Hulse, Russell, 194
Human eye, abilities of, 232–34. *See also* Vision
Human species, 680
Hunter's moon, 67
Huygens, Christian, 111, 366
Huygens probe, 375
Hyades cluster, xxix, 449, 549
Hydra galaxy cluster, 612
Hydrogen. *See also* Nuclear fusion
 absorption spectra, 447, 451
 atomic structure, 27, 150
 attraction to developing planets, 256
 availability for fusion reactions, 491–92
 Bohr's model, 169
 consumption in high-mass stars, 476
 electron orbitals, 150
 escape from Venus's atmosphere, 310–11, 339, 341
 fusion reactions in Sun, 413, 414–16
 in giant planets' composition, 342–43, 350, 356, 357
 ionization in interstellar clouds, 578
 in Jupiter's atmosphere, 348
 in molecular clouds, 481
 as percentage of interstellar clouds, 252
 as percentage of stars' mass, 451
 as percentage of Sun, 174, 175, 246
 presence or absence from supernovae, 527
 in Saturn's atmosphere, 349
 spectral signature, 169, 170–71
 21-centimeter radiation, 576, 602, 695

Hydrosphere of Earth, 281. *See also* Oceans; Water
Hydrostatic equilibrium, 404–6, 489, 490
Hydroxyl molecules, 695
Hydrus constellation, xxxiii
Hyperbolas, 133
Hyperion, 363, 364
Hypernovae, 530
Hypotenuse, 430
Hypotheses, 25–26, 27

I

Iapetus, 363, 364
Ice. *See also* Water
 on asteroids, 332
 climate data from, 287
 in comet nuclei, 386
 in giant planets, 342
 on Mars, 319–20, 322, 324, 325
 on Mercury, 307
 on outer planets' moons, 344–45, 363, 364, 371, 373–74
 possible presence on Moon, 299
 in Saturn's rings, 367
 in solar nebula, 254, 256
 in Solar System, 246, 342
Ice core data, 287
Ice giants, 355
Ice worlds, 344–45, 371
Ida, 332
Ideal gas law, 405
Idunn Mons, 314
Impact features. *See also* Craters
 on asteroids, 330, 331, 393
 on Earth, 396–99, 400
 on Moon, 2
Impact heating, 256–57, 273
Impacts. *See also* Collisions
 as cause of planets' dissimilarities, 344
 of Comet Shoemaker-Levy 9 with Jupiter, 369, 396
 crater formation from, 295
 energy of, 395–96
 evidence on Earth, 396–99
 mare formation from, 298
 mass extinctions from, 399–400
 Moon formation from, 293
 of probes with comets, 387
 suggested as cause of Uranus's tilt, 359
 suggested in Mercury's early history, 304, 305
Ina, 299
Inertia, 110, 111, 121, 122. *See also* Mass
Inferior conjunction, 88, 89
Inflation, 635, 654, 665–69, 677
Infrared and optical wavelength interferometers, 222
Infrared detectors
 arrays of, 222
 brown dwarf discoveries, 554
 overcoming interstellar dust effects, 575, 576
 space-based, 229
 technologies used for, 198

Infrared radiation. *See also* Greenhouse effect; Heat
 atmospheric absorption, 225, 226, 285
 discovery, 158
 from interstellar clouds, 484
 measuring asteroids by, 332
 thermal, 162
 visible light versus, 158, 159
Initial mass function, 553–55
Inner core of Earth, 272
Inner planets. *See* Terrestrial planets
Instability strip, 505, 506
Instrumentation, 196–97. *See also* Telescopes
Intelligent life on other planets, 692, 694–96
Interference of light waves, 148, 219
Interferometers, 220–22, 460
Internal heat. *See* Heat
International Astronomical Union, 243, 378–79
International date line, 54, 55
International Space Station, 129, 633
Interstellar clouds. *See also* Dust (interstellar); Gases (interstellar)
 around blue stars, 493
 collisions between, 567, 569, 586
 collisions within, 577–78
 continuing discoveries, 616
 detectors used to observe, 160, 198–99
 formation of, 477
 heating and cooling, 577–78
 in Milky Way, 11
 as obstacle to Milky Way observations, 557, 559, 561, 562, 591
 organic molecules in, 683
 properties of, 572–74
 searching for planets in, 261–62
 Solar System development from, 252–53
 spectra, 160, 173, 175
 star clusters from, 548
 stellar evolution from, 10, 474–75, 477, 481–85
Interstellar grains, 252. *See also* Dust (interstellar)
Interstellar medium. *See also* Interstellar clouds
 continuing discoveries, 616
 heating and cooling, 577–78
 properties of, 572–74
 radio waves from, 576–77
Inverse-square law, 149, 441–42, 443, 553–54
Inverse-square proportionality, 122
Io
 comparative size, 361, 362, 363, 371
 major features, 372
 volcanoes, 364, 372
Ionization
 by absorption, 152
 energy required for, 174–75
 in hot stars, 451, 452
 in Sun, 404, 409
 by UV radiation, 578

Ionosphere, 282–83
Ion tails, 384–85
Iridium, 399
Iris of eye, 232
Iron
  in asteroids, 332, 333
  condensation, 254–56
  in cores of gas giants, 350
  in cores of high-mass stars, 477, 522, 524, 526
  in Earth's core, 270–71, 278
  in Mars's core, 316–17
  on Mercury, 305
  Moon's lack of, 293, 297
  in neutron stars, 536, 540
  in planets' composition, 247, 248
  in stars with extrasolar planets, 266
Iron meteorites, 393, 394, 395
Irregular galaxies, 599, 602, 611
Irregular variables, 503–4
Ishtar, 312
Islamic astronomy, 87
Islamic calendar, 70, 74
Islands on Mars, 320
Isotopes, 250, 414, 679
Itokawa, 331, 334

J

Jansky, Karl, 160
Jets
  from active galactic nuclei, 619, 620, 622, 623–24
  from protostars, 485
Jet streams, 290, 351
Jewel Box, xxxii
Jewish calendar, 70–71, 74
Joule, James, 142
Joules, 142
Jove, 70
Jovian-mass exoplanets, 263
Jovian planets, 242–43, 256, 342–45. *See also* Gas giants; Outer planets
Julian calendar, 72
Jupiter
  appearance of, 348–49
  atmosphere, 242–43, 343, 348, 351–52
  axial tilt, 343
  comet impact, 369, 396
  core, 248, 342, 350
  days, 343
  density, 343, 349
  distance from Sun, 90, 95, 343
  effects on planet formation, 257
  equatorial bulge, 343, 344, 349
  escape velocity, 343
  Galileo's observations, 97
  gravitational effects on asteroid belt, 257, 330, 333–34
  interior structure, 248, 349–51
  magnetic field, 353
  mass, 241, 343
  orbital period, 95, 343
  radius, 3, 343
  ring, 367, 368, 369
  satellites, 244, 361, 362, 363, 372–74

K

Kaguya satellite, 63
Kant, Immanuel, 252, 558, 591, 636
Kapteyn, Jacobus, 559–60, 561
Kapteyn's Universe, 560, 592
Karnak temple, 47
Keck telescopes, 196, 206, 215, 222
Kelvin (William Thomson), 164
Kelvin scale, 164
Kepler, Johannes
  contributions to distance measures, 430
  discovery of elliptical orbits, 94–95
  guess on Mars's moons, 326
  ideas on finite universe, 645
  ideas preceding gravity discoveries, 120
  naming of Galilean satellites, 361
Kepler's laws
  conservation of angular momentum and, 145
  formulation of, 95–96
  measuring stellar masses with, 455–57
  Newton's modifications, 127, 456
  varying day lengths explained, 106–7
Kepler spacecraft, 267
Kepler's supernova, 520
Keratin, 682
Kilograms, 19, 110
Kiloparsecs, 560
Kinetic energy
  of asteroid impacts, 295
  calculating, 142, 395–96
  escape velocity as, 143–44
  of expanding universe, 656
Kirchhoff, Gustav, 173
Kirchhoff's laws, 173
Kirkwood, Daniel, 333, 367
Kirkwood gaps, 333–34
K stars, 449, 452
Kuiper belt
  captured objects, 245
  comets from, 388
  formation of, 257, 265
  location in Solar System, 5, 243
  Pluto's apparent origin, 378

L

Lada Terra, 313
Lag of the seasons, 48
Lagoon Nebula, xxxi
Lakes on Mars, 320, 321
Lambda, 658
Laminated terrain, 319–20
Landers, 321–22, 331, 375. *See also* Spacecraft
Laplace, Pierre-Simon, 252, 540
Large Binocular Telescope, 204
Large Hadron Collider, 30, 31, 665
Large Magellanic Cloud
  in Local Group, 610
  location in night sky, xxxiii, 237
  Milky Way collision with, 570
  SN 1987A in, 527
  visibility of, 589, 590
Large Millimeter Telescope, 205
Large-scale structures, 614–15
Larissa, 363
Laser Interferometer Gravitational-wave Observatory, 194
Last scattering epoch, 647, 660
Latitude, 37
Launching platforms, 133
Lava
  magnetic field evidence from, 279
  on Mercury, 303, 304
  on Moon, 295
  on Venus, 313, 314
  water in, 288
Lavinia Planitia, 313
Law of action-reaction, 117–18
Law of gravity, 121–22
Law of inertia, 111
Laws, defined, 26
Laws of conservation, 141–45, 533–34
Leaning Tower of Pisa, 123
Leap seconds, 56
Leap years, 72–73
Leavitt, Henrietta, 592
Lemaître, Georges, 641
Lenses
  chromatic aberration, 216
  color corrected, 235
  gravitational, 630–32
  in reflecting telescopes, 214, 216
  in refracting telescopes, 210–12, 214
Leo I galaxy, 601
Leo II group, 15
Leo constellation, 34, 35, 41, 42
Leonid meteor shower, 390
Leucippus, 150
Leverrier, Urbain, 356
Libra constellation, 42, 102
Libration of Moon, 300
Life
  anthropic principle, 686
  chemical elements, 681–82
  Gaia hypothesis, 686
  history on Earth, 679–81
  impact on Earth's development, 286, 340
  Mars studies, 316, 688–90
  mass extinctions, 399–400
  origins on Earth, 683–85
  potential on extra-solar planets, 690–96
Light. *See also* Light speed; Luminosity; Photons
  atmospheric interactions with, 284–85
  from Big Bang, 647–48
  deflection by gravity, 191–92
  detectors, 196–201, 203–8 (see also Observatories; Telescopes)
  diffraction, 218–19, 459
  dispersion, 216
  distance effects on, 149
  Doppler shift, 177–79
  dual nature, 147–49
  elementary particles, 28
  filtering, 207
  identifying atoms from, 170–73
  interactions with matter, 151–53
  inverse-square law, 441–42, 443
  luminosity, 164, 166, 440–44
  from moving bodies, 181–83
  Olbers' paradox, 644–46
  refraction, 210–12
  relation to temperature, 163–66
  speed, 10, 147
  visible spectrum, 155–56, 157–58
  white, 157–58
Lightbulbs, 449
Light curves
  for eclipsing binaries, 461–62
  for galaxies, 628
  supernova, 520, 527
  for variable stars, 503, 504
Light echoes, 518
Lightning on Jupiter, 352
Lightning on Venus, 314
Light pollution, 226
Light speed
  as absolute limit, 27, 147
  black holes and, 540, 542–43
  distance measures based on, 10–11
  expansion of universe and, 637–39, 667
  Michelson-Morley experiment on, 183–84
  relation to frequency and wavelength, 156
  relative motion and, 181–83
  special relativity and, 185–87
  through matter, 211–12
  through space, 147
Light travel time distance, 642
Light-years, 10–11, 20, 21–22, 434–35, 604
LIGO, 194
Linear size, 81
Lippershey, Hans, 210
Liquid core of Earth, 272
Liquid metallic hydrogen, 350, 353
Liters, 19
Lithium, 663
Little Dipper, xxv, xxvi, 101, 443–44
Little ice age, 427
Local Group
  basic features, 12–13, 14, 610–11
  density, 655
  future of, 675–76
Local Supercluster, 14, 613, 614
Loners, 691–92
Longitude, 37
Long-period comets, 388
Lorentz factor, 184, 185, 186, 192
Lovelock, James, 686
Lowell, Percival, 688
Low-mass black holes, 546
Low-mass stars
  comparative dimness, 554–55
  death, 509–14
  density, 489–90
  evolution, 475, 476, 477, 479
  helium fusion in, 499–501
  as oldest in Milky Way, 567–68
  shell fusion in, 497–98

Index

size limits, 487
structure, 490–91
Low surface brightness galaxies, 600, 601
Low tides, 136
L stars, 450, 452, 487
Lucretius, 688
Luminosity. *See also* H-R diagrams; Stefan-Boltzmann law
  absolute magnitude versus, 444
  of active galactic nuclei, 618–19, 620–21
  changes during stellar evolution, 494, 497–98, 501
  classification, 470–71
  overview, 440–41
  relation to distance, 440–44, 553–54, 592
  relation to main-sequence lifetime, 492–93
  relation to star mass, 469–70, 492, 554
  relation to star size, 467
  relation to temperature, 164, 166, 465
  Stefan-Boltzmann law and, 462–63
  Sun's variations, 426–27
  supernovae, 519
  of variable stars, 503–7
Lunar cycles, 58–67, 69, 70
Lunar eclipses
  ancient discoveries from, 76–77
  causes, 61, 62
  frequency of occurrence, 65–66
  total, 62, 63
Lunar months, 59
Lunar Prospector satellite, 299
Lunar Reconnaissance Orbiter, 298, 299
Lunisolar calendars, 70–71
Lutetia, 331
Lynden-Bell, Donald, 566
Lyra constellation, xxviii, 102

M

M3, 552
M13 globular cluster, 550, 696
M22 globular cluster, xxxi
M31. *See* Andromeda Galaxy (M31)
M33, 15
M51, xxvi
M52, xxv
M57, xxviii
M61, 562
M63, 587
M67, 549, 610
M81, xxv, 587, 602
M82, xxv, 605
M87, 612
M97, xxvi
M100, 593
M101, xxv
M104, 572
MACHOs, 632–33
Magellanic Clouds, 15. *See also* Large Magellanic Cloud; Small Magellanic Cloud

Magellan spacecraft, 312
Magnetars, 537
Magnetic dynamos, 278
Magnetic fields
  of accretion disks, 623–24
  of Earth, 278–79, 283, 426
  of Jupiter and Saturn, 353
  of Mars, 317, 338
  of Mercury, 306
  Moon's lack of, 297
  of outer planets' moons, 374
  in protostars, 485
  of pulsars, 533, 534–35, 537
  reversals, 426
  of solar wind, 384–85
  suggested effects on planetary nebulae, 513
  suggested importance to life, 694
  of Sun, 408–10, 420–26
  in supernova remnants, 520
  of Uranus and Neptune, 359
  of Venus, 314
Magnetic force, 29, 148, 278
Magnetic poles, 278–79
Magnetograms, 421
Magnifying power of telescopes, 235–36
Magnitude system, 443–44
Main-sequence lifetime, 492–94, 522–23
Main-sequence stars
  basic features, 466, 467
  in clusters, 551–52
  entry into giant phase, 496–98
  evolution, 475, 492–94, 522–24
  mass-luminosity relation, 469–70, 492
  properties and structure, 489–92
Major axes of ellipses, 94
Makemake, 243, 371, 380
Manicouagan Crater, 398
Mantle, 272, 297, 313–14
Many-worlders, 691–94
Mare Imbrium, 298
Mare Orientale, 298
Margulis, Lynn, 686
Maria, 294, 295, 298
Mariner missions, 689
Mariner 4 spacecraft, 316, 317
Mariner 10 spacecraft, 302, 303, 304
Mars
  atmosphere, 3, 319, 324–26, 338, 339, 340
  density, 338, 341
  distance from Earth, 430–31
  distance from Sun, 90, 95, 316, 341
  escape velocity, 338
  evolutionary stage, 337
  interior structure, 248, 316–17, 337
  Kepler's studies, 94–95
  magnetic fields, 317
  mass, 242, 338
  orbit, 319
  orbital period, 95, 341
  radius, 3, 338
  retrograde motion, 86, 88
  satellites, 244, 326
  search for life on, 316, 688–90
  surface features, 317–23, 337–38

Mars *Climate Orbiter*, 18
Mars Express spacecraft, 316, 320
Mars Global Surveyor spacecraft, 316, 320, 322
Mars Odyssey spacecraft, 316, 318, 323, 325
Mars Reconnaissance Orbiter, 316, 322, 323, 325
Mason, Charles, 431
Mass. *See also* High-mass stars; Low-mass stars
  acceleration and, 116–18, 122–23
  amount and distribution in Milky Way, 580–83
  atomic, 28
  Chandrasekhar limit, 519, 520
  conservation of, 144
  curved space and, 541–43
  defined, 110
  of Earth, 2
  Earth's water and atmosphere, 281
  effects on giant planets, 349–51
  in English system, 19
  escape velocity and, 131
  impact on stellar evolution, 476–77, 486–87
  limits for stars, 487
  measuring for galaxies, 627–29
  measuring from orbital motion, 125–27, 246, 247, 455–57
  of Moon, 2
  of neutrinos, 31
  in Newton's laws of motion, 116–18
  of Pluto, 377
  of quarks, 30
  relation to luminosity of stars, 469–70, 492, 554
  relation to terrestrial planets' properties, 337–38
  relativistic, 185
  role in stellar evolution, 474
  special measurement units, 22
  Sun's conversion to light, 412
  Sun's percentage in Solar System, 5
  in universal law of gravity, 121–22
  variations among stars, 9
  weight versus, 110, 111–12, 123
Mass-energy equation, 144, 413, 414, 664
Mass excess, 525–26
Mass extinctions, 399–400
Massive compact halo objects, 632–33
Massive stars. *See* High-mass stars
Mass-luminosity relation, 469–70, 492, 493, 554
Mass transfers, 516–18
2MASSWJ1207334-393254, 266
Mathilde, 331
Matter
  basic nature of, 27–28, 150–51
  creation in young universe, 661–65
  effects on light speed, 211–12
  elementary particles, 30–31
  fundamental forces, 29
  interactions with light, 151–53

Mauna Kea, 318
Mauna Kea observatory, 196, 205, 215
Mauna Loa, 288
Maunder, E. W., 426–27
Maunder minimum, 427
Maury, Antonia, 470
Maxwell, James C., 160, 312, 366
Maxwell Montes, 312
Mayan calendar, 74
Mean solar day, 52, 56, 106
Measurement
  acceleration, 114–16
  by ancient astronomers, 77–80, 430
  approximation in, 17, 23
  astronomical units, 6
  Brahe's precision, 93, 94
  calendars, 69–74
  density and composition, 246–48
  diameter, 80–81
  distances to galaxies, 592–96
  Doppler shift, 177–79, 437
  Earth radii, 1–2, 3
  energy, 142
  informal scale for night sky, 100, 101
  inverse-square law and, 441–42, 443
  latitude and longitude, 37
  light-years, 10–11
  of mass, 110, 125–27, 246, 247, 455–57
  metric system, 2, 17, 18–19
  parallax, 432–36
  scales compared, 10, 14
  scientific notation, 17, 20
  sizes of stars, 9, 459–63
  special units, 21–22
  standard-candles method, 442–43, 505, 592–93
  of Sun's properties, 404
  thermal radiation, 163–66, 332
  tidal forces, 137–38
  triangulation, 429–31
  of volume, 246, 247
Medium-size satellites, 363
Megaparsecs, 595, 596
Melting point, pressure and, 273
Membranes of cells, 685
Mensa constellation, xxxiii
Mercury
  conjunctions, 89
  density, 305, 338, 341
  deviation in orbit, 193
  difficulty of observing, 105
  distance from Sun, 6, 90, 95, 341
  escape velocity, 338
  evolutionary stage, 337
  greatest elongation, 89, 302
  interior structure, 248, 305–6, 337
  lack of atmosphere, 258–59
  mass, 242, 338
  orbital period, 95, 341
  probes of, 133
  radius, 3, 338
  rotation, 306
  Sun's tidal effects, 139
  surface features, 302–4, 337, 345
  temperature extremes, 307, 341

Mergers of galaxies, 568–69, 570, 604–7, 611–12, 625
Meridian, 51, 52
Messenger spacecraft, 302, 303, 304, 307
Messier, Charles, 200, 590
Metallic liquid hydrogen, 350, 353
Metals, 564, 567, 570. *See also* Elements; Iron
Meteorites
  age, 252
  defined, 392
  organic molecules in, 683
  potential evidence of Martian life, 689
  types, 393–95
Meteoroids, 392–93, 395–400
Meteors, 389–90, 392–93
Meteor showers, 389–90
Meters, 19, 21
Methane, 345, 356–57, 375, 386
Method of standard candles. *See* Standard candles
Metric prefixes, 18–19
Metric system, 2, 17, 18–19
The Mice, 605
Michell, John, 540
Michelson, Albert, 183–84
Microorganisms, 679–80
Microwaves, 175
Mid-Atlantic Ridge, 275
Midnight sun phenomenon, 53
Migrating planets, 265–66
Milankovitch cycles, 287
Milky Way Galaxy
  age and future, 570, 649, 675, 676
  center and edge features, 562, 583–85, 624
  compared to universe, 14
  formation, 566–70
  Galileo's observations, 98
  gas and dust distribution in, 572–73
  general features, 9, 11, 562
  growth of understanding about, 558–61, 591–92
  infrared image, 576
  in Local Group, 12–13, 15, 610–11
  in night sky, 9, 557
  position in universe, 636–37
  potential life-bearing planets, 691–94
  rotation and mass, 580–83
  search for other life in, 694–96
  spiral arms, 578, 585–87
  stellar populations, 564–66
  zone of avoidance, 591
Millennia, 74
Miller, Stanley, 683
Miller-Urey experiments, 683, 692
Millisecond pulsars, 536
Mimas, 363, 367, 368
Minerals, 321, 322–23, 332. *See also* Elements; Rock
Minor axes of ellipses, 94
Minor planets, 329
Mira, 503, 512
Miranda, 363, 365
Mira variables, 504, 507, 510, 511–12

Mirrors, telescope, 204, 206, 212–15
MIR space station, 62
Mitochondria, 685
Mizar, xxvi, 454, 455
MKS system, 19
Mnemonics, 450
Models
  celestial sphere, 34
  defined, 26
  evolving concepts of universe, 635–36
  galaxy formation, 568–69, 570, 585–87
  geocentric, 86–87, 94, 97–98
  heliocentric, 78, 88–91, 97–98
  inflationary, 665–69
  large-scale structure of universe, 615
  of light, 147–49
  Standard Model of particle physics, 30, 31
  stellar, 473–74
  of Sun, 405–6, 410
Molecular clouds, 481–85, 576–77. *See also* Interstellar clouds
Molecules, spectroscopic identification, 172–73
Monoceros, 518
Months, 58, 70–71, 72
Moon. *See also* Lunar cycles
  age, 252
  ancient measurement, 77–79
  angular size, 80–81
  comparative size, 362, 371
  crust and interior, 248, 292–93, 297–99, 337
  density, 293, 297, 338, 341, 342
  diameter, 81
  distance from Earth, 2
  early distance calculations, 430
  eclipses, 61–66
  escape velocity, 131, 338
  evolutionary stage, 337
  Galileo's studies, 97
  gravitational force on Earth, 118
  lack of atmosphere, 258–59, 299
  lore concerning, 66–67
  mass, 2, 338
  months based on, 70–71
  orbital motion explained, 120–21
  origins, 258, 292–93
  phases, 58–61
  radius, 338
  rotation, 300
  surface features, 2–3, 294–96
  surface gravity, 123
  tidal effects on Earth, 135–39
  tracking daily motions, 104–5
Moon illusion, 82
Moons. *See* Satellites
Morley, Edward, 183–84
Morning Star, 105, 309
Motion. *See also* Expansion of space; Orbits; Rotation
  acceleration, 22, 114–18, 122–23
  annual motions of Earth around Sun, 41–48
  within atoms, 27–28
  daily motions in night sky, 35–36

defined, 114
Doppler shift, 177–79
of Earth on axis, 36, 43–44, 51–53
of galaxies, 179
Galileo's studies, 97–98, 109, 110
inability to sense on Earth, 91
Kepler's laws, 95–96, 106–7
measuring mass from, 125–27, 246, 247
of Moon around Earth, 58–61
Newton's laws, 111, 116–18, 121
of planets, 84–91
special relativity and, 185–87
of stars, 267, 436–37
Motor drives, 237, 239
Mountain ranges on Earth, 274, 275
Mountains, 298, 312, 323. *See also* Volcanoes
Mount Everest, 318
Mount Fuji, 288
Mounts for telescopes, 235, 236–37
Mount Sharp, 323
Mount Wilson Observatory, 226, 564
M stars
  comparative dimness, 554
  emission spectra, 450
  on H-R diagram, 466
  temperature, 449, 452
Multicellular organisms, 685
Multimirror instruments, 215
Multiverse, 677
Muons, 30, 186
Musca constellation, xxxiii
Music of the spheres, 96
Mutations, 682

N

Nadir, 37
Naming of asteroids, 329
Nanometers, 19, 169
National Aeronautics and Space Administration (NASA), 316, 331
National Optical Astronomical Observatory, 203
National Radio Astronomy Observatory, 203
Natural selection, 685
Neap tides, 138
Near-Earth objects, 334
NEAR spacecraft, 331
Nebulae
  Butterfly, 513
  California, xxvii
  Carina, 485
  Crab, xxix, 200–201, 528, 533, 534, 535
  Dumbbell, xxviii
  Eagle, xxxi, 484
  Helix, 513
  Horsehead, xxx
  Hourglass, xxxiii, 230, 513
  Lagoon, xxxi
  Orion, xxx, 261, 262, 482–83, 573
  Owl, xxvi
  planetary, 476, 509, 512–14

  Ring, xxviii, 513
  Snake, 575
  Solar System origins from, 252–57
  Swan, xxxi
  Tarantula, xxxiii
  telescopic viewing, 208, 239
  Thumbprint, xxxiii
  Trifid, xxxi
  types, 573–74
Negative curvature, 653
Neptune
  atmosphere, 242–43, 343, 356–57
  axial tilt, 343
  core, 248, 343, 357
  days, 343
  density, 343, 357
  discovery, 356
  distance from Sun, 5, 6, 343, 357
  escape velocity, 343
  interior structure, 248, 357
  magnetic field, 359
  mass, radius, and orbital period, 343
  rings, 367, 368
  satellites, 244, 245, 361, 362, 363, 376
  suggested "migration," 265
Nereid, 363, 376
Net force, 111
Neutrinos
  benefits of detecting, 412, 416–18
  from Big Bang, 665
  discovery, 30
  ejection during fusion, 414, 524
  extraordinary nature of, 30, 416, 633
  mass, 31
  in supernova explosions, 201, 526, 527
  types, 30
Neutrons
  in atomic structure, 27
  components of, 30
  creation in proton-proton chain, 418
  degenerate, 526, 533–34
  neutrino interaction with, 417
  strong force on, 413
Neutron stars
  Crab Nebula, 201
  discovery, 532–34
  emissions, 534–37
  escape velocity, 540
  evolution of, 477
  formation of, 529, 532
  gravitational wave evidence from, 194
New General Catalog, 590
New moon, 59, 61, 65, 66
Newton, Isaac. *See also* Newton's laws
  modification of Kepler's third law, 127, 456
  overview of contributions, 109, 120
  reflecting telescope concept, 212
  white light experiments, 157
Newtonian telescopes, 213, 234
Newtons, 116, 123

Newton's laws
  continued validity, 26–27
  determining masses of distant objects with, 125–27
  first law of motion, 111, 121
  second law of motion, 116–17
  third law of motion, 117–18
  universal law of gravity, 121–23
NGC 1288, 599
NGC 1365, 600, 623
NGC 2264, 486, 493
NGC 3293, 549
NGC 383, 619
NGC 4261, 622
NGC 4303, 562
NGC 4414, 599
NGC 4458, 599
NGC 4565, 599
NGC 4595, 572
NGC 4660, 599
NGC 4755, xxxii
NGC 5236, 600
NGC 5532, 619
NGC 5907, 606
NGC 6822, 599
NGC 7582, 15
NGC 7742, 618
NGC 7788, 552
Nickel, 271, 519
Night sky
  amateur astronomy equipment, 234–39
  celestial sphere, 33–38
  coordinate system for, 38
  daily motions, 35–36
  effect of location on appearance, 37
  human eye perception abilities, 232–34
  learning constellations, 100–102
  lunar cycles, 58–67
  Milky Way in, 9, 557
  Olbers' paradox, 644–46
  origins of constellations, 34–35
  planetary motions, 84–91
  selected features, xxv–xxxiii
  tracking motions of celestial objects, 103–5
  visibility of giant stars in, 553–54
Nitrogen, 309, 340
Nitrogen oxides, 399
Nonvisible wavelengths, 158–59, 198–99
Northern circumpolar constellations, xxv
Northern Hemisphere (Earth)
  night sky observations, 100–102
  seasonal changes, 43–44, 46, 53
North Star. See Polaris
Novae, 503, 516–18. See also Supernovae
Nuclear fusion
  of deuterium, 485, 487
  following Big Bang, 661–63
  in giant stars, 496–501
  heavy elements from, 524–26
  in high-mass stars, 490, 491, 522–26
  in low-mass stars, 490, 491

  in main-sequence phase, 475, 489–94
  measurement from solar neutrinos, 416–18
  overview, 413–16
  weak force in, 418
Nuclei
  of active galaxies, 618–25
  in atomic structure, 27, 150
  of cells, 685
  of comets, 383–84, 386–87
  forces within, 29
  of Milky Way, 562, 584–85, 624
Nucleosynthesis, 524–25
Nulling interferometers, 221

O

Oberon, 363
Observatories. See also Telescopes
  atmospheric effects on, 224–28
  effects of light pollution on, 226
  for gravitational waves, 194
  Greenwich, 37
  overview of modern types, 203–5
  prehistoric, 40, 46–47
  siting of, 230
  space-based, 197, 229–30
Occam's razor, 87
Ocean currents of Earth, 289
Ocean floor of Earth, 275, 276, 279
Oceans. See also Water
  on Earth, 275, 286, 310
  on Europa, 373, 690
  formation of, 258, 259
  on Mars, 321
Ocean temperatures, solar cycle and, 427
Ocean tides, 135–38
Octans constellation, xxxiii
Off-axis foci, 214
Olbers, Heinrich, 645
Olbers' paradox, 644–46
Olivine, 272
Olympus Mons, 318
Omega Centauri, xxxii, 550
Omega quantity, 656–57, 671–72
On the Origin of Species by Natural Selection (Darwin), 685
On the Revolutions of the Celestial Orbs (Copernicus), 90
Oort cloud, 243–44, 257, 387–88
Open clusters, 549, 566
Ophiuchus constellation, 42, 102
Opportunity lander, 316, 321–22, 689
Opposition of planets, 88–89, 90, 105
Optical illusions, 82
Orbital motion of stars, 182, 454–57
Orbital periods, 95, 96, 343
Orbital perturbations, 356
Orbital planes, 244, 245
Orbitals
  basic principles, 150, 151
  relation to energy levels, 152–53, 168–69
  spectral signatures, 168, 169, 170–73
Orbital velocity, 129–30

Orbits
  of asteroids, 328, 333–34
  Bode's rule, 329–30
  comets, 5, 387–88
  Copernicus's calculations, 88–91
  Earth, 6, 22, 41–48, 106–7, 126
  of electrons, 28, 150, 151
  exoplanet, 262–64, 265, 266
  in galaxies, 580, 584–85
  gravity and, 112, 120–21
  Kepler's discoveries, 94–96
  Mars, 316, 319
  measuring mass from, 125–27, 246, 247
  Mercury, 193, 306, 307
  Moon, 59, 60–61, 65–66, 300
  observed planetary motions based on, 84, 85
  Roche limit, 369
  of satellites, 362, 376
  shapes, 132–33
  Solar System structure, 5, 6, 242, 244–45
  trans-Neptunian objects, 379–80
  velocity calculations, 129–30
  Venus, 309
Orcus, 371, 379, 380
Organic molecules, 393, 682, 683–85
Orion constellation
  amateur photos, 238, 239
  annual motions, 41
  location in night sky, xxx
  star colors, 448
Orion Nebula
  amateur photos, 239
  as emission nebula, 573
  location in night sky, xxx
  protostars in, 482–83
  young stars near, 261, 262
Ortelius, Abraham, 276
O stars
  effects on interstellar medium, 578
  on H-R diagram, 466
  in spiral arms of galaxies, 566, 586
  temperature, 449, 452
Outer core of Earth, 272
Outer planets. See also Gas giants
  basic features, 242–43
  compared, 342–45
  density and composition, 248
  formation of, 256
  satellites, 361–65
Overall expansion of the universe, 596
Owens Valley radio telescope, 221
Owl Nebula (M97), xxvi
Oxygen
  absorption of electromagnetic radiation, 225, 284–85
  breakdown of organic molecules, 684
  detecting in exoplanet atmospheres, 690
  fusion in stars, 519, 525
  origins on Earth, 286, 340, 685
  production in stars, 501, 512
Ozone, 225, 284–85, 685
Ozone layer, 685

P

Pacini, Franco, 533
Pallas, 329
Palomar observatory, 214–15
Pancake domes, 312
Pangea, 276–77
Panspermia hypothesis, 690
Parabolas, 133
Parallax
  anti-Copernican views based on, 90
  Brahe's inability to detect, 93
  early knowledge of, 78–79
  limits as distance measure, 435–36, 592
  principles and calculation, 432–36
Parsecs, 22, 23, 434–36
Parsons, William, 200
Partial eclipses, 62, 77
Particle accelerators, 30, 31, 665
Particle-antiparticle pairs, 664–65
Particle colliders, 661, 665
Particle physics, 30–31
Pathfinder spacecraft, 321
Pattern speed, 586
Payne, Cecilia, 451
Pegasus constellation, 102
Penumbra, 62
Penumbral eclipses, 62
Penzias, Arno, 648
Perihelion of Mercury, 306, 307
Periodic table of elements, 28
Period-luminosity relation, 507, 592–93
Periods, orbital, 95
Periods of variable stars, 504, 505, 507
Perseid meteor shower, 389
Perseus constellation, xxvii, 102, 461
Perseus galaxy cluster, 612
Perturbations, orbital, 356
PG 0052 251 quasar, 620
Phases of Moon, 58–61
Phases of Venus, 98
Philosophiae Naturalis Principia Mathematica (Newton), 121
Phobos, 326
Phoenix lander, 322, 689
Photodissociation, 310
Photoelectric effect, 198
Photography, 104, 197–98, 208, 237–39
Photometers, 442
Photons
  absorption, 152
  black hole effects, 542–43
  collecting area and, 206
  deflection by gravity, 191–92
  as elementary particles, 28
  energy carried by, 29, 157, 490–91, 666
  in expanding space, 638–39
  following Big Bang, 647–48, 650, 665
  interactions with matter in stars, 446–47
  inverse-square law and, 441–42
  ionization by, 174–75

mass equivalent in young universe, 660
movement in curved space, 652–53
movement in Sun, 406–7
from moving bodies, 182–83
nature of, 148–49
wavelength, 157
Photopigments, 232, 233–34
Photosphere, 403, 407, 446–47
Photosynthesis, 682, 685, 686
Piazzi, Giuseppe, 329
Pisces constellation, 42
Pixels, 198, 233
Planck's constant, 157
Planck time, 665–66
Planetary nebulae, 476, 509, 512–14
Planetesimals
  asteroids from, 333
  formation of, 255–56
  formation of planets from, 256–58
  late-stage impacts from, 258–59
  Mercury's suggested collision with, 305
  organic molecules in, 683
Planets
  Bode's rule, 329–30
  days named for, 69–70
  defining, 243, 378–79
  density and composition, 246–48
  difficulty of detecting near stars, 219
  formation of, 256–58
  general features, 3
  Kepler's discoveries, 94–96
  mass compared to Sun, 241
  motion in night sky, 84–91, 105
  in other systems, 261–68, 690
  outer planets compared, 342–45
  in Solar System structure, 5, 6, 7, 241–48
  terrestrial planets compared, 242, 336–42
Plants, 286, 686
Plasmas, 404, 409
Plate tectonics. See also Tectonic activity
  apparent absence from Venus, 313
  on Earth, 274–77
  on Mars, 318, 319
Plato, 688
Pleiades, xxvii, xxix, 549, 574
Plumes, 313
Plutinos, 380
Pluto
  compared to other planets, 243
  discovery, 377
  distance from Sun, 6, 7, 96, 378
  major features, 377–78
  orbit, 96, 245, 377
  reclassification, 5, 378–79
  satellites, 244, 371
Plutoids, 5, 243, 379
Pointer stars, xxv, xxvi, 101
Polaris
  as current circumpolar star, 49
  as giant star, 467

lack of apparent motion, 103
location in night sky, xxv, xxvi, 36, 37
magnitude, 444
pointer stars for, xxv, xxvi, 101
Polarity of Sun's magnetic field, 421, 425
Polar regions
  Earth, 48
  Mars, 319–20
  Mercury, 307
  Moon, 299
  Uranus, 359
  Venus, 310
Poles (celestial), 36, 37
Poles (magnetic), 278–79
Ponnamperuma, Cyril, 683
Poor clusters, 611–12
Population I stars
  basic properties, 564, 565
  formation in Milky Way, 566
  galaxy types favoring, 602
  heavy elements in, 567
Population II stars
  basic properties, 565
  formation in Milky Way, 565, 566, 568
  galaxy types favoring, 602
Population III stars, 567–68
Positive curvature, 653
Positrons, 414, 415, 490, 664, 665
Post meridian, 51
Potassium-40, 251
Potential energy, 142, 143–44
Potholes, 295
Pounds, 19
Power, special measurement units, 22
Powers of 10, 18, 20, 23
Precession, 49, 66, 320
Precision in measurement, 23
Prefixes, metric, 18–19
Prehistoric observatories, 40, 46–47. See also Ancient astronomy
Pre-main-sequence phase of protostars, 483, 484–85, 486–87
Pressure
  atmospheric, 281–82, 310
  of gases, 403–6
  impact on melting point, 273
  radiation, 384, 385
Primary mirrors, 212–13
Prime meridian, 37
Primeval Atom, 641
Principia (Newton), 121
Principle of equivalence, 189–90
Prisms, 157, 158, 168, 216
Procyon, 52
Project Ozma, 695
Prokaryotes, 684–85
Prominences, 63, 423
Proper motion method, 267, 436–37, 560
Proportionality, 122
Proteinoids, 683
Proteins, 682, 683
Proteus, 363, 364
Proton decay, 676
Proton-proton chain, 414–16, 418, 490

Protons
  antiparticles, 664
  atomic number, 150
  in atomic structure, 27, 28, 150–51
  components of, 30
  electric charge, 413
  forces between, 29
Proton storms, 424
Protoplanetary disks, xxx, 261–62
Protostars, 474, 475, 482–87
Proxima Centauri
  comparative dimness, 554
  distance from Earth, 23, 181, 436
  distance from Sun, 435
  on H-R diagram, 467
  location in night sky, xxxii
Ptolemy, 87
Pulsars, 264, 532–37
Pulsating variables, 504, 505–7, 592–93
Pupil of eye, 232, 233
Purkinje effect, 234
P waves, 271–72
Pythagoras, 76

Q

Quadratures, 90
Quakes. See also Tectonic activity
  on Earth, 271–72, 275
  on Moon, 297
  of Sun, 410
Quantized energy, 148
Quantized orbitals, 151, 152
Quantum fluctuations, 545, 666, 669
Quantum mechanics, 151
Quaoar, 371, 379
Quarks, 30, 31, 664, 665
Quarter moon, 59
Quasars
  absorption studies with, 616
  causes of activity, 622–24
  gravitational lensing of, 630, 631
  luminosity, 620–21
  redshifts, 620, 625

R

Radar guns, 177
Radar maps, 311–12, 313, 332
Radar measurement, 429
Radial velocity, 178, 437
Radiants, 390
Radiation, matter from, 661–65
Radiation era, 660–61
Radiation pressure, 384, 385
Radiative zone of Sun, 403, 406–7
Radioactive dating, 250–52
Radioactive decay, 29, 250–51, 273–74
Radioactive elements, 273
Radioactivity in Earth's interior, 273–74
Radio galaxies, 199, 619
Radio lobes, 619

Radio telescopes
  arrays of, 221
  basic features, 205
  importance, 160, 198
  resolution, 213, 220
  search for life via, 695–96
Radio waves
  from active galactic nuclei, 618
  atmospheric window, 224, 225
  cosmic microwave background, 647–48
  from Crab Nebula, 200, 201
  detectors, 198
  discovery and overview, 155, 160
  from interstellar clouds, 482, 576–77, 578
  from pulsars, 532–33, 534–35
  sample spectrum, 175
  search for life via, 694–96
  transmitting to create images, 179
Radius
  Earth as reference, 1–2, 3
  escape velocity and, 131, 539–40
  indicating on H-R diagram, 468
  measuring for stars, 459, 460, 461–63
  outer planets compared, 343
  scales compared, 15
  special measurement units, 22
  terrestrial planets compared, 338
Rainfall, carbon dioxide in, 286, 339
Ramadan, 70
Rays of Mercury, 303
Rays of Moon, 295
Recession velocities of galaxies, 595–96, 636–37
Recombination epoch, 647
Recurrent novae, 517–18
Reddening, 575
Red giant branch, 498
Red giants. See also Giant stars
  Antares as, 467
  evolution of, 475, 476, 498, 501
  mass transfer from, 516–17
  stellar winds, 511–12, 513
  Sun as, 509, 510
Red sequence of galaxies, 606, 607
Redshifts
  of cosmic background radiation, 648, 649
  defined, 178
  determining galaxy distances from, 594–96
  gravitational, 193
  measuring universe expansion from, 636–37, 639–41
  Omega quantity data from, 671–72
  of quasars, 620, 625
Red stars, Pop II stars as, 564, 565
Red supergiants, 523
Rees, Martin, 688
Reflected light, 165, 332
Reflecting telescopes, 212–15, 234
Reflection nebulae, 574
Refracting telescopes, 210–12, 214, 216, 235
Refraction, 62, 63, 210–12, 228
Regolith, 297, 331

Regular orbits of satellites, 362
Relativistic mass, 185
Relativity theory. *See* General relativity; Special relativity
Religions, 70–71, 73–74, 91, 98
Reproduction, 682, 685
Reproductive advantage, 685
Resolution of telescopes, 197, 218–22, 459–60
Resonances, 334, 367, 378
Rest frames, 181–83, 187, 189
Retina of eye, 232
Retrograde motion of planets
   Copernicus's explanation, 88–91
   early theories, 86–87
   Kepler's third law and, 96
   tracking in night sky, 105
Retrograde spin of Venus, 314
Revolution of Earth, 41
Rhea, 362, 363
Ribonucleic acid, 684
Rich clusters, 611–12
Ridges, mid-ocean, 275, 277, 279
Rifting, 274–75
Rigel, xxx, 448, 450, 471
Right ascension, 38, 45, 49, 236–37
Rilles, 295
Ring galaxies, 605
Ringlets, 367, 368
Ring Nebula, xxviii, 513
Rings, 97, 344, 366–69
Ritter, Johann, 158–59
Rivers on Mars, 320
Rivers on Titan, 375
RNA (ribonucleic acid), 684
Roche, Edouard, 369, 517
Roche limit, 369
Roche lobe, 516–17
Rock
   as carbon sink, 286, 340
   in cores of gas giants, 350
   on Mars, 321, 324
   in meteorites, 393–95
   of Moon, 292–93, 297
   radioactive decay, 250–52
   sedimentary, 286
   in Solar System, 246
   on Venus, 310, 312, 313
   water's effects on, 288
Rocket engines, 118
Rods of retina, 232, 233–34
Roman calendar, 69, 71–72
Rosh Hashanah, 70–71, 74
Rotation
   Earth, 35–36, 43–44, 51–53, 289–90
   effects on Doppler shift, 179
   galaxies, 580–83
   gas giants, 343
   interstellar clouds, 252
   Jupiter, 349
   Mars, 316
   Mercury, 306
   Milky Way, 580–83, 585–86
   Moon, 300
   protostars, 484–85
   pulsars, 533–34
   Saturn, 349
   Sun, 97, 425–26

   tidal forces on, 139
   Venus, 314
Rotation axes
   angular momentum and, 145
   of Mars, 316
   of planets, 244, 245
   seasons and, 341
   of Uranus and Neptune, 358–59
Rotation axis (Earth)
   Earth days, 51–53, 106–7
   effects on zodiac's appearance, 85
   impact on seasons, 43–44, 45
   precession, 49
Rotation curves of galaxies, 580–81, 627, 628, 629
Round numbers, 17, 23
Rovers, 316, 321–22, 323, 689
Royal Observatory (Greenwich), 37
RR Lyrae variables
   fixed luminosity, 507
   formation, 506
   light curve, 504
   Shapley's use of, 560
   as standard candles, 505
Rubin, Vera, 628
Runaway greenhouse effect, 310, 311, 686
Russell, Henry Norris, 465
Russian meteoroid impacts, 395–96, 398
Rutherford, Ernest, 150
RX J1856.5-3754, 536

S

S0 galaxies, 600, 607, 611
Sagan, Carl, 683
Sagittarius, xxxi, 560
Sagittarius A*, 584–85
Sagittarius constellation, 42, 102
Sandage, Allan R., 566
Saros, 66
Satellites. *See also* Moon
   of asteroids, 332
   craters on, 345
   effects on ring particles, 367–68
   formation of, 257–58
   Galilean satellite features, 372–74
   ice worlds, 344–45, 371
   of Jupiter, 97, 361, 362, 363, 372–74
   major features, 363–65
   of Mars, 326
   Moon as, 2
   of Neptune, 361, 362, 363, 376
   orbital velocity, 129–30
   orbits of, 245, 362
   of Pluto, 377
   of Saturn, 361, 362, 363, 364–65, 368, 375–76
   as systems orbiting gas giants, 361–62
   of trans-Neptunian objects, 380
   of Uranus, 358–59, 362, 363, 365
Saturn
   appearance, 348–49
   atmosphere, 242–43, 343, 349, 351, 352–53

   axial tilt, 343
   core, 248, 342, 350
   days, 343
   density, 343, 349
   distance from Sun, 90, 95, 343
   equatorial bulge, 349
   escape velocity, 343
   Galileo's observations, 97
   interior structure, 248, 349–51
   magnetic field, 353
   mass, radius, and orbital period, 95, 343
   rings, 366–68, 369
   satellites, 244, 361, 362, 363, 364–65, 367, 375–76
SB galaxies, 600, 601
Scales of universe, 15
Scarps, 303
Scattering, 574–75
Schiaparelli, Giovanni, 688
Schmidt, Maarten, 620
Schmitt, Harrison, 292
Schwarzschild, Karl, 540
Schwarzschild radius, 540, 543, 544, 545
Scientific method, 25–27
Scientific notation, 17, 20, 23
Scintillation, 226–27
Scorpius constellation, 41, 42, 102
Sea floor vents, 683, 684
Search for Extraterrestrial Intelligence, 694–96
Seasons
   causes on Earth, 40, 43–44
   causes on terrestrial planets, 341
   on Mars, 316, 319–20
   precession's effects, 49
   solstices and equinoxes, 46–47, 79–80, 107
Secondary mirrors, 213, 214, 235
Seconds, 56
Sedimentary rock, 286, 323
Sedna, 244, 371, 379, 380
Seeing, 227
Segmented mirrors, 215
Seismic waves, 271–72, 410
Selection effect, 554
Self-propagating star formation model, 587
Semimajor axes of ellipses, 94, 95, 96
Separate universes, 677
Serpens constellation, 102
SETI, 694–96
SETI@home, 696
Seyfert, Carl, 618–19
Seyfert galaxies, 618–19
Shapley, Harlow, 560–61, 591–92
Shell fusion, 497–98, 501, 523–24
Shepherding satellites, 368
Shielding, 577
Short-period comets, 388
Shorty Crater, 292
The Sickle, 34
Sidereal days, 52, 61
Sidereal months, 60–61
Significant digits, 23
Silicates
   in asteroids, 332
   condensation, 254–56

   in Earth's crust, 270
   in interstellar dust, 574
   on Moon, 294, 297
Silicon
   fusion into iron, 524, 525
   as fusion product, 519, 520
   potential life based on, 691
Simulations. *See* Models
Singularities, 541
Sirius, 219, 493
Sirius A, 516
Sirius B, 468, 516
Size, measurement methods, 80–81
Slipher, Vesto, 594
Sloan Digital Sky Survey, 601, 603
Small-angle formula, 81
Small Magellanic Cloud
   as irregular galaxy, 599
   in Local Group, 610
   location in night sky, xxxiii
   Milky Way collision with, 570
   visibility of, 589, 590
SN 1987A, 527
Snake Nebula, 575
Snider-Pellegrini, Antonio, 276
Snow line, 342
SO-2, 585
Soils of Mars, 322
Solar activity, defined, 420
Solar and Heliospheric Observatory, 403, 422, 423, 424
Solar cycle, 425–27
Solar days, 51, 52, 56, 106
Solar eclipses, 61–66
Solar flares, 423, 425
Solar masses, 9
Solar nebula theory, 252–57, 261, 332–33, 393
Solar neutrinos, 416–18. *See also* Neutrinos
Solar prominences, 63, 423
Solar seismology, 410
Solar System
   age, 250–52
   Bode's rule, 329–30
   center of mass, 456
   compared to universe, 14
   composition of bodies in, 246–48
   formation of, 252–59
   future of, 675
   geocentric models, 86–87, 94, 97–98
   heliocentric model's advance, 78, 88–91, 97–98
   location in Milky Way, 11, 562
   major features, 1–7, 241–44
   motion through space, 179
   orbital patterns, 5, 7, 242, 244–45
   orbital shifts suggested, 265
   outer planets compared, 342–45
   terrestrial planets compared, 336–42
Solar tides, 138
Solar wind, 384–85, 410
Solid core of Earth, 272
Solstices
   day length near, 107
   Eratosthenes's use of, 79–80

Solstices—cont.
  Sun's motions near, 46, 47, 104
  varying times of occurrence, 107
Sombrero galaxy, 230, 572
Sonograms, 271
Sound, 147, 160, 177
South celestial pole, xxxiii
Southern circumpolar constellations, xxxiii
Southern Cross, xxxii, xxxiii
Southern Hemisphere (Earth), 43–44, 46
Space-based observatories, 197, 229–30. See also Hubble Space Telescope
Spacecraft
  Apollo, 123, 292, 296
  Cassini, 352, 363, 364, 366, 375
  Clementine, 296
  Deep Impact, 387
  EPOXI, 386
  Galileo, 348, 350, 351, 363, 372, 373, 374
  Giotto, 383–84
  Hayabusa, 331
  Huygens, 375
  Kaguya, 63
  Kepler, 267
  launching platform concept, 133
  Lunar Reconnaissance Orbiter, 298, 299
  Magellan, 312
  Mariner, 302, 303, 304, 316, 317, 689
  Mars Climate Orbiter, 18
  Mars Express, 316, 320
  Mars Global Surveyor, 316, 320, 322
  Mars Odyssey, 316, 318, 323, 325
  Mars Reconnaissance Orbiter, 316, 323, 325
  Messenger, 302, 303, 304, 307
  NEAR, 331
  Opportunity, 316, 321–22, 689
  Pathfinder, 321
  Phoenix, 322, 689
  potential orbits, 132–33
  Solar and Heliospheric Observatory, 403, 422, 423, 424
  Solar System exploration with, 6–7
  Spirit, 321–22, 689
  Stardust, 386, 387
  Venera, 311
  Venus Express, 309, 310
  Viking, 320, 321, 324, 689
  Voyager 1, 6–7, 352, 367, 410
  Voyager 2, 6–7, 355, 358, 363, 376, 410
Space travel, 186–87
Space velocity, 437
Special measurement units, 21–22. See also Measurement
Special relativity, 27, 181–87
Speckle imaging techniques, 460
Spectra
  application to astronomy, 174–75
  of binary stars, 455
  classification, 449–50, 451–52

effects of interstellar medium on, 616
  electromagnetic, 155–60
  formation, 170–73
  hydrogen, 168–69
  interstellar clouds, 573
  luminosity class and, 470–71
  from planetary nebulae, 512–13
  principles, 216
  redshifts, 594–96
  star data from, 446
  temperature's relation to, 447–52
  of Type Ia supernovae, 520
  types, 173
  visible, 157–58 (see also Visible light)
  Zeeman effect on, 420–21
Spectral types of stars, 449–50
Spectroscopic binaries, 455, 457
Spectroscopy, 168, 216, 268, 449–50
Speed, special measurement units, 22
Speed of light. See Light speed
Spherical form of Earth, 76–77
Spicules, 408
Spindown of pulsars, 535
Spin of subatomic particles, 499, 576
Spiral arms of galaxies, 562, 566, 578, 585–87
Spiral density waves, 368
Spiral galaxies
  basic features, 598–99
  clusters containing, 611
  formation, 607
  Milky Way as, 566, 578
  rotation curves, 627, 628, 629
  subdivisions, 600, 601
  types of stars in, 602
Spirit lander, 321–22, 689
Spitzer Infrared Space Telescope, 229
Spring tides, 138
Square Kilometer Array, 205
Stadia, 80
Standard candles
  method described, 442–43
  supernovae as, 520, 593, 672
  variable stars as, 505, 592–93
Standard Model of particle physics, 30, 31
Standard time, 54
Starburst galaxies, 605
Star charts, 38, 100, 105
Star clusters
  formation of, 548
  initial mass function, 553–55
  testing stellar evolution with, 551–52
  types, 548–50
Stardust spacecraft, 386, 387
Starfire Optical Range, 227
Star lore, 102
Stars. See also Binary stars; Constellations; Variable stars
  aberration of light from, 437–38
  annual motions, 41
  apparent motions, 103
  binary system types, 454–55
  circumpolar, xxv, xxxiii, 36, 49
  composition, 174, 175

death of, 509–14, 522–30
  density in Milky Way, 583
  evolutionary forces, 9–10, 473–79
  exploding white dwarfs, 517–20, 672
  formation, 481–87
  giant phase, 496–501
  gravity within, 9, 474
  main-sequence features, 489–94
  mass limits, 487
  measuring days by, 52
  measuring distance to, 433–37, 440–44
  measuring mass, 126, 455–57
  measuring sizes of, 459–63
  Milky Way population, 564–66
  night sky observations, xxv–xxxiii, 33–38, 100–103
  origins of names, 35
  parallax, 78–79, 90, 93, 433–36
  photon-matter interactions in, 446–47
  placing on H-R diagrams, 465–71
  proper motion, 267, 436–37, 560
  spectral classes and temperatures, 449–52
  studies of light from, 182–83
  Sun as, 4, 9
  surface brightness, 208
  surface temperature measurement, 165, 447–49
  types in different galaxy classes, 602
  variable types, 503–7
Stars (by name)
  Alberio, xxviii
  Alcor, xxvi, 454
  Aldebaran, xxix
  Algol, 461
  Alpha Centauri, xxxii, 435, 436, 440, 443, 456, 457
  Altair, xxviii, 102
  Antares, 467
  Arcturus, 101
  Barnard's star, 436
  Beta Centauri, 440
  Beta Pictoris, 262
  Betelgeuse, xxx, 443, 448–49, 450, 459, 460, 501
  55 Cancri, 263
  Canopus, 77
  Capella, xxvii
  Delta Cepheus, xxv, 505
  Deneb, xxviii, 102
  Denebola, 35
  Epsilon Lyra, xxviii
  Eta Carina, 487
  Eta Carinae, xxxii
  Fomalhaut, 262
  Gliese 581, 264
  HD 209458, 268
  HR 4796A, 262
  HR 8799, 266
  2MASSWJ1207334-393254, 266
  Mira, 503, 512
  Mizar, xxvi, 454, 455
  Polaris, xxv, xxvi, 36, 37, 49, 101, 103
  Procyon, 52

Proxima Centauri, xxxii, 23, 181, 435, 436, 467, 554
  Rigel, xxx, 448, 450, 471
  Sirius, 219, 468, 493, 516
  Thuban, xxv
  Vega, xxviii, 49, 102
  Zeta Tauri, xxix
Star trails, 238
Steady state cosmology, 648
Stefan-Boltzmann constant, 166
Stefan-Boltzmann law
  basic principles, 164
  calculations, 166
  H-R diagram applications, 466–67, 468
  measuring star sizes with, 459, 462–63
Steins, 331
Stellar associations, 549–50
Stellar evolution
  death of high-mass stars, 522–30
  death of low-mass stars, 509–14
  giant stars, 496–501
  main-sequence stars, 492–94
  overview, 9–10, 473–79
  star formation, 481–87
  testing theories of, 551–52
Stellar models, 473–74
Stellar parallax. See Parallax
Stellar remnants, 474, 477. See also Black holes; Neutron stars
Stellar winds, 476, 477, 511–12, 513
Stereovision, 432–33
Stonehenge, 46–47
Stony asteroids, 333
Stony-iron meteorites, 393, 394, 395
Stony meteorites, 393–95
Storms
  Coriolis effect and, 290
  on gas giants, 343, 351–53
  on Mars, 324
  on Sun, 402, 403, 420
  on Venus, 309
Strange quarks, 30
Stratosphere, 282
String theories, 677
Stromatolites, 679–80, 685
Strong force
  basic features, 29
  following unification, 666–67
  on quarks, 30
  role in fusion reactions, 413–14, 498, 524, 526
Subatomic energy, 413
Subatomic forces, 28, 29
Subatomic particles, 27–28, 29–31, 144, 150–51
Subdivision of galaxy classes, 600, 601
Subduction, 274–75
Sublimation, 384, 389
Subscripts, 127
Sudbury iron mines, 398
Sudbury Neutrino Observatory, 417, 418
Sulfur, 306, 372, 684
Sulfuric acid, 309, 339
Summer Triangle, xxviii, 102

Sun. *See also* Solar System
  ancient measurement, 78
  annual motions, 40, 41–48
  atmosphere, 407–10
  charged particles from, 283
  composition, 174, 175, 246
  convection zone, 491
  energy generation, 4, 166, 412–16
  energy transport within, 406–7
  escape velocity, 132, 133, 540
  expected lifetime, 4, 10, 492–93
  future of, 509–11
  Galileo's studies, 97
  general features, 4, 241, 565
  gravitational force on Earth, 118
  hydrostatic equilibrium, 404–6
  interior structure, 404–6
  luminosity, 440–41, 494
  magnetic field effects, 408–10, 420–26
  mass and luminosity, 126, 404, 555
  in Milky Way, 11, 580, 582, 583
  planetary distances from, 90
  radius, 404, 463
  as reference for star properties, 9
  relation to Moon in night sky, 59–61
  Schwarzschild radius, 540
  solar cycle, 425–27
  spectral class, 450
  structure and physical properties, 402–4
  temperature, 165, 166, 404, 406, 408
  tidal effects on Earth, 138
  tracking daily motions, 104, 106–7
Sundials, 51, 106–7
Sunlight, 284–85, 341–42, 412–16
Sun-like stars. *See* Low-mass stars
Sunrise and sunset, 46
Sunshine, 412–16
Sunspots, 97, 420–22, 425–27
Superclusters, 13, 15, 609, 613–14
Supergiants, 471
Superior conjunction, 88, 89
Superior planets, 242. *See also* Outer planets
Super-Kamiokande detector, 417
Supermassive black holes, 622–25
Supernovae
  Brahe's observations, 93, 503
  Crab Nebula, 200–201
  evolution of, 477, 526–27
  neutron stars from, 532
  Omega quantity data from, 671–73
  as standard candles, 520, 593, 672
  Type Ia, 519–20, 527, 671–73
Supernova remnants, 520, 528–29
Surface brightness, 208, 590
Surface gravity, 123
Surface temperatures. *See also* Temperatures
  Mars, 316, 324, 338, 341
  Mercury, 307, 338, 341
  Moon, 299, 338
  of stars, 447–49

Sun, 403, 404
Venus, 310, 338, 341
Survival factor, 696
Swan Nebula, xxxi
S waves, 271–72
Syene, 79–80
Symbiosis, 681
Synchronous rotation, 300
Synchrotron radiation, 535, 619
Synodic periods, 105

T

Tails of comets, 383, 384–85
Tarantula Nebula, xxxiii
Tauons, 30
Taurus constellation, xxix, 41, 42
Taurus molecular cloud, 481
Taylor, F. B., 276
Taylor, Joseph, 194
Tectonic activity
  apparent absence from Venus, 313
  on Earth, 274–77
  on Mars, 318, 319
  on outer planets' moons, 345, 365, 374
  terrestrial planets compared, 337–38
Telescopes. *See also* Hubble Space Telescope; Observatories; Radio telescopes
  for amateur astronomy, 234–37
  arrays of, 220–22, 695
  atmospheric effects on, 197, 224–28
  basic features, 196–97, 234–37
  collecting power, 206
  diffraction in, 218–19, 459
  filters with, 207
  Galileo's use, 97–98
  gamma ray, 199, 529, 530
  overview of modern types, 203–5
  radio, 160
  reflectors, 212–15, 234
  refractors, 210–12, 214, 216, 235
  resolution, 197, 218–22, 459–60
  space-based, 197, 229–30
  using, 197–98
Temperatures. *See also* Luminosity; Stefan-Boltzmann law; Surface temperatures
  condensation, 254, 255
  cosmic background radiation, 648
  determining for stars, 447–49
  determining from emitted radiation, 163–66
  of heavy element fusion, 524
  hydrostatic equilibrium and, 405
  of Jupiter and Saturn, 349
  low-mass versus high-mass stars, 490
  luminosity versus, 164, 166, 465
  of molecular clouds, 482
  Moon, 299
  relation to stellar spectra, 447–52

role in nuclear fusion, 413–14
Sun, 165, 166, 403, 404, 406, 408
terrestrial planets compared, 338, 341
of Uranus and Neptune, 357
in young universe, 647, 661–62, 665
Terrestrial planets
  basic features, 242
  density and composition, 248
  effects of mass and radius, 337–38
  effects of water, biology, and sunlight, 339–42
  interior structure compared, 337
  surface features compared, 336
Tethys, 362, 363
Tharsis bulge, 317–19
Theories, 26–27
Thermal energy
  defined, 142, 164
  in interstellar medium, 577–78
  measuring asteroids by, 332
Thermal radiation, 162–66, 535. *See also* Stefan-Boltzmann law
Thermonuclear runaways, 517
Third-quarter moon, 59
30-Meter Telescope, 205
Thor, 70
Thorium, 273, 299
Thought experiments, 120–21
Thuban, xxv, 49
Thumbprint Nebula, xxxiii
Tidal bores, 136
Tidal braking, 56, 139, 145, 300
Tidal bulges, 135–36, 138, 139
Tidal forces
  on asteroids, 332, 333
  black holes, 546
  conservation of angular momentum with, 145
  defined, 135
  on Io, 372
  measurement, 137–38
  on satellites, 369
  of Sun on Mercury, 306
  on Venus, 314
Tidal lock of Mercury, 306
Tides, 56, 135–38, 300
Tilt of Earth, 43–44, 53
Tilt of ecliptic, 45
Tilt of Uranus, 358–59
Time
  calendars, 69–74
  distortion by gravity, 192–93, 543
  Earth days, 51–56
  Earth years, 41–48
  equation of, 106
  special measurement units, 22
  special relativity and, 185–87
Time zones, 54–55
Titan
  atmosphere, 345
  comparative size, 344, 361, 362, 363, 371
  major features, 375–76
  potential for life on, 691
Titania, 363
Titius, Johann, 329

Tiw, 70
Tombaugh, Clyde, 377
Topographic map of Mars, 317
Top quarks, 30
Torque, 145
Total eclipses, 62–64, 191, 196. *See also* Eclipses
Toutatis, 179
Trade winds, 289–90
Transition probabilities, 174
Transit method, 267, 268, 431
Trans-Neptunian objects, 243, 344, 378–80
Transverse velocity of stars, 437
Trenches, 275
Triangulation, 429–31
Trifid Nebula, xxxi
Trigonometry, 430
Triple alpha process, 498–99
Triton
  comparative size, 361, 362, 363, 371
  major features, 376
  orbit, 245, 300
Trojan asteroids, 328, 334
Tropic of Cancer, 47–48, 49
Tropic of Capricorn, 48
Tropics of Earth, 47–48
Troposphere, 282
True binaries, 454
T stars, 450, 452, 487
Tsunamis, 398
T Tauri stars, 485, 486, 503–4
Tunguska event, 397–98
Tuning-fork diagram, 600, 601
Turnoff point, 551, 552
21-centimeter radiation, 576, 602, 695
Twinkling of stars, 226–27
Twin paradox, 187
2 Micron All Sky Survey (2MASS), 576, 609
Two-stage collapse model, 566–68, 569
Tycho crater, 295
Tycho's supernova, 520
Type Ia supernovae, 519–20, 527, 671–73
Type Ib supernovae, 527
Type Ic supernovae, 527
Type II supernovae, 527

U

Ultraviolet radiation
  atmospheric absorption, 225, 284–85
  detectors, 229
  discovery, 158–59
  in emission nebulae, 573
  fluorescence and, 385–86
  ionization of hydrogen, 578
  from planetary nebulae, 511, 512
  Sun image, 403
Umbra, 62
Umbriel, 363
Uncompressed density, 342

Undersea vents, 683, 684
Unification hypotheses, 667
Units, 6, 17. *See also* Measurement
Universality hypothesis, 27, 28
Universal law of gravity, 121–23
Universal time, 55
Universe
   age, 13–14, 641–42, 673
   curvature, 541–43, 652–54
   distance scales, 10, 15
   evolving concepts of, 635–36
   expansion, 14, 596, 636–39 (*see also* Expansion of space)
   first moments, 660–69
   geocentric models, 86–87, 94, 97–98
   heliocentric models, 78, 88–91, 97–98
   possible fates, 654–58
   scales of measurement, 15
   unknown matter and energy, 14–15
Updike, John, 418
Up quarks, 30, 31
Uranium, 273
Uranus
   atmosphere, 242–43, 343, 356–57
   axial tilt, 343
   core, 248, 343, 357
   days, 343
   density, 343, 357
   discovery, 329, 356
   distance from Sun, 343, 357
   escape velocity, 343
   interior structure, 248, 357
   magnetic field, 359
   mass, radius, and orbital period, 343
   rings, 367
   rotation axis, 244, 358–59
   satellites, 244, 258, 358–59, 362, 363, 365
Urey, Harold, 683
Ursa Major, xxvi, 34, 35, 101. *See also* Big Dipper
Ursa Major group, 15, 548–49
Ursa Minor, 35, 101, 444

V

Valles Marineris, 317–18, 689
Valve mechanism, 506
Van Allen radiation belts, 283
van Maanen, Adriaan, 592
Vaporization temperatures, 254
Variable stars
   classification, 503–4
   evolution of, 476
   giant stars as, 505–6, 507
   measuring distances with, 505, 507, 592–93
   perceiving brightness of, 234
   period-luminosity relation, 507
   pulsating types, 504, 505–7, 592–93

Shapley's use of, 560
T Tauri stars, 485, 486, 503–4
Vectors, 114, 115, 137
Vega, xxviii, 49, 102
Vela constellation, 49
Velocity. *See also* Escape velocity; Light speed
   acceleration and, 114–16
   in angular momentum, 145
   defined, 111
   escape, 130–32, 143–44
   orbital, 22, 126, 129–30
   radial, 178, 437
Venera spacecraft, 311
Venus
   apparent magnitude, 444
   atmosphere, 3, 309–11, 339, 340
   conjunctions, 89
   density, 338, 341
   distance from Earth, 430–31
   distance from Sun, 90, 95, 309, 341
   escape velocity, 338
   evolutionary stage, 337
   Galileo's observations, 98
   greatest elongation, 89
   interior structure, 248, 337
   mass, 338
   orbital period, 95, 341
   phases, 98
   position in night sky, 309
   radius, 3, 338
   rotation, 244, 314
   Sun's tidal effects, 139
   surface features, 311–14, 337
   tracking motions in night sky, 105
Venus Express satellite, 309, 310
Vernal equinox, 46, 73
Very Large Telescope (VLT)
   as array, 215, 222
   basic features, 204
   exoplanet image, 266
   mirror technology, 215
   molecular cloud images, 484, 485
   neutron star images, 536
Very Long Baseline Array (VLBA), 221
Vesta, 329, 330, 331
Victoria Crater, 322
Viking spacecraft, 320, 321, 324, 689
Virgo Cluster, 13, 15, 612, 613, 675–76
Virgo constellation, 42
Virgo III groups, 15
Virgo Supercluster, 13, 15
Virtual particles, 545
Visible light
   atmospheric window, 224
   contrast with radio emissions, 618
   detectors, 197–98 (*see also* Telescopes)
   nonvisible versus, 159
   spectrum, 155–56, 157–58
Visible universe, 646
Vision
   basic structures, 232–33
   color perception, 157–58, 165, 232–33, 234

dark adaptation, 233–34
   limitations, 197
   parallax and, 432–33
   perception of brightness, 206
Visual binaries, 454–55, 456
Voids, 614, 615
Volans constellation, xxxiii
Volatility of ice, 246
Volcanoes
   on asteroids, 333
   atmospheric gases from, 258, 259
   causes of, 275, 276
   on Earth, 2, 275, 276, 288
   on Mars, 317, 318–19
   on Mercury, 303, 304
   on outer planets' moons, 345, 364, 372, 376
   on Venus, 312, 313–14, 338
Volume, 110, 246, 247
Vortices, 310, 352, 358
Voyager spacecraft
   detection of ice volcano on Triton, 376
   exit from Solar System, 410
   Jupiter lightning image, 352
   Jupiter ring images, 367
   satellite images, 363
   Solar System view, 7
   Uranus and Neptune studies, 355, 358

W

Waning moon, 59
Water. *See also* Ice
   in comets, 386
   condensation, 254
   Earth's hydrosphere, 281
   effects on rock formation, 288
   on Europa, 373, 690
   heavy, 417
   impact on Earth's development, 339–40
   infrared absorption, 225
   in Mars ice, 320, 322
   in outer planets, 348, 357, 359
   on outer planets' moons, 373
   possible presence on Moon, 299
   suggested by Mars surface features, 320–23, 689
   suggested loss in Mars's history, 325
   on Venus, 310–11
Waterhole, 695
Water ice. *See also* Ice
   on asteroids, 332
   on Mars, 320, 322, 324, 325
   on outer planets' moons, 344–45, 371, 373–74
   possible presence on Moon, 299
   in Saturn's rings, 367
Water line, 342
Water vapor, 285, 310
Wavelengths. *See also* Spectra
   atmospheric absorption and, 224–26

blackbody radiation, 162
   colors and, 155–56
   cosmic background radiation, 648
   dispersion, 216
   Doppler shift, 177–79, 437
   filtering, 207
   gravity's effects, 192–93
   high-energy, 160
   interference, 148, 219
   observing nonvisible, 198–99
   redshifts, 594–96, 639–40
   relation to energy transmitted, 157
   relation to scattering effects, 575
   relation to telescope resolution, 220
   relation to temperature, 163–64, 165, 448–49
   for seeking extraterrestrial life, 695
   telescope sensitivity to, 213
Wave-particle duality, 149, 151, 152–53
Waves
   diffraction, 218–19
   Doppler shift, 177–79
   gravitational, 194
   light as, 148–49
   seismic, 271–72, 410
Waxing moon, 59
Weak force
   asymmetry, 665
   basic features, 29, 525
   as fusion "bottleneck," 418
   on quarks, 30, 31
   relation to electromagnetic force, 666
Weakly interacting massive particles, 633
Weather, 290
Weeks, 69–70, 74
Wegener, Alfred, 276–77
Weight, mass versus, 110, 111–12, 123
Weizsäcker, Carl F. von, 413
Whirlpool Galaxy (M51), xxvi
White dwarfs
   accretion by, 516–17
   density, 516, 518
   evolution of, 514
   exploding, 517–20, 672
   formation, 476, 509, 510, 514
   gravitational redshift, 192–93
   on H-R diagram, 468
   pulsating, 506–7
   range of sizes, 468, 469
   Sun as, 510, 511
   temperature, 514
White holes, 621
White light, 157–58, 216
Wien's law, 163–64, 165, 166, 448–49, 545
Wilkinson, David T., 668
Wilkinson Microwave Anisotropy Probe, 668
Wilson, Robert, 648
WIMPs, 633

Winding problem, 586
Winds
    on Jupiter and Saturn, 351–53
    on Mars, 324
    on Neptune, 358
    in protostars, 485
    solar, 410
    stellar, 476, 477, 511–12, 513
Wind shear, 351
Woden, 70
Wolf Creek Crater, 398
W particles, 666
Wright, Thomas, 558

X

X-ray pulsars, 537
X-rays. *See also* Wavelength
    atmospheric absorption, 225
    from Crab Nebula, 200, 201
    from galaxy clusters, 612
    identifying black holes from, 544–45
    sample spectrum, 175
    from supernova remnants, 520
    wavelength, 155, 160
X-ray telescopes, 199, 213–14, 229–30, 584

Y

Years
    calendars, 69–74
    on Mars, 316
    on Mercury, 306
    Sun's annual motions, 41–48
Yellow giants, 476, 505, 509
Yellow supergiants, 505
Yerkes telescope, 212
Young, John, 123
Young planetary systems, 261–62

Z

Zeeman effect, 420–21
Zenith, 37
Zero-age main sequence, 487, 494
Zeta Tauri, xxix
Zodiac, 42, 84, 85
Zone of avoidance, 591
Zones of Jupiter, 348
Z particles, 666
Zwicky, Fritz, 532, 630

Spring Night Sky

When to Use This Star Map

Early March:	2 a.m.
Late March:	1 a.m.
Early April:	midnight
Late April:	11 p.m.
Early May:	10 p.m.
Late May:	Dusk

This star chart is most accurate if used within an hour or so of the times listed and is plotted for observers located between 30° and 50° north latitude. All times are standard time; if daylight-saving time is in effect, add one hour.

To use this chart, hold it in front of you and rotate it so that the yellow label corresponding to the direction you are facing is positioned at the bottom, right-side up. The stars in the sky should match those depicted on the chart. The center of the chart is the zenith, the point in the sky directly overhead.

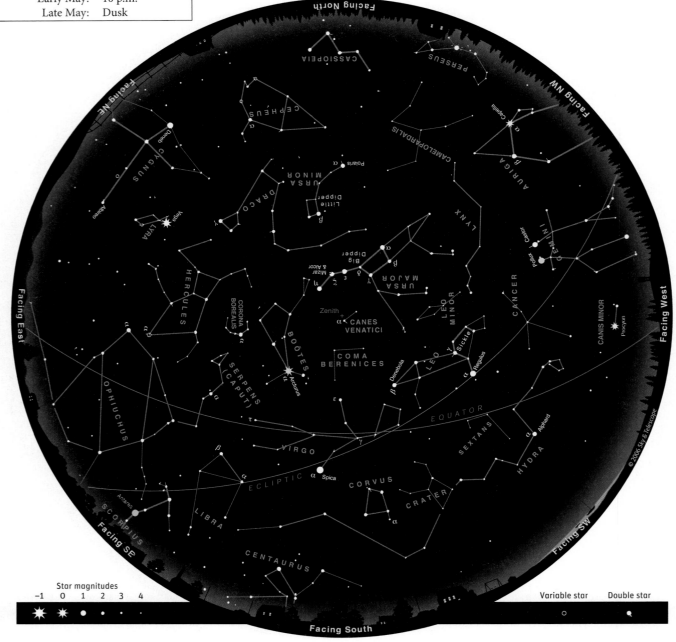

Summer Night Sky

When to Use This Star Map

Early June:	1 a.m.
Late June:	midnight
Early July:	11 p.m.
Late July:	10 p.m.
Early August:	9 p.m.
Late August:	Dusk

This star chart is most accurate if used within an hour or so of the times listed and is plotted for observers located between 30° and 50° north latitude. All times are standard time; if daylight-saving time is in effect, add one hour.

To use this chart, hold it in front of you and rotate it so that the yellow label corresponding to the direction you are facing is positioned at the bottom, right-side up. The stars in the sky should match those depicted on the chart. Ignore all the parts of the map above horizons you are not facing.

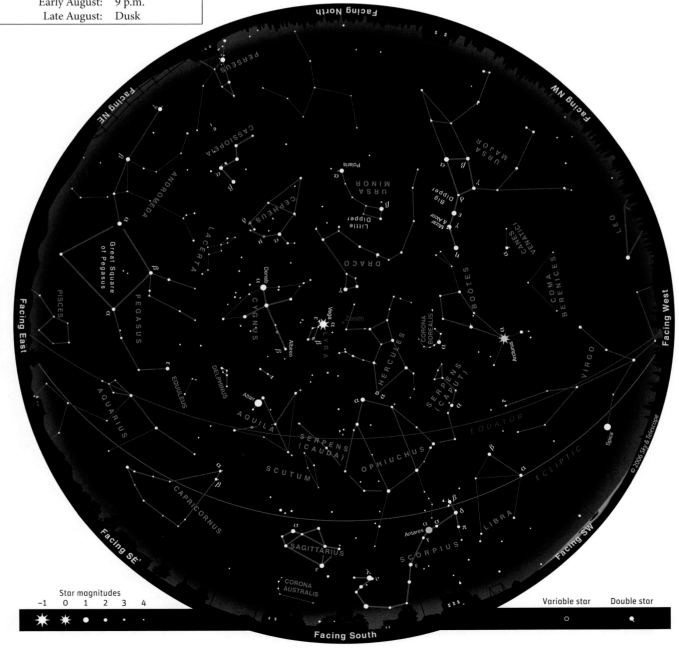

Autumn Night Sky

When to Use This Star Map

Early September:	midnight
Late September:	11 p.m.
Early October:	10 p.m.
Late October:	9 p.m.
Early November:	8 p.m.
Late November:	7 p.m.

This star chart is most accurate if used within an hour or so of the times listed and is plotted for observers located between 30° and 50° north latitude. All times are standard time: if daylight-saving time is in effect, add one hour.

To use this chart, hold it in front of you and rotate it so that the yellow label corresponding to the direction you are facing is positioned at the bottom, right-side up. The stars in the sky should match those depicted on the chart. The farther up from the map's edge they appear, the higher they'll be shining in your sky.

Winter Night Sky

When to Use This Star Map

Early December:	midnight
Late December:	11 p.m.
Early January:	10 p.m.
Late January:	9 p.m.
Early February:	8 p.m.
Late February:	Dusk

This star chart is most accurate if used within an hour or so of the times listed and is plotted for observers located between 30° and 50° north latitude. All times are standard time.

To use this chart, hold it in front of you and rotate it so that the yellow label corresponding to the direction you are facing is positioned at the bottom, right-side up. The stars in the sky should match those depicted on the chart. Remember that star patterns will look much larger in the sky than they do here on paper.

Using the Foldout Star Charts ➤

These star charts cover the entire sky and can be used to identify stars and constellations from any location at any time. They were produced at Stephen F. Austin State University by Professor Dan Bruton and his students, with additions by the publisher. The large rectangular chart covers a wide region around the celestial equator, and the two circular charts are centered on the northern and southern celestial poles. The charts are labeled with Right Ascension and Declination coordinates, which are like longitude and latitude on the celestial sphere (see Unit 5). Stars visible to the naked eye have sizes on the charts that depend on their brightness (or apparent magnitude—see Unit 54). Lines connecting the stars help to identify the pattern and primary stars in each constellation. The pale blue region shows the approximate location of the Milky Way.

For viewers in the northern hemisphere, to observe the northern half of the sky, use the **Northern Region** chart. If you are looking at the sky around 8 P.M., you should locate the closest date along the outer perimeter of the northern region chart. Rotate the chart so that the correct date is at the top. For each hour earlier/later that you are outside, you should rotate the map so that a Right Ascension an hour earlier/later is at the top. For example, if it is 8 P.M., on September 18, you would rotate the map so that "Sep 20" (Right Ascension of 20^h) is at the top. On the same date at 9 P.M., you would rotate the map so that a Right Ascension of 21^h is at the top. Held with this orientation, the chart should match what you see over the northern sky. The stars that are due north will match what you see at the center of the chart. Stars at the bottom of the chart will be below the horizon and stars near the top of the chart will be overhead.

Looking southward from the northern hemisphere, use the **Equatorial Region** chart. Find the date along the top of the chart, and adjust to an earlier or later Right Ascension according to how much earlier or later than 8 P.M. you are observing the sky as explained for the Northern Region chart. Stars located at the Right Ascension you have identified will lie straight up from the point on the horizon that is due south of you, along the "meridian" (see Unit 7). For observers north of latitude 30°, some stars along the bottom of the chart will never rise above the horizon.

The curving "sine wave" line running through the middle of the Equatorial Region chart is the ecliptic (Unit 6), which marks the path of the Sun among the stars. The dates along this line indicate where the Sun is located throughout the year. The Moon and planets also remain close to this line as they move around the celestial sphere.

The star charts can be used from the southern hemisphere using the same instructions but swapping the words "north" and "south" wherever they occur. Looking northward from the southern hemisphere, you will also need to hold the Equatorial Region chart upside down.

Using the Moon and Planet Finder Tables

Along the bottom of the foldout chart, information is provided to help you find where the Moon and five bright planets will be located each month. The Moon will be in its new phase and will lie close to the ecliptic at the Sun's position on the date given in the table. It then shifts about 13° to the east each subsequent day. If the date of the new moon is in a black circle, there will be a total solar eclipse on that date. The dates of total lunar eclipses of the *full moon* are also provided, listed in red circles.

The locations of Venus, Mars, Jupiter, and Saturn each month are indicated by the abbreviations for the constellations they are in. The abbreviation is listed in green when the planet is primarily in the morning sky and listed in blue for the evening sky. The listing is in black when the planet is in opposition to the Sun, indicating that it rises at sunset and sets at sunrise. The word *Sun* is listed when the planet is nearly in conjunction and therefore difficult to see. For Venus, the month when it is at its greatest elongation from the Sun is marked by an asterisk.

Mercury is always fairly close to the Sun and can be seen only shortly after sunset or shortly before sunrise. The date of its greatest elongation is given in green when the planet is in the morning sky and in blue when it is in the evening sky. The best opportunity for seeing Mercury is generally within about one week of this date.